Student Solutions Manual

for

Precalculus with Unit-Circle Trigonometry

Fourth Edition

by

David Cohen

with Theodore B. Lee

and David Sklar

Prepared by

Ross Rueger

Department of Mathematics

College of the Sequoias

Visalia, California

Australia • Canada • Mexico • Singapore • Spain • United Kingdom • United States

Printed in Canada
2 3 4 5 6 7 09 08 07

Printer: Transcontinental Printing

ISBN-13: 978-0-534-40232-7
ISBN-10: 0-534-40232-1

Thomson Higher Education
10 Davis Drive
Belmont, CA 94002-3098
USA

Asia (including India)
Thomson Learning
5 Shenton Way
#01-01 UIC Building
Singapore 068808

Australia/New Zealand
Thomson Learning Australia
102 Dodds Street
Southbank, Victoria 3006
Australia

Canada
Thomson Nelson
1120 Birchmount Road
Toronto, Ontario M1K 5G4
Canada

UK/Europe/Middle East/Africa
Thomson Learning
High Holborn House
50–51 Bedford Road
London WC1R 4LR
United Kingdom

Latin America
Thomson Learning
Seneca, 53
Colonia Polanco
11560 Mexico
D.F. Mexico

Spain (including Portugal)
Thomson Paraninfo
Calle Magallanes, 25
28015 Madrid, Spain

Contents

Preface

This *Student Solutions Manual* contains complete solutions to all odd-numbered exercises (and all chapter test exercises) of *Precalculus, With Unit-Circle Trigonometry* by David Cohen, Theodore B. Lee, and David Sklar. I have attempted to format solutions for readability and accuracy, and apologize to you for any errors that you may encounter. If you have any comments, suggestions, error corrections, or alternative solutions please feel free to drop me a note or send an email (address below).

Please use this manual with some degree of caution. Be sure that you have attempted a solution, and re-attempted it, before you look it up in this manual. Mathematics can only be learned by doing (not by reading), and finding your own mistake in a solution is part of the learning process. As you use this manual, do not just read the solution, rather work it along with the manual, using my solution to check your work. If you use this manual in that fashion then it should be helpful to you in your studying and preparation for calculus.

I would like to thank a number of people for their assistance in preparing this manual. Thanks go to my editors Katherine Brayton, Rachael Sturgeon, and John-Paul Ramin at Thomson Brooks/Cole Publishing for their valuable assistance and support, and for keeping me on schedule. Special thanks go to Matt Bourez of College of the Sequoias for his meticulous error-checking of my solutions, and prompt return of my manuscript under tight deadlines. Also thanks go to various instructors who have sent me alternative solutions and corrections to this and my student manual.

Finally, I would like to thank Theodore Lee and David Sklar, not only for their suggestions, but for continuing the tradition of excellence of David Cohen's classic textbook on Precalculus. I'm sure the changes they have made would be met with approval by Dave.

This manual is dedicated to David Cohen, my mentor, teacher, and friend.

Ross Rueger
College of the Sequoias
matmanross@aol.com

March, 2005

Chapter 1
Fundamentals

1.1 Sets of Real Numbers

1. **a.** integer, rational number **b.** rational number

3. **a.** natural number, integer, rational number **b.** rational number

5. **a.** rational number **b.** rational number

7. irrational number

9. irrational number

11. Since $\frac{11}{4} = 2.75$, sketch the graph:

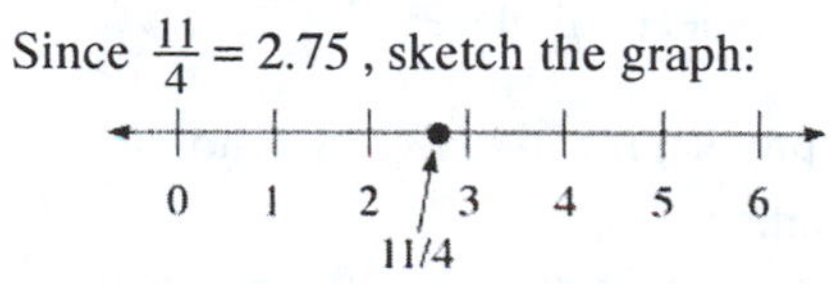

13. Since $1+\sqrt{2} \approx 2.4$, sketch the graph:

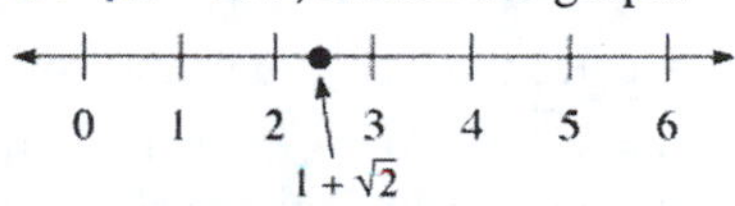

15. Since $\sqrt{2}-1 \approx 0.4$, sketch the graph:

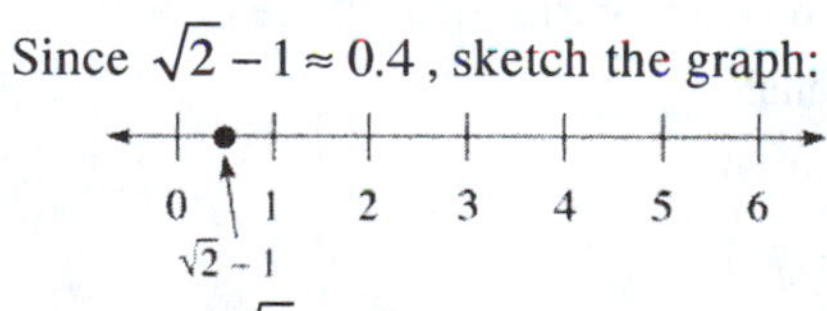

17. Since $\sqrt{2}+\sqrt{3} \approx 3.1$, sketch the graph:

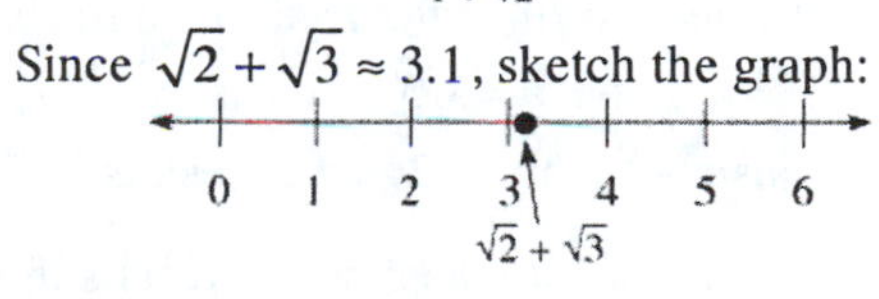

19. Since $\frac{1+\sqrt{2}}{2} \approx \frac{2.4}{2} \approx 1.2$, sketch the graph:

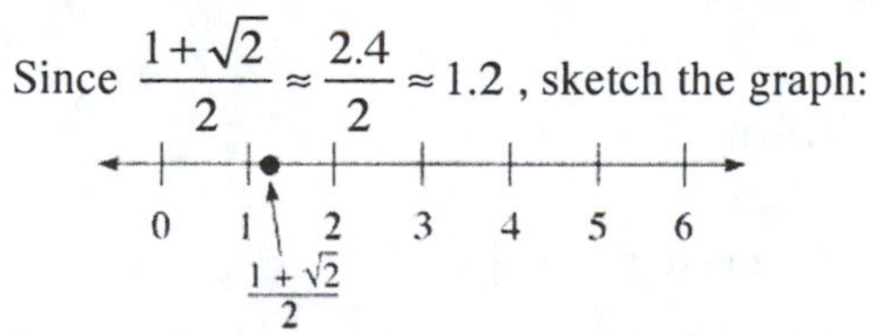

21. Sketching the graph:

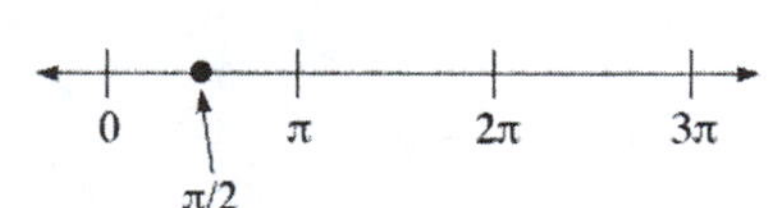

23. Sketching the graph:

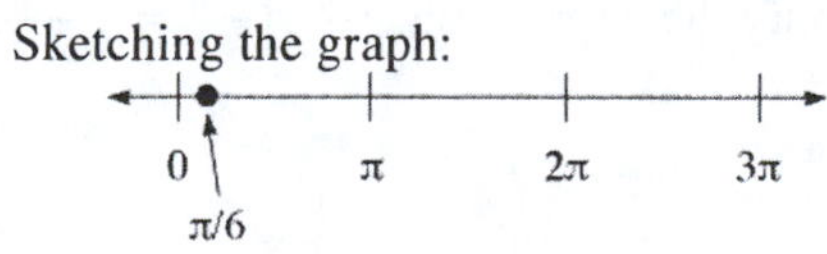

25. Sketching the graph:

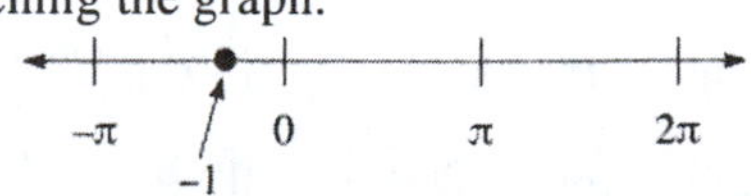

27. Sketching the graph:

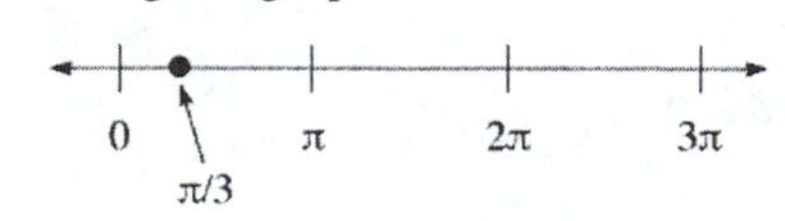

29. Since $1 \approx \frac{\pi}{3}$, sketch the following graph:

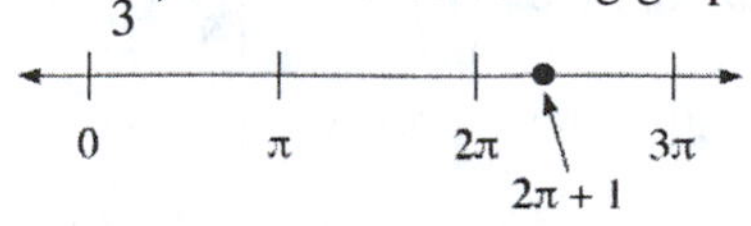

31. False

33. True (since $-2 = -2$, it is also true that $-2 \le -2$)

35. False

37. False (since $2\pi \approx 6.2$)

39. True (since $2\sqrt{2} \approx 2.8$)

41. The inequality notation is $2 < x < 5$:

43. The inequality notation is $1 \le x \le 4$:

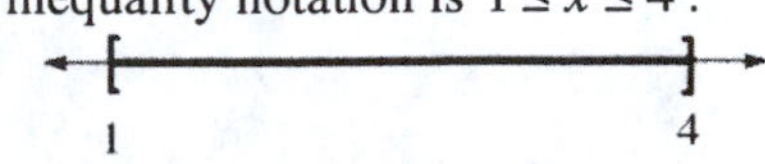

45. The inequality notation is $0 \le x < 3$:

47. The inequality notation is $-3 < x < \infty$:

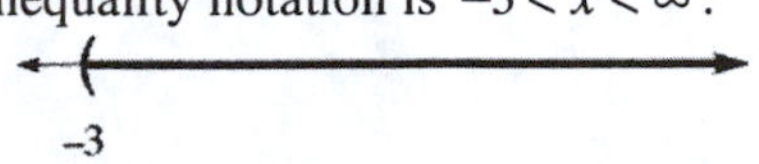

49. The inequality notation is $-1 \le x < \infty$:

51. The inequality notation is $-\infty < x < 1$:

53. The inequality notation is $-\infty < x \le \pi$:

55. **a.** Since $\left(\frac{4}{3}\right)^4 \approx 3.16$, it agrees with π to one decimal place.

b. Since $\frac{22}{7} \approx 3.142$, it agrees with π to two decimal places.

c. Since $\frac{355}{113} \approx 3.1415929$, it agrees with π to six decimal places.

d. Since $\frac{63}{25}\left(\frac{17+15\sqrt{5}}{7+15\sqrt{5}}\right) \approx 3.1415926538$, it agrees with π to nine decimal places.

57. **a.** We need to find two irrational numbers a and b such that their product is rational. If we choose $a = \sqrt{2}$ and $b = \sqrt{8}$, then $ab = \sqrt{2} \bullet \sqrt{8} = \sqrt{16} = 4$, which is rational.

b. We need to find two irrational numbers a and b such that their product is irrational. If we choose $a = \sqrt{2}$ and $b = \sqrt{3}$, then $ab = \sqrt{2} \bullet \sqrt{3} = \sqrt{6}$, which is irrational.

59. **a.** Raising 2 to the 1/2 power results in $2^{1/2} = \sqrt{2}$, which is irrational.

b. Raising $\sqrt{2}$ to the 2 power results in $\left(\sqrt{2}\right)^2 = 2$, which is rational.

1.2 Absolute Value

1. Computing: $|3| = 3$

3. Computing: $|-6| = 6$

5. Computing: $|-1+3| = |2| = 2$

7. Simplify the expression: $\left|-\frac{4}{5}\right| - \frac{4}{5} = \frac{4}{5} - \frac{4}{5} = 0$

9. Simplify the expression: $|-6+2| - |4| = |-4| - |4| = 4 - 4 = 0$

11. Simplify the expression: $\big||-8| + |-9|\big| = |8+9| = |17| = 17$

13. Simplify the expression: $\left|\frac{27-5}{5-27}\right| = \left|\frac{22}{-22}\right| = |-1| = 1$

15. Simplify the expression: $|7(-8)| - |7||-8| = |-56| - 7(8) = 56 - 56 = 0$

17. Substituting $a = -2$ and $b = 3$: $|a-b|^2 = |-2-3|^2 = |-5|^2 = (5)^2 = 25$

19. Substituting $a = -2$, $b = 3$, and $c = -4$: $|c| - |b| - |a| = |-4| - |3| - |-2| = 4 - 3 - 2 = -1$

21. Substituting $a = -2$, $b = 3$, and $c = -4$:

$$|a+b|^2 - |b+c|^2 = |-2+3|^2 - |3+(-4)|^2 = |1|^2 - |-1|^2 = (1)^2 - (1)^2 = 1 - 1 = 0$$

23. Substituting $a=-2$ and $b=3$: $\dfrac{a+b+|a-b|}{2}=\dfrac{-2+3+|-2-3|}{2}=\dfrac{1+|-5|}{2}=\dfrac{1+5}{2}=\dfrac{6}{2}=3$

25. Since $\sqrt{2}-1>0$: $\left|\sqrt{2}-1\right|-1=\left(\sqrt{2}-1\right)-1=\sqrt{2}-2$

27. Since $x\geq 3$, $x-3\geq 0$ and thus: $|x-3|=x-3$

29. Since $t^2+1>0$: $\left|t^2+1\right|=t^2+1$

31. Since $-\sqrt{3}-4<0$: $\left|-\sqrt{3}-4\right|=-\left(-\sqrt{3}-4\right)=\sqrt{3}+4$

33. Since $x<3$, $x-3<0$ and $x-4<0$, and thus: $|x-3|+|x-4|=-(x-3)+[-(x-4)]=-x+3-x+4=-2x+7$

35. Since $3<x<4$, $x-3>0$ and $x-4<0$, and thus: $|x-3|+|x-4|=(x-3)+[-(x-4)]=x-3-x+4=1$

37. Since $-\frac{5}{2}<x<-\frac{3}{2}$, $x+1<0$ and $x+3>0$, and thus:

$$|x+1|+4|x+3|=[-(x+1)]+4(x+3)=-x-1+4x+12=3x+11$$

39. The absolute value equality can be written as $|x-1|=\frac{1}{2}$.

41. The absolute value inequality can be written as $|x-1|\geq\frac{1}{2}$.

43. The absolute value inequality can be written as $|y-(-4)|<1$, or $|y+4|<1$.

45. The absolute value inequality can be written as $|y-0|<3$, or $|y|<3$.

47. The absolute value inequality can be written as $\left|x^2-a^2\right|<M$.

49. Graphing the interval $|x|<4$:

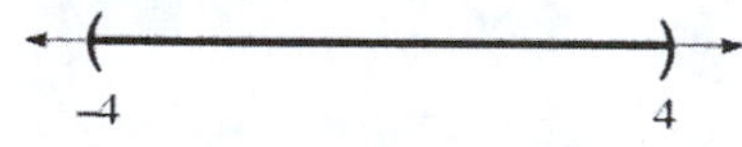

51. Graphing the interval $|x|>1$:

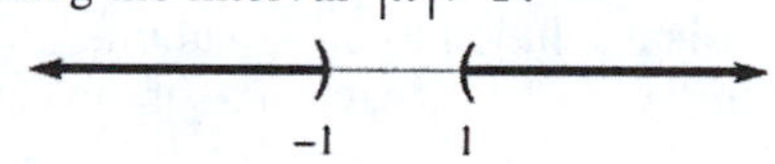

53. Graphing the interval $|x-5|<3$:

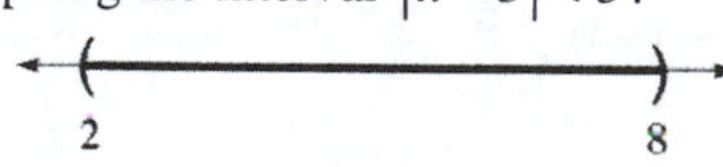

55. Graphing the interval $|x-3|\leq 4$:

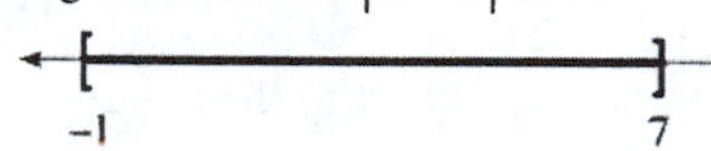

57. Graphing the interval $\left|x+\frac{1}{3}\right|<\frac{3}{2}$:

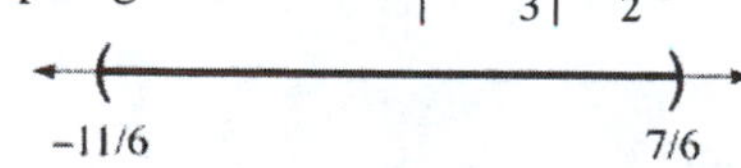

59. Graphing the interval $|x-5|\geq 2$:

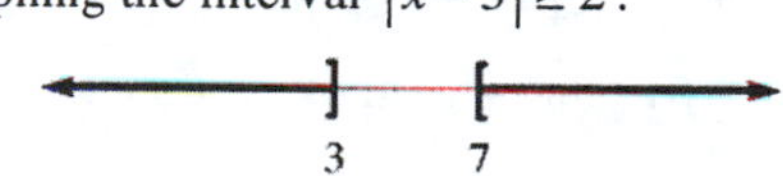

61. **a.** Graphing the interval $|x-2|<1$:

b. Graphing the interval $0<|x-2|<1$, noting that $x=2$ is excluded:

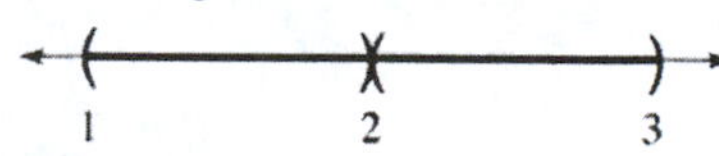

c. The interval in part **b** does not include 2.

63. Using the triangle inequality twice: $|a+b+c|=|a+(b+c)|\leq|a|+|b+c|\leq|a|+|b|+|c|$

65. Consider three cases: $a = b$, $a > b$, and $a < b$.

case 1: If $a = b$, then max $(a, b) = a$, now verifying: $\frac{a+b+|a-b|}{2} = \frac{a+a+|a-a|}{2} = \frac{2a}{2} = a$

Thus the equation is verified.

case 2: If $a > b$, then max $(a, b) = a$, and since $a - b > 0$, we have: $\frac{a+b+|a-b|}{2} = \frac{a+b+a-b}{2} = \frac{2a}{2} = a$

Thus the equation is verified.

case 3: If $a < b$, then max $(a, b) = b$, and since $a - b < 0$, we have:

$$\frac{a+b+|a-b|}{2} = \frac{a+b+-(a-b)}{2} = \frac{a+b-a+b}{2} = \frac{2b}{2} = b$$

Thus the equation is verified.

67. **a.** Property 1(b). **b.** $a+b \le |a|+|b|$

c. Since $(-a)+(-b) \le |a|+|b|, -(a+b) \le |a|+|b|$.

d. Since $a+b \le |a|+|b|$ and $-(a+b) \le |a|+|b|$, $|a+b| \le |a|+|b|$, since $|a+b|$ is either $a + b$ or $-(a + b)$.

1.3 Solving Equations (Review and Preview)

1. Substituting $x = -2$ into the equation: $4x - 5 = 4(-2) - 5 = -8 - 5 = -13$

So $x = -2$ is a solution to the equation.

3. Substituting $y = -3$ into each side of the equation:

$$\frac{2}{y-1} - \frac{3}{y} = \frac{2}{-3-1} - \frac{3}{-3} = -\frac{1}{2} + 1 = \frac{1}{2}$$

$$\frac{7}{y^2 - y} = \frac{7}{(-3)^2 - (-3)} = \frac{7}{9+3} = \frac{7}{12}$$

So $y = -3$ is not a solution to the equation.

5. Substituting $m = \frac{1}{4}$ into the equation: $m^2 + m - \frac{5}{16} = \left(\frac{1}{4}\right)^2 + \left(\frac{1}{4}\right) - \left(\frac{5}{16}\right) = \frac{1}{16} + \frac{1}{4} - \frac{5}{16} = \frac{5}{16} - \frac{5}{16} = 0$

So $m = \frac{1}{4}$ is a solution to the equation.

7. Solving for x:

$$\begin{aligned} 2x - 3 &= -5 \\ 2x &= -2 \\ x &= -1 \end{aligned}$$

9. Solving for m:

$$\begin{aligned} 1-(2m+5) &= -3m \\ 1-2m-5 &= -3m \\ m &= 4 \end{aligned}$$

11. Solving for t:

$$\begin{aligned} t-\{4-[t-(4+t)]\} &= 6 \\ t-\{4-[t-4-t]\} &= 6 \\ t-[4-(-4)] &= 6 \\ t-8 &= 6 \\ t &= 14 \end{aligned}$$

13. Multiplying by 3 and then solving for x:

$$\begin{aligned} 1-\frac{y}{3} &= 6 \\ 3(1)-3\left(\frac{y}{3}\right) &= 3(6) \\ 3-y &= 18 \\ -y &= 15 \\ y &= -15 \end{aligned}$$

15. Multiplying by x and then solving for x:

$$\begin{aligned} \frac{1}{x} &= \frac{4}{x} - 1 \\ 1 &= 4 - x \\ x &= 3 \end{aligned}$$

Check: Replacing x by 3 in the original equation yields:

$\frac{1}{3} = \frac{4}{3} - 1$

$\frac{1}{3} = \frac{4}{3} - \frac{3}{3}$, which is true

17. Multiplying by $x^2-9=(x+3)(x-3)$ and then solving for x:

$$\begin{aligned}\frac{1}{x-3}-\frac{2}{x+3}&=\frac{1}{x^2-9}\\ \frac{(x-3)(x+3)}{x-3}-\frac{2(x-3)(x+3)}{x+3}&=\frac{(x-3)(x+3)}{(x-3)(x+3)}\\ x+3-2(x-3)&=1\\ x+3-2x+6&=1\\ -x&=-8\\ x&=8\end{aligned}$$

Check: Replacing x by 8 in the original equation yields:

$$\begin{aligned}&\tfrac{1}{5}-\tfrac{2}{11}=\tfrac{1}{55}\\ &\tfrac{11}{55}-\tfrac{10}{55}=\tfrac{1}{55}, \text{ which is true}\end{aligned}$$

19. Multiplying by $x^2-4=(x+2)(x-2)$ and then solving for x:

$$\begin{aligned}\frac{4}{x+2}+\frac{1}{x-2}&=\frac{4}{x^2-4}\\ \frac{4(x-2)(x+2)}{x+2}+\frac{(x-2)(x+2)}{x-2}&=\frac{4(x-2)(x+2)}{(x-2)(x+2)}\\ 4(x-2)+(x+2)&=4\\ 4x-8+x+2&=4\\ 5x&=10\\ x&=2\end{aligned}$$

This shows that if the original equation has a solution, it must be $x=2$. However, the value $x=2$ does not check in the original equation (it produces zeros in two of the denominators). Consequently the original equation has no solutions.

21. Multiplying by $2x^2-5x-12=(x-4)(2x+3)$ and then solving for x:

$$\begin{aligned}\frac{5}{x-4}-\frac{3}{2x^2-5x-12}&=\frac{1}{2x+3}\\ \frac{5(x-4)(2x+3)}{x-4}-\frac{3(x-4)(2x+3)}{(x-4)(2x+3)}&=\frac{1(x-4)(2x+3)}{2x+3}\\ 5(2x+3)-3&=1(x-4)\\ 10x+15-3&=x-4\\ 10x+12&=x-4\\ 9x&=-16\\ x&=-\tfrac{16}{9}\end{aligned}$$

Check: Replacing x by $-\frac{16}{9}$ in the original equation yields:

$$\begin{aligned}\frac{5}{-\frac{16}{9}-4}-\frac{3}{2\left(\frac{256}{81}\right)-5\left(-\frac{16}{9}\right)-12}&=\frac{1}{2\left(-\frac{16}{9}\right)+3}\\ \frac{45}{-16-36}-\frac{243}{512+720-972}&=\frac{9}{-32+27}\\ -\tfrac{45}{52}-\tfrac{243}{260}&=-\tfrac{9}{5}\\ -\tfrac{9}{5}&=-\tfrac{9}{5}, \text{ which is true}\end{aligned}$$

23. **a.** Multiplying by $9x(x-2)$ and then solving for x:

$$9x(x-2)\bullet\frac{3}{x-2}=9x(x-2)\bullet\frac{5}{9x}$$
$$27x=5x-10$$
$$22x=-10$$
$$x=-\tfrac{5}{11}$$

Check: Replacing x by $-\frac{5}{11}$ in the original equation yields:

$$\frac{3}{-\frac{5}{11}-2}=\frac{5}{-\frac{45}{11}}$$
$$-\tfrac{11}{9}=-\tfrac{11}{9}\text{, which is true}$$

b. Multiplying by $(x-2)(9x-2)$ and then solving for x:

$$(x-2)(9x-2)\bullet\frac{3}{x-2}=(x-2)(9x-2)\bullet\frac{5}{9x-2}$$
$$27x-6=5x-10$$
$$22x=-4$$
$$x=-\tfrac{2}{11}$$

Check: Replacing x by $-\frac{2}{11}$ in the original equation yields:

$$\frac{3}{-\frac{2}{11}-2}=\frac{5}{-\frac{18}{11}-2}$$
$$-\tfrac{11}{8}=-\tfrac{11}{8}\text{, which is true}$$

c. Multiplying by $(x-2)\left(\frac{5}{3}x-2\right)$ and then solving for x:

$$(x-2)\left(\tfrac{5}{3}x-2\right)\bullet\frac{3}{x-2}=(x-2)\left(\tfrac{5}{3}x-2\right)\bullet\frac{5}{\frac{5}{3}x-2}$$
$$5x-6=5x-10$$
$$-6=-10$$

Since there are no values of x which can make this last statement true, there is no solution.

25. Factoring then solving for x:

$$x^2-5x+6=0$$
$$(x-3)(x-2)=0$$
$$x=3 \text{ or } x=2$$

27. Factoring then solving for t:

$$3t^2-t-4=0$$
$$(3t-4)(t+1)=0$$
$$t=\tfrac{4}{3} \text{ or } t=-1$$

29. Factoring then solving for x:

$$x^2+3x-40=0$$
$$(x+8)(x-5)=0$$
$$x=-8 \text{ or } x=5$$

31. Factoring then solving for x:

$$x(3x-23)=8$$
$$3x^2-23x-8=0$$
$$(3x+1)(x-8)=0$$
$$x=-\tfrac{1}{3} \text{ or } x=8$$

33. Factoring then solving for x:

$$x^2+\left(2\sqrt{5}\right)x+5=0$$
$$\left(x+\sqrt{5}\right)^2=0$$
$$x+\sqrt{5}=0$$
$$x=-\sqrt{5}\ \text{(double root)}$$

35. Using the quadratic formula with $a = 4$, $b = -3$, and $c = -9$:

$$x = \frac{-(-3) \pm \sqrt{(-3)^2 - 4(4)(-9)}}{2(4)} = \frac{3 \pm \sqrt{153}}{8} = \frac{-3 \pm 3\sqrt{17}}{8} \approx -1.17, 1.92$$

37. The equation is equivalent to $3x^2 + 8x = -2$, or $3x^2 + 8x + 2 = 0$. Using the quadratic formula with $a = 3$, $b = 8$, and $c = 2$: $x = \frac{-8 \pm \sqrt{(8)^2 - 4(3)(2)}}{2(3)} = \frac{-8 \pm \sqrt{40}}{6} = \frac{-8 \pm 2\sqrt{10}}{6} = \frac{-4 \pm \sqrt{10}}{3} \approx -2.39, -0.28$

39. The equation is equivalent to $\sqrt{3}\,x^2 - 6x + \sqrt{3} = 0$. Using the quadratic formula with $a = \sqrt{3}$, $b = -6$, and $c = \sqrt{3}$:

$$x = \frac{-(-6) \pm \sqrt{(-6)^2 - 4\left(\sqrt{3}\right)\left(\sqrt{3}\right)}}{2\left(\sqrt{3}\right)} = \frac{6 \pm \sqrt{24}}{2\sqrt{3}} = \frac{6 \pm 2\sqrt{6}}{2\sqrt{3}} = \frac{3 \pm \sqrt{6}}{\sqrt{3}} = \frac{3\sqrt{3} \pm 3\sqrt{2}}{3} = \sqrt{3} \pm \sqrt{2} \approx 0.32, 3.15$$

41. The equation is equivalent to $24x^2 + 23x + 5 = 0$. Using the quadratic formula with $a = 24$, $b = 23$, and $c = 5$:

$$x = \frac{-23 \pm \sqrt{(23)^2 - 4(24)(5)}}{2(24)} = \frac{-23 \pm \sqrt{49}}{48} = \frac{-23 \pm 7}{48} = -\tfrac{5}{8}, -\tfrac{1}{3}$$

43. Taking square roots yields:

$$\begin{aligned} 2y^2 - 50 &= 0 \\ 2y^2 &= 50 \\ y^2 &= 25 \\ y &= \pm\sqrt{25} \\ y &= \pm 5 \end{aligned}$$

45. Taking square roots yields:

$$\begin{aligned} x^2 - \sqrt{5} &= 0 \\ x^2 &= \sqrt{5} \\ x &= \pm\sqrt{\sqrt{5}} \\ x &= \pm\left(5^{1/2}\right)^{1/2} \\ x &= \pm 5^{1/4} \\ x &= \pm\sqrt[4]{5} \end{aligned}$$

47. **a.** Applying the quadratic formula: $x = \frac{-156 \pm \sqrt{156^2 - 4(1)(5963)}}{2(1)} = \frac{-156 \pm \sqrt{484}}{2} = \frac{-156 \pm 22}{2} = -89, -67$

b. Write the equation as $144y^2 - 54y - 13 = 0$, now apply the quadratic formula:

$$y = \frac{-(-54) \pm \sqrt{(-54)^2 - 4(144)(-13)}}{2(144)} = \frac{54 \pm \sqrt{10404}}{288} = \frac{54 \pm 102}{288} = -\tfrac{1}{6}, \tfrac{13}{24}$$

49. Solving for x:

$$\begin{aligned} 3ax - 2b &= b + 3 \\ 3ax &= 3b + 3 \\ ax &= b + 1 \\ x &= \frac{b+1}{a} \end{aligned}$$

51. Solving for x:

$$\begin{aligned} ax + b &= bx + a \\ ax - bx &= a - b \\ x(a - b) &= a - b \\ x &= \frac{a-b}{a-b} \\ x &= 1 \end{aligned}$$

Note that the last step required $a \neq b$.

53. Multiplying by x and solving for x yields:

$$\begin{aligned} \frac{1}{x} &= a + b \\ 1 &= x(a + b) \\ x &= \frac{1}{a+b} \end{aligned}$$

55. Multiplying by abx and solving for x yields:

$$\begin{aligned} \frac{1}{a} - \frac{1}{x} &= \frac{1}{x} - \frac{1}{b} \\ bx - ab &= ab - ax \\ bx + ax &= ab + ab \\ x(b + a) &= 2ab \\ x &= \frac{2ab}{b+a} \end{aligned}$$

57. Solving for x by factoring:

$$(ax+b)^2-(bx+a)^2=0$$
$$[ax+b-(bx+a)][(ax+b)+(bx+a)]=0$$
$$(ax-bx+b-a)(ax+bx+b+a)=0$$

Setting each factor equal to 0, we have:

$$x(a-b)=a-b \quad \text{or} \quad x(a+b)=-(a+b)$$
$$x=\frac{a-b}{a-b}=1 \quad \text{or} \quad x=\frac{-(a+b)}{a+b}=-1$$
$$x=1 \quad \text{or} \quad x=-1$$

59. Solving for x:

$$a^2(a-x)=b^2(b+x)-2abx$$
$$a^3-a^2x=b^3+b^2x-2abx$$
$$a^3-b^3=a^2x+b^2x-2abx$$
$$a^3-b^3=x\left(a^2+b^2-2ab\right)$$
$$a^3-b^3=x(a-b)^2$$
$$x=\frac{(a-b)\left(a^2+ab+b^2\right)}{(a-b)^2}$$
$$x=\frac{a^2+ab+b^2}{a-b}$$

61. Multiplying by $(a-b)(b-c)$ and solving for x:

$$\frac{a-x}{a-b}-2=\frac{c-x}{b-c}$$
$$(a-x)(b-c)-2(a-b)(b-c)=(c-x)(a-b)$$
$$ab-bx-ac+cx-2ab+2b^2+2ac-2bc=ac-ax-bc+bx$$
$$-ab-bx+ac+cx+2b^2-2bc=ac-ax-bc+bx$$
$$ax-2bx+cx=ab-2b^2+bc$$
$$x(a-2b+c)=b(a-2b+c)$$
$$x=b \qquad \text{(assuming } a-2b+c\neq 0\text{)}$$

63. Multiplying by $(x-a)(x-b)$ and solving for x:

$$\frac{x-a}{x-b}=\frac{b-x}{a-x}=\frac{x-b}{x-a}$$
$$(x-a)^2=(x-b)^2$$
$$x^2-2ax+a^2=x^2-2bx+b^2$$
$$-2ax+a^2=-2bx+b^2$$
$$2bx-2ax=b^2-a^2$$
$$2x(b-a)=b^2-a^2$$
$$x=\frac{b^2-a^2}{2(b-a)}$$
$$x=\frac{(b-a)(b+a)}{2(b-a)}$$
$$x=\frac{a+b}{2} \qquad \text{(assuming } a\neq b\text{)}$$

65. Solving for h:

$$S=2\pi r^2+2\pi rh$$
$$S-2\pi r^2=2\pi rh$$
$$h=\frac{S-2\pi r^2}{2\pi r}$$

67. Multiplying by $1+rt$ and solving for r:

$$\begin{aligned}
d &= \frac{r}{1+rt} \\
d(1+rt) &= r \\
d + drt &= r \\
drt - r &= -d \\
r(dt-1) &= -d \\
r &= \frac{-d}{dt-1} \\
r &= \frac{d}{1-dt}
\end{aligned}$$

69. Multiplying through by the least common denominator $x(x+5)$ yields:

$$\begin{aligned}
3x + 4(x+5) &= 2x(x+5) \\
7x + 20 &= 2x^2 + 10x \\
-2x^2 - 3x + 20 &= 0 \\
2x^2 + 3x - 20 &= 0 \\
(2x-5)(x+4) &= 0 \\
x = \tfrac{5}{2} &\text{ or } x = -4
\end{aligned}$$

Both of these values check in the original equation.

71. Multiplying through by the least common denominator $6x + 1$ yields:

$$\begin{aligned}
-6x^2 + 5x - 1 &= 0 \\
6x^2 - 5x + 1 &= 0 \\
(3x-1)(2x-1) &= 0 \\
x = \tfrac{1}{3} &\text{ or } x = \tfrac{1}{2}
\end{aligned}$$

Both values check in the original equation.

73. Multiplying through by the least common denominator $(x+2)(x-2)$ yields:

$$\begin{aligned}
x(x+2) + x(x-2) &= 8 \\
2x^2 - 8 &= 0 \\
x^2 - 4 &= 0 \\
(x-2)(x+2) &= 0 \\
x = 2 &\text{ or } x = -2
\end{aligned}$$

However, neither value satisfies the original equation. Thus the original equation has no solution.

75. **a.** Multiplying each side of the equation by abx:

$$\begin{aligned}
abx\left(\frac{1}{x}\right) &= abx\left(\frac{1}{a} + \frac{1}{b}\right) \\
ab &= bx + ax \\
ab &= x(a+b) \\
x &= \frac{ab}{a+b}
\end{aligned}$$

b. Replacing x by $\frac{ab}{a+b}$ yields:

$$\frac{1}{\frac{ab}{a+b}} = \frac{1}{a} + \frac{1}{b}$$
$$\frac{a+b}{ab} = \frac{1}{a} + \frac{1}{b}$$
$$\frac{a}{ab} + \frac{b}{ab} = \frac{1}{a} + \frac{1}{b}$$
$$\frac{1}{b} + \frac{1}{a} = \frac{1}{a} + \frac{1}{b}, \text{ which is true}$$

1.4 Rectangular Coordinates. Visualizing Data

1. Plotting the points:

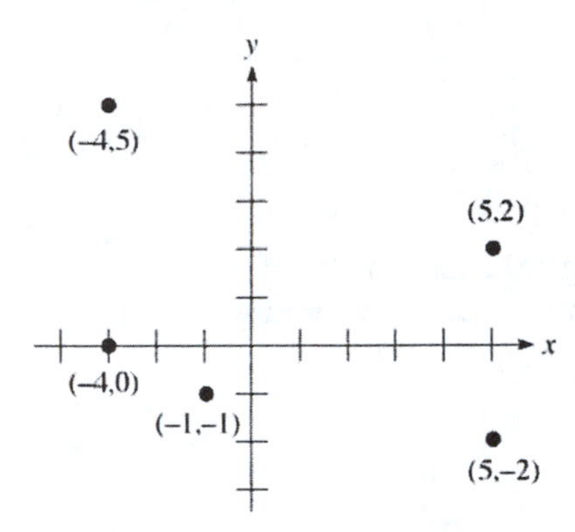

3. **a.** Draw the right triangle PQR:

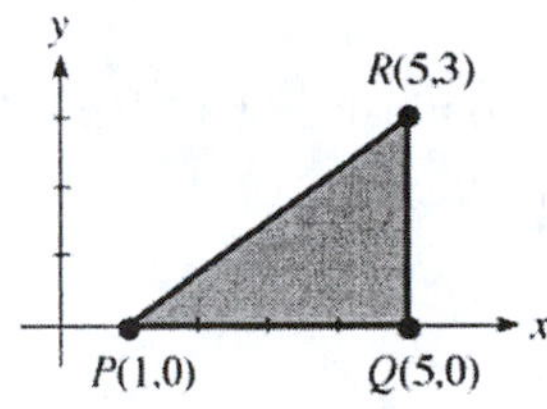

b. Since the base is $b = 5 - 1 = 4$ and the height is $h = 3 - 0 = 3$, the area is given by:

$$A = \tfrac{1}{2}bh = \tfrac{1}{2}(4)(3) = 6 \text{ square units}$$

5. **a.** Here $(x_1, y_1) = (0, 0)$ and $(x_2, y_2) = (-3, 4)$, so by the distance formula:

$$d = \sqrt{(-3-0)^2 + (4-0)^2} = \sqrt{9+16} = \sqrt{25} = 5$$

b. Here $(x_1, y_1) = (2, 1)$ and $(x_2, y_2) = (7, 13)$, so: $d = \sqrt{(7-2)^2 + (13-1)^2} = \sqrt{25+144} = \sqrt{169} = 13$

7. **a.** Here $(x_1, y_1) = (-5, 0)$ and $(x_2, y_2) = (5, 0)$, so: $d = \sqrt{[5-(-5)]^2 + (0-0)^2} = \sqrt{100+0} = \sqrt{100} = 10$

b. Here $(x_1, y_1) = (0, -8)$ and $(x_2, y_2) = (0, 1)$, so: $d = \sqrt{(0-0)^2 + [1-(-8)]^2} = \sqrt{0+81} = \sqrt{81} = 9$

9. Here $(x_1, y_1) = (1, \sqrt{3})$ and $(x_2, y_2) = (-1, -\sqrt{3})$, so:

$$d = \sqrt{(-1-1)^2 + \left(-\sqrt{3}-\sqrt{3}\right)^2} = \sqrt{(-2)^2 + \left(-2\sqrt{3}\right)^2} = \sqrt{4+12} = 4$$

11. **a.** Calculate the distance of each point from the origin:

$$(3,-2): \quad d=\sqrt{(3-0)^2+(-2-0)^2}=\sqrt{9+4}=\sqrt{13}$$

$$(4,\tfrac{1}{2}): \quad d=\sqrt{(4-0)^2+\left(\tfrac{1}{2}-0\right)^2}=\sqrt{16+\tfrac{1}{4}}=\sqrt{16.25}$$

So $\left(4,\frac{1}{2}\right)$ is farther from the origin.

b. Calculate the distance of each point from the origin:

$$(-6,7): \quad d=\sqrt{(-6-0)^2+(7-0)^2}=\sqrt{36+49}=\sqrt{85}$$

$$(9,0): \quad d=\sqrt{(9-0)^2+(0-0)^2}=\sqrt{81+0}=\sqrt{81}$$

So $(-6,7)$ is farther from the origin.

13. We will determine (using the converse of the Pythagorean theorem) whether $a^2+b^2=c^2$.

a. Calculate the distances:

$$a=\sqrt{(-3-7)^2+[5-(-1)]^2}=\sqrt{100+36}=\sqrt{136}$$

$$b=\sqrt{[-12-(-3)]^2+(-10-5)^2}=\sqrt{81+225}=\sqrt{306}$$

$$c=\sqrt{(-12-7)^2+[-10-(-1)]^2}=\sqrt{361+81}=\sqrt{442}$$

Now check the converse of the Pythagorean theorem: $a^2+b^2=136+306=442=c^2$
So the triangle is a right triangle.

b. Calculate the distances:

$$a=\sqrt{(-3-1)^2+(9-3)^2}=\sqrt{16+36}=\sqrt{52}$$

$$b=\sqrt{(4-1)^2+(5-3)^2}=\sqrt{9+4}=\sqrt{13}$$

$$c=\sqrt{(-3-4)^2+(9-5)^2}=\sqrt{49+16}=\sqrt{65}$$

Now check the converse of the Pythagorean theorem: $a^2+b^2=52+13=65=c^2$
So the triangle is a right triangle.

c. Calculate the distances:

$$a=\sqrt{(-8-1)^2+[-2-(-1)]^2}=\sqrt{81+1}=\sqrt{82}$$

$$b=\sqrt{(10-1)^2+[19-(-1)]^2}=\sqrt{81+400}=\sqrt{481}$$

$$c=\sqrt{[10-(-8)]^2+[19-(-2)]^2}=\sqrt{324+441}=\sqrt{765}$$

Now check the converse of the Pythagorean theorem: $a^2+b^2=82+481=563\neq 765=c^2$
So the triangle is not a right triangle.

15. Let $(x_1,y_1)=(1,-4)$, $(x_2,y_2)=(5,3)$, and $(x_3,y_3)=(13,17)$, so using the formula from Exercise 14(b) we have:

$$\text{Area}=\tfrac{1}{2}\left|1(3)-5(-4)+5(17)-13(3)+13(-4)-1(17)\right|=\tfrac{1}{2}\left|3+20+85-39-52-17\right|=\tfrac{1}{2}\left|0\right|=0$$

For the area of the triangle to be 0 it must be that these three points do not form a triangle. The only way that could occur is if the three points are collinear; that is, they all lie on the same line.

17. **a.** Here $(x_1, y_1) = (3,2)$ and $(x_2, y_2) = (9,8)$, so by the midpoint formula: $M = \left(\frac{3+9}{2}, \frac{2+8}{2}\right) = \left(\frac{12}{2}, \frac{10}{2}\right) = (6,5)$

b. Here $(x_1, y_1) = (-4,0)$ and $(x_2, y_2) = (5,-3)$, so by the midpoint formula: $M = \left(\frac{-4+5}{2}, \frac{0-3}{2}\right) = \left(\frac{1}{2}, -\frac{3}{2}\right)$

c. Here $(x_1, y_1) = (3,-6)$ and $(x_2, y_2) = (-1,-2)$, so by the midpoint formula:

$$M = \left(\frac{3-1}{2}, \frac{-6-2}{2}\right) = \left(\frac{2}{2}, -\frac{8}{2}\right) = (1,-4)$$

19. **a.** Since $\overline{PQ}$ is the diameter of the circle, its midpoint must be the center of the circle. Using the midpoint formula:

$$\text{center} = \left(\frac{-4+6}{2}, \frac{-2+4}{2}\right) = \left(\frac{2}{2}, \frac{2}{2}\right) = (1,1)$$

b. The radius of the circle is the distance from this center $(1,1)$ to point Q (or point P), so using the distance formula:

$$r = \sqrt{(6-1)^2 + (4-1)^2} = \sqrt{25+9} = \sqrt{34}$$

21. **a.** Plotting the data:

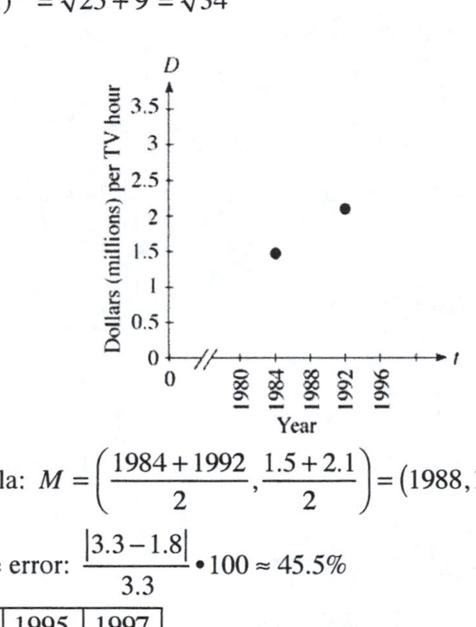

b. Using the midpoint formula: $M = \left(\frac{1984+1992}{2}, \frac{1.5+2.1}{2}\right) = (1988, 1.8)$. The estimate is \$1.8 million.

c. Computing the percentage error: $\frac{|3.3-1.8|}{3.3} \bullet 100 \approx 45.5\%$

23. **a.** Completing the table:

t	1995	1997
n	14	30

Using the midpoint formula: $M = \left(\frac{1995+1997}{2}, \frac{14+30}{2}\right) = (1996, 22)$

For 1996, the number of global Internet host computers was 22 million.

b. Completing the table:

t	1985	1987
n	2	28

Using the midpoint formula: $M = \left(\frac{1985+1987}{2}, \frac{2+28}{2}\right) = (1986, 15)$

For 1986, the number of global Internet host computers was 15 thousand.

c. For 1996, the percentage error is: $\frac{|21.819-22|}{21.819} \bullet 100 \approx 0.83\%$

For 1986, the percentage error is: $\frac{|5.089-15|}{5.089} \bullet 100 \approx 194.75\%$

25. **a.** Sketch the parallelogram $ABCD$:

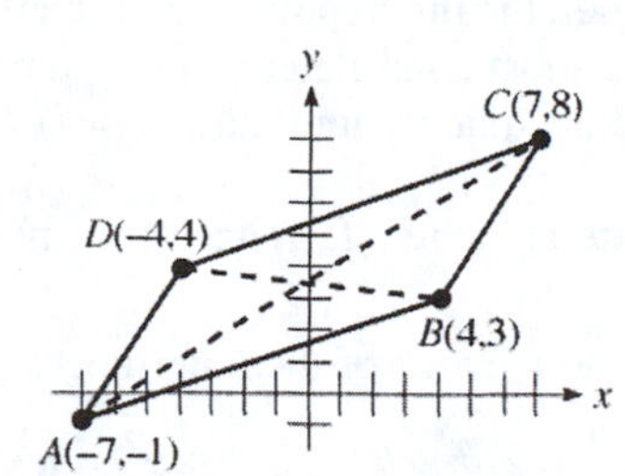

b. For AC, let $(x_1, y_1) = (-7, -1)$ and $(x_2, y_2) = (7, 8)$, so the midpoint is: $M_1 = \left(\frac{-7+7}{2}, \frac{-1+8}{2}\right) = \left(0, \frac{7}{2}\right)$

For BD, let $(x_1, y_1) = (4, 3)$ and $(x_2, y_2) = (-4, 4)$, so the midpoint is: $M_2 = \left(\frac{4-4}{2}, \frac{3+4}{2}\right) = \left(0, \frac{7}{2}\right)$

c. It appears that the midpoints of the two diagonals of a parallelogram are the same. Stated more concisely, the diagonals of a parallelogram bisect each other.

27. Using the Pythagorean theorem:

$a^2 = 1^2 + 1^2 = 2$, so $a = \sqrt{2}$ $\qquad$ $b^2 = 1^2 + \left(\sqrt{2}\right)^2 = 1 + 2 = 3$, so $b = \sqrt{3}$

$c^2 = 1^2 + \left(\sqrt{3}\right)^2 = 1 + 3 = 4$, so $c = 2$ $\qquad$ $d^2 = 1^2 + 2^2 = 1 + 4 = 5$, so $d = \sqrt{5}$

$e^2 = 1^2 + \left(\sqrt{5}\right)^2 = 1 + 5 = 6$, so $e = \sqrt{6}$ $\qquad$ $f^2 = 1^2 + \left(\sqrt{6}\right)^2 = 1 + 6 = 7$, so $f = \sqrt{7}$

$g^2 = 1^2 + \left(\sqrt{7}\right)^2 = 1 + 7 = 8$, so $g = \sqrt{8} = 2\sqrt{2}$

29. **a.** Call (x, y) the coordinates of B. Clearly $y = c$, since the top and bottom of the parallelogram must be parallel, and thus since the bottom is horizontal the top must be also. Now compute the lengths OC and OB using the distance formula:

$$OC = \sqrt{(b-0)^2 + (c-0)^2} = \sqrt{b^2 + c^2} \qquad AB = \sqrt{(x-a)^2 + (c-0)^2} = \sqrt{(x-a)^2 + c^2}$$

Since $OC = AB$, we have:

$$\sqrt{b^2 + c^2} = \sqrt{(x-a)^2 + c^2}$$
$$b^2 + c^2 = (x-a)^2 + c^2$$
$$b^2 = (x-a)^2$$

Taking roots, we have $x - a = \pm b$, or $x = a \pm b$. But clearly $x = a - b$ doesn't make sense, as $x \ge a$ from the figure, so $x = a + b$.

b. Find the midpoints of the two diagonals $\overline{OB}$ and $\overline{AC}$:

For $O(0,0)$ and $B(a+b, c)$, $M_{\overline{OB}} = \left(\frac{a+b}{2}, \frac{c}{2}\right)$ $\qquad$ For $A(a,c)$ and $C(b,c)$, $M_{\overline{AC}} = \left(\frac{a+b}{2}, \frac{c}{2}\right)$

c. Clearly our two midpoints from part **b** are equal. Since the two midpoints on the diagonals are equal they must bisect each other.

31. **a.** Using the distance formula:

$$PM = \sqrt{\left(\frac{x_1 + x_2}{2} - x_1\right)^2 + \left(\frac{y_1 + y_2}{2} - y_1\right)^2} = \sqrt{\left(\frac{x_2 - x_1}{2}\right)^2 + \left(\frac{y_2 - y_1}{2}\right)^2} = \frac{\sqrt{\left(x_2 - x_1\right)^2 + \left(y_2 - y_1\right)^2}}{2}$$

$$MQ = \sqrt{\left(x_2 - \frac{x_1 + x_2}{2}\right)^2 + \left(y_2 - \frac{y_1 + y_2}{2}\right)^2} = \sqrt{\left(\frac{x_2 - x_1}{2}\right)^2 + \left(\frac{y_2 - y_1}{2}\right)^2} = \frac{\sqrt{\left(x_2 - x_1\right)^2 + \left(y_2 - y_1\right)^2}}{2}$$

Thus $PM = MQ$.

b. Adding the two distances together: $PM + MQ = 2(PM) = \sqrt{\left(x_2 - x_1\right)^2 + \left(y_2 - y_1\right)^2} = PQ$

33. **a.** Each side of the outermost quadrilateral is the hypotenuse of the triangle in Figure A, so each side has a length of c. Also, since each angle of the outermost quadrilateral is the sum of the two acute angles from the triangle, the measure of each angle is 90°. Since the quadrilateral has angles of 90° and sides of equal length, it must be a square.

b. Each side of the innermost quadrilateral is the difference of the two legs of the triangle in Figure A, so the length is $b - a$.

c. The area of each triangle is $\frac{1}{2}ab$, and the area of the innermost square is $(b-a)^2$. Since there are four triangles, the figure area is: $4\left(\frac{1}{2}ab\right)+(b-a)^2 = 2ab+b^2-2ab+a^2 = a^2+b^2$

Since the figure has an area of c^2, we have the result $a^2+b^2=c^2$.

1.5 Graphs and Graphing Utilities

1. Substituting $x = 8$: $y=\frac{1}{2}(8)+3=4+3=7\neq 6$. The point (8, 6) does not lie on the graph.

3. Substituting $(x, y) = (4, 3)$: $3(4)^2+(3)^2=3(16)+9=48+9=57\neq 52$. The point (4, 3) does not lie on the graph.

5. Substituting $x = a$: $y = 4a$. The point $(a, 4a)$ lies on the graph.

7. **a.** Solving for y:

$$2x-3y=-3$$
$$-3y=-2x-3$$
$$y=\frac{2}{3}x+1$$

Completing the table:

x	–6	–3	0	3	6
y	–3	–1	1	3	5

b. Graphing the equation $2x-3y=-3$:

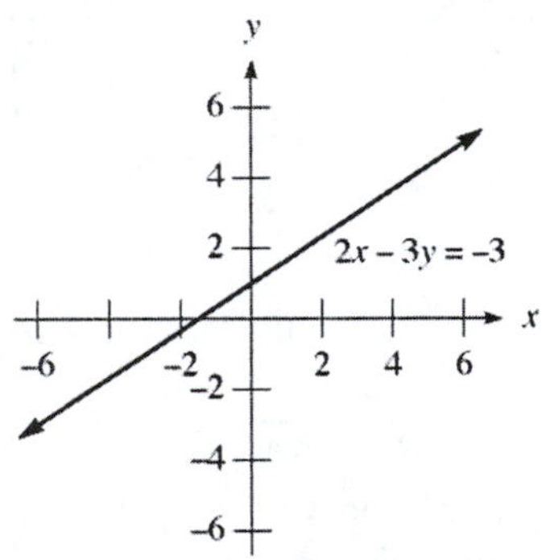

9. To find the x-intercept, substitute $y = 0$ into the equation $3x + 4y = 12$:

$$3x+4(0)=12$$
$$3x=12$$
$$x=4$$

To find the y-intercept, substitute $x = 0$ into the equation $3x + 4y = 12$:

$$3(0)+4y=12$$
$$4y=12$$
$$y=3$$

Now graph the line $3x+4y=12$:

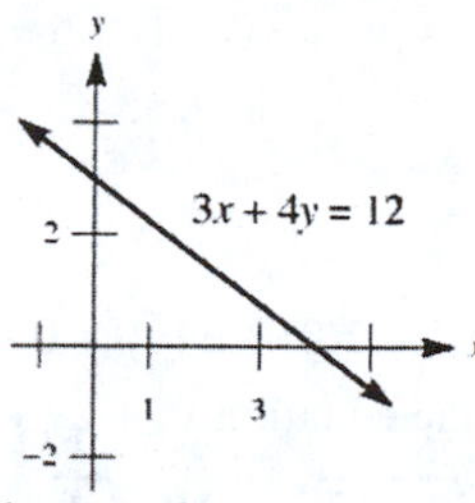

11. To find the x-intercept, substitute $y = 0$ into the equation $y = 2x - 4$:

$$0 = 2x - 4$$
$$-2x = -4$$
$$x = 2$$

To find the y-intercept, substitute $x = 0$ into the equation $y = 2x - 4$:

$$y = 2(0) - 4$$
$$y = -4$$

Now graph the line $y = 2x - 4$:

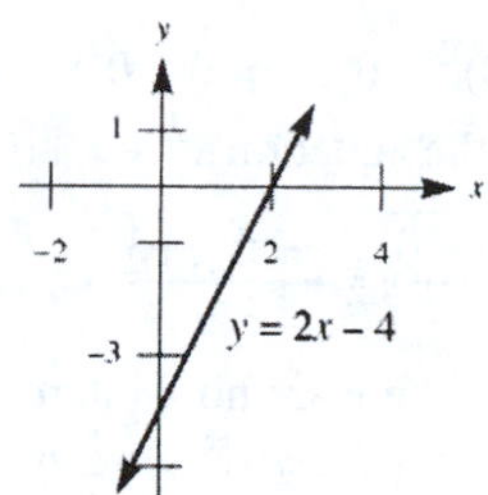

13. To find the x-intercept, substitute $y = 0$ into the equation $x + y = 1$:

$$x + 0 = 1$$
$$x = 1$$

To find the y-intercept, substitute $x = 0$ into the equation $x + y = 1$:

$$0 + y = 1$$
$$y = 1$$

Now graph the line $x + y = 1$:

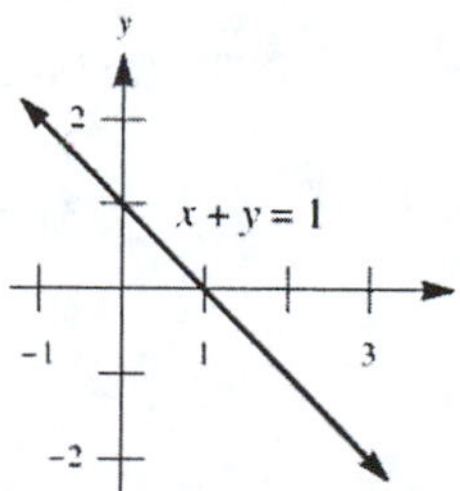

15. **a.** The viewing rectangle can be represented as Xmin = –3, Xmax = 3, Xscl = 1 and Ymin = –200, Ymax = 200, Yscl = 100. The bracket notation describes the rectangle as [–3,3,1] by [–200,200,100].

b. The viewing rectangle can be represented as Xmin = –1, Xmax = 1.25, Xscl = 0.25 and Ymin = –0.2, Ymax = 0.1, Yscl = 0.1. The bracket notation describes the rectangle as [–1,1.25,0.25] by [–0.2,0.1,0.1].

17. **a.** To find the y-intercepts, let $x = 0$: $y = (0)^2 + 3(0) + 2 = 0 + 0 + 2 = 2$
To find the x-intercepts, let $y = 0$:

$$\begin{aligned} x^2 + 3x + 2 &= 0 \\ (x+2)(x+1) &= 0 \\ x &= -1, -2 \end{aligned}$$

b. To find the y-intercepts, let $x = 0$: $y = (0)^2 + 2(0) + 3 = 0 + 0 + 3 = 3$
To find the x-intercepts, we must solve the equation $x^2 + 2x + 3 = 0$. Using the quadratic formula:

$$x = \frac{-2 \pm \sqrt{(2)^2 - 4(1)(3)}}{2(1)} = \frac{-2 \pm \sqrt{4-12}}{2} = \frac{-2 \pm \sqrt{-8}}{2}$$

Since this equation has no real solutions, there are no x-intercepts.

19. **a.** To find the y-intercepts, let $x = 0$: $y = (0)^2 + 0 - 1 = 0 + 0 - 1 = -1$
To find the x-intercepts, we must solve the equation $x^2 + x - 1 = 0$. Using the quadratic formula:

$$x = \frac{-1 \pm \sqrt{(1)^2 - 4(1)(-1)}}{2(1)} = \frac{-1 \pm \sqrt{1+4}}{2} = \frac{-1 \pm \sqrt{5}}{2}$$

b. To find the y-intercepts, let $x = 0$: $y = (0)^2 + 0 + 1 = 0 + 0 + 1 = 1$
To find the x-intercepts, we must solve the equation $x^2 + x + 1 = 0$. Using the quadratic formula:

$$x = \frac{-1 \pm \sqrt{(1)^2 - 4(1)(1)}}{2(1)} = \frac{-1 \pm \sqrt{1-4}}{2} = \frac{-1 \pm \sqrt{-3}}{2}$$

Since this equation has no real solutions, there are no x-intercepts.

21. To find the y-intercepts, let $x = 0$: $y = 11(0) - 2(0)^2 - (0)^3 = 0 - 0 - 0 = 0$
To find the x-intercepts, let $y = 0$:

$$\begin{aligned} 0 &= 11x - 2x^2 - x^3 \\ 0 &= x\left(11 - 2x - x^2\right) \end{aligned}$$

So $x = 0$ is clearly one x-intercept. Find the other two using the quadratic formula:

$$x = \frac{2 \pm \sqrt{(-2)^2 - 4(11)(-1)}}{2(-1)} = \frac{2 \pm \sqrt{48}}{-2} = -1 \pm 2\sqrt{3} \approx -4.46, 2.46$$

23. To find the y-intercepts, let $x = 0$:

$$\begin{aligned} y^2 - 4y - 8 &= 0 \\ (y-6)(y+2) &= 0 \\ y &= -2, 6 \end{aligned}$$

To find the x-intercepts, let $y = 0$:

$$\begin{aligned} 3x &= (0)^2 - 4(0) - 8 \\ 3x &= -8 \\ x &= -\tfrac{8}{3} \end{aligned}$$

25. **a.** Graphing the equation $y = x^2 - 2x - 2$:

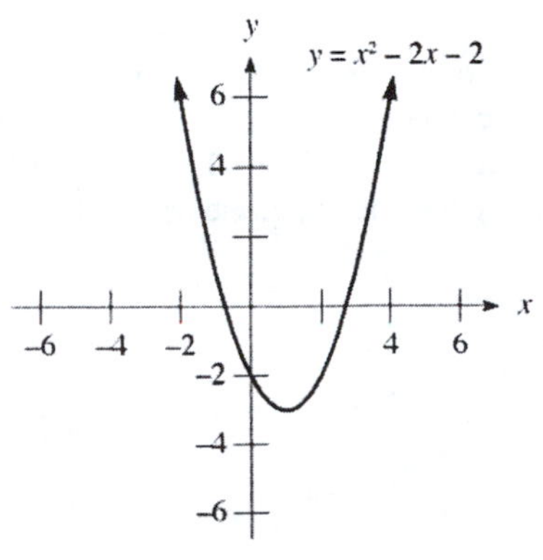

b. The x-intercepts are approximately –0.7 and 2.7.

c. Using the quadratic formula: $x = \dfrac{2 \pm \sqrt{(-2)^2 - 4(1)(-2)}}{2} = \dfrac{2 \pm \sqrt{12}}{2} = \dfrac{2 \pm 2\sqrt{3}}{2} = 1 \pm \sqrt{3} \approx -0.7, 2.7$

27. **a.** Graphing the equation $y = 2x^3 - 5x$:

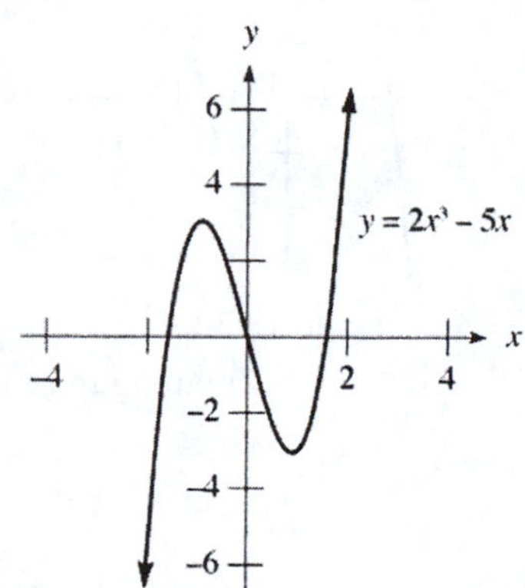

b. The x-intercepts are approximately –1.6, 0, and 1.6.

c. Factoring results in $x(2x^2 - 5) = 0$, so one x-intercept is $x = 0$. Finding the other two intercepts:

$$\begin{aligned} 2x^2 - 5 &= 0 \\ x^2 &= \tfrac{5}{2} \\ x &= \pm\sqrt{\tfrac{5}{2}} \approx \pm 1.6 \end{aligned}$$

29. **a.** Graphing the equation $2xy - x^3 - 5 = 0$:

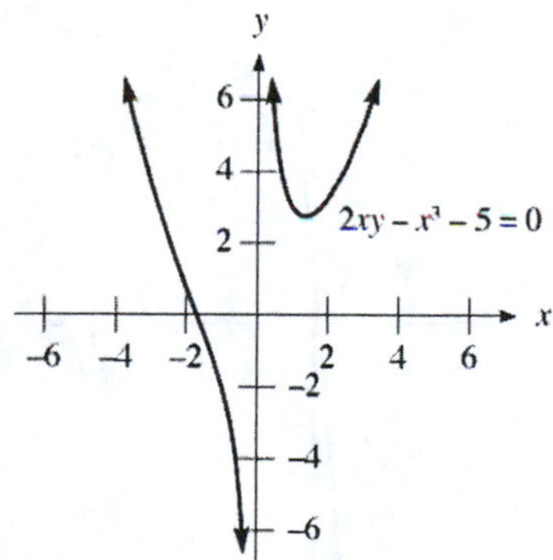

b. The x-intercept is approximately –1.7.

c. Set $y = 0$ to find the x-intercept:

$$\begin{aligned} -x^3 - 5 &= 0 \\ x^3 &= -5 \\ x &= \sqrt[3]{-5} \approx -1.7 \end{aligned}$$

31. Graphing the equation $y = x^3 - 3x + 1$:

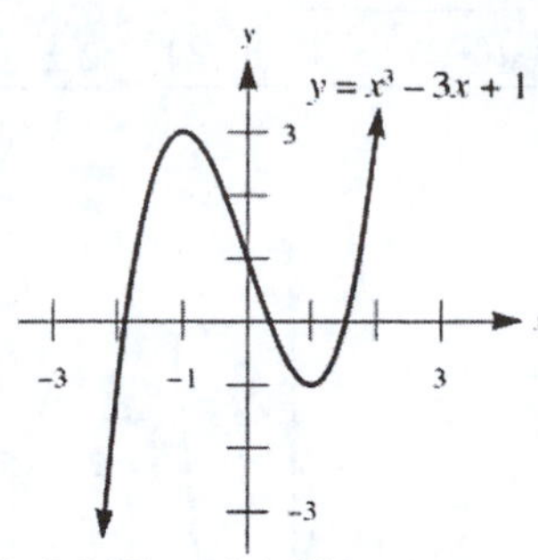

The x-intercepts are approximately –1.879, 0.347, and 1.532.

33. Graphing the equation $y = x^5 - 6x^4 + 3$: Zooming in near the origin:

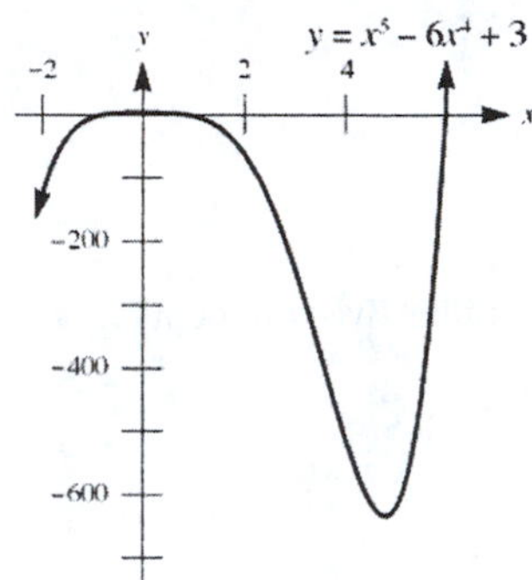

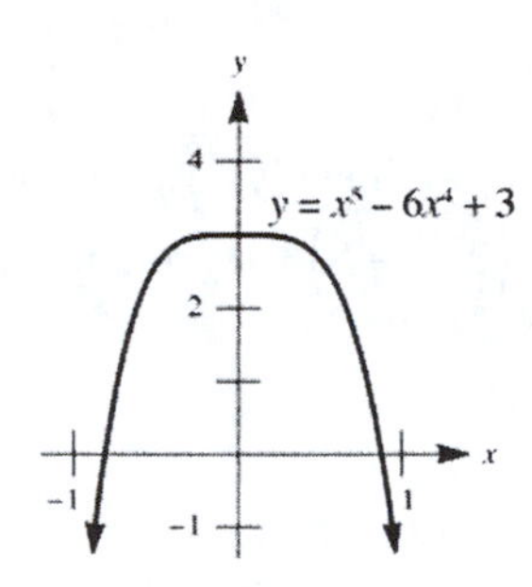

The x-intercepts are approximately –0.815, 0.875, and 5.998.

35. **a.** Graphing in the standard viewing rectangle:

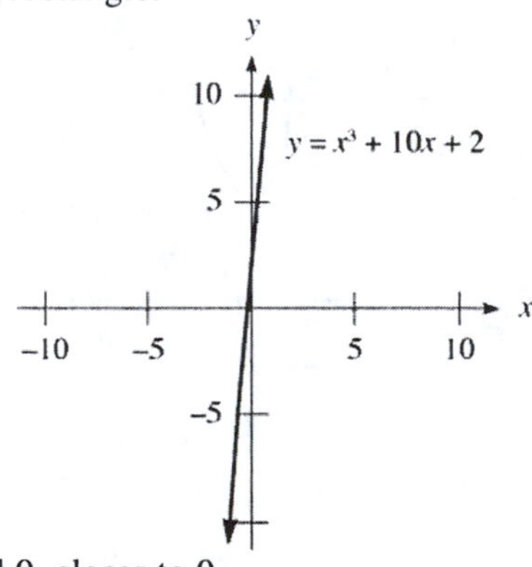

The x-intercept is between –1 and 0, closer to 0.

c. Any viewing rectangle not including –0.2 (such as [0,10]) will not reveal the x-intercept near –0.2.

d. Generally, start with a large viewing rectangle, then zoom in near the x-intercepts to find a more specific viewing rectangle.

37. **a.** The Celsius temperature is approximately –20°. This is the C-intercept of the graph.

b. Substituting $F = 0$ in the equation:

$$\frac{9}{5}C + 32 = 0$$
$$\frac{9}{5}C = -32$$
$$C = -\frac{160}{9} = -17\frac{7}{9}$$

39. **a.** Tracing up to the curve from $x = 2$, $y = \sqrt{2} \approx 1.4$.

b. Tracing up to the curve from $x = 3$, $y = \sqrt{3} \approx 1.7$.

c. Since $\sqrt{ab} = \sqrt{a} \bullet \sqrt{b}$, $\sqrt{6} = \sqrt{2 \bullet 3} = \sqrt{2} \bullet \sqrt{3} \approx (1.4)(1.7) \approx 2.4$.

41. **a.** Tracing up to the curve from $t = 0$, $N = 500$ bacteria.

b. Since $(0, 500)$ lies on the curve, find where $(t, 1000)$ would be on the curve, as the population would now be double. Tracing down from $N = 1000$, $t = 1.5$ hours. So the population will double in 1.5 hours.

c. As in **b**, find where $(t, 2500)$ would be on the curve. Tracing down from $N = 2500$, $t = 3.5$ hours.

d. Between $t = 0$ and $t = 1$, the population has grown from $N = 500$ to $N = 800$, so it has increased at an average rate of 300 bacteria per hour. Between $t = 3$ and $t = 4$, the population has grown from $N = 2000$ to $N = 3000$, so it has increased at an average rate of 1000 bacteria per hour. So the population has increased more rapidly between $t = 3$ and $t = 4$.

43. Since the y-coordinate for point A is $0.4 = \frac{2}{5}$, substitute to find x:

$$\frac{1}{2}x^2 = \frac{2}{5}$$
$$x^2 = \frac{4}{5}$$
$$x = \sqrt{\frac{4}{5}} = \frac{2}{5}\sqrt{5} \approx 0.894$$

Since the y-coordinate for point B is $0.6 = \frac{3}{5}$, substitute to find x:

$$\frac{1}{2}x^2 = \frac{3}{5}$$
$$x^2 = \frac{6}{5}$$
$$x = \sqrt{\frac{6}{5}} = \frac{1}{5}\sqrt{30} \approx 1.095$$

1.6 Equations of Lines

1. **a.** Here $(x_1, y_1) = (-3, 2)$ and $(x_2, y_2) = (1, -6)$, so: slope $= \dfrac{-6-2}{1-(-3)} = -\frac{8}{4} = -2$

b. Here $(x_1, y_1) = (2, -5)$ and $(x_2, y_2) = (4, 1)$, so: slope $= \dfrac{1-(-5)}{4-2} = \frac{6}{2} = 3$

c. Here $(x_1, y_1) = (-2, 7)$ and $(x_2, y_2) = (1, 0)$, so: slope $= \dfrac{0-7}{1-(-2)} = -\frac{7}{3}$

d. Here $(x_1, y_1) = (4, 5)$ and $(x_2, y_2) = (5, 8)$, so: slope $= \dfrac{8-5}{5-4} = \frac{3}{1} = 3$

3. **a.** Here $(x_1, y_1) = (1, 1)$ and $(x_2, y_2) = (-1, -1)$, so: slope $= \dfrac{-1-1}{-1-1} = \dfrac{-2}{-2} = 1$

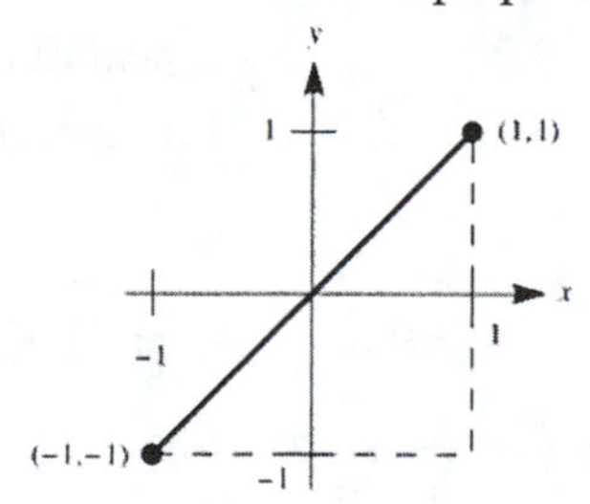

b. Here $(x_1, y_1) = (0, 5)$ and $(x_2, y_2) = (-8, 5)$, so: slope $= \frac{5-5}{-8-0} = \frac{0}{-8} = 0$

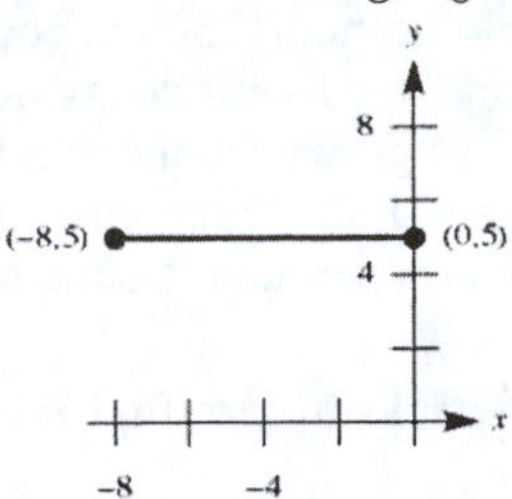

5. a. From $x = 10$ seconds to $x = 15$ seconds:

$\Delta x = x_2 - x_1 = 15 \text{ sec} - 10 \text{ sec} = 5 \text{ sec}$

$\Delta y = y_2 - y_1 = 45 \text{ ft} - 30 \text{ ft} = 15 \text{ ft}$

$\frac{\Delta y}{\Delta x} = \frac{15 \text{ ft}}{5 \text{ sec}} = 3 \text{ ft/sec}$

b. From $x = 10$ seconds to $x = 25$ seconds:

$\Delta x = x_2 - x_1 = 25 \text{ sec} - 10 \text{ sec} = 15 \text{ sec}$

$\Delta y = y_2 - y_1 = 75 \text{ ft} - 30 \text{ ft} = 45 \text{ ft}$

$\frac{\Delta y}{\Delta x} = \frac{45 \text{ ft}}{15 \text{ sec}} = 3 \text{ ft/sec}$

c. From $x = 5$ seconds to $x = 30$ seconds:

$\Delta x = x_2 - x_1 = 30 \text{ sec} - 5 \text{ sec} = 25 \text{ sec}$

$\Delta y = y_2 - y_1 = 90 \text{ ft} - 15 \text{ ft} = 75 \text{ ft}$

$\frac{\Delta y}{\Delta x} = \frac{75 \text{ ft}}{25 \text{ sec}} = 3 \text{ ft/sec}$

7. a. From $x = 1970$ to $x = 1980$:

$\Delta x = x_2 - x_1 = 1980 - 1970 = 10$ years

$\Delta y = y_2 - y_1 = 225{,}000{,}000 - 200{,}000{,}000 = 25{,}000{,}000$ people

b. From $x = 1970$ to $x = 1990$:

$\Delta x = x_2 - x_1 = 1990 - 1970 = 20$ years

$\Delta y = y_2 - y_1 = 250{,}000{,}000 - 200{,}000{,}000 = 50{,}000{,}000$ people

c. From $x = 1980$ to $x = 1990$:

$\Delta x = x_2 - x_1 = 1990 - 1980 = 10$ years

$\Delta y = y_2 - y_1 = 250{,}000{,}000 - 225{,}000{,}000 = 25{,}000{,}000$ people

9. From $t = 1990$ to $t = 1993$:

$\Delta t = t_2 - t_1 = 1993 - 1990 = 3$ years

$\Delta N = N_2 - N_1 = 5.300 - 5.419 = -0.119$ trillion cigarettes

11. m_3 is smallest (since it is negative)

m_4 is next (it is positive but not as steep as m_1)

m_2 is next (it appears to be near zero)

m_1 is largest (it is steeper than m_4)

13. Compute the slopes:

$\text{slope}_{AB} = \frac{\frac{1}{2} - (-2)}{2 - (-8)} = \frac{\frac{5}{2}}{10} = \frac{1}{4}$

$\text{slope}_{BC} = \frac{-1 - \frac{1}{2}}{11 - 2} = \frac{-\frac{3}{2}}{9} = -\frac{1}{6}$

Since these slopes are different the three points cannot be collinear.

15. **a.** Here $(x_1, y_1) = (-2, 1)$ and $m = -5$, so by the point-slope formula:

$$\begin{aligned} y-1 &= -5[x-(-2)] \\ y-1 &= -5(x+2) \\ y-1 &= -5x-10 \\ y &= -5x-9 \end{aligned}$$

b. Here $(x_1, y_1) = \left(-6, -\frac{2}{3}\right)$ and $m = \frac{1}{3}$, so by the point-slope formula:

$$\begin{aligned} y-\left(-\tfrac{2}{3}\right) &= \left(\tfrac{1}{3}\right)[x-(-6)] \\ y+\tfrac{2}{3} &= \left(\tfrac{1}{3}\right)(x+6) \\ y+\tfrac{2}{3} &= \tfrac{1}{3}x+2 \\ y &= \tfrac{1}{3}x+\tfrac{4}{3} \end{aligned}$$

17. **a.** First find the slope: $m = \dfrac{-6-8}{-3-4} = \dfrac{-14}{-7} = 2$

Using $(x_1, y_1) = (4, 8)$ in the point-slope formula:

$$\begin{aligned} y-8 &= 2(x-4) \\ y-8 &= 2x-8 \\ y &= 2x \end{aligned}$$

b. First find the slope: $m = \dfrac{-10-0}{3-(-2)} = \dfrac{-10}{5} = -2$

Using $(x_1, y_1) = (-2, 0)$ in the point-slope formula:

$$\begin{aligned} y-0 &= -2[x-(-2)] \\ y &= -2(x+2) \\ y &= -2x-4 \end{aligned}$$

c. First find the slope: $m = \dfrac{-1-(-2)}{4-(-3)} = \frac{1}{7}$

Using $(x_1, y_1) = (4, -1)$ in the point-slope formula:

$$\begin{aligned} y-(-1) &= \tfrac{1}{7}(x-4) \\ y+1 &= \tfrac{1}{7}x-\tfrac{4}{7} \\ y &= \tfrac{1}{7}x-\tfrac{11}{7} \end{aligned}$$

19. **a.** The vertical line is $x = -3$.

b. The horizontal line is $y = 4$.

21. The y-axis is vertical, so its equation must have the form x = constant. Since $(0, 0)$ is on the y-axis, the equation is $x = 0$.

23. **a.** Using the slope-intercept formula with $m = -4$ and $b = 7$, we have $y = -4x + 7$.

b. Using the slope-intercept formula with $m = 2$ and $b = \frac{3}{2}$, we have $y = 2x + \frac{3}{2}$.

25. **a.** Use the point-slope formula:

$$\begin{aligned} y-(-1) &= 4[x-(-3)] \\ y+1 &= 4(x+3) \\ y+1 &= 4x+12 \\ y &= 4x+11 \end{aligned}$$

Now draw the graph:

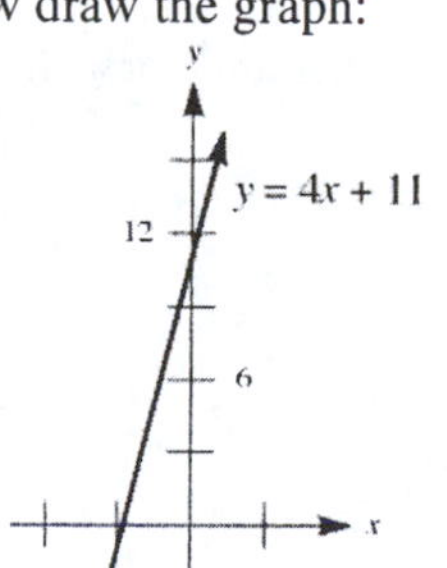

b. Use the point-slope formula:

$$\begin{aligned} y-0 &= \tfrac{1}{2}\left(x-\tfrac{5}{2}\right) \\ y &= \tfrac{1}{2}x-\tfrac{5}{4} \end{aligned}$$

Now draw the graph:

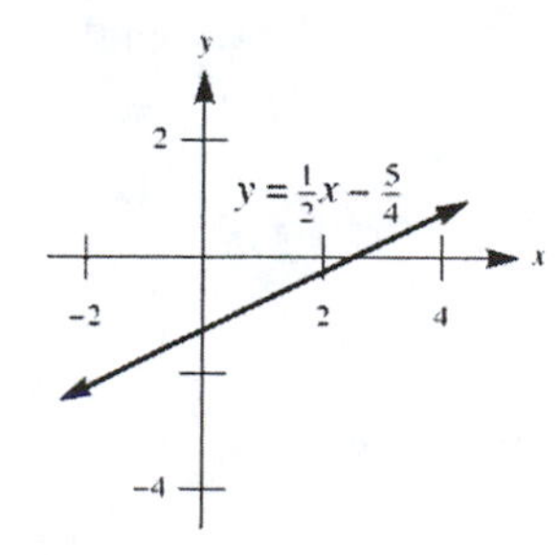

c. Find the slope between the points (6,0) and (0,5): $m=\dfrac{5-0}{0-6}=-\frac{5}{6}$

Since 5 is the y-intercept, by the slope-intercept formula: $y=-\frac{5}{6}x+5$. Now draw the graph:

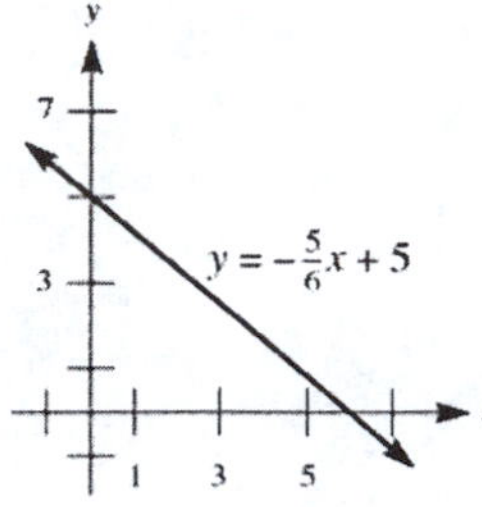

d. Use the point-slope formula with the point (–2,0):

$$y-0=\tfrac{3}{4}[x-(-2)]$$
$$y=\tfrac{3}{4}(x+2)$$
$$y=\tfrac{3}{4}x+\tfrac{3}{2}$$

Now draw the graph:

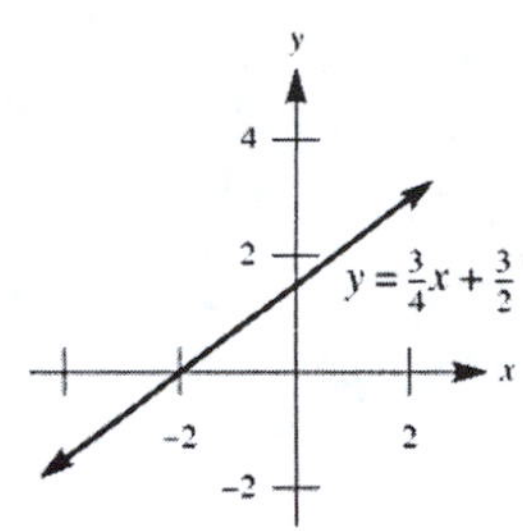

e. First find the slope: $m=\dfrac{6-2}{2-1}=\frac{4}{1}=4$. Using the point (1,2) in the point-slope formula:

$$y-2=4(x-1)$$
$$y-2=4x-4$$
$$y=4x-2$$

Now draw the graph:

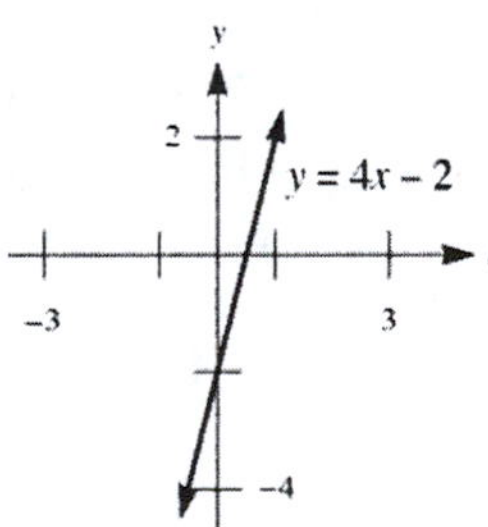

27. If the line is parallel to the x-axis, then its slope is 0. Since this line is of the form y = constant, for (–3,4) to lie on the line the equation is $y=4$, or $y-4=0$ in the desired form. Draw the graph:

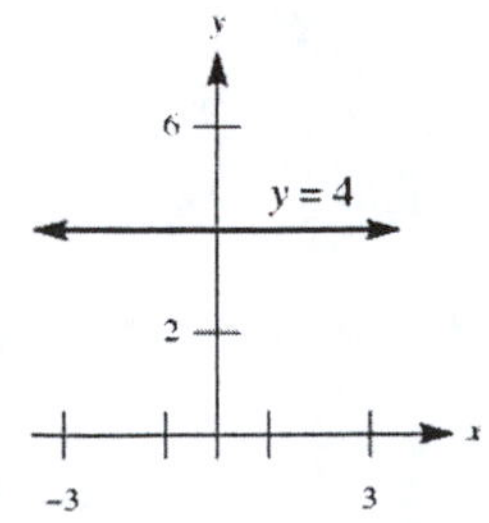

29. **a.** If $x = 0$, then $y = 3$ and if $y = 0$, then $x = 5$. So the x-intercept is 5 and the y-intercept is 3. Thus:

$$\text{area} = \tfrac{1}{2}(5)(3) = \tfrac{15}{2} \qquad \text{perimeter} = 5 + 3 + \sqrt{5^2 + 3^2} = 8 + \sqrt{34}$$

b. If $x = 0$, then $y = -3$ and if $y = 0$, then $x = 5$. So the x-intercept is 5 and the y-intercept is -3. Thus:

$$\text{area} = \tfrac{1}{2}(5)(3) = \tfrac{15}{2} \qquad \text{perimeter} = 5 + 3 + \sqrt{5^2 + 3^2} = 8 + \sqrt{34}$$

31. **a.** Find the slopes of each:

$$\begin{aligned} 3x - 4y &= 12 \\ -4y &= -3x + 12 \\ y &= \tfrac{3}{4}x - 3 \\ m &= \tfrac{3}{4} \end{aligned} \qquad \begin{aligned} 4x - 3y &= 12 \\ -3y &= -4x + 12 \\ y &= \tfrac{4}{3}x - 4 \\ m &= \tfrac{4}{3} \end{aligned}$$

The lines are not parallel (slopes aren't the same), the lines are not perpendicular $\left(\frac{3}{4} \bullet \frac{4}{3} = 1, \text{ not } -1\right)$, so they are neither.

b. Find the slopes of each:

$$\begin{aligned} y &= 5x - 16 \\ m &= 5 \end{aligned} \qquad \begin{aligned} y &= 5x + 2 \\ m &= 5 \end{aligned}$$

Since these slopes are the same, the lines are parallel.

c. Find the slopes of each:

$$\begin{aligned} 5x - 6y &= 25 \\ -6y &= -5x + 25 \\ y &= \tfrac{5}{6}x - \tfrac{25}{6} \\ m &= \tfrac{5}{6} \end{aligned} \qquad \begin{aligned} 6x + 5y &= 0 \\ 5y &= -6x \\ y &= -\tfrac{6}{5}x \\ m &= -\tfrac{6}{5} \end{aligned}$$

Since $\left(\frac{5}{6}\right)\left(-\frac{6}{5}\right) = -1$, the lines are perpendicular.

d. Find the slopes of each:

$$\begin{aligned} y &= -\tfrac{2}{3}x - 1 \\ m &= -\tfrac{2}{3} \end{aligned} \qquad \begin{aligned} y &= \tfrac{3}{2}x - 1 \\ m &= \tfrac{3}{2} \end{aligned}$$

Since $\left(-\frac{2}{3}\right)\left(\frac{3}{2}\right) = -1$, the lines are perpendicular.

33. First find the slope:

$$\begin{aligned} 2x - 5y &= 10 \\ -5y &= -2x + 10 \\ y &= \tfrac{2}{5}x - 2 \end{aligned}$$

So the slope is $\frac{2}{5}$. Using the point $(-1,2)$ in the point-slope formula:

$$\begin{aligned} y - 2 &= \tfrac{2}{5}[x - (-1)] \\ y - 2 &= \tfrac{2}{5}(x + 1) \\ y - 2 &= \tfrac{2}{5}x + \tfrac{2}{5} \\ y &= \tfrac{2}{5}x + \tfrac{12}{5} && [\text{form } y = mx + b] \\ 5y &= 2x + 12 \\ 2x - 5y + 12 &= 0 && [\text{form } Ax + By + C = 0] \end{aligned}$$

35. First find the slope of $4y - 3x = 1$:

$$4y = 3x + 1$$
$$y = \frac{3}{4}x + \frac{1}{4}$$

Since this slope is $\frac{3}{4}$, the perpendicular line slope is $-\frac{4}{3}$. Now use the point (4,0) in the point-slope formula:

$$y - 0 = -\frac{4}{3}(x - 4)$$
$$y = -\frac{4}{3}x + \frac{16}{3} \quad [\text{form } y = mx + b]$$
$$3y = -4x + 16$$
$$4x + 3y - 16 = 0 \quad [\text{form } Ax + By + C = 0]$$

37. The three lines appear parallel as long as the viewing rectangle is square. Sketching the graphs:

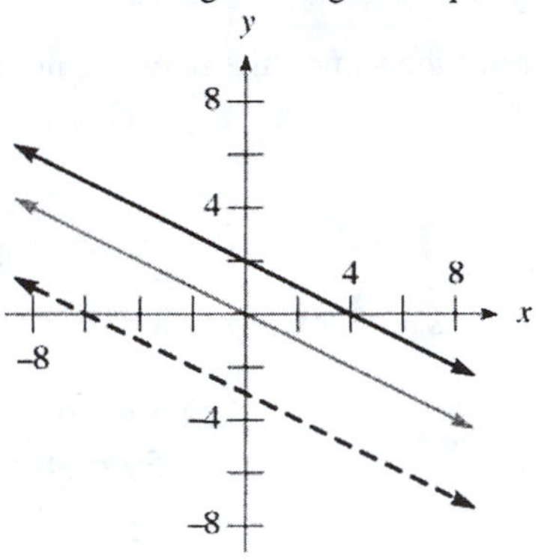

39. **a.** Solve for y to find the slope:

$$3x + 4y = 12$$
$$4y = -3x + 12$$
$$y = -\frac{3}{4}x + 3$$

Thus the perpendicular slope is $\frac{4}{3}$, and the line through the origin is $y = \frac{4}{3}x$.

b. Sketching the two graphs using a square viewing rectangle:

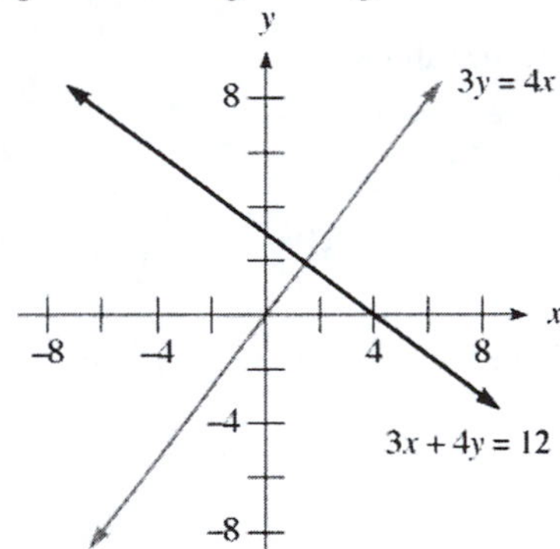

41. **a.** First find the slope: $m = \frac{230 - 260}{270 - 225} = \frac{-30}{45} = -\frac{2}{3}$. Now using the point-slope formula:

$$y - 230 = -\frac{2}{3}(p - 270)$$
$$y - 230 = -\frac{2}{3}p + 180$$
$$y = -\frac{2}{3}p + 410$$

b. Substituting $p = 303$: $y = -\frac{2}{3}(303) + 410 = -202 + 410 = 208$

So 208 units can be sold when the price is \$303 per unit.

c. Substituting $y = 288$:

$$288 = -\tfrac{2}{3}p + 410$$
$$\tfrac{2}{3}p = 122$$
$$p = 183$$

The price should be \$183 to sell 288 units per month.

43. Finding the slope: $m = \dfrac{x^2 - a^2}{x - a} = \dfrac{(x+a)(x-a)}{x-a} = x + a$

45. Finding the slope: $m = \dfrac{x^3 - a^3}{x - a} = \dfrac{(x-a)(x^2 + ax + a^2)}{x-a} = x^2 + ax + a^2$

47. Use the point-slope formula to find the equation of the line:

$$y - 6 = -5(x - 3)$$
$$y - 6 = -5x + 15$$
$$y = -5x + 21$$

To find the x-intercept, let $y = 0$:

$$-5x + 21 = 0$$
$$5x = 21$$
$$x = \tfrac{21}{5}$$

To find the y-intercept, let $x = 0$, so $y = 21$. The area of the triangle is given by:

$$\text{Area} = \tfrac{1}{2}(\text{base})(\text{height}) = \tfrac{1}{2} \cdot \tfrac{21}{5} \cdot 21 = 44.1 \text{ square units}$$

49. a. Draw the graph:

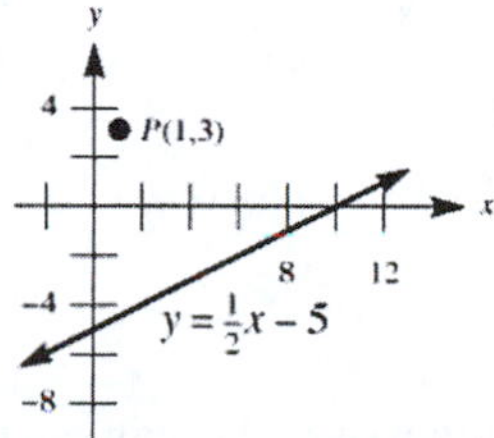

b. Since the slope must be –2, use (1,3) in the point-slope formula:

$$y - 3 = -2(x - 1)$$
$$y - 3 = -2x + 2$$
$$y = -2x + 5$$

c. Set the two equations equal:

$$-2x + 5 = \tfrac{1}{2}x - 5$$
$$-4x + 10 = x - 10$$
$$-5x = -20$$
$$x = 4$$
$$y = -2(4) + 5 = -3$$

The intersection point is (4,–3).

d. Find the distance from (1,3) to (4,–3): $d = \sqrt{(4-1)^2 + (-3-3)^2} = \sqrt{9 + 36} = \sqrt{45} = 3\sqrt{5}$

51. a. Draw the triangle with vertices $A(-4, 0)$, $B(2, 0)$, and $C(0, 6)$:

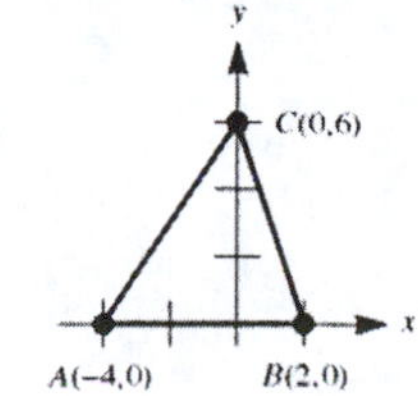

b. Using the midpoint formula:

$$M_{AB} = \left(\frac{-4+2}{2}, \frac{0+0}{2}\right) = \left(-\frac{2}{2}, \frac{0}{2}\right) = (-1, 0)$$
$$M_{BC} = \left(\frac{2+0}{2}, \frac{0+6}{2}\right) = \left(\frac{2}{2}, \frac{6}{2}\right) = (1, 3)$$
$$M_{AC} = \left(\frac{-4+0}{2}, \frac{0+6}{2}\right) = \left(-\frac{4}{2}, \frac{6}{2}\right) = (-2, 3)$$

Now add the medians to the graph of the triangle:

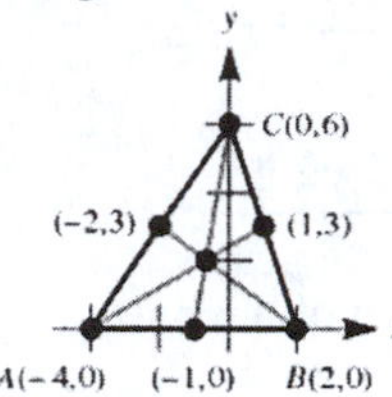

The medians appear to intersect at the point (–1,2).

c. For the median from *A* to side *BC*, the line passes through the points (–4,0) and (1,3). Finding the slope:

$$m = \frac{3-0}{1+4} = \frac{3}{5}$$

Using the point-slope formula with the point (1,3):

$$y - 3 = \frac{3}{5}(x-1)$$
$$y - 3 = \frac{3}{5}x - \frac{3}{5}$$
$$y = \frac{3}{5}x + \frac{12}{5}$$

For the median from *B* to side *AC*, the line passes through the points (2,0) and (–2,3). Finding the slope:

$$m = \frac{3-0}{-2-2} = -\frac{3}{4}$$

Using the point-slope formula with the point (2,0):

$$y - 0 = -\frac{3}{4}(x-2)$$
$$y = -\frac{3}{4}x + \frac{3}{2}$$

For the median from *C* to side *AB*, the line passes through the points (0,6) and (–1,0). Finding the slope:

$$m = \frac{0-6}{-1-0} = 6$$

Since the *y*-intercept is 6, its equation is $y = 6x + 6$. Solving any pair of simultaneous equations will yield the coordinates of the centroid. Using $y = -\frac{3}{4}x + \frac{3}{2}$ and $y = 6x + 6$, solve the system:

$$-\frac{3}{4}x + \frac{3}{2} = 6x + 6$$
$$-3x + 6 = 24x + 24$$
$$-27x = 18$$
$$x = -\frac{2}{3}$$
$$y = 6\left(-\frac{2}{3}\right) + 6 = 2$$

The exact coordinates of the centroid are $\left(-\frac{2}{3}, 2\right)$, which agrees with our estimate from part **b**, to the nearest integer.

53. **a.** Since A and C lie on the line $y = m_1x + b_1$, and the x-coordinates of A and C are 0 and 1, respectively, the y-coordinates will be b_1 and $m_1 + b_1$, so $A = (0, b_1)$ and $C = (1, m_1 + b_1)$. Similarly, the points B and D are $B = (0, b_2)$ and $D = (1, m_2 + b_2)$.

b. Verify the results:

$$AB = \sqrt{(0-0)^2 + (b_1 - b_2)^2} = \sqrt{(b_1 - b_2)^2} = |b_1 - b_2|, \text{ which is } b_1 - b_2 \text{ since } b_1 > b_2$$

$$\begin{aligned} CD &= \sqrt{(1-1)^2 + [(m_1 + b_1) - (m_2 + b_2)]^2} \\ &= \sqrt{[(m_1 + b_1) - (m_2 + b_2)]^2} \\ &= |(m_1 + b_1) - (m_2 + b_2)| \\ &= m_1 + b_1 - (m_2 + b_2) \text{ since } m_1 + b_1 > m_2 + b_2 \end{aligned}$$

c. Since $AB = CD$:

$$\begin{aligned} b_1 - b_2 &= m_1 + b_1 - m_2 - b_2 \\ 0 &= m_1 - m_2 \\ m_1 &= m_2 \end{aligned}$$

55. Multiplying both numerator and denominator by –1: $\dfrac{y_2 - y_1}{x_2 - x_1} \bullet \dfrac{-1}{-1} = \dfrac{y_1 - y_2}{x_1 - x_2}$

This identity tells us that, in calculating slope, either point can be chosen as the starting point.

1.7 Symmetry and Graphs. Circles

1. **a.** The reflection of $\overline{AB}$ about the x-axis is given by the graph:

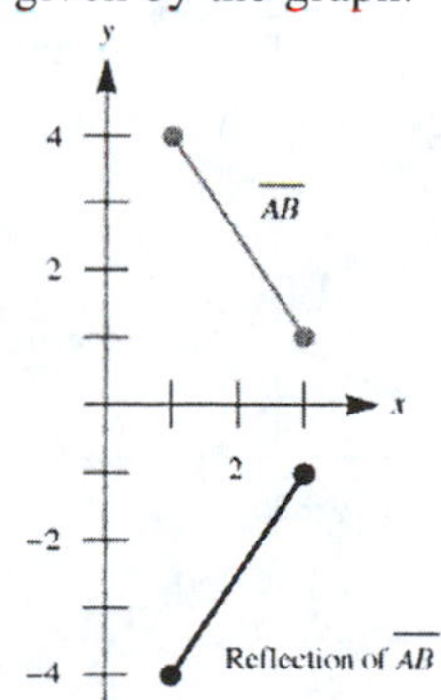

b. The reflection of $\overline{AB}$ about the y-axis is given by the graph:

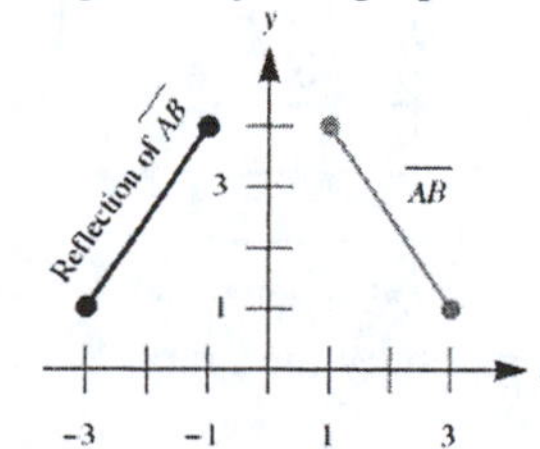

c. The reflection of $\overline{AB}$ about the origin is given by the graph:

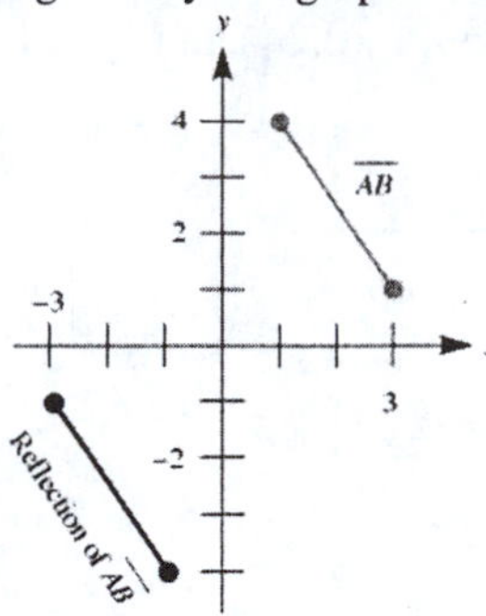

3. a. The reflection of $\overline{AB}$ about the x-axis is given by the graph:

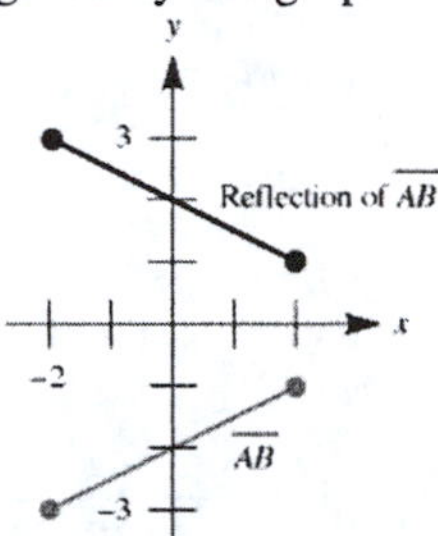

b. The reflection of $\overline{AB}$ about the y-axis is given by the graph:

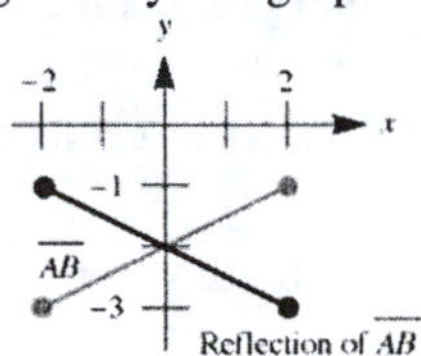

c. The reflection of $\overline{AB}$ about the origin is given by the graph:

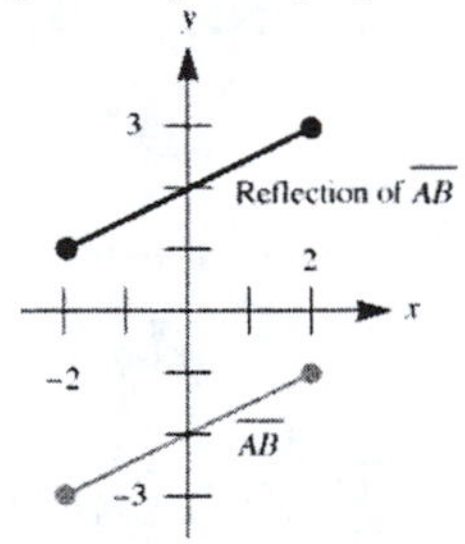

5. a. The reflection of $\overline{AB}$ about the x-axis is given by the graph:

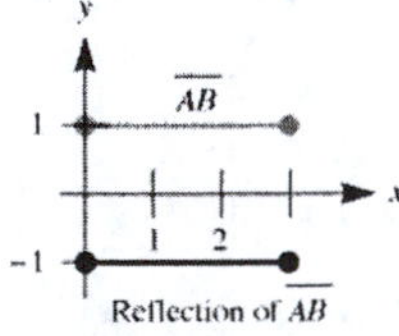

b. The reflection of $\overline{AB}$ about the y-axis is given by the graph:

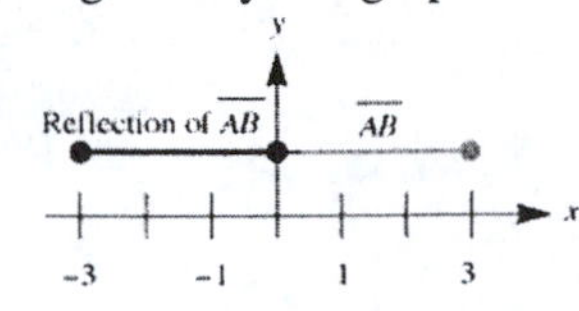

c. The reflection of $\overline{AB}$ about the origin is given by the graph:

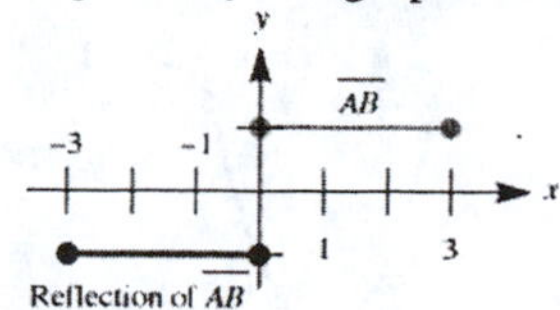

7. The x-intercepts are 2 and -2, and the y-intercept is 4. The graph is symmetric about the y-axis.

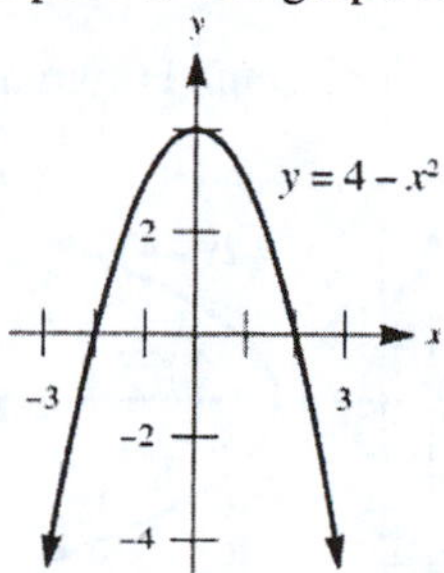

9. There are no x- or y-intercepts. The graph is symmetric about the origin.

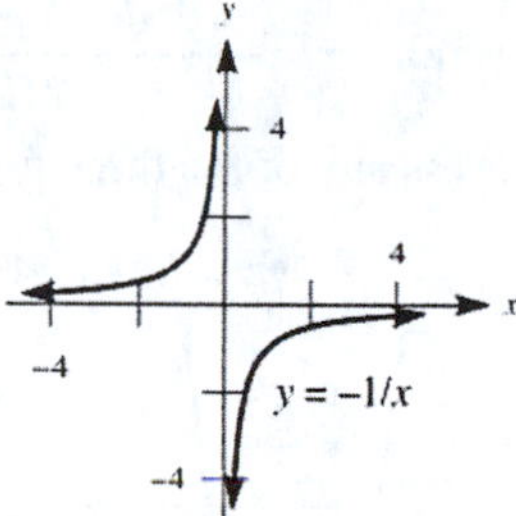

11. The x- and y-intercepts are both 0. The graph is symmetric about the y-axis.

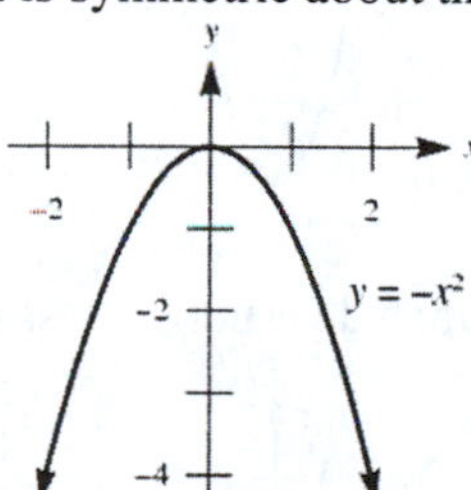

13. There are no x- or y-intercepts. The graph is symmetric about the origin.

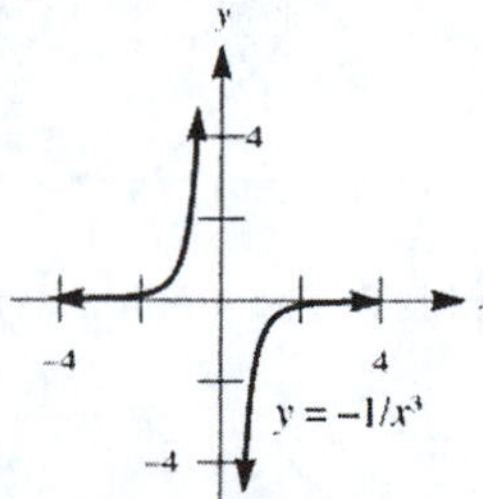

15. The x- and y-intercepts are both 0. The graph is symmetric about the y-axis.

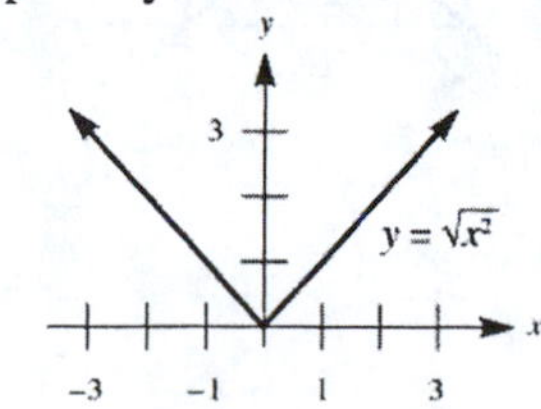

17. The x- and y-intercepts are both 1. The graph does not possess any of the three types of symmetry.

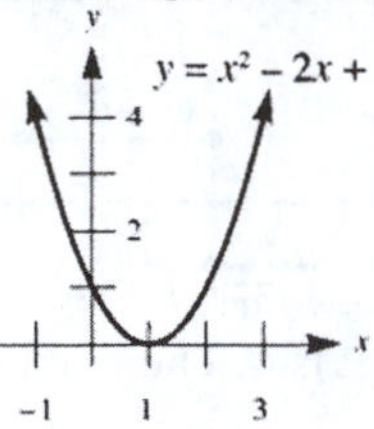

19. The x-intercept is 2 and there is no y-intercept. The graph is symmetric about the x-axis.

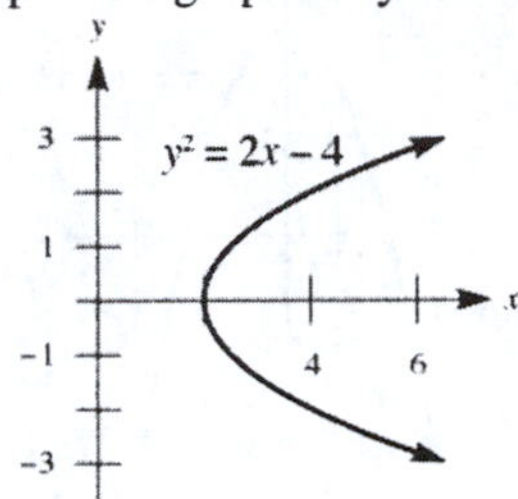

21. To find the x-intercepts, use the quadratic formula: $x = \dfrac{-1 \pm \sqrt{(1)^2 - 4(2)(-4)}}{2(2)} = \dfrac{-1 \pm \sqrt{33}}{4} \approx -1.69, 1.19$

The y-intercept is –4. The graph does not possess any of the three types of symmetry.

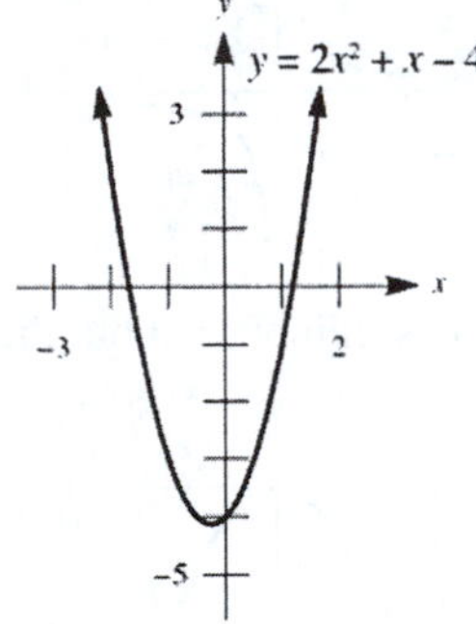

23. **a.** The x- and y-intercepts are both 2. The graph does not possess any of the three types of symmetry.

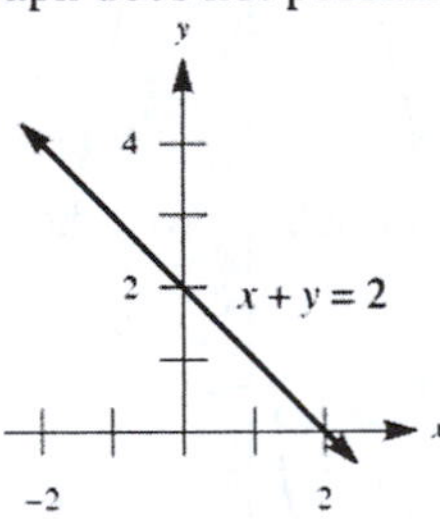

b. The x-intercepts are ±2 and the y-intercept is 2. The graph is symmetric about the y-axis.

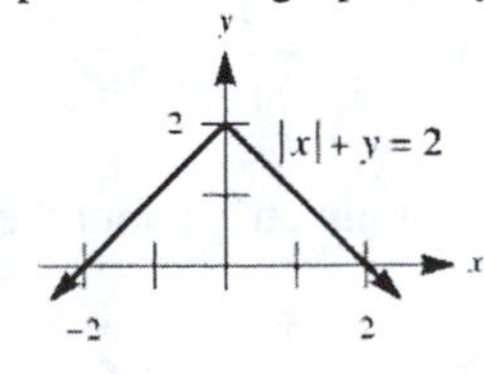

25. **a.** Graphing $y = x^2 - 3x$ in the standard viewing rectangle:

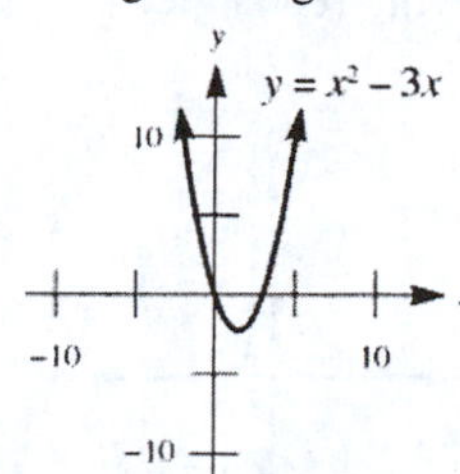

b. No, the graph does not appear to possess any of the three types of symmetry.

c. Now graphing $y = x^2 - 3x$ using the suggested settings:

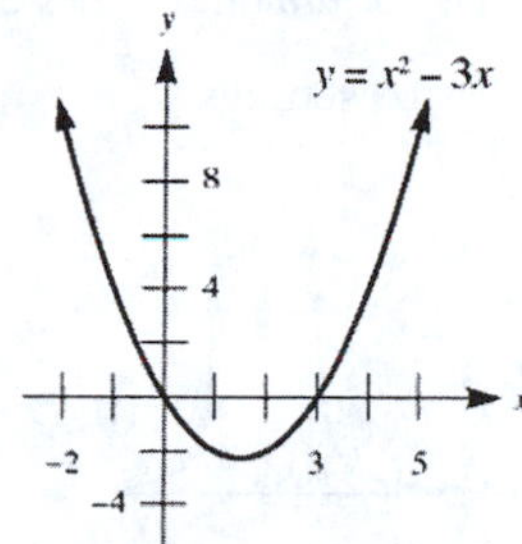

d. The graph does not possess any of the three types of symmetry.

27. **a.** Graphing $y = 2^x$ in the standard viewing rectangle:

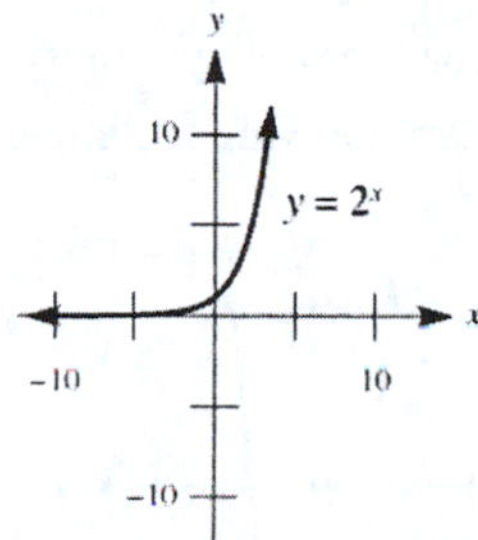

b. No, the graph does not appear to possess any of the three types of symmetry.

c. Now graphing $y = 2^x$ using the suggested settings:

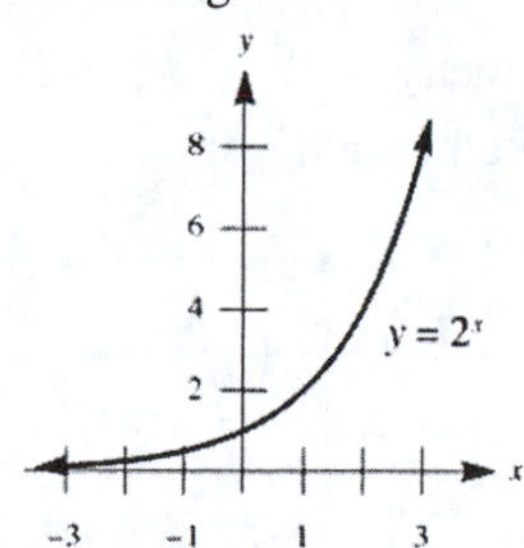

d. The graph does not possess any of the three types of symmetry.

29. **a.** Graphing $y = \dfrac{1}{x^2 - x}$ in the standard viewing rectangle:

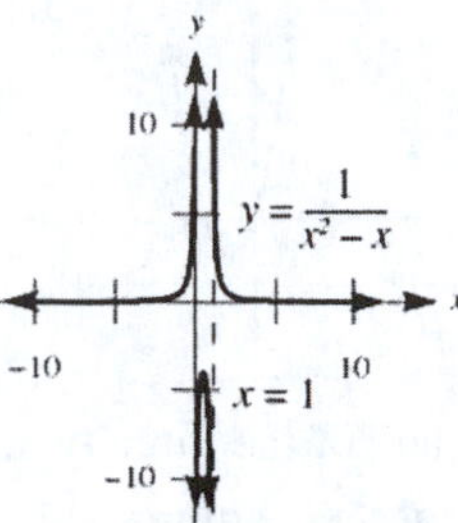

b. No, the graph does not appear to possess any of the three types of symmetry.

c. Now graphing $y = \dfrac{1}{x^2 - x}$ using the suggested settings:

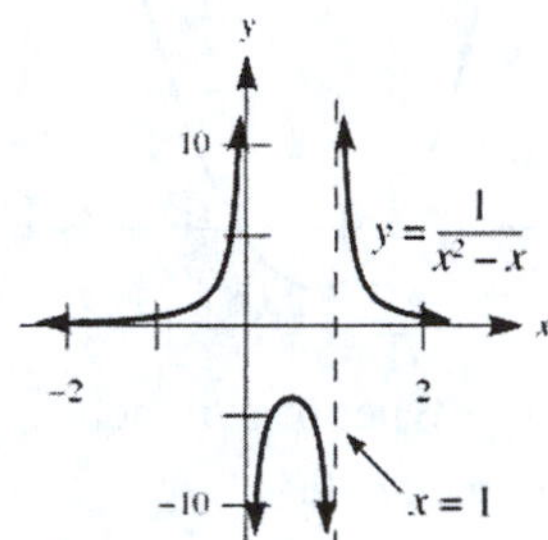

d. The graph does not possess any of the three types of symmetry.

31. **a.** Graphing $y = x^2 - 0.2x - 15$ in the standard viewing rectangle:

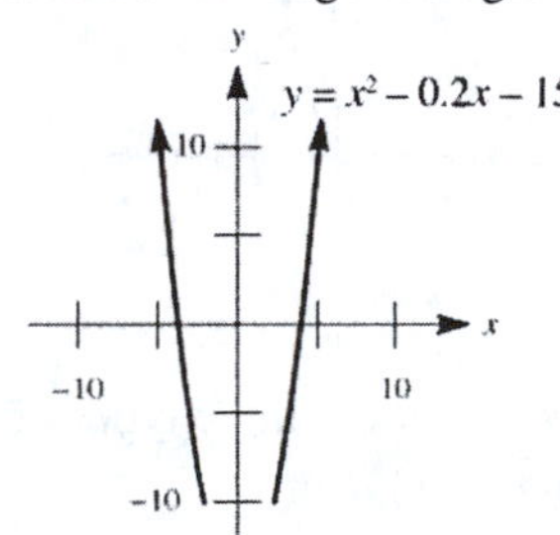

b. The graph appears to possess y-axis symmetry.

c. Now graphing $y = x^2 - 0.2x - 15$ using other settings:

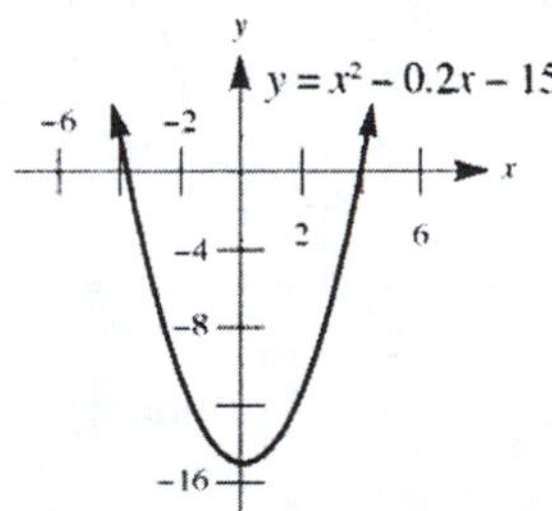

d. The graph does not possess any of the three types of symmetry (note the x-intercepts are not opposites).

33. **a.** Graphing $y=\sqrt{|x|}$ in the standard viewing rectangle:

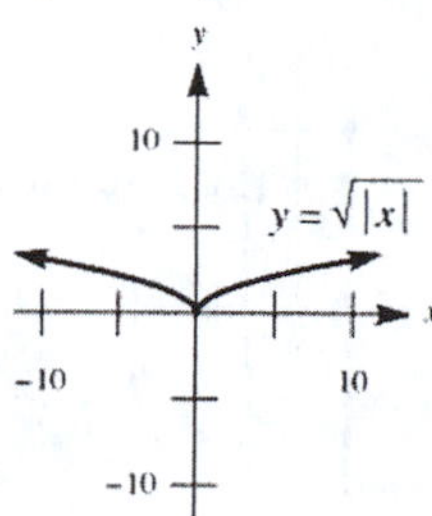

b. The graph appears to possess y-axis symmetry.

c. Now graphing $y=\sqrt{|x|}$ using other settings:

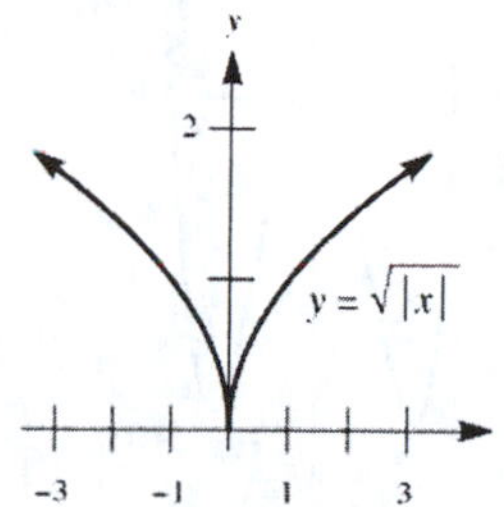

d. The graph appears to possess y-axis symmetry. Replacing (x, y) with $(-x, y)$: $y=\sqrt{|-x|}=\sqrt{|x|}$

So the graph does possess y-axis symmetry.

35. **a.** Graphing $y=2x-x^3-x^5+x^7$ in the standard viewing rectangle:

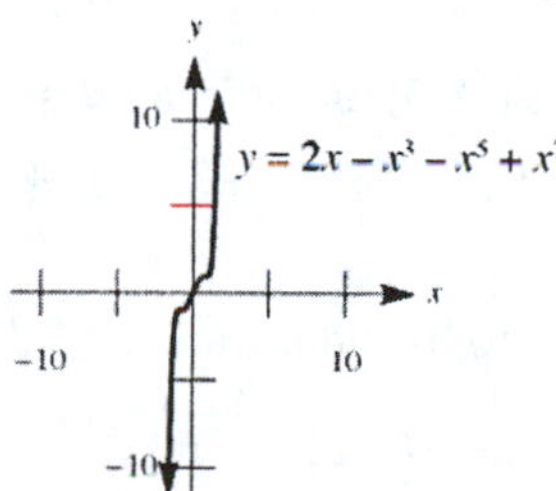

b. The graph appears to possess origin symmetry.

c. Now graphing $y=2x-x^3-x^5+x^7$ using other settings:

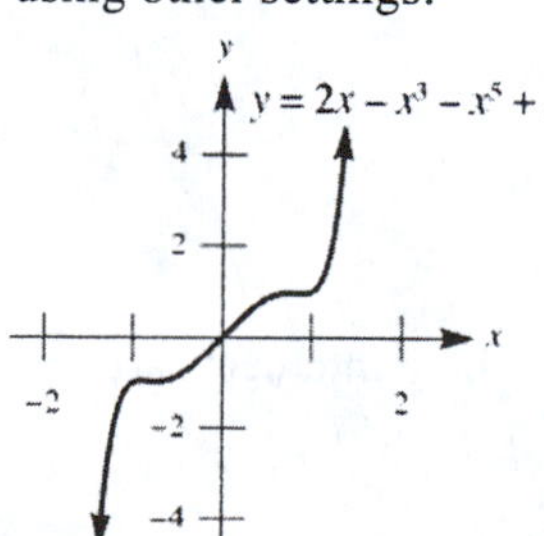

d. The graph appears to have origin symmetry. Replacing (x, y) with $(-x, -y)$:

$$-y=2(-x)-(-x)^3-(-x)^5+(-x)^7$$
$$-y=-2x+x^3+x^5-x^7$$
$$y=2x-x^3-x^5+x^7$$

So the graph does possess origin symmetry.

37. a. Graphing $y = x^4 - 10x^2 + \frac{1}{4}x$ in the standard viewing rectangle:

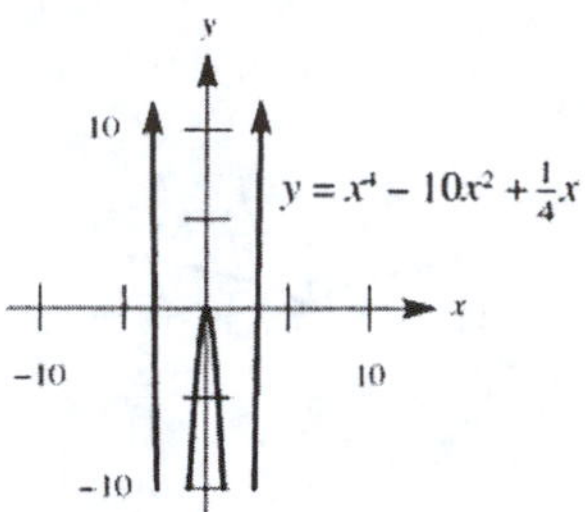

b. The graph appears to possess y-axis symmetry.

c. Now graphing $y = x^4 - 10x^2 + \frac{1}{4}x$ using other settings:

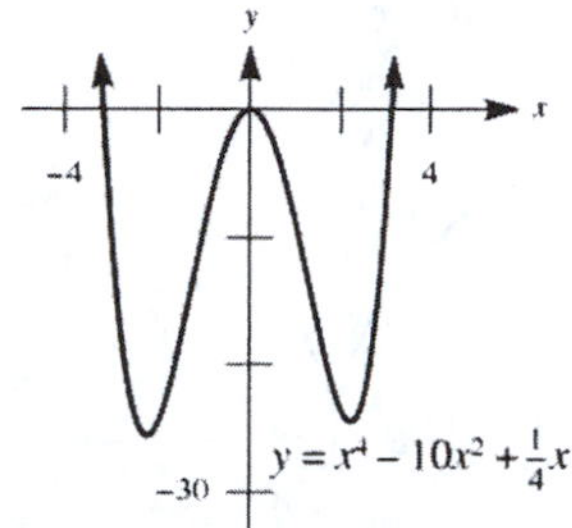

d. The graph does not possess any of the three types of symmetry.

39. The circle is in standard form, so its center is (1,5) and its radius is $\sqrt{169} = 13$. To determine if the point (6,–7) lies on the circle, substitute into the equation: $(x-1)^2 + (y-5)^2 = (6-1)^2 + (-7-5)^2 = (5)^2 + (-12)^2 = 25 + 144 = 169$
So (6,–7) lies on the circle.

41. The circle is in standard form, so its center is (–8,5) and its radius is $\sqrt{13}$. To determine if the point (–5,2) lies on the circle, substitute into the equation: $(x+8)^2 + (y-5)^2 = (-5+8)^2 + (2-5)^2 = (3)^2 + (-3)^2 = 9 + 9 = 18 \neq 13$
So (–5,2) does not lie on the circle.

43. The circle is in standard form, so its center is (0,0) and its radius is $\sqrt{\sqrt{2}} = \sqrt[4]{2}$. The y-intercepts are found by setting $x = 0$:

$$y^2 = \sqrt{2}$$
$$y = \pm\sqrt{\sqrt{2}} = \pm\sqrt[4]{2}$$

45. Complete the square:

$$x^2 + 8x + y^2 - 6y = -24$$
$$\left(x^2 + 8x + 16\right) + \left(y^2 - 6y + 9\right) = -24 + 16 + 9$$
$$(x+4)^2 + (y-3)^2 = 1$$

The center is (–4,3) and the radius is $\sqrt{1} = 1$. The y-intercepts are found by setting $x = 0$:

$$16 + (y-3)^2 = 1$$
$$(y-3)^2 = -15$$

Since this equation has no real solution, there are no y-intercepts.

47. Divide by 9 and complete the square:

$$x^2+6x+y^2-\tfrac{2}{3}y=-\tfrac{64}{9}$$
$$\left(x^2+6x+9\right)+\left(y^2-\tfrac{2}{3}y+\tfrac{1}{9}\right)=-\tfrac{64}{9}+9+\tfrac{1}{9}$$
$$(x+3)^2+\left(y-\tfrac{1}{3}\right)^2=2$$

The center is $\left(-3,\tfrac{1}{3}\right)$ and the radius is $\sqrt{2}$. The y-intercepts are found by setting $x=0$:

$$9+\left(y-\tfrac{1}{3}\right)^2=2$$
$$\left(y-\tfrac{1}{3}\right)^2=-7$$

Since this equation has no real solution, there are no y-intercepts.

49. **a.** Completing the square:

$$16x^2-64x+16y^2+48y-69=0$$
$$x^2-4x+y^2+3y-\tfrac{69}{16}=0$$
$$\left(x^2-4x+4\right)+\left(y^2+3y+\tfrac{9}{4}\right)=\tfrac{69}{16}+4+\tfrac{9}{4}$$
$$(x-2)^2+\left(y+\tfrac{3}{2}\right)^2=\tfrac{169}{16}$$

The center is $\left(2,-\tfrac{3}{2}\right)$ and the radius is $\sqrt{\tfrac{169}{16}}=\tfrac{13}{4}$.

b. Solving for y: $y=-\tfrac{3}{2}\pm\sqrt{\tfrac{169}{16}-(x-2)^2}$. Sketching the graph:

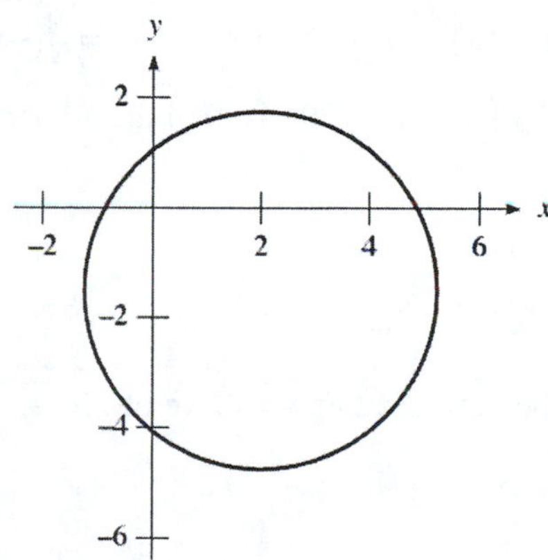

The x-intercepts are approximately -0.88 and 4.88.

c. The x-intercepts are found by setting $y=0$:

$$(x-2)^2+\left(\tfrac{3}{2}\right)^2=\tfrac{169}{16}$$
$$(x-2)^2+\tfrac{9}{4}=\tfrac{169}{16}$$
$$(x-2)^2=\tfrac{133}{16}$$
$$x-2=\pm\frac{\sqrt{133}}{4}$$
$$x=\frac{8\pm\sqrt{133}}{4}\approx-0.88, 4.88$$

51. Graphing $y=-\frac{4}{x}$ for the given viewing rectangle:

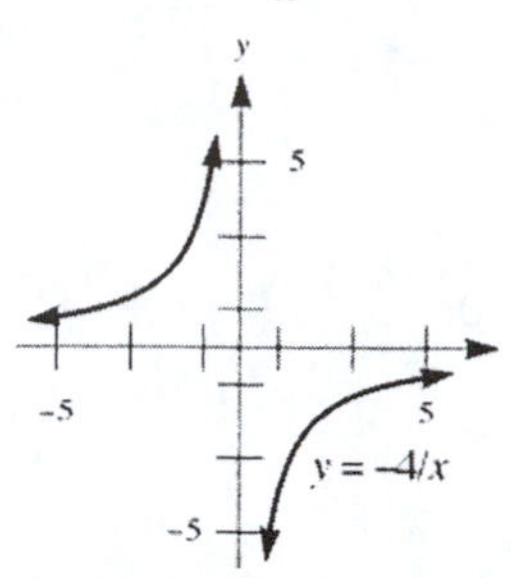

Now expanding the viewing rectangle:

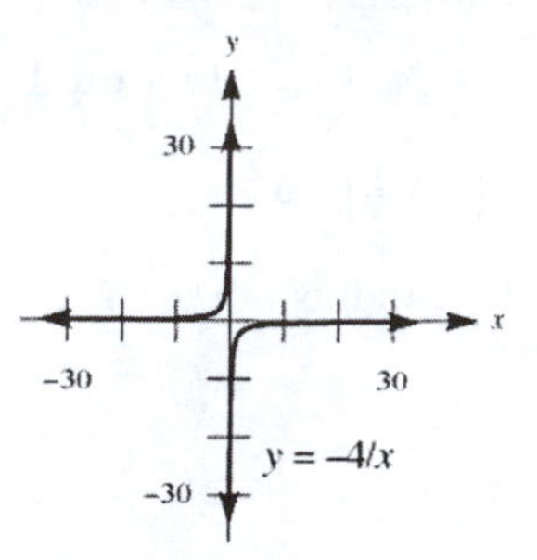

Note how closely the axes approximate the curve.

53. Since the center of the circle is (3,2), we know the equation will take the form $(x-3)^2+(y-2)^2=r^2$, where r is the radius of the circle. We can find r since the point (–2,–10) must satisfy the equation for the circle:

$$(-2-3)^2+(-10-2)^2=r^2$$
$$25+144=r^2$$
$$169=r^2$$

The equation of the circle is $(x-3)^2+(y-2)^2=169$.

55. Since the radius is 3, the equation of the circle is $(x-3)^2+(y-5)^2=9$.

57. The center of the circle is the midpoint of the diameter, which is: $\left(\frac{-1+3}{2},\frac{6-2}{2}\right)=(1,2)$

The radius is the distance from the center to the point (–1,6), thus: $r=\sqrt{(-1-1)^2+(6-2)^2}=\sqrt{4+16}=\sqrt{20}$

So the equation of the circle is $(x-1)^2+(y-2)^2=20$. To find the y-intercepts, let $x=0$:

$$(-1)^2+(y-2)^2=20$$
$$(y-2)^2=19$$
$$y-2=\pm\sqrt{19}$$
$$y=2\pm\sqrt{19}$$

59. **a.** The x-intercept for each graph is the same. Setting $y=0$, we obtain:

$$\tfrac{3}{4}x-2=0$$
$$\tfrac{3}{4}x=2$$
$$x=\tfrac{8}{3}$$

The y-intercept for $y=\frac{3}{4}x-2$ is –2, so the y-intercept for $y=\left|\frac{3}{4}x-2\right|$ is $|-2|=2$.

b. The two graphs are identical on the interval $\left[\frac{8}{3},\infty\right)$.

c. The graph of $y=\left|\frac{3}{4}x-2\right|$ can be obtained by reflecting $y=\frac{3}{4}x-2$ about the x-axis for the interval $\left(-\infty,\frac{8}{3}\right)$. For the interval $\left[\frac{8}{3},\infty\right)$, no reflection is necessary.

61. a. Using the quadratic formula: $y=\dfrac{0.1\pm\sqrt{(-0.1)^2-4(1)(-(15+x))}}{2}=\dfrac{0.1\pm\sqrt{0.01+60+4x}}{2}=\dfrac{0.1\pm\sqrt{60.01+4x}}{2}$

b. Sketching the graphs in the given viewing rectangle [–20,5,5] by [–10,10,5]:

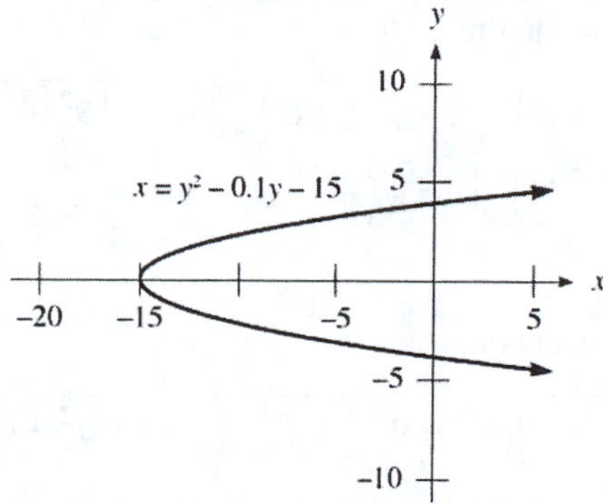

c. Sketching the graphs in the viewing rectangle [–15.01,–14.99,0.005] by [–0.1,0.2,0.1]:

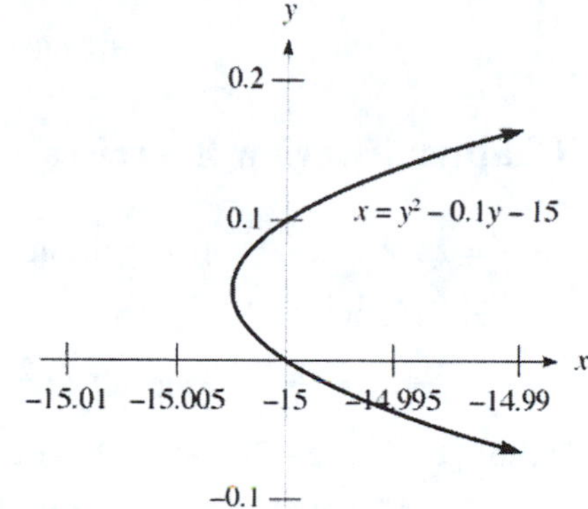

Note that the graph is not symmetric about the x-axis.

63. Since $(x_1,0)$ and $(x_2,0)$ satisfy the equation of the circle:

$$x_1^2+2Ax_1+0^2+2B(0)=C,\text{ so } x_1^2+2Ax_1=C$$
$$x_2^2+2Ax_2+0^2+2B(0)=C,\text{ so } x_2^2+2Ax_2=C$$

Similarly, since $(0,y_1)$ and $(0,y_2)$ satisfy the equation of the circle:

$$0^2+2A(0)+y_1^2+2By_1=C,\text{ so } y_1^2+2By_1=C$$
$$0^2+2A(0)+y_2^2+2By_2=C,\text{ so } y_2^2+2By_2=C$$

Since each of these four equations is quadratic, solve for x_1, x_2, y_1, and y_2 using the quadratic formula. Assuming x_1 and y_1 are the larger of the intercepts:

$$x_1=\frac{-(2A)+\sqrt{(2A)^2-4(1)(-C)}}{2(1)}=\frac{-2A+2\sqrt{A^2+C}}{2}=-A+\sqrt{A^2+C}$$
$$x_2=\frac{-(2A)-\sqrt{(2A)^2-4(1)(-C)}}{2(1)}=\frac{-2A-2\sqrt{A^2+C}}{2}=-A-\sqrt{A^2+C}$$
$$y_1=\frac{-(2B)+\sqrt{(2B)^2-4(1)(-C)}}{2(1)}=\frac{-2B+2\sqrt{B^2+C}}{2}=-B+\sqrt{B^2+C}$$
$$y_2=\frac{-(2B)-\sqrt{(2B)^2-4(1)(-C)}}{2(1)}=\frac{-2B-2\sqrt{B^2+C}}{2}=-B-\sqrt{B^2+C}$$

Using each of these expressions, now prove the statements.

a. Substitute for x_1, x_2, y_1, and y_2 to obtain: $\dfrac{x_1+x_2}{y_1+y_2}=\dfrac{-A+\sqrt{A^2+C}-A-\sqrt{A^2+C}}{-B+\sqrt{B^2+C}-B-\sqrt{B^2+C}}=\dfrac{-2A}{-2B}=\dfrac{A}{B}$

b. Substitute for x_1, x_2, y_1, and y_2 to obtain:

$$\begin{aligned} x_1x_2-y_1y_2 &= \left(-A+\sqrt{A^2+C}\right)\left(-A-\sqrt{A^2+C}\right)-\left(-B+\sqrt{B^2+C}\right)\left(-B-\sqrt{B^2+C}\right) \\ &= \left[A^2-\left(A^2+C\right)\right]-\left[B^2-\left(B^2+C\right)\right] \\ &= -C+C \\ &= 0 \end{aligned}$$

c. Substitute for x_1, x_2, y_1, and y_2 to obtain:

$$\begin{aligned} x_1x_2+y_1y_2 &= \left(-A+\sqrt{A^2+C}\right)\left(-A-\sqrt{A^2+C}\right)+\left(-B+\sqrt{B^2+C}\right)\left(-B-\sqrt{B^2+C}\right) \\ &= \left[A^2-\left(A^2+C\right)\right]+\left[B^2-\left(B^2+C\right)\right] \\ &= -C-C \\ &= -2C \end{aligned}$$

Chapter 1 Review Exercises

1. Using absolute value, the equality is $|x-6|=2$.

3. Using absolute value, the equality is $|a-b|=3$.

5. Using absolute value, the inequality is $|x-0|>10$, or $|x|>10$.

7. Since $\sqrt{6}-2>0$: $\left|\sqrt{6}-2\right|=\sqrt{6}-2$

9. Since $x^4+x^2+1>0$: $\left|x^4+x^2+1\right|=x^4+x^2+1$

11. a. If $x<2$, then $x-2<0$ and $x-3<0$, so: $|x-2|+|x-3|=-(x-2)-(x-3)=-2x+5$

b. If $2<x<3$, then $x-2>0$ and $x-3<0$, so: $|x-2|+|x-3|=(x-2)-(x-3)=1$

c. If $x>3$, then $x-2>0$ and $x-3>0$, so: $|x-2|+|x-3|=x-2+x-3=2x-5$

13. In inequality notation, the interval is $3<x<5$:

15. In inequality notation, the interval is $-5\le x\le 2$:

17. In inequality notation, the interval is $-1\le x<\infty$:

19. Sketch the interval (3,9):

21. Sketch the intervals $(-\infty,-2]$ and $[0,\infty)$:

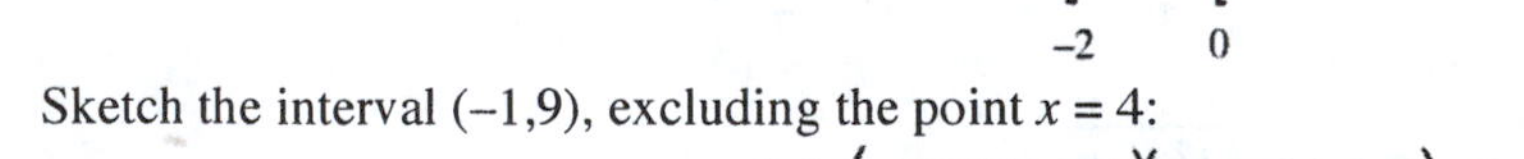

23. a. Sketch the interval (–1,9), excluding the point $x=4$:

b. Sketch the interval (–1,9):

25. Solving the equation:

$$\begin{aligned} 5-9x&=2\\ -9x&=-3\\ x&=\tfrac{1}{3}\end{aligned}$$

27. Solving the equation:

$$\begin{aligned} (t-4)(t+3)&=(t+5)^2\\ t^2-t-12&=t^2+10t+25\\ -11t&=37\\ t&=-\tfrac{37}{11}\end{aligned}$$

29. Solving the equation:

$$\begin{aligned} \frac{2t-1}{t+2}&=5\\ 2t-1&=5t+10\\ -3t&=11\\ t&=-\tfrac{11}{3}\end{aligned}$$

31. Solving the equation:

$$\begin{aligned} \frac{2y-5}{4y+1}&=\frac{y-1}{2y+5}\\ 4y^2-25&=4y^2-3y-1\\ 3y&=24\\ y&=8\end{aligned}$$

This value checks in the original equation.

33. Solving the equation:

$$\begin{aligned} 12x^2+2x-2&=0\\ 6x^2+x-1&=0\\ (3x-1)(2x+1)&=0\\ x&=\tfrac{1}{3},-\tfrac{1}{2}\end{aligned}$$

35. Multiplying by 2, solve the equation:

$$\begin{aligned} \tfrac{1}{2}x^2+x-12&=0\\ x^2+2x-24&=0\\ (x+6)(x-4)&=0\\ x&=-6,4\end{aligned}$$

37. Solving the equation:

$$\begin{aligned} \frac{x}{5-x}&=\frac{-2}{11-x}\\ 11x-x^2&=-10+2x\\ -x^2+9x+10&=0\\ x^2-9x-10&=0\\ (x-10)(x+1)&=0\\ x&=10,-1\end{aligned}$$

39. Multiply by 2 to obtain $2t^2+2t-1=0$. Now using the quadratic formula:

$$t=\frac{-2\pm\sqrt{4-4(2)(-1)}}{2(2)}=\frac{-2\pm\sqrt{12}}{4}=\frac{-2\pm2\sqrt{3}}{4}=\frac{-1\pm\sqrt{3}}{2}$$

41. Find the slope: $m=\dfrac{6-2}{-6-(-4)}=\dfrac{4}{-2}=-2$. Using the point (–4,2) in the point-slope formula:

$$\begin{aligned} y-2&=-2[x-(-4)]\\ y-2&=-2(x+4)\\ y-2&=-2x-8\\ y&=-2x-6\end{aligned}$$

43. Using the point-slope formula:

$$\begin{aligned} y-(-3)&=\tfrac{1}{4}[x-(-2)]\\ y+3&=\tfrac{1}{4}(x+2)\\ y+3&=\tfrac{1}{4}x+\tfrac{1}{2}\\ y&=\tfrac{1}{4}x-\tfrac{5}{2}\end{aligned}$$

45. Find the slope between the points (–4,0) and (0,8): $m=\dfrac{8-0}{0-(-4)}=\dfrac{8}{4}=2$

Using the slope-intercept formula, we have $y = 2x + 8$.

47. Since the line is parallel to the x-axis, its equation will be of the form y = constant (horizontal line). Since (0,–2) is on the line, its equation is $y = -2$.

49. Find the slope of $x + y + 1 = 0$ by writing it in slope-intercept form, and obtain $y = -x - 1$. So its slope is -1, and thus the perpendicular slope would be 1. Use the point (1,2) in the point-slope formula:

$$\begin{aligned} y - 2 &= 1(x - 1) \\ y - 2 &= x - 1 \\ y &= x + 1 \end{aligned}$$

51. The centers of the circles are $(-2,-1)$ and $(2,8)$. Find the slope between $(-2,-1)$ and $(2,8)$: $m = \dfrac{8 - (-1)}{2 - (-2)} = \frac{9}{4}$

Use the point (2,8) in the point-slope formula:

$$\begin{aligned} y - 8 &= \tfrac{9}{4}(x - 2) \\ 4y - 32 &= 9x - 18 \\ -9x + 4y - 14 &= 0 \\ 9x - 4y + 14 &= 0 \end{aligned}$$

53. Find the midpoint of the line segment joining $(-2,-3)$ and $(6,-5)$: $M = \left(\dfrac{-2+6}{2}, \dfrac{-3-5}{2}\right) = \left(\frac{4}{2}, -\frac{8}{2}\right) = (2, -4)$

Now find the slope between (0,0) and $(2,-4)$: $m = \dfrac{-4 - 0}{2 - 0} = \dfrac{-4}{2} = -2$

Using the slope-intercept formula, we obtain $y = -2x$, or $2x + y = 0$.

55. The center is $(3,-4)$. Now find the slope of the radius drawn from $(3,-4)$ to (0,0): $m = \dfrac{0 - (-4)}{0 - 3} = \dfrac{4}{-3} = -\frac{4}{3}$

So the slope of the perpendicular tangent line is $\frac{3}{4}$. Using the slope-intercept formula, we obtain $y = \frac{3}{4}x$, or $3x - 4y = 0$.

57. Call the x-intercept a. Then the y-intercept is $2 - a$, and we find the slope from each to $(2,-1)$:

$$m = \frac{-1 - 0}{2 - a} = \frac{-1}{2 - a} \qquad m = \frac{-1 - (2 - a)}{2 - 0} = \frac{-1 - 2 + a}{2} = \frac{a - 3}{2}$$

Since these two slopes must be equal:

$$\begin{aligned} \frac{-1}{2 - a} &= \frac{a - 3}{2} \\ -2 &= (2 - a)(a - 3) \\ -2 &= -a^2 + 5a - 6 \\ 0 &= a^2 - 5a + 4 \\ 0 &= (a - 4)(a - 1) \\ a &= 1 \text{ or } a = 4 \end{aligned}$$

When $a = 1$, we have $m = \dfrac{1 - 3}{2} = \dfrac{-2}{2} = -1$. When $a = 4$, we have $m = \dfrac{4 - 3}{2} = \frac{1}{2}$. Using the point $(2,-1)$ and each of these slopes in the point-slope formula:

$$\begin{aligned} y - (-1) &= -1(x - 2) \\ y + 1 &= -x + 2 \\ x + y - 1 &= 0 \end{aligned} \qquad \begin{aligned} y - (-1) &= \tfrac{1}{2}(x - 2) \\ 2y + 2 &= x - 2 \\ x - 2y - 4 &= 0 \end{aligned}$$

Both of these lines satisfy the given conditions.

59. For x-axis symmetry, replace (x, y) with $(x, -y)$:

$$-y = x^4 - 2x^2$$
$$y = -x^4 + 2x^2$$

The graph is not symmetric about the x-axis.

For y-axis symmetry, replace (x, y) with $(-x, y)$: $y = (-x)^4 - 2(-x)^2 = x^4 - 2x^2$

The graph is symmetric about the y-axis. For origin symmetry, replace (x, y) with $(-x, -y)$:

$$-y = (-x)^4 - 2(-x)^2$$
$$-y = x^4 - 2x^2$$
$$y = -x^4 + 2x^2$$

The graph is not symmetric about the origin.

61. For x-axis symmetry, replace (x, y) with $(x, -y)$:

$$-y = 2^x + 2^{-x}$$
$$y = -2^x - 2^{-x}$$

The graph is not symmetric about the x-axis. For y-axis symmetry, replace (x, y) with $(-x, y)$:

$$y = 2^{-x} + 2^{-(-x)} = 2^{-x} + 2^x$$

The graph is symmetric about the y-axis. For origin symmetry, replace (x, y) with $(-x, -y)$:

$$-y = 2^{-x} + 2^{-(-x)}$$
$$-y = 2^{-x} + 2^x$$
$$y = -2^{-x} - 2^x$$

The graph is not symmetric about the origin.

63. The graph is symmetric about the x-axis only.

65. The graph is symmetric about the x-axis, the y-axis, and the origin.

67. The graph is symmetric about the x-axis only.

69. Draw the graph:

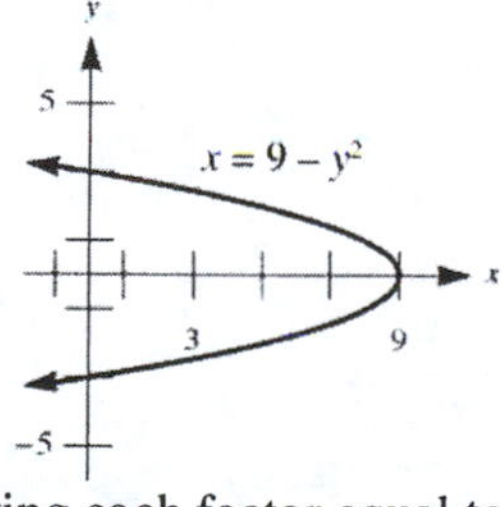

71. Draw the graph:

73. Setting each factor equal to 0, we have $4x - y + 4 = 0$ and thus $y = 4x + 4$, or $4x + y - 4 = 0$ and thus $y = -4x + 4$. So the graph consists of two lines:

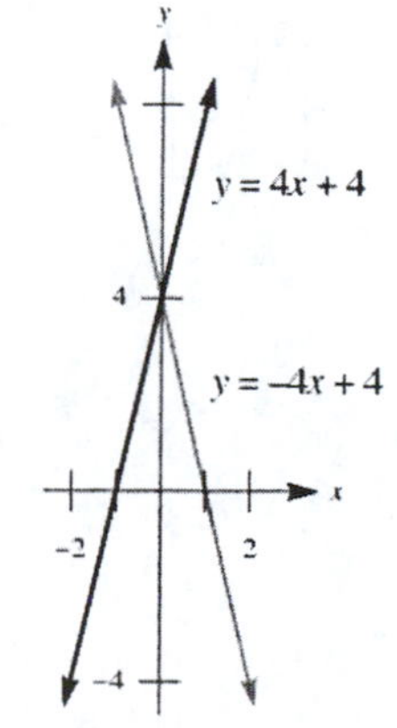

75. **a.** Graphing the ellipse:

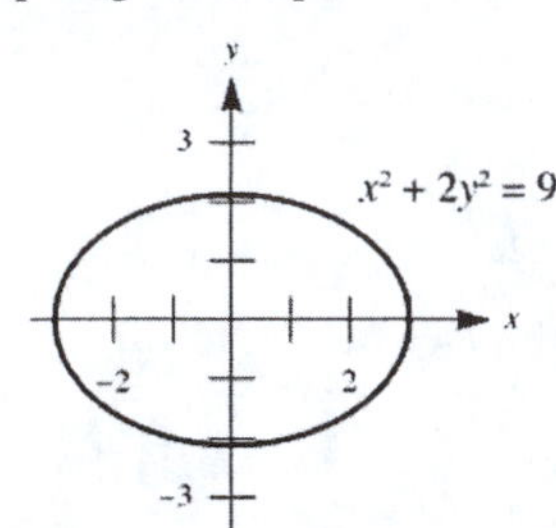

b. The curves appear to intersect at (–3,0) and (3,0):

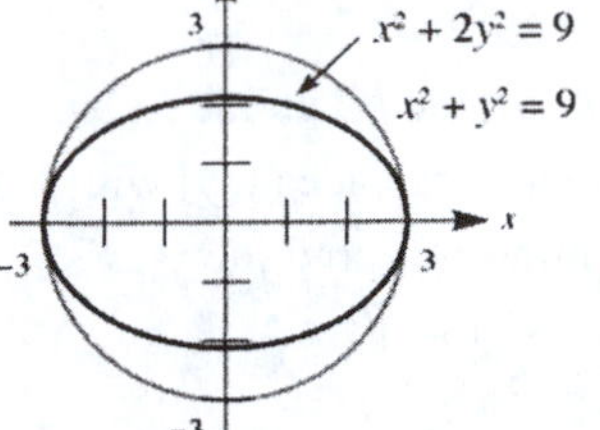

c. Substituting the points (±3,0) into the two equations:

$$(\pm 3)^2 + (0)^2 = 9 \qquad (\pm 3)^2 + 2(0)^2 = 9$$

Thus the intersection points are indeed (–3,0) and (3,0).

77. **a.** Graphing the hyperbola:

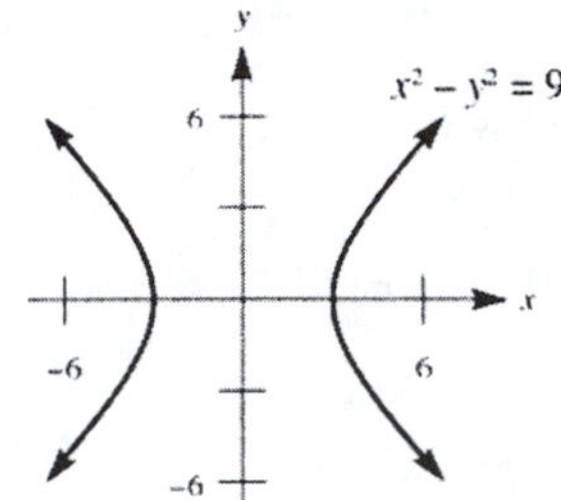

b. Graphing the two lines $y = x$ and $y = -x$:

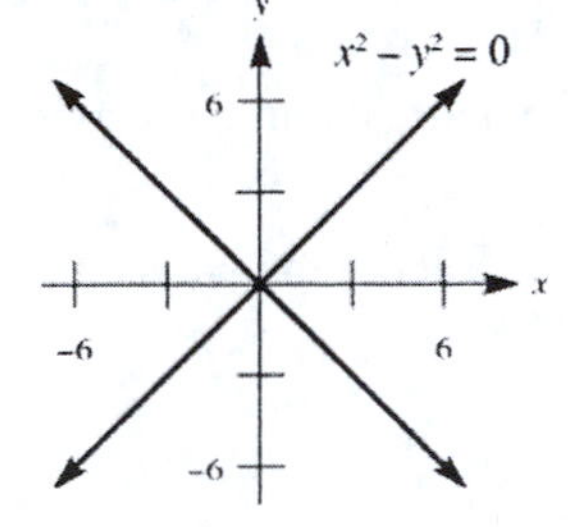

c. Putting the two graphs together:

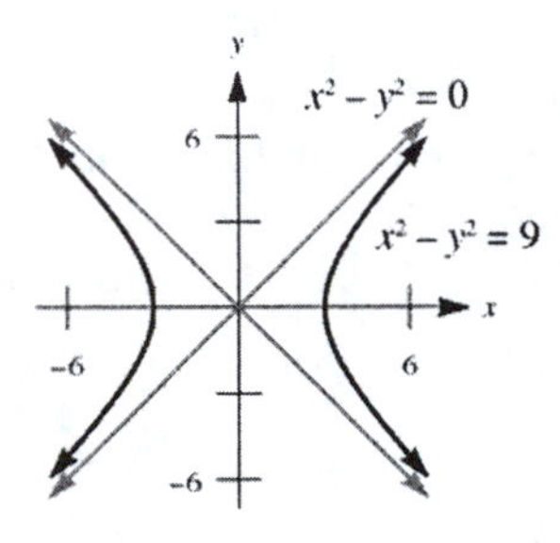

There are no intersection points.

79. Solving for *y* yields $y = \pm\sqrt{x^3 - 9}$. Graphing the curve:

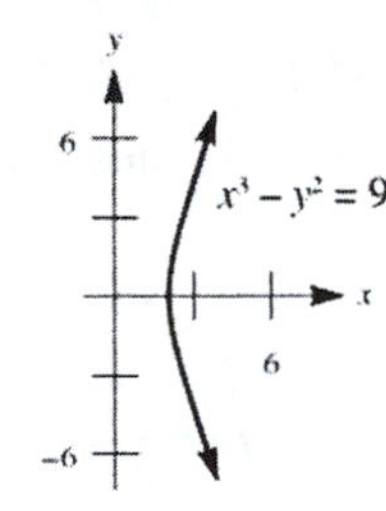

81. a. Solving for y yields $y = \pm\sqrt[4]{9 - x^4}$.

b. Solving for y yields $y = \pm\sqrt[6]{9 - x^6}$. Graphing:

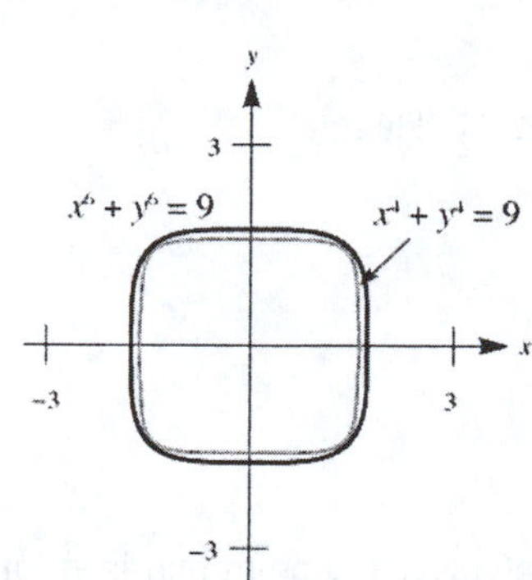

c. Solving for y yields $y = \pm\sqrt[8]{9 - x^8}$. Graphing:

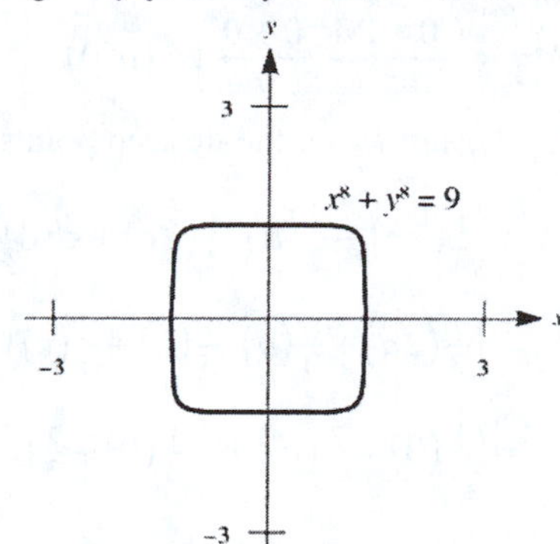

83. The midpoint of the hypotenuse BC would be: $M = \left(\frac{0+2c}{2}, \frac{2b+0}{2}\right) = \left(\frac{2c}{2}, \frac{2b}{2}\right) = (c, b)$. Compute the distances:

$$MA = \sqrt{(c-0)^2 + (b-0)^2} = \sqrt{c^2 + b^2}$$
$$MB = \sqrt{(c-0)^2 + (b-2b)^2} = \sqrt{c^2 + b^2}$$
$$MC = \sqrt{(c-2c)^2 + (b-0)^2} = \sqrt{c^2 + b^2}$$

These distances are all the same.

85. a. Compute the distance: $d = \sqrt{(5-2)^2 + (-6-5)^2} = \sqrt{9+121} = \sqrt{130}$

b. Compute the slope: $m = \frac{-6-5}{5-2} = -\frac{11}{3}$

c. Compute the midpoint: $M = \left(\frac{2+5}{2}, \frac{5-6}{2}\right) = \left(\frac{7}{2}, -\frac{1}{2}\right)$

87. Find the slope: $m = \frac{1-2}{4-1} = -\frac{1}{3}$. Using (1,2) in the point-slope formula:

$$y - 2 = -\frac{1}{3}(x-1)$$
$$y - 2 = -\frac{1}{3}x + \frac{1}{3}$$
$$y = -\frac{1}{3}x + \frac{7}{3}$$

So the x- and y-intercepts are $\frac{7}{3}$ (if $x = 0$) and 7 (if $y = 0$). Then the area is given by:

$$\text{Area} = \tfrac{1}{2}(\text{base})(\text{height}) = \tfrac{1}{2} \bullet 7 \bullet \tfrac{7}{3} = \tfrac{49}{6}$$

89. a. Find each midpoint:

$$M_1 = \left(\frac{-5+7}{2}, \frac{3+7}{2}\right) = (1,5) \qquad M_2 = \left(\frac{3+7}{2}, \frac{1+7}{2}\right) = (5,4) \qquad M_3 = \left(\frac{-5+3}{2}, \frac{3+1}{2}\right) = (-1,2)$$

Let P_1, P_2, and P_3 be the desired points on medians CM_1, AM_2, and BM_3. Then:

$$P_1 = \left(\tfrac{1}{3}(3) + \tfrac{2}{3}(1), \tfrac{1}{3}(1) + \tfrac{2}{3}(5)\right) = \left(1 + \tfrac{2}{3}, \tfrac{1}{3} + \tfrac{10}{3}\right) = \left(\tfrac{5}{3}, \tfrac{11}{3}\right)$$
$$P_2 = \left(\tfrac{1}{3}(-5) + \tfrac{2}{3}(5), \tfrac{1}{3}(3) + \tfrac{2}{3}(4)\right) = \left(-\tfrac{5}{3} + \tfrac{10}{3}, 1 + \tfrac{8}{3}\right) = \left(\tfrac{5}{3}, \tfrac{11}{3}\right)$$
$$P_3 = \left(\tfrac{1}{3}(7) + \tfrac{2}{3}(-1), \tfrac{1}{3}(7) + \tfrac{2}{3}(2)\right) = \left(\tfrac{7}{3} - \tfrac{2}{3}, \tfrac{7}{3} + \tfrac{4}{3}\right) = \left(\tfrac{5}{3}, \tfrac{11}{3}\right)$$

They all intersect at the same point.

b. Find each midpoint:

$$M_1 = \left(\frac{0+2b}{2}, \frac{0+2c}{2}\right) = (b,c) \qquad M_1 = \left(\frac{2b+2a}{2}, \frac{2c+0}{2}\right) = (a+b,c)$$

$$M_3 = \left(\frac{0+2a}{2}, \frac{0+0}{2}\right) = (a,0)$$

Let P_1, P_2, and P_3 be the desired points on medians CM_3, BM_1, and AM_2. Then:

$$P_1 = \left(\tfrac{1}{3}(2b) + \tfrac{2}{3}(a), \tfrac{1}{3}(2c) + \tfrac{2}{3}(0)\right) = \left(\frac{2b}{3} + \frac{2a}{3}, \frac{2c}{3}\right)$$

$$P_2 = \left(\tfrac{1}{3}(2a) + \tfrac{2}{3}(b), \tfrac{1}{3}(0) + \tfrac{2}{3}(c)\right) = \left(\frac{2a}{3} + \frac{2b}{3}, \frac{2c}{3}\right)$$

$$P_3 = \left(\tfrac{1}{3}(0) + \tfrac{2}{3}(a+b), \tfrac{1}{3}(0) + \tfrac{2}{3}(c)\right) = \left(\frac{2a}{3} + \frac{2b}{3}, \frac{2c}{3}\right)$$

Again all these intersect at the same point. All medians of a triangle intersect at a point that is $\frac{2}{3}$ of the distance from each vertex to the midpoint of the opposite side.

91. Using the hint:

$$m^2 = (a+c-0)^2 + (b-0)^2 = a^2 + 2ac + c^2 + b^2$$
$$s^2 = (2a-0)^2 + (2b-0)^2 = 4a^2 + 4b^2$$
$$t^2 = (2c)^2 = 4c^2$$
$$u^2 = (2a-2c)^2 + (2b-0)^2 = 4a^2 - 8ac + 4c^2 + 4b^2$$

Working from the right-hand side:

$$\begin{aligned}\tfrac{1}{2}\left(s^2 + t^2\right) - \tfrac{1}{4}u^2 &= \tfrac{1}{2}\left(4a^2 + 4b^2 + 4c^2\right) - \tfrac{1}{4}\left(4a^2 - 8ac + 4c^2 + 4b^2\right) \\ &= 2a^2 + 2b^2 + 2c^2 - a^2 + 2ac - c^2 - b^2 \\ &= a^2 + b^2 + c^2 + 2ac \\ &= m^2\end{aligned}$$

93. **a.** The reflection of $\overline{AB}$ about the origin is given by the graph:

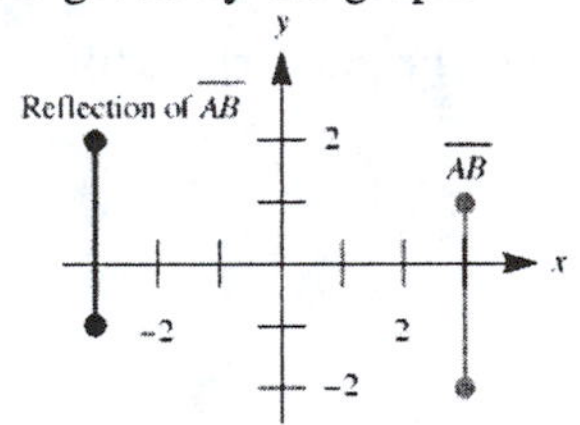

b. The reflection of $\overline{AB}$ about the x-axis is given by the graph:

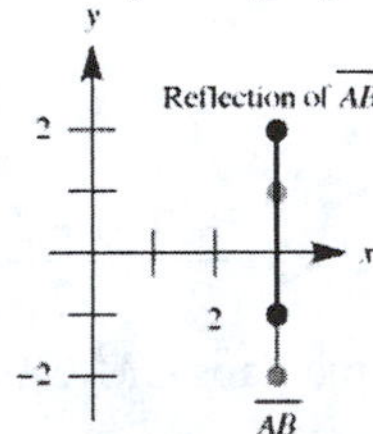

c. The reflection of $\overline{AB}$ about the y-axis is given by the graph:

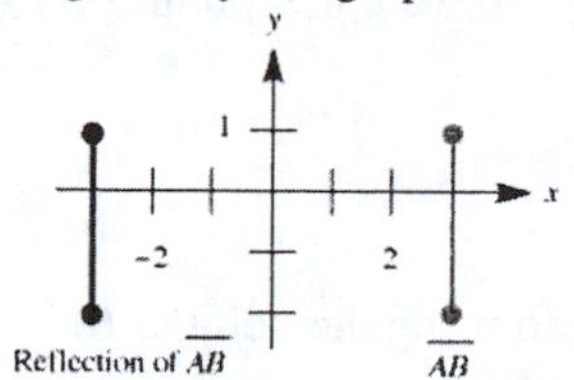

95. Here $(x_0, y_0) = (1,2)$, $m = \frac{1}{2}$, and $b = -5$: $d = \dfrac{2 - \frac{1}{2}(1) - (-5)}{\sqrt{1 + \left(\frac{1}{2}\right)^2}} = \dfrac{2 - \frac{1}{2} + 5}{\sqrt{1 + \frac{1}{4}}} = \dfrac{\frac{13}{2}}{\sqrt{\frac{5}{4}}} = \dfrac{13\sqrt{5}}{5}$

97. Here $(x_0, y_0) = (-1,-3)$, $A = 2$, $B = 3$ and $C = -6$: $d = \dfrac{|2(-1) + 3(-3) + (-6)|}{\sqrt{2^2 + 3^2}} = \dfrac{|-2-9-6|}{\sqrt{13}} = \dfrac{17}{\sqrt{13}} = \dfrac{17\sqrt{13}}{13}$

99. a. If the center is (h, k) and the circle is tangent to the coordinate axes, then the points of tangency are $(h, 0)$ and $(0, k)$. Since the distances from the center to the coordinate axes are the same, $h = k = r$ (the radius) and the center must be (r, r). Write the line in slope-intercept form:

$$3x + 4y = 12$$
$$4y = -3x + 12$$
$$y = -\frac{3}{4}x + 3$$

Find the distance from (r, r) to this line, using the fact that this distance is also r: $r = \dfrac{\left|r + \frac{3}{4}r - 3\right|}{\sqrt{1 + (3/4)^2}} = \dfrac{\left|\frac{7}{4}r - 3\right|}{5/4}$

So we have the equation $\frac{5}{4}r = \left|\frac{7}{4}r - 3\right|$. Thus we have:

$$\frac{5}{4}r = \frac{7}{4}r - 3 \qquad \text{or} \qquad -\frac{5}{4}r = \frac{7}{4}r - 3$$
$$5r = 7r - 12 \qquad\qquad -5r = 7r - 12$$
$$-2r = -12 \qquad\qquad -12r = -12$$
$$r = 6 \qquad\qquad r = 1$$

Clearly $r = 6$ corresponds to a circle above the line, so $r = 1$. Thus the equation of the circle is $(x - 1)^2 + (y - 1)^2 = 1$.

b. We have the coordinates $S(1,0)$ and $U(0,1)$. We must find the coordinates of T. We know the slope of the center to $3x + 4y = 12$ is 4/3, so use the point (1,1) in the point-slope formula:

$$y - 1 = \frac{4}{3}(x - 1)$$
$$y - 1 = \frac{4}{3}x - \frac{4}{3}$$
$$y = \frac{4}{3}x - \frac{1}{3}$$

Now find the intersection point of this line with our given line:

$$\frac{4}{3}x - \frac{1}{3} = -\frac{3}{4}x + 3$$
$$16x - 4 = -9x + 36$$
$$25x = 40$$
$$x = \frac{8}{5}$$

Substituting, we find $y = 9/5$, so we have the point $T(8/5, 9/5)$. We now can find the required equations. For $\overline{AT}$ the slope is $m = \dfrac{9/5 - 0}{8/5 - 0} = \frac{9}{8}$, so using the point (0,0) in the point-slope equation yields:

$$y - 0 = \frac{9}{8}(x - 0)$$
$$y = \frac{9}{8}x$$

For $\overline{BU}$ the slope is $m=\dfrac{0-1}{4-0}=-\frac{1}{4}$, so using the point (0,1) in the point-slope equation yields:

$$y-1=-\tfrac{1}{4}(x-0)$$
$$y=-\tfrac{1}{4}x+1$$

For $\overline{CS}$ the slope is $m=\dfrac{3-0}{0-1}=-3$, so using the point (1,0) in the point-slope equation yields:

$$y-0=-3(x-1)$$
$$y=-3x+3$$

c. To find the required intersection points, set the equations equal. For $\overline{AT}$ and $\overline{CS}$ we have:

$$\tfrac{9}{8}x=-3x+3$$
$$\tfrac{33}{8}x=3$$
$$x=\tfrac{24}{33}=\tfrac{8}{11}$$
$$y=\tfrac{9}{11}$$

The intersection point is (8/11, 9/11). For $\overline{AT}$ and $\overline{BU}$ we have:

$$\tfrac{9}{8}x=-\tfrac{1}{4}x+1$$
$$9x=-2x+8$$
$$11x=8$$
$$x=\tfrac{8}{11}$$
$$y=\tfrac{9}{11}$$

The intersection point is (8/11,9/11). Observe that the point of intersection is the same in both cases.

Chapter 1 Test

1. Since $x>6$, $x-6>0$ and thus $|x-6|=x-6$. Since $x<7$, $x-7<0$ and thus $|x-7|=-(x-7)=7-x$. Therefore:

$$|x-6|+|x-7|=x-6+7-x=1$$

2. a. The inequality is equivalent to:

$$|x-4|<\tfrac{1}{10}$$
$$-\tfrac{1}{10}<x-4<\tfrac{1}{10}$$
$$\tfrac{39}{10}<x<\tfrac{41}{10}$$

The interval is $\left(\frac{39}{10},\frac{41}{10}\right)$.

b. The interval is $[2,\infty)$.

3. Multiplying by $(x+4)(x-4)$:

$$\frac{2}{x+4}-\frac{1}{x-4}=\frac{-7}{x^2-16}$$
$$2(x-4)-(x+4)=-7$$
$$2x-8-x-4=-7$$
$$x=5$$

4. Multiplying by $6(1-x)(2-x)$:

$$\frac{1}{1-x}+\frac{4}{2-x}=\frac{11}{6}$$
$$6(2-x)+4(6)(1-x)=11(1-x)(2-x)$$
$$12-6x+24-24x=22-33x+11x^2$$
$$-11x^2+3x+14=0$$
$$11x^2-3x-14=0$$
$$(11x-14)(x+1)=0$$
$$x=\tfrac{14}{11},-1$$

5. **a.** Solve by factoring:

$$\begin{aligned} x^2+4x&=5 \\ x^2+4x-5&=0 \\ (x+5)(x-1)&=0 \\ x&=-5,1 \end{aligned}$$

b. Solve by writing the equation as $x^2+4x-1=0$, then using the quadratic formula:

$$x=\frac{-4\pm\sqrt{(4)^2-4(1)(-1)}}{2(1)}=\frac{-4\pm\sqrt{20}}{2}=\frac{-4\pm2\sqrt{5}}{2}=-2\pm\sqrt{5}$$

6. Multiplying by $cx+d$:

$$\begin{aligned} \frac{ax+b}{cx+d}&=e \\ ax+b&=cex+de \\ ax-cex&=de-b \\ x(a-ce)&=de-b \\ x&=\frac{de-b}{a-ce} \end{aligned}$$

7. **a.** The corresponding value is: $\Delta N=11565-10788=777$

b. The slope is: $\dfrac{\Delta N}{\Delta t}=\dfrac{777 \text{ stations}}{4 \text{ years}}\approx 194$ stations per year

c. In 1995, we would expect 194 more stations, which is 11,565 + 194 = 11,759 stations.

8. **a.** Plotting the points:

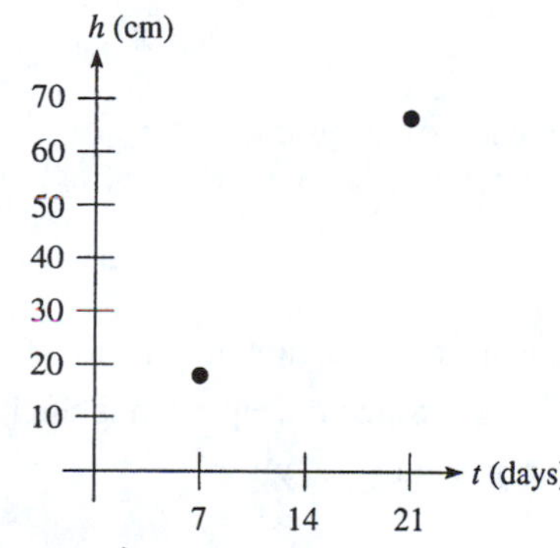

b. The midpoint is: $\left(\dfrac{7+21}{2},\dfrac{17.93+67.76}{2}\right)\approx(14,42.8)$

At $t=14$ days, we would expect the average height of the sunflowers to be 42.8 cm.

c. Computing the percentage error: $\dfrac{|36.36-42.8|}{36.36}\bullet100\approx18\%$

9. First find the slope: $m=\dfrac{8-(-2)}{3-1}=\frac{10}{2}=5$. Now using the point (3,8) in the point-slope formula:

$$\begin{aligned} y-8&=5(x-3) \\ y-8&=5x-15 \\ y&=5x-7 \end{aligned}$$

10. First write $5x + 6y = 30$ in slope-intercept form:

$$5x+6y=30$$
$$6y=-5x+30$$
$$y=-\tfrac{5}{6}x+5$$

So the perpendicular slope is $\frac{6}{5}$. Now using the point (2,–1) in the point-slope formula:

$$y-(-1)=\tfrac{6}{5}(x-2)$$
$$y+1=\tfrac{6}{5}x-\tfrac{12}{5}$$
$$y=\tfrac{6}{5}x-\tfrac{17}{5}$$

11. Compute the distance from each point to the origin:

$$d_1=\sqrt{(3-0)^2+(9-0)^2}=\sqrt{9+81}=\sqrt{90} \qquad d_2=\sqrt{(5-0)^2+(8-0)^2}=\sqrt{25+64}=\sqrt{89}$$

So (3,9) is farther from the origin.

12. **a.** To test for symmetry about the x-axis, replace (x, y) with $(x, -y)$:

$$-y=x^3+5x$$
$$y=-x^3-5x$$

Since the equation is changed, there is no x-axis symmetry. To test for symmetry about the y-axis, replace (x, y) with $(-x, y)$: $y=(-x)^3+5(-x)=-x^3-5x$. Since the equation is changed, there is no y-axis symmetry. To test for symmetry about the origin, replace (x, y) with $(-x, -y)$:

$$-y=(-x)^3+5(-x)$$
$$-y=-x^3-5x$$
$$y=x^3+5x$$

Since the equation is unchanged, there is origin symmetry.

b. To test for symmetry about the x-axis, replace (x, y) with $(x, -y)$:

$$-y=3^x+3^{-x}$$
$$y=-3^x-3^{-x}$$

Since the equation is changed, there is no x-axis symmetry. To test for symmetry about the y-axis, replace (x, y) with $(-x, y)$: $y=3^{-x}+3^{-(-x)}=3^{-x}+3^x$. Since the equation is unchanged, there is y-axis symmetry. To test for symmetry about the origin, replace (x, y) with $(-x, -y)$:

$$-y=3^{-x}+3^{-(-x)}$$
$$-y=3^{-x}+3^x$$
$$y=-3^{-x}-3^x$$

Since the equation is changed, there is no origin symmetry.

c. To test for symmetry about the x-axis, replace (x, y) with $(x, -y)$:

$$(-y)^2=5x^2+x$$
$$y^2=5x^2+x$$

Since the equation is unchanged, there is x-axis symmetry. To test for symmetry about the y-axis, replace (x, y) with $(-x, y)$: $y^2=5(-x)^2+(-x)=5x^2-x$. Since the equation is changed, there is no y-axis symmetry. To test for symmetry about the origin, replace (x, y) with $(-x, -y)$:

$$(-y)^2=5(-x)^2+(-x)$$
$$y^2=5x^2-x$$

Since the equation is changed, there is no origin symmetry.

13. To find the x-intercept, let $y = 0$:

$$3x = 15$$
$$x = 5$$

To find the y-intercept, let $x = 0$:

$$-5y = 15$$
$$y = -3$$

Now draw the graph:

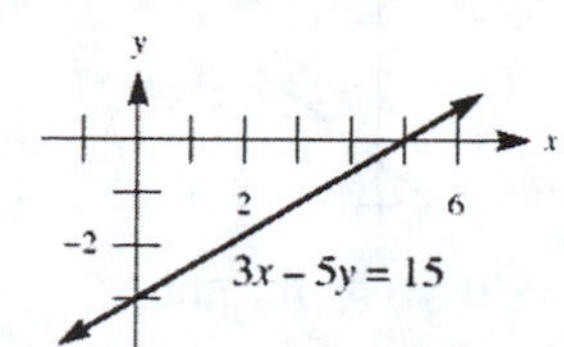

14. To find the x-intercepts, let $y = 0$:

$$(x-1)^2 + (0+2)^2 = 9$$
$$(x-1)^2 + 4 = 9$$
$$(x-1)^2 = 5$$
$$x - 1 = \pm\sqrt{5}$$
$$x = 1 \pm \sqrt{5}$$

To find the y-intercepts, let $x = 0$:

$$(0-1)^2 + (y+2)^2 = 9$$
$$1 + (y+2)^2 = 9$$
$$(y+2)^2 = 8$$
$$y + 2 = \pm\sqrt{8} = \pm 2\sqrt{2}$$
$$y = -2 \pm 2\sqrt{2}$$

Now draw the graph:

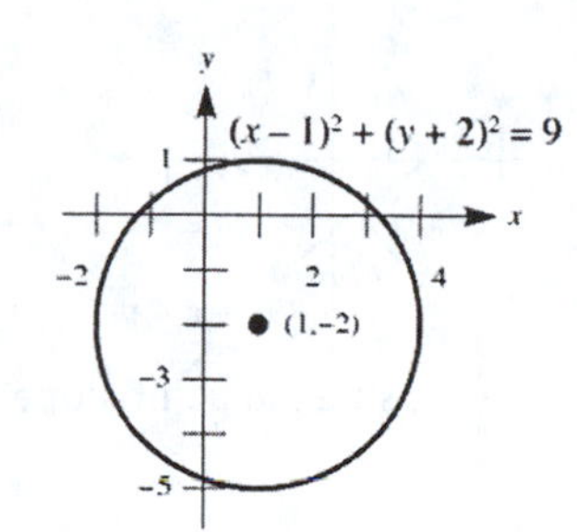

15. **a.** Graphing the equation $y = x^3 - 2x^2 - 9x$:

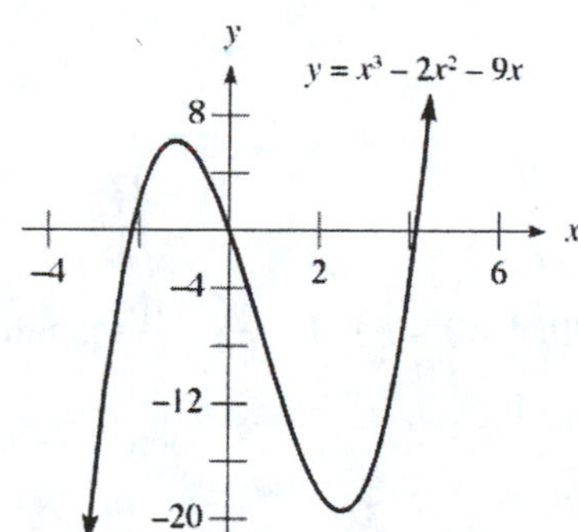

b. By tracing and zooming, the x-intercepts are approximately -2.2, 0, and 4.2.

c. Factoring as $y = x\left(x^2 - 2x - 9\right)$, one x-intercept is 0. Using the quadratic formula:

$$x = \frac{2 \pm \sqrt{(-2)^2 - 4(1)(-9)}}{2(1)} = \frac{2 \pm \sqrt{40}}{2} = \frac{2 \pm 2\sqrt{10}}{2} = 1 \pm \sqrt{10} \approx -2.2, 4.2$$

16. Substituting $x = -\frac{1}{2}$ into $y = 4x^2 - 8x$: $y = 4\left(-\frac{1}{2}\right)^2 - 8\left(-\frac{1}{2}\right) = 4\left(\frac{1}{4}\right) + 4 = 1 + 4 = 5$

So $\left(-\frac{1}{2}, 5\right)$ lies on the graph of $y = 4x^2 - 8x$.

17. **a.** The reflection of $\overline{AB}$ about the x-axis is given by the graph:

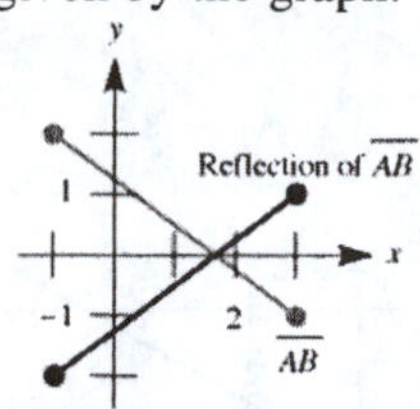

b. The reflection of $\overline{AB}$ about the y-axis is given by the graph:

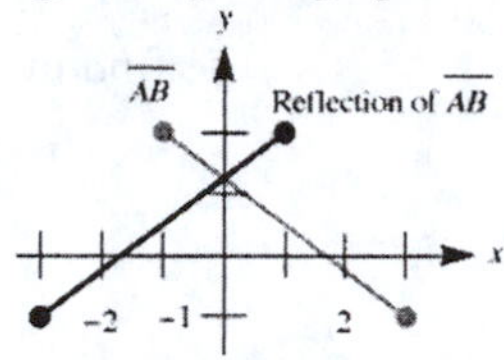

c. The reflection of $\overline{AB}$ about the origin is given by the graph:

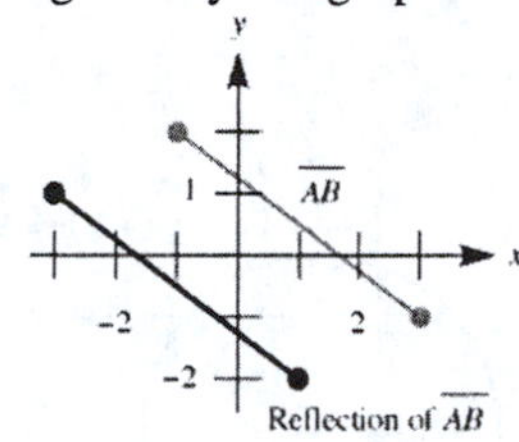

18. First find the slope of the line: $m = \dfrac{4-2}{1-6} = -\frac{2}{5}$. Using the point-slope formula with the point (1,4):

$$y - 4 = -\tfrac{2}{5}(x-1)$$
$$5y - 20 = -2x + 2$$
$$2x + 5y = 22$$

Now find the x- and y-intercepts:

Let $x = 0$	Let $y = 0$
$5y = 22$	$2x = 22$
$y = \frac{22}{5}$	$x = 11$

So the area is given by: Area $= \frac{1}{2}(\text{base})(\text{height}) = \frac{1}{2} \bullet 11 \bullet \frac{22}{5} = \frac{121}{5}$ sq. units

Chapter 2
Equations and Inequalities

2.1 Quadratic Equations: Theory and Examples

1. Solve by completing the square:

$$x^2+8x-2=0$$
$$x^2+8x=2$$
$$x^2+8x+16=2+16$$
$$(x+4)^2=18$$
$$x+4=\pm\sqrt{18}=\pm 3\sqrt{2}$$
$$x=-4\pm 3\sqrt{2}\approx -8.24,\ 0.24$$

3. Solve by completing the square:

$$x^2+4x+1=0$$
$$x^2+4x=-1$$
$$x^2+4x+4=-1+4$$
$$(x+2)^2=3$$
$$x+2=\pm\sqrt{3}$$
$$x=-2\pm\sqrt{3}\approx -3.73,\ -0.27$$

5. Solve by completing the square:

$$2y^2-5y-2=0$$
$$y^2-\tfrac{5}{2}y-1=0$$
$$y^2-\tfrac{5}{2}y=1$$
$$y^2-\tfrac{5}{2}y+\tfrac{25}{16}=1+\tfrac{25}{16}$$
$$\left(y-\tfrac{5}{4}\right)^2=\tfrac{41}{16}$$
$$y-\tfrac{5}{4}=\pm\sqrt{\tfrac{41}{16}}=\pm\frac{\sqrt{41}}{4}$$
$$y=\frac{5\pm\sqrt{41}}{4}\approx -0.35,\ 2.85$$

7. Solve by completing the square:

$$4y^2+8y+5=0$$
$$y^2+2y+\tfrac{5}{4}=0$$
$$y^2+2y=-\tfrac{5}{4}$$
$$y^2+2y+1=-\tfrac{5}{4}+1$$
$$(y+1)^2=-\tfrac{1}{4}$$

The equation has no real roots.

9. Solve by completing the square:

$$4s^2 - 20s + 25 = 0$$
$$s^2 - 5s + \frac{25}{4} = 0$$
$$s^2 - 5s = -\frac{25}{4}$$
$$s^2 - 5s + \frac{25}{4} = -\frac{25}{4} + \frac{25}{4}$$
$$\left(s - \frac{5}{2}\right)^2 = 0$$
$$s - \frac{5}{2} = 0$$
$$s = \frac{5}{2}$$

11. First write the equation as $x^2 - 8x + 6 = 0$. Using the quadratic formula with $a = 1$, $b = -8$, and $c = 6$:

$$x = \frac{-(-8) \pm \sqrt{(-8)^2 - 4(1)(6)}}{2(1)} = \frac{8 \pm \sqrt{64-24}}{2} = \frac{8 \pm \sqrt{40}}{2} = \frac{8 \pm 2\sqrt{10}}{2} = 4 \pm \sqrt{10} \approx 0.84, 7.16$$

13. First write the equation as $-3x^2 + x + 3 = 0$. Using the quadratic formula with $a = -3$, $b = 1$, and $c = 3$:

$$x = \frac{-1 \pm \sqrt{(1)^2 - 4(-3)(3)}}{2(-3)} = \frac{-1 \pm \sqrt{1+36}}{-6} = \frac{-1 \pm \sqrt{37}}{-6} = \frac{1 \pm \sqrt{37}}{6} \approx -0.85, 1.18$$

15. Rewrite the equation as $y^2 + 8y + 1 = 0$. Now using the quadratic formula with $a = 1$, $b = 8$, and $c = 1$:

$$y = \frac{-8 \pm \sqrt{(8)^2 - 4(1)(1)}}{2(1)} = \frac{-8 \pm \sqrt{64-4}}{2} = \frac{-8 \pm \sqrt{60}}{2} = \frac{-8 \pm 2\sqrt{15}}{2} = -4 \pm \sqrt{15} \approx -7.87, -0.13$$

17. Rewrite the equation as $t^2 + 3t + 4 = 0$. Now using the quadratic formula with $a = 1$, $b = 3$, and $c = 4$:

$$t = \frac{-3 \pm \sqrt{(3)^2 - 4(1)(4)}}{2(1)} = \frac{-3 \pm \sqrt{9-16}}{2} = \frac{-3 \pm \sqrt{-7}}{2}$$

The equation has no real roots.

19. **a.** Substituting $y = 100$ into the equation yields:

$$0.006609x^2 - 23.771x + 21382 = 100$$
$$0.006609x^2 - 23.771x + 21282 = 0$$

Using $a = 0.006609$, $b = -23.771$, and $c = 21282$ in the quadratic formula:

$$x = \frac{23.771 \pm \sqrt{(-23.771)^2 - 4(0.006609)(21282)}}{2(0.006609)} \approx 1680, 1917$$

The year 1917 is approximately the year the population will reach 100 million.

b. Our estimate is between 1910 and 1920, so it is reasonable.

21. **a.** Substituting $y = 1620$ into the equation yields:

$$-0.09781t^2 + 385.8336t - 378850.4046 = 1620$$
$$-0.09781t^2 + 385.8336t - 380470.4046 = 0$$
$$0.09781t^2 - 385.8336t + 380470.4046 = 0$$

Using $a = 0.09781$, $b = -385.8336$, and $c = 380470.4046$ in the quadratic formula:

$$t = \frac{385.8336 \pm \sqrt{(-385.8336)^2 - 4(0.09781)(380470.4046)}}{2(0.09781)} \approx 1954, 1990$$

The estimate is the year 1990. Our estimate is 3 years off the actual value.

b. Substituting $y = 1590$ into the equation yields:

$$-0.09781t^2 + 385.8336t - 378850.4046 = 1590$$
$$-0.09781t^2 + 385.8336t - 380440.4046 = 0$$
$$0.09781t^2 - 385.8336t + 380440.4046 = 0$$

Using $a = 0.09781$, $b = -385.8336$, and $c = 380440.4046$ in the quadratic formula:

$$t = \frac{385.8336 \pm \sqrt{(-385.8336)^2 - 4(0.09781)(380440.4046\ )}}{2(0.09781)} \approx 1947, 1997$$

The estimate is the year 1997. Our estimate hits the year exactly.

23. **a.** Graphing the equation $y = x^2 - 4x + 1$:

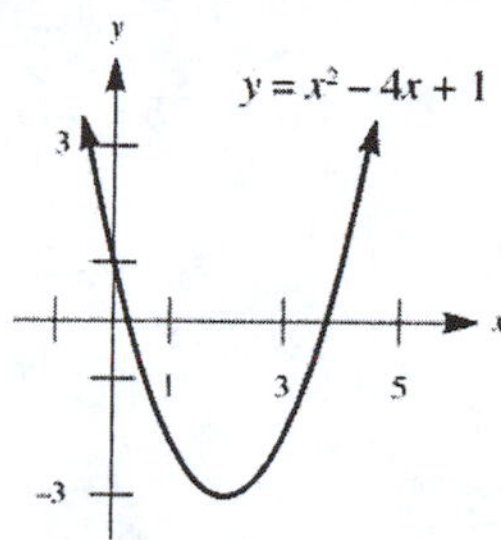

The x-intercepts are approximately 0.268 and 3.732.

b. Using the quadratic formula:

$$x = \frac{-(-4) \pm \sqrt{(-4)^2 - 4(1)(1)}}{2(1)} = \frac{4 \pm \sqrt{16-4}}{2} = \frac{4 \pm \sqrt{12}}{2} = \frac{4 \pm 2\sqrt{3}}{2} = 2 \pm \sqrt{3} \approx 0.2679, 3.7321$$

These intercepts are consistent with those found in **a**.

25. **a.** Graphing the equation $y = 0.5x^2 + 8x - 3$:

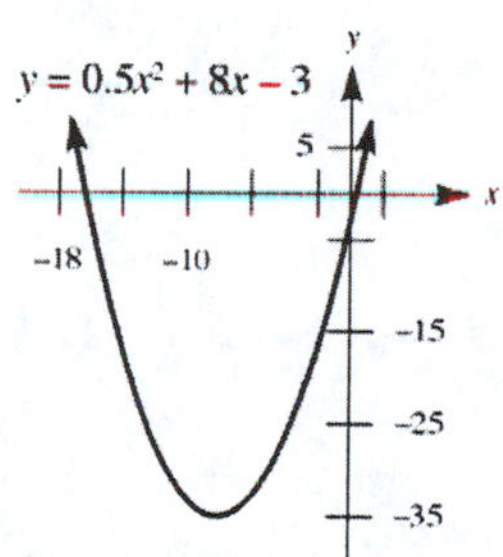

The x-intercepts are approximately –16.367 and 0.367.

b. Using the quadratic formula: $x = \dfrac{-8 \pm \sqrt{(8)^2 - 4(0.5)(-3)}}{2(0.5)} = \dfrac{-8 \pm \sqrt{64+6}}{1} = -8 \pm \sqrt{70} \approx -16.3666, 0.3666$

These intercepts are consistent with those found in part **a**.

27. **a.** Graphing the equation $y = 2x^2 + 2\sqrt{26}\,x + 13$:

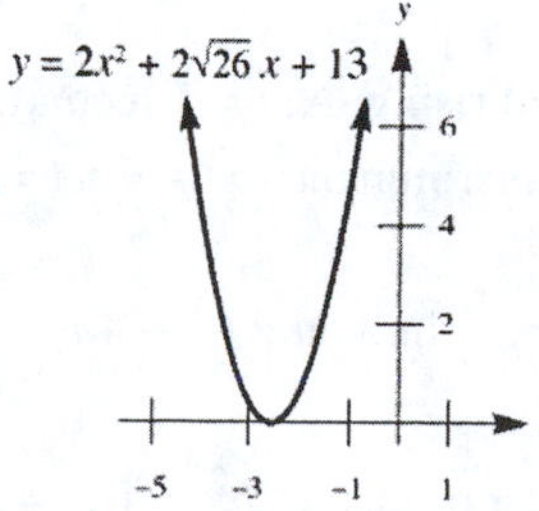

The x-intercept is approximately –2.550.

b. Using the quadratic formula:

$$x=\frac{-2\sqrt{26}\pm\sqrt{\left(2\sqrt{26}\right)^2-4(2)(13)}}{2(2)}=\frac{-2\sqrt{26}\pm\sqrt{104-104}}{4}=\frac{-2\sqrt{26}}{4}=\frac{-\sqrt{26}}{2}\approx-2.5495$$

This intercept is consistent with that found in part **a**.

29. Finding the sum and the product of the roots:

$$r_1+r_2=-b=-8 \qquad r_1\bullet r_2=c=-20$$

31. First divide by 4 to write the equation as $y^2-7y+\frac{9}{4}=0$. Now finding the sum and the product of the roots:

$$r_1+r_2=-b=-(-7)=7 \qquad r_1\bullet r_2=c=\tfrac{9}{4}$$

33. If $a = 1$, we can find c and b:

$$c=r_1\bullet r_2=3\bullet 11=33 \qquad b=-\left(r_1+r_2\right)=-(3+11)=-14$$

Thus the equation is $x^2-14x+33=0$.

35. If $a = 1$, we can find c and b:

$$c=r_1\bullet r_2=\left(1-\sqrt{2}\right)\left(1+\sqrt{2}\right)=1-2=-1 \qquad b=-\left(r_1+r_2\right)=-\left(1-\sqrt{2}+1+\sqrt{2}\right)=-2$$

Thus the equation is $x^2-2x-1=0$.

37. If $a = 1$, we can find c and b:

$$c=r_1\bullet r_2=\tfrac{1}{2}\left(2+\sqrt{5}\right)\bullet\tfrac{1}{2}\left(2-\sqrt{5}\right)=\tfrac{1}{4}(4-5)=-\tfrac{1}{4}$$
$$b=-\left(r_1+r_2\right)=-\left(1+\tfrac{1}{2}\sqrt{5}+1-\tfrac{1}{2}\sqrt{5}\right)=-2$$

Thus the equation is $x^2-2x-\frac{1}{4}=0$. Multiplying by 4 (to produce integer coefficients) results in the equation $4x^2-8x-1=0$.

39. If $r_1=\sqrt{2}$ and $r_2=\sqrt{5}$, note that:

$$c=r_1r_2=\sqrt{2}\bullet\sqrt{5}=\sqrt{10} \qquad b=-\left(r_1+r_2\right)=-\left(\sqrt{2}+\sqrt{5}\right)$$

So the roots are $\sqrt{2}$ and $\sqrt{5}$.

41. **a.** When the ball lands $h = 0$. Therefore:

$$\begin{aligned}-16t^2+96t&=0\\-16t(t-6)&=0\\t=0 \text{ or } t&=6\end{aligned}$$

The value $t = 0$ gives the time when the ball is first thrown. Consequently the ball lands after 6 seconds.

b. With $h = 80$, the equation becomes:

$$\begin{aligned}80&=-16t^2+96t\\16t^2-96t+80&=0\\t^2-6t+5&=0\\(t-5)(t-1)&=0\\t=5 \text{ or } t&=1\end{aligned}$$

At $t = 1$ second the ball is 80 ft. high and rising. At $t = 5$ seconds the ball is 80 ft high and falling.

43. Given $a = 1, b = -12, c = 16$, compute the discriminant: $b^2-4ac=144-4(1)(16)=80>0$

The equation has two real roots.

45. Given $a = 4, b = -5, c=-\frac{1}{2}$, compute the discriminant: $b^2-4ac=25-4(4)\left(-\frac{1}{2}\right)=33>0$

The equation has two real roots.

47. Given $a = 1, b=\sqrt{3}, c=\frac{3}{4}$, compute the discriminant: $b^2-4ac=3-4(1)\left(\frac{3}{4}\right)=0$

The equation has one real root.

49. Given $a = 1, b = -\sqrt{5}, c = 1$, compute the discriminant: $b^2 - 4ac = 5 - 4(1)(1) = 1 > 0$
The equation has two real roots.

51. Set the discriminant equal to 0:
$$b^2 - 4ac = 0$$
$$144 - 4(1)(k) = 0$$
$$4k = 144$$
$$k = 36$$

53. Set the discriminant equal to 0:
$$b^2 - 4ac = 0$$
$$k^2 - 4(1)(5) = 0$$
$$k = \pm\sqrt{20} = \pm 2\sqrt{5}$$

55. Solving for r using the quadratic formula:
$$2\pi r^2 + 2\pi rh = 20\pi$$
$$2\pi r^2 + 2\pi rh - 20\pi = 0$$
$$2\pi\left(r^2 + rh - 10\right) = 0$$
$$r^2 + rh - 10 = 0$$
$$r = \frac{-h \pm \sqrt{h^2 - 4(1)(-10)}}{2(1)} = \frac{-h \pm \sqrt{h^2 + 40}}{2}$$

57. Solving for t by factoring:
$$-16t^2 + v_0 t = 0$$
$$t\left(-16t + v_0\right) = 0$$
$$t = 0 \text{ or } t = \frac{v_0}{16}$$

59. **a.** Graphing the two equations:

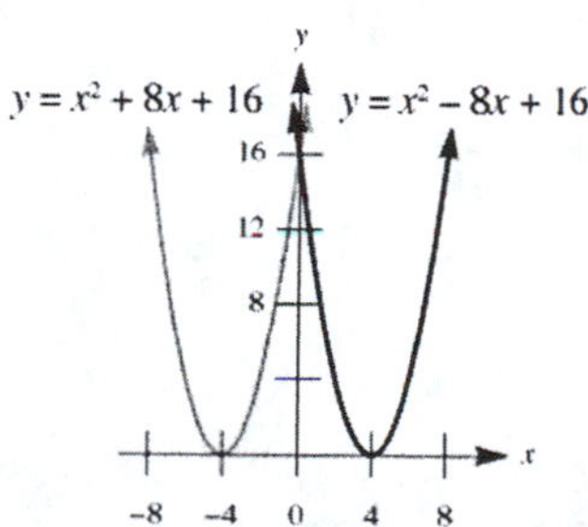

b. The root of $x^2 + 8x + 16 = 0$ appears to be –4, while the root of $x^2 - 8x + 16 = 0$ appears to be 4. The roots are opposites.

c. Solving each equation:
$$x^2 + 8x + 16 = 0$$
$$(x + 4)^2 = 0$$
$$x = -4$$
$$x^2 - 8x + 16 = 0$$
$$(x - 4)^2 = 0$$
$$x = 4$$
The results support the response for part **b**.

61. **a.** Using the quadratic formula with $a = 1$, $b = 3$, and $c = 1$:
$$x = \frac{-3 \pm \sqrt{(3)^2 - 4(1)(1)}}{2(1)} = \frac{-3 \pm \sqrt{9 - 4}}{2} = \frac{-3 \pm \sqrt{5}}{2} = \tfrac{1}{2}\left(-3 \pm \sqrt{5}\right)$$

b. Rationalizing the denominator: $\dfrac{1}{\frac{1}{2}\left(-3 - \sqrt{5}\right)} \bullet \dfrac{-3 + \sqrt{5}}{-3 + \sqrt{5}} = \dfrac{-3 + \sqrt{5}}{\frac{1}{2}(9 - 5)} = \dfrac{-3 + \sqrt{5}}{2} = \tfrac{1}{2}\left(-3 + \sqrt{5}\right)$

c. Yes. Since $r_1 r_2 = c$, and $c = 1$, then $r_1 r_2 = 1$ and thus $r_1 = \dfrac{1}{r_2}$. Thus the roots of the equation $x^2 + 3x + 1 = 0$ are reciprocals.

63. Using the quadratic formula: $x = \dfrac{-b \pm \sqrt{b^2 - 4(a)(-a)}}{2a} = \dfrac{-b \pm \sqrt{b^2 + 4a^2}}{2a}$

Since $a \neq 0$, the quantity $b^2 + 4a^2 > 0$, thus the equation has two real roots.

65. First rewrite the equation as $x^2 - 2(3k + 1)x + 7(2k + 3) = 0$. Since the roots are equal, set the discriminant equal to 0:

$$\begin{aligned} [-2(3k+1)]^2 - 4(1)[7(2k+3)] &= 0 \\ 4(9k^2 + 6k + 1) - 28(2k+3) &= 0 \\ 36k^2 + 24k + 4 - 56k - 84 &= 0 \\ 36k^2 - 32k - 80 &= 0 \\ 9k^2 - 8k - 20 &= 0 \\ (9k+10)(k-2) &= 0 \\ k = -\tfrac{10}{9} \text{ or } k &= 2 \end{aligned}$$

67. **a.** Following the instructions:

$$\begin{aligned} ax^2 + bx &= -c \\ 4a^2x^2 + 4abx &= -4ac \\ 4a^2x^2 + 4abx + b^2 &= b^2 - 4ac \end{aligned}$$

b. Factoring and taking roots:

$$\begin{aligned} (2ax+b)^2 &= b^2 - 4ac \\ 2ax + b &= \pm\sqrt{b^2 - 4ac} \\ 2ax &= -b \pm \sqrt{b^2 - 4ac} \\ x &= \frac{-b \pm \sqrt{b^2 - 4ac}}{2a} \end{aligned}$$

69. Making the substitution $x = y + k$:

$$\begin{aligned} 2(y+k)^2 - 3(y+k) + 1 &= 0 \\ 2y^2 + 4ky + 2k^2 - 3y - 3k + 1 &= 0 \\ 2y^2 + (4k-3)y &= -1 + 3k - 2k^2 \end{aligned}$$

When $k = \frac{3}{4}$:

$$\begin{aligned} 2y^2 + (3-3)y &= -1 + \tfrac{9}{4} - \tfrac{9}{8} \\ 2y^2 &= \tfrac{1}{8} \\ y^2 &= \tfrac{1}{16} \\ y &= \pm\tfrac{1}{4} \end{aligned}$$

Since $x = y + k$, $x = \frac{3}{4} \pm \frac{1}{4}$, so $x = 1$ or $x = \frac{1}{2}$.

71. **a.** Graphing the desired equations:

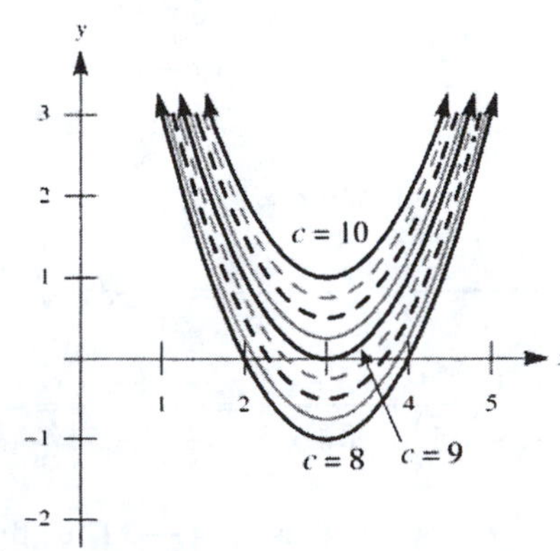

b. It appears when $c = 9$ the two x-intercepts merge into one.

c. Solving the equation when $c = 9$:

$$x^2 - 6x + 9 = 0$$
$$(x-3)^2 = 0$$
$$x = 3$$

This confirms our observation from part **b**.

d. When $c > 9$ the graphs appear to have no x-intercepts. Picking $c = 10$, we have:

$$x = \frac{-(-6) \pm \sqrt{(-6)^2 - 4(1)(10)}}{2(1)} = \frac{6 \pm \sqrt{36-40}}{2} = \frac{6 \pm \sqrt{-4}}{2}$$

Note that this is not a real number.

2.2 Other Types of Equations

1. If $x - 5 \geq 0$, the equation becomes:

$$x - 5 = 1$$
$$x = 6$$

If $x - 5 < 0$, the equation becomes:

$$-(x-5) = 1$$
$$-x + 5 = 1$$
$$-x = -4$$
$$x = 4$$

The solutions of the original equation are 6 and 4.

3. If $x + 6 \geq 0$, the equation becomes:

$$x + 6 = \tfrac{1}{2}$$
$$x = -\tfrac{11}{2}$$

If $x + 6 < 0$, the equation becomes:

$$-(x+6) = \tfrac{1}{2}$$
$$-x - 6 = \tfrac{1}{2}$$
$$-x = \tfrac{13}{2}$$
$$x = -\tfrac{13}{2}$$

The solutions of the original equation are $-\frac{11}{2}$ and $-\frac{13}{2}$.

5. If $6x - 5 \geq 0$, the equation becomes:

$$6x - 5 = 25$$
$$6x = 30$$
$$x = 5$$

If $6x - 5 < 0$, the equation becomes:

$$-(6x-5) = 25$$
$$-6x + 5 = 25$$
$$-6x = 20$$
$$x = -\tfrac{10}{3}$$

The solutions of the original equation are 5 and $-\frac{10}{3}$.

7. Squaring each side of the equation:

$$|x+3| = 2x - 2$$
$$(x+3)^2 = (2x-2)^2$$
$$x^2 + 6x + 9 = 4x^2 - 8x + 4$$
$$0 = 3x^2 - 14x - 5$$
$$0 = (3x+1)(x-5)$$
$$x = -\tfrac{1}{3}, 5$$

After checking in the original equation, note that $x = 5$ is the only solution.

9. Squaring each side of the equation:

$$|2x-1| = 1-\tfrac{1}{2}x$$
$$(2x-1)^2 = \left(1-\tfrac{1}{2}x\right)^2$$
$$4x^2-4x+1 = 1-x+\tfrac{1}{4}x^2$$
$$16x^2-16x+4 = 4-4x+x^2$$
$$15x^2-12x = 0$$
$$3x(5x-4) = 0$$
$$x = 0, \tfrac{4}{5}$$

Both values check in the original equation.

11. a. Graphing the two equations:

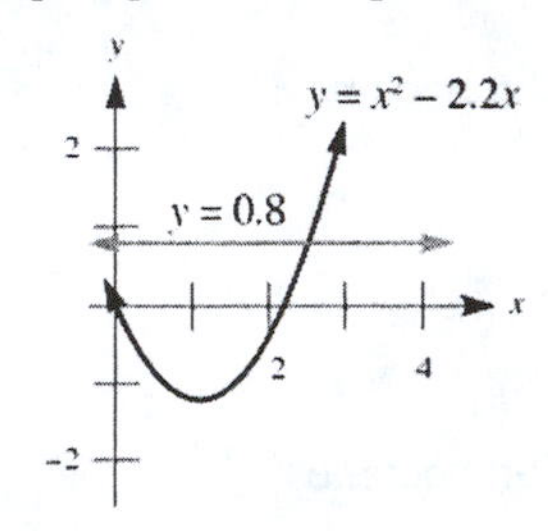

b. Zooming in with a new viewing rectangle:

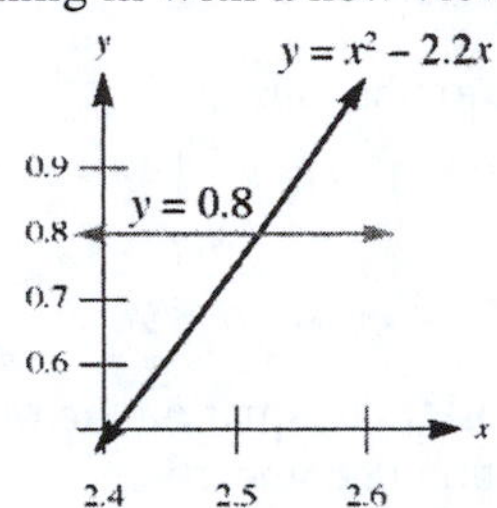

To three decimal places the root is 2.518.

c. Using the quadratic formula to find the positive root for the equation $x^2-2.2x-0.8=0$:

$$x = \frac{-(-2.2)\pm\sqrt{(-2.2)^2-4(1)(-0.8)}}{2(1)} = \frac{2.2\pm\sqrt{4.84+3.2}}{2} = \frac{2.2\pm\sqrt{8.04}}{2} \approx 2.5177$$

13. a. Graphing the equation:

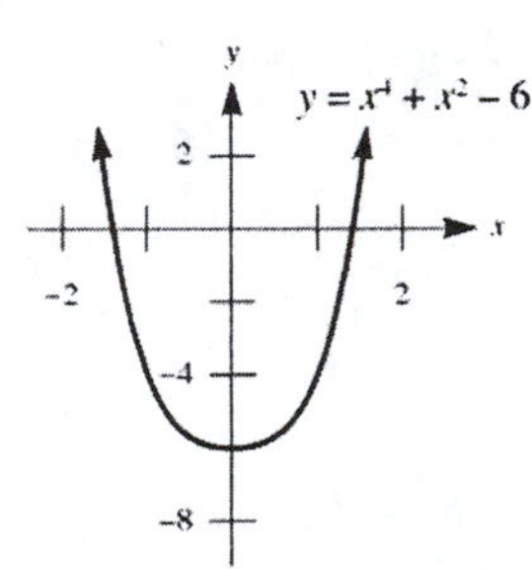

b. There are two roots at ±1.414. These values agree with those found in the example.

15. Factoring the equation:

$$3x^2-48x = 0$$
$$3x(x-16) = 0$$
$$x = 0, 16$$

The solutions are 0 and 16.

17. Factoring the equation:

$$t^3-125 = 0$$
$$(t-5)\left(t^2+5t+25\right) = 0$$

Setting the first factor equal to zero yields $t=5$. Setting the second factor equal to zero yields a quadratic equation with no real roots. The only solution is 5.

19. Factoring the equation:

$$\begin{aligned} 7x^4 - 28x^2 &= 0 \\ 7x^2\left(x^2 - 4\right) &= 0 \\ 7x^2(x-2)(x+2) &= 0 \\ x &= 0, 2, -2 \end{aligned}$$

The solutions are 0, 2, and –2.

21. Factoring the equation:

$$\begin{aligned} t^4 + 2t^3 - 3t^2 &= 0 \\ t^2\left(t^2 + 2t - 3\right) &= 0 \\ t^2(t+3)(t-1) &= 0 \\ t &= 0, -3, 1 \end{aligned}$$

The solutions are 0, –3, and 1.

23. Factoring the equation:

$$\begin{aligned} 6x &= 23x^2 + 4x^3 \\ 4x^3 + 23x^2 - 6x &= 0 \\ x\left(4x^2 + 23x - 6\right) &= 0 \\ x(4x-1)(x+6) &= 0 \\ x &= 0, \tfrac{1}{4}, -6 \end{aligned}$$

The solutions are 0, $\frac{1}{4}$, and –6.

25. Factoring the equation:

$$\begin{aligned} x^4 - x^2 &= 6 \\ x^4 - x^2 - 6 &= 0 \\ \left(x^2 - 3\right)\left(x^2 + 2\right) &= 0 \\ x &= \pm\sqrt{3} \end{aligned}$$

Note that $x^2 + 2 \neq 0$. The solutions are $-\sqrt{3}$ and $\sqrt{3}$.

27. Factoring the equation:

$$\begin{aligned} 4y^2 &= 5 - y^4 \\ y^4 + 4y^2 - 5 &= 0 \\ \left(y^2 + 5\right)\left(y^2 - 1\right) &= 0 \\ \left(y^2 + 5\right)(y+1)(y-1) &= 0 \\ y &= -1, 1 \end{aligned}$$

Note that $y^2 + 5 \neq 0$. The solutions are –1 and 1.

29. Factoring the equation:

$$\begin{aligned} 3t^2 + 2 &= 9t^4 \\ 9t^4 - 3t^2 - 2 &= 0 \\ \left(3t^2 + 1\right)\left(3t^2 - 2\right) &= 0 \\ 3t^2 &= 2 \\ t &= \pm\sqrt{\tfrac{2}{3}} = \pm\tfrac{\sqrt{6}}{3} \end{aligned}$$

Note that $3t^2 + 1 \neq 0$. The solutions are $\pm\frac{\sqrt{6}}{3}$.

31. Taking roots:

$$\begin{aligned} (x-2)^3 - 5 &= 0 \\ (x-2)^3 &= 5 \\ x - 2 &= \sqrt[3]{5} \\ x &= 2 + \sqrt[3]{5} \approx 3.71 \end{aligned}$$

The only solution is $2 + \sqrt[3]{5} \approx 3.71$.

33. Taking roots:

$$\begin{aligned} (x+4)^5 + 16 &= 0 \\ (x+4)^5 &= -16 \\ x + 4 &= -\sqrt[5]{16} \\ x &= -4 - \sqrt[5]{16} \approx -5.74 \end{aligned}$$

The only solution is $-4 - \sqrt[5]{16} \approx -5.74$.

35. **a.** Taking roots:

$$\begin{aligned} (x-3)^4 - 30 &= 0 \\ (x-3)^4 &= 30 \\ x - 3 &= \pm\sqrt[4]{30} \\ x &= 3 \pm \sqrt[4]{30} \approx 5.34, 0.66 \end{aligned}$$

The solutions are $3 + \sqrt[4]{30} \approx 5.34$ and $3 - \sqrt[4]{30} \approx 0.66$.

b. Solving for x:

$$(x-3)^4+30=0$$
$$(x-3)^4=-30$$

Since $(x-3)^4 \geq 0$ for real values of x, there are no solutions to this equation.

37. Factoring the equation:

$$x^6-10x^4+24x^2=0$$
$$x^2\left(x^4-10x^2+24\right)=0$$
$$x^2\left(x^2-6\right)\left(x^2-4\right)=0$$
$$x^2\left(x^2-6\right)(x-2)(x+2)=0$$
$$x=0,\pm\sqrt{6},\pm 2$$

The solutions are 0, $\pm\sqrt{6}$, and ± 2.

39. Letting $x^2=t$ and $x^4=t^2$ yields:

$$t^2+t-1=0$$
$$t=\frac{-1\pm\sqrt{1-4(-1)}}{2}=\frac{-1\pm\sqrt{5}}{2}$$

Since t must be non-negative (because $t=x^2$) choose the value: $t=\dfrac{-1+\sqrt{5}}{2}$. Therefore:

$$x^2=\frac{-1+\sqrt{5}}{2}$$
$$x=\pm\sqrt{\frac{-1+\sqrt{5}}{2}}$$

The solutions are $\pm\sqrt{\dfrac{-1+\sqrt{5}}{2}}$.

41. Letting $x^2=t$ and $x^4=t^2$ yields:

$$t^2+3t-2=0$$
$$t=\frac{-3\pm\sqrt{9-4(-2)}}{2}=\frac{-3\pm\sqrt{17}}{2}$$

Since t must be non-negative (because $t=x^2$) choose the value: $t=\dfrac{-3+\sqrt{17}}{2}$

Therefore:

$$x^2=\frac{-3+\sqrt{17}}{2}$$
$$x=\pm\sqrt{\frac{-3+\sqrt{17}}{2}}$$

The solutions are $\pm\sqrt{\dfrac{-3+\sqrt{17}}{2}}$.

43. Letting $x^3=t$ and $x^6=t^2$, the equation becomes:

$$t^2+7t-8=0$$
$$(t+8)(t-1)=0$$
$$t=-8,1$$

So $x^3=-8$ or $x^3=1$, thus $x=-2$ or $x=1$. The solutions are -2 and 1.

45. Let $t^{-1} = x$ and $t^{-2} = x^2$. Then the equation becomes:

$$\begin{aligned} x^2 - 7x + 12 &= 0 \\ (x-3)(x-4) &= 0 \\ x &= 3, 4 \end{aligned}$$

So $t^{-1} = 3$ or $t^{-1} = 4$, thus $t = \frac{1}{3}$ or $t = \frac{1}{4}$. The solutions are $\frac{1}{3}$ and $\frac{1}{4}$.

47. Letting $y^{-1} = t$ and $y^{-2} = t^2$, we have the equation:

$$\begin{aligned} 12t^2 - 23t + 5 &= 0 \\ (3t-5)(4t-1) &= 0 \\ t &= \frac{5}{3}, \frac{1}{4} \end{aligned}$$

So $y^{-1} = \frac{5}{3}$ or $y^{-1} = \frac{1}{4}$, thus $y = \frac{3}{5}$ or $y = 4$. The solutions are $\frac{3}{5}$ and 4.

49. Letting $x^{-2} = t$ and $x^{-4} = t^2$, we have the equation:

$$\begin{aligned} 4t^2 - 33t - 27 &= 0 \\ (4t+3)(t-9) &= 0 \\ t &= -\frac{3}{4}, 9 \end{aligned}$$

So $x^{-2} = -\frac{3}{4}$ or $x^{-2} = 9$. If $x^{-2} = -\frac{3}{4}$ then $x^2 = -\frac{4}{3}$, which is impossible for real values of x since $x^2 \ge 0$. If $x^{-2} = 9$ then $x^2 = \frac{1}{9}$ and thus $x = \pm\frac{1}{3}$. The solutions are $\pm\frac{1}{3}$.

51. Raising both sides of the given equation to the power $\frac{3}{2}$ yields: $t = \pm 9^{3/2} = \pm\left(\sqrt{9}\right)^3 = \pm 27$

Since $t = \pm 27$ both satisfy the original equation, the solutions are ± 27.

53. Taking cube roots:

$$\begin{aligned} (y-1)^3 &= 7 \\ y-1 &= \sqrt[3]{7} \\ y &= 1+\sqrt[3]{7} \end{aligned}$$

The solution is $1+\sqrt[3]{7}$.

55. Taking fifth roots:

$$\begin{aligned} (t+1)^5 &= -243 \\ t+1 &= \sqrt[5]{-243} \\ t+1 &= -3 \\ t &= -4 \end{aligned}$$

The solution is -4.

57. Letting $x^{2/3} = t$ and $x^{4/3} = t^2$, the equation becomes:

$$\begin{aligned} 9t^2 - 10t + 1 &= 0 \\ (t-1)(9t-1) &= 0 \\ t &= 1, \frac{1}{9} \end{aligned}$$

So $x^{2/3} = 1$ or $x^{2/3} = \frac{1}{9}$. Thus we have either of the two equations:

$$\begin{aligned} x^{2/3} &= 1 \\ \left(\sqrt[3]{x}\right)^2 &= 1 \\ \sqrt[3]{x} &= \pm 1 \\ x &= \pm 1 \end{aligned}$$

$$\begin{aligned} x^{2/3} &= \frac{1}{9} \\ \left(\sqrt[3]{x}\right)^2 &= \frac{1}{9} \\ \sqrt[3]{x} &= \pm\frac{1}{3} \\ x &= \pm\frac{1}{27} \end{aligned}$$

The solutions are ± 1 and $\pm\frac{1}{27}$.

59. Set $y = 0$ to produce the equation $x^4 - 2x^2 - 1 = 0$. Using the quadratic formula:

$$x^2 = \frac{-(-2) \pm \sqrt{(-2)^2 - 4(1)(-1)}}{2(1)} = \frac{2 \pm \sqrt{4+4}}{2} = \frac{2 \pm \sqrt{8}}{2} = \frac{2 \pm 2\sqrt{2}}{2} = 1 \pm \sqrt{2}$$

Since $1 - \sqrt{2} < 0$, $x^2 = 1 - \sqrt{2}$ is impossible. Thus:

$$\begin{aligned} x^2 &= 1 + \sqrt{2} \\ x &= \pm\sqrt{1+\sqrt{2}} \\ x &\approx \pm 1.55 \end{aligned}$$

Note that these two intercepts are consistent with the graph.

61. Squaring both sides yields:

$$\begin{aligned} \sqrt{1-3x} &= 2 \\ 1 - 3x &= 4 \\ -3x &= 3 \\ x &= -1 \end{aligned}$$

Upon checking, –1 is the solution.

63. Isolating the radical and squaring yields:

$$\begin{aligned} \sqrt{x} + 6 &= x \\ \sqrt{x} &= x - 6 \\ x &= x^2 - 12x + 36 \\ 0 &= x^2 - 13x + 36 \\ 0 &= (x-9)(x-4) \\ x &= 4, 9 \end{aligned}$$

Upon checking, 9 is the solution ($x = 4$ does not check).

65. Isolating the radical and squaring yields:

$$\begin{aligned} x - \sqrt{3-x} &= -3 \\ x + 3 &= \sqrt{3-x} \\ x^2 + 6x + 9 &= 3 - x \\ x^2 + 7x + 6 &= 0 \\ (x+6)(x+1) &= 0 \\ x &= -1, -6 \end{aligned}$$

Upon checking, –1 is the solution ($x = -6$ does not check).

67. Isolating the radical and squaring yields:

$$\begin{aligned} 4x + \sqrt{2x+5} &= 0 \\ \sqrt{2x+5} &= -4x \\ 2x + 5 &= 16x^2 \\ 0 &= 16x^2 - 2x - 5 \\ 0 &= (8x-5)(2x+1) \\ x &= -\tfrac{1}{2}, \tfrac{5}{8} \end{aligned}$$

Upon checking, $-\frac{1}{2}$ is the solution ($x = \frac{5}{8}$ does not check).

69. Letting $x^2 = t$ and $x^4 = t^2$:

$$\begin{aligned} \sqrt{t^2 - 13t + 37} &= 1 \\ t^2 - 13t + 37 &= 1 \\ t^2 - 13t + 36 &= 0 \\ (t-9)(t-4) &= 0 \\ t &= 4, 9 \end{aligned}$$

Since $x^2 = t$, $x^2 = 4$ or $x^2 = 9$, thus $x = \pm 2$ or $x = \pm 3$. Upon checking, ±2 and ±3 are solutions.

71. Isolating $\sqrt{1-2x}$ and squaring yields:

$$\begin{aligned}\sqrt{1-2x} &= 4-\sqrt{x+5}\\ 1-2x &= 16-8\sqrt{x+5}+x+5\\ 8\sqrt{x+5} &= 3x+20\\ 64(x+5) &= 9x^2+120x+400\\ 0 &= 9x^2+56x+80\\ 0 &= (x+4)(9x+20)\\ x &= -4, -\tfrac{20}{9}\end{aligned}$$

Upon checking, -4 and $-\frac{20}{9}$ are solutions.

73. Isolating $\sqrt{3+2t}$ and squaring yields:

$$\begin{aligned}\sqrt{3+2t} &= 1-\sqrt{-1+4t}\\ 3+2t &= 1-2\sqrt{-1+4t}-1+4t\\ 2\sqrt{-1+4t} &= 2t-3\\ 4(-1+4t) &= 4t^2-12t+9\\ 0 &= 4t^2-28t+13\\ 0 &= (2t-1)(2t-13)\\ t &= \tfrac{1}{2}, \tfrac{13}{2}\end{aligned}$$

Upon checking, neither of these values are solutions.

75. Isolating $\sqrt{3y+3}$ and squaring yields:

$$\begin{aligned}\sqrt{2y-3}+\sqrt{3y-2} &= \sqrt{3y+3}\\ 2y-3+2\sqrt{2y-3}\sqrt{3y-2}+3y-2 &= 3y+3\\ 2\sqrt{6y^2-13y+6} &= -2y+8\\ \sqrt{6y^2-13y+6} &= -y+4\\ 6y^2-13y+6 &= y^2-8y+16\\ 5y^2-5y-10 &= 0\\ y^2-y-2 &= 0\\ (y-2)(y+1) &= 0\\ y &= -1, 2\end{aligned}$$

Upon checking, 2 is the solution ($y=-1$ does not check).

77. **a.** **a.** Graphing the curve $y=x^4-5x^2+6.2$:

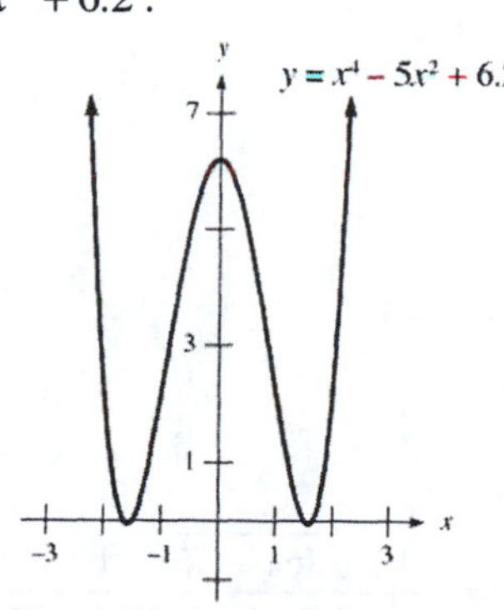

The roots are approximately -1.65, -1.51, 1.51, and 1.65.

b. Using the quadratic formula: $x^2=\dfrac{-(-5)\pm\sqrt{(-5)^2-4(1)(6.2)}}{2(1)}=\dfrac{5\pm\sqrt{0.2}}{2}$

Therefore there are four roots:

$$x=-\sqrt{\frac{5-\sqrt{0.2}}{2}}\approx -1.51 \qquad x=\sqrt{\frac{5-\sqrt{0.2}}{2}}\approx 1.51$$

$$x=-\sqrt{\frac{5+\sqrt{0.2}}{2}}\approx -1.65 \qquad x=\sqrt{\frac{5+\sqrt{0.2}}{2}}\approx 1.65$$

These roots are consistent with those found in part **a**.

b. **a.** Graphing the curve $y = x^4 - 5x^2 + 6.3$:

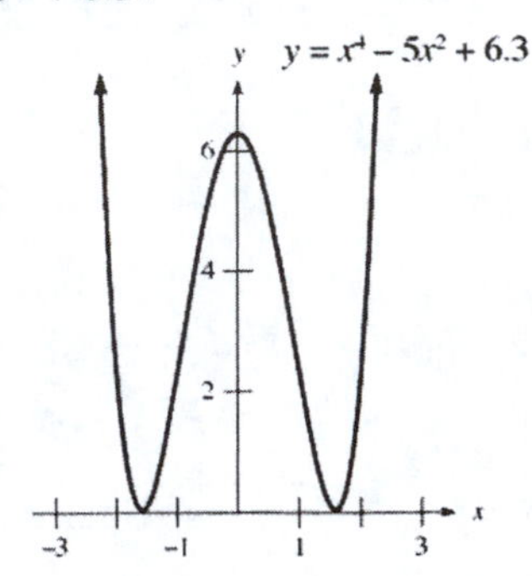

There are no real roots.

b. Using the quadratic formula: $x^2 = \frac{-(-5) \pm \sqrt{(-5)^2 - 4(1)(6.3)}}{2(1)} = \frac{5 \pm \sqrt{-0.2}}{2}$

Therefore there are no real roots.

79. **a.** Graphing the curve $y = x - 5\sqrt{x} + 3$:

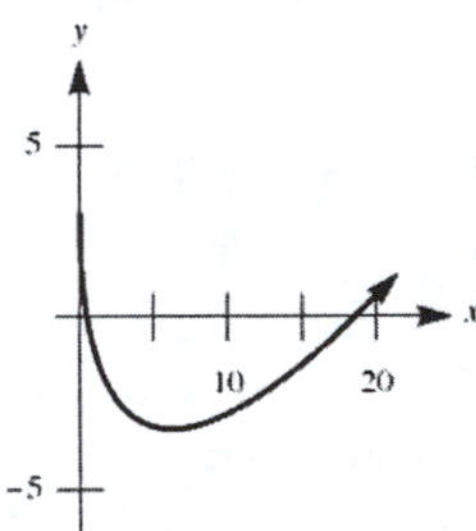

The roots are approximately 0.49 and 18.51.

b. Solving the equation:

$$x - 5\sqrt{x} = -3$$
$$x + 3 = 5\sqrt{x}$$
$$(x+3)^2 = \left(5\sqrt{x}\right)^2$$
$$x^2 + 6x + 9 = 25x$$
$$x^2 - 19x + 9 = 0$$

Using the quadratic formula:

$$x = \frac{-(-19) \pm \sqrt{(-19)^2 - 4(1)(9)}}{2(1)} = \frac{19 \pm \sqrt{361 - 36}}{2} = \frac{19 \pm \sqrt{325}}{2} = \frac{19 \pm 5\sqrt{13}}{2} \approx 0.4861, 18.5139$$

These roots are consistent with those found in part **a**.

81. **a.** Graphing the curve $y = \sqrt{2x-1} - \sqrt{x-2} - 2$:

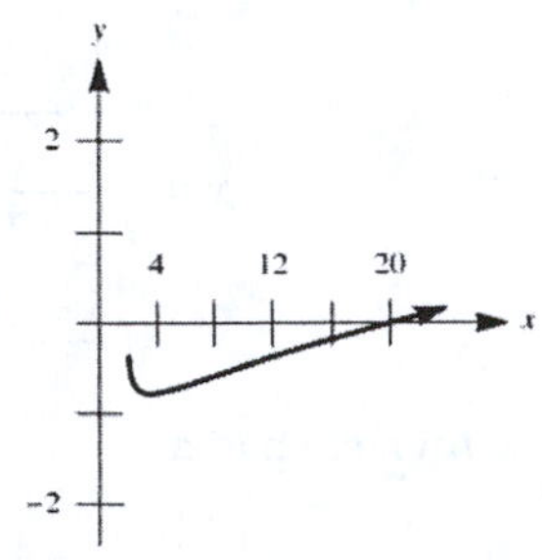

The root is approximately 19.94.

b. Solving the equation:

$$\begin{aligned}
\sqrt{2x-1}-\sqrt{x-2} &= 2\\
\sqrt{2x-1} &= 2+\sqrt{x-2}\\
2x-1 &= 4+4\sqrt{x-2}+x-2\\
2x-1 &= x+2+4\sqrt{x-2}\\
x-3 &= 4\sqrt{x-2}\\
(x-3)^2 &= 16(x-2)\\
x^2-6x+9 &= 16x-32\\
x^2-22x+41 &= 0
\end{aligned}$$

Using the quadratic formula:

$$\begin{aligned}
x &= \frac{-(-22)\pm\sqrt{(-22)^2-4(1)(41)}}{2(1)}\\
&= \frac{22\pm\sqrt{484-164}}{2}\\
&= \frac{22\pm\sqrt{320}}{2}\\
&= \frac{22\pm 8\sqrt{5}}{2}\\
&= 11\pm 4\sqrt{5}\\
&\approx 2.0557,\ 19.9443
\end{aligned}$$

Since $x \approx 2.0557$ is extraneous, the root is $x \approx 19.9443$. This root is consistent with that found in part **a**.

83. **a.** Graphing the curve $y=\dfrac{\sqrt{x}-4}{\sqrt{x}+3}+\dfrac{1}{2}$:

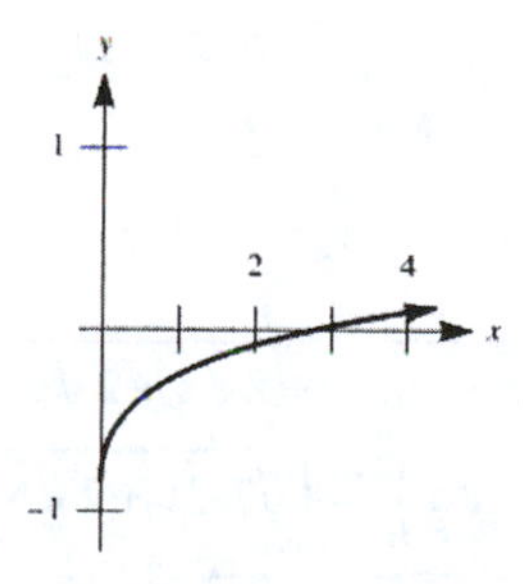

The root is approximately 2.78.

b. Solving the equation:

$$\begin{aligned}
\frac{\sqrt{x}-4}{\sqrt{x}+3}+\frac{1}{2} &= 0\\
\frac{\sqrt{x}-4}{\sqrt{x}+3} &= -\frac{1}{2}\\
2\sqrt{x}-8 &= -\sqrt{x}-3\\
3\sqrt{x} &= 5\\
9x &= 25\\
x &= \tfrac{25}{9}
\end{aligned}$$

This root is consistent with that found in part **a**.

85. **a.** Graphing the curve $y = x^{2/3} - x^{1/3} - 1$:

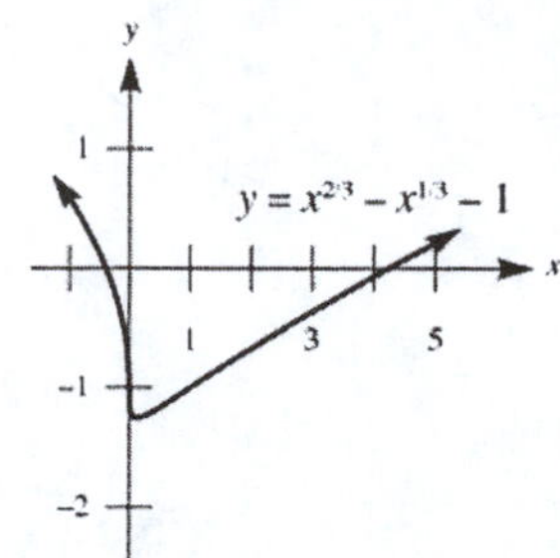

The roots are approximately –0.24 and 4.24.

b. Solving the equation $x^{2/3} - x^{1/3} - 1 = 0$, use the quadratic formula:

$$x^{1/3} = \frac{-(-1) \pm \sqrt{(-1)^2 - 4(1)(-1)}}{2(1)} = \frac{1 \pm \sqrt{1+4}}{2} = \frac{1 \pm \sqrt{5}}{2}$$

Cubing each side, we have two roots:

$$x = \left(\frac{1-\sqrt{5}}{2}\right)^3 \approx -0.2361 \qquad x = \left(\frac{1+\sqrt{5}}{2}\right)^3 \approx 4.2361$$

These roots are consistent with those found in part **a**.

87. We'll show the verification for $x = \sqrt{3+\sqrt{5}}$. (The other three cases are similar.)

With $x = \sqrt{3+\sqrt{5}}$ the equation becomes:

$$\left(\sqrt{3+\sqrt{5}}\right)^4 - 6\left(\sqrt{3+\sqrt{5}}\right)^2 + 4 = 0$$
$$\left(3+\sqrt{5}\right)^2 - 6\left(3+\sqrt{5}\right) + 4 = 0$$
$$9 + 6\sqrt{5} + 5 - 18 - 6\sqrt{5} + 4 = 0$$
$$(9+5-18+4) + 6\sqrt{5} - 6\sqrt{5} = 0$$
$$0 = 0$$

89. Squaring both sides yields:

$$\sqrt{\sqrt{x}+\sqrt{a}} + \sqrt{\sqrt{x}-\sqrt{a}} = \sqrt{2\sqrt{x}+2\sqrt{b}}$$
$$\left(\sqrt{\sqrt{x}+\sqrt{a}} + \sqrt{\sqrt{x}-\sqrt{a}}\right)^2 = \left(\sqrt{2\sqrt{x}+2\sqrt{b}}\right)^2$$
$$\sqrt{x} + \sqrt{a} + 2\sqrt{\sqrt{x}+\sqrt{a}}\sqrt{\sqrt{x}-\sqrt{a}} + \sqrt{x} - \sqrt{a} = 2\sqrt{x} + 2\sqrt{b}$$
$$\sqrt{\sqrt{x}+\sqrt{a}}\sqrt{\sqrt{x}-\sqrt{a}} = \sqrt{b}$$
$$\left(\sqrt{x}+\sqrt{a}\right)\left(\sqrt{x}-\sqrt{a}\right) = b$$
$$x - a = b$$
$$x = a + b$$

Upon checking, we find that this value of x satisfies the original equation. *Note*: The following fact is useful in carrying out the check. Two non-negative quantities are equal if and only if their squares are equal.

91. With $t = \frac{x-a}{x}$ the given equation becomes:

$$\begin{aligned}\sqrt{t} + \frac{4}{\sqrt{t}} &= 5\\ t+4 &= 5\sqrt{t}\\ t^2+8t+16 &= 25t\\ t^2-17t+16 &= 0\\ (t-1)(t-16) &= 0\\ t &= 1, 16\end{aligned}$$

Note that both $t = 1$ and $t = 16$ satisfy the original equation involving t. If $t = 1$:

$$\begin{aligned}1 &= \frac{x-a}{x}\\ x &= x-a\\ 0 &= -a\end{aligned}$$

Since $a \neq 0$, we discard this case. If $t = 16$:

$$\begin{aligned}16 &= \frac{x-a}{x}\\ 16x &= x-a\\ 15x &= -a\\ x &= -\frac{a}{15}\end{aligned}$$

2.3 Inequalities

1. Solving the inequality:

$$\begin{aligned}2x-7 &< 11\\ 2x &< 18\\ x &< 9\end{aligned}$$

The solution set is $(-\infty, 9)$.

3. Solving the inequality:

$$\begin{aligned}4x+6 &< 3(x-1)-x\\ 4x+6 &< 2x-3\\ 2x &< -9\\ x &< -\tfrac{9}{2}\end{aligned}$$

The solution set is $\left(-\infty, -\tfrac{9}{2}\right)$.

5. Multiplying by 15 then solving the inequality:

$$\begin{aligned}\frac{3x}{5} - \frac{x-1}{3} &< 1\\ 9x-5(x-1) &< 15\\ 4x+5 &< 15\\ 4x &< 10\\ x &< \tfrac{5}{2}\end{aligned}$$

The solution set is $\left(-\infty, \tfrac{5}{2}\right)$.

7. Multiplying by 6 then solving the inequality:

$$\begin{aligned}\frac{2x+1}{2} - \frac{x-1}{3} &< x+\tfrac{1}{2}\\ 3(2x+1)-2(x-1) &< 6x+3\\ 4x+5 &< 6x+3\\ -2x &< -2\\ x &> 1\end{aligned}$$

The solution set is $(1, \infty)$.

9. Multiplying by 3 then solving the double inequality:

$$\begin{aligned}-1 \le \frac{1-4t}{3} &\le 1\\ -3 \le 1-4t &\le 3\\ -4 \le -4t &\le 2\\ 1 \ge t &\ge -\tfrac{1}{2}\\ -\tfrac{1}{2} \le t &\le 1\end{aligned}$$

The solution set is $\left[-\tfrac{1}{2}, 1\right]$.

11. Multiplying by 2 then solving the double inequality:

$$\begin{aligned}0.99 < \frac{x}{2} - 1 &< 0.999\\ 1.98 < x-2 &< 1.998\\ 3.98 < x &< 3.998\end{aligned}$$

The solution set is $(3.98, 3.998)$.

13. Solving each inequality separately:

$$x-5\le 2x+3 \qquad -x\le 8 \qquad x\ge -8$$

$$2x+3<10-3x \qquad 5x<7 \qquad x<\tfrac{7}{5}$$

Since both inequalities must be satisfied, the solution set is $\left[-8,\frac{7}{5}\right)$.

15. **a.** The inequality is equivalent to $-\frac{1}{2}\le x\le\frac{1}{2}$, so the solution set is $\left[-\frac{1}{2},\frac{1}{2}\right]$.

b. The inequality is equivalent to $x>\frac{1}{2}$ or $x<-\frac{1}{2}$, so the solution set is $\left(-\infty,-\frac{1}{2}\right)\cup\left(\frac{1}{2},\infty\right)$.

17. **a.** The inequality is equivalent to $x>0$ or $x<0$, so the solution set is $(-\infty,0)\cup(0,\infty)$.

b. Since $|x|\ge 0$ for all x, there is no solution to this inequality.

19. **a.** Solving the inequality:

$$x-2<1 \qquad x<3$$

The solution set is $(-\infty,3)$.

b. Solving the inequality:

$$|x-2|<1 \qquad -1<x-2<1 \qquad 1<x<3$$

The solution set is $(1,3)$.

c. Solving the inequality:

$$|x-2|>1$$

$$x-2>1 \text{ or } x-2<-1$$

$$x>3 \text{ or } x<1$$

The solution set is $(-\infty,1)\cup(3,\infty)$.

21. **a.** Solving the inequality:

$$1-x\le 5 \qquad -x\le 4 \qquad x\ge -4$$

The solution set is $[-4,\infty)$.

b. Solving the inequality:

$$|1-x|\le 5 \qquad -5\le 1-x\le 5 \qquad -6\le -x\le 4 \qquad 6\ge x\ge -4$$

The solution set is $[-4,6]$.

c. Solving the inequality:

$$|1-x|>5$$

$$1-x>5 \text{ or } 1-x<-5$$

$$-x>4 \text{ or } -x<-6$$

$$x<-4 \text{ or } x>6$$

The solution set is $(-\infty,-4)\cup(6,\infty)$.

23. **a.** Solving the inequality:

$$a-x<c \qquad -x<c-a \qquad x>a-c$$

The solution set is $(a-c,\infty)$.

b. Solving the inequality:

$$|a-x|<c \qquad -c<a-x<c \qquad -c-a<-x<c-a \qquad a+c>x>a-c$$

The solution set is $(a-c,a+c)$.

c. Solving the inequality:

$$|a-x|\ge c$$

$$a-x\ge c \text{ or } a-x\le -c$$

$$-x\ge c-a \text{ or } -x\le -c-a$$

$$x\le a-c \text{ or } x\ge a+c$$

The solution set is $(-\infty,a-c]\cup[a+c,\infty)$.

25. Solving the inequality:

$$\left|\frac{x-2}{3}\right| < 4$$
$$-4 < \frac{x-2}{3} < 4$$
$$-12 < x-2 < 12$$
$$-10 < x < 14$$

The solution set is $(-10, 14)$.

27. Solving the inequality:

$$\left|\frac{x+1}{2} - \frac{x-1}{3}\right| < 1$$
$$-1 < \frac{x+1}{2} - \frac{x-1}{3} < 1$$
$$-6 < 3(x+1) - 2(x-1) < 6$$
$$-6 < x+5 < 6$$
$$-11 < x < 1$$

The solution set is $(-11, 1)$.

29. **a.** Solving the inequality:

$$\left|(x+h)^2 - x^2\right| < 3h^2$$
$$\left|x^2 + 2xh + h^2 - x^2\right| < 3h^2$$
$$\left|2xh + h^2\right| < 3h^2$$
$$-3h^2 < 2xh + h^2 < 3h^2$$
$$-4h^2 < 2xh < 2h^2$$

Since $h > 0$, we can divide by $2h$ to obtain $-2h < x < h$. The solution set is $(-2h, h)$.

b. Since $h < 0$, when we divide the inequality in **a** by $2h$, we reverse the inequalities to obtain $-2h > x > h$. The solution set is $(h, -2h)$.

31. **a.** The solution set will be of the form $[a, \infty)$.

b. Sketching the graph:

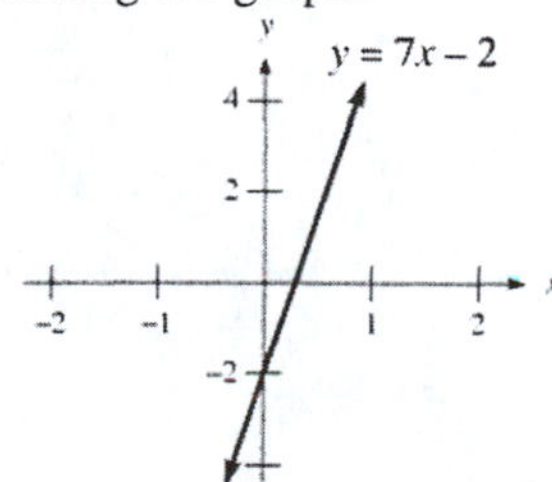

The value of a is approximately 0.3.

c. Solving the inequality:

$$7x - 2 \geq 0$$
$$7x \geq 2$$
$$x \geq \tfrac{2}{7}$$

The solution set is $\left[\tfrac{2}{7}, \infty\right)$, where $a = \tfrac{2}{7}$.

33. **a.** The solution set will be of the form $[a, b]$.

b. Sketching the graph:

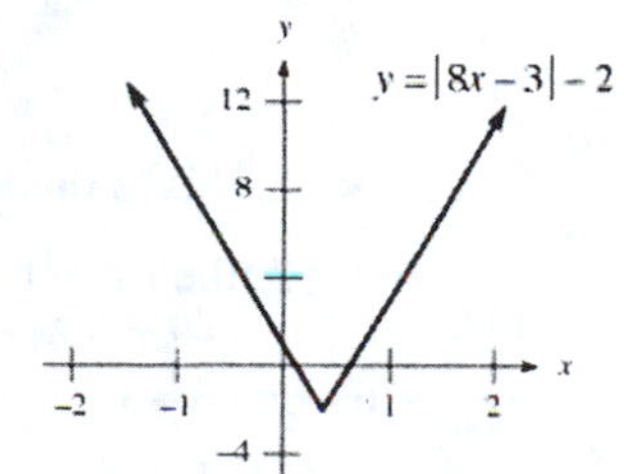

The values are $a \approx 0.1$ and $b \approx 0.6$.

c. Solving the inequality:

$$|8x-3| - 2 \leq 0$$
$$|8x-3| \leq 2$$
$$-2 \leq 8x - 3 \leq 2$$
$$1 \leq 8x \leq 5$$
$$\tfrac{1}{8} \leq x \leq \tfrac{5}{8}$$

The solution set is $\left[\tfrac{1}{8}, \tfrac{5}{8}\right]$, where $a = \tfrac{1}{8}$ and $b = \tfrac{5}{8}$.

35. First solve the equation $F = \frac{9}{5}C + 32$ for C. This yields: $C = \dfrac{5F-160}{9}$

The following inequalities are then equivalent:

$$-183 \le C \le 112$$
$$-183 \le \frac{5F-160}{9} \le 112$$
$$-1647 \le 5F - 160 \le 1008$$
$$-1487 \le 5F \le 1168$$
$$-\tfrac{1487}{5} \le F \le \tfrac{1168}{5}$$

After carrying out the arithmetic and rounding off to the nearest integer, we obtain $-297° \le F \le 234°$, or $[-297°, 234°]$.

37. Using the expression for C:

$$-170 \le C \le 430$$
$$-170 \le \frac{5F-160}{9} \le 430$$
$$-1530 \le 5F - 160 \le 3870$$
$$-1370 \le 5F \le 4030$$
$$-274 \le F \le 806$$

Rounding to the nearest 10°, the corresponding range is $[-270°, 810°]$.

39. Consider three cases.

case 1: If $x < -3$, then $x + 3 < 0$ and $x - 2 < 0$. Therefore:

$$6 + (x+3) + (x-2) < 0$$
$$2x + 7 < 0$$
$$2x < -7$$
$$x < -\tfrac{7}{2}$$

Since $x < -3$, this is the interval $\left(-\infty, -\frac{7}{2}\right)$.

case 2: If $-3 \le x < 2$, then $x + 3 \ge 0$ and $x - 2 < 0$. Therefore:

$$6 - (x+3) + (x-2) < 0$$
$$6 - x - 3 + x - 2 < 0$$
$$1 < 0$$

Since this is impossible, there are no solutions for this case.

case 3: If $x \ge 2$, then $x + 3 \ge 0$ and $x - 2 \ge 0$. Therefore:

$$6 - (x+3) - (x-2) < 0$$
$$6 - x - 3 - x + 2 < 0$$
$$-2x + 5 < 0$$
$$-2x < -5$$
$$x > \tfrac{5}{2}$$

Since $x \ge 2$, this is the interval $\left(\frac{5}{2}, \infty\right)$.

Putting our solutions together, the solution is $\left(-\infty, -\frac{7}{2}\right) \cup \left(\frac{5}{2}, \infty\right)$.

41. **a.** Using the result of the previous exercise: $\sqrt{pqr} = \sqrt{(pq)(r)} \le \dfrac{pq+r}{2}$

b. Using the result of the previous exercise: $\sqrt{pqrs} = \sqrt{(pq)(rs)} \le \dfrac{pq+rs}{2}$

43. **a.** Sketching the graph:

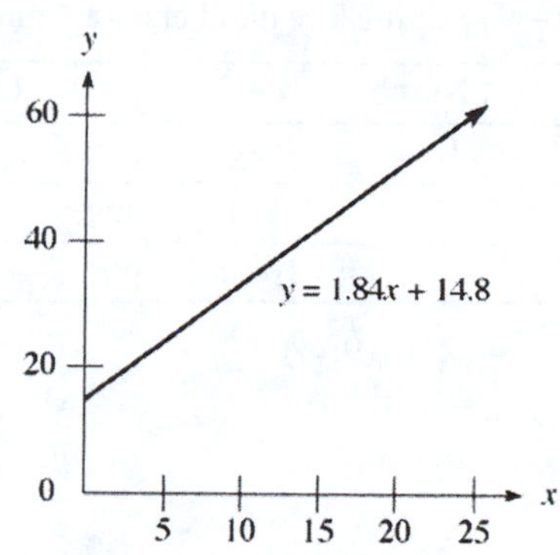

The positive slope tells us the sulfur dioxide emissions are increasing.

b. Solving the inequality:

$$1.84x + 14.8 > 65$$
$$1.84x > 50.2$$
$$x > 27.3$$

The sulfer dioxide emissions might exceed 65 million tons when $x = 28$, which is the year 2008.

2.4 More on Inequalities

1. **a.** Using the graph, the solution is $(-\infty,-1]\cup[4,\infty)$. **b.** Using the graph, the solution is $[-1,4]$.

3. **a.** From the graph, there is no solution.

b. From the graph, the solution is all real numbers, or $(-\infty,\infty)$.

5. **a.** Using the graph, the solution is $[-1,1]\cup[3,\infty)$. **b.** Using the graph, the solution is $(-\infty,-1)\cup(1,3)$.

7. Graphing the equation $y = x^3 - 2x^2 - 3x$:

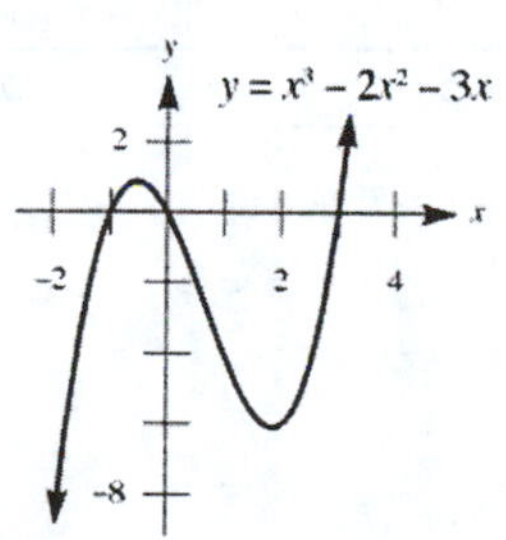

Note that $y > 0$ when $-1 < x < 0$ and when $x > 3$. Thus the curve is positive on the set $(-1,0)\cup(3,\infty)$, which represents the solution set for the inequality.

9. First factor $x^2 + x - 6 = (x+3)(x-2)$, so the key numbers are –3 and 2. Now draw the sign chart:

Interval	Test Number	$x+3$	$x-2$	$(x+3)(x-2)$
$(-\infty,-3)$	–4	neg.	neg.	pos.
$(-3,2)$	0	pos.	neg.	neg.
$(2,\infty)$	3	pos.	pos.	pos.

So $x^2 + x - 6 < 0$ on the interval $(-3, 2)$.

11. First factor $x^2 - 11x + 18 = (x-2)(x-9)$, so the key numbers are 2 and 9. Now draw the sign chart:

Interval	Test Number	$x-2$	$x-9$	$(x-2)(x-9)$
$(-\infty, 2)$	0	neg.	neg.	pos.
$(2, 9)$	3	pos.	neg.	neg.
$(9, \infty)$	10	pos.	pos.	pos.

So $x^2 - 11x + 18 > 0$ on the intervals $(-\infty, 2) \cup (9, \infty)$.

13. First simplify and factor the inequality:

$$9x - x^2 \le 20$$
$$-x^2 + 9x - 20 \le 0$$
$$x^2 - 9x + 20 \ge 0$$
$$(x-4)(x-5) \ge 0$$

So the key numbers are 4 and 5. Now draw the sign chart:

Interval	Test Number	$x-4$	$x-5$	$(x-4)(x-5)$
$(-\infty, 4)$	0	neg.	neg.	pos.
$(4, 5)$	$\frac{9}{2}$	pos.	neg.	neg.
$(5, \infty)$	6	pos.	pos.	pos.

So $x^2 - 9x + 20 \ge 0$ (and consequently the original inequality is satisfied) on the intervals $(-\infty, 4] \cup [5, \infty)$.

15. First factor $x^2 - 16 = (x+4)(x-4)$, so the key numbers are –4 and 4. Now draw the sign chart:

Interval	Test Number	$x-4$	$x+4$	$(x-4)(x+4)$
$(-\infty, -4)$	–5	neg.	neg.	pos.
$(-4, 4)$	0	neg.	pos.	neg.
$(4, \infty)$	5	pos.	pos.	pos.

So $x^2 - 16 \ge 0$ on the intervals $(-\infty, -4] \cup [4, \infty)$.

17. First simplify and factor the inequality:

$$16x^2 + 24x < -9$$
$$16x^2 + 24x + 9 < 0$$
$$(4x+3)^2 < 0$$

Since $(4x+3)^2 \ge 0$ for all real values of x, this inequality has no solution.

19. First factor $x^3 + 13x^2 + 42x = x(x^2 + 13x + 42) = x(x+7)(x+6)$, so the key numbers are 0, –7 and –6. Now draw the sign chart:

Interval	Test Number	x	$x+7$	$x+6$	$x(x+7)(x+6)$
$(-\infty, -7)$	–8	neg.	neg.	neg.	neg.
$(-7, -6)$	$-\frac{13}{2}$	neg.	pos.	neg.	pos.
$(-6, 0)$	–1	neg.	pos.	pos.	neg.
$(0, \infty)$	1	pos.	pos.	pos.	pos.

So $x^3 + 13x^2 + 42x > 0$ on the intervals $(-7, -6) \cup (0, \infty)$.

21. Since $2x^2 + 1 \ge 1$ for all real numbers x, the inequality is satisfied for the interval $(-\infty, \infty)$.

23. First factor $12x^3+17x^2+6x=x\left(12x^2+17x+6\right)=x(4x+3)(3x+2)$, so the key numbers are $0,-\frac{3}{4}$ and $-\frac{2}{3}$.
Now draw the sign chart:

Interval	Test Number	x	$4x+3$	$3x+2$	$x(4x+3)(3x+2)$
$\left(-\infty,-\frac{3}{4}\right)$	-1	neg.	neg.	neg.	neg.
$\left(-\frac{3}{4},-\frac{2}{3}\right)$	$-\frac{17}{24}$	neg.	pos.	neg.	pos.
$\left(-\frac{2}{3},0\right)$	$-\frac{1}{3}$	neg.	pos.	pos.	neg.
$(0,\infty)$	1	pos.	pos.	pos.	pos.

So $12x^3+17x^2+6x<0$ on the intervals $\left(-\infty,-\frac{3}{4}\right)\cup\left(-\frac{2}{3},0\right)$.

25. The key numbers are found by solving the equation $x^2+x-1=0$. Using the quadratic formula we have:

$$x=\frac{-1\pm\sqrt{1-4(-1)}}{2}=\frac{-1\pm\sqrt{5}}{2}$$

For purposes of picking appropriate test numbers, note that: $\frac{-1+\sqrt{5}}{2}\approx 0.6$ and $\frac{-1-\sqrt{5}}{2}\approx -1.6$

Draw the sign chart:

Interval	Test Number	$x+1.6$	$x-0.6$	x^2+x-1
$\left(-\infty,\frac{-1-\sqrt{5}}{2}\right)$	-2	neg.	neg.	pos.
$\left(\frac{-1-\sqrt{5}}{2},\frac{-1+\sqrt{5}}{2}\right)$	0	pos.	neg.	neg.
$\left(\frac{-1+\sqrt{5}}{2},\infty\right)$	1	pos.	pos.	pos.

So $x^2+x-1>0$ on the intervals $\left(-\infty,\frac{-1-\sqrt{5}}{2}\right)\cup\left(\frac{-1+\sqrt{5}}{2},\infty\right)$.

27. The key numbers are found by using the quadratic formula to solve $x^2-8x+2=0$. We have:

$$x=\frac{8\pm\sqrt{64-4(2)}}{2}=\frac{8\pm\sqrt{56}}{2}=\frac{8\pm 2\sqrt{14}}{2}=4\pm\sqrt{14}$$

For purposes of picking appropriate test numbers, note that $4+\sqrt{14}\approx 7.7$ and $4-\sqrt{14}\approx 0.3$. Draw the sign chart:

Interval	Test Number	$x-7.7$	$x-0.3$	x^2-8x+2
$(-\infty,4-\sqrt{14})$	0	neg.	neg.	pos.
$(4-\sqrt{14},4+\sqrt{14})$	4	neg.	pos.	neg.
$(4+\sqrt{14},\infty)$	8	pos.	pos.	pos.

So $x^2-8x+2\le 0$ on the interval $\left[4-\sqrt{14},4+\sqrt{14}\right]$.

29. The key numbers are –4, –3, and 1. Draw the sign chart:

Interval	Test Number	$x-1$	$x+3$	$x+4$	$(x-1)(x+3)(x+4)$
$(-\infty,-4)$	-5	neg.	neg.	neg.	neg.
$(-4,-3)$	$-\frac{7}{2}$	neg.	neg.	pos.	pos.
$(-3,1)$	0	neg.	pos.	pos.	neg.
$(1,\infty)$	2	pos.	pos.	pos.	pos.

So $(x-1)(x+3)(x+4) \geq 0$ on the intervals $[-4,-3]\cup[1,\infty)$.

31. The key numbers are –4, –5, and –6. Draw the sign chart:

Interval	Test Number	$x+4$	$x+5$	$x+6$	$(x+4)(x+5)(x+6)$
$(-\infty,-6)$	-7	neg.	neg.	neg.	neg.
$(-6,-5)$	$-\frac{11}{2}$	neg.	neg.	pos.	pos.
$(-5,-4)$	$-\frac{9}{2}$	neg.	pos.	pos.	neg.
$(-4,\infty)$	0	pos.	pos.	pos.	pos.

So $(x+4)(x+5)(x+6) < 0$ on the intervals $(-\infty,-6)\cup(-5,-4)$.

33. The key numbers are $-\frac{1}{3}, \frac{1}{3}$ and 2. Draw the sign chart:

Interval	Test Number	$(x-2)^2$	$(3x+1)^3$	$3x-1$	product
$\left(-\infty,-\frac{1}{3}\right)$	-1	pos.	neg.	neg.	pos.
$\left(-\frac{1}{3},\frac{1}{3}\right)$	0	pos.	pos.	neg.	neg.
$\left(\frac{1}{3},2\right)$	1	pos.	pos.	pos.	pos.
$(2,\infty)$	3	pos.	pos.	pos.	pos.

So $(x-2)^2(3x+1)^3(3x-1) > 0$ on the intervals $\left(-\infty,-\frac{1}{3}\right)\cup\left(\frac{1}{3},2\right)\cup(2,\infty)$.

35. The key numbers are $3, -1, -\frac{1}{2}$ and $-\frac{2}{3}$. Draw the sign chart:

Interval	Test Number	$(x-3)^2$	$(x+1)^4$	$(2x+1)^4$	$3x+2$	product
$(-\infty,-1)$	-2	pos.	pos.	pos.	neg.	neg.
$\left(-1,-\frac{2}{3}\right)$	$-\frac{3}{4}$	pos.	pos.	pos.	neg.	neg.
$\left(-\frac{2}{3},-\frac{1}{2}\right)$	$-\frac{7}{12}$	pos.	pos.	pos.	pos.	pos.
$\left(-\frac{1}{2},3\right)$	0	pos.	pos.	pos.	pos.	pos.
$(3,\infty)$	4	pos.	pos.	pos.	pos.	pos.

So $(x-3)^2(x+1)^4(2x+1)^4(3x+2) \leq 0$ on the intervals $(-\infty,-1]\cup\left[-1,-\frac{2}{3}\right]$, which is equivalent to the interval $\left(-\infty,-\frac{2}{3}\right]$. It is also zero at 3 and –1/2, so the inequality is satisfied for the values $\left(-\infty,-\frac{2}{3}\right]\cup\left\{3,-\frac{1}{2}\right\}$.

37. First simplify and factor the inequality:

$$20 \ge x^2\left(9 - x^2\right)$$
$$20 \ge 9x^2 - x^4$$
$$x^4 - 9x^2 + 20 \ge 0$$
$$\left(x^2 - 4\right)\left(x^2 - 5\right) \ge 0$$
$$(x-2)(x+2)\left(x-\sqrt{5}\right)\left(x+\sqrt{5}\right) \ge 0$$

The key numbers are -2, 2, $-\sqrt{5}$, and $\sqrt{5}$. Draw the sign chart:

Interval	Test Number	$x-\sqrt{5}$	$x-2$	$x+2$	$x+\sqrt{5}$	x^4-9x^2+20
$\left(-\infty,-\sqrt{5}\right)$	–3	neg.	neg.	neg.	neg.	pos.
$\left(-\sqrt{5},-2\right)$	–2.1	neg.	neg.	neg.	pos.	neg.
$(-2,2)$	0	neg.	neg.	pos.	pos.	pos.
$\left(2,\sqrt{5}\right)$	2.1	neg.	pos.	pos.	pos.	neg.
$(\sqrt{5},\infty)$	3	pos.	pos.	pos.	pos.	pos.

So $x^4 - 9x^2 + 20 \ge 0$ (and consequently the original inequality is satisfied) on the intervals $\left(-\infty,-\sqrt{5}\right] \cup [-2,2] \cup \left[\sqrt{5},\infty\right)$.

39. First simplify and factor the inequality:

$$9(x-4) - x^2(x-4) < 0$$
$$(x-4)\left(9-x^2\right) < 0$$
$$(x-4)(3+x)(3-x) < 0$$

The key numbers are 4, –3 and 3. Draw the sign chart:

Interval	Test Number	$x-4$	$3+x$	$3-x$	$(x-4)(3+x)(3-x)$
$(-\infty,-3)$	-4	neg.	neg.	pos.	pos.
$(-3,3)$	0	neg.	pos.	pos.	neg.
$(3,4)$	3.5	neg.	pos.	neg.	pos.
$(4,\infty)$	6	pos.	pos.	neg.	neg.

So $(x-4)(3+x)(3-x) < 0$ (and consequently the original inequality is satisfied) on the intervals $(-3,3) \cup (4,\infty)$.

41. First simplify and factor the inequality:

$$4\left(x^2-9\right) - \left(x^2-9\right)^2 > -5$$
$$4x^2 - 36 - x^4 + 18x^2 - 81 > -5$$
$$-x^4 + 22x^2 - 112 > 0$$
$$x^4 - 22x^2 + 112 < 0$$
$$\left(x^2-8\right)\left(x^2-14\right) < 0$$
$$\left(x-2\sqrt{2}\right)\left(x+2\sqrt{2}\right)\left(x-\sqrt{14}\right)\left(x+\sqrt{14}\right) < 0$$

The key numbers are $2\sqrt{2}, -2\sqrt{2}, \sqrt{14}$, and $-\sqrt{14}$. Draw the sign chart:

Interval	Test Number	x^2-8	x^2-14	x^4-22x^2+112
$(-\infty,-\sqrt{14})$	-5	pos.	pos.	pos.
$(-\sqrt{14},-2\sqrt{2})$	-3	pos.	neg.	neg.
$(-2\sqrt{2},2\sqrt{2})$	0	neg.	neg.	pos.
$(2\sqrt{2},\sqrt{14})$	3	pos.	neg.	neg.
$(\sqrt{14},\infty)$	5	pos.	pos.	pos.

So $x^4-22x^2+112<0$ (and consequently the original inequality is satisfied) on the intervals $\left(-\sqrt{14},-2\sqrt{2}\right)\cup\left(2\sqrt{2},\sqrt{14}\right)$.

43. The key numbers are -1 and 1. Draw the sign chart:

Interval	Test Number	$x-1$	$x+1$	$\frac{x-1}{x+1}$
$(-\infty,-1)$	-2	neg.	neg.	pos.
$(-1,1)$	0	neg.	pos.	neg.
$(1,\infty)$	2	pos.	pos.	pos.

Thus the quotient is negative on $(-1, 1)$ and it is zero when $x = 1$. So $\dfrac{x-1}{x+1}\le 0$ on the interval $(-1,1]$.

45. The key numbers are 2 and $\frac{3}{2}$. Draw the sign chart:

Interval	Test Number	$2-x$	$3-2x$	$\frac{2-x}{3-2x}$
$(-\infty,\frac{3}{2})$	0	pos.	pos.	pos.
$(\frac{3}{2},2)$	$\frac{7}{4}$	pos.	neg.	neg.
$(2,\infty)$	3	neg.	neg.	pos.

Thus the quotient is positive on $\left(-\infty,\frac{3}{2}\right)\cup(2,\infty)$ and it is zero when $x = 2$. So $\dfrac{2-x}{3-2x}\ge 0$ on the intervals $\left(-\infty,\frac{3}{2}\right)\cup[2,\infty)$.

47. Factoring, we have $\dfrac{(x+1)(x-9)}{x}<0$. So the key numbers are -1, 0, and 9. Draw the sign chart:

Interval	Test Number	$x-9$	x	$x+1$	$\frac{(x-9)(x+1)}{x}$
$(-\infty,-1)$	-2	neg.	neg.	neg.	neg.
$(-1,0)$	$-\frac{1}{2}$	neg.	neg.	pos.	pos.
$(0,9)$	1	neg.	pos.	pos.	neg.
$(9,\infty)$	10	pos.	pos.	pos.	pos.

Thus the quotient is negative on the intervals $(-\infty,-1)\cup(0,9)$.

49. Factoring the inequality yields:

$$\frac{2x^3+5x^2-7x}{3x^2+7x+4}>0$$
$$\frac{x(2x^2+5x-7)}{(3x+4)(x+1)}>0$$
$$\frac{x(2x+7)(x-1)}{(3x+4)(x+1)}>0$$

So the key numbers are $0, -\frac{7}{2}, 1, -\frac{4}{3}$ and -1. Draw the sign chart:

Interval	Test Number	$x-1$	x	$x+1$	$3x+4$	$2x+7$	$\frac{x(2x+7)(x-1)}{(3x+4)(x+1)}$
$\left(-\infty,-\frac{7}{2}\right)$	-4	neg.	neg.	neg.	neg.	neg.	neg.
$\left(-\frac{7}{2},-\frac{4}{3}\right)$	-2	neg.	neg.	neg.	neg.	pos.	pos.
$\left(-\frac{4}{3},-1\right)$	$-\frac{7}{6}$	neg.	neg.	neg.	pos.	pos.	neg.
$(-1,0)$	$-\frac{1}{2}$	neg.	neg.	pos.	pos.	pos.	pos.
$(0,1)$	$\frac{1}{2}$	neg.	pos.	pos.	pos.	pos.	neg.
$(1,\infty)$	2	pos.	pos.	pos.	pos.	pos.	pos.

Thus the quotient is positive on the intervals $\left(-\frac{7}{2},-\frac{4}{3}\right)\cup(-1,0)\cup(1,\infty)$.

51. First simplify the inequality:

$$\frac{x}{x+1}-1>0$$
$$\frac{x-x-1}{x+1}>0$$
$$\frac{-1}{x+1}>0$$

This is true whenever $x+1<0$, so $x<-1$. So the solution is $(-\infty,-1)$.

53. First simplify the inequality:

$$\frac{1}{x}-\frac{1}{x+1}\le 0$$
$$\frac{x+1-x}{x(x+1)}\le 0$$
$$\frac{1}{x(x+1)}\le 0$$

So the key numbers are 0 and -1. Draw the sign chart:

Interval	Test Number	x	$x+1$	$\frac{1}{x(x+1)}$
$(-\infty,-1)$	-2	neg.	neg.	pos.
$(-1,0)$	-0.5	neg.	pos.	neg.
$(0,\infty)$	1	pos.	pos.	pos.

So $\frac{1}{x(x+1)}\le 0$ (and consequently the original inequality is satisfied) on the interval $(-1,0)$.

55. First simplify the inequality:

$$\frac{1}{x-2}-\frac{1}{x-1}-\frac{1}{6}\geq 0$$
$$\frac{6(x-1)-6(x-2)-(x-2)(x-1)}{6(x-2)(x-1)}\geq 0$$
$$\frac{-x^2+3x+4}{6(x-2)(x-1)}\geq 0$$
$$\frac{x^2-3x-4}{6(x-2)(x-1)}\leq 0$$
$$\frac{(x-4)(x+1)}{6(x-2)(x-1)}\leq 0$$

So the key numbers are –1, 1, 2, and 4. Draw the sign chart:

Interval	Test Number	$x-4$	$x-2$	$x-1$	$x+1$	$\frac{(x-4)(x+1)}{6(x-2)(x-1)}$
$(-\infty,-1)$	–2	neg.	neg.	neg.	neg.	pos.
$(-1,1)$	0	neg.	neg.	neg.	pos.	neg.
$(1,2)$	$\frac{3}{2}$	neg.	neg.	pos.	pos.	pos.
$(2,4)$	3	neg.	pos.	pos.	pos.	neg.
$(4,\infty)$	5	pos.	pos.	pos.	pos.	pos.

So the quotient is negative on the intervals $(-1,1)\cup(2,4)$ and zero at $x=4$ and $x=-1$, so our original inequality is satisfied on the intervals $[-1,1)\cup(2,4]$.

57. First simplify the inequality:

$$\frac{1+x}{1-x}-\frac{1-x}{1+x}+1<0$$
$$\frac{(1+x)^2-(1-x)^2+(1-x^2)}{(1-x)(1+x)}<0$$
$$\frac{-x^2+4x+1}{(1-x)(1+x)}<0$$
$$\frac{x^2-4x-1}{(x-1)(x+1)}<0$$

The denominator is zero when $x=\pm 1$. Using the quadratic formula, we find the numerator is zero when $x=2\pm\sqrt{5}$. Thus, the key numbers are ± 1 and $2\pm\sqrt{5}$. For purposes of picking appropriate test numbers, note that $2+\sqrt{5}\approx 4.2$ and that $2-\sqrt{5}\approx -0.2$. Draw the sign chart:

Interval	Test Number	$x-4.2$	$x-1$	$x+0.2$	$x+1$	$\frac{x^2-4x-1}{(x-1)(x+1)}$
$(-\infty,-1)$	–2	neg.	neg.	neg.	neg.	pos.
$(-1,-0.2)$	– 0.5	neg.	neg.	neg.	pos.	neg.
$(-0.2,1)$	0	neg.	neg.	pos.	pos.	pos.
$(1,4.2)$	2	neg.	pos.	pos.	pos.	neg.
$(4.2,\infty)$	5	pos.	pos.	pos.	pos.	pos.

So $\frac{x^2-4x-1}{(x-1)(x+1)}<0$ (and consequently the original inequality is satisfied) on the intervals $\left(-1,2-\sqrt{5}\right)\cup\left(1,2+\sqrt{5}\right)$.

59. First factor the inequality as $\frac{(x-2)(x+1)}{(x-2)(x-1)} > 0$. The key numbers are –1, 1, and 2. Draw the sign chart:

Interval	Test Number	$x-2$	$x-1$	$x+1$	$\frac{(x-2)(x+1)}{(x-2)(x-1)}$
$(-\infty,-1)$	–2	neg.	neg.	neg.	pos.
$(-1,1)$	0	neg.	neg.	pos.	neg.
$(1,2)$	1.5	neg.	pos.	pos.	pos.
$(2,\infty)$	3	pos.	pos.	pos.	pos.

So $\frac{(x-2)(x+1)}{(x-2)(x-1)} > 0$ on the intervals $(-\infty,-1)\cup(1,2)\cup(2,\infty)$.

61. **a.** Graphing the curve $y = x^2 - 5x + 3$:

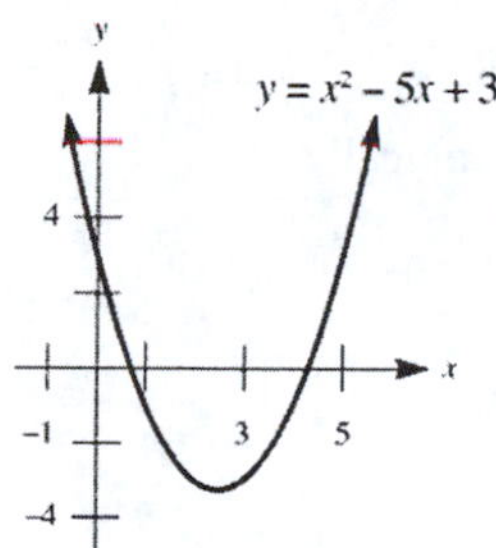

Thus $y \le 0$ on the interval $[0.697, 4.303]$.

b. Finding the x-intercepts algebraically, use the quadratic formula:

$$x = \frac{-(-5) \pm \sqrt{(-5)^2 - 4(1)(3)}}{2(1)} = \frac{5 \pm \sqrt{25-12}}{2} = \frac{5 \pm \sqrt{13}}{2} \approx 0.697, 4.303$$

These values agree with those found graphically.

63. **a.** Graphing the curve $y = 0.25x^2 - 6x - 2$:

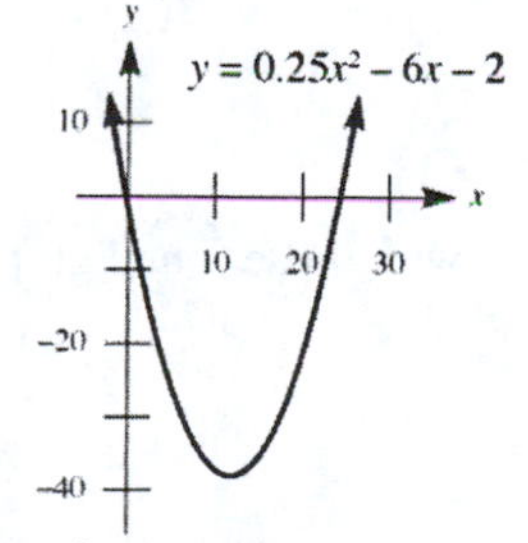

Thus $y < 0$ on the interval $(-0.329, 24.329)$.

b. Finding the x-intercepts algebraically, use the quadratic formula:

$$x = \frac{-(-6) \pm \sqrt{(-6)^2 - 4(0.25)(-2)}}{2(0.25)} = \frac{6 \pm \sqrt{36+2}}{0.5} = \frac{6 \pm \sqrt{38}}{0.5} = 12 \pm 2\sqrt{38} \approx -0.329, 24.329$$

These values agree with those found graphically.

65. a. Graphing the curve $y = x^4 - 2x^2 - 1$:

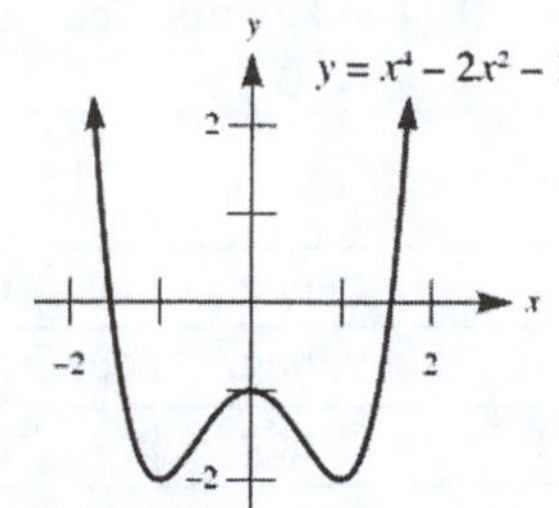

Thus $y > 0$ on the set $(-\infty, -1.554) \cup (1.554, \infty)$.

b. Finding the x-intercepts algebraically, use the quadratic formula:

$$x^2 = \frac{-(-2) \pm \sqrt{(-2)^2 - 4(1)(-1)}}{2(1)} = \frac{2 \pm \sqrt{4+4}}{2} = \frac{2 \pm 2\sqrt{2}}{2} = 1 \pm \sqrt{2}$$

Since $1 - \sqrt{2} < 0$, we have $x^2 = 1 + \sqrt{2}$, thus: $x = \pm\sqrt{1+\sqrt{2}} \approx \pm 1.554$

These values agree with those found graphically.

67. a. Graphing the curve $y = \dfrac{x^2 - 5}{x^2 + 1}$:

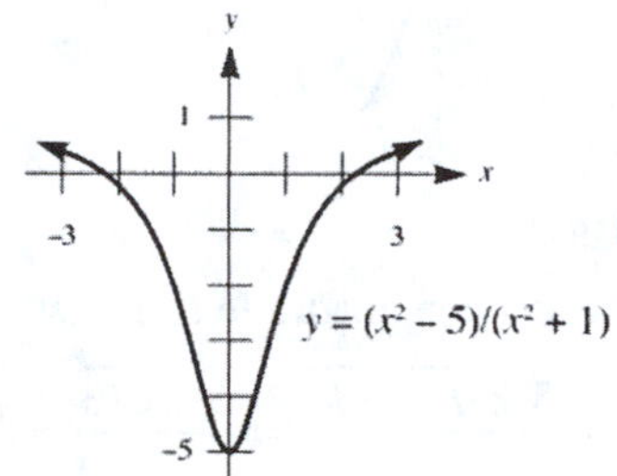

Thus $y \le 0$ on the interval $[-2.236, 2.236]$.

b. To find the x-intercepts algebraically, set the numerator equal to 0:

$$x^2 - 5 = 0$$
$$x^2 = 5$$
$$x = \pm\sqrt{5} \approx \pm 2.236$$

Note that $x^2 + 1 \ne 0$. These values agree with those found graphically.

69. a. Graphing the curve $y = \dfrac{x^2 + 1}{x^2 - 5}$:

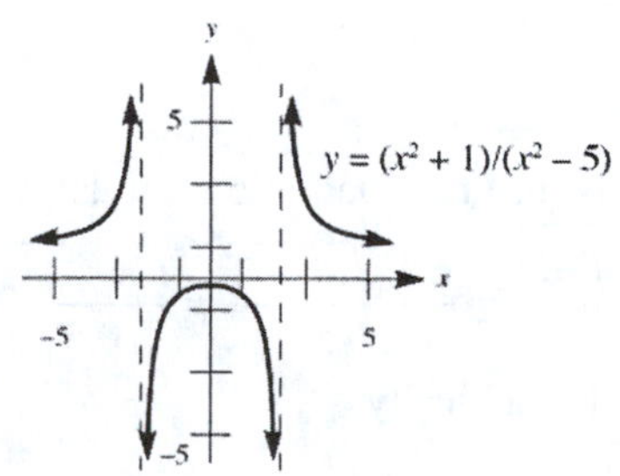

Thus $y \le 0$ on the interval $(-2.236, 2.236)$.

b. To find the asymptotes algebraically, set the denominator equal to 0:

$$x^2 - 5 = 0$$
$$x^2 = 5$$
$$x = \pm\sqrt{5} \approx \pm 2.236$$

Note that $x^2 + 1 \neq 0$. These values agree with those found graphically.

71. Graphing the curve $y = x^3 + 2x + 1$:

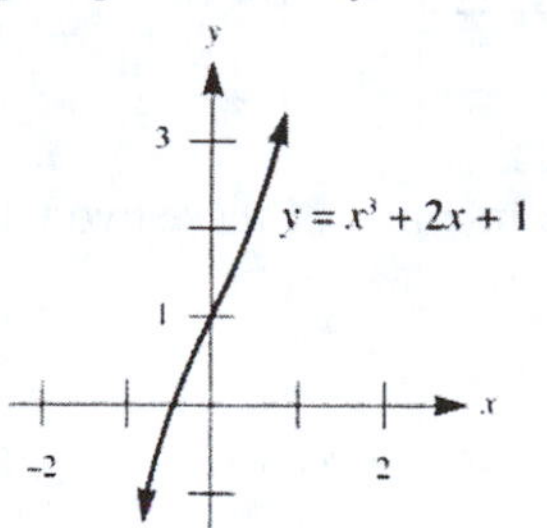

Thus $y \geq 0$ on the interval $[-0.453, \infty)$.

73. Graphing the curve $y = x^4 - 2x + 1$:

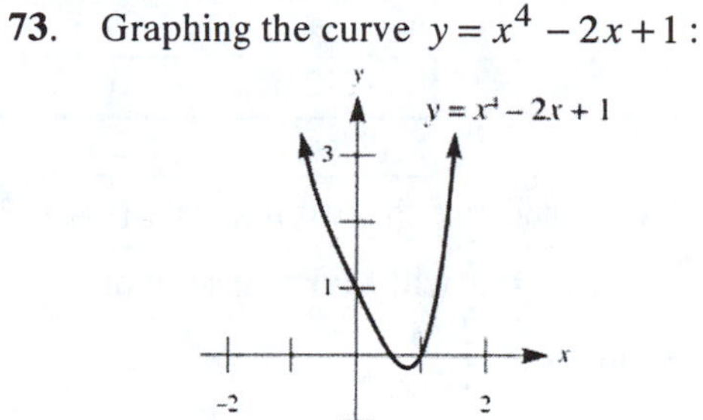

Thus $y > 0$ on the set $(-\infty, 0.544) \cup (1, \infty)$.

75. Graphing the curve $y = x - \frac{x^3}{3!} + \frac{x^5}{5!} - \frac{x^7}{7!}$:

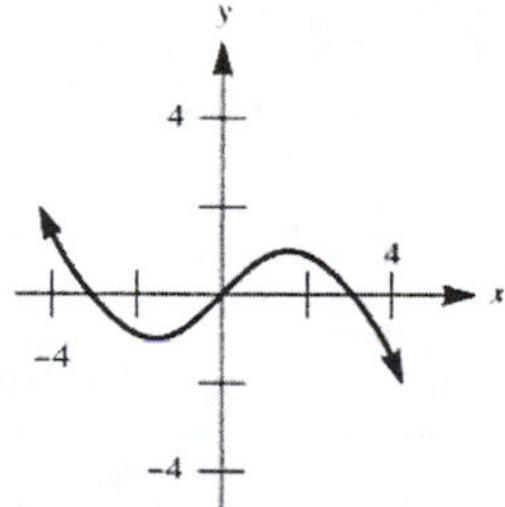

Thus $y < 0$ on the set $(-3.079, 0) \cup (3.079, \infty)$.

77. The following is a graph of R:

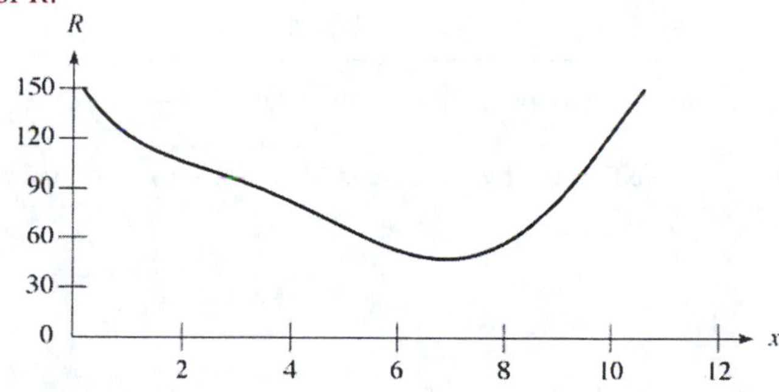

a. Using the solver feature, we find $R \leq 80$ when $4 \leq x \leq 9$, which is the months from April through September.

b. Using the solver feature, we find $R \geq 120$ when $x \leq 1$ and $x \geq 10$, which is the months January, October, November, and December.

79. The solutions will be real provided the discriminant $b^2 - 4ac$ is non-negative. Thus:

$$b^2 - 4 \geq 0$$
$$(b-2)(b+2) \geq 0$$

The key numbers are ±2. Draw the sign chart:

Interval	Test Number	$b-2$	$b+2$	$(b-2)(b+2)$
$(-\infty,-2)$	-3	neg.	neg.	pos.
$(-2,2)$	0	neg.	pos.	neg.
$(2,\infty)$	3	pos.	pos.	pos.

The solution set consists of the two intervals $(-\infty,-2]\cup[2,\infty)$. For values of b in either of these two intervals, the equation $x^2 + bx + 1 = 0$ will have real solutions.

81. If $x = 1$ is a solution of $\frac{2a+x}{x-2a} < 1$, then:

$$\frac{2a+1}{1-2a} < 1$$
$$\frac{2a+1}{1-2a} - 1 < 0$$
$$\frac{2a+1-(1-2a)}{1-2a} < 0$$
$$\frac{4a}{1-2a} < 0$$

The key numbers are 0 and $\frac{1}{2}$. Draw the sign chart:

Interval	Test Number	$4a$	$1-2a$	$\frac{4a}{1-2a}$
$(-\infty,0)$	-1	neg.	pos.	neg.
$\left(0,\frac{1}{2}\right)$	$\frac{1}{4}$	pos.	pos.	pos.
$\left(\frac{1}{2},\infty\right)$	1	pos.	neg.	neg.

The allowable values of a are those numbers in either of the intervals $(-\infty,0)\cup\left(\frac{1}{2},\infty\right)$.

83. Using the Pythagorean theorem, we find the hypotenuse is $\sqrt{x^2 + (1-x)^2}$, or $\sqrt{2x^2 - 2x + 1}$. If this is less than $\frac{\sqrt{17}}{5}$, then:

$$\sqrt{2x^2 - 2x + 1} < \frac{\sqrt{17}}{5}$$
$$\left(5\sqrt{2x^2 - 2x + 1}\right)^2 < \left(\sqrt{17}\right)^2$$
$$50x^2 - 50x + 25 < 17$$
$$50x^2 - 50x + 8 < 0$$
$$25x^2 - 25x + 4 < 0$$
$$(5x-1)(5x-4) < 0$$

The key numbers here are $\frac{1}{5}$ and $\frac{4}{5}$. Draw the sign chart:

Interval	Test Number	$5x-1$	$5x-4$	$(5x-1)(5x-4)$
$(-\infty,\frac{1}{5})$	0	neg.	neg.	pos.
$\left(\frac{1}{5},\frac{4}{5}\right)$	$\frac{2}{5}$	pos.	neg.	neg.
$\left(\frac{4}{5},\infty\right)$	1	pos.	pos.	pos.

The solution set is the interval $\left(\frac{1}{5},\frac{4}{5}\right)$.

85. Computing the key numbers by the quadratic formula:

$$x=\frac{-2c\pm\sqrt{4c^2-4(-6c)}}{2}=\frac{-2c\pm 2\sqrt{c^2+6c}}{2}=-c\pm\sqrt{c^2+6c}$$

Since these key numbers must be equal to $-3c$ and c, respectively:

$$-c-\sqrt{c^2+6c}=-3c \quad \text{and} \quad -c+\sqrt{c^2+6c}=c$$

$$\sqrt{c^2+6c}=2c \qquad\qquad \sqrt{c^2+6c}=2c$$

Squaring:

$$\begin{aligned} c^2+6c&=4c^2\\ 3c^2-6c&=0\\ 3c(c-2)&=0\\ c&=0,\ c=2 \end{aligned}$$

Thus a (non-zero) value is $c = 2$.

Chapter 2 Review Exercises

1. True

3. False (For instance: $x = 5$)

5. False (For instance: $x = 1$)

7. True

9. False (For instance: $x = 0, y = -1$)

11. False (For instance: $a = -5, b = 1$)

13. Solve the equation:

$$\begin{aligned} 12x^2+2x-2&=0\\ 6x^2+x-1&=0\\ (3x-1)(2x+1)&=0\\ x&=\tfrac{1}{3},-\tfrac{1}{2} \end{aligned}$$

15. Multiplying by 2, solve the equation:

$$\begin{aligned} \tfrac{1}{2}x^2+x-12&=0\\ x^2+2x-24&=0\\ (x+6)(x-4)&=0\\ x&=-6,4 \end{aligned}$$

17. Solve the equation:

$$\begin{aligned} \frac{x}{5-x}&=\frac{-2}{11-x}\\ 11x-x^2&=-10+2x\\ -x^2+9x+10&=0\\ x^2-9x-10&=0\\ (x-10)(x+1)&=0\\ x&=10,-1 \end{aligned}$$

Both values check in the original equation.

19. Solve the equation:

$$\begin{aligned} \frac{1}{3x-7}-\frac{2}{5x-5}-\frac{3}{3x+1}&=0 \\ (5x-5)(3x+1)-2(3x-7)(3x+1)-3(3x-7)(5x-5)&=0 \\ 15x^2-10x-5-18x^2+36x+14-45x^2+150x-105&=0 \\ -48x^2+176x-96&=0 \\ 3x^2-11x+6&=0 \\ (3x-2)(x-3)&=0 \\ x&=\tfrac{2}{3}, 3 \end{aligned}$$

Both values check in the original equation.

21. Multiply by 2 to obtain $2t^2+2t-1=0$. Now using the quadratic formula:

$$t=\frac{-2\pm\sqrt{4-4(2)(-1)}}{2(2)}=\frac{-2\pm\sqrt{12}}{4}=\frac{-2\pm 2\sqrt{3}}{4}=\frac{-1\pm\sqrt{3}}{2}$$

23. Multiplying by x^2:

$$\begin{aligned} x^2+14x+48&=0 \\ (x+6)(x+8)&=0 \\ x&=-6, -8 \end{aligned}$$

25. Let $t=x^{1/4}$ and $t^2=x^{1/2}$. Then:

$$\begin{aligned} t^2-13t+36&=0 \\ (t-9)(t-4)&=0 \\ t&=9 \text{ or } t=4 \end{aligned}$$

If $t=9$, then $x^{1/4}=9$ and $x=9^4=6561$. If $t=4$, then $x^{1/4}=4$ and $x=4^4=256$. So the solutions are 6561 and 256.

27. Squaring:

$$\begin{aligned} \sqrt{4x+3}&=\sqrt{11-8x}-1 \\ 4x+3&=11-8x-2\sqrt{11-8x}+1 \\ 12x-9&=-2\sqrt{11-8x} \\ 144x^2-216x+81&=44-32x \\ 144x^2-184x+37&=0 \\ (4x-1)(36x-37)&=0 \\ x&=\tfrac{1}{4} \text{ or } x=\tfrac{37}{36} \end{aligned}$$

Upon checking, we find only $x=\frac{1}{4}$ checks in the original equation.

29. Multiplying by $\sqrt{x+6}\sqrt{x-6}$:

$$\begin{aligned} \frac{2}{\sqrt{x^2-36}}+\frac{1}{\sqrt{x+6}}-\frac{1}{\sqrt{x-6}}&=0 \\ 2+\sqrt{x-6}-\sqrt{x+6}&=0 \\ 2+\sqrt{x-6}&=\sqrt{x+6} \\ 4+4\sqrt{x-6}+x-6&=x+6 \\ 4\sqrt{x-6}&=8 \\ \sqrt{x-6}&=2 \\ x-6&=4 \\ x&=10 \end{aligned}$$

This value checks in the original equation.

31. Squaring:

$$\begin{aligned}
\sqrt{x+7}-\sqrt{x+2}&=\sqrt{x-1}-\sqrt{x-2}\\
x+7-2\sqrt{x+7}\sqrt{x+2}+x+2&=x-1-2\sqrt{x-1}\sqrt{x-2}+x-2\\
-2\sqrt{x+7}\sqrt{x+2}+9&=-2\sqrt{x-1}\sqrt{x-2}-3\\
-2\sqrt{x+7}\sqrt{x+2}&=-2\sqrt{x-1}\sqrt{x-2}-12\\
\sqrt{x+7}\sqrt{x+2}&=\sqrt{x-1}\sqrt{x-2}+6\\
(x+7)(x+2)&=(x-1)(x-2)+12\sqrt{x-1}\sqrt{x-2}+36\\
x^2+9x+14&=x^2-3x+2+12\sqrt{x-1}\sqrt{x-2}+36\\
12x-24&=12\sqrt{x-1}\sqrt{x-2}\\
x-2&=\sqrt{x-1}\sqrt{x-2}\\
x^2-4x+4&=x^2-3x+2\\
-x&=-2\\
x&=2
\end{aligned}$$

This value checks in the original equation.

33. Solve the equation for x:

$$\begin{aligned}
4x^2y^2-4xy&=-1\\
4x^2y^2-4xy+1&=0\\
(2xy-1)^2&=0\\
2xy-1&=0\\
x&=\frac{1}{2y}
\end{aligned}$$

35. Multiplying through by a^2bx yields:

$$\begin{aligned}
a^2bx^2+abx-a^2x&=2b+2a\\
a^2bx^2+(ab-a^2)x-2(b+a)&=0
\end{aligned}$$

Applying the quadratic formula:

$$\begin{aligned}
x&=\frac{-\left(ab-a^2\right)\pm\sqrt{\left(ab-a^2\right)^2-4\left(a^2b\right)[-2(b+a)]}}{2\left(a^2b\right)}\\
&=\frac{-ab+a^2\pm\sqrt{a^4+6a^3b+9a^2b^2}}{2a^2b}\\
&=\frac{-ab+a^2\pm\sqrt{\left(a^2+3ab\right)^2}}{2a^2b}\\
&=\frac{-ab+a^2\pm\left(a^2+3ab\right)}{2a^2b}
\end{aligned}$$

So the two solutions are:

$$x=\frac{-ab+a^2+a^2+3ab}{2a^2b}=\frac{2a^2+2ab}{2a^2b}=\frac{2a(a+b)}{2a^2b}=\frac{a+b}{ab}$$

$$x=\frac{-ab+a^2-a^2-3ab}{2a^2b}=\frac{-4ab}{2a^2b}=-\frac{2}{a}$$

Both of these values satisfy the original equation.

37. Multiplying by $abx(x+a+b)$ yields:

$$\frac{1}{x+a+b}=\frac{1}{x}+\frac{1}{a}+\frac{1}{b}$$
$$abx=ab(x+a+b)+bx(x+a+b)+ax(x+a+b)$$
$$abx=abx+a^2b+ab^2+bx^2+abx+b^2x+ax^2+a^2x+abx$$
$$0=(a+b)x^2+\left(a^2+2ab+b^2\right)x+\left(a^2b+ab^2\right)$$
$$0=(a+b)x^2+(a+b)^2x+ab(a+b)$$
$$0=x^2+(a+b)x+ab \qquad \text{(since } a+b\neq 0\text{)}$$
$$0=(x+a)(x+b)$$
$$x=-a \text{ or } x=-b$$

Both of these values satisfy the original equation.

39. Solve the inequality:

$$-1<\frac{1-2(1+x)}{3}<1$$
$$-3<1-2-2x<3$$
$$-2<-2x<4$$
$$1>x>-2$$

The solution set is (–2, 1).

41. The inequality is equivalent to $-\frac{1}{2}\leq x\leq\frac{1}{2}$, so the solution set is $\left[-\frac{1}{2},\frac{1}{2}\right]$.

43. Solve the inequality $|2x-1|\geq 5$:

$$2x-1\geq 5 \quad \text{or} \quad -(2x-1)\geq 5$$
$$2x\geq 6 \qquad\qquad 2x-1\leq -5$$
$$x\geq 3 \qquad\qquad 2x\leq -4$$
$$x\geq 3 \qquad\qquad x\leq -2$$

The solution set is $(-\infty,-2]\cup[3,\infty)$.

45. Factor $x^2+3x-40=(x-5)(x+8)$, so the key numbers are 5 and –8. Now draw the sign chart:

Interval	Test Number	$x+8$	$x-5$	$(x+8)(x-5)$
$(-\infty,-8)$	–9	neg.	neg.	pos.
$(-8,5)$	0	pos.	neg.	neg.
$(5,\infty)$	6	pos.	pos.	pos.

So the product is negative on the interval $(-8,5)$.

47. The key numbers are found by using the quadratic formula to solve $x^2-6x-1=0$. The result is $x=3\pm\sqrt{10}$. For purposes of picking appropriate test numbers, note that $3+\sqrt{10}\approx 6.2$ and that $3-\sqrt{10}\approx -0.2$. Now draw the sign chart:

Interval	Test Number	$x-6.2$	$x+0.2$	x^2-6x-1
$\left(-\infty,3-\sqrt{10}\right)$	–1	neg.	neg.	pos.
$\left(3-\sqrt{10},3+\sqrt{10}\right)$	0	neg.	pos.	neg.
$\left(3+\sqrt{10},\infty\right)$	7	pos.	pos.	pos.

So the product is negative on the interval $\left(3-\sqrt{10},3+\sqrt{10}\right)$.

49. First factor the inequality:

$$x^4 - 34x^2 + 225 < 0$$
$$\left(x^2 - 9\right)\left(x^2 - 25\right) < 0$$
$$(x-3)(x+3)(x-5)(x+5) < 0$$

The key numbers are –3, 3, –5, and 5. Now draw the sign chart:

Interval	Test Number	$x-5$	$x-3$	$x+3$	$x+5$	$\left(x^2-9\right)\left(x^2-25\right)$
$(-\infty,-5)$	–6	neg.	neg.	neg.	neg.	pos.
$(-5,-3)$	–4	neg.	neg.	neg.	pos.	neg.
$(-3,3)$	0	neg.	neg.	pos.	pos.	pos.
$(3,5)$	4	neg.	pos.	pos.	pos.	neg.
$(5,\infty)$	6	pos.	pos.	pos.	pos.	pos.

So the product is negative on the intervals $(-5,-3)\cup(3,5)$.

51. The key numbers are 7 and –2. Now draw the sign chart:

Interval	Test Number	$(x+2)^3$	$(x-7)^2$	$\frac{(x-7)^2}{(x+2)^3}$
$(-\infty,-2)$	–3	neg.	pos.	neg.
$(-2,7)$	0	pos.	pos.	pos.
$(7,\infty)$	8	pos.	pos.	pos.

The quotient is positive on the intervals $(-2,7)\cup(7,\infty)$, and zero at $x = 7$. The single interval $(-2, \infty)$ satisfies the inequality.

53. First simplify the inequality:

$$\frac{3x+1}{x-4} - 1 < 0$$
$$\frac{3x+1-(x-4)}{x-4} < 0$$
$$\frac{2x+5}{x-4} < 0$$

The key numbers are $-\frac{5}{2}$ and 4. Draw the sign chart:

Interval	Test Number	$2x+5$	$x-4$	$\frac{2x+5}{x-4}$
$\left(-\infty,-\frac{5}{2}\right)$	–3	neg.	neg.	pos.
$\left(-\frac{5}{2},4\right)$	0	pos.	neg.	neg.
$(4,\infty)$	5	pos.	pos.	pos.

The quotient is negative on the interval $\left(-\frac{5}{2},4\right)$.

55. First simplify the inequality:

$$x^2 + \frac{1}{x^2} - 3 > 0$$

$$\frac{x^4 + 1 - 3x^2}{x^2} > 0$$

$$\frac{x^4 - 3x^2 + 1}{x^2} > 0$$

One key number is $x = 0$. To find the others, let $t = x^2$ then $t^2 = x^4$ to obtain $t^2 - 3t + 1 = 0$. Using the quadratic formula: $t = \frac{3 \pm \sqrt{(-3)^2 - 4(1)}}{2(1)} = \frac{3 \pm \sqrt{5}}{2}$. Therefore: $x = \pm\sqrt{\frac{3 \pm \sqrt{5}}{2}} \approx \pm 0.62, \pm 1.62$

Using the key numbers 0, ±0.62, and ±1.62, draw the sign chart:

Interval	Test Number	$x^4 - 3x^2 + 1$	x^2	$\frac{x^4 - 3x^2 + 1}{x^2}$
$(-\infty, -1.62)$	-2	pos.	pos.	pos.
$(-1.62, -0.62)$	-1	neg.	pos.	neg.
$(-0.62, 0)$	-0.5	pos.	pos.	pos.
$(0, 0.62)$	0.5	pos.	pos.	pos.
$(0.62, 1.62)$	1	neg.	pos.	neg.
$(1.62, \infty)$	2	pos.	pos.	pos.

The quotient is positive on the intervals: $\left(-\infty, -\sqrt{\frac{3+\sqrt{5}}{2}}\right) \cup \left(-\sqrt{\frac{3-\sqrt{5}}{2}}, 0\right) \cup \left(0, \sqrt{\frac{3-\sqrt{5}}{2}}\right) \cup \left(\sqrt{\frac{3+\sqrt{5}}{2}}, \infty\right)$

57. The discriminant is: $b^2 - 4ac = (k+1)^2 - 4(2k) = k^2 + 2k + 1 - 8k = k^2 - 6k + 1$

To find the key numbers for $k^2 - 6k + 1 \geq 0$, use the quadratic formula:

$$k = \frac{6 \pm \sqrt{36 - 4}}{2} = \frac{6 \pm \sqrt{32}}{2} = \frac{6 \pm 4\sqrt{2}}{2} = 3 \pm 2\sqrt{2}$$

For purposes of picking appropriate test numbers, note that $3 + 2\sqrt{2} \approx 5.8$ and that $3 - 2\sqrt{2} \approx 0.2$. Now draw the sign chart:

Interval	Test Number	$k - 5.8$	$k - 0.2$	$k^2 - 6k + 1$
$\left(-\infty, 3 - 2\sqrt{2}\right)$	0	neg.	neg.	pos.
$\left(3 - 2\sqrt{2}, 3 + 2\sqrt{2}\right)$	3	neg.	pos.	neg.
$\left(3 + 2\sqrt{2}, \infty\right)$	7	pos.	pos.	pos.

The expression is positive on the interval $\left(-\infty, 3 - 2\sqrt{2}\right) \cup \left(3 + 2\sqrt{2}, \infty\right)$, and zero at $k = 3 - 2\sqrt{2}$ and $k = 3 + 2\sqrt{2}$.

So the values of k are chosen from the interval $\left(-\infty, 3 - 2\sqrt{2}\right] \cup \left[3 + 2\sqrt{2}, \infty\right)$.

59. If $a = 1$, we can find c and b:

$$c = r_1 \bullet r_2 = (-3)(-4) = 12 \qquad b = -(r_1 + r_2) = -(-3 - 4) = -(-7) = 7$$

Thus the equation is $x^2 + 7x + 12 = 0$.

61. Let x and y represent the two numbers. So $x^3 + y^3 = 2071$ while $x + y = 19$. Solving for y yields $y = 19 - x$, now substitute:

$$\begin{aligned} x^3 + y^3 &= 2071 \\ x^3 + (19 - x)^3 &= 2071 \\ x^3 + 6859 - 1083x + 57x^2 - x^3 &= 2071 \\ 57x^2 - 1083x + 4788 &= 0 \\ x^2 - 19x + 84 &= 0 \\ (x - 12)(x - 7) &= 0 \\ x &= 7, 12 \end{aligned}$$

If $x = 7$ then $y = 19 - x = 12$, and if $x = 12$ then $y = 19 - x = 7$. The two numbers are 7 and 12.

63. First re-draw the figure:

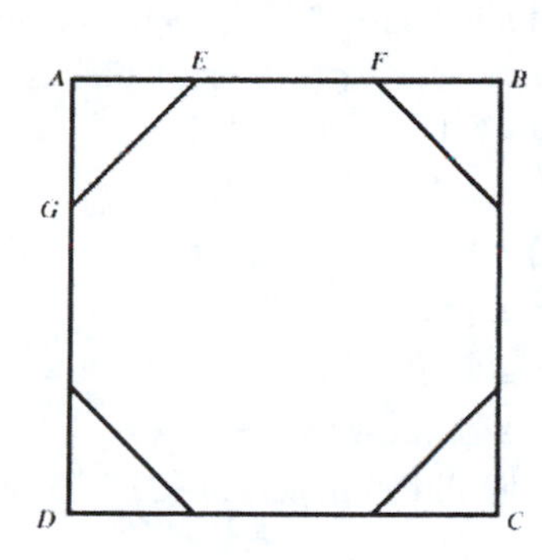

Since $AE = FB$:

$$\begin{aligned} AE + EF + FB &= 1 \\ EF + 2AE &= 1 \\ AE &= \frac{1 - EF}{2} \end{aligned}$$

By the Pythagorean theorem and the fact that $AG = AE$ and $EF = GE$:

$$\begin{aligned} (AG)^2 + (AE)^2 &= (GE)^2 \\ (AE)^2 + (AE)^2 &= (EF)^2 \\ 2(AE)^2 &= (EF)^2 \\ 2\left(\frac{1 - EF}{2}\right)^2 &= EF^2 \end{aligned}$$

Now, for convenience, let $EF = x$. Then:

$$\begin{aligned} 2\left(\frac{1 - 2x + x^2}{4}\right) &= x^2 \\ 1 - 2x + x^2 &= 2x^2 \\ 0 &= x^2 + 2x - 1 \end{aligned}$$

The quadratic formula then gives us: $x = \dfrac{-2 \pm \sqrt{4+4}}{2} = \dfrac{-2 \pm \sqrt{8}}{2} = \dfrac{-2 \pm 2\sqrt{2}}{2} = -1 \pm \sqrt{2}$

Choose the positive root, since $x > 0$. Thus $x = EF = -1 + \sqrt{2}$ cm.

65. Given that the length of a side is $\frac{x}{4}$, compute the area and the perimeter:

$$\text{area} = \left(\frac{x}{4}\right)^2 = \frac{x^2}{16} \qquad \text{perimeter} = 4\left(\frac{x}{4}\right) = x$$

If the area is numerically greater than the perimeter, then:

$$\frac{x^2}{16} > x$$
$$x^2 > 16x$$
$$x^2 - 16x > 0$$
$$x(x-16) > 0$$

Since x must be positive, this is satisfied only if $x - 16 > 0$, and hence $x > 16$. We conclude that the area will be numerically greater than the perimeter when $x > 16$ cm, which is the interval $(16 \text{ cm}, \infty)$.

67. Denoting the integers by $x - 1$, x and $x + 1$:

$$(x-1)^2 + x^2 + (x+1)^2 = 1454$$
$$x^2 - 2x + 1 + x^2 + x^2 + 2x + 1 = 1454$$
$$3x^2 = 1452$$
$$x^2 = 484$$
$$x = \sqrt{484} = 22$$

Thus $x - 1 = 21$ and $x + 1 = 23$, so the three integers are 21, 22, and 23.

69. First draw a figure:

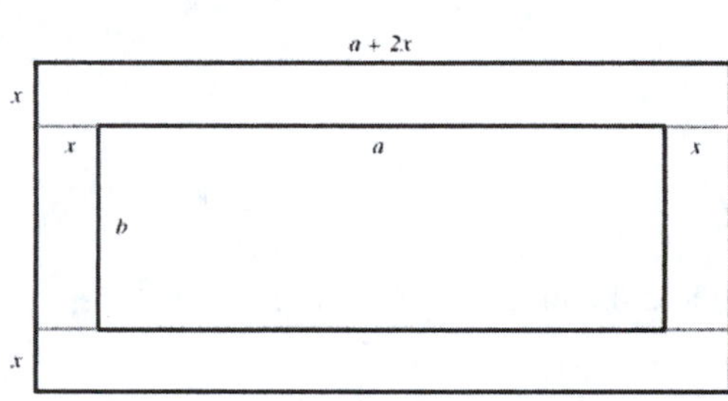

Let x denote the width of the path, as indicated in the figure. To compute the area of the path, divide it into four rectangular portions, as indicated by the dotted lines in the figure. Then the area of the path is:

$$2[x(a+2x)] + 2(bx) = 4x^2 + 2ax + 2bx$$

Since the area of the path and the flower garden are equal:

$$4x^2 + 2ax + 2bx = ab$$
$$4x^2 + 2(a+b)x - ab = 0$$
$$x = \frac{-2(a+b) \pm \sqrt{4(a+b)^2 + 16ab}}{8}$$

Choosing the positive root here (since $x > 0$): $x = \frac{-2(a+b) + 2\sqrt{(a+b)^2 + 4ab}}{8} = \frac{-(a+b) + \sqrt{a^2 + 6ab + b^2}}{4}$

71. For each ball we will find at what time the height is 0. For the first ball we have $h=-16t^2+40t+50$, so setting $h=0$ yields $-16t^2+40t+50=0$, or $8t^2-20t-25=0$. Using the quadratic formula and choosing the positive root yields: $t=\dfrac{20+\sqrt{400+32(25)}}{2(8)}=\dfrac{20+20\sqrt{3}}{16}=\dfrac{5+5\sqrt{3}}{4}\approx 3.4$ sec

For the second ball we have $h=-16t^2+5t+100$, so setting $h=0$ yields $-16t^2+5t+100=0$, or $16t^2-5t-100=0$. Using the quadratic formula and choosing the positive root yields:

$$t=\frac{5+\sqrt{25+6400}}{2(16)}=\frac{5+\sqrt{6425}}{32}=\frac{5+5\sqrt{257}}{32}\approx 2.7 \text{ sec}$$

Therefore the second ball (the one thrown from 100 ft) hits the ground first.

73. Let y be the height of the rectangle, so the area of the rectangle (using the Pythagorean theorem) is:

$$\text{area}=2xy=2x\sqrt{1-x^2}$$

If the area is 1 cm^2, then:

$$\begin{aligned}2x\sqrt{1-x^2}&=1\\4x^2\left(1-x^2\right)&=1\\4x^2-4x^4-1&=0\\4x^4-4x^2+1&=0\\\left(2x^2-1\right)^2&=0\\2x^2-1&=0\\x^2&=\tfrac{1}{2}\\x&=\sqrt{\tfrac{1}{2}}=\tfrac{\sqrt{2}}{2}\end{aligned}$$

This value checks in the original equation.

75. Call x the radius of the inscribed circle. Then note the diagonal (drawn from the center of this circle to the right angle) is: $\sqrt{x^2+x^2}=\sqrt{2x^2}=x\sqrt{2}$. But tracing along this diagonal, we know the full length is equal to r, thus:

$$\begin{aligned}x\sqrt{2}+x&=r\\x\left(1+\sqrt{2}\right)&=r\\x&=\frac{r}{1+\sqrt{2}}\end{aligned}$$

So the radius of the inscribed circle is $\dfrac{r}{1+\sqrt{2}}$.

Chapter 2 Test

1. **a.** Solve by factoring:

$$\begin{aligned}x^2+4x&=5\\x^2+4x-5&=0\\(x+5)(x-1)&=0\\x&=-5,1\end{aligned}$$

b. Solve by writing the equation as $x^2+4x-1=0$, then using the quadratic formula:

$$x=\frac{-4\pm\sqrt{(4)^2-4(1)(-1)}}{2(1)}=\frac{-4\pm\sqrt{20}}{2}=\frac{-4\pm 2\sqrt{5}}{2}=-2\pm\sqrt{5}$$

2. Solve by factoring:

$$x^2\left(x^2-7\right)+12=0$$
$$x^4-7x^2+12=0$$
$$\left(x^2-4\right)\left(x^2-3\right)=0$$

So $x^2=4$ or $x^2=3$, and thus $x=\pm 2$ or $x=\pm\sqrt{3}$.

3. Isolating $\sqrt{5-2x}$ and squaring:

$$1+\sqrt{2-x}-\sqrt{5-2x}=0$$
$$\sqrt{5-2x}=1+\sqrt{2-x}$$
$$5-2x=1+2\sqrt{2-x}+2-x$$
$$2-x=2\sqrt{2-x}$$
$$(2-x)^2=4(2-x)$$
$$(2-x)^2-4(2-x)=0$$
$$(2-x)(-x-2)=0$$
$$x=-2,2$$

Both values check in the original equation.

4. Given $|3x-1|=2$, either $3x-1=2$ or $3x-1=-2$. Therefore:

$$3x-1=2 \quad\text{or}\quad 3x-1=-2$$
$$3x=3 \qquad\qquad 3x=-1$$
$$x=1 \qquad\qquad x=-\tfrac{1}{3}$$

The solutions are 1 and $-\frac{1}{3}$.

5. First divide by 2 to write the equation as $x^2+4x-\frac{9}{2}=0$. Then the sum of the roots is given by $r_1+r_2=-b=-4$.

6. If $a=1$, we can find c and b:

$$c=r_1\bullet r_2=\left(2+3\sqrt{7}\right)\left(2-3\sqrt{7}\right)=4-63=-59 \qquad b=-\left(r_1+r_2\right)=-\left(2+3\sqrt{7}+2-3\sqrt{7}\right)=-4$$

Thus the equation is $x^2-4x-59=0$.

7. Using the quadratic formula: $x^2=\dfrac{-(-3)\pm\sqrt{(-3)^2-4(1)(-1)}}{2(1)}=\dfrac{3\pm\sqrt{9+4}}{2}=\dfrac{3\pm\sqrt{13}}{2}$

Since $\dfrac{3-\sqrt{13}}{2}<0$, we choose $x^2=\dfrac{3+\sqrt{13}}{2}$. Solving for x: $x=\pm\sqrt{\dfrac{3+\sqrt{13}}{2}}\approx\pm 1.817$

These values are consistent with the given graph.

8. **a.** Sketching the graph of $y=2x^{4/3}-x^{2/3}-6$:

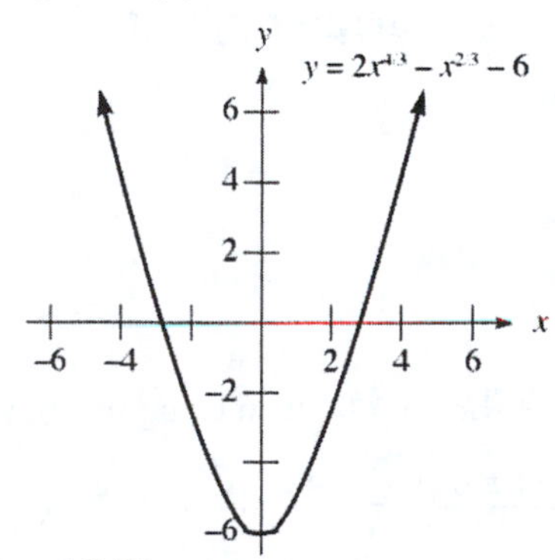

The roots are approximately –2.83 and 2.83.

b. Factoring:

$$2x^{4/3} - x^{2/3} - 6 = 0$$
$$\left(2x^{2/3} + 3\right)\left(x^{2/3} - 2\right) = 0$$
$$x^{2/3} = -\frac{3}{2}, 2$$
$$x^2 = -\frac{27}{8}, 8$$
$$x = \pm 2\sqrt{2} \approx \pm 2.83$$

9. Solve the inequality:

$$4(1+x) - 3(2x-1) \ge 1$$
$$4 + 4x - 6x + 3 \ge 1$$
$$-2x \ge -6$$
$$x \le 3$$

The solution set is $(-\infty, 3]$.

10. Solving the double inequality:

$$\frac{3}{5} < \frac{3-2x}{-4} < \frac{4}{5}$$
$$-12 > 15 - 10x > -16$$
$$-27 > -10x > -31$$
$$\tfrac{27}{10} < x < \tfrac{31}{10}$$

The solution set is $\left(\frac{27}{10}, \frac{31}{10}\right)$.

11. Solving $|3x-8| \le 1$:

$$-1 \le 3x - 8 \le 1$$
$$7 \le 3x \le 9$$
$$\tfrac{7}{3} \le x \le 3$$

The solution set is $\left[\frac{7}{3}, 3\right]$.

12. The key numbers are 4 and –8. Draw the sign chart:

Interval	Test Number	$(x+8)^3$	$(x-4)^2$	$(x-4)^2(x+8)^3$
$(-\infty, -8)$	–9	neg.	pos.	neg.
$(-8, 4)$	0	pos.	pos.	pos.
$(4, \infty)$	5	pos.	pos.	pos.

The product is positive on $(-8,4) \cup (4,\infty)$ and zero at $x = -8$ and $x = 4$. So the inequality is satisfied on $[-8,4] \cup [4,\infty)$, which combine as $[-8,\infty)$.

13. Simplifying the inequality and factoring:

$$\frac{(x+1)(x+2) + x(x+2) + x(x+1)}{x(x+1)(x+2)} \ge 0$$
$$\frac{3x^2 + 6x + 2}{x(x+1)(x+2)} \ge 0$$

Three of the key numbers are 0, –1 and –2. The other two are found by using the quadratic formula to solve $3x^2 + 6x + 2 = 0$. The roots are found to be $\frac{-3 \pm \sqrt{3}}{3}$. For purposes of picking appropriate test numbers, note that $\frac{-3+\sqrt{3}}{3} \approx -0.4$ and that $\frac{-3-\sqrt{3}}{3} \approx -1.6$.

Draw the sign chart:

Interval	Test Number	x	$x+0.4$	$x+1$	$x+1.6$	$x+2$	$\frac{3x^2+6x+2}{x(x+1)(x+2)}$
$(-\infty,-2)$	-3	neg.	neg.	neg.	neg.	neg.	neg.
$(-2,-1.6)$	-1.8	neg.	neg.	neg.	neg.	pos.	pos.
$(-1.6,-1)$	-1.5	neg.	neg.	neg.	pos.	pos.	neg.
$(-1,-0.4)$	-0.5	neg.	neg.	pos.	pos.	pos.	pos.
$(-0.4,0)$	-0.1	neg.	pos.	pos.	pos.	pos.	neg.
$(0,\infty)$	2	pos.	pos.	pos.	pos.	pos.	pos.

So the solution set is $\left(-2,\frac{-3-\sqrt{3}}{3}\right]\cup\left(-1,\frac{-3+\sqrt{3}}{3}\right]\cup(0,\infty)$.

14. The discriminant is $b^2-4ac=(3)^2-4(1)(k^2)=9-4k^2$. For the equation to have real solutions, this discriminant must be non-negative:

$$9-4k^2\geq 0$$
$$(3+2k)(3-2k)\geq 0$$

The key numbers are $-\frac{3}{2}$ and $\frac{3}{2}$. Draw the sign chart:

Interval	Test Number	$3+2k$	$3-2k$	$9-4k^2$
$\left(-\infty,-\frac{3}{2}\right)$	-2	neg.	pos.	neg.
$\left(-\frac{3}{2},\frac{3}{2}\right)$	0	pos.	pos.	pos.
$\left(\frac{3}{2},\infty\right)$	2	pos.	neg.	neg.

So the product is positive on $\left(-\frac{3}{2},\frac{3}{2}\right)$, and zero at $k=-\frac{3}{2}$ and $k=\frac{3}{2}$. Thus the equation will have real solutions if k is in the interval $\left[-\frac{3}{2},\frac{3}{2}\right]$.

15. Since (a, b) lies in the third quadrant, $a<0$ and $b<0$. Also, since (a, b) lies on the line $y=2x+1$, we have $b=2a+1$. Substituting into the distance formula:

$$\sqrt{a^2+b^2}=\sqrt{65}$$
$$a^2+b^2=65$$
$$a^2+(2a+1)^2=65$$
$$a^2+4a^2+4a+1=65$$
$$5a^2+4a-64=0$$
$$(5a-16)(a+4)=0$$
$$a=-4,\ \frac{16}{5}$$

Since $a<0$ (third quadrant), choose $a=-4$. Then $b=2(-4)+1=-7$. The point $(a,b)=(-4,-7)$.

16. Solve by isolating the radical and squaring each side of the equation:

$$\sqrt{3x+7}=x-1$$
$$3x+7=x^2-2x+1$$
$$0=x^2-5x-6$$
$$0=(x-6)(x+1)$$
$$x=-1,\ 6$$

Note that $x=6$ is indeed a solution to the original equation. However, $x=-1$ results in the statement $\sqrt{4}=-2$, which is false. Thus $x=-1$ is an extraneous solution of the original equation, even though it is a solution to the subsequent equation $x^2-5x-6=0$.

Chapter 3
Functions

3.1 The Definition of a Function

1. **a.** Finding the value: $g(1975) = \$2.00$

b. Finding the value: $g(1995) - g(1975) = \$4.25 - \$2.00 = \$2.25$

The minimum wage increased \$2.25 from 1975 to 1995.

3. **a.** The range of h consists of real numbers.

b. Finding the value: $h(\text{Mars}) = 2$

c. Since $h(\text{Neptune}) = 8$ and $h(\text{Pluto}) = 1$, $h(\text{Neptune})$ is larger. Neptune has more moons (8) than does Pluto (1).

5. **a.** Rules f and g are functions. Rule h is not a function since $h(x) = 1$ and $h(x) = 2$, which violates the definition of a function

b. The range of f is $\{1,2,3\}$, and the range of g is $\{2,3\}$.

7. **a.** Rule g is a function. Rule f is not a function since $f(b) = j$ and $f(b) = k$, which violates the definition of a function.

b. The range of g is $\{i,j\}$.

9. **a.** The domain is all real numbers, or $(-\infty,\infty)$.

b. Find where the denominator is 0:

$$-5x+1=0$$
$$-5x=-1$$
$$x=\tfrac{1}{5}$$

The domain is all real numbers except $\frac{1}{5}$, or $\left(-\infty,\frac{1}{5}\right)\cup\left(\frac{1}{5},\infty\right)$.

c. Find where the radical is defined:

$$-5x+1\geq 0$$
$$-5x\geq -1$$
$$x\leq \tfrac{1}{5}$$

The domain is $x\leq\frac{1}{5}$, or $\left(-\infty,\frac{1}{5}\right]$.

d. The domain is all real numbers, or $(-\infty,\infty)$.

11. a. The domain is all real numbers, or $(-\infty,\infty)$.

b. Find where the denominator is 0:

$$x^2-9=0$$
$$(x+3)(x-3)=0$$
$$x=-3,\ 3$$

The domain is all real numbers except ±3, or $(-\infty,-3)\cup(-3,3)\cup(3,\infty)$.

c. Find where the radical is defined:

$$x^2-9\ge 0$$
$$(x+3)(x-3)\ge 0$$

The key numbers are –3 and 3. Construct the sign chart:

Interval	Test Number	$x+3$	$x-3$	$(x+3)(x-3)$
$(-\infty,-3)$	–4	neg.	neg.	pos.
$(-3,3)$	0	pos.	neg.	neg.
$(3,\infty)$	4	pos.	pos.	pos.

So $x^2-9\ge 0$ on the set $(-\infty,-3]\cup[3,\infty)$. Thus the domain is $(-\infty,-3]\cup[3,\infty)$.

d. The domain is all real numbers, or $(-\infty,\infty)$.

13. a. The domain is all real numbers, or $(-\infty,\infty)$.

b. Find where the denominator is 0:

$$t^2-8t+15=0$$
$$(t-5)(t-3)=0$$
$$t=3,\ 5$$

The domain is all real numbers except 3 and 5, or $(-\infty,3)\cup(3,5)\cup(5,\infty)$.

c. Find where the radical is defined:

$$t^2-8t+15\ge 0$$
$$(t-5)(t-3)\ge 0$$

The key numbers are 3 and 5. Construct the sign chart:

Interval	Test Number	$t-3$	$t-5$	$t^2-8t+15$
$(-\infty,3)$	2	neg.	neg.	pos.
$(3,5)$	4	pos.	neg.	neg.
$(5,\infty)$	6	pos.	pos.	pos.

So $t^2-8t+15\ge 0$ on the set $(-\infty,3]\cup[5,\infty)$. Thus the domain is $(-\infty,3]\cup[5,\infty)$.

d. The domain is all real numbers, or $(-\infty,\infty)$.

15. **a.** The denominator is 0 when $x = -3$, so the domain is all real numbers except -3, or $(-\infty,-3)\cup(-3,\infty)$.

b. The radical is defined when $\dfrac{x-2}{2x+6} \geq 0$. The key numbers are -3 and 2. Construct the sign chart:

Interval	Test Number	$x-2$	$2x+6$	$\dfrac{x-2}{2x+6}$
$(-\infty,-3)$	-4	neg.	neg.	pos.
$(-3,2)$	0	neg.	pos.	neg.
$(2,\infty)$	4	pos.	pos.	pos.

So $\dfrac{x-2}{2x+6} \geq 0$ on the set $(-\infty,-3)\cup[2,\infty)$. Notice that $x = -3$ results in a 0 denominator, so that endpoint is not included. Thus the domain is $(-\infty,-3)\cup[2,\infty)$.

c. The denominator is 0 when $x = -3$, so the domain is all real numbers except -3, or $(-\infty,-3)\cup(-3,\infty)$.

17. The domain is all real numbers, or $(-\infty,\infty)$. The range is all real numbers, or $(-\infty,\infty)$.

19. The domain is all real numbers, or $(-\infty,\infty)$. The range is all real numbers, or $(-\infty,\infty)$.

21. Find where the denominator is 0:

$$\begin{aligned} 3x-18 &= 0 \\ 3x &= 18 \\ x &= 6 \end{aligned}$$

The domain is all real numbers except 6, or $(-\infty,6)\cup(6,\infty)$. To find the range, let $y = g(x)$ and solve for x:

$$\begin{aligned} y &= \frac{4x-20}{3x-18} \\ 3xy-18y &= 4x-20 \\ 3xy-4x &= 18y-20 \\ x(3y-4) &= 18y-20 \\ x &= \frac{18y-20}{3y-4} \end{aligned}$$

Since the denominator is 0 when $y = \frac{4}{3}$, the range is all real numbers except $\frac{4}{3}$, or $\left(-\infty,\frac{4}{3}\right)\cup\left(\frac{4}{3},\infty\right)$.

23. **a.** The denominator is 0 when $x = 5$, so the domain is all real numbers except 5, or $(-\infty,5)\cup(5,\infty)$. To find the range, let $y = f(x)$ and solve for x:

$$\begin{aligned} y &= \frac{x+3}{x-5} \\ xy-5y &= x+3 \\ xy-x &= 5y+3 \\ x(y-1) &= 5y+3 \\ x &= \frac{5y+3}{y-1} \end{aligned}$$

Since the denominator is 0 when $y = 1$, the range is all real numbers except 1, or $(-\infty,1)\cup(1,\infty)$.

b. Find where the denominator is 0:

$$x^3 - 5 = 0$$
$$x^3 = 5$$
$$x = \sqrt[3]{5}$$

The domain is all real numbers except $\sqrt[3]{5}$, or $(-\infty, \sqrt[3]{5}) \cup (\sqrt[3]{5}, \infty)$. To find the range, let $y = F(x)$ and solve for x:

$$y = \frac{x^3 + 3}{x^3 - 5}$$
$$x^3 y - 5y = x^3 + 3$$
$$x^3 y - x^3 = 5y + 3$$
$$x^3 (y-1) = 5y + 3$$
$$x^3 = \frac{5y+3}{y-1}$$
$$x = \sqrt[3]{\frac{5y+3}{y-1}}$$

Since the denominator is 0 when $y = 1$, the range is all real numbers except 1, or $(-\infty, 1) \cup (1, \infty)$.

25. The domain is all real numbers, or $(-\infty, \infty)$. Since $t^2 \geq 0$, $t^2 + 4 \geq 4$ and thus the range is $s \geq 4$, or $[4, \infty)$.

27. a. The rule can be written as $y = (x-3)^2$.

b. The rule can be written as $y = x^2 - 3$.

c. The rule can be written as $y = (3x)^2$.

d. The rule can be written as $y = 3x^2$.

29. a. Compute $f(1)$: $f(1) = (1)^2 - 3(1) + 1 = 1 - 3 + 1 = -1$

b. Compute $f(0)$: $f(0) = (0)^2 - 3(0) + 1 = 0 - 0 + 1 = 1$

c. Compute $f(-1)$: $f(-1) = (-1)^2 - 3(-1) + 1 = 1 + 3 + 1 = 5$

d. Compute $f\left(\frac{3}{2}\right)$: $f\left(\frac{3}{2}\right) = \left(\frac{3}{2}\right)^2 - 3\left(\frac{3}{2}\right) + 1 = \frac{9}{4} - \frac{9}{2} + 1 = -\frac{5}{4}$

e. Compute $f(z)$: $f(z) = (z)^2 - 3(z) + 1 = z^2 - 3z + 1$

f. Compute $f(x+1)$: $f(x+1) = (x+1)^2 - 3(x+1) + 1 = x^2 + 2x + 1 - 3x - 3 + 1 = x^2 - x - 1$

g. Compute $f(a+1)$: $f(a+1) = (a+1)^2 - 3(a+1) + 1 = a^2 + 2a + 1 - 3a - 3 + 1 = a^2 - a - 1$

h. Compute $f(-x)$: $f(-x) = (-x)^2 - 3(-x) + 1 = x^2 + 3x + 1$

i. Using our result from part **a**: $|f(1)| = |-1| = 1$

j. Compute $f(\sqrt{3})$: $f(\sqrt{3}) = (\sqrt{3})^2 - 3(\sqrt{3}) + 1 = 3 - 3\sqrt{3} + 1 = 4 - 3\sqrt{3}$

k. Compute $f(1+\sqrt{2})$: $f(1+\sqrt{2}) = (1+\sqrt{2})^2 - 3(1+\sqrt{2}) + 1 = 1 + 2\sqrt{2} + 2 - 3 - 3\sqrt{2} + 1 = 1 - \sqrt{2}$

l. Compute $|1 - f(2)|$: $|1 - f(2)| = \left|1 - \left[(2)^2 - 3(2) + 1\right]\right| = |1 - [4 - 6 + 1]| = |1 - (-1)| = |1 + 1| = 2$

31. **a.** Compute $f(2x)$: $f(2x)=3(2x)^2=3(4x^2)=12x^2$

b. Compute $2f(x)$: $2f(x)=2(3x^2)=6x^2$

c. Compute $f(x^2)$: $f(x^2)=3(x^2)^2=3x^4$

d. Compute $[f(x)]^2$: $[f(x)]^2=(3x^2)^2=9x^4$

e. Compute $f\left(\frac{x}{2}\right)$: $f\left(\frac{x}{2}\right)=3\left(\frac{x}{2}\right)^2=3\bullet\frac{x^2}{4}=\frac{3}{4}x^2$

f. Compute $\frac{f(x)}{2}$: $\frac{f(x)}{2}=\frac{3x^2}{2}=\frac{3}{2}x^2$

33. **a.** Compute $H(\sqrt{2})$: $H(\sqrt{2})=1-2(\sqrt{2})^2=1-2(2)=1-4=-3$

b. Compute $H\left(\frac{5}{6}\right)$: $H\left(\frac{5}{6}\right)=1-2\left(\frac{5}{6}\right)^2=1-2\left(\frac{25}{36}\right)=1-\frac{25}{18}=-\frac{7}{18}$

c. Compute $H(x+1)$: $H(x+1)=1-2(x+1)^2=1-2(x^2+2x+1)=1-2x^2-4x-2=-2x^2-4x-1$

d. Compute $H(x+h)$: $H(x+h)=1-2(x+h)^2=1-2(x^2+2xh+h^2)=1-2x^2-4xh-2h^2$

35. **a.** We must have $g(0)=2$, since $g(x)=2$ for all x.

b. We must have $g(5)=2$, since $g(x)=2$ for all x.

c. We must have $g(x+h)=2$, since $g(x)=2$ for all x.

37. **a.** Solving the equation $f(x)=16$:

$$\begin{aligned} x^2-6x&=16\\ x^2-6x-16&=0\\ (x-8)(x+2)&=0\\ x&=-2,8 \end{aligned}$$

b. Solving the equation $f(x)=-10$:

$$\begin{aligned} x^2-6x&=-10\\ x^2-6x+10&=0 \end{aligned}$$

Using the quadratic formula: $x=\frac{6\pm\sqrt{(-6)^2-4(1)(10)}}{2(1)}=\frac{6\pm\sqrt{36-40}}{2}=\frac{6\pm\sqrt{-4}}{2}$

There are no real solutions.

c. Solving the equation $f(x)=-9$:

$$\begin{aligned} x^2-6x&=-9\\ x^2-6x+9&=0\\ (x-3)^2&=0\\ x&=3 \end{aligned}$$

39. Solving the equation $p(n)=8$:

$$\begin{aligned} -0.012n+20.49&=8\\ -0.012n&=-12.49\\ n&\approx 1041 \end{aligned}$$

When the price is set at $8 per tee shirt, the store can expect to sell 1041 tee shirts per month.

41. Solving the equation $p(n)=19$:

$$\begin{aligned}-0.012n+20.49&=19\\-0.012n&=-1.49\\n&\approx 124\end{aligned}$$

Now solving the equation $f(n)=19$:

$$\begin{aligned}4+\frac{3000}{n+100}&=19\\\frac{3000}{n+100}&=15\\3000&=15n+1500\\15n&=1500\\n&=100\end{aligned}$$

When the price is set at \$19 per tee shirt, the first model predicts 124 sales while the second model predicts 100 sales.

43. a. Simplifying: $T(x+2)=2(x+2)^2-3(x+2)=2x^2+8x+8-3x-6=2x^2+5x+2$

b. Simplifying: $T(x-2)=2(x-2)^2-3(x-2)=2x^2-8x+8-3x+6=2x^2-11x+14$

c. Subtracting the results:

$$T(x+2)-T(x-2)=\left(2x^2+5x+2\right)-\left(2x^2-11x+14\right)=2x^2+5x+2-2x^2+11x-14=16x-12$$

45. Let a = 1 and b = 2. Then:

$$f(a+b)=f(3)=3^2-1=8$$

$$f(a)=f(1)=1^2-1=0 \qquad f(b)=f(2)=2^2-1=3$$

So $f(a+b)\neq f(a)+f(b)$.

47. Let a = 2. Then:

$$f\left(\tfrac{1}{a}\right)=f\left(\tfrac{1}{2}\right)=\left(\tfrac{1}{2}\right)^2-1=\tfrac{1}{4}-1=-\tfrac{3}{4} \qquad \frac{1}{f(a)}=\frac{1}{f(2)}=\frac{1}{2^2-1}=\tfrac{1}{3}$$

So $f\left(\tfrac{1}{a}\right)\neq\dfrac{1}{f(a)}$.

49. a. Compute $f(a)$, $f(2a)$ and $f(3a)$:

$$f(a)=\frac{a-a}{a+a}=\frac{0}{2a}=0 \qquad f(2a)=\frac{2a-a}{2a+a}=\frac{a}{3a}=\tfrac{1}{3}$$

$$f(3a)=\frac{3a-a}{3a+a}=\frac{2a}{4a}=\tfrac{1}{2}$$

Since $\tfrac{1}{2}\neq 0+\tfrac{1}{3}$, $f(3a)\neq f(a)+f(2a)$.

b. Compute $f(5a)$: $f(5a)=\dfrac{5a-a}{5a+a}=\dfrac{4a}{6a}=\tfrac{2}{3}$

Since $f(2a)=\tfrac{1}{3}$, $f(5a)=2f(2a)$.

51. Simplify $f(ax+b)=2(ax+b)+3=2ax+2b+3$. Since $f(ax+b)=x$, we have $2ax+2b+3=x$. Since a and b are constants, we must have:

$$\begin{aligned}2ax&=x\\2a&=1\\a&=\tfrac{1}{2}\end{aligned} \qquad \begin{aligned}2b+3&=0\\2b&=-3\\b&=-\tfrac{3}{2}\end{aligned}$$

So $a=\tfrac{1}{2}$ and $b=-\tfrac{3}{2}$.

53. Compute $f\left(\frac{3z-4}{5z-3}\right)$: $f\left(\frac{3z-4}{5z-3}\right)=\dfrac{3\left(\frac{3z-4}{5z-3}\right)-4}{5\left(\frac{3z-4}{5z-3}\right)-3}=\dfrac{3(3z-4)-4(5z-3)}{5(3z-4)-3(5z-3)}=\dfrac{9z-12-20z+12}{15z-20-15z+9}=\dfrac{-11z}{-11}=z$

55. Since $f(0)=-2(0)^2+6(0)+k=0+0+k=k$, if $f(0)=-1$, we have $k=-1$.

57. Finding where $h(c)=c$:

$$\begin{aligned} h(c)&=c\\ c^2-4c-c&=c\\ c^2-6c&=0\\ c(c-6)&=0\\ c&=0,6 \end{aligned}$$

Since c is nonzero, we have $c=6$.

59. Actually, we already know the answer to this question. Since $\dfrac{-b+\sqrt{b^2-4ac}}{2a}$ is one of the roots to the quadratic equation $q(x)=0$, we know $q\left(\dfrac{-b+\sqrt{b^2-4ac}}{2a}\right)=0$. Let's check our answer manually:

$$\begin{aligned} q\left(\frac{-b+\sqrt{b^2-4ac}}{2a}\right)&=a\left(\frac{-b+\sqrt{b^2-4ac}}{2a}\right)^2+b\left(\frac{-b+\sqrt{b^2-4ac}}{2a}\right)+c\\ &=a\left(\frac{b^2-2b\sqrt{b^2-4ac}+b^2-4ac}{4a^2}\right)+\left(\frac{-b^2+b\sqrt{b^2-4ac}}{2a}\right)+c\\ &=\frac{2b^2-4ac-2b\sqrt{b^2-4ac}}{4a}+\frac{-b^2+b\sqrt{b^2-4ac}}{2a}+c\\ &=\frac{b^2-2ac-b\sqrt{b^2-4ac}-b^2+b\sqrt{b^2-4ac}}{2a}+c\\ &=\frac{-2ac}{2a}+c\\ &=-c+c\\ &=0 \end{aligned}$$

61. For the rule F, each input has exactly one output. That is, every person has a mother (so each value of x has been assigned), and no person has two (natural) mothers (so no value of x is assigned twice). The rule G, however, fails on both accounts. Not every person has an aunt (in the case that both parents are only children), and so not every value of x has been assigned. Furthermore, two aunts could be assigned to a person, which would violate the definition of a function.

63. Compute the values:

$f(8)=4$, since the primes ≤ 8 are $\{2,3,5,7\}$

$f(10)=4$, since the primes ≤ 10 are $\{2,3,5,7\}$

$f(50)=15$, since the primes ≤ 50 are $\{2,3,5,7,11,13,17,19,23,29,31,37,41,43,47\}$

65. **a.** From the expression given, $G(10)=5$ and $G(14)=9$.

b. The values are $G(100)=9$, $G(750)=0$, and $G(1000)=9$.

67. Simplifying the right-hand side of the desired inequality:

$$f\left(\frac{x_1+x_2}{2}\right)=a\left(\frac{x_1+x_2}{2}\right)^2+b\left(\frac{x_1+x_2}{2}\right)+c$$
$$=\frac{ax_1^2+2ax_1x_2+ax_2^2}{4}+\frac{bx_1+bx_2}{2}+c$$
$$=\frac{ax_1^2+2ax_1x_2+ax_2^2+2bx_1+2bx_2+4c}{4}$$

Simplifying the left-hand side of the desired inequality:

$$\frac{f(x_1)+f(x_2)}{2}=\frac{ax_1^2+bx_1+c+ax_2^2+bx_2+c}{2}=\frac{ax_1^2+bx_1+ax_2^2+bx_2+2c}{2}$$

Therefore we need to show:

$$\frac{ax_1^2+bx_1+ax_2^2+bx_2+2c}{2}\le\frac{ax_1^2+2ax_1x_2+ax_2^2+2bx_1+2bx_2+4c}{4}$$
$$2ax_1^2+2bx_1+2ax_2^2+2bx_2+4c\le ax_1^2+2ax_1x_2+ax_2^2+2bx_1+2bx_2+4c$$
$$ax_1^2-2ax_1x_2+ax_2^2\le 0$$
$$a(x_1-x_2)^2\le 0$$

Since $(x_1-x_2)^2\ge 0$ and $a<0$, then $a(x_1-x_2)^2<0$. Thus the desired inequality is valid.

3.2 The Graph of a Function

1. Substituting $x = 3$: $y=\sqrt{x}=\sqrt{3}\approx 1.732$

3. Substituting $x=\sqrt{5}$: $y=\frac{1}{x}=\frac{1}{\sqrt{5}}=\frac{\sqrt{5}}{5}\approx 0.447$

5.
- **a.** Since any vertical line drawn intersects the graph in at most one point, this is the graph of a function.
- **b.** Since a vertical line can be drawn which intersects the graph in two points, this cannot be the graph of a function.
- **c.** Since a vertical line can be drawn which intersects the graph in two points, this cannot be the graph of a function.
- **d.** Since any vertical line drawn intersects the graph in only one point, this is the graph of a function.

7. The domain is $[-4,2]$ and the range is $[-3,3]$.

9. The domain is $[-3,4]$ and the range is $[-2,2]$.

11. The domain is $[-4,-1)\cup(-1,4]$ and the range is $[-2, 3)$.

13. The domain is $[-4, 3]$ and the range is $\{2\}$.

15.
- **a.** Since $(-5,1)$ lies on the graph of F, $F(-5) = 1$.
- **b.** Since $(2,-3)$ lies on the graph of F, $F(2) = -3$.
- **c.** Since $F(1)\approx -2$, $F(1)$ is negative, not positive.
- **d.** Since $(2,-3)$ lies on the graph of F, $F(x) = -3$ when $x = 2$.
- **e.** Since $F(2) = -3$ and $F(-2) = -2$, $F(2) - F(-2) = -3 - (-2) = -1$.

17.
- **a.** positive
- **b.** $f(-2) = 4; f(1) = 1; f(2) = 2; f(3) = 0$
- **c.** $f(2)$, since $f(2) > 0$ and $f(4) < 0$
- **d.** $f(4) - f(1) = -2 - 1 = -3$
- **e.** $f(4) - f(1) \ = \ -3 \ = 3$
- **f.** domain = $[-2,4]$; range = $[-2,4]$

19.
- **a.** $f(-2) = 0$ and $g(-2) = 1$, so $g(-2)$ is larger.
- **b.** $f(0) - g(0) = 2 - (-3) = 2 + 3 = 5$
- **c.** Compute the three values:

  $f(1)-g(1)=1-(-1)=2$ $\qquad$ $f(2)-g(2)=1-0=1$

  $f(3)-g(3)=4-1=3$

  So $f(2) - g(2)$ is the smallest.
- **d.** Since $f(1) = 1$, we look for where $g(x) = 1$. This occurs at $x = -2$ or $x = 3$.
- **e.** Since $(3, 4)$ is a point on the graph of f, 4 is in the range of f.

21. **a.** Set up a table of values:

x	-3	-2	-1	0	1	2	3
$y=\|x\|$	3	2	1	0	1	2	3

Now graph the equation $y=|x|$:

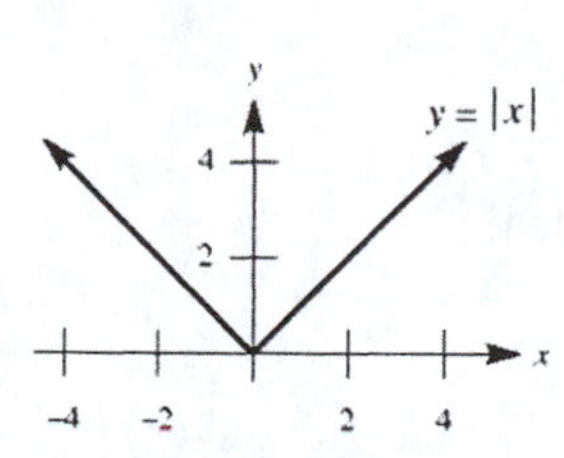

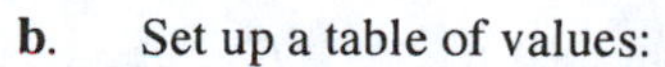

b. Set up a table of values:

x	-3	-2	-1	0	1	2	3
$y=x^2$	9	4	1	0	1	4	9

Now graph the equation $y=x^2$:

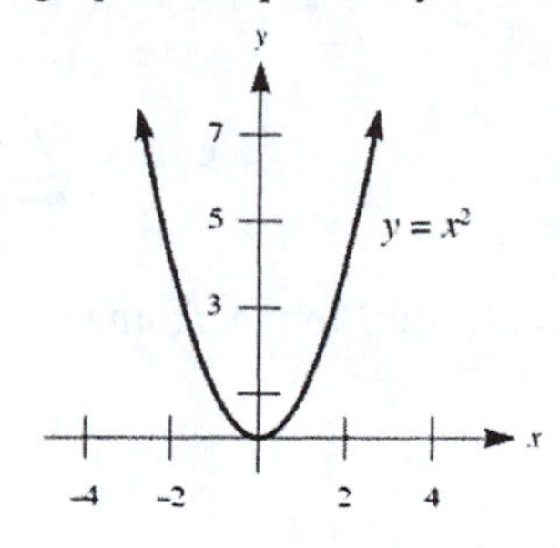

c. Set up a table of values $y=x^3$:

x	-2	-1	0	1	2
$y=x^3$	-8	-1	0	1	8

Now graph the equation $y=x^3$:

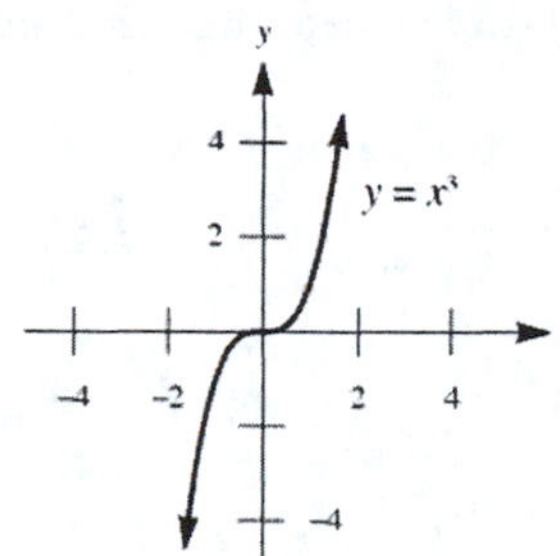

23. Graphing the function:

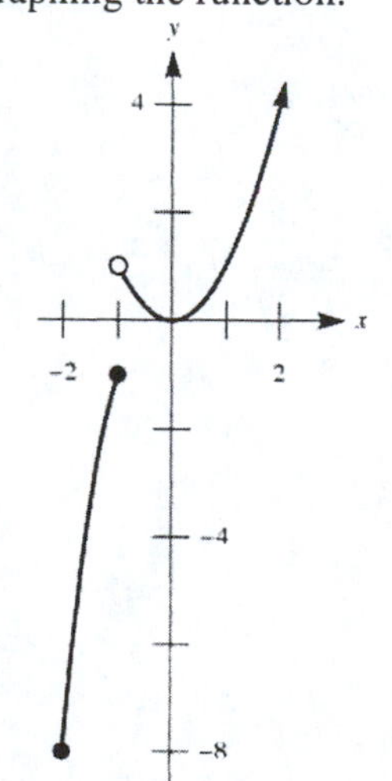

25. Graphing the function:

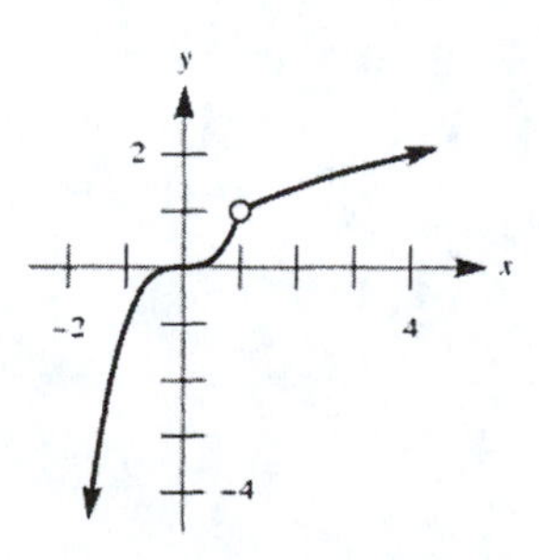

27. **a.** We "piece together" the curves $y=\sqrt{x}$ when $0 \le x \le 1$ and $y=\frac{1}{x}$ when $1 < x < 2$:

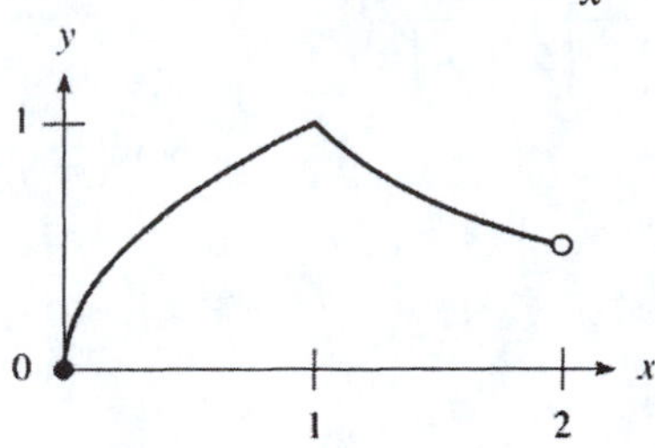

b. This graph is identical to that from part **a**, except the point (1,1) is excluded:

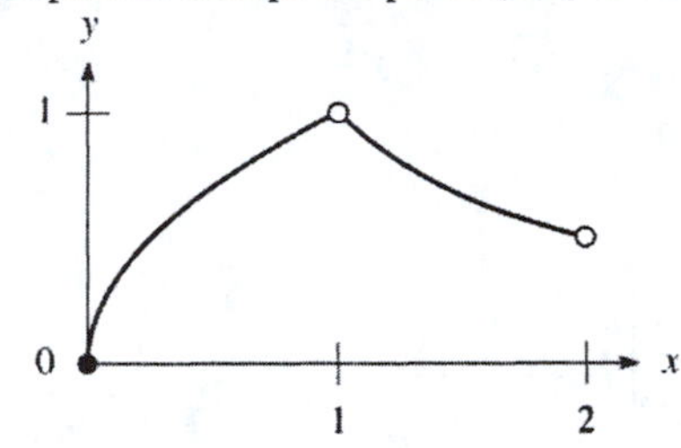

29. **a.** We "piece together" the curves $y=\sqrt{1-x^2}$ when $-1 \le x \le 0$ and $y=x^2$ when $0 < x \le 2$:

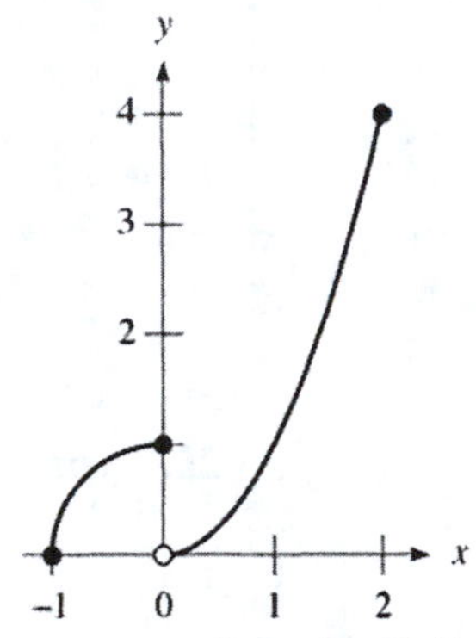

b. This graph is identical to that from part **a**, except the point (0,1) is excluded:

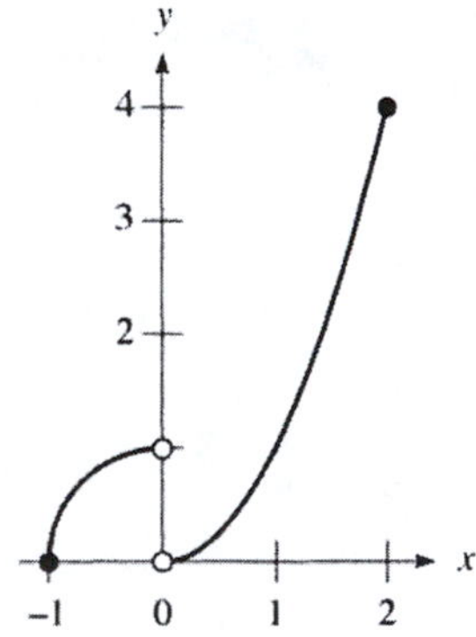

31. **a.** Sketching the graph:

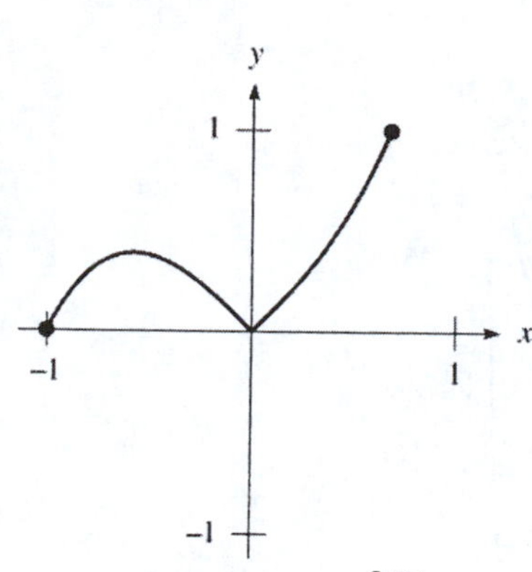

b. Sketching the graph:

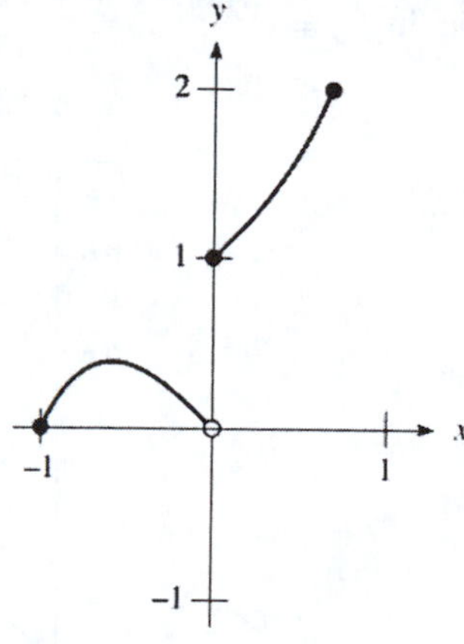

33. For point P, $x = 4$ and thus $y = \sqrt[3]{4}$. For point Q, $y = \sqrt[3]{4}$ and thus, since Q lies on the line $y = x$, $x = \sqrt[3]{4}$. For point R, $x = \sqrt[3]{4}$ and thus $y = \sqrt[3]{\sqrt[3]{4}} = \sqrt[9]{4}$. Thus the coordinates of each point are given by $P\left(4, \sqrt[3]{4}\right) \approx P(4, 1.587)$, $Q\left(\sqrt[3]{4}, \sqrt[3]{4}\right) \approx Q(1.587, 1.587)$, and $R\left(\sqrt[3]{4}, \sqrt[9]{4}\right) \approx R(1.587, 1.167)$.

35. For point P, $x = \sqrt{2}$ and thus: $y = \left(\sqrt{2}\right)^3 - 3\sqrt{2} = 2\sqrt{2} - 3\sqrt{2} = -\sqrt{2}$

For point Q, $y = -\sqrt{2}$ and thus, since Q lies on the line $y = x$, $x = -\sqrt{2}$.

For point R, $x = -\sqrt{2}$ and thus: $y = \left(-\sqrt{2}\right)^3 - 3\left(-\sqrt{2}\right) = -2\sqrt{2} + 3\sqrt{2} = \sqrt{2}$

Thus the coordinates of each point are given by $P\left(\sqrt{2}, -\sqrt{2}\right) \approx P(1.414, -1.414)$, $Q\left(-\sqrt{2}, -\sqrt{2}\right) \approx Q(-1.414, -1.414)$, and $R\left(-\sqrt{2}, \sqrt{2}\right) \approx R(-1.414, 1.414)$.

37. **a.** Sketching the graph:

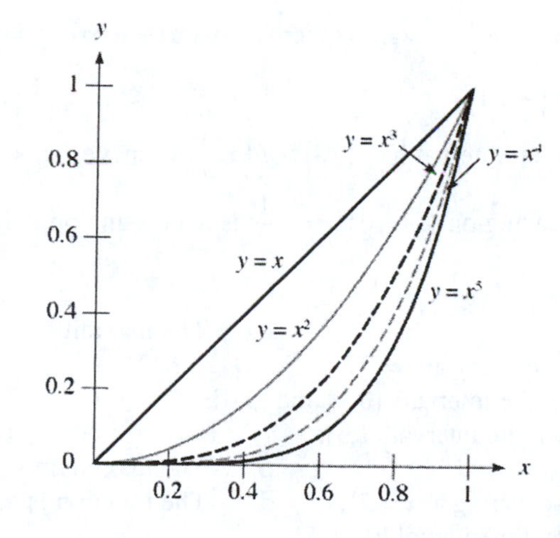

b. As the power on x increases, the curve "hugs" the x-axis closer, before rising to the point (1,1). Sketching the graph including $y = x^{100}$:

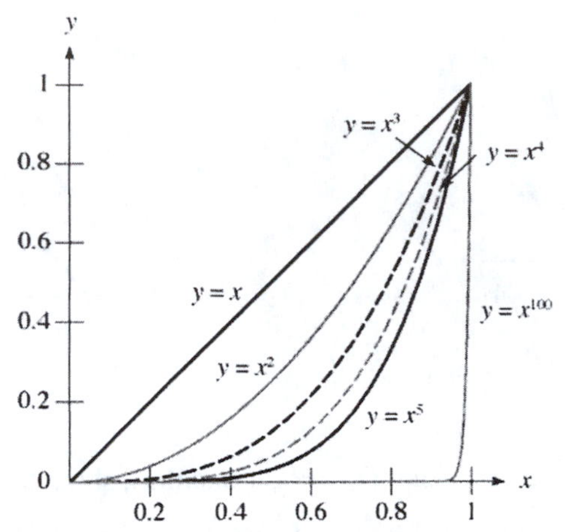

3.3 Shapes of Graphs. Average Rate of Change

1. a. A turning point is a point where the function changes from increasing to decreasing, or from decreasing to increasing. For example, $y = |x|$, $y = x^2$, and $y = \sqrt{1 - x^2}$ all have turning points at $x = 0$.

   b. A maximum value is a y-value corresponding to the highest point on the graph. For example, $y = 1$ is a maximum value for the function $y = \sqrt{1 - x^2}$ occurring at $x = 0$.

   c. A minimum value is a y-value corresponding to the lowest point on the graph. For example, $y = 0$ is a minimum value for the functions $y = |x|$ and $y = x^2$.

   d. A function f is increasing on an interval if $f(x_1) < f(x_2)$ whenever $x_1 < x_2$ on the interval. For example, $y = |x|$ and $y = x^2$ are increasing on $[0, \infty)$, $y = x^3$ is increasing on $(-\infty, \infty)$, $y = \sqrt{x}$ is increasing on $[0, \infty)$, and $y = \sqrt{1 - x^2}$ is increasing on $[-1, 0]$.

   e. A function f is decreasing on an interval if $f(x_1) > f(x_2)$ whenever $x_1 < x_2$ on the interval. For example, $y = |x|$ and $y = x^2$ are decreasing on $(-\infty, 0]$, $y = \frac{1}{x}$ is decreasing on either $(-\infty, 0)$ or $(0, \infty)$, and $y = \sqrt{1 - x^2}$ is decreasing on $[0, 1]$.

3. a. The range is [–1,1]. b. The maximum value is 1 (occurring at $x = 1$).

   c. The minimum value is –1 (occurring at $x = 3$).

   d. The function is increasing on the intervals [0,1] and [3,4].

   e. The function is decreasing on the interval [1,3].

5. a. The range is [–3,0]. b. The maximum value is 0 (occurring at $x = 0$ and $x = 4$).

   c. The minimum value is –3 (occurring at $x = 2$). d. The function is increasing on the interval [2,4].

   e. The function is decreasing on the interval [0,2].

7. The turning points are approximately (–1.15,1.08) and (1.15,–5.08). The function is increasing on the intervals $(-\infty, -1.15] \cup [1.15, \infty)$ and it is decreasing on the interval $[-1.15, 1.15]$.

9. Since $f(3) = 3^2 + 2(3) = 15$ and $f(5) = 5^2 + 2(5) = 35$, the average rate of change is given by:

$$\frac{f(5) - f(3)}{5 - 3} = \frac{35 - 15}{5 - 3} = \frac{20}{2} = 10$$

11. Since $g(-1)=2(-1)^2-4(-1)=6$ and $g(3)=2(3)^2-4(3)=6$, the average rate of change is given by:

$$\frac{g(3)-g(-1)}{3-(-1)}=\frac{6-6}{3+1}=0$$

13. Since $h(5) = 2(5) - 6 = 4$ and $h(12) = 2(12) - 6 = 18$, the average rate of change is given by:

$$\frac{h(12)-h(5)}{12-5}=\frac{18-4}{12-5}=\frac{14}{7}=2$$

15. a. Since $G(0) = 22$ and $G(3) = 23$, the average rate of change is given by: $\frac{G(3)-G(0)}{3-0}=\frac{23-22}{3-0}=\frac{1}{3}$ °C/min

b. Since $G(3) = 23$ and $G(6) = 27$, the average rate of change is given by: $\frac{G(6)-G(3)}{6-3}=\frac{27-23}{6-3}=\frac{4}{3}$ °C/min

c. Since $G(6) = 27$ and $G(8) = 28$, the average rate of change is given by: $\frac{G(8)-G(6)}{8-6}=\frac{28-27}{8-6}=\frac{1}{2}$ °C/min

17. a. $\frac{\Delta P}{\Delta t}$ appears to be greater over the period 1984-1990. Calculating $\frac{\Delta P}{\Delta t}$ over each period:

1978-1984: $\frac{\Delta P}{\Delta t}=\frac{11\%-0\%}{6 \text{ years}}\approx 1.8\% / \text{year}$ 1984-1990: $\frac{\Delta P}{\Delta t}=\frac{69\%-11\%}{6 \text{ years}}\approx 9.7\% / \text{year}$

b. Using the given data:

1978-1984: $\frac{\Delta P}{\Delta t}=\frac{10.6\%-0.3\%}{6 \text{ years}}\approx 1.7\% / \text{year}$ 1984-1990: $\frac{\Delta P}{\Delta t}=\frac{68.6\%-10.6\%}{6 \text{ years}}\approx 9.7\% / \text{year}$

19. a. Calculating $\frac{\Delta f}{\Delta t}$ over the period: $\frac{\Delta f}{\Delta t}=\frac{20-19 \text{ million tons}}{16 \text{ years}}=\frac{1}{16}\approx 0.06$ million tons / year

b. Using the exact values: $\frac{\Delta f}{\Delta t}=\frac{20.32-18.68 \text{ million tons}}{16 \text{ years}}\approx 0.10$ million tons / year

21. a. Finding the difference quotient: $\frac{f(x)-f(3)}{x-3}=\frac{2x^2-18}{x-3}=\frac{2(x+3)(x-3)}{x-3}=2x+6$

b. Finding the difference quotient: $\frac{f(x)-f(a)}{x-a}=\frac{2x^2-2a^2}{x-a}=\frac{2(x+a)(x-a)}{x-a}=2x+2a$

23. a. Finding the difference quotient: $\frac{f(2+h)-f(2)}{h}=\frac{(2+h)^2-4}{h}=\frac{4+4h+h^2-4}{h}=\frac{h(4+h)}{h}=4+h$

b. Finding the difference quotient: $\frac{f(x+h)-f(x)}{h}=\frac{(x+h)^2-x^2}{h}=\frac{x^2+2xh+h^2-x^2}{h}=\frac{h(2x+h)}{h}=2x+h$

25. a. Finding the difference quotient: $\frac{f(x)-f(a)}{x-a}=\frac{(8x-3)-(8a-3)}{x-a}=\frac{8x-8a}{x-a}=\frac{8(x-a)}{x-a}=8$

b. Finding the difference quotient: $\frac{f(x+h)-f(x)}{h}=\frac{(8x+8h-3)-(8x-3)}{h}=\frac{8h}{h}=8$

27. a. Finding the difference quotient:

$$\frac{f(x)-f(a)}{x-a}=\frac{\left(x^2-2x+4\right)-\left(a^2-2a+4\right)}{x-a}=\frac{x^2-a^2-2x+2a}{x-a}=\frac{(x+a)(x-a)-2(x-a)}{x-a}=x+a-2$$

b. Finding the difference quotient:

$$\begin{aligned}\frac{f(x+h)-f(x)}{h}&=\frac{(x+h)^2-2(x+h)+4-\left(x^2-2x+4\right)}{h}\\&=\frac{x^2+2xh+h^2-2x-2h+4-x^2+2x-4}{h}\\&=\frac{2xh+h^2-2h}{h}\\&=2x+h-2\end{aligned}$$

29. a. Finding the difference quotient: $\frac{f(x)-f(a)}{x-a}=\frac{\frac{1}{x}-\frac{1}{a}}{x-a}\bullet\frac{ax}{ax}=\frac{a-x}{ax(x-a)}=-\frac{1}{ax}$

b. Finding the difference quotient: $\frac{f(x+h)-f(x)}{h}=\frac{\frac{1}{x+h}-\frac{1}{x}}{h}\bullet\frac{x(x+h)}{x(x+h)}=\frac{x-(x+h)}{hx(x+h)}=\frac{-h}{hx(x+h)}=-\frac{1}{x(x+h)}$

31. a. Finding the difference quotient: $\frac{f(x)-f(a)}{x-a}=\frac{2x^3-2a^3}{x-a}=\frac{2(x-a)\left(x^2+ax+a^2\right)}{x-a}=2x^2+2ax+2a^2$

b. Finding the difference quotient:

$$\begin{aligned}\frac{f(x+h)-f(x)}{h}&=\frac{2(x+h)^3-2x^3}{h}\\&=\frac{2x^3+6hx^2+6h^2x+2h^3-2x^3}{h}\\&=\frac{6hx^2+6h^2x+2h^3}{h}\\&=6x^2+6hx+2h^2\end{aligned}$$

33. Finding the average velocity: $\frac{\Delta s}{\Delta t}=\frac{s(2)-s(1)}{2-1}=\frac{16(2)^2-16(1)^2}{2-1}=\frac{64-16}{1}=48$ feet/second

35. a. Finding the average velocity:

$$\frac{\Delta s}{\Delta t}=\frac{s(2+h)-s(2)}{2+h-2}=\frac{16(2+h)^2-16(2)^2}{h}=\frac{64+64h+16h^2-64}{h}=\frac{64h+16h^2}{h}=(64+16h)\text{ feet/second}$$

b. Completing the table:

h (seconds)	0.1	0.01	0.001	0.0001	0.00001
Average velocity $\frac{\Delta s}{\Delta t}$ on interval $[2,2+h]$	65.6	64.16	64.016	64.0016	64.00016

c. The values seem to be approaching 64 feet/second.

37. Computing $\frac{\Delta p}{\Delta x}$ over each of the intervals:

$$0 \le x \le 100: \frac{\Delta p}{\Delta x} = \frac{p(100) - p(0)}{100 - 0} = \frac{\frac{24}{2^1} - \frac{24}{1}}{100} = \frac{-24}{200} = -\frac{3}{25} \text{ dollars/item}$$

$$300 \le x \le 400: \frac{\Delta p}{\Delta x} = \frac{p(400) - p(300)}{400 - 300} = \frac{\frac{24}{2^4} - \frac{24}{2^3}}{100} = \frac{-24}{1600} = -\frac{3}{200} \text{ dollars/item}$$

The answers make sense since, as more items are produced, the price should decrease.

39. Complete the table:

Function	$\lvert x \rvert$	x^2	x^3
Domain	$(-\infty, \infty)$	$(-\infty, \infty)$	$(-\infty, \infty)$
Range	$[0, \infty)$	$[0, \infty)$	$(-\infty, \infty)$
Turning Point	$(0,0)$	$(0,0)$	none
Maximum Value	none	none	none
Minimum Value	0	0	none
Interval(s) where Increasing	$(0, \infty)$	$(0, \infty)$	$(-\infty, \infty)$
Intervals(s) where Decreasing	$(-\infty, 0)$	$(-\infty, 0)$	none

41. Solving the equation:

$$\frac{\frac{1}{b} - 1}{b - 1} = -\frac{1}{5}$$

$$\frac{1 - b}{b(b - 1)} = -\frac{1}{5}$$

$$-\frac{1}{b} = -\frac{1}{5}$$

$$b = 5$$

43. Finding the difference quotient:

$$\begin{aligned}\frac{f(x+h) - f(x)}{h} &= \frac{a(x+h)^2 + b(x+h) + c - (ax^2 + bx + c)}{h} \\ &= \frac{ax^2 + 2axh + ah^2 + bx + bh + c - ax^2 - bx - c}{h} \\ &= \frac{2axh + ah^2 + bh}{h} \\ &= 2ax + ah + b\end{aligned}$$

45. Using a new viewing rectangle:

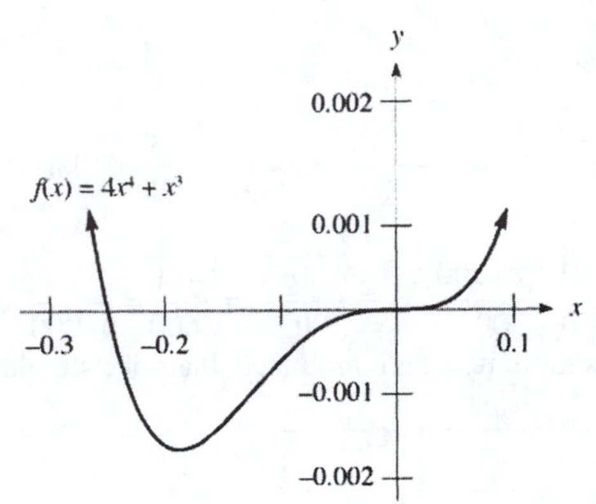

The turning point is (–0.1875,–0.0016). The interval of increase is $[-0.1875,\infty)$ and the interval of decrease is $(-\infty,-0.1875]$.

47. **a.** The functions have the same graph:

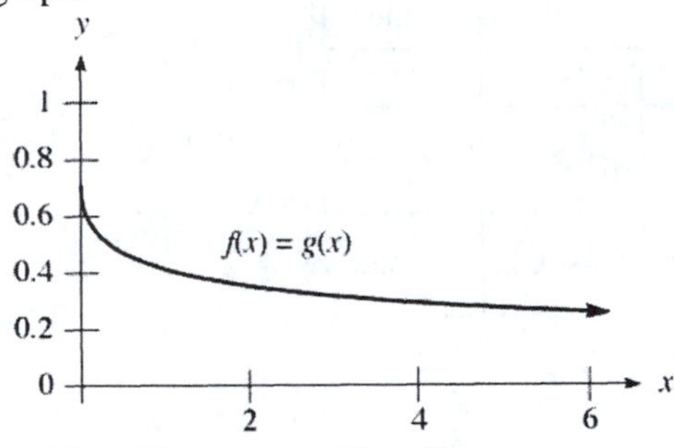

b. Verifying the identity: $f(x)=\dfrac{\sqrt{x}-\sqrt{2}}{x-2}=\dfrac{\sqrt{x}-\sqrt{2}}{\left(\sqrt{x}-\sqrt{2}\right)\left(\sqrt{x}+\sqrt{2}\right)}=\dfrac{1}{\sqrt{x}+\sqrt{2}}=g(x)$

This identity is not valid when $x = 2$.

3.4 Techniques in Graphing

1. **a.** C **b.** F **c.** I **d.** A **e.** J **f.** K
g. D **h.** B **i.** E **j.** H **k.** G

3. This graph will be that of $y=x^3$ translated down 3 units:

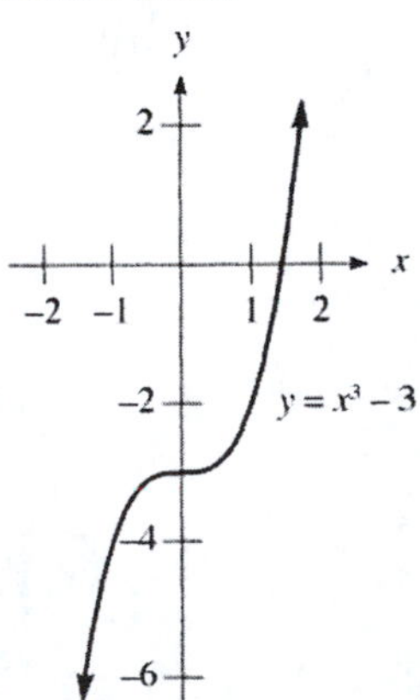

5. This graph will be that of $y = x^2$ translated to the left 4 units:

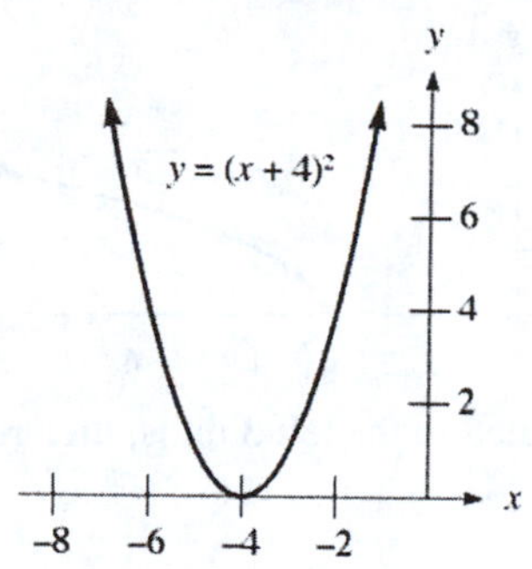

7. This graph will be that of $y = x^2$ translated to the right 4 units:

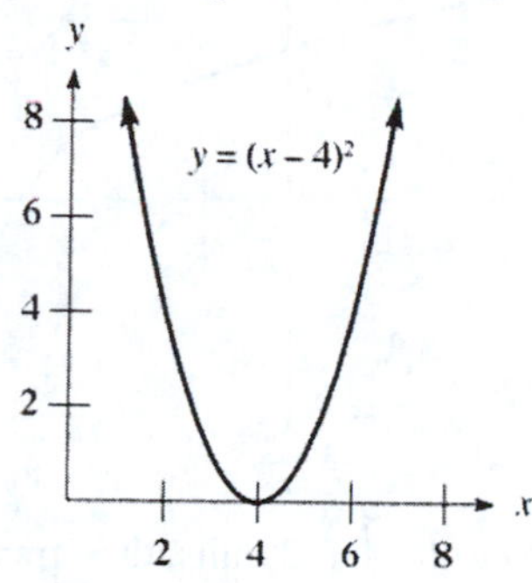

9. This graph will be that of $y = x^2$ reflected across the x-axis:

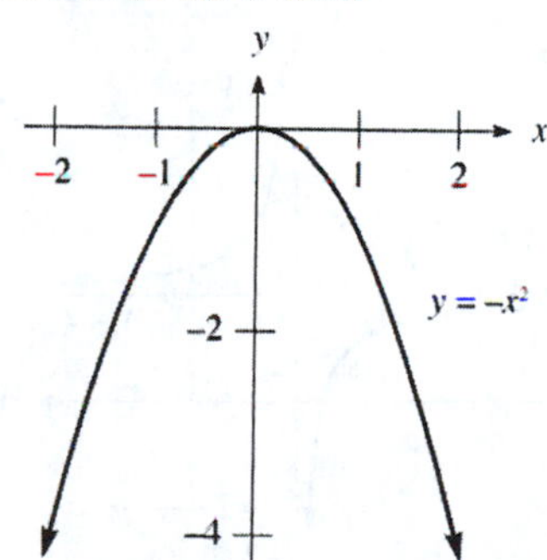

11. This graph will be that of $y = x^2$ translated to the right 3 units, then reflected across the x-axis:

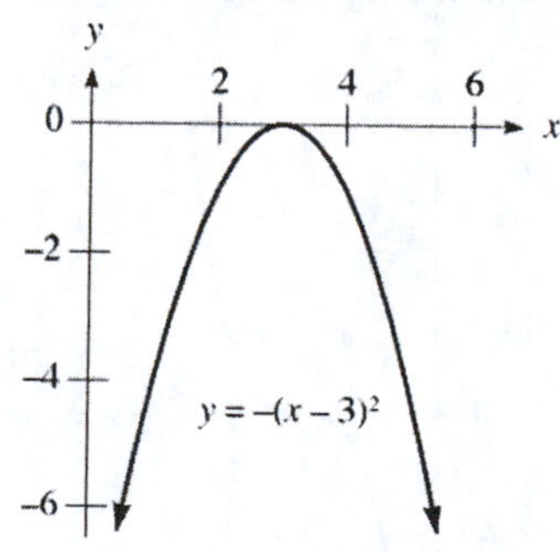

13. This graph will be that of $y=\sqrt{x}$ translated to the right 3 units:

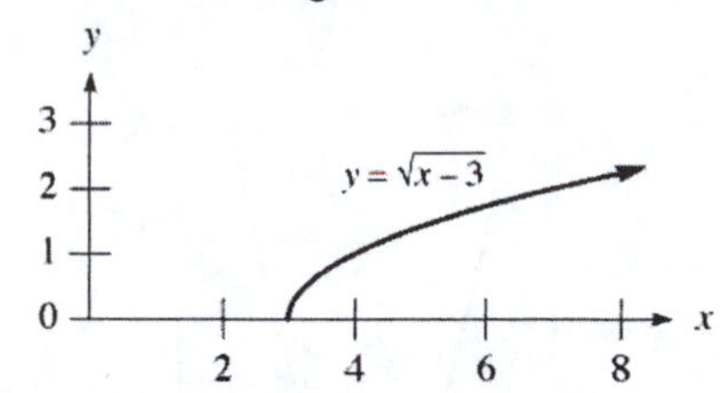

15. This graph will be that of $y=\sqrt{x}$ translated to the left 3 units, then reflected across the y-axis:

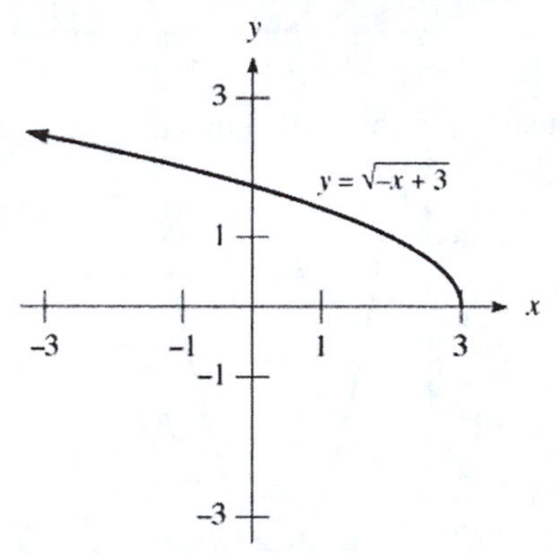

17. This graph will be that of $y=\dfrac{1}{x}$ translated to the left 2 units, then translated up 2 units:

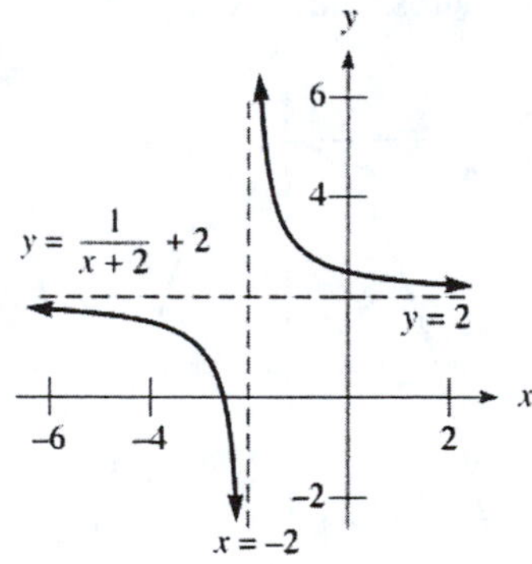

19. This graph will be that of $y=x^3$ translated to the right 2 units:

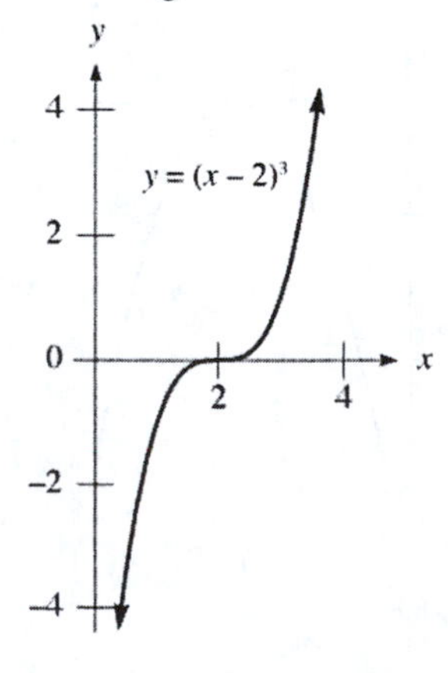

21. This graph will be that of $y = x^3$ reflected across the x-axis, then translated up 4 units:

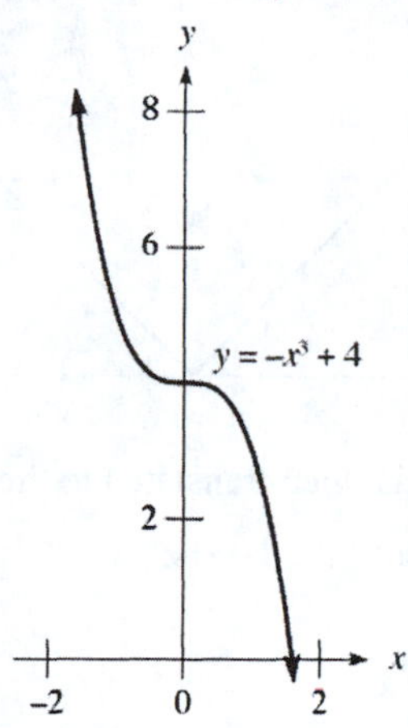

23. **a.** This graph will be that of $y = |x|$ translated to the left 4 units:

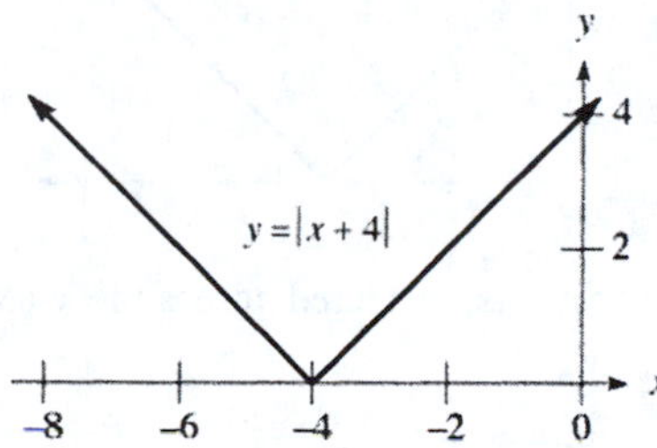

b. This graph will be that of $y = |x|$ reflected across the y-axis, then translated to the right 4 units. Note that the reflection has no effect on the graph, since $y = |x|$ is symmetric about the y-axis:

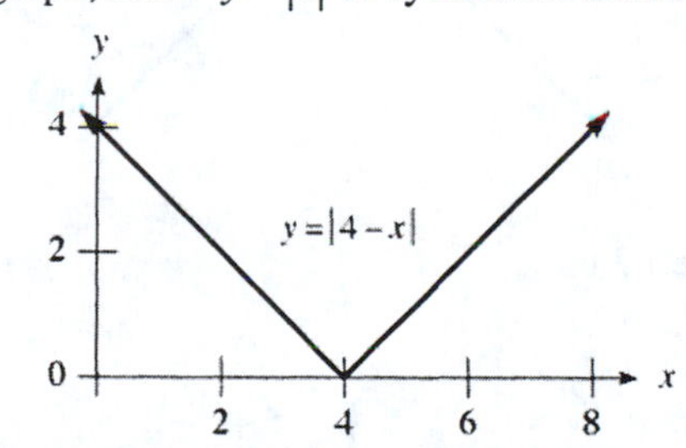

c. This graph will be that of $y = |x|$ reflected across the y-axis, translated to the right 4 units, reflected across the x-axis, then translated up 1 unit:

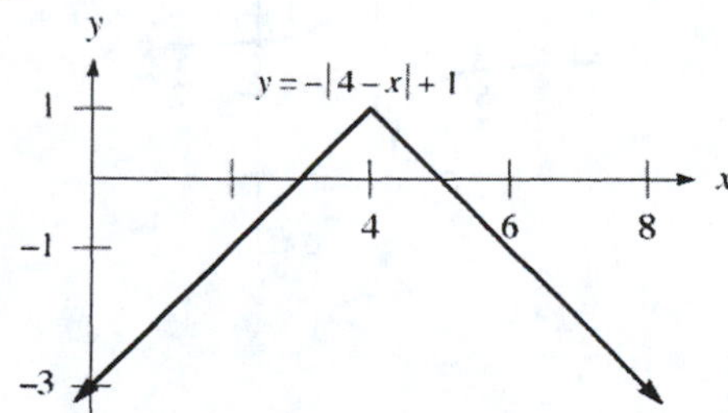

25. This is $f(x)=|x|$ translated to the right 5 units:

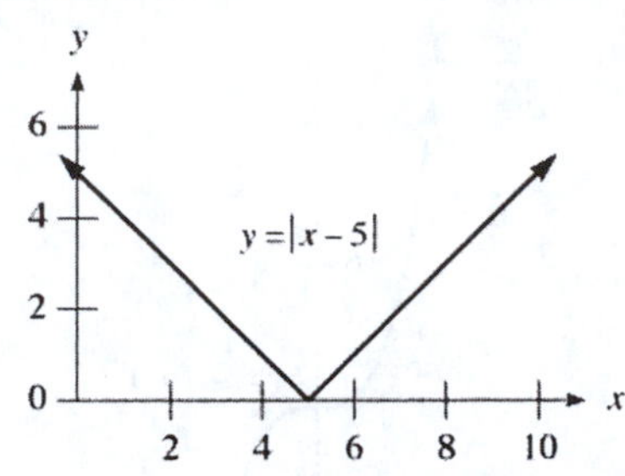

27. This is $f(x)=|x|$ reflected across the y-axis, then translated to the right 5 units. Note that the reflection has no effect on the graph, since $y=|x|$ is symmetric about the y-axis:

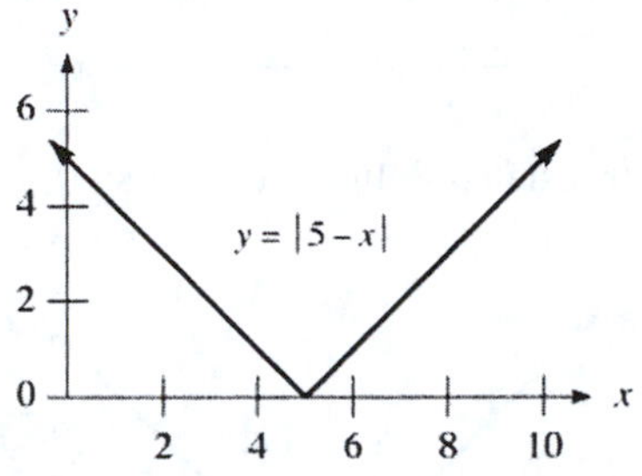

29. This is $f(x)=|x|$ translated to the right 5 units, reflected across the x-axis, then translated up 1 unit:

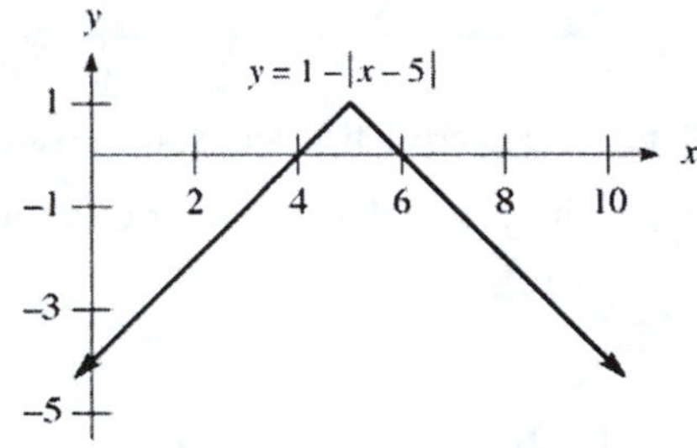

31. This is $F(x)=\dfrac{1}{x}$ translated to the left 3 units:

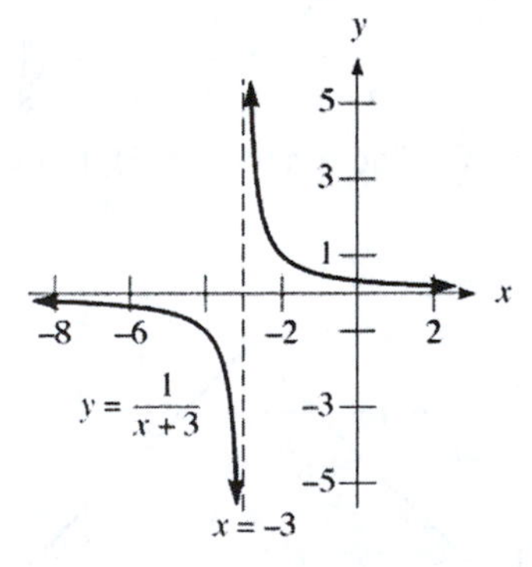

33. This is $F(x)=\frac{1}{x}$ translated to the left 3 units, then reflected across the x-axis:

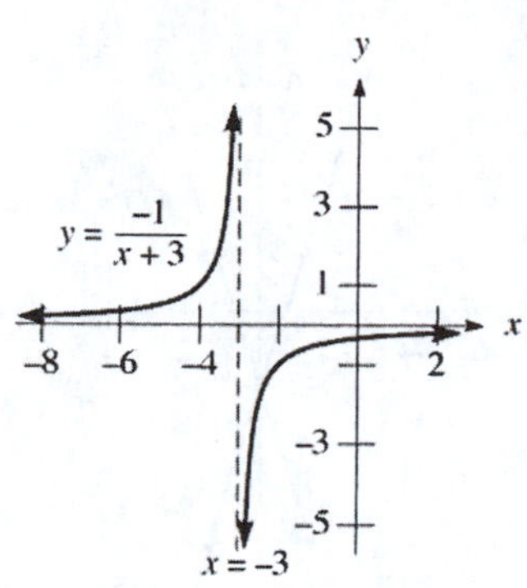

35. This is $g(x)=\sqrt{1-x^2}$ translated to the right 2 units:

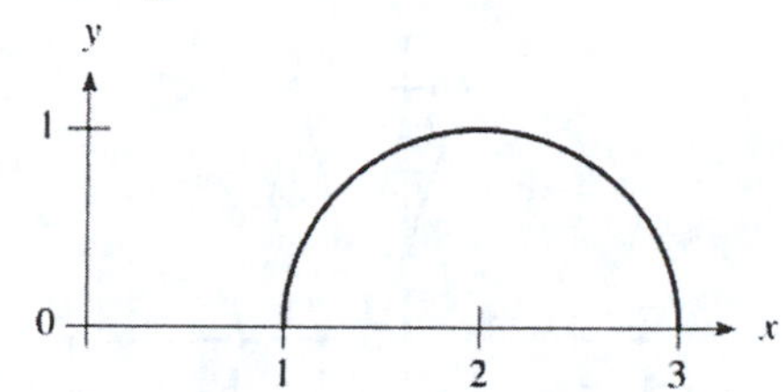

37. This is $g(x)=\sqrt{1-x^2}$ translated to the right 2 units, reflected across the x-axis, then translated up 1 unit:

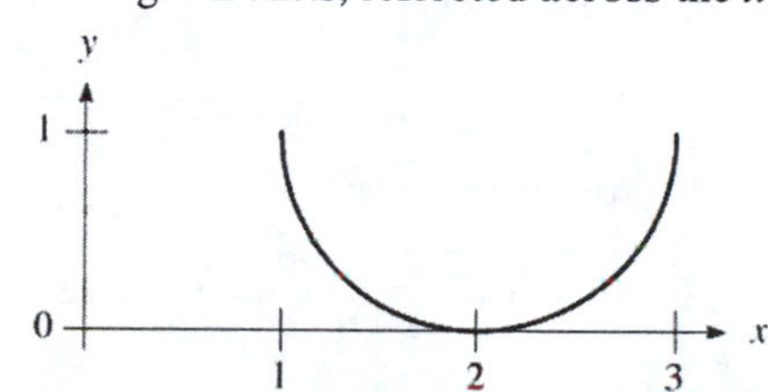

39. This is $g(x)=\sqrt{1-x^2}$ reflected across the y-axis, then translated to the right 2 units. Note that the graph is the same as in Exercise 35, since $(2-x)^2=(x-2)^2$:

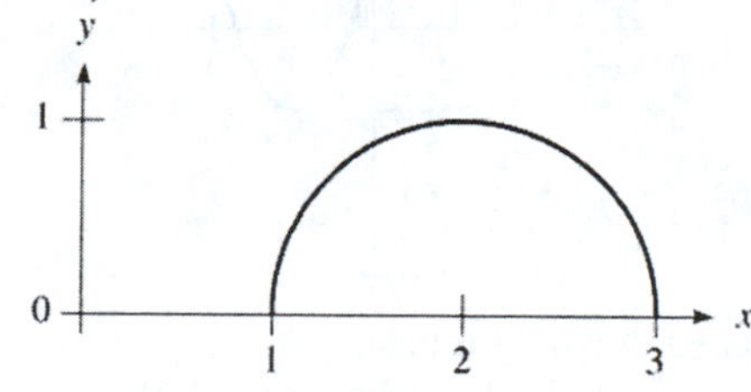

41. **a.** This is the graph of $f(x)$ reflected across the y-axis:

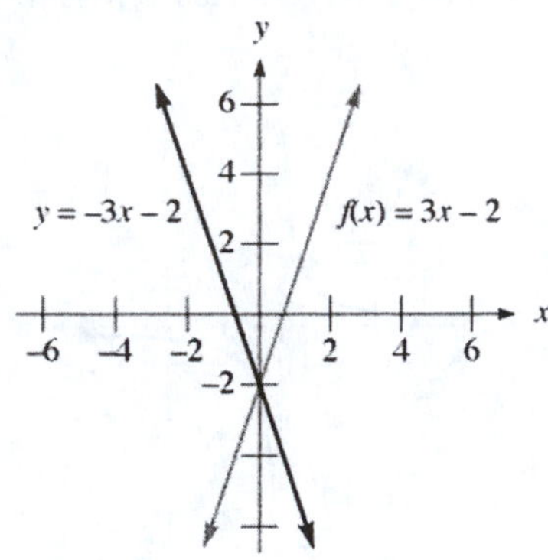

b. This is the graph of $f(x)$ reflected across the x-axis:

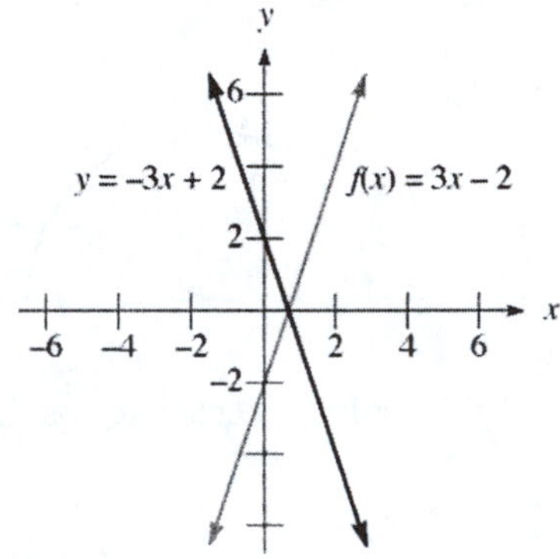

43. **a.** This is the graph of $f(x)$ reflected across the y-axis:

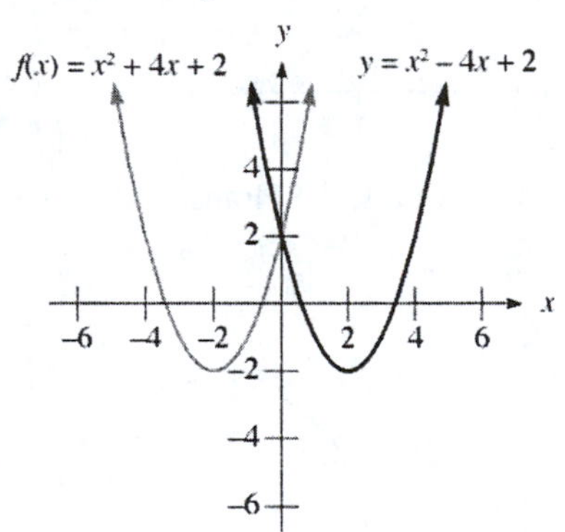

b. This is the graph of $f(x)$ displaced down 2 units:

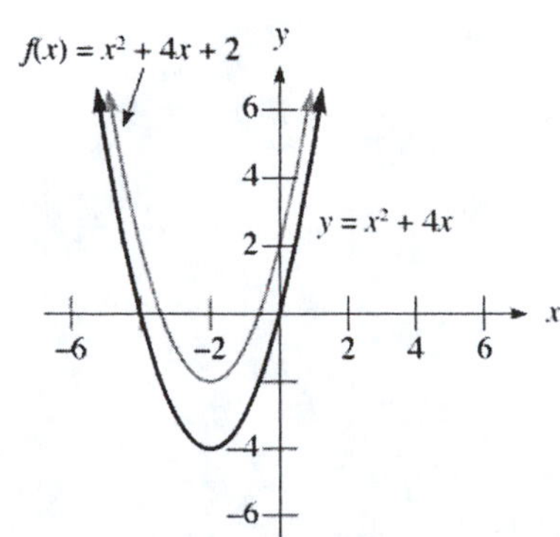

45. **a.** This is the graph of $f(x)$ reflected across the y-axis:

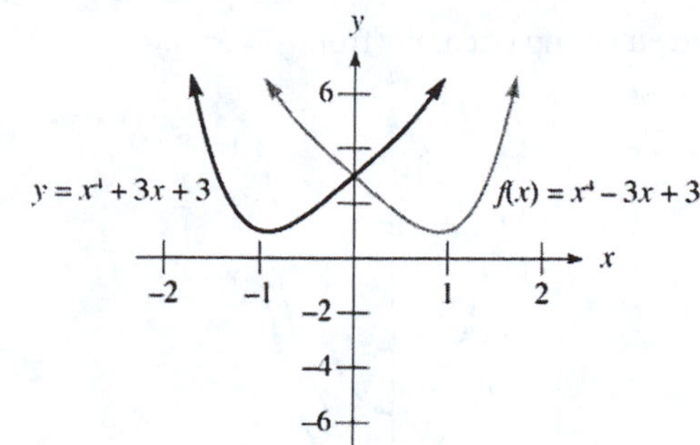

b. This is the graph of $f(x)$ reflected across the x-axis:

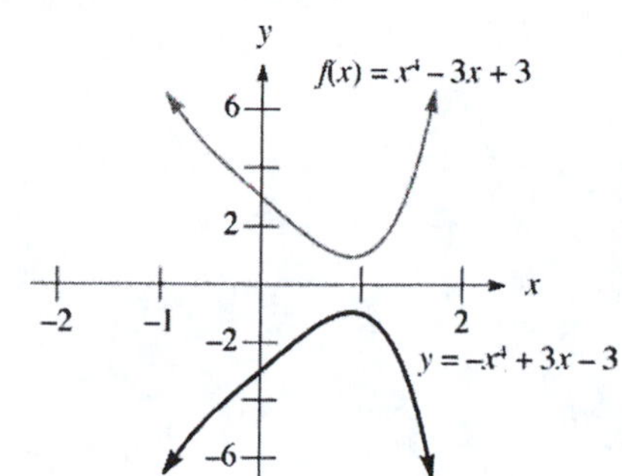

c. This is the graph of $f(x)$ reflected across the x-axis, then displaced up 3 units:

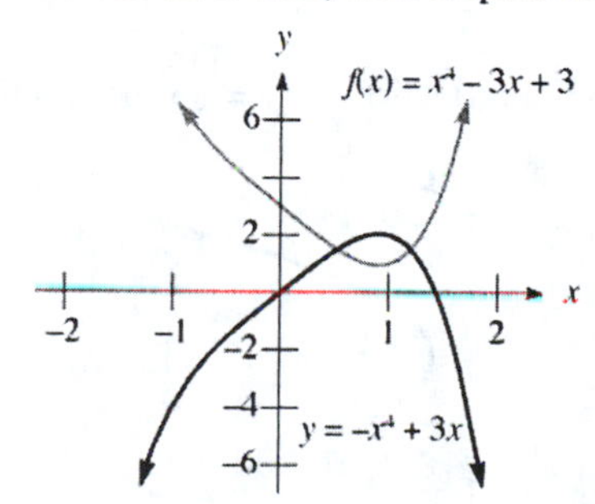

47. **a.** Complete the table:

x	x^2	x^2-1	x^2+1
0	0	−1	1
±1	1	0	2
±2	4	3	5
±3	9	8	10

b. Notice that the graph of $y = x^2 - 1$ is a vertical displacement down one unit (from $y = x^2$), while the graph of $y = x^2 + 1$ is a vertical displacement up one unit (from $y = x^2$):

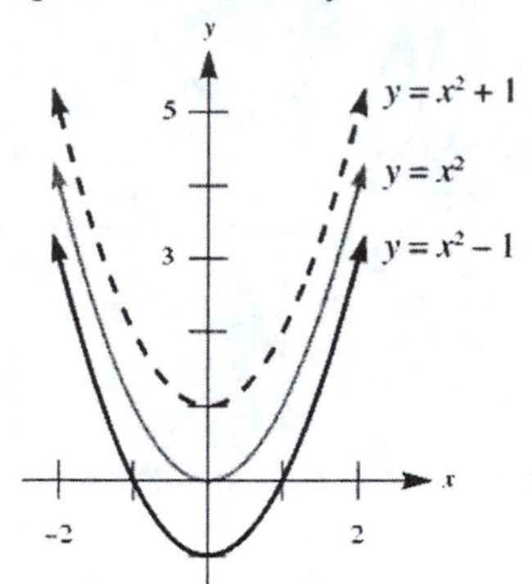

49. **a.** Complete the table:

x	$\sqrt{x}$	$-\sqrt{x}$
0	0.0	0.0
1	1.0	−1.0
2	1.4	−1.4
3	1.7	−1.7
4	2.0	−2.0
5	2.2	−2.2

b. Notice that the graph of $y = -\sqrt{x}$ is a reflection of $y = \sqrt{x}$ across the x-axis:

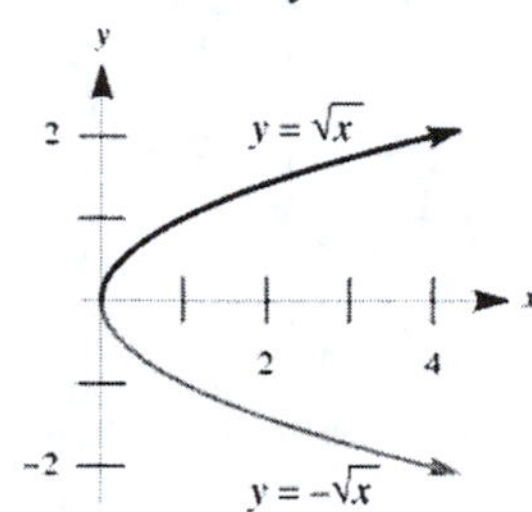

51. **a.** Graphing the two functions:

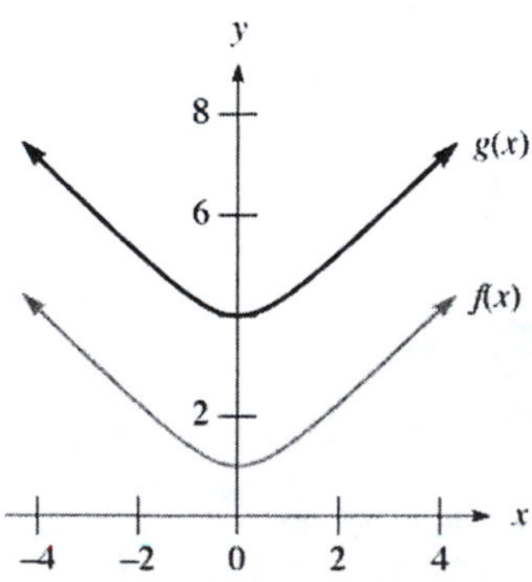

Notice that the graph of g is the graph of f translated up three units.

b. Graphing the two functions:

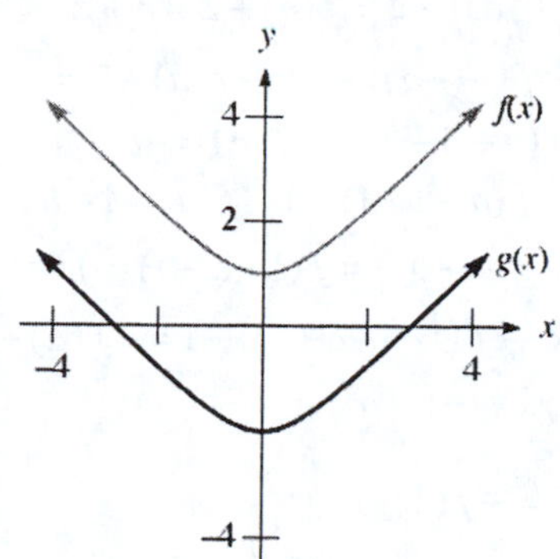

Notice that the graph of g is the graph of f translated down three units.

53. a. Graphing the two functions:

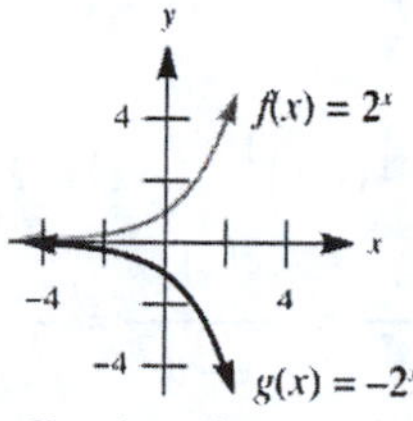

Notice that the graph of g is obtained by reflecting the graph of f in the x-axis.

b. Graphing the two functions:

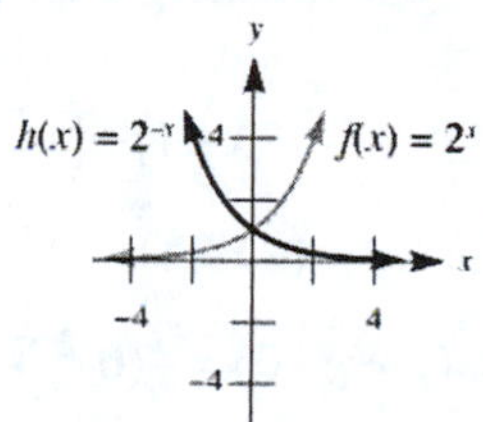

Notice that the graph of h is obtained by reflecting the graph of f in the y-axis.

55. a. The coordinates of Q are $Q(a+c,b)$, so the coordinates of R are $R(a+c,b+d)$.

b. The coordinates of S are $S(a,b+d)$, so the coordinates of T are $T(a+c,b+d)$.

c. The results demonstrate that the order of translations do not affect the final point.

57. a. Graphing the function:

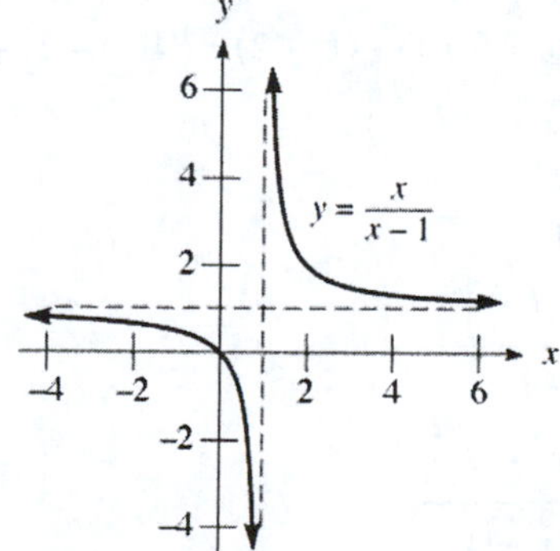

b. Simplifying the right side: $\frac{1}{x-1}+1=\frac{1}{x-1}+\frac{x-1}{x-1}=\frac{1+x-1}{x-1}=\frac{x}{x-1}$

c. Starting with $y=\frac{1}{x}$, translate to the right 1 unit and 1 unit up. Our graph from part **a** verifies this result.

59. **a.** The point is $(-a, b+2)$, since $f(-(-a))+2 = f(a)+2 = b+2$.

b. The point is $(-a, -b+2)$, since $-f(-(-a))+2 = -f(a)+2 = -b+2$.

c. The point is $(a+3, -b)$, since $-f(a+3-3) = -f(a) = -b$.

d. The point is $(a-1, 1-b)$, since $1-f(a-1+1) = 1-f(a) = 1-b$.

e. The point is $(-a+1, b)$, since $f(1-(-a+1)) = f(1+a-1) = f(a) = b$.

f. The point is $(-a+1, -b+1)$, since $-f(1-(-a+1))+1 = -f(a)+1 = -b+1$.

61. **a.** Call $y = f(x)$. Replace x by $-x$ and y by $-y$:

$$-y = f(-x)$$
$$-y = -f(x) \qquad [\text{since } f(-x) = -f(x)]$$
$$y = f(x)$$

So the resulting equation is identical to the original equation, and thus the graph of $y = f(x)$ is symmetric about the origin.

b. (i) Compute $f(-x)$: $f(-x) = (-x)^3 = -x^3 = -f(x)$

(ii) Compute $f(-x)$: $f(-x) = -2(-x)^5 + 4(-x)^3 - (-x) = 2x^5 - 4x^3 + x = -(-2x^5 + 4x^3 - x) = -f(x)$

(iii) Compute $f(-x)$: $f(-x) = \dfrac{|-x|}{(-x)+(-x)^7} = \dfrac{|x|}{-x-x^7} = -\dfrac{|x|}{x+x^7} = -f(x)$

3.5 Methods of Combining Functions. Iteration

1. **a.** Compute $(f+g)(x)$: $(f+g)(x) = f(x)+g(x) = (2x-1)+(x^2-3x-6) = x^2-x-7$

b. Compute $(f-g)(x)$: $(f-g)(x) = f(x)-g(x) = (2x-1)-(x^2-3x-6) = -x^2+5x+5$

c. Using our answer from part **b**, we have: $(f-g)(0) = -(0)^2+5(0)+5 = 0+0+5 = 5$

3. **a.** Compute $(m-f)(x)$: $(m-f)(x) = m(x)-f(x) = (x^2-9)-(2x-1) = x^2-9-2x+1 = x^2-2x-8$

b. Compute $(f-m)(x)$: $(f-m)(x) = f(x)-m(x) = (2x-1)-(x^2-9) = 2x-1-x^2+9 = -x^2+2x+8$

5. **a.** Compute $(fk)(x)$: $(fk)(x) = f(x)k(x) = (2x-1)(2) = 4x-2$

b. Compute $(kf)(x)$: $(kf)(x) = k(x)f(x) = 2(2x-1) = 4x-2$

c. Using our results from parts **a** and **b**: $(fk)(1)-(kf)(2) = [4(1)-2]-[4(2)-2] = 2-6 = -4$

7. **a.** Compute $(f/m)(x) - (m/f)(x)$:

$$\frac{f}{m}(x) - \frac{m}{f}(x) = \frac{f(x)}{m(x)} - \frac{m(x)}{f(x)}$$
$$= \frac{[f(x)]^2 - [m(x)]^2}{f(x)m(x)}$$
$$= \frac{(2x-1)^2 - (x^2-9)^2}{(x^2-9)(2x-1)}$$
$$= \frac{(4x^2-4x+1)-(x^4-18x^2+81)}{2x^3-x^2-18x+9}$$
$$= \frac{-x^4+22x^2-4x-80}{2x^3-x^2-18x+9}$$

b. Using our result from part **a**: $\frac{f}{m}(0)-\frac{m}{f}(0)=\frac{-0^4+22(0)^2-4(0)-80}{2(0)^3-0^2-18(0)+9}=-\frac{80}{9}$

9. **a.** Compute $(f\circ g)(x)$: $(f\circ g)(x)=f[g(x)]=f(-2x-5)=3(-2x-5)+1=-6x-15+1=-6x-14$

b. Using our result from part **a**: $(f\circ g)(10)=-6(10)-14=-60-14=-74$

c. Compute $(g\circ f)(x)$: $(g\circ f)(x)=g[f(x)]=g(3x+1)=-2(3x+1)-5=-6x-2-5=-6x-7$

d. Using our result from part **a**: $(g\circ f)(10)=-6(10)-7=-60-7=-67$

11. **a.** Finding the indicated compositions:

$$(f\circ g)(x)=f[g(x)]=f(2-3x)=(2-3x)^2-3(2-3x)-4=4-12x+9x^2-6+9x-4=9x^2-3x-6$$
$$(f\circ g)(-2)=9(-2)^2-3(-2)-6=36+6-6=36$$
$$(g\circ f)(x)=g[f(x)]=g\left(x^2-3x-4\right)=2-3\left(x^2-3x-4\right)=2-3x^2+9x+12=-3x^2+9x+14$$
$$(g\circ f)(-2)=-2(-2)^2+9(-2)+14=-12-18+14=-16$$

b. Finding the indicated compositions:

$$(f\circ g)(x)=f[g(x)]=f\left(x^2+1\right)=2^{x^2+1}\qquad (f\circ g)(-2)=2^{(-2)^2+1}=2^{4+1}=2^5=32$$
$$(g\circ f)(x)=g[f(x)]=g\left(2^x\right)=\left(2^x\right)^2+1=2^{2x}+1\qquad (g\circ f)(-2)=2^{2(-2)}+1=2^{-4}+1=\tfrac{1}{16}+1=\tfrac{17}{16}$$

c. Finding the indicated compositions:

$$(f\circ g)(x)=f[g(x)]=f\left(3x^5-4x^2\right)=3x^5-4x^2\qquad (f\circ g)(-2)=3(-2)^5-4(-2)^2=-96-16=-112$$
$$(g\circ f)(x)=g[f(x)]=g(x)=3x^5-4x^2\qquad (g\circ f)(-2)=3(-2)^5-4(-2)^2=-96-16=-112$$

d. Finding the indicated compositions:

$$(f\circ g)(x)=f[g(x)]=f\left(\frac{x+4}{3}\right)=3\left(\frac{x+4}{3}\right)-4=x+4-4=x$$
$$(f\circ g)(-2)=-2$$
$$(g\circ f)(x)=g[f(x)]=g(3x-4)=\frac{3x-4+4}{3}=\frac{3x}{3}=x$$
$$(g\circ f)(-2)=-2$$

13. **a.** Compute $(F\circ G)(x)$:

$$(F\circ G)(x)=F[G(x)]=F\left(\tfrac{x+1}{x-1}\right)=\frac{3\left(\frac{x+1}{x-1}\right)-4}{3\left(\frac{x+1}{x-1}\right)+3}=\frac{3(x+1)-4(x-1)}{3(x+1)+3(x-1)}=\frac{3x+3-4x+4}{3x+3+3x-3}=\frac{-x+7}{6x}$$

b. Using our result from part **a**: $F[G(t)]=\frac{-t+7}{6t}$

c. Using our result from part **a**: $(F\circ G)(2)=F[G(2)]=\frac{-2+7}{6(2)}=\frac{5}{12}$

d. Compute $(G\circ F)(x)$:

$$(G\circ F)(x)=G[F(x)]=G\left(\tfrac{3x-4}{3x+3}\right)=\frac{\frac{3x-4}{3x+3}+1}{\frac{3x-4}{3x+3}-1}=\frac{(3x-4)+1(3x+3)}{(3x-4)-1(3x+3)}=\frac{3x-4+3x+3}{3x-4-3x-3}=\frac{6x-1}{-7}=\frac{1-6x}{7}$$

e. Using our result from part **d**: $G[F(y)]=\frac{1-6y}{7}$

f. Using our result from part **d**: $(G\circ F)(2)=G[F(2)]=\frac{1-6(2)}{7}=-\frac{11}{7}$

15. **a.** Compute $M(7)$ and $M[M(7)]$:

$$M(7)=\frac{2(7)-1}{7-2}=\frac{14-1}{5}=\frac{13}{5} \qquad M\left[M(7)\right]=M\left(\frac{13}{5}\right)=\frac{2\left(\frac{13}{5}\right)-1}{\frac{13}{5}-2}=\frac{\frac{26}{5}-1}{\frac{3}{5}}=\frac{\frac{21}{5}}{\frac{3}{5}}=7$$

b. Compute $(M\circ M)(x)$:

$$(M\circ M)(x)=M\left[M(x)\right]=M\left(\frac{2x-1}{x-2}\right)=\frac{2\left(\frac{2x-1}{x-2}\right)-1}{\left(\frac{2x-1}{x-2}\right)-2}=\frac{2(2x-1)-1(x-2)}{(2x-1)-2(x-2)}=\frac{4x-2-x+2}{2x-1-2x+4}=\frac{3x}{3}=x$$

c. Using our result from part **b**, we have $(M\circ M)(7)=M\left[M(7)\right]=7$. This agrees with our answer from part **a**.

17. **a.** $f[g(3)]=f(0)=1$ **b.** $g[f(3)]=g(4)=-3$

c. $f[h(3)]=f(2)=-1$ **d.** $(h\circ g)(2)=h[g(2)]=h(1)=2$

e. $h\{f[g(3)]\}=h[f(0)]=h(1)=2$

f. $(g\circ f\circ h\circ f)(2)=(g\circ f\circ h)(-1)=(g\circ f)(3)=g(4)=-3$

19. Compute the compositions:

$(f\circ g)(0)=f[g(0)]=f(3)=1$ $(f\circ g)(1)=f[g(1)]=f(2)=3$

$(f\circ g)(2)=f[g(2)]=f(0)=2$ $(f\circ g)(3)=f[g(3)]=f(4)=\text{undefined}$

$(f\circ g)(4)=f[g(4)]=f(-1)=2$

Thus we have the table:

x	0	1	2	3	4
$(f\circ g)(x)$	1	3	2	undef.	2

Compute the compositions:

$(g\circ f)(-1)=g[f(-1)]=g(2)=0$ $(g\circ f)(0)=g[f(0)]=g(2)=0$

$(g\circ f)(1)=g[f(1)]=g(0)=3$ $(g\circ f)(2)=g[f(2)]=g(3)=4$

$(g\circ f)(3)=g[f(3)]=g(1)=2$ $(g\circ f)(4)=g[f(4)]=\text{undefined}$

Thus we have the table:

x	−1	0	1	2	3	4
$(g\circ f)(x)$	0	0	3	4	2	undef.

21. **a.** The graph of $f\circ g$ is the graph of f displaced 4 units to the left.

b. Graphing the two functions to confirm our answer:

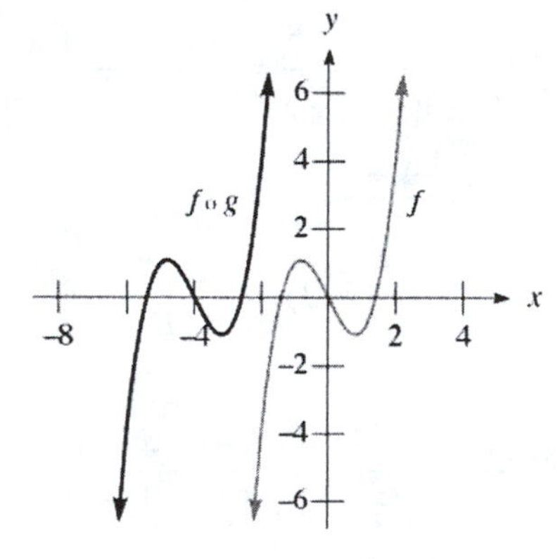

23. a. The domain is $[0, \infty)$ and the range is $[-3, \infty)$:

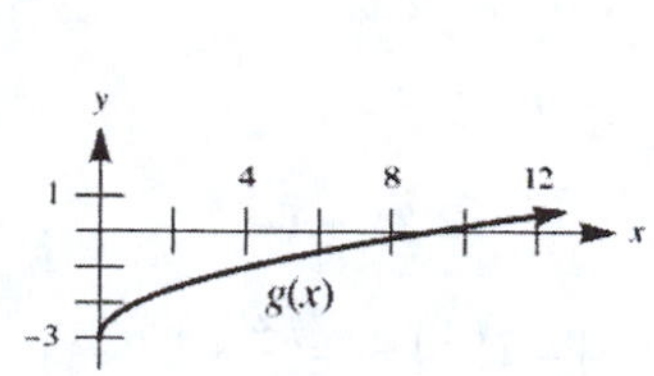

b. The domain is $(-\infty, \infty)$ and the range is $(-\infty, \infty)$:

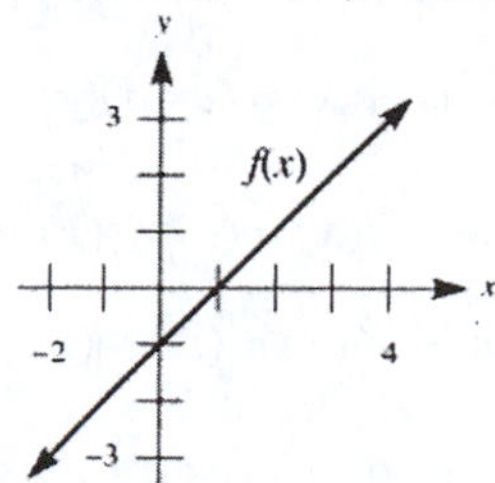

c. First compute $(f \circ g)(x)$: $(f \circ g)(x) = f[g(x)] = f\left(\sqrt{x} - 3\right) = \left(\sqrt{x} - 3\right) - 1 = \sqrt{x} - 4$

The domain is $[0, \infty)$ and the range is $[-4, \infty)$:

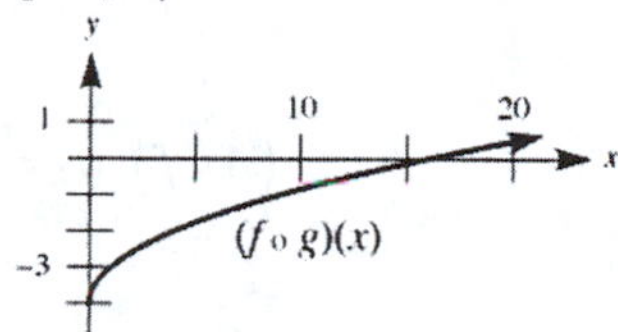

d. First compute $g[f(x)]$: $g[f(x)] = g(x-1) = \sqrt{x-1} - 3$

The domain is $[1, \infty)$.

e. Draw the graph:

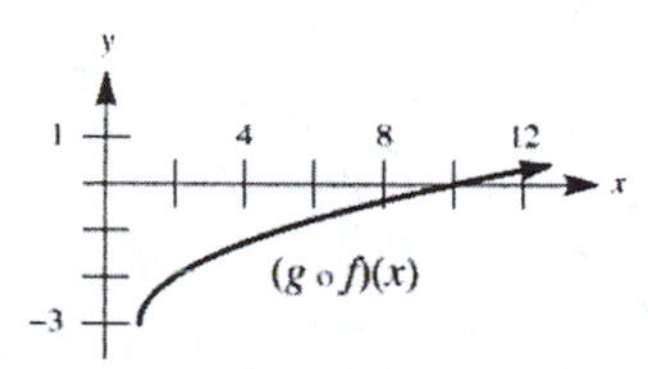

25. a. The area function is given by: $A(r) = A\left(f(t)\right) = \pi\left(15 + t^{1.65}\right)^2$

Completing the table (rounding to the nearest whole number):

t	0	0.5	1	1.5	2	2.5	3	3.5	4	4.5	5
A	707	737	804	903	1034	1199	1402	1648	1940	2284	2685

b. After one hour the area is approximately 800 m^2.

c. Initially the area of the spill was 707 m^2. After approximately three hours this area has doubled.

d. Computing the average rates of change:

0 to 2.5: $\dfrac{1199 - 707}{2.5} = 196.8 \text{ m}^2/\text{hr}$ 2.5 to 5: $\dfrac{2685 - 1199}{2.5} = 594.4 \text{ m}^2/\text{hr}$

The area is increasing faster over the interval from $t = 2.5$ to $t = 5$.

27. a. Compute $(C \circ f)(t)$: $(C \circ f)(t) = C[f(t)] = C(5t) = 100 + 90(5t) - (5t)^2 = 100 + 450t - 25t^2$

b. When $t = 3$ hr: $C[f(3)] = 100 + 450(3) - 25(3)^2 = \$1225$

c. When $t = 6$ hr: $C[f(6)] = 100 + 450(6) - 25(6)^2 = \$1900$

No, the cost is not twice as much for 6 hours.

29. a. Let $f(x) = \sqrt[3]{x}$ and $g(x) = 3x + 4$. Then $F(x) = (f \circ g)(x)$, since: $(f \circ g)(x) = f[g(x)] = f(3x+4) = \sqrt[3]{3x+4}$

b. Let $f(x) = |x|$ and $g(x) = 2x - 3$. Then $G(x) = (f \circ g)(x)$, since: $(f \circ g)(x) = f[g(x)] = f(2x-3) = |2x-3|$

c. Let $f(x) = x^5$ and $g(x) = ax + b$. Then $H(x) = (f \circ g)(x)$, since: $(f \circ g)(x) = f[g(x)] = f(ax+b) = (ax+b)^5$

d. Let $f(x) = \frac{1}{x}$ and $g(x) = \sqrt{x}$. Then $T(x) = (f \circ g)(x)$, since: $(f \circ g)(x) = f[g(x)] = f\left(\sqrt{x}\right) = \dfrac{1}{\sqrt{x}}$

31. a. The composition is $f(x) = (b \circ c)(x)$, since: $(b \circ c)(x) = b[c(x)] = b(2x+1) = \sqrt[3]{2x+1}$

b. The composition is $g(x) = (a \circ d)(x)$, since: $(a \circ d)(x) = a[d(x)] = a\left(x^2\right) = \frac{1}{x^2}$

c. The composition is $h(x) = (c \circ d)(x)$, since: $(c \circ d)(x) = c[d(x)] = c\left(x^2\right) = 2x^2 + 1$

d. The composition is $K(x) = (c \circ b)(x)$, since: $(c \circ b)(x) = c[b(x)] = c\left(\sqrt[3]{x}\right) = 2\sqrt[3]{x} + 1$

e. The composition is $l(x) = (c \circ a)(x)$, since: $(c \circ a)(x) = c[a(x)] = c\left(\frac{1}{x}\right) = 2\left(\frac{1}{x}\right) + 1 = \frac{2}{x} + 1$

f. The composition is $m(x) = (a \circ c)(x)$, since: $(a \circ c)(x) = a[c(x)] = a(2x+1) = \frac{1}{2x+1}$

33. a. Computing the values:

$$(A \circ f)(0) = \pi\left(\frac{0}{0+4}\right)^2 = 0 \qquad (A \circ f)(6) = \pi\left(\frac{6}{12+4}\right)^2 = \pi\left(\frac{3}{8}\right)^2 = \frac{9\pi}{64}$$

b. Sketching the graph:

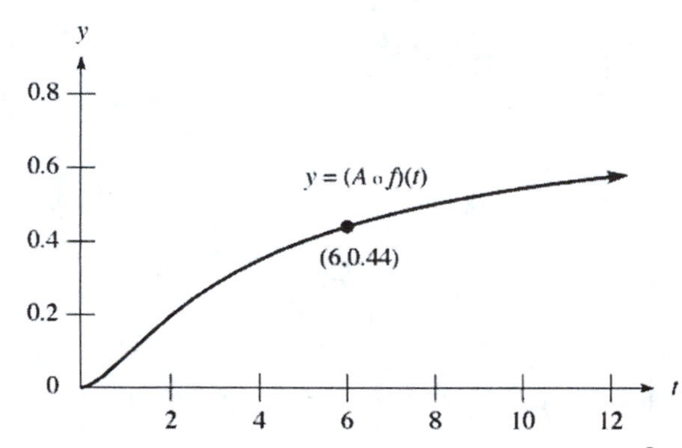

c. The y-coordinate is approximately 0.44. This corresponds to $(A \circ f)(6) = \frac{9\pi}{64} \approx 0.44$.

35. a. The first six iterates for $x_0 = 1$ are:

$x_1 = 2 \qquad x_2 = 4 \qquad x_3 = 8$
$x_4 = 16 \qquad x_5 = 32 \qquad x_6 = 64$

b. For $x_0 = 0$, all six iterates are 0.

c. The first six iterates for $x_0 = -1$ are:

$x_1 = -2 \qquad x_2 = -4 \qquad x_3 = -8$
$x_4 = -16 \qquad x_5 = -32 \qquad x_6 = -64$

37. a. The first six iterates for $x_0 = -2$ are:

$x_1 = -3 \qquad x_2 = -5 \qquad x_3 = -9$
$x_4 = -17 \qquad x_5 = -33 \qquad x_6 = -65$

b. For $x_0 = -1$, all six iterates are -1.

c. The first six iterates for $x_0 = 1$ are:

$x_1 = 3 \qquad x_2 = 7 \qquad x_3 = 15$
$x_4 = 31 \qquad x_5 = 63 \qquad x_6 = 127$

39. **a.** The first six iterates for $x_0 = 0.9$ are:

$x_1 = 0.81$ $x_2 \approx 0.656$ $x_3 \approx 0.430$
$x_4 \approx 0.185$ $x_5 \approx 0.034$ $x_6 \approx 0.001$

b. For $x_0 = 1$, all six iterates are 1.

c. The first six iterates for $x_0 = 1.1$ are:

$x_1 = 1.21$ $x_2 \approx 1.464$ $x_3 \approx 2.144$
$x_4 \approx 4.595$ $x_5 \approx 21.114$ $x_6 \approx 445.792$

41. The first four iterates for $x_0 = 0.1$ are:

$x_1 \approx 0.316$ $x_2 \approx 0.562$ $x_3 \approx 0.750$ $x_4 \approx 0.866$

These values are consistent with the graph.

43. **a.** Compute the function values:

$f(1) = 3(1)+1 = 4$ $f(2) = 2/2 = 1$
$f(3) = 3(3)+1 = 10$ $f(4) = 4/2 = 2$
$f(5) = 3(5)+1 = 16$ $f(6) = 6/2 = 3$

b. The first three iterates of $x_0 = 1$ are:

$f(1) = 3(1)+1 = 4$ $f(4) = 4/2 = 2$
$f(2) = 2/2 = 1$

c. Compute the iterates of $x_0 = 3$:

$f(3) = 3(3)+1 = 10$ $f(10) = 10/2 = 5$
$f(5) = 3(5)+1 = 16$ $f(16) = 16/2 = 8$
$f(8) = 8/2 = 4$ $f(4) = 4/2 = 2$
$f(2) = 2/2 = 1$

d. The iterates of $x_0 = 2$ are: $f(2) = 2/2 = 1$

The iterates of $x_0 = 4$ are:

$f(4) = 4/2 = 2$ $f(2) = 2/2 = 1$

The iterates of $x_0 = 5$ are:

$f(5) = 3(5)+1 = 16$ $f(16) = 16/2 = 8$
$f(8) = 8/2 = 4$ $f(4) = 4/2 = 2$
$f(2) = 2/2 = 1$

The iterates of $x_0 = 6$ are:

$f(6) = 6/2 = 3$ $f(3) = 3(3)+1 = 10$
$f(10) = 10/2 = 5$ $f(5) = 3(5)+1 = 16$
$f(16) = 16/2 = 8$ $f(8) = 8/2 = 4$
$f(4) = 4/2 = 2$ $f(2) = 2/2 = 1$

The iterates of $x_0 = 7$ are:

$f(7) = 3(7)+1 = 22$ $f(22) = 22/2 = 11$
$f(11) = 3(11)+1 = 34$ $f(34) = 34/2 = 17$
$f(17) = 3(17)+1 = 52$ $f(52) = 52/2 = 26$
$f(26) = 26/2 = 13$ $f(13) = 3(13)+1 = 40$
$f(40) = 40/2 = 20$ $f(20) = 20/2 = 10$
$f(10) = 10/2 = 5$ $f(5) = 3(5)+1 = 16$
$f(16) = 16/2 = 8$ $f(8) = 8/2 = 4$
$f(4) = 4/2 = 2$ $f(2) = 2/2 = 1$

The conjecture is valid for each of the given values of x_0.

45. Call $y = f(x)$. Since $(g \circ f)(x) = g[f(x)] = g(y)$:

$$\begin{aligned} g(y) &= x+5 \\ 4y-1 &= x+5 \\ 4y &= x+6 \\ y &= \frac{x+6}{4} \end{aligned}$$

So $f(x) = \dfrac{x+6}{4}$.

47. Set $f[g(x)] = x$:

$$\begin{aligned} f(ax+b) &= x \\ -2(ax+b)+1 &= x \\ -2ax-2b+1 &= x \end{aligned}$$

Since a and b are constants, we can equate components:

$$-2a = 1 \quad\Rightarrow\quad a = -\tfrac{1}{2} \qquad \text{and} \qquad -2b+1 = 0 \quad\Rightarrow\quad -2b = -1 \quad\Rightarrow\quad b = \tfrac{1}{2}$$

So $a = -\frac{1}{2}$ and $b = \frac{1}{2}$.

49. **a.** Compute the difference quotient:

$$\begin{aligned} \frac{f[g(x)]-f[g(a)]}{g(x)-g(a)} &= \frac{f(2x-1)-f(2a-1)}{(2x-1)-(2a-1)} \\ &= \frac{(2x-1)^2-(2a-1)^2}{2x-1-2a+1} \\ &= \frac{[(2x-1)+(2a-1)][(2x-1)-(2a-1)]}{2x-2a} \\ &= \frac{(2x+2a-2)(2x-2a)}{2x-2a} \\ &= 2x+2a-2 \end{aligned}$$

b. Compute the difference quotient:

$$\frac{f[g(x)]-f[g(a)]}{x-a} = \frac{(2x+2a-2)(2x-2a)}{x-a} = \frac{4(x+a-1)(x-a)}{x-a} = 4x+4a-4$$

51. **a.** Computing the first ten iterates for $x_0 = 1$:

$x_1 = 3$ | $x_2 \approx 2.259259259$
$x_3 \approx 1.963308018$ | $x_4 \approx 1.914212754$
$x_5 \approx 1.912932041$ | $x_6 = \ldots = x_{10} \approx 1.912931183$

Notice that the iterates converge to a number which is approximately 1.912931183.

b. Note that $\sqrt[3]{7} \approx 1.912931183$. They are the same.

c. The fifth iterate agrees with $\sqrt[3]{7}$ through the first three decimal places. The sixth iterate agrees with $\sqrt[3]{7}$ through the first eight decimal places.

3.6 Inverse Functions

1. **a.** $h[k(x)] = x$ for every x in the domain of k. **b.** $k[h(x)] = x$ for every x in the domain of h.

3. **a.** We must show that $f[g(x)] = x$ and $g[f(x)] = x$:

$$f[g(x)] = f\left(\tfrac{x}{3}\right) = 3\left(\tfrac{x}{3}\right) = x \qquad g[f(x)] = g(3x) = \tfrac{3x}{3} = x$$

So $f(x)$ and $g(x)$ are inverse functions.

b. We must show that $f[g(x)] = x$ and $g[f(x)] = x$:

$$f[g(x)] = f\left(\tfrac{x+1}{4}\right) = 4\left(\tfrac{x+1}{4}\right) - 1 = x + 1 - 1 = x \qquad g[f(x)] = g(4x-1) = \tfrac{(4x-1)+1}{4} = \tfrac{4x}{4} = x$$

So $f(x)$ and $g(x)$ are inverse functions.

c. We must show that $g[h(x)] = x$ and $h[g(x)] = x$:

$$g[h(x)] = g\left(x^2\right) = \sqrt{x^2} = x, \quad \text{since } x \ge 0 \qquad h[g(x)] = h\left(\sqrt{x}\right) = \left(\sqrt{x}\right)^2 = x$$

So $g(x)$ and $h(x)$ are inverse functions.

5. **a.** Graphing the three functions:

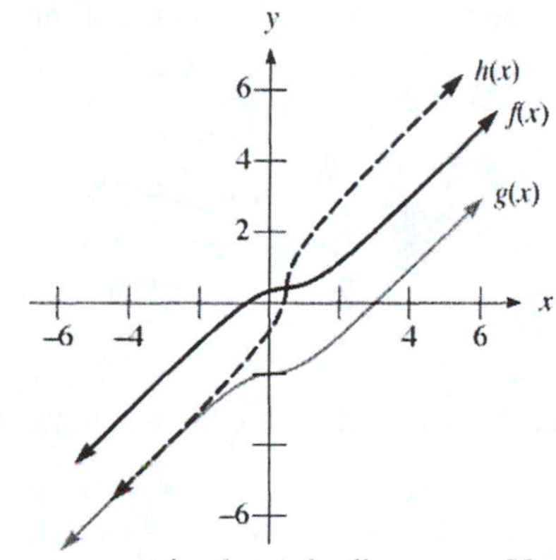

b. The graphs of f and h appear to be symmetric about the line $y = x$. Verifying these are inverses:

$$(f \circ h)(x) = f\left(\sqrt[3]{(x+1)^3 - 3}\right) = \sqrt[3]{\left(\sqrt[3]{(x+1)^3 - 3}\right)^3 + 3} - 1 = \sqrt[3]{(x+1)^3 - 3 + 3} - 1 = (x+1) - 1 = x$$

$$(h \circ f)(x) = h\left(\sqrt[3]{x^3 + 3} - 1\right) = \sqrt[3]{\left(\sqrt[3]{x^3 + 3} - 1 + 1\right)^3 - 3} = \sqrt[3]{\left(\sqrt[3]{x^3 + 3}\right)^3 - 3} = \sqrt[3]{x^3 + 3 - 3} = \sqrt[3]{x^3} = x$$

7. Since $f(7) = 12$ and g is the inverse function for f, $g(12) = 7$.

9. **a.** Since $f\left[f^{-1}(x)\right] = x$ for all x in the domain of f^{-1}, $f\left[f^{-1}(4)\right] = 4$ since the domain of f^{-1} is $(-\infty, \infty)$.

b. Since $f^{-1}[f(x)] = x$ as long as $f(x)$ is in the domain of f^{-1}, $f^{-1}[f(-1)] = -1$ since the domain of f^{-1} is $(-\infty, \infty)$.

c. Since $f\left[f^{-1}(x)\right] = x$ for all x in the domain of f^{-1}, $f\left[f^{-1}\left(\sqrt{2}\right)\right] = \sqrt{2}$ since the domain of f^{-1} is $(-\infty, \infty)$.

d. Since $f\left[f^{-1}(x)\right] = x$ for all x in the domain of f^{-1}, $f\left[f^{-1}(t+1)\right] = t + 1$ since the domain of f^{-1} is $(-\infty, \infty)$.

11. **a.** Let $y = 2x + 1$. Switch the roles of x and y and solve the resulting equation for y:

$$x = 2y + 1$$
$$2y = x - 1$$
$$y = \frac{x-1}{2}$$

So the inverse is $f^{-1}(x) = \dfrac{x-1}{2}$.

b. Calculate the required values:

$$f^{-1}(5)=\frac{5-1}{2}=\frac{4}{2}=2 \qquad \frac{1}{f(5)}=\frac{1}{2(5)+1}=\frac{1}{11}$$

Note that the two answers are not the same. Remember that f^{-1} does not mean the reciprocal of f.

13. **a.** Let $y = 3x - 1$. Switch the roles of x and y and solve the resulting equation for y:

$$x = 3y-1$$
$$3y = x+1$$
$$y=\frac{x+1}{3}$$

So the inverse is $f^{-1}(x)=\frac{x+1}{3}$.

b. Verify that $f\left[f^{-1}(x)\right]=x$ and $f^{-1}\left[f(x)\right]=x$:

$$f\left[f^{-1}(x)\right]=f\left(\tfrac{x+1}{3}\right)=3\left(\tfrac{x+1}{3}\right)-1=x+1-1=x \qquad f^{-1}\left[f(x)\right]=f^{-1}(3x-1)=\frac{(3x-1)+1}{3}=\frac{3x}{3}=x$$

c. The graphs of each line are given below. Note the symmetry of the two lines about the line $y = x$:

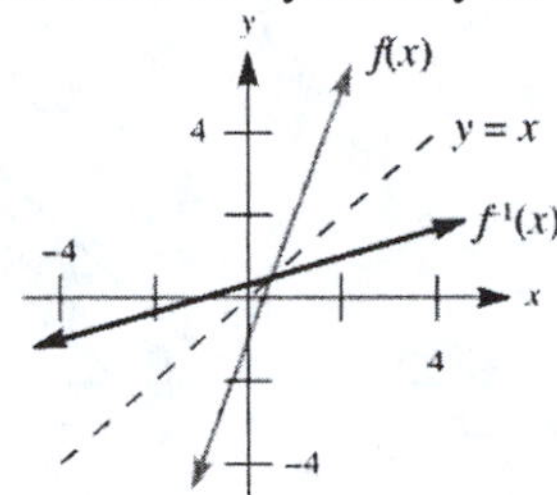

15. **a** Let $y=\sqrt{x-1}$. Switch the roles of x and y and solve the resulting equation for y:

$$x=\sqrt{y-1}$$
$$x^2 = y-1$$
$$y = x^2+1$$

So the inverse is $f^{-1}(x)=x^2+1$ for $x \ge 0$.

b. Verify that $f\left[f^{-1}(x)\right]=x$ and $f^{-1}\left[f(x)\right]=x$:

$$f\left[f^{-1}(x)\right]=f\left(x^2+1\right)=\sqrt{\left(x^2+1\right)-1}=\sqrt{x^2}=x \quad (\text{since } x \ge 0)$$
$$f^{-1}\left[f(x)\right]=f^{-1}\left(\sqrt{x-1}\right)=\left(\sqrt{x-1}\right)^2+1=x-1+1=x$$

c. The graphs of each curve are given below. Note the symmetry of the curves about the line $y = x$:

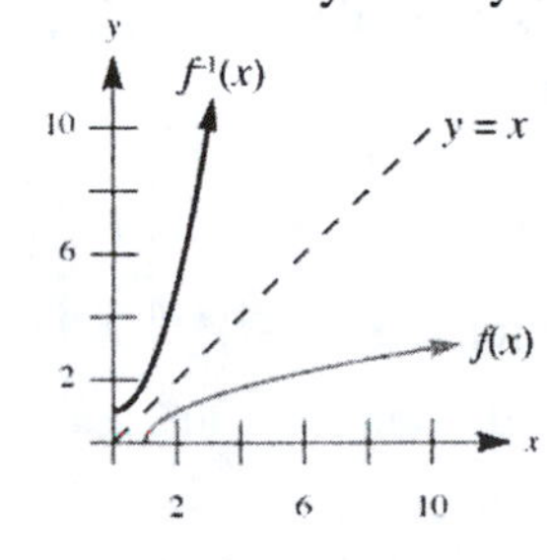

17. **a.** Since the denominator is 0 when $x = 3$, the domain of f is the set of all real numbers except 3, or $(-\infty, 3) \cup (3, \infty)$. To find the range of f, first solve for x:

$$\begin{aligned} y &= \frac{x+2}{x-3} \\ y(x-3) &= x+2 \\ yx - 3y &= x+2 \\ yx - x &= 3y+2 \\ x(y-1) &= 3y+2 \\ x &= \frac{3y+2}{y-1} \end{aligned}$$

Since the denominator is 0 when $y = 1$, the range of f is the set of all real numbers except 1, or $(-\infty, 1) \cup (1, \infty)$.

b. Let $y = \dfrac{x+2}{x-3}$. Switch the roles of x and y and solve the resulting equation for y:

$$\begin{aligned} x &= \frac{y+2}{y-3} \\ x(y-3) &= y+2 \\ xy - 3x &= y+2 \\ xy - y &= 3x+2 \\ y(x-1) &= 3x+2 \\ y &= \frac{3x+2}{x-1} \end{aligned}$$

So the inverse is $f^{-1}(x) = \dfrac{3x+2}{x-1}$.

c. Since the denominator is 0 when $x = 1$, the domain of f^{-1} is the set of all real numbers except 1, or $(-\infty, 1) \cup (1, \infty)$. To find the range of f^{-1}, first solve for x:

$$\begin{aligned} y &= \frac{3x+2}{x-1} \\ y(x-1) &= 3x+2 \\ yx - y &= 3x+2 \\ yx - 3x &= y+2 \\ x(y-3) &= y+2 \\ x &= \frac{y+2}{y-3} \end{aligned}$$

Since the denominator is 0 when $y = 3$, the range of f^{-1} is the set of all real numbers except 3, or $(-\infty, 3) \cup (3, \infty)$. Observe that the domain of f is equal to the range of f^{-1}, and that the range of f is equal to the domain of f^{-1}.

19. Let $y = 2x^3 + 1$. Switch the roles of x and y and solve the resulting equation for y:

$$\begin{aligned} x &= 2y^3 + 1 \\ 2y^3 &= x - 1 \\ y^3 &= \tfrac{x-1}{2} \\ y &= \sqrt[3]{\frac{x-1}{2}} \end{aligned}$$

So the inverse is $f^{-1}(x) = \sqrt[3]{\dfrac{x-1}{2}}$.

21. Finding the compositions:

$$f\left[f^{-1}(x)\right]=f\left(\frac{-5x-1}{3x-1}\right)=\frac{\frac{-5x-1}{3x-1}-1}{3\left(\frac{-5x-1}{3x-1}\right)+5}\bullet\frac{3x-1}{3x-1}=\frac{-5x-1-3x+1}{-15x-3+15x-5}=\frac{-8x}{-8}=x$$

$$f^{-1}\left[f(x)\right]=f^{-1}\left(\frac{x-1}{3x+5}\right)=\frac{-5\left(\frac{x-1}{3x+5}\right)-1}{3\left(\frac{x-1}{3x+5}\right)-1}\bullet\frac{3x+5}{3x+5}=\frac{-5x+5-3x-5}{3x-3-3x-5}=\frac{-8x}{-8}=x$$

This shows that the two functions are inverse functions.

23. **a.** The graph of $y=f^{-1}(x)$ is a reflection of $f(x)$ across the line $y=x$:

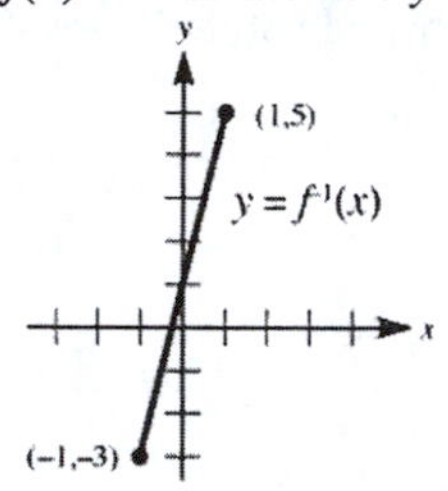

b. The graph of $y=f^{-1}(x-2)$ is a displacement of $f^{-1}(x)$ to the right 2 units:

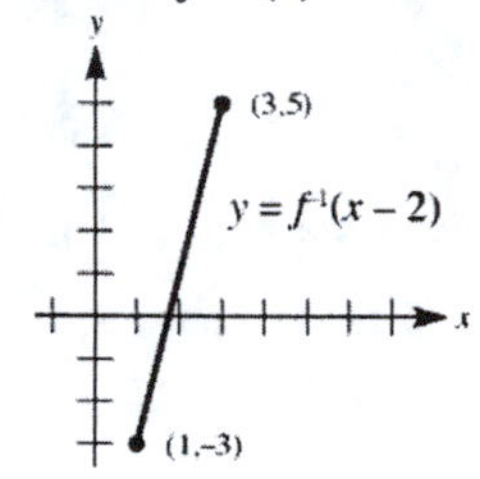

c. The graph of $y=f^{-1}(x)-2$ is a displacement of $f^{-1}(x)$ down 2 units:

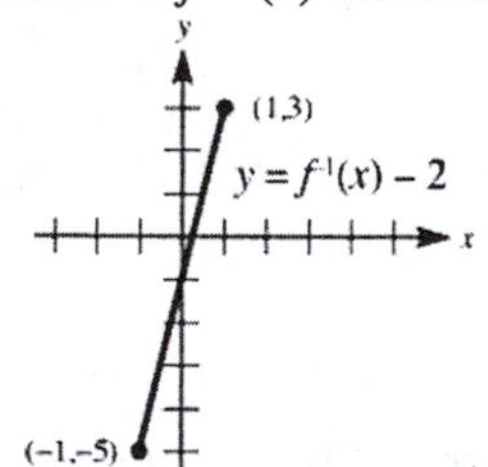

d. The graph of $y=f^{-1}(x-2)-2$ is a displacement of $f^{-1}(x)$ to the right 2 units and down 2 units:

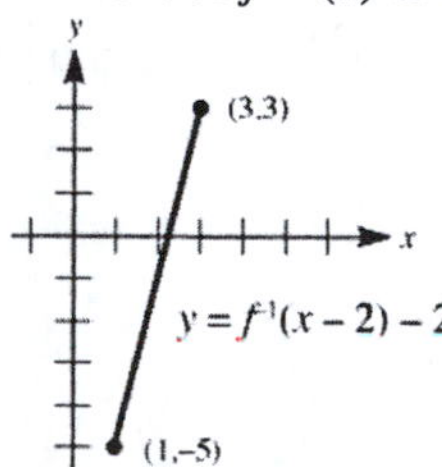

25. **a.** Solve the equation:

$$7+f^{-1}(x-1)=9$$
$$f^{-1}(x-1)=2$$
$$x-1=f(2)$$
$$x-1=6$$
$$x=7$$

b. Solve the equation:

$$4+f(x+3)=-3$$
$$f(x+3)=-7$$
$$x+3=f^{-1}(-7)$$
$$x+3=0$$
$$x=-3$$

27. Solve the equation:

$$f^{-1}\left(\frac{t+1}{t-2}\right)=12$$
$$\frac{t+1}{t-2}=f(12)$$
$$\frac{t+1}{t-2}=13$$
$$t+1=13t-26$$
$$27=12t$$
$$t=\frac{27}{12}=\frac{9}{4}$$

29. The graph of $f(x)=-x^2+1$ fails the horizontal line test, so it is not one-to-one:

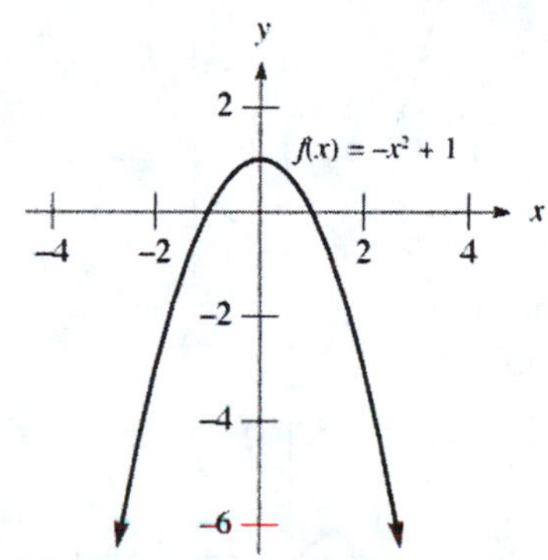

31. The graph of $g(x) = 5$ fails the horizontal line test (it *is* a horizontal line), so it is not one-to-one:

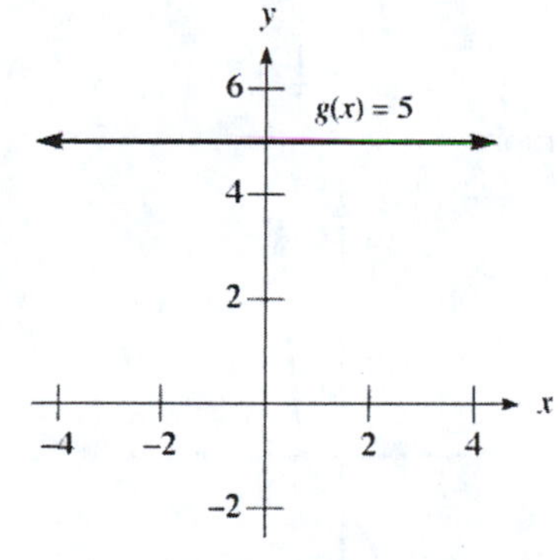

33. The graph of $g(x)$ fails the horizontal line test, since $g(-1) = g(0) = 1$, so it is not one-to-one:

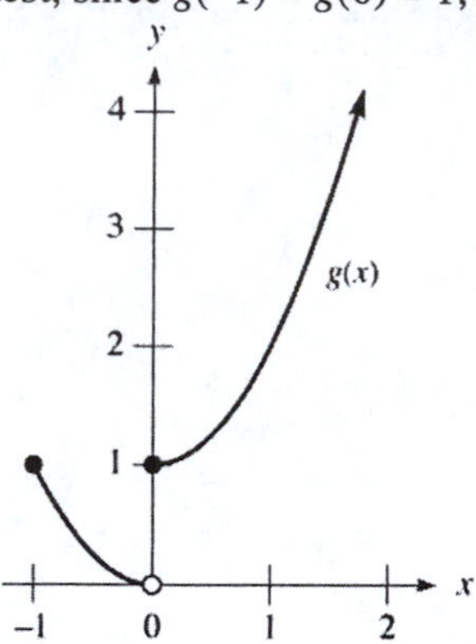

35. The function is not one-to-one. Draw the graph:

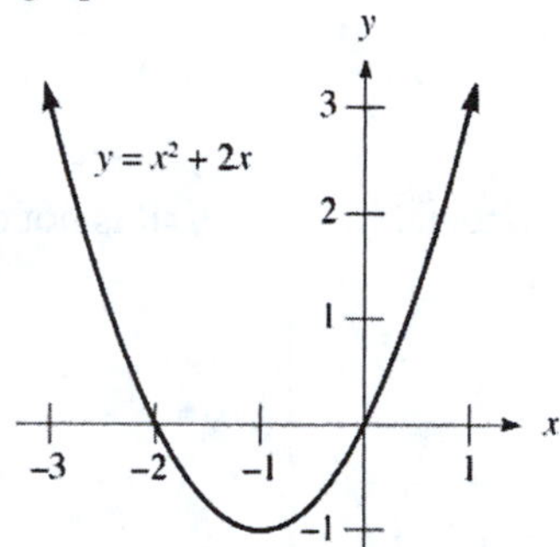

37. The function is not one-to-one. Draw the graph:

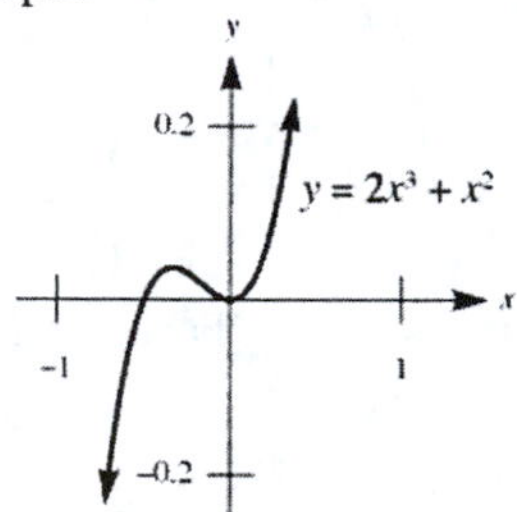

39. The function is one-to-one. Draw the graph:

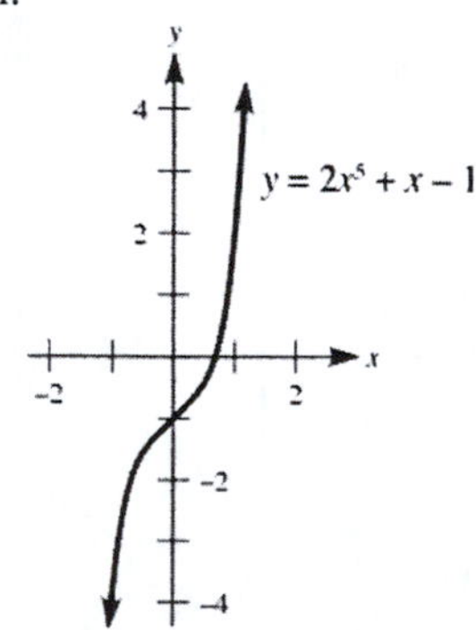

41. **a.** In the standard viewing rectangle, both graphs appear to be one-to-one. However, graphing the two functions in the viewing rectangle [–0.8,–0.4] x [1.78,1.82]:

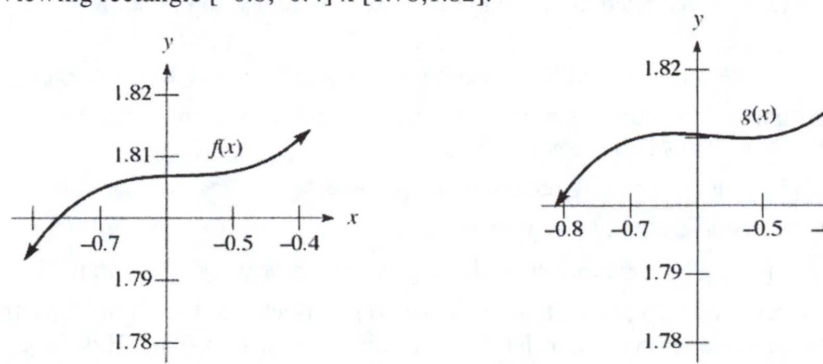

b. Note that f appears to be one-to-one, while g does not.

43. The coordinates of A are $(a, f(a))$. Since B lies on the line $y = x$, and its y-coordinate is $f(a)$, the coordinates of B are $(f(a), f(a))$. The y-coordinate of C is also $f(a)$, and so its x-coordinate is $f(f(a))$, since $f^{-1}(f(f(a))) = f(a)$, so the coordinates of C are $(f(f(a)), f(a))$. Since D is the reflection of C across $y = x$, the coordinates of D are $(f(a), f(f(a)))$.

45. **a.** Computing the average rate of change: $\dfrac{f(9)-f(4)}{9-4} = \dfrac{\frac{3}{8}-1}{5} = \dfrac{-\frac{5}{8}}{5} = -\frac{1}{8}$

b. To find $f^{-1}(x)$, call $y = f(x)$. Switch the roles of x and y and solve the resulting equation for y:

$$x = \frac{3}{y-1}$$
$$xy - x = 3$$
$$xy = x + 3$$
$$y = \frac{x+3}{x}$$

So $f^{-1}(x) = \dfrac{x+3}{x}$. Now compute the average rate of change of $f^{-1}(x)$ on the interval $[f(4), f(9)] = \left[1, \frac{3}{8}\right]$:

$$\frac{f^{-1}\left(\frac{3}{8}\right) - f^{-1}(1)}{\frac{3}{8}-1} = \frac{9-4}{-\frac{5}{8}} = \frac{5}{-\frac{5}{8}} = -8$$

Notice that our answer is the reciprocal of that found in part **a**.

47. **a.** Since the graph of $y = f(x) + 1$ is obtained by translating the graph of $y = f(x)$ up one unit, the point (a, b) is translated to the point $(a, b + 1)$. Thus E is the correct answer.

b. Since the graph of $y = f(x + 1)$ is obtained by translating the graph of $y = f(x)$ to the left one unit, the point (a, b) is translated to the point $(a - 1, b)$. Thus C is the correct answer.

c. Since the graph of $y = f(x - 1) + 1$ is obtained by translating the graph of $y = f(x)$ to the right one unit and up one unit, the point (a, b) is translated to the point $(a + 1, b + 1)$. Thus L is the correct answer.

d. Since the graph of $y = f(-x)$ is obtained by reflecting the graph of $y = f(x)$ across the y-axis, the point (a, b) is reflected to the point $(-a, b)$. Thus A is the correct answer.

e. Since the graph of $y = -f(x)$ is obtained by reflecting the graph of $y = f(x)$ across the x-axis, the point (a, b) is reflected to the point $(a, -b)$. Thus J is the correct answer.

f. Since the graph of $y = -f(-x)$ is obtained by reflecting the graph of $y = f(x)$ across the x-axis and across the y-axis, the point (a, b) is reflected to the point $(-a, -b)$. Thus G is the correct answer.

g. Since the graph of $y = f^{-1}(x)$ is obtained by reflecting the graph of $y = f(x)$ across the line $y = x$, the point (a, b) is reflected to the point (b, a). Thus B is the correct answer.

h. Since the graph of $y = f^{-1}(x) + 1$ is obtained by reflecting the graph of $y = f(x)$ across the line $y = x$, translating up one unit, then the point (a, b) is reflected to the point (b, a) then translated to the point $(b, a + 1)$. Thus M is the correct answer.

i. Since the graph of $y = f^{-1}(x-1)$ is obtained by reflecting the graph of $y = f(x)$ across the line $y = x$, translating to the right one unit, the point (a, b) is reflected to the point (b, a) then translated to the point $(b + 1, a)$. Thus K is the correct answer.

j. Since the graph of $y = f^{-1}(-x)+1$ is obtained by reflecting the graph of $y = f(x)$ across the line $y = x$, across the y-axis, then translating up one unit, the point (a, b) is reflected to the point (b, a), reflected to the point $(-b, a)$, then translated to the point $(-b, a + 1)$. Thus D is the correct answer.

k. Since the graph of $y = -f^{-1}(x)$ is obtained by reflecting the graph of $y = f(x)$ across the line $y = x$ then across the x-axis, the point (a, b) is reflected to the point (b, a) then reflected to the point $(b, -a)$. Thus I is the correct answer.

l. Since the graph of $y = -f^{-1}(-x)+1$ is obtained by reflecting the graph of $y = f(x)$ across the line $y = x$, the x-axis, and the y-axis, then translating up one unit, the point (a, b) is reflected to the point (b, a), then reflected to the point $(-b, a)$ and then $(-b, -a)$, and finally translated to the point $(-b, -a + 1)$. Thus H is the correct answer.

m. Since the graph of $y = 1 - f^{-1}(x)$ is obtained by reflecting the graph of $y = f(x)$ across the line $y = x$ and across the x-axis, then translated up one unit, the point (a, b) is reflected to the point (b, a), then reflected to the point $(b, -a)$, then translated to the point $(b, -a + 1)$. Thus N is the correct answer.

n. Since the graph of $y = f(1 - x)$ is obtained by reflecting the graph of $y = f(x)$ across the y-axis, then translating to the right one unit, the point (a, b) is reflected to the point $(-a, b)$ then translated to the point $(-a + 1, b)$. Thus F is the correct answer.

49. a. Finding the composition: $f\left[f(x)\right] = f\left(-\frac{2x+2}{x}\right) = -\frac{2\left(-\frac{2x+2}{x}\right)+2}{-\frac{2x+2}{x}} \bullet \frac{x}{x} = \frac{-4x-4+2x}{2x+2} = \frac{-2x-4}{2x+2} = -\frac{x+2}{x+1}$

b. Sketching the graph:

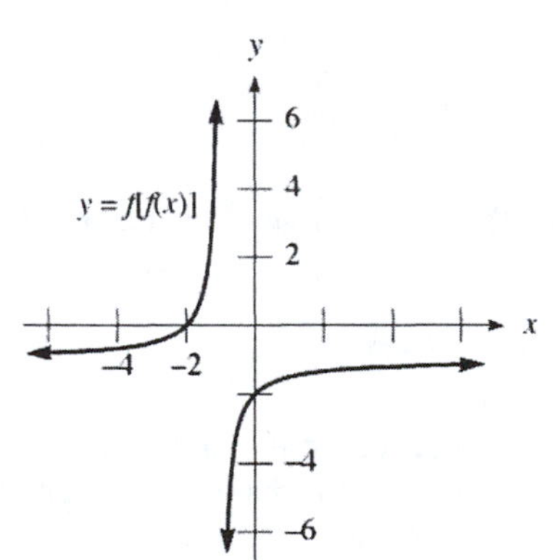

The graph appears to have symmetry about the line $y = x$.

c. Finding the composition:

$$(f \circ f)(f \circ f)(x) = (f \circ f)\left(-\frac{x+2}{x+1}\right) = -\frac{-\frac{x+2}{x+1}+2}{-\frac{x+2}{x+1}+1} \bullet \frac{x+1}{x+1} = -\frac{-x-2+2x+2}{-x-2+x+1} = -\frac{x}{-1} = x$$

Thus $f \circ f$ is the inverse of itself.

51. Let the points $P = (a, b)$ and $Q = (b, a)$. We must first show that the line segment $\overline{PQ}$ is perpendicular to the line $y = x$. Line segment $\overline{PQ}$ has a slope of: $m = \dfrac{a-b}{b-a} = -1$. Since $y = x$ has a slope of 1, and $1(-1) = -1$, the two lines are perpendicular. Next, show that P and Q are equidistant from $y = x$. We find the midpoint of line segment $\overline{PQ}$:

$$M = \left(\frac{a+b}{2}, \frac{b+a}{2}\right) = \left(\frac{a+b}{2}, \frac{a+b}{2}\right)$$

Since $\left(\dfrac{a+b}{2}, \dfrac{a+b}{2}\right)$ lies on the line $y = x$, and since $\overline{PQ}$ is perpendicular to $y = x$, and $PM = QM$, by the definition of symmetry P and Q are symmetric about the line $y = x$.

Chapter 3 Review Exercises

1. **a.** We must guarantee that the quantity within the radical is non-negative:

$$\begin{aligned} 15 - 5x &\geq 0 \\ -5x &\geq -15 \\ x &\leq 3 \end{aligned}$$

So the domain is $(-\infty, 3]$.

b. Solve $y = \dfrac{3+x}{2x-5}$ for x:

$$\begin{aligned} y(2x-5) &= 3+x \\ 2xy - 5y &= 3+x \\ 2xy - x &= 5y+3 \\ x(2y-1) &= 5y+3 \\ x &= \frac{5y+3}{2y-1} \end{aligned}$$

So the range is all real numbers except $\frac{1}{2}$, or $\left(-\infty, \frac{1}{2}\right) \cup \left(\frac{1}{2}, \infty\right)$.

3. Yes. Let $f(x) = ax + b$ and $g(x) = cx + d$, then: $(f \circ g)(x) = f(cx+d) = a(cx+d) + b = acx + (ad+b)$
So $(f \circ g)(x)$ is a linear function.

5. **a.** Compute the difference quotient: $\dfrac{F(x) - F(a)}{x-a} = \dfrac{\frac{1}{x} - \frac{1}{a}}{x-a} \bullet \dfrac{ax}{ax} = \dfrac{a-x}{ax(x-a)} = -\dfrac{1}{ax}$

b. First compute: $g(x+h) = (x+h) - 2(x+h)^2 = x + h - 2x^2 - 4xh - 2h^2$. So we have:

$$\begin{aligned} \frac{g(x+h) - g(x)}{h} &= \frac{\left(x + h - 2x^2 - 4xh - 2h^2\right) - \left(x - 2x^2\right)}{h} \\ &= \frac{x + h - 2x^2 - 4xh - 2h^2 - x + 2x^2}{h} \\ &= \frac{h - 4xh - 2h^2}{h} \\ &= 1 - 4x - 2h \end{aligned}$$

7. a. Switch the roles of x and y and solve the resulting equation for y:

$$x = \frac{1-5y}{3y}$$
$$3xy = 1-5y$$
$$3xy+5y = 1$$
$$y(3x+5) = 1$$
$$y = \frac{1}{3x+5}$$

So $g^{-1}(x) = \dfrac{1}{3x+5}$.

b. Graph f^{-1}:

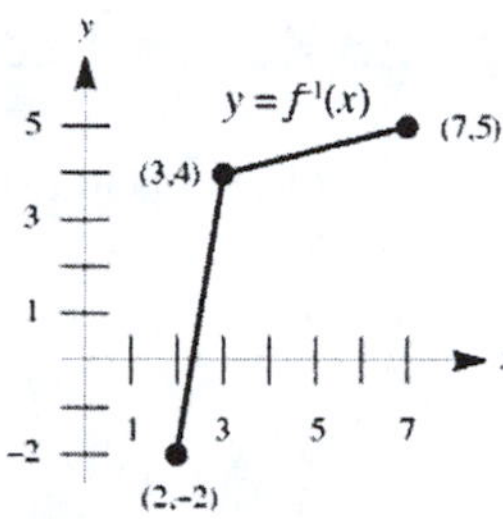

9. a. Displace $y = |x|$ two units to the left and 3 units down. The x-intercepts are –5 and 1, and the y-intercept is –1:

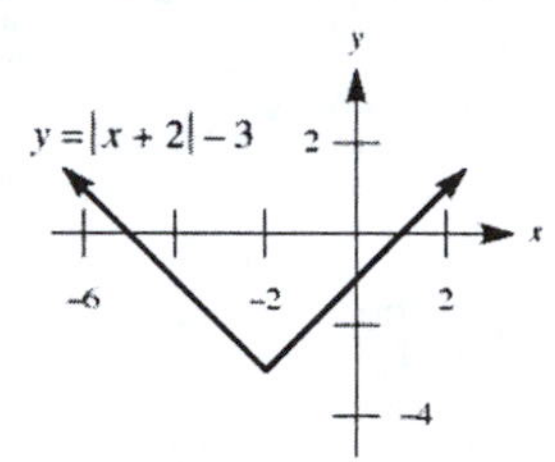

b. Displace $y = \dfrac{1}{x}$ two units to the left and 1 unit down. The x-intercept is –1 and the y-intercept is $-\frac{1}{2}$:

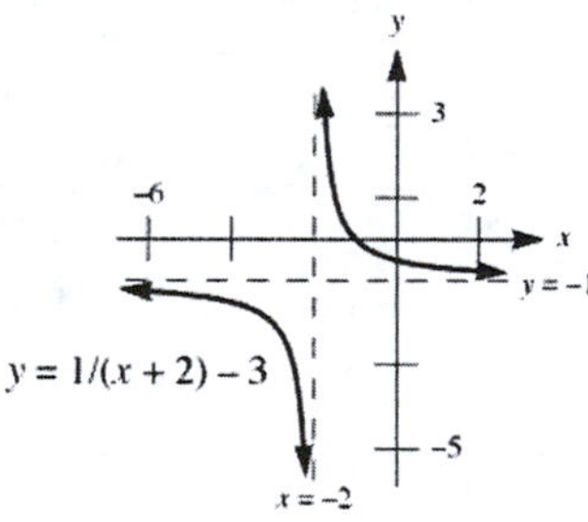

11. a. Compute $f(-1)$: $f(-1) = 3(-1)^2 - 2(-1) = 3+2 = 5$

b. Compute $f\left(1-\sqrt{2}\right)$:

$$f\left(1-\sqrt{2}\right) = 3\left(1-\sqrt{2}\right)^2 - 2\left(1-\sqrt{2}\right) = 3\left(1-2\sqrt{2}+2\right) - 2 + 2\sqrt{2} = 9 - 6\sqrt{2} - 2 + 2\sqrt{2} = 7 - 4\sqrt{2}$$

13. The slope is given by: $m = \dfrac{(5+h)^2 - 25}{5+h-5} = \dfrac{25+10h+h^2-25}{h} = \dfrac{10h+h^2}{h} = 10+h$

Using functional notation, we have $m(h) = 10 + h$.

15. The graph of $y = f(-x)$ will result in a reflection across the y-axis:

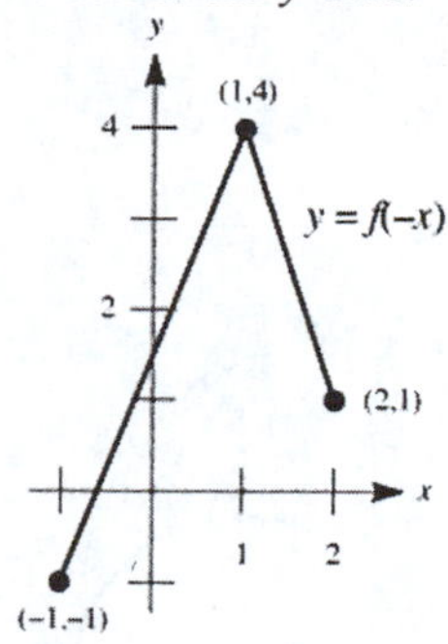

17.
a. The first three iterates are 2.5, 3.25, and 3.625.
b. The y-coordinate of point A is $f(1) = 2.5$. The x-coordinate of point B is 2.5, so the y-coordinate of point B is $f(2.5) = 3.25$. The x-coordinate of point C is 3.25, so the y-coordinate of point C is $f(3.25) = 3.625$.
c. At the intersection point the y-coordinates must be equal, so:

$$x = \frac{1}{2}x + 2$$
$$2x = x + 4$$
$$x = 4$$

Since $y = x$, the y-coordinate is also 4.
d. The fourth through tenth iterates are 3.813, 3.906, 3.953, 3.977, 3.988, 3.994, and 3.997. These iterates appear to be approaching 4.

19. To find the x-intercepts, let $y = 0$:

$$\frac{1}{x+1} = 0$$
$$1 = 0$$

Clearly this is impossible, so there are no x-intercepts. To find the y-intercept, let $x = 0$: $y = \frac{1}{0+1} = 1$

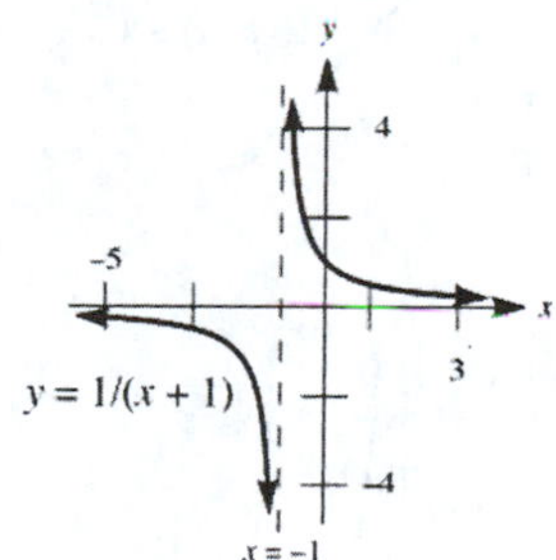

21. To find the x-intercept, let $y = 0$:

$$|x+3| = 0$$
$$x = -3$$

To find the y-intercept, let $x = 0$: $y = |0+3| = 3$

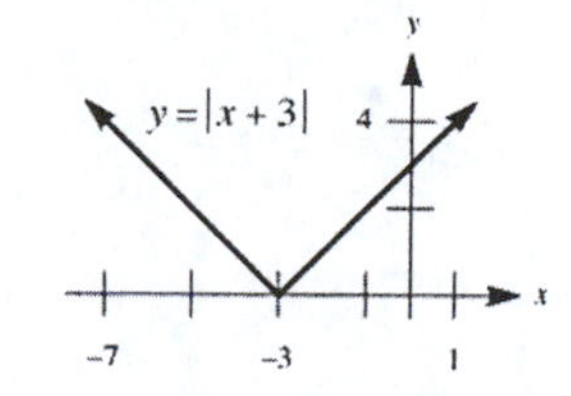

23. To find the x-intercepts, let $y = 0$:

$$\begin{aligned}\sqrt{1-x^2} &= 0\\ 1-x^2 &= 0\\ -x^2 &= -1\\ x &= \pm 1\end{aligned}$$

To find the y-intercept, let $x = 0$: $y = \sqrt{1-0} = 1$

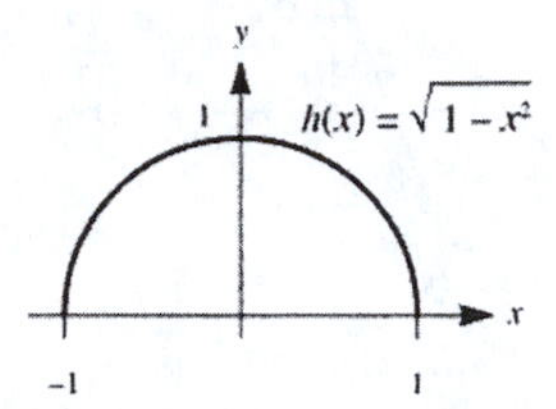

25. The x-intercept is 0 and the y-intercept is 0:

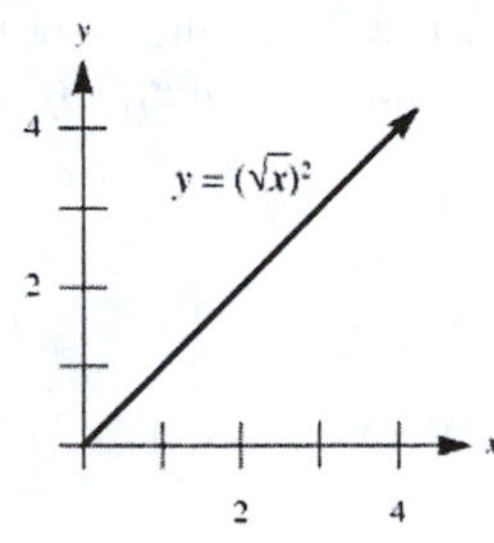

27. First compute $(f \circ g)(x)$: $(f \circ g)(x) = f\left(\sqrt{x-1}\right) = -\left(\sqrt{x-1}\right)^2 = 1 - x$ for $x \geq 1$

The x-intercept is 1 and there is no y-intercept:

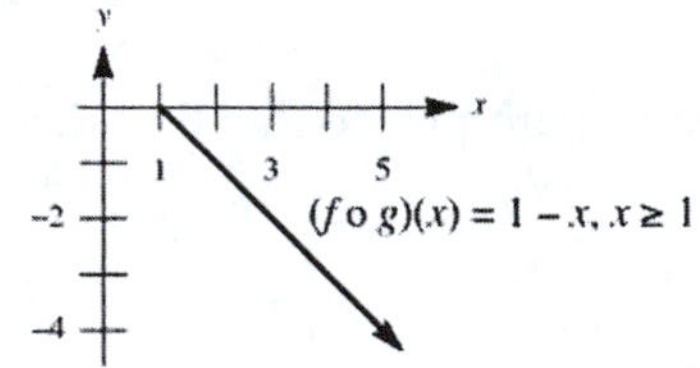

29. There are no x- or y-intercepts:

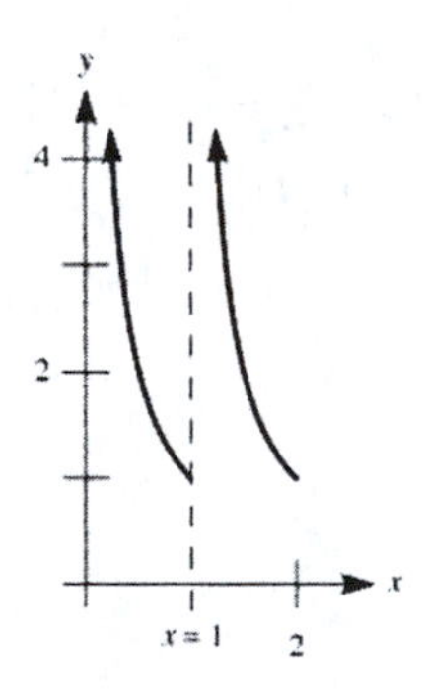

31. To find the inverse function, switch the roles of x and y and solve the resulting equation for y:

$$x = \frac{y+1}{2}$$
$$2x = y+1$$
$$2x-1 = y$$

So $f^{-1}(x) = 2x-1$. To find the x-intercept, let $y = 0$:

$$2x-1 = 0$$
$$2x = 1$$
$$x = \tfrac{1}{2}$$

To find the y-intercept, let $x = 0$: $y = 2(0)-1 = -1$

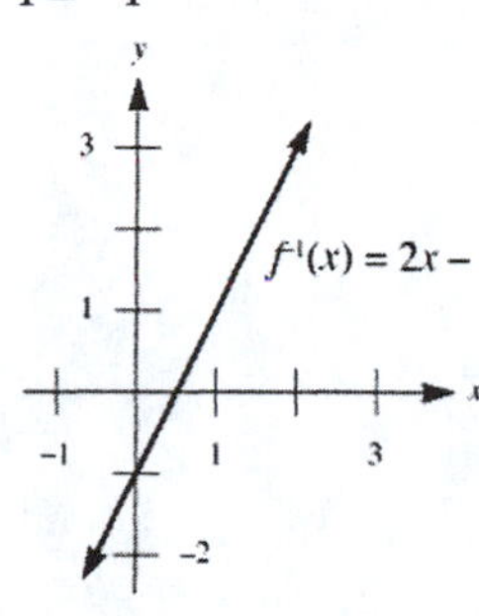

33. Compute $\left(f \circ f^{-1}\right)(x) = x$ for $x \geq 0$. The x-intercept is 0 and the y-intercept is 0:

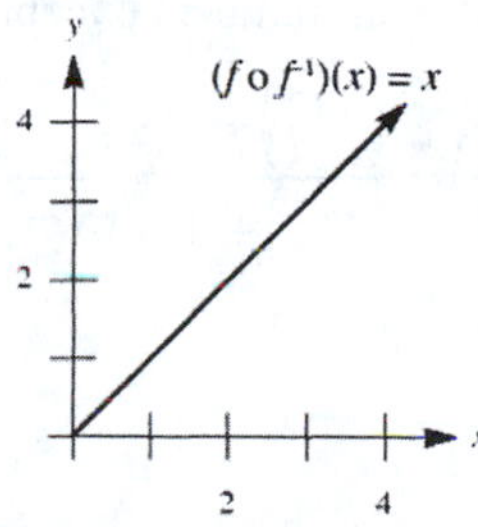

35. We must make sure that $x^2 - 9 \neq 0$. Find the points to exclude:

$$x^2 - 9 = 0$$
$$x^2 = 9$$
$$x = \pm 3$$

So the domain is all real numbers except 3 and –3, or $(-\infty,-3) \cup (-3,3) \cup (3,\infty)$.

37. We must be sure that the quantity inside the radical is non-negative:

$$8 - 2x \geq 0$$
$$-2x \geq -8$$
$$x \leq 4$$

So the domain is $(-\infty, 4]$.

39. We must be sure that the quantity inside the radical is non-negative:

$$x^2 - 2x - 3 \geq 0$$
$$(x-3)(x+1) \geq 0$$

The key numbers are –1 and 3. Draw the sign chart:

Interval	Test Number	$x-3$	$x+1$	$(x-3)(x+1)$
$(-\infty,-1)$	–2	neg.	neg.	pos.
$(-1,3)$	0	neg.	pos.	neg.
$(3,\infty)$	4	pos.	pos.	pos.

So the product is positive on the intervals $(-\infty,-1)\cup(3,\infty)$, and it is zero when $x = -1$ and $x = 3$. So the domain is $(-\infty,-1]\cup[3,\infty)$.

41. Solve for x:

$$y = \frac{x+4}{3x-1}$$
$$y(3x-1) = x+4$$
$$3xy - y = x+4$$
$$3xy - x = y+4$$
$$x(3y-1) = y+4$$
$$x = \frac{y+4}{3y-1}$$

Now $3y - 1 = 0$ when $y = \frac{1}{3}$, so the range is all real numbers except $\frac{1}{3}$, or $\left(-\infty,\frac{1}{3}\right)\cup\left(\frac{1}{3},\infty\right)$.

43. First compute $(g\circ f)(x)$: $(g\circ f)(x) = g\left(\frac{x+2}{x-1}\right) = \dfrac{\left(\frac{x+2}{x-1}\right)+1}{\left(\frac{x+2}{x-1}\right)+4} = \dfrac{(x+2)+1(x-1)}{(x+2)+4(x-1)} = \dfrac{x+2+x-1}{x+2+4x-4} = \dfrac{2x+1}{5x-2}$

Solve $y = \dfrac{2x+1}{5x-2}$ for x:

$$y = \frac{2x+1}{5x-2}$$
$$y(5x-2) = 2x+1$$
$$5xy - 2y = 2x+1$$
$$5xy - 2x = 2y+1$$
$$x(5y-2) = 2y+1$$
$$x = \frac{2y+1}{5y-2}$$

Now $5y - 2 = 0$ when $y = \frac{2}{5}$, so the range is all real numbers except $\frac{2}{5}$, or $\left(-\infty,\frac{2}{5}\right)\cup\left(\frac{2}{5},\infty\right)$.

45. The composition is $a(x) = (f\circ g)(x)$, since: $(f\circ g)(x) = f[g(x)] = f(x-1) = \dfrac{1}{x-1}$

47. The composition is $c(x) = (G\circ g)(x)$, since: $(G\circ g)(x) = G[g(x)] = G(x-1) = \sqrt{x-1}$

49. The composition is $A(x) = (g\circ f\circ G)(x)$, since: $(g\circ f\circ G)(x) = (g\circ f)\left(\sqrt{x}\right) = g\left(\dfrac{1}{\sqrt{x}}\right) = \dfrac{1}{\sqrt{x}} - 1$

51. The composition is $C(x) = (g\circ G\circ G)(x)$, since: $(g\circ G\circ G)(x) = (g\circ G)\left(\sqrt{x}\right) = g\left(\sqrt{\sqrt{x}}\right) = g\left(\sqrt[4]{x}\right) = \sqrt[4]{x} - 1$

53. Using $f(x)$, compute $f(-3)$: $f(-3) = (-3)^2 - (-3) = 9 + 3 = 12$

55. Using $F(x)$, compute $F\left(\frac{3}{4}\right)$: $F\left(\frac{3}{4}\right)=\dfrac{\frac{3}{4}-3}{\frac{3}{4}+4}=\dfrac{3-3(4)}{3+4(4)}=\dfrac{3-12}{3+16}=-\frac{9}{19}$

57. Using $f(x)$, compute $f(-t)$: $f(-t)=(-t)^2-(-t)=t^2+t$

59. Using $f(x)$, compute $f(x-2)$: $f(x-2)=(x-2)^2-(x-2)=x^2-4x+4-x+2=x^2-5x+6$

61. Using $f(x)$, compute $f\left(x^2\right)$: $f\left(x^2\right)=\left(x^2\right)^2-x^2=x^4-x^2$

63. Compute $[f(x)][g(x)]$: $[f(x)][g(x)]=\left(x^2-x\right)(1-2x)=x^2-2x^3-x+2x^2=-2x^3+3x^2-x$

65. Compute $f[g(x)]$: $f[g(x)]=f(1-2x)=(1-2x)^2-(1-2x)=1-4x+4x^2-1+2x=4x^2-2x$

67. Compute $(g\circ f)(x)$: $(g\circ f)(x)=g[f(x)]=g\left(x^2-x\right)=1-2\left(x^2-x\right)=2-2x^2+2x=-2x^2+2x+1$

69. Compute $(F\circ g)(x)$: $(F\circ g)(x)=F[g(x)]=F(1-2x)=\dfrac{(1-2x)-3}{(1-2x)+4}=\dfrac{-2x-2}{-2x+5}$ or $\dfrac{2x+2}{2x-5}$

71. Find the difference quotient:

$$\frac{f(x+h)-f(x)}{h}=\frac{(x+h)^2-(x+h)-\left(x^2-x\right)}{h}=\frac{2xh+h^2-h}{h}=\frac{h(2x+h-1)}{h}=2x+h-1$$

73. Let $y=\dfrac{x-3}{x+4}$. To find $F^{-1}(x)$, switch the roles of x and y and solve the resulting equation for y:

$$\begin{aligned}x&=\frac{y-3}{y+4}\\x(y+4)&=y-3\\xy+4x&=y-3\\xy-y&=-4x-3\\y(x-1)&=-4x-3\\y&=\frac{-4x-3}{x-1}\end{aligned}$$

So $F^{-1}(x)=\dfrac{-4x-3}{x-1}=\dfrac{4x+3}{1-x}$.

75. The value is x, since $\left(g\circ g^{-1}\right)(x)=x$ for all x.

77. Since $g^{-1}(x)=\dfrac{1-x}{2}$, we have: $g^{-1}(-x)=\dfrac{1-(-x)}{2}=\dfrac{1+x}{2}$

79. Since $\frac{22}{7}$ is in the domain of $F(x)$, $F^{-1}\left[F\left(\frac{22}{7}\right)\right]=\frac{22}{7}$.

81. Since $f(0)=-2$, it is negative.
83. Since the point $(-3,-1)$ lies on the graph, $f(-3)=-1$.
85. Since $f(0)=-2$ and $f(8)=-1$, $f(0)-f(8)=-2-(-1)=-2+1=-1$.
87. The coordinates of the turning points are $(0,-2)$ and $(5,1)$.
89. $f(x)$ is decreasing on the intervals $[-6,0]$ and $[5,8]$.
91. Since $|x|\le 2$ corresponds to the interval $-2\le x\le 2$, the largest value of $f(x)$ is 0, occurring at $x=2$.
93. Since $f(x)$ is not a one-to-one function (it does not pass the horizontal line test) it does not possess an inverse function.
95. From the graph, note that $f(x)=g(x)$ when $x=4$.
97. **a.** From the graph, note that $f(x)=0$ when $x=10$. **b.** From the graph, note that $g(x)=0$ when $x=0$.

99. a. Computing the value: $(f+g)(8)=f(8)+g(8)=1+4=5$
b. Computing the value: $(f-g)(8)=f(8)-g(8)=1-4=-3$
c. Computing the value: $fg(8)=f(8)\bullet g(8)=1\bullet 4=4$
d. Computing the value: $(f/g)(8)=\dfrac{f(8)}{g(8)}=\frac{1}{4}$

101. Compute each value:
$(f\circ f)(10)=f[f(10)]=f(0)=5$ $\qquad$ $(g\circ g)(10)=g[g(10)]=g(3)=1$
So $(f\circ f)(10)$ is larger.

103. Note that $f(x)\geq 3$ for $0\leq x\leq 4$, which is the interval $[0,4]$.

105. The maximum point of $g(x)$ is $(6,5)$, so the largest number in the range of g is 5.

107. Note that $g(x)$ is decreasing when $1<x<3$ and $6<x<10$, which are the intervals $(1,3)\cup(6,10)$.

109. Consider two cases:

If $4<x<5$, then $f(x)>f(5)$ and $x<5$, thus $f(x)-f(5)>0$ while $x-5<0$. Thus $\dfrac{f(x)-f(5)}{x-5}<0$.

If $5<x<7$, then $f(x)<f(5)$ and $x>5$, thus $f(x)-f(5)<0$ while $x-5>0$. Thus $\dfrac{f(x)-f(5)}{x-5}<0$.

Therefore, the quantity is negative for all x-values in the interval $(4,7)$.

Chapter 3 Test

1. We must be sure the quantity inside the radical is non-negative:

$$x^2-5x-6\geq 0$$
$$(x-6)(x+1)\geq 0$$

The key numbers are -1 and 6. Draw the sign chart:

Interval	Test Number	$x-6$	$x+1$	$(x-6)(x+1)$
$(-\infty,-1)$	-2	neg.	neg.	pos.
$(-1,6)$	0	neg.	pos.	neg.
$(6,\infty)$	8	pos.	pos.	pos.

The product is positive on the intervals $(-\infty,-1)\cup(6,\infty)$ and 0 at $x=-1$ and $x=6$. So the domain is $(-\infty,-1]\cup[6,\infty)$.

2. Let $y=\dfrac{2x-8}{3x+5}$. Solve for x:

$$\frac{2x-8}{3x+5}=y$$
$$2x-8=3yx+5y$$
$$2x-3yx=5y+8$$
$$x(2-3y)=5y+8$$
$$x=\frac{5y+8}{2-3y}$$

So $y\neq\frac{2}{3}$. The range is $\left(-\infty,\frac{2}{3}\right)\cup\left(\frac{2}{3},\infty\right)$.

3. **a.** Compute $(f-g)(x)$: $(f-g)(x)=f(x)-g(x)=(2x^2-3x)-(2-x)=2x^2-3x-2+x=2x^2-2x-2$

b. Compute $(f\circ g)(x)$:

$$(f\circ g)(x)=f[g(x)]=f(2-x)=2(2-x)^2-3(2-x)=8-8x+2x^2-6+3x=2x^2-5x+2$$

c. Using our result from part **b**: $f[g(-4)]=2(-4)^2-5(-4)+2=32+20+2=54$

4. Compute the difference quotient: $\dfrac{f(t)-f(a)}{t-a}=\dfrac{\frac{2}{t}-\frac{2}{a}}{t-a}\bullet\dfrac{at}{at}=\dfrac{2a-2t}{at(t-a)}=\dfrac{-2(t-a)}{at(t-a)}=-\dfrac{2}{at}$

5. First compute $g(x+h)$: $g(x+h)=2(x+h)^2-5(x+h)=2x^2+4xh+2h^2-5x-5h$

Now compute the difference quotient:

$$\frac{g(x+h)-g(x)}{h}=\frac{2x^2+4xh+2h^2-5x-5h-2x^2+5x}{h}=\frac{4xh+2h^2-5h}{h}=4x+2h-5$$

6. Let $y=\dfrac{-4x}{6x+1}$. Switch the roles of x and y, then solve the resulting equation for y:

$$\begin{aligned}\frac{-4y}{6y+1}&=x\\ -4y&=6xy+x\\ -x&=6xy+4y\\ -x&=y(6x+4)\\ \frac{-x}{6x+4}&=y\end{aligned}$$

So $g^{-1}(x)=-\dfrac{x}{6x+4}$.

7. Since $f^{-1}(x)$ will be the line segment joining the points (1,–3) and (6,5), $y=-f^{-1}(x)$ will be the line segment joining the points (1,3) and (6,–5). Compute the slope: $m=\dfrac{-5-3}{6-1}=-\frac{8}{5}$

8. **a.** The x-intercepts are 4 and 2, and the y-intercept is –2:

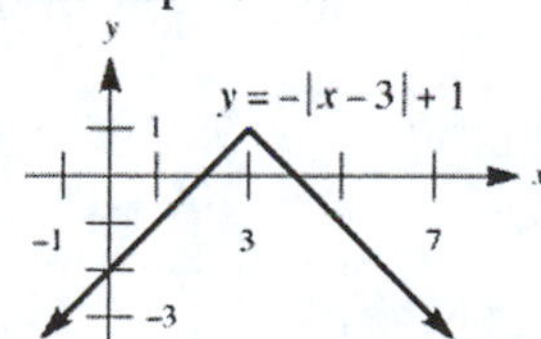

b. The x-intercept is $-\frac{5}{2}$ and the y-intercept is $-\frac{5}{3}$:

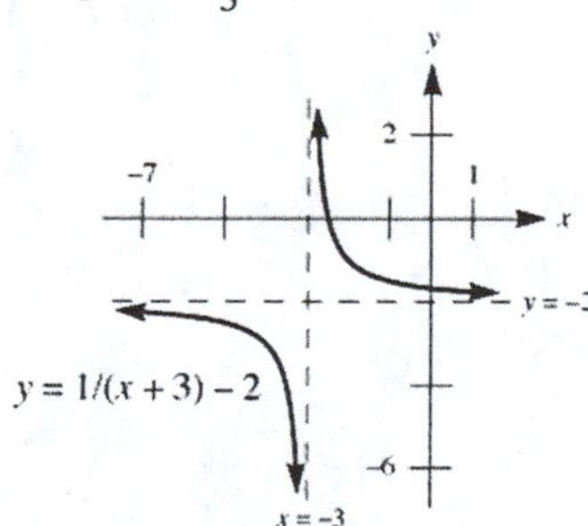

9. **a.** The range of g is [–3,1].

b. The turning point has coordinates (–2,1).

c. The minimum value of g is –3, which occurs at $x = 2$.

d. The maximum value of g is 1, occurring at $x = -2$.

e. For $-2 \le x \le 2$, g is decreasing. So g is decreasing on [–2,2].

f. Since g fails the horizontal line test, it is not a one-to-one function.

10. **a.** Compute $f\left(-\frac{3}{2}\right)$: $f\left(-\frac{3}{2}\right)=\left(-\frac{3}{2}\right)^2-3\left(-\frac{3}{2}\right)-1=\frac{9}{4}+\frac{9}{2}-1=\frac{23}{4}$

b. Compute $f\left(\sqrt{3}-2\right)$: $f\left(\sqrt{3}-2\right)=\left(\sqrt{3}-2\right)^2-3\left(\sqrt{3}-2\right)-1=3-4\sqrt{3}+4-3\sqrt{3}+6-1=12-7\sqrt{3}$

11. The domain is all real numbers except -1, or $(-\infty,-1)\cup(-1,\infty)$. Draw the graph:

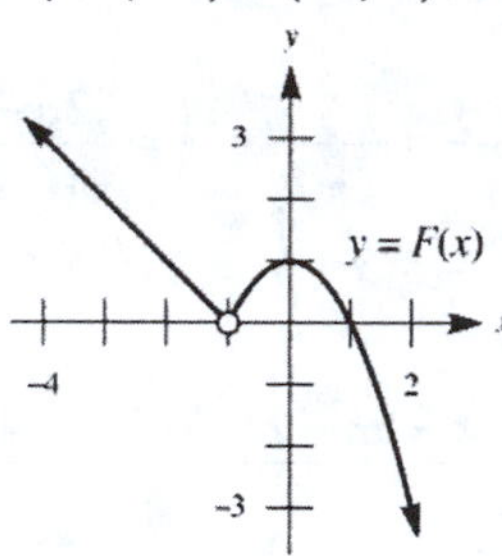

12. **a.** The first six iterates are 0.4, 0.8, 0.4, 0.8, 0.4, and 0.8.

b. The calculated six iterates are 0.4, 0.8, 0.4, 0.8, 0.4, and 0.8. These results are consistent with those found in part **a**.

13. Since $y=f(-x)$ is a reflection of $y=f(x)$ across the y-axis, it will be the line segment joining the points $(-1,3)$ and $(-5,-2)$:

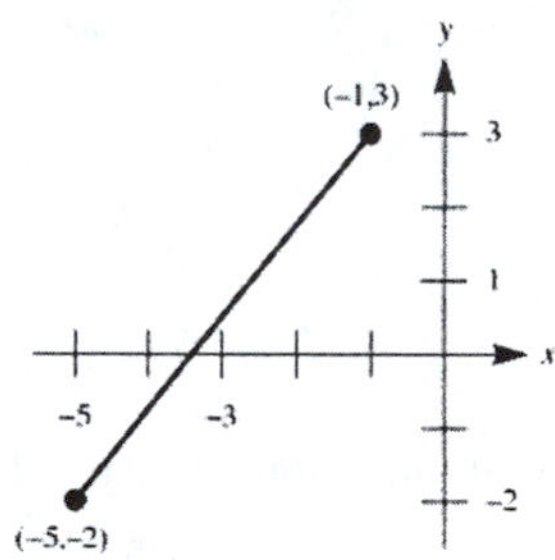

14. Solve the equation:

$$\begin{aligned}5+f(4t-3)&=2\\ f(4t-3)&=-3\\ 4t-3&=f^{-1}(-3)\\ 4t-3&=1\\ 4t&=4\\ t&=1\end{aligned}$$

15. Since $f(1)=1$ and $f(b)=\frac{1}{b}$: $\frac{\Delta f}{\Delta x}=\frac{f(b)-f(1)}{b-1}=\frac{\frac{1}{b}-1}{b-1}\bullet\frac{b}{b}=\frac{1-b}{b(b-1)}=-\frac{1}{b}$. Therefore:

$$\begin{aligned}-\frac{1}{b}&=-\frac{1}{10}\\ b&=10\end{aligned}$$

16. **a.** Since $F(a) = 0.16a^2 - 1.6a + 35$ and $F(5) = 0.16(5)^2 - 1.6(5) + 35 = 31$, the average value is given by:

$$\begin{aligned}\frac{\Delta F}{\Delta t} &= \frac{F(5) - F(a)}{5 - a} \\ &= \frac{31 - \left(0.16a^2 - 1.6a + 35\right)}{5 - a} \\ &= \frac{-0.16a^2 + 1.6a - 4}{5 - a} \\ &= \frac{0.16a^2 - 1.6a + 4}{a - 5} \\ &= \frac{(a - 5)(0.16a - 0.8)}{a - 5} \\ &= 0.16a - 0.8\end{aligned}$$

b. Using the result from part **a** where $a = 4$: $\frac{\Delta F}{\Delta t} = 0.16(4) - 0.8 = -0.16$ °C/hour

Chapter 4
Polynomial and Rational Functions. Applications to Optimization

4.1 Linear Functions

1. First find the slope between the points (–1,0) and (5,4): $m = \dfrac{4-0}{5-(-1)} = \frac{4}{6} = \frac{2}{3}$

Now use the point (–1,0) in the point-slope formula:

$$\begin{aligned} y - 0 &= \tfrac{2}{3}[x-(-1)] \\ y &= \tfrac{2}{3}(x+1) \\ y &= \tfrac{2}{3}x + \tfrac{2}{3} \end{aligned}$$

Using functional notation we have $f(x) = \frac{2}{3}x + \frac{2}{3}$.

3. First find the slope between the points (0,0) and $\left(1, \sqrt{2}\right)$: $m = \dfrac{\sqrt{2}-0}{1-0} = \sqrt{2}$. Now, since (0,0) is the y-intercept, we use the slope-intercept formula to write $y = \sqrt{2}x$. Using functional notation we have $g(x) = \sqrt{2}x$.

5. Find the slope of $x - y = 1$:

$$\begin{aligned} -y &= -x + 1 \\ y &= x - 1 \end{aligned}$$

So the parallel slope is 1. Use the point $\left(\frac{1}{2}, -3\right)$ in the point-slope formula:

$$\begin{aligned} y - (-3) &= 1\left(x - \tfrac{1}{2}\right) \\ y + 3 &= x - \tfrac{1}{2} \\ y &= x - \tfrac{7}{2} \end{aligned}$$

Using functional notation we have $f(x) = x - \frac{7}{2}$.

7. Since the points $(-1, 2)$ and $(0, 4)$ lie on the graph of the inverse function, (2,–1) and (4,0) must lie on the graph of the function. We find the slope: $m = \frac{0-(-1)}{4-2} = \frac{1}{2}$. Use the point (4,0) in the point-slope formula:

$$y - 0 = \frac{1}{2}(x-4)$$
$$y = \frac{1}{2}x - 2$$

Using functional notation we have $f(x) = \frac{1}{2}x - 2$.

9. Compute $(f \circ g)(x)$: $(f \circ g)(x) = f[g(x)] = f(1-2x) = 3(1-2x) - 4 = 3 - 6x - 4 = -6x - 1$
So $f \circ g$ is a linear function (it is in standard form).

11. Call $V(t)$ the value of the machine after t years. When $t = 0$ we have $V = 20{,}000$ and when $t = 8$ we have $V = 1{,}000$.
Find the slope of the line between the points (0,20000) and (8,1000): $m = \frac{1000 - 20000}{8-0} = \frac{-19000}{8} = -2375$
Since (0,20000) is the y-intercept, we use the slope-intercept formula to write $V = -2375t + 20000$. Using functional notation we have $V(t) = -2375t + 20000$.

13. a. Call $V(t)$ the value of the machine after t years. Now $V = 60{,}000$ when $t = 0$ and $V = 0$ when $t = 5$. Find the slope of the line between (0,60000) and (5,0): $m = \frac{0 - 60000}{5-0} = \frac{-60000}{5} = -12000$. Since (0,60000) is the y-intercept, we use the slope-intercept formula to write $V = -12000t + 60000$. Using functional notation we have $V(t) = -12000t + 60000$.

b. The completed schedule is:

End of Year	Yearly Depreciation	Accumulated Depreciation	Value V
0	0	0	60,000
1	12,000	12,000	48,000
2	12,000	24,000	36,000
3	12,000	36,000	24,000
4	12,000	48,000	12,000
5	12,000	60,000	0

15. a. Let $x = 10$, so $C(10) = 450 + 8(10) = \$530$.   b. Let $x = 11$, so $C(11) = 450 + 8(11) = \$538$.

c. There are two ways to find the marginal cost. One way is to recognize that the marginal cost will be the slope of the line, which is \$8/fan. Another way would be to use the definition of the marginal cost: it is cost of producing the ***next*** unit. Since $C(10)$ is the cost of producing 10 fans and $C(11)$ is the cost of producing 11 fans, the marginal cost would be $C(11) - C(10) = 538 - 530 = \$8$/fan. We get the same answer using either approach.

17. a. Graph $C(x) = 0.5x + 500$:

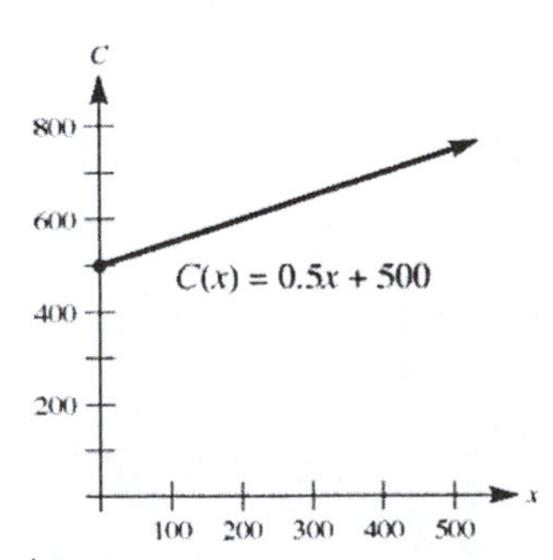

b. Compute $C(150) = 0.5(150) + 500 = \$575$.

c. Adding the marginal cost: $\$575 + \$0.50 = \$575.50$. This amount represents the cost of producing 151 units.
As a check: $C(151) = 0.5(151) + 500 = 75.50 + 500 = \$575.50$

19. **a.** Since the velocity of A is 3 units/sec and the velocity of B is 20 units/sec, B is traveling faster.

b. When $t = 0$, A is at $x = 100$ and B is at $x = -36$. So A is farther to the right.

c. Set the two x-coordinates equal:

$$\begin{aligned} 3t + 100 &= 20t - 36 \\ 136 &= 17t \\ 8 &= t \end{aligned}$$

When $t = 8$ sec, A and B have the same x-coordinate.

21. **a.** First find the slope: $m = \dfrac{32,217,708 - 31,493,525}{1997 - 1995} = \dfrac{724183}{2} = 362091.5$

Using the point (1995,31493525) in the point-slope formula:

$$\begin{aligned} y - 31,493,525 &= 362091.5(x - 1995) \\ y - 31,493,525 &= 362091.5x - 722,372,542.5 \\ y &= 362091.5x - 690,879,017.5 \end{aligned}$$

b. Substituting $x = 2000$: $y = 362091.5(2000) - 690,879,017.5 \approx 33,304,000$

The population projection is 33,304,000.

c. The linear function yields a projection that is too low. Calculating the percentage error:

$$\text{percentage error} = \frac{|33,871,648 - 33,304,000|}{33,871,648} \times 100 \approx 1.7\%$$

23. **a.** First find the slope: $m = \dfrac{57,184 - 53,504}{1995 - 1994} = \dfrac{3680}{1} = 3680$

Using the point (1994,53504) in the point-slope formula:

$$\begin{aligned} y - 53,504 &= 3680(x - 1994) \\ y - 53,504 &= 3680x - 7,337,920 \\ y &= 3680x - 7,284,416 \end{aligned}$$

b. Sketching the graph:

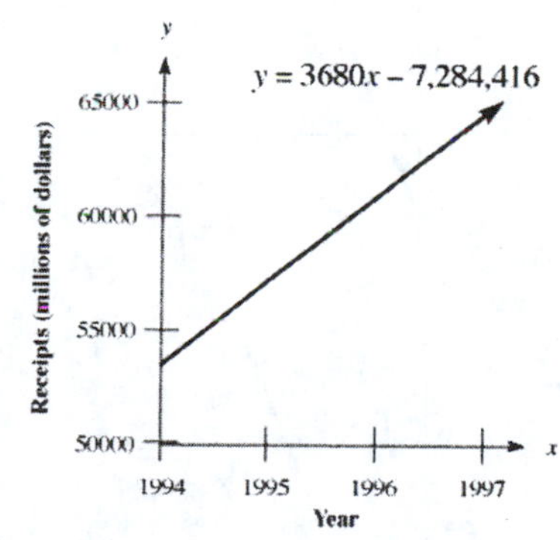

The approximate receipts (in millions of dollars) for 1997 are 64,544.

c. Calculating the percentage error: $\text{percentage error} = \dfrac{|63,010 - 64,544|}{63,010} \times 100 \approx 2.4\%$

25. **a.** First find the slope: $m = \dfrac{205.50 - 67.76}{49 - 21} = \dfrac{137.74}{28} \approx 4.92$. Using the point (21,67.76) in the point-slope formula:

$$\begin{aligned} y - 67.76 &= 4.92(x - 21) \\ y - 67.76 &= 4.92x - 103.31 \\ y &= 4.92x - 35.55 \end{aligned}$$

b. Substituting $x = 28$: $y = 4.92(28) - 35.55 = 102.21$. The projection for average height is 102.21 cm.

c. The estimate from part **b** is too high. Calculating the percentage error:

$$\text{percentage error} = \frac{|98.10 - 102.21|}{98.10} \times 100 \approx 4.2\%$$

d. Substituting $x = 14$: $y = 4.92(14) - 35.55 = 33.33$. The projection for average height is 33.33 cm. This estimate is too low. Calculating the percentage error: percentage error $= \frac{|36.36 - 33.33|}{36.36} \times 100 \approx 8.3\%$

e. Substituting $x = 84$: $y = 4.92(84) - 35.55 = 377.73$. The projection for average height is 377.73 cm. This estimate is too high. Calculating the percentage error: percentage error $= \frac{|254.50 - 377.73|}{254.50} \times 100 \approx 48.4\%$

Note the percentage error is fairly large for this longer time period.

27. **a.** Plotting the points:

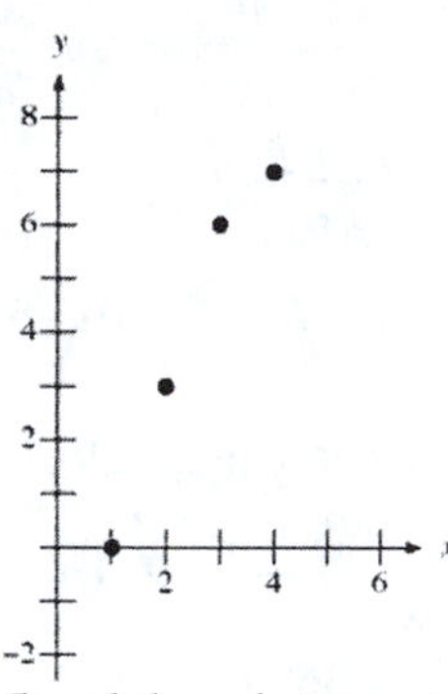

b. The slope appears to be approximately 2.5 and the y-intercept is -2:

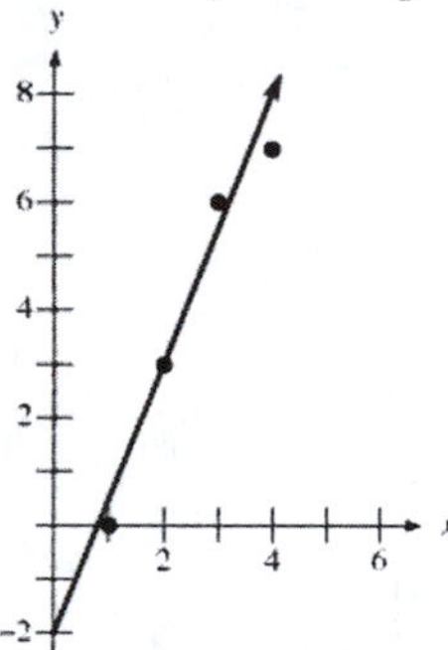

c. Graphing the regression line $y = 2.4x - 2$:

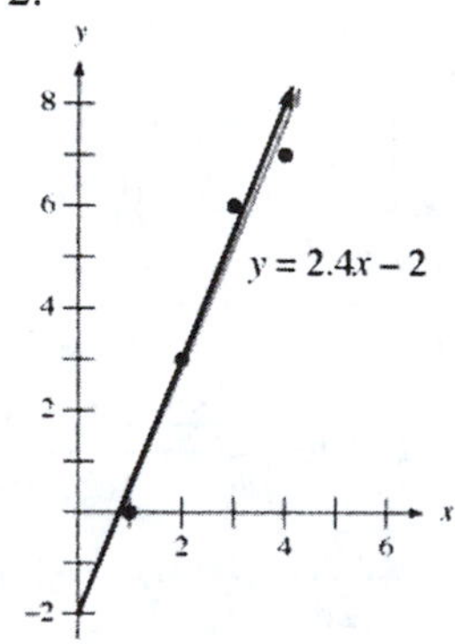

29. **a.** Substituting $x = 2000$ into the regression line: $f(2000) = 37546.068(2000) - 71238863.429 \approx 3{,}853{,}000$
The projected population is 3,853,000 for the year 2000. Calculating the percentage error:

$$\text{percentage error} = \frac{|3{,}823{,}000 - 3{,}853{,}000|}{3{,}823{,}000} \times 100 \approx 0.8\%$$

b. Let $y = f(x)$. Switching the roles of x and y and solving the resulting equation for y:

$$\begin{aligned} x &= 37546.068y - 71238863.429 \\ x + 71238863.429 &= 37546.068y \\ y &= \frac{x + 71238863.429}{37546.068} \end{aligned}$$

So $f^{-1}(x) = \dfrac{x + 71238863.429}{37546.068}$.

c. Substituting $x = 4{,}000{,}000$ into the inverse function: $f^{-1}(4000000) = \dfrac{4000000 + 71238863.429}{37546.068} \approx 2004$
The population of Los Angeles might reach 4 million in the year 2004.

31. **a.** The equation of the regression line is $y = 28.82x - 55628.8$. Graphing the data and the regression line:

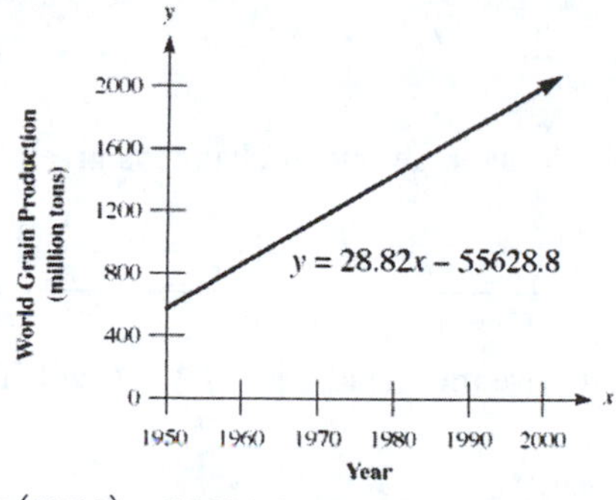

b. Substituting $x = 1993$: $y = 28.82(1993) - 55628.8 \approx 1809$ million tons

Computing the percentage error: $\text{percentage error} = \dfrac{|1714 - 1809|}{1714} \times 100 \approx 5.5\%$

c. Substituting $x = 1998$: $y = 28.82(1998) - 55628.8 \approx 1954$

Computing the percentage error: $\text{percentage error} = \dfrac{|1844 - 1954|}{1844} \times 100 \approx 6.0\%$

33. **a.** Let $x = 1990$: $y = (28632.69)(1990) - 52928780 \approx 4050273$
The population of California in 1990 would be approximately 4,050,273 people.

b. No.

35. **a.** Substituting $x = 1954$: $y = -0.318(1954) + 858.955 = 237.583$
Our estimate is 237.6 minutes, or 3:57.6. This estimate is too low. Calculating the percentage error:

$$\text{percentage error} = \frac{|238 - 237.6|}{238} \times 100 \approx 0.17\%$$

b. Substituting $x = 1911$: $y = -0.318(1911) + 858.955 = 251.257$. Our estimate is 251.3 minutes, or 4:11.3. This estimate is too low. Calculating the percentage error: $\text{percentage error} = \dfrac{|255.4 - 251.3|}{255.4} \times 100 \approx 1.61\%$
This percent error is much larger. That result is to be expected, since the regression line is based on data from 1957-1993. Since 1954 is close in year to this data, the regression line should be a good predictor of the actual value. Since 1911 is much farther away, we would expect the regression line to be a much poorer predictor of the actual value.

37. a. The average rate of change is given by: $\dfrac{f(b)-f(a)}{b-a}=\dfrac{(Ab+B)-(Aa+B)}{b-a}=\dfrac{Ab-Aa}{b-a}=\dfrac{A(b-a)}{b-a}=A$

b. Since $(f\circ f)(x)=f(Ax+B)=A(Ax+B)+B=A^2x+AB+B$, the average rate of change is given by:

$$\frac{(f\circ f)(b)-(f\circ f)(a)}{b-a}=\frac{\left(A^2b+AB+B\right)-\left(A^2a+AB+B\right)}{b-a}=\frac{A^2b-A^2a}{b-a}=\frac{A^2(b-a)}{b-a}=A^2$$

c. Since $(g\circ f)(x)=g(Ax+B)=C(Ax+B)+D=ACx+BC+D$, the average rate of change is given by:

$$\frac{(g\circ f)(b)-(g\circ f)(a)}{b-a}=\frac{(ACb+BC+D)-(ACa+BC+D)}{b-a}=\frac{ACb-ACa}{b-a}=\frac{AC(b-a)}{b-a}=AC$$

d. Since $(f\circ g)(x)=f(Cx+D)=A(Cx+D)+B=ACx+AD+B$, the average rate of change is given by:

$$\frac{(f\circ g)(b)-(f\circ g)(a)}{b-a}=\frac{(ACb+AD+B)-(ACa+AD+B)}{b-a}=\frac{ACb-ACa}{b-a}=\frac{AC(b-a)}{b-a}=AC$$

39. a. Since $(f\circ g)(x)=ACx+AD+B$, we must find $(f\circ g)^{-1}$ by switching the roles of *x* and *y*, then solving for *y*:

$$ACy+AD+B=x$$
$$ACy=x-AD-B$$
$$y=\frac{x}{AC}-\frac{D}{C}-\frac{B}{AC}$$

So $(f\circ g)^{-1}=\dfrac{x}{AC}-\dfrac{D}{C}-\dfrac{B}{AC}$. The average rate of change is given by:

$$\frac{(f\circ g)^{-1}(b)-(f\circ g)^{-1}(a)}{b-a}=\frac{\left(\frac{b}{AC}-\frac{D}{C}-\frac{B}{AC}\right)-\left(\frac{a}{AC}-\frac{D}{C}-\frac{B}{AC}\right)}{b-a}=\frac{\frac{b}{AC}-\frac{a}{AC}}{b-a}=\frac{\frac{1}{AC}(b-a)}{b-a}=\frac{1}{AC}$$

b. Since $(g\circ f)(x)=ACx+BC+D$, we must find $(g\circ f)^{-1}$ by switching the roles of *x* and *y*, then solving for *y*:

$$ACy+BC+D=x$$
$$ACy=x-BC-D$$
$$y=\frac{x}{AC}-\frac{B}{A}-\frac{D}{AC}$$

So $(g\circ f)^{-1}=\dfrac{x}{AC}-\dfrac{B}{A}-\dfrac{D}{AC}$. The average rate of change is given by:

$$\frac{(g\circ f)^{-1}(b)-(g\circ f)^{-1}(a)}{b-a}=\frac{\left(\frac{b}{AC}-\frac{B}{A}-\frac{D}{AC}\right)-\left(\frac{a}{AC}-\frac{B}{A}-\frac{D}{AC}\right)}{b-a}=\frac{\frac{b}{AC}-\frac{a}{AC}}{b-a}=\frac{\frac{1}{AC}(b-a)}{b-a}=\frac{1}{AC}$$

41. a. Using $f(x)=mx$, we have: $f(a+b)=m(a+b)=ma+mb=f(a)+f(b)$

b. Using $f(x)=mx$, we have: $f(ax)=m(ax)=a(mx)=af(x)$

43. First simplify $(f\circ f)(x)$: $(f\circ f)(x)=f(mx+b)=m(mx+b)+b=m^2x+(mb+b)$

Since $(f\circ f)(x)=9x+4$, and since *m* and *b* are constants, we must have the two equations (equating *x*-coefficients and constants): $m^2=9$ and $mb+b=4$. Since $m>0$, $m=3$, so:

$$3b+b=4$$
$$4b=4$$
$$b=1$$

So $f(x)=3x+1$.

45. **a.** Compute the sums:

$$\sum x = 1+2+3+4 = 10 \qquad \sum y = 0+3+6+7 = 16$$

b. Compute the sums:

$$\sum x^2 = 1+4+9+16 = 30 \qquad \sum xy = 0+6+18+28 = 52$$

c. Multiply the first equation by –3:

$$-12b - 30m = -48$$
$$10b + 30m = 52$$

Adding, we obtain:

$$-2b = 4$$
$$b = -2$$

Substituting into the first equation:

$$4b + 10m = 16$$
$$-8 + 10m = 16$$
$$10m = 24$$
$$m = 2.4$$

So the regression line is $y = 2.4x - 2$.

47. First do the computations:

$$\sum x = 1+2+3+4+5 = 15 \qquad \sum y = 2+3+9+9+11 = 34$$
$$\sum x^2 = 1+4+9+16+25 = 55 \qquad \sum xy = 2+6+27+36+55 = 126$$

So the system of equations becomes:

$$5b + 15m = 34$$
$$15b + 55m = 126$$

We multiply the first equation by –3:

$$-15b - 45m = -102$$
$$15b + 55m = 126$$

Adding, we obtain:

$$10m = 24$$
$$m = 2.4$$

Substituting into the first equation:

$$5b + 36 = 34$$
$$5b = -2$$
$$b = -0.4$$

So the regression line is $y = 2.4x - 0.4$.

49. First do the computations:

$$\sum x = 520 + 740 + 560 + 610 + 650 = 3080$$
$$\sum y = 81 + 98 + 83 + 88 + 95 = 445$$
$$\sum x^2 = 270400 + 547600 + 313600 + 372100 + 422500 = 1926200$$
$$\sum xy = 42120 + 72520 + 46480 + 53680 + 61750 = 276550$$

So the system of equations becomes:

$$5b + 3080m = 445$$
$$3080b + 1926200m = 276550$$

Divide the first equation by 5 and the second equation by 10:

$$b + 616m = 89$$
$$308b + 192620m = 27655$$

Multiply the first equation by –308:

$$-308b - 189728m = -27412$$
$$308b + 192620m = 27655$$

Adding, we obtain:

$$2892m = 243$$
$$m \approx 0.084$$

Substituting into the first equation:

$$5b + 258.8 = 445$$
$$5b = 186.2$$
$$b \approx 37.241$$

So the regression line is $y = 0.084x + 37.241$. This is verified using a graphing calculator.

51. **a.** Since f is a linear function, let $f(x) = ax + b$. Therefore: $f(f(x)) = a(ax+b)+b = a^2x+(ab+b)$

Since $f(f(x)) = 2x+1$, we have $a^2 = 2$, thus $a = \pm\sqrt{2}$. Also $ab+b=1$:

$$ab+b=1$$
$$b(a+1)=1$$
$$b=\frac{1}{a+1}$$

If $a=\sqrt{2}$, then $b=\frac{1}{\sqrt{2}+1}=\sqrt{2}-1$ and if $a=-\sqrt{2}$, then $b=\frac{1}{-\sqrt{2}+1}=-\sqrt{2}-1$.

The possible linear functions are $f(x)=\sqrt{2}\,x+\left(-1+\sqrt{2}\right)$ and $f(x)=-\sqrt{2}\,x+\left(-1-\sqrt{2}\right)$.

b. Since f is a linear function, let $f(x)=ax+b$. From part **a** we have $f(f(x))=a^2x+(ab+b)$, so:

$$f(f(f(x)))=f\left(a^2x+(ab+b)\right)=a^3x+\left(a^2b+ab\right)+b=a^3x+\left(a^2b+ab+b\right)$$

Since $f(f(f(x)))=2x+1$, we have $a^3 = 2$, so $a=\sqrt[3]{2}$. Also $a^2b+ab+b=1$, so:

$$a^2b+ab+b=1$$
$$\sqrt[3]{4}\,b+\sqrt[3]{2}\,b+b=1$$
$$b\left(\sqrt[3]{4}+\sqrt[3]{2}+1\right)=1$$
$$b=\frac{1}{1+\sqrt[3]{2}+\sqrt[3]{4}}\bullet\frac{1-\sqrt[3]{2}}{1-\sqrt[3]{2}}=\frac{1-\sqrt[3]{2}}{1-2}=-1+\sqrt[3]{2}$$

The only possible linear function is $f(x)=\sqrt[3]{2}\,x+\left(-1+\sqrt[3]{2}\right)$.

53. If $f^{-1}(x)=f(f(x))$, then $f(f(f(x)))=x$. From Exercise 51b, we know $f(f(f(x)))=a^3x+\left(a^2b+ab+b\right)$. Thus $a^3 = 1$, so $a = 1$. Also $a^2b+ab+b=0$, so:

$$a^2b+ab+b=0$$
$$b+b+b=0$$
$$3b=0$$
$$b=0$$

Thus $f(x)=x$ is the only linear function satisfying the conditions.

4.2 Quadratic Functions

1. a. The linear model is $y = 23.8288x - 113.7508$ and the quadratic model is $y = 0.7499x^2 - 2.4181x + 17.4838$. Sketching the data and models:

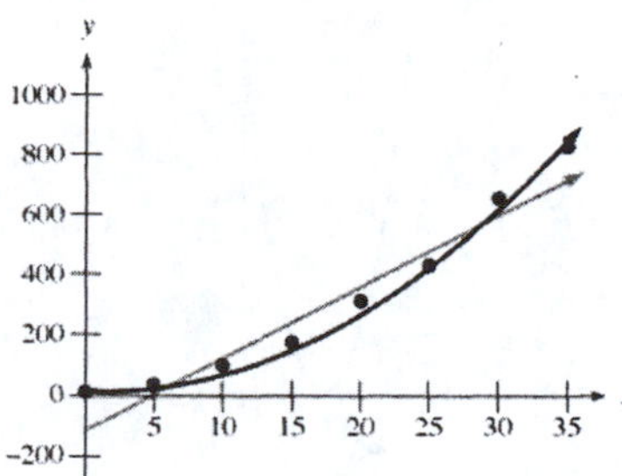

b. For 1990 ($t = 40$), the projections are:

linear: $y = 23.8288(40) - 113.7508 \approx 839.40$ billion

quadratic: $y = 0.7499(40)^2 - 2.4181(40) + 17.4838 \approx 1314.05$ billion

For 1998 ($t = 48$), the projections are:

linear: $y = 23.8288(48) - 113.7508 \approx 1030.03$ billion

quadratic: $y = 0.7499(48)^2 - 2.4181(48) + 17.4838 \approx 1861.32$ billion

c. For 1990, the percentage errors are:

linear: percentage error $= \dfrac{|1145.94 - 839.40|}{1145.94} \times 100 \approx 26.8\%$

quadratic: percentage error $= \dfrac{|1145.94 - 1314.05|}{1145.94} \times 100 \approx 14.7\%$

For 1998, the percentage errors are:

linear: percentage error $= \dfrac{|1585.74 - 1030.03|}{1585.74} \times 100 \approx 35.0\%$

quadratic: percentage error $= \dfrac{|1585.74 - 1861.32|}{1585.74} \times 100 \approx 17.4\%$

In both cases the quadratic model produces the smaller percent error.

3. Completing the table:

	Projected U.S. population (millions)	Actual U.S. population (millions)	Projection too high or too low?	% Error
1970	191.12	203.30	too low	5.99%
2000	258.35	275.60	too low	6.26%

5. The vertex is (–2,0), the axis of symmetry is $x = -2$, the minimum value is 0, the x-intercept is –2, and the y-intercept is 4. Sketching the graph:

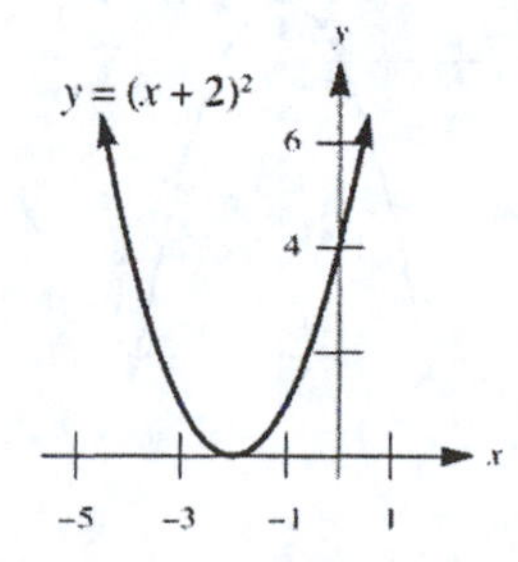

7. The vertex is (–2,0), the axis of symmetry is $x = -2$, the minimum value is 0, the x-intercept is –2, and the y-intercept is 8. Sketching the graph:

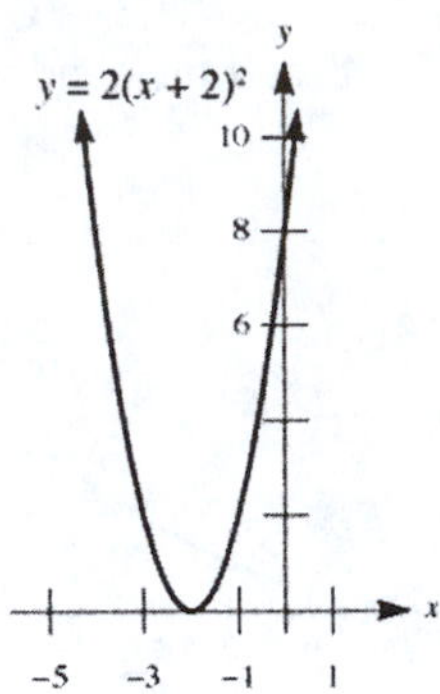

9. The vertex is (–2,4), the axis of symmetry is $x = -2$, the maximum value is 4, the x-intercepts are $-2 \pm \sqrt{2}$, and the y-intercept is –4. Sketching the graph:

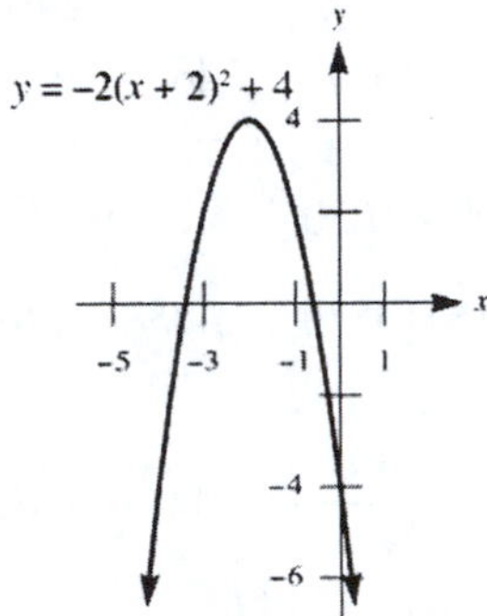

11. Completing the square: $f(x) = x^2 - 4x = \left(x^2 - 4x + 4\right) - 4 = (x-2)^2 - 4$. The vertex is (2,–4), the axis of symmetry is $x = 2$, the minimum value is –4, the x-intercepts are 0 and 4, and the y-intercept is 0. Sketching the graph:

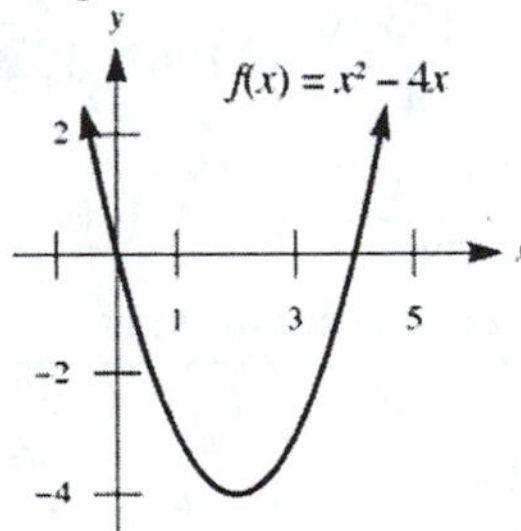

13. The vertex is (0,1), the axis of symmetry is $x = 0$, the maximum value is 1, the x-intercepts are ±1, and the y-intercept is 1. Sketching the graph:

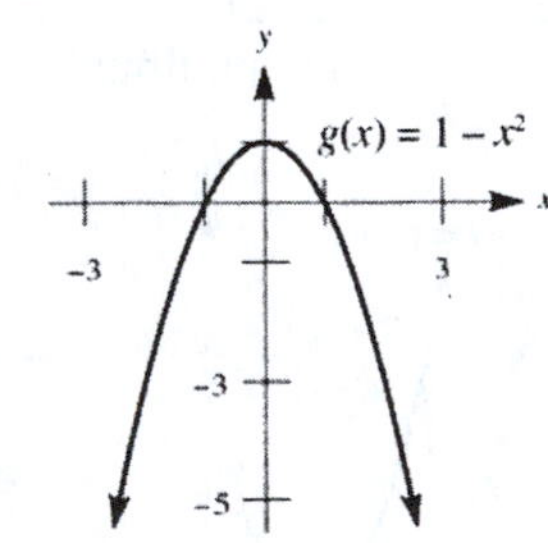

15. Completing the square: $y = x^2 - 2x - 3 = (x^2 - 2x + 1) - 4 = (x-1)^2 - 4$. The vertex is (1,–4), the axis of symmetry is $x = 1$, the minimum value is –4, the x-intercepts are 3 and –1, and the y-intercept is –3. Sketching the graph:

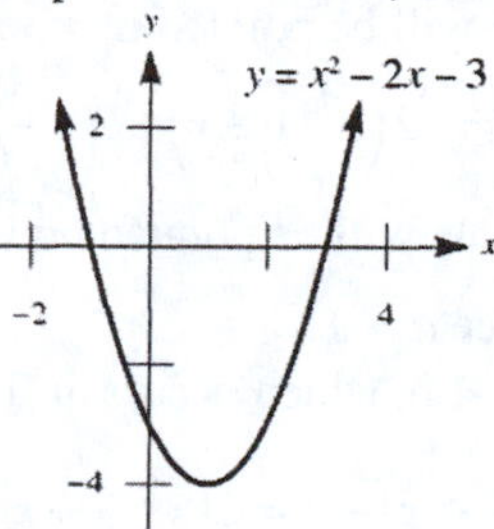

17. Completing the square: $y = -x^2 + 6x + 2 = -(x^2 - 6x) + 2 = -(x^2 - 6x + 9) + 11 = -(x-3)^2 + 11$

The vertex is (3,11), the axis of symmetry is $x = 3$, the maximum value is 11, the x-intercepts are $3 \pm \sqrt{11}$, and the y-intercept is 2. Sketching the graph:

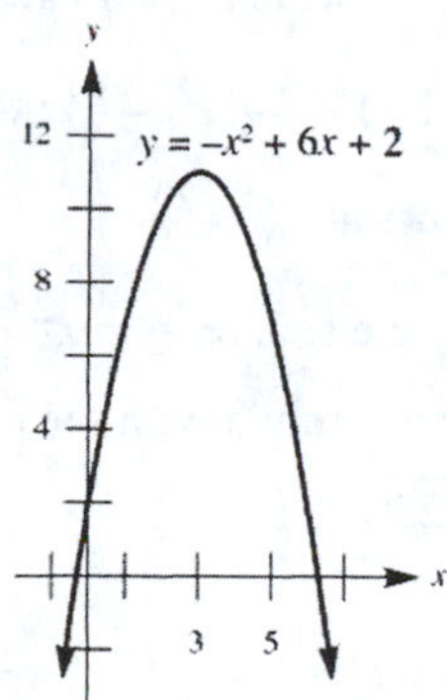

19. Completing the square: $s = -9t^2 + 3t + 2 = -9\left(t^2 - \frac{1}{3}t\right) + 2 = -9\left(t^2 - \frac{1}{3}t + \frac{1}{36}\right) + \frac{9}{4} = -9\left(t - \frac{1}{6}\right)^2 + \frac{9}{4}$

The vertex is $\left(\frac{1}{6}, \frac{9}{4}\right)$, the axis of symmetry is $t = \frac{1}{6}$, the maximum value is $\frac{9}{4}$, the t-intercepts are $-\frac{1}{3}$ and $\frac{2}{3}$, and the s-intercept is 2. Sketching the graph:

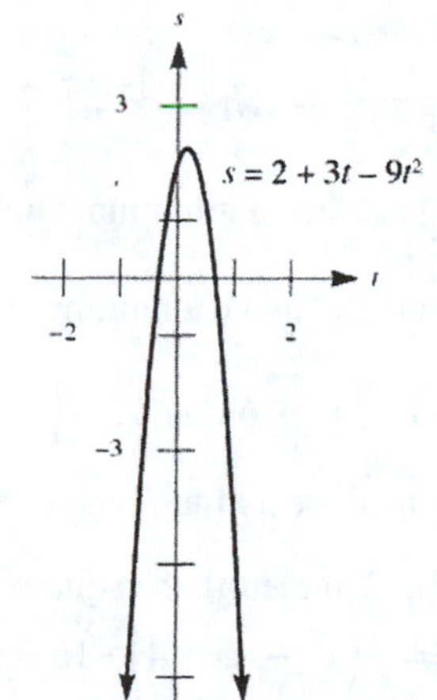

21. Completing the square: $y = 2x^2 - 4x + 11 = 2(x^2 - 2x) + 11 = 2(x^2 - 2x + 1) + 9 = 2(x-1)^2 + 9$

Since the vertex is (1,9) and the parabola will be pointed up, $x = 1$ will yield a minimum output value.

23. Completing the square: $g(x) = -6x^2 + 18x = -6(x^2 - 3x) = -6\left(x^2 - 3x + \frac{9}{4}\right) + \frac{27}{2} = -6\left(x - \frac{3}{2}\right)^2 + \frac{27}{2}$

Since the vertex is $\left(\frac{3}{2}, \frac{27}{2}\right)$ and the parabola will be pointed down, $x = \frac{3}{2}$ will yield a maximum output value.

25. Since the vertex is (0,–10) and the parabola will be pointed up, $x = 0$ will yield a minimum output value.

27. Completing the square: $y = x^2 - 8x + 3 = \left(x^2 - 8x + 16\right) - 13 = (x-4)^2 - 13$

Since the vertex is (4,–13) and the parabola will be pointed up, the function has a minimum value of –13.

29. Completing the square: $y = -2x^2 - 3x + 2 = -2\left(x^2 + \frac{3}{2}x\right) + 2 = -2\left(x^2 + \frac{3}{2}x + \frac{9}{16}\right) + 2 + \frac{9}{8} = -2\left(x + \frac{3}{4}\right)^2 + \frac{25}{8}$

Since the vertex is $\left(-\frac{3}{4}, \frac{25}{8}\right)$ and the parabola will be pointed down, the function has a maximum value of $\frac{25}{8}$.

31. **a.** The function has a minimum value since $a > 0$.

b. The minimum value is approximately –4.0, which occurs at $x_0 \approx 0.2$.

c. Completing the square: $f(x) = 6\left(x^2 - \frac{1}{6}x\right) - 4 = 6\left(x^2 - \frac{1}{6}x + \frac{1}{144}\right) - 4 - \frac{1}{24} = 6\left(x - \frac{1}{12}\right)^2 - \frac{97}{24}$

The minimum value is $-\frac{97}{24}$ which occurs at $x_0 = \frac{1}{12}$.

33. **a.** The function has a maximum value since $a < 0$.

b. The maximum value is approximately –43.4 which occurs at $t_0 \approx 2.2$.

c. Completing the square: $y = -9\left(t^2 - \frac{40}{9}t\right) + 1 = -9\left(t^2 - \frac{40}{9}t + \frac{400}{81}\right) + 1 - \frac{400}{9} = -9\left(t - \frac{20}{9}\right)^2 - \frac{391}{9}$

The maximum value is $-\frac{391}{9}$ which occurs at $t_0 = \frac{20}{9}$.

35. Find the vertex of the parabola by completing the square: $y = x^2 - 6x + 13 = \left(x^2 - 6x + 9\right) + 4 = (x-3)^2 + 4$

So the vertex is (3,4). We now use the distance formula with the points (0,0) and (3,4):

$$d = \sqrt{(3-0)^2 + (4-0)^2} = \sqrt{9+16} = \sqrt{25} = 5$$

So the vertex is 5 units from the origin.

37. Compute $(f \circ g)(x)$: $(f \circ g)(x) = 2\left(x^2 + 4x + 1\right) - 3 = 2x^2 + 8x + 2 - 3 = 2x^2 + 8x - 1$

So $f \circ g$ is a quadratic function.

39. Compute $(g \circ h)(x)$: $(g \circ h)(x) = \left(1 - 2x^2\right)^2 + 4\left(1 - 2x^2\right) + 1 = 1 - 4x^2 + 4x^4 + 4 - 8x^2 + 1 = 4x^4 - 12x^2 + 6$

So $g \circ h$ is neither linear nor quadratic.

41. Compute $(f \circ f)(x)$: $(f \circ f)(x) = 2(2x-3) - 3 = 4x - 6 - 3 = 4x - 9$. So $f \circ f$ is a linear function.

43. **a.** First complete the square on $x^2 - 6x + 73$: $x^2 - 6x + 73 = \left(x^2 - 6x + 9\right) + 73 - 9 = (x-3)^2 + 64$

So $f(x) = \sqrt{(x-3)^2 + 64}$. This would achieve a minimum value at $\left(3, \sqrt{64}\right) = (3,8)$.

b. Here $g(x) = \sqrt[3]{(x-3)^2 + 64}$, which would achieve a minimum value at $\left(3, \sqrt[3]{64}\right) = (3,4)$.

c. Completing the square on $x^4 - 6x^2 + 73$: $x^4 - 6x^2 + 73 = \left(x^4 - 6x^2 + 9\right) + 73 - 9 = \left(x^2 - 3\right)^2 + 64$

So $h(x) = (x^2 - 3)^2 + 64$, which would achieve a minimum value at $\left(\pm\sqrt{3}, 64\right)$.

45. **a.** The maximum value occurs at (2.0,4.0). Completing the square on $-x^2 + 4x + 12$:

$$-x^2 + 4x + 12 = -\left(x^2 - 4x\right) + 12 = -\left(x^2 - 4x + 4\right) + 16 = -(x-2)^2 + 16$$

So $f(x) = \sqrt{-(x-2)^2 + 16}$, which has a maximum value at $\left(2, \sqrt{16}\right) = (2,4)$.

b. The maximum value occurs at (2.0,2.5). Now $g(x) = \sqrt[3]{-(x-2)^2 + 16}$, which has a maximum value at $\left(2, \sqrt[3]{16}\right) = \left(2, 2\sqrt[3]{2}\right)$.

c. The maximum value occurs at (±1.4,16.0). Here $h(x) = -\left(x^2 - 2\right)^2 + 16$, which has a maximum value at $\left(\pm\sqrt{2}, 16\right)$.

47. The first differences are –22, –7, 11, 26, so the function is not linear. The second differences are 15, 18, 15, so the function is not quadratic. Therefore it is neither.

49. The first differences are 18, 10, 2, –6, so the function is not linear. The second differences are –8, –8, –8, so the function is quadratic.

51. Graphing all four functions:

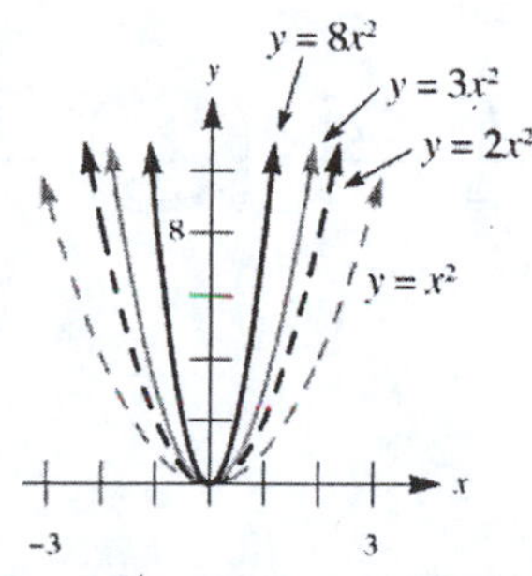

As the coefficient of x^2 increases from 1 to 8, the graph "narrows" and produces larger y-coordinates. The graph of $y = 50x^2$ should be "narrower" than the others:

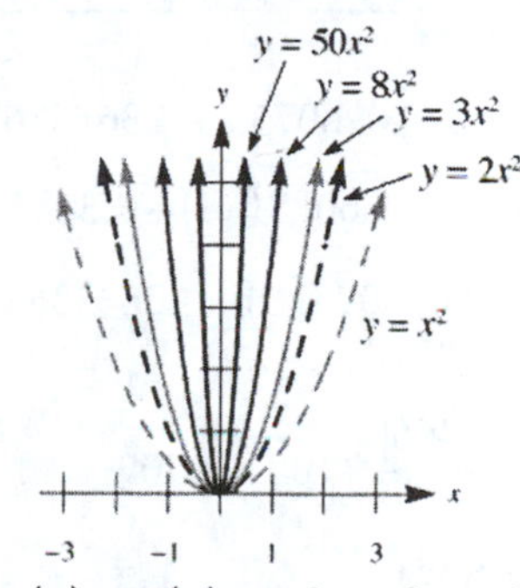

53. The average rate of change is given by: $\dfrac{f(x)-f(a)}{x-a} = \dfrac{x^2-a^2}{x-a} = \dfrac{x^2-a^2}{x-a} = \dfrac{(x+a)(x-a)}{x-a} = x+a$

55. Since a and b will be the x-intercepts of the parabola, and all points with the same y-coordinate will be symmetric about the axis of symmetry, it follows that the midpoint $\left(\dfrac{a+b}{2}, 0\right)$ must lie on the axis of symmetry. Thus the x-coordinate of the vertex (which also lies on this axis) is $\dfrac{a+b}{2}$.

57. Since the vertex is (2,2), its equation will be $y = A(x-2)^2 + 2$. Since (0,0) must lie on this parabola, substitute the points to find A:

$$\begin{aligned} y &= A(x-2)^2 + 2 \\ 0 &= A(0-2)^2 + 2 \\ 0 &= 4A + 2 \\ -2 &= 4A \\ A &= -\tfrac{1}{2} \end{aligned}$$

So the parabola is $y = -\frac{1}{2}(x-2)^2 + 2$.

59. Since the vertex is (3,–1), its equation will be $y = A(x-3)^2 - 1$. Since (1,0) must lie on this parabola, substitute the points to find A:

$$\begin{aligned} y &= A(x-3)^2 - 1 \\ 0 &= A(1-3)^2 - 1 \\ 0 &= 4A - 1 \\ 1 &= 4A \\ A &= \tfrac{1}{4} \end{aligned}$$

So the parabola is $y = \frac{1}{4}(x-3)^2 - 1$.

61. Completing the square: $g(x) = \left(x^2 + bx + \frac{b^2}{4}\right) - \frac{b^2}{4} = \left(x + \frac{b}{2}\right)^2 - \frac{b^2}{4}$

For the minimum value to equal –1, we must have:

$$\begin{aligned} -\frac{b^2}{4} &= -1 \\ b^2 &= 4 \\ b &= \pm 2 \end{aligned}$$

63. **a.** Completing the table:

x	a	$a+h$	$a+2h$
$f(x)$	$ma+b$	$ma+mh+b$	$ma+2mh+b$

b. All of the first differences are mh.

65. **a.** The quadratic model is $y = 0.015089x^2 - 0.086075x + 3.364536$. Computing the estimates:

1989 ($x = -1$): $y = 0.015089(-1)^2 - 0.086075(-1) + 3.364536 \approx 3.466$ billion tons

1996 ($x = 6$): $y = 0.015089(6)^2 - 0.086075(6) + 3.364536 \approx 3.391$ billion tons

b. Computing the percentage errors:

1989: percentage error $= \dfrac{|3.408 - 3.466|}{3.408} \times 100 \approx 1.70\%$

1996: percentage error $= \dfrac{|3.428 - 3.391|}{3.428} \times 100 \approx 1.08\%$

c. In 1998 ($x = 8$), the projection is: $y = 0.015089(8)^2 - 0.086075(8) + 3.364536 \approx 3.642$ billion tons

Computing the percentage error: percentage error $= \dfrac{|3.329 - 3.642|}{3.329} \times 100 \approx 9.40\%$

67. Completing the square:

$$\begin{aligned} y &= ax^2 + bx + c \\ &= a\left(x^2 + \frac{b}{a}x\right) + c \\ &= a\left(x^2 + \frac{b}{a}x + \frac{b^2}{4a^2}\right) + c - \frac{ab^2}{4a^2} \\ &= a\left(x + \frac{b}{2a}\right)^2 + \frac{4ac - b^2}{4a} \\ &= a\left(x + \frac{b}{2a}\right)^2 - \frac{D}{4a}, \text{ where } D = b^2 - 4ac \end{aligned}$$

So the vertex is $\left(-\frac{b}{2a}, -\frac{D}{4a}\right)$, where $D = b^2 - 4ac$.

69. **a.** First complete the square to find the vertex: $y = p\left(x^2 + x\right) + r = p\left(x^2 + x + \frac{1}{4}\right) + r - \frac{p}{4} = p\left(x + \frac{1}{2}\right)^2 + \frac{4r - p}{4}$

If the vertex lies on the x-axis, then $\frac{4r - p}{4} = 0$, so $4r - p = 0$, thus $p = 4r$.

b. If $p = 4r$, then $y = 4rx^2 + 4rx + r$. Completing the square:

$$y = 4r\left(x^2 + x\right) + r = 4r\left(x^2 + x + \frac{1}{4}\right) + r - r = 4r\left(x + \frac{1}{2}\right)^2$$

But then the vertex is $\left(-\frac{1}{2}, 0\right)$, and thus it lies on the x-axis.

4.3 Using Iteration to Model Population Growth (Optional Section)

1. Finding where $f(x) = x$:

$$\begin{aligned} -4x + 5 &= x \\ 5 &= 5x \\ x &= 1 \end{aligned}$$

The fixed point is $x = 1$.

3. Finding where $G(x) = x$:

$$\begin{aligned} \tfrac{1}{2} + x &= x \\ \tfrac{1}{2} &= 0 \end{aligned}$$

Since this last equation is false, there are no fixed points.

5. Finding where $h(x) = x$:

$$\begin{aligned} x^2 - 3x - 5 &= x \\ x^2 - 4x - 5 &= 0 \\ (x - 5)(x + 1) &= 0 \\ x &= -1, 5 \end{aligned}$$

The fixed points are $x = -1$ and $x = 5$.

7. Finding where $f(t) = t$:

$$\begin{aligned} t^2 - t + 1 &= t \\ t^2 - 2t + 1 &= 0 \\ (t - 1)^2 &= 0 \\ t &= 1 \end{aligned}$$

The fixed point is $t = 1$.

9. Finding where $k(t) = t$:

$$\begin{aligned} t^2 - 12 &= t \\ t^2 - t - 12 &= 0 \\ (t - 4)(t + 3) &= 0 \\ t &= -3, 4 \end{aligned}$$

The fixed points are $t = -3$ and $t = 4$.

11. Finding where $T(x) = x$:

$$\begin{aligned} 1.8x(1 - x) &= x \\ 1.8x - 1.8x^2 &= x \\ 0.8x - 1.8x^2 &= 0 \\ 0.2x(4 - 9x) &= 0 \\ x &= 0, \tfrac{4}{9} \end{aligned}$$

The fixed points are $x = 0$ and $x = \frac{4}{9}$.

13. Finding where $g(u) = u$:

$$\begin{aligned} 2u^2 + 3u - 4 &= u \\ 2u^2 + 2u - 4 &= 0 \\ 2\left(u^2 + u - 2\right) &= 0 \\ 2(u + 2)(u - 1) &= 0 \\ u &= -2, 1 \end{aligned}$$

The fixed points are $u = -2$ and $u = 1$.

15. Finding where $f(x) = x$:

$$\begin{aligned} 7+\sqrt{x-1} &= x \\ \sqrt{x-1} &= x-7 \\ x-1 &= x^2 - 14x + 49 \\ 0 &= x^2 - 15x + 50 \\ 0 &= (x-5)(x-10) \\ x &= 5, 10 \end{aligned}$$

Note that $x = 5$ does not check in the original equation. The fixed point is $x = 10$.

17. **a.** There is one fixed point for the function:

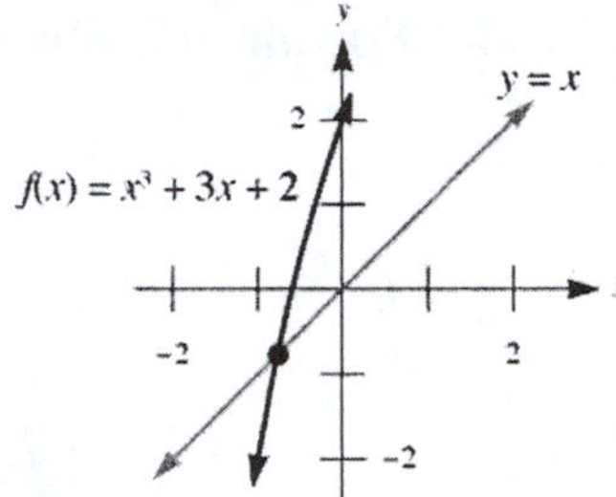

b. The fixed point is approximately –0.771.

19. **a.** There are three fixed points for the function:

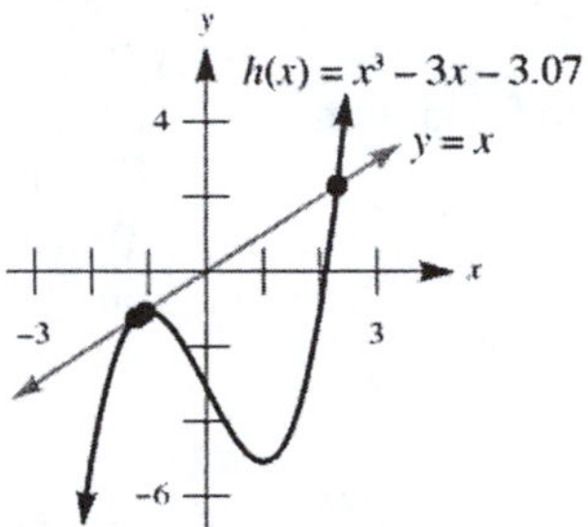

b. The fixed points are approximately –1.206, –1.103, and 2.309.

21. **a.** There are two fixed points for the function:

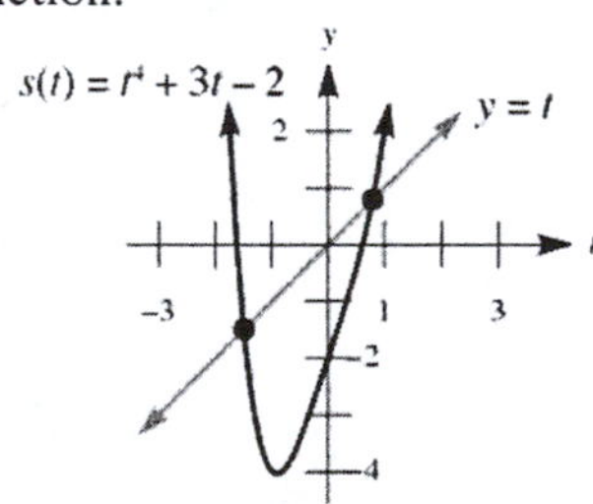

b. The fixed points are approximately –1.495 and 0.798.

23. **a.** Finding where $g(x) = x$:

$$\begin{aligned} x^2 - 0.5 &= x \\ 2x^2 - 1 &= 2x \\ 2x^2 - 2x - 1 &= 0 \end{aligned}$$

Using the quadratic formula: $x = \dfrac{-(-2) \pm \sqrt{(-2)^2 - 4(2)(-1)}}{2(2)} = \dfrac{2 \pm \sqrt{12}}{4} = \dfrac{2 \pm 2\sqrt{3}}{4} = \dfrac{1 \pm \sqrt{3}}{2}$

Thus one of the fixed points is $x = \dfrac{1-\sqrt{3}}{2} \approx -0.366$.

b. Calculating the iterates of $x_0 = -1$:

$x_1 = -0.49$ $\quad$ $x_2 = -0.2599$

$x_3 \approx -0.4325$ $\quad$ $x_4 \approx -0.3130$

The fourth iterate is the first iterate to have the digit 3 in the first decimal place.

c. Continuing the iterates:

$x_5 \approx -0.4020$ $\quad$ $x_6 \approx -0.3384$ $\quad$ $x_7 \approx -0.3855$

$x_8 \approx -0.3514$ $\quad$ $x_9 \approx -0.3765$ $\quad$ $x_{10} \approx -0.3582$

$x_{11} \approx -0.3717$ $\quad$ $x_{12} \approx -0.3619$

The twelfth iterate is the first iterate to have the digit 6 in the second decimal place.

25. **a.** Completing the table:

	x_1	x_2	x_3	x_4	x_5	x_6	x_7	x_8
From graph	1.7	0.8	1.4	1.0	1.3	1.1	1.2	1.1
From calculator	1.72	0.796	1.443	0.990	1.307	1.085	1.240	1.132

b. Finding where $f(x) = x$:

$$-0.7x + 2 = x$$
$$2 = 1.7x$$
$$x = \frac{2}{1.7} = \frac{20}{17}$$

The fixed point is $x = \frac{20}{17} \approx 1.176$.

c. The eighth iterate has the same digit in the first decimal place.

27. Completing the table:

	x_1	x_2	x_3	x_4	x_5	x_6	x_7	x_8	x_9
From graph	0.36	0.92	0.29	0.82	0.58	0.97	0.11	0.40	0.96
From calculator	0.36	0.922	0.289	0.822	0.585	0.971	0.113	0.402	0.962

29. **a.** Since $x_{20} \approx 0.64594182$ and $f(x) = 2.9x(1-x)$, we can compute the required iterates:

$x_{21} \approx 0.6632$ $\quad$ $x_{22} \approx 0.6477$ $\quad$ $x_{23} \approx 0.6617$

$x_{24} \approx 0.6492$ $\quad$ $x_{25} \approx 0.6605$

b. Multiplying each of these iterates by 500 results in the population of catfish in the pond:

n	20	21	22	23	24	25
Number of fish after n breeding seasons	323	332	324	331	325	330

Note that although the population is oscillating, the changes (sizes) of the oscillations are decreasing.

c. Using $f(x) = 0.75x(1-x)$ and $x_0 = 0.1$, compute the iterates:

$x_1 = 0.0675$ $\quad$ $x_2 \approx 0.04721$ $\quad$ $x_3 \approx 0.03373$

$x_4 \approx 0.02445$ $\quad$ $x_5 \approx 0.01789$

The iterates appear to be approaching 0. Now calculate the fixed point of the function:

$$f(x) = x$$
$$0.75x(1-x) = x$$
$$0.75x - 0.75x^2 = x$$
$$-0.25x - 0.75x^2 = 0$$
$$-0.25x(1+3x) = 0$$
$$x = 0, -\tfrac{1}{3}$$

Since our iterates seem to be approaching 0, they are approaching a fixed point of the function. Since this fixed point is 0, eventually the population will decrease to 0.

31. **a.** Using $f(x)=3.1x(1-x)$ and $x_0=0.1$, complete the tables:

n	0	1	2	3	4	5
x_n	0.1	0.279	0.6236	0.7276	0.6143	0.7345
Number of fish after n breeding seasons	50	140	312	364	307	367

n	6	7	8	9	10
x_n	0.6046	0.7411	0.5948	0.7471	0.5857
Number of fish after n breeding seasons	302	371	297	374	293

Using $x_{20}\approx 0.56140323$, complete the table:

n	21	22	23	24	25	26
x_n	0.7633	0.5601	0.7638	0.5592	0.7641	0.5587
Number of fish after n breeding seasons	382	280	382	280	382	279

b. Let $f(x)=x$:

$$\begin{aligned} 3.1x(1-x)&=x\\ 3.1x-3.1x^2&=x\\ 2.1x-3.1x^2&=0\\ 0.1x(21-31x)&=0\\ x&=0,\ \tfrac{21}{31}\end{aligned}$$

The nonzero fixed point is $x=\frac{21}{31}\approx 0.6774$. The iterates are not approaching this fixed point.

c. Evaluating these expressions when $k=3.1$:

$$a=\frac{1+3.1+\sqrt{(3.1-3)(3.1+1)}}{2(3.1)}=\frac{4.1+\sqrt{(0.1)(4.1)}}{6.2}\approx 0.7646$$

$$b=\frac{1+3.1-\sqrt{(3.1-3)(3.1+1)}}{2(3.1)}=\frac{4.1-\sqrt{(0.1)(4.1)}}{6.2}\approx 0.5580$$

These iterates are consistent with the table from part **a**.

d. These correspond to the populations of 382 and 279 fish.

33. **a.** The first six iterates of $x_0=c$ are:

$$x_1=g(c)=d \qquad x_2=g(x_1)=g(d)=c \qquad x_3=g(x_2)=g(c)=d$$
$$x_4=g(x_3)=g(d)=c \qquad x_5=g(x_4)=g(c)=d \qquad x_6=g(x_5)=g(d)=c$$

Similarly, the first six iterates of $x_0=d$ are:

$$x_1=g(d)=c \qquad x_2=g(x_1)=g(c)=d \qquad x_3=g(x_2)=g(d)=c$$
$$x_4=g(x_3)=g(c)=d \qquad x_5=g(x_4)=g(d)=c \qquad x_6=g(x_5)=g(c)=d$$

With an initial input of either c or d, the sequence of iterates alternates between c and d. This is what is meant by $\{c,d\}$ being a 2-cycle for the function g.

b. The first six iterates of 0.4 are 0.8, 0.4, 0.8, 0.4, 0.8, and 0.4.

c. The work in part **b** shows that $\{0.4, 0.8\}$ is a 2-cycle for the function T.

d. Computing the first two iterates:

$$x_1=T(0.4)=1-|0.8-1|=1-0.2=0.8 \qquad x_2=T(0.8)=1-|1.6-1|=1-0.6=0.4$$

These answers are consistent with the work in part **b**.

35. **a.** Subtracting the two equations:

$$4a(1-a)-4b(1-b)=b-a$$
$$4a-4a^2-4b+4b^2=b-a$$
$$4b^2-4a^2+4a-4b=b-a$$
$$4(b+a)(b-a)-4(b-a)=b-a$$
$$4(b-a)(b+a-1)=b-a$$

b. Since $a \neq b$, we can divide by $b-a$ to obtain:

$$4(b+a-1)=1$$
$$b+a-1=\tfrac{1}{4}$$
$$b+a=\tfrac{5}{4}$$
$$b=\tfrac{5}{4}-a$$

c. Substituting $b=\frac{5}{4}-a$:

$$4a(1-a)=\tfrac{5}{4}-a$$
$$16a-16a^2=5-4a$$
$$0=16a^2-20a+5$$

d. Using the quadratic formula: $a=\dfrac{-(-20)\pm\sqrt{(-20)^2-4(16)(5)}}{2(16)}=\dfrac{20\pm\sqrt{80}}{32}=\dfrac{20\pm4\sqrt{5}}{32}=\dfrac{5\pm\sqrt{5}}{8}$

e. Choosing the positive root, we have $a=\frac{1}{8}\left(5+\sqrt{5}\right)$. Substituting to find b:

$$b=\tfrac{5}{4}-\tfrac{1}{8}\left(5+\sqrt{5}\right)=\tfrac{5}{4}-\tfrac{5}{8}-\tfrac{1}{8}\sqrt{5}=\tfrac{5}{8}-\tfrac{1}{8}\sqrt{5}=\tfrac{1}{8}\left(5-\sqrt{5}\right)$$

Now checking these values:

$$\begin{aligned}f\left[\tfrac{1}{8}\left(5+\sqrt{5}\right)\right]&=4\bullet\tfrac{1}{8}\left(5+\sqrt{5}\right)\left[1-\tfrac{1}{8}\left(5+\sqrt{5}\right)\right]\\&=\tfrac{1}{2}\left(5+\sqrt{5}\right)\left(1-\tfrac{5}{8}-\tfrac{1}{8}\sqrt{5}\right)\\&=\tfrac{1}{2}\left(5+\sqrt{5}\right)\left(\tfrac{3}{8}-\tfrac{1}{8}\sqrt{5}\right)\\&=\tfrac{1}{16}\left(5+\sqrt{5}\right)\left(3-\sqrt{5}\right)\\&=\tfrac{1}{16}\left(10-2\sqrt{5}\right)\\&=\tfrac{1}{8}\left(5-\sqrt{5}\right)\end{aligned}$$

$$\begin{aligned}f\left[\tfrac{1}{8}\left(5-\sqrt{5}\right)\right]&=4\bullet\tfrac{1}{8}\left(5-\sqrt{5}\right)\left[1-\tfrac{1}{8}\left(5-\sqrt{5}\right)\right]\\&=\tfrac{1}{2}\left(5-\sqrt{5}\right)\left(1-\tfrac{5}{8}+\tfrac{1}{8}\sqrt{5}\right)\\&=\tfrac{1}{2}\left(5-\sqrt{5}\right)\left(\tfrac{3}{8}+\tfrac{1}{8}\sqrt{5}\right)\\&=\tfrac{1}{16}\left(5-\sqrt{5}\right)\left(3+\sqrt{5}\right)\\&=\tfrac{1}{16}\left(10+2\sqrt{5}\right)\\&=\tfrac{1}{8}\left(5+\sqrt{5}\right)\end{aligned}$$

Therefore $f(a)=b$ and $f(b)=a$.

37. **a.** Substituting $k=3.5$ yields:

$$a=\frac{1+3.5+\sqrt{(3.5+1)(3.5-3)}}{2(3.5)}=\frac{4.5+\sqrt{(4.5)(0.5)}}{7}=\frac{\frac{9}{2}+\sqrt{\frac{9}{4}}}{7}=\frac{\frac{9}{2}+\frac{3}{2}}{7}=\frac{6}{7}$$

$$b=\frac{\frac{9}{2}-\frac{3}{2}}{7}=\frac{3}{7}$$

b. The coordinates are $P\left(\frac{3}{7},\frac{3}{7}\right)$, $Q\left(\frac{3}{7},\frac{6}{7}\right)$, $R\left(\frac{6}{7},\frac{6}{7}\right)$, and $S\left(\frac{6}{7},\frac{3}{7}\right)$.

4.4 Setting Up Equations That Define Functions

1. **a.** Call P the perimeter and A the area. We are asked to come up with a formula for A in terms of x. Since $P = 16$ and $P = 2x + 2l$:

$$2x+2l=16$$
$$2l=16-2x$$
$$l=8-x$$

Now $A = xl = x(8-x) = 8x - x^2$. Using functional notation we have $A(x) = 8x - x^2$. The domain is $(0,8)$.

b. Since $A = 85$ and $A = xl$:

$$xl=85$$
$$l=\frac{85}{x}$$

Now $P=2x+2l=2x+2\left(\frac{85}{x}\right)=2x+\frac{170}{x}$. Using functional notation we have $P(x)=2x+\frac{170}{x}$.

The domain is $(0,\infty)$.

3. **a.** First draw the figure:

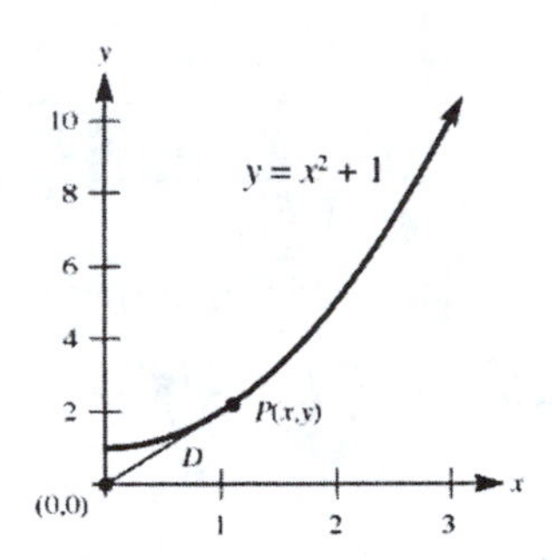

Call D the distance from $P(x, y)$ to the origin. We are asked to come up with a formula for D in terms of x. By the distance formula: $D=\sqrt{(x-0)^2+(y-0)^2}=\sqrt{x^2+y^2}$. Since $P(x, y)$ lies on the curve, $y = x^2 + 1$. Substituting this for y in our equation for D: $D=\sqrt{x^2+y^2}=\sqrt{x^2+\left(x^2+1\right)^2}=\sqrt{x^2+x^4+2x^2+1}=\sqrt{x^4+3x^2+1}$

Using functional notation we have $D(x)=\sqrt{x^4+3x^2+1}$. The domain is $(-\infty,\infty)$.

b. Let m denote the slope of the line segment from the origin to $P(x, y)$. Then: $m=\frac{y-0}{x-0}=\frac{y}{x}$. Substituting $y=x^2+1$ in for y in this equation: $m=\frac{y}{x}=\frac{x^2+1}{x}$. Using functional notation we have $m(x)=\frac{x^2+1}{x}$.

The domain is $(-\infty,0)\cup(0,\infty)$.

5. **a.** First draw the figure:

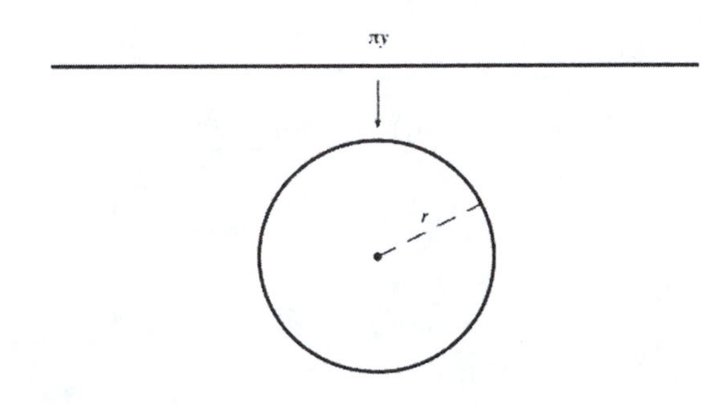

Call A the area of the circle and r its radius. We are asked to come up with a formula for A in terms of y. Since πy is the circumference C of the circle, and $C = 2\pi r$:

$$2\pi r = \pi y$$
$$r = \frac{y}{2}$$

Now $A = \pi r^2 = \pi\left(\frac{y}{2}\right)^2 = \frac{\pi y^2}{4}$. Using functional notation we have $A(y) = \frac{\pi y^2}{4}$. The domain is $(0, \infty)$.

b. Draw the figure:

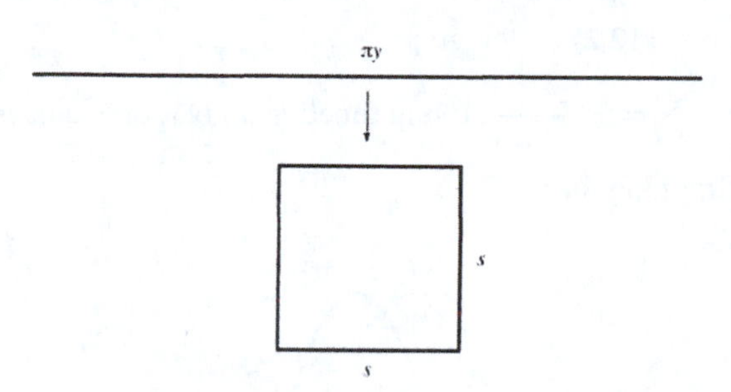

Let A denote the area of the square, and s the length of the side. Since the perimeter P is πy and $P = 4s$:

$$4s = \pi y$$
$$s = \frac{\pi y}{4}$$

Now $A = s^2 = \left(\frac{\pi y}{4}\right)^2 = \frac{\pi^2 y^2}{16}$. Using functional notation we have $A(y) = \frac{\pi^2 y^2}{16}$. The domain is $(0, \infty)$.

7. **a.** Let the two numbers be x and $16 - x$. Then the product P would be: $P = x(16 - x) = 16x - x^2$

Using functional notation we have $P(x) = 16x - x^2$. The domain is $(-\infty, \infty)$.

b. Since the two numbers are x and $16 - x$, the sum of squares S would be:

$$S = (x)^2 + (16 - x)^2 = x^2 + 256 - 32x + x^2 = 2x^2 - 32x + 256$$

Using functional notation we have $S(x) = 2x^2 - 32x + 256$. The domain is $(-\infty, \infty)$.

c. There are two ways to set this up. Since the two numbers are x and $16 - x$, the difference of the cubes D could be:

$$D = x^3 - (16 - x)^3 \text{ or } D = (16 - x)^3 - x^3$$

Using functional notation we have $D(x) = x^3 - (16 - x)^3$ or $D(x) = (16 - x)^3 - x^3$. The domain is $(-\infty, \infty)$.

d. Let A denote the average of the two numbers. Since the two numbers are x and $16 - x$: $A = \frac{x + 16 - x}{2} = \frac{16}{2} = 8$

So $A(x) = 8$. Notice that the average does not depend on what the two numbers are.

9. **a.** Graph $p = 5 - \frac{x}{4} = -\frac{1}{4}x + 5$ (note the domain is $[0, 20]$.

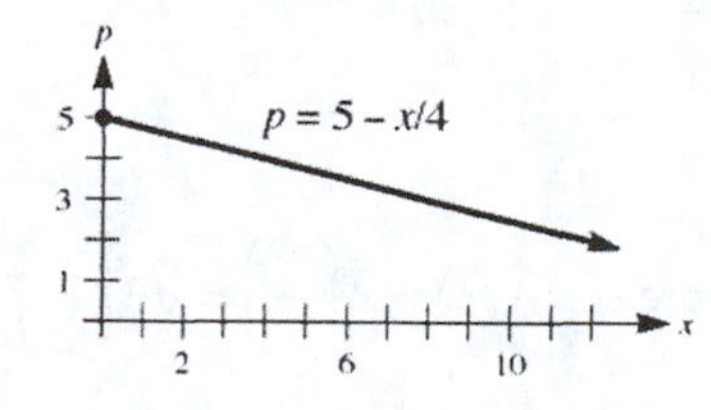

b. When $p = 3$:

$$3 = 5 - \frac{x}{4}$$
$$-2 = -\frac{x}{4}$$
$$x = 8$$

Thus 8 units can be sold. This corresponds to the point (8,3) on the graph.

c. When $x = 12$, we have $p = 5 - \frac{12}{4} = 5 - 3 = 2$. The unit price should be set at \$2.

This corresponds to the point (12,2) on the graph.

d. Substitute: $R = xp = x\left(5 - \frac{x}{4}\right) = 5x - \frac{x^2}{4}$. Using functional notation we have $R(x) = 5x - \frac{x^2}{4}$.

The domain is $[0, 20]$. Graphing $R(x)$:

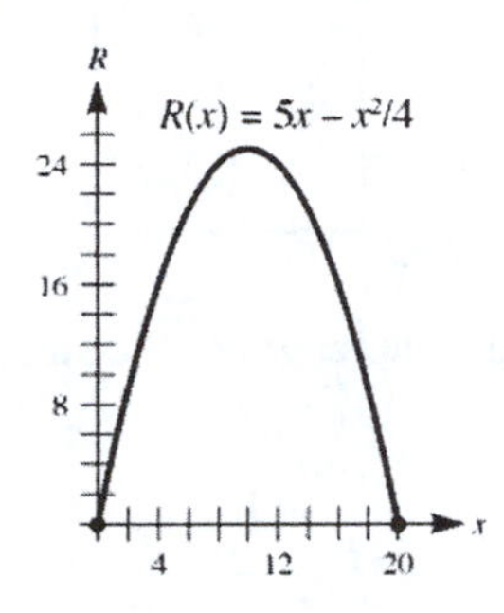

e. Compute the revenue values:

$$R(2) = 5(2) - \frac{2^2}{4} = 10 - 1 = \$9 \qquad R(8) = 5(8) = \frac{8^2}{4} = 40 - 16 = \$24$$

$$R(14) = 5(14) - \frac{14^2}{4} = 70 - 49 = \$21$$

f. We know that $x = 10$ (10 units sold) will generate the largest revenue. The revenue when $x = 10$ is \$25. The corresponding unit price is $p = 5 - \frac{10}{4} = 5 - 2.5 = \$2.50$.

11. **a.** Complete the table:

x	1	2	3	4	5	6	7
$P(x)$	17.88	19.49	20.83	21.86	22.49	22.58	21.75

b. The largest value for $P(x)$ is 22.58, corresponding to $x = 6$.

c. Compute $P\left(4\sqrt{2}\right)$:

$$P\left(4\sqrt{2}\right) = 2\left(4\sqrt{2}\right) + 2\sqrt{64 - \left(4\sqrt{2}\right)^2} = 8\sqrt{2} + 2\sqrt{64 - 32} = 8\sqrt{2} + 2\sqrt{32} = 8\sqrt{2} + 8\sqrt{2} = 16\sqrt{2} \approx 22.63$$

This is indeed larger than any of our table values.

13. If x is the length of a side of the triangle, then:

$$2s = x$$
$$s = \frac{x}{2}$$

Using our function from the previous exercise: $A(s) = A\left(\frac{x}{2}\right) = \sqrt{3}\left(\frac{x}{2}\right)^2 = \frac{\sqrt{3}}{4}x^2$. The domain is $[0, \infty)$.

15. Since $V = 2\pi r^3$, solve for r:

$$2\pi r^3 = V$$
$$r^3 = \frac{V}{2\pi}$$
$$r = \sqrt[3]{\frac{V}{2\pi}}$$

Using functional notation we have $r(V) = \sqrt[3]{\frac{V}{2\pi}}$. The domain is $[0, \infty)$.

17. Solve $S = 4\pi r^2$ for r:

$$4\pi r^2 = S$$
$$r^2 = \frac{S}{4\pi}$$
$$r = \sqrt{\frac{S}{4\pi}}$$

Now since $V = \frac{4}{3}\pi r^3$: $V = \frac{4}{3}\pi\left(\sqrt{\frac{S}{4\pi}}\right)^3 = \frac{4\pi S\sqrt{s}}{3(4\pi)\sqrt{4\pi}} = \frac{S\sqrt{S}}{3\sqrt{4\pi}} = \frac{S\sqrt{S\pi}}{6\pi}$

Using functional notation we have $V(S) = \frac{S\sqrt{S\pi}}{6\pi}$. The domain is $[0, \infty)$.

19. Draw a figure:

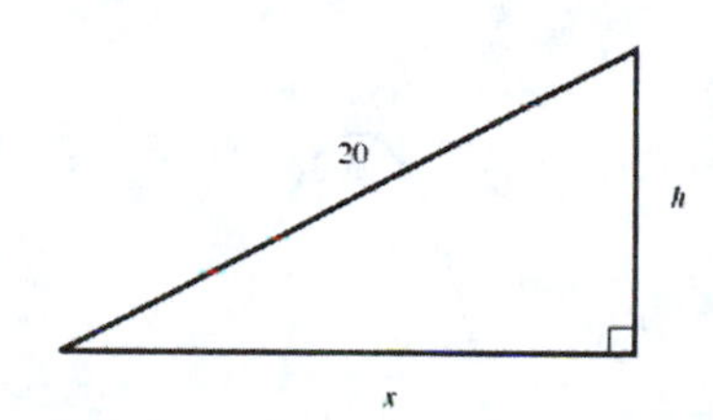

Let A be the area of the triangle and let x and h be its two legs. By the Pythagorean theorem:

$$x^2 + h^2 = 20^2$$
$$h^2 = 400 - x^2$$
$$h = \sqrt{400 - x^2}$$

Therefore: $A = \frac{1}{2}(\text{base})(\text{height}) = \frac{1}{2}(x)(h) = \frac{1}{2}x\sqrt{400 - x^2}$

Using functional notation we have $A(x) = \frac{1}{2}x\sqrt{400 - x^2}$. The domain is $(0, 20)$.

21. Draw the figure:

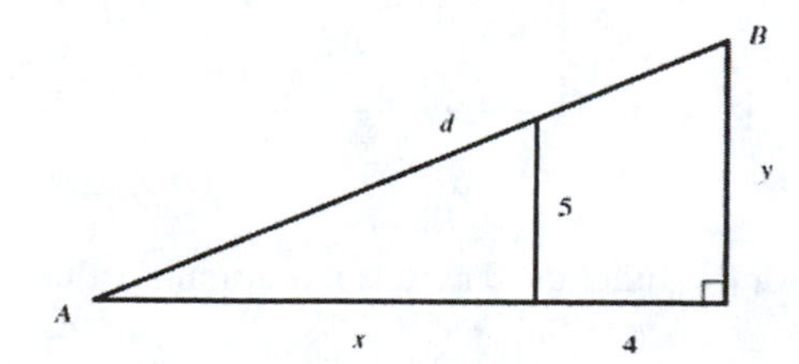

Using similar triangles:

$$\frac{x}{5} = \frac{x+4}{y}$$
$$xy = 5(x+4)$$
$$y = \frac{5(x+4)}{x}$$

Therefore:

$$d^2 = (x+4)^2 + y^2$$
$$= (x+4)^2 + \left(\frac{5(x+4)}{x}\right)^2$$
$$= (x+4)^2 + \frac{25(x+4)^2}{x^2}$$
$$= \frac{x^2(x+4)^2 + 25(x+4)^2}{x^2}$$
$$= \frac{(x+4)^2(x^2+25)}{x^2}$$

So $d = \frac{(x+4)\sqrt{x^2+25}}{x}$. Using functional notation we have $d(x) = \frac{(x+4)\sqrt{x^2+25}}{x}$. The domain is $(0, \infty)$.

23. **a.** Yes, this is a quadratic function. Graphing the function:

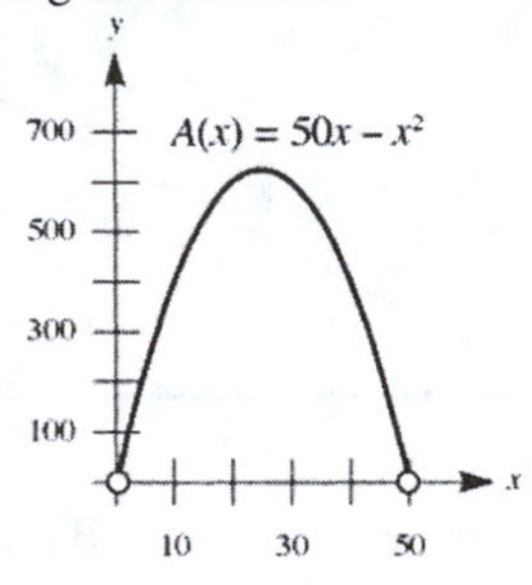

b. There is one turning point.

c. There is a maximum value of 625, which occurs at $x = 25$. There is no minimum value for the function.

25. **a.** No, this is not a quadratic function. Graphing the function:

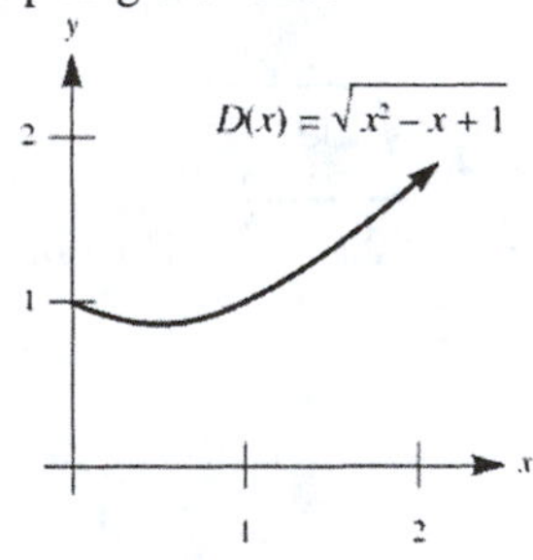

b. There is one turning point.

c. There is no maximum value for the function. There is a minimum value of $\sqrt{0.75} \approx 0.87$, which occurs at $x = 1/2$.

27. **a.** No, this is not a quadratic function. Graphing the function:

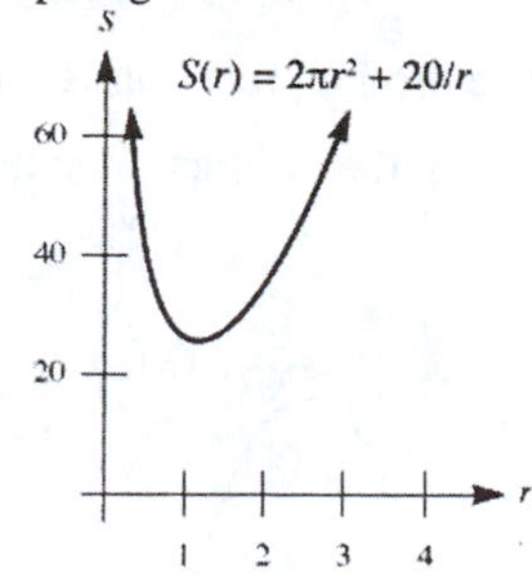

b. There is one turning point.

c. There is no maximum value for the function. There is a minimum value of approximately 25.69, which occurs at approximately $x \approx 1.17$.

29. **a.** **a.** No, this is not a quadratic function. Graphing the function:

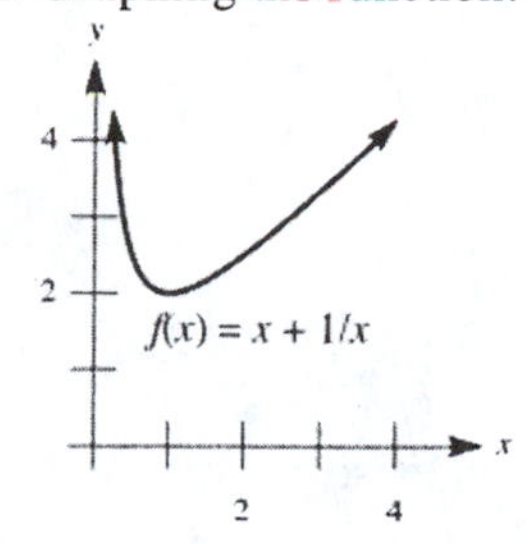

b. There is one turning point.

c. There is no maximum value for the function. There is a minimum value of 2, which occurs at $x = 1$.

b. **a.** No, this is not a quadratic function. Graphing the function:

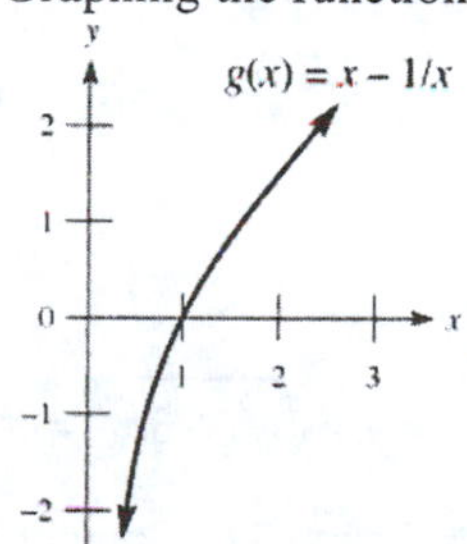

b. There are no turning points.

c. There are no maximum or minimum values for the function.

31. **a.** Let x and y represent the two numbers. Then $xy = \sqrt{11}$, so $y = \dfrac{\sqrt{11}}{x}$. The sum is given by: $S = x + y = x + \dfrac{\sqrt{11}}{x}$

The sum is given by $S(x) = x + \dfrac{\sqrt{11}}{x}$. The domain is $(0, \infty)$. Graphing the function:

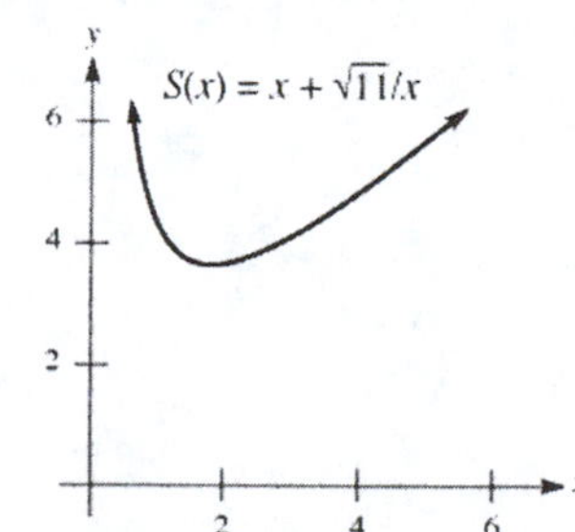

The sum has a minimum value of approximately 3.642.

b. Let x and y represent the two numbers. Then $x+y=\sqrt{11}$, so $y=\sqrt{11}-x$. The product is given by: $P=xy=x\left(\sqrt{11}-x\right)=-x^2+\sqrt{11}\,x$. The product function is thus $P(x)=-x^2+\sqrt{11}\,x$. The domain is $\left(0,\sqrt{11}\right)$. Since this is a quadratic function with $a<0$, the product must have a maximum value. Graphing the function:

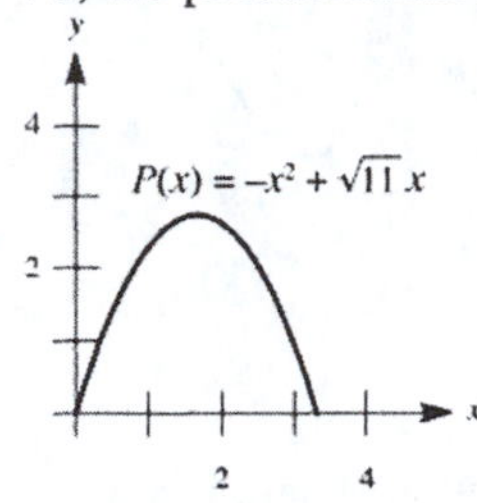

The product has a maximum value of 2.75.

33. For the square, the side is $\frac{x}{4}$. For the rectangle which has a perimeter of $3-x$:

$$\begin{aligned}2w+2l&=3-x\\ 2\bullet\tfrac{1}{2}l+2l&=3-x\\ 3l&=3-x\\ l&=1-\tfrac{1}{3}x\\ w&=\tfrac{1}{2}-\tfrac{1}{6}x\end{aligned}$$

So the combined area of the square and rectangle is:

$$A=\left(\tfrac{x}{4}\right)^2+\left(1-\tfrac{1}{3}x\right)\left(\tfrac{1}{2}-\tfrac{1}{6}x\right)=\tfrac{1}{16}x^2+\tfrac{1}{2}-\tfrac{1}{3}x+\tfrac{1}{18}x^2=\tfrac{17}{144}x^2-\tfrac{1}{3}x+\tfrac{1}{2}$$

Using functional notation we have $A(x)=\frac{17}{144}x^2-\frac{1}{3}x+\frac{1}{2}$. The domain is $(0,3)$. Yes, this is a quadratic function.

35. **a.** Since $V=\frac{1}{3}\pi r^2h$ and $h=\sqrt{3}r$: $V=\frac{1}{3}\pi r^2\left(\sqrt{3}r\right)=\frac{\sqrt{3}}{3}\pi r^3$. Using functional notation we have $V(r)=\frac{\sqrt{3}}{3}\pi r^3$. The domain is $(0,\infty)$.

b. Since $S=\pi r\sqrt{r^2+h^2}$ and $h=\sqrt{3}r$:

$$S=\pi r\sqrt{r^2+h^2}=\pi r\sqrt{r^2+\left(\sqrt{3}r\right)^2}=\pi r\sqrt{r^2+3r^2}=\pi r\sqrt{4r^2}=\pi r(2r)=2\pi r^2$$

Using functional notation we have $S(r)=2\pi r^2$. The domain is $(0,\infty)$.

37. **a.** Since $V=\frac{1}{3}\pi r^2h$ and $S=\pi r\sqrt{r^2+h^2}$, and $V=S$:

$$\begin{aligned}\tfrac{1}{3}\pi r^2h&=\pi r\sqrt{r^2+h^2}\\ rh&=3\sqrt{r^2+h^2}\\ r^2h^2&=9\left(r^2+h^2\right)\\ r^2h^2&=9r^2+9h^2\\ r^2h^2-9r^2&=9h^2\\ r^2\left(h^2-9\right)&=9h^2\\ r^2&=\frac{9h^2}{h^2-9}\\ r&=\sqrt{\frac{9h^2}{h^2-9}}=\frac{3h}{\sqrt{h^2-9}}\end{aligned}$$

Using functional notation we have $r(h)=\dfrac{3h}{\sqrt{h^2-9}}$. The domain is $(3,\infty)$.

b. After squaring in part **a**, we had:

$$r^2h^2 = 9r^2 + 9h^2$$
$$r^2h^2 - 9h^2 = 9r^2$$
$$h^2\left(r^2 - 9\right) = 9r^2$$
$$h^2 = \frac{9r^2}{r^2 - 9}$$

Taking roots: $h = \sqrt{\dfrac{9r^2}{r^2 - 9}} = \dfrac{3r}{\sqrt{r^2 - 9}}$

Using functional notation we have $h(r) = \dfrac{3r}{\sqrt{r^2 - 9}}$. The domain is $(3, \infty)$.

39. Let x be the length of wire used for the circle, so $14 - x$ is the length of wire used for the square. For the circle, we have $2\pi r = x$, so $r = \frac{x}{2\pi}$ and thus the area is given by: $\pi r^2 = \pi\left(\dfrac{x}{2\pi}\right)^2 = \dfrac{x^2}{4\pi}$

For the square, we have $4s = 14 - x$, so $s = \dfrac{14 - x}{4}$ and thus the area is given by: $\left(\dfrac{14 - x}{4}\right)^2 = \dfrac{(14 - x)^2}{16}$

So the total combined area is: $A = \dfrac{x^2}{4\pi} + \dfrac{(14 - x)^2}{16} = \dfrac{4x^2 + \pi(14 - x)^2}{16\pi}$

Using functional notation we have $A(x) = \dfrac{4x^2 + \pi(14 - x)^2}{16\pi}$. The domain is $(0, 14)$.

41. The perimeter of each semi-circle is $\frac{1}{2}(2\pi r) = \pi r$, so the total perimeter P is given by:

$P = \pi r + \pi r + l + l = 2\pi r + 2l$, where l is the length of the rectangle

Since $P = \frac{1}{4}$:

$$2\pi r + 2l = \frac{1}{4}$$
$$2l = \frac{1}{4} - 2\pi r$$
$$2l = \frac{1 - 8\pi r}{4}$$
$$l = \frac{1 - 8\pi r}{8}$$

Now find the area A. The area of each semicircle is $\frac{1}{2}\pi r^2$, and the area of the rectangle is length • width, where $w = 2r$, so: $A = \frac{1}{2}\pi r^2 + \frac{1}{2}\pi r^2 + lw = \pi r^2 + \left(\dfrac{1 - 8\pi r}{8}\right)(2r) = \pi r^2 + \dfrac{r - 8\pi r^2}{4} = \dfrac{4\pi r^2 + r - 8\pi r^2}{4} = \dfrac{r - 4\pi r^2}{4}$

Using functional notation we have $A(r) = \dfrac{r(1 - 4\pi r)}{4}$. The domain is $\left(0, \dfrac{1}{4\pi}\right)$.

43. Draw the figure:

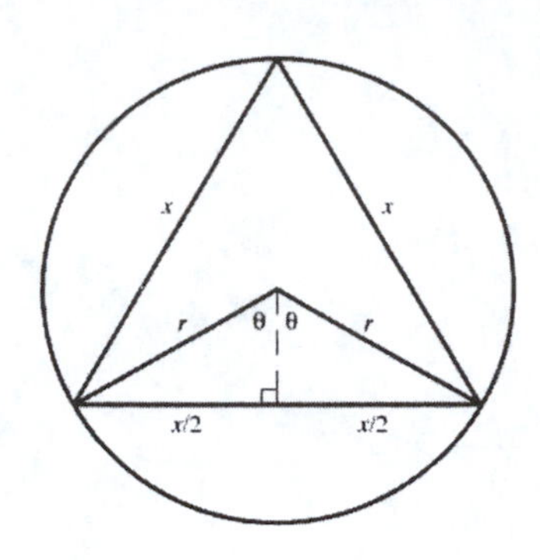

Using geometry, we see that:

$$2\theta = \tfrac{1}{3}(360°)$$
$$2\theta = 120°$$
$$\theta = 60°$$

Since $\theta = 60°$:

$$\frac{x}{2} = \frac{\sqrt{3}}{2}r$$
$$x = \sqrt{3}r$$
$$r = \frac{x}{\sqrt{3}}$$

So the area of the circle A is: $A = \pi r^2 = \pi\left(\frac{x}{\sqrt{3}}\right)^2 = \frac{\pi x^2}{3}$. Using functional notation we have $A(x) = \frac{\pi}{3}x^2$.

The domain is $(0, \infty)$.

45. **a.** Since $V = l \bullet w \bullet h$ and $l = 8 - 2w$, $w = 6 - 2x$, $h = x$:

$$V = (8-2x)(6-2x)(x) = \left(48 - 28x + 4x^2\right)(x) = 4x^3 - 28x^2 + 48x$$

Using functional notation we have $V(x) = 4x^3 - 28x^2 + 48x$. The domain is $(0, 3)$.

b. Completing the table:

x (in.)	0	0.5	1.0	1.5	2.0	2.5	3.0
volume $\left(\text{in.}^3\right)$	0	17.5	24	22.5	16	7.5	0

c. The value $x = 1.0$ appears to yield the largest volume.

d. Completing the table:

x (in.)	0.8	0.9	1.0	1.1	1.2	1.3	1.4
volume $\left(\text{in.}^3\right)$	22.5	23.4	24	24.2	24.2	23.9	23.3

e. The value $x = 1.1$ appears to yield the largest volume.

47. **a.** The area of the window would be: $A = \frac{1}{2}\left(\pi r^2\right) + lw$. It remains to find l and w in terms of r. We see that $w = 2r$, and the perimeter $P = 32$, so: $P = \frac{1}{2}\left(2\pi r\right) + 2l + w$. Therefore:

$$\tfrac{1}{2}\left(2\pi r\right) + 2l + w = 32$$
$$\pi r + 2l + 2r = 32$$
$$2l = 32 - \pi r - 2r$$
$$l = \frac{32 - \pi r - 2r}{2}$$

Now find the area:

$$A = \tfrac{1}{2}\left(\pi r^2\right) + lw = \tfrac{1}{2}\left(\pi r^2\right) + \left(\frac{32 - \pi r - 2r}{2}\right)(2r) = \frac{\pi r^2}{2} + 32r - \pi r^2 - 2r^2 = 32r - 2r^2 - \frac{\pi r^2}{2}$$

Using functional notation we write $A(r) = 32r - 2r^2 - \dfrac{\pi r^2}{2}$. The domain is $\left(0, \dfrac{32}{\pi + 2}\right)$.

b. We have $A(r) = -\left(\frac{4+\pi}{2}\right)r^2 + 32r$, which will open downward. Since $A(0) = 0$, it does pass through the origin. Complete the square:

$$\begin{aligned} A(r) &= -\left(\frac{4+\pi}{2}\right)\left(r^2 - \frac{64}{4+\pi}r\right) \\ &= -\left(\frac{4+\pi}{2}\right)\left[r^2 - \frac{64}{4+\pi}r + \left(\frac{32}{4+\pi}\right)^2\right] + \left(\frac{4+\pi}{2}\right)\left(\frac{32}{4+\pi}\right)^2 \\ &= -\left(\frac{4+\pi}{2}\right)\left(r - \frac{32}{4+\pi}\right)^2 + \frac{512}{4+\pi} \end{aligned}$$

So the vertex is $\left(\frac{32}{4+\pi}, \frac{512}{4+\pi}\right)$.

49. a. Use the Pythagorean theorem: $3^2 + y^2 = z^2$

Taking roots: $z = \sqrt{y^2 + 9}$. Now $s = \dfrac{y}{z} = \dfrac{y}{\sqrt{y^2 + 9}}$, so $s\sqrt{y^2 + 9} = y$. Squaring yields:

$$\begin{aligned} s^2\left(y^2 + 9\right) &= y^2 \\ s^2 y^2 + 9s^2 &= y^2 \\ y^2 - s^2 y^2 &= 9s^2 \\ y^2\left(1 - s^2\right) &= 9s^2 \\ y^2 &= \frac{9s^2}{1 - s^2} \\ y &= \frac{3s}{\sqrt{1 - s^2}} \end{aligned}$$

Using functional notation we have $y(s) = \dfrac{3s}{\sqrt{1 - s^2}}$. The domain is $(0, 1)$.

b. From part **a** we have $s(y) = \dfrac{y}{\sqrt{y^2 + 9}}$. The domain is $(0, \infty)$.

c. Since $s = \frac{y}{z}$, then $z = \frac{y}{s}$. Using our result from part **a**: $z = \dfrac{y}{s} = \dfrac{\frac{3s}{\sqrt{1-s^2}}}{s} = \dfrac{3}{\sqrt{1 - s^2}}$

Using functional notation we have $z(s) = \dfrac{3}{\sqrt{1 - s^2}}$. The domain is $(0, 1)$.

d. Using our answer from part **c**:

$$z = \frac{3}{\sqrt{1-s^2}}$$

$$z\sqrt{1-s^2} = 3$$

Squaring each side, we obtain:

$$z^2\left(1-s^2\right) = 9$$

$$z^2 - z^2s^2 = 9$$

$$-z^2s^2 = 9 - z^2$$

$$s^2 = \frac{z^2-9}{z^2}$$

$$s = \frac{\sqrt{z^2-9}}{z}$$

Using functional notation we have $s(z) = \dfrac{\sqrt{z^2-9}}{z}$. The domain is $(3,\infty)$.

51. **a.** Using the points (0,–1) and (a, a^2): $m(a) = \dfrac{a^2-(-1)}{a-0} = \dfrac{a^2+1}{a}$. The domain is $(-\infty,0)\cup(0,\infty)$.

b. If x_0 is the x-intercept, then the area of the triangle A is: $A = \frac{1}{2}(\text{base})(\text{height}) = \frac{1}{2}\left(a-x_0\right)\left(a^2\right)$

To find the x-intercept, we must find the equation of the line. Use $m = \dfrac{a^2+1}{a}$ (from part **a**) and (0,–1) in the slope-intercept formula to obtain: $y = \dfrac{a^2+1}{a}x - 1$. Find x_0 by letting $y = 0$:

$$\frac{a^2+1}{a}x_0 - 1 = 0$$

$$\frac{a^2+1}{a}x_0 = 1$$

$$x_0 = \frac{a}{a^2+1}$$

Therefore:

$$A = \tfrac{1}{2}\left(a-x_0\right)a^2 = \tfrac{1}{2}\left(a-\frac{a}{a^2+1}\right)a^2 = \frac{a^2}{2}\left[\frac{a\left(a^2+1\right)-a}{a^2+1}\right] = \frac{a^2\left(a^3+a-a\right)}{2\left(a^2+1\right)} = \frac{a^2\left(a^3\right)}{2\left(a^2+1\right)} = \frac{a^5}{2\left(a^2+1\right)}$$

53. Re-draw the figure (differently):

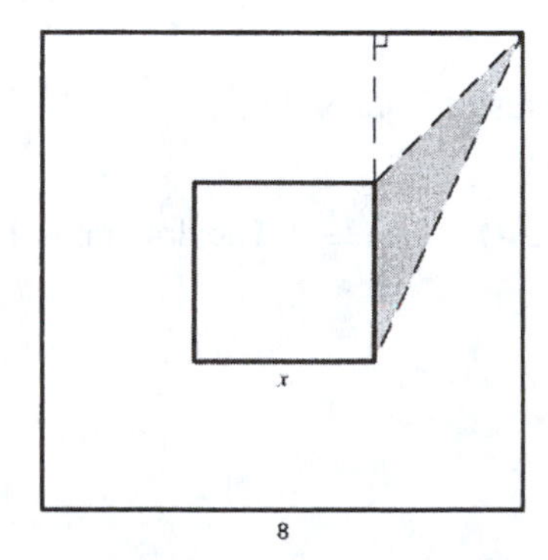

Extend the triangle to form a right triangle as pictured. Now find the areas of the large and small right triangles:

$$A_{\text{large}} = \tfrac{1}{2}(\text{base})(\text{height}) = \tfrac{1}{2}\left(x + \frac{8-x}{2}\right)\left(\frac{8-x}{2}\right) = \tfrac{1}{2}\left(\frac{2x+8-x}{2}\right)\left(\frac{8-x}{2}\right) = \frac{(x+8)(8-x)}{8} = \frac{64-x^2}{8}$$

$$A_{\text{small}} = \tfrac{1}{2}(\text{base})(\text{height}) = \tfrac{1}{2}\left(\frac{8-x}{2}\right)\left(\frac{8-x}{2}\right) = \frac{64-16x+x^2}{8}$$

Therefore: $A = A_{\text{large}} - A_{\text{small}} = \dfrac{64-x^2}{8} - \dfrac{64-16x+x^2}{8} = \dfrac{64-x^2-64+16x-x^2}{8} = \dfrac{16x-2x^2}{8} = \dfrac{8x-x^2}{4}$

Using functional notation we write $A(x) = \dfrac{8x-x^2}{4}$. The domain is $(0,8)$.

Note: A much easier approach is to realize that the altitude of the triangle need not lie on the triangle. That is:

$$\text{Area} = \tfrac{1}{2}(\text{base})(\text{altitude}) = \tfrac{1}{2}(x)\left(\frac{8-x}{2}\right) = \frac{8x-x^2}{4}$$

Both approaches are correct.

55. Draw the figure:

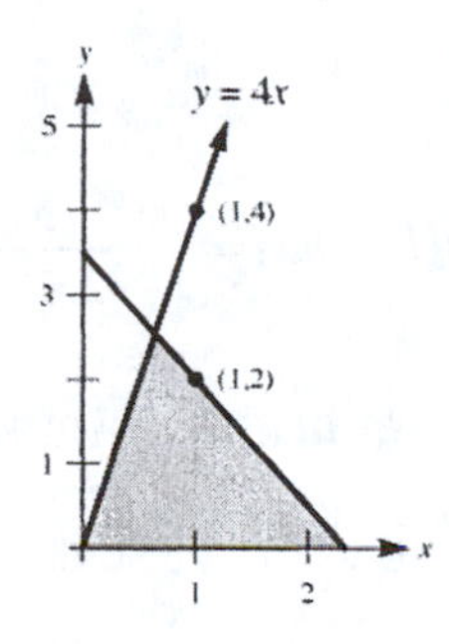

We are asked to find the area A of the shaded triangle. Since the line has slope m and passes through (1,2), by the point-slope formula:

$$y-2 = m(x-1)$$
$$y-2 = mx-m$$
$$y = mx+(2-m)$$

Its x-intercept is where $y = 0$:

$$0 = mx+(2-m)$$
$$mx = m-2$$
$$x = \frac{m-2}{m}$$

This is the base of a triangle. To find its height, we must find the value of y where this line and $y = 4x$ intersect. Set the two y-values equal:

$$mx+2-m = 4x$$
$$mx-4x = m-2$$
$$x(m-4) = m-2$$
$$x = \frac{m-2}{m-4}$$

Since this point lies on $y = 4x$, its y-coordinate is: $y = 4x = 4\left(\dfrac{m-2}{m-4}\right)$. Finally, find the area:

$$A = \tfrac{1}{2}(\text{base})(\text{height}) = \tfrac{1}{2}\left(\frac{m-2}{m}\right)(4)\left(\frac{m-2}{m-4}\right) = 2\frac{(m-2)^2}{m(m-4)} = \frac{2\left(m^2-4m+4\right)}{m^2-4m} = \frac{2m^2-8m+8}{m^2-4m}$$

Using functional notation we have $A(m) = \dfrac{2m^2-8m+8}{m^2-4m}$. The domain is $(-\infty,0)$.

57. Draw a figure:

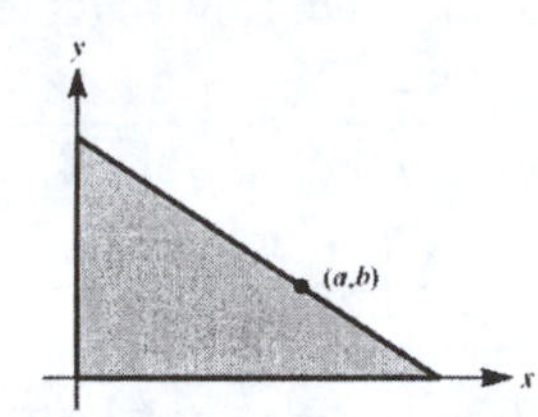

The line has the equation $y - b = m(x - a)$. Since the base and height are the x- and y-intercepts, respectively, find each intercept:

$$\begin{aligned} \text{base:} \quad & y = 0 \\ -b &= m(x-a) \\ -b &= mx - ma \\ mx &= ma - b \\ x &= \frac{ma-b}{m} \\ \text{height:} \quad & x = 0 \\ y - b &= m(-a) \\ y &= b - ma \end{aligned}$$

So the area of the triangle is $A = \frac{1}{2}\left(\frac{ma-b}{m}\right)(b - ma) = \frac{(ma-b)^2}{-2m}$. The domain is $(-\infty, 0)$.

4.5 Maximum and Minimum Problems

1. Call the two numbers x and y. Then $x + y = 5$, so $y = 5 - x$. So the product can be written as:

$$P = xy = x(5-x) = 5x - x^2$$

Completing the square: $P = -\left(x^2 - 5x\right) = -\left(x^2 - 5x + \frac{25}{4}\right) + \frac{25}{4} = -\left(x - \frac{5}{2}\right)^2 + \frac{25}{4}$

Since this is a parabola opening downward, it will have a maximum value of $\frac{25}{4}$.

3. Call the two numbers x and y. Then $y - x = 1$, so $y = x + 1$. The sum of their squares can be written as:

$$S = x^2 + y^2 = x^2 + (x+1)^2 = x^2 + x^2 + 2x + 1 = 2x^2 + 2x + 1$$

Completing the square: $S = 2\left(x^2 + x\right) + 1 = 2\left(x^2 + x + \frac{1}{4}\right) + 1 - \frac{1}{2} = 2\left(x + \frac{1}{2}\right)^2 + \frac{1}{2}$

Since this is a parabola opening upward, it will have a minimum value of $\frac{1}{2}$.

5. Let w and l be the width and length, respectively. Since $P = 2w + 2l$:

$$\begin{aligned} 2w + 2l &= 25 \\ 2l &= 25 - 2w \\ l &= \frac{25-2w}{2} \end{aligned}$$

So the area is given by: $A = wl = w\left(\frac{25-2w}{2}\right) = \frac{1}{2}\left(-2w^2 + 25w\right)$

Completing the square: $A = -\left(w^2 - \frac{25}{2}w\right) = -\left(w^2 - \frac{25}{2}w + \frac{625}{16}\right) = -\left(w - \frac{25}{4}\right)^2 + \frac{625}{16}$

This is a parabola opening downward, so it will achieve a maximum value when $w = \frac{25}{4}$.

Substitute to find l: $l = \frac{25 - 2\left(\frac{25}{4}\right)}{2} = \frac{25 - \frac{25}{2}}{2} = \frac{25}{4}$

So the largest such rectangle is a square of dimensions $\frac{25}{4}$ m by $\frac{25}{4}$ m.

7. Let x and y be the lengths of the two shorter sides, so $x + y = 100$, and $y = 100 - x$. Then the area is given by:

$$A = \frac{1}{2}xy = \frac{1}{2}x(100 - x) = \frac{1}{2}(-x^2 + 100x)$$

Completing the square: $A = -\frac{1}{2}\left(x^2 - 100x\right) = -\frac{1}{2}\left(x^2 - 100x + 2500\right) + 1250 = -\frac{1}{2}(x - 50)^2 + 1250$

This is a parabola opening downward, so it will achieve a maximum value of 1250 in.^{2}.

9. Let x and y be the two numbers, so $x + y = 6$ and thus $y = 6 - x$.

a. Simplifying: $T = x^2 + y^2 = x^2 + (6 - x)^2 = x^2 + 36 - 12x + x^2 = 2x^2 - 12x + 36$

Completing the square: $T = 2\left(x^2 - 6x\right) + 36 = 2\left(x^2 - 6x + 9\right) + 36 - 18 = 2(x - 3)^2 + 18$

This is a parabola opening upward, so it will have a minimum value of 18.

b. Simplifying: $S = x + y^2 = x + (6 - x)^2 = x + 36 - 12x + x^2 = x^2 - 11x + 36$

Completing the square: $S = \left(x^2 - 11x\right) + 36 = \left(x^2 - 11x + \frac{121}{4}\right) + 36 - \frac{121}{4} = \left(x - \frac{11}{2}\right)^2 + \frac{23}{4}$

This is a parabola opening upward, so it will have a minimum value of $\frac{23}{4}$.

c. Simplifying: $U = x + 2y^2 = x + 2(6 - x)^2 = x + 72 - 24x + 2x^2 = 2x^2 - 23x + 72$

Completing the square: $U = 2\left(x^2 - \frac{23}{2}x\right) + 72 = 2\left(x^2 - \frac{23}{2}x + \frac{529}{16}\right) + 72 - \frac{529}{8} = 2\left(x - \frac{23}{4}\right)^2 + \frac{47}{8}$

This is a parabola opening upward, so it will have a minimum value of $\frac{47}{8}$.

d. Substituting: $V = x + (2y)^2 = x + 4y^2 = x + 4(6 - x)^2 = x + 144 - 48x + 4x^2 = 4x^2 - 47x + 144$

Completing the square: $V = 4\left(x^2 - \frac{47}{4}x\right) + 144 = 4\left(x^2 - \frac{47}{4}x + \frac{2209}{64}\right) + 144 - \frac{2209}{16} = 4\left(x - \frac{47}{8}\right)^2 + \frac{95}{16}$

This is a parabola opening upward, so it will have a minimum value of $\frac{95}{16}$.

11. a. Compute the heights:

$$h(1) = -16(1)^2 + 32(1) = -16 + 32 = 16 \text{ ft}$$

$$h\left(\frac{3}{2}\right) = -16\left(\frac{3}{2}\right)^2 + 32\left(\frac{3}{2}\right) = -16\left(\frac{9}{4}\right) + 48 = -36 + 48 = 12 \text{ ft}$$

b. Completing the square: $h = -16t^2 + 32t = -16\left(t^2 - 2t\right) = -16\left(t^2 - 2t + 1\right) + 16 = -16(t - 1)^2 + 16$

This is a parabola opening downward, so it will have a maximum height of 16 ft, attained after 1 second.

c. Set $h = 7$ and solve for t:

$$7 = -16t^2 + 32t$$
$$16t^2 - 32t + 7 = 0$$
$$(4t - 7)(4t - 1) = 0$$
$$t = \frac{7}{4}, \frac{1}{4}$$

So $h = 7$ when $t = \frac{7}{4}$ sec or $t = \frac{1}{4}$ sec.

13. Every point on the given curve has coordinates of the form $\left(x, \sqrt{x-2}+1\right)$, and using the distance formula gives:

$$d=\sqrt{(4-x)^2+\left(1-\sqrt{x-2}-1\right)^2}=\sqrt{16-8x+x^2+x-2}=\sqrt{x^2-7x+14}$$

Completing the square: $\left(x^2-7x\right)+14=\left(x^2-7x+\frac{49}{4}\right)+14-\frac{49}{4}=\left(x-\frac{7}{2}\right)^2+\frac{7}{4}$

This is a parabola opening upward which will achieve a minimum value of $\sqrt{\frac{7}{4}}=\dfrac{\sqrt{7}}{2}$ at $x=\frac{7}{2}$. Then:

$$y=\sqrt{\tfrac{7}{2}-2}+1=\sqrt{\tfrac{3}{2}}+1=\tfrac{2+\sqrt{6}}{2}$$

So the point is $\left(\frac{7}{2}, \frac{2+\sqrt{6}}{2}\right)$ and the distance is $\dfrac{\sqrt{7}}{2}$.

15. **a.** We must find the value of x such that $x-x^2$ is as large as possible. Call $f(x)=-x^2+x$. Completing the square:

$$f(x)=-\left(x^2-x\right)=-\left(x^2-x+\tfrac{1}{4}\right)+\tfrac{1}{4}=-\left(x-\tfrac{1}{2}\right)^2+\tfrac{1}{4}$$

This is a parabola opening downward, so it will achieve a maximum value when $x=\frac{1}{2}$. So the number is $\frac{1}{2}$.

b. We must find the value of x such that $x-2x^2$ is as large as possible. Call $f(x)=-2x^2+x$. Completing the square:

$$f(x)=-2\left(x^2-\tfrac{1}{2}x\right)=-2\left(x^2-\tfrac{1}{2}x+\tfrac{1}{16}\right)+\tfrac{1}{8}=-2\left(x-\tfrac{1}{4}\right)^2+\tfrac{1}{8}$$

This is a parabola opening downward, so it will achieve a maximum value when $x=\frac{1}{4}$. So the number is $\frac{1}{4}$.

17. If we choose x for the depth of the pasture, then $500-2x$ is the length parallel to the river. The area of the pasture will then be given by: $A=x(500-2x)=-2x^2+500x$

Completing the square: $A=-2\left(x^2-250x\right)=-2\left(x^2-250x+125^2\right)+2(125)^2=-2(x-125)^2+31250$

This is a parabola opening downward, so it will achieve a maximum value at $x=125$. Then the length is $500-2(125)=500-250=250$. So the dimensions are 125 ft by 250 ft.

19. Simplifying:

$$R-C=\left(0.4x^2+10x+5\right)-\left(0.5x^2+2x+101\right)=0.4x^2+10x+5-0.5x^2-2x-101=-0.1x^2+8x-96$$

Completing the square: $R-C=-0.1\left(x^2-80x\right)-96=-0.1\left(x^2-80x+1600\right)-96+160=-0.1(x-40)^2+64$

This is a parabola opening downward, so it will achieve a maximum value when $x=40$.

21. Recall that revenue R is $x \bullet p$. So: $R=x\left(-\frac{1}{4}x+30\right)=-\frac{1}{4}x^2+30x$

Completing the square: $R=-\frac{1}{4}x^2+30x=-\frac{1}{4}\left(x^2-120x\right)=-\frac{1}{4}\left(x^2-120x+3600\right)+900=-\frac{1}{4}(x-60)^2+900$

This is a parabola opening downward, so it will achieve a maximum value at $x=60$. The maximum revenue is \$900. The corresponding unit price p is: $p=-\frac{1}{4}(60)+30=-15+30=\$15$

23. **a.** To use max/min methods, we need to substitute in the quantity $x^2 + y^2$ and write it strictly in terms of x or y. So take $2x + 3y = 6$ and solve for y:

$$3y = 6 - 2x$$
$$y = \frac{6-2x}{3}$$

Then substitute, and the quantity $x^2 + y^2$ becomes: $Q = x^2 + \left(\frac{6-2x}{3}\right)^2 = x^2 + \frac{36-24x+4x^2}{9} = \frac{13}{9}x^2 - \frac{8}{3}x + 4$

Completing the square: $Q = \frac{13}{9}\left(x^2 - \frac{24}{13}x\right) + 4 = \frac{13}{9}\left(x^2 - \frac{24}{13}x + \frac{144}{169}\right) + 4 - \frac{13}{9}\left(\frac{144}{169}\right) = \frac{13}{9}\left(x - \frac{12}{13}\right)^2 + \frac{36}{13}$

This is a parabola opening up, so it will achieve a minimum value of $\frac{36}{13}$.

b. The equation of a circle with its center at the origin is $x^2 + y^2 = r^2$ where r is the radius. The line $2x + 3y = 6$ will intersect the circle in two points wherever r is sufficiently large. As we reduce r, we gradually reach a position where the circle and line are tangent and this is the minimum value of r or $\sqrt{x^2 + y^2}$. In this case, it is $\sqrt{\frac{36}{13}} = \frac{6\sqrt{13}}{13}$. This is the square root of the answer from part **a**.

25. **a.** Sketching the graph of the two functions:

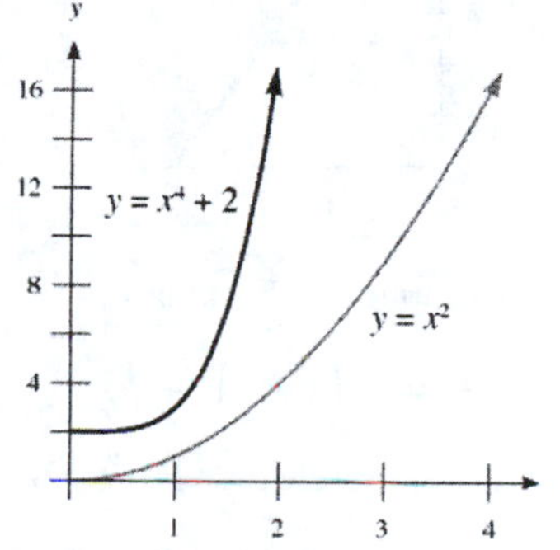

b. The minimum length of road is approximately 1.5 miles.

c. Let $y = x^4 + 2 - x^2 = x^4 - x^2 + 2$. Substituting $t = x^2$, this is a quadratic function $y = t^2 - t + 2$ which has a minimum value when $t = -\frac{b}{2a} = \frac{1}{2}$, so $x = \sqrt{\frac{1}{2}} = \frac{\sqrt{2}}{2}$. The two points the road connects are $\left(\frac{\sqrt{2}}{2}, \frac{1}{2}\right)$ and $\left(\frac{\sqrt{2}}{2}, \frac{9}{4}\right)$, so the minimum length of the road is $\frac{9}{4} - \frac{1}{2} = \frac{7}{4}$ miles.

27. **a.** Substitute $y = 15 - x$: $Q = x^2 + y^2 = x^2 + (15-x)^2 = x^2 + 225 - 30x + x^2 = 2x^2 - 30x + 225$

Completing the square:

$$Q = 2x^2 - 30x + 225 = 2\left(x^2 - 15x\right) + 225 = 2\left(x^2 - 15x + \frac{225}{4}\right) + 225 - \frac{225}{2} = 2\left(x - \frac{15}{2}\right)^2 + \frac{225}{2}$$

This is a parabola opening upward, so it will achieve a minimum value of $\frac{225}{2}$.

b. Substitute $y = C - x$: $Q = x^2 + y^2 = x^2 + (C-x)^2 = x^2 + C^2 - 2Cx + x^2 = 2x^2 - 2Cx + C^2$

Completing the square:

$$Q = 2x^2 - 2Cx + C^2 = 2\left(x^2 - Cx\right) + C^2 = 2\left(x^2 - Cx + \frac{C^2}{4}\right) + C^2 - \frac{C^2}{2} = 2\left(x - \frac{C}{2}\right)^2 + \frac{C^2}{2}$$

This is a parabola opening upward, so it will achieve a minimum value of $\frac{C^2}{2}$. When $C = 15$, the result from part **a** is verified.

29. Let the other two sides of each of the four triangles be t and $1-t$, respectively. Then the area of the square will be a minimum when the area of these triangles is a maximum. Now write an expression for the total area of the four triangles: $A = 4\left(\frac{1}{2}\right)(t)(1-t) = 2t - 2t^2 = -2t^2 + 2t$

Completing the square: $A = -2t^2 + 2t = -2\left(t^2 - t\right) = -2\left(t^2 - t + \frac{1}{4}\right) + \frac{1}{2} = -2\left(t - \frac{1}{2}\right)^2 + \frac{1}{2}$

This is a parabola opening downward, so it will achieve a maximum area of $\frac{1}{2}$ when $t = \frac{1}{2}$. Since the large square has area = 1, the smaller square has area $1 - A = 1 - \frac{1}{2} = \frac{1}{2}$. Using the Pythagorean theorem to find x:

$$x^2 = t^2 + (1-t)^2 = \frac{1}{4} + \frac{1}{4} = \frac{1}{2}$$

So $x = \sqrt{\frac{1}{2}} = \frac{\sqrt{2}}{2}$.

31. Draw the figure:

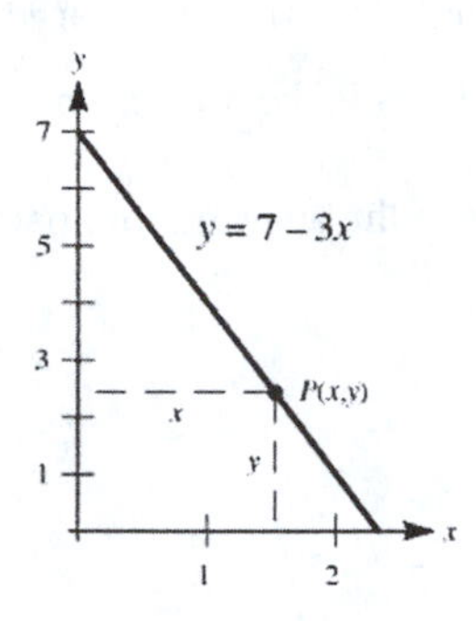

We have $A = xy = x(7-3x) = -3x^2 + 7x$. Completing the square:

$$A = -3\left(x^2 - \frac{7}{3}x\right) = -3\left(x^2 - \frac{7}{3}x + \frac{49}{36}\right) + \frac{49}{12} = -3\left(x - \frac{7}{6}\right)^2 + \frac{49}{12}$$

Since this parabola opens downward, the largest possible area is $\frac{49}{12}$.

33. Draw the figure:

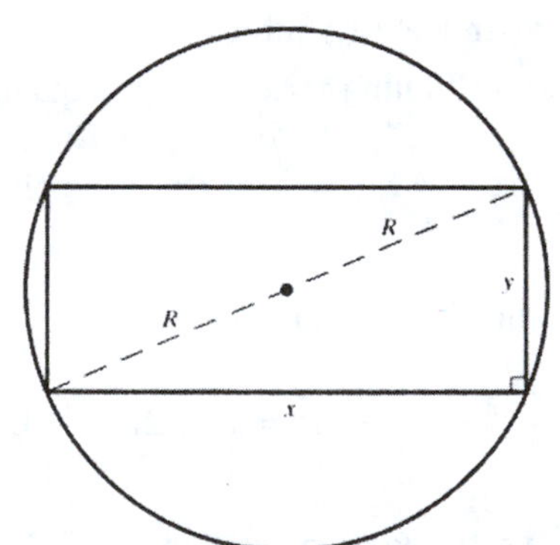

Using the Pythagorean theorem, we obtain $y = \sqrt{4R^2 - x^2}$. The area A of the rectangle is: $A = xy = x\sqrt{4R^2 - x^2}$

Therefore: $A^2 = x^2\left(4R^2 - x^2\right) = 4R^2x^2 - x^4$. In order to maximize the expression $4R^2x^2 - x^4$, first let $t = x^2$, so that the expression becomes $4R^2t - t^2$. Completing the square:

$$A^2 = -t^2 + 4R^2t = -\left(t^2 - 4R^2t\right) = -\left(t^2 - 4R^2t + 4R^4\right) + 4R^4 = -\left(t - 2R^2\right)^2 + 4R^4$$

This parabola opens downward, so the maximum value of A^2 is $4R^4$. Thus the maximum area is $\sqrt{4R^4} = 2R^2$.

35. Let x = east-west dimension, and y = north-south dimension. So the cost is given by $C = 12(2x) + 8(2y) = 24x + 16y$. Since this cost is \$4800, we have $24x + 16y = 4800$, so: $y = \dfrac{4800 - 24x}{16} = \dfrac{600 - 3x}{2}$

Now the area is $A = xy = x\left(\dfrac{600 - 3x}{2}\right) = -\frac{3}{2}x^2 + 300x$. So $A(x) = -\frac{3}{2}x^2 + 300x$.

This will be a parabola opening downward, so it will have a maximum value. Completing the square:

$$A(x) = -\tfrac{3}{2}x^2 + 300x = -\tfrac{3}{2}\left(x^2 - 200x\right) = -\tfrac{3}{2}\left(x^2 - 200x + 100^2\right) + 15000 = -\tfrac{3}{2}(x - 100)^2 + 15000$$

So $x = 100$ will maximize area, which is 15,000 yd^2. We find y: $y = \dfrac{600 - 3(100)}{2} = \dfrac{600 - 300}{2} = 150$ yd

So the dimensions are 100 yd by 150 yd.

37. The given function can be rewritten: $y = \left(a_1 + a_2\right)x^2 - 2\left(a_1x_1 + a_2x_2\right)x + \left(a_1x_1^2 + a_2x_2^2\right)$

Completing the square:

$$y = \left(a_1 + a_2\right)\left[x^2 - \frac{2\left(a_1x_1 + a_2x_2\right)}{a_1 + a_2}\right] + \left(a_1x_1^2 + a_2x_2^2\right)$$

$$= \left(a_1 + a_2\right)\left(x - \frac{a_1x_1 + a_2x_2}{a_1 + a_2}\right)^2 + \left(a_1x_1^2 + a_2x_2^2\right) - \frac{\left(a_1x_1 + a_2x_2\right)^2}{a_1 + a_2}$$

Since a_1 and a_2 are both positive, $a_1 + a_2 > 0$ and thus this parabola opens upward. So the minimum must occur where $x = \dfrac{a_1x_1 + a_2x_2}{a_1 + a_2}$.

39. **a.** We have $\dfrac{\Delta p}{\Delta x} = \dfrac{10}{-5} = -2$. Also $p = 200$ when $x = 150$. Using the point-slope formula with the point (150,200):

$$p - 200 = -2(x - 150)$$
$$p - 200 = -2x + 300$$
$$p = -2x + 500$$

Using functional notation we have $p(x) = -2x + 500$.

b. Since $R = xp$, we have $R = x(-2x + 500) = -2x^2 + 500x$. Completing the square:

$$R = -2x^2 + 500x = -2\left(x^2 - 250x\right) = -2\left(x^2 - 250x + 15625\right) + 31250 = -2(x - 125)^2 + 31250$$

Using functional notation we have $R(x) = -2(x - 125)^2 + 31250$. Since this parabola opens downward, we have a maximum revenue of \$31,250 when $x = 125$. We find $p = -2(125) + 500 = \$250$.

41. Let $x = t^2$. Then $y = -t^4 + 6t^2 - 6 = -x^2 + 6x - 6$. Completing the square:

$$y = -x^2 + 6x - 6 = -\left(x^2 - 6x\right) - 6 = -\left(x^2 - 6x + 9\right) - 6 + 9 = -(x - 3)^2 + 3$$

So $x = 3$ will yield the largest output. Since $x = t^2$, we have:

$$t^2 = 3$$
$$t = \pm\sqrt{3}$$

So $t = \sqrt{3}$ or $t = -\sqrt{3}$ will yield the largest output.

43. **a.** Since $x^2 - x + 1 = \left(x - \frac{1}{2}\right)^2 + \frac{3}{4}$, the quantity must be positive. We know that $\left(x - \frac{1}{2}\right)^2 \geq 0$, thus $\left(x - \frac{1}{2}\right)^2 + \frac{3}{4} \geq \frac{3}{4} > 0$.

b. Computing the discriminant: $D = (-1)^2 - 4(1)(1) = 1 - 4 = -3$
Since $D < 0$, there are no x-intercepts (because no real solutions exist). Since the graph is a parabola opening up, and there are no x-intercepts, the y-values must be positive for all values of x.

45. **a.** For the circle, we have $2\pi r = x$, so $r = \frac{x}{2\pi}$, and thus the area is: $A = \pi r^2 = \pi\left(\frac{x}{2\pi}\right)^2 = \frac{x^2}{4\pi}$

For the square, we have $4s = 16 - x$, so $s = \frac{16 - x}{4}$, and thus the area is:

$$A = s^2 = \left(\frac{16 - x}{4}\right)^2 = \frac{256 - 32x + x^2}{16} = 16 - 2x + \frac{1}{16}x^2$$

So the total combined area is given by $A(x) = \frac{x^2}{4\pi} + 16 - 2x + \frac{1}{16}x^2 = \frac{4+\pi}{16\pi}x^2 - 2x + 16$. The domain is $(0, \infty)$.

b. This is a parabola opening upward, so it will have a minimum value at: $x = \frac{2}{2 \bullet \frac{4+\pi}{16\pi}} = \frac{16\pi}{4 + \pi}$

c. This ratio is given by: $\frac{\frac{16\pi}{4+\pi}}{16 - \frac{16\pi}{4+\pi}} \bullet \frac{4+\pi}{4+\pi} = \frac{16\pi}{64 + 16\pi - 16\pi} = \frac{16\pi}{64} = \frac{\pi}{4}$

47. **a.** Call r the radius of the base and h the height. Then $\pi r^2 h = 500$, and thus $h = \frac{500}{\pi r^2}$. The amount of material is given by: $A = \pi r^2 + 2\pi r h = \pi r^2 + 2\pi r\left(\frac{500}{\pi r^2}\right) = \pi r^2 + \frac{1000}{r}$

b. Graphing the function:

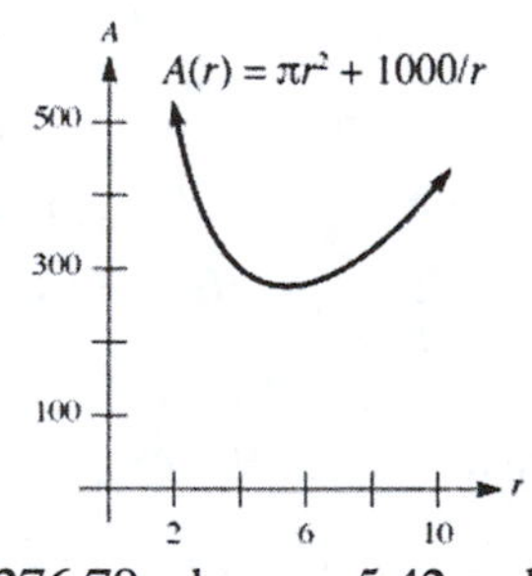

The minimum value is approximately 276.79 when $r \approx 5.42$ and $h \approx 5.42$. The height and radius are both approximately 5.42 cm.

49. **a.** Let $\left(x, x^2\right)$ represent a point on the parabola. The distance from this point to the point $(3, 0)$ is given by the distance formula: $d = \sqrt{(x-3)^2 + \left(x^2 - 0\right)^2} = \sqrt{x^2 - 6x + 9 + x^4} = \sqrt{x^4 + x^2 - 6x + 9}$

b. Graphing the function:

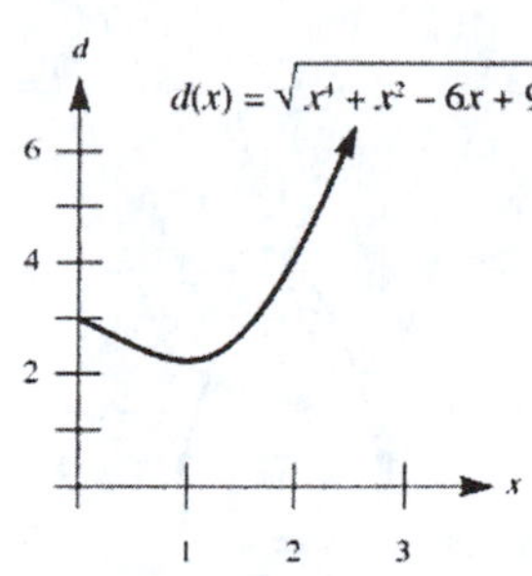

The minimum value is $\sqrt{5} \approx 2.236$ when $x = 1$. The closest point on the parabola is $(1, 1)$.

51. **a.** First draw the figure:

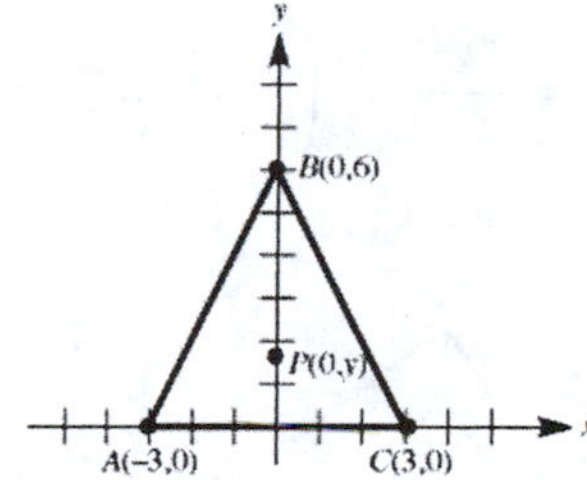

Call P the point $P(0, y)$. The sum of the distances between P and the vertices is given by:

$$\begin{aligned} S &= \sqrt{(0+3)^2 + (y-0)^2} + \sqrt{(y-6)^2} + \sqrt{(0-3)^2 + (y-0)^2} \\ &= \sqrt{9+y^2} + 6 - y + \sqrt{9+y^2} \\ &= 2\sqrt{9+y^2} + 6 - y \end{aligned}$$

Note that $\sqrt{(y-6)^2} = 6 - y$, since $y - 6 < 0$.

b. Graphing the function:

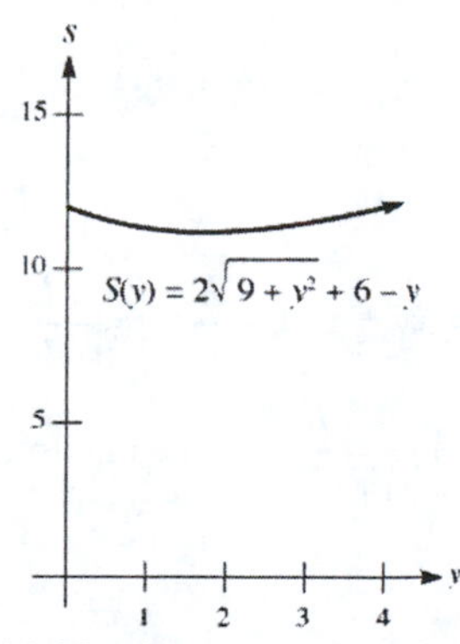

The minimum value is approximately 11.20 when $y \approx 1.73$. The position of P is approximately $P(0,1.73)$.

53. **a.** Graphing the curves $y = x^2 + kx$ for $k = 2, -2, 3, -3$:

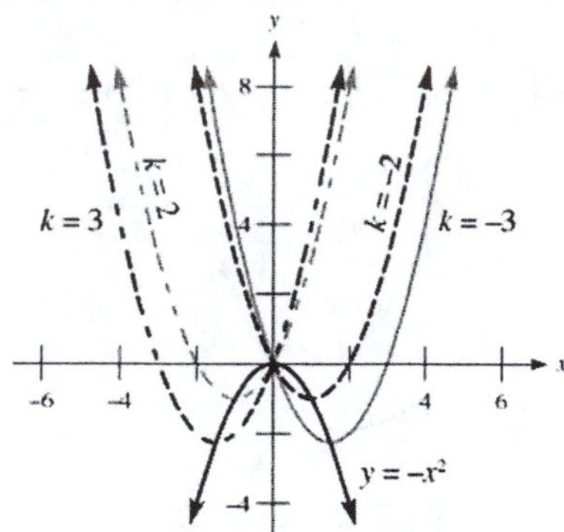

Note that the vertex appears to lie on the curve $y = -x^2$.

b. The x-coordinate of the vertex of $y = x^2 + kx$ is $x = -\frac{k}{2}$, and the y-coordinate is:

$$y = \left(-\frac{k}{2}\right)^2 + k\left(-\frac{k}{2}\right) = \frac{k^2}{4} - \frac{k^2}{2} = -\frac{k^2}{4} = -\left(-\frac{k}{2}\right)^2 = -x^2$$

This proves the vertex lies on the curve $y = -x^2$.

55. Re-draw the figure and label additional sides:

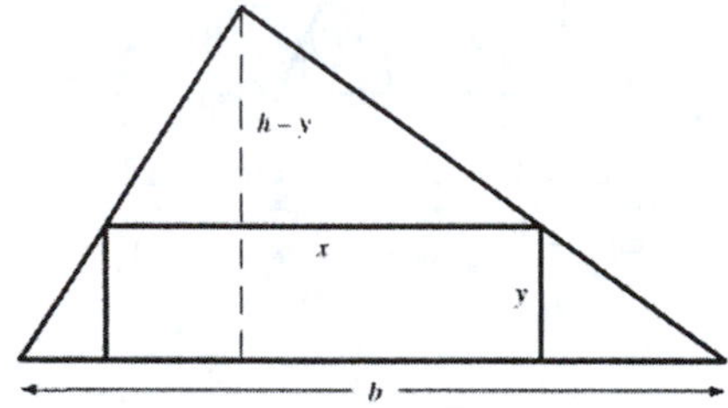

By similar triangles:

$$\frac{h-y}{h} = \frac{x}{b}$$
$$bh - by = xh$$
$$by = bh - xh$$
$$y = h - \frac{h}{b}x$$

So the area of the rectangle is given by: $A = xy = x\left(h - \frac{h}{b}x\right) = -\frac{h}{b}x^2 + hx$. This is a parabola opening downward, so it will have a maximum value when $x = \frac{-h}{2\left(-\frac{h}{b}\right)} = \frac{b}{2}$. So the maximum area is given by:

$$A\left(\frac{b}{2}\right) = -\frac{h}{b} \bullet \frac{b^2}{4} + h \bullet \frac{b}{2} = -\frac{hb}{4} + \frac{hb}{2} = \frac{hb}{4}$$

Since the triangle has an area of $\frac{hb}{2}$, the desired ratio is: $\frac{\frac{hb}{2}}{\frac{hb}{4}} = 2$

57. Call h the height of the triangle (see figure), and call y the indicated value:

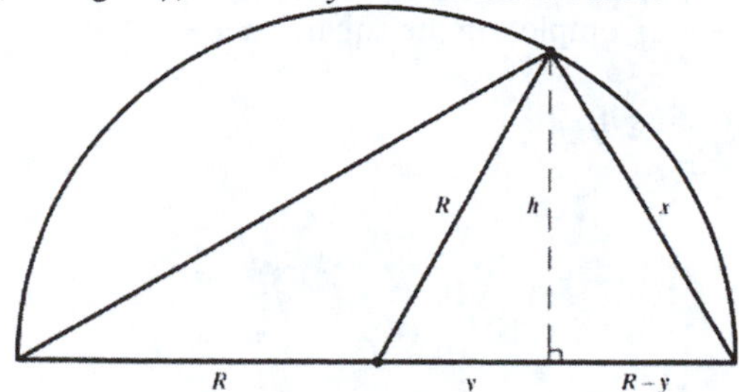

By the Pythagorean theorem, we have the following relationships:

$$y^2 + h^2 = R^2 \qquad \text{and} \qquad (R-y)^2 + h^2 = x^2$$
$$y^2 = R^2 - h^2$$
$$y = \sqrt{R^2 - h^2}$$

Substituting:

$$\left(R - \sqrt{R^2 - h^2}\right)^2 + h^2 = x^2$$
$$R^2 - 2R\sqrt{R^2 - h^2} + R^2 - h^2 + h^2 = x^2$$
$$2R^2 - 2R\sqrt{R^2 - h^2} = x^2$$

Solve for h:

$$-2R\sqrt{R^2 - h^2} = x^2 - 2R^2$$
$$\sqrt{R^2 - h^2} = R - \frac{x^2}{2R}$$
$$R^2 - h^2 = R^2 - x^2 + \frac{x^4}{4R^2}$$
$$-h^2 = \frac{x^4}{4R^2} - x^2$$
$$h^2 = -\frac{x^4}{4R^2} + x^2$$

Now use the hint. Since the area of the triangle is given by $A = \frac{1}{2}(2R)(h) = Rh$, we maximize:

$$A^2 = R^2h^2 = R^2\left(\frac{-x^4}{4R^2} + x^2\right) = -\tfrac{1}{4}x^4 + R^2x^2$$

This will have a maximum value when: $x^2 = \dfrac{-b}{2a} = \dfrac{-R^2}{-\frac{1}{2}} = 2R^2$

Therefore we have: $A^2 = -\frac{1}{4}\left(2R^2\right)^2 + R^2\left(2R^2\right) = -R^4 + 2R^4 = R^4$

So the maximum value is $A^2 = R^4$, or $A = R^2$. Thus the shaded area is: $\frac{1}{2}\left(\pi R^2\right) - R^2 = \dfrac{\pi R^2}{2} - R^2 = R^2\left(\dfrac{\pi - 2}{2}\right)$

59. We wish to minimize the sum $x+\frac{1}{x}$. Completing the square: $x+\frac{1}{x}=\left(\sqrt{x}-\sqrt{\frac{1}{x}}\right)^2+2$

The smallest possible value of 2 occurs if:

$$\sqrt{x}=\sqrt{\frac{1}{x}}$$
$$x=\frac{1}{x}$$
$$x^2=1$$
$$x=1$$

If both numbers are 1, the minimum sum of 2 occurs.

4.6 Polynomial Functions

1. The functions in (a) through (d) are polynomial functions. Graph (e) is not the graph of a polynomial function, since it has a break or discontinuity in it. Graph (f) is not the graph of a polynomial function, since its domain is not all real numbers.

3. **a.** Sketching the graphs:

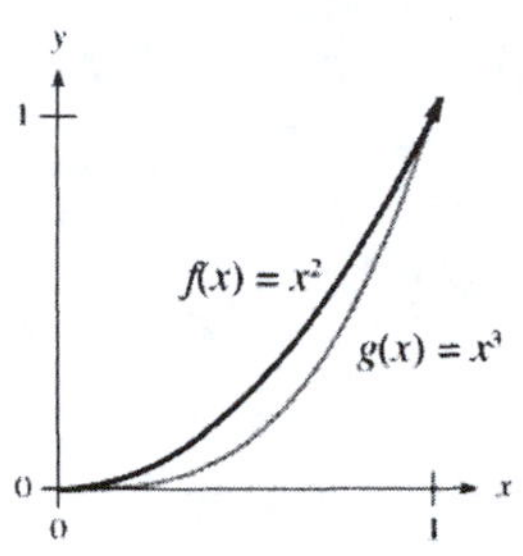

b. On the interval $[0,1]$, the average rates of change are:

$$\frac{f(1)-f(0)}{1-0}=\frac{1-0}{1}=1 \qquad \frac{g(1)-g(0)}{1-0}=\frac{1-0}{1}=1$$

On the interval $\left[0,\frac{1}{2}\right]$, the average rates of change are:

$$\frac{f\left(\frac{1}{2}\right)-f(0)}{\frac{1}{2}-0}=\frac{\frac{1}{4}-0}{\frac{1}{2}}=\frac{1}{2} \qquad \frac{g\left(\frac{1}{2}\right)-g(0)}{\frac{1}{2}-0}=\frac{\frac{1}{8}-0}{\frac{1}{2}}=\frac{1}{4}$$

5. This is $y=x^2$ translated 2 units to the right and 1 unit up. There are no x-intercepts and the y-intercept is 5. Sketching the graph:

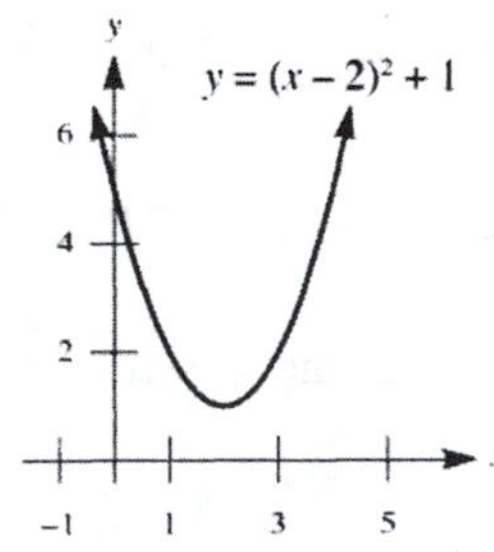

7. This is $y = x^4$ reflected across the x-axis and translated 1 unit to the right. The x-intercept is 1 and the y-intercept is -1. Sketching the graph:

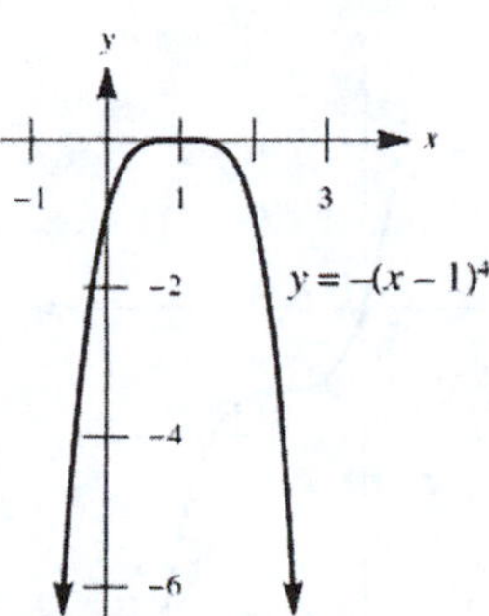

9. This is $y = x^3$ translated 4 units to the right and 2 units down. The x-intercept is $4+\sqrt[3]{2}$ and the y-intercept is -66. Sketching the graph:

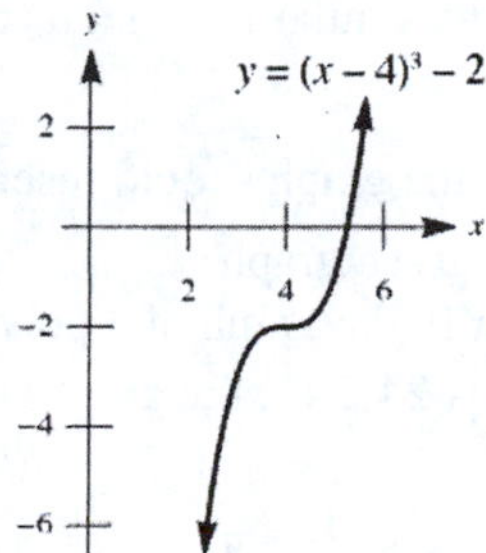

11. This is $y = 2x^4$ translated 5 units to the right and reflected across the x-axis. The x-intercept is -5 and the y-intercept is -1250. Sketching the graph:

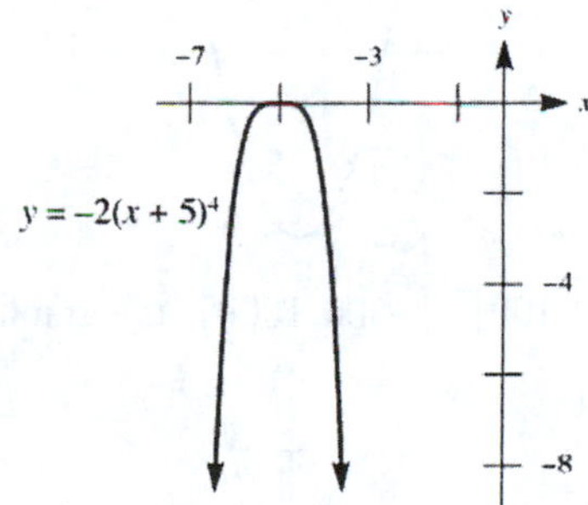

13. This is $y = \frac{1}{2}x^5$ translated 1 unit to the left. The x-intercept is -1 and the y-intercept is $\frac{1}{2}$. Sketching the graph:

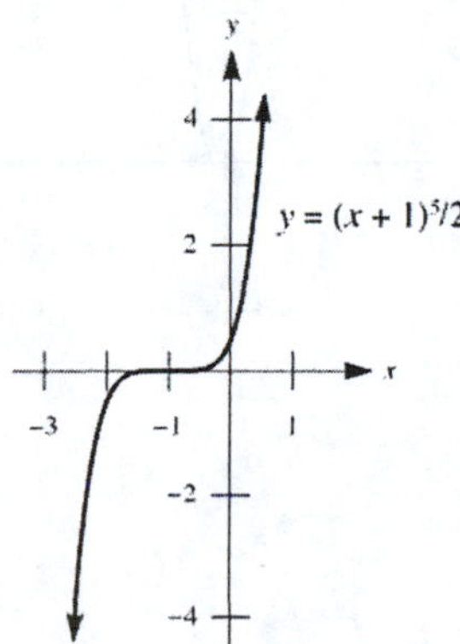

15. This is $y = x^3$ translated 1 unit to the right and 1 unit down, then reflected across the x-axis. The x- and y-intercepts are both 0. Sketching the graph:

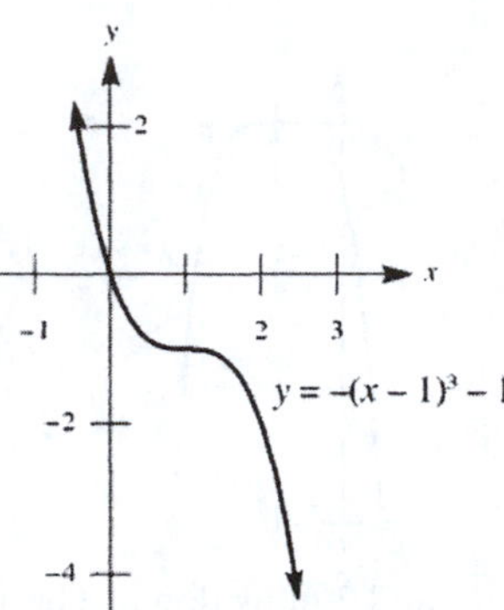

17. This graph has 4 turning points, but a polynomial function of degree 3 can have at most 2 turning points.

19. As $|x|$ gets very large, our function should be similar to $f(x) = a_3x^3$. But $f(x)$ does not have a parabolic shape like the given graph.

21. As $|x|$ gets very large with x negative, then the graph should resemble $2x^5$. But the y-values of $2x^5$ are always negative when x is negative, contrary to the given graph.

23. This graph has a corner, which cannot occur in the graph of a polynomial function.

25. **a.** Using the viewing rectangle $[-6,12]\times[-24,24]$, the graphs appear as:

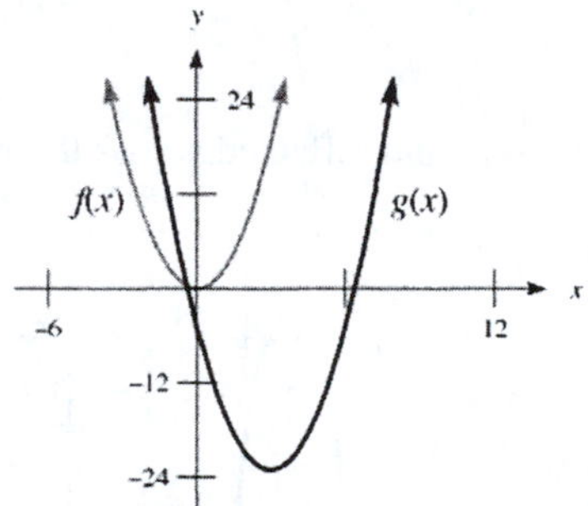

b. Using the viewing rectangle $[-100,100]\times[-500,1000]$, the graphs appear as:

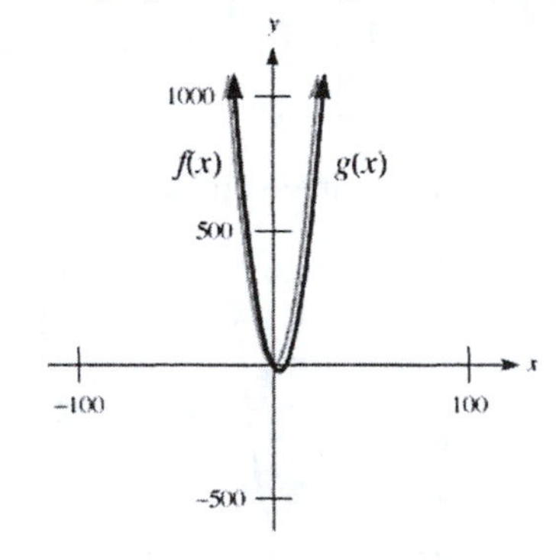

27. **a.** The x-intercepts are 2, 1, and -1, and the y-intercept is 2. The signs of y are given by:

Interval	$y=(x-2)(x-1)(x+1)$
$(-\infty,-1)$	neg.
$(-1,1)$	pos.
$(1,2)$	neg.
$(2,\infty)$	pos.

Sketching the excluded regions:

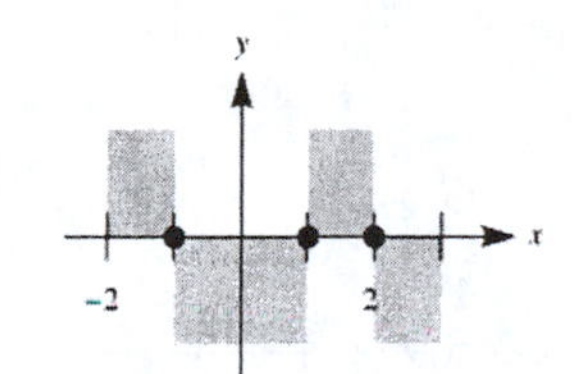

b. Now graph the function:

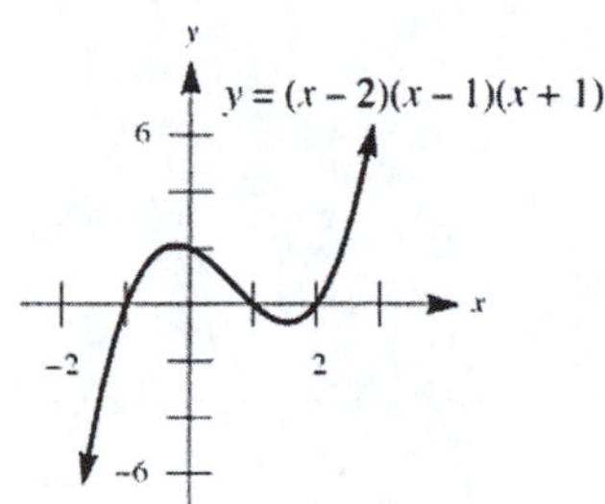

29. **a.** The x-intercepts are 0, 2, and 1, and the y-intercept is 0. The signs of y are given by:

Interval	$y=2x(x-2)(x-1)$
$(-\infty,0)$	neg.
$(0,1)$	pos.
$(1,2)$	neg.
$(2,\infty)$	pos.

Sketching the excluded regions:

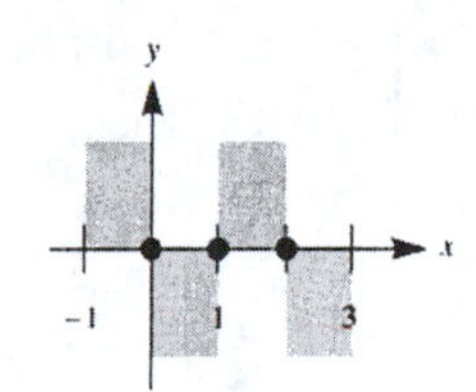

b. Now graph the function:

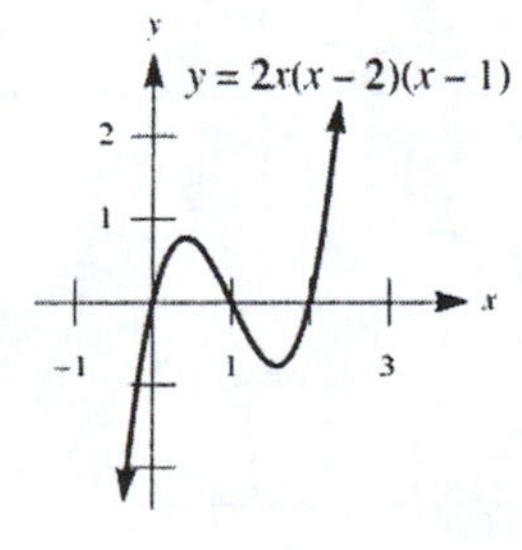

31. a. First factor to find the x-intercepts: $y = x\left(x^2 - 4x - 5\right) = x(x-5)(x+1)$

The x-intercepts are 0, 5, and –1, and the y-intercept is 0. The signs of y are given by:

Interval	$y = x(x-5)(x+1)$
$(-\infty, -1)$	neg.
$(-1, 0)$	pos.
$(0, 5)$	neg.
$(5, \infty)$	pos.

Sketching the excluded regions:

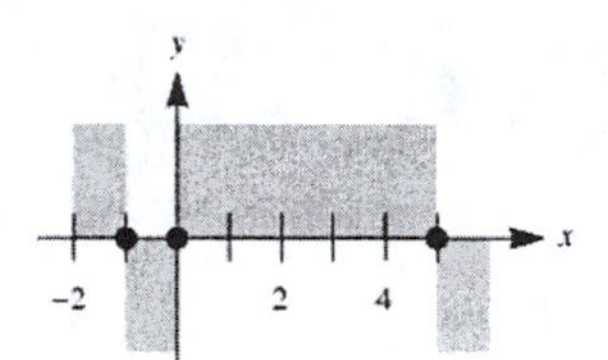

b. Now graph the function:

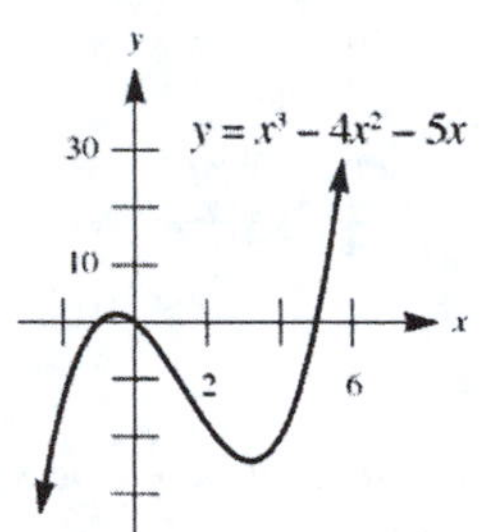

33. a. First factor (by grouping) to find the x-intercepts:

$$y = x^2(x+3) - 4(x+3) = (x+3)\left(x^2 - 4\right) = (x+3)(x+2)(x-2)$$

The x-intercepts are –3, –2, and 2, and the y-intercept is –12. The signs of y are given by:

Interval	$y = (x+3)(x+2)(x-2)$
$(-\infty, -3)$	neg.
$(-3, -2)$	pos.
$(-2, 2)$	neg.
$(2, \infty)$	pos.

Sketching the excluded regions:

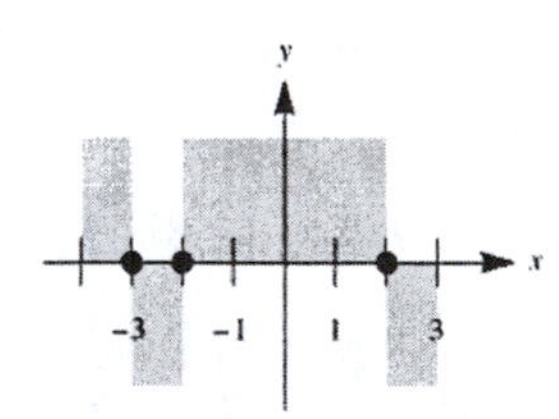

b. Now graph the function:

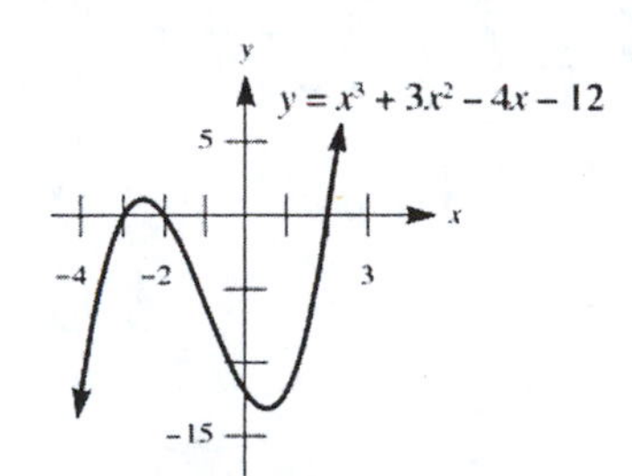

35. **a.** The x-intercepts are 0 and -2, and the y-intercept is 0. The signs of y are given by:

Interval	$y=x^3(x+2)$
$(-\infty,-2)$	pos.
$(-2,0)$	neg.
$(0,\infty)$	pos.

Sketching the excluded regions:

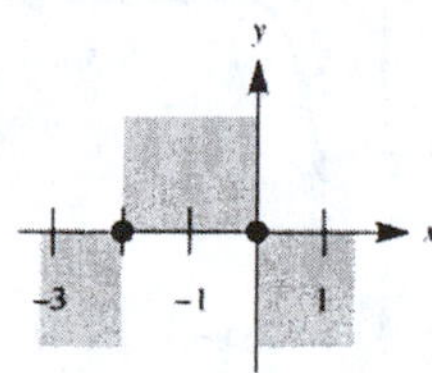

b. Near $x=0$, we have the approximation $y \approx x^3(0+2) = 2x^3$. So in the immediate vicinity of $x=0$, the graph of y resembles $y=2x^3$:

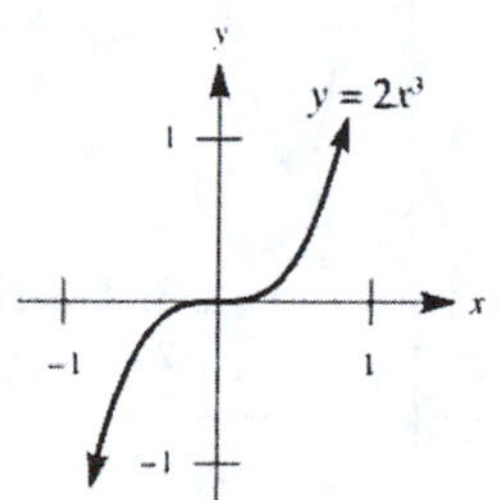

c. Now graph the function:

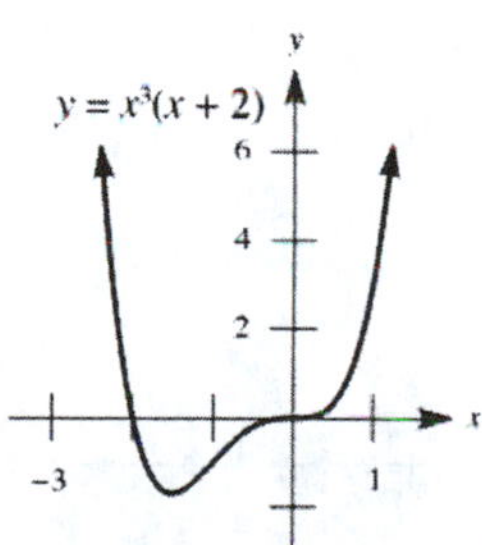

37. **a.** The x-intercepts are 1 and 4, and the y-intercept is 128. The signs of y are given by:

Interval	$y=2(x-1)(x-4)^3$
$(-\infty,1)$	pos.
$(1,4)$	neg.
$(4,\infty)$	pos.

Sketching the excluded regions:

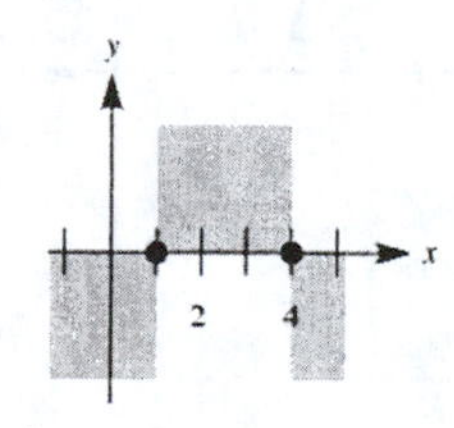

b. Near $x = 4$, we have the approximation $y \approx 2(4-1)(x-4)^3 = 6(x-4)^3$. So in the immediate vicinity of $x = 4$, the graph of y resembles $y = 6(x-4)^3$:

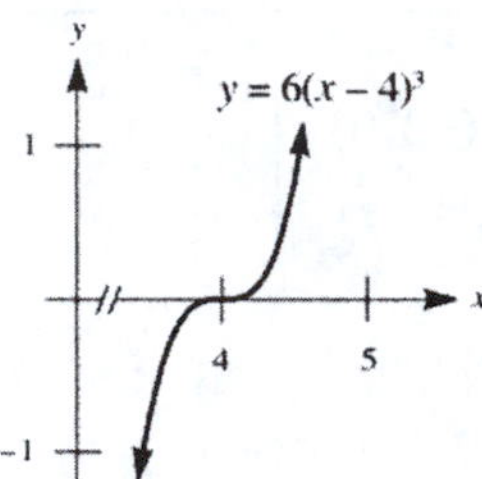

c. Now graph the function:

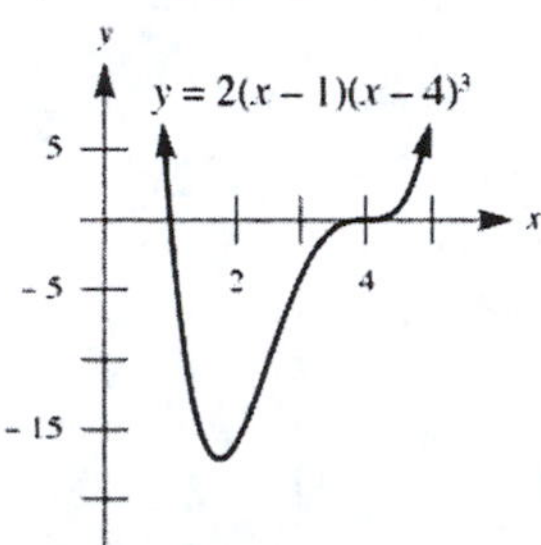

39. **a.** The x-intercepts are -1, 1, and 3, and the y-intercept is 3. The signs of y are given by:

Interval	$y=(x+1)^2(x-1)(x-3)$
$(-\infty,-1)$	pos.
$(-1,1)$	pos.
$(1,3)$	neg.
$(3,\infty)$	pos.

Sketching the excluded regions:

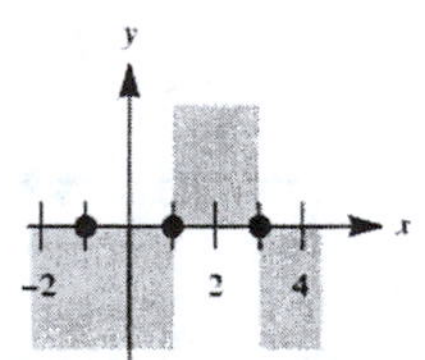

b. Near $x = -1$, we have the approximation $y \approx (x+1)^2(-2)(-4) = 8(x+1)^2$. So in the immediate vicinity of $x = -1$, the graph of y resembles $y = 8(x+1)^2$:

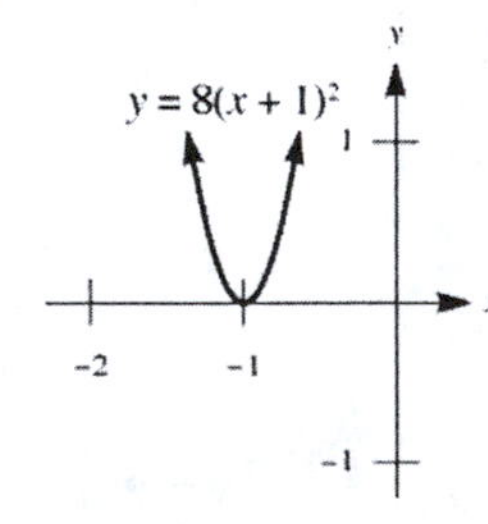

c. Now graph the function:

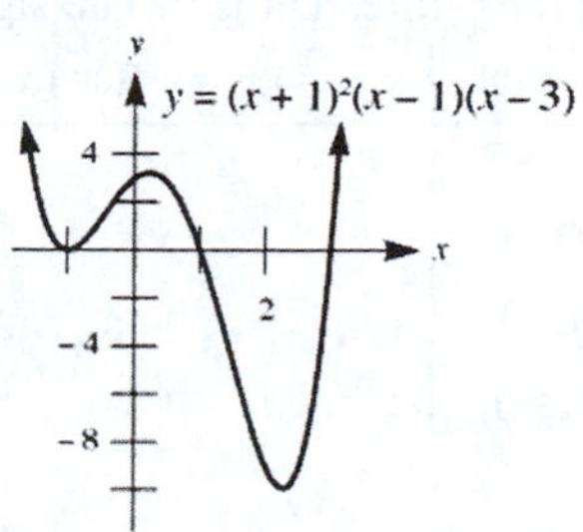

41. a. The x-intercepts are 0, 4, and –2, and the y-intercept is 0. The signs of y are given by:

Interval	$y=-x^3(x-4)(x+2)$
$(-\infty,-2)$	pos.
$(-2,0)$	neg.
$(0,4)$	pos.
$(4,\infty)$	neg.

Sketching the excluded regions:

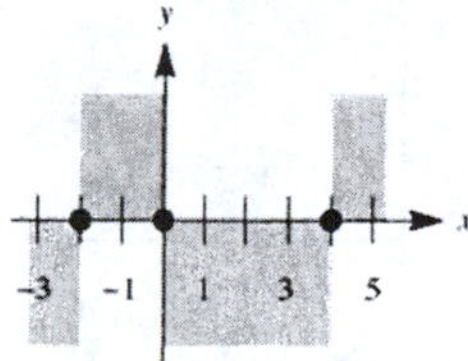

b. Near $x = 0$, we have the approximation $y \approx -x^3(0-4)(0+2) = 8x^3$. So in the immediate vicinity of $x = 0$, the graph resembles $y = 8x^3$:

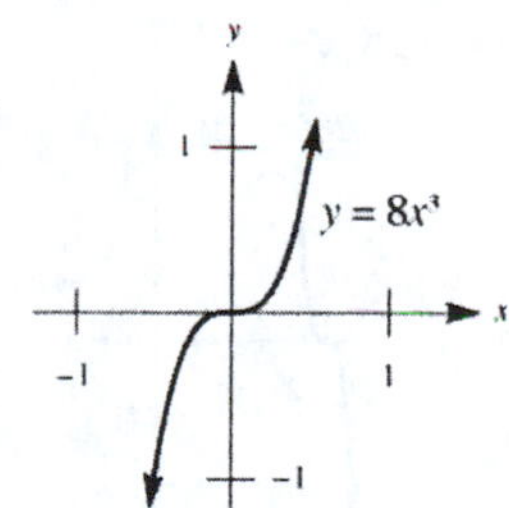

c. Now graph the function:

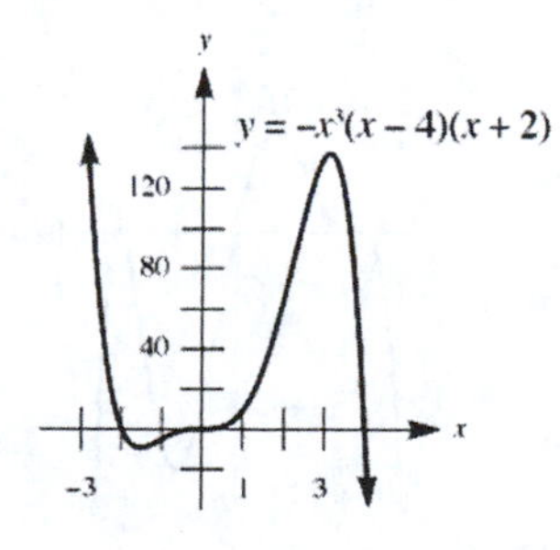

43. **a.** The x-intercepts are 0, 2, and –2, and the y-intercept is 0. The signs of y are given by:

Interval	$y=-4x(x-2)^2(x+2)^3$
$(-\infty,-2)$	neg.
$(-2,0)$	pos.
$(0,2)$	neg.
$(2,\infty)$	neg.

Sketching the excluded regions:

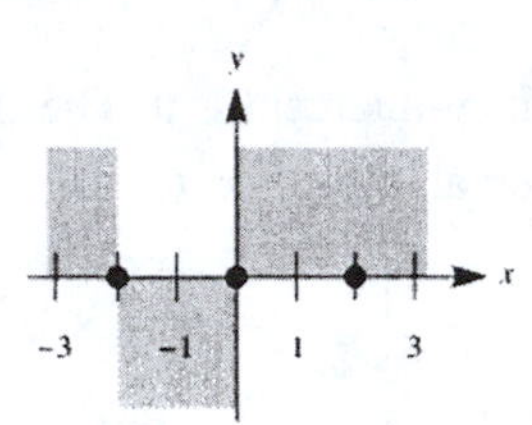

b. Near $x = 2$, we have the approximation $y \approx -4(2)(x-2)^2(2+2)^3 = -512(x-2)^2$. So in the immediate vicinity of $x = 2$, the graph resembles $y = -512(x-2)^2$:

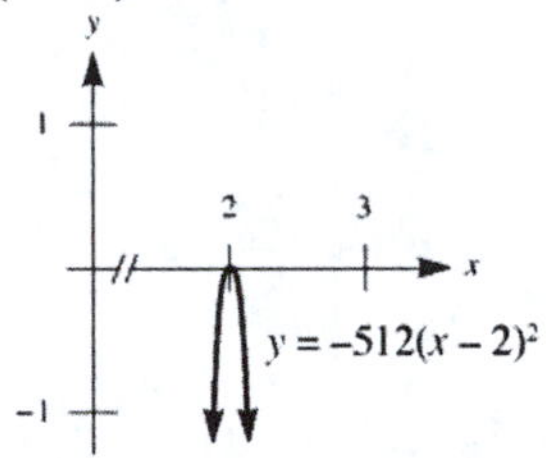

Near $x = -2$, we have the approximation $y \approx -4(-2)(-2-2)^2(x+2)^3 = 128(x+2)^3$. So in the immediate vicinity of $x = -2$, the graph resembles $y = 128(x+2)^3$:

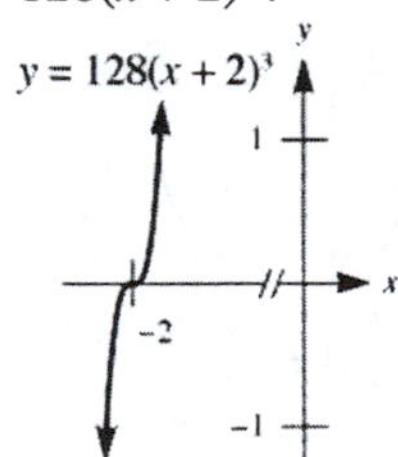

c. Now graph the function:

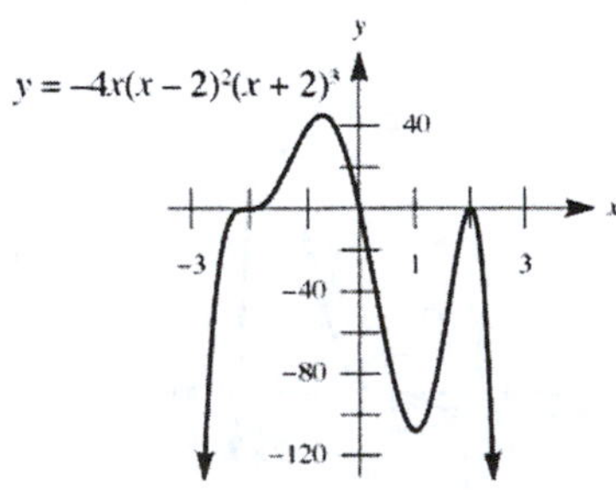

45. From the figure, the average rates of change are:

$$\frac{\Delta g}{\Delta x}=\frac{g(1.25)-g(0)}{1.25-0}=\frac{5-0}{1.25}\approx 4 \qquad \frac{\Delta f}{\Delta x}=\frac{f(1.25)-f(0)}{1.25-0}=\frac{2.5-0}{1.25}\approx 2$$

47. Finding the x-intercepts:

$$x^3 - 3x^2 - 5x = 0$$
$$x\left(x^2 - 3x - 5\right) = 0$$

So $x = 0$ is one x-intercept. Find the other two using the quadratic formula:

$$x = \frac{-(-3) \pm \sqrt{(-3)^2 - 4(1)(-5)}}{2(1)} = \frac{3 \pm \sqrt{29}}{2} \approx -1.193, 4.193$$

Note that the x-intercepts are consistent with the given graph.

49. Finding the x-intercepts:

$$x^3 + 6x^2 - 3x - 18 = 0$$
$$x^2(x+6) - 3(x+6) = 0$$
$$(x+6)\left(x^2 - 3\right) = 0$$
$$x = -6, \pm\sqrt{3} \approx \pm 1.732$$

Note that the x-intercepts are consistent with the given graph.

51. **a.** Graphing the function:

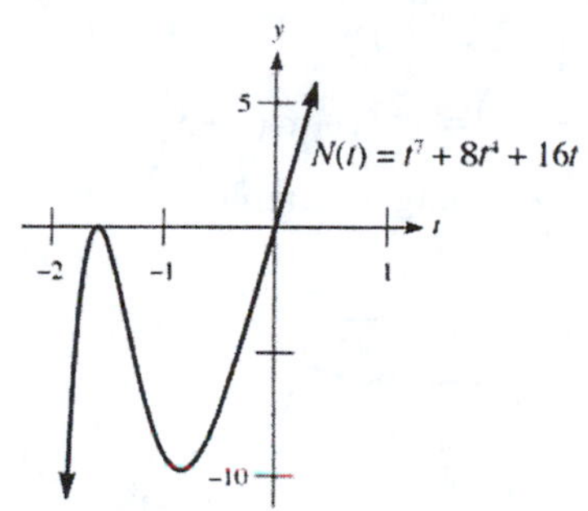

b. The t-intercepts are approximately 0.0 and −1.6.

c. Finding the t-intercepts:

$$t^7 + 8t^4 + 16t = 0$$
$$t\left(t^6 + 8t^3 + 16\right) = 0$$
$$t\left(t^3 + 4\right)^2 = 0$$
$$t = 0, -\sqrt[3]{4} \approx -1.587$$

Note that the t-intercepts are consistent with the graph.

53. From left to right, they are $f(x) = x$, $g(x) = x^2$, $h(x) = x^3$, $F(x) = x^4$, $G(x) = x^5$, and $H(x) = x^6$.

55. Find where $0 \le H(x) < 0.1$:

$$0 \le x^6 < 0.1$$
$$0 \le x < 0.68$$

When x lies in the interval $[0,0.68)$, then $H(x)$ will lie in the interval $[0, 0.1)$.

57. Find where:

$$g(t) - F(t) = 0.26$$
$$t^2 - t^4 = 0.26$$
$$t^4 - t^2 + 0.26 = 0$$

Using the quadratic formula: $t^2 = \dfrac{1 \pm \sqrt{1 - 4(0.26)}}{2} = \dfrac{1 \pm \sqrt{1 - 1.04}}{2} = \dfrac{1 \pm \sqrt{-0.04}}{2}$

Since this equation has no real solutions, there is no such value of t.

59. **a.** Set the two y-coordinates equal:

$$4x = \frac{1}{100}x^2$$
$$0 = \frac{1}{100}x^2 - 4x$$
$$0 = \frac{1}{100}\left(x^2 - 400x\right)$$
$$0 = \frac{1}{100}x(x - 400)$$

The graphs intersect at the origin but also at the point (400,1600).

b. Graphing the two functions:

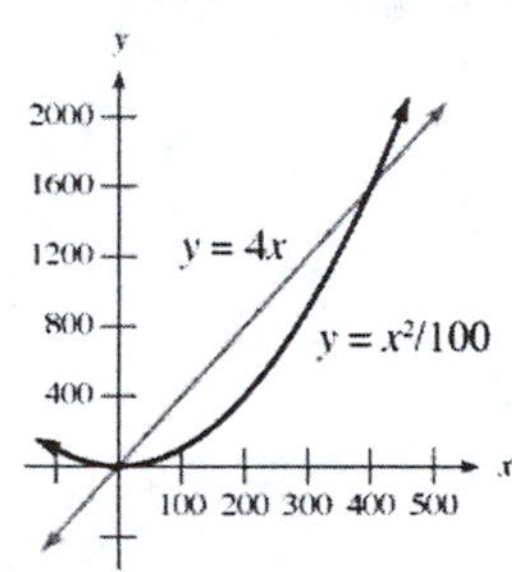

Note that there are two intersection points.

61. **a.** First factor to obtain: $D(x) = x^2\left(1 - x^2\right) = x^2(1 + x)(1 - x)$

So the x-intercepts are 0, –1, and 1. Sketching the graph:

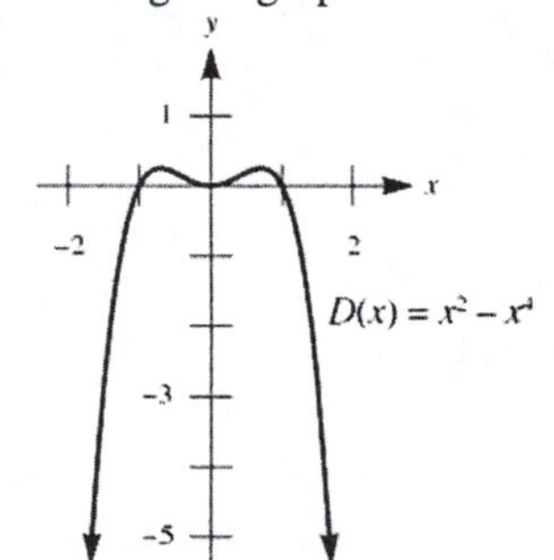

b. Complete the square: $D(x) = x^2 - x^4 = -\left(x^4 - x^2\right) = -\left(x^4 - x^2 + \frac{1}{4}\right) + \frac{1}{4} = -\left(x^2 - \frac{1}{2}\right)^2 + \frac{1}{4}$

The turning points are at $\left(\pm\frac{1}{\sqrt{2}}, \frac{1}{4}\right) = \left(\pm\frac{\sqrt{2}}{2}, \frac{1}{4}\right)$, which yield maximum values. Note that the graph also has a turning point at (0,0), which is a minimum value.

c. Graph the two functions:

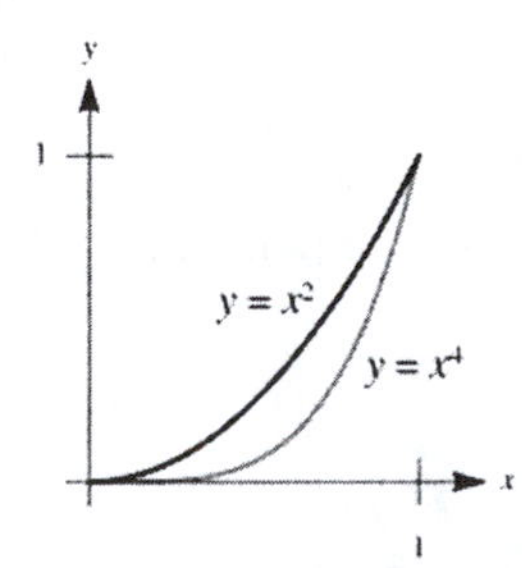

From part **b**, we know that the maximum vertical distance between the two curves is $\frac{1}{4}$.

63. a. Let $(x, 1 - x^4)$ represent the point in the first quadrant where the rectangle touches the curve. Since the base of the rectangle is x and the height is $1 - x^4$, the area is given by: $A(x) = x\left(1 - x^4\right) = x - x^5$

b. Graphing the area function:

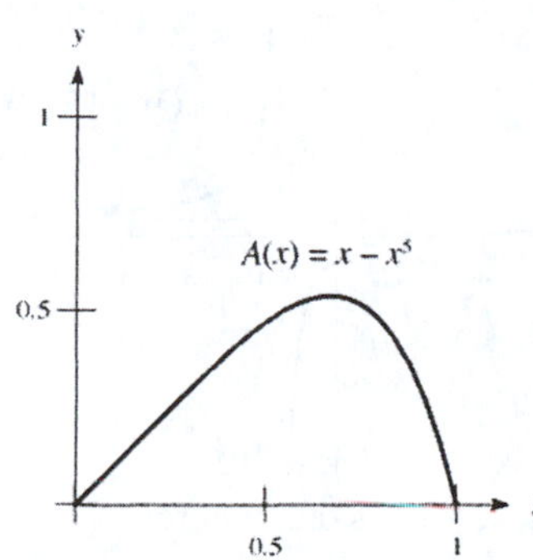

The maximum point on the resulting curve occurs when $x \approx 0.669$, with a maximum possible area of 0.53 square units.

65. a. If $0 \le x \le 1$ and $n > m$, then $0 \le x^n \le x^m$, since x^n has more multiplications of a number less than or equal to 1. If $x > 1$, then $x^n > x^m$, since x^n has more multiplications of a number greater than 1.

b. From part **a**, if $|x| < 1$, then $x^n < x^m$, so the graph of $y = x^n$ lies below the graph of $y = x^m$. If $|x| > 1$, then $x^n > x^m$, so the graph of $y = x^n$ lies above the graph of $y = x^m$. These are valied since m and n are even, thus ensuring $x^m \ge 0$ and $x^n \ge 0$.

c. If m and n are positive odd integers, from part **a** the graph of $y = x^m$ lies above the graph of $y = x^n$ for $0 < x < 1$, and it lies below the graph of $y = x^n$ for $x > 1$. If $-1 < x < 0$, however, $x^n > x^m$ (since x is negative), and so the graph of $y = x^n$ lies above the graph of $y = x^m$. If $x < 1$, then $x^n < x^m$, so the graph of $y = x^n$ lies below the graph of $y = x^m$.

4.7 Rational Functions

1. a. For the domain, we must exclude those values of x where:

$$4x - 12 = 0$$
$$4x = 12$$
$$x = 3$$

So the domain is all real numbers except 3, or $(-\infty, 3) \cup (3, \infty)$. For the x-intercepts, we must find where:

$$3x + 15 = 0$$
$$3x = -15$$
$$x = -5$$

For the y-intercept, let $x = 0$ to obtain: $y = \dfrac{0 + 15}{0 - 12} = -\frac{5}{4}$. The vertical asymptote occurs where the denominator is 0, which is the line $x = 3$. For the horizontal asymptote: $y = \dfrac{3x + 15}{4x - 12} = \dfrac{x\left(3 + \frac{15}{x}\right)}{x\left(4 - \frac{12}{x}\right)} = \dfrac{3 + \frac{15}{x}}{4 - \frac{12}{x}} \approx \dfrac{3}{4}$.

b. Sketching the graph of $y = \dfrac{3x+15}{4x-12}$:

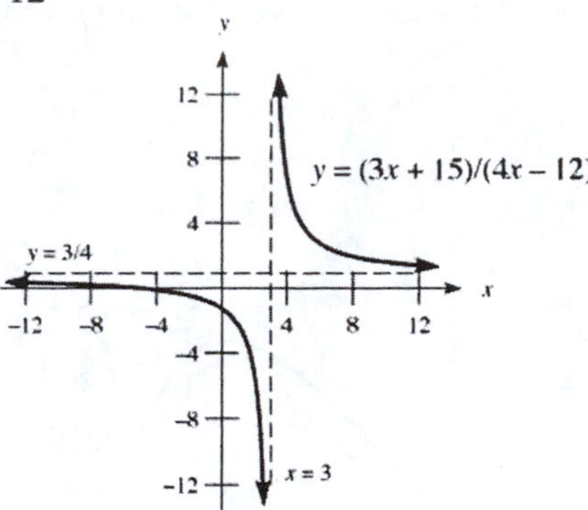

3. **a.** For the domain, we must exclude those values of x where:

$$2x^2 = 0$$
$$x = 0$$

So the domain is all real numbers except 0, or $(-\infty, 0) \cup (0, \infty)$. For the x-intercepts, we must find where:

$$6x^2 - 5x + 1 = 0$$
$$(3x-1)(2x-1) = 0$$
$$x = \tfrac{1}{3}, \tfrac{1}{2}$$

There is no y-intercept, since $x = 0$ is not in the domain of the function. The vertical asymptotes occur where the denominator is 0, which is the line $x = 0$. For the horizontal asymptote:

$$y = \frac{6x^2 - 5x + 1}{2x^2} = \frac{x^2\left(6 - \frac{5}{x} + \frac{1}{x^2}\right)}{x^2(2)} = \frac{6 - \frac{5}{x} + \frac{1}{x^2}}{2} \approx 3$$

b. Sketching the graph of $y = \dfrac{6x^2 - 5x + 1}{2x^2}$:

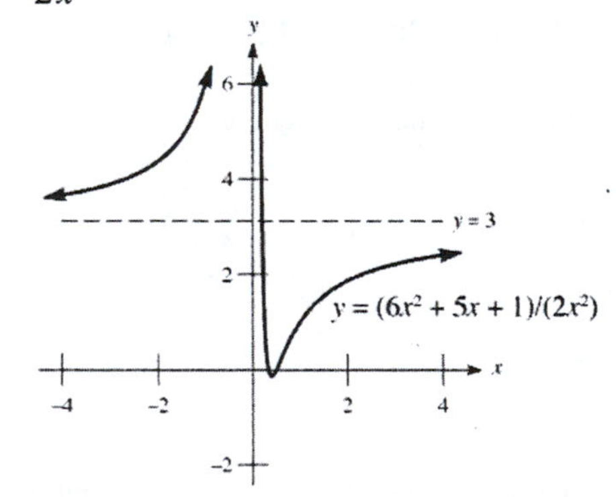

5. a. For the domain, we must exclude those values of x where:

$$4x^2 - 1 = 0$$
$$x^2 = \frac{1}{4}$$
$$x = \pm\frac{1}{2}$$

So the domain is all real numbers except $\pm\frac{1}{2}$, or $\left(-\infty,-\frac{1}{2}\right)\cup\left(-\frac{1}{2},\frac{1}{2}\right)\cup\left(\frac{1}{2},\infty\right)$. For the x-intercepts, we must find where:

$$x^2 - 9 = 0$$
$$x^2 = 9$$
$$x = \pm 3$$

For the y-intercept, let $x = 0$ to obtain: $y = \frac{-9}{-1} = 9$. The vertical asymptotes occur where the denominator is 0, which are the lines $x = \frac{1}{2}$ and $x = -\frac{1}{2}$. For the horizontal asymptote: $y = \frac{x^2 - 9}{4x^2 - 1} = \frac{x^2\left(1 - \frac{9}{x^2}\right)}{x^2\left(4 - \frac{1}{x^2}\right)} = \frac{1 - \frac{9}{x^2}}{4 - \frac{1}{x^2}} \approx \frac{1}{4}$

b. Sketching the graph of $y = \frac{x^2 - 9}{4x^2 - 1}$:

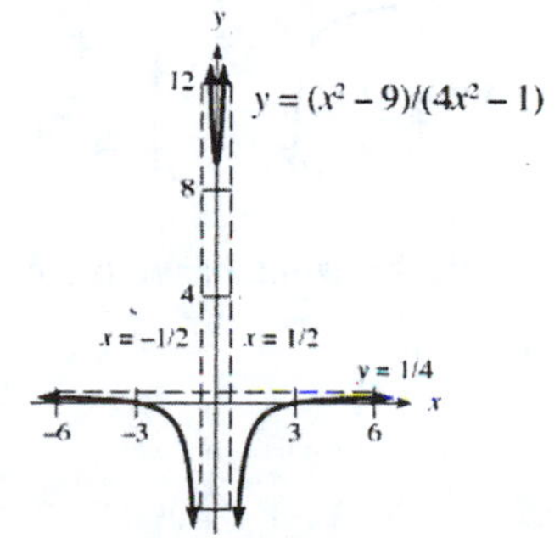

7. a. For the domain, we must exclude those values of x where:

$$2x^3 + x^2 - 3x = 0$$
$$x\left(2x^2 + x - 3\right) = 0$$
$$x(2x + 3)(x - 1) = 0$$
$$x = -\frac{3}{2}, 0, 1$$

So the domain is all real numbers except $-\frac{3}{2}, 0, 1$, or $\left(-\infty,-\frac{3}{2}\right)\cup\left(-\frac{3}{2},0\right)\cup(0,1)\cup(1,\infty)$. For the x-intercepts, we must find where:

$$3x^2 - 2x - 8 = 0$$
$$(3x + 4)(x - 2) = 0$$
$$x = -\frac{4}{3}, 2$$

There is no y-intercept, since $x = 0$ is not in the domain of the function. The vertical asymptotes occur where the denominator is 0, which are the lines $x = -\frac{3}{2}$, $x = 0$, and $x = 1$. For the horizontal asymptote:

$$y = \frac{3x^2 - 2x - 8}{2x^3 + x^2 - 3x} = \frac{x^3\left(\frac{3}{x} - \frac{2}{x^2} - \frac{8}{x^3}\right)}{x^3\left(2 + \frac{1}{x} - \frac{3}{x^2}\right)} = \frac{\frac{3}{x} - \frac{2}{x^2} - \frac{8}{x^3}}{2 + \frac{1}{x} - \frac{3}{x^2}} \approx 0$$

b. Sketching the graph of $y=\dfrac{3x^2-2x-8}{2x^3+x^2-3x}$:

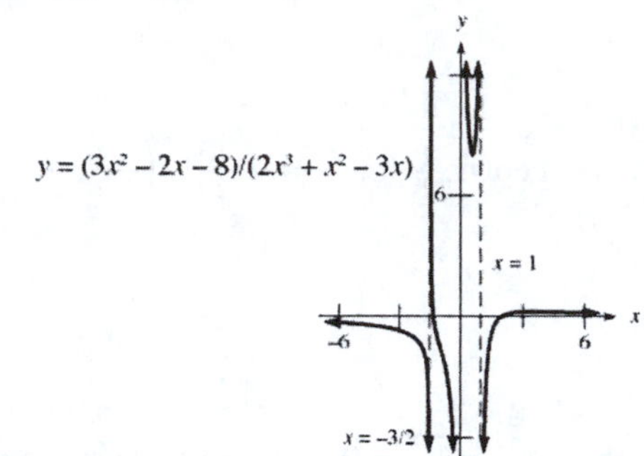

9. There is no x-intercept, the y-intercept is $\frac{1}{4}$, the horizontal asymptote is $y=0$, and the vertical asymptote is $x=-4$.

Sketching the graph of $y=\dfrac{1}{x+4}$:

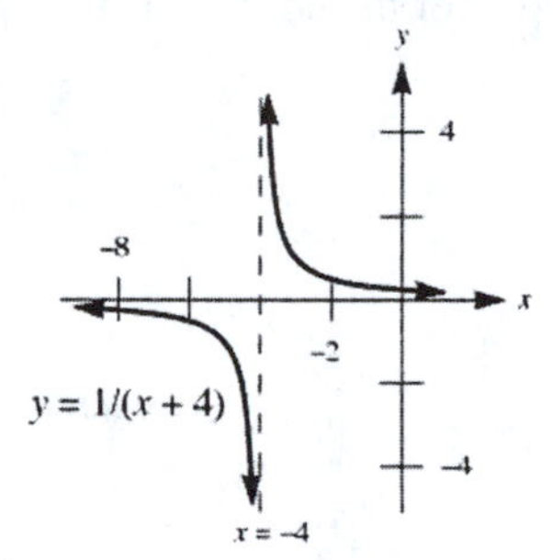

11. There is no x-intercept, the y-intercept is $\frac{3}{2}$, the horizontal asymptote is $y=0$, and the vertical asymptote is $x=-2$.

Sketching the graph of $y=\dfrac{3}{x+2}$:

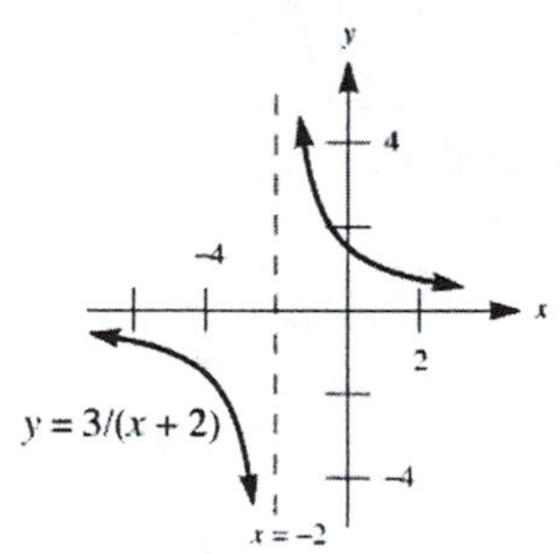

13. There is no x-intercept, the y-intercept is $\frac{2}{3}$, the horizontal asymptote is $y=0$, and the vertical asymptote is $x=3$.

Sketching the graph of $y=\dfrac{-2}{x-3}$:

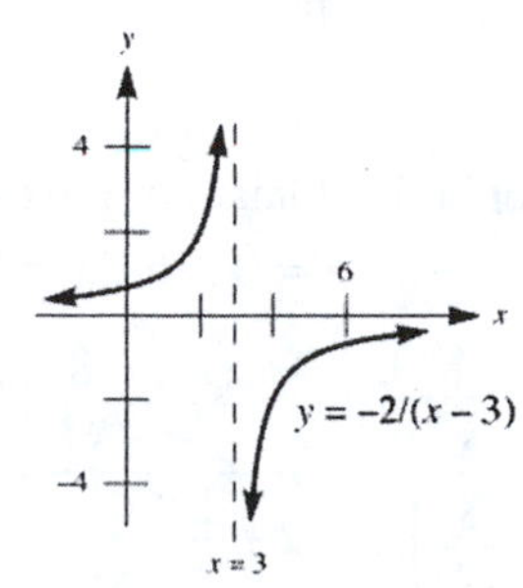

15. Using long division: $\frac{x-3}{x-1}=1-\frac{2}{x-1}$. The x-intercept is 3, the y-intercept is 3, the horizontal asymptote is $y=1$, and the vertical asymptote is $x=1$. Sketching the graph of $y=\frac{x-3}{x-1}$:

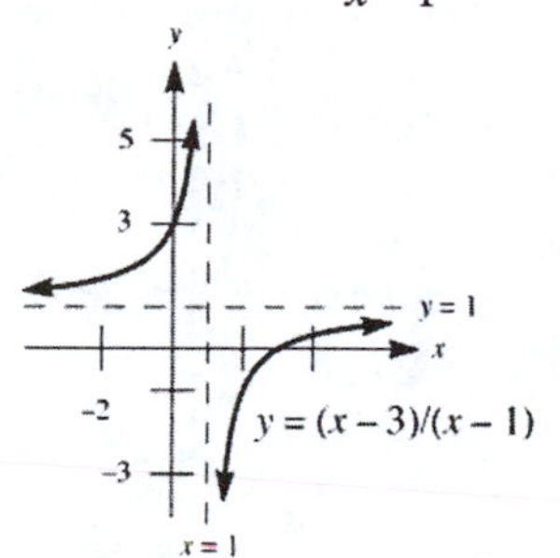

17. Using long division: $\frac{4x-2}{2x+1}=2-\frac{4}{2x+1}$. The x-intercept is $\frac{1}{2}$, the y-intercept is –2, the horizontal asymptote is $y=2$, and the vertical asymptote is $x=-\frac{1}{2}$. Sketching the graph of $y=\frac{4x-2}{2x+1}$:

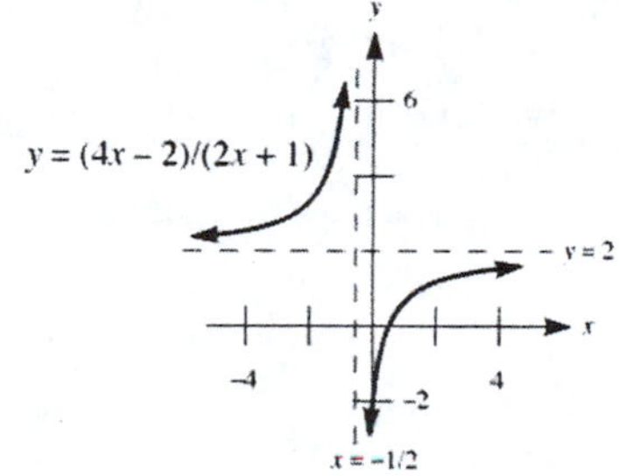

19. There is no x-intercept, the y-intercept is $\frac{1}{4}$, the horizontal asymptote is $y=0$, and the vertical asymptote is $x=2$. Sketching the graph of $y=\frac{1}{(x-2)^2}$:

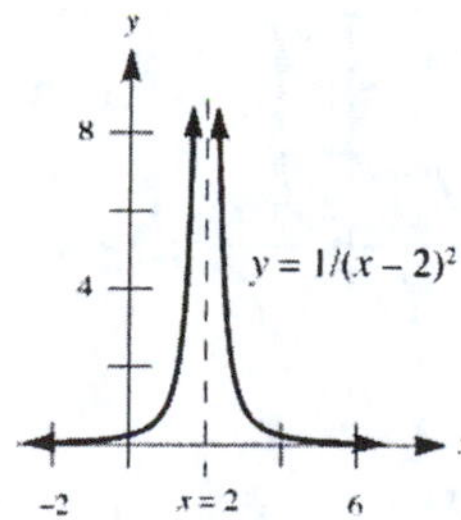

21. There is no x-intercept, the y-intercept is 3, the horizontal asymptote is $y=0$, and the vertical asymptote is $x=-1$. Sketching the graph of $y=\frac{3}{(x+1)^2}$:

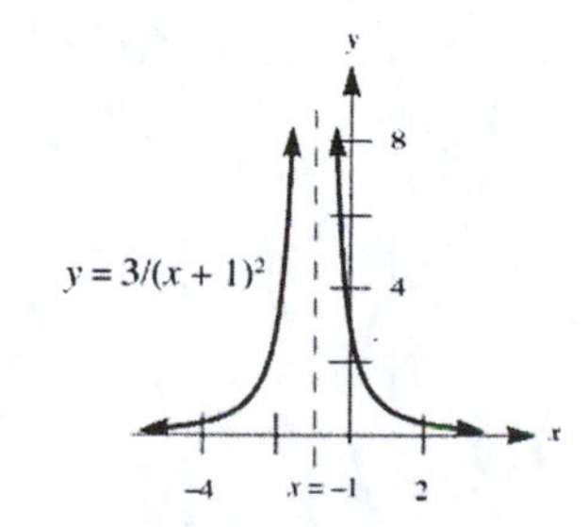

23. There is no x-intercept, the y-intercept is $\frac{1}{8}$, the horizontal asymptote is $y = 0$, and the vertical asymptote is $x = -2$.

Sketching the graph of $y = \dfrac{1}{(x+2)^3}$:

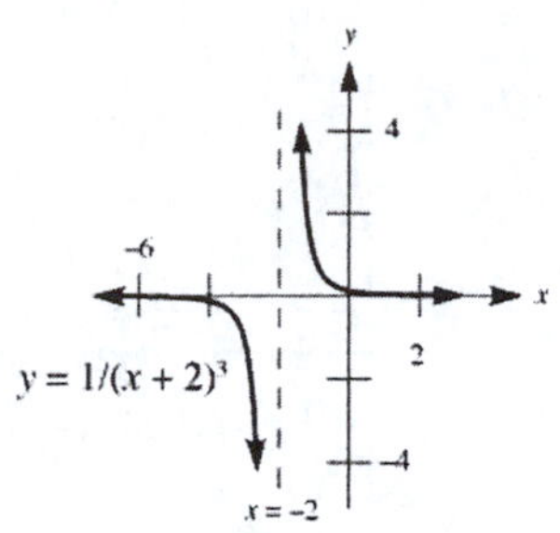

25. There is no x-intercept, the y-intercept is $-\frac{4}{125}$, the horizontal asymptote is $y = 0$, and the vertical asymptote is $x = -5$.

Sketching the graph of $y = \dfrac{-4}{(x+5)^3}$:

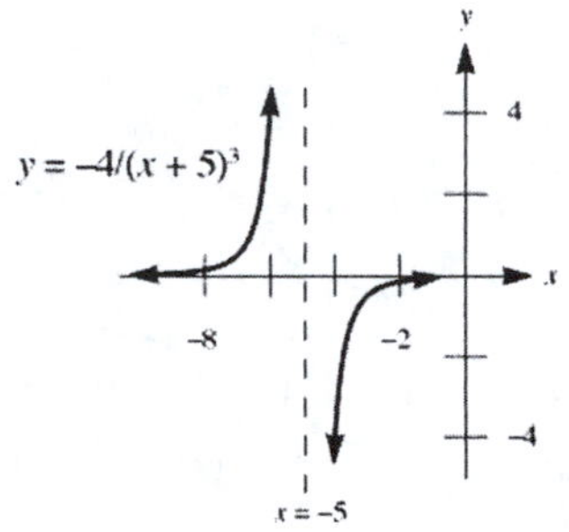

27. The x- and y-intercepts are 0, the horizontal asymptote is $y = 0$, and the vertical asymptotes are $x = -2$ and $x = 2$.

Sketching the graph of $y = \dfrac{-x}{(x+2)(x-2)}$:

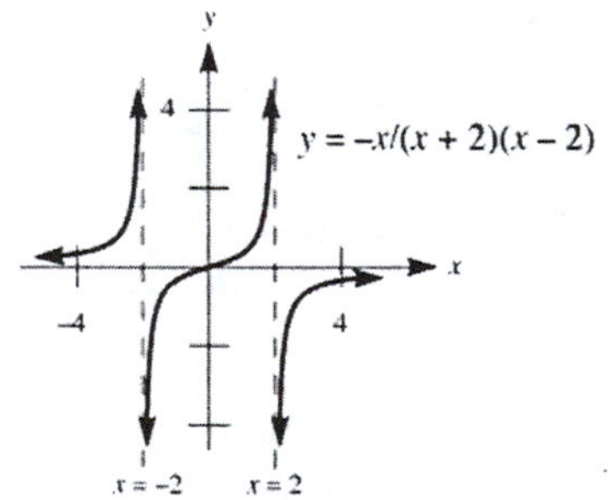

29. **a.** The x-intercept is 0, the y-intercept is 0, the horizontal asymptote is $y = 0$, and the vertical asymptotes are $x = 1$ and $x = -3$. Sketching the graph of $y = \dfrac{3x}{(x-1)(x+3)}$:

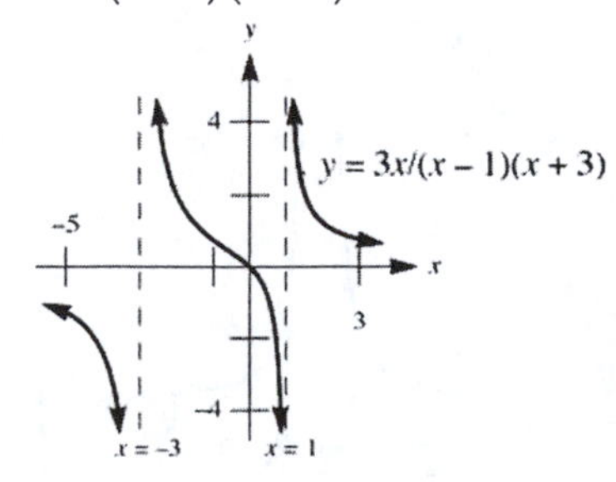

b. The x-intercept is 0, the y-intercept is 0, the horizontal asymptote is $y = 3$, and the vertical asymptotes are $x = 1$ and $x = -3$. Sketching the graph of $y = \dfrac{3x^2}{(x-1)(x+3)}$:

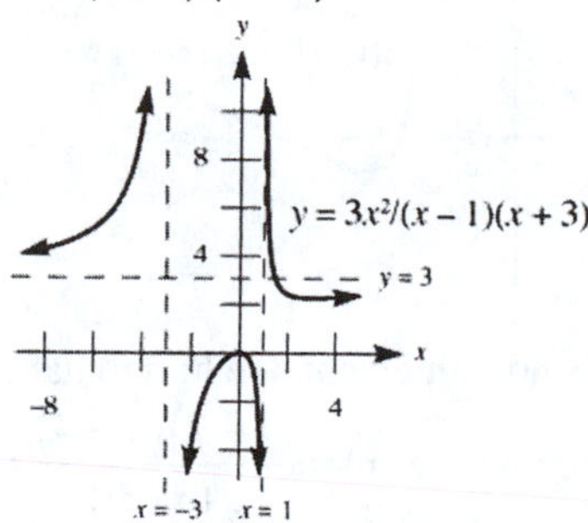

31. Factor the expression as $y = \dfrac{(4x+5)(x-1)}{(2x-5)(x+1)}$. The x-intercepts are $-\frac{5}{4}, 1$, the y-intercept is 1, the horizontal asymptote is $y = 2$, and the vertical asymptotes are $x = \frac{5}{2}$ and $x = -1$. Sketching the graph of $y = \dfrac{4x^2 + x - 5}{2x^2 - 3x - 5}$:

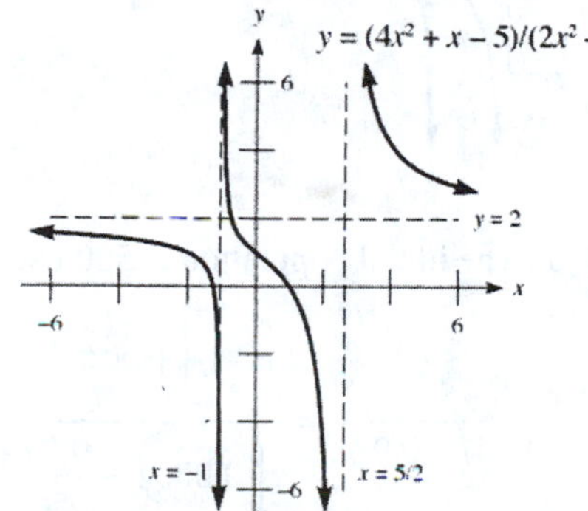

33. **a.** The x-intercepts are 2 and 4, there is no y-intercept, the horizontal asymptote is $y = 1$, and the vertical asymptotes are $x = 0$ and $x = 1$. Sketching the graph of $f(x) = \dfrac{(x-2)(x-4)}{x(x-1)}$:

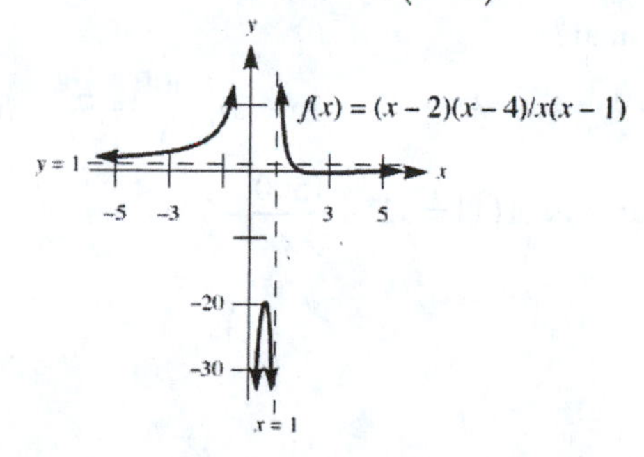

An expanded view of the graph near $x = 3$ illustrates the behavior of the curve for $x \geq 2$:

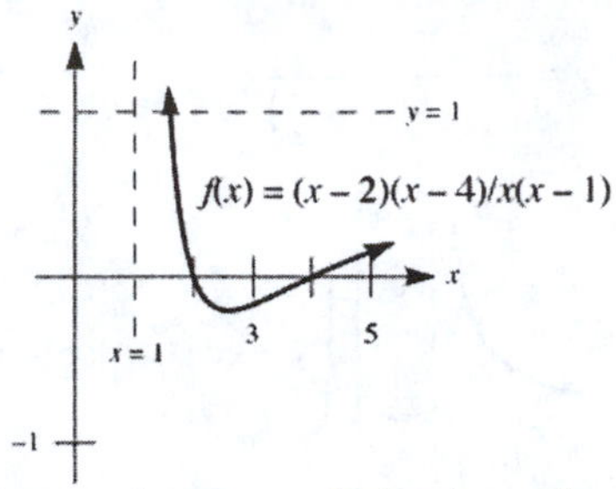

b. The x-intercepts are 2 and 4, there is no y-intercept, the horizontal asymptote is $y = 1$, and the vertical asymptotes are $x = 0$ and $x = 3$. Sketching the graph of $g(x) = \dfrac{(x-2)(x-4)}{x(x-3)}$:

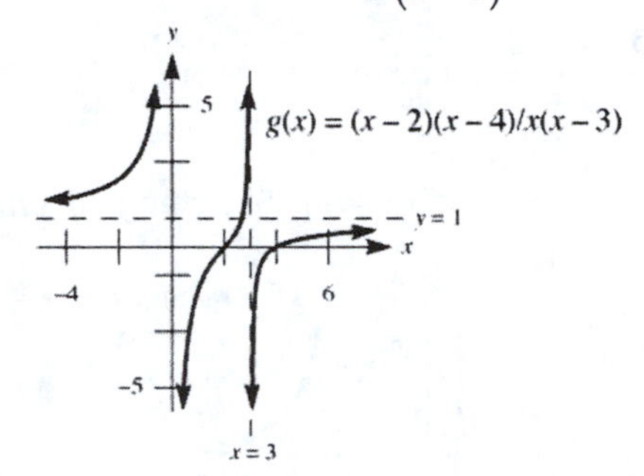

35. **a.** Substituting $t = 0$: $y = \dfrac{0+12}{0+0.024} = 500$. The initial population is 500 bacteria.

b. Finding the long term behavior: $y = \dfrac{t+12}{0.0004t+0.024} = \dfrac{t\left(1+\frac{12}{t}\right)}{t\left(0.0004+\frac{0.024}{t}\right)} = \dfrac{1+\frac{12}{t}}{0.0004+\frac{0.024}{t}} \approx \dfrac{1}{0.0004} = 2500$

The population is approaching 2500 bacteria.

37. **a.** Let x represent the entire poster width, and let y represent the entire poster length. Then $xy = 500$, so $y = \dfrac{500}{x}$.

Finding the area of the printed matter:

$$A = (x-1.5-1.5)(y-3-4) = (x-3)(y-7) = (x-3)\left(\frac{500}{x}-7\right) = 500 - \frac{1500}{x} - 7x + 21 = 521 - \frac{1500}{x} - 7x$$

Using functional notation, the area is $A(x) = 521 - \dfrac{1500}{x} - 7x$.

b. Graphing the area function $A(x)=521-\dfrac{1500}{x}-7x$:

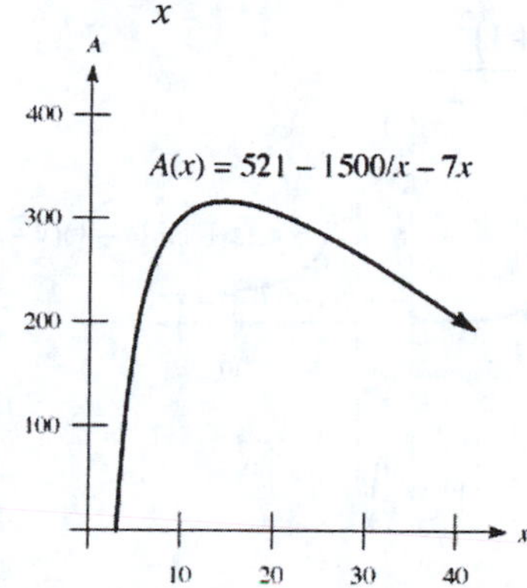

c. The area is a maximum when $x\approx 14.6$ and $y\approx 34.2$. A width of approximately 14.6 inches and a length of approximately 34.2 inches will maximize the area.

39. Finding where the curve crosses the horizontal asymptote:

$$\frac{(x-4)(x+2)}{(x-1)(x-3)}=1$$
$$(x-4)(x+2)=(x-1)(x-3)$$
$$x^2-2x-8=x^2-4x+3$$
$$-2x-8=-4x+3$$
$$2x=11$$
$$x=\tfrac{11}{2}$$

The x-intercepts are –2 and 4, the y-intercept is $-\frac{8}{3}$, the horizontal asymptote is $y = 1$, and the vertical asymptotes are $x = 1$ and $x = 3$. Sketching the graph of $y=\dfrac{(x-4)(x+2)}{(x-1)(x-3)}$:

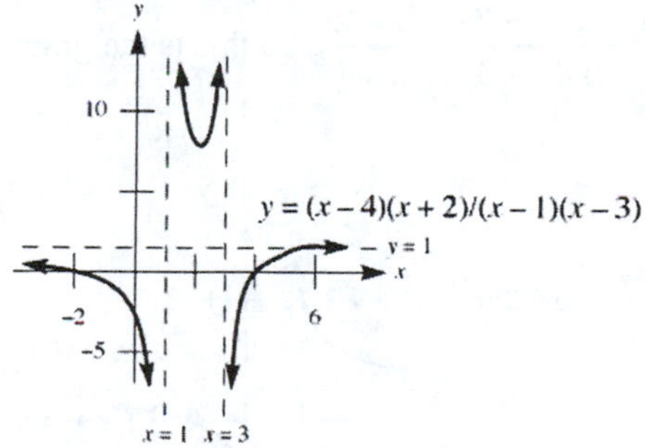

41. Finding where the curve crosses the horizontal asymptote:

$$\frac{(x+1)^2}{(x-1)(x-3)}=1$$
$$(x+1)^2=(x-1)(x-3)$$
$$x^2+2x+1=x^2-4x+3$$
$$2x+1=-4x+3$$
$$6x=2$$
$$x=\tfrac{1}{3}$$

The x-intercept is -1, the y-intercept is $\frac{1}{3}$, the horizontal asymptote is $y = 1$, and the vertical asymptotes are $x = 1$ and $x = 3$. Sketching the graph of $y = \dfrac{(x+1)^2}{(x-1)(x-3)}$:

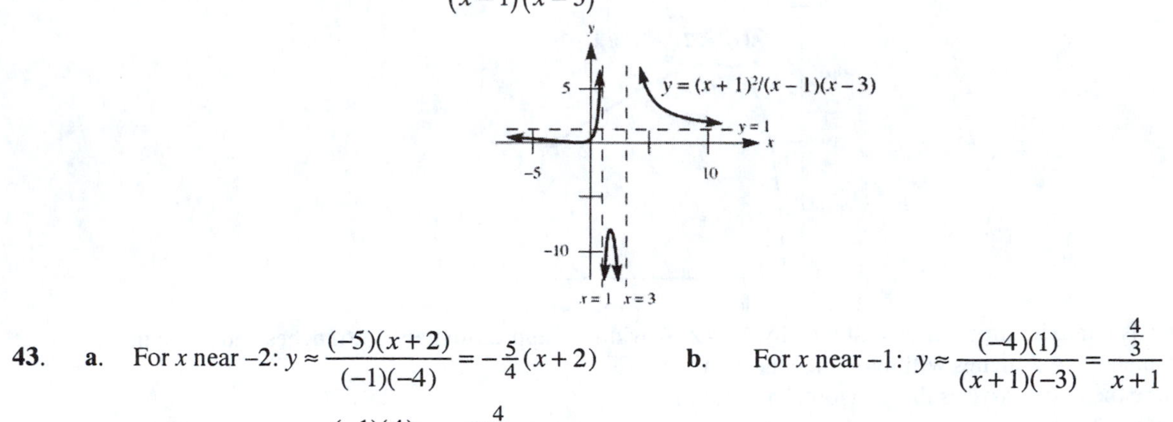

43. **a.** For x near -2: $y \approx \dfrac{(-5)(x+2)}{(-1)(-4)} = -\frac{5}{4}(x+2)$ **b.** For x near -1: $y \approx \dfrac{(-4)(1)}{(x+1)(-3)} = \dfrac{\frac{4}{3}}{x+1}$

c. For x near 2: $y \approx \dfrac{(-1)(4)}{(3)(x-2)} = \dfrac{-\frac{4}{3}}{x-2}$

45. **a.** For $x \neq -3$: $y = \dfrac{x^2-9}{x+3} = \dfrac{(x+3)(x-3)}{x+3} = x-3$. So this is the graph of $y = x - 3$, without the point at (–3,–6):

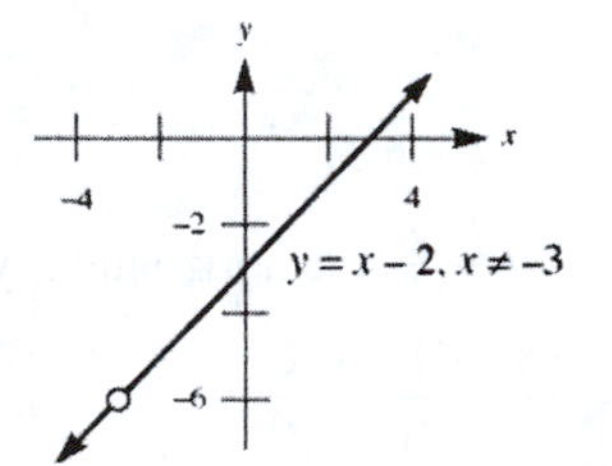

b. For $x \neq 3$: $y = \dfrac{x^2-5x+6}{x^2-2x-3} = \dfrac{(x-2)(x-3)}{(x+1)(x-3)} = \dfrac{x-2}{x+1}$. So this is the graph of $y = \dfrac{x-2}{x+1}$, without the point at $\left(3, \frac{1}{4}\right)$:

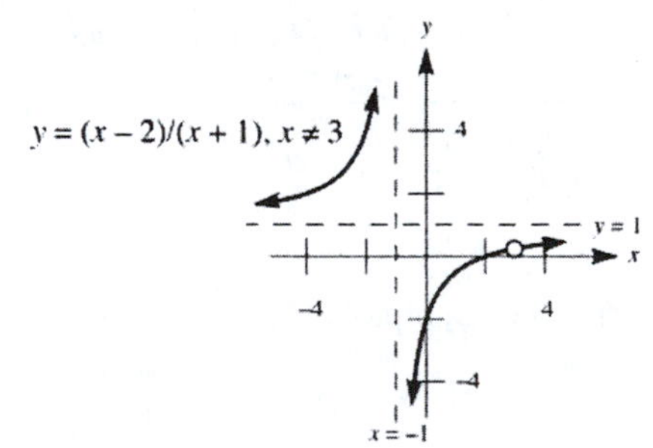

c. For $x \neq 1, 2, 3$: $y = \dfrac{(x-1)(x-2)(x-3)}{(x-1)(x-2)(x-3)(x-4)} = \dfrac{1}{x-4}$

So this is the graph of $y = \dfrac{1}{x-4}$, without the points at $\left(1, -\frac{1}{3}\right), \left(2, -\frac{1}{2}\right), (3, -1)$:

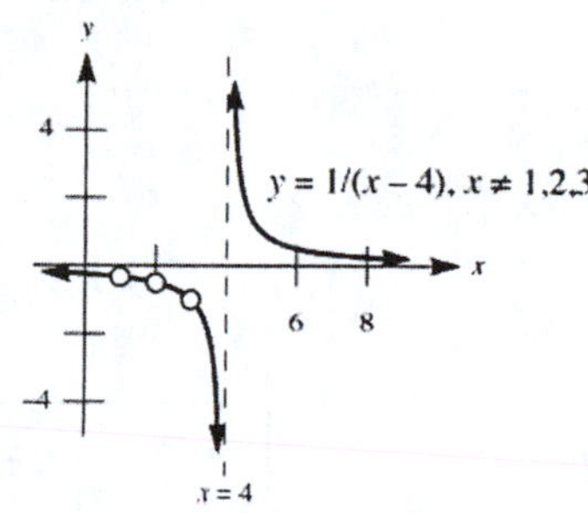

47. The horizontal asymptote is $y = 0$ and the vertical asymptote is $x = 3$. We find the value of k where $k = \dfrac{x}{(x-3)^2}$:

$$k(x-3)^2 = x$$
$$kx^2 - 6kx + 9k = x$$
$$kx^2 - (6k+1)x + 9k = 0$$

Since this equation must have only one solution, we set the discriminant equal to zero:

$$[-(6k+1)]^2 - 4k(9k) = 0$$
$$(6k+1)^2 - 36k^2 = 0$$
$$(6k+1+6k)(6k+1-6k) = 0$$
$$(12k+1)(1) = 0$$
$$k = -\tfrac{1}{12}$$

Thus $y = -\frac{1}{12}$. Find x:

$$-\tfrac{1}{12} = \frac{x}{(x-3)^2}$$
$$-(x-3)^2 = 12x$$
$$(x-3)^2 = -12x$$
$$x^2 - 6x + 9 = -12x$$
$$x^2 + 6x + 9 = 0$$
$$(x+3)^2 = 0$$
$$x = -3$$

So the low point is $\left(-3, -\frac{1}{12}\right)$. Sketching the graph:

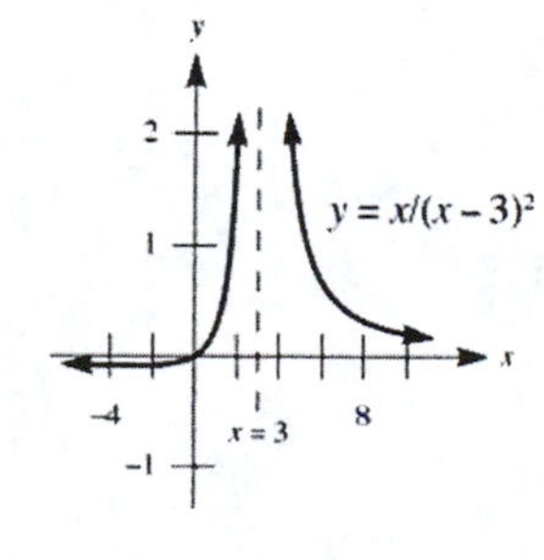

49. **a.** Use long division:

$$\begin{array}{r} x+4 \\ x-3\overline{)x^2+\ x-6} \\ \underline{x^2-3x} \\ 4x-6 \\ \underline{4x-12} \\ 6 \end{array}$$

So $\dfrac{x^2+x-6}{x-3}=(x+4)+\dfrac{6}{x-3}$.

b. Complete the tables:

x	$x+4$	$\dfrac{x^2+x-6}{x-3}$
10	14	14.8571
100	104	104.0619
1000	1004	1004.0060

x	$x+4$	$\dfrac{x^2+x-6}{x-3}$
−10	−6	−6.4615
−100	−96	−96.0583
−1000	−996	−996.0600

c. The vertical asymptote is $x=3$, the x-intercepts are −3 and 2, and the y-intercept is 2.

d. Sketching the graph:

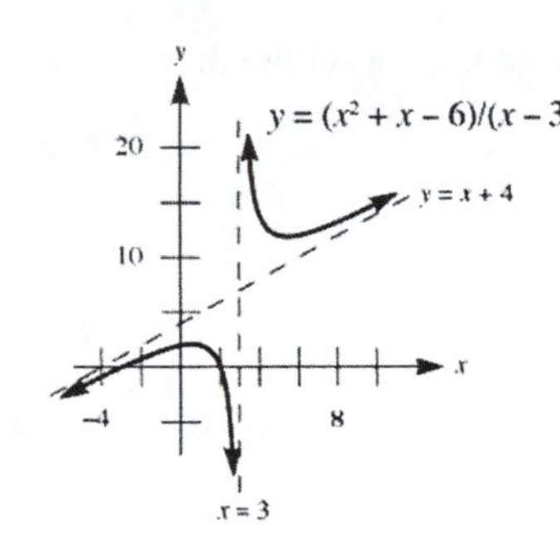

e. Find where:

$$\begin{aligned} \frac{x^2+x-6}{x-3}&=k \\ x^2+x-6&=kx-3k \\ x^2+(1-k)x+(3k-6)&=0 \end{aligned}$$

Setting the discriminant equal to 0:

$$\begin{aligned} (1-k)^2-4(1)(3k-6)&=0 \\ 1-2k+k^2-12k+24&=0 \\ k^2-14k+25&=0 \\ (k-7)^2&=-25+49 \\ (k-7)^2&=24 \\ k-7&=\pm\sqrt{24} \\ k&=7\pm2\sqrt{6} \end{aligned}$$

So either $y = 7 + 2\sqrt{6}$ or $y = 7 - 2\sqrt{6}$. For each of these values, we find x:

$$\frac{x^2 + x - 6}{x - 3} = 7 + 2\sqrt{6}$$
$$x^2 + x - 6 = \left(7 + 2\sqrt{6}\right)x - 21 - 6\sqrt{6}$$
$$x^2 + \left(-6 - 2\sqrt{6}\right)x + \left(15 + 6\sqrt{6}\right) = 0$$
$$\left(x - \left(3 + \sqrt{6}\right)\right)^2 = 0$$
$$x = 3 + \sqrt{6}$$

So one point is $\left(3 + \sqrt{6}, 7 + 2\sqrt{6}\right)$. For $y = 7 - 2\sqrt{6}$, we find x:

$$\frac{x^2 + x - 6}{x - 3} = 7 - 2\sqrt{6}$$
$$x^2 + x - 6 = \left(7 - 2\sqrt{6}\right)x - 21 + 6\sqrt{6}$$
$$x^2 + \left(-6 + 2\sqrt{6}\right)x + \left(15 - 6\sqrt{6}\right) = 0$$
$$\left(x - \left(3 - \sqrt{6}\right)\right)^2 = 0$$
$$x = 3 - \sqrt{6}$$

So the other point is $\left(3 - \sqrt{6}, 7 - 2\sqrt{6}\right)$.

51. Note that: $\dfrac{-x^2 + 1}{x} = -x + \dfrac{1}{x}$. Thus $y = -x$ is a slant asymptote. Sketching the graph:

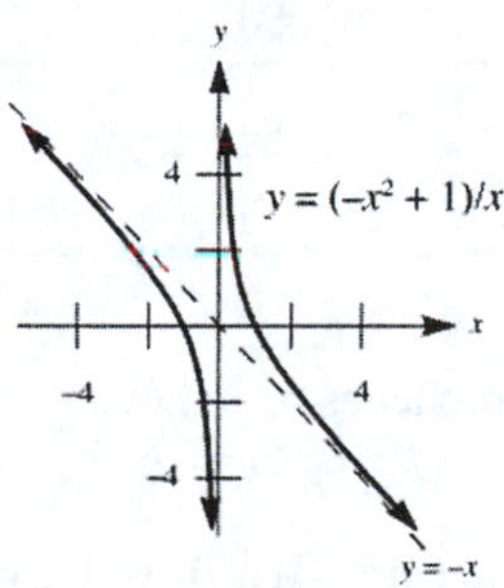

53. **a.** Graphing the function:

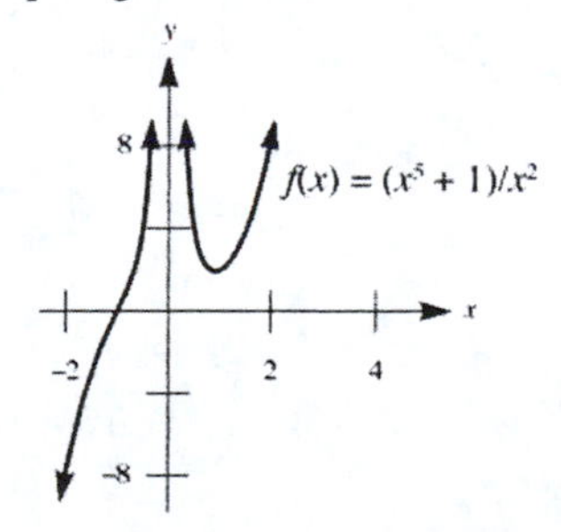

b. Adding the curve $y = x^3$ to the graph:

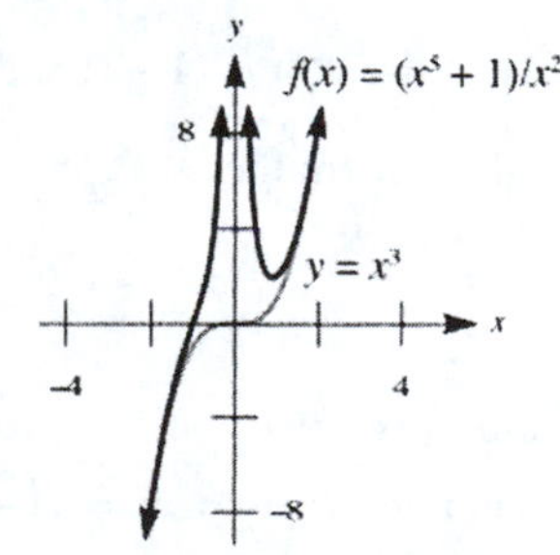

Using the viewing rectangle $-4 \le x \le 4$ and $-20 \le y \le 20$:

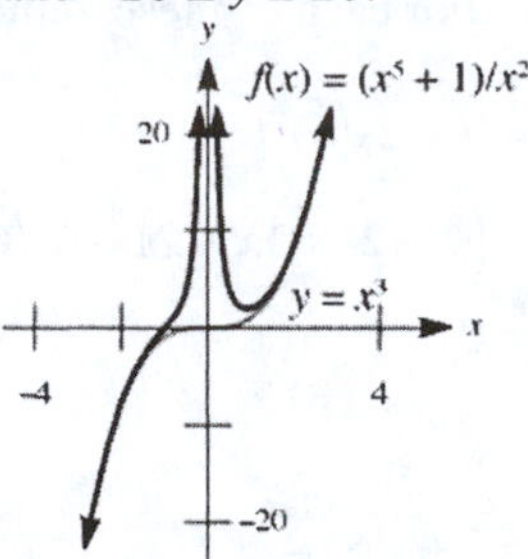

Using the viewing rectangle $-10 \le x \le 10$ and $-100 \le y \le 100$:

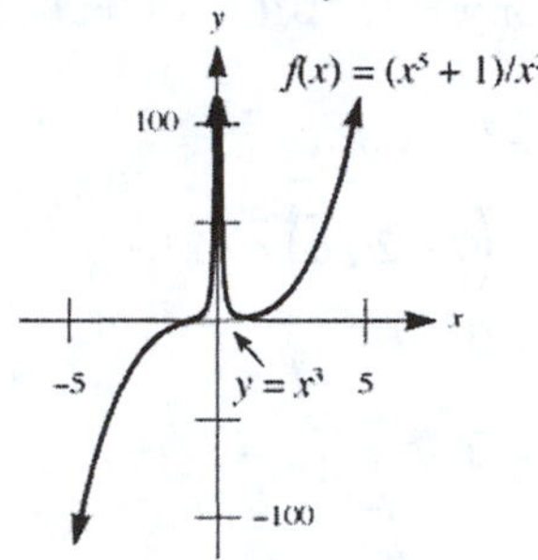

Note that as $|x|$ increases, the curve $f(x)$ approaches the curve $y = x^3$.

c. Completing the tables:

x	5	10	50	100	500
d	0.04	0.01	0.0004	0.0001	0.000004

x	–5	–10	–50	–100	–500
d	0.04	0.01	0.0004	0.0001	0.000004

Note that the values of d decrease as $|x|$ increases.

d. Working from the left-hand side: $\dfrac{x^5+1}{x^2} = \dfrac{x^5}{x^2} + \dfrac{1}{x^2} = x^3 + \dfrac{1}{x^2}$

As $|x|$ increases, the quantity $\dfrac{1}{x^2}$ approaches 0, and thus $f(x) \approx x^3$ when $|x|$ gets very large.

Chapter 4 Review Exercises

1. Find the slope between the points (1,–2) and (–2,–11): $m = \dfrac{-11-(-2)}{-2-1} = \dfrac{-9}{-3} = 3$

Now use the point-slope formula:

$$y-(-2) = 3(x-1)$$
$$y+2 = 3x-3$$
$$y = 3x-5$$

So $G(x) = 3x - 5$, and thus $G(0) = -5$.

3. First compute the revenue: $R(x) = xp = x\left(-\frac{1}{8}x+100\right) = -\frac{1}{8}x^2+100x$

Now complete the square: $R(x) = -\frac{1}{8}\left(x^2-800x\right) = -\frac{1}{8}\left(x^2-800x+160000\right)+20000 = -\frac{1}{8}(x-400)^2+20000$

So the maximum possible revenue is \$20,000.

5. Graphing the function:

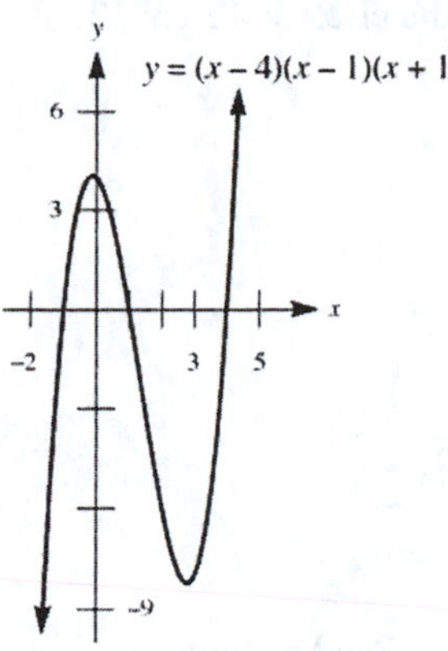

7. We have the points (0,1000) and (5,100). Find the slope: $m = \dfrac{100-1000}{5-0} = \dfrac{-900}{5} = -180$. So $V(t) = -180t + 1000$.

9. The x-intercept is $-\frac{5}{3}$, the y-intercept is $\frac{5}{2}$, the horizontal asymptote is $y = 3$, and the vertical asymptote is $x = -2$. Graphing the function:

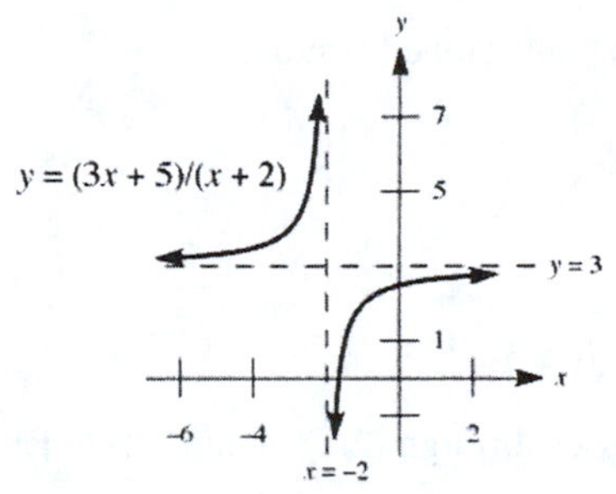

11. Graphing the function:

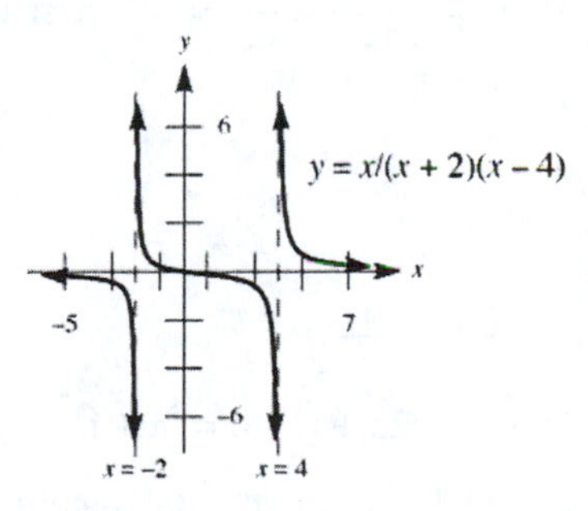

13. Drawing the figure:

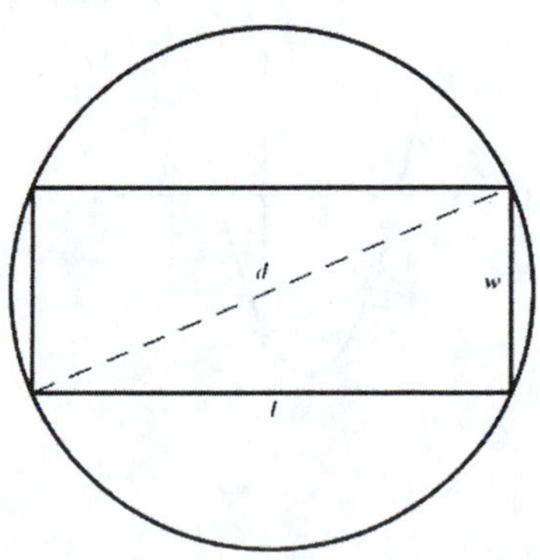

We can find r, since the circumference of the circle is 12 cm: $2\pi r = 12$, so $r = \frac{6}{\pi}$ and thus $d = \frac{12}{\pi}$

Using the Pythagorean theorem:

$$w^2 + l^2 = \left(\tfrac{12}{\pi}\right)^2$$
$$w^2 + l^2 = \frac{144}{\pi^2}$$
$$l^2 = \frac{144}{\pi^2} - w^2$$
$$l = \sqrt{\frac{144 - \pi^2 w^2}{\pi^2}} = \frac{\sqrt{144 - \pi^2 w^2}}{\pi}$$

Since the perimeter is given by $P = 2w + 2l$: $P(w) = 2w + \frac{2\sqrt{144 - \pi^2 w^2}}{\pi}$

15. Find the slope of the line $3x - 8y = 16$:

$$-8y = -3x + 16$$
$$y = \tfrac{3}{8}x - 2$$

Use $m = \frac{3}{8}$ and the point (4,–1) in the point-slope formula:

$$y - (-1) = \tfrac{3}{8}(x - 4)$$
$$y + 1 = \tfrac{3}{8}x - \tfrac{3}{2}$$
$$y = \tfrac{3}{8}x - \tfrac{5}{2}$$

Using functional notation we have $f(x) = \frac{3}{8}x - \frac{5}{2}$.

17. If the graph of the inverse function passes through (2,1), then (1,2) must lie on the graph of the function. Find the slope between the points (1,2) and (–3,5): $m = \frac{5-2}{-3-1} = \frac{3}{-4} = -\frac{3}{4}$. Use the point (1,2) in the point-slope formula:

$$y - 2 = -\tfrac{3}{4}(x - 1)$$
$$y - 2 = -\tfrac{3}{4}x + \tfrac{3}{4}$$
$$y = -\tfrac{3}{4}x + \tfrac{11}{4}$$

Using functional notation we have $f(x) = -\frac{3}{4}x + \frac{11}{4}$.

19. Completing the square: $y = x^2 + 2x - 3 = \left(x^2 + 2x + 1\right) - 4 = (x+1)^2 - 4$

The vertex is (–1,–4), the x-intercepts are 1 and –3, and the y-intercept is –3. Graphing the function:

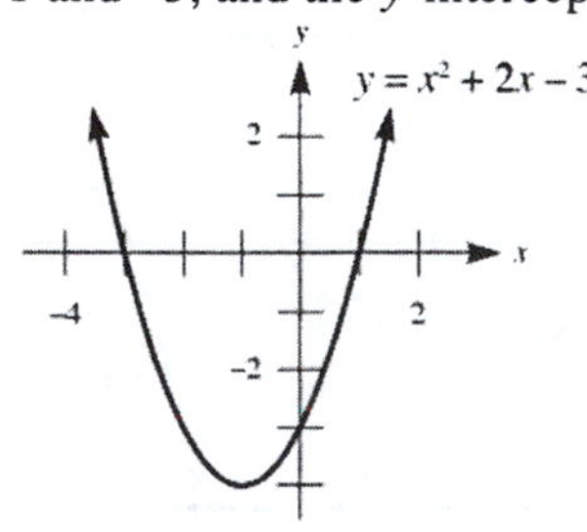

21. Complete the square: $y = 2x^2 - 2x + 1 = 2\left(x^2 - x\right) + 1 = 2\left(x^2 - x + \frac{1}{4}\right) + 1 - \frac{1}{2} = 2\left(x - \frac{1}{2}\right)^2 + \frac{1}{2}$

The vertex is $\left(\frac{1}{2}, \frac{1}{2}\right)$, there is no x-intercept, and the y-intercept is 1. Draw the graph:

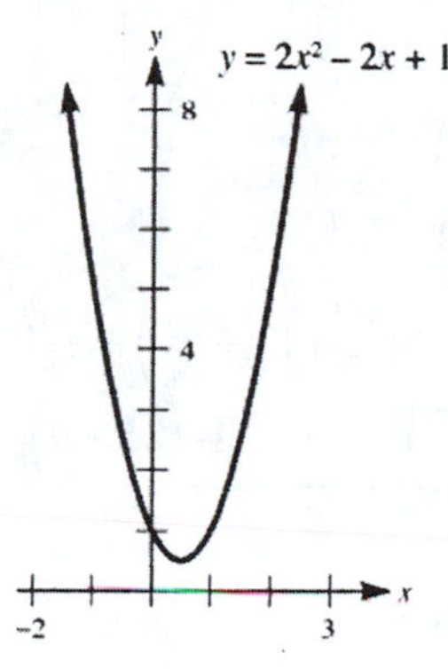

23. Find the two vertices:

$$\begin{aligned} y &= x^2 - 4x + 6 \\ y - 6 &= x^2 - 4x \\ y - 6 + 4 &= x^2 - 4x + 4 \\ y - 2 &= (x-2)^2 \end{aligned}$$

vertex: $(2,2)$

$$\begin{aligned} y &= -x^2 - 4x - 5 \\ y + 5 &= -\left(x^2 + 4x\right) \\ y + 5 - 4 &= -\left(x^2 + 4x + 4\right) \\ y + 1 &= -(x+2)^2 \end{aligned}$$

vertex: $(-2,-1)$

Now find the distance between (2,2) and (–2,–1): $d = \sqrt{(-2-2)^2 + (-1-2)^2} = \sqrt{16+9} = \sqrt{25} = 5$

25. **a.** Completing the square: $h = -16t^2 + v_0 t = -16\left(t^2 - \frac{v_0}{16}t\right) = -16\left(t^2 - \frac{v_0}{16}t + \frac{v_0^2}{1024}\right) + \frac{v_0^2}{64} = -16\left(t - \frac{v_0}{32}\right)^2 + \frac{v_0^2}{64}$

So the maximum height of $\frac{v_0^2}{64}$ ft. is obtained when $t = \frac{v_0}{32}$ sec.

b. The object will strike the ground when $h(t) = 0$:

$$\begin{aligned} -16t^2 + v_0 t &= 0 \\ t\left(-16t + v_0\right) &= 0 \\ t = 0 \text{ or } t &= \frac{v_0}{16} \end{aligned}$$

So the object will strike the ground when $t = \frac{v_0}{16}$ sec.

27. Find the distance between (2,0) and $\left(x, \frac{4}{3}x+b\right)$:

$$d=\sqrt{(x-2)^2+\left(\frac{4}{3}x+b\right)^2}=\sqrt{x^2-4x+4+\frac{16}{9}x^2+\frac{8b}{3}x+b^2}=\sqrt{\frac{25}{9}x^2+\left(\frac{8b}{3}-4\right)x+\left(4+b^2\right)}$$

Completing the square on d^2:

$$\begin{aligned}
d^2&=\frac{25}{9}\left[x^2+\frac{9}{25}\left(\frac{8b}{3}-4\right)x\right]+\left(4+b^2\right)\\
&=\frac{25}{9}\left[x^2+\frac{12}{25}(2b-3)x\right]+\left(4+b^2\right)\\
&=\frac{25}{9}\left[x^2+\frac{12}{25}(2b-3)x+\frac{36}{625}(2b-3)^2\right]+\left(4+b^2\right)-\frac{4}{25}(2b-3)^2\\
&=\frac{25}{9}\left[x+\frac{6}{25}(2b-3)\right]^2+\frac{25\left(4+b^2\right)-4\left(4b^2-12b+9\right)}{25}\\
&=\frac{25}{9}\left[x+\frac{6}{25}(2b-3)\right]^2+\frac{(3b+8)^2}{25}
\end{aligned}$$

Since the minimum distance is 5, $5=\left|\frac{3b+8}{5}\right|$. Now solve for b:

$$|3b+8|=25$$

$$\begin{aligned}
3b+8&=25 &\quad\text{or}\quad&& 3b+8&=-25\\
3b&=17 &&& 3b&=-33\\
b&=\frac{17}{3} &&& b&=-11
\end{aligned}$$

29. Since $x+y=\sqrt{2}$, $y=\sqrt{2}-x$. Therefore: $s=x^2+y^2=x^2+\left(\sqrt{2}-x\right)^2=x^2+2-2\sqrt{2}x+x^2=2x^2-2\sqrt{2}x+2$

Completing the square: $s=2\left(x^2-\sqrt{2}x\right)+2=2\left(x^2-\sqrt{2}x+\frac{1}{2}\right)+2-1=2\left(x-\frac{\sqrt{2}}{2}\right)^2+1$

So the minimum value of s is 1.

31. Let x and h be the two legs. We have:

$$\begin{aligned}
x^2+h^2&=15^2\\
h^2&=225-x^2\\
h&=\sqrt{225-x^2}
\end{aligned}$$

So $A=\frac{1}{2}$(base)(height) $=\frac{1}{2}x\sqrt{225-x^2}$. We find $A^2=\frac{1}{4}x^2\left(225-x^2\right)=-\frac{1}{4}x^4+\frac{225}{4}x^2$.

Now complete the square on A^2:

$$A^2=-\frac{1}{4}\left(x^4-225x^2\right)=-\frac{1}{4}\left[x^4-225x^2+\left(\frac{225}{2}\right)^2\right]+\left(\frac{225}{4}\right)^2=-\frac{1}{4}\left(x^2-\frac{225}{2}\right)^2+\left(\frac{225}{4}\right)^2$$

So the maximum of $A^2=\left(\frac{225}{4}\right)^2$, thus $A=\frac{225}{4}$ cm^2.

33. Factor $f(x)$ as: $f(x)=x^2-\left(a^2+2a\right)x+2a^3=\left(x-a^2\right)(x-2a)$

So the x-intercepts are a^2 and $2a$. Since $2a>a^2$ when $0<a<2$, the distance between $(a^2,0)$ and $(2a,0)$ will be $D=2a-a^2=-a^2+2a$. Now complete the square: $D=-\left(a^2-2a\right)=-\left(a^2-2a+1\right)+1=-(a-1)^2+1$

So D is maximum when $a=1$.

35. For the first piece:

$$\begin{aligned} 2(w)+2(2w) &= x \\ 6w &= x \\ w &= \frac{x}{6} \end{aligned}$$

The area is given by: $A = w \bullet 2w = \frac{x}{6} \bullet \frac{x}{3} = \frac{x^2}{18}$. For the second piece:

$$\begin{aligned} 2w+2(3w) &= 16-x \\ 8w &= 16-x \\ w &= 2-\tfrac{1}{8}x \end{aligned}$$

The area is given by: $A = w \bullet 3w = \left(2-\frac{1}{8}x\right)\left(6-\frac{3}{8}x\right) = \frac{3}{64}x^2 - \frac{3}{2}x + 12$

The total area is given by: $A(x) = \frac{x^2}{18} + \frac{3}{64}x^2 - \frac{3}{2}x + 12 = \frac{59}{576}x^2 - \frac{3}{2}x + 12$

Now complete the square: $A = \frac{59}{576}\left(x^2 - \frac{864}{59}x\right) + 12 = \frac{59}{576}\left(x - \frac{432}{59}\right)^2 + \frac{384}{59}$

The total area will be a minimum when $x = \frac{432}{59}$ cm ≈ 7.32 cm.

37. Since the x-coordinate of point A is a, the coordinates of P are $(a, f(a))$. Similarly, the coordinates of point R are $(b, f(b))$. But, since $PQRS$ is a square with diagonal $y = x$, the points P and R are reflections of one another about the line $y = x$. So their coordinates are just interchanges of each other, thus $a = f(b)$ and $b = f(a)$.

39. The x-intercepts are –4 and 2 and the y-intercept is –8. Drawing the graph:

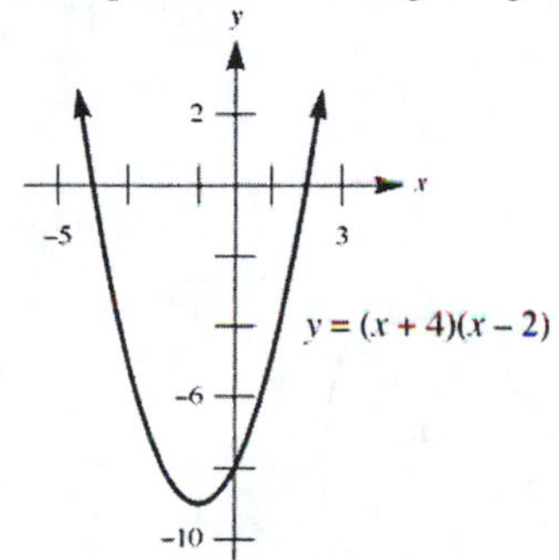

41. The x-intercepts are –1 and 0, and the y-intercept is 0. Drawing the graph:

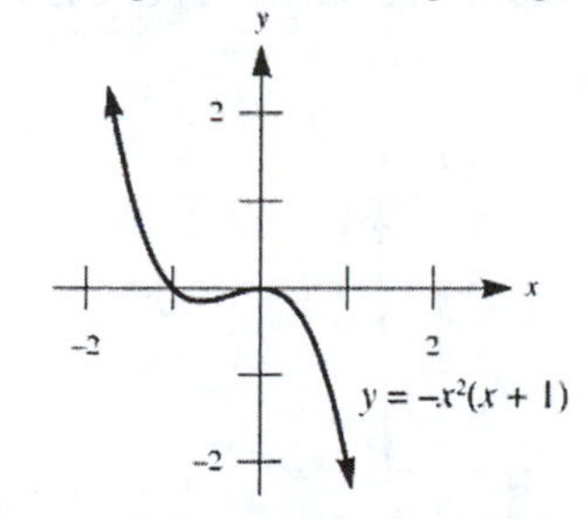

43. The x-intercepts are 0, 2, and –2, and the y-intercept is 0. Drawing the graph:

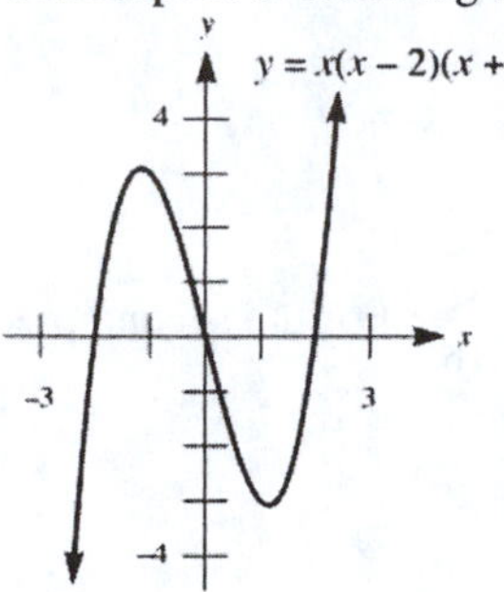

45. There is no x-intercept, the y-intercept is –1, the horizontal asymptote is $y = 0$, and the vertical asymptote is $x = 1$. Drawing the graph:

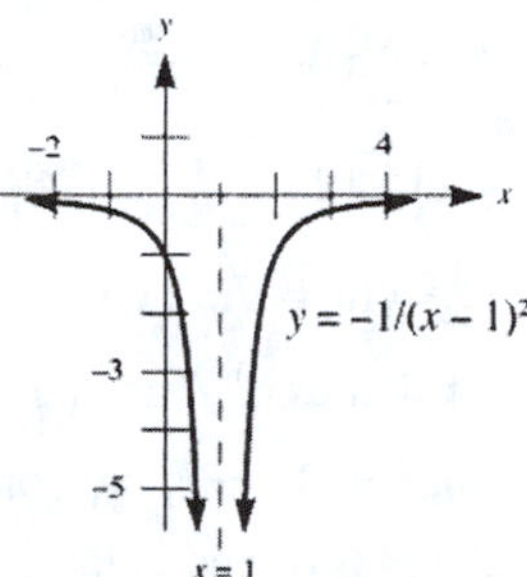

47. The x-intercept is 2, the y-intercept is $\frac{2}{3}$, the horizontal asymptote is $y = 1$, and the vertical asymptote is $x = 3$. Drawing the graph:

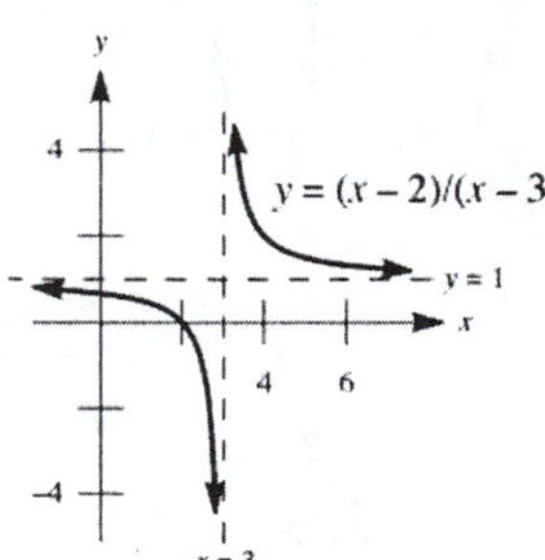

49. The x-intercept is 1, the y-intercept is $\frac{1}{4}$, the horizontal asymptote is $y = 1$, and the vertical asymptote is $x = 2$. Drawing the graph:

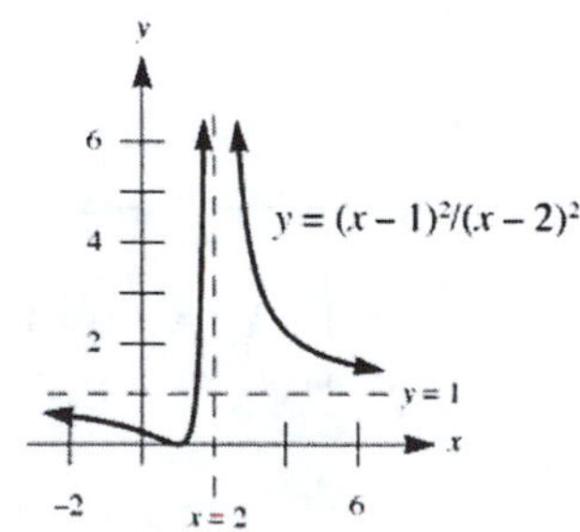

51. We need to find the vertex of the parabola. Completing the square:

$$\begin{aligned} y &= x^2 - 2x + k \\ y - k &= x^2 - 2x \\ y - k + 1 &= x^2 - 2x + 1 \\ y &= (x-1)^2 + (k-1) \end{aligned}$$

Since the vertex is $(1, k-1)$ and the parabola is opening upward, $k - 1 = 5$, so $k = 6$.

53. Solve for x:

$$\begin{aligned} y &= \frac{(x-1)(x-3)}{x-4} \\ y(x-4) &= (x-1)(x-3) \\ yx - 4y &= x^2 - 4x + 3 \\ 0 &= x^2 - (4+y)x + (4y+3) \end{aligned}$$

Using the quadratic formula:

$$x = \frac{4+y \pm \sqrt{(4+y)^2 - 4(4y+3)}}{2} = \frac{4+y \pm \sqrt{16+8y+y^2-16y-12}}{2} = \frac{4+y \pm \sqrt{y^2-8y+4}}{2}$$

So we must make sure that $y^2 - 8y + 4 \geq 0$. Find the key numbers by using the quadratic formula:

$$y = \frac{8 \pm \sqrt{64-16}}{2} = \frac{8 \pm 4\sqrt{3}}{2} = 4 \pm 2\sqrt{3}$$

From a sign chart, we see that the range is $y \leq 4 - 2\sqrt{3}$ or $y \geq 4 + 2\sqrt{3}$. We write this as $\left(-\infty, 4-2\sqrt{3}\right] \cup \left[4+2\sqrt{3}, \infty\right)$.

55. First draw a figure:

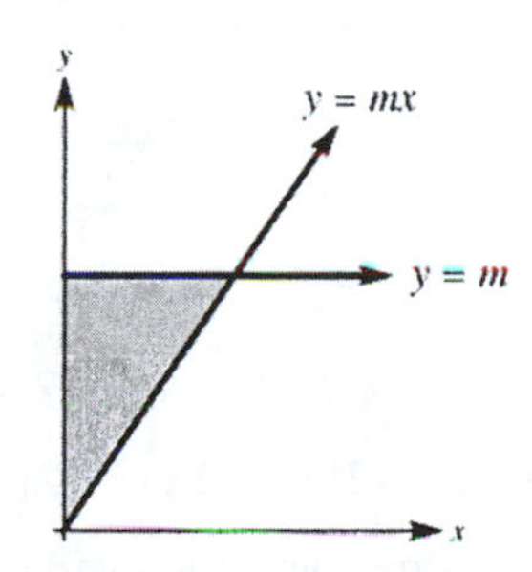

Find the intersection point of these two lines:

$$\begin{aligned} mx &= m \\ x &= 1 \\ y &= m \end{aligned}$$

So the point is $(1, m)$. Since these are the base and height, respectively, of the triangle, we have: $A = \frac{1}{2}(1)(m) = \frac{m}{2}$

Using functional notation we have $A(m) = \frac{m}{2}$.

57. Re-draw the figure and label essential parts:

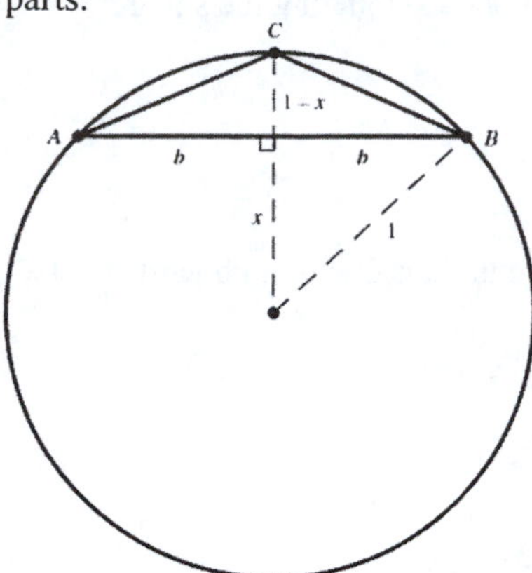

By the labeled parts of the figure, we have: $x^2 + b^2 = 1^2$, so $b = \sqrt{1-x^2}$

Thus the base of the triangle is $2b = 2\sqrt{1-x^2}$ and the height is $1 - x$, thus the area is given by:

$$A = \frac{1}{2} \bullet 2\sqrt{1-x^2} \bullet (1-x) = (1-x)\sqrt{1-x^2}$$

Using functional notation we have $A(x) = (1-x)\sqrt{1-x^2}$.

Chapter 4 Test

1. First find the slope between (–2,–4) and (5,1): $m = \dfrac{-4-1}{-2-5} = \dfrac{-5}{-7} = \frac{5}{7}$. Using the point-slope formula:

$$y - 1 = \tfrac{5}{7}(x-5)$$
$$y - 1 = \tfrac{5}{7}x - \tfrac{25}{7}$$
$$y = \tfrac{5}{7}x - \tfrac{18}{7}$$

So $L(x) = \frac{5}{7}x - \frac{18}{7}$, and thus $L(0) = -\frac{18}{7}$.

2. **a.** Completing the square: $F(x) = -2x^2 + 4x = -2\left(x^2 - 2x\right) = -2\left(x^2 - 2x + 1\right) + 2 = -2(x-1)^2 + 2$

So the graph of $F(x)$ is a parabola pointing downward with vertex $(1, 2)$, and thus it is increasing on the interval $(-\infty, 1)$. The maximum value of $F(x)$ is 2.

b. Since $9t^4 \ge 0$ and $6t^2 \ge 0$, $G(t)$ will achieve a minimum value of 2 when $t = 0$.

3. **a.** The x-intercepts are 3 and –4, and the y-intercept is –48. The signs of $f(x)$ are given by:

Interval	$f(x) = (x-3)(x+4)^2$
$(-\infty, -4)$	neg.
$(-4, 3)$	neg.
$(3, \infty)$	pos.

Sketch the excluded regions for $f(x)$:

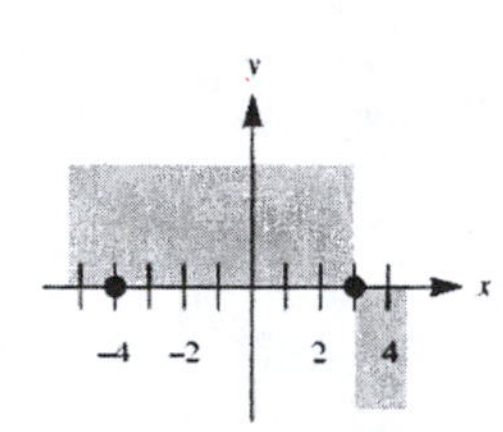

b. For $x \approx -4$, we have $f(x) \approx (-4-3)(x+4)^2 = -7(x+4)^2$. So $f(x)$ is approximately $-7(x+4)^2$ when x is close to -4. Sketch $y = -7(x+4)^2$:

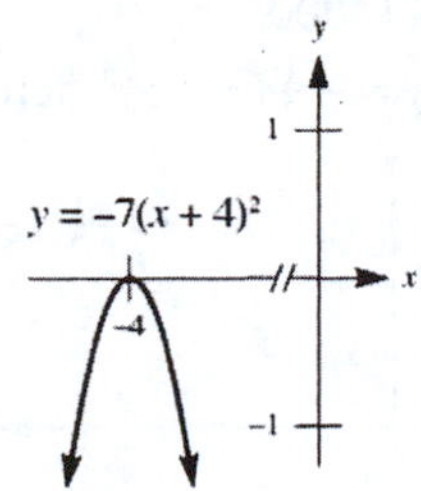

c. Graph the function:

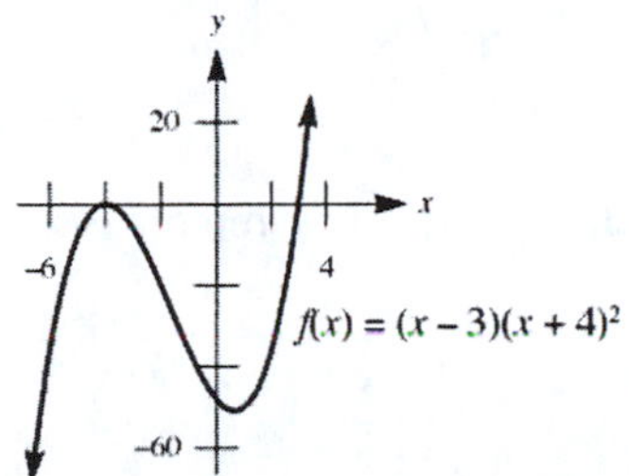

4. **a.** Sketch the graph:

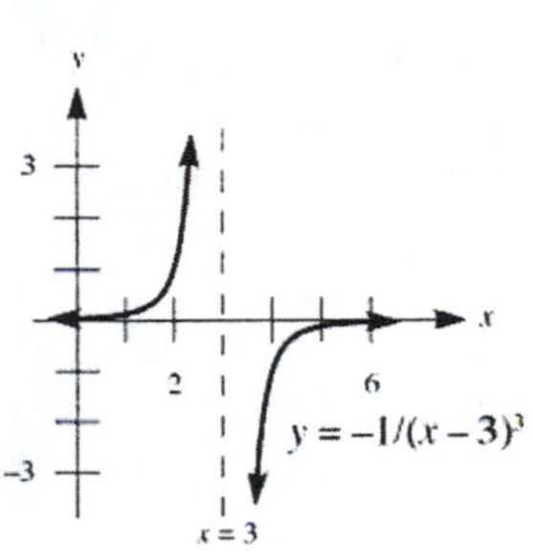

b. Sketch the graph:

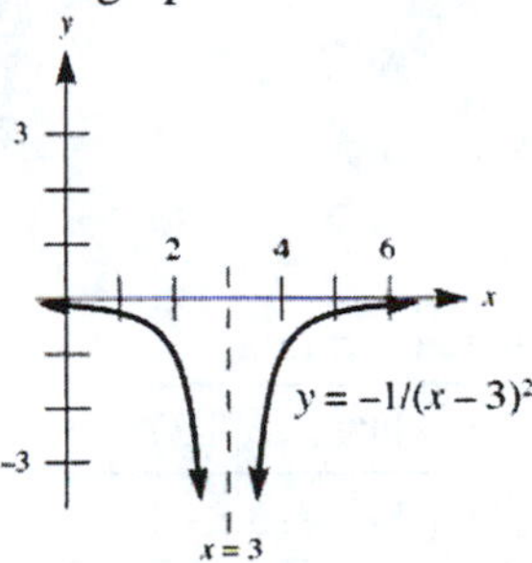

5. First complete the square: $y = -x^2 + 7x + 6 = -\left(x^2 - 7x\right) + 6 = -\left(x^2 - 7x + \frac{49}{4}\right) + \frac{73}{4} = -\left(x - \frac{7}{2}\right)^2 + \frac{73}{4}$

The turning point (vertex) is $\left(\frac{7}{2}, \frac{73}{4}\right)$, the x-intercepts are $\dfrac{7 \pm \sqrt{73}}{2}$, the y-intercept is 6, and the axis of symmetry is $x = \frac{7}{2}$. Draw the graph:

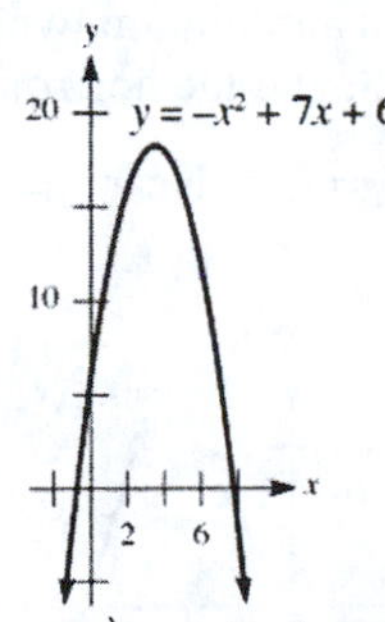

6. Set up the revenue function: $R = xp = x\left(-\frac{1}{6}x + 80\right) = -\frac{1}{6}x^2 + 80x$

Now complete the square: $R = -\frac{1}{6}\left(x^2 - 480x\right) = -\frac{1}{6}\left(x^2 - 480x + 57600\right) + 9600 = -\frac{1}{6}(x - 240)^2 + 9600$

The maximum possible revenue is \$9600, which occurs when $x = 240$. Thus $p = -\frac{1}{6}(240) + 80 = \$40 / \text{unit}$.

7. Find the slope between the points (0,14000) and (10,750): $m=\dfrac{750-14000}{10-0}=\dfrac{-13250}{10}=-1325$

So the value of the machine is $V(t) = -1325t + 14000$.

8. Since $y=-\frac{1}{2}(3-x)^3$ will be a reflection of $y=-\frac{1}{2}(x-3)^3$ across the x-axis, we have the graph:

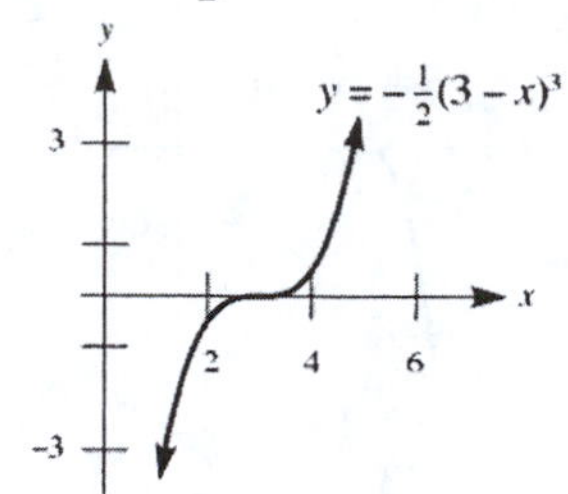

The x-intercept is 3 and the y-intercept is $-\frac{27}{2}$.

9. The x-intercept is $\frac{3}{2}$, the y-intercept is -3, the vertical asymptote is $x = -1$, and the horizontal asymptote is $y = 2$. Draw the graph:

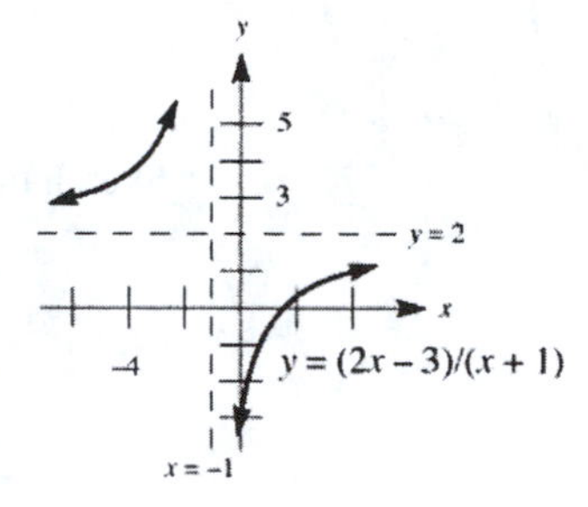

10. **a.** Using the distance formula:

$$\begin{aligned} L(x) &= \sqrt{(x+1)^2+(y-3)^2} \\ &= \sqrt{(x+1)^2+(3x-1-3)^2} \\ &= \sqrt{x^2+2x+1+9x^2-24x+16} \\ &= \sqrt{10x^2-22x+17} \end{aligned}$$

b. Complete the square inside the radical:

$$10x^2-22x+17=10\left(x^2-2.2x\right)+17=10\left(x^2-2.2x+1.21\right)+4.9=10(x-1.1)^2+4.9$$

So the length will be a minimum when L^2 is a minimum, which occurs when $x = 1.1$.

11. **a.** The vertical asymptotes are $x = 3$ and $x = -3$, and the horizontal asymptote is $y = 1$.

b. Near $x = 0$, $f(x)$ will look like: $y=\dfrac{x(0-2)}{0-9}=\frac{2}{9}x$. Near $x = 2$, $f(x)$ will look like: $y=\dfrac{2(x-2)}{4-9}=-\frac{2}{5}(x-2)$

Sketch the graph:

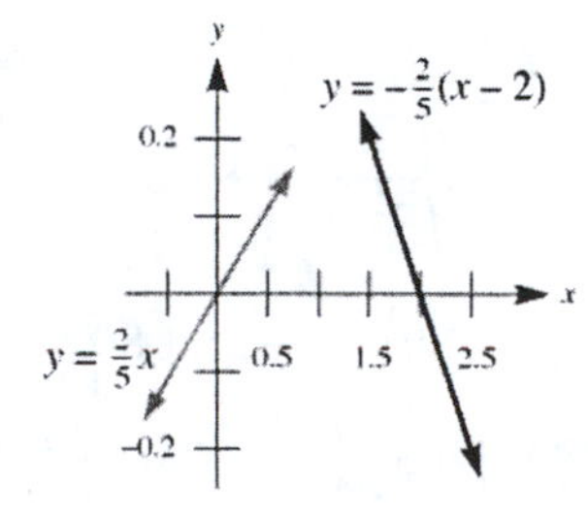

c. Near $x = 3$, $f(x)$ will look like: $y = \dfrac{3(3-2)}{(3+3)(x-3)} = \dfrac{1}{2(x-3)}$

Near $x = -3$, $f(x)$ will look like: $y = \dfrac{-3(-3-2)}{(x+3)(-3-3)} = -\dfrac{5}{2(x+3)}$

Sketch the graph:

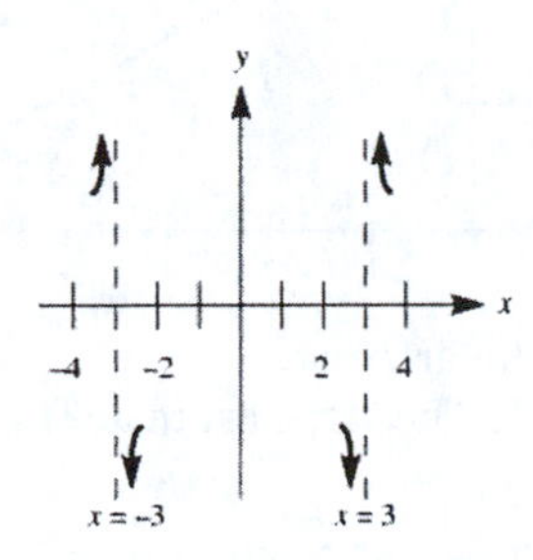

d. Graph the function:

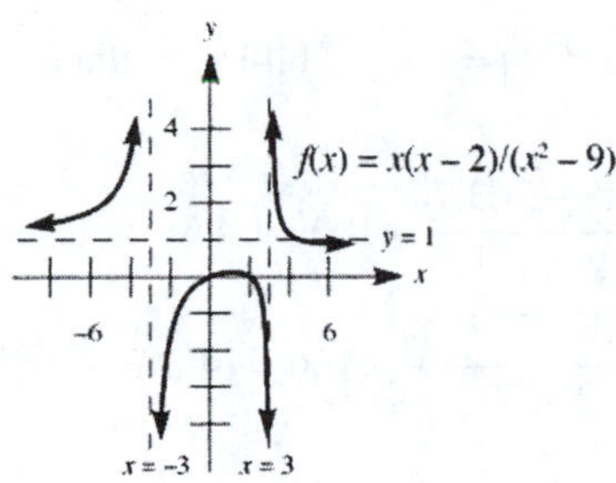

12. Let w and l represent the width and length, respectively. Draw the figure:

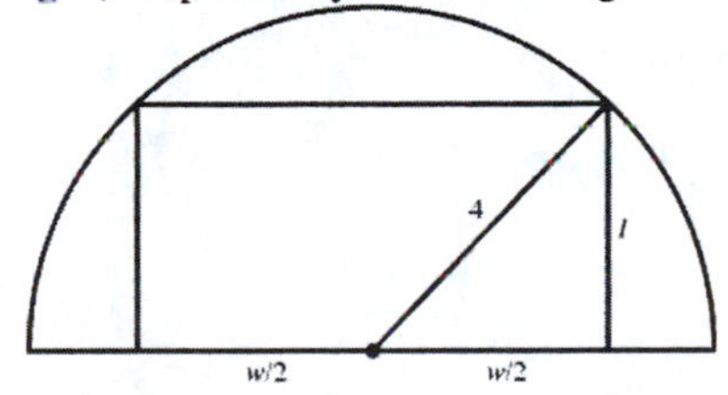

From the Pythagorean theorem:

$$\left(\frac{w}{2}\right)^2 + l^2 = 4^2$$

$$\frac{w^2}{4} + l^2 = 16$$

$$l^2 = \frac{64 - w^2}{4}$$

$$l = \frac{\sqrt{64 - w^2}}{2}$$

So the area of the rectangle is given by: $A(w) = w \bullet l = w \bullet \dfrac{\sqrt{64 - w^2}}{2} = \frac{1}{2} w\sqrt{64 - w^2}$

13. a. As $|x|$ increases in size for x positive, $-x^3$ increases in the negative direction, which the pictured function does not.

b. This graph has four turning points, and a polynomial function with highest degree term $-x^3$ can have at most two turning points.

14. a. Sketching the data and the best-fit line:

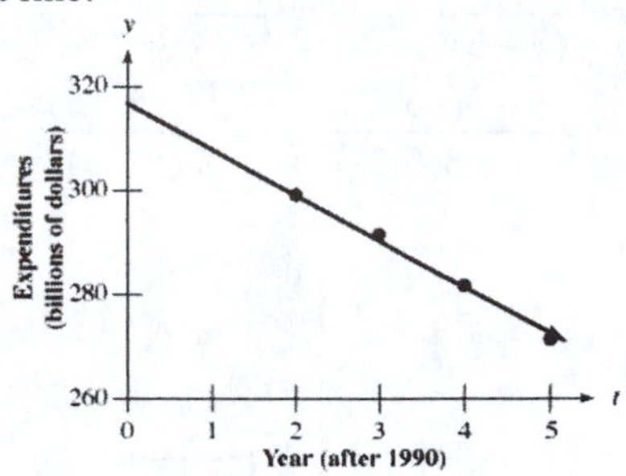

The slope is approximately –9 billion dollars/year.

b. The best-fit line is $y = -8.84x + 316.74$. The slope estimate from part **a** is fairly close to the actual slope.

c. Projecting the estimates:

$$1996\ (t = 6):\ \ y = -8.84(6) + 316.74 = 263.7 \text{ billion dollars}$$

$$1999\ (t = 9):\ \ y = -8.84(9) + 316.74 = 237.18 \text{ billion dollars}$$

Computing the percentage errors:

$$1996:\ \text{percentage error} = \frac{|265.8 - 263.7|}{265.8} \times 100 \approx 0.8\%$$

$$1999:\ \text{percentage error} = \frac{|276.7 - 237.18|}{276.7} \times 100 \approx 14.3\%$$

15. a. Finding the fixed points:

$$\begin{aligned} f(x) &= x \\ 2x(1-x) &= x \\ 2x - 2x^2 &= x \\ x - 2x^2 &= 0 \\ x(1-2x) &= 0 \\ x &= 0, \tfrac{1}{2} \end{aligned}$$

b. Finding the fixed points:

$$\begin{aligned} g(x) &= x \\ \frac{4x-1}{3x+6} &= x \\ 4x - 1 &= 3x^2 + 6x \\ 0 &= 3x^2 + 2x + 1 \end{aligned}$$

Using the quadratic formula: $x = \dfrac{-2 \pm \sqrt{(2)^2 - 4(3)(1)}}{2(3)} = \dfrac{-2 \pm \sqrt{4-12}}{6} = \dfrac{-2 \pm \sqrt{-8}}{6}$

Since there are no real solutions, there are no fixed points for the function.

16. **a.** Completing the table:

	x_1	x_2	x_3	x_4	x_5	x_6
From graph	0.56	0.28	0.52	0.33	0.49	0.36
From calculator	0.56	0.286	0.518	0.332	0.490	0.360

b. Finding the fixed points:

$$\begin{aligned} f(x) &= x \\ 0.6 - x^2 &= x \\ 0 &= x^2 + x - 0.6 \\ 0 &= 10x^2 + 10x - 6 \\ 0 &= 5x^2 + 5x - 3 \end{aligned}$$

Using the quadratic formula: $x = \dfrac{-5 \pm \sqrt{(5)^2 - 4(5)(-3)}}{2(5)} = \dfrac{-5 \pm \sqrt{25+60}}{10} = \dfrac{-5 \pm \sqrt{85}}{10}$

The fixed points are $\dfrac{-5-\sqrt{85}}{10} \approx -1.4220$ and $\dfrac{-5+\sqrt{85}}{10} \approx 0.4220$. The iterates from part **a** are approaching 0.4220.

c. Computing the first six iterates of $x_0 = 1$:

$x_1 = -0.4$ $\quad x_2 = 0.44$ $\quad x_3 = 0.4064$

$x_4 \approx 0.4348$ $\quad x_5 \approx 0.4109$ $\quad x_6 \approx 0.4311$

Yes, these iterates are approaching the fixed point 0.4220.

Chapter 5
Exponential and Logarithmic Functions

5.1 Exponential Functions

1. **a.** Since $2^{10} \approx 10^3$, $2^{30} = \left(2^{10}\right)^3 \approx \left(10^3\right)^3 = 10^9$.

b. Since $2^{10} \approx 10^3$, $2^{50} = \left(2^{10}\right)^5 \approx \left(10^3\right)^5 = 10^{15}$.

3. Using the property $\left(b^m\right)^n = b^{mn}$: $\left(5^{\sqrt{3}}\right)^{\sqrt{3}} = 5^{\sqrt{3}\cdot\sqrt{3}} = 5^3 = 125$

5. Using the property $b^n b^m = b^{n+m}$: $\left(4^{1+\sqrt{2}}\right)\left(4^{1-\sqrt{2}}\right) = 4^{1+\sqrt{2}+1-\sqrt{2}} = 4^2 = 16$

7. Using the property $\dfrac{b^m}{b^n} = b^{m-n}$: $\dfrac{2^{4+\pi}}{2^{1+\pi}} = 2^{4+\pi-1-\pi} = 2^3 = 8$

9. Using the property $\left(b^m\right)^n = b^{mn}$: $\left(\sqrt{5}^{\sqrt{2}}\right)^2 = \sqrt{5}^{2\sqrt{2}} = \left(5^{1/2}\right)^{2\sqrt{2}} = 5^{\sqrt{2}}$

11. **a.** Solve for x:

$$3^x = 27$$
$$3^x = 3^3$$
$$x = 3$$

b. Solve for t:

$$9^t = 27$$
$$\left(3^2\right)^t = 3^3$$
$$3^{2t} = 3^3$$
$$2t = 3$$
$$t = \tfrac{3}{2}$$

c. Solve for y:

$$3^{1-2y} = \sqrt{3}$$
$$3^{1-2y} = 3^{1/2}$$
$$1 - 2y = \tfrac{1}{2}$$
$$-2y = -\tfrac{1}{2}$$
$$y = \tfrac{1}{4}$$

d. Solve for z:

$$3^z = 9\sqrt{3}$$
$$3^z = 3^2 \bullet 3^{1/2}$$
$$3^z = 3^{5/2}$$
$$z = \tfrac{5}{2}$$

13. The domain is all real numbers, or $(-\infty, \infty)$.

15. Since $2^{x-1} \neq 0$ (even if $x = 1$), the domain is all real numbers, or $(-\infty, \infty)$.

17. Graph $y = 2^x$ and $y = 2^{-x}$:

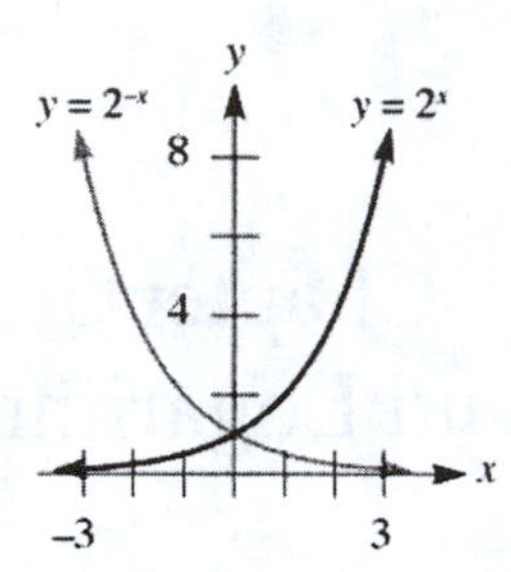

19. Graph $y = 3^x$ and $y = -3^x$ (note that $y = -3^x$ is a reflection of $y = 3^x$ about the x-axis):

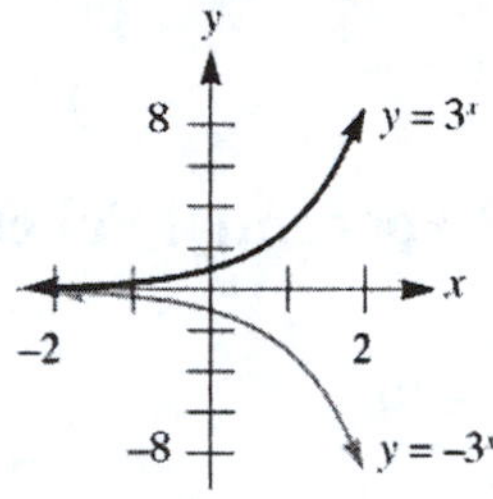

21. Graph $y = 2^x$ and $y = 3^x$:

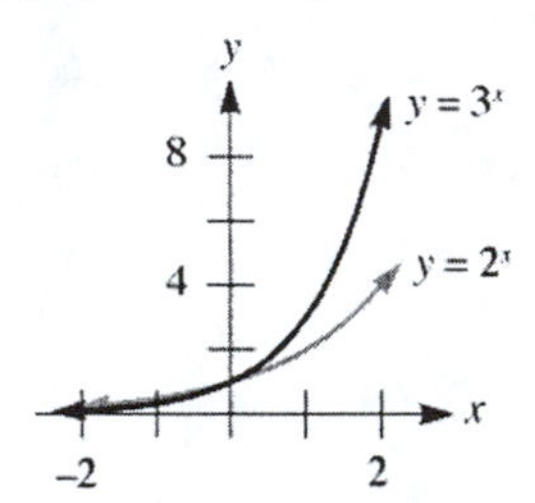

23. Graph $y = \left(\frac{1}{2}\right)^x = 2^{-x}$ and $y = \left(\frac{1}{3}\right)^x = 3^{-x}$:

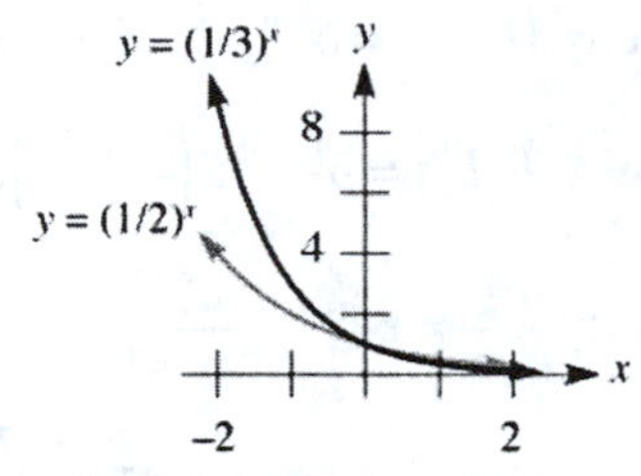

25. The domain is $(-\infty, \infty)$, the range is $(-\infty, 1)$, the x- and y-intercepts are 0, and the asymptote is $y = 1$:

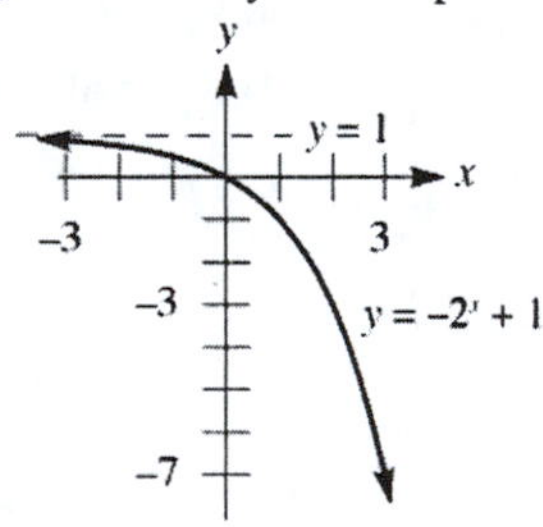

27. The domain is $(-\infty, \infty)$, the range is $(1, \infty)$, there is no x-intercept, the y-intercept is 2, and the asymptote is $y = 1$:

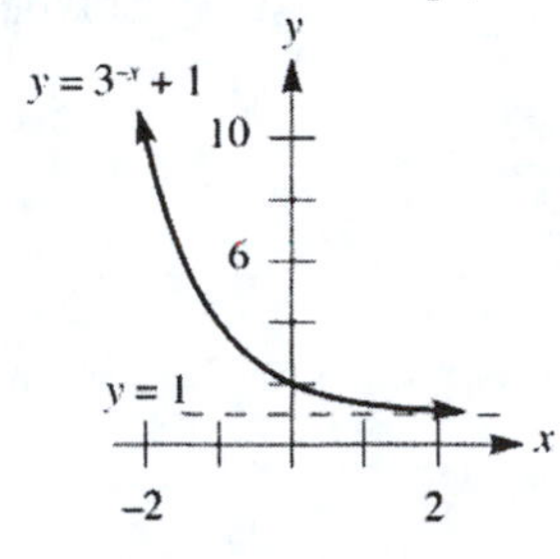

29. The domain is $(-\infty, \infty)$, the range is $(0, \infty)$, there is no x-intercept, the y-intercept is $\frac{1}{2}$, and the asymptote is $y = 0$:

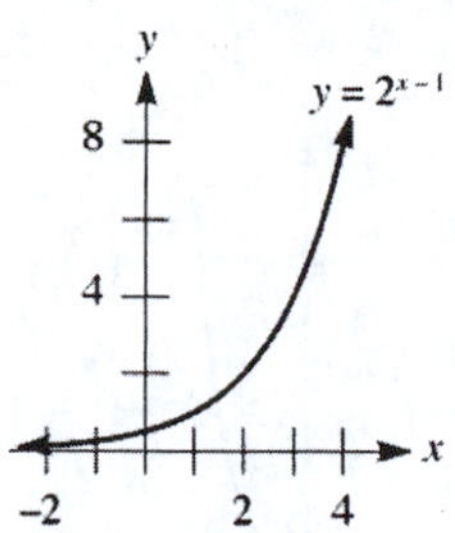

31. The domain is $(-\infty, \infty)$, the range is $(1, \infty)$, there is no x-intercept, the y-intercept is 4, and the asymptote is $y = 1$:

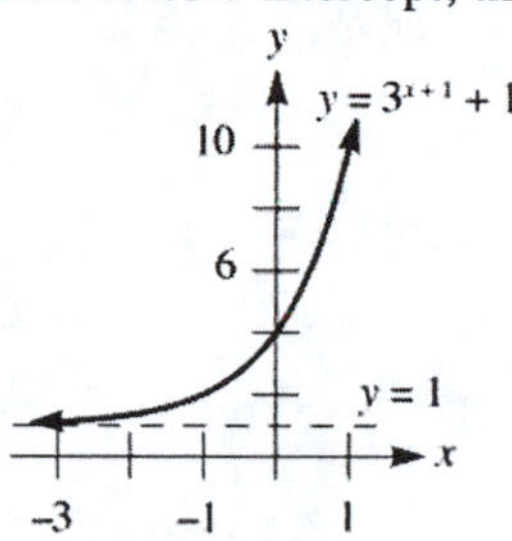

33. **a.** There is no x-intercept, the y-intercept is $-\frac{1}{9}$, and the asymptote is $y = 0$.

b. Graph the function:

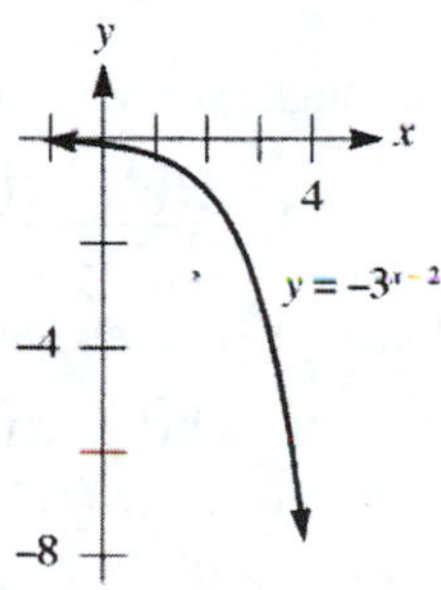

35. **a.** The x-intercept is -1, the y-intercept is -3, and the asymptote is $y = -4$.

b. Graph the function:

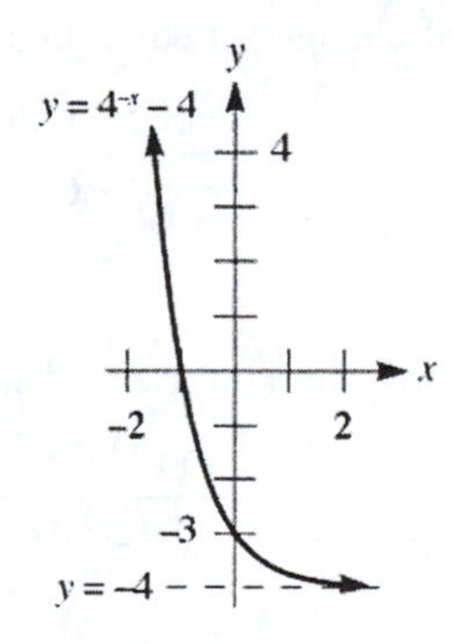

37. a. There is no x-intercept, the y-intercept is $\frac{1}{10}$, and the asymptote is $y = 0$.

b. Graph the function:

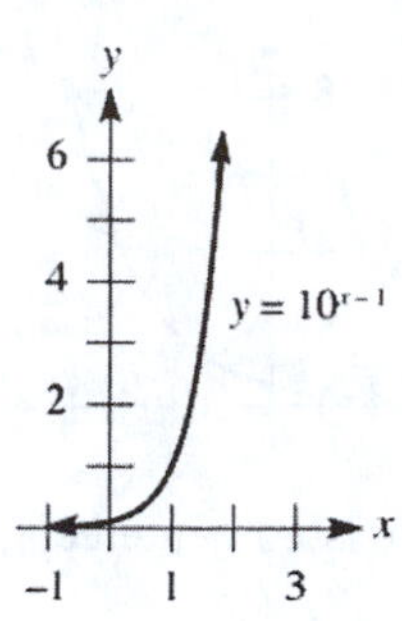

39. Solve for x:

$$3x\left(10^x\right) + 10^x = 0$$
$$10^x\left(3x+1\right) = 0$$
$$x = -\tfrac{1}{3} \text{ (since } 10^x \neq 0)$$

41. Solve for x:

$$3\left(3^x\right) - 5x\left(3^x\right) + 2x^2\left(3^x\right) = 0$$
$$3^x\left(3 - 5x + 2x^2\right) = 0$$
$$3^x\left(3-2x\right)\left(1-x\right) = 0$$
$$x = \tfrac{3}{2}, 1 \text{ (since } 3^x \neq 0)$$

43. a. For the interval $[0,2]$, the average rate of change for each function is:

$$\frac{f(2)-f(0)}{2-0} = \frac{4-1}{2} = 1.5 \qquad \frac{g(2)-g(0)}{2-0} = \frac{4-1}{2} = 1.5$$

For the interval $[2,4]$, the average rate of change for each function is:

$$\frac{f(4)-f(2)}{4-2} = \frac{16-4}{2} = 6 \qquad \frac{g(4)-g(2)}{4-2} = \frac{16-4}{2} = 6$$

b. For the interval $[4,6]$, the average rate of change for each function is:

$$\frac{f(6)-f(4)}{6-4} = \frac{37-16}{2} = 10.5 \qquad \frac{g(6)-g(4)}{6-4} = \frac{64-16}{2} = 24$$

Note that $\frac{\Delta g}{\Delta x}$ is more than twice $\frac{\Delta f}{\Delta x}$.

For the interval $[6,8]$, the average rate of change for each function is:

$$\frac{f(8)-f(6)}{8-6} = \frac{67-37}{2} = 15 \qquad \frac{g(8)-g(6)}{8-6} = \frac{256-64}{2} = 96$$

Note that $\frac{\Delta g}{\Delta x}$ is more than six times $\frac{\Delta f}{\Delta x}$.

c. For the interval $[10,12]$, the average rate of change for each function is:

$$\frac{f(12)-f(10)}{12-10} = \frac{154-106}{2} = 24 \qquad \frac{g(12)-g(10)}{12-10} = \frac{4096-1024}{2} = 1536$$

Note that $\frac{\Delta g}{\Delta x}$ is 64 times $\frac{\Delta f}{\Delta x}$.

45. Simplify the difference quotient: $\frac{f(x+h)-f(x)}{h} = \frac{2^{x+h}-2^x}{h} = \frac{2^x 2^h - 2^x}{h} = \frac{2^x\left(2^h-1\right)}{h} = 2^x\left(\frac{2^h-1}{h}\right)$

47. a. We know the graph of $y = 2^x$ and the graph of g, the inverse of f, should contain these points with the x and y coordinates interchanged. So $g(x)$ will be the reflection of $f(x)$ across the line $y = x$:

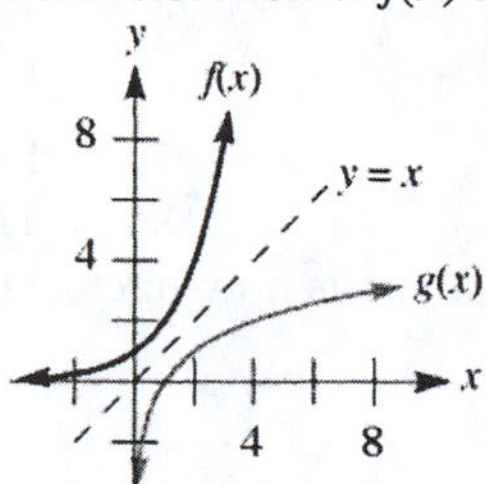

b. The domain is $(0, \infty)$, the range is $(-\infty, \infty)$, the x-intercept is 1, there is no y-intercept, and the asymptote is $x = 0$.

49. a. From the graph, we have $2^{1/2} \approx 1.4$.
b. Using a calculator, we have $\sqrt{2} \approx 1.41$.

51. a. From the graph, we have $2^{3/5} = 2^{0.6} \approx 1.5$.
b. Using a calculator, we have $2^{3/5} \approx 1.52$.

53. a. From the graph, we have $\sqrt{3} = 3^{1/2} \approx 1.7$.
b. Using a calculator, we have $\sqrt{3} \approx 1.73$.

55. a. From the graph, we have $5^{3/10} = 5^{0.3} \approx 1.6$.
b. Using a calculator, we have $5^{3/10} \approx 1.62$.

57. From the graph, we have $x \approx 0.3$.

59. From the graph, we have $x \approx 0.7$.

61. a. Draw the graph:

b. Adding $y = x^2$, draw the graph:

c. Sketch the graph of both functions:

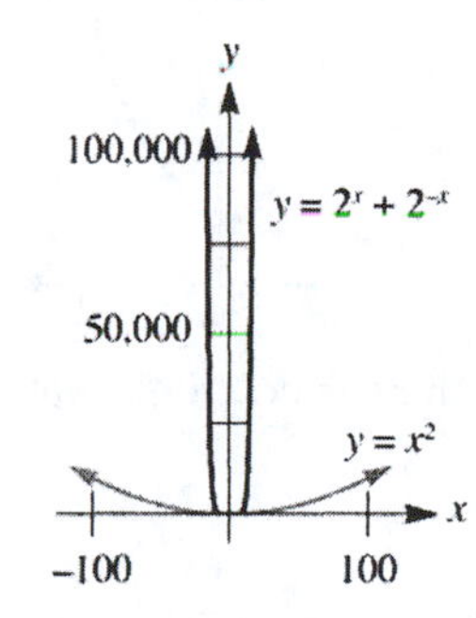

63. a. Graph the two functions:

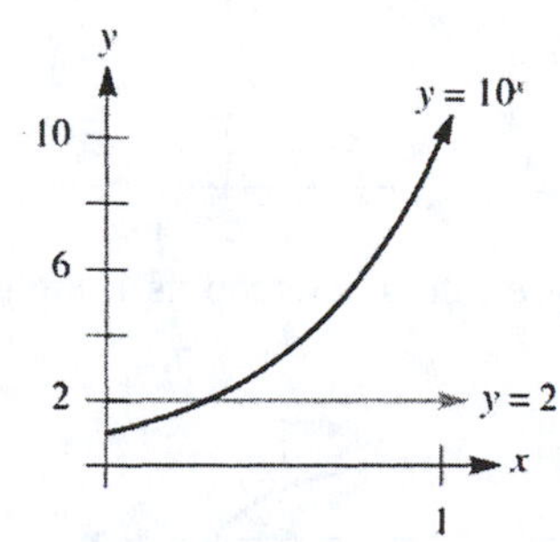

b. The point (0.3,2) is the approximate point of intersection, and thus $10^{0.3} \approx 2$.

5.2 The Exponential Function $y = e^x$

1. False (since $e \approx 2.7$)
3. False (since $\sqrt{2.7} > 1$)
5. True
7. False (since $2.7^{-1} > 0$)
9. **a.** These agree to two decimal places. **b.** These agree to two decimal places.
11. The domain is $(-\infty, \infty)$, the range is $(0, \infty)$, there is no x-intercept, the y-intercept is 1, and the asymptote is $y = 0$:

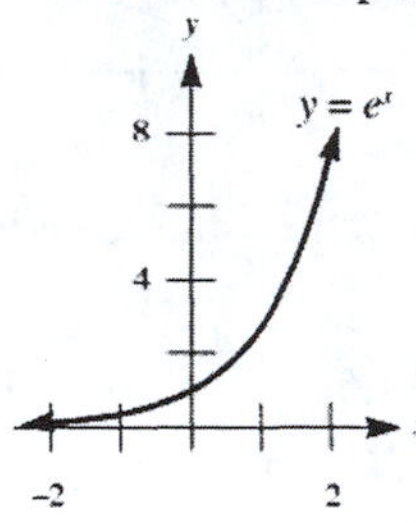

13. The domain is $(-\infty, \infty)$, the range is $(-\infty, 0)$, there is no x-intercept, the y-intercept is -1, and the asymptote is $y = 0$:

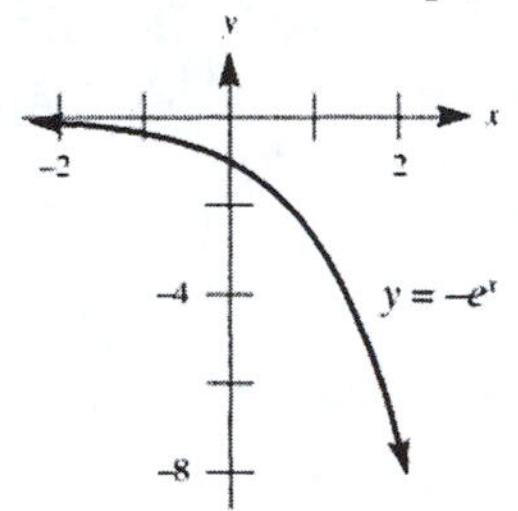

15. The domain is $(-\infty, \infty)$, the range is $(1, \infty)$, there is no x-intercept, the y-intercept is 2, and the asymptote is $y = 1$:

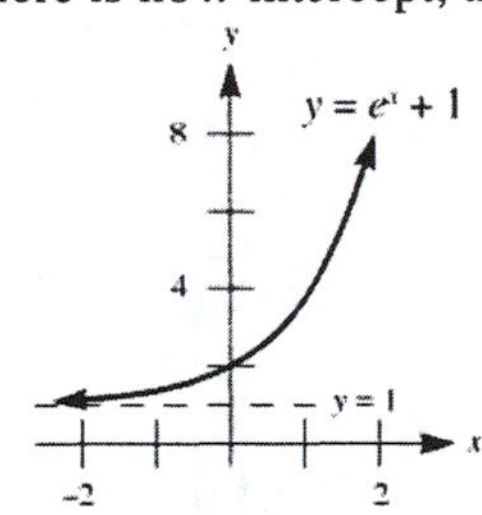

17. The domain is $(-\infty, \infty)$, the range is $(1, \infty)$, there is no x-intercept, the y-intercept is $e + 1$, and the asymptote is $y = 1$:

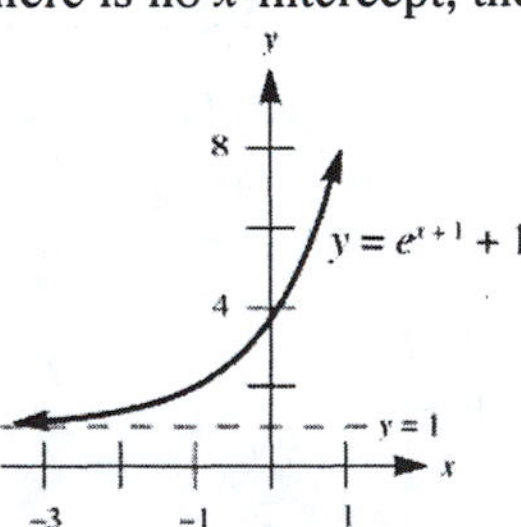

19. The domain is $(-\infty, \infty)$, the range is $(-\infty, e)$, the x-intercept is 1, the y-intercept is $e - 1$, and the asymptote is $y = e$:

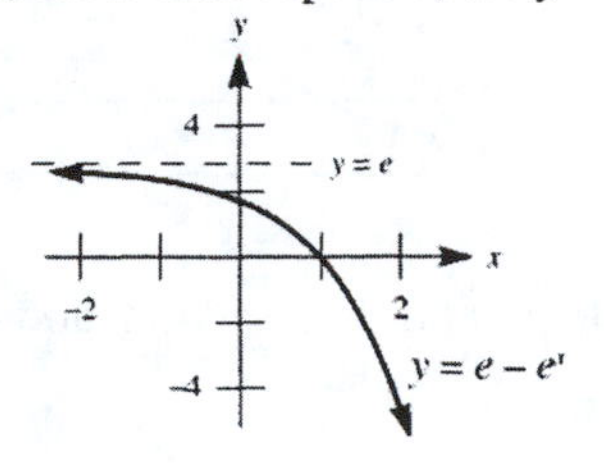

21. **a.** This is the graph of $y = e^x$ displaced to the left 2 units:

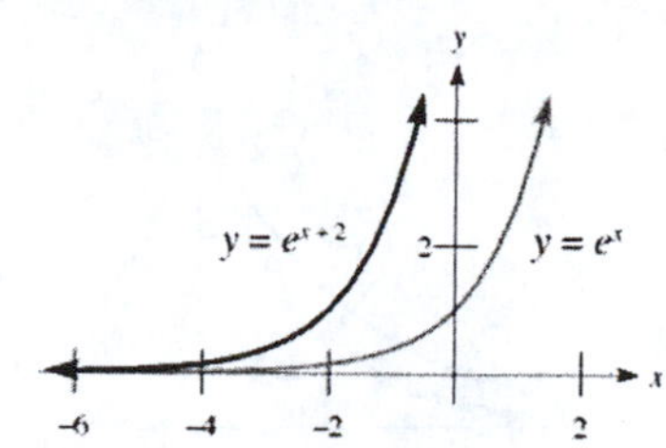

b. This is the graph of $y = e^x$ displaced to the left 2 units, then reflected across the y-axis:

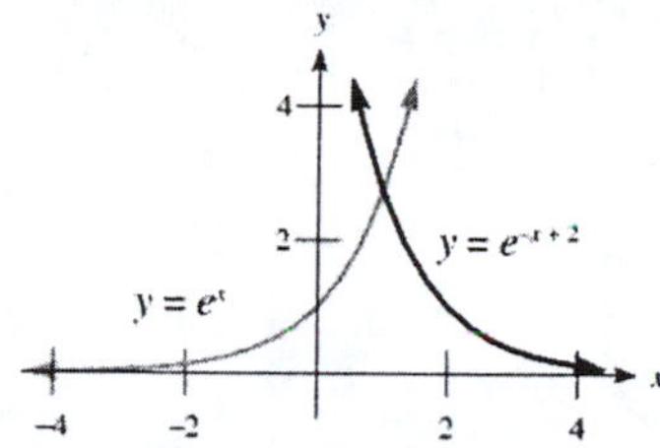

23. For the interval $[0,1]$, the average rate of change for each function is:

$$\frac{f(1)-f(0)}{1-0} = \frac{2-0}{1} = 1 \qquad \frac{g(1)-g(0)}{1-0} = \frac{2-1}{1} = 1 \qquad \frac{h(1)-h(0)}{1-0} = \frac{e-1}{1} \approx 1.7$$

The results are verified.

25. **a.** Sketch the graphs using the standard viewing rectangle:

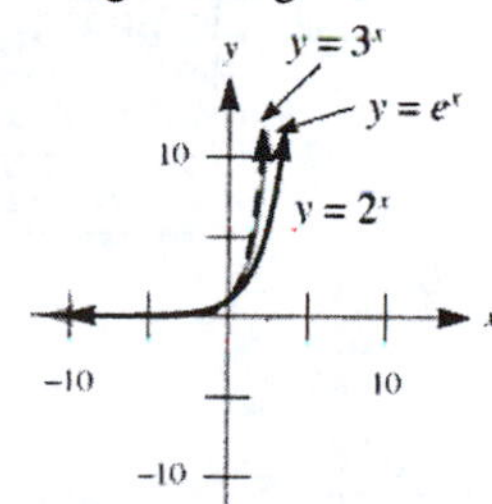

b. Sketch the graphs using the indicated settings:

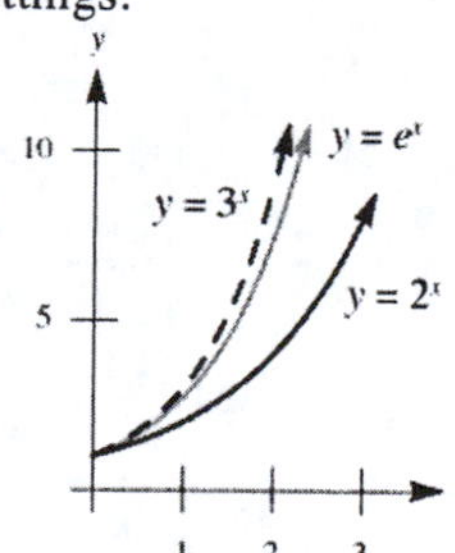

Note that $2^x < e^x < 3^x$ when $x > 0$.

c. Sketch the graphs using the indicated settings:

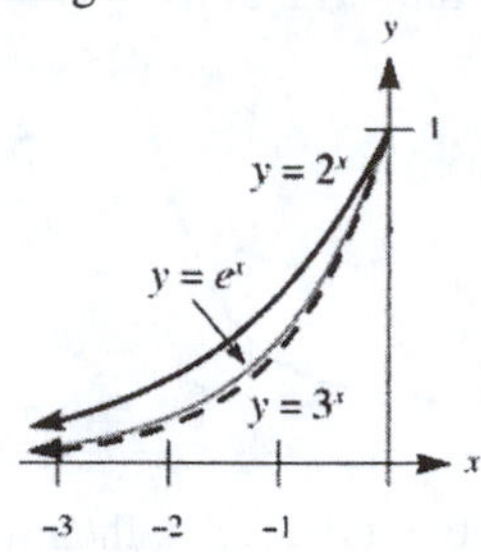

Note that $3^x < e^x < 2^x$ when $x < 0$.

d. If $x < 0$, then $-x > 0$, so by part **b**:

$$2^{-x} < e^{-x} < 3^{-x}$$

$$\frac{1}{2^x} < \frac{1}{e^x} < \frac{1}{3^x}$$

$$2^x > e^x > 3^x$$

27. Completing the tables:

Interval	$\Delta f/\Delta x$ for $f(x)=x^2$
$[3,3.1]$	6.1
$[3,3.01]$	6.01
$[3,3.001]$	6.001
$[3,3.0001]$	6.0001
$[3,3.00001]$	6.00001

Interval	$\Delta f/\Delta x$ for $f(x)=x^2$
$[2.9,3]$	5.9
$[2.99,3]$	5.99
$[2.999,3]$	5.999
$[2.9999,3]$	5.9999
$[2.99999,3]$	5.99999

The instantaneous rate of change appears to be 6.

29. Completing the table:

Interval	$\Delta f/\Delta x$ for $f(x)=e^x$
$[1.9,2]$	7.031617
$[1.99,2]$	7.352234
$[1.999,2]$	7.385363
$[1.9999,2]$	7.388687
$[1.99999,2]$	7.389019
$[1.999999,2]$	7.389053
$[1.9999999,2]$	7.389056

The value is close to the value of e^2.

31. **a.** Finding the average rate of change: $\dfrac{N(6)-N(5)}{6-5}=\dfrac{e^6-e^5}{1}\approx 255$ bacteria per hour

b. At $t = 5$ hours, the instantaneous rate of change is $e^5 \approx 148$ bacteria per hour. At $t = 5.5$ hours, the instantaneous rate of change is $e^{5.5} \approx 245$ bacteria per hour.

33. The instantaneous rates of change at each point are:

$x=-1$: $e^{-1} \approx 0.37$ $x=0$: $e^0 = 1$ $x=0.5$: $e^{0.5} \approx 1.65$

35. The function possesses properties A, D, E, and G.

37. The function possesses properties B, D, E, and G.

39. The function possesses properties A, D, F, G, and H.

41. The function possesses properties A, D, F, G, and H.

43. From the graph, we have $e^{0.1} \approx 1.1$. Using a calculator, we have $e^{0.1} \approx 1.105$.

45. From the graph, we have $e^{-0.3} \approx 0.75$. Using a calculator, we have $e^{-0.3} \approx 0.741$.

47. From the graph, we have $e^{-1} \approx 0.35$. Using a calculator, we have $e^{-1} \approx 0.368$.

49. From the graph, we have $\dfrac{1}{\sqrt{e}} = e^{-0.5} \approx 0.6$. Using a calculator, we have $\dfrac{1}{\sqrt{e}} = e^{-0.5} \approx 0.607$.

51. **a.** When $y = 1.5$, the value of x is $x \approx 0.4$.
b. The value is $\ln 1.5 \approx 0.405$, which is consistent with the answer from part **a**.

53. **a.** When $y = 1.8$, the value of x is $x \approx 0.6$.
b. The value is $\ln 1.8 \approx 0.588$, which is consistent with the answer from part **a**.

55. **a.** Evaluating when $x = 0$: $\cosh(0) = \dfrac{e^0 + e^0}{2} = \dfrac{1+1}{2} = 1$

Evaluating when $x = 1$: $\cosh(1) = \dfrac{e^1 + e^{-1}}{2} \approx 1.54$

Evaluating when $x = -1$: $\cosh(-1) = \dfrac{e^{-1} + e^1}{2} \approx 1.54$

b. The domain is all real numbers, or $(-\infty, \infty)$.

c. Compute $\cosh(-x)$: $\cosh(-x) = \dfrac{e^{-x} + e^x}{2} = \cosh(x)$. The graph must be symmetric about the y-axis.

d. Drawing the graph:

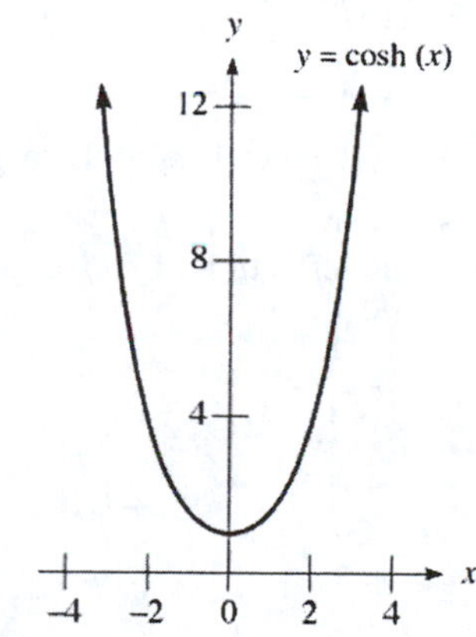

57. **a.** Evaluating when $x = 0$: $\sinh(0) = \dfrac{e^0 - e^0}{2} = 0$ Evaluating when $x = 1$: $\sinh(1) = \dfrac{e^1 - e^{-1}}{2} \approx 1.18$

Evaluating when $x = -1$: $\sinh(-1) = \dfrac{e^{-1} - e^1}{2} \approx -1.18$

b. The domain is all real numbers, or $(-\infty, \infty)$.

c. Compute $\sinh(-x)$: $\sinh(-x) = \dfrac{e^{-x} - e^x}{2} = -\dfrac{e^x - e^{-x}}{2} = -\sinh(x)$

The graph must be symmetric about the origin.

d. Draw the graph:

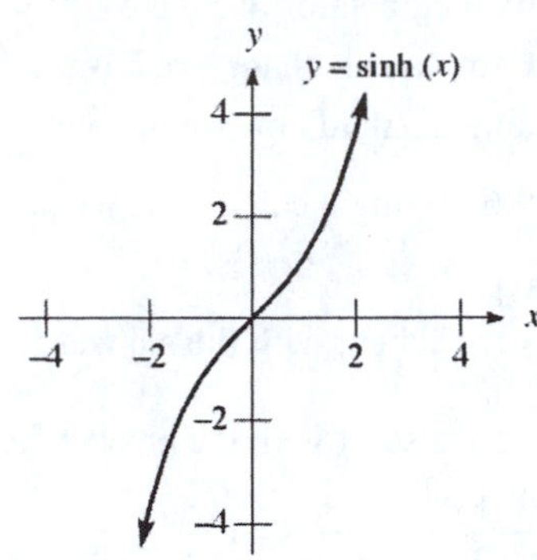

e. The two curves intersect in three points. See the graph:

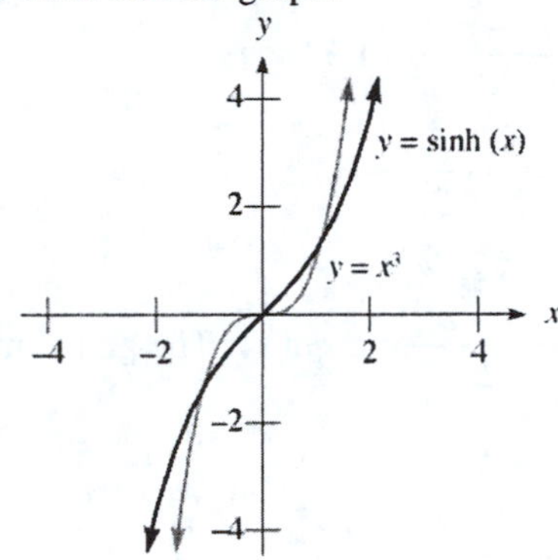

59. **a.** Based on the graph, $e^{\pi} \approx 23.4$ while $\pi^{e} \approx 22.2$, so e^{π} is slightly larger.

b. Since $e^{\pi} \approx 23.14$ while $\pi^{e} \approx 22.46$, e^{π} is larger.

61. **a.** Draw the graphs, noting that $L(x)$ will be the reflection of $f(x)$ across $y = x$:

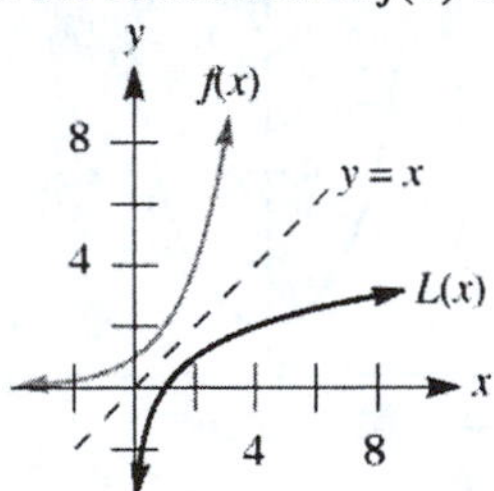

b. The domain is $(0, \infty)$, the range is $(-\infty, \infty)$, the x-intercept is 1, and the asymptote is $x = 0$.

c. (i) This will be a reflection of $L(x)$ across the x-axis. The x-intercept is 1 and the asymptote is $x = 0$:

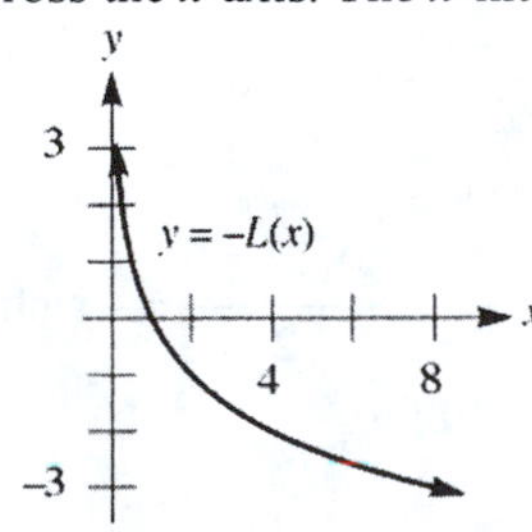

(ii) This will be a reflection of $L(x)$ across the y-axis. The x-intercept is -1 and the asymptote is $x = 0$:

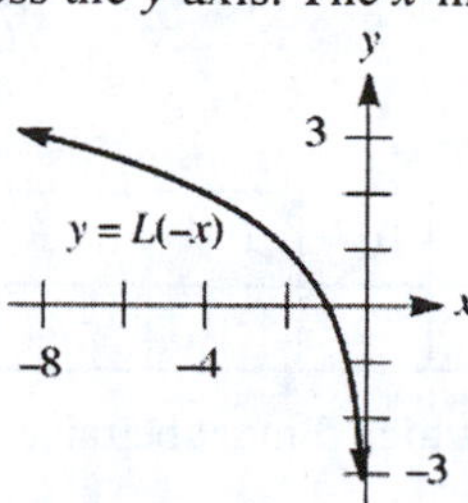

(iii) This will be a displacement of $L(x)$ to the right one unit. The x-intercept is 2 and the asymptote is $x = 1$:

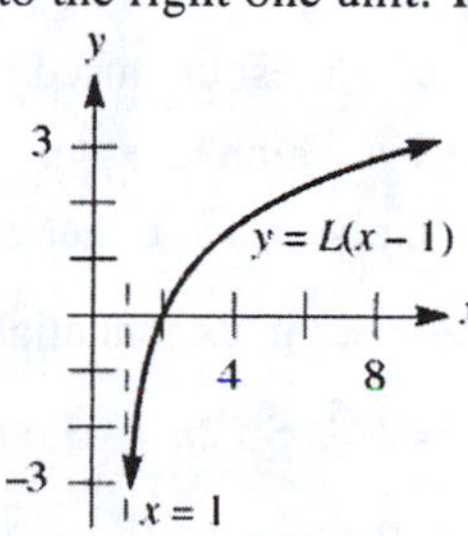

63. **a.** The equation $[\cosh(x)]^2 - 2[\sinh(x)]^2 = 1$ is not an identity. The only solution to it is $x = 0$.

b. Proving the identity:

$$
\begin{aligned}
[\cosh(x)]^2 - [\sinh(x)]^2 &= \left(\frac{e^x + e^{-x}}{2}\right)^2 - \left(\frac{e^x - e^{-x}}{2}\right)^2 \\
&= \frac{e^{2x} + 2 + e^{-2x}}{4} - \frac{e^{2x} - 2 + e^{-2x}}{4} \\
&= \frac{e^{2x} + 2 + e^{-2x} - e^{2x} + 2 - e^{-2x}}{4} \\
&= \frac{4}{4} \\
&= 1
\end{aligned}
$$

5.3 Logarithmic Functions

1. **a.** $f[g(x)] = x$ for each x in the domain of g. **b.** $g[f(x)] = x$ for each x in the domain of f.

3. **a.** The functions $f(x) = 2^x$ and $g(x) = \log_2 x$ are inverse functions.

b. $2^{\log_2 x} = x$ for each positive number x, and $\log_2\left(2^x\right) = x$ for all real numbers x.

c. $2^{\log_2 99} = 99$ and also $\log_2\left(2^{-\pi}\right) = -\pi$.

5. **a.** $y = x + 1$ is one-to-one, since any horizontal line intersects the line only once.

b. $y = (x+1)^2$ is not one-to-one, since a horizontal line can intersect the parabola twice.

c. $y = (x+1)^3$ is one-to-one, since any horizontal line intersects the curve only once.

7. $f\left[f^{-1}(6)\right] = 6$

9. **a.** $\log_3 9 = 2$ **b.** $\log_{10} 1000 = 3$

c. $\log_7 343 = 3$ **d.** $\log_2 \sqrt{2} = \frac{1}{2}$

11. a. $2^5 = 32$ b. $10^0 = 1$

c. $e^{1/2} = \sqrt{e}$ d. $e^{-1} = \frac{1}{e}$

13. Complete the table:

x	1	10	10^2	10^3	10^{-1}	10^{-2}	10^{-3}
$\log_{10} x$	0	1	2	3	−1	−2	−3

15. a. Since $\log_5 30$ represents the power to which 5 must be raised to get 30, it is clearly greater than 2, since $5^2 = 25$. But $\log_8 60$ is less than 2, since $8^2 = 64$. Hence $\log_5 30$ is larger.

b. Since $\ln 17$ represents the power to which e must be raised to get 17, it is greater than 2. But $\log_{10} 17$ is the power to which 10 must be raised to get 17, which is less than 2. Hence $\ln 17$ is larger.

17. a. Since $\log_9 27$ is the power to which 9 must be raised to get 27, we can see it is between 1 and 2, since $9^1 = 9$ while $9^2 = 81$. To find it let $\log_9 27 = n$ then $9^n = 27$ in exponential form and $3^{2n} = 3^3$. So $2n = 3$ and $n = \frac{3}{2}$.

b. If $\log_4 \frac{1}{32} = n$ then $4^n = \frac{1}{32}$, thus $2^{2n} = 2^{-5}$. So $2n = -5$, and $n = -\frac{5}{2}$. So $\log_4 \frac{1}{32} = -\frac{5}{2}$.

c. Follow the same steps. If $\log_5 5\sqrt{5} = n$, then:

$$5^n = 5\sqrt{5}$$
$$5^n = 5^{3/2}$$
$$n = \frac{3}{2}$$

19. a. Writing $\log_4 x = -2$ in exponential form, we have $x = 4^{-2} = \frac{1}{16}$.

b. Writing $\ln x = -2$ in exponential form, we have $x = e^{-2} \approx 0.14$.

21. a. We must have $5x > 0$, so $x > 0$. So the domain is $(0, \infty)$.

b. We must have $3 - 4x > 0$, so $3 > 4x$ and $x < \frac{3}{4}$. So the domain is $\left(-\infty, \frac{3}{4}\right)$.

c. We must have $x^2 > 0$, so $x \neq 0$. So the domain is all real numbers except 0, or $(-\infty, 0) \cup (0, \infty)$.

d. We must have $x > 0$. So the domain is $(0, \infty)$.

e. Solve the inequality:

$$x^2 - 25 > 0$$
$$(x+5)(x-5) > 0$$

The key numbers are −5 and 5. Construct the sign chart:

Inteval	Test Number	$x+5$	$x-5$	$(x+5)(x-5)$
$(-\infty, -5)$	−6	neg.	neg.	pos.
$(-5, 5)$	0	pos.	neg.	neg.
$(5, \infty)$	6	pos.	pos.	pos.

So the product is positive on the intervals $(-\infty, -5) \cup (5, \infty)$, which is the domain.

Note: This inequality can also be solved without using a sign chart:

$$x^2 - 25 > 0$$
$$x^2 > 25$$
$$|x| > 5$$
$$x > 5 \text{ or } x < -5$$

23. The coordinates for each point are:

A: (0,1) B: (1,0) C: $(4, \log_2 4) = (4, 2)$ D: (2,4)

25. **a.** The domain is $(0, \infty)$, the range is $(-\infty, \infty)$, the x-intercept is 1, there is no y-intercept, and the asymptote is $x = 0$:

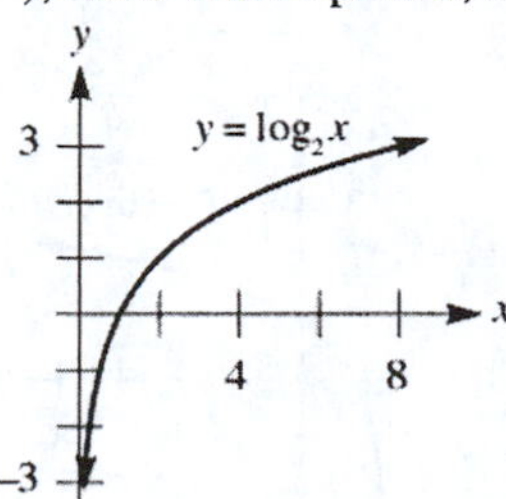

b. The domain is $(0, \infty)$, the range is $(-\infty, \infty)$, the x-intercept is 1, there is no y-intercept, and the asymptote is $x = 0$:

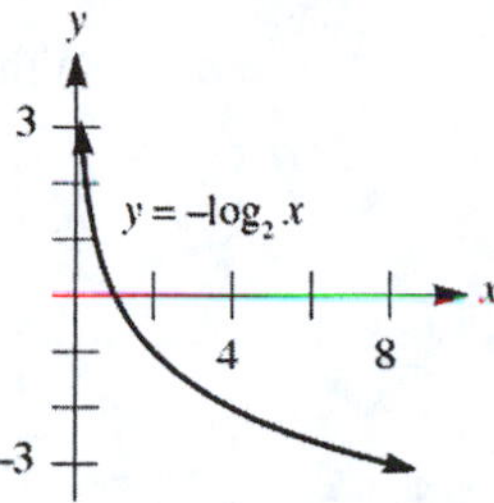

c. The domain is $(-\infty, 0)$, the range is $(-\infty, \infty)$, the x-intercept is -1, there is no y-intercept, and the asymptote is $x = 0$:

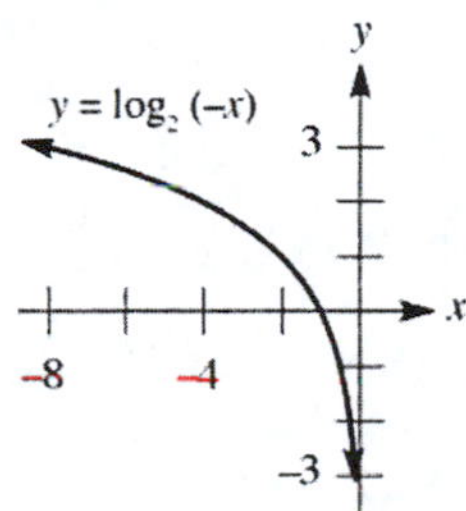

d. The domain is $(-\infty, 0)$, the range is $(-\infty, \infty)$, the x-intercept is -1, there is no y-intercept, and the asymptote is $x = 0$:

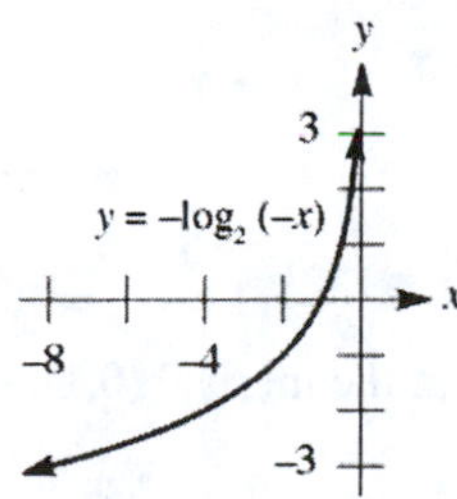

27. Translate $y = \log_3 x$ two units to the right, reflect across the x-axis, and translate 1 unit up. The domain is $(2, \infty)$, the range is $(-\infty, \infty)$, the x-intercept is 5, there is no y-intercept, and the asymptote is $x = 2$:

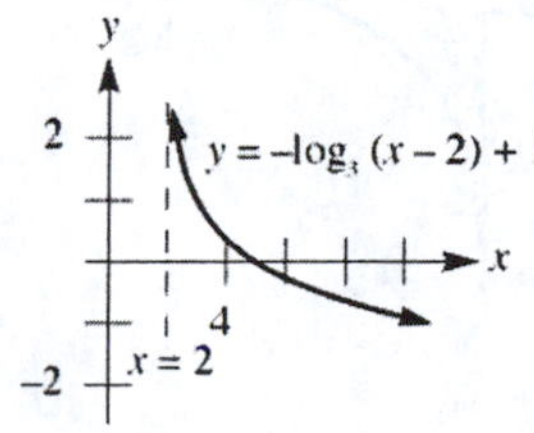

29. Translate $y = \ln x$ to the left e units. The domain is $(-e, \infty)$, the range is $(-\infty, \infty)$, the x-intercept is $-e + 1$, the y-intercept is 1, and the asymptote is $x = -e$:

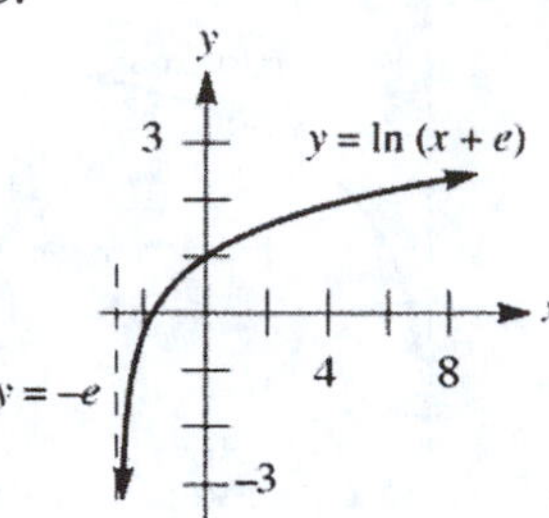

31. **a.** Writing $x = \ln e^4$ in exponential form:

$$e^x = e^4$$
$$x = 4$$

b. Writing $x = \ln \frac{1}{e}$ in exponential form:

$$e^x = \frac{1}{e}$$
$$e^x = e^{-1}$$
$$x = -1$$

c. Writing $x = \ln \sqrt{e}$ in exponential form:

$$e^x = \sqrt{e}$$
$$e^x = e^{1/2}$$
$$x = \frac{1}{2}$$

33. **a.** The solution is approximately 1.4.

b. Solve for x:

$$10^x = 25$$
$$x = \log_{10} 25 \approx 1.398$$

35. **a.** The solutions are approximately ±1.3.

b. Solve for x:

$$10^{(x^2)} = 40$$
$$x^2 = \log_{10} 40$$
$$x = \pm\sqrt{\log_{10} 40} = \pm\sqrt{1+\log_{10} 4} \approx \pm 1.266$$

37. **a.** The solution is approximately –0.3.

b. Solve for t:

$$e^{2t+3} = 10$$
$$2t + 3 = \ln 10$$
$$2t = -3 + \ln 10$$
$$t = \frac{-3+\ln 10}{2} \approx -0.349$$

39. **a.** The solution is approximately –0.4.

b. Solve for t:

$$e^{1-4t} = 12.405$$
$$1 - 4t = \ln 12.405$$
$$-4t = -1 + \ln 12.405$$
$$t = \frac{1-\ln 12.405}{4} \approx -0.380$$

41. **a.** From the graphs, note that $\ln x < \log_{10} x$ on the interval (0,1):

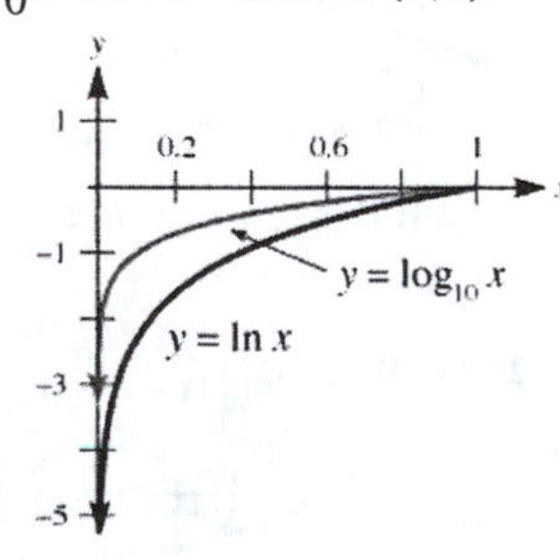

b. From the graphs, note that $\ln x > \log_{10} x$ on the interval $(1,\infty)$:

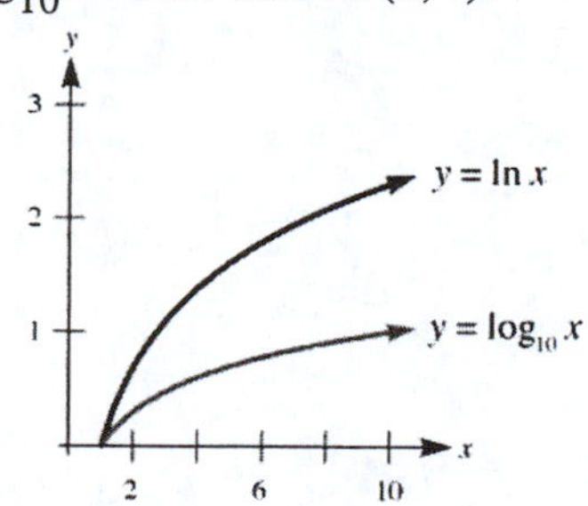

c. From the graphs, note that $e^x < 10^x$ on the interval $(0,\infty)$:

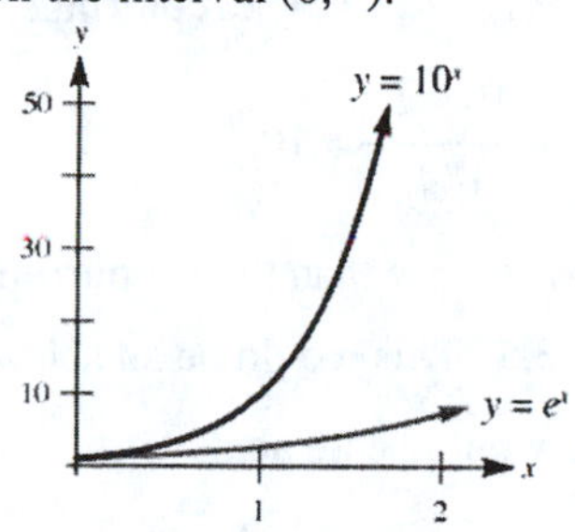

d. From the graphs, note that $e^x > 10^x$ on the interval $(-\infty, 0)$:

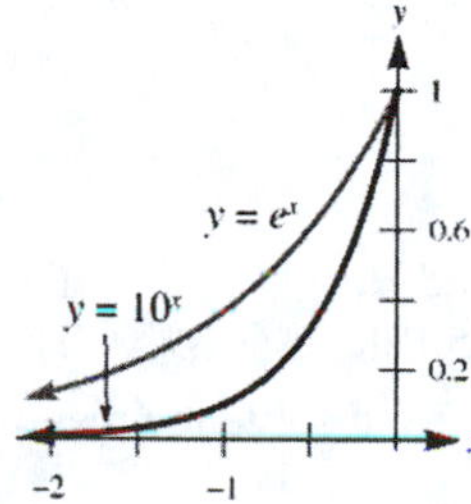

43. The intensities for Bombay and San Salvador are:

Bombay: $A_0 10^{6.4}$ San Salvador: $A_0 10^{6.6}$

Computing the ratio: $\dfrac{\text{San Salvador}}{\text{Bombay}} = \dfrac{A_0 10^{6.6}}{A_0 10^{6.4}} = 10^{0.2} \approx 1.6$

The San Salvador quake was approximately 1.6 times stronger than the Bombay quake.

45. The intensities for the first quake and the second quake are:

First: $A_0 10^{M_0}$ Second: $A_0 10^{M_0+d}$

Computing the ratio: $\dfrac{\text{Second}}{\text{First}} = \dfrac{A_0 10^{M_0+d}}{A_0 10^{M_0}} = 10^d$

The second quake is 10^d times stronger than the first quake.

47. **a.** Solving the equation:

$$\beta = 10\log_{10}\frac{I}{I_0}$$
$$\frac{\beta}{10} = \log_{10}\frac{I}{I_0}$$
$$\frac{I}{I_0} = 10^{\beta/10}$$
$$I = 10^{\beta/10} I_0$$

b. Computing the intensities for each:

power mower: $10^{100/10} I_0 = 10^{10} I_0$ cat purring: $10^{10/10} I_0 = 10 I_0$

Computing the ratio: $\frac{\text{power mower}}{\text{cat purring}} = \frac{10^{10} I_0}{10 I_0} = 10^9$

The power mower is 10^9 times more intense than the cat purring.

49. **a.** Compute $\text{pH} = -\log_{10}\left(3\times 10^{-4}\right) \approx 3.5$. This would be an acid.

b. Compute $\text{pH} = -\log_{10}(1) = 0$. This would be an acid.

51. Solving the equation:

$$-\log_{10}\left[\text{H}^+\right] = 5.9$$
$$\log_{10}\left[\text{H}^+\right] = -5.9$$
$$\left[\text{H}^+\right] = 10^{-5.9}$$

53. The function possesses properties A, D, E, and H.

55. The function possesses properties B, D, E, and H.

57. The function possesses properties A, D, E, and H.

59. The function possesses properties A, D, E, and H.

61. Let $y = e^{x+1}$. Interchange the roles of x and y and solve the resulting equation for y:

$$x = e^{y+1}$$
$$\ln x = \ln\left(e^{y+1}\right)$$
$$\ln x = y+1$$
$$y = -1+\ln x$$
$$f^{-1}(x) = -1+\ln x$$

The x-intercept is e and the asymptote is $x = 0$:

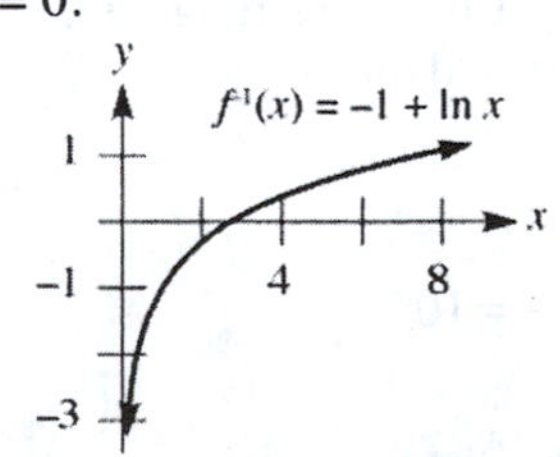

63. Using the approximation $2^{10} \approx 10^3$, then writing $\log_2 x = 100$ in exponential form:

$$x = 2^{100} = \left(2^{10}\right)^{10} \approx \left(10^3\right)^{10} = 10^{30}$$

65. **a.** Complete the table:

Planet	x	y	$\ln x$	$\ln y$
Mercury	0.387	0.241	–0.95	–1.42
Venus	0.723	0.615	–0.32	–0.49
Earth	1.000	1.000	0.00	0.00
Mars	1.523	1.881	0.42	0.63
Jupiter	5.202	11.820	1.65	2.47

b. Entering the data as points ($\ln x$, $\ln y$), we obtain $a = 0.00$ and $b = 1.50$. So the relationship between $\ln x$ and $\ln y$ is $\ln y = 1.50 \ln x$.

c. Taking the exponential of each side of the equation:

$$e^{\ln y} = e^{0.00} \bullet e^{1.50 \ln x}$$
$$y = 1 \bullet e^{\ln x^{1.50}}$$
$$y = x^{1.5}$$

d. Complete the table:

Planet	x	y (calculated)	y (observed)
Saturn	9.555	29.54	29.46
Uranus	19.22	84.26	84.01
Neptune	30.11	165.22	164.79
Pluto	39.44	247.69	248.5

67. **a.** The logarithmic regression equation is $y = -24.795459 + 7.453496 \ln x$.

b. The linear regression equation is $y = 0.1086x - 0.8890$. Sketching both lines with the data:

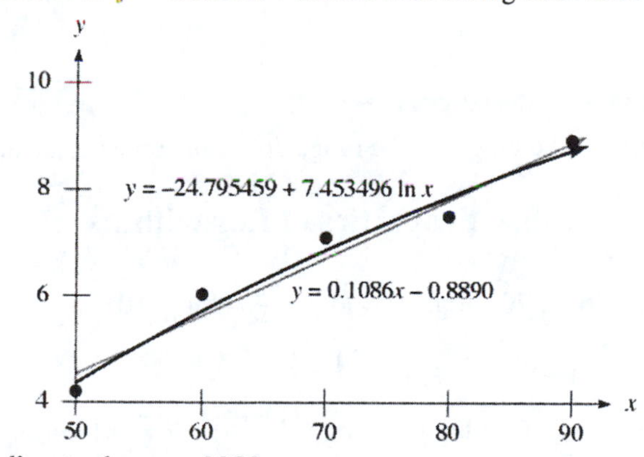

c. Using x = 100 corresponding to the year 2000:

logarithmic: $y = -24.795459 + 7.453496 \ln(100) \approx 9.529$ million

linear: $y = 0.1086(100) - 0.8890 = 9.971$ million

d. The logarithmic model is closer to this. Computing the pertentage errors:

logarithmic: percentage error $= \dfrac{|9.519 - 9.529|}{9.519} \times 100 \approx 0.11\%$

linear: percentage error $= \dfrac{|9.519 - 9.971|}{9.519} \times 100 \approx 4.75\%$

69. a. First, for $\ln x$ to be defined we must have $x > 0$. Now, for $\ln(\ln x)$ to be defined we must have $\ln x > 0$, so $x > 1$. Thus the domain is $(1, \infty)$.

b. Interchange the roles of x and y, and solve the resulting equation for y:

$$\ln(\ln y) = x$$
$$\ln y = e^x$$
$$y = e^{e^x}$$

So $f^{-1}(x) = e^{e^x}$.

71. a. Graphing the function:

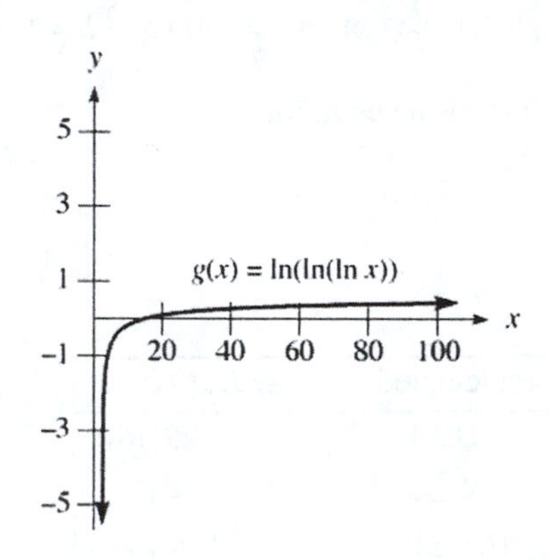

b. Solving for x:

$$\ln(\ln(\ln x)) = y$$
$$\ln(\ln x) = e^y$$
$$\ln x = e^{e^y}$$
$$x = e^{e^{e^y}}$$

Since there is no restriction on y, the range is $(-\infty, \infty)$.

c. No, the graph in part **a** appears bounded, so the range does not appear to be all real numbers.

5.4 Properties of Logarithms

1. Using properties of logarithms: $\log_{10} 70 - \log_{10} 7 = \log_{10} \frac{70}{7} = \log_{10} 10 = 1$

3. Using properties of logarithms: $\log_7 \sqrt{7} = \log_7 \left(7^{1/2}\right) = \frac{1}{2}$

5. Using properties of logarithms: $\log_3 108 + \log_3 \frac{3}{4} = \log_3 \left(\frac{108 \bullet 3}{4}\right) = \log_3 81 = \log_3 3^4 = 4$

7. Using properties of logarithms: $-\frac{1}{2} + \ln\sqrt{e} = -\frac{1}{2} + \ln e^{1/2} = -\frac{1}{2} + \frac{1}{2} = 0$

9. Using properties of logarithms: $2^{\log_2 5} - 3\log_5 \sqrt[3]{5} = 5 - 3\log_5 5^{1/3} = 5 - 3\left(\frac{1}{3}\right) = 5 - 1 = 4$

11. Using properties of logarithms: $\log_{10} 30 + \log_{10} 2 = \log_{10}(30 \bullet 2) = \log_{10} 60$

13. Using properties of logarithms: $\log_5 6 + \log_5 \frac{1}{3} + \log_5 10 = \log_5\left(6 \bullet \frac{1}{3} \bullet 10\right) = \log_5 20$

15. **a.** Using properties of logarithms: $\ln 3 - 2\ln 4 + \ln 32 = \ln 3 - \ln\left(4^2\right) + \ln 32 = \ln\left(\frac{3 \bullet 32}{4^2}\right) = \ln 6$

b. Using properties of logarithms:

$$\ln 3 - 2(\ln 4 + \ln 32) = \ln 3 - 2\left[\ln(4 \bullet 32)\right] = \ln 3 - 2\ln 128 = \ln 3 - \ln\left(128^2\right) = \ln\left(\frac{3}{128^2}\right) = \ln\left(\tfrac{3}{16384}\right)$$

17. Using properties of logarithms:

$$\begin{aligned}\log_b 4 + 3\left[\log_b(1+x) - \tfrac{1}{2}\log_b(1-x)\right] &= \log_b 4 + 3\left(\log_b(1+x) - \log_b\sqrt{1-x}\right)\\ &= \log_b 4 + \log_b(1+x)^3 - \log_b(1-x)^{3/2}\\ &= \log_b\left[\frac{4(1+x)^3}{(1-x)^{3/2}}\right]\end{aligned}$$

19. Using properties of logarithms:

$$\begin{aligned}&4\log_{10} 3 - 6\log_{10}\left(x^2+1\right) + \tfrac{1}{2}\left[\log_{10}(x+1) - 2\log_{10} 3\right]\\ &= \log_{10} 3^4 - \log_{10}\left(x^2+1\right)^6 + \tfrac{1}{2}\left[\log_{10}(x+1) - \log_{10} 3^2\right]\\ &= \log_{10} 81 - \log_{10}\left(x^2+1\right)^6 + \tfrac{1}{2}\log_{10}\left(\frac{x+1}{9}\right)\\ &= \log_{10} 81 - \log_{10}\left(x^2+1\right)^6 + \log_{10}\left(\frac{x+1}{9}\right)^{1/2}\\ &= \log_{10}\left[\frac{81\frac{\sqrt{x+1}}{3}}{\left(x^2+1\right)^6}\right]\\ &= \log_{10}\left[\frac{27\sqrt{x+1}}{\left(x^2+1\right)^6}\right]\end{aligned}$$

21. **a.** Using properties of logarithms: $\log_{10}\left(\frac{x^2}{1+x^2}\right) = \log_{10} x^2 - \log_{10}\left(1+x^2\right) = 2\log_{10} x - \log_{10}\left(1+x^2\right)$

b. Using properties of logarithms: $\ln\left(\frac{x^2}{\sqrt{1+x^2}}\right) = \ln x^2 - \ln\sqrt{1+x^2} = 2\ln x - \frac{1}{2}\ln\left(1+x^2\right)$

23. **a.** Using properties of logarithms:

$$\log_{10}\sqrt{9-x^2} = \tfrac{1}{2}\log_{10}\left(9-x^2\right) = \tfrac{1}{2}\log_{10}\left[(3+x)(3-x)\right] = \tfrac{1}{2}\log_{10}(3+x) + \tfrac{1}{2}\log_{10}(3-x)$$

b. Using properties of logarithms:

$$\begin{aligned}\ln\left[\frac{\sqrt{4-x^2}}{(x-1)(x+1)^{3/2}}\right] &= \tfrac{1}{2}\ln\left(4-x^2\right) - \ln(x-1) - \ln(x+1)^{3/2}\\ &= \tfrac{1}{2}\ln\left[(2+x)(2-x)\right] - \ln(x-1) - \tfrac{3}{2}\ln(x+1)\\ &= \tfrac{1}{2}\ln(2+x) + \tfrac{1}{2}\ln(2-x) - \ln(x-1) - \tfrac{3}{2}\ln(x+1)\end{aligned}$$

25. **a.** Using properties of logarithms: $\log_b\sqrt{\frac{x}{b}} = \frac{1}{2}\log_b\frac{x}{b} = \frac{1}{2}\log_b x - \frac{1}{2}\log_b b = \frac{1}{2}\log_b x - \frac{1}{2}$

b. Using properties of logarithms:

$$2\ln\sqrt{\left(1+x^2\right)\left(1+x^4\right)\left(1+x^6\right)} = \ln\left[\left(1+x^2\right)\left(1+x^4\right)\left(1+x^6\right)\right] = \ln\left(1+x^2\right) + \ln\left(1+x^4\right) + \ln\left(1+x^6\right)$$

27. a. Using properties of logarithms: $\log_b 6 = \log_b (2 \cdot 3) = \log_b 2 + \log_b 3 = A + B$

b. Using properties of logarithms: $\log_b \frac{1}{6} = \log_b (6^{-1}) = -\log_b 6 = -(A+B) = -A - B$

c. Using properties of logarithms: $\log_b 27 = \log_b (3^3) = 3\log_b 3 = 3B$

d. Using properties of logarithms: $\log_b \frac{1}{27} = \log_b (27^{-1}) = -\log_b 27 = -3B$

29. a. Using properties of logarithms: $\log_b \frac{5}{3} = \log_b 5 - \log_b 3 = C - B$

b. Using properties of logarithms: $\log_b 0.6 = \log_b \frac{3}{5} = \log_b 3 - \log_b 5 = B - C$

c. Using properties of logarithms: $\log_b \frac{5}{9} = \log_b 5 - \log_b (3^2) = \log_b 5 - 2\log_b 3 = C - 2B$

d. Using properties of logarithms: $\log_b \frac{5}{16} = \log_b 5 - \log_b (2^4) = \log_b 5 - 4\log_b 2 = C - 4A$

31. a. Using the change of base formula: $\log_3 b = \dfrac{\log_b b}{\log_b 3} = \dfrac{1}{B}$

b. Using the change of base formula:

$$\log_3 (10b) = \frac{\log_b (10b)}{\log_b 3} = \frac{\log_b (2 \cdot 5b)}{\log_b 3} = \frac{\log_b 2 + \log_b 5 + \log_b b}{\log_b 3} = \frac{A+C+1}{B}$$

33. a. Using the change of base formula: $\log_{3b} 2 = \dfrac{\log_b 2}{\log_b (3b)} = \dfrac{\log_b 2}{\log_b 3 + \log_b b} = \dfrac{A}{B+1}$

b. Using the change of base formula: $\log_{3b} 15 = \dfrac{\log_b 15}{\log_b (3b)} = \dfrac{\log_b 3 + \log_b 5}{\log_b 3 + \log_b b} = \dfrac{B+C}{B+1}$

35. a. Using the change of base formula: $(\log_b 5)(\log_5 b) = \log_b 5 \cdot \dfrac{\log_b b}{\log_b 5} = \log_b b = 1$

b. Using the change of base formula: $(\log_b 6)(\log_6 b) = \log_b 6 \cdot \dfrac{\log_b b}{\log_b 6} = \log_b b = 1$

37. a. Using properties of logarithms:

$$\log_{10} (AB^2C^3) = \log_{10} A + \log_{10} B^2 + \log_{10} C^3 = \log_{10} A + 2\log_{10} B + 3\log_{10} C = a + 2b + 3c$$

b. Using properties of logarithms: $\log_{10} 10\sqrt{A} = \log_{10} 10 + \log_{10} \sqrt{A} = 1 + \frac{1}{2}\log_{10} A = 1 + \frac{1}{2}a$

c. Using properties of logarithms:

$$\begin{aligned}\log_{10} \sqrt{10ABC} &= \tfrac{1}{2}\log_{10} (10ABC) \\ &= \tfrac{1}{2}\log_{10} 10 + \tfrac{1}{2}\log_{10} A + \tfrac{1}{2}\log_{10} B + \tfrac{1}{2}\log_{10} C \\ &= \tfrac{1}{2}(1) + \tfrac{1}{2}a + \tfrac{1}{2}b + \tfrac{1}{2}c \\ &= \tfrac{1}{2}(1 + a + b + c)\end{aligned}$$

d. Using properties of logarithms:

$$\log_{10}\left(\frac{10A}{\sqrt{BC}}\right) = \log_{10} (10A) - \tfrac{1}{2}\log_{10} (BC) = \log_{10} 10 + \log_{10} A - \tfrac{1}{2}\log_{10} B - \tfrac{1}{2}\log_{10} C = 1 + a - \tfrac{1}{2}b - \tfrac{1}{2}c$$

39. **a.** Using properties of logarithms: $\ln(ex)=\ln e+\ln x=1+t$

b. Using properties of logarithms: $\ln(xy)-\ln\left(x^2\right)=\ln\frac{xy}{x^2}=\ln\frac{y}{x}=\ln y-\ln x=u-t$

c. Using properties of logarithms: $\ln\sqrt{xy}+\ln\frac{x}{e}=\frac{1}{2}(\ln x+\ln y)+\ln x-\ln e=\frac{1}{2}(t+u)+t-1=\frac{3}{2}t+\frac{1}{2}u-1$

d. Using properties of logarithms: $\ln\left(e^2x\sqrt{y}\right)=\ln e^2+\ln x+\ln\sqrt{y}=2\ln e+\ln x+\frac{1}{2}\ln y=2+t+\frac{1}{2}u$

41. **a.** To find the x-intercept, set $y=0$:

$$\begin{aligned}0&=2^x-5\\2^x&=5\\\ln 2^x&=\ln 5\\x\ln 2&=\ln 5\\x&=\frac{\ln 5}{\ln 2}\approx 2.32\end{aligned}$$

Now sketch the graph:

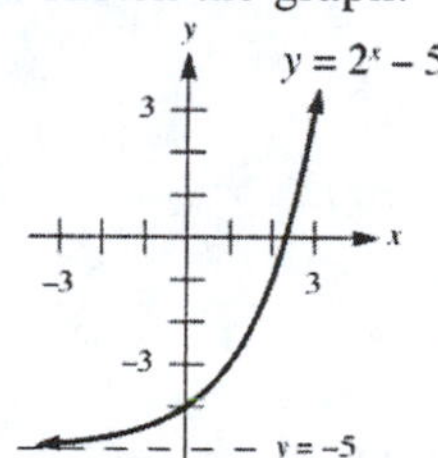

b. To find the x-intercept, set $y=0$:

$$\begin{aligned}0&=2^{x/2}-5\\2^{x/2}&=5\\\ln 2^{x/2}&=\ln 5\\\frac{x}{2}\ln 2&=\ln 5\\x&=\frac{2\ln 5}{\ln 2}\approx 4.64\end{aligned}$$

Now sketch the graph:

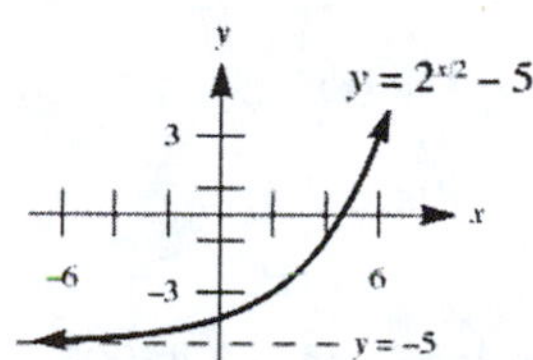

43. Taking the natural log of each side:

$$\begin{aligned}\ln 5&=\ln 2+\ln e^{2x-1}\\\ln 5&=\ln 2+(2x-1)\\\ln 5&=\ln 2+2x-1\\2x&=\ln 5-\ln 2+1\\x&=\frac{\ln 5-\ln 2+1}{2}\end{aligned}$$

45. Taking the natural logarithm of each side:

$$\begin{aligned}2^x&=13\\\ln 2^x&=\ln 13\\x\ln 2&=\ln 13\\x&=\frac{\ln 13}{\ln 2}\end{aligned}$$

47. Taking the natural logarithm of each side:

$$\begin{aligned}10^x&=e\\\ln 10^x&=\ln e\\x\ln 10&=1\\x&=\frac{1}{\ln 10}\end{aligned}$$

49. Taking the natural logarithm of each side:

$$\begin{aligned}3^{x^2-1}&=12\\\log 3^{x^2-1}&=\log 12\\\left(x^2-1\right)\log 3&=\log 12\\x^2-1&=\frac{\log 12}{\log 3}\\x^2&=1+\frac{\log 12}{\log 3}\\x&=\pm\sqrt{1+\frac{\log 12}{\log 3}}\\x&\approx\pm 1.806\end{aligned}$$

51. Using the change of base formula: $\log_2 5=\frac{\log_{10} 5}{\log_{10} 2}$

53. Using the change of base formula: $\ln 3=\log_e 3=\frac{\log_{10} 3}{\log_{10} e}$

55. Using the change of base formula: $\log_b 2 = \dfrac{\log_{10} 2}{\log_{10} b}$

57. Using the change of base formula: $\log_{10} 6 = \dfrac{\ln 6}{\ln 10}$

59. Using the change of base formula: $\log_{10} e = \dfrac{\ln e}{\ln 10} = \dfrac{1}{\ln 10}$

61. Using the change of base formula: $\log_{10}\left(\log_{10} x\right) = \log_{10}\left(\dfrac{\ln x}{\ln 10}\right) = \dfrac{\ln\left(\frac{\ln x}{\ln 10}\right)}{\ln 10} = \dfrac{\ln(\ln x) - \ln(\ln 10)}{\ln 10}$

63.
- **a.** true
- **b.** true
- **c.** true
- **d.** false, since $\ln x^3 = 3\ln x$, not $\ln 3x$
- **e.** true
- **f.** false, since $\ln 2x^3 = \ln 2 + 3\ln x \neq 3\ln 2x = \ln(2x)^3$
- **g.** true
- **h.** false, since $\log_5 24$ is between 1 and 2, not 5^1 and 5^2
- **i.** true
- **j.** false, since $\log_5 24$ is close to 2 ($2 = \log_5 25$)
- **k.** false, since the domain is $(0, \infty)$
- **l.** true
- **m.** true

65.
a. Computing:

$\log_{10} \pi^7 \approx 3.48005$ $\qquad$ $7\log_{10} \pi \approx 7(0.49715) = 3.48005$

So $\log_{10} \pi^7 = 7\log_{10} \pi$.

b. Computing:

$\log_b P^n = \log_{10} \pi^7 \approx 3.48005$ $\qquad$ $\left(\log_b P\right)^n = \left(\log_{10} \pi\right)^7 \approx 0.00751$

So $\log_b \pi^7 \neq \left(\log_b \pi\right)^7$.

c. Computing: $b^{\log_b P} = 10^{\log_{10} 1776} = 10^{3.24944} = 1776$. So $10^{\log_{10} 1776} = 1776$.

d. Computing:

$\ln 2 + \ln 3 + \ln 4 \approx 0.69315 + 1.09861 + 1.38629 \approx 3.17805$

$\ln 24 \approx 3.17805$

So $\ln 2 + \ln 3 + \ln 4 = \ln 24$.

e. Computing:

$\log_{10} A + \log_{10} B + \log_{10} C = \log_{10} 11 + \log_{10} 12 + \log_{10} 13 \approx 1.04139 + 1.07918 + 1.11394 = 3.23451$

$\log_{10}(ABC) = \log_{10}(11 \bullet 12 \bullet 13) = \log_{10} 1716 \approx 3.23451$

So $\log_{10} A + \log_{10} B + \log_{10} C = \log_{10}(ABC)$.

f. Compute $f[g(2345.6)]$: $f[g(2345.6)] = f(\ln 2345.6) \approx f(7.76030) \approx e^{7.76030} \approx 2345.6$

g. Compute $g[f(0.123456)]$: $g[f(0.123456)] = g\left(10^{0.123456}\right) = \log_{10}(1.32879) = 0.123456$

67. Complete the table:

x	0.1	0.05	0.005	0.0005
$\ln(1+x)$	0.095310	0.048790	0.004987	0.000499

Note that the values of $\ln(1 + x)$ are close to x, when x is close to 0.

69. Since $(-2,324)$ and $\left(\frac{1}{2},\frac{4}{3}\right)$ lie on the curve $y=ae^{bx}$: $324=ae^{-2b}$ and $\frac{4}{3}=ae^{\frac{1}{2}b}$

Dividing:

$$\frac{324}{\frac{4}{3}}=\frac{ae^{-2b}}{ae^{\frac{1}{2}b}}$$
$$243=e^{-2.5b}$$
$$\ln 243=-2.5b$$
$$b=-\frac{\ln 243}{2.5}$$

Substituting into $324=ae^{-2b}$:

$$324=ae^{(2\ln 243)/2.5}$$
$$324=ae^{0.8\ln 243}$$
$$324=a\left(e^{\ln 243}\right)^{0.8}$$
$$324=a\bullet 243^{0.8}$$
$$a=\frac{324}{243^{0.8}}=\frac{324}{81}=4$$

71. **a.** Solve the equation:

$$2^x-3=0$$
$$2^x=3$$
$$\log_{10}2^x=\log_{10}3$$
$$x\log_{10}2=\log_{10}3$$
$$x=\frac{\log_{10}3}{\log_{10}2}$$

b. We have $\frac{\log_{10}3}{\log_{10}2}\approx 1.585$ and $\frac{\ln 3}{\ln 2}\approx 1.585$.

c. Using the change of base formula: $\frac{\log_{10}3}{\log_{10}2}=\frac{\frac{\ln 3}{\ln 10}}{\frac{\ln 2}{\ln 10}}=\frac{\ln 3}{\ln 2}$

73. Working from the left-hand side:

$$\log_b\left(\frac{\sqrt{3}+\sqrt{2}}{\sqrt{3}-\sqrt{2}}\right)=\log_b\left(\frac{\sqrt{3}+\sqrt{2}}{\sqrt{3}-\sqrt{2}}\bullet\frac{\sqrt{3}+\sqrt{2}}{\sqrt{3}+\sqrt{2}}\right)=\log_b\left[\frac{\left(\sqrt{3}+\sqrt{2}\right)^2}{3-2}\right]=\log_b\left(\sqrt{3}+\sqrt{2}\right)^2=2\log_b\left(\sqrt{3}+\sqrt{2}\right)$$

75. Using properties of logarithms: $b^{3\log_b x}=b^{\log_b x^3}=x^3$

77. Using the change of base formula: $\log_b a=\frac{\log_a a}{\log_a b}=\frac{1}{\log_a b}$

79. a. Using the hint: $\frac{1}{\log_2 \pi}+\frac{1}{\log_5 \pi}=\frac{1}{\frac{\log_\pi \pi}{\log_\pi 2}}+\frac{1}{\frac{\log_\pi \pi}{\log_\pi 5}}=\log_\pi 2+\log_\pi 5=\log_\pi 10$

Since $\pi^2<10$, $\log_\pi 10>\log_\pi \pi^2=2$, which completes the proof.

b. Since $\log_\pi 2>\log_\pi 1=0$, $\log_\pi 2$ is positive. Using the inequality $a+b\geq 2\sqrt{ab}$ with $a=\log_\pi 2$ and $b=\frac{1}{\log_\pi 2}$: $\log_\pi 2+\frac{1}{\log_\pi 2}\geq 2\sqrt{1}=2$. The equality occurs only if $a=b$, which occurs when:

$$\log_\pi 2=\frac{1}{\log_\pi 2}$$

$$\left(\log_\pi 2\right)^2=1, \text{ which is impossible}$$

Thus $\log_\pi 2+\frac{1}{\log_\pi 2}>2$.

c. The two quantities are:

$$\frac{1}{\log_2 \pi}+\frac{1}{\log_5 \pi}=\log_\pi 10=\frac{\ln 10}{\ln \pi}\approx 2.0115$$

$$\log_\pi 2+\frac{1}{\log_\pi 2}=\frac{\ln 2}{\ln \pi}+\frac{\ln \pi}{\ln 2}\approx 2.2570$$

Thus $\log_\pi 2+\frac{1}{\log_\pi 2}$ is the larger quantity.

81. Using the change of base formula: $\log_{ab} x=\frac{\log_a x}{\log_a ab}=\frac{\log_a x}{\log_a a+\log_a b}=\frac{\log_a x}{1+\log_a b}$

Therefore: $\frac{\log_a x}{\log_{ab} x}=\frac{\log_a x}{\frac{\log_a x}{1+\log_a b}}=1+\log_a b$

83. Work from the right-hand side: $\frac{1}{2}(\log a+\log b)=\frac{1}{2}\log(ab)=\log\sqrt{ab}$

Proving the desired equality is equivalent to proving: $\frac{1}{3}(a+b)=\sqrt{ab}$, since log is a one-to-one function.

If a and b are both positive, we can square each side: $\frac{1}{9}\left(a^2+2ab+b^2\right)=ab$

Now work with the left-hand side:

$$\begin{aligned}\tfrac{1}{9}\left(a^2+2ab+b^2\right)&=\tfrac{1}{9}(7ab+2ab) \text{ by our assumption } a^2+b^2=7ab\\&=\tfrac{1}{9}(9ab)\\&=ab\end{aligned}$$

5.5 Equations and Inequalities with Logs and Exponents

1. Taking natural logarithms:

$$\begin{aligned}5^x&=3^{2x-1}\\ \ln 5^x&=\ln 3^{2x-1}\\ x\ln 5&=(2x-1)\ln 3\\ x\ln 5&=2x\ln 3-\ln 3\\ x(\ln 5-2\ln 3)&=-\ln 3\\ x&=\frac{-\ln 3}{\ln 5-2\ln 3}\\ x&=\frac{\ln 3}{2\ln 3-\ln 5}\approx 1.869\end{aligned}$$

3. Converting to exponential form:

$$\begin{aligned}\ln(\ln x)&=1.5\\ \ln x&=e^{1.5}\\ x&=e^{e^{1.5}}\approx 88.384\end{aligned}$$

5. Converting to exponential form:

$$\begin{aligned}\log_{10}\left(x^2+36\right)&=2\\ x^2+36&=10^2\\ x^2+36&=100\\ x^2&=64\\ x&=\pm 8\end{aligned}$$

7. Converting to exponential form:

$$\begin{aligned}\log_{10}\left(2x^2-3x\right)&=2\\ 2x^2-3x&=10^2\\ 2x^2-3x&=100\\ 2x^2-3x-100&=0\end{aligned}$$

Using the quadratic formula: $x=\dfrac{-(-3)\pm\sqrt{(-3)^2-4(2)(-100)}}{2(2)}=\dfrac{3\pm\sqrt{9+800}}{4}=\dfrac{3\pm\sqrt{809}}{4}$

The solutions are $x=\dfrac{3-\sqrt{809}}{4}\approx -6.361$ and $x=\dfrac{3+\sqrt{809}}{4}\approx 7.861$.

9. Let $u=10^x$. Then the equation becomes:

$$\begin{aligned}\left(10^x\right)^2+3\left(10^x\right)-10&=0\\ u^2+3u-10&=0\\ (u+5)(u-2)&=0\\ u&=-5, 2\end{aligned}$$

So either $10^x=-5$, which is impossible, or $10^x=2$, so $x=\log 2\approx 0.301$.

11. **a.** Since $\ln\left(x^3\right)=3\ln x$, the equation is true for all $x>0$.

b. Let $u=\ln x$. Then the equation becomes:

$$\begin{aligned}(\ln x)^3&=3\ln x\\ u^3&=3u\\ u^3-3u&=0\\ u\left(u^2-3\right)&=0\\ u&=0,-\sqrt{3},\sqrt{3}\end{aligned}$$

So either $\ln x=0$, so $x=1$, $\ln x=-\sqrt{3}$, so $x=e^{-\sqrt{3}}\approx 0.177$, or $\ln x=\sqrt{3}$, so $x=e^{\sqrt{3}}\approx 5.652$.

13. **a.** Since $\log_3 6x=\log_3 6+\log_3 x$, the equation is true for all $x>0$.

b. Using properties of logarithms:

$$\begin{aligned}\log_3 6x&=6\log_3 x\\ \log_3 6+\log_3 x&=6\log_3 x\\ \log_3 6&=5\log_3 x\\ \tfrac{1}{5}\log_3 6&=\log_3 x\\ \log_3 6^{1/5}&=\log_3 x\\ x&=6^{1/5}\approx 1.431\end{aligned}$$

15. Since $7^{\log_7 2x}=2x$, the equation is true for all $x>0$.

17. Converting to exponential form:

$$\begin{aligned} \log_2\left(\log_3 x\right) &= -1 \\ \log_3 x &= 2^{-1} \\ \log_3 x &= \tfrac{1}{2} \\ x &= 3^{1/2} \\ x &= \sqrt{3} \approx 1.732 \end{aligned}$$

19. Solving the equation:

$$\begin{aligned} \ln 4 - \ln x &= \frac{\ln 4}{\ln x} \\ (\ln 4)(\ln x) - (\ln x)^2 &= \ln 4 \end{aligned}$$

Let $u = \ln x$. Then the equation becomes:

$$\begin{aligned} (\ln 4)u - u^2 &= \ln 4 \\ 0 &= u^2 - (\ln 4)u + \ln 4 \end{aligned}$$

Using the quadratic formula: $u = \dfrac{\ln 4 \pm \sqrt{(\ln 4)^2 - 4\ln 4}}{2} = \dfrac{\ln 4 \pm \sqrt{(\ln 4)(\ln 4 - 4)}}{2}$

Since $\ln 4 > 0$ and $\ln 4 - 4 < 0$, the quantity inside the radical is negative and thus the equation has no real solutions.

21. Using properties of logarithms:

$$\begin{aligned} \ln\left(3x^2\right) &= 2\ln(3x) \\ \ln\left(3x^2\right) &= \ln(3x)^2 \\ 3x^2 &= (3x)^2 \\ 3x^2 &= 9x^2 \\ 0 &= 6x^2 \\ x &= 0 \end{aligned}$$

Since ln 0 is undefined, the equation has no real solutions.

23. Converting to exponential form:

$$\begin{aligned} \log_{16}\frac{x+3}{x-1} &= \tfrac{1}{2} \\ \frac{x+3}{x-1} &= 16^{1/2} \\ \frac{x+3}{x-1} &= 4 \\ x+3 &= 4x-4 \\ 7 &= 3x \\ x &= \tfrac{7}{3} \end{aligned}$$

25. a. Let $u = e^x$. Then the equation becomes:

$$\begin{aligned} \left(e^x\right)^2 + 2e^x + 1 &= 0 \\ u^2 + 2u + 1 &= 0 \\ (u+1)^2 &= 0 \\ u &= -1 \end{aligned}$$

So $e^x = -1$, which is impossible. The equation has no real solutions.

b. Let $u = e^x$. Then the equation becomes:

$$\begin{aligned}(e^x)^2 - 2e^x + 1 &= 0\\ u^2 - 2u + 1 &= 0\\ (u-1)^2 &= 0\\ u &= 1\end{aligned}$$

So $e^x = 1$, thus $x = 0$.

c. Let $u = e^x$. Then the equation becomes:

$$\begin{aligned}(e^x)^2 - 2e^x - 3 &= 0\\ u^2 - 2u - 3 &= 0\\ (u-3)(u+1) &= 0\\ u &= -1, 3\end{aligned}$$

So either $e^x = -1$, which is impossible, or $e^x = 3$, so $x = \ln 3 \approx 1.099$.

d. Let $u = e^x$. Then the equation becomes:

$$\begin{aligned}(e^x)^2 - 2e^x - 4 &= 0\\ u^2 - 2u - 4 &= 0\end{aligned}$$

Using the quadratic formula: $u = \dfrac{-(-2) \pm \sqrt{(-2)^2 - 4(1)(-4)}}{2(1)} = \dfrac{2 \pm \sqrt{4+16}}{2} = \dfrac{2 \pm 2\sqrt{5}}{2} = 1 \pm \sqrt{5}$

So either $e^x = 1 - \sqrt{5}$, which is impossible (since $1 - \sqrt{5} < 0$), or $e^x = 1 + \sqrt{5}$, so $x = \ln(1+\sqrt{5}) \approx 1.174$.

27. Using the hint:

$$\begin{aligned}e^x - e^{-x} &= 1\\ e^{2x} - 1 &= e^x\\ e^{2x} - e^x - 1 &= 0\end{aligned}$$

Let $u = e^x$. Then the equation becomes $u^2 - u - 1 = 0$. Using the quadratic formula:

$$u = \frac{-(-1) \pm \sqrt{(-1)^2 - 4(1)(-1)}}{2(1)} = \frac{1 \pm \sqrt{1+4}}{2} = \frac{1 \pm \sqrt{5}}{2}$$

So either $e^x = \dfrac{1-\sqrt{5}}{2}$, which is impossible, or $e^x = \dfrac{1+\sqrt{5}}{2}$, so $x = \ln\left(\dfrac{1+\sqrt{5}}{2}\right) \approx 0.481$.

29. Taking natural logarithms:

$$\begin{aligned}2^{5x} &= 3^x \bullet 5^{x+3}\\ \ln(2^{5x}) &= \ln(3^x \bullet 5^{x+3})\\ (5x)\ln 2 &= \ln(3^x) + \ln(5^{x+3})\\ (5\ln 2)x &= (\ln 3)x + (x+3)\ln 5\\ (5\ln 2)x &= (\ln 3)x + (\ln 5)x + 3\ln 5\\ (5\ln 2 - \ln 3 - \ln 5)x &= 3\ln 5\\ x &= \frac{3\ln 5}{5\ln 2 - \ln 3 - \ln 5} \approx 6.372\end{aligned}$$

31. Using properties of logarithms:

$$\begin{aligned} \log_6 x + \log_6 (x+1) &= 1 \\ \log_6 \left(x(x+1)\right) &= 1 \\ x(x+1) &= 6 \\ x^2 + x &= 6 \\ x^2 + x - 6 &= 0 \\ (x+3)(x-2) &= 0 \\ x &= -3, 2 \end{aligned}$$

But $x = -3$ is an extraneous root $\left(\log_6 (-3) \text{ is undefined}\right)$, so $x = 2$ is the only solution.

33. Solving for x:

$$\begin{aligned} \log_9 (x+1) &= \tfrac{1}{2} + \log_9 x \\ \log_9 (x+1) - \log_9 x &= \tfrac{1}{2} \\ \log_9 \left(\frac{x+1}{x}\right) &= \tfrac{1}{2} \\ 9^{1/2} &= \frac{x+1}{x} \\ 3 &= \frac{x+1}{x} \\ 3x &= x+1 \\ 2x &= 1 \\ x &= \tfrac{1}{2} \end{aligned}$$

35. Solving for x:

$$\begin{aligned} \log_{10} (2x+4) + \log_{10} (x-2) &= 1 \\ \log_{10} \left[(2x+4)(x-2)\right] &= 1 \\ (2x+4)(x-2) &= 10^1 \\ 2x^2 - 8 &= 10 \\ 2x^2 &= 18 \\ x^2 &= 9 \\ x &= \pm 3 \end{aligned}$$

Since $\log_{10} (-2)$ is undefined, $x = -3$ is an extraneous root. So the solution is $x = 3$.

37. Solving for x:

$$\begin{aligned} \log_{10} (x+3) - \log_{10} (x-2) &= 2 \\ \log_{10} \left(\frac{x+3}{x-2}\right) &= 2 \\ 10^2 &= \frac{x+3}{x-2} \\ 100(x-2) &= x+3 \\ 100x - 200 &= x+3 \\ 99x &= 203 \\ x &= \tfrac{203}{99} \end{aligned}$$

39. Solving for x:

$$\log_b(x+1) = 2\log_b(x-1)$$
$$\log_b(x+1) = \log_b(x-1)^2$$
$$x+1 = (x-1)^2$$
$$x+1 = x^2 - 2x + 1$$
$$x^2 - 3x = 0$$
$$x(x-3) = 0$$
$$x = 0, 3$$

Since $\log_b(-1)$ is undefined, $x = 0$ is an extraneous root. So the solution is $x = 3$.

41. Solving for x:

$$\log_{10}(x-6) + \log_{10}(x+3) = 1$$
$$\log_{10}[(x-6)(x+3] = 1$$
$$10^1 = (x-6)(x+3)$$
$$x^2 - 3x - 18 = 10$$
$$x^2 - 3x - 28 = 0$$
$$(x-7)(x+4) = 0$$
$$x = 7, -4$$

Since $\log_{10}(-10)$ is undefined, $x = -4$ is an extraneous root. So the solution is $x = 7$.

43. **a.** Solving for x:

$$\log_{10} x - y = \log_{10}(3x-1)$$
$$\log_{10} x - \log_{10}(3x-1) = y$$
$$\log_{10}\left(\frac{x}{3x-1}\right) = y$$
$$10^y = \frac{x}{3x-1}$$
$$10^y(3x-1) = x$$
$$3\left(10^y\right)x - 10^y = x$$
$$3\left(10^y\right)x - x = 10^y$$
$$x\left[3\left(10^y\right) - 1\right] = 10^y$$
$$x = \frac{10^y}{3\left(10^y\right) - 1}$$

b. Solving for x:

$$\log_{10}(x-y) = \log_{10}(3x-1)$$
$$x - y = 3x - 1$$
$$-2x = y - 1$$
$$x = \frac{y-1}{-2} \text{ or } \frac{1-y}{2} \qquad \left(y < \tfrac{1}{3}\right)$$

45. **a.** The two graphs intersect in one point, so the equation has one solution.

b. Sketching the graph:

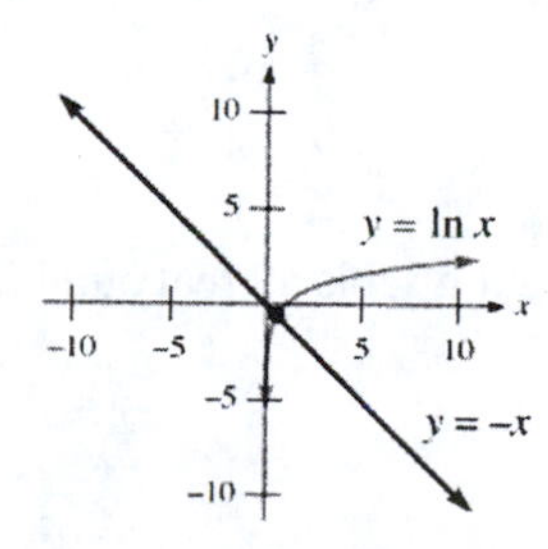

The solution is $x \approx 0.57$.

47. **a.** The two graphs intersect in one point, so the equation has one solution.
b. Sketching the graph:

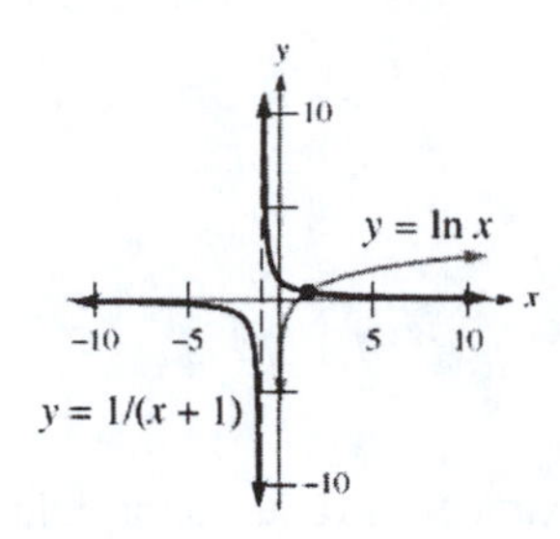

The solution is $x \approx 1.49$.

49. **a.** The two graphs intersect in two points, so the equation has two solutions.
b. Sketching the graph:

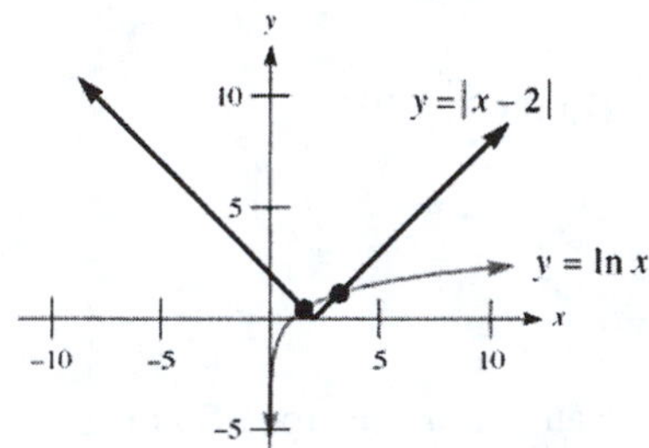

The solutions are $x \approx 1.56$ and $x \approx 3.15$.

51. Solving the inequality:

$$\begin{aligned} 3\left(2-0.6^x\right) &\le 1 \\ 2-0.6^x &\le \tfrac{1}{3} \\ -0.6^x &\le -\tfrac{5}{3} \\ 0.6^x &\ge \tfrac{5}{3} \\ \left(\tfrac{3}{5}\right)^x &\ge \tfrac{5}{3} \\ \left(\tfrac{5}{3}\right)^{-x} &\ge \tfrac{5}{3} \\ -x &\ge 1 \\ x &\le -1 \end{aligned}$$

53. Solving the inequality:

$$\begin{aligned} 4\left(10-e^x\right) &\le -3 \\ 10-e^x &\le -\tfrac{3}{4} \\ -e^x &\le -\tfrac{43}{4} \\ e^x &\ge \tfrac{43}{4} \\ x &\ge \ln\tfrac{43}{4} \\ x &\ge 2.375 \end{aligned}$$

55. Solving the inequality:

$$\begin{aligned} \ln(2-5x) &> 2 \\ 2-5x &> e^2 \\ -5x &> e^2-2 \\ 5x &< 2-e^2 \\ x &< \frac{2-e^2}{5} \\ x &< -1.078 \end{aligned}$$

57. Solving the inequality:

$$\begin{aligned} e^{2+x} &\ge 100 \\ 2+x &\ge \ln 100 \\ x &\ge \ln 100 - 2 \\ x &\ge 2.605 \end{aligned}$$

59. Since $2^x > 0$ for all real numbers x, the solution set is all real numbers.

61. Solving the inequality:

$$\log_2 \frac{2x-1}{x-2} < 0$$
$$0 < \frac{2x-1}{x-2} < 1$$

Now $\frac{2x-1}{x-2} > 0$ for $x < \frac{1}{2}$ or $x > 2$ (using a sign chart). Solving the second inequality:

$$\frac{2x-1}{x-2} < 1$$
$$\frac{2x-1}{x-2} - \frac{x-2}{x-2} < 0$$
$$\frac{2x-1-x+2}{x-2} < 0$$
$$\frac{x+1}{x-2} < 0$$

This inequality is satisfied (using a sign chart) when $-1 < x < 2$. Combining answers, the solution set is $-1 < x < \frac{1}{2}$.

63. Solving the inequality:

$$e^{x^2-4x} \geq e^5$$
$$x^2 - 4x \geq 5$$
$$x^2 - 4x - 5 \geq 0$$
$$(x-5)(x+1) \geq 0$$

This inequality is satisfied (using a sign chart) when $x \leq -1$ or $x \geq 5$.

65. Solving the inequality:

$$e^{1/x-1} > 1$$
$$\frac{1}{x} - 1 > 0$$
$$\frac{1}{x} > 1$$
$$0 < x < 1$$

67. **a.** We must have both $x > 0$ and $x - 4 > 0$, which occurs when $x > 4$. The domain is $(4, \infty)$.

b. Solving the inequality:

$$\ln x + \ln(x-4) \leq \ln 21$$
$$\ln\left(x^2 - 4x\right) \leq \ln 21$$
$$x^2 - 4x \leq 21$$
$$x^2 - 4x - 21 \leq 0$$
$$(x-7)(x+3) \leq 0$$

The inequality is satisfied (using a sign chart) when $-3 \leq x \leq 7$. Combined with the domain from part **a**, the solution set is $4 < x \leq 7$.

69. First note that the domain requires $x > 0$, $x + 1 > 0$, and $2x + 6 > 0$, which occurs when $x > 0$. Solving the inequality:

$$\log_2 x + \log_2 (x+1) - \log_2 (2x+6) < 0$$
$$\log_2 \frac{x(x+1)}{2x+6} < 0$$
$$\frac{x(x+1)}{2x+6} < 1$$
$$\frac{x^2+x}{2x+6} - \frac{2x+6}{2x+6} < 0$$
$$\frac{x^2-x-6}{2x+6} < 0$$
$$\frac{(x-3)(x+2)}{2x+6} < 0$$

This inequality is satisfied (using a sign chart) when $x < -3$ or $-2 < x < 3$. Combined with the domain $x > 0$, the solution set is $0 < x < 3$.

71. Use the change of base formula:

$$\log_2 x = \log_x 2$$
$$\log_2 x = \frac{\log_2 2}{\log_2 x}$$
$$\left(\log_2 x\right)^2 = 1$$
$$\log_2 x = -1, 1$$
$$x = 2^{-1}, 2^1$$
$$x = \tfrac{1}{2}, 2$$

73. Graphing the two functions:

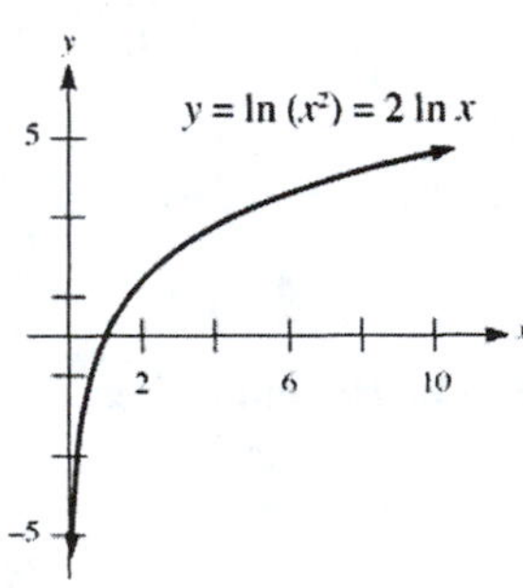

The two graphs are identical. This indicates that $\ln\left(x^2\right) = 2\ln x$ is an identity for $x > 0$.

75. Let $u = \ln x$. Then the equation becomes:

$$3(\ln x)^2 - \ln\left(x^2\right) - 8 = 0$$
$$3(\ln x)^2 - 2\ln x - 8 = 0$$
$$3u^2 - 2u - 8 = 0$$
$$(3u+4)(u-2) = 0$$
$$u = -\tfrac{4}{3}, 2$$

Either $\ln x = -\frac{4}{3}$, so $x = e^{-4/3} \approx 0.264$, or $\ln x = 2$, so $x = e^2 \approx 7.389$.

77. Changing all logarithms to natural logarithms:

$$\log_6 x = \frac{1}{\frac{1}{\log_2 x} + \frac{1}{\log_3 x}}$$

$$\frac{\ln x}{\ln 6} = \frac{1}{\frac{\ln 2}{\ln x} + \frac{\ln 3}{\ln x}}$$

$$\frac{\ln x}{\ln 6} = \frac{\ln x}{\ln 2 + \ln 3}$$

$$\ln 6 = \ln 2 + \ln 3$$

Since this last equation is true for all real x, the solution set is $x > 0$ (note that the domain of $\ln x$ requires that $x > 0$) and $x \neq 1$ (to avoid dividing by 0), which is $(0,1) \cup (1,\infty)$.

79. Solving for x:

$$\alpha \ln x + \ln \beta = 0$$
$$\alpha \ln x = -\ln \beta$$
$$\ln x = -\frac{\ln \beta}{\alpha}$$
$$x = e^{-(\ln \beta)/\alpha}$$
$$x = e^{\ln \beta^{-1/\alpha}}$$
$$x = \beta^{-1/\alpha}$$

81. Solving for x:

$$y = Ae^{kx}$$
$$\frac{y}{A} = e^{kx}$$
$$\ln \frac{y}{A} = kx$$
$$x = \frac{1}{k} \ln \frac{y}{A}$$

83. Solving for x:

$$y = \frac{a}{1 + be^{-kx}}$$
$$y + bye^{-kx} = a$$
$$bye^{-kx} = a - y$$
$$e^{-kx} = \frac{a-y}{by}$$
$$-kx = \ln\left(\frac{a-y}{by}\right)$$
$$x = -\frac{1}{k} \ln\left(\frac{a-y}{by}\right)$$

85. **a.** Converting to exponential form:

$$\ln(\ln x) = 2$$
$$\ln x = e^2$$
$$x = e^{e^2}$$

b. The solution is approximately 1618.178, which checks in the original equation.

87. **a.** Let $u = 3^x$. Then the equation becomes:

$$2\left(3^{2x}\right) - 3^x - 3 = 0$$
$$2u^2 - u - 3 = 0$$
$$(2u - 3)(u + 1) = 0$$
$$u = \tfrac{3}{2}, -1$$

Since $3^x \neq -1$, we have $3^x = \frac{3}{2}$, so $x = \log_3 1.5 = \dfrac{\ln 1.5}{\ln 3}$.

b. The solution is approximately 0.369, which checks in the original equation.

89. Since $e > 0$, $2 - e < 2$, thus $\frac{2-e}{3} < \frac{2}{3}$.

91. Since $\log_\pi \left[\log_4\left(x^2-5\right)\right] < 0$, we must have:

$$\begin{aligned} 0 &< \log_4\left(x^2-5\right) < 1 \\ 4^0 &< x^2-5 < 4^1 \\ 1 &< x^2-5 < 4 \\ 6 &< x^2 < 9 \end{aligned}$$

This inequality is satisfied on the set $\left(-3,-\sqrt{6}\right)\cup\left(\sqrt{6},3\right)$. Note that the original domain requires that $x^2 > 5$, which is satisfied by our solution set.

93. Taking natural logarithms of each side:

$$\begin{aligned} x^{\left(x^x\right)} &= \left(x^x\right)^x \\ \ln x^{\left(x^x\right)} &= \ln\left(x^x\right)^x \\ x^x \ln x &= x\ln\left(x^x\right) \\ x^x \ln x &= x^2 \ln x \\ (\ln x)\left(x^x - x^2\right) &= 0 \end{aligned}$$

So either $\ln x = 0$, thus $x = 1$, or $x^x - x^2 = 0$:

$$\begin{aligned} x^x - x^2 &= 0 \\ x^x &= x^2 \\ \ln x^x &= \ln x^2 \\ x\ln x &= 2\ln x \\ (\ln x)(x-2) &= 0 \\ x &= 1, 2 \end{aligned}$$

The solutions are $x = 1$ or $x = 2$.

95. Solving for x:

$$\begin{aligned} \left(a^4 - 2a^2b^2 + b^4\right)^{x-1} &= (a-b)^{2x}(a+b)^{-2} \\ \left(a^2-b^2\right)^{2x-2} &= \frac{(a-b)^{2x}}{(a+b)^2} \\ (a+b)^{2x-2}(a-b)^{2x-2} &= \frac{(a-b)^{2x}}{(a+b)^2} \\ (a+b)^{2x}(a-b)^{-2} &= 1 \\ (a+b)^{2x} &= (a-b)^2 \\ 2x\ln(a+b) &= 2\ln(a-b) \\ x &= \frac{\ln(a-b)}{\ln(a+b)} \end{aligned}$$

97. Solving for x:

$$6^x = \tfrac{10}{3} - 6^{-x}$$
$$6^x + 6^{-x} = \tfrac{10}{3}$$
$$6^{2x} + 1 = \tfrac{10}{3}6^x$$
$$3 \cdot 6^{2x} - 10 \cdot 6^x + 3 = 0$$
$$\left(3 \cdot 6^x - 1\right)\left(6^x - 3\right) = 0$$

Setting each factor equal to 0, we have:

$$3 \cdot 6^x - 1 = 0$$
$$6^x = \tfrac{1}{3}$$
$$x\log_{10} 6 = -\log_{10} 3$$
$$x = -\frac{\log_{10} 3}{\log_{10} 6}$$

$$6^x - 3 = 0$$
$$6^x = 3$$
$$x\log_{10} 6 = \log_{10} 3$$
$$x = \frac{\log_{10} 3}{\log_{10} 6}$$

Now, since $\log_{10} 6 = \log_{10}(2 \cdot 3) = \log_{10} 2 + \log_{10} 3$, and since $\log_{10} 2 = a$ and $\log_{10} 3 = b$, we have:

$$x = \frac{-b}{a+b} \quad \text{or} \quad x = \frac{b}{a+b}$$

So $x = \dfrac{\pm b}{a+b}$.

5.6 Compound Interest

1. For annual compounding of money use the formula $A = P(1 + r)^t$. Now find A when $P = \$800$, $r = 0.06$ and $t = 4$ yrs:

$$A = 800(1+0.06)^4 = 800(1.06)^4 \approx \$1009.98$$

3. For annual compounding of money use the formula $A = P(1 + r)^t$. Now find r when $A = \$6000$, $P = \$4000$ and $t = 5$:

$$6000 = 4000(1+r)^5$$
$$1.5 = (1+r)^5$$
$$(1.5)^{1/5} = 1+r$$
$$r = (1.5)^{1/5} - 1 \approx 0.0845$$

So the interest rate is 8.45%.

5. In the first bank, we have $P = \$500$, $r = 0.05$, and $t = 4$: $A = 500(1+0.05)^4 = 500(1.05)^4 \approx 607.75$

Now deposit $P = \$607.75$, $r = 0.06$, and $t = 4$: $A = 607.75(1+0.06)^4 = 607.75(1.06)^4 \approx 767.27$

So the new balance will be \$767.27.

7. **a.** Use $A = P(1 + r)^t$ with $P = 1000$, $r = 0.07$, and $t = 20$: $A = 1000(1+0.07)^{20} = 1000(1.07)^{20} \approx 3869.68$

The new balance is \$3869.68.

b. Use $A = P\left(1+\frac{r}{N}\right)^t$ with $P = 1000$, $r = 0.07$, $N = 4$, and $t = 20$:

$$A = 1000\left(1+\frac{0.07}{4}\right)^{4(20)} = 1000(1.0175)^{80} \approx 4006.39$$

The new balance is \$4006.39.

9. For compounding quarterly use the formula: $A = P\left(1+\frac{r}{N}\right)^{Nt}$

So here $P = \$100$, $r = 6\%$ and we have $N = 4$ compoundings per year. Find the value of t for which $A \geq \$120$:

$$\begin{aligned} 120 &\leq 100\left(1+\tfrac{0.06}{4}\right)^{4t} \\ 1.2 &\leq (1.015)^{4t} \\ \ln 1.2 &\leq 4t \ln 1.015 \\ 4t &\geq \frac{\ln 1.2}{\ln 1.015} \\ t &\geq \frac{\ln 1.2}{4 \ln 1.015} \approx 3.06 \end{aligned}$$

This is slightly over 3 years, and so 13 quarters will be required.

11. Use $A = P\left(1+\frac{r}{N}\right)^{Nt}$ where $r = 0.055$, $N = 2$, $A = 6000$, and $t = 10$:

$$\begin{aligned} 6000 &= P\left(1+\tfrac{0.55}{2}\right)^{2(10)} \\ 6000 &= P(1.0275)^{20} \\ P &= \frac{6000}{(1.0275)^{20}} \approx 3487.50 \end{aligned}$$

You must deposit a principal of \$3487.50.

13. **a.** The balance will reach \$800 in approximately 4.4 years.

b. Use $A = Pe^{rt}$ where $P = \$600$, $r = 0.065$, and $A = \$800$:

$$\begin{aligned} 800 &= 600e^{(0.065)t} \\ \tfrac{4}{3} &= e^{0.065t} \\ \ln \tfrac{4}{3} &= 0.065t \\ t &= \frac{\ln \frac{4}{3}}{0.065} \approx 4.43 \end{aligned}$$

The balance will reach \$800 in approximately 4.43 years.

15. Account #1 grows according to the equation $A = 3500e^{0.07t}$ while Account #2 grows according to the equation $A = 3500\left(1+\frac{0.07}{2}\right)^{2t} = 3500(1.035)^{2t}$. Let $f(t) = 3500e^{0.07t} - 3500(1.035)^{2t}$. We want to find when $f(t) = \$200$. Using the table feature, the difference reaches \$200 when $t = 16$. It will take 16 years for the accounts to differ by at least \$200.

17. Use $A = Pe^{rt}$ where $A = 5000$, $t = 10$ and $r = 0.065$:

$$\begin{aligned} 5000 &= Pe^{(0.065)(10)} \\ 5000 &= Pe^{0.65} \\ P &= \frac{5000}{e^{0.65}} \approx 2610.23 \end{aligned}$$

A principal of \$2610.23 will grow to \$5000 in 10 yrs.

19. Since the effective rate is $r = 0.06$: $A = P(1+0.06)^1 = 1.06P$

The nominal rate r would yield a balance of: $A = Pe^{r(1)} = Pe^r$

Setting these equal:

$$\begin{aligned} Pe^r &= P(1.06) \\ e^r &= 1.06 \\ r &= \ln 1.06 \approx 0.0583 \end{aligned}$$

So the nominal rate is 5.83%.

21. Compute the 6% investment first: $A = 10000(1+0.06)^5 = 10000(1.3382) \approx \$13382.26$

The second choice will be: $A = 10000e^{0.05(5)} = 10000e^{0.25} \approx \$12840.25$, considerably less.

23. **a.** We have $T_2 \approx \frac{70}{5} = 14$ yrs. **b.** We have $T_2 = \frac{\ln 2}{r} = \frac{\ln 2}{0.05} = 13.86$ yrs.

c. Here $d_1 = 13.86$, $d_2 = 14$, so $d = |13.86 - 14| = 0.14$.

This represents $\frac{0.14}{13.86}(100) \approx 1.01\%$ of the actual doubling time.

25. Compute: $A = 1000e^{(0.08)(300)} = 1000e^{24} = 1000\left(2.65 \times 10^{10}\right) = \$2.65 \times 10^{13}$

That's $26.5 trillion, a nice inheritance.

27. **a.** We have $T_2 \approx \frac{0.7}{0.05} = 14$ yrs.

b. Sketch the graph:

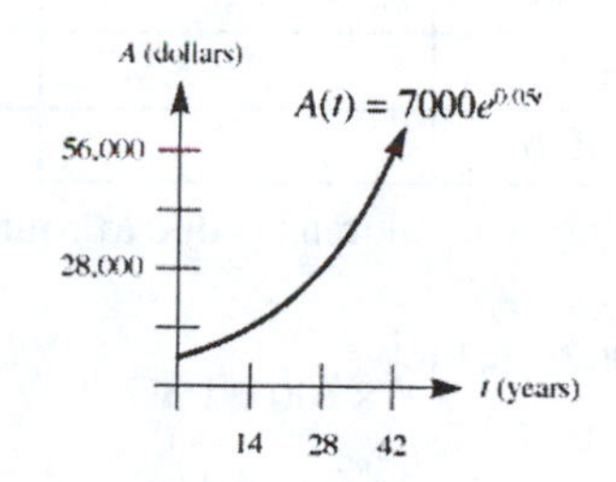

5.7 Exponential Growth and Decay

1. **a.** Since 2000 bacteria are present initially, we have $N_0 = 2000$. Now $N(2) = 3800$, so:

$$3800 = 2000e^{k \cdot 2}$$
$$1.9 = e^{2k}$$
$$\ln 1.9 = 2k$$
$$k = \frac{\ln 1.9}{2} \approx 0.3209$$

b. Using the formula $N(t) = 2000e^{0.3209t}$, find $N(5)$: $N(5) = 2000e^{(0.3209)(5)} \approx 9951$ bacteria.

c. Find when $N(t) = 10000$:

$$10000 = 2000e^{0.3209t}$$
$$5 = e^{0.3209t}$$
$$\ln 5 = 0.3209t$$
$$t = \frac{\ln 5}{0.3209} \approx 5.0 \text{ hours}$$

3. Since $N(0) = 2000$, we have $N_0 = 2000$. Using $N(t) = 2000e^{kt}$, find when $N(3) = 3400$:

$$3400 = 2000e^{k \cdot 3}$$
$$1.7 = e^{3k}$$
$$\ln 1.7 = 3k$$
$$k = \frac{\ln 1.7}{3} \approx 0.1769$$

5. **a.** The growth problems are solved by assuming an exponential growth rate: $N = N_0 e^{kt}$

First we find the percent distribution by dividing the total population into each of the two regions:

$$\frac{1.169}{5.702} \approx 20.5\% \text{ for more developed regions}$$

Applying the growth formula to the total world population:

$$N = N_0 e^{kt} = 5.702e^{0.015(5)} = 5.702e^{0.075} \approx 6.146 \text{ billion}$$

The other calculations are similar to these two. The completed table is:

Region	1995 Population (billions)	Percent of Population in 1995	Relative Growth Rate (percent per year)	Year 2000 Population (billions)	Percent of World Population in 2000
World	5.702	100	1.5	6.146	100
More dev.	1.169	20.5	0.2	1.181	19.22
Less dev.	4.533	79.5	1.9	4.985	81.11

Note that some discrepancy in the last two columns is due to round-off error.

b. Computing the percentage errors:

$$\text{world: percentage error} = \frac{|6.067 - 6.146|}{6.067} \times 100 \approx 1.3\%$$

$$\text{more developed: percentage error} = \frac{|1.184 - 1.181|}{1.184} \times 100 \approx 0.3\%$$

$$\text{less developed: percentage error} = \frac{|4.883 - 4.985|}{4.883} \times 100 \approx 2.1\%$$

7. **a.** For Chad, the exponential growth law is: $N(t) = 8.0e^{0.033t}$

For United Kingdom, the exponential growth law is: $N(t) = 59.8e^{0.001t}$

b. Setting the two expressions equal:

$$8.0e^{0.033t} = 59.8e^{0.001t}$$

$$\frac{e^{0.033t}}{e^{0.001t}} = \frac{59.8}{8.0}$$

$$e^{0.032t} = 7.475$$

$$0.032t = \ln 7.475$$

$$t = \frac{\ln 7.475}{0.032} \approx 60 \text{ years}$$

9. **a.** Finding the population for each country in 2015:

$$\text{Niger: } N(15) = 10.1e^{0.030(15)} \approx 15.8 \text{ million}$$

$$\text{Portugal: } N(15) = 10.0e^{0.001(15)} \approx 10.2 \text{ million}$$

b. For Niger, finding when the population reaches 15 million:

$$10.1e^{0.030t} = 15$$

$$e^{0.030t} = \frac{15}{10.1}$$

$$0.030t = \ln\left(\frac{15}{10.1}\right)$$

$$t = \frac{\ln\left(\frac{15}{10.1}\right)}{0.030} \approx 13 \text{ years}$$

After 13 years, the population of Portugal would be: $N(13) = 10.0e^{0.001(13)} \approx 10.1$ million

11. **a.** Using the function $N(t) = 4.088e^{0.0175t}$: $N(25) = 4.088e^{0.0175(25)} \approx 6.332$ billion

b. Our projection is higher than the actual population. Computing the percentage error:

$$\text{percentage error} = \frac{|6.067 - 6.332|}{6.067} \times 100 \approx 4\%$$

13. **a.** Finding the doubling time: $T = \frac{\ln 2}{0.006} \approx 116$ years **b.** Finding the doubling time: $T = \frac{\ln 2}{0.016} \approx 43$ years

c. Finding the doubling time: $T = \frac{\ln 2}{0.026} \approx 27$ years **d.** Finding the doubling time: $T = \frac{\ln 2}{0.037} \approx 19$ years

15. **a.** Completing the table:

Year	1998 $(t = 10)$	2000 $(t = 12)$
Concentration of Carbon Dioxide (ppm)	365.6	368.6

b. For 1998, the projection is too low: $\text{percentage error} = \frac{|366.7 - 365.6|}{366.7} \times 100 \approx 0.3\%$

For 2000, the projection is too high: $\text{percentage error} = \frac{|368.4 - 368.6|}{368.4} \times 100 \approx 0.1\%$

17. **a.** For New York, the model is: $N(t) = 18.976e^{0.006t}$. For Arizona, the model is: $N(t) = 5.131e^{0.04t}$

b. Equating the two expressions:

$$5.131e^{0.04t} = 18.976e^{0.006t}$$
$$\frac{e^{0.04t}}{e^{0.006t}} = \frac{18.976}{5.131}$$
$$e^{0.034t} = \frac{18.976}{5.131}$$
$$0.034t = \ln\left(\frac{18.976}{5.131}\right)$$
$$t = \frac{\ln\left(\frac{18.976}{5.131}\right)}{0.034} \approx 40 \text{ years}$$

The two states will have equal populations in the year 2040.

19. **a.** Complete the table:

Region	1990 Population (millions)	Growth Rate (%)	2025 Population
North America	275.2	0.7	351.6
Soviet Union	291.3	0.7	372.2
Europe	499.5	0.2	535.7
Nigeria	113.3	3.1	335.3

b. It will be 335.3 mil – 113.3 mil = 222.0 million.

c. The net increases are given by:

North America: 351.6 mil – 275.2 mil = 76.4 mil
Soviet Union: 372.2 mil – 291.3 mil = 80.9 mil
Europe: 535.7 mil – 499.5 mil = 36.2 mil
combined: 193.5 mil

d. For Nigeria, our results support this projection.

21. Computing the value: $N(0.5) = 99.6e^{0.02(0.5)} \approx 100.6$ million

By the end of the year 2000, the population of Mexico exceeded 100 million people.

23. a. Complete the table:

t(seconds)	0	550	1100	1650	2200
N(grams)	8	4	2	1	0.5

b. Complete the table:

t(years)	0	4.9×10^9	9.8×10^9	14.7×10^9	19.6×10^9
N(grams)	10	5	2.5	1.25	0.625

25. Given the decay law $N = N_0 e^{kt}$, and $N = \frac{1}{2}N_0$ when $t = 8$:

$$\frac{1}{2}N_0 = N_0 e^{8k}$$
$$\frac{1}{2} = e^{8k}$$
$$\ln\frac{1}{2} = 8k$$
$$k = \frac{\ln\frac{1}{2}}{8} \approx -0.0866$$

So when $t = 7$ and $N_0 = 1$: $N(7) = 1e^{(-0.0866)(7)} \approx 0.55$ g

27. a. Using $N = \frac{1}{2}N_0$ when $t = 14.9$ hours:

$$\frac{1}{2}N_0 = N_0 e^{14.9k}$$
$$\frac{1}{2} = e^{14.9k}$$
$$\ln\frac{1}{2} = 14.9k$$
$$k = \frac{\ln\frac{1}{2}}{14.9} \approx -0.0465$$

Now using $t = 48$ hours and $N_0 = 40$ g: $N(48) = 40e^{(-0.0465)(48)} \approx 4.29$ g

Approximately 4.29 g of the sample will remain after 48 hours.

b. Find when $N(t) = 1$:

$$1 = 40e^{-0.0465t}$$
$$\frac{1}{40} = e^{-0.0465t}$$
$$\ln\frac{1}{40} = -0.0465t$$
$$t = \frac{\ln\frac{1}{40}}{-0.0465} \approx 79 \text{ hours}$$

The isotope will decay to 1 gram in approximately 79 hours.

29. a. Draw the graph:

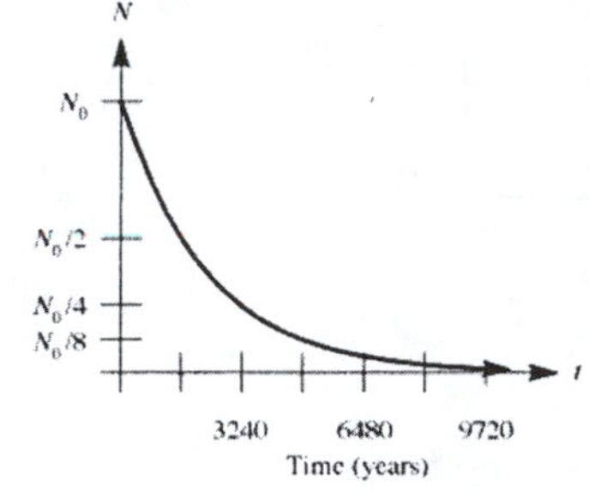

b. Draw the graph:

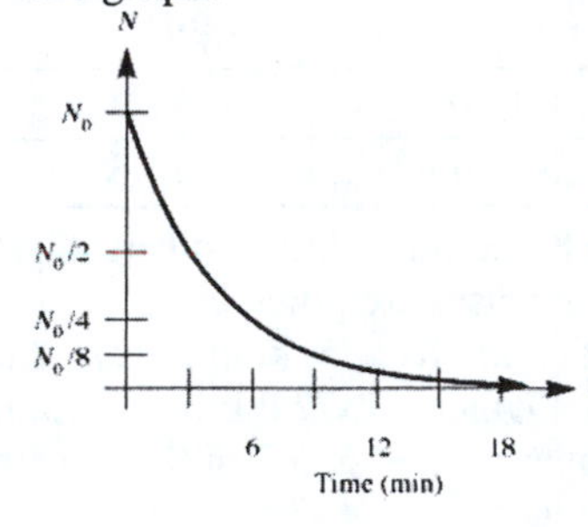

31. **a.** First find the decay constant k:

$$\tfrac{1}{2}N_0 = N_0 e^{13k}$$
$$\tfrac{1}{2} = e^{13k}$$
$$\ln\tfrac{1}{2} = 13k$$
$$k = \frac{\ln\frac{1}{2}}{13} \approx -0.0533$$

Now using $N_0 = 2$ g and $t = 5$ years: $N(5) = 2e^{(-0.0533)(5)} \approx 1.53$ grams

b. Find when $N(t) = 0.2$ g:

$$0.2 = 2e^{-0.0533t}$$
$$0.1 = e^{-0.0533t}$$
$$\ln 0.1 = -0.0533t$$
$$t = \frac{\ln 0.1}{-0.0533} \approx 43 \text{ years}$$

33. **a.** The decay law is $N = N_0 e^{kt}$, where $k = \frac{\ln 0.5}{7340}$. We want to determine t when $\frac{N_0}{1000} = N_0 e^{kt}$. After dividing by N_0, and then taking the logarithm of both sides, we obtain $\ln 0.001 = kt$. Thus $t = \frac{\ln 0.001}{k}$, where $k = \frac{\ln 0.5}{7340}$. A calculator shows that it is approximately 73,000 yrs.

b. The answer 73,000 yrs is approximately 10 half-lives.

c. After one half-life, we have $N = \frac{N_0}{2}$. After two half-lives, we have $N = \frac{N_0}{2^2}$. After three half-lives, we have $N = \frac{N_0}{2^3}$. Continuing the pattern, we see that after 10 half-lives, $N = \frac{N_0}{2^{10}} \approx \frac{N_0}{1000}$. This agrees with the result in part **b**.

35. **a.** We have $k = \frac{\ln 0.5}{28} \approx -0.0248$.

b. Solve for t:

$$\frac{N_0}{1000} = N_0 e^{kt}$$
$$0.001 = e^{-0.0248t}$$
$$\ln 0.001 = -0.0248t$$
$$t = \frac{\ln 0.001}{-0.0248} \approx 279 \text{ years}$$

c. Since $2^{10} \approx 1000$, after 10 half-lives, it should be reduced to $\frac{N_0}{1000}$. Since each half-life is 28 years, this is approximately 280 years. Note that this is close to our answer from part **b**.

37. First compute the decay constant:

$$\tfrac{1}{2}N_0 = N_0 e^{\left(7.1\times10^8\right)k}$$
$$\tfrac{1}{2} = e^{\left(7.1\times10^8\right)k}$$
$$\ln\tfrac{1}{2} = \left(7.1\times10^8\right)k$$
$$k = \frac{\ln\frac{1}{2}}{7.1\times10^8} \approx -9.76\times10^{-10}$$

Now using the decay law: $N = 0.028e^{\left(-9.76\times10^{-10}\right)(1000)} \approx 0.02799997$ ounces

39. a. Using $N_0 = 5.7$, $k = 0.016$, and $t = 25$: $N(t) = 5.7e^{(0.016)(25)} \approx 8.5$ billion

The projected world population in the year 2020 is 8.5 billion. This is greater than the target value of 7.2 billion.

b. Using $N_0 = 5.7$, $N = 7.2$, and $t = 25$:

$$\begin{aligned} 7.2 &= 5.7e^{k(25)} \\ \frac{7.2}{5.7} &= e^{25k} \\ 25k &= \ln\frac{7.2}{5.7} \\ k &= \tfrac{1}{25}\ln\frac{7.2}{5.7} \approx 0.00934 \end{aligned}$$

The growth constant would need to be 0.934% per year.

41. a. First multiply both sides by $\dfrac{k}{A_0}$ to get:

$$\begin{aligned} \frac{Ak}{A_0} &= e^{kT} - 1 \\ e^{kT} &= \frac{Ak}{A_0} + 1 \end{aligned}$$

Now taking the natural logarithm of both sides:

$$\begin{aligned} kT &= \ln\left(\frac{Ak}{A_0} + 1\right) \\ T &= \frac{\ln\left(\frac{Ak}{A_0} + 1\right)}{k} \end{aligned}$$

b. We have: $T = \dfrac{\ln\left[\frac{(700)(0.07)}{18.7} + 1\right]}{0.07} \approx 18$ yrs. The depletion year would have been 1990.

Using $k = 1\%$ instead: $T = \dfrac{\ln\left[\frac{(700)(0.01)}{18.7} + 1\right]}{0.01} \approx 32$ yrs. The depletion year would have been 2004.

c. The equation for the regression line is $y = 0.4606x + 24.0385$.

d. We have: $T = \dfrac{-24.0385 + \sqrt{24.0385^2 + 2(0.4606)(1007)}}{0.4606} \approx 32$ yrs. The depletion year is 2025.

43. a. Computing the life expectancy: $T = \dfrac{\ln\left[\frac{(650)(0.024)}{12.6} + 1\right]}{0.024} \approx 34$ yrs. The depletion year is 2033.

b. Computing the life expectancy: $T = \dfrac{\ln\left[\frac{(1300)(0.012)}{12.6} + 1\right]}{0.012} \approx 67$ yrs. The depletion year is 2066.

c. Computing the life expectancy: $T = \dfrac{650}{12.6} \approx 52$ yrs. The depletion year is 2051.

45. **a.** Solving for N_0:

$$N_r = N_0 e^{kT}$$
$$\frac{N_r}{e^{kT}} = N_0$$
$$N_0 = N_r e^{-kT}$$

So $N_s = N_0 - N_r = N_r e^{-kT} - N_r$.

b. Solving for T:

$$N_s + N_r = N_r e^{-kT}$$
$$\frac{N_s}{N_r} + 1 = e^{-kT}$$
$$\ln\left(\frac{N_s}{N_r} + 1\right) = -kT$$
$$T = \frac{\ln\left(\frac{N_s}{N_r} + 1\right)}{-k}$$

47. Compute T: $T = \dfrac{\ln\left(\frac{N_s}{N_r} + 1\right)}{-k} = \dfrac{\ln(0.0636 + 1)}{-\left(-1.4748 \times 10^{-11}\right)} \approx 4.181 \times 10^9$ yrs

The rock is approximately 4.181 billion years old.

49. Compute T: $T = \dfrac{5730 \ln \frac{N}{920}}{\ln \frac{1}{2}} = \dfrac{5730 \ln \frac{141}{920}}{\ln \frac{1}{2}} \approx 15505$ yrs. The two paintings are 15,505 years old.

51. Compute T: $T = \dfrac{5730 \ln \frac{N}{920}}{\ln \frac{1}{2}} = \dfrac{5730 \ln \frac{348}{920}}{\ln \frac{1}{2}} \approx 8000$ yrs. The site is older than which was originally estimated.

53. Compute T: $T = \dfrac{5730 \ln\left(\frac{N}{920}\right)}{\ln\left(\frac{1}{2}\right)} = \dfrac{5730 \ln\left(\frac{723}{920}\right)}{\ln\left(\frac{1}{2}\right)} \approx 1992$

This would correspond to the years 10-5 B.C., which fit in the historical range.

55. **a.** Completing the table:

	$N(-1)$	$N(0)$	$N(1)$	$N(4)$	$N(5)$
From graph	0.25	0.5	1.0	3.5	3.75
From calculator	0.176	0.444	1.014	3.489	3.795

b. Computing the values:

$$N(10) = \frac{4}{1 + 8e^{-10}} \approx 3.99854773 \qquad N(15) = \frac{4}{1 + 8e^{-15}} \approx 3.99999021$$
$$N(20) = \frac{4}{1 + 8e^{-20}} \approx 3.99999993$$

Note that the values are approaching $N = 4$.

c. The value is approximately $t \approx 3$.

d. Let $N = 3$ and solve for t:

$$3 = \frac{4}{1 + 8e^{-t}}$$
$$1 + 8e^{-t} = \frac{4}{3}$$
$$8e^{-t} = \frac{1}{3}$$
$$e^{-t} = \frac{1}{24}$$
$$e^t = 24$$
$$t = \ln 24 \approx 3.178$$

The answer is consistent with the answer from part **c**.

57. **a.** Since $N = 1$ when $t = 0$:

$$1 = \frac{5}{1+ae^0}$$
$$1 = \frac{5}{1+a}$$
$$a+1 = 5$$
$$a = 4$$

b. Since $N = 2$ when $t = 1$:

$$2 = \frac{5}{1+4e^{-b(1)}}$$
$$2 = \frac{5}{1+4e^{-b}}$$
$$1+4e^{-b} = \frac{5}{2}$$
$$4e^{-b} = \frac{3}{2}$$
$$e^{-b} = \frac{3}{8}$$
$$-b = \ln\frac{3}{8}$$
$$b = -\ln\frac{3}{8}$$
$$b = \ln\frac{8}{3} \approx 0.9808$$

59. **a.** Since $N = 1$ when $t = 0$:

$$1 = \frac{162}{1+ae^0}$$
$$1 = \frac{162}{1+a}$$
$$1+a = 162$$
$$a = 161$$

Thus the logistic curve is $N(t) = \dfrac{162}{1+161e^{-bt}}$. Since $N = 77.9$ when $t = 10$:

$$77.9 = \frac{162}{1+161e^{-b(10)}}$$
$$1+161e^{-10b} = \frac{162}{77.9}$$
$$161e^{-10b} \approx 1.080$$
$$e^{-10b} \approx 0.00671$$
$$-10b \approx \ln 0.00671$$
$$b \approx \frac{\ln 0.00671}{-10} \approx 0.50$$

Thus the logistic curve is $N(t) = \dfrac{162}{1+161e^{-0.50t}}$.

b. Computing the function values:

$$N(4) = \frac{162}{1+161e^{-0.50(4)}} \approx 7.1 \qquad N(8) = \frac{162}{1+161e^{-0.50(8)}} \approx 41.0$$
$$N(12) = \frac{162}{1+161e^{-0.50(12)}} \approx 115.8 \qquad N(16) = \frac{162}{1+161e^{-0.50(16)}} \approx 153.7$$

$N(4)$ and $N(8)$ are lower than the actual values, while $N(12)$ and $N(16)$ are higher than the actual values.

c. Solving the equation $N(t) = 81$:

$$\begin{aligned}
\frac{162}{1+161e^{-0.50t}} &= 81 \\
2 &= 1+161e^{-0.50t} \\
161e^{-0.50t} &= 1 \\
e^{-0.50t} &= \frac{1}{161} \\
-0.50t &= \ln\left(\frac{1}{161}\right) \\
t &= \frac{\ln\left(\frac{1}{161}\right)}{-0.50} \approx 10.16 \text{ days} = 10 \text{ days } 4 \text{ hours}
\end{aligned}$$

61. **a.** Simplifying: $\frac{N(t+1)-N(t)}{N(t)} = \frac{N_0e^{k(t+1)} - N_0e^{kt}}{N_0e^{kt}} = \frac{N_0e^{kt}e^k - N_0e^{kt}}{N_0e^{kt}} = \frac{N_0e^{kt}\left(e^k - 1\right)}{N_0e^{kt}} = e^k - 1$

b. If k is close to 0, then $e^k - 1 \approx (k+1) - 1 = k$.

63. **a.** Compute $\frac{\Delta N}{\Delta t}$ over the interval $[t, t+\Delta t]$:

$$\frac{\Delta N}{\Delta t} = \frac{N(t+\Delta t) - N(t)}{t+\Delta t - t} = \frac{N_0e^{k(t+\Delta t)} - N_0e^{kt}}{\Delta t} = \frac{N_0e^{kt}e^{k\Delta t} - N_0e^{kt}}{\Delta t} = \frac{N_0e^{kt}\left(e^{k\Delta t} - 1\right)}{\Delta t} = \frac{N\left(e^{k\Delta t} - 1\right)}{\Delta t}$$

b. Substituting: $\frac{\Delta N}{\Delta t} = \frac{N\left(e^{k\Delta t} - 1\right)}{\Delta t} \approx \frac{N(k\Delta t + 1 - 1)}{\Delta t} = \frac{kN\Delta t}{\Delta t} = kN$

Thus $\frac{\Delta N}{\Delta t} \approx kN$ when Δt is close to 0.

Chapter 5 Review Exercises

1. If $x = \log_5 126$, then $5^x = 126$. Since $5^3 = 125$, $x > 3$.

If $y = \log_{10} 999$, then $10^y = 999$. Since $10^3 = 1000$, $y < 3$. So $\log_5 126$ is larger.

3. Since $N(t) = N_0e^{kt}$ and $N_0 = 8000$, $N(t) = 8000e^{kt}$. Since $N(4) = 10000$, we can find k:

$$\begin{aligned}
N(4) &= 10000 \\
8000e^{4k} &= 10000 \\
e^{4k} &= 1.25 \\
k &= \frac{\ln 1.25}{4}
\end{aligned}$$

Now find t such that $N(t) = 12000$:

$$\begin{aligned}
N(t) &= 12000 \\
8000e^{kt} &= 12000 \\
e^{kt} &= 1.5 \\
kt &= \ln 1.5 \\
t &= \frac{\ln 1.5}{k} = \frac{\ln 1.5}{\frac{\ln 1.25}{4}} = \frac{4\ln 1.5}{\ln 1.25} \text{ hours}
\end{aligned}$$

5. Graph $f(x)$:

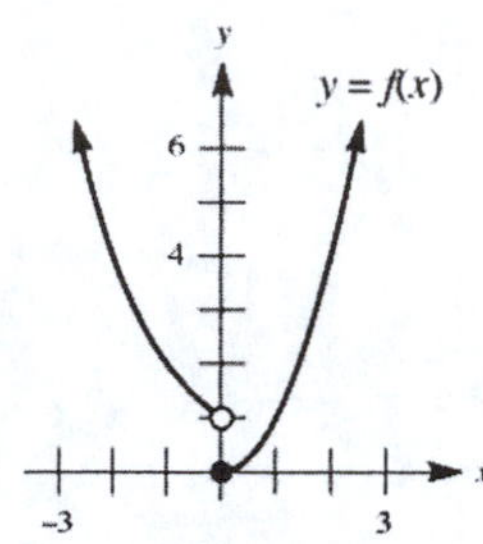

f is not one-to-one since it does not pass the horizontal line test.

7. Solve for x:

$$\ln(x+1)-1=\ln(x-1)$$
$$\ln(x+1)-\ln(x-1)=1$$
$$\ln\frac{x+1}{x-1}=1$$
$$\frac{x+1}{x-1}=e$$
$$x+1=ex-e$$
$$e+1=ex-x$$
$$e+1=x(e-1)$$
$$x=\frac{e+1}{e-1}$$

9. For $y=e^x$, the domain is $(-\infty,\infty)$ and the range is $(0,\infty)$. For $y=\ln x$, the domain is $(0,\infty)$ and the range is $(-\infty,\infty)$:

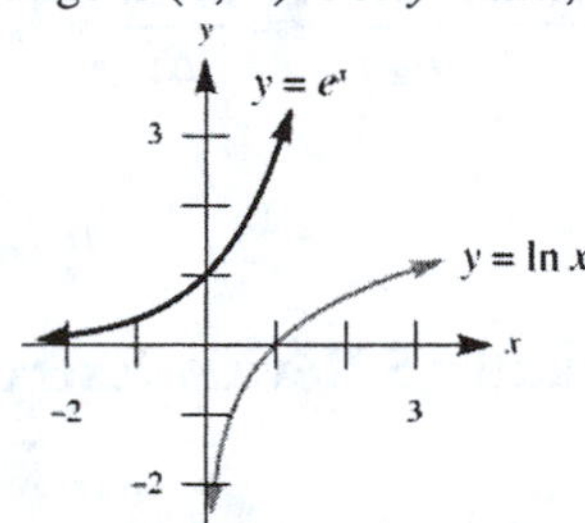

11. Simplify the logarithm: $\log_9\frac{1}{27}=\log_9\left(3^{-3}\right)=\log_9\left(9^{1/2}\right)^{-3}=\log_9 9^{-3/2}=-\frac{3}{2}$

13. Using the decay equation $N=N_0e^{kt}$, find k when $t=13$ and $N=\frac{1}{2}N_0$:

$$\tfrac{1}{2}N_0=N_0e^{13k}$$
$$\tfrac{1}{2}=e^{13k}$$
$$\ln\tfrac{1}{2}=13k$$
$$k=\frac{\ln\frac{1}{2}}{13}\approx-0.05$$

15. Solve for x:

$$5e^{2-x}=12$$
$$e^{2-x}=\tfrac{12}{5}$$
$$2-x=\ln\tfrac{12}{5}$$
$$-x=-2+\ln\tfrac{12}{5}$$
$$x=2-\ln\tfrac{12}{5}$$

17. Use $N = N_0e^{kt}$ where $N_0 = 2$ and $k = 0.02$, so $N(t) = 2e^{0.02t}$. Find t when $N = 3$:

$$\begin{aligned} 3 &= 2e^{0.02t} \\ 1.5 &= e^{0.02t} \\ \ln 1.5 &= 0.02t \\ t &= \frac{\ln 1.5}{0.02} \approx 20 \text{ yrs} \end{aligned}$$

We would expect the population to reach 3 million in the year 2015.

19. Use $A = Pe^{rt}$ where $P = \$1000$ and $r = 0.10$, so $A(t) = 1000e^{0.1t}$. The doubling time is approximately $\frac{70}{10} = 7$ years. Sketching the graph:

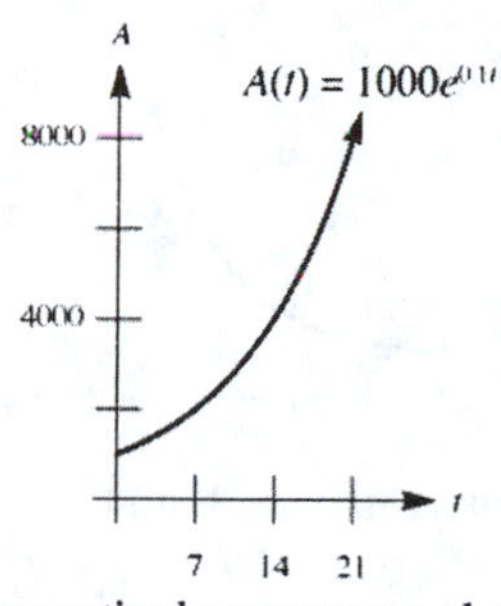

21. The horizontal asymptote is $y = 0$, there is no vertical asymptote, there is no x-intercept, and the y-intercept is 1:

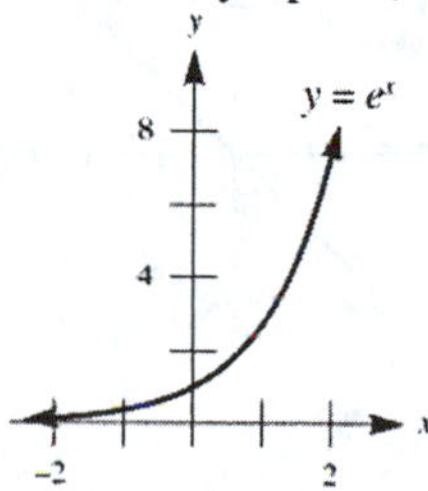

23. There is no horizontal asymptote, the vertical asymptote is $x = 0$, the x-intercept is 1, and there is no y-intercept:

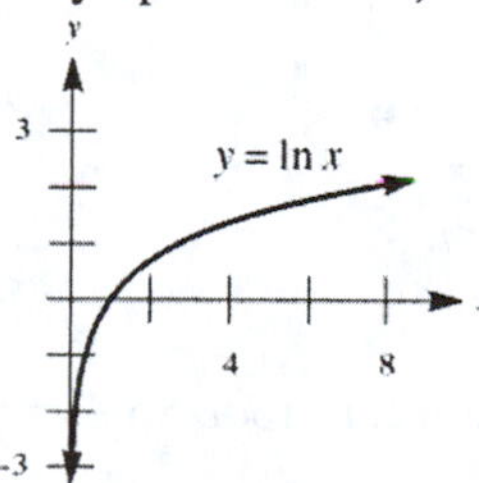

25. The horizontal asymptote is $y = 1$, there is no vertical asymptote, there is no x-intercept, and the y-intercept is 3:

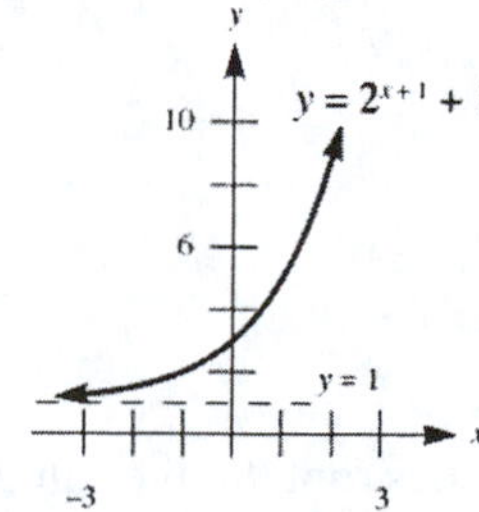

27. The horizontal asymptote is $y = 0$, there is no vertical asymptote, there is no x-intercept, and the y-intercept is 1:

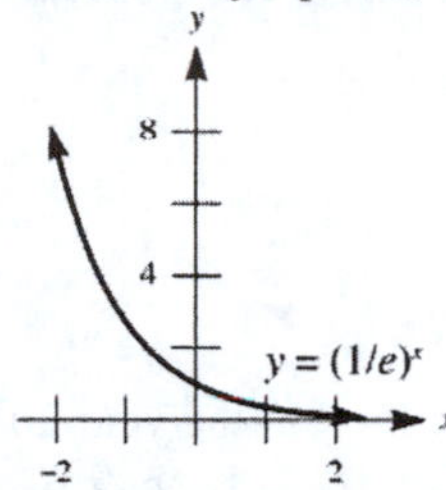

29. The horizontal asymptote is $y = 1$, there is no vertical asymptote, there is no x-intercept, and the y-intercept is $e + 1$:

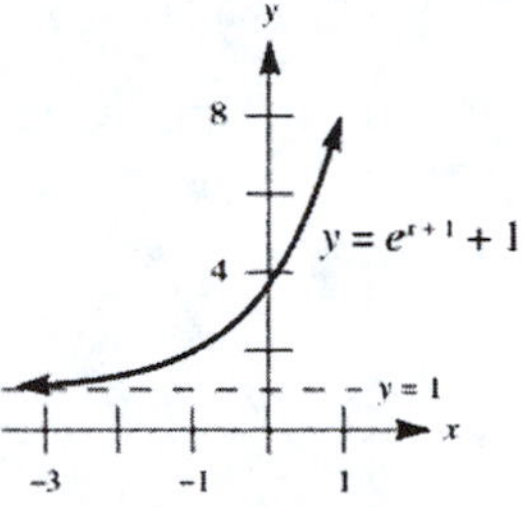

31. There are no asymptotes, and the x- and y-intercepts are both 0:

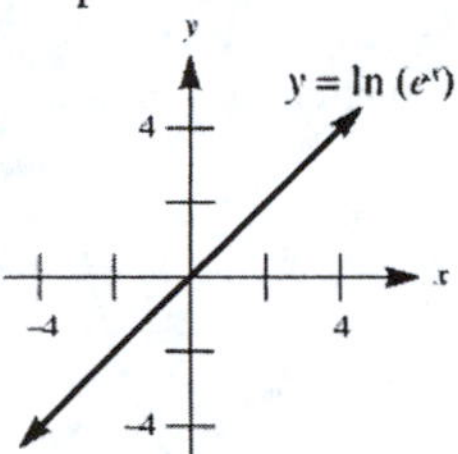

33. Using properties of logarithms:

$$\begin{aligned}
\log_4 x + \log_4 (x-3) &= 1 \\
\log_4 [x(x-3)] &= 1 \\
x(x-3) &= 4^1 \\
x^2 - 3x &= 4 \\
x^2 - 3x - 4 &= 0 \\
(x-4)(x+1) &= 0 \\
x &= 4, -1
\end{aligned}$$

Since $\log_4(-1)$ is undefined, $x = -1$ is an extraneous root. So the solution is $x = 4$.

35. Using properties of logarithms:

$$\begin{aligned}
\ln x + \ln(x+2) &= \ln 15 \\
\ln[x(x+2)] &= \ln 15 \\
x(x+2) &= 15 \\
x^2 + 2x &= 15 \\
x^2 + 2x - 15 &= 0 \\
(x+5)(x-3) &= 0 \\
x &= -5, 3
\end{aligned}$$

Since $\ln(-5)$ is undefined, $x = -5$ is an extraneous root. So the solution is $x = 3$.

37. Using properties of logarithms:

$$\begin{aligned}\log_2 x+\log_2(3x+10)-3&=0\\ \log_2\left[x(3x+10)\right]&=3\\ x(3x+10)&=2^3\\ 3x^2+10x&=8\\ 3x^2+10x-8&=0\\ (3x-2)(x+4)&=0\\ x&=\tfrac{2}{3},-4\end{aligned}$$

Since $\log_2(-4)$ is undefined, $x=-4$ is an extraneous root. So the solution is $x=\frac{2}{3}$.

39. Solve for x:

$$\begin{aligned}3\log_9 x&=\tfrac{1}{2}\\ \log_9 x&=\tfrac{1}{6}\\ x&=9^{1/6}=\sqrt[6]{9}=\sqrt[3]{3}\end{aligned}$$

41. Solve for x:

$$\begin{aligned}e^{1-5x}&=3\sqrt{e}\\ 1-5x&=\ln 3\sqrt{e}\\ 1-5x&=\ln 3+\tfrac{1}{2}\ln e\\ 1-5x&=\ln 3+\tfrac{1}{2}\\ -5x&=\ln 3-\tfrac{1}{2}\\ x&=\frac{1-2\ln 3}{10}\end{aligned}$$

43. Using properties of logarithms:

$$\begin{aligned}\log_{10}x-2&=\log_{10}(x-2)\\ \log_{10}x-\log_{10}(x-2)&=2\\ \log_{10}\frac{x}{x-2}&=2\\ \frac{x}{x-2}&=100\\ x&=100(x-2)\\ x&=100x-200\\ 200&=99x\\ x&=\tfrac{200}{99}\end{aligned}$$

45. Using properties of logarithms:

$$\begin{aligned}\ln(x+2)&=\ln x+\ln 2\\ \ln(x+2)&=\ln 2x\\ x+2&=2x\\ x&=2\end{aligned}$$

47. Since $\ln\left(x^4\right)=4\ln x$ is an identity, the solution is all real numbers $x>0$, or $(0,\infty)$.

49. Using properties of logarithms:

$$\begin{aligned}\log_{10}x&=\ln x\\ \frac{\ln x}{\ln 10}&=\ln x\\ \ln x&=(\ln 10)(\ln x)\\ \ln x-(\ln 10)(\ln x)&=0\\ \ln x(1-\ln 10)&=0\\ \ln x&=0,\text{ so }x=1\end{aligned}$$

51. Using properties of logarithms: $\log_{10}\sqrt{10}=\log_{10}\left(10^{1/2}\right)=\frac{1}{2}$

53. Using properties of logarithms: $\ln\left(\sqrt[5]{e}\right)=\ln\left(e^{1/5}\right)=\frac{1}{5}$

55. Using properties of logarithms:

$$\log_{10}\pi-\log_{10}10\pi=\log_{10}\pi-\left(\log_{10}10+\log_{10}\pi\right)=\log_{10}\pi-\log_{10}10-\log_{10}\pi=-\log_{10}10=-1$$

57. Using properties of logarithms: $10^{\log_{10}16}=16$

59. Evaluate: $\ln\left(e^4\right)=4$

61. Using properties of logarithms: $\log_{12} 2+\log_{12} 18+\log_{12} 4=\log_{12}(2\bullet 18\bullet 4)=\log_{12} 144=\log_{12}\left(12^2\right)=2$

63. Using properties of logarithms: $\dfrac{\ln 100}{\ln 10}=\dfrac{\ln\left(10^2\right)}{\ln 10}=\dfrac{2\ln 10}{\ln 10}=2$

65. Using properties of logarithms:

$$\begin{aligned}\log_2 \sqrt[7]{16\sqrt[3]{2\sqrt{2}}} &= \tfrac{1}{7}\log_2 16\sqrt[3]{2\sqrt{2}}\\ &= \tfrac{1}{7}\left(\log_2 16+\log_2 \sqrt[3]{2\sqrt{2}}\right)\\ &= \tfrac{1}{7}\left[\log_2\left(2^4\right)+\tfrac{1}{3}\log_2 2\sqrt{2}\right]\\ &= \tfrac{1}{7}\left[4+\tfrac{1}{3}\left(\log_2 2+\log_2 \sqrt{2}\right)\right]\\ &= \tfrac{1}{7}\left[4+\tfrac{1}{3}\left(1+\log_2\left(2^{1/2}\right)\right)\right]\\ &= \tfrac{1}{7}\left(4+\tfrac{1}{3}\bullet\tfrac{3}{2}\right)\\ &= \tfrac{1}{7}\left(4+\tfrac{1}{2}\right)\\ &= \tfrac{1}{7}\bullet\tfrac{9}{2}\\ &= \tfrac{9}{14}\end{aligned}$$

67. Using properties of logarithms:

$$\log_{10}\left(A^2B^3\sqrt{C}\right)=\log_{10} A^2+\log_{10} B^3+\log_{10}\sqrt{C}=2\log_{10} A+3\log_{10} B+\tfrac{1}{2}\log_{10} C=2a+3b+\tfrac{c}{2}$$

69. Using properties of logarithms:

$$16\log_{10}\sqrt{A}\sqrt[4]{B}=16\left(\log_{10}\sqrt{A}+\log_{10}\sqrt[4]{B}\right)=16\left(\tfrac{1}{2}\log_{10} A+\tfrac{1}{4}\log_{10} B\right)=8\log_{10} A+4\log_{10} B=8a+4b$$

71. Since $\log_{10} 100=2$ and $\log_{10} 1000=3$, $\log_{10} 209$ lies between 2 and 3.

73. Since $\log_6 36=2$ and $\log_6 216=3$, $\log_6 100$ lies between 2 and 3.

75. Since $\log_{10} 0.01=-2$ and $\log_{10} 0.001=-3$, $\log_{10} 0.003$ lies between -2 and -3.

77. **a.** Graph $y=\ln(x+2)$ and $y=\ln(-x)-1$:

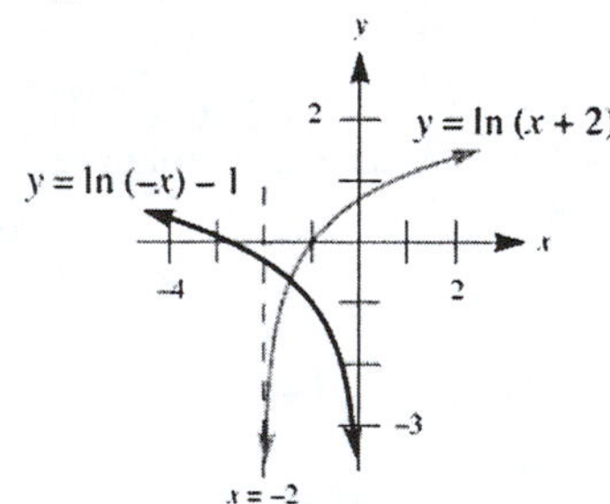

The graph shows these two curves intersect in the third quadrant.

b. Setting the y-coordinates equal:

$$\begin{aligned}
\ln(x+2) &= \ln(-x)-1 \\
\ln(x+2)-\ln(-x) &= -1 \\
\ln\frac{x+2}{-x} &= -1 \\
\frac{x+2}{-x} &= e^{-1} \\
e(x+2) &= -x \\
ex+2e &= -x \\
ex+x &= -2e \\
x(e+1) &= -2e \\
x &= \frac{-2e}{e+1} \approx -1.46
\end{aligned}$$

79. We know $N = N_0 e^{kt}$, and $N = \frac{1}{2}N_0$ when $t = T$:

$$\begin{aligned}
\tfrac{1}{2}N_0 &= N_0 e^{kT} \\
\tfrac{1}{2} &= e^{kT} \\
\ln\tfrac{1}{2} &= kT \\
k &= \frac{\ln\frac{1}{2}}{T}
\end{aligned}$$

81. Since 4 half-lives have passed, we will have $\left(\frac{1}{2}\right)^4 N_0 = \frac{1}{16}N_0$ of the substance left, or 6.25% remaining.

83. We have $k = \dfrac{\ln\frac{1}{2}}{d}$ and $N_0 = b$. We want to find the value of t when $N = c$:

$$\begin{aligned}
N &= N_0 e^{kt} \\
c &= be^{kt} \\
\frac{c}{b} &= e^{kt} \\
\ln\frac{c}{b} &= kt \\
\ln\frac{c}{b} &= \frac{\ln\frac{1}{2}}{d} \bullet t \\
t &= \frac{d\ln\frac{c}{b}}{\ln\frac{1}{2}} \text{ days}
\end{aligned}$$

85. Using properties of logarithms: $\log_{10} 8 + \log_{10} 3 - \log_{10} 12 = \log_{10}\left(\dfrac{8 \bullet 3}{12}\right) = \log_{10} 2$

87. Using properties of logarithms: $\ln 5 - 3\ln 2 + \ln 16 = \ln 5 - \ln 2^3 + \ln 16 = \ln 5 - \ln 8 + \ln 16 = \ln\left(\dfrac{5 \bullet 16}{8}\right) = \ln 10$

89. Using properties of logarithms: $a\ln x + b\ln y = \ln x^a + \ln y^b = \ln x^a y^b$

91. Using properties of logarithms: $\ln\sqrt{(x-3)(x+4)} = \frac{1}{2}\ln\left[(x-3)(x+4)\right] = \frac{1}{2}\ln(x-3) + \frac{1}{2}\ln(x+4)$

93. Using properties of logarithms:

$$\log_{10}\frac{x^3}{\sqrt{1+x}} = \log_{10} x^3 - \log_{10}\sqrt{1+x} = \log_{10} x^3 - \log_{10}(1+x)^{1/2} = 3\log_{10} x - \tfrac{1}{2}\log_{10}(1+x)$$

95. Using properties of logarithms: $\ln\left(\frac{1+2e}{1-2e}\right)^3 = 3\ln\frac{1+2e}{1-2e} = 3\ln(1+2e) - 3\ln(1-2e)$

97. Use $B = P(1+r)^t$, where $P = A$, $B = 2A$, and $r = \frac{R}{100}$:

$$\begin{aligned} 2A &= A\left(1+\tfrac{R}{100}\right)^t \\ 2 &= \left(1+\tfrac{R}{100}\right)^t \\ \ln 2 &= t\ln\left(1+\tfrac{R}{100}\right) \\ t &= \frac{\ln 2}{\ln\left(1+\frac{R}{100}\right)} \text{ years} \end{aligned}$$

99. The balance after 1 year will be $A = P\left(1+\frac{0.095}{12}\right)^{12}$. We must find the effective interest rate r where $A = P(1+r)^1$. Set these equal:

$$\begin{aligned} P(1+r) &= P\left(1+\tfrac{0.095}{12}\right)^{12} \\ 1+r &= \left(1+\tfrac{0.095}{12}\right)^{12} \\ r &= \left(1+\tfrac{0.095}{12}\right)^{12} - 1 \approx 0.0992 \end{aligned}$$

So the effective interest rate is 9.92%.

101. Use $A = Pe^{rt}$ where $P = D$, $A = 2D$, and $r = \frac{R}{100}$:

$$\begin{aligned} 2D &= De^{(R/100)t} \\ 2 &= e^{(R/100)t} \\ \ln 2 &= \frac{R}{100}(t) \\ t &= \frac{100\ln 2}{R} \text{ years} \end{aligned}$$

103. **a.** Use $A = P\left(1+\frac{r}{N}\right)^{Nt}$ where $P = \$660$, $r = 0.055$, $N = 4$, and $A = \$1000$:

$$\begin{aligned} 1000 &= 660\left(1+\tfrac{0.055}{4}\right)^{4t} \\ \tfrac{50}{33} &= (1.01375)^{4t} \\ \ln\tfrac{50}{33} &= 4t\ln 1.01375 \\ t &= \frac{\ln\frac{50}{33}}{4\ln 1.01375} \approx 7.61 \text{ yrs} \end{aligned}$$

So the balance will reach \$1000 after $7\frac{3}{4}$ years.

b. Use $A = P\left(1+\frac{r}{N}\right)^{Nt}$ where $P = D$, $r = \frac{R}{100}$, $N = 4$, and $A = nD$:

$$\begin{aligned} nD &= D\left(1+\tfrac{R}{400}\right)^t \\ n &= \left(1+\tfrac{R}{400}\right)^t \\ \ln n &= t\ln\left(1+\tfrac{R}{400}\right) \\ t &= \frac{\ln n}{\ln\left(1+\frac{R}{400}\right)} \text{ years} \end{aligned}$$

105. **a.** The domain is $x > 0$, or $(0, \infty)$.

b. We must make sure $\log_{10} x \ge 0$, so $x \ge 1$. So the domain is $x \ge 1$, or $[1, \infty)$.

107. We must have $x > 0$, and also must exclude where $\ln x = 2$, which is $x = e^2$. So the domain is $\left(0, e^2\right) \cup \left(e^2, \infty\right)$.

Chapter 5 Test

1. The domain is $(-\infty, \infty)$, the range is $(-3, \infty)$, the x-intercept is $-\frac{\ln 3}{\ln 2}$, the y-intercept is -2, and the asymptote is $y = -3$. Draw the graph:

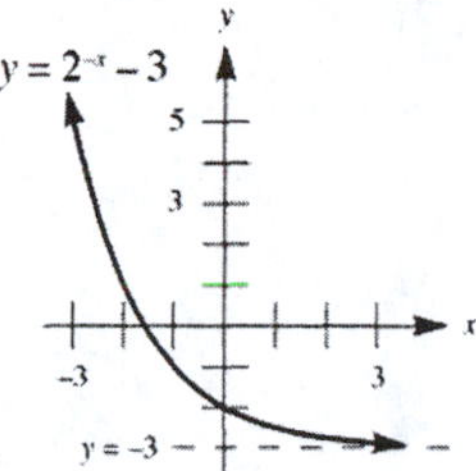

2. Using $N_0 = 6000$, $N = 6200$, and $t = 1$, find the growth constant k:

$$6200 = 6000e^{k \cdot 1}$$
$$1.033 = e^k$$
$$k = \ln 1.033 \approx 0.03279$$

Now find the value of t when $N = 10000$:

$$10000 = 6000e^{kt}$$
$$\frac{5}{3} = e^{kt}$$
$$kt = \ln\frac{5}{3}$$
$$t = \frac{\ln\frac{5}{3}}{\ln 1.033} \approx 16 \text{ hrs}$$

3. Since $2^4 = 16$, $\log_2 17 > 4$. Since $3^4 = 81$, $\log_3 80 < 4$. So $\log_2 17$ is larger.

4. Using the change-of-base formula: $\log_2 15 = \dfrac{\ln 15}{\ln 2}$

5. Since $2^{10} \approx 10^3$: $2^{40} = \left(2^{10}\right)^4 \approx \left(10^3\right)^4 = 10^{12}$

6. Using $A = P(1 + r)^t$ for interest compounded annually, with $P = \$9500$, $A = \$12{,}000$ and $r = 0.06$:

$$12000 = 9500(1.06)^t$$
$$1.2632 = (1.06)^t$$
$$\ln 1.2632 = t \ln 1.06$$
$$t = \frac{\ln 1.2632}{\ln 1.06} \approx 4 \text{ yrs}$$

7. **a.** The identity is valid only if $\ln x$ is defined, which is the interval $(0, \infty)$.

b. Graph the two functions:

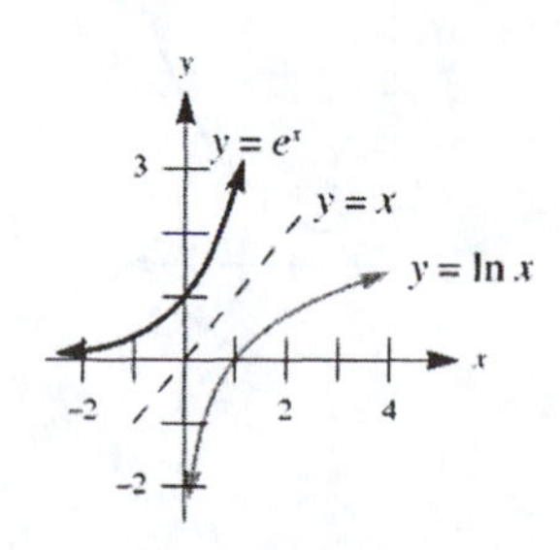

8. Using properties of logarithms: $\log_{10}\dfrac{A^3}{\sqrt{B}} = \log_{10}A^3 - \log_{10}\sqrt{B} = 3\log_{10}A - \frac{1}{2}\log_{10}B = 3a - \frac{1}{2}b$

9. **a.** Using properties of logarithms: $\log_5\dfrac{1}{\sqrt{5}} = \log_5 5^{-1/2} = -\frac{1}{2}$

b. Using properties of logarithms: $\ln e^2 + \ln 1 - e^{\ln 3} = 2 + 0 - 3 = -1$

10. Factoring:

$$\begin{aligned} e^x\left(x^3 - 4x\right) &= 0 \\ xe^x\left(x^2 - 4\right) &= 0 \\ xe^x(x+2)(x-2) &= 0 \\ x &= 0, -2, 2 \quad (\text{since } e^x \neq 0) \end{aligned}$$

11. **a.** Using the formula $T = \dfrac{\ln\frac{1}{2}}{k}$:

$$\begin{aligned} 4 &= \frac{\ln\frac{1}{2}}{k} \\ k &= \frac{\ln\frac{1}{2}}{4} \approx -0.1733 \end{aligned}$$

b. Using $N = N_0e^{kt}$ with $N_0 = 2$, $k = -0.1733$, and $t = 10$: $N = 2e^{-0.1733(10)} \approx 0.35$ grams

12. Using properties of logarithms: $2\ln x - \ln\sqrt[3]{x^2+1} = \ln x^2 - \ln\left(x^2+1\right)^{1/3} = \ln\dfrac{x^2}{\left(x^2+1\right)^{1/3}}$

13. Solve for x:

$$\begin{aligned} -2e^{3x-1} &= 9 \\ e^{3x-1} &= -4.5 \end{aligned}$$

But this is impossible $\left(e^{3x-1} > 0 \text{ for all } x\right)$, so there is no solution.

14. Let $y = \log_{10}(x-1)$. Switching the roles of x and y and solving the resulting equation for y yields:

$$\begin{aligned} x &= \log_{10}(y-1) \\ 10^x &= y-1 \\ y &= 10^x + 1 \end{aligned}$$

So $g^{-1}(x) = 10^x + 1$. The range of $g^{-1}(x)$ is $(1, \infty)$.

15. **a.** Using the formula $T = \frac{70}{r}$: $T = \dfrac{70}{6} \approx 12$ years

b. Graph $A = 12000e^{0.06t}$:

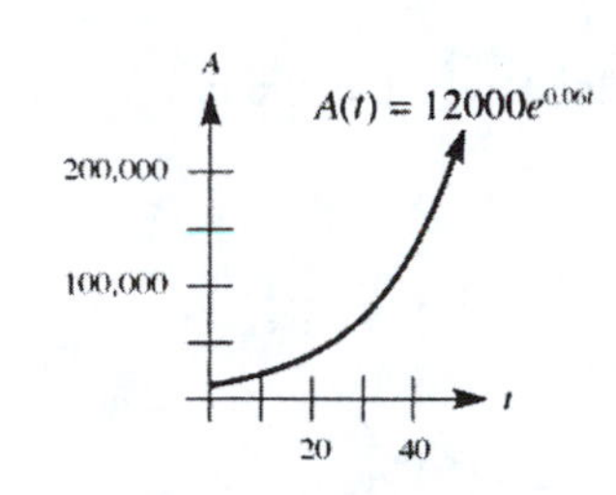

16. **a.** Since $\ln x$ must be defined, this is true on the interval $(0, \infty)$. Note that $\ln\left(x^2\right) = 2\ln x$ is an identity on the interval.

b. Solve for x:

$$\begin{aligned}(\ln x)^2 &= 2\ln x\\ (\ln x)^2 - 2\ln x &= 0\\ (\ln x)(\ln x - 2) &= 0\\ \ln x = 0 \quad &\text{or} \quad \ln x = 2\\ x = 1 \quad &\quad\quad x = e^2\end{aligned}$$

17. **a.** Computing the average rate of change: $\dfrac{f(\ln 4) - f(\ln 3)}{\ln 4 - \ln 3} = \dfrac{e^{\ln 4} - e^{\ln 3}}{\ln \frac{4}{3}} = \dfrac{4-3}{\ln \frac{4}{3}} = \dfrac{1}{\ln \frac{4}{3}} \approx 3.48$

b. Recall that the instantaneous rate of change of $f(x) = e^x$ is e^x, so the instantaneous rate of change is $e^{\ln 4} = 4$.

18. Using properties of logarithms, solve for x:

$$\begin{aligned}\tfrac{1}{2} - \log_{16}(x-3) &= \log_{16} x\\ \tfrac{1}{2} &= \log_{16} x + \log_{16}(x-3)\\ \tfrac{1}{2} &= \log_{16}[x(x-3)]\\ 16^{1/2} &= x(x-3)\\ 4 &= x^2 - 3x\\ 0 &= x^2 - 3x - 4\\ 0 &= (x-4)(x+1)\\ x &= -1, 4\end{aligned}$$

Since $\log_{10}(-1)$ is undefined, $x = -1$ is an extraneous root. So the solution is $x = 4$.

19. Let $u = e^x$. Then the equation becomes:

$$\begin{aligned}6\left(e^x\right)^2 - 5e^x &= 6\\ 6u^2 - 5u - 6 &= 0\\ (3u+2)(2u-3) &= 0\\ u &= -\tfrac{2}{3}, \tfrac{3}{2}\end{aligned}$$

So either $e^x = -\frac{2}{3}$, which is impossible, or $e^x = \frac{3}{2}$, so $x = \ln\frac{3}{2} \approx 0.405$.

20. a. Solving the inequality:

$$\begin{aligned} 5\left(4-0.3^x\right) &> 12 \\ 4-0.3^x &> 2.4 \\ -0.3^x &> -1.6 \\ 0.3^x &< 1.6 \\ x\ln 0.3 &< \ln 1.6 \\ x &> \frac{\ln 1.6}{\ln 0.3} \approx -0.39 \end{aligned}$$

(since ln 0.3<0)

b. Solving the inequality:

$$\begin{aligned} \ln x + \ln(x-3) &\le \ln 4 \\ \ln\left(x^2-3x\right) &\le \ln 4 \\ x^2-3x &\le 4 \\ x^2-3x-4 &\le 0 \\ (x-4)(x+1) &\le 0 \end{aligned}$$

Using a sign chart, the solution is $-1 \le x \le 4$. For the original domains, we must have $x > 0$ and $x > 3$, so the combined solution is $3 < x \le 4$, or $(3,4]$.

Chapter 6
The Trigonometric Functions

6.1 Radian Measure

1. Using $\theta = \frac{s}{r}$, the radian measure is: $\theta = \frac{5 \text{ cm}}{2 \text{ cm}} = 2.5$ radians

3. Using $\theta = \frac{s}{r}$, the radian measure is: $\theta = \frac{200 \text{ cm}}{1 \text{ m}} = \frac{200 \text{ cm}}{100 \text{ cm}} = 2$ radians

5. **a.** Multiplying by the conversion factor $\frac{\pi}{180°}$: $45° \bullet \frac{\pi}{180°} = \frac{\pi}{4} \approx 0.79$ radians

   b. Multiplying by the conversion factor $\frac{\pi}{180°}$: $90° \bullet \frac{\pi}{180°} = \frac{\pi}{2} \approx 1.57$ radians

   c. Multiplying by the conversion factor $\frac{\pi}{180°}$: $135° \bullet \frac{\pi}{180°} = \frac{3\pi}{4} \approx 2.36$ radians

7. **a.** Multiplying by the conversion factor $\frac{\pi}{180°}$: $0° \bullet \frac{\pi}{180°} = 0$ radians

   b. Multiplying by the conversion factor $\frac{\pi}{180°}$: $360° \bullet \frac{\pi}{180°} = 2\pi \approx 6.28$ radians

   c. Multiplying by the conversion factor $\frac{\pi}{180°}$: $450° \bullet \frac{\pi}{180°} = \frac{5\pi}{2} \approx 7.85$ radians

9. **a.** Multiplying by the conversion factor $\frac{180°}{\pi}$: $\frac{\pi}{12} \bullet \frac{180°}{\pi} = 15°$

   b. Multiplying by the conversion factor $\frac{180°}{\pi}$: $\frac{\pi}{6} \bullet \frac{180°}{\pi} = 30°$

   c. Multiplying by the conversion factor $\frac{180°}{\pi}$: $\frac{\pi}{4} \bullet \frac{180°}{\pi} = 45°$

11. **a.** Multiplying by the conversion factor $\frac{180°}{\pi}$: $\frac{\pi}{3} \bullet \frac{180°}{\pi} = 60°$

    b. Multiplying by the conversion factor $\frac{180°}{\pi}$: $\frac{5\pi}{3} \bullet \frac{180°}{\pi} = 300°$

    c. Multiplying by the conversion factor $\frac{180°}{\pi}$: $4\pi \bullet \frac{180°}{\pi} = 720°$

13. **a.** Multiplying by the conversion factor $\frac{180°}{\pi}$: $2 \bullet \frac{180°}{\pi} = \frac{360°}{\pi} \approx 114.59°$

    b. Multiplying by the conversion factor $\frac{180°}{\pi}$: $3 \bullet \frac{180°}{\pi} = \frac{540°}{\pi} \approx 171.89°$

    c. Multiplying by the conversion factor $\frac{180°}{\pi}$: $\pi^2 \bullet \frac{180°}{\pi} = 180\pi° \approx 565.49°$

15. Since a right angle has radian measure of $\frac{\pi}{2}$, which is larger than $\frac{3}{2}$ (since π is larger than 3), this angle is smaller than a right angle.

17. Multiplying by the conversion factor $\frac{\pi}{180°}$:

$30° = 30° \bullet \frac{\pi}{180°} = \frac{\pi}{6}$ radians

$45° = 45° \bullet \frac{\pi}{180°} = \frac{\pi}{4}$ radians

$60° = 60° \bullet \frac{\pi}{180°} = \frac{\pi}{3}$ radians

$120° = 120° \bullet \frac{\pi}{180°} = \frac{2\pi}{3}$ radians

$135° = 135° \bullet \frac{\pi}{180°} = \frac{3\pi}{4}$ radians

$150° = 150° \bullet \frac{\pi}{180°} = \frac{5\pi}{6}$ radians

19. Using the formula $s = r\theta$ where $r = 3$ ft and $\theta = \frac{4\pi}{3}$: $s = r\theta = 3 \bullet \frac{4\pi}{3} = 4\pi$ ft

21. First convert 45° to radian measure: $45° = 45° \bullet \frac{\pi}{180°} = \frac{\pi}{4}$ radians

Now using the formula $s = r\theta$ where $r = 2$ cm and $\theta = \frac{\pi}{4}$: $s = r\theta = 2 \bullet \frac{\pi}{4} = \frac{\pi}{2}$ cm

23. a. The area is given by: $A = \frac{1}{2}r^2\theta = \frac{1}{2}(6)^2\left(\frac{2\pi}{3}\right) = 12\pi \text{ cm}^2 \approx 37.70 \text{ cm}^2$

b. First convert θ to radian measure: $\theta = 80° \bullet \frac{\pi}{180°} = \frac{4\pi}{9}$ radians

The area is given by: $A = \frac{1}{2}r^2\theta = \frac{1}{2}(5)^2\left(\frac{4\pi}{9}\right) = \frac{50\pi}{9} \text{ m}^2 \approx 17.45 \text{ m}^2$

c. The area is given by: $A = \frac{1}{2}r^2\theta = \frac{1}{2}(24)^2\left(\frac{\pi}{20}\right) = \frac{72\pi}{5} \text{ m}^2 \approx 45.24 \text{ m}^2$

d. First convert θ to radian measure: $\theta = 144° \bullet \frac{\pi}{180°} = \frac{4\pi}{5}$ radians

The area is given by: $A = \frac{1}{2}r^2\theta = \frac{1}{2}(1.8)^2\left(\frac{4\pi}{5}\right) = \frac{12.96\pi}{10} = 1.296\pi \text{ cm}^2 \approx 4.07 \text{ cm}^2$

25. We have $r = 1$ cm and $A = \frac{\pi}{5} \text{ cm}^2$. Substituting into $A = \frac{1}{2}r^2\theta$:

$$\frac{\pi}{5} = \frac{1}{2}(1)^2\theta$$
$$\frac{\pi}{5} = \frac{\theta}{2}$$
$$\theta = \frac{2\pi}{5} \text{ radians}$$

27. a. First convert $30° = \frac{\pi}{6}$ radians. The arc length is given by: $s = r\theta = (5 \text{ in.})\left(\frac{\pi}{6}\right) = \frac{5\pi}{6}$ in.

Therefore the perimeter is: $5 \text{ in.} + 5 \text{ in.} + \frac{5\pi}{6} \text{ in.} = \left(10 + \frac{5\pi}{6}\right) \text{ in.} \approx 12.62 \text{ in.}$

b. The area is given by: $A = \frac{1}{2}r^2\theta = \frac{1}{2}(5 \text{ in.})^2\left(\frac{\pi}{6}\right) = \frac{25\pi}{12} \text{ in.}^2 \approx 6.54 \text{ in.}^2$

29. a. Each revolution of the wheel is 2π radians, so in 6 revolutions there are $\theta = 6(2\pi) = 12\pi$ radians.

Consequently: $\omega = \frac{\theta}{t} = \frac{12\pi \text{ radians}}{1 \text{ sec}} = 12\pi \frac{\text{radians}}{\text{sec}}$

b. Using the formula $s = r\theta$, where $r = 12$ cm and $\theta = 12\pi$ radians,

$s = (12 \text{ cm})(12\pi \text{ radians}) = 144\pi$ cm. The linear speed, therefore, is: $v = \frac{d}{t} = \frac{144\pi \text{ cm}}{1 \text{ sec}} = 144\pi \frac{\text{cm}}{\text{sec}}$

c. Using the formula $s = r\theta$, where $r = 6$ cm and $\theta = 12\pi$ radians,

$s = (6 \text{ cm})(12\pi \text{ radians}) = 72\pi$ cm. Thus: $v = \frac{d}{t} = \frac{72\pi \text{ cm}}{1 \text{ sec}} = 72\pi \frac{\text{cm}}{\text{sec}}$

31. a. In 1 second, the wheel has rotated $1080° \bullet \frac{\pi \text{ radians}}{180°} = 6\pi$ radians. Consequently: $\omega = \frac{\theta}{t} = \frac{6\pi \text{ radians}}{1 \text{ sec}} = 6\pi \frac{\text{radians}}{\text{sec}}$

b. Using $\theta = 6\pi$ radians and $r = 25$ cm, then $s = r\theta = (25 \text{ cm})(6\pi \text{ radians}) = 150\pi$ cm.

Thus: $v = \frac{150\pi \text{ cm}}{1 \text{ sec}} = 150\pi \frac{\text{cm}}{\text{sec}}$

c. Using $\theta = 6\pi$ radians and $r = \frac{25}{2}$ cm, then $s = r\theta = \left(\frac{25}{2} \text{ cm}\right)(6\pi \text{ radians}) = 75\pi$ cm. Thus: $v = \frac{75\pi \text{ cm}}{1 \text{ sec}} = 75\pi \frac{\text{cm}}{\text{sec}}$

33. **a.** In 1 minute, the wheel has rotated 500 rev. Since each revolution is equal to 2π radians:

$$\theta = (500)(2\pi) = 1000\pi \text{ radians}$$

Consequently: $\omega = \frac{\theta}{t} = \frac{1000\pi \text{ radians}}{60 \text{ sec}} = \frac{50\pi}{3} \frac{\text{radians}}{\text{sec}}$

b. Using $\theta = 1000\pi$ radians and $r = 45$ cm, then: $s = r\theta = (45 \text{ cm})(1000\pi \text{ radians}) = 45000\pi \text{ cm}$

Thus: $v = \frac{45000\pi \text{ cm}}{60 \text{ sec}} = 750\pi \frac{\text{cm}}{\text{sec}}$

c. Using $\theta = 1000\pi$ radians and $r = \frac{45}{2}$ cm, then: $s = r\theta = \left(\frac{45}{2} \text{ cm}\right)(1000\pi \text{ radians}) = 22500\pi \text{ cm}$

Thus: $v = \frac{22500\pi \text{ cm}}{60 \text{ sec}} = 375\pi \frac{\text{cm}}{\text{sec}}$

35. **a.** Each revolution is 2π radians, and 24 hr = 86400 sec, so: $\omega = \frac{\theta}{t} = \frac{2\pi \text{ radians}}{86400 \text{ sec}} \approx 0.000073 \frac{\text{radians}}{\text{sec}}$

b. Using $\theta = 2\pi$ radians and $r = 3960$ mi, then: $s = r\theta = (3960 \text{ mi})(2\pi \text{ radians}) = 7920\pi \text{ mi}$

Thus: $v = \frac{s}{t} = \frac{7920\pi \text{ mi}}{24 \text{ hr}} = 330\pi \frac{\text{mi}}{\text{hr}} \approx 1040 \text{ mph}$

37. Since the triangle is equilateral with sides 1 ft, the circle formed will have a radius of 1 ft and arc length of 1 ft, so $\theta = \frac{s}{r} = \frac{1}{1} = 1$ radian. Converting to degrees, the angle at A will have a measure of: $1 \bullet \frac{180°}{\pi} = \frac{180°}{\pi} \approx 57.30°$

39. Since $x° = \frac{\pi x}{180}$ radians, the following equation must hold:

$$x = \frac{\pi x}{180}$$

$$180x = \pi x$$

The only solution to this equation is $x = 0$. The real number $x = 0$ satisfies the property that x degrees equals x radians.

41. **a.** The angular speed of the larger wheel is given by: $\omega = 100 \frac{\text{rev}}{\text{min}} \bullet 2\pi \frac{\text{rad}}{\text{rev}} = 200\pi \frac{\text{rad}}{\text{min}}$

b. In 1 minute, a point on the larger wheel has traveled: $s = R\theta = (10 \text{ cm})(200\pi \text{ rad}) = 2000\pi \text{ cm}$

So the linear speed is given by: $v = \frac{2000\pi \text{ cm}}{1 \text{ min}} = 2000\pi \frac{\text{cm}}{\text{min}}$

c. Since the linear speed on the smaller wheel must also be $2000\pi \frac{\text{cm}}{\text{min}}$, and the radius of the smaller wheel is $r = 6$ cm, its angular speed is: $\omega = \frac{v}{r} = \frac{2000\pi \text{ cm/min}}{6 \text{ cm}} = \frac{1000}{3}\pi \frac{\text{radians}}{\text{min}}$

d. In rpm, the angular speed of the smaller wheel is: $\frac{1 \text{ rev}}{2\pi \text{ rad}} \bullet \frac{1000}{3}\pi \frac{\text{rad}}{\text{min}} = \frac{500}{3}$ rpm

43. First convert θ to radian measure: $\theta = 71°23' = 71\frac{23}{60}° \bullet \frac{\pi}{180°} = \frac{4283\pi}{10800}$

Now compute the arc length: $s = r\theta = 3960 \text{ mi} \bullet \frac{4283\pi}{10800} = \frac{47113\pi}{30} \text{ mi} \approx 4930 \text{ mi}$

45. First convert θ to radian measure: $\theta = 21°19' = 21\frac{19}{60}° \bullet \frac{\pi}{180°} = \frac{1279\pi}{10800}$

Now compute the arc length: $s = r\theta = 3960 \text{ mi} \bullet \frac{1279\pi}{10800} = \frac{14069\pi}{30} \text{ mi} \approx 1470 \text{ mi}$

47. First convert θ to radian measure: $\theta = 38°54' = 38\frac{54}{60}° \bullet \frac{\pi}{180°} = \frac{2334\pi}{10800}$

Now compute the arc length: $s = r\theta = 3960 \text{ mi} \bullet \frac{2334\pi}{10800} = \frac{8558\pi}{10} \text{ mi} \approx 2690 \text{ mi}$

49. Since the altitude bisects the base of the triangle, we can use the following figure:

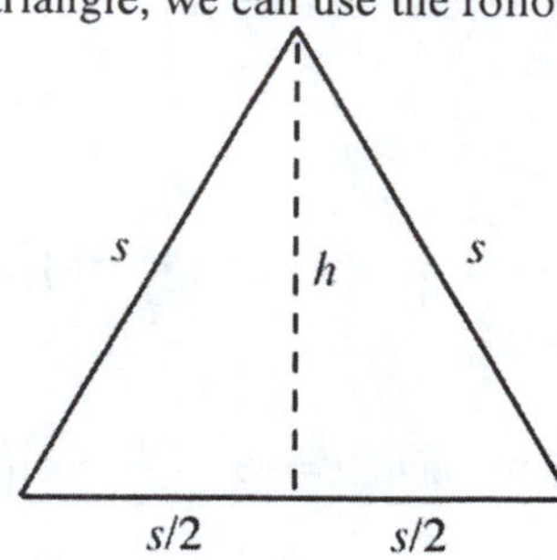

Find h using the Pythagorean theorem:

$$\left(\frac{s}{2}\right)^2 + h^2 = s^2$$
$$\frac{s^2}{4} + h^2 = s^2$$
$$h^2 = \frac{3}{4}s^2$$
$$h = \frac{\sqrt{3}}{2}s$$

Thus the area of the triangle is given by: $\frac{1}{2}(\text{base})(\text{height}) = \frac{1}{2}(s)\left(\frac{\sqrt{3}}{2}s\right) = \frac{\sqrt{3}}{4}s^2$

51. **a.** Since each angle of the equilateral triangle is $60° = \frac{\pi}{3}$ radians, each arc length is given by $r\theta = \frac{\pi}{3}s$.

Therefore the perimeter is given by: $P = \frac{\pi}{3}s + \frac{\pi}{3}s + s = \left(\frac{2\pi}{3} + 1\right)s$

b. The area of sectors ABC and CAB are both given by: $A_s = \frac{1}{2}r^2\theta = \frac{1}{2}(s)^2 \bullet \frac{\pi}{3} = \frac{\pi}{6}s^2$

The area of the triangle is given by: $A_t = \frac{1}{2}(s)(s)\sin\frac{\pi}{3} = \left(\frac{1}{2}s^2\right)\left(\frac{\sqrt{3}}{2}\right) = \frac{\sqrt{3}}{4}s^2$

Therefore the area of the equilateral arch is given by:

$$A = A_s + A_s - A_t = \frac{\pi}{6}s^2 + \frac{\pi}{6}s^2 - \frac{\sqrt{3}}{4}s^2 = \left(\frac{\pi}{3} - \frac{\sqrt{3}}{4}\right)s^2$$

53. **a.** From Exercise 51b, the area of equilateral arch ABC is given by: $A_{ABC} = \left(\frac{\pi}{3} - \frac{\sqrt{3}}{4}\right)s^2$

For arch ADF, use $\frac{s}{2}$ as the radius of the circle. The area of sectors DAF and AFD are each given by: $A_s = \frac{1}{2}r^2\theta = \frac{1}{2}\left(\frac{s}{2}\right)^2 \bullet \frac{\pi}{3} = \frac{\pi}{24}s^2$

The area of the triangle is given by: $A_t = \frac{1}{2}\left(\frac{s}{2}\right)\left(\frac{s}{2}\right)\sin\frac{\pi}{3} = \left(\frac{1}{8}s^2\right)\left(\frac{\sqrt{3}}{2}\right) = \frac{\sqrt{3}}{16}s^2$

Therefore the area of arch ADF is given by: $A_{ADF} = A_s + A_s - A_t = \frac{\pi}{24}s^2 + \frac{\pi}{24}s^2 - \frac{\sqrt{3}}{16}s^2 = \left(\frac{\pi}{12} - \frac{\sqrt{3}}{16}\right)s^2$

Finally, the required ratio is given by: $\frac{A_{ADF}}{A_{ABC}} = \frac{\left(\frac{\pi}{12} - \frac{\sqrt{3}}{16}\right)s^2}{\left(\frac{\pi}{3} - \frac{\sqrt{3}}{4}\right)s^2} = \frac{\frac{\pi}{12} - \frac{\sqrt{3}}{16}}{\frac{\pi}{3} - \frac{\sqrt{3}}{4}} \bullet \frac{48}{48} = \frac{4\pi - 3\sqrt{3}}{16\pi - 12\sqrt{3}} = \frac{1}{4}$

b. From Exercise 52a, use $\frac{s}{2}$ as the radius of the circle. The area of each sector is given by:

$$A_s = \tfrac{1}{2}r^2\theta = \tfrac{1}{2}\left(\frac{s}{2}\right)^2\left(\tfrac{\pi}{3}\right) = \tfrac{\pi}{24}s^2$$

The area of the triangle is given by: $A_t = \tfrac{1}{2}\left(\frac{s}{2}\right)\left(\frac{s}{2}\right)\sin\tfrac{\pi}{3} = \left(\tfrac{1}{8}s^2\right)\left(\frac{\sqrt{3}}{2}\right) = \frac{\sqrt{3}}{16}s^2$

Therefore the area of the equilateral curved triangle *DBE* is given by:

$$A_{DBE} = 3A_s - 2A_t = 3\left(\tfrac{\pi}{24}s^2\right) - 2\left(\frac{\sqrt{3}}{16}s^2\right) = \left(\frac{\pi}{8} - \frac{\sqrt{3}}{8}\right)s^2$$

c. The area of triangle *DEF* is given by: $A_t = \tfrac{1}{2}\left(\frac{s}{2}\right)\left(\frac{s}{2}\right)\sin\tfrac{\pi}{3} = \left(\tfrac{1}{8}s^2\right)\left(\frac{\sqrt{3}}{2}\right) = \frac{\sqrt{3}}{16}s^2$

The area of each sector to be removed is given by: $A_s = \tfrac{1}{2}\left(\frac{s}{2}\right)^2\left(\tfrac{\pi}{3}\right) - \frac{\sqrt{3}}{16}s^2 = \tfrac{\pi}{24}s^2 - \frac{\sqrt{3}}{16}s^2 = \left(\frac{\pi}{24} - \frac{\sqrt{3}}{16}\right)s^2$

Therefore the area of the curved figure *DEF* is given by:

$$A_t - 3A_s = \frac{\sqrt{3}}{16}s^2 - 3\left(\frac{\pi}{24} - \frac{\sqrt{3}}{16}\right)s^2 = \left(\frac{\sqrt{3}}{16} - \frac{\pi}{8} + \frac{3\sqrt{3}}{16}\right)s^2 = \left(\frac{\sqrt{3}}{4} - \frac{\pi}{8}\right)s^2$$

d. From part **c**, the sectors *BE* and *EC* each have areas of $A_s = \left(\frac{\pi}{24} - \frac{\sqrt{3}}{16}\right)s^2$. The large sector *BAC* has an area of:

$$A_{BAC} = \tfrac{1}{2}(s)^2\left(\tfrac{\pi}{3}\right) - \tfrac{1}{2}(s)^2\sin\tfrac{\pi}{3} = \tfrac{\pi}{6}s^2 - \frac{\sqrt{3}}{4}s^2 = \left(\frac{\pi}{6} - \frac{\sqrt{3}}{4}\right)s^2$$

Therefore the area of the curved figure *BEC* is given by:

$$A_{BAC} - 2A_s = \left(\frac{\pi}{6} - \frac{\sqrt{3}}{4}\right)s^2 - 2\left(\frac{\pi}{24} - \frac{\sqrt{3}}{16}\right)s^2 = \left(\frac{\pi}{6} - \frac{\sqrt{3}}{4} - \frac{\pi}{12} + \frac{\sqrt{3}}{8}\right)s^2 = \left(\frac{\pi}{12} - \frac{\sqrt{3}}{8}\right)s^2$$

55. **a.** The arc length is $s = r\theta$. Since the perimeter is 12 cm:

$$r + r + r\theta = 12$$
$$2r + r\theta = 12$$
$$r(2+\theta) = 12$$
$$r = \frac{12}{2+\theta}$$

Using functional notation, the function is $r(\theta) = \frac{12}{2+\theta}$.

b. The area is given by: $A = \tfrac{1}{2}r^2(\theta) = \tfrac{1}{2}\left(\frac{12}{2+\theta}\right)^2(\theta) = \frac{72\theta}{(2+\theta)^2}$

Using functional notation, the function is $A(\theta) = \frac{72\theta}{(2+\theta)^2}$. This is not a quadratic function.

c. Solving the expression in part **a** for θ:

$$r = \frac{12}{2+\theta}$$
$$2+\theta = \frac{12}{r}$$
$$\theta = \frac{12}{r} - 2$$

Using functional notation, the function is $\theta(r) = \frac{12}{r} - 2$.

d. The area is given by: $A = \frac{1}{2}r^2(\theta) = \frac{1}{2}r^2\left(\frac{12}{r} - 2\right) = 6r - r^2$

Using functional notation, the function is $A(r) = 6r - r^2$. Yes, this is a quadratic function.

e. Completing the square on A: $A = 6r - r^2 = -\left(r^2 - 6r\right) = -\left(r^2 - 6r + 9\right) + 9 = -(r-3)^2 + 9$

The area will be a maximum of 9 cm^2 when $r = 3$ cm. The corresponding value of θ is $\theta = 2$ radians.

6.2 Trigonometric Functions of Angles

1. **a.** The reference number for $\frac{\pi}{4}$ is $\frac{\pi}{4}$.

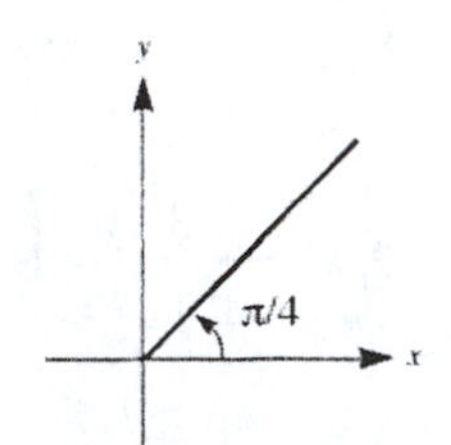

b. The reference number for $-\frac{\pi}{4}$ is $\frac{\pi}{4}$.

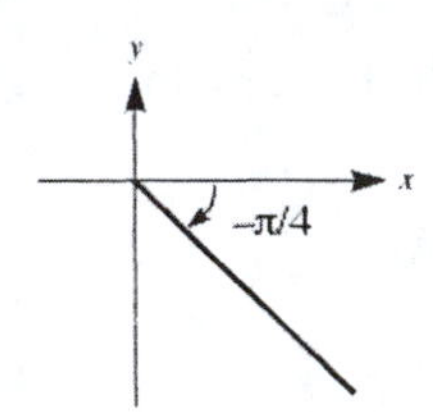

c. The reference number for $\frac{3\pi}{4}$ is $\frac{\pi}{4}$.

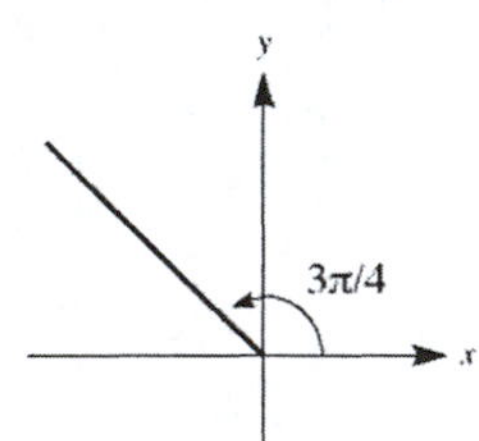

3. **a.** The reference number for $\frac{\pi}{3}$ is $\frac{\pi}{3}$.

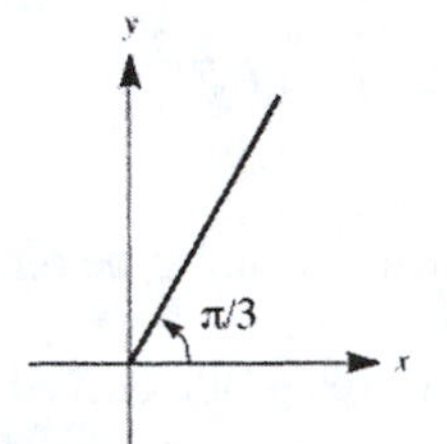

b. The reference number for $-\frac{5\pi}{3}$ is $\frac{\pi}{3}$.

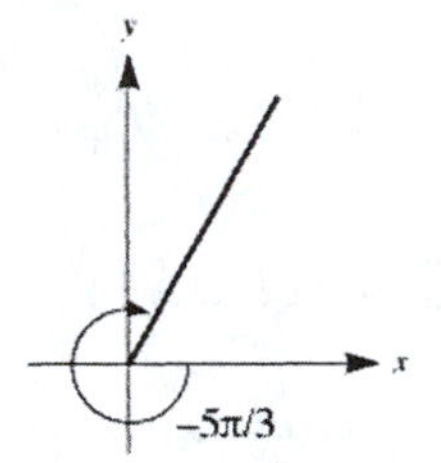

c. The reference number for $-\frac{7\pi}{3}$ is $\frac{\pi}{3}$.

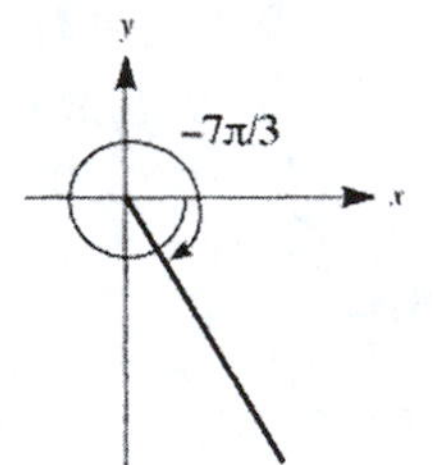

5. **a.** The reference angle for 210° is 30°.

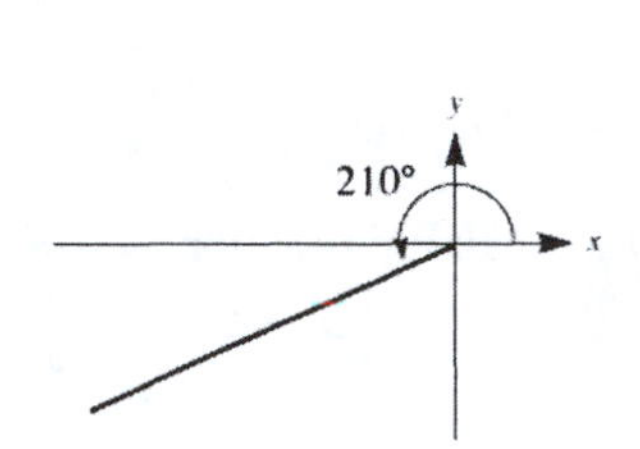

7. **a.** The reference angle for 120° is 60°.

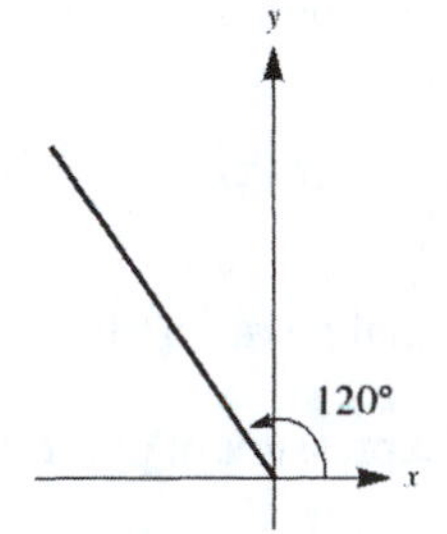

b. The reference angle for –210° is 30°.

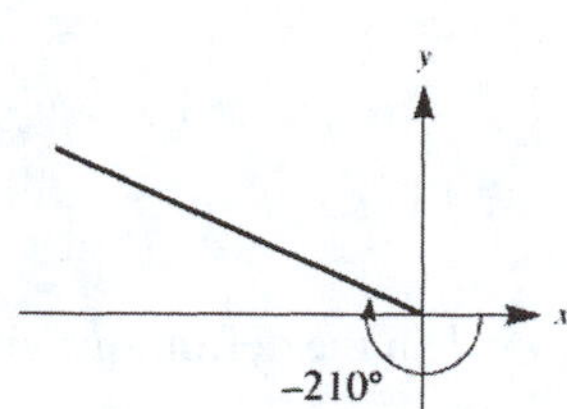

c. The reference angle for –570° is 30°.

b. The reference angle for –120° is 60°.

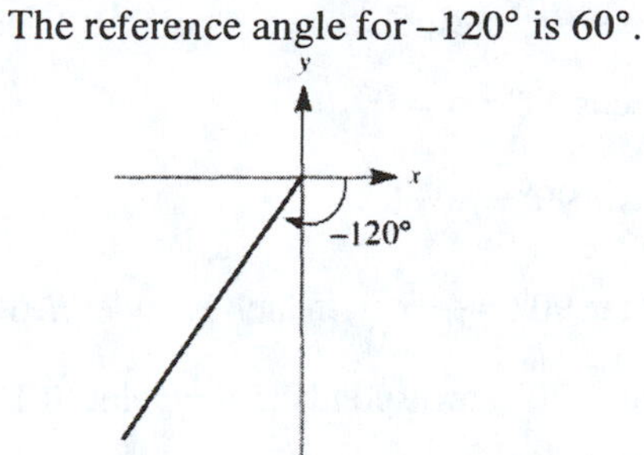

c. The reference angle for 300° is 60°.

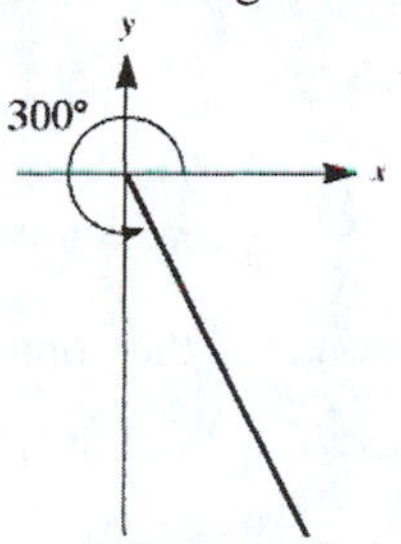

9. Since π corresponds to the point (–1,0) on the unit circle, using $x = -1$ and $y = 0$ in the definitions yields:

$\cos\pi = x = -1$ $\qquad$ $\sec\pi = \frac{1}{x} = \frac{1}{-1} = -1$

$\sin\pi = y = 0$ $\qquad$ $\csc\pi = \frac{1}{y} = \frac{1}{0}$, which is undefined

$\tan\pi = \frac{y}{x} = \frac{0}{-1} = 0$ $\qquad$ $\cot\pi = \frac{x}{y} = \frac{-1}{0}$, which is undefined

11. Since –2π corresponds to the point (1,0) on the unit circle, using $x = 1$ and $y = 0$ in the definitions yields:

$\cos(-2\pi) = x = 1$ $\qquad$ $\sec(-2\pi) = \frac{1}{x} = \frac{1}{1} = 1$

$\sin(-2\pi) = y = 0$ $\qquad$ $\csc(-2\pi) = \frac{1}{y} = \frac{1}{0}$, which is undefined

$\tan(-2\pi) = \frac{y}{x} = \frac{0}{1} = 0$ $\qquad$ $\cot(-2\pi) = \frac{x}{y} = \frac{1}{0}$, which is undefined

13. Since $-\frac{3\pi}{2}$ corresponds to the point (0,1) on the unit circle, using $x = 0$ and $y = 1$ in the definitions yields:

$\cos\left(-\frac{3\pi}{2}\right) = x = 0$ $\qquad$ $\sec\left(-\frac{3\pi}{2}\right) = \frac{1}{x} = \frac{1}{0}$, which is undefined

$\sin\left(-\frac{3\pi}{2}\right) = y = 1$ $\qquad$ $\csc\left(-\frac{3\pi}{2}\right) = \frac{1}{y} = \frac{1}{1} = 1$

$\tan\left(-\frac{3\pi}{2}\right) = \frac{y}{x} = \frac{1}{0}$, which is undefined $\qquad$ $\cot\left(-\frac{3\pi}{2}\right) = \frac{x}{y} = \frac{0}{1} = 0$

15. Since 0 corresponds to the point (1,0) on the unit circle, using $x = 1$ and $y = 0$ in the definitions yields:

$\cos 0 = x = 1$ $\qquad$ $\sec 0 = \frac{1}{x} = \frac{1}{1} = 1$

$\sin 0 = y = 0$ $\qquad$ $\csc 0 = \frac{1}{y} = \frac{1}{0}$, which is undefined

$\tan 0 = \frac{y}{x} = \frac{0}{1} = 0$ $\qquad$ $\cot 0 = \frac{x}{y} = \frac{1}{0}$, which is undefined

17. Since 90° corresponds to the point (0,1) on the unit circle, using $x = 0$ and $y = 1$ in the definitions yields:

$\cos 90° = x = 0$ $\qquad$ $\sec 90° = \frac{1}{x} = \frac{1}{0}$, which is undefined

$\sin 90° = y = 1$ $\qquad$ $\csc 90° = \frac{1}{y} = \frac{1}{1} = 1$

$\tan 90° = \frac{y}{x} = \frac{1}{0}$, which is undefined $\qquad$ $\cot 90° = \frac{x}{y} = \frac{0}{1} = 0$

19. Since –270° corresponds to the point (0,1) on the unit circle, using $x = 0$ and $y = 1$ in the definitions yields:

$\cos(-270°) = x = 0$ $\qquad$ $\sec(-270°) = \frac{1}{x} = \frac{1}{0}$, which is undefined

$\sin(-270°) = y = 1$ $\qquad$ $\csc(-270°) = \frac{1}{y} = \frac{1}{1} = 1$

$\tan(-270°) = \frac{y}{x} = \frac{1}{0}$, which is undefined $\qquad$ $\cot(-270°) = \frac{x}{y} = \frac{0}{1} = 0$

21. Since 180° corresponds to the point (–1,0) on the unit circle, using $x = -1$ and $y = 0$ in the definitions yields:

$\cos 180° = x = -1$ $\qquad$ $\sec 180° = \frac{1}{x} = \frac{1}{-1} = -1$

$\sin 180° = y = 0$ $\qquad$ $\csc 180° = \frac{1}{y} = \frac{1}{0}$, which is undefined

$\tan 180° = \frac{y}{x} = \frac{0}{-1} = 0$ $\qquad$ $\cot 180° = \frac{x}{y} = \frac{-1}{0}$, which is undefined

23. **a.** Verifying that $\left(-\frac{3}{5}, \frac{4}{5}\right)$ lies on the curve $x^2 + y^2 = 1$: $\left(-\frac{3}{5}\right)^2 + \left(\frac{4}{5}\right)^2 = \frac{9}{25} + \frac{16}{25} = \frac{25}{25} = 1$

b. Evaluating the six trigonometric functions:

$\sin\theta = y = \frac{4}{5}$ $\qquad$ $\cos\theta = x = -\frac{3}{5}$ $\qquad$ $\tan\theta = \frac{y}{x} = -\frac{4}{3}$

$\sec\theta = \frac{1}{x} = -\frac{5}{3}$ $\qquad$ $\csc\theta = \frac{1}{y} = \frac{5}{4}$ $\qquad$ $\cot\theta = \frac{x}{y} = -\frac{3}{4}$

25. Complete the table:

θ	$\cos\theta$	$\sin\theta$	$\tan\theta$	$\sec\theta$	$\csc\theta$	$\cot\theta$
0	1	0	0	1	undefined	undefined
$\frac{\pi}{2}$	0	1	undefined	undefined	1	0
π	–1	0	0	–1	undefined	undefined
$\frac{3\pi}{2}$	0	–1	undefined	undefined	–1	0
2π	1	0	0	1	undefined	undefined

27. **a.** Since sin 2 = y in the second quadrant, sin 2 is positive.
b. Since cos 2 = x in the second quadrant, cos 2 is negative.
c. Since $\tan 2 = \frac{y}{x}$ in the second quadrant, tan 2 is negative.

29. **a.** Since cos 1 > 0 and cos 6 > 0, cos 1 + cos 6 is positive.
b. Since cos 6 > cos 1, cos 1 – cos 6 is negative.

31. The y-coordinate at 2 is larger than the y-coordinate at 3, so sin 2 is larger.

33. The x-coordinate at 2 is larger than the x-coordinate at 3 (it is less negative), so cos 2 is larger.

35. Since tan 2 < 0 while tan 4 > 0, tan 4 is larger.

37. From the figure we have $0.5 < \cos 1 < 0.6$ and $0.8 < \sin 1 < 0.9$.
Using a calculator, we have $\cos 1 \approx 0.54$ and $\sin 1 \approx 0.84$.

39. From the figure we have $0.5 < \cos(-1) < 0.6$ and $-0.9 < \sin(-1) < -0.8$.
Using a calculator, we have $\cos(-1) \approx 0.54$ and $\sin(-1) \approx -0.84$.

41. From the figure we have $-0.7 < \cos 4 < -0.6$ and $-0.8 < \sin 4 < -0.7$.
Using a calculator, we have $\cos 4 \approx -0.65$ and $\sin 4 \approx -0.76$.

43. From the figure, we have $-0.7 < \cos(-4) < -0.6$ and $0.7 < \sin(-4) < 0.8$.
Using a calculator, we have $\cos(-4) \approx -0.65$ and $\sin(-4) \approx 0.76$.

45. From the figure we have $\sin 10° \approx 0.2$ and $\sin(-10°) \approx -0.2$.
Using a calculator, we have $\sin 10° \approx 0.17$ and $\sin(-10°) \approx -0.17$.

47. From the figure we have $\cos 80° \approx 0.2$ and $\cos(-80°) \approx 0.2$.
Using a calculator, we have $\cos 80° \approx 0.17$ and $\cos(-80°) \approx 0.17$.

49. From the figure we have $\sin 120° \approx 0.9$ and $\sin(-120°) \approx -0.9$.
Using a calculator, we have $\sin 120° \approx 0.87$ and $\sin(-120°) \approx -0.87$.

51. From the figure we have $\sin 150° = 0.5$ and $\sin(-150°) = -0.5$.
Using a calculator, we have $\sin 150° = 0.5$ and $\sin(-150°) = -0.5$.

53. From the figure we have $\cos 220° \approx -0.8$ and $\cos(-220°) \approx -0.8$.
Using a calculator, we have $\cos 220° \approx -0.77$ and $\cos(-220°) \approx -0.77$.

55. From the figure we have $\cos 310° \approx 0.6$ and $\cos(-310°) \approx 0.6$.
Using a calculator, we have $\cos 310° \approx 0.64$ and $\cos(-310°) \approx 0.64$.

57. Since $\sin(1 + 2\pi) = \sin 1$, we can use the figure to obtain $0.8 < \sin(1 + 2\pi) < 0.9$.
Using a calculator, we have $\sin(1 + 2\pi) \approx 0.84$.

59. Since $P(x,y)$ lies on the unit circle $x^2 + y^2 = 1$ and $y > 0$ in the first quadrant:

$$y = \sqrt{1-x^2} = \sqrt{1-\left(\tfrac{1}{3}\right)^2} = \sqrt{1-\tfrac{1}{9}} = \sqrt{\tfrac{8}{9}} = \frac{2\sqrt{2}}{3}$$

Therefore:

$$\sin\theta = y = \frac{2\sqrt{2}}{3} \qquad \cos\theta = x = \tfrac{1}{3} \qquad \tan\theta = \frac{y}{x} = 2\sqrt{2}$$

$$\csc\theta = \frac{1}{y} = \frac{3}{2\sqrt{2}} = \frac{3\sqrt{2}}{4} \qquad \sec\theta = \frac{1}{x} = 3 \qquad \cot\theta = \frac{x}{y} = \frac{1}{2\sqrt{2}} = \frac{\sqrt{2}}{4}$$

61. Since $P(x,y)$ lies on the unit circle $x^2 + y^2 = 1$ and $y < 0$ in the third quadrant:

$$y = -\sqrt{1-x^2} = -\sqrt{1-\left(-\tfrac{3}{5}\right)^2} = -\sqrt{1-\tfrac{9}{25}} = -\sqrt{\tfrac{16}{25}} = -\tfrac{4}{5}$$

Therefore:

$$\sin\theta = y = -\tfrac{4}{5} \qquad \cos\theta = x = -\tfrac{3}{5} \qquad \tan\theta = \frac{y}{x} = \tfrac{4}{3}$$

$$\csc\theta = \frac{1}{y} = -\tfrac{5}{4} \qquad \sec\theta = \frac{1}{x} = -\tfrac{5}{3} \qquad \cot\theta = \frac{x}{y} = \tfrac{3}{4}$$

63. Since $P(x,y)$ lies on the unit circle $x^2 + y^2 = 1$ and $x < 0$ in the second quadrant:

$$x = -\sqrt{1-y^2} = -\sqrt{1-\left(\tfrac{5}{13}\right)^2} = -\sqrt{1-\tfrac{25}{169}} = -\sqrt{\tfrac{144}{169}} = -\tfrac{12}{13}$$

Therefore:

$$\sin\theta = y = \tfrac{5}{13} \qquad \cos\theta = x = -\tfrac{12}{13} \qquad \tan\theta = \frac{y}{x} = -\tfrac{5}{12}$$

$$\csc\theta = \frac{1}{y} = \tfrac{13}{5} \qquad \sec\theta = \frac{1}{x} = -\tfrac{13}{12} \qquad \cot\theta = \frac{x}{y} = -\tfrac{12}{5}$$

65. Since $P(x,y)$ lies on the unit circle $x^2 + y^2 = 1$ and $x < 0$ in the third quadrant:

$$x = -\sqrt{1-y^2} = -\sqrt{1-\left(-\tfrac{3}{4}\right)^2} = -\sqrt{1-\tfrac{9}{16}} = -\sqrt{\tfrac{7}{16}} = -\frac{\sqrt{7}}{4}$$

Therefore:

$$\sin\theta = y = -\tfrac{3}{4} \qquad \cos\theta = x = -\frac{\sqrt{7}}{4} \qquad \tan\theta = \frac{y}{x} = -\frac{3}{\sqrt{7}} = -\frac{3\sqrt{7}}{7}$$

$$\csc\theta = \frac{1}{y} = -\tfrac{4}{3} \qquad \sec\theta = \frac{1}{x} = -\frac{4}{\sqrt{7}} = -\frac{4\sqrt{7}}{7} \qquad \cot\theta = \frac{x}{y} = -\frac{\sqrt{7}}{3}$$

67. Since $P(x,y)$ lies on the unit circle $x^2 + y^2 = 1$ and $y > 0$ in the second quadrant:

$$y = \sqrt{1-x^2} = \sqrt{1-\left(-\tfrac{8}{15}\right)^2} = \sqrt{1-\tfrac{64}{225}} = \sqrt{\tfrac{161}{225}} = \frac{\sqrt{161}}{15}$$

Therefore:

$$\sin\theta = y = \frac{\sqrt{161}}{15} \qquad \cos\theta = x = -\tfrac{8}{15} \qquad \tan\theta = \frac{y}{x} = -\frac{\sqrt{161}}{8}$$

$$\csc\theta = \frac{1}{y} = \frac{15}{\sqrt{161}} = \frac{15\sqrt{161}}{161} \qquad \sec\theta = \frac{1}{x} = -\tfrac{15}{8} \qquad \cot\theta = \frac{x}{y} = -\frac{8}{\sqrt{161}} = -\frac{8\sqrt{161}}{161}$$

69. Since $P(x,y)$ lies on the unit circle $x^2 + y^2 = 1$ and $x > 0$ in the fourth quadrant:

$$x = \sqrt{1-y^2} = \sqrt{1-\left(-\tfrac{2}{9}\right)^2} = \sqrt{1-\tfrac{4}{81}} = \sqrt{\tfrac{77}{81}} = \frac{\sqrt{77}}{9}$$

Therefore:

$$\sin\theta = y = -\tfrac{2}{9} \qquad \cos\theta = x = \frac{\sqrt{77}}{9} \qquad \tan\theta = \frac{y}{x} = -\frac{2}{\sqrt{77}} = -\frac{2\sqrt{77}}{77}$$

$$\csc\theta = \frac{1}{y} = -\tfrac{9}{2} \qquad \sec\theta = \frac{1}{x} = \frac{9}{\sqrt{77}} = \frac{9\sqrt{77}}{77} \qquad \cot\theta = \frac{x}{y} = -\frac{\sqrt{77}}{2}$$

71. Since $P(x,y)$ lies on the unit circle $x^2 + y^2 = 1$ and $y < 0$ in the fourth quadrant:

$$y = -\sqrt{1-x^2} = -\sqrt{1-\left(\tfrac{7}{25}\right)^2} = -\sqrt{1-\tfrac{49}{625}} = -\sqrt{\tfrac{576}{625}} = -\tfrac{24}{25}$$

Therefore:

$$\sin\theta = y = -\tfrac{24}{25} \qquad \cos\theta = x = \tfrac{7}{25} \qquad \tan\theta = \frac{y}{x} = -\tfrac{24}{7}$$

$$\csc\theta = \frac{1}{y} = -\tfrac{25}{24} \qquad \sec\theta = \frac{1}{x} = \tfrac{25}{7} \qquad \cot\theta = \frac{x}{y} = -\tfrac{7}{24}$$

73. Since $P(x,y)$ lies on the unit circle $x^2 + y^2 = 1$ and $y > 0$ in the first quadrant:

$$y = \sqrt{1-x^2} = \sqrt{1-\left(\tfrac{1}{2}\right)^2} = \sqrt{1-\tfrac{1}{4}} = \sqrt{\tfrac{3}{4}} = \frac{\sqrt{3}}{2}$$

Therefore:

$$\sin\theta = y = \frac{\sqrt{3}}{2} \qquad \cos\theta = x = \tfrac{1}{2} \qquad \tan\theta = \frac{y}{x} = \sqrt{3}$$

$$\csc\theta = \frac{1}{y} = \frac{2}{\sqrt{3}} = \frac{2\sqrt{3}}{3} \qquad \sec\theta = \frac{1}{x} = 2 \qquad \cot\theta = \frac{x}{y} = \frac{1}{\sqrt{3}} = \frac{\sqrt{3}}{3}$$

75. Using a calculator in radian mode, we find $\sec 2.06 \approx -2.13$, $\csc 2.06 \approx 1.13$, and $\cot 2.06 \approx -0.53$.

77. Using a calculator in radian mode, we find $\sec 9 \approx -1.10$, $\csc 9 \approx 2.43$, and $\cot 9 \approx -2.21$.

79. Using a calculator in radian mode, we find $\sec(-0.55) \approx 1.17$, $\csc(-0.55) \approx -1.91$, and $\cot(-0.55) \approx -1.63$.

81. Using a calculator in radian mode, we find $\sec\frac{\pi}{6} \approx 1.15$, $\csc\frac{\pi}{6} = 2$ (exact), and $\cot\frac{\pi}{6} \approx 1.73$.

83. Using a calculator in radian mode, we find $\sec 1400 \approx 2.45$, $\csc 1400 \approx -1.10$, and $\cot 1400 \approx -0.45$.

85. Using a calculator in degree mode, we find $\sec 33° \approx 1.19$, $\csc 33° \approx 1.84$, and $\cot 33° \approx 1.54$.

87. Using a calculator in degree mode, we find $\sec(-125°) \approx -1.74$, $\csc(-125°) \approx -1.22$, and $\cot(-125°) \approx 0.70$.

89. Using a calculator in degree mode, we find $\sec 225° \approx -1.41$, $\csc 225° \approx -1.41$, and $\cot 225° = 1$ (exact).

91.
- **a.** Since $\frac{\pi}{2}$ results in the point (0,1) on the unit circle, $\sin\frac{\pi}{2} = 1$. Since π results in the point (–1,0) on the unit circle, $\sin\pi = 0$.
- **b.** Note that $\sin\frac{\pi}{2} = 1$ while $\dfrac{\sin\pi}{2} = 0$.

6.3 Evaluating the Trigonometric Functions

1. a. The reference angle for 110° is 70°:

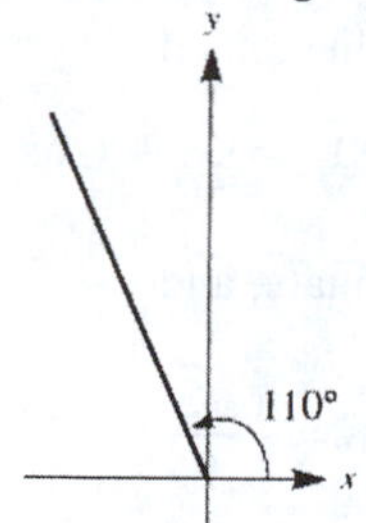

b. The reference angle for 240° is 60°:

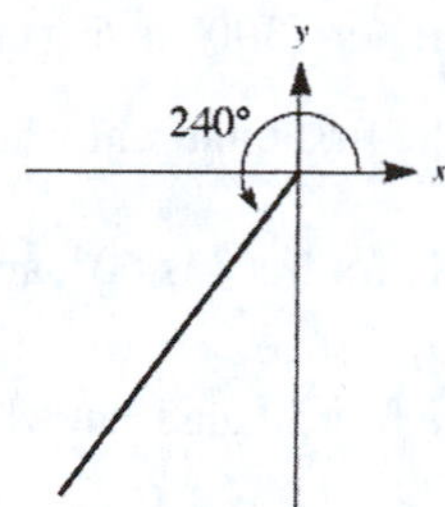

c. The reference angle for 60° is 60°:

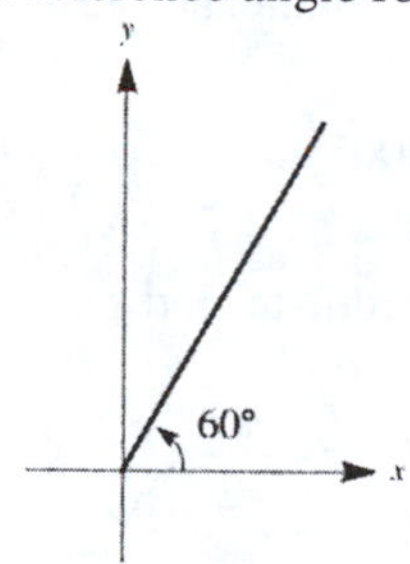

d. The reference angle for – 60° is 60°:

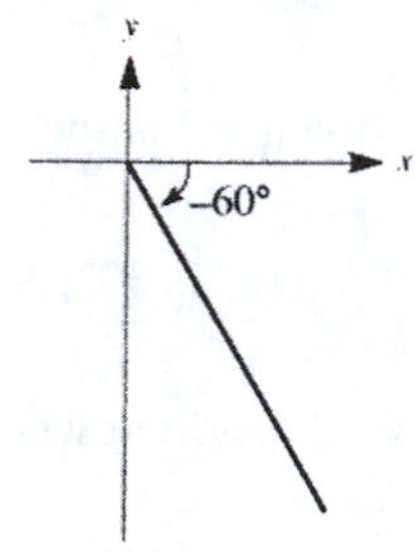

3. a. The reference number for $\frac{3\pi}{4}$ is $\frac{\pi}{4}$:

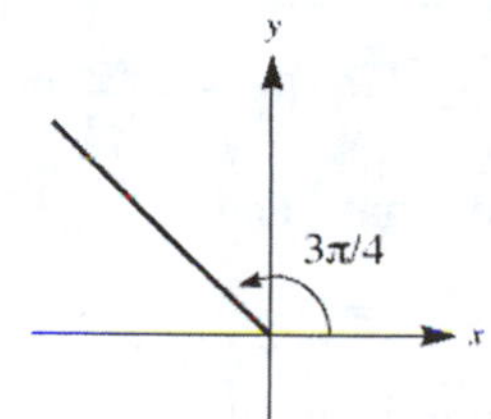

b. The reference number for $-\frac{5\pi}{6}$ is $\frac{\pi}{6}$:

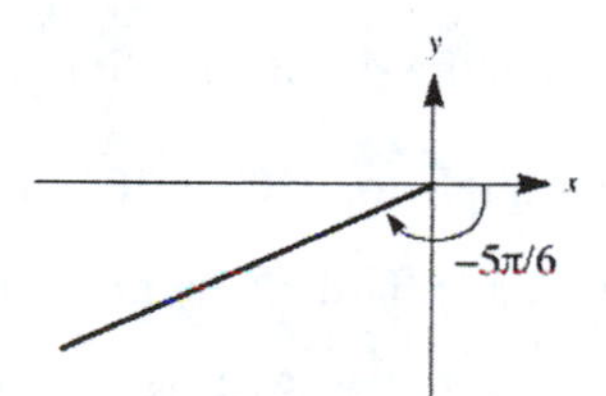

c. The reference number for $\frac{5\pi}{3}$ is $\frac{\pi}{3}$:

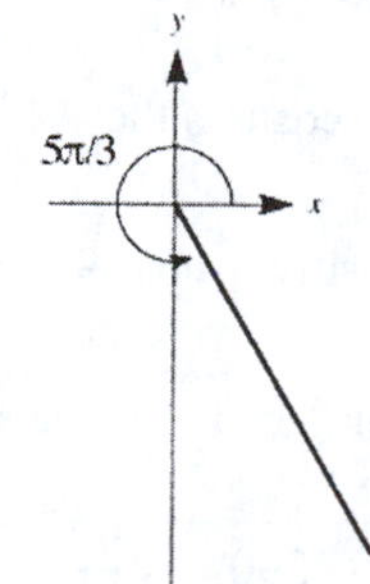

d. The reference number for $\frac{7\pi}{6}$ is $\frac{\pi}{6}$:

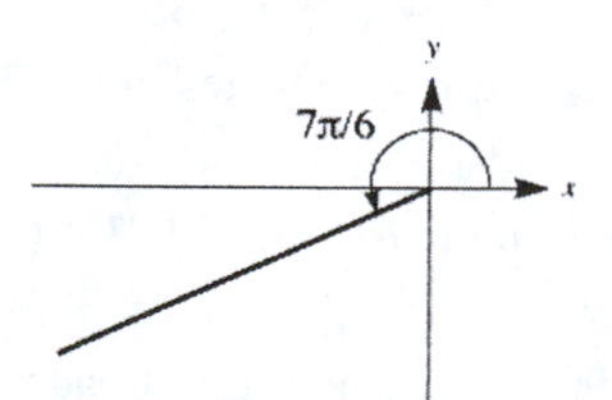

5. a. Since $\cos\frac{\pi}{3} = \cos 60° = \frac{1}{2}$, the correct answer is B.

b. Since $\sin\frac{\pi}{3} = \sin 60° = \frac{\sqrt{3}}{2}$, the correct answer is A.

c. Since $\tan\frac{\pi}{3} = \tan 60° = \sqrt{3}$, the correct answer is D.

d. Since $\cos\frac{\pi}{6} = \cos 30° = \frac{\sqrt{3}}{2}$, the correct answer is A.

e. Since $\sin\frac{\pi}{6} = \sin 30° = \frac{1}{2}$, the correct answer is B.

f. Since $\tan\frac{\pi}{6} = \tan 30° = \frac{\sqrt{3}}{3}$, the correct answer is C.

7. a. The reference angle for 300° is 60°, and $\cos 60° = \frac{1}{2}$. Since cos θ is the x-coordinate, and $\theta = 300°$ lies in the fourth quadrant where the x-coordinates are positive, $\cos 300° = \frac{1}{2}$.

b. The reference angle for –300° is 60°, and $\cos 60° = \frac{1}{2}$. Since cos θ is the x-coordinate, and $\theta = -300°$ lies in the first quadrant where the x-coordinates are positive, $\cos(-300°) = \frac{1}{2}$.

c. The reference angle for 300° is 60°, and $\sin 60° = \frac{\sqrt{3}}{2}$. Since sin θ is the y-coordinate, and $\theta = 300°$ lies in the fourth quadrant where the y-coordinates are negative, $\sin 300° = -\frac{\sqrt{3}}{2}$.

d. The reference angle for –300° is 60°, and $\sin 60° = \frac{\sqrt{3}}{2}$. Since sin θ is the y-coordinate, and $\theta = -300°$ lies in the first quadrant where the y-coordinates are positive, $\sin(-300°) = \frac{\sqrt{3}}{2}$.

9. a. The reference angle for 210° is 30°, and $\cos 30° = \frac{\sqrt{3}}{2}$. Since cos θ is the x-coordinate, and $\theta = 210°$ lies in the third quadrant where the x-coordinates are negative, $\cos 210° = -\frac{\sqrt{3}}{2}$.

b. The reference angle for –210° is 30°, and $\cos 30° = \frac{\sqrt{3}}{2}$. Since cos θ is the x-coordinate, and $\theta = -210°$ lies in the second quadrant where the x-coordinates are negative, $\cos(-210°) = -\frac{\sqrt{3}}{2}$.

c. The reference angle for 210° is 30°, and $\sin 30° = \frac{1}{2}$. Since sin θ is the y-coordinate, and $\theta = 210°$ lies in the third quadrant where the y-coordinates are negative, $\sin 210° = -\frac{1}{2}$.

d. The reference angle for –210° is 30°, and $\sin 30° = \frac{1}{2}$. Since sin θ is the y-coordinate, and $\theta = -210°$ lies in the second quadrant where the y-coordinates are positive, $\sin(-210°) = \frac{1}{2}$.

11. a. The reference angle for 390° is 30° (390° = 30° + 360°), and $\cos 30° = \frac{\sqrt{3}}{2}$. Since cos θ is the x-coordinate, and $\theta = 390°$ lies in the first quadrant where the x-coordinates are positive, $\cos 390° = \frac{\sqrt{3}}{2}$.

b. The reference angle for –390° is 30° (–390° = –30° – 360°), and $\cos 30° = \frac{\sqrt{3}}{2}$. Since cos θ is the x-coordinate, and $\theta = -390°$ lies in the fourth quadrant where the x-coordinates are positive, $\cos(-390°) = \frac{\sqrt{3}}{2}$.

c. The reference angle for 390° is 30°, and $\sin 30° = \frac{1}{2}$. Since sin θ is the y-coordinate, and $\theta = 390°$ lies in the first quadrant where the y-coordinates are positive, $\sin 390° = \frac{1}{2}$.

d. The reference angle for –390° is 30°, and $\sin 30° = \frac{1}{2}$. Since sin θ is the y-coordinate, and $\theta = -390°$ lies in the fourth quadrant where the y-coordinates are negative, $\sin(-390°) = -\frac{1}{2}$.

13. **a.** The reference angle for 600° is 60° (600° = 240° + 360°), and sec 60° = 2. Since sec θ is the reciprocal of the x-coordinate, and $\theta = 600°$ lies in the third quadrant where the x-coordinates are negative, sec 600° = –2.

b. The reference angle for – 600° is 60° (– 600° = –240° – 360°), and $\csc 60° = \frac{2\sqrt{3}}{3}$. Since csc θ is the reciprocal of the y-coordinate, and $\theta = -600°$ lies in the second quadrant where the y-coordinates are positive, $\csc(-600°) = \frac{2\sqrt{3}}{3}$.

c. The reference angle for 600° is 60°, and $\tan 60° = \sqrt{3}$. Since $\tan\theta = \frac{y}{x}$, and $\theta = 600°$ lies in the third quadrant where both the x- and y-coordinates are negative, $\tan 600° = \sqrt{3}$.

d. The reference angle for – 600° is 60°, and $\cot 60° = \frac{\sqrt{3}}{3}$. Since $\cot\theta = \frac{x}{y}$ and $\theta = -600°$ lies in the second quadrant where the y-coordinates are positive and the x-coordinates are negative, $\cot(-600°) = -\frac{\sqrt{3}}{3}$.

15. **a.** The reference angle for $\frac{4\pi}{3}$ is $\frac{\pi}{3}$, and $\cos\frac{\pi}{3} = \frac{1}{2}$. The terminal side of $\frac{4\pi}{3}$ lies in the third quadrant where the x-coordinates are negative, so: $\cos\frac{4\pi}{3} = -\cos\frac{\pi}{3} = -\frac{1}{2}$

b. The reference angle for $-\frac{4\pi}{3}$ is $\frac{\pi}{3}$, and $\cos\frac{\pi}{3} = \frac{1}{2}$. The terminal side of $-\frac{4\pi}{3}$ lies in the second quadrant where the x-coordinates are negative, so: $\cos\left(-\frac{4\pi}{3}\right) = -\cos\frac{\pi}{3} = -\frac{1}{2}$

c. The reference angle for $\frac{4\pi}{3}$ is $\frac{\pi}{3}$, and $\sin\frac{\pi}{3} = \frac{\sqrt{3}}{2}$. The terminal side of $\frac{4\pi}{3}$ lies in the third quadrant where the y-coordinates are negative, so: $\sin\frac{4\pi}{3} = -\sin\frac{\pi}{3} = -\frac{\sqrt{3}}{2}$

d. The reference angle for $-\frac{4\pi}{3}$ is $\frac{\pi}{3}$, and $\sin\frac{\pi}{3} = \frac{\sqrt{3}}{2}$. The terminal side of $-\frac{4\pi}{3}$ lies in the second quadrant where the y-coordinates are positive, so: $\sin\left(-\frac{4\pi}{3}\right) = \sin\frac{\pi}{3} = \frac{\sqrt{3}}{2}$

17. **a.** The reference angle for $\frac{5\pi}{4}$ is $\frac{\pi}{4}$, and $\cos\frac{\pi}{4} = \frac{\sqrt{2}}{2}$. The terminal side of $\frac{5\pi}{4}$ lies in the third quadrant where the x-coordinates are negative, so: $\cos\frac{5\pi}{4} = -\cos\frac{\pi}{4} = -\frac{\sqrt{2}}{2}$

b. The reference angle for $-\frac{5\pi}{4}$ is $\frac{\pi}{4}$, and $\cos\frac{\pi}{4} = \frac{\sqrt{2}}{2}$. The terminal side of $-\frac{5\pi}{4}$ lies in the second quadrant where the x-coordinates are negative, so: $\cos\left(-\frac{5\pi}{4}\right) = -\cos\frac{\pi}{4} = -\frac{\sqrt{2}}{2}$

c. The reference angle for $\frac{5\pi}{4}$ is $\frac{\pi}{4}$, and $\sin\frac{\pi}{4} = \frac{\sqrt{2}}{2}$. The terminal side of $\frac{5\pi}{4}$ lies in the third quadrant where the y-coordinates are negative, so: $\sin\frac{5\pi}{4} = -\sin\frac{\pi}{4} = -\frac{\sqrt{2}}{2}$

d. The reference angle for $-\frac{5\pi}{4}$ is $\frac{\pi}{4}$, and $\sin\frac{\pi}{4} = \frac{\sqrt{2}}{2}$. The terminal side of $-\frac{5\pi}{4}$ lies in the second quadrant where the y-coordinates are positive, so: $\sin\left(-\frac{5\pi}{4}\right) = \sin\frac{\pi}{4} = \frac{\sqrt{2}}{2}$

19. **a.** The reference angle for $\frac{4\pi}{3}$ is $\frac{\pi}{3}$, and $\sec\frac{\pi}{3}=2$. The terminal side of $\frac{4\pi}{3}$ lies in the third quadrant where the x-coordinates are negative, so: $\sec\frac{4\pi}{3}=-\sec\frac{\pi}{3}=-2$

b. The reference angle for $-\frac{4\pi}{3}$ is $\frac{\pi}{3}$, and $\csc\frac{\pi}{3}=\frac{2\sqrt{3}}{3}$. The terminal side of $-\frac{4\pi}{3}$ lies in the second quadrant where the y-coordinates are positive, so: $\csc\left(-\frac{4\pi}{3}\right)=\csc\frac{\pi}{3}=\frac{2\sqrt{3}}{3}$

c. The reference angle for $\frac{4\pi}{3}$ is $\frac{\pi}{3}$, and $\tan\frac{\pi}{3}=\sqrt{3}$. The terminal side of $\frac{4\pi}{3}$ lies in the third quadrant where both the x- and y-coordinates are negative, so: $\tan\frac{4\pi}{3}=\tan\frac{\pi}{3}=\sqrt{3}$

d. The reference angle for $-\frac{4\pi}{3}$ is $\frac{\pi}{3}$, and $\cot\frac{\pi}{3}=\frac{\sqrt{3}}{3}$. The terminal side of $-\frac{4\pi}{3}$ lies in the second quadrant where the x-coordinates are negative while the y-coordinates are positive, so: $\cot\left(-\frac{4\pi}{3}\right)=-\cot\frac{\pi}{3}=-\frac{\sqrt{3}}{3}$

21. **a.** The reference angle for $\frac{17\pi}{6}$ is $\frac{\pi}{6}$, and $\sec\frac{\pi}{6}=\frac{2\sqrt{3}}{3}$. The terminal side of $\frac{17\pi}{6}$ lies in the second quadrant where the x-coordinates are negative, so: $\sec\frac{17\pi}{6}=-\sec\frac{\pi}{6}=-\frac{2\sqrt{3}}{3}$

b. The reference angle for $-\frac{17\pi}{6}$ is $\frac{\pi}{6}$, and $\csc\frac{\pi}{6}=2$. The terminal side of $-\frac{17\pi}{6}$ lies in the third quadrant where the y-coordinates are negative, so: $\csc\left(-\frac{17\pi}{6}\right)=-\csc\frac{\pi}{6}=-2$

c. The reference angle for $\frac{17\pi}{6}$ is $\frac{\pi}{6}$, and $\tan\frac{\pi}{6}=\frac{\sqrt{3}}{3}$. The terminal side of $\frac{17\pi}{6}$ lies in the second quadrant where the x-coordinates are negative while the y-coordinates are positive, so: $\tan\frac{17\pi}{6}=-\tan\frac{\pi}{6}=-\frac{\sqrt{3}}{3}$

d. The reference angle for $-\frac{17\pi}{6}$ is $\frac{\pi}{6}$, and $\cot\frac{\pi}{6}=\sqrt{3}$. The terminal side of $-\frac{17\pi}{6}$ lies in the third quadrant where both the x- and y-coordinates are negative, so: $\cot\left(-\frac{17\pi}{6}\right)=\cot\frac{\pi}{6}=\sqrt{3}$

23. Three radian angles with a cosine of $-\frac{1}{2}$ are $-\frac{2\pi}{3}$, $\frac{2\pi}{3}$, and $\frac{4\pi}{3}$. Other answers are possible.

25. Three degree angles with a cosine of $\frac{\sqrt{3}}{2}$ are –30°, 30°, and 330°. Other answers are possible.

27. **a.** Since the x-coordinate of P is $\frac{3}{4}$, and this is the same as $\cos\theta$, $\cos\theta=\frac{3}{4}$.

b. Since P is in the first quadrant, $y>0$ and thus:

$$\sin\theta=y=\sqrt{1-x^2}=\sqrt{1-\left(\tfrac{3}{4}\right)^2}=\sqrt{1-\tfrac{9}{16}}=\sqrt{\tfrac{7}{16}}=\frac{\sqrt{7}}{4}$$

$$\tan\theta=\frac{y}{x}=\frac{\sqrt{7}}{3}$$

c. Since $\beta-\frac{\pi}{2}$ lies in the first quadrant $\cos\left(\beta-\frac{\pi}{2}\right)>0$. Since $\theta+\frac{\pi}{2}$ lies in the second quadrant $\cos\left(\theta+\frac{\pi}{2}\right)<0$. Therefore $\cos\left(\beta-\frac{\pi}{2}\right)$ is larger than $\cos\left(\theta+\frac{\pi}{2}\right)$.

29. **a.** Compute the quantity: $f\left(\frac{5\pi}{6}+\frac{\pi}{6}\right)=\sin\left(\frac{5\pi}{6}+\frac{\pi}{6}\right)=\sin\pi=0$

b. Compute the quantity: $f\left(\frac{5\pi}{6}\right)+f\left(\frac{\pi}{6}\right)=\sin\left(\frac{5\pi}{6}\right)+\sin\left(\frac{\pi}{6}\right)=\frac{1}{2}+\frac{1}{2}=1$

c. Compute the quantity: $g\left[k\left(\frac{3\pi}{4}\right)\right]=\cos\left[2\left(\frac{3\pi}{4}\right)\right]=\cos\frac{3\pi}{2}=0$

d. Compute the quantity: $k\left[g\left(\frac{3\pi}{4}\right)\right] = 2\cos\frac{3\pi}{4} = 2\left(-\frac{\sqrt{2}}{2}\right) = -\sqrt{2}$

e. Compute the quantity: $\frac{\Delta f}{\Delta x} = \frac{\sin\frac{\pi}{2} - \sin\frac{\pi}{4}}{\frac{\pi}{2} - \frac{\pi}{4}} = \frac{1 - \frac{\sqrt{2}}{2}}{\frac{\pi}{4}} = \frac{4 - 2\sqrt{2}}{\pi}$

f. Compute the quantity: $\frac{\Delta g}{\Delta x} = \frac{\cos\frac{\pi}{2} - \cos\frac{\pi}{4}}{\frac{\pi}{2} - \frac{\pi}{4}} = \frac{0 - \frac{\sqrt{2}}{2}}{\frac{\pi}{4}} = -\frac{2\sqrt{2}}{\pi}$

g. Compute the quantity: $\frac{\Delta f}{\Delta x} = \frac{\sin\frac{7\pi}{6} - \sin\frac{5\pi}{6}}{\frac{7\pi}{6} - \frac{5\pi}{6}} = \frac{-\frac{1}{2} - \frac{1}{2}}{\frac{\pi}{3}} = -\frac{3}{\pi}$

h. Compute the quantity: $\frac{\Delta g}{\Delta x} = \frac{\cos\frac{7\pi}{6} - \cos\frac{5\pi}{6}}{\frac{7\pi}{6} - \frac{5\pi}{6}} = \frac{-\frac{\sqrt{3}}{2} - \left(-\frac{\sqrt{3}}{2}\right)}{\frac{\pi}{3}} = 0$

31. **a.** Complete the table:

θ	$\sin\theta$	larger?
0.1	0.0998	θ
0.2	0.1987	θ
0.3	0.2955	θ
0.4	0.3894	θ
0.5	0.4794	θ

b. In right ΔPQR the hypotenuse is $\overline{PR}$ while $\overline{PQ}$ is a leg, and thus $PQ < PR$. Assuming that the shortest path from point P to point R is a straight line, then $PR < \theta$. (Recall that θ is the arc length). Thus $PQ < PR < \theta$.

c. When $0 < \theta < \frac{\pi}{2}$ we have $PQ = \sin\theta$, and so $\sin\theta < \theta$.

33. **a.** The slope of line segment OP is $\frac{b}{a}$, and since OQ is perpendicular to OP, its slope must be $-\frac{a}{b}$. Since the line passing through O and Q has a y-intercept of 0, its equation must be $y = -\frac{a}{b}x$.

b. Substituting $y = -\frac{a}{b}x$ into the second equation:

$$x^2 + \left(-\frac{a}{b}x\right)^2 = 1$$

$$x^2 + \frac{a^2}{b^2}x^2 = 1$$

$$\left(1 + \frac{a^2}{b^2}\right)x^2 = 1$$

$$\frac{b^2 + a^2}{b^2}x^2 = 1$$

$$x^2 = \frac{b^2}{b^2 + a^2}$$

$$x = \frac{-b}{\sqrt{b^2 + a^2}} \quad (\text{since } x < 0)$$

Substituting into $y=-\frac{a}{b}x$ to find y: $y=-\frac{a}{b}\bullet\frac{-b}{\sqrt{b^2+a^2}}=\frac{a}{\sqrt{b^2+a^2}}$

So the coordinates of Q are $\left(\frac{-b}{\sqrt{b^2+a^2}},\frac{a}{\sqrt{b^2+a^2}}\right)$.

c. Since $P(a,b)$ lies on the unit circle $x^2+y^2=1$, its coordinates must satisfy the equation. Thus $a^2+b^2=1$. Thus $\sqrt{a^2+b^2}=1$, and so the coordinates of Q are $(-b,a)$.

6.4 Algebra and the Trigonometric Functions

1. a. Combining like terms: $-SC+12SC=11SC$
   b. Combining as in part **a**: $-\sin\theta\cos\theta+12\sin\theta\cos\theta=11\sin\theta\cos\theta$

3. a. Combining like terms: $4C^3S-12C^3S=-8C^3S$
   b. Combining as in part **a**: $4\cos^3\theta\sin\theta-12\cos^3\theta\sin\theta=-8\cos^3\theta\sin\theta$

5. a. Squaring: $(1+T)^2=1+2T+T^2$
   b. Squaring as in part **a**: $(1+\tan\theta)^2=1+2\tan\theta+\tan^2\theta$

7. a. Using the distributive property: $(T+3)(T-2)=T^2+3T-2T-6=T^2+T-6$
   b. Using the distributive property as in part **a**:
   $$(\tan\theta+3)(\tan\theta-2)=\tan^2\theta+3\tan\theta-2\tan\theta-6=\tan^2\theta+\tan\theta-6$$

9. a. Factoring then simplifying: $\frac{S-C}{C-S}=\frac{-1(C-S)}{C-S}=-1$
   b. Factoring then simplifying as in part **a**: $\frac{\sin\theta-\cos\theta}{\cos\theta-\sin\theta}=\frac{-1(\cos\theta-\sin\theta)}{\cos\theta-\sin\theta}=-1$

11. a. Obtaining common denominators then adding: $C+\frac{2}{S}=\frac{CS}{S}+\frac{2}{S}=\frac{CS+2}{S}$
    b. Obtaining common denominators then adding as in part **a**:
    $$\cos A+\frac{2}{\sin A}=\frac{\cos A\sin A}{\sin A}+\frac{2}{\sin A}=\frac{\cos A\sin A+2}{\sin A}$$

13. a. The expression factors as: $T^2+8T-9=(T-1)(T+9)$
    b. Factoring as in part **a**: $\tan^2\beta+8\tan\beta-9=(\tan\beta-1)(\tan\beta+9)$

15. a. Factoring as a difference of squares: $4C^2-1=(2C+1)(2C-1)$
    b. Factoring as in part **a**: $4\cos^2 B-1=(2\cos B+1)(2\cos B-1)$

17. a. Factoring the greatest common factor: $9S^2T^3+6ST^2=3ST^2(3ST+2)$
    b. Factoring as in part **a**: $9\sec^2 B\tan^3 B+6\sec B\tan^2 B=3\sec B\tan^2 B(3\sec B\tan B+2)$

19. By factoring the numerator: $\frac{\sin^2 A-\cos^2 A}{\sin A-\cos A}=\frac{(\sin A-\cos A)(\sin A+\cos A)}{\sin A-\cos A}=\sin A+\cos A$

21. By using the identities for $\csc\theta$ and $\sec\theta$:
    $$\sin^2\theta\cos\theta\csc^3\theta\sec\theta=\sin^2\theta\bullet\cos\theta\bullet\frac{1}{\sin^3\theta}\bullet\frac{1}{\cos\theta}=\frac{1}{\sin\theta}=\csc\theta$$

23. By using the identity for $\cot B$: $\cot B\sin^2 B\cot B=\frac{\cos B}{\sin B}\bullet\sin^2 B\bullet\frac{\cos B}{\sin B}=\cos^2 B$

25. By factoring the numerator: $\frac{\cos^2 A+\cos A-12}{\cos A-3}=\frac{(\cos A-3)(\cos A+4)}{\cos A-3}=\cos A+4$

27. By using the identities for $\tan\theta$ and $\sec\theta$:

$$\frac{\tan\theta}{\sec\theta-1}+\frac{\tan\theta}{\sec\theta+1}=\frac{\frac{\sin\theta}{\cos\theta}}{\frac{1}{\cos\theta}-1}+\frac{\frac{\sin\theta}{\cos\theta}}{\frac{1}{\cos\theta}+1}$$
$$=\frac{\sin\theta}{1-\cos\theta}+\frac{\sin\theta}{1+\cos\theta}$$
$$=\frac{\sin\theta(1+\cos\theta)+\sin\theta(1-\cos\theta)}{1-\cos^2\theta}$$
$$=\frac{2\sin\theta}{\sin^2\theta}$$
$$=\frac{2}{\sin\theta}$$
$$=2\csc\theta$$

29. By using the identities for $\sec A$, $\csc A$, $\tan A$ and $\cot A$:

$$\sec A\csc A-\tan A-\cot A=\frac{1}{\cos A}\bullet\frac{1}{\sin A}-\frac{\sin A}{\cos A}-\frac{\cos A}{\sin A}$$
$$=\frac{1-\sin^2 A-\cos^2 A}{\sin A\cos A}$$
$$=\frac{1-\left(\sin^2 A+\cos^2 A\right)}{\sin A\cos A}$$
$$=\frac{1-1}{\sin A\cos A}$$
$$=0$$

31. By using the identities for $\cot\theta$, $\tan\theta$, $\csc\theta$ and $\sec\theta$: $\dfrac{\cot^2\theta}{\csc^2\theta}+\dfrac{\tan^2\theta}{\sec^2\theta}=\dfrac{\frac{\cos^2\theta}{\sin^2\theta}}{\frac{1}{\sin^2\theta}}+\dfrac{\frac{\sin^2\theta}{\cos^2\theta}}{\frac{1}{\cos^2\theta}}=\cos^2\theta+\sin^2\theta=1$

33. Since $\pi<\theta<\frac{3\pi}{2}$ (third quadrant), $\cos\theta<0$ and thus:

$$\cos\theta=-\sqrt{1-\sin^2\theta}=-\sqrt{1-\left(-\tfrac{3}{5}\right)^2}=-\sqrt{1-\tfrac{9}{25}}=-\sqrt{\tfrac{16}{25}}=-\tfrac{4}{5}$$
$$\tan\theta=\frac{\sin\theta}{\cos\theta}=\frac{-\frac{3}{5}}{-\frac{4}{5}}=\frac{3}{4}$$

35. Since $\frac{\pi}{2}<t<\pi$ (second quadrant), $\cos t<0$ and thus:

$$\cos t=-\sqrt{1-\sin^2 t}=-\sqrt{1-\left(\tfrac{\sqrt{3}}{4}\right)^2}=-\sqrt{1-\tfrac{3}{16}}=-\sqrt{\tfrac{13}{16}}=-\frac{\sqrt{13}}{4}$$
$$\tan t=\frac{\sin t}{\cos t}=\frac{\frac{\sqrt{3}}{4}}{-\frac{\sqrt{13}}{4}}=-\frac{\sqrt{3}}{\sqrt{13}}=-\frac{\sqrt{39}}{13}$$

37. Since $\sec\beta=-\frac{17}{15}$ we must have $\cos\beta=-\frac{15}{17}$. Since $\frac{\pi}{2}<\beta<\pi$ (second quadrant), $\sin\beta>0$ and thus:

$$\sin\beta=\sqrt{1-\cos^2\beta}=\sqrt{1-\left(-\tfrac{15}{17}\right)^2}=\sqrt{1-\tfrac{225}{289}}=\sqrt{\tfrac{64}{289}}=\tfrac{8}{17}$$
$$\csc\beta=\frac{1}{\sin\beta}=\frac{17}{8}$$
$$\cot\beta=\frac{\cos\beta}{\sin\beta}=\frac{-\frac{15}{17}}{\frac{8}{17}}=-\frac{15}{8}$$

39. Substituting $\sin\theta = \frac{1}{5}$ into $\cos^2\theta = 1 - \sin^2\theta$ yields: $\cos^2\theta = 1 - \left(\frac{1}{5}\right)^2 = 1 - \frac{1}{25} = \frac{24}{25}$

Since the terminal side of θ lies in quadrant 2, $\cos\theta < 0$ and thus: $\cos\theta = -\sqrt{\frac{24}{25}} = -\frac{2\sqrt{6}}{5}$

Now compute the remaining four trigonometric values:

$$\tan\theta = \frac{\sin\theta}{\cos\theta} = \frac{\frac{1}{5}}{-\frac{2\sqrt{6}}{5}} = -\frac{1}{2\sqrt{6}} = -\frac{\sqrt{6}}{12} \qquad \cot\theta = \frac{\cos\theta}{\sin\theta} = \frac{-\frac{2\sqrt{6}}{5}}{\frac{1}{5}} = -2\sqrt{6}$$

$$\sec\theta = \frac{1}{\cos\theta} = \frac{1}{-\frac{2\sqrt{6}}{5}} = -\frac{5}{2\sqrt{6}} = -\frac{5\sqrt{6}}{12} \qquad \csc\theta = \frac{1}{\sin\theta} = \frac{1}{\frac{1}{5}} = 5$$

41. Substituting $\cos\theta = -\frac{3}{5}$ into $\sin^2\theta = 1 - \cos^2\theta$ yields: $\sin^2\theta = 1 - \left(-\frac{3}{5}\right)^2 = 1 - \frac{9}{25} = \frac{16}{25}$

Since the terminal side of θ lies in quadrant 3, $\sin\theta < 0$ and thus: $\sin\theta = -\sqrt{\frac{16}{25}} = -\frac{4}{5}$

Now compute the remaining four trigonometric values:

$$\tan\theta = \frac{\sin\theta}{\cos\theta} = \frac{-\frac{4}{5}}{-\frac{3}{5}} = \frac{4}{3} \qquad \cot\theta = \frac{\cos\theta}{\sin\theta} = \frac{-\frac{3}{5}}{-\frac{4}{5}} = \frac{3}{4}$$

$$\sec\theta = \frac{1}{\cos\theta} = \frac{1}{-\frac{3}{5}} = -\frac{5}{3} \qquad \csc\theta = \frac{1}{\sin\theta} = \frac{1}{-\frac{4}{5}} = -\frac{5}{4}$$

43. Since $\sec\beta = 3$, $\cos\beta = \frac{1}{3}$. Substituting into $\sin^2\beta = 1 - \cos^2\beta$ yields: $\sin^2\beta = 1 - \left(\frac{1}{3}\right)^2 = 1 - \frac{1}{9} = \frac{8}{9}$

Since the terminal side of β lies in quadrant 1, $\sin\beta > 0$ and thus: $\sin\beta = \sqrt{\frac{8}{9}} = \frac{2\sqrt{2}}{3}$

Now compute the remaining three trigonometric values:

$$\tan\beta = \frac{\sin\beta}{\cos\beta} = \frac{\frac{2\sqrt{2}}{3}}{\frac{1}{3}} = 2\sqrt{2} \qquad \cot\beta = \frac{\cos\beta}{\sin\beta} = \frac{\frac{1}{3}}{\frac{2\sqrt{2}}{3}} = \frac{1}{2\sqrt{2}} = \frac{\sqrt{2}}{4}$$

$$\csc\beta = \frac{1}{\sin\beta} = \frac{1}{\frac{2\sqrt{2}}{3}} = \frac{3}{2\sqrt{2}} = \frac{3\sqrt{2}}{4}$$

45. Using the identities for $\sec\theta$ and $\csc\theta$: $\sin\theta\cos\theta\sec\theta\csc\theta = \frac{\sin\theta\cos\theta}{\sin\theta\cos\theta} = 1$

47. Using the identities for $\sec\theta$ and $\tan\theta$: $\frac{\sin\theta\sec\theta}{\tan\theta} = \frac{\sin\theta \bullet \frac{1}{\cos\theta}}{\frac{\sin\theta}{\cos\theta}} = \frac{\frac{\sin\theta}{\cos\theta}}{\frac{\sin\theta}{\cos\theta}} = 1$

49. Working from the right-hand side and using the identities for $\sec x$ and $\tan x$:

$$\sec x - 5\tan x = \frac{1}{\cos x} - \frac{5\sin x}{\cos x} = \frac{1 - 5\sin x}{\cos x}$$

51. Using the identity for $\sec A$: $\cos A(\sec A - \cos A) = \cos A\left(\frac{1}{\cos A} - \cos A\right) = 1 - \cos^2 A = \sin^2 A$

53. Multiplying out parentheses and using the identities for sec θ and tan θ:

$$\begin{aligned}(1-\sin\theta)(\sec\theta+\tan\theta) &= \sec\theta+\tan\theta-\sin\theta\sec\theta-\sin\theta\tan\theta \\ &= \frac{1}{\cos\theta}+\frac{\sin\theta}{\cos\theta}-\sin\theta\bullet\frac{1}{\cos\theta}-\sin\theta\bullet\frac{\sin\theta}{\cos\theta} \\ &= \frac{1+\sin\theta-\sin\theta-\sin^2\theta}{\cos\theta} \\ &= \frac{1-\sin^2\theta}{\cos\theta} \\ &= \frac{\cos^2\theta}{\cos\theta} \\ &= \cos\theta\end{aligned}$$

55. Multiplying out parentheses and using the identities for sec α and tan α:

$$(\sec\alpha-\tan\alpha)^2=\left(\frac{1}{\cos\alpha}-\frac{\sin\alpha}{\cos\alpha}\right)^2=\frac{(1-\sin\alpha)^2}{\cos^2\alpha}=\frac{(1-\sin\alpha)^2}{1-\sin^2\alpha}=\frac{(1-\sin\alpha)^2}{(1+\sin\alpha)(1-\sin\alpha)}=\frac{1-\sin\alpha}{1+\sin\alpha}$$

57. Working from the right-hand side and using identities for cot A and tan A:

$$\begin{aligned}\frac{\sin A}{1-\cot A}-\frac{\cos A}{\tan A-1} &= \frac{\sin A}{1-\frac{\cos A}{\sin A}}-\frac{\cos A}{\frac{\sin A}{\cos A}-1} \\ &= \frac{\sin^2 A}{\sin A-\cos A}-\frac{\cos^2 A}{\sin A-\cos A} \\ &= \frac{(\sin A-\cos A)(\sin A+\cos A)}{\sin A-\cos A} \\ &= \sin A+\cos A\end{aligned}$$

59. Working from the left-hand side and using identities for csc θ and sec θ:

$$\csc^2\theta+\sec^2\theta=\frac{1}{\sin^2\theta}+\frac{1}{\cos^2\theta}=\frac{\cos^2\theta+\sin^2\theta}{\sin^2\theta\cos^2\theta}=\frac{1}{\sin^2\theta\cos^2\theta}=\frac{1}{\sin^2\theta}\bullet\frac{1}{\cos^2\theta}=\csc^2\theta\sec^2\theta$$

61. Using the identity for tan A and $\sin^2 A = 1-\cos^2 A$: $\sin A\tan A=\sin A\bullet\frac{\sin A}{\cos A}=\frac{\sin^2 A}{\cos A}=\frac{1-\cos^2 A}{\cos A}$

63. Working from the right-hand side:

$$\begin{aligned}-\cot^4 A+\csc^4 A &= \frac{-\cos^4 A}{\sin^4 A}+\frac{1}{\sin^4 A} \\ &= \frac{(1-\cos^2 A)(1+\cos^2 A)}{\sin^4 A} \\ &= \frac{\sin^2 A(1+\cos^2 A)}{\sin^4 A} \\ &= \frac{1+\cos^2 A}{\sin^2 A} \\ &= \frac{1}{\sin^2 A}+\frac{\cos^2 A}{\sin^2 A} \\ &= \csc^2 A+\cot^2 A\end{aligned}$$

65. Working from the left-hand side:

$$\begin{aligned}\frac{\sin A-\cos A}{\sin A}+\frac{\cos A-\sin A}{\cos A}&=\frac{\cos A(\sin A-\cos A)+\sin A(\cos A-\sin A)}{\sin A\cos A}\\&=\frac{\cos A\sin A-\cos^2 A+\sin A\cos A-\sin^2 A}{\sin A\cos A}\\&=\frac{2\cos A\sin A-\left(\cos^2 A+\sin^2 A\right)}{\sin A\cos A}\\&=2-\frac{1}{\sin A\cos A}\\&=2-\sec A\csc A\end{aligned}$$

67. Since the height of the triangle is the *y*-coordinate of point *P*, the height is $\sin\theta$. Thus the area is:

$$\text{area}=\tfrac{1}{2}(\text{base})(\text{height})=\tfrac{1}{2}(2)(\sin\theta)=\sin\theta$$

69. **a.** Label point *P* as $P(x,mx)$. Now draw the triangle:

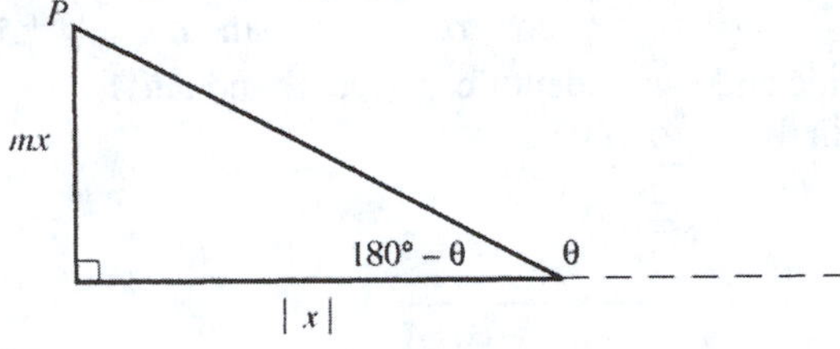

Using the right triangle: $\tan(180°-\theta)=\dfrac{mx}{|x|}=-m$, since $x<0$

But $180°-\theta$ is the reference angle for θ when θ is in the second quadrant, so: $\tan\theta=-\tan(180°-\theta)=m$

Thus, using functional notation, $m(\theta)=\tan\theta$.

b. **i.** Using our function from part **a**: $m(120°)=\tan 120°=\dfrac{\sin 120°}{\cos 120°}=\dfrac{\sqrt{3}/2}{-1/2}=-\sqrt{3}$

ii. Using our function from part **a**: $m(1)=\tan 1\approx 1.6$

71. **a.** Start by testing the value $\alpha=30°$. Since $\cos 30°=\dfrac{\sqrt{3}}{2}$ and $\csc 30°=2$, the equation states that:

$$\begin{aligned}\frac{2^2-1}{2^2}&=\frac{\sqrt{3}}{2}\\\frac{3}{4}&=\frac{\sqrt{3}}{2}\end{aligned}$$

Since the left-hand side does not equal the right-hand side, this is not an identity.

b. Proceeding as in part **a**, test the value $\alpha=30°$. Since $\sec 30°=\dfrac{2}{\sqrt{3}}$ and $\csc 30°=2$, the equation states that:

$$\begin{aligned}\left[\left(\frac{2}{\sqrt{3}}\right)^2-1\right]\left[(2)^2-1\right]&=1\\\left[\tfrac{4}{3}-1\right][4-1]&=1\\\tfrac{1}{3}\bullet 3&=1\end{aligned}$$

Since this is true, proceed by proving the identity:

$$\left(\sec^2\alpha-1\right)\left(\csc^2\alpha-1\right)=\left(\frac{1}{\cos^2\alpha}-1\right)\left(\frac{1}{\sin^2\alpha}-1\right)=\frac{1-\cos^2\alpha}{\cos^2\alpha}\bullet\frac{1-\sin^2\alpha}{\sin^2\alpha}=\frac{\sin^2\alpha}{\cos^2\alpha}\bullet\frac{\cos^2\alpha}{\sin^2\alpha}=1$$

73. **a.** Starting with the left-hand side, multiply the numerator and denominator by $1 + \cos\theta$ to obtain: $\dfrac{\sin\theta}{1-\cos\theta}\bullet\dfrac{1+\cos\theta}{1+\cos\theta}=\dfrac{\sin\theta(1+\cos\theta)}{1-\cos^2\theta}=\dfrac{\sin\theta(1+\cos\theta)}{\sin^2\theta}=\dfrac{1+\cos\theta}{\sin\theta}$

b. Starting with the left-hand side, multiply the numerator and denominator by $\sin\theta$ to obtain:

$$\frac{\sin\theta}{1-\cos\theta}\bullet\frac{\sin\theta}{\sin\theta}=\frac{\sin^2\theta}{\sin\theta(1-\cos\theta)}=\frac{1-\cos^2\theta}{\sin\theta(1-\cos\theta)}=\frac{(1+\cos\theta)(1-\cos\theta)}{\sin\theta(1-\cos\theta)}=\frac{1+\cos\theta}{\sin\theta}$$

75. Starting with the left-hand side, use the identities for $\sec\theta$ and $\csc\theta$, then multiply the resulting fraction by $\sin\theta$ to obtain: $\dfrac{\sec\theta-\csc\theta}{\sec\theta+\csc\theta}=\dfrac{\frac{1}{\cos\theta}-\frac{1}{\sin\theta}}{\frac{1}{\cos\theta}+\frac{1}{\sin\theta}}=\dfrac{\frac{\sin\theta}{\cos\theta}-\frac{\sin\theta}{\sin\theta}}{\frac{\sin\theta}{\cos\theta}+\frac{\sin\theta}{\sin\theta}}=\dfrac{\tan\theta-1}{\tan\theta+1}$

77. Starting with the left-hand side, use the identity for $\cot\theta$ to obtain:

$$\sin^2\theta\left(1+n\cot^2\theta\right)=\sin^2\theta\left(1+n\bullet\frac{\cos^2\theta}{\sin^2\theta}\right)=\sin^2\theta+n\cos^2\theta=\cos^2\theta\left(\frac{\sin^2\theta}{\cos^2\theta}+n\right)=\cos^2\theta\left(n+\tan^2\theta\right)$$

79. **a.** Using the difference of cubes formula $A^3-B^3=(A-B)\left(A^2+AB+B^2\right)$:

$$\cos^3\theta-\sin^3\theta=(\cos\theta-\sin\theta)\left(\cos^2\theta+\cos\theta\sin\theta+\sin^2\theta\right)=(\cos\theta-\sin\theta)(1+\cos\theta\sin\theta)$$

b. Starting with the left-hand side, use the identities for $\cot\phi$, $\tan\phi$, $\sec\phi$ and $\csc\phi$ to obtain:

$$\begin{aligned}\frac{\cos\phi\cot\phi-\sin\phi\tan\phi}{\csc\phi-\sec\phi}&=\frac{\cos\phi\bullet\frac{\cos\phi}{\sin\phi}-\sin\phi\bullet\frac{\sin\phi}{\cos\phi}}{\frac{1}{\sin\phi}-\frac{1}{\cos\phi}}\\&=\frac{\frac{\cos^2\phi}{\sin\phi}-\frac{\sin^2\phi}{\cos\phi}}{\frac{\cos\phi-\sin\phi}{\sin\phi\cos\phi}}\\&=\frac{\cos^3\phi-\sin^3\phi}{\cos\phi-\sin\phi}\\&=\frac{(\cos\phi-\sin\phi)(1+\cos\phi\sin\phi)}{\cos\phi-\sin\phi}\\&=1+\sin\phi\cos\phi\end{aligned}$$

81. Since $\tan\alpha\tan\beta=1$, $\tan\alpha=\dfrac{1}{\tan\beta}$. Using the identities from Exercise 80:

$$\sec^2\alpha=1+\tan^2\alpha=1+\frac{1}{\tan^2\beta}=1+\cot^2\beta=\csc^2\beta$$

Since α and β are acute angles, $\sec\alpha>0$ and $\csc\beta>0$, so $\sec^2\alpha=\csc^2\beta$ implies $\sec\alpha=\csc\beta$.

83. Simplify the right-hand side of the equality:

$$\frac{a+b}{a-b}=\frac{(\sin\alpha+\cos\alpha)+(\sin\alpha-\cos\alpha)}{(\sin\alpha+\cos\alpha)-(\sin\alpha-\cos\alpha)}=\frac{\sin\alpha+\cos\alpha+\sin\alpha-\cos\alpha}{\sin\alpha+\cos\alpha-\sin\alpha+\cos\alpha}=\frac{2\sin\alpha}{2\cos\alpha}=\tan\alpha$$

85. **a.** Using $\theta=20°$, calculate:

$\log_{10}\left(\sin^2 20°\right)\approx-0.9319$ $\qquad$ $2\log_{10}(\sin 20°)\approx-0.9319$

The two expressions are equal.

b. They are equal as long as $\sin\theta>0$, thus ensuring $\log_{10}(\sin\theta)$ is defined. Thus they are equal on the interval $0°<\theta<180°$.

87. Since $(L\circ S)(\theta)=\ln(\sin\theta)$, we must have $\sin\theta>0$. This occurs on the interval $0°<\theta<180°$, which (in interval notation) is $(0°,180°)$.

89. a. Using $\theta = 20°$, calculate:

$$\ln\sqrt{1-\cos 20°} + \ln\sqrt{1+\cos 20°} \approx -1.0729 \qquad \ln(\sin 20°) \approx -1.0729$$

The two expressions are equal.

b. Working from the left-hand side: $\ln\sqrt{1-\cos\theta} + \ln\sqrt{1+\cos\theta} = \ln\sqrt{1-\cos^2\theta} = \ln\sqrt{\sin^2\theta} = \ln(\sin\theta)$

c. For the equation to be valid, we must ensure $\sin\theta > 0$. This occurs on the interval $0° < \theta < 180°$, which (in interval notation) is $(0°, 180°)$.

91. Following the hint, let $P(\cos\theta, \sin\theta)$ be any point on the unit circle. By the distance formula:

$$PA = \sqrt{(\cos\theta - 1)^2 + (\sin\theta - 0)^2} = \sqrt{\cos^2\theta - 2\cos\theta + 1 + \sin^2\theta} = \sqrt{2 - 2\cos\theta}$$

$$PB = \sqrt{\left(\cos\theta + \tfrac{1}{2}\right)^2 + \left(\sin\theta - \tfrac{\sqrt{3}}{2}\right)^2} = \sqrt{\cos^2\theta + \cos\theta + \tfrac{1}{4} + \sin^2\theta - \sqrt{3}\sin + \tfrac{3}{4}} = \sqrt{2 + \cos\theta - \sqrt{3}\sin\theta}$$

$$PC = \sqrt{\left(\cos\theta + \tfrac{1}{2}\right)^2 + \left(\sin\theta + \tfrac{\sqrt{3}}{2}\right)^2} = \sqrt{\cos^2\theta + \cos\theta + \tfrac{1}{4} + \sin^2\theta + \sqrt{3}\sin\theta + \tfrac{3}{4}} = \sqrt{2 + \cos\theta + \sqrt{3}\sin\theta}$$

Now adding up the squares of the distances:

$$(PA)^2 + (PB)^2 + (PC)^2 = 2 - 2\cos\theta + 2 + \cos\theta - \sqrt{3}\sin\theta + 2 + \cos\theta + \sqrt{3}\sin\theta = 6$$

6.5 Right-Triangle Trigonometry

1. a. Use the definitions:

$$\sin\theta = \frac{\text{opposite}}{\text{hypotenuse}} = \tfrac{15}{17} \qquad \cos\theta = \frac{\text{adjacent}}{\text{hypotenuse}} = \tfrac{8}{17}$$

$$\tan\theta = \frac{\text{opposite}}{\text{adjacent}} = \tfrac{15}{8} \qquad \cot\theta = \frac{\text{adjacent}}{\text{opposite}} = \tfrac{8}{15}$$

$$\sec\theta = \frac{\text{hypotenuse}}{\text{adjacent}} = \tfrac{17}{8} \qquad \csc\theta = \frac{\text{hypotenuse}}{\text{opposite}} = \tfrac{17}{15}$$

b. Use the definitions:

$$\sin\beta = \frac{\text{opposite}}{\text{hypotenuse}} = \tfrac{8}{17} \qquad \cos\beta = \frac{\text{adjacent}}{\text{hypotenuse}} = \tfrac{15}{17}$$

$$\tan\beta = \frac{\text{opposite}}{\text{adjacent}} = \tfrac{8}{15} \qquad \cot\beta = \frac{\text{adjacent}}{\text{opposite}} = \tfrac{15}{8}$$

$$\sec\beta = \frac{\text{hypotenuse}}{\text{adjacent}} = \tfrac{17}{15} \qquad \csc\beta = \frac{\text{hypotenuse}}{\text{opposite}} = \tfrac{17}{8}$$

3. a. Use the definitions:

$$\sin\theta = \frac{\text{opposite}}{\text{hypotenuse}} = \frac{3}{3\sqrt{5}} = \frac{1}{\sqrt{5}} = \frac{\sqrt{5}}{5} \qquad \cos\theta = \frac{\text{adjacent}}{\text{hypotenuse}} = \frac{6}{3\sqrt{5}} = \frac{2}{\sqrt{5}} = \frac{2\sqrt{5}}{5}$$

$$\tan\theta = \frac{\text{opposite}}{\text{adjacent}} = \frac{3}{6} = \tfrac{1}{2} \qquad \cot\theta = \frac{\text{adjacent}}{\text{opposite}} = \frac{6}{3} = 2$$

$$\sec\theta = \frac{\text{hypotenuse}}{\text{adjacent}} = \frac{3\sqrt{5}}{6} = \frac{\sqrt{5}}{2} \qquad \csc\theta = \frac{\text{hypotenuse}}{\text{opposite}} = \frac{3\sqrt{5}}{3} = \sqrt{5}$$

b. Use the definitions:

$$\sin\beta = \frac{\text{opposite}}{\text{hypotenuse}} = \frac{6}{3\sqrt{5}} = \frac{2}{\sqrt{5}} = \frac{2\sqrt{5}}{5} \qquad \cos\beta = \frac{\text{adjacent}}{\text{hypotenuse}} = \frac{3}{3\sqrt{5}} = \frac{1}{\sqrt{5}} = \frac{\sqrt{5}}{5}$$

$$\tan\beta = \frac{\text{opposite}}{\text{adjacent}} = \frac{6}{3} = 2 \qquad \cot\beta = \frac{\text{adjacent}}{\text{opposite}} = \frac{3}{6} = \tfrac{1}{2}$$

$$\sec\beta = \frac{\text{hypotenuse}}{\text{adjacent}} = \frac{3\sqrt{5}}{3} = \sqrt{5} \qquad \csc\beta = \frac{\text{hypotenuse}}{\text{opposite}} = \frac{3\sqrt{5}}{6} = \frac{\sqrt{5}}{2}$$

5. First draw ΔABC and label $AC = 3$ and $BC = 2$:

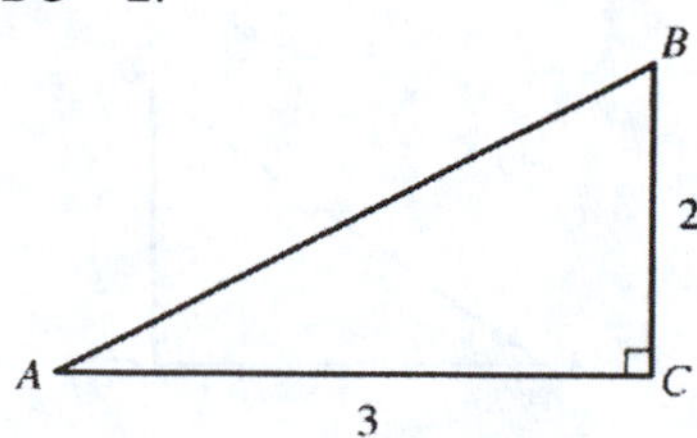

Now find AB by using the Pythagorean theorem:

$$(AC)^2 + (BC)^2 = (AB)^2$$
$$(3)^2 + (2)^2 = (AB)^2$$
$$9 + 4 = (AB)^2$$
$$\sqrt{13} = AB$$

a. Find $\cos A$, $\sin A$ and $\tan A$:

$$\cos A = \frac{\text{adjacent}}{\text{hypotenuse}} = \frac{3}{\sqrt{13}} = \frac{3\sqrt{13}}{13}$$
$$\sin A = \frac{\text{opposite}}{\text{hypotenuse}} = \frac{2}{\sqrt{13}} = \frac{2\sqrt{13}}{13}$$
$$\tan A = \frac{\text{opposite}}{\text{adjacent}} = \frac{2}{3}$$

b. Find $\sec B$, $\csc B$ and $\cot B$:

$$\sec B = \frac{\text{hypotenuse}}{\text{adjacent}} = \frac{\sqrt{13}}{2}$$
$$\csc B = \frac{\text{hypotenuse}}{\text{opposite}} = \frac{\sqrt{13}}{3}$$
$$\cot B = \frac{\text{adjacent}}{\text{opposite}} = \frac{2}{3}$$

7. Draw a sketch with $AB = 13$ and $BC = 5$:

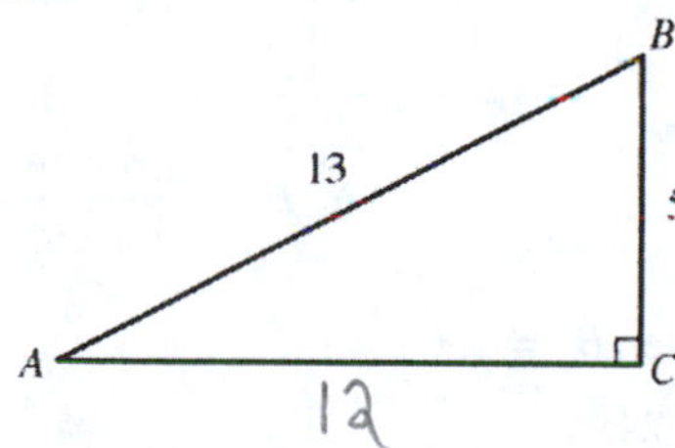

Now find AC by using the Pythagorean theorem:

$$(AC)^2 + (BC)^2 = (AB)^2$$
$$(AC)^2 + (5)^2 = (13)^2$$
$$(AC)^2 + 25 = 169$$
$$(AC)^2 = 144$$
$$AC = 12$$

Now find the six trigonometric functions of angle B:

$$\sin B = \frac{\text{opposite}}{\text{hypotenuse}} = \frac{12}{13} \qquad \cos B = \frac{\text{adjacent}}{\text{hypotenuse}} = \frac{5}{13}$$
$$\tan B = \frac{\text{opposite}}{\text{adjacent}} = \frac{12}{5} \qquad \cot B = \frac{\text{adjacent}}{\text{opposite}} = \frac{5}{12}$$
$$\sec B = \frac{\text{hypotenuse}}{\text{adjacent}} = \frac{13}{5} \qquad \csc B = \frac{\text{hypotenuse}}{\text{opposite}} = \frac{13}{12}$$

9. Draw a sketch where $AC = 1$ and $BC = \frac{3}{4}$:

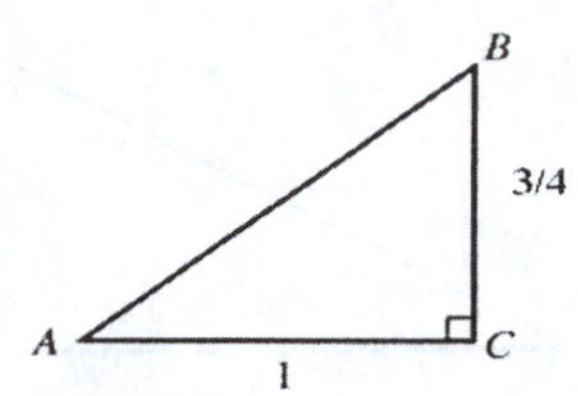

Now find AB by using the Pythagorean theorem:

$$(AC)^2 + (BC)^2 = (AB)^2$$
$$(1)^2 + \left(\tfrac{3}{4}\right)^2 = (AB)^2$$
$$1 + \tfrac{9}{16} = (AB)^2$$
$$\tfrac{25}{16} = (AB)^2$$
$$\tfrac{5}{4} = AB$$

a. Find $\sin B$ and $\cos A$:

$$\sin B = \frac{\text{opposite}}{\text{hypotenuse}} = \frac{1}{\frac{5}{4}} = \frac{4}{5} \qquad \cos A = \frac{\text{adjacent}}{\text{hypotenuse}} = \frac{1}{\frac{5}{4}} = \frac{4}{5}$$

b. Find $\sin A$ and $\cos B$:

$$\sin A = \frac{\text{opposite}}{\text{hypotenuse}} = \frac{\frac{3}{4}}{\frac{5}{4}} = \frac{3}{5} \qquad \cos B = \frac{\text{adjacent}}{\text{hypotenuse}} = \frac{\frac{3}{4}}{\frac{5}{4}} = \frac{3}{5}$$

c. First find $\tan A$ and $\tan B$:

$$\tan A = \frac{\text{opposite}}{\text{adjacent}} = \frac{\frac{3}{4}}{1} = \frac{3}{4} \qquad \tan B = \frac{\text{opposite}}{\text{adjacent}} = \frac{1}{\frac{3}{4}} = \frac{4}{3}$$

Now find $(\tan A)(\tan B)$: $(\tan A)(\tan B) = \frac{3}{4} \bullet \frac{4}{3} = 1$

11. Draw a sketch where $AB = 25$ and $AC = 24$:

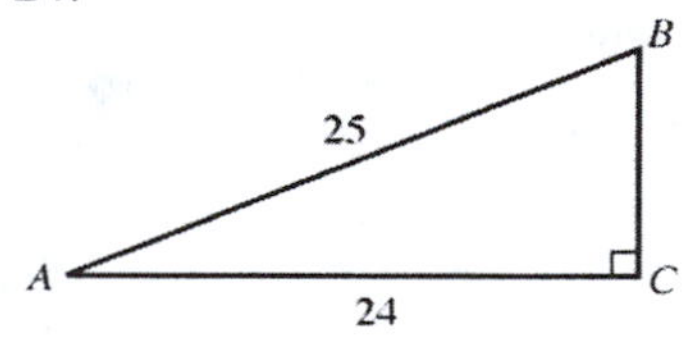

Now find BC by using the Pythagorean theorem:

$$(AC)^2 + (BC)^2 = (AB)^2$$
$$(24)^2 + (BC)^2 = (25)^2$$
$$576 + (BC)^2 = 625$$
$$(BC)^2 = 49$$
$$BC = 7$$

a. Find $\cos A$, $\sin A$ and $\tan A$:

$$\cos A = \frac{\text{adjacent}}{\text{hypotenuse}} = \frac{24}{25}$$
$$\sin A = \frac{\text{opposite}}{\text{hypotenuse}} = \frac{7}{25}$$
$$\tan A = \frac{\text{opposite}}{\text{adjacent}} = \frac{7}{24}$$

b. Find $\cos B$, $\sin B$ and $\tan B$:

$$\cos B = \frac{\text{adjacent}}{\text{hypotenuse}} = \frac{7}{25}$$
$$\sin B = \frac{\text{opposite}}{\text{hypotenuse}} = \frac{24}{25}$$
$$\tan B = \frac{\text{opposite}}{\text{adjacent}} = \frac{24}{7}$$

c. Using the values obtained in parts **a** and **b**: $(\tan A)(\tan B) = \frac{7}{24} \bullet \frac{24}{7} = 1$

13. Compute each side of the equation:

$$\cos 60° = \tfrac{1}{2} \qquad \cos^2 30° - \sin^2 30° = \left(\tfrac{\sqrt{3}}{2}\right)^2 - \left(\tfrac{1}{2}\right)^2 = \tfrac{3}{4} - \tfrac{1}{4} = \tfrac{2}{4} = \tfrac{1}{2}$$

So $\cos 60° = \cos^2 30° - \sin^2 30°$.

15. Compute the left-hand side of the equation:

$$\sin^2 30° + \sin^2 45° + \sin^2 60° = \left(\tfrac{1}{2}\right)^2 + \left(\tfrac{\sqrt{2}}{2}\right)^2 + \left(\tfrac{\sqrt{3}}{2}\right)^2 = \tfrac{1}{4} + \tfrac{2}{4} + \tfrac{3}{4} = \tfrac{6}{4} = \tfrac{3}{2}$$

So $\sin^2 30° + \sin^2 45° + \sin^2 60° = \tfrac{3}{2}$.

17. Compute each side of the equation:

$$2\sin 30° \cos 30° = 2\left(\tfrac{1}{2}\right)\left(\tfrac{\sqrt{3}}{2}\right) = \tfrac{2\sqrt{3}}{4} = \tfrac{\sqrt{3}}{2} \qquad \sin 60° = \tfrac{\sqrt{3}}{2}$$

So $2\sin 30° \cos 30° = \sin 60°$.

19. Compute each side of the equation:

$$\sin 30° = \tfrac{1}{2} \qquad \sqrt{\tfrac{1}{2}(1 - \cos 60°)} = \sqrt{\tfrac{1}{2}\left(1 - \tfrac{1}{2}\right)} = \sqrt{\tfrac{1}{2}\left(\tfrac{1}{2}\right)} = \sqrt{\tfrac{1}{4}} = \tfrac{1}{2}$$

So $\sin 30° = \sqrt{\tfrac{1}{2}(1 - \cos 60°)}$.

21. Compute each side of the equation:

$$\tan 30° = \frac{1}{\sqrt{3}} = \frac{\sqrt{3}}{3} \qquad \frac{\sin 60°}{1 + \cos 60°} = \frac{\frac{\sqrt{3}}{2}}{1 + \frac{1}{2}} = \frac{\frac{\sqrt{3}}{2}}{\frac{3}{2}} = \frac{\sqrt{3}}{3}$$

So $\tan 30° = \dfrac{\sin 60°}{1 + \cos 60°}$.

23. Compute each side of the equation:

$$1 + \tan^2 45° = 1 + (1)^2 = 1 + 1 = 2 \qquad \sec^2 45° = \left(\tfrac{\sqrt{2}}{1}\right)^2 = 2$$

So $1 + \tan^2 45° = \sec^2 45°$.

25. Using the Pythagorean theorem, the hypotenuse is given by: $\sqrt{(3)^2 + (2x)^2} = \sqrt{4x^2 + 9}$

a. Using the right-triangle definitions:

$$\sin\theta = \frac{\text{opposite}}{\text{hypotenuse}} = \frac{2x}{\sqrt{4x^2+9}} = \frac{2x\sqrt{4x^2+9}}{4x^2+9}$$

$$\cos\theta = \frac{\text{adjacent}}{\text{hypotenuse}} = \frac{3}{\sqrt{4x^2+9}} = \frac{3\sqrt{4x^2+9}}{4x^2+9}$$

$$\tan\theta = \frac{\text{opposite}}{\text{adjacent}} = \frac{2x}{3}$$

b. Squaring the results in part **a**:

$$\sin^2\theta = \left(\frac{2x}{\sqrt{4x^2+9}}\right)^2 = \frac{4x^2}{4x^2+9}$$

$$\cos^2\theta = \left(\frac{3}{\sqrt{4x^2+9}}\right)^2 = \frac{9}{4x^2+9}$$

$$\tan^2\theta = \left(\frac{2x}{3}\right)^2 = \frac{4x^2}{9}$$

c. Using the right-triangle definitions:

$$\sin(90° - \theta) = \sin\alpha = \frac{\text{opposite}}{\text{hypotenuse}} = \frac{3}{\sqrt{4x^2+9}} = \frac{3\sqrt{4x^2+9}}{4x^2+9}$$

$$\cos(90° - \theta) = \cos\alpha = \frac{\text{adjacent}}{\text{hypotenuse}} = \frac{2x}{\sqrt{4x^2+9}} = \frac{2x\sqrt{4x^2+9}}{4x^2+9}$$

$$\tan(90° - \theta) = \tan\alpha = \frac{\text{opposite}}{\text{adjacent}} = \frac{3}{2x}$$

27. Using the Pythagorean theorem, the other leg is given by: $\sqrt{(4x)^2 - (1)^2} = \sqrt{16x^2 - 1}$

a. Using the right-triangle definitions:

$$\sin\beta = \frac{\text{opposite}}{\text{hypotenuse}} = \frac{\sqrt{16x^2-1}}{4x} \qquad \cos\beta = \frac{\text{adjacent}}{\text{hypotenuse}} = \frac{1}{4x}$$

$$\tan\beta = \frac{\text{opposite}}{\text{adjacent}} = \frac{\sqrt{16x^2-1}}{1} = \sqrt{16x^2-1}$$

b. Inverting the results from part **a**:

$$\csc\beta = \frac{\text{hypotenuse}}{\text{opposite}} = \frac{4x}{\sqrt{16x^2-1}} = \frac{4x\sqrt{16x^2-1}}{16x^2-1}$$

$$\sec\beta = \frac{\text{hypotenuse}}{\text{adjacent}} = \frac{4x}{1} = 4x$$

$$\cot\beta = \frac{\text{adjacent}}{\text{opposite}} = \frac{1}{\sqrt{16x^2-1}} = \frac{\sqrt{16x^2-1}}{16x^2-1}$$

c. Using the right-triangle definitions:

$$\sin(90° - \beta) = \frac{\text{opposite}}{\text{hypotenuse}} = \frac{1}{4x} \qquad \cos(90° - \beta) = \frac{\text{adjacent}}{\text{hypotenuse}} = \frac{\sqrt{16x^2-1}}{4x}$$

$$\tan(90° - \beta) = \frac{\text{opposite}}{\text{adjacent}} = \frac{1}{\sqrt{16x^2-1}} = \frac{\sqrt{16x^2-1}}{16x^2-1}$$

29. Construct the right triangle:

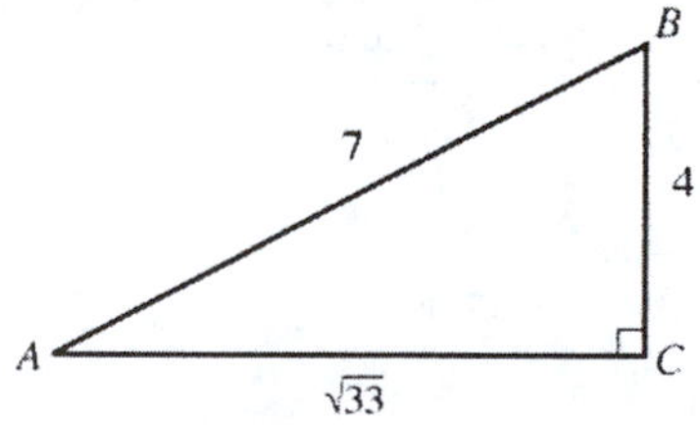

The remaining side was found using the Pythagorean theorem: $\sqrt{7^2 - 4^2} = \sqrt{49-16} = \sqrt{33}$

Now find the remaining five trigonometric functions of B:

$$\sin B = \frac{\text{opposite}}{\text{hypotenuse}} = \frac{\sqrt{33}}{7} \qquad \tan B = \frac{\text{opposite}}{\text{adjacent}} = \frac{\sqrt{33}}{4}$$

$$\sec B = \frac{\text{hypotenuse}}{\text{adjacent}} = \frac{7}{4} \qquad \csc B = \frac{\text{hypotenuse}}{\text{opposite}} = \frac{7}{\sqrt{33}} = \frac{7\sqrt{33}}{33}$$

$$\cot B = \frac{\text{adjacent}}{\text{opposite}} = \frac{4}{\sqrt{33}} = \frac{4\sqrt{33}}{33}$$

31. Construct the right triangle:

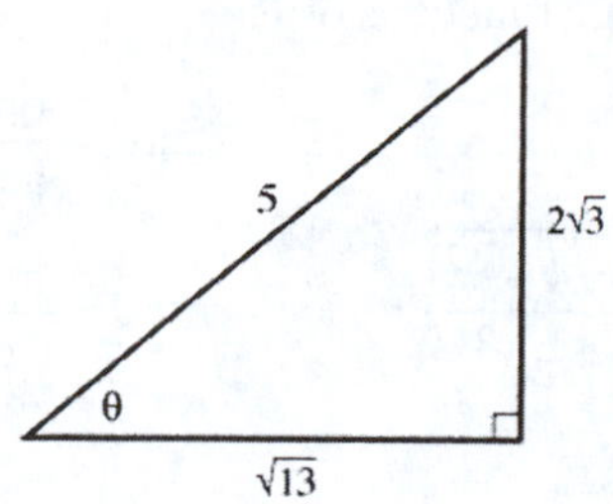

The remaining side was found using the Pythagorean theorem: $\sqrt{5^2-\left(2\sqrt{3}\right)^2}=\sqrt{25-12}=\sqrt{13}$

Now find the remaining five trigonometric functions of θ:

$$\cos\theta=\frac{\text{adjacent}}{\text{hypotenuse}}=\frac{\sqrt{13}}{5} \qquad \tan\theta=\frac{\text{opposite}}{\text{adjacent}}=\frac{2\sqrt{3}}{\sqrt{13}}=\frac{2\sqrt{39}}{13}$$

$$\sec\theta=\frac{\text{hypotenuse}}{\text{adjacent}}=\frac{5}{\sqrt{13}}=\frac{5\sqrt{13}}{13} \qquad \csc\theta=\frac{\text{hypotenuse}}{\text{opposite}}=\frac{5}{2\sqrt{3}}=\frac{5\sqrt{3}}{6}$$

$$\cot\theta=\frac{\text{adjacent}}{\text{opposite}}=\frac{\sqrt{13}}{2\sqrt{3}}=\frac{\sqrt{39}}{6}$$

33. Construct the right triangle:

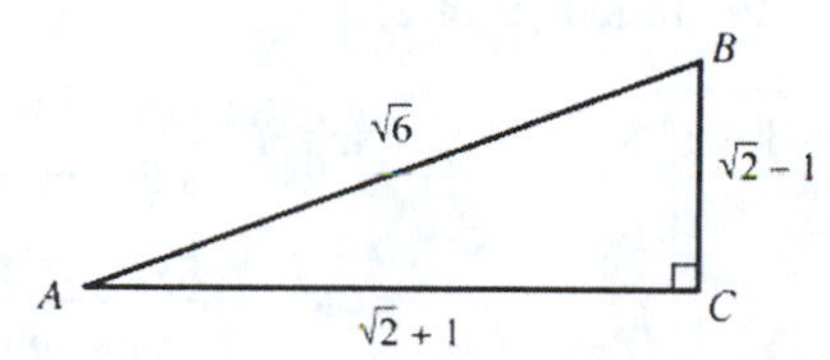

The hypotenuse was found using the Pythagorean theorem: $\sqrt{\left(\sqrt{2}+1\right)^2+\left(\sqrt{2}-1\right)^2}=\sqrt{\left(3+2\sqrt{2}\right)+\left(3-2\sqrt{2}\right)}=\sqrt{6}$

Now find the remaining five trigonometric functions of A:

$$\sin A=\frac{\text{opposite}}{\text{hypotenuse}}=\frac{\sqrt{2}-1}{\sqrt{6}}=\frac{2\sqrt{3}-\sqrt{6}}{6}$$

$$\cos A=\frac{\text{adjacent}}{\text{hypotenuse}}=\frac{\sqrt{2}+1}{\sqrt{6}}=\frac{2\sqrt{3}+\sqrt{6}}{6}$$

$$\sec A=\frac{\text{hypotenuse}}{\text{adjacent}}=\frac{\sqrt{6}}{\sqrt{2}+1}\bullet\frac{\sqrt{2}-1}{\sqrt{2}-1}=\frac{2\sqrt{3}-\sqrt{6}}{1}=2\sqrt{3}-\sqrt{6}$$

$$\csc A=\frac{\text{hypotenuse}}{\text{opposite}}=\frac{\sqrt{6}}{\sqrt{2}-1}\bullet\frac{\sqrt{2}+1}{\sqrt{2}+1}=\frac{2\sqrt{3}+\sqrt{6}}{1}=2\sqrt{3}+\sqrt{6}$$

$$\cot A=\frac{\text{adjacent}}{\text{opposite}}=\frac{\sqrt{2}+1}{\sqrt{2}-1}\bullet\frac{\sqrt{2}+1}{\sqrt{2}+1}=\frac{3+2\sqrt{2}}{2}=3+2\sqrt{2}$$

35. Construct the right triangle:

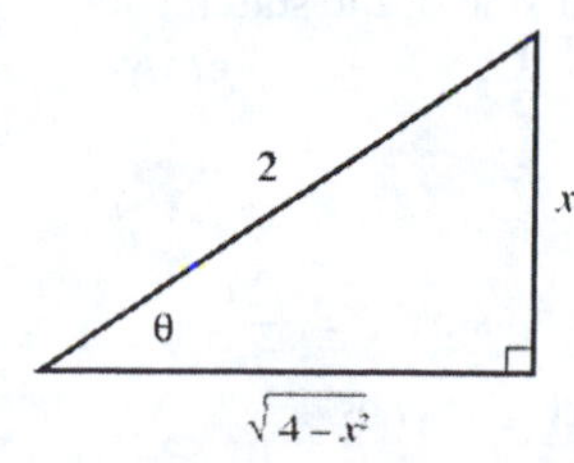

The remaining side was found using the Pythagorean theorem: $\sqrt{2^2-x^2}=\sqrt{4-x^2}$

Now find the remaining five trigonometric functions of θ:

$$\cos\theta = \frac{\text{adjacent}}{\text{hypotenuse}} = \frac{\sqrt{4-x^2}}{2} \qquad \tan\theta = \frac{\text{opposite}}{\text{adjacent}} = \frac{x}{\sqrt{4-x^2}} = \frac{x\sqrt{4-x^2}}{4-x^2}$$

$$\sec\theta = \frac{\text{hypotenuse}}{\text{adjacent}} = \frac{2}{\sqrt{4-x^2}} = \frac{2\sqrt{4-x^2}}{4-x^2} \qquad \csc\theta = \frac{\text{hypotenuse}}{\text{opposite}} = \frac{2}{x}$$

$$\cot\theta = \frac{\text{adjacent}}{\text{opposite}} = \frac{\sqrt{4-x^2}}{x}$$

37. Construct the right triangle:

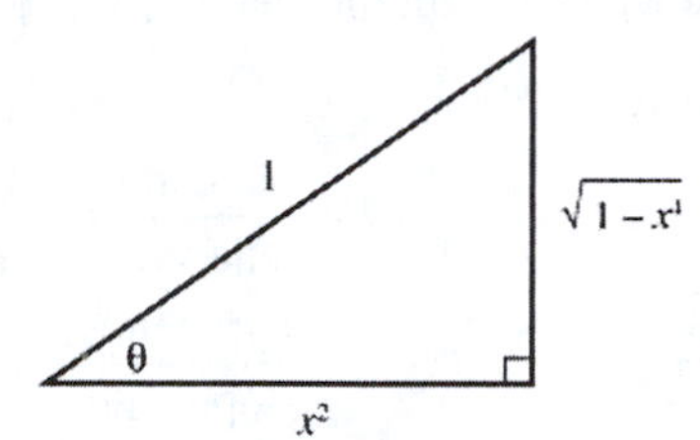

The remaining side was found using the Pythagorean theorem: $\sqrt{1^2 - \left(x^2\right)^2} = \sqrt{1-x^4}$

Now find the remaining five trigonometric functions of θ:

$$\sin\theta = \frac{\text{opposite}}{\text{hypotenuse}} = \frac{\sqrt{1-x^4}}{1} = \sqrt{1-x^4} \qquad \tan\theta = \frac{\text{opposite}}{\text{adjacent}} = \frac{\sqrt{1-x^4}}{x^2}$$

$$\sec\theta = \frac{\text{hypotenuse}}{\text{adjacent}} = \frac{1}{x^2} \qquad \csc\theta = \frac{\text{hypotenuse}}{\text{opposite}} = \frac{1}{\sqrt{1-x^4}} = \frac{\sqrt{1-x^4}}{1-x^4}$$

$$\cot\theta = \frac{\text{adjacent}}{\text{opposite}} = \frac{x^2}{\sqrt{1-x^4}} = \frac{x^2\sqrt{1-x^4}}{1-x^4}$$

39. Using the values in Table 1 to evaluate each side of the statement:

$$2\left[S(30°)\right] = 2\sin 30° = 2\left(\tfrac{1}{2}\right) = 1 \qquad S(60°) = \sin 60° = \frac{\sqrt{3}}{2}$$

The statement is false.

41. Using the values in Table 1 to evaluate the left-hand side of the statement:

$$(T \circ D)(30°) = \tan\left[2 \bullet 30°\right] = \tan 60° = \sqrt{3} > 1$$

The statement is true.

43. Using the values in Table 1 to evaluate the left-hand side of the statement:

$$S(45°) - C(45°) = \sin 45° - \cos 45° = \frac{\sqrt{2}}{2} - \frac{\sqrt{2}}{2} = 0$$

The statement is true.

45. Using the values in Table 1 to evaluate each side of the statement:

$$(C \circ D)(30°) = \cos\left[2 \bullet 30°\right] = \cos 60° = \tfrac{1}{2} \qquad S(30°) = \sin 30° = \tfrac{1}{2}$$

The statement is true.

47. **a.** The expressions can be written as:

$$\sin 20° = \frac{RC}{r}, \qquad \sin 40° = \frac{QB}{r}, \qquad \sin 60° = \frac{PA}{r}$$

But $RC < QB < PA$, so $\sin 20° < \sin 40° < \sin 60°$.

b. Using a calculator:

$$\sin 20° \approx 0.3420, \qquad \sin 40° \approx 0.6428, \qquad \sin 60° \approx 0.8660$$

Thus $\sin 20° < \sin 40° < \sin 60°$.

49. **a.** Since the coordinates of C are $(\cos\theta, \sin\theta)$, $OA = \cos\theta$.

b. Since the coordinates of C are $(\cos\theta, \sin\theta)$, $AC = \sin\theta$.

c. Using similar triangles ΔOAC and ΔOBF:

$$\frac{BF}{OB} = \frac{AC}{OA}$$
$$\frac{BF}{1} = \frac{\sin\theta}{\cos\theta}$$
$$BF = \tan\theta$$

d. Using similar triangles ΔOAC and ΔDEO:

$$\frac{ED}{OE} = \frac{OA}{AC}$$
$$\frac{ED}{1} = \frac{\cos\theta}{\sin\theta}$$
$$ED = \cot\theta$$

e. Using ΔOBF:

$$\cos\theta = \frac{OB}{OF}$$
$$\cos\theta = \frac{1}{OF}$$
$$OF = \frac{1}{\cos\theta} = \sec\theta$$

f. Using ΔDEO:

$$\sin\theta = \frac{OE}{OD}$$
$$\sin\theta = \frac{1}{OD}$$
$$OD = \frac{1}{\sin\theta} = \csc\theta$$

51. Starting with the left-hand side and dividing each term by $\cos\theta$:

$$\frac{p\sin\theta - q\cos\theta}{p\sin\theta + q\cos\theta} = \frac{p\dfrac{\sin\theta}{\cos\theta} - q}{p\dfrac{\sin\theta}{\cos\theta} + q} = \frac{p \bullet \dfrac{p}{q} - q}{p \bullet \dfrac{p}{q} + q} = \frac{p^2 - q^2}{p^2 + q^2}$$

53. **a.** Call the length of a side of the polygon s. Drawing a perpendicular as suggested, note that the acute angle at θ is $\frac{1}{2} \bullet \frac{360°}{11} = \frac{360°}{22}$. Thus:

$$\sin\left(\tfrac{360°}{22}\right) = \frac{s/2}{r}$$
$$\sin\left(\tfrac{360°}{22}\right) = \frac{s}{2r}$$
$$s = 2r\sin\left(\tfrac{360°}{22}\right)$$
$$s \approx 0.5635r$$

b. Translating the quote into symbols: $s = \frac{1}{4}(2r) + \frac{1}{8} \bullet \frac{1}{4}(2r) = \frac{1}{2}r + \frac{1}{16}r = \frac{9}{16}r = 0.5625r$

c. Calculating the percentage error: $\left|\dfrac{0.5635r - 0.5625r}{0.5635r}\right| \times 100 = \dfrac{0.001}{0.5635} \times 100 \approx 0.1775\%$ error

This is a very low percentage error.

55. **a.** Substituting $x = \sin 18° = \frac{1}{4}\left(\sqrt{5} - 1\right)$ into the equation:

$$4\left[\tfrac{1}{4}\left(\sqrt{5}-1\right)\right]^2 + 2\left[\tfrac{1}{4}\left(\sqrt{5}-1\right)\right] - 1 = 4 \bullet \tfrac{1}{16}\left(6 - 2\sqrt{5}\right) + \tfrac{1}{2}\left(\sqrt{5}-1\right) - 1 = \tfrac{3}{2} - \tfrac{1}{2}\sqrt{5} + \tfrac{1}{2}\sqrt{5} - \tfrac{1}{2} - 1 = 0$$

Thus $\sin 18°$ is a root of the quadratic equation $4x^2 + 2x - 1 = 0$.

b. Substituting $x = \sin 15° = \frac{1}{4}\left(\sqrt{6} - \sqrt{2}\right)$ into the equation:

$$\begin{aligned}
16\left[\tfrac{1}{4}\left(\sqrt{6}-\sqrt{2}\right)\right]^4 - 16\left[\tfrac{1}{4}\left(\sqrt{6}-\sqrt{2}\right)\right]^2 + 1 &= 16 \bullet \tfrac{1}{256}\left(6 - 4\sqrt{3} + 2\right)^2 - 16 \bullet \tfrac{1}{16}\left(6 - 4\sqrt{3} + 2\right) + 1 \\
&= \tfrac{1}{16}\left(8 - 4\sqrt{3}\right)^2 - 1\left(8 - 4\sqrt{3}\right) + 1 \\
&= \tfrac{1}{16}\left(64 - 64\sqrt{3} + 48\right) - 8 + 4\sqrt{3} + 1 \\
&= 7 - 4\sqrt{3} - 8 + 4\sqrt{3} + 1 \\
&= 0
\end{aligned}$$

Thus $\sin 15°$ is a root of the equation $16x^4 - 16x^2 + 1 = 0$.

57. We have $\sin\alpha = p\sin\beta$ and $\cos\alpha = q\cos\beta$, so:

$$(p\sin\beta)^2 + (q\cos\beta)^2 = 1$$
$$p^2\sin^2\beta + q^2\cos^2\beta = 1$$

Letting $\sin^2\beta = 1 - \cos^2\beta$:

$$p^2 - p^2\cos^2\beta + q^2\cos^2\beta = 1$$
$$\left(q^2 - p^2\right)\cos^2\beta = 1 - p^2$$
$$\cos^2\beta = \frac{p^2 - 1}{p^2 - q^2}$$

Similarly, letting $\cos^2\beta = 1 - \sin^2\beta$:

$$p^2\sin^2\beta + q^2 - q^2\sin^2\beta = 1$$
$$\left(p^2 - q^2\right)\sin^2\beta = 1 - q^2$$
$$\sin^2\beta = \frac{1 - q^2}{p^2 - q^2}$$

Thus $\tan^2\beta = \dfrac{\sin^2\beta}{\cos^2\beta} = \dfrac{1-q^2}{p^2-1}$, and thus $\tan\beta = \sqrt{\dfrac{1-q^2}{p^2-1}}$.

We can find $\tan\alpha$ by the following relationships: $\sin\alpha = p\sin\beta$ and $\cos\alpha = q\cos\beta$
Therefore:

$$\frac{\sin\alpha}{\cos\alpha} = \frac{p}{q}\bullet\frac{\sin\beta}{\cos\beta}$$
$$\tan\alpha = \frac{p}{q}\bullet\tan\beta$$

Thus $\tan\alpha = \dfrac{p}{q}\sqrt{\dfrac{1-q^2}{p^2-1}}$.

Chapter 6 Review Exercises

1. Complete the table:

θ	$\cos\theta$	$\sin\theta$	$\tan\theta$	$\sec\theta$	$\csc\theta$	$\cot\theta$
0	1	0	0	1	undef.	undef.
$\pi/6$	$\sqrt{3}/2$	1/2	$\sqrt{3}/3$	$2\sqrt{3}/3$	2	$\sqrt{3}$
$\pi/4$	$\sqrt{2}/2$	$\sqrt{2}/2$	1	$\sqrt{2}$	$\sqrt{2}$	1
$\pi/3$	1/2	$\sqrt{3}/2$	$\sqrt{3}$	2	$2\sqrt{3}/3$	$\sqrt{3}/3$
$\pi/2$	0	1	undef.	undef.	1	0
$2\pi/3$	$-1/2$	$\sqrt{3}/2$	$-\sqrt{3}$	-2	$2\sqrt{3}/3$	$-\sqrt{3}/3$
$3\pi/4$	$-\sqrt{2}/2$	$\sqrt{2}/2$	-1	$-\sqrt{2}$	$\sqrt{2}$	-1
$5\pi/6$	$-\sqrt{3}/2$	1/2	$-\sqrt{3}/3$	$-2\sqrt{3}/3$	2	$-\sqrt{3}$
π	-1	0	0	-1	undef.	undef.

3. Draw the triangle:

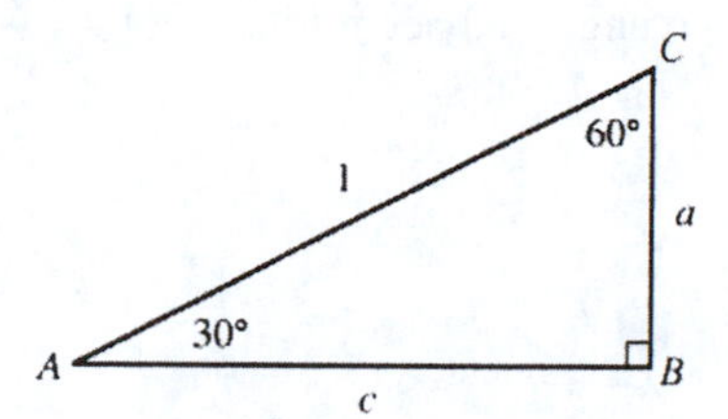

Now compute a and c:

$$\sin 30° = \frac{a}{1}, \text{ so } a = \sin 30° = \tfrac{1}{2} \qquad \cos 30° = \frac{c}{1}, \text{ so } c = \cos 30° = \frac{\sqrt{3}}{2}$$

5. Draw the triangle:

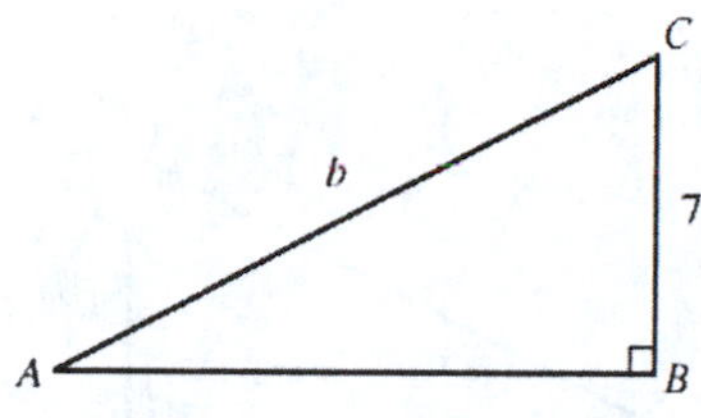

Now compute b: $\sin A = \dfrac{7}{b}$, so $b = \dfrac{7}{\sin A} = \dfrac{7}{\frac{2}{5}} = \frac{35}{2}$

7. Draw the triangle:

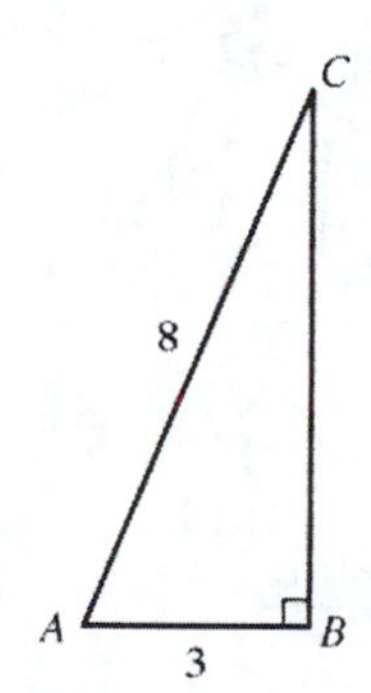

Using the Pythagorean theorem:

$$\begin{aligned} 3^2 + b^2 &= 8^2 \\ 9 + b^2 &= 64 \\ b^2 &= 55 \\ b &= \sqrt{55} \end{aligned}$$

From the triangle: $\sin A = \dfrac{\sqrt{55}}{8}$ and $\cot A = \dfrac{3}{\sqrt{55}} = \dfrac{3\sqrt{55}}{55}$

9. Letting z be the side opposite β: $\tan(\alpha + \beta) = \dfrac{y + z}{x}$ and $\tan \beta = \dfrac{z}{x}$. Therefore:

$$\begin{aligned} \tan(\alpha + \beta) - \tan \beta &= \frac{y + z}{x} - \frac{z}{x} \\ \tan(\alpha + \beta) - \tan \beta &= \frac{y}{x} \\ x\left[\tan(\alpha + \beta) - \tan \beta\right] &= y \end{aligned}$$

11. Letting c be the side opposite the 90° - α angle (adjacent to α): $\cot\theta = \dfrac{a+c}{b}$ and $\cot\alpha = \dfrac{c}{b}$. Therefore:

$$\cot\theta - \cot\alpha = \frac{a+c}{b} - \frac{c}{b}$$
$$\cot\theta - \cot\alpha = \frac{a}{b}$$
$$\cot\theta = \frac{a}{b} + \cot\alpha$$

13. **a.** Start with the identity $\sin^2\alpha + \cos^2\alpha = 1$. Dividing by $\cos^2\alpha$:

$$\frac{\sin^2\alpha}{\cos^2\alpha} + 1 = \frac{1}{\cos^2\alpha}$$
$$\tan^2\alpha + 1 = \sec^2\alpha$$

b. Using the hint, draw the triangle:

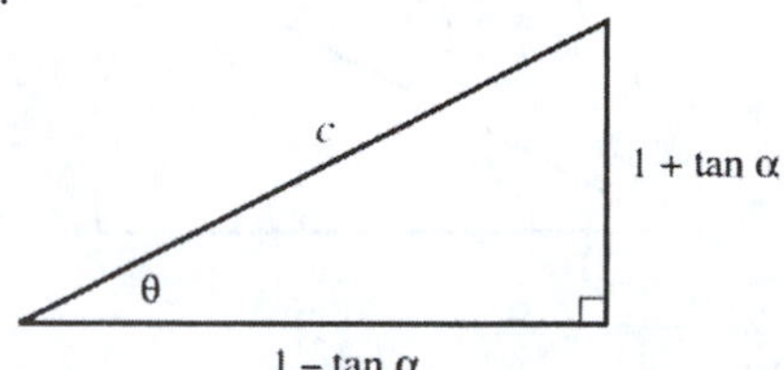

Use the Pythagorean theorem to find c:

$$(1-\tan\alpha)^2 + (1+\tan\alpha)^2 = c^2$$
$$1 - 2\tan\alpha + \tan^2\alpha + 1 + 2\tan\alpha + \tan^2\alpha = c^2$$
$$2\left(1+\tan^2\alpha\right) = c^2$$
$$2\sec^2\alpha = c^2$$
$$c = \sqrt{2}\sec\alpha$$

Now compute $\sin\theta$ and $\cos\theta$:

$$\sin\theta = \frac{1+\tan\alpha}{\sqrt{2}\sec\alpha} = \frac{1+\frac{\sin\alpha}{\cos\alpha}}{\frac{\sqrt{2}}{\cos\alpha}} = \frac{\cos\alpha + \sin\alpha}{\sqrt{2}}$$

$$\cos\theta = \frac{1-\tan\alpha}{\sqrt{2}\sec\alpha} = \frac{1-\frac{\sin\alpha}{\cos\alpha}}{\frac{\sqrt{2}}{\cos\alpha}} = \frac{\cos\alpha - \sin\alpha}{\sqrt{2}}$$

15. Using the identities for $\sec A$ and $\csc A$: $\dfrac{\sin A + \cos A}{\sec A + \csc A} = \dfrac{\sin A + \cos A}{\frac{1}{\cos A} + \frac{1}{\sin A}} = \dfrac{\sin A\cos A(\sin A + \cos A)}{\sin A + \cos A} = \sin A\cos A$

17. Using the identities for $\sec A$, $\tan A$ and $\cot A$: $\dfrac{\sin A\sec A}{\tan A + \cot A} = \dfrac{\frac{\sin A}{\cos A}}{\frac{\sin A}{\cos A} + \frac{\cos A}{\sin A}} = \dfrac{\sin^2 A}{\sin^2 A + \cos^2 A} = \sin^2 A$

19. Using the identities for tan A and cot A:

$$\begin{aligned}\frac{\cos A}{1-\tan A}+\frac{\sin A}{1-\cot A} &= \frac{\cos A}{1-\frac{\sin A}{\cos A}}+\frac{\sin A}{1-\frac{\cos A}{\sin A}}\\ &= \frac{\cos^2 A}{\cos A-\sin A}+\frac{\sin^2 A}{\sin A-\cos A}\\ &= \frac{\cos^2 A}{\cos A-\sin A}-\frac{\sin^2 A}{\cos A-\sin A}\\ &= \frac{\cos^2 A-\sin^2 A}{\cos A-\sin A}\\ &= \frac{(\cos A+\sin A)(\cos A-\sin A)}{\cos A-\sin A}\\ &= \cos A+\sin A\end{aligned}$$

21. Using the identities for sec A and csc A:

$$\begin{aligned}(\sec A+\csc A)^{-1}\left[(\sec A)^{-1}+(\csc A)^{-1}\right] &= \frac{1}{\sec A+\csc A}\bullet\left[\frac{1}{\sec A}+\frac{1}{\csc A}\right]\\ &= \frac{1}{\sec A+\csc A}\bullet\frac{\csc A+\sec A}{\sec A\csc A}\\ &= \frac{1}{\sec A\csc A}\\ &= \frac{1}{\frac{1}{\cos A}\bullet\frac{1}{\sin A}}\\ &= \sin A\cos A\end{aligned}$$

23. **a.** Using the figure we have $0.9 < \cos 6 < 1.0$. Using a calculator we have $\cos 6 \approx 0.96$.
b. Using the figure we have $0.9 < \cos(-6) < 1.0$. Using a calculator we have $\cos(-6) \approx 0.96$.

25. **a.** Using the figure we have $-0.8 < \cos 140° < -0.7$. Using a calculator we have $\cos 140° \approx -0.77$.
b. Using the figure we have $-0.8 < \cos(-140°) < -0.7$. Using a calculator we have $\cos(-140°) \approx -0.77$.

27. **a.** Using the figure we have $0.7 < \sin\frac{\pi}{4} < 0.8$. Using a calculator we have $\sin\frac{\pi}{4} \approx 0.71$.
b. Using the figure we have $-0.8 < \sin\left(-\frac{\pi}{4}\right) < -0.7$. Using a calculator we have $\sin\left(-\frac{\pi}{4}\right) \approx -0.71$.

29. **a.** Using the figure we have $-1.0 < \sin 250° < -0.9$. Using a calculator we have $\sin 250° \approx -0.94$.
b. Using the figure we have $0.9 < \sin(-250°) < 1.0$. Using a calculator we have $\sin(-250°) \approx 0.94$.

31. **a.** Using the figure we have $-0.7 < \cos 4 < -0.6$. Using a calculator we have $\cos 4 \approx -0.65$.
b. Using the figure we have $-0.7 < \cos(-4) < -0.6$. Using a calculator we have $\cos(-4) \approx -0.65$.

33. **a.** Since $4 + 2\pi$ will have the same trigonometric values as 4, we have $-0.7 < \cos(4 + 2\pi) < -0.6$ from the figure and $\cos(4 + 2\pi) \approx -0.65$ from a calculator.
b. Since $-4 + 2\pi$ will have the same trigonometric values as -4, we have $-0.7 < \cos(-4 + 2\pi) < -0.6$ from the figure and $\cos(-4 + 2\pi) \approx -0.65$ from a calculator.

35. Since 20° lies in the first quadrant, $\sin 20° > 0$ and thus: $\sin 20° = \sqrt{1-\cos^2 20°} = \sqrt{1-a^2}$

37. Since $\cos 70° = \sin(90° - 70°) = \sin 20°$: $\cos 70° = \sin 20° = \sqrt{1-a^2}$

39. Since the reference angle for 160° is 20°, and 160° lies in the second quadrant: $\cos 160° = -\cos 20° = -a$

41. Since the reference angle for −160° is 20°, and −160° lies in the third quadrant: $\cos(-160°) = -\cos 20° = -a$

43. Since $\cos 200° = -\cos 20° = -a$ and 200° lies in the third quadrant, $\sin 200° < 0$ and thus:

$$\sin 200° = -\sqrt{1-\cos^2 200°} = -\sqrt{1-(-a)^2} = -\sqrt{1-a^2}$$

45. Converting to multiplication and factoring:

$$\frac{\sin^4\theta-\cos^4\theta}{\sin^2\theta-\cos^2\theta}\div\frac{1+\sin\theta\cos\theta}{\sin^3\theta-\cos^3\theta}$$
$$=\frac{(\sin^2\theta+\cos^2\theta)(\sin^2\theta-\cos^2\theta)}{\sin^2\theta-\cos^2\theta}\bullet\frac{(\sin\theta-\cos\theta)(\sin^2\theta+\sin\theta\cos\theta+\cos^2\theta)}{1+\sin\theta\cos\theta}$$
$$=1\bullet\frac{(\sin\theta-\cos\theta)(1+\sin\theta\cos\theta)}{1+\sin\theta\cos\theta}$$
$$=\sin\theta-\cos\theta$$

47. Using the identities for sec θ and csc θ:

$$\frac{1-\sin\theta\cos\theta}{\cos\theta(\sec\theta-\csc\theta)}\bullet\frac{\sin^2\theta-\cos^2\theta}{\sin^3\theta+\cos^3\theta}=\frac{1-\sin\theta\cos\theta}{\cos\theta\left(\frac{1}{\cos\theta}-\frac{1}{\sin\theta}\right)}\bullet\frac{(\sin\theta+\cos\theta)(\sin\theta-\cos\theta)}{(\sin\theta+\cos\theta)(\sin^2\theta-\sin\theta\cos\theta+\cos^2\theta)}$$
$$=\frac{(1-\sin\theta\cos\theta)(\sin\theta\cos\theta)}{\cos\theta(\sin\theta-\cos\theta)}\bullet\frac{\sin\theta-\cos\theta}{1-\sin\theta\cos\theta}$$
$$=\sin\theta$$

49. Using the identity for cot A: $\dfrac{\cot A-1}{\cot A+1}=\dfrac{\frac{\cos A}{\sin A}-1}{\frac{\cos A}{\sin A}+1}=\dfrac{\cos A-\sin A}{\cos A+\sin A}$

51. Using the identity $\sin^2\theta=1-\cos^2\theta$: $\cos^2\theta-\sin^2\theta=\cos^2\theta-(1-\cos^2\theta)=2\cos^2\theta-1$

53. Using the identity for tan A and $\sin^2 A=1-\cos^2 A$: $\sin A\tan A=\sin A\bullet\dfrac{\sin A}{\cos A}=\dfrac{\sin^2 A}{\cos A}=\dfrac{1-\cos^2 A}{\cos A}$

55. Working from the right-hand side and using identities for tan A and tan B:

$$\frac{\tan A+\tan B}{\cot A+\cot B}=\frac{\frac{\sin A}{\cos A}+\frac{\sin B}{\cos B}}{\frac{\cos A}{\sin A}+\frac{\cos B}{\sin B}}=\frac{\frac{\sin A\cos B+\cos A\sin B}{\cos A\cos B}}{\frac{\cos A\sin B+\sin A\cos B}{\sin A\sin B}}=\frac{\sin A\sin B}{\cos A\cos B}=\frac{\sin A}{\cos A}\bullet\frac{\sin B}{\cos B}=\tan A\tan B$$

57. Simplifying the left-hand side and using the identity $\sin^2 A=1-\cos^2 A$:

$$\tan A-\frac{\sec A\sin^3 A}{1+\cos A}=\frac{\sin A}{\cos A}-\frac{\sin^3 A}{\cos A(1+\cos A)}$$
$$=\frac{(\sin A)(1+\cos A)-\sin^3 A}{\cos A(1+\cos A)}$$
$$=\frac{\sin A(1+\cos A)-\sin A(1-\cos^2 A)}{\cos A(1+\cos A)}$$
$$=\frac{\sin A-\sin A(1-\cos A)}{\cos A}$$
$$=\frac{\sin A-\sin A+\sin A\cos A}{\cos A}$$
$$=\frac{\sin A\cos A}{\cos A}$$
$$=\sin A$$

59. Simplifying the left-hand side and using the identities for $\cot A$ and $\csc A$:

$$\begin{aligned}\frac{1}{\csc A-\cot A}-\frac{1}{\csc A+\cot A}&=\frac{\csc A+\cot A-\csc A+\cot A}{(\csc A-\cot A)(\csc A+\cot A)}\\&=\frac{2\cot A}{\csc^2 A-\cot^2 A}\\&=\frac{2\cot A}{\frac{1}{\sin^2 A}-\frac{\cos^2 A}{\sin^2 A}}\\&=\frac{2\cot A\bullet\sin^2 A}{1-\cos^2 A}\\&=\frac{2\cot A\bullet\sin^2 A}{\sin^2 A}\\&=2\cot A\end{aligned}$$

61. Using the identities for $\sec A$ and $\csc A$: $\dfrac{\sec A-\csc A}{\sec A+\csc A}=\dfrac{\frac{1}{\cos A}-\frac{1}{\sin A}}{\frac{1}{\cos A}+\frac{1}{\sin A}}=\dfrac{\sin A-\cos A}{\sin A+\cos A}$

63. Working from the right-hand side and using the identity for $\tan A$:

$$\frac{\tan A}{1+\tan^2 A}=\frac{\frac{\sin A}{\cos A}}{1+\frac{\sin^2 A}{\cos^2 A}}=\frac{\sin A\cos A}{\cos^2 A+\sin^2 A}=\sin A\cos A$$

65. Using $s=r\theta$ and $A=\frac{1}{2}r^2\theta$:

$$s=r\theta=16\text{ cm}\bullet\tfrac{\pi}{8}=2\pi\text{ cm}$$
$$A=\tfrac{1}{2}r^2\theta=\tfrac{1}{2}(16\text{ cm})^2\bullet\tfrac{\pi}{8}=16\pi\text{ cm}^2$$

67. Since $s=r\theta$, $\theta=\frac{s}{r}=\frac{1\text{ cm}}{1\text{ cm}}=1$. Therefore: $A=\frac{1}{2}r^2\theta=\frac{1}{2}(1\text{ cm})^2\bullet 1=\frac{1}{2}\text{ cm}^2$

69. First convert 36° to radians: $36°=36°\bullet\frac{\pi}{180°}=\frac{\pi}{5}$ radians

Since $s=r\theta$: $r=\dfrac{s}{\theta}=\dfrac{4\text{ cm}}{\frac{\pi}{5}}=\dfrac{20}{\pi}$ radians. Now using $A=\frac{1}{2}r^2\theta$: $A=\frac{1}{2}r^2\theta=\frac{1}{2}\left(\frac{20}{\pi}\text{ cm}\right)^2\left(\frac{\pi}{5}\right)=\frac{40}{\pi}\text{ cm}^2$

71. Since $s=r\theta$:

$$\begin{aligned}s&=r\theta\\12&=r(r+1)\\r^2+r-12&=0\\(r+4)(r-3)&=0\\r&=-4,3\end{aligned}$$

Since r cannot be negative, we must have $r=3$ cm. Then $\theta=\dfrac{s}{r}=\dfrac{12\text{ cm}}{3\text{ cm}}=4$ radians.

73. Since the three angles must sum to 180°, the third angle is 180° – 40° – 70° = 70°. Now convert 70° to radians: $70°=70°\bullet\frac{\pi}{180°}=\frac{7\pi}{18}$ radians

75. Since P is in the second quadrant, $x < 0$ and thus: $x = -\sqrt{1-y^2} = -\sqrt{1-\left(\frac{2}{5}\right)^2} = -\sqrt{1-\frac{4}{25}} = -\sqrt{\frac{21}{25}} = -\frac{\sqrt{21}}{5}$

Using the definitions of the trigonometric functions:

$$\sin\theta = y = \frac{2}{5} \qquad \cos\theta = x = -\frac{\sqrt{21}}{5}$$

$$\tan\theta = \frac{y}{x} = \frac{\frac{2}{5}}{-\frac{\sqrt{21}}{5}} = -\frac{2}{\sqrt{21}} = -\frac{2\sqrt{21}}{21} \qquad \cot\theta = \frac{x}{y} = \frac{-\frac{\sqrt{21}}{5}}{\frac{2}{5}} = -\frac{\sqrt{21}}{2}$$

$$\sec\theta = \frac{1}{x} = \frac{1}{-\frac{\sqrt{21}}{5}} = -\frac{5}{\sqrt{21}} = -\frac{5\sqrt{21}}{21} \qquad \csc\theta = \frac{1}{y} = \frac{1}{\frac{2}{5}} = \frac{5}{2}$$

77. Since P is in the third quadrant, $y < 0$ and thus: $y = -\sqrt{1-x^2} = -\sqrt{1-\left(-\frac{5}{7}\right)^2} = -\sqrt{1-\frac{25}{49}} = -\sqrt{\frac{24}{49}} = -\frac{2\sqrt{6}}{7}$

Using the definitions of the trigonometric functions:

$$\sin\theta = y = -\frac{2\sqrt{6}}{7} \qquad \cos\theta = x = -\frac{5}{7}$$

$$\tan\theta = \frac{y}{x} = \frac{-\frac{2\sqrt{6}}{7}}{-\frac{5}{7}} = \frac{2\sqrt{6}}{5} \qquad \cot\theta = \frac{x}{y} = \frac{-\frac{5}{7}}{-\frac{2\sqrt{6}}{7}} = \frac{5}{2\sqrt{6}} = \frac{5\sqrt{6}}{12}$$

$$\sec\theta = \frac{1}{x} = \frac{1}{-\frac{5}{7}} = -\frac{7}{5} \qquad \csc\theta = \frac{1}{y} = \frac{1}{-\frac{2\sqrt{6}}{7}} = -\frac{7}{2\sqrt{6}} = -\frac{7\sqrt{6}}{12}$$

79. Since $\frac{\pi}{2} < \theta < \pi$, P is in the second quadrant and thus $y > 0$: $y = \sqrt{1-x^2} = \sqrt{1-\left(-\frac{15}{17}\right)^2} = \sqrt{1-\frac{225}{289}} = \sqrt{\frac{64}{289}} = \frac{8}{17}$

Using the definitions of the trigonometric functions:

$$\sin\theta = y = \frac{8}{17} \qquad \cos\theta = x = -\frac{15}{17}$$

$$\tan\theta = \frac{y}{x} = \frac{\frac{8}{17}}{-\frac{15}{17}} = -\frac{8}{15} \qquad \cot\theta = \frac{x}{y} = \frac{-\frac{15}{17}}{\frac{8}{17}} = -\frac{15}{8}$$

$$\sec\theta = \frac{1}{x} = \frac{1}{-\frac{15}{17}} = -\frac{17}{15} \qquad \csc\theta = \frac{1}{y} = \frac{1}{\frac{8}{17}} = \frac{17}{8}$$

81. Since $\frac{3\pi}{2} < \theta < 2\pi$, P is in the fourth quadrant and thus $x > 0$: $x = \sqrt{1-y^2} = \sqrt{1-\left(-\frac{7}{25}\right)^2} = \sqrt{1-\frac{49}{625}} = \sqrt{\frac{576}{625}} = \frac{24}{25}$

Using the definitions of the trigonometric functions:

$$\sin\theta = y = -\frac{7}{25} \qquad \cos\theta = x = \frac{24}{25}$$

$$\tan\theta = \frac{y}{x} = \frac{-\frac{7}{25}}{\frac{24}{25}} = -\frac{7}{24} \qquad \cot\theta = \frac{x}{y} = \frac{\frac{24}{25}}{-\frac{7}{25}} = -\frac{24}{7}$$

$$\sec\theta = \frac{1}{x} = \frac{1}{\frac{24}{25}} = \frac{25}{24} \qquad \csc\theta = \frac{1}{y} = \frac{1}{-\frac{7}{25}} = -\frac{25}{7}$$

83. Three such angles are $-\frac{5\pi}{6}$, $\frac{5\pi}{6}$ and $\frac{7\pi}{6}$ (other answers are possible).

85. Four such angles are –450°, –90°, 270°, and 630° (other answers are possible).

87. Since $\pi < \theta < \frac{3\pi}{2}$, $\sin\theta < 0$ and thus:

$$\sin\theta = -\sqrt{1-\cos^2\theta} = -\sqrt{1-\left(-\tfrac{5}{13}\right)^2} = -\sqrt{1-\tfrac{25}{169}} = -\sqrt{\tfrac{144}{169}} = -\tfrac{12}{13}$$

$$\tan\theta = \frac{\sin\theta}{\cos\theta} = \frac{-\frac{12}{13}}{-\frac{5}{13}} = \frac{12}{5}$$

89. The x-coordinate of P is approximately 0.9848, so $\cos 10° \approx 0.9848$.

91. The y-coordinate of P is approximately 0.1736, so $\sin 10° \approx 0.1736$.

Chapter 6 Test

1. **a.** Since $-\frac{\pi}{2}$ corresponds to the point $(0,-1)$ on the unit circle, $\sin\left(-\frac{\pi}{2}\right) = -1$.

b. Since 540° corresponds to the point $(-1,0)$ on the unit circle, $\cos 540° = -1$.

c. Since 450° corresponds to the point $(0, 1)$ on the unit circle: $\cot 450° = \frac{x}{y} = \frac{0}{1} = 0$

2. **a.** $\cos\frac{\pi}{6} = \frac{\sqrt{3}}{2}$

b. $\sin 45° = \frac{\sqrt{2}}{2}$

c. Since $\sin^2\theta + \cos^2\theta = 1$ for any value of θ, $\sin^2 7° + \cos^2 7° = 1$.

3. Since $\frac{3\pi}{2} < \theta < 2\pi$ is the fourth quadrant, $\cos\theta > 0$ and thus:

$$\cos\theta = \sqrt{1-\sin^2\theta} = \sqrt{1-\left(-\tfrac{1}{5}\right)^2} = \sqrt{1-\tfrac{1}{25}} = \sqrt{\tfrac{24}{25}} = \frac{2\sqrt{6}}{5}$$

The other four trigonometric functions are:

$$\tan\theta = \frac{\sin\theta}{\cos\theta} = \frac{-\frac{1}{5}}{\frac{2\sqrt{6}}{5}} = -\frac{1}{2\sqrt{6}} = -\frac{\sqrt{6}}{12} \qquad \cot\theta = \frac{\cos\theta}{\sin\theta} = \frac{\frac{2\sqrt{6}}{5}}{-\frac{1}{5}} = -2\sqrt{6}$$

$$\sec\theta = \frac{1}{\cos\theta} = \frac{1}{\frac{2\sqrt{6}}{5}} = \frac{5}{2\sqrt{6}} = \frac{5\sqrt{6}}{12} \qquad \csc\theta = \frac{1}{\sin\theta} = \frac{1}{-\frac{1}{5}} = -5$$

4. Since $90° < \theta < 180°$ is the second quadrant, $\sin\theta > 0$ and thus:

$$\sin\theta = \sqrt{1-\cos^2\theta} = \sqrt{1-\left(-\frac{\sqrt{5}}{6}\right)^2} = \sqrt{1-\tfrac{5}{36}} = \sqrt{\tfrac{31}{36}} = \frac{\sqrt{31}}{6}$$

The other four trigonometric functions are:

$$\tan\theta = \frac{\sin\theta}{\cos\theta} = \frac{\frac{\sqrt{31}}{6}}{-\frac{\sqrt{5}}{6}} = -\frac{\sqrt{31}}{\sqrt{5}} = -\frac{\sqrt{155}}{5} \qquad \cot\theta = \frac{\cos\theta}{\sin\theta} = \frac{-\frac{\sqrt{5}}{6}}{\frac{\sqrt{31}}{6}} = -\frac{\sqrt{5}}{\sqrt{31}} = -\frac{\sqrt{155}}{31}$$

$$\sec\theta = \frac{1}{\cos\theta} = \frac{1}{-\frac{\sqrt{5}}{6}} = -\frac{6}{\sqrt{5}} = -\frac{6\sqrt{5}}{5} \qquad \csc\theta = \frac{1}{\sin\theta} = \frac{1}{\frac{\sqrt{31}}{6}} = \frac{6}{\sqrt{31}} = \frac{6\sqrt{31}}{31}$$

5. Since the terminal side of $\frac{5\pi}{3}$ lies in the fourth quadrant, $\sin\frac{5\pi}{3} < 0$. Using a reference angle of $\frac{\pi}{3}$:

$$\sin\tfrac{5\pi}{3} = -\sin\tfrac{\pi}{3} = -\frac{\sqrt{3}}{2}$$

6. Since the terminal side of $-\frac{5\pi}{4}$ lies in the second quadrant, $\cot\left(-\frac{5\pi}{4}\right) < 0$. Using a reference angle of $\frac{\pi}{4}$:

$$\cot\left(-\tfrac{5\pi}{4}\right) = -\cot\tfrac{\pi}{4} = -1$$

7. Since the terminal side of 300° lies in the fourth quadrant, $\cos 300° > 0$. Using a reference angle of 60°:

$$\cos 300° = \cos 60° = \frac{1}{2}$$

8. Since the terminal side of –135° lies in the third quadrant, $\csc(-135°) < 0$. Using a reference angle of 45°:

$$\csc(-135°) = -\csc 45° = -\frac{1}{\sin 45°} = -\frac{1}{\frac{\sqrt{2}}{2}} = -\sqrt{2}$$

9. Since 2 radians lies in the second quadrant, $\sin 2 > 0$ while $\cos 2 < 0$. So $\sin 2$ is larger than $\cos 2$.

10. a. Converting 165° to radian measure: $165° = 165° \bullet \frac{\pi}{180°} = \frac{11\pi}{12}$ radians

b. Converting 3 radians to degree measure: $3 \text{ radians} = 3 \bullet \frac{180°}{\pi} = \frac{540°}{\pi}$

11. First convert 75° to radian measure: $75° = 75° \bullet \frac{\pi}{180°} = \frac{5\pi}{12}$ radians

Now using the arc length formula: $s = r\theta = 5 \text{ cm} \bullet \frac{5\pi}{12} = \frac{25\pi}{12}$ cm

12. Using the formula for the area of a sector: $A = \frac{1}{2}r^2\theta = \frac{1}{2}(5 \text{ cm})^2 \bullet \frac{5\pi}{12} = \frac{125\pi}{24}$ cm^2

13. Multiplying both numerator and denominator by $\tan\theta$:

$$\frac{1+\frac{\tan\theta+1}{\tan\theta}}{-1+\frac{\tan\theta-1}{\tan\theta}} \bullet \frac{\tan\theta}{\tan\theta} = \frac{\tan\theta+\tan\theta+1}{-\tan\theta+\tan\theta-1} = \frac{2\tan\theta+1}{-1} = -2\tan\theta-1$$

14. Working from the left-hand side and using identities for $\csc\theta$ and $\cot\theta$:

$$\frac{\cos\theta+1}{\csc\theta+\cot\theta} = \frac{\cos\theta+1}{\frac{1}{\sin\theta}+\frac{\cos\theta}{\sin\theta}} \bullet \frac{\sin\theta}{\sin\theta} = \frac{\sin\theta(\cos\theta+1)}{\cos\theta+1} = \sin\theta$$

15. a. The angular speed of the smaller wheel is given by: $\omega = 600 \frac{\text{rev}}{\text{min}} \bullet 2\pi \frac{\text{rad}}{\text{rev}} = 1200\pi \frac{\text{radians}}{\text{min}}$

b. Since $v = r\omega$ and $r = 15$ cm: $v = r\omega = 15 \text{ cm} \bullet 1200\pi \frac{\text{rad}}{\text{min}} = 18000\pi \frac{\text{cm}}{\text{min}}$

c. Since the linear speed on the larger wheel must also be $18000\pi \frac{\text{cm}}{\text{min}}$, and the radius of the larger wheel is $R = 25$ cm, its angular speed is: $\omega = \frac{v}{r} = \frac{18000\pi \text{ cm/min}}{25 \text{ cm}} = 720\pi \frac{\text{radians}}{\text{min}}$

d. In rpm, the angular speed of the larger wheel is: $\frac{1 \text{ rev}}{2\pi \text{ rad}} \bullet 720\pi \frac{\text{rad}}{\text{min}} = 360$ rpm

16. Given $\tan\theta = t$, we can construct the right triangle:

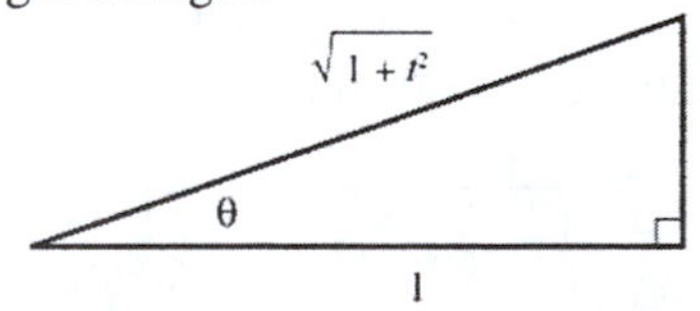

The hypotenuse was found using the Pythagorean theorem. Using the right-triangle definitions for the trigonometric functions:

$$\sin\theta = \frac{t}{\sqrt{1+t^2}} = \frac{t\sqrt{1+t^2}}{1+t^2} \qquad \cos\theta = \frac{1}{\sqrt{1+t^2}} = \frac{\sqrt{1+t^2}}{1+t^2}$$

$$\sec\theta = \frac{\sqrt{1+t^2}}{1} = \sqrt{1+t^2} \qquad \csc\theta = \frac{\sqrt{1+t^2}}{t}$$

$$\cot\theta = \frac{1}{t}$$

17. Draw ΔABC, labeling appropriate sides:

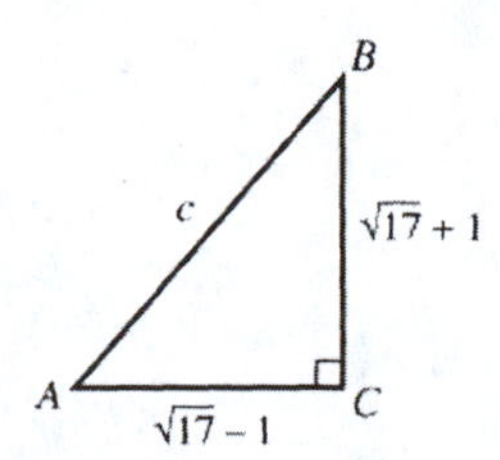

Find c using the Pythagorean theorem:

$$c^2 = \left(\sqrt{17}-1\right)^2 + \left(\sqrt{17}+1\right)^2 = \left(17-2\sqrt{17}+1\right)+\left(17+2\sqrt{17}+1\right) = 36$$
$$c = \sqrt{36} = 6$$

Therefore $\cos A = \dfrac{\sqrt{17}-1}{6}$.

18. **a.** Using the right-triangle definitions, we know $\tan(\angle DAB) = \dfrac{DB}{AB}$ while $\tan(\angle CAB) = \dfrac{CB}{AB}$. Clearly $CB > DB$, so $\tan(\angle CAB)$ is larger than $\tan(\angle DAB)$.

b. Using the right-triangle definitions, we know $\cos(\angle DAB) = \dfrac{AB}{AD}$ while $\cos(\angle CAB) = \dfrac{AB}{AC}$. But $AD < AC$, so $\cos(\angle DAB)$ is larger than $\cos(\angle CAB)$.

Chapter 7
Graphs of the Trigonometric Functions

7.1 Trigonometric Functions of Real Numbers

1. **a.** The reference angle for $\frac{11\pi}{6}$ is $\frac{\pi}{6}$, and $\cos\frac{\pi}{6} = \frac{\sqrt{3}}{2}$. The terminal side of $\frac{11\pi}{6}$ lies in the fourth quadrant where the x-coordinate is positive, so: $\cos\frac{11\pi}{6} = \cos\frac{\pi}{6} = \frac{\sqrt{3}}{2}$

   b. The reference angle for $-\frac{11\pi}{6}$ is $\frac{\pi}{6}$, and $\cos\frac{\pi}{6} = \frac{\sqrt{3}}{2}$. The terminal side of $-\frac{11\pi}{6}$ lies in the first quadrant where the x-coordinate is positive, so: $\cos\left(-\frac{11\pi}{6}\right) = \cos\frac{\pi}{6} = \frac{\sqrt{3}}{2}$

   c. The reference angle for $\frac{11\pi}{6}$ is $\frac{\pi}{6}$, and $\sin\frac{\pi}{6} = \frac{1}{2}$. The terminal side of $\frac{11\pi}{6}$ lies in the fourth quadrant where the y-coordinate is negative, so: $\sin\frac{11\pi}{6} = -\sin\frac{\pi}{6} = -\frac{1}{2}$

   d. The reference angle for $-\frac{11\pi}{6}$ is $\frac{\pi}{6}$, and $\sin\frac{\pi}{6} = \frac{1}{2}$. The terminal side of $-\frac{11\pi}{6}$ lies in the first quadrant where the y-coordinate is positive, so: $\sin\left(-\frac{11\pi}{6}\right) = \sin\frac{\pi}{6} = \frac{1}{2}$

3. **a.** The reference angle for $\frac{\pi}{6}$ is $\frac{\pi}{6}$, and $\cos\frac{\pi}{6} = \frac{\sqrt{3}}{2}$. The terminal side of $\frac{\pi}{6}$ lies in the first quadrant where the x-coordinate is positive, so: $\cos\frac{\pi}{6} = \frac{\sqrt{3}}{2}$

   b. The reference angle for $-\frac{\pi}{6}$ is $\frac{\pi}{6}$, and $\cos\frac{\pi}{6} = \frac{\sqrt{3}}{2}$. The terminal side of $-\frac{\pi}{6}$ lies in the fourth quadrant where the x-coordinate is positive, so: $\cos\left(-\frac{\pi}{6}\right) = \cos\frac{\pi}{6} = \frac{\sqrt{3}}{2}$

   c. The reference angle for $\frac{\pi}{6}$ is $\frac{\pi}{6}$, and $\sin\frac{\pi}{6} = \frac{1}{2}$. The terminal side of $\frac{\pi}{6}$ lies in the first quadrant where the y-coordinate is positive, so: $\sin\frac{\pi}{6} = \frac{1}{2}$

d. The reference angle for $-\frac{\pi}{6}$ is $\frac{\pi}{6}$, and $\sin\frac{\pi}{6}=\frac{1}{2}$. The terminal side of $-\frac{\pi}{6}$ lies in the fourth quadrant where the y-coordinate is negative, so: $\sin\left(-\frac{\pi}{6}\right)=-\sin\frac{\pi}{6}=-\frac{1}{2}$

5. **a.** The reference angle for $\frac{5\pi}{4}$ is $\frac{\pi}{4}$, and $\cos\frac{\pi}{4}=\frac{\sqrt{2}}{2}$. The terminal side of $\frac{5\pi}{4}$ lies in the third quadrant where the x-coordinate is negative, so: $\cos\frac{5\pi}{4}=-\cos\frac{\pi}{4}=-\frac{\sqrt{2}}{2}$

b. The reference angle for $-\frac{5\pi}{4}$ is $\frac{\pi}{4}$, and $\cos\frac{\pi}{4}=\frac{\sqrt{2}}{2}$. The terminal side of $-\frac{5\pi}{4}$ lies in the second quadrant where the x-coordinate is negative, so: $\cos\left(-\frac{5\pi}{4}\right)=-\cos\frac{\pi}{4}=-\frac{\sqrt{2}}{2}$

c. The reference angle for $\frac{5\pi}{4}$ is $\frac{\pi}{4}$, and $\sin\frac{\pi}{4}=\frac{\sqrt{2}}{2}$. The terminal side of $\frac{5\pi}{4}$ lies in the third quadrant where the y-coordinate is negative, so: $\sin\frac{5\pi}{4}=-\sin\frac{\pi}{4}=-\frac{\sqrt{2}}{2}$

d. The reference angle for $-\frac{5\pi}{4}$ is $\frac{\pi}{4}$, and $\sin\frac{\pi}{4}=\frac{\sqrt{2}}{2}$. The terminal side of $-\frac{5\pi}{4}$ lies in the second quadrant where the y-coordinate is positive, so: $\sin\left(-\frac{5\pi}{4}\right)=\sin\frac{\pi}{4}=\frac{\sqrt{2}}{2}$

7. **a.** The reference angle for $\frac{5\pi}{3}$ is $\frac{\pi}{3}$, and $\sec\frac{\pi}{3}=2$. The terminal side of $\frac{5\pi}{3}$ lies in the fourth quadrant where the x-coordinate is positive, so: $\sec\frac{5\pi}{3}=\sec\frac{\pi}{3}=2$

b. The reference angle for $-\frac{5\pi}{3}$ is $\frac{\pi}{3}$, and $\csc\frac{\pi}{3}=\frac{2\sqrt{3}}{3}$. The terminal side of $-\frac{5\pi}{3}$ lies in the first quadrant where the y-coordinate is positive, so: $\csc\left(-\frac{5\pi}{3}\right)=\csc\frac{\pi}{3}=\frac{2\sqrt{3}}{3}$

c. The reference angle for $\frac{5\pi}{3}$ is $\frac{\pi}{3}$, and $\tan\frac{\pi}{3}=\sqrt{3}$. The terminal side of $\frac{5\pi}{3}$ lies in the fourth quadrant where the x-coordinate is positive while the y-coordinate is negative, so: $\tan\frac{5\pi}{3}=-\tan\frac{\pi}{3}=-\sqrt{3}$

d. The reference angle for $-\frac{5\pi}{3}$ is $\frac{\pi}{3}$, and $\cot\frac{\pi}{3}=\frac{\sqrt{3}}{3}$. The terminal side of $-\frac{5\pi}{3}$ lies in the first quadrant where both the x- and y-coordinates are positive, so: $\cot\left(-\frac{5\pi}{3}\right)=\cot\frac{\pi}{3}=\frac{\sqrt{3}}{3}$

9. **a.** Choose positive radian values such that the x-coordinate of the point on the unit circle is 0. Four such values are $t=\frac{\pi}{2},\frac{3\pi}{2},\frac{5\pi}{2}$, and $\frac{7\pi}{2}$ (other answers are possible).

b. Choose negative radian values such that the x-coordinate of the point on the unit circle is 0. Four such values are $t=-\frac{\pi}{2},-\frac{3\pi}{2},-\frac{5\pi}{2}$, and $-\frac{7\pi}{2}$ (other answers are possible).

11. **a.** Using a calculator to two decimal place accuracy:

$\sin 2.06\approx 0.88$	$\cos 2.06\approx -0.47$	$\tan 2.06\approx -1.88$
$\sec 2.06\approx -2.13$	$\csc 2.06\approx 1.13$	$\cot 2.06\approx -0.53$

b. Using a calculator to two decimal place accuracy:

$\sin(-2.06)\approx -0.88$	$\cos(-2.06)\approx -0.47$	$\tan(-2.06)\approx 1.88$
$\sec(-2.06)\approx -2.13$	$\csc(-2.06)\approx -1.13$	$\cot(-2.06)\approx 0.53$

13. **a.** Using a calculator to two decimal place accuracy:

$\sin\frac{\pi}{6}\approx 0.50$	$\cos\frac{\pi}{6}\approx 0.87$	$\tan\frac{\pi}{6}\approx 0.58$
$\sec\frac{\pi}{6}\approx 1.15$	$\csc\frac{\pi}{6}\approx 2.00$	$\cot\frac{\pi}{6}\approx 1.73$

b. Since $\frac{\pi}{6}+2\pi$ will intersect the unit circle at the same location as in part **a**, all six trigonometric functions will have the same values as in part **a**:

$$\sin\left(\tfrac{\pi}{6}+2\pi\right)\approx 0.50 \qquad \cos\left(\tfrac{\pi}{6}+2\pi\right)\approx 0.87 \qquad \tan\left(\tfrac{\pi}{6}+2\pi\right)\approx 0.58$$

$$\sec\left(\tfrac{\pi}{6}+2\pi\right)\approx 1.15 \qquad \csc\left(\tfrac{\pi}{6}+2\pi\right)\approx 2.00 \qquad \cot\left(\tfrac{\pi}{6}+2\pi\right)\approx 1.73$$

15. **a.** Checking the identity $\sin^2 t+\cos^2 t=1$: $\sin^2\frac{\pi}{3}+\cos^2\frac{\pi}{3}=\left(\frac{\sqrt{3}}{2}\right)^2+\left(\frac{1}{2}\right)^2=\frac{3}{4}+\frac{1}{4}=1$

b. Checking the identity $\sin^2 t+\cos^2 t=1$: $\sin^2\frac{5\pi}{4}+\cos^2\frac{5\pi}{4}=\left(-\frac{\sqrt{2}}{2}\right)^2+\left(-\frac{\sqrt{2}}{2}\right)^2=\frac{2}{4}+\frac{2}{4}=1$

c. Checking the identity $\sin^2 t+\cos^2 t=1$ and using a calculator:

$$\sin^2(-53)+\cos^2(-53)\approx(-0.3959)^2+(-0.9183)^2\approx 0.1568+0.8432=1$$

17. **a.** Checking the identity $\cot^2 t+1=\csc^2 t$:

$$\cot^2\left(-\tfrac{\pi}{6}\right)+1=\left(-\sqrt{3}\right)^2+1=3+1=4 \qquad \csc^2\left(-\tfrac{\pi}{6}\right)=(-2)^2=4$$

b. Checking the identity $\cot^2 t+1=\csc^2 t$:

$$\cot^2\left(\tfrac{7\pi}{4}\right)+1=(-1)^2+1=1+1=2 \qquad \csc^2\left(\tfrac{7\pi}{4}\right)=\left(-\sqrt{2}\right)^2=2$$

c. Checking the identity $\cot^2 t+1=\csc^2 t$ and using a calculator:

$$\cot^2(0.12)+1\approx 68.7787+1=69.7787 \qquad \csc^2(0.12)\approx 69.7787$$

19. **a.** Checking the identity $\sin(-t)=-\sin t$:

$$\sin\left(-\tfrac{3\pi}{2}\right)=1 \qquad -\sin\tfrac{3\pi}{2}=-(-1)=1$$

b. Checking the identity $\sin(-t)=-\sin t$:

$$\sin\tfrac{5\pi}{6}=\tfrac{1}{2} \qquad -\sin\left(-\tfrac{5\pi}{6}\right)=-\left(-\tfrac{1}{2}\right)=\tfrac{1}{2}$$

c. Checking the identity $\sin(-t)=-\sin t$, and using a calculator:

$$\sin(-13.24)\approx -0.6238 \qquad -\sin 13.24\approx-(0.6238)=-0.6238$$

21. **a.** Checking the identity $\sin(t+2\pi)=\sin t$:

$$\sin\left(\tfrac{5\pi}{3}+2\pi\right)=\sin\left(\tfrac{11\pi}{3}\right)=-\frac{\sqrt{3}}{2} \qquad \sin\tfrac{5\pi}{3}=-\frac{\sqrt{3}}{2}$$

b. Checking the identity $\sin(t+2\pi)=\sin t$:

$$\sin\left(-\tfrac{3\pi}{2}+2\pi\right)=\sin\tfrac{\pi}{2}=1 \qquad \sin\left(-\tfrac{3\pi}{2}\right)=1$$

c. Checking the identity $\sin(t+2\pi)=\sin t$ and using a calculator:

$$\sin\left(\sqrt{19}+2\pi\right)\approx\sin(10.6421)\approx -0.9382 \qquad \sin\left(\sqrt{19}\right)\approx\sin(4.3589)\approx -0.9382$$

23. Evaluating each side of the "equality" when $t=\frac{\pi}{6}$:

$$\cos\left(2\bullet\tfrac{\pi}{6}\right)=\cos\tfrac{\pi}{3}=\tfrac{1}{2} \qquad 2\cos\tfrac{\pi}{6}=2\bullet\tfrac{\sqrt{3}}{2}=\sqrt{3}$$

Since these results are unequal, $\cos 2t=2\cos t$ is not an identity.

25. Using $\sin^2 t+\cos^2 t=1$: $\cos t=\pm\sqrt{1-\sin^2 t}=\pm\sqrt{1-\frac{9}{25}}=\pm\frac{4}{5}$

Since $\pi<t<\frac{3\pi}{2}$, $\cos t=-\frac{4}{5}$ and thus: $\tan t=\dfrac{\sin t}{\cos t}=\dfrac{-\frac{3}{5}}{-\frac{4}{5}}=\dfrac{3}{4}$

27. Since $\frac{\pi}{2} < t < \pi$, $\cos t < 0$ and thus: $\cos t = -\sqrt{1-\sin^2 t} = -\sqrt{1-\frac{3}{16}} = -\sqrt{\frac{13}{16}} = -\frac{\sqrt{13}}{4}$

Thus: $\tan t = \frac{\sin t}{\cos t} = \frac{\frac{\sqrt{3}}{4}}{-\frac{\sqrt{13}}{4}} = -\frac{\sqrt{3}}{\sqrt{13}} = -\frac{\sqrt{39}}{13}$

29. Using the identity $1+\tan^2\alpha = \sec^2\alpha$:

$$\begin{aligned} 1+\left(\tfrac{12}{5}\right)^2 &= \sec^2\alpha \\ 1+\tfrac{144}{25} &= \sec^2\alpha \\ \tfrac{169}{25} &= \sec^2\alpha \\ \pm\tfrac{13}{5} &= \sec\alpha \end{aligned}$$

Since $\cos\alpha > 0$, $\sec\alpha > 0$, pick $\sec\alpha = \frac{13}{5}$. Thus $\cos\alpha = \frac{5}{13}$, and use $\sin^2\alpha + \cos^2\alpha = 1$ to get:

$$\sin^2\alpha = 1-\left(\tfrac{5}{13}\right)^2 = \tfrac{144}{169}, \text{ thus } \sin\alpha = \pm\tfrac{12}{13}$$

Pick the positive value since, if both the tangent and cosine are positive, so is the sine. Thus $\sin\alpha = \frac{12}{13}$.

31. For $0 < \theta < \frac{\pi}{2}$: $\sqrt{9-x^2} = \sqrt{9-(3\sin\theta)^2} = \sqrt{9\left(1-\sin^2\theta\right)} = \sqrt{9\cos^2\theta} = 3\cos\theta$

Choose the positive root since $\cos\theta > 0$ for $0 < \theta < \frac{\pi}{2}$.

33. Since $0 < \theta < \frac{\pi}{2}$, $\tan\theta > 0$. Thus: $\frac{1}{\left(u^2-25\right)^{3/2}} = \frac{1}{\left(25\sec^2\theta - 25\right)^{3/2}} = \frac{1}{125\tan^3\theta} = \frac{\cot^3\theta}{125}$

35. Since $\sec\theta > 0$: $\frac{1}{\sqrt{u^2+7}} = \frac{1}{\sqrt{7\tan^2\theta+7}} = \frac{1}{\sqrt{7}\sec\theta} = \frac{\cos\theta}{\sqrt{7}} = \frac{\sqrt{7}\cos\theta}{7}$

37.
a. Since $\sin(-t) = -\sin t$: $\sin(-t) = -\sin t = -\frac{2}{3}$

b. Since $\sin(-\phi) = -\sin\phi$: $\sin(-\phi) = -\sin\phi = -\left(-\frac{1}{4}\right) = \frac{1}{4}$

c. Since $\cos(-\alpha) = \cos\alpha$: $\cos(-\alpha) = \cos\alpha = \frac{1}{5}$

d. Since $\cos(-s) = \cos s$: $\cos(-s) = \cos s = -\frac{1}{5}$

39.
a. Since $\cos t = -\frac{1}{3}$ and $\frac{\pi}{2} < t < \pi$, $\sin t > 0$ and thus: $\sin t = \sqrt{1-\frac{1}{9}} = \sqrt{\frac{8}{9}} = \frac{2\sqrt{2}}{3}$

Therefore the values are:

$$\sin(-t) = -\sin t = -\frac{2\sqrt{2}}{3} \qquad \cos(-t) = \cos t = -\tfrac{1}{3}$$

Thus: $\sin(-t)+\cos(-t) = -\frac{2\sqrt{2}}{3} - \frac{1}{3} = -\frac{1+2\sqrt{2}}{3}$

b. Note that $\sin^2(-t)+\cos^2(-t) = 1$, regardless of the value of t.

41.
a. Using the identity $\cos(t+2\pi k) = \cos t$: $\cos\left(\frac{\pi}{4}+2\pi\right) = \cos\frac{\pi}{4} = \frac{\sqrt{2}}{2}$

b. Using the identity $\sin(t+2\pi k) = \sin t$: $\sin\left(\frac{\pi}{3}+2\pi\right) = \sin\frac{\pi}{3} = \frac{\sqrt{3}}{2}$

c. Using the identity $\sin(t+2\pi k) = \sin t$: $\sin\left(\frac{\pi}{2}-6\pi\right) = \sin\frac{\pi}{2} = 1$

43. Using the Pythagorean identities: $\frac{\sin^2 t+\cos^2 t}{\tan^2 t+1} = \frac{1}{\sec^2 t} = \cos^2 t$

45. Using the Pythagorean identities: $\dfrac{\sec^2\theta-\tan^2\theta}{1+\cot^2\theta}=\dfrac{\tan^2\theta+1-\tan^2\theta}{\csc^2\theta}=\dfrac{1}{\csc^2\theta}=\sin^2\theta$

47. Working from the right-hand side and using the identity for $\cot t$:

$$\sin t+\cot t\cos t=\sin t+\frac{\cos t}{\sin t}(\cos t)=\frac{\sin^2 t}{\sin t}+\frac{\cos^2 t}{\sin t}=\frac{\sin^2 t+\cos^2 t}{\sin t}=\frac{1}{\sin t}=\csc t$$

49. Combining fractions on the left-hand side:

$$\frac{1}{1+\sec s}+\frac{1}{1-\sec s}=\frac{(1-\sec s)+(1+\sec s)}{(1-\sec s)(1+\sec s)}=\frac{2}{1-\sec^2 s}=\frac{-2}{\tan^2 s}=-2\cot^2 s$$

51. Using the identities for $\sec s$, $\cot s$ and $\csc s$:

$$\frac{\sec s+\cot s\csc s}{\cos s}=\frac{\frac{1}{\cos s}+\frac{\cos s}{\sin s}\bullet\frac{1}{\sin s}}{\cos s}=\frac{1}{\cos^2 s}+\frac{1}{\sin^2 s}=\frac{\sin^2 s+\cos^2 s}{\cos^2 s\sin^2 s}=\frac{1}{\cos^2 s\sin^2 s}=\sec^2 s\csc^2 s$$

53. The expression on the left-hand side becomes:

$$\begin{aligned}\cos^2\alpha\cos^2\beta-\sin^2\alpha\sin^2\beta&=\left(\cos^2\alpha\right)\left(1-\sin^2\beta\right)-\left(1-\cos^2\alpha\right)\left(\sin^2\beta\right)\\&=\cos^2\alpha-\cos^2\alpha\sin^2\beta-\sin^2\beta+\cos^2\alpha\sin^2\beta\\&=\cos^2\alpha-\sin^2\beta\end{aligned}$$

55. If $\sec t=\frac{13}{5}$, then $\cos t=\frac{5}{13}$. Given $\frac{3\pi}{2}<t<2\pi$, then $\sin t<0$ (fourth quadrant), thus:

$$\sin t=-\sqrt{1-\cos^2 t}=-\sqrt{1-\tfrac{25}{169}}=-\sqrt{\tfrac{144}{169}}=-\tfrac{12}{13}$$

Therefore: $\dfrac{2\sin t-3\cos t}{4\sin t-9\cos t}=\dfrac{2\left(-\frac{12}{13}\right)-3\left(\frac{5}{13}\right)}{4\left(-\frac{12}{13}\right)-9\left(\frac{5}{13}\right)}=\dfrac{-24-15}{-48-45}=\dfrac{-39}{-93}=\dfrac{13}{31}$

57. Assuming that the radius of the circle is 1, the coordinates of the point labeled (x, y) are $(\cos t, \sin t)$, and the coordinates of the point labeled $(-x, -y)$ are $(\cos(t+\pi),\sin(t+\pi))$. So $y=\sin t$ and $-y=\sin(t+\pi)$, from which it follows that $\sin(t+\pi)=-\sin t$. Similarly, $-x=\cos(t+\pi)$ and $x=\cos t$, from which it follows that $\cos(t+\pi)=-\cos t$. Since $t-\pi$ results in the same intersection point with the unit circle as $t+\pi$, identities (ii) and (iv) follow in a similar manner.

59. The substitutions are $x=\frac{\sqrt{2}}{2}(X-Y)$ and $y=\frac{\sqrt{2}}{2}(X+Y)$. Thus:

$$x^2=\tfrac{1}{2}\left(X^2-2XY+Y^2\right)\text{ and }y^2=\tfrac{1}{2}\left(X^2+2XY+Y^2\right)$$

Squaring again, we obtain:

$$\begin{aligned}x^4&=\tfrac{1}{4}\left(X^4-4X^3Y+6X^2Y^2-4XY^3+Y^4\right)\\y^4&=\tfrac{1}{4}\left(X^4+4X^3Y+6X^2Y^2+4XY^3+Y^4\right)\end{aligned}$$

Also find $x^2y^2=\frac{1}{4}\left(X^2-Y^2\right)^2=\frac{1}{4}\left(X^4-2X^2Y^2+Y^4\right)$. Now use the expressions that have been found for x^2, y^2, x^4, and y^4 to substitute in the expression $x^4+6x^2y^2+y^4$ to obtain:

$$\tfrac{X^4}{4}-X^3Y+\tfrac{3}{2}X^2Y^2-XY^3+\tfrac{Y^4}{4}+\tfrac{3}{2}X^4-3X^2Y^2+\tfrac{3}{2}Y^4+\tfrac{X^4}{4}+X^3Y+\tfrac{3}{2}X^2Y^2+XY^3+\tfrac{Y^4}{4}$$

which is $2X^4+2Y^4$. In light of this result, the equation $x^4+6x^2y^2+y^4=32$ is equivalent to $2X^4+2Y^4=32$, or $X^4+Y^4=16$, as required.

61. **a.** Complete the table:

t	0.2	0.4	0.6	0.8	1.0	1.2	1.4
$f(t)$	219.07	50.53	19.70	9.55	6.14	7.98	33.88

b. The smallest output is 6.14, which occurs at $t = 1.0$.

c. Work from the right-hand side:

$$\begin{aligned}(\tan t - 3\cot t)^2 + 6 &= \tan^2 t - 6\tan t\cot t + 9\cot^2 t + 6\\ &= \tan^2 t - 6 + 9\cot^2 t + 6\\ &= \tan^2 t + 9\cot^2 t\end{aligned}$$

d. Since $(\tan t - 3\cot t)^2 \geq 0$: $\tan^2 t + 9\cot^2 t = (\tan t - 3\cot t)^2 + 6 \geq 0 + 6 \geq 6$

e. The minimum will occur when:

$$\begin{aligned}\tan t - 3\cot t &= 0\\ \tan t &= 3\cot t\\ \tan t &= \frac{3}{\tan t}\\ \tan^2 t &= 3\\ \tan t &= \sqrt{3} \text{ (since } 0 < t < \pi/2\text{, then } \tan t > 0)\\ t &= \tfrac{\pi}{3} \approx 1.05\end{aligned}$$

The answer from part **b** is consistent with this result.

63. **a.** When $t = \frac{\pi}{6}$:

$$2\sin^2\tfrac{\pi}{6} - \sin\tfrac{\pi}{6} = 2\left(\tfrac{1}{2}\right)^2 - \tfrac{1}{2} = \tfrac{1}{2} - \tfrac{1}{2} = 0$$

$$2\sin\tfrac{\pi}{6}\cos\tfrac{\pi}{6} - \cos\tfrac{\pi}{6} = 2\bullet\tfrac{1}{2}\bullet\frac{\sqrt{3}}{2} - \frac{\sqrt{3}}{2} = \frac{\sqrt{3}}{2} - \frac{\sqrt{3}}{2} = 0$$

b. When $t = \frac{\pi}{4}$:

$$2\sin^2\tfrac{\pi}{4} - \sin\tfrac{\pi}{4} = 2\left(\frac{\sqrt{2}}{2}\right)^2 - \frac{\sqrt{2}}{2} = 1 - \frac{\sqrt{2}}{2} = \frac{2-\sqrt{2}}{2}$$

$$2\sin\tfrac{\pi}{4}\cos\tfrac{\pi}{4} - \cos\tfrac{\pi}{4} = 2\bullet\frac{\sqrt{2}}{2}\bullet\frac{\sqrt{2}}{2} - \frac{\sqrt{2}}{2} = 1 - \frac{\sqrt{2}}{2} = \frac{2-\sqrt{2}}{2}$$

c. No. Using the value $t = 0$:

$$2\sin^2 0 - \sin 0 = 2(0)^2 - 0 = 0 - 0 = 0$$

$$2\sin 0\cos 0 - \cos 0 = 2(0)(1) - 1 = 0 - 1 = -1$$

Since the two sides of the equation are not equal, the given equation is not an identity.

65. **a.** Complete the table:

t	$1-\frac{1}{2}t^2$	$\cos\theta$
0.02	0.9998	0.999800
0.05	0.99875	0.998750
0.1	0.995	0.995004
0.2	0.980	0.980067
0.3	0.955	0.955336

67. **a.** Complete the table:

x	$\frac{1}{3}x^3+x$	$\frac{2}{15}x^5+\frac{1}{3}x^3+x$	$\tan x$
0.1	0.100333	0.100335	0.100335
0.2	0.202667	0.202709	0.202710
0.3	0.309	0.309324	0.309336
0.4	0.421333	0.422699	0.422793
0.5	0.541667	0.545833	0.546302

b. Sketching the graphs:

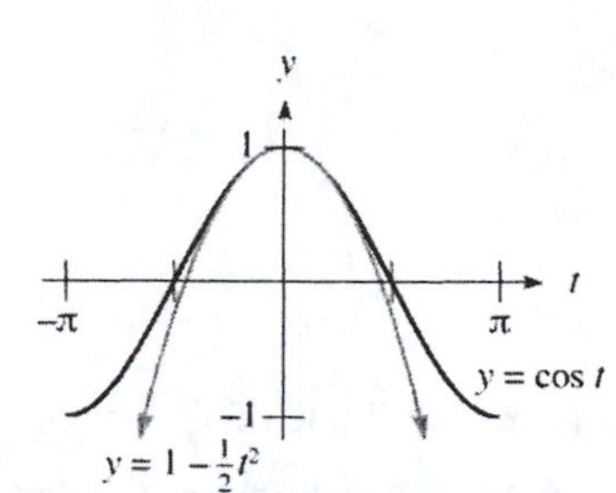

Both curves are similar when t is near 0.

b. Sketching the graphs:

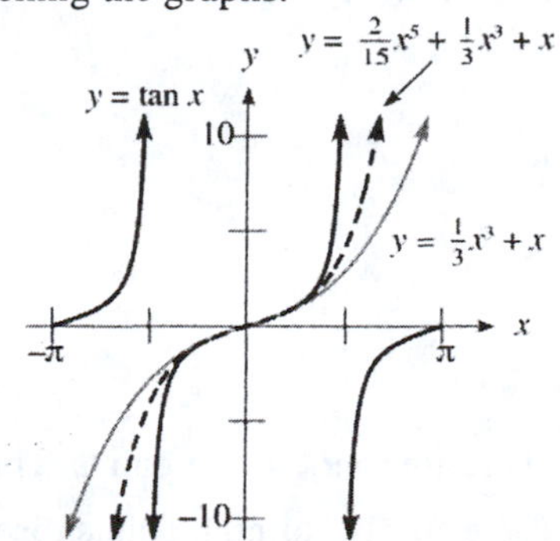

All three curves are similar when x is near 0.

69. The y-coordinates of points P, Q, R, S, and T are just the sines of the radian measures of the corresponding arcs. Thus the y-coordinates are:

$P: \sin\frac{\pi}{12} \approx 0.259$ $\qquad$ $Q: \sin\frac{\pi}{6} = \frac{1}{2} = 0.5$

$R: \sin\frac{\pi}{4} = \frac{\sqrt{2}}{2} \approx 0.707$ $\qquad$ $S: \sin\frac{\pi}{3} = \frac{\sqrt{3}}{2} \approx 0.866$

$T: \sin\frac{5\pi}{12} \approx 0.966$

7.2 Graphs of the Sine and Cosine Functions

1. A cycle is completed every 2 units, so the period is 2. The curve has high and low points of 1 and –1, respectively, so the amplitude is 1.

3. A cycle is completed every 4 units, so the period is 4. The curve has high and low points of 6 and – 6, respectively, so the amplitude is 6.

5. A cycle is completed every 4 units, so the period is 4. The curve has high and low points of 6 and 2, respectively, so the amplitude is $\frac{6-2}{2} = 2$.

7. A cycle is completed every 6 units, so the period is 6. The curve has high and low points of –3 and –6, respectively, so the amplitude is $\frac{-3-(-6)}{2} = \frac{3}{2}$.

9. The coordinates of point C are $\left(-\frac{7\pi}{2}, 1\right) \approx (-10.996, 1)$.

11. The coordinates of point G are $\left(\frac{5\pi}{2}, 1\right) \approx (7.854, 1)$.

13. The coordinates of point B are $(-4\pi, 0) \approx (-12.566, 0)$.

15. The coordinates of point D are $(-3\pi, 0) \approx (-9.425, 0)$.

17. The coordinates of point E are $(-\pi, 0) \approx (-3.142, 0)$.

19. Referring to the graph of $y = \sin x$, note that $y = \sin x$ is increasing on the interval $\frac{3\pi}{2} < x < 2\pi$.

21. Referring to the graph of $y = \sin x$, note that $y = \sin x$ is decreasing on the interval $\frac{5\pi}{2} < x < \frac{7\pi}{2}$.

23. The coordinates of point J are $\left(\frac{9\pi}{2}, 0\right) \approx (14.137, 0)$.

25. The coordinates of point A are $(-4\pi, 1) \approx (-12.566, 1)$.

27. The coordinates of point E are $\left(\frac{\pi}{2}, 0\right) \approx (1.571, 0)$.

29. The coordinates of point I are $(4\pi, 1) \approx (12.566, 1)$.

31. The coordinates of point B are $\left(-\frac{5\pi}{2}, 0\right) \approx (-7.854, 0)$.

33. Referring to the graph of $y = \cos x$, note that $y = \cos x$ is decreasing on the interval $0 < x < \pi$.

35. Referring to the graph of $y = \cos x$, note that $y = \cos x$ is increasing on the interval $-\frac{\pi}{2} < x < 0$.

37. **a.** The x-intercept is π.

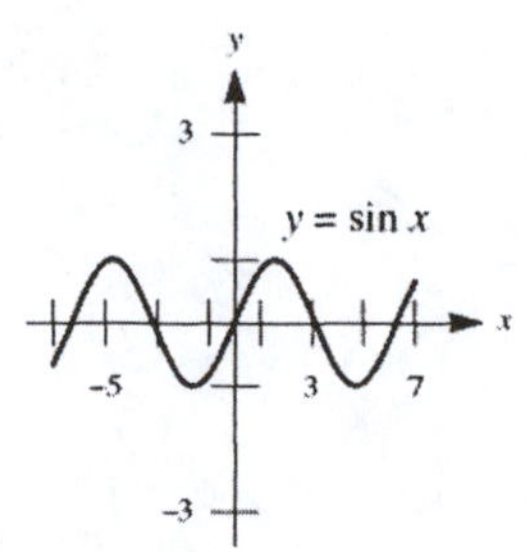

b. There are four turning points. The turning point between 1 and 2 has an x-coordinate of $\frac{\pi}{2}$.

c. There are five turning points for $y = \cos x$. The turning point between 6 and 7 has an x-coordinate of 2π.

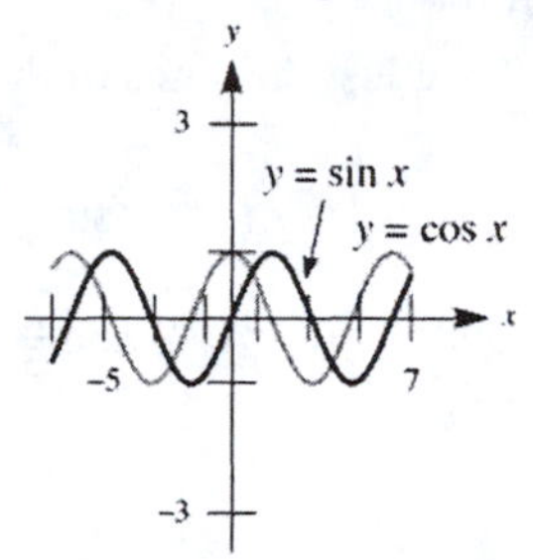

d. If $y = \sin x$ is shifted $\frac{\pi}{2}$ units to the left, its graph will coincide with $y = \cos x$.

e. The results from parts **a** through **c** are confirmed.

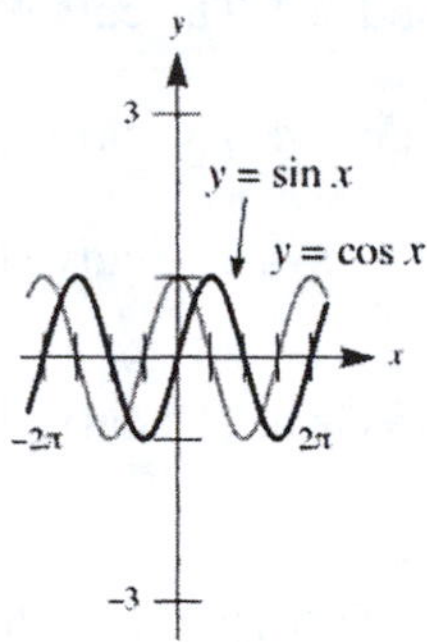

39. **a.** Between $-\frac{\pi}{2}$ and $\frac{\pi}{2}$, the curve does appear similar to that of a parabola:

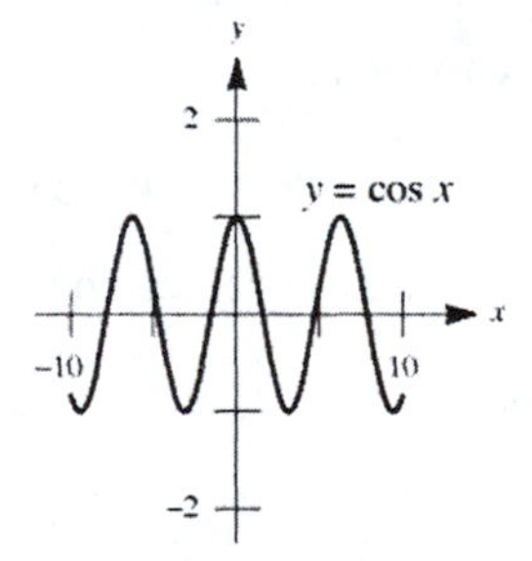

b. Graphing the two curves:

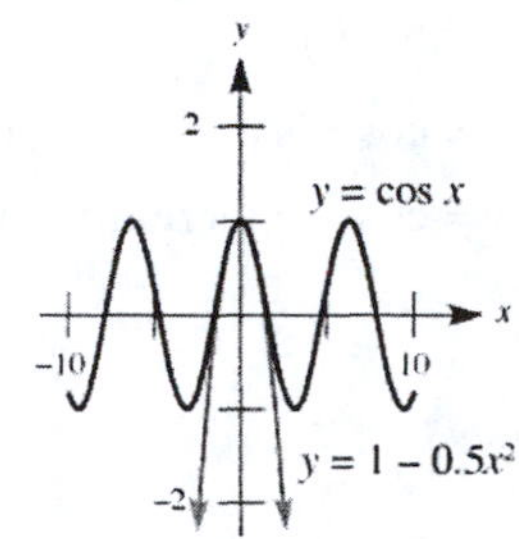

In the vicinity of $x = 0$, the graphs of $y = \cos x$ and $y = 1 - 0.5x^2$ are very similar.

c. Completing the table:

x	1	0.5	0.1	0.01	0.001
$\cos x$	0.54	0.8776	0.9950	0.99995	0.9999995
$1-0.5x^2$	0.50	0.8750	0.9950	0.99995	0.9999995

41. a. From the figure, the root is $x \approx 0.45$.
b. Using a calculator, the root is $x \approx 0.4510$.
c. Another root would be $2\pi - x \approx 5.8322$.
d. These roots would be $\pi - x \approx 2.6906$ and $\pi + x \approx 3.5926$.

43. a. From the figure, the root is $x \approx 1.25$.
b. Using a calculator, the root is $x \approx 1.2661$.
c. Another root would be $2\pi - x \approx 5.0171$.
d. These roots would be $\pi - x \approx 1.8755$ and $\pi + x \approx 4.4077$.

45. a. From the figure, the root is $x \approx 1.0$.
b. Using a calculator, the root is $x \approx 0.9884$.
c. Another root would be $2\pi - x \approx 5.2948$.
d. These roots would be $\pi - x \approx 2.1532$ and $\pi + x \approx 4.1300$.

47. a. From the figure, the root is $x \approx 0.65$.
b. Using a calculator, the root is $x \approx 0.6435$.
c. Another root would be $\pi - x \approx 2.4981$.
d. These roots would be $\pi + x \approx 3.7851$ and $2\pi - x \approx 5.6397$.

49. a. From the figure, the root is $x \approx 0.2$.
b. Using a calculator, the root is $x \approx 0.2014$.
c. Another root would be $\pi - x \approx 2.9402$.
d. These roots would be $\pi + x \approx 3.3430$ and $2\pi - x \approx 6.0818$.

51. a. From the figure, the root is $x \approx 0.8$.
b. Using a calculator, the root is $x \approx 0.7754$.
c. Another root would be $\pi - x \approx 2.3662$.
d. These roots would be $\pi + x \approx 3.9170$ and $2\pi - x \approx 5.5078$.

53. Graphing the two curves:

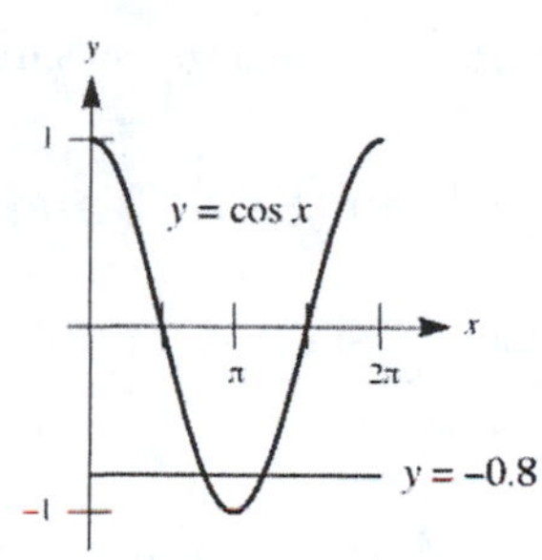

The x-coordinates of the intersection points are $x_3 \approx 2.498$ and $x_4 \approx 3.785$, matching the results in the text.

55. a. The roots to $\sin x = 0.687$ are approximately $x \approx 0.7574$ and $x \approx 2.3842$.

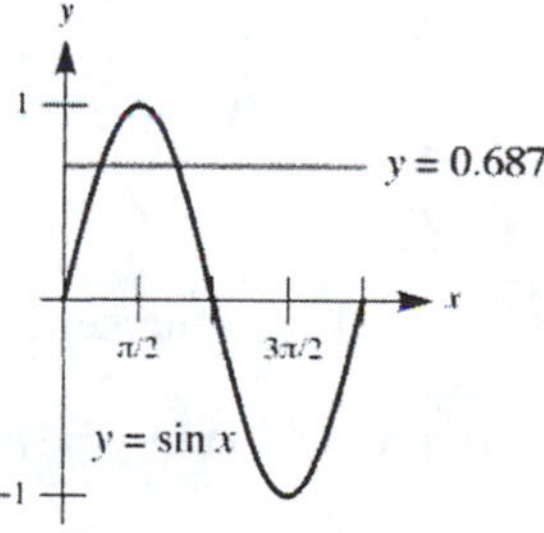

b. The roots to $\sin x = -0.687$ are approximately $x \approx 3.8989$ and $x \approx 5.5258$.

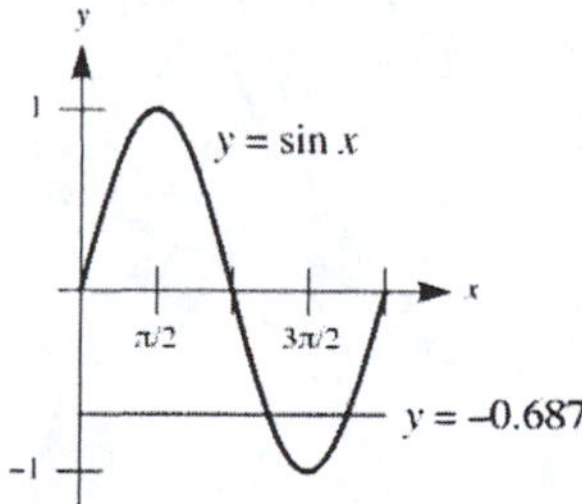

57. a. The corresponding value of $\cos x$ is 0.
b. The corresponding value of $\sin x$ is 0.

59. Both functions are decreasing on the open interval $\left(\frac{\pi}{2}, \pi\right)$.

61. **a.** The completed table values using $x_0 = \frac{50}{500} = 0.1$ are:

	x_1	x_2	x_3	x_4
From graph	1.0	0.55	0.85	0.65
From calculator	0.99500	0.54450	0.85539	0.65593

	x_5	x_6	x_7
From graph	0.80	0.70	0.75
From calculator	0.79248	0.70208	0.76350

b. Multiplying the results by 500, the completed table is:

n	0	1	2	3	4	5	6	7
Number of fish after n breeding seasons	50	498	272	428	328	396	351	382

c. The corresponding equilibrium population is $(0.7391)(500) \approx 370$ fish.

63. **a.** Since C lies on the unit circle, its coordinates are $C(\cos\theta, \sin\theta)$.

b. Since $\angle AOB = \theta = \angle COD$ and $AO = CO$, the two triangles are congruent.

c. Since $AB = CD$ and $OB = OD$, the coordinates of A are $A(-\sin\theta, \cos\theta)$.

d. Matching up x- and y-coordinates: $\cos\left(\theta + \frac{\pi}{2}\right) = -\sin\theta$ and $\sin\left(\theta + \frac{\pi}{2}\right) = \cos\theta$

7.3 Graphs of $y = A \sin(Bx - C)$ and $y = A \cos(Bx - C)$

1. **a.** The amplitude is 2, the period is 2π, and the x-intercepts are $0, \pi, 2\pi$. The function is increasing on the intervals $\left(0, \frac{\pi}{2}\right)$ and $\left(\frac{3\pi}{2}, 2\pi\right)$.

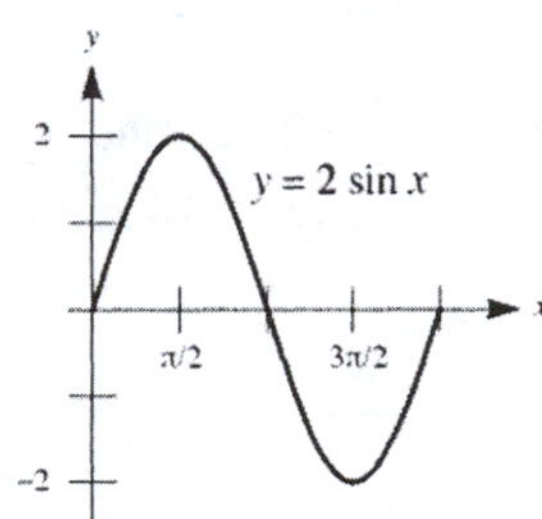

b. The amplitude is 1 and the period is $\frac{2\pi}{2} = \pi$. The x-intercepts are $0, \frac{\pi}{2}, \pi$, and the function is increasing on the interval $\left(\frac{\pi}{4}, \frac{3\pi}{4}\right)$. Notice that the graph is a reflection of $y = \sin 2x$ across the x-axis.

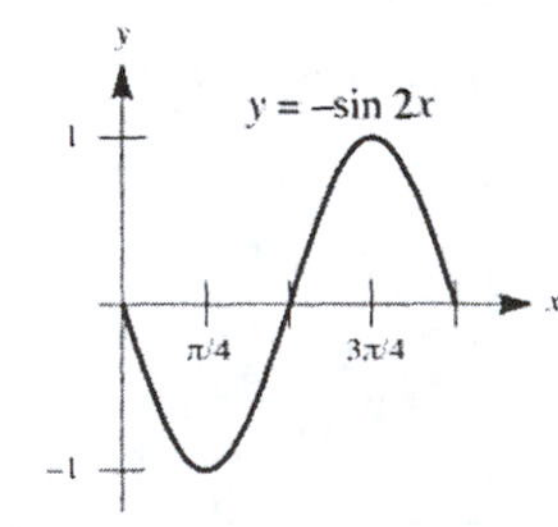

3. a. The amplitude is 1 and the period is $\frac{2\pi}{2} = \pi$. The x-intercepts are $\frac{\pi}{4}, \frac{3\pi}{4}$, and the function is increasing on the interval $\left(\frac{\pi}{2}, \pi\right)$.

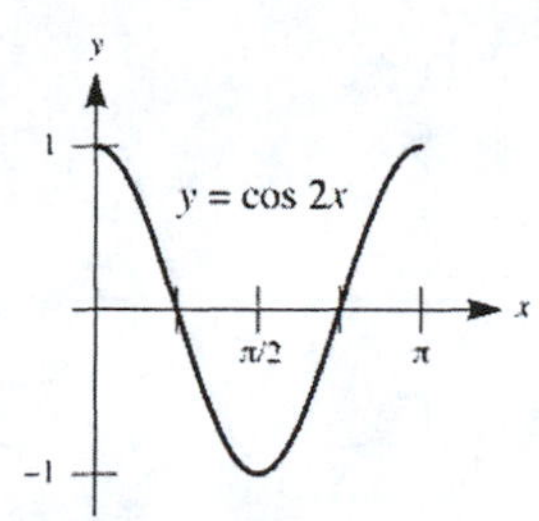

b. The amplitude is 2 and the period is $\frac{2\pi}{2} = \pi$. The x-intercepts are $\frac{\pi}{4}, \frac{3\pi}{4}$, and the function is increasing on the interval $\left(\frac{\pi}{2}, \pi\right)$.

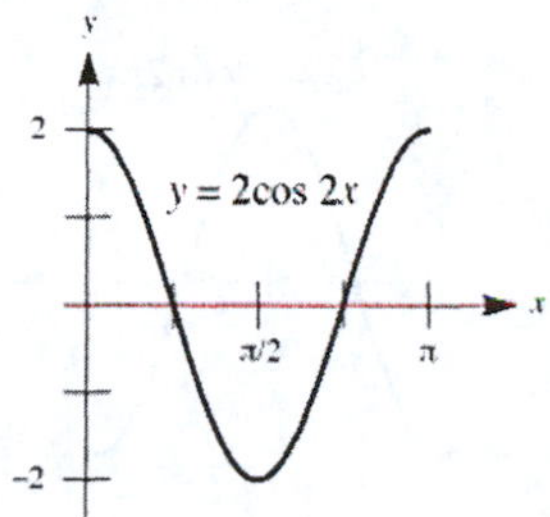

5. a. The amplitude is 3 and the period is $\frac{2\pi}{\pi/2} = 4$. The x-intercepts are 0, 2, 4, and the function is increasing on the intervals (0,1) and (3,4).

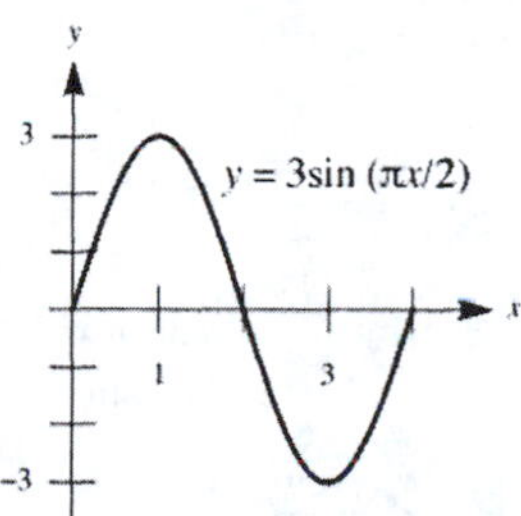

b. The amplitude is 3 and the period is $\frac{2\pi}{\pi/2} = 4$. The x-intercepts are 0, 2, 4, and the function is increasing on the interval (1,3). Notice that the graph is a reflection of $y = 3\sin\frac{\pi x}{2}$ across the x-axis.

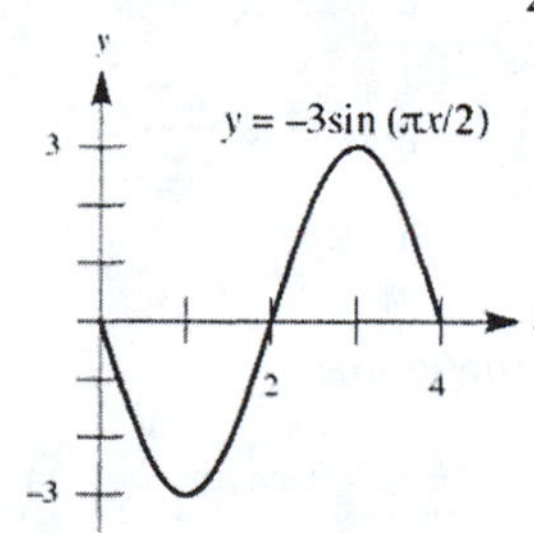

7. **a.** The amplitude is 1 and the period is $\frac{2\pi}{2\pi} = 1$. The x-intercepts are $\frac{1}{4}, \frac{3}{4}$, and the function is increasing on the interval $\left(\frac{1}{2}, 1\right)$.

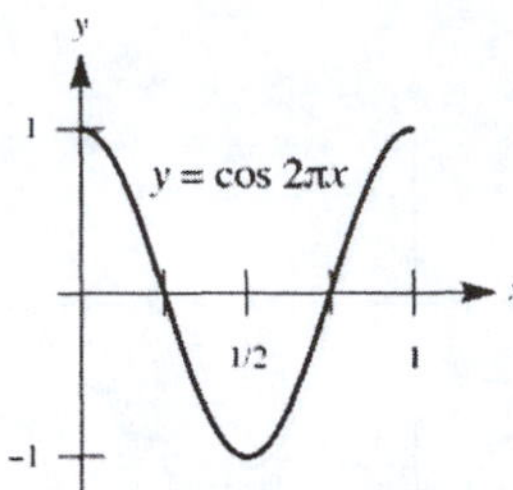

b. The amplitude is 4 and the period is $\frac{2\pi}{2\pi} = 1$. The x-intercepts are $\frac{1}{4}, \frac{3}{4}$, and the function is increasing on the interval $\left(0, \frac{1}{2}\right)$. Notice that the graph is a reflection of $y = 4\cos 2\pi x$ across the x-axis.

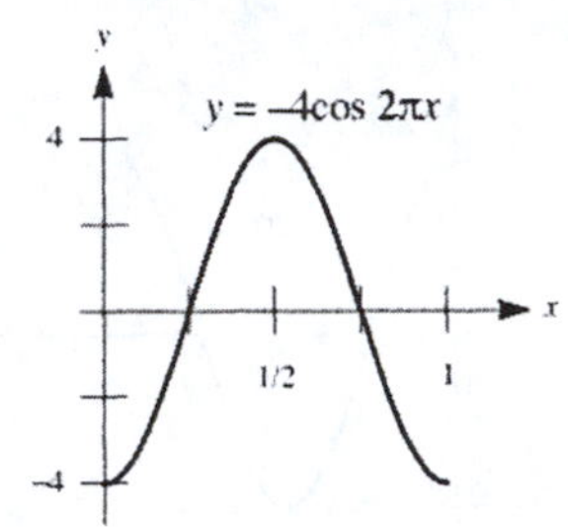

9. The amplitudes and periods are:

$y = \sin x$: amplitude = 1, period = 2π
$y = 2\sin x$: amplitude = 2, period = 2π
$y = 3\sin x$: amplitude = 3, period = 2π
$y = 4\sin x$: amplitude = 4, period = 2π

The graphs of the four functions are:

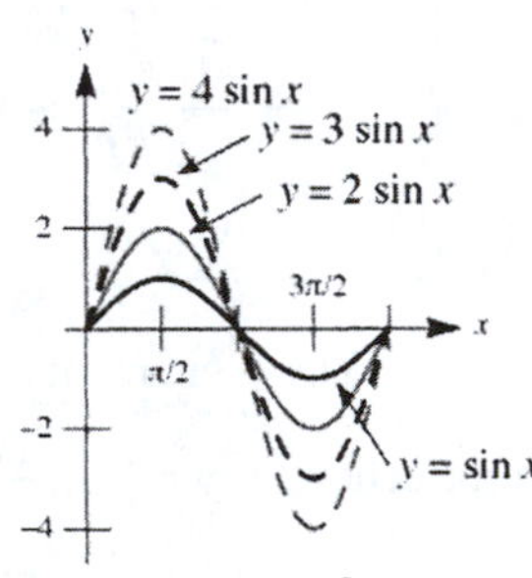

11. **a.** For $y = 2\sin \pi x$, the amplitude is 2 and the period is $\frac{2\pi}{\pi} = 2$. For $y = \sin 2\pi x$, the amplitude is 1 and the period is $\frac{2\pi}{2\pi} = 1$.

b. The amplitudes and periods agree with the graphs:

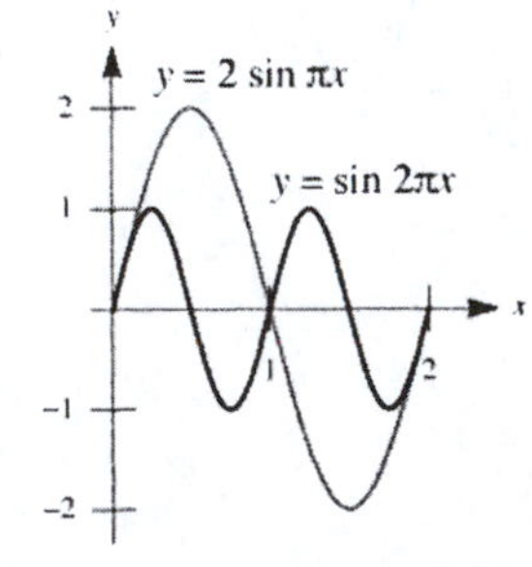

13. The amplitude is 1 and the period is $\frac{2\pi}{1/2} = 4\pi$. There are no x-intercepts, and the function is increasing on the intervals $(0, \pi)$ and $(3\pi, 4\pi)$. Notice that the graph is a displacement of $y = \sin\frac{x}{2}$ down 2 units.

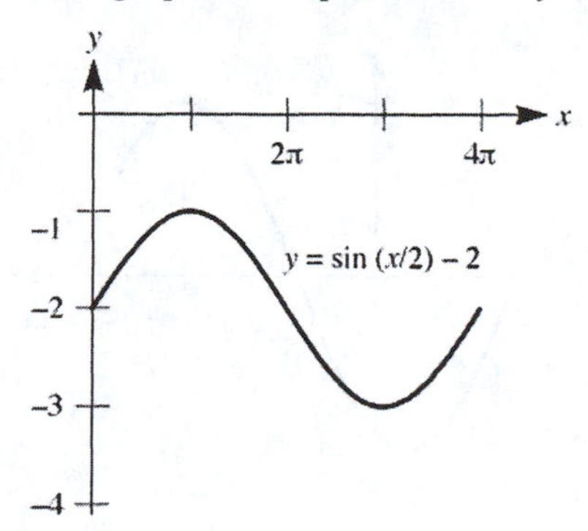

15. The amplitude is 2 and the period is $\frac{2\pi}{3\pi} = \frac{2}{3}$. The x-intercept is $\frac{1}{3}$, and the function is increasing on the interval $\left(0, \frac{1}{3}\right)$. Notice that the graph is a reflection of $y = 2\cos 3\pi x$ across the x-axis, then a displacement down 2 units.

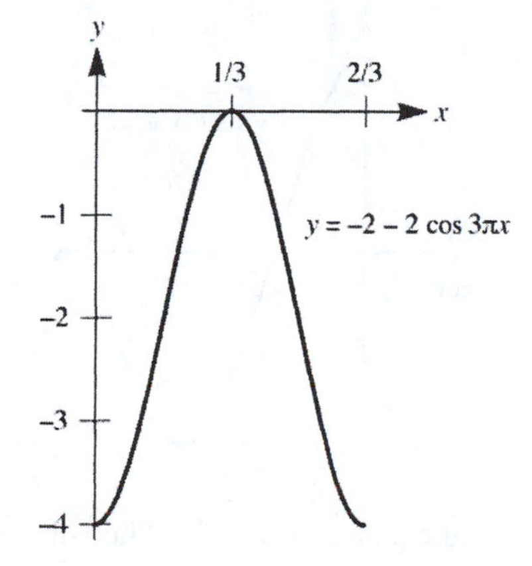

17. The amplitude is 1, the period is 2π, and the phase shift is $-\frac{\pi}{3}$. The x-intercepts are $\frac{\pi}{6}, \frac{7\pi}{6}$, the high points are $\left(-\frac{\pi}{3}, 1\right)$ and $\left(\frac{5\pi}{3}, 1\right)$, and the low point is $\left(\frac{2\pi}{3}, -1\right)$.

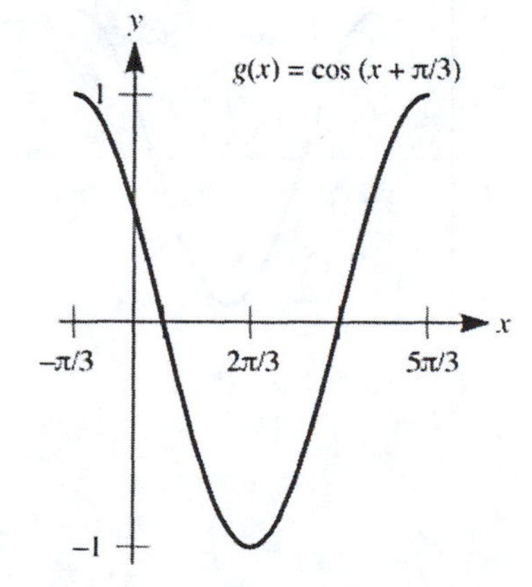

19. The amplitude is 1, the period is 2π, and the phase shift is –2. The x-intercepts are –2, $\pi - 2$, $2\pi - 2$, the high point is $\left(\frac{3\pi}{2} - 2, 1\right)$, and the low point is $\left(\frac{\pi}{2} - 2, -1\right)$.

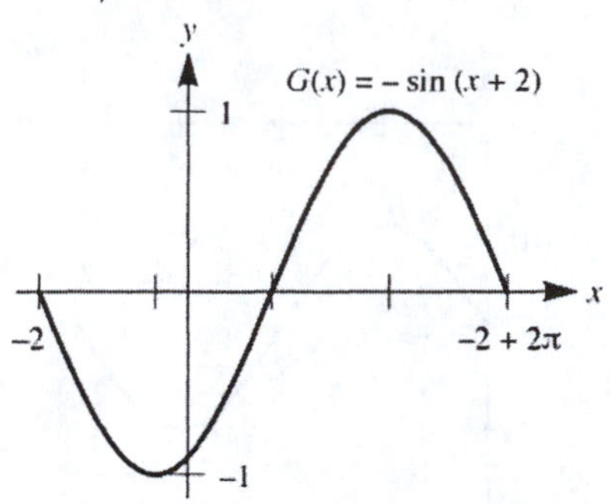

21. The amplitude is 1, the period is $\frac{2\pi}{3}$, and the phase shift is $\frac{-\pi/2}{3} = -\frac{\pi}{6}$. The x-intercepts are $-\frac{\pi}{6}, \frac{\pi}{6}, \frac{\pi}{2}$, the high point is (0,1), and the low point is $\left(\frac{\pi}{3}, -1\right)$.

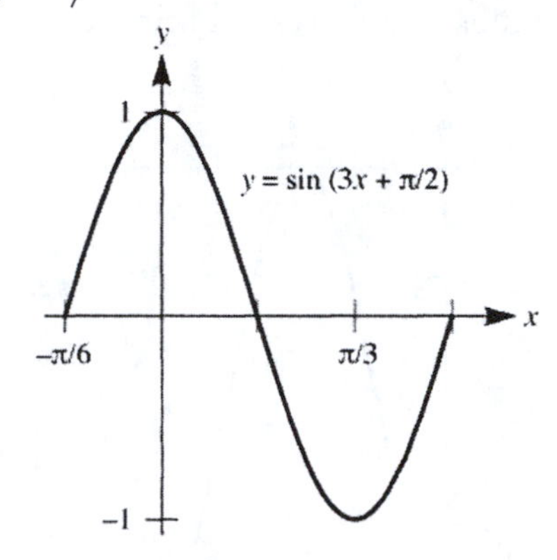

23. The amplitude is 1, the period is 2π, and the phase shift is $\frac{\pi}{2}$. The x-intercepts are π and 2π, the high points are $\left(\frac{\pi}{2}, 1\right)$ and $\left(\frac{5\pi}{2}, 1\right)$, and the low point is $\left(\frac{3\pi}{2}, -1\right)$.

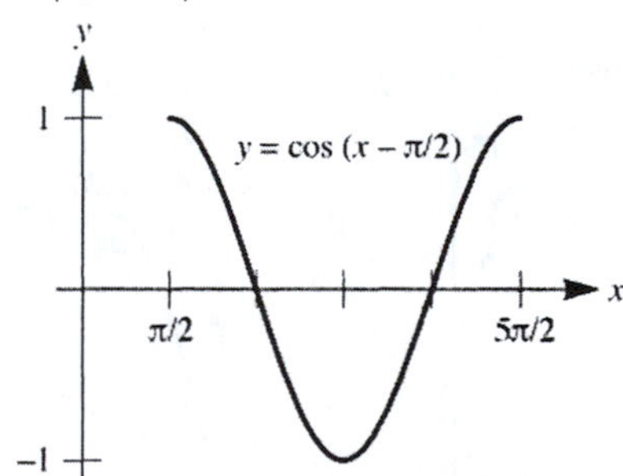

25. The amplitude is 2, the period is $\frac{2\pi}{\pi} = 2$, and the phase shift is $\frac{-\pi}{\pi} = -1$. The x-intercepts are –1, 0, 1, the high point is $\left(\frac{1}{2}, 2\right)$, and the low point is $\left(-\frac{1}{2}, -2\right)$. Notice that the graph is a reflection of $y = 2\sin(\pi x + \pi)$ across the x-axis.

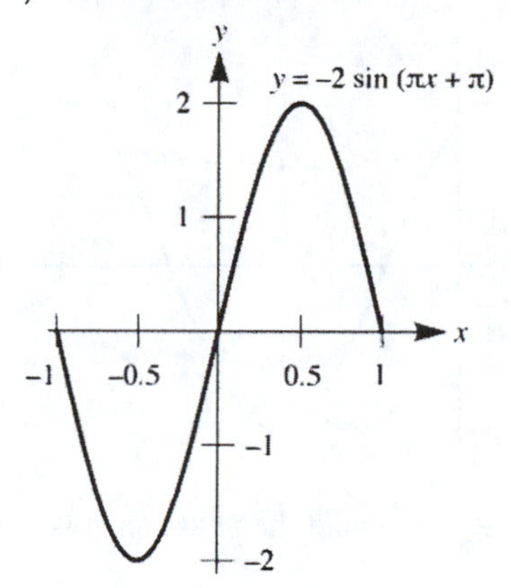

27. The amplitude is 1, the period is 2π, and the phase shift is –1. The x-intercepts are $\frac{\pi}{2} - 1$ and $\frac{3\pi}{2} - 1$, the high points are $(-1, 1)$ and $(2\pi - 1, 1)$, and the low point is $(\pi - 1, -1)$.

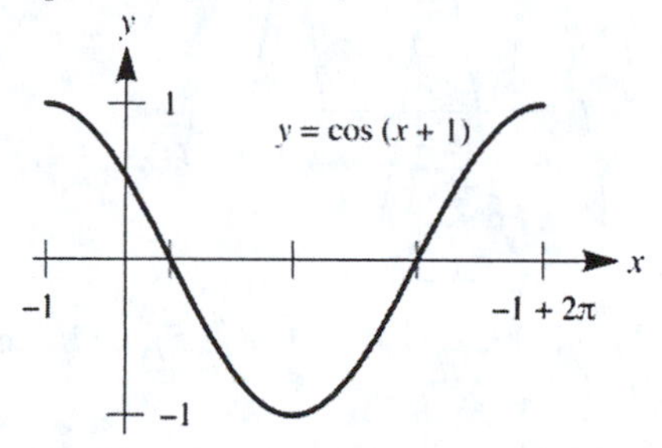

29. The amplitude is 1, the period is $\frac{2\pi}{2} = \pi$, and the phase shift is $\frac{\pi/3}{2} = \frac{\pi}{6}$. The x-intercept is $\frac{2\pi}{3}$, the high points are $\left(\frac{\pi}{6}, 2\right)$ and $\left(\frac{7\pi}{6}, 2\right)$, and the low point is $\left(\frac{2\pi}{3}, 0\right)$. Notice that the graph is a displacement of $y = \cos\left(2x - \frac{\pi}{3}\right)$ up 1 unit.

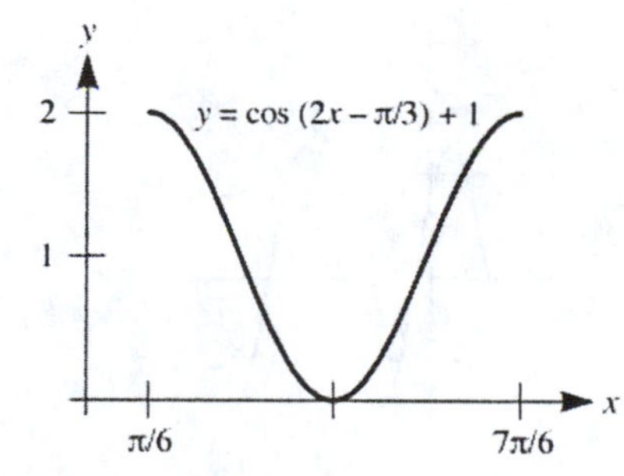

31. The amplitude is 3, the period is $\frac{2\pi}{2/3} = 3\pi$, and the phase shift is $\frac{-\pi/6}{2/3} = -\frac{\pi}{4}$. The x-intercepts are $\frac{\pi}{2}$ and 2π, the high points are $\left(-\frac{\pi}{4}, 3\right)$ and $\left(\frac{11\pi}{4}, 3\right)$, and the low point is $\left(\frac{5\pi}{4}, -3\right)$.

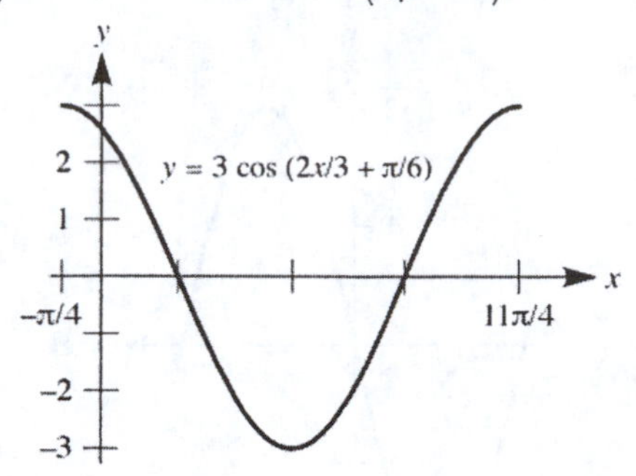

33. **a.** The amplitude is 2.5, the period is $\frac{2\pi}{3\pi} = \frac{2}{3}$, and the phase shift is $-\frac{4}{3\pi}$.

b. Graphing the curve:

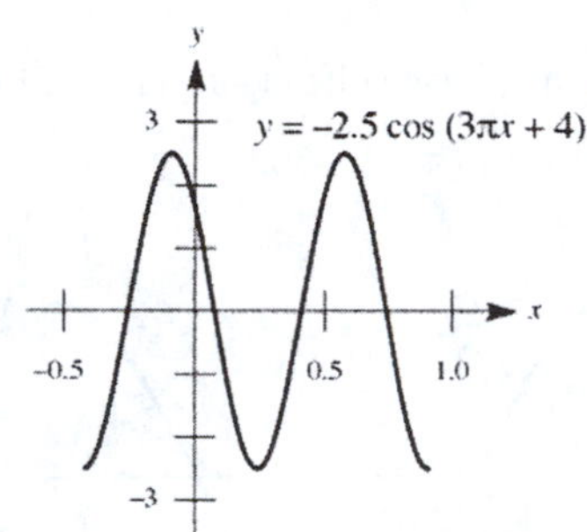

c. The high points are approximately (–0.09, 2.5) and (0.58, 2.5), while the low points are approximately (–0.42, –2.5), (0.24, –2.5), and (0.91, –2.5).

d. The high points are $\left(-\frac{4}{3\pi}+\frac{1}{3}, 2.5\right)$ and $\left(-\frac{4}{3\pi}+1, 2.5\right)$, while the low points are $\left(-\frac{4}{3\pi}, -2.5\right)$, $\left(-\frac{4}{3\pi}+\frac{2}{3}, -2.5\right)$, and $\left(-\frac{4}{3\pi}+\frac{4}{3}, -2.5\right)$.

35. **a.** The amplitude is 2.5, the period is 6, and the phase shift is $-\frac{12}{\pi}$.

b. Graphing the curve:

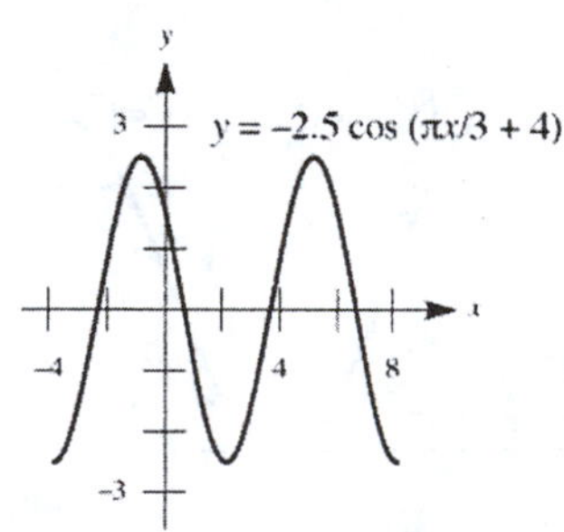

c. The high points are approximately (–0.82, 2.5) and (5.18, 2.5), while the low points are approximately (–3.82, –2.5), (2.18, –2.5), and (8.18, –2.5).

d. The high points are $\left(-\frac{12}{\pi}+3, 2.5\right)$ and $\left(-\frac{12}{\pi}+9, 2.5\right)$, while the low points are $\left(-\frac{12}{\pi}, -2.5\right)$, $\left(-\frac{12}{\pi}+6, -2.5\right)$, and $\left(-\frac{12}{\pi}+12, -2.5\right)$.

37. **a.** The amplitude is 1, the period is $\frac{2\pi}{0.5} = 4\pi$, and the phase shift is $\frac{-0.75}{0.5} = -1.5$.

b. Graphing the curve:

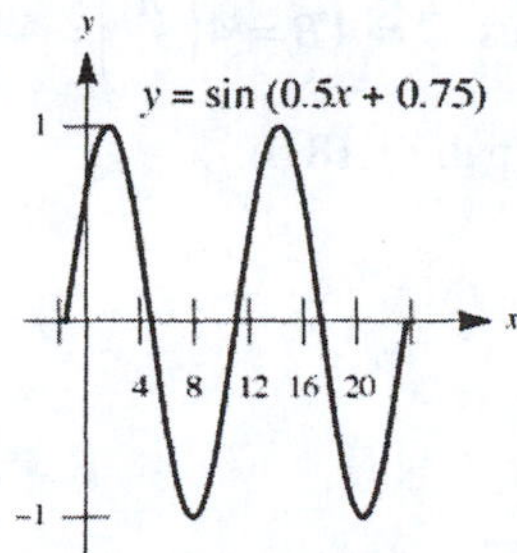

c. The high points are approximately (1.64, 1) and (14.21, 1), while the low points are approximately (7.92, –1) and (20.49, –1).

d. The high points are $(-1.5+\pi, 1)$ and $(-1.5+5\pi, 1)$, while the low points are $(-1.5+3\pi, -1)$ and $(-1.5+7\pi, -1)$.

39. **a.** The amplitude is 0.02, the period is $\frac{2\pi}{0.01\pi} = 200$, and the phase shift is $\frac{4\pi}{0.01\pi} = 400$.

b. Graphing the curve:

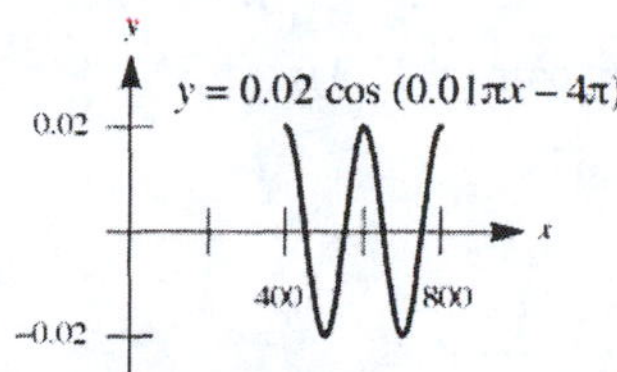

c. The high points are (400,0.02), (600,0.02), and (800,0.02), while the low points are (500,–0.02) and (700,–0.02).

d. The high points are (400,0.02), (600,0.02), and (800,0.02), while the low points are (500,–0.02) and (700,–0.02).

41. This is a sine function where the amplitude is 1.5, so $A = 1.5$. Since the period is $\frac{4\pi}{3}$:

$$\frac{2\pi}{B} = \frac{4\pi}{3}$$
$$6\pi = 4\pi B$$
$$\frac{3}{2} = B$$

The equation is $y = 1.5\sin\frac{3}{2}x$.

43. This is a cosine function where the amplitude is 1, so $A = 1$. Since the period is 5:

$$\frac{2\pi}{B} = 5$$
$$2\pi = 5B$$
$$\frac{2\pi}{5} = B$$

The equation is $y = \cos\frac{2\pi}{5}x$.

45. This is a cosine function where the amplitude is π, so $A = \pi$. Since the period is 8:

$$\frac{2\pi}{B} = 8$$
$$2\pi = 8B$$
$$\frac{\pi}{4} = B$$

The equation is $y = \pi\cos\frac{\pi}{4}x$.

47. First find A: $A=\frac{M-m}{2}=\frac{68.5-18.0}{2}=25.25$. Assuming $\frac{2\pi}{B}=12$, then $B=\frac{2\pi}{12}=\frac{\pi}{6}$. The maximum occurs when $t=7$, so the phase shift is 4. So $\frac{C}{B}=4$, thus $C=4B=4\left(\frac{\pi}{6}\right)=\frac{2\pi}{3}$.

Using $y=A\sin(Bt-C)+D$ and the data pair (1,18.0):

$$18.0=25.25\sin\left(\frac{\pi}{6}\bullet 1-\frac{2\pi}{3}\right)+D$$
$$18.0=25.25\sin\left(-\frac{\pi}{2}\right)+D$$
$$18.0=-25.25+D$$
$$43.25=D$$

The function is $y=25.25\sin\left(\frac{\pi}{6}t-\frac{2\pi}{3}\right)+43.25$.

49. First find A: $A=\frac{M-m}{2}=\frac{81.3-68.7}{2}=6.3$. Assuming $\frac{2\pi}{B}=12$, then $B=\frac{2\pi}{12}=\frac{\pi}{6}$. The maximum occurs when $t=1$ or $t=2$, so using $t=1.5$ the phase shift is $-\frac{3}{2}$. So $\frac{C}{B}=-\frac{3}{2}$, thus $C=-\frac{3}{2}B=-\frac{3}{2}\left(\frac{\pi}{6}\right)=-\frac{\pi}{4}$.

Using $y=A\sin(Bt-C)+D$ and the data pair (1,81.3):

$$81.3=6.3\sin\left(\frac{\pi}{6}\bullet 1+\frac{\pi}{4}\right)+D$$
$$81.3=6.3\sin\left(\frac{5\pi}{12}\right)+D$$
$$81.3=6.085+D$$
$$75.215=D$$

The function is $y=6.3\sin\left(\frac{\pi}{6}t+\frac{\pi}{4}\right)+75.215$.

51. First find A: $A=\frac{M-m}{2}=\frac{115-65}{2}=25$. Assuming $\frac{2\pi}{B}=12$, then $B=\frac{2\pi}{12}=\frac{\pi}{6}$. The maximum occurs when $t=7$, so the phase shift is 4. So $\frac{C}{B}=4$, thus $C=4B=4\left(\frac{\pi}{6}\right)=\frac{2\pi}{3}$.

Using $y=A\sin(Bt-C)+D$ and the data pair (1,65):

$$65=25\sin\left(\frac{\pi}{6}\bullet 1-\frac{2\pi}{3}\right)+D$$
$$65=25\sin\left(-\frac{\pi}{2}\right)+D$$
$$65=-25+D$$
$$90=D$$

The function is $y=25\sin\left(\frac{\pi}{6}t-\frac{2\pi}{3}\right)+90$.

53. Graph $y=\frac{1}{2}+\frac{1}{2}\cos 2x$, which has an amplitude of $\frac{1}{2}$ and a period of $\frac{2\pi}{2}=\pi$. The graph will be a displacement of $y=\frac{1}{2}\cos 2x$ up $\frac{1}{2}$ unit.

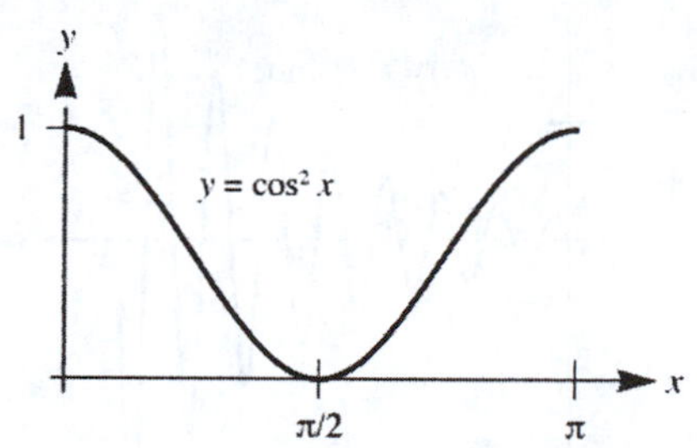

55. Graph $y = \cos 2x$, which has an amplitude of 1 and a period of $\frac{2\pi}{2}=\pi$.

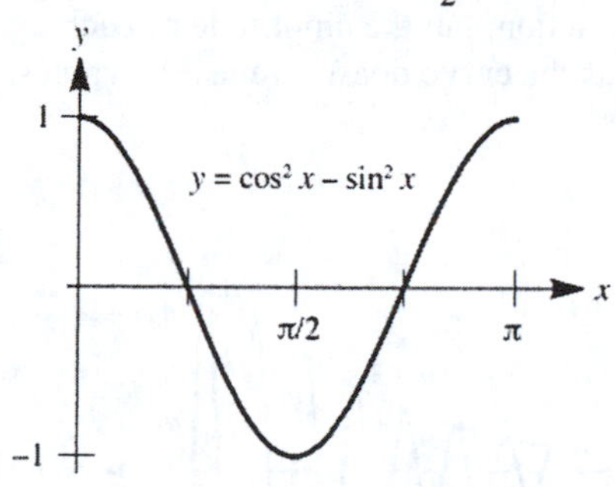

57. First compute each of the composite functions:

$$G[F(x)] = G(\sin x) = (\sin x)^2 \qquad F[G(x)] = F(x^2) = \sin(x^2)$$
$$H[F(x)] = H(\sin x) = (\sin x)^3 \qquad F[H(x)] = F(x^3) = \sin(x^3)$$

The first composite function $G[F(x)]$ has a graph which will not go below the x-axis, since regardless of the value of $\sin x$, the value of $(\sin x)^2 \geq 0$. The second composite function $F[G(x)]$ has a graph which will go below the x-axis, for example $F[G(2)] = \sin 4 \approx -0.76$. The third composite function $H[F(x)]$ has a graph which will go below the x-axis, for example $H[F(4)] = (\sin 4)^3 \approx -0.43$. The fourth composite function $F[H(x)]$ has a graph which will go below the x-axis, for example $F[H(1.5)] = \sin 3.375 \approx -0.23$.

59. **a.** Graphing the curve:

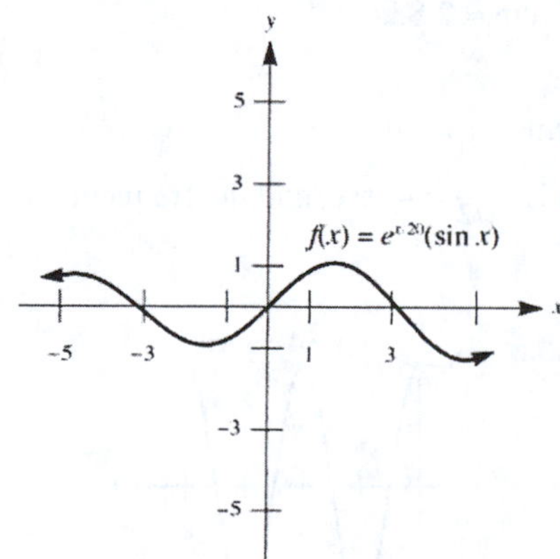

The curve is similar to a sine curve.

b. Using the revised settings:

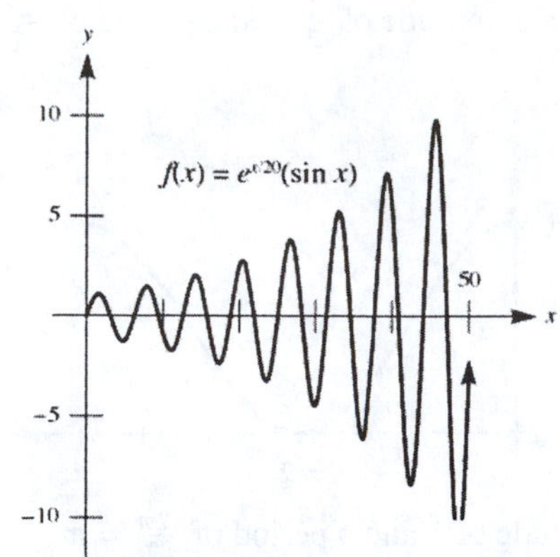

The graph oscillates like a sine function, but the amplitude of each cycle increases as x increases. No, the function is not periodic, as the curve doesn't repeat its values.

c. Including the other two functions:

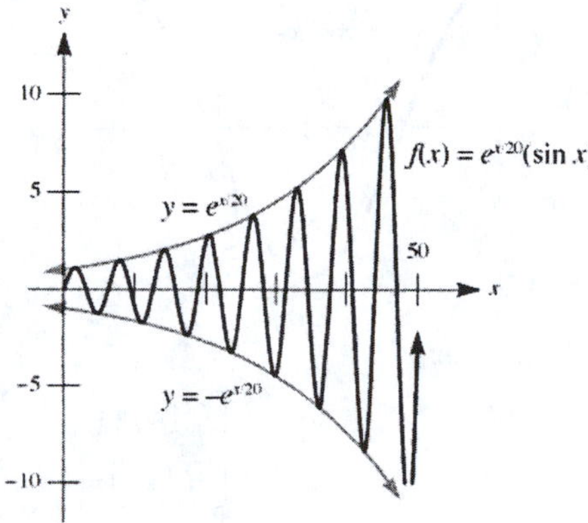

The graph of $y = e^{x/20}$ touches the top of each cycle of the sine curve, while the graph of $y = -e^{x/20}$ touches the bottom of each cycle.

7.4 Simple Harmonic Motion

1. **a.** Evaluating $s = 4\cos\dfrac{\pi t}{2}$ at the given values of t:

$t = 0$: $s = 4\cos 0 = 4$ cm

$t = 0.5$: $s = 4\cos\frac{\pi}{4} = 2\sqrt{2}$ cm ≈ 2.83 cm

$t = 1$: $s = 4\cos\frac{\pi}{2} = 0$ cm

$t = 2$: $s = 4\cos\pi = -4$ cm

b. The amplitude is 4 cm, the period is $\frac{2\pi}{\pi/2} = 4$ sec, and the frequency is $\frac{1}{4}$ cycles/sec. Sketch the graph over the interval $0 \le t \le 8$:

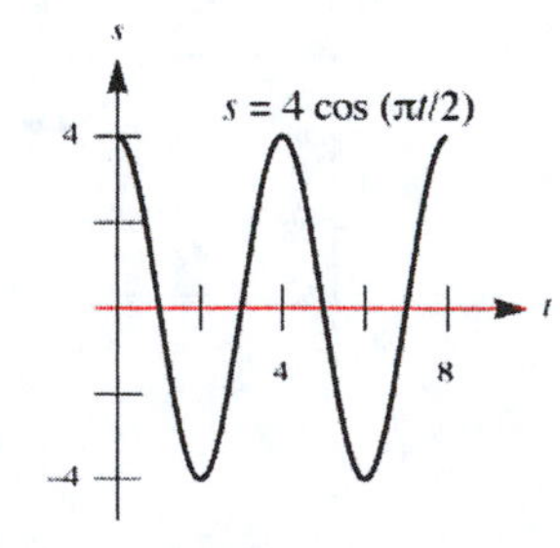

c. The mass is farthest from the origin at high and low points, which occur at $t = 0$, $t = 2$, $t = 4$, $t = 6$ and $t = 8$ sec.

d. The mass is passing through the origin when $s = 0$, which occurs at $t = 1$, $t = 3$, $t = 5$ and $t = 7$ sec.

e. The mass is moving to the right when the s-coordinate is increasing, which occurs during the intervals $2 < t < 4$ and $6 < t < 8$.

3. a. The amplitude is 3 feet, the period is $\frac{2\pi}{\pi/3} = 6$ sec, and the frequency is $\frac{1}{6}$ cycles/sec. Sketch the graph over the interval $0 \le t \le 12$:

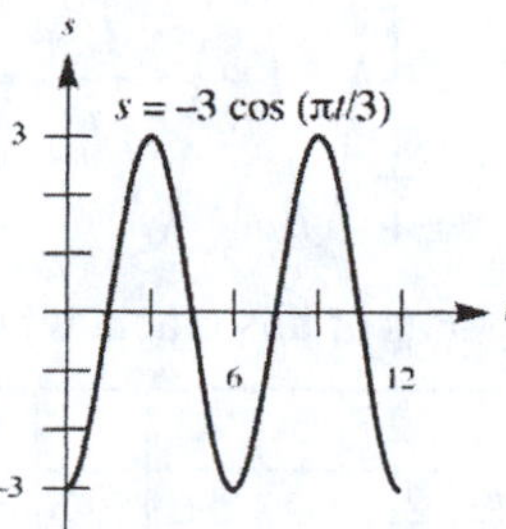

b. The mass is moving upward when the s-coordinate is increasing, which occurs during the intervals $0 < t < 3$ and $6 < t < 9$.

c. The mass is moving downward when the s-coordinate is decreasing, which occurs during the intervals $3 < t < 6$ and $9 < t < 12$.

d. Graph the velocity function for $0 \le t \le 12$, noting its period is $\frac{2\pi}{\pi/3} = 6$ sec and its amplitude is π feet/sec.

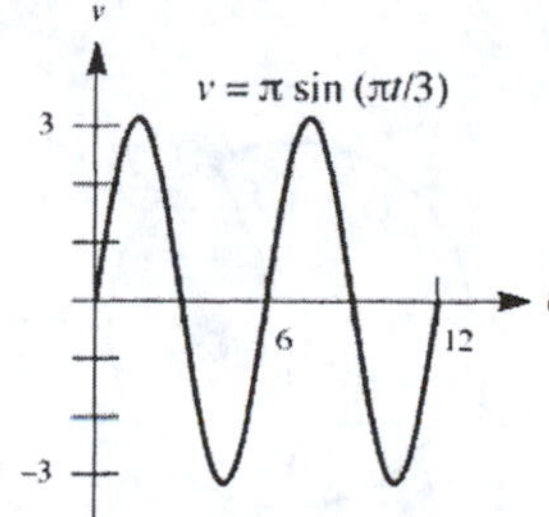

e. Note that $v = 0$ when $t = 0$, $t = 3$, $t = 6$, $t = 9$, and $t = 12$ sec. At these times the mass (s-coordinate) is at –3, 3, –3, 3 and –3 feet, respectively.

f. The velocity is a maximum at the high points of this graph, which occur at $t = 1.5$ and $t = 7.5$ sec. At these times the mass is at 0 feet.

g. The velocity is a minimum at the low points of this graph, which occur at $t = 4.5$ and $t = 10.5$ sec. At these times the mass is at 0 feet.

h. Graph the velocity function and position function on the same set of axes for $0 \le t \le 12$:

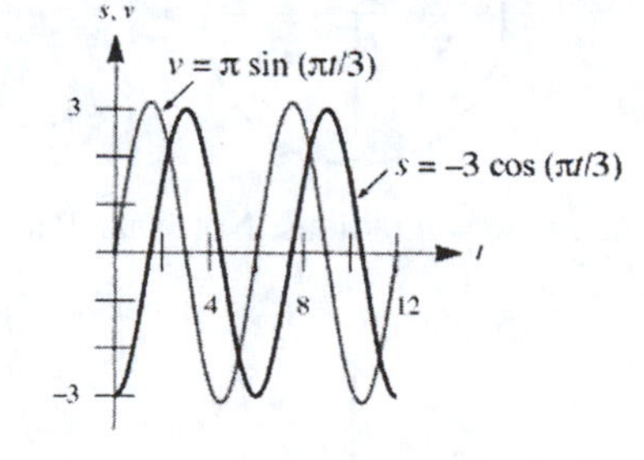

5. a. The amplitude is 170 volts and the period is $\frac{2\pi}{120\pi} = \frac{1}{60}$ sec, so the frequency is 60 cycles/sec.

   b. Graph the voltage for $0 \le t \le \frac{1}{30}$ sec:

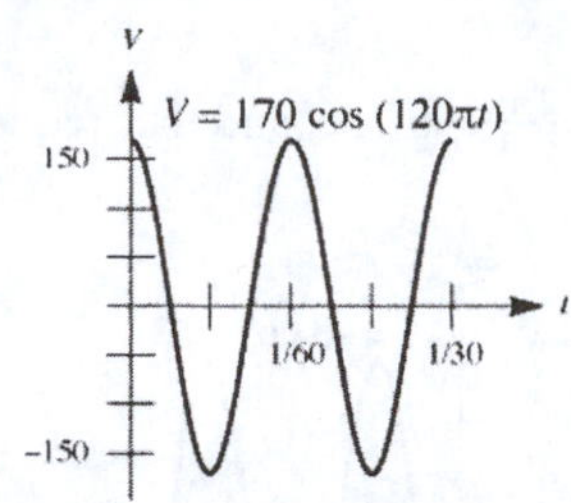

   c. The voltage is a maximum at the high points of this graph, which occur at $t = 0$, $t = \frac{1}{60}$ and $t = \frac{1}{30}$ sec.

7. a. Complete the table:

t(sec)	0	1	2	3	4	5	6	7
θ(radians)	0	$\frac{\pi}{3}$	$\frac{2\pi}{3}$	π	$\frac{4\pi}{3}$	$\frac{5\pi}{3}$	2π	$\frac{7\pi}{3}$

   b. Since the x-coordinate of the point Q is $\cos\theta$, the corresponding x-coordinates are $\cos 0 = 1$, $\cos\frac{\pi}{3} = \frac{1}{2}$, $\cos\frac{2\pi}{3} = -\frac{1}{2}$, $\cos\pi = -1$, $\cos\frac{4\pi}{3} = -\frac{1}{2}$, $\cos\frac{5\pi}{3} = \frac{1}{2}$, $\cos 2\pi = 1$, and $\cos\frac{7\pi}{3} = \frac{1}{2}$.

   c. Sketch P and Q when $t = 1$ second, thus $\theta = \frac{\pi}{3}$ radians:

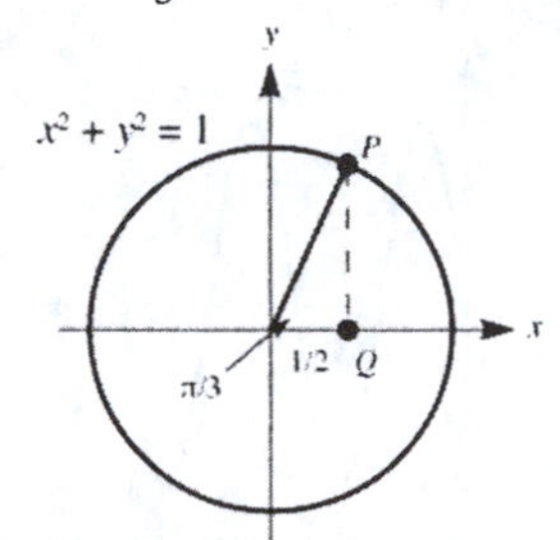

   d. Sketch P and Q when $t = 2$ seconds, thus $\theta = \frac{2\pi}{3}$ radians:

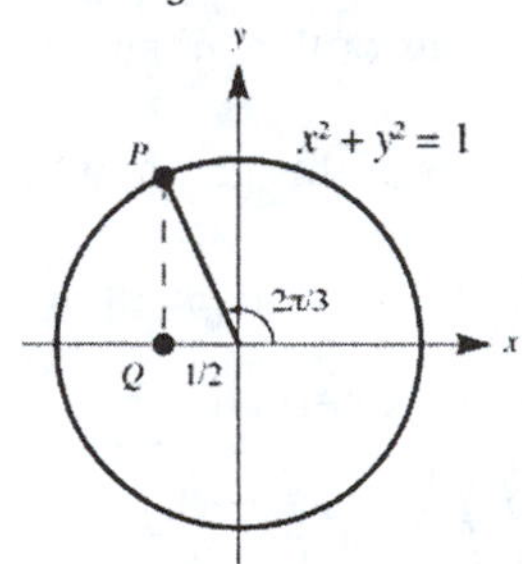

   Sketch P and Q when $t = 3$ seconds, thus $\theta = \pi$ radians. Note that P and Q coincide at the same point, since P lies on the x-axis:

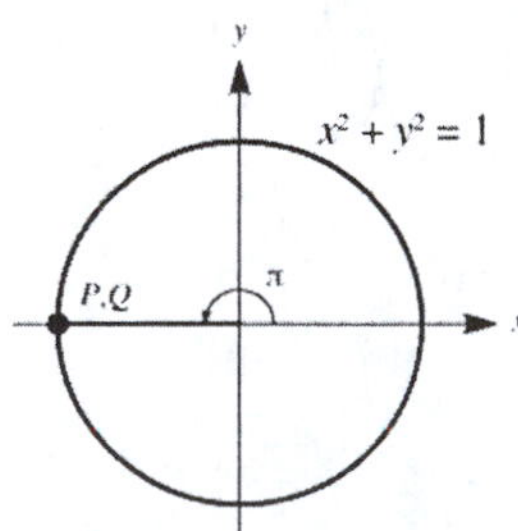

Sketch P and Q when $t = 4$ seconds, thus $\theta = \frac{4\pi}{3}$ radians:

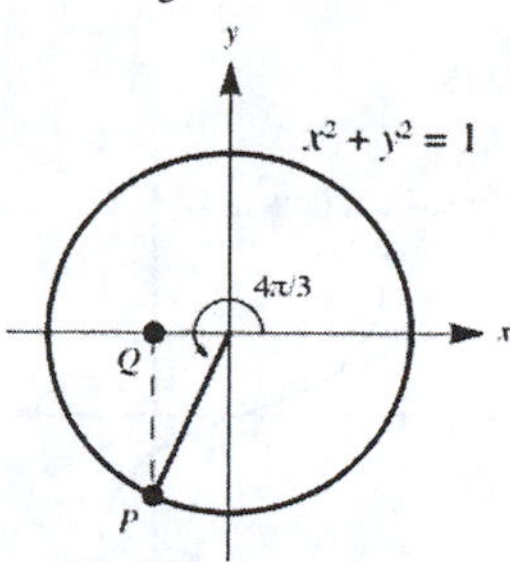

e. The x-coordinate of Q is the same as the x-coordinate of P. But P is a point on the unit circle and θ is the radian measure to that point, thus the x-coordinate is $\cos\theta$.

f. The period of this function is $\frac{2\pi}{\pi/3} = 6$, the amplitude is 1, and the frequency is $\frac{1}{6}$. Graph the function for two complete cycles, or $0 \le t \le 12$:

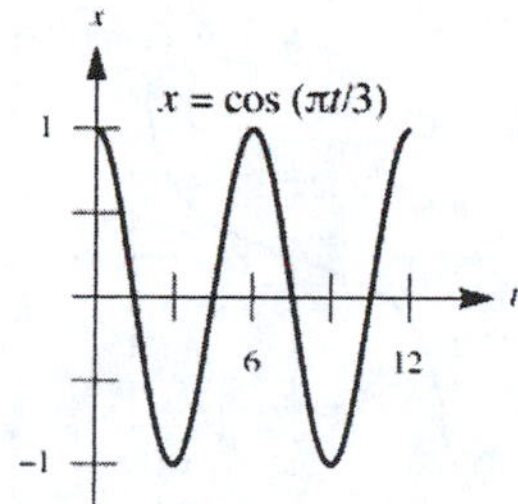

g. The period is $\frac{2\pi}{\pi/3} = 6$ and the amplitude is $\frac{\pi}{3}$. Graph the velocity for two complete cycles, or $0 \le t \le 12$:

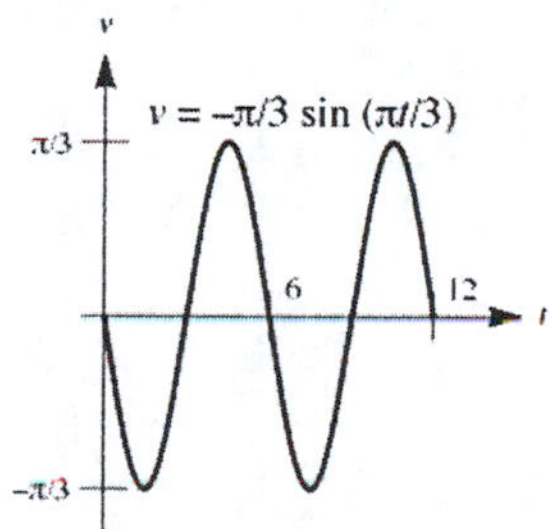

h. The velocity is 0 when $t = 0, 3, 6, 9$ and 12 sec. The corresponding x-coordinates are $1, -1, 1, -1$ and 1, respectively.

i. The velocity is a maximum at the high points of the curve, which occur at $t = 4.5$ and $t = 10.5$ sec. The x-coordinate of Q is 0 at each of these times, so Q is located at the origin.

j. The velocity is a minimum at the low points of the curve, which occur at $t = 1.5$ and $t = 7.5$ sec. The x-coordinate of Q is 0 at each of these times, so Q is located at the origin.

7.5 Graphs of the Tangent and the Reciprocal Functions

1. **a.** The x-intercept is $-\frac{\pi}{4}$, the y-intercept is 1, and the asymptotes are $x = -\frac{3\pi}{4}$ and $x = \frac{\pi}{4}$. Notice that this graph is $y = \tan x$ displaced $\frac{\pi}{4}$ units to the left.

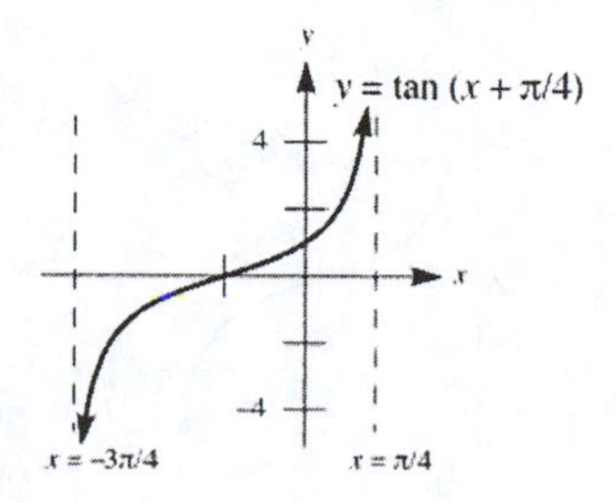

b. The x-intercept is $-\frac{\pi}{4}$, the y-intercept is -1, and the asymptotes are $x = -\frac{3\pi}{4}$ and $x = \frac{\pi}{4}$. Notice that this graph is $y = \tan\left(x + \frac{\pi}{4}\right)$ reflected across the x-axis.

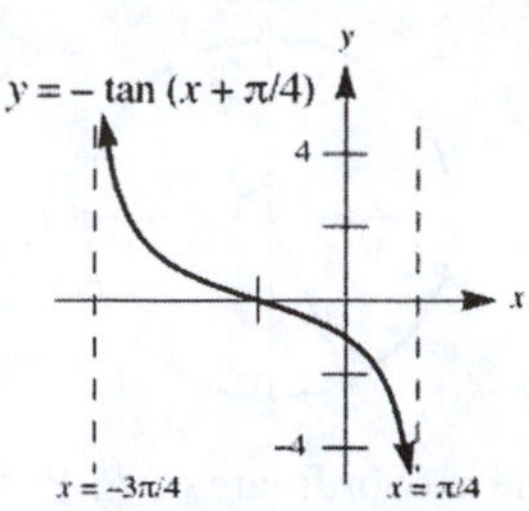

3. **a.** The x- and y-intercepts are both 0, and the asymptotes are $x = -\frac{3\pi}{2}$ and $x = \frac{3\pi}{2}$. Notice that the period is $\frac{\pi}{1/3} = 3\pi$.

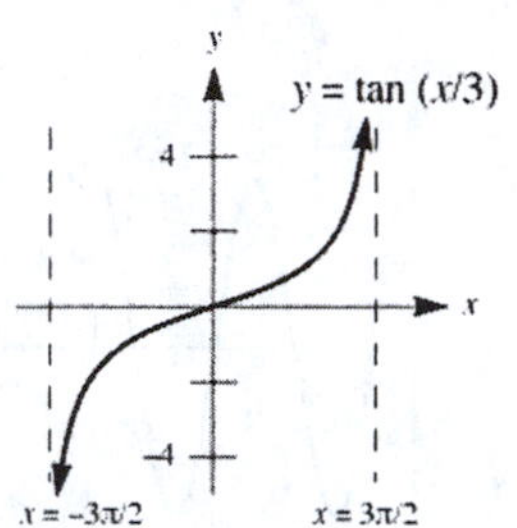

b. The x- and y-intercepts are both 0, and the asymptotes are $x = -\frac{3\pi}{2}$ and $x = \frac{3\pi}{2}$. Notice that this graph is $y = \tan\frac{x}{3}$ reflected across the x-axis.

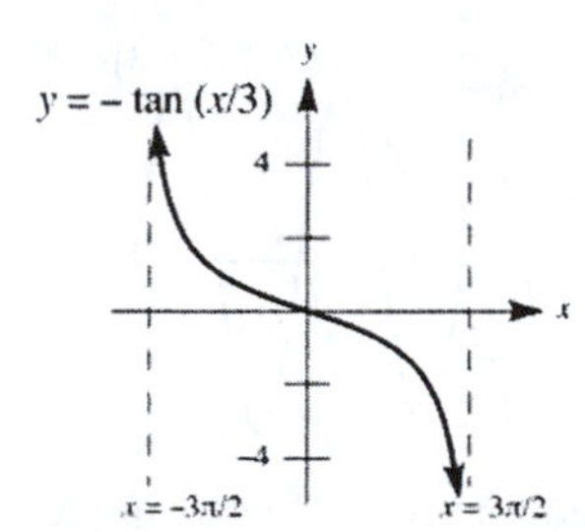

5. The x- and y-intercepts are both 0, and the asymptotes are $x = -1$ and $x = 1$. Notice that the period is $\frac{\pi}{\pi/2} = 2$.

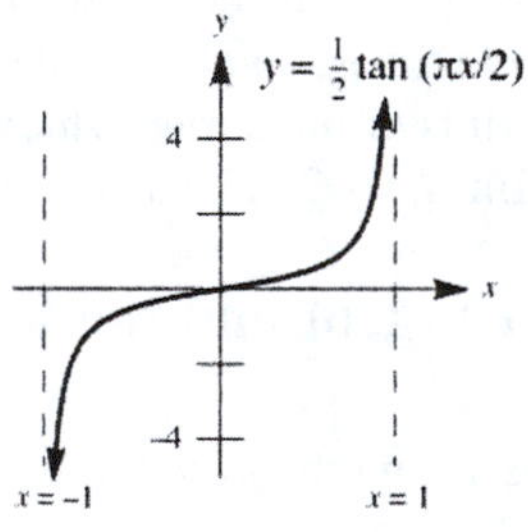

7. The x-intercept is 1, there is no y-intercept, and the asymptotes are $x = 0$ and $x = 2$. Notice that the period is $\frac{\pi}{\pi/2} = 2$.

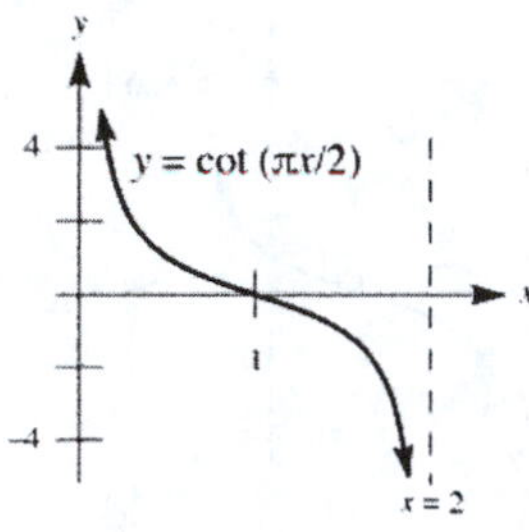

9. The x-intercept is $\frac{3\pi}{4}$, the y-intercept is 1, and the asymptotes are $x=\frac{\pi}{4}$ and $x=\frac{5\pi}{4}$. Notice that this graph is $y=\cot x$ displaced $\frac{\pi}{4}$ units to the right, then reflected across the x-axis.

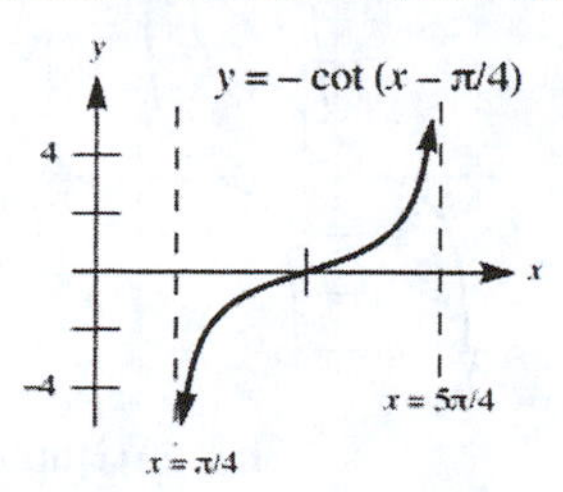

11. The x-intercept is $\frac{\pi}{4}$, there is no y-intercept, and the asymptotes are $x = 0$ and $x=\frac{\pi}{2}$. Notice that the period is $\frac{\pi}{2}$.

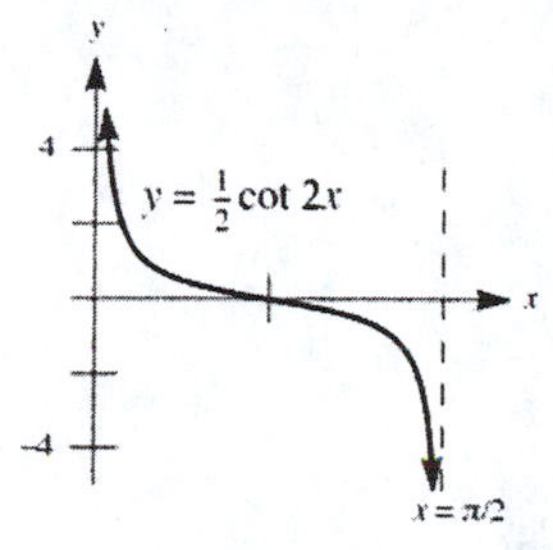

13. The x-intercepts are $-\pi$, 0, and π:

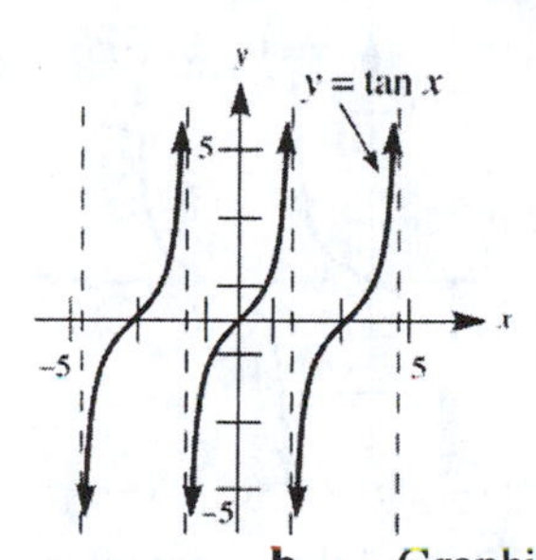

15. **a.** Graphing the function:

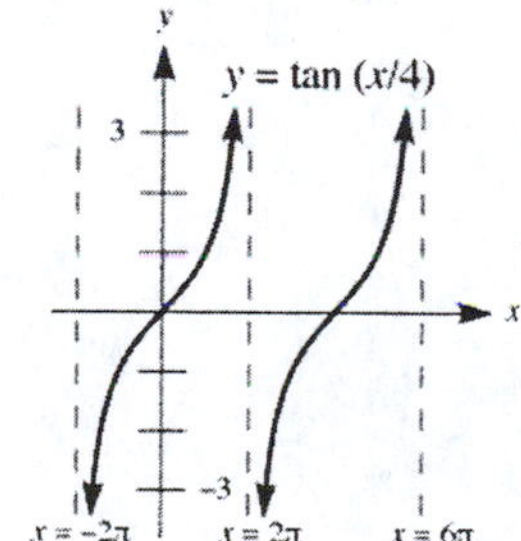

b. Graphing the function:

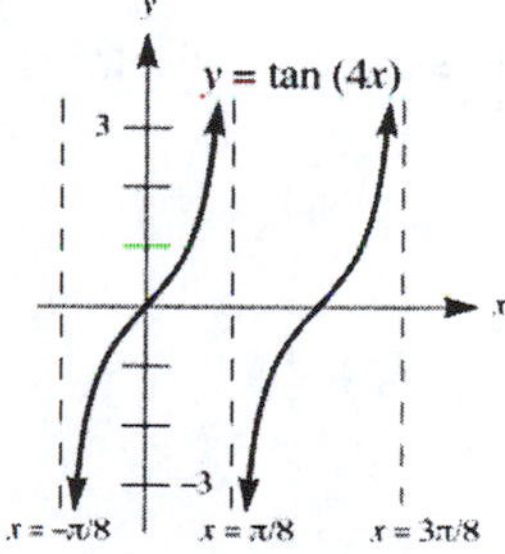

17. **a.** Graphing the function:

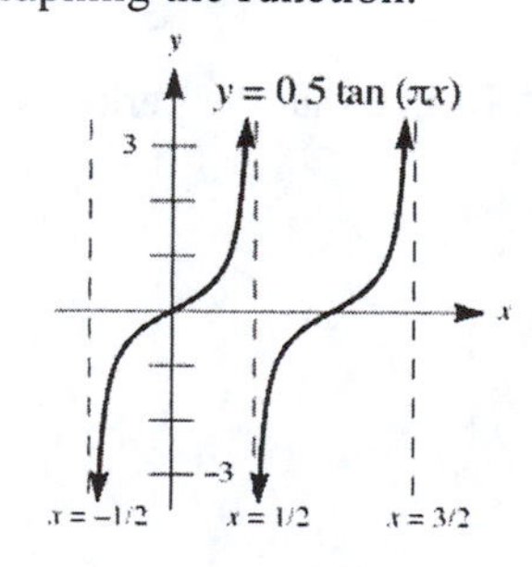

b. Graphing the function:

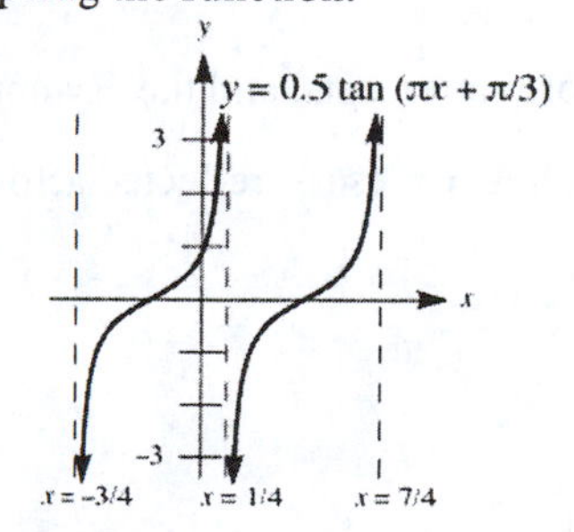

c. Graphing the function:

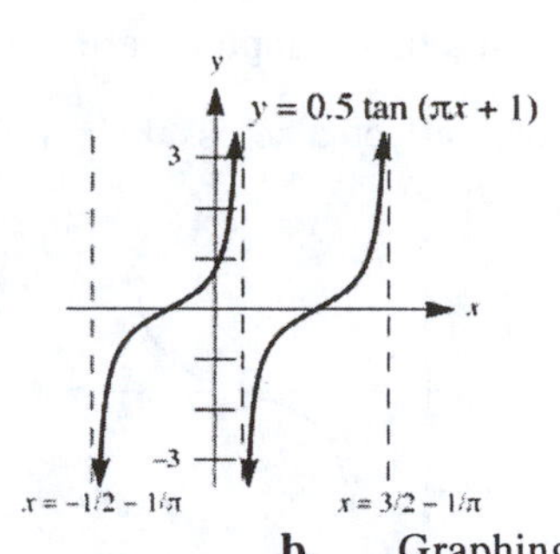

19. a. Graphing the function:

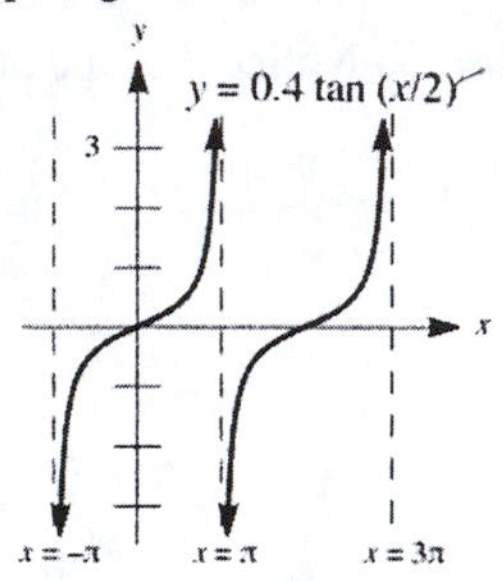

b. Graphing the function:

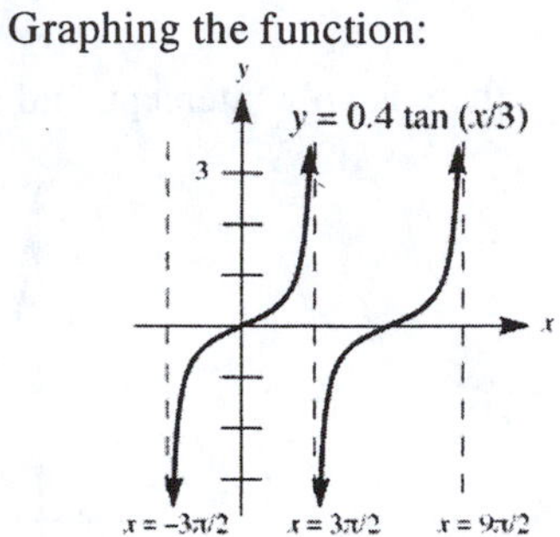

c. Graphing the function:

y = 0.4 tan (x/5)

x = –5π/2 x = 5π/2 x = 15π/2

21. There is no x-intercept, the y-intercept is $-\sqrt{2}$, and the asymptotes are $x=-\frac{3\pi}{4}$, $x=\frac{\pi}{4}$, and $x=\frac{5\pi}{4}$.

Notice that this graph is $y = \csc x$ displaced $\frac{\pi}{4}$ units to the right.

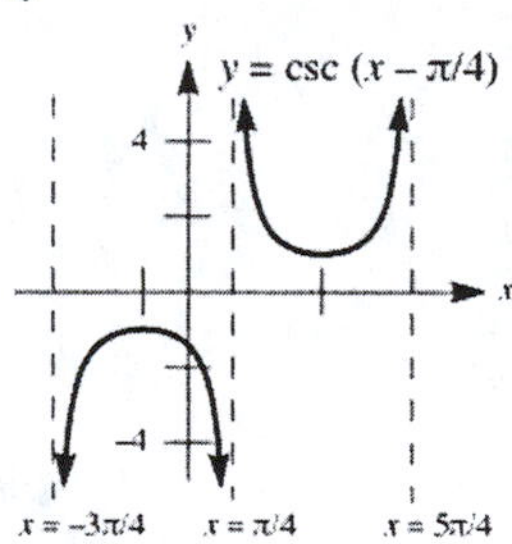

23. There are no x- or y-intercepts, and the asymptotes are $x = -2\pi$, $x = 0$ and $x = 2\pi$. Notice that the period is $\frac{2\pi}{1/2} = 4\pi$, and that this graph is $y = \csc\frac{x}{2}$ reflected across the x-axis.

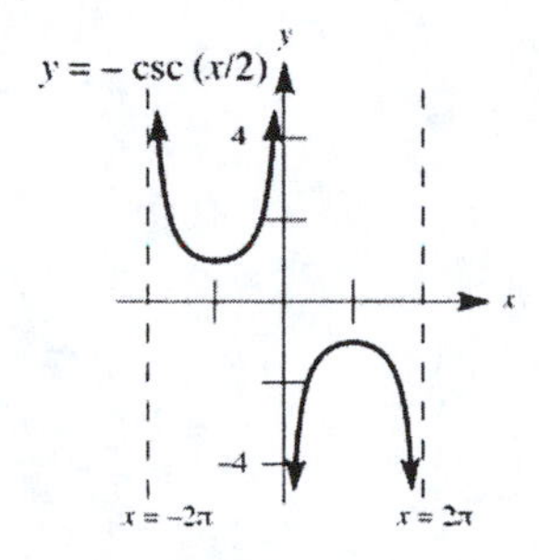

25. There are no x- or y-intercepts, and the asymptotes are $x = -1$, $x = 0$, and $x = 1$. Notice that the period is $\frac{2\pi}{\pi} = 2$.

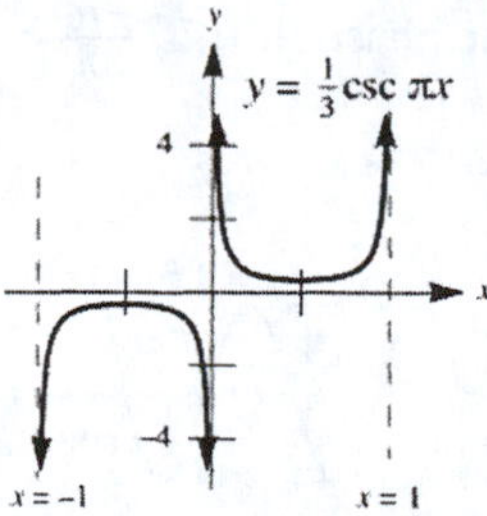

27. There is no x-intercept, the y-intercept is -1, and the asymptotes are $x = -\frac{\pi}{2}$, $x = \frac{\pi}{2}$, and $x = \frac{3\pi}{2}$. Notice that this graph is $y = \sec x$ reflected across the x-axis.

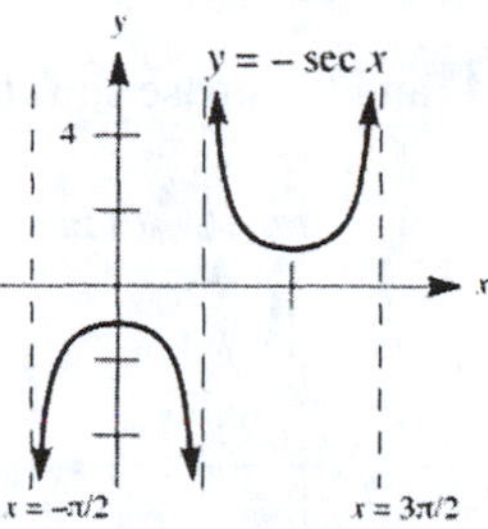

29. There is no x-intercept, the y-intercept is -1, and the asymptotes are $x = \frac{\pi}{2}$, $x = \frac{3\pi}{2}$, and $x = \frac{5\pi}{2}$. Notice that this graph is $y = \sec x$ displaced π units to the right.

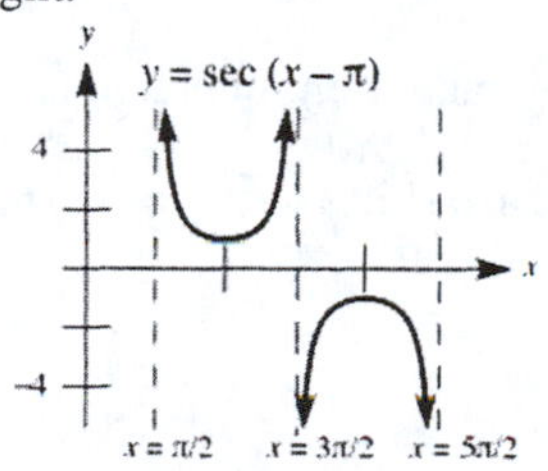

31. There is no x-intercept, the y-intercept is 3, and the asymptotes are $x = -1$, $x = 1$, and $x = 3$. Notice that the period is $\frac{2\pi}{\pi/2} = 4$.

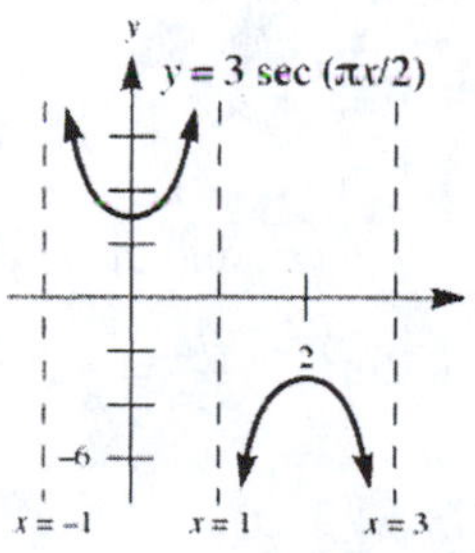

33. It appears that $\sin x = \csc x$ at odd multiples of $\frac{\pi}{2}$, such as $-\frac{3\pi}{2}, -\frac{\pi}{2}, \frac{\pi}{2}$, and $\frac{3\pi}{2}$. Since $\sin x$ and $\csc x$ have the same sign, there are no points in which $\sin x = -\csc x$. Using the standard viewing rectangle, the graphs appear as:

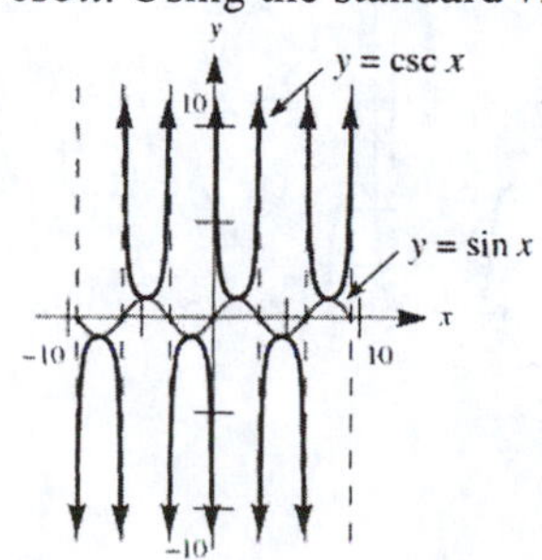

35. **a.** The x-intercepts are $-\frac{11}{18}, -\frac{5}{18}, \frac{1}{18}, \frac{7}{18}$ and $\frac{13}{18}$, and the y-intercept is -1, and there are no asymptotes. Notice that the period is $\frac{2\pi}{3\pi} = \frac{2}{3}$ and the phase shift is $\frac{\pi/6}{3\pi} = \frac{1}{18}$.

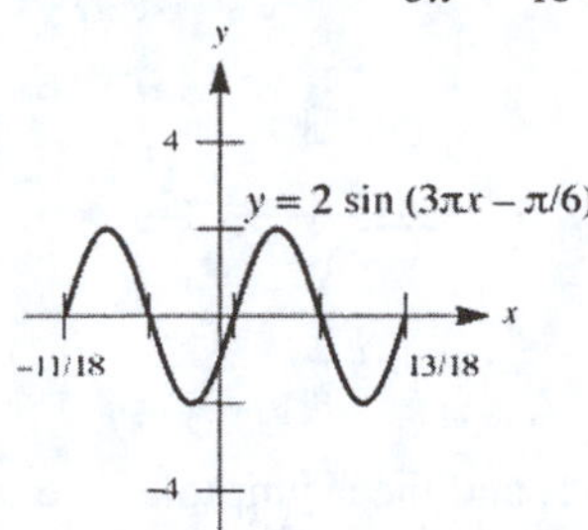

b. There are no x-intercepts, the y-intercept is -4, and the asymptotes are $x = -\frac{11}{18}$, $x = -\frac{5}{18}$, $x = \frac{1}{18}$, $x = \frac{7}{18}$ and $x = \frac{13}{18}$. Notice that the period is $\frac{2\pi}{3\pi} = \frac{2}{3}$ and the phase shift is $\frac{\pi/6}{3\pi} = \frac{1}{18}$.

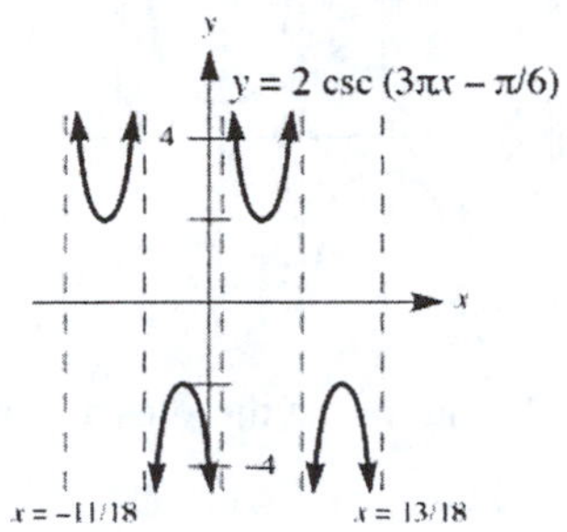

37. **a.** The x-intercepts are $-\frac{5}{8}, -\frac{1}{8}, \frac{3}{8}$, and $\frac{7}{8}$, and the y-intercept is $-\frac{3\sqrt{2}}{2} \approx -2.12$, and there are no asymptotes. Notice that the period is $\frac{2\pi}{2\pi} = 1$, the phase shift is $\frac{\pi/4}{2\pi} = \frac{1}{8}$, and that this graph is $y = 3\cos\left(2\pi x - \frac{\pi}{4}\right)$ reflected across the x-axis.

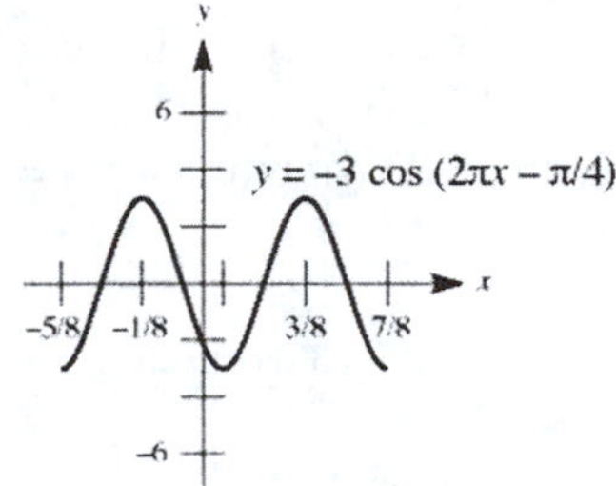

b. There are no x-intercepts, the y-intercept is $-3\sqrt{2}$, and the asymptotes are $x = -\frac{5}{8}$, $x = -\frac{1}{8}$, $x = \frac{3}{8}$ and $x = \frac{7}{8}$. Notice that the period is $\frac{2\pi}{2\pi} = 1$, the phase shift is $\frac{\pi/4}{2\pi} = \frac{1}{8}$, and that this graph is $y = 3\sec\left(2\pi x - \frac{\pi}{4}\right)$ reflected across the x-axis.

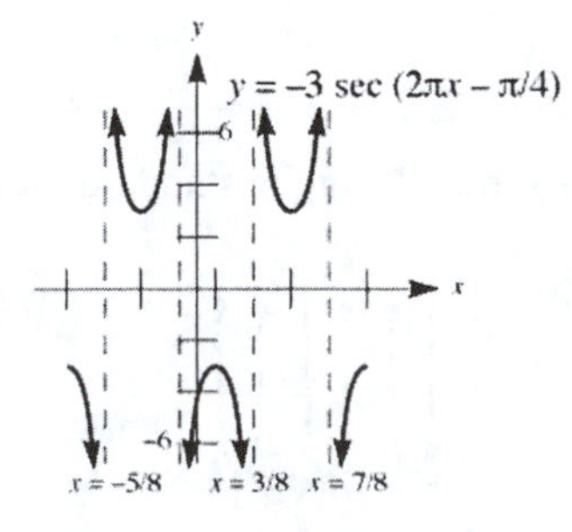

39. Graphing the two functions:

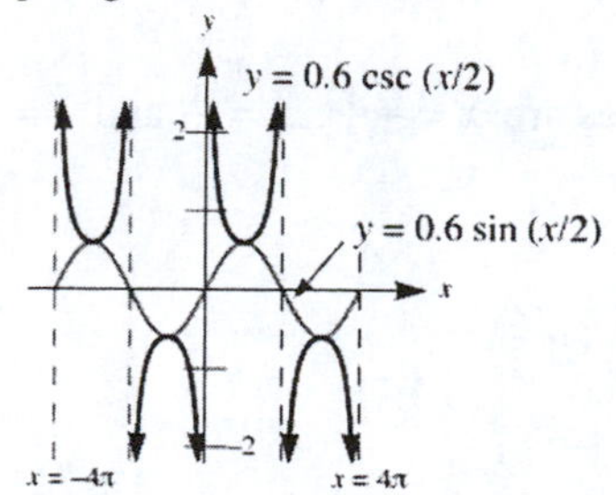

41. Graphing the two functions:

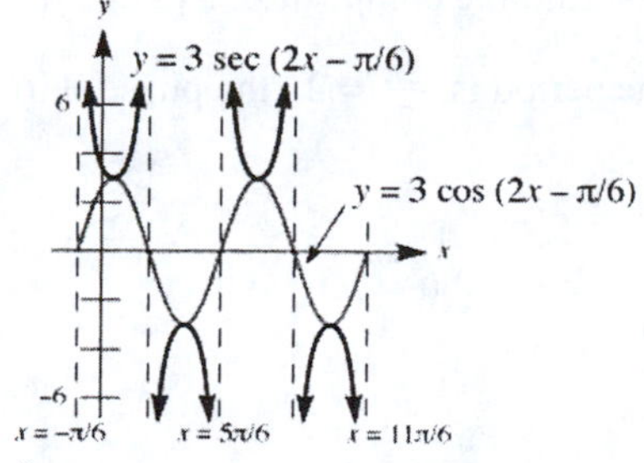

43. **a.** First find the composition: $(f \circ h)(x) = f\left(\pi x - \frac{\pi}{6}\right) = \sin\left(\pi x - \frac{\pi}{6}\right)$

The period is $\frac{2\pi}{\pi} = 2$, the amplitude is 1, and the phase shift is $\frac{\pi/6}{\pi} = \frac{1}{6}$.

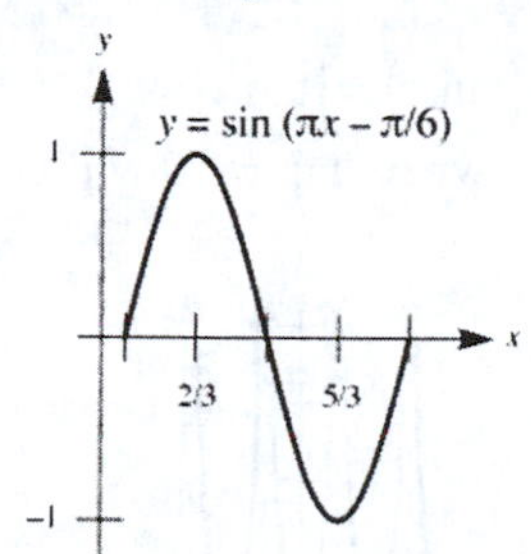

b. First find the composition: $(g \circ h)(x) = g\left(\pi x - \frac{\pi}{6}\right) = \csc\left(\pi x - \frac{\pi}{6}\right)$

The period is $\frac{2\pi}{\pi} = 2$, the phase shift is $\frac{\pi/6}{\pi} = \frac{1}{6}$, and the asymptotes are $x = \frac{1}{6}$, $x = \frac{7}{6}$ and $x = \frac{13}{6}$.

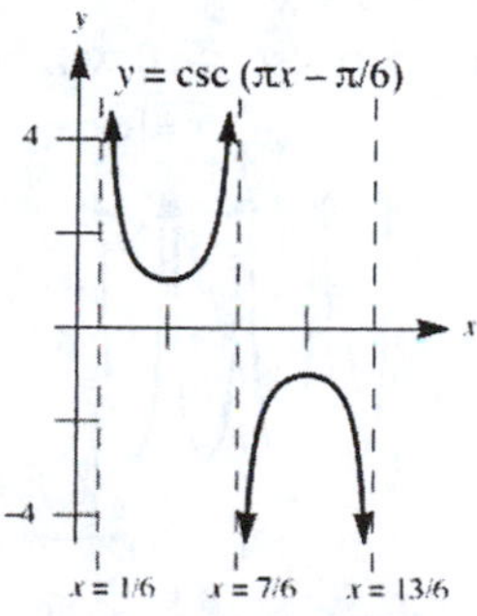

45. **a.** First find the composition: $(f \circ H)(x) = f\left(\pi x + \frac{\pi}{4}\right) = \sin\left(\pi x + \frac{\pi}{4}\right)$

The period is $\frac{2\pi}{\pi} = 2$, the amplitude is 1, and the phase shift is $\frac{-\pi/4}{\pi} = -\frac{1}{4}$.

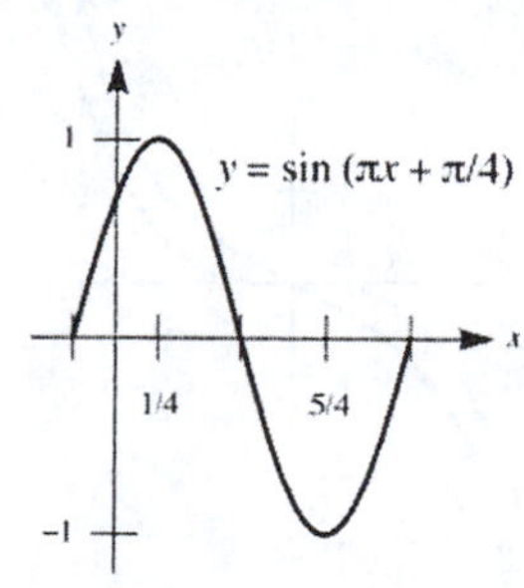

b. First find the composition: $(g \circ H)(x) = g\left(\pi x + \frac{\pi}{4}\right) = \csc\left(\pi x + \frac{\pi}{4}\right)$

The period is $\frac{2\pi}{\pi} = 2$, the phase shift is $\frac{-\pi/4}{\pi} = -\frac{1}{4}$, and the asymptotes are $x = -\frac{1}{4}$, $x = \frac{3}{4}$ and $x = \frac{7}{4}$.

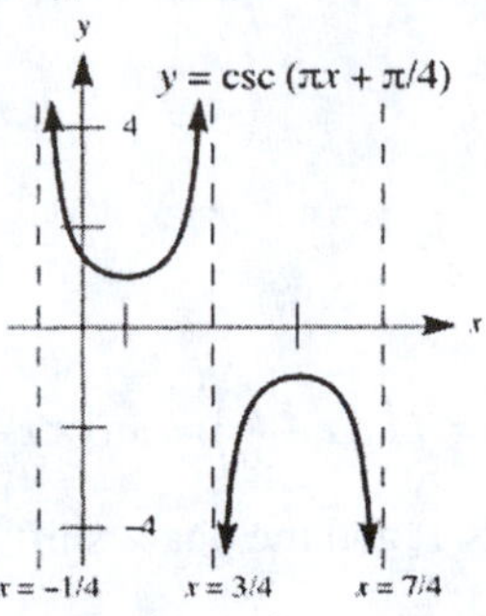

47. First find the composition: $(A \circ T)(x) = A(\tan x) = |\tan x|$

Now draw the graph, noting that values of x where $\tan x < 0$ will be reflected across the x-axis.

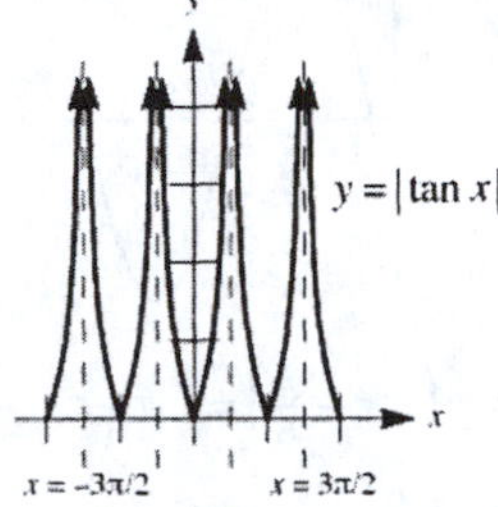

49. First find the composition: $(A \circ f)(x) = A(\csc x) = |\csc x|$

Now draw the graph, noting that values of x where $\csc x < 0$ will be reflected across the x-axis.

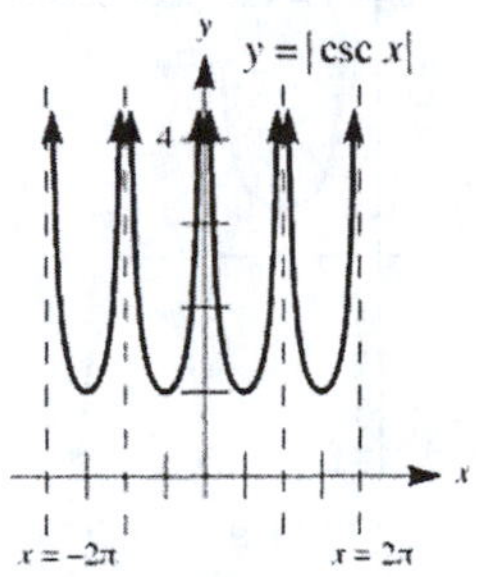

51. Graphing the two functions:

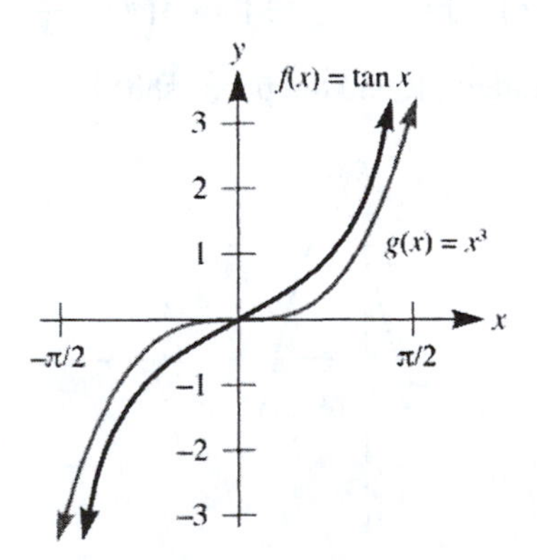

a. The graph of $y = x^3$ appears to be horizontal as it passes through the origin.

b. In the interval $0 < x < \frac{\pi}{2}$, the values of $\tan x$ appear to be larger. Also note that x^3 is finite at $x = \frac{\pi}{2}$, while $\tan x$ is positive infinite at $x = \frac{\pi}{2}$.

c. In the interval $-\frac{\pi}{2} < x < 0$, the values of $y = x^3$ appear to be larger. Also note that x^3 is finite at $x = -\frac{\pi}{2}$, while $\tan x$ is negative infinite at $x = -\frac{\pi}{2}$.

53. The two graphs are identical. This demonstrates that $\cot^2 x = \csc^2 x - 1$ is a trigonometric identity:

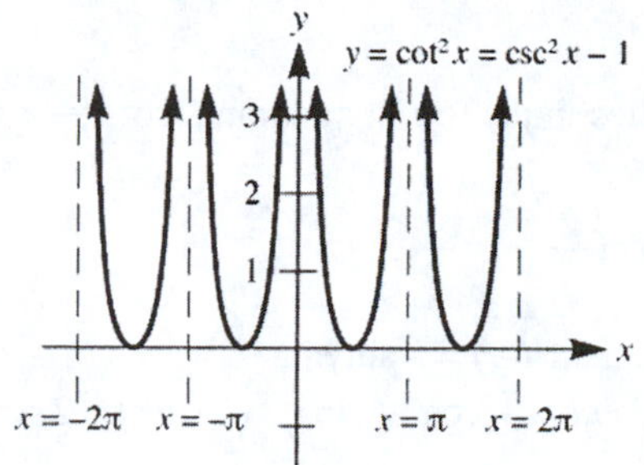

55. a. Graphing the two functions:

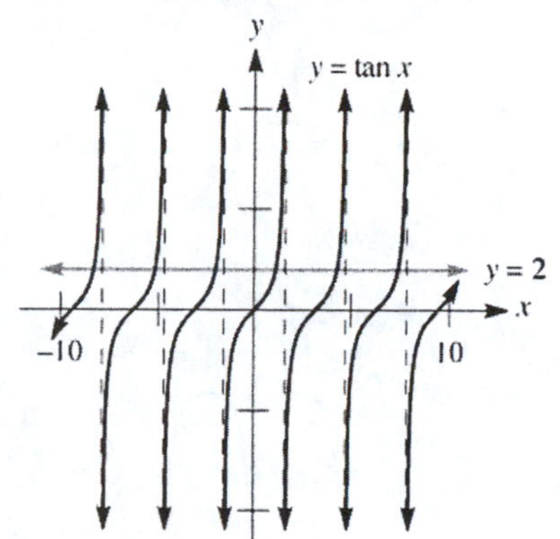

b. The two graphs intersect at approximately $x \approx 1.1$.
c. More accurately, the root is $x \approx 1.1071$.
d. Yes, since $\tan r = 2$, we must have $\tan(r + \pi) = 2$ also.

57. a. Since P and Q are both points on the unit circle, the coordinates are $P(\cos s, \sin s)$ and $Q\left(\cos\left(s - \frac{\pi}{2}\right), \sin\left(s - \frac{\pi}{2}\right)\right)$.

b. Since ΔOAP is congruent to ΔOBQ (labeling the third vertex B), $OA = OB$ and $AP = BQ$. Because the y-coordinate at Q is negative, we have concluded what was required.

c. Restating part **b**, we have shown that: $\cos\left(s - \frac{\pi}{2}\right) = \sin s$ and $\sin\left(s - \frac{\pi}{2}\right) = -\cos s$

d. Compute $\cot s$: $\cot s = \dfrac{\cos s}{\sin s} = \dfrac{-\sin\left(s - \frac{\pi}{2}\right)}{\cos\left(s - \frac{\pi}{2}\right)} = -\tan\left(s - \frac{\pi}{2}\right)$

59. a. $\angle AOE = \frac{\pi}{2} - s = \angle COD$, and since both triangles are right triangles, $\Delta AOE \approx \Delta COD$. Since $AO = OC = 1$ (both are radii of the unit circle), ΔAOE is congruent to ΔCOD.

b. $\angle COD = \frac{\pi}{2} - s$, so the y-coordinate of C is $\sin\left(\frac{\pi}{2} - s\right)$. $\angle AOD = \frac{\pi}{2} + s$, so the y-coordinate of A is $\sin\left(\frac{\pi}{2} + s\right)$

c. Since the two triangles are congruent, the y-coordinates of A and C are the same. Thus $\sin\left(\frac{\pi}{2} - s\right) = \sin\left(\frac{\pi}{2} + s\right)$. Thus $\csc\left(\frac{\pi}{2} - s\right) = \csc\left(\frac{\pi}{2} + s\right)$.

Chapter 7 Review Exercises

1. **a.** Since the terminal side of $\frac{5\pi}{3}$ lies in the fourth quadrant, $\sin\frac{5\pi}{3} < 0$. Using a reference angle of $\frac{\pi}{3}$:

$$\sin\frac{5\pi}{3} = -\sin\frac{\pi}{3} = -\frac{\sqrt{3}}{2}$$

b. Since the terminal side of $\frac{11\pi}{6}$ lies in the fourth quadrant, $\cot\frac{11\pi}{6} < 0$. Using a reference angle of $\frac{\pi}{6}$:

$$\cot\frac{11\pi}{6} = -\cot\frac{\pi}{6} = -\frac{\cos\frac{\pi}{6}}{\sin\frac{\pi}{6}} = -\frac{\frac{\sqrt{3}}{2}}{\frac{1}{2}} = -\sqrt{3}$$

3. Using the identities $\cos(-t) = \cos t$ and $\sin(-t) = -\sin t$:

$$\cos t - \cos(-t) + \sin t - \sin(-t) = \cos t - \cos t + \sin t - (-\sin t) = 2\sin t$$

5. The period is $\frac{2\pi}{2\pi} = 1$ and the phase shift is $\frac{3}{2\pi}$. The asymptotes will occur at $x = -\frac{1}{4} + \frac{3}{2\pi}$, $x = \frac{1}{4} + \frac{3}{2\pi}$ and $x = \frac{3}{4} + \frac{3}{2\pi}$.

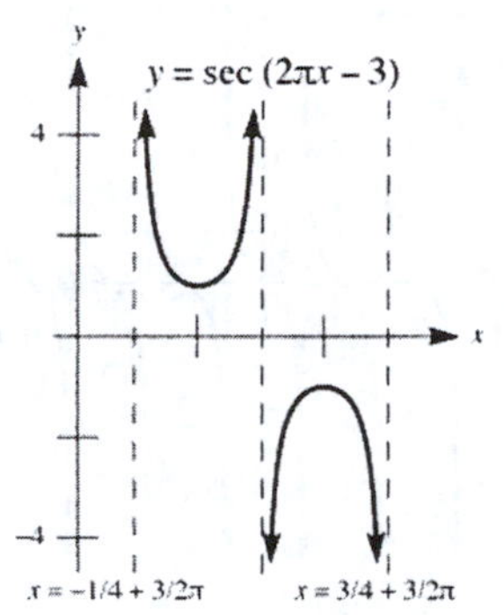

7. The amplitude is 1, the period is $\frac{2\pi}{2} = \pi$, and the phase shift is $\frac{\pi}{2}$. Notice that this graph is $y = \sin(2x - \pi)$ reflected across the x-axis.

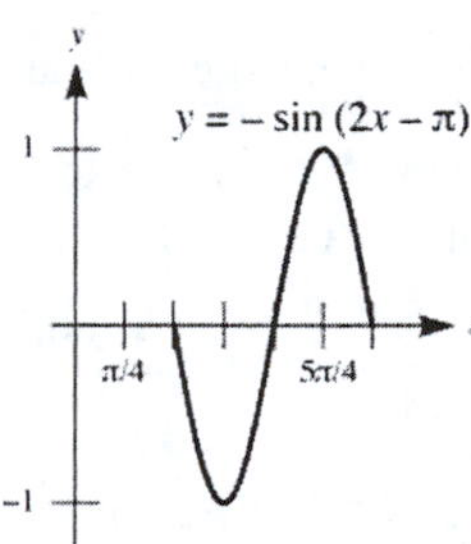

9. Since the terminal side of π radians intersects the unit circle at the point $(-1, 0)$, $\cos\pi = -1$.

11. Since the terminal side of $\frac{2\pi}{3}$ radians lies in the second quadrant, $\csc\frac{2\pi}{3} > 0$. Using a reference angle of $\frac{\pi}{3}$ radians:

$$\csc\frac{2\pi}{3} = \csc\frac{\pi}{3} = \frac{1}{\sin\frac{\pi}{3}} = \frac{2}{\sqrt{3}} = \frac{2\sqrt{3}}{3}$$

13. Since the terminal side of $\frac{11\pi}{6}$ radians lies in the fourth quadrant, $\tan\frac{11\pi}{6} < 0$. Using a reference angle of $\frac{\pi}{6}$ radians:

$$\tan\frac{11\pi}{6} = -\tan\frac{\pi}{6} = -\frac{\sin\frac{\pi}{6}}{\cos\frac{\pi}{6}} = -\frac{\frac{1}{2}}{\frac{\sqrt{3}}{2}} = -\frac{1}{\sqrt{3}} = -\frac{\sqrt{3}}{3}$$

15. Since the terminal side of $\frac{\pi}{6}$ radians lies in the first quadrant, $\sin\frac{\pi}{6} > 0$. Thus $\sin\frac{\pi}{6} = \frac{1}{2}$.

17. Since the terminal side of $\frac{5\pi}{4}$ radians lies in the third quadrant, $\cot\frac{5\pi}{4} > 0$. Using a reference angle of $\frac{\pi}{4}$ radians:

$$\cot\frac{5\pi}{4} = \cot\frac{\pi}{4} = \frac{\cos\frac{\pi}{4}}{\sin\frac{\pi}{4}} = \frac{\frac{\sqrt{2}}{2}}{\frac{\sqrt{2}}{2}} = 1$$

19. Since the terminal side of $-\frac{5\pi}{6}$ radians lies in the third quadrant, $\csc\left(-\frac{5\pi}{6}\right) < 0$. Using a reference angle of $\frac{\pi}{6}$ radians: $\csc\left(-\frac{5\pi}{6}\right) = -\csc\frac{\pi}{6} = -\frac{1}{\sin\frac{\pi}{6}} = -\frac{1}{\frac{1}{2}} = -2$

21. Using a calculator, we find $\sin 1 \approx 0.841$.

23. Since $\frac{3\pi}{2}$ radians intersects the unit circle at the point $(0,-1)$, $\sin\frac{3\pi}{2} = -1$. A calculator also verifies this value.

25. Using a calculator, we find $\sin(\sin 0.0123) \approx 0.0123$.

27. Since $\sin 1776 \approx -0.842$ and $\cos 1776 \approx -0.540$: $\sin^2 1776 + \cos^2 1776 \approx (-0.842)^2 + (-0.540)^2 = 1$

29. Since $\cos(0.25) \approx 0.969$ and $\sin(0.25) \approx 0.247$: $\cos^2(0.25) - \sin^2(0.25) \approx (0.969)^2 - (0.247)^2 \approx 0.878$
Since $\cos(0.5) \approx 0.878$, the equality is verified.

31. Since $0 < \theta < \frac{\pi}{2}$, $\cos\theta > 0$ and thus: $\sqrt{25 - x^2} = \sqrt{25 - 25\sin^2\theta} = 5\sqrt{1 - \sin^2\theta} = 5\sqrt{\cos^2\theta} = 5\cos\theta$

33. Since $0 < \theta < \frac{\pi}{2}$, $\tan\theta > 0$ and thus:

$$\left(x^2 - 100\right)^{1/2} = \left(100\sec^2\theta - 100\right)^{1/2} = 10\left(\sec^2\theta - 1\right)^{1/2} = 10\left(\tan^2\theta\right)^{1/2} = 10\tan\theta$$

35. Since $0 < \theta < \frac{\pi}{2}$, $\sec\theta > 0$ and thus: $\left(x^2 + 5\right)^{-1/2} = \left(5\tan^2\theta + 5\right)^{-1/2} = \frac{\sqrt{5}}{5}\left(\sec^2\theta\right)^{-1/2} = \frac{\sqrt{5}}{5}\cos\theta$

37. Since $\sin\theta$ is negative: $\sin\theta = -\sqrt{1 - \cos^2\theta} = -\sqrt{1 - \left(\frac{8}{17}\right)^2} = -\sqrt{1 - \frac{64}{289}} = -\sqrt{\frac{225}{289}} = -\frac{15}{17}$

Thus: $\tan\theta = \frac{\sin\theta}{\cos\theta} = \frac{-\frac{15}{17}}{\frac{8}{17}} = -\frac{15}{8}$

39. Since $\angle BPA = \theta$, $\angle APC = \pi - \theta$, and the area of the sector formed by $\angle APC$ is: $\frac{1}{2}r^2\theta = \frac{1}{2}\left(\sqrt{2}\right)^2(\pi - \theta) = \pi - \theta$

Now using the area formula for ΔAPC: $\text{Area}_{\Delta APC} = \frac{1}{2}ab\sin(\pi - \theta) = \frac{1}{2}\left(\sqrt{2}\right)\left(\sqrt{2}\right)\sin\theta = \sin\theta$

Thus, the area of the shaded region is given by: $A(\theta) = \pi - \theta - \sin\theta$

41. Using the hint, we have two congruent shaded regions each with $r = 1$ cm and $\theta = \frac{\pi}{2}$, and thus the total area is:

$$2 \bullet \frac{1}{2}\left(\frac{\pi}{2} - 1\right)\text{ cm}^2 = \frac{\pi - 2}{2}\text{ cm}^2$$

43. This is a sine function where the amplitude is 4, so $A = 4$. Since the period is 2π:

$$\frac{2\pi}{B} = 2\pi$$
$$2\pi = 2\pi B$$
$$1 = B$$

So the equation is $y = 4\sin x$.

45. This is a cosine function (reflected across the x-axis) where the amplitude is 2, so $A = -2$. Since the period is $\frac{\pi}{2}$:

$$\frac{2\pi}{B} = \frac{\pi}{2}$$
$$4\pi = \pi B$$
$$4 = B$$

So the equation is $y = -2\cos 4x$.

47. The x-intercepts are $\frac{\pi}{8}$ and $\frac{3\pi}{8}$, the high point is $\left(\frac{\pi}{4}, 3\right)$, and the low points are $(0, -3)$ and $\left(\frac{\pi}{2}, -3\right)$. Notice that the period is $\frac{2\pi}{4} = \frac{\pi}{2}$, and that this graph is $y = 3\cos 4x$ reflected across the x-axis.

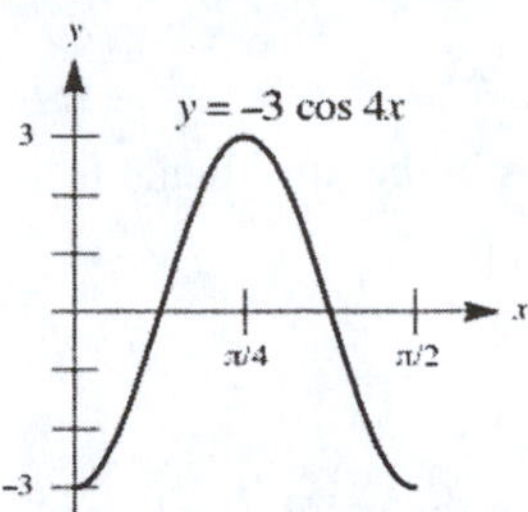

49. The x-intercepts are $\frac{1}{2}, \frac{5}{2}$ and $\frac{9}{2}$, the high point is $\left(\frac{3}{2}, 2\right)$, and the low point is $\left(\frac{7}{2}, -2\right)$. Notice that the period is $\frac{2\pi}{\pi/2} = 4$, and the phase shift is $\frac{\pi/4}{\pi/2} = \frac{1}{2}$.

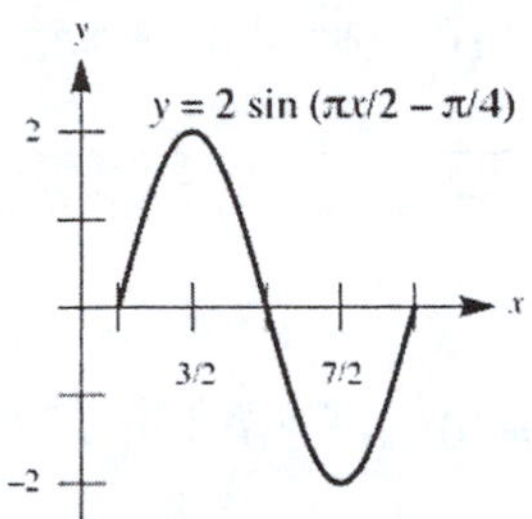

51. The x-intercepts are $\frac{5}{2}$ and $\frac{11}{2}$, the high points are $(1, 3)$ and $(7, 3)$, and the low point is $(4, -3)$. Notice that the period is $\frac{2\pi}{\pi/6} = 6$, and the phase shift is $\frac{\pi/3}{\pi/3} = 1$.

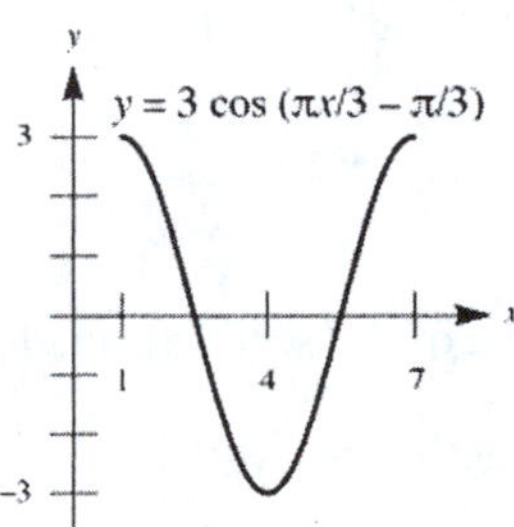

53. **a.** Notice that the period is $\frac{\pi}{\pi/4} = 4$, and the asymptotes occur at $x = 2$ and $x = -2$.

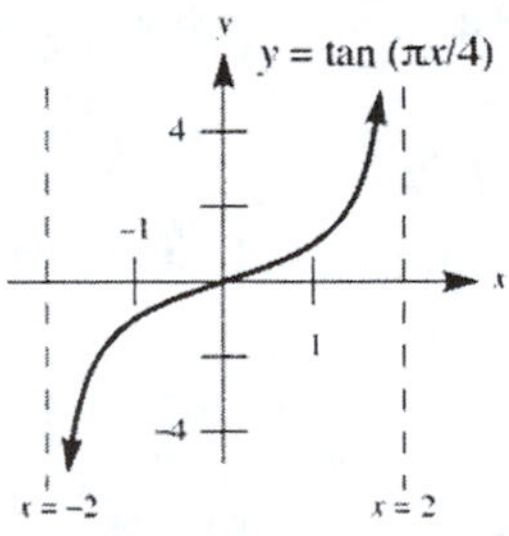

b. Notice that the period is $\frac{\pi}{\pi/4} = 4$, and the asymptotes occur at $x = 0$ and $x = 4$.

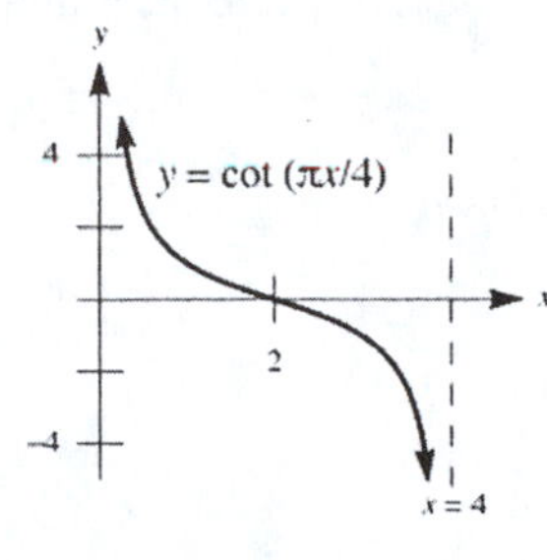

55. **a.** Notice that the period is $\frac{2\pi}{1/4} = 8\pi$, and the asymptotes occur at $x = -2\pi$, $x = 2\pi$ and $x = 6\pi$.

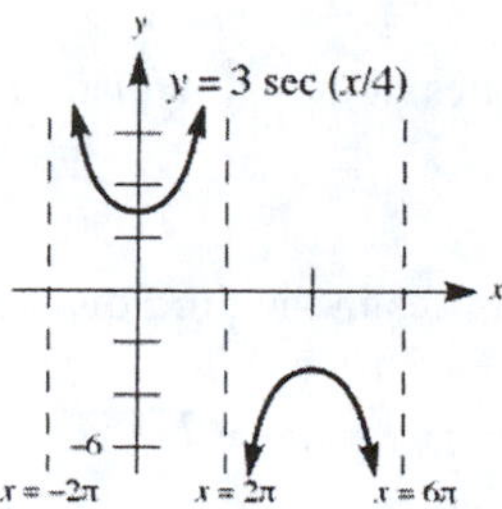

b. Notice that the period is $\frac{2\pi}{1/4} = 8\pi$, and the asymptotes occur at $x = -4\pi$, $x = 0$ and $x = 4\pi$.

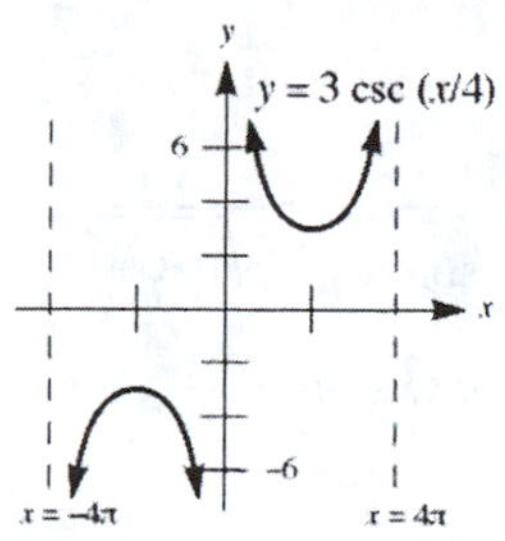

57. **a.** The amplitude is 2.5 cm, the period is $\frac{2\pi}{\pi/8} = 16$ sec, and the frequency is $\frac{1}{16}$ cycles/sec. Sketch the graph over the interval $0 \le t \le 32$:

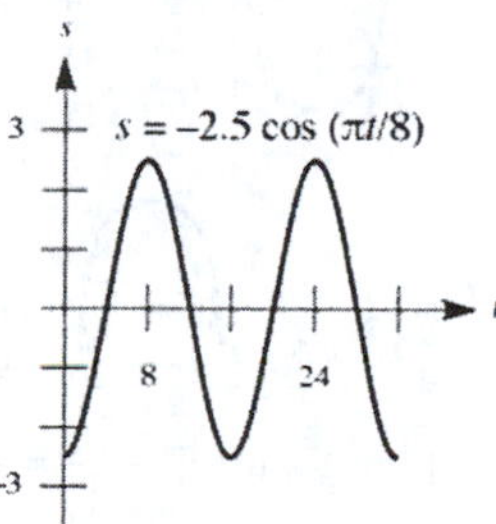

b. The mass is farthest from the origin at high and low points, which occur at $t = 0$, $t = 8$, $t = 16$, $t = 24$ and $t = 32$ sec.

c. The mass is passing through the equilibrium position when $s = 0$, which occurs at $t = 4$, $t = 12$, $t = 20$ and $t = 28$ sec.

59. First find the area of each region, noting that $AD = \tan\alpha$ and $AE = \tan\beta$:

$$A_{ABDA} = \tfrac{1}{2}(1)(\tan\alpha) - \tfrac{1}{2}(1)^2\alpha = \tfrac{1}{2}(\tan\alpha - \alpha)$$
$$A_{ABCEDA} = \tfrac{1}{2}(1)(\tan\beta) - \tfrac{1}{2}(1)^2\beta = \tfrac{1}{2}(\tan\beta - \beta)$$

Since $A_{ABDA} < A_{ABCEDA}$, the resulting inequality is:

$$\tfrac{1}{2}(\tan\alpha - \alpha) < \tfrac{1}{2}(\tan\beta - \beta)$$
$$\tan\alpha - \alpha < \tan\beta - \beta$$
$$\tan\alpha - \tan\beta < \alpha - \beta$$
$$\tan\beta - \tan\alpha > \beta - \alpha$$

This proves the desired inequality.

Chapter 7 Test

1. a. Since the terminal side of $\frac{4\pi}{3}$ radians lies in the third quadrant, $\cos\frac{4\pi}{3} < 0$. Using a reference angle of $\frac{\pi}{3}$ radians: $\cos\frac{4\pi}{3} = -\cos\frac{\pi}{3} = -\frac{1}{2}$

   b. Since the terminal side of $-\frac{5\pi}{6}$ radians lies in the third quadrant, $\csc\left(-\frac{5\pi}{6}\right) < 0$. Using a reference angle of $\frac{\pi}{6}$ radians: $\csc\left(-\frac{5\pi}{6}\right) = -\csc\frac{\pi}{6} = -\frac{1}{\sin\frac{\pi}{6}} = -\frac{1}{\frac{1}{2}} = -2$

   c. Since $\sin^2 t + \cos^2 t = 1$ for all values of t: $\sin^2\frac{3\pi}{4} + \cos^2\frac{3\pi}{4} = 1$

2. Substituting $t = 4\sin u$ and noting that $\sin u > 0$ when $0 < u < \frac{\pi}{2}$:

$$\frac{1}{\sqrt{16-t^2}} = \frac{1}{\sqrt{16-16\sin^2 u}} = \frac{1}{\sqrt{16\left(1-\sin^2 u\right)}} = \frac{1}{\sqrt{16\cos^2 u}} = \frac{1}{4\cos u} = \frac{1}{4}\sec u$$

3. Noting that the period is $\frac{2\pi}{4\pi} = \frac{1}{2}$, and the phase shift is $\frac{1}{4\pi}$, then the asymptotes are $x = -\frac{1}{8} + \frac{1}{4\pi}$, $x = \frac{1}{8} + \frac{1}{4\pi}$, and $x = \frac{3}{8} + \frac{1}{4\pi}$.

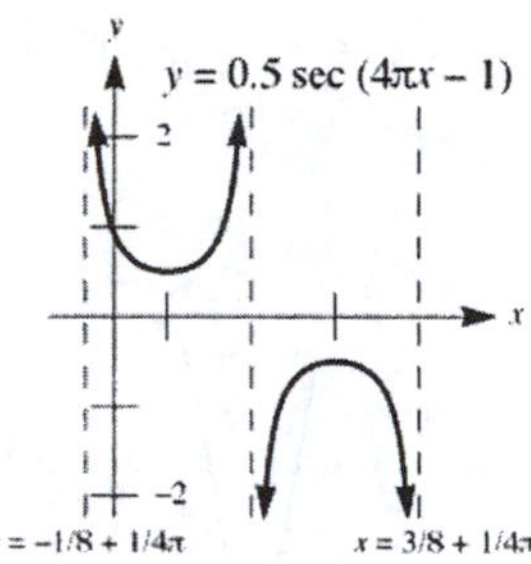

4. The amplitude is 1, the period is $\frac{2\pi}{3}$, and the phase shift is $\frac{\pi/4}{3} = \frac{\pi}{12}$. Notice that this graph is $y = \sin\left(3x - \frac{\pi}{4}\right)$ reflected across the x-axis.

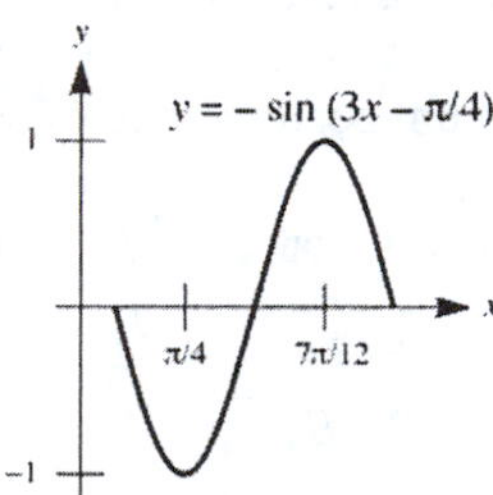

5. Notice that the period is $\frac{\pi}{\pi/4} = 4$, and that the asymptote occurs at $x = 2$.

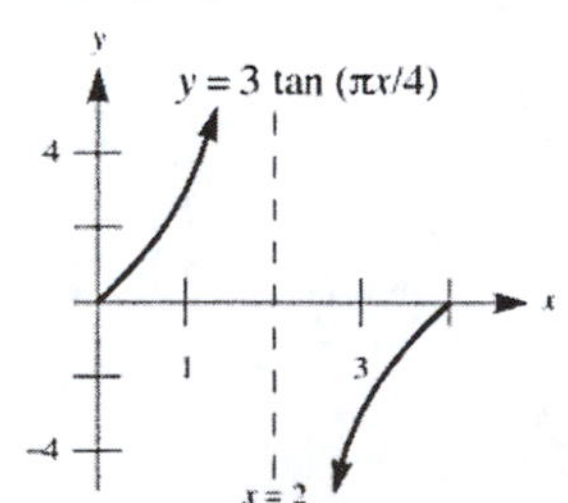

6. a. Converting the angular speed to radians: $\omega = 25\,\frac{\text{rev}}{\text{sec}} \bullet 2\pi\,\frac{\text{rad}}{\text{rev}} = 50\pi\,\frac{\text{rad}}{\text{sec}}$

   b. In 1 second, the wheel has rotated 50π radians, so the total distance traveled by a point 5 cm from the center is $d = (5\text{ cm})(50\pi\text{ rad}) = 250\pi\text{ cm}$. Thus the linear velocity is given by: $v = \frac{d}{t} = \frac{250\pi\text{ cm}}{1\text{ sec}} = 250\pi\,\frac{\text{cm}}{\text{sec}}$

7. Simplify each side of the equation, using the identities $\sin(-\theta) = -\sin\theta$ and $\cos(-\theta) = \cos\theta$:

$$\frac{\cot\theta}{1+\tan(-\theta)}+\frac{\tan\theta}{1+\cot(-\theta)}=\frac{\frac{\cos\theta}{\sin\theta}}{1+\frac{\sin(-\theta)}{\cos(-\theta)}}+\frac{\frac{\sin\theta}{\cos\theta}}{1+\frac{\cos(-\theta)}{\sin(-\theta)}}$$

$$=\frac{\frac{\cos\theta}{\sin\theta}}{1-\frac{\sin\theta}{\cos\theta}}+\frac{\frac{\sin\theta}{\cos\theta}}{1-\frac{\cos\theta}{\sin\theta}}$$

$$=\frac{\cos^2\theta}{\sin\theta\cos\theta-\sin^2\theta}+\frac{\sin^2\theta}{\sin\theta\cos\theta-\cos^2\theta}$$

$$=\frac{\cos^2\theta}{\sin\theta(\cos\theta-\sin\theta)}+\frac{\sin^2\theta}{\cos\theta(\sin\theta-\cos\theta)}$$

$$=\frac{-\cos^3\theta}{\sin\theta\cos\theta(\sin\theta-\cos\theta)}+\frac{\sin^3\theta}{\sin\theta\cos\theta(\sin\theta-\cos\theta)}$$

$$=\frac{\sin^3\theta-\cos^3\theta}{\sin\theta\cos\theta(\sin\theta-\cos\theta)}$$

$$=\frac{(\sin\theta-\cos\theta)(\sin^2\theta+\sin\theta\cos\theta+\cos^2\theta)}{\sin\theta\cos\theta(\sin\theta-\cos\theta)}$$

$$=\frac{1+\sin\theta\cos\theta}{\sin\theta\cos\theta}$$

$$\cot\theta+\tan\theta+1=\frac{\cos\theta}{\sin\theta}+\frac{\sin\theta}{\cos\theta}+1=\frac{\cos^2\theta+\sin^2\theta+\sin\theta\cos\theta}{\sin\theta\cos\theta}=\frac{1+\sin\theta\cos\theta}{\sin\theta\cos\theta}$$

Since both sides of the equation simplify to the same quantity, the equation is an identity.

8. **a.** The period is $\frac{2\pi}{\pi/3} = 6$ sec, and the amplitude is 10 cm. Graph the function over the interval $0 \le t \le 12$:

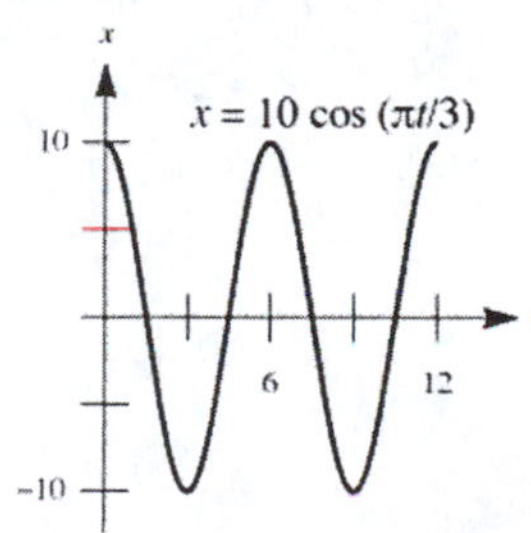

b. The point is passing through the origin when $x = 0$, which occurs when $t = 1.5$, $t = 4.5$, $t = 7.5$ and $t = 10.5$ sec. The point is farthest from the origin at the high and low points of this graph, which occur when $t = 0$, $t = 3$, $t = 6$, $t = 9$ and $t = 12$ sec.

9. **a.** $\sin^2 13 + \cos^2 13 = 1$, since $\sin^2 x + \cos^2 x = 1$ regardless of the value of x.

b. $\sin 5 + \sin(-5) = \sin 5 - \sin 5 = 0$

c. $\tan 1 + \tan(-1 - 2\pi) = \tan 1 + \tan(-1) = \tan 1 - \tan 1 = 0$

10. **a.** Using the graph, a root is $x \approx 1.1$.

b. Using a calculator, a root is $x \approx 1.1198$.

c. Another root would be $\pi - x \approx 2.0218$.

11. First find A: $A=\dfrac{M-m}{2}=\dfrac{15.27-9.08}{2}=3.095$. Assuming $\dfrac{2\pi}{B}=366$, then $B=\dfrac{2\pi}{366}=\dfrac{\pi}{183}$. The maximum occurs when t = 167 (and normally it would occur at $\dfrac{366}{4}=91.5$, so the phase shift is 75.5. So $\dfrac{C}{B}=75.5$, thus $C=75.5B=75.5\left(\dfrac{\pi}{183}\right)=\dfrac{151\pi}{366}$. Using $y=A\sin(Bt-C)+D$ and the data pair (15,9.42):

$$\begin{aligned}9.42&=3.095\sin\left(\frac{\pi}{183}\bullet 15-\frac{151\pi}{366}\right)+D\\ 9.42&=3.095\sin\left(-\frac{121\pi}{366}\right)+D\\ 9.42&=-2.667+D\\ 12.087&=D\end{aligned}$$

The function is $y=3.095\sin\left(\dfrac{\pi}{183}t-\dfrac{151\pi}{366}\right)+12.087$.

Chapter 8
Analytical Trigonometry

8.1 The Addition Formulas

1. Using the identity $\sin(s+t) = \sin s \cos t + \cos s \sin t$ where $s = \theta$ and $t = 2\theta$:
$\sin\theta\cos 2\theta + \cos\theta\sin 2\theta = \sin(\theta + 2\theta) = \sin 3\theta$

3. Using the identity $\sin(s-t) = \sin s \cos t - \cos s \sin t$ where $s = 3\theta$ and $t = \theta$:
$\sin 3\theta\cos\theta - \cos 3\theta\sin\theta = \sin(3\theta - \theta) = \sin 2\theta$

5. Using the identity $\cos(s+t) = \cos s \cos t - \sin s \sin t$ where $s = 2u$ and $t = 3u$:
$\cos 2u\cos 3u - \sin 2u\sin 3u = \cos(2u + 3u) = \cos 5u$

7. Using the identity $\cos(s-t) = \cos s \cos t + \sin s \sin t$ where $s = \frac{2\pi}{9}$ and $t = \frac{\pi}{18}$:
$$\cos\tfrac{2\pi}{9}\cos\tfrac{\pi}{18} + \sin\tfrac{2\pi}{9}\sin\tfrac{\pi}{18} = \cos\left(\tfrac{2\pi}{9} - \tfrac{\pi}{18}\right) = \cos\tfrac{3\pi}{18} = \cos\tfrac{\pi}{6} = \frac{\sqrt{3}}{2}$$

9. Using the identity $\sin(s-t) = \sin s \cos t - \cos s \sin t$ where $s = A + B$ and $t = A$:
$\sin(A+B)\cos A - \cos(A+B)\sin A = \sin(A + B - A) = \sin B$

11. Using the identity $\sin(s-t) = \sin s \cos t - \cos s \sin t$ with $s = \theta$ and $t = \frac{3\pi}{2}$:
$$\sin\left(\theta - \tfrac{3\pi}{2}\right) = \sin\theta\cos\tfrac{3\pi}{2} - \cos\theta\sin\tfrac{3\pi}{2} = \sin\theta \bullet 0 - \cos\theta \bullet (-1) = \cos\theta$$

13. Using the identity $\cos(s+t) = \cos s \cos t - \sin s \sin t$ with $s = \theta$ and $t = \pi$:
$\cos(\theta + \pi) = \cos\theta\cos\pi - \sin\theta\sin\pi = \cos\theta \bullet (-1) - \sin\theta \bullet 0 = -\cos\theta$

15. Using the identity $\sin(s+t) = \sin s \cos t + \cos s \sin t$ with $s = t$ and $t = 2\pi$:
$\sin(t + 2\pi) = \sin t\cos 2\pi + \cos t\sin 2\pi = \sin t \bullet 1 + \cos t \bullet 0 = \sin t$

Notice that this verifies the formula $\sin(t + 2\pi) = \sin t$.

17. Using the identity $\cos(s+t) = \cos s \cos t - \sin s \sin t$:
$$\cos 75^\circ = \cos(45^\circ + 30^\circ) = \cos 45^\circ\cos 30^\circ - \sin 45^\circ\sin 30^\circ = \frac{\sqrt{2}}{2} \bullet \frac{\sqrt{3}}{2} - \frac{\sqrt{2}}{2} \bullet \frac{1}{2} = \frac{\sqrt{6} - \sqrt{2}}{4}$$

19. Using the identity $\sin(s+t) = \sin s \cos t + \cos s \sin t$:
$$\sin\tfrac{7\pi}{12} = \sin\left(\tfrac{\pi}{3} + \tfrac{\pi}{4}\right) = \sin\tfrac{\pi}{3}\cos\tfrac{\pi}{4} + \sin\tfrac{\pi}{4}\cos\tfrac{\pi}{3} = \frac{\sqrt{3}}{2} \bullet \frac{\sqrt{2}}{2} + \frac{\sqrt{2}}{2} \bullet \frac{1}{2} = \frac{\sqrt{6} + \sqrt{2}}{4}$$

21. Using the identities $\sin(s+t) = \sin s \cos t + \cos s \sin t$ and $\sin(s-t) = \sin s \cos t - \cos s \sin t$:

$$\begin{aligned}\sin\left(\tfrac{\pi}{4}+s\right)-\sin\left(\tfrac{\pi}{4}-s\right) &= \left(\sin\tfrac{\pi}{4}\cos s+\cos\tfrac{\pi}{4}\sin s\right)-\left(\sin\tfrac{\pi}{4}\cos s-\cos\tfrac{\pi}{4}\sin s\right)\\ &= \frac{\sqrt{2}}{2}\cos s+\frac{\sqrt{2}}{2}\sin s-\frac{\sqrt{2}}{2}\cos s+\frac{\sqrt{2}}{2}\sin s\\ &= \sqrt{2}\sin s\end{aligned}$$

23. Using the identities $\cos(s+t) = \cos s \cos t - \sin s \sin t$ and $\cos(s-t) = \cos s \cos t + \sin s \sin t$:

$$\begin{aligned}\cos\left(\tfrac{\pi}{3}-\theta\right)-\cos\left(\tfrac{\pi}{3}+\theta\right) &= \left(\cos\tfrac{\pi}{3}\cos\theta+\sin\tfrac{\pi}{3}\sin\theta\right)-\left(\cos\tfrac{\pi}{3}\cos\theta-\sin\tfrac{\pi}{3}\sin\theta\right)\\ &= \tfrac{1}{2}\cos\theta+\frac{\sqrt{3}}{2}\sin\theta-\tfrac{1}{2}\cos\theta+\frac{\sqrt{3}}{2}\sin\theta\\ &= \sqrt{3}\sin\theta\end{aligned}$$

25. First compute $\cos\alpha$ and $\sin\beta$. Since $\frac{\pi}{2}<\alpha<\pi$, $\cos\alpha<0$ and thus:

$$\cos\alpha = -\sqrt{1-\sin^2\alpha} = -\sqrt{1-\left(\tfrac{12}{13}\right)^2} = -\sqrt{1-\tfrac{144}{169}} = -\sqrt{\tfrac{25}{169}} = -\tfrac{5}{13}$$

Since $\pi<\beta<\frac{3\pi}{2}$, $\sin\beta<0$ and thus:

$$\sin\beta = -\sqrt{1-\cos^2\beta} = -\sqrt{1-\left(-\tfrac{3}{5}\right)^2} = -\sqrt{1-\tfrac{9}{25}} = -\sqrt{\tfrac{16}{25}} = -\tfrac{4}{5}$$

a. Using the identity for $\sin(\alpha+\beta)$:

$$\sin(\alpha+\beta) = \sin\alpha\cos\beta+\cos\alpha\sin\beta = \left(\tfrac{12}{13}\right)\cdot\left(-\tfrac{3}{5}\right)+\left(-\tfrac{5}{13}\right)\cdot\left(-\tfrac{4}{5}\right) = -\tfrac{36}{65}+\tfrac{20}{65} = -\tfrac{16}{65}$$

b. Using the identity for $\cos(\alpha+\beta)$:

$$\cos(\alpha+\beta) = \cos\alpha\cos\beta-\sin\alpha\sin\beta = \left(-\tfrac{5}{13}\right)\cdot\left(-\tfrac{3}{5}\right)-\left(\tfrac{12}{13}\right)\cdot\left(-\tfrac{4}{5}\right) = \tfrac{15}{65}+\tfrac{48}{65} = \tfrac{63}{65}$$

27. Use the value for $\cos\alpha$ computed in Exercise 26. Since $-2\pi<\theta<-\frac{3\pi}{2}$, $\sin\theta>0$ and thus:

$$\sin\theta = \sqrt{1-\cos^2\theta} = \sqrt{1-\left(\tfrac{7}{25}\right)^2} = \sqrt{1-\tfrac{49}{625}} = \sqrt{\tfrac{576}{625}} = \tfrac{24}{25}$$

a. Using the identity for $\sin(\theta-\beta)$:

$$\sin(\theta-\beta) = \sin\theta\cos\beta-\cos\theta\sin\beta = \left(\tfrac{24}{25}\right)\cdot\left(-\tfrac{3}{5}\right)-\left(\tfrac{7}{25}\right)\cdot\left(-\tfrac{4}{5}\right) = -\tfrac{72}{125}+\tfrac{28}{125} = -\tfrac{44}{125}$$

b. Using the identity for $\sin(\theta+\beta)$:

$$\sin(\theta+\beta) = \sin\theta\cos\beta+\cos\theta\sin\beta = \left(\tfrac{24}{25}\right)\cdot\left(-\tfrac{3}{5}\right)+\left(\tfrac{7}{25}\right)\cdot\left(-\tfrac{4}{5}\right) = -\tfrac{72}{125}-\tfrac{28}{125} = -\tfrac{100}{125} = -\tfrac{4}{5}$$

29. **a.** Since $0<\theta<\frac{\pi}{2}$, $\cos\theta>0$ and thus: $\cos\theta = \sqrt{1-\sin^2\theta} = \sqrt{1-\left(\tfrac{1}{5}\right)^2} = \sqrt{1-\tfrac{1}{25}} = \sqrt{\tfrac{24}{25}} = \dfrac{2\sqrt{6}}{5}$

b. Since $\sin 2\theta = \sin(\theta+\theta)$, use the addition formula for $\sin(s+t)$ to obtain:

$$\sin 2\theta = \sin\theta\cos\theta+\sin\theta\cos\theta = 2\sin\theta\cos\theta = 2\cdot\frac{1}{5}\cdot\frac{2\sqrt{6}}{5} = \frac{4\sqrt{6}}{25}$$

31. First find $\sin\theta$, $\cos\theta$, $\sin\beta$ and $\cos\beta$. Since $\frac{\pi}{2}<\theta<\pi$, $\sec\theta<0$, so using the identity $\sec^2\theta = 1+\tan^2\theta$:

$$\sec\theta = -\sqrt{1+\tan^2\theta} = -\sqrt{1+\left(-\tfrac{2}{3}\right)^2} = -\sqrt{1+\tfrac{4}{9}} = -\sqrt{\tfrac{13}{3}} = -\frac{\sqrt{13}}{3}$$

Thus $\cos\theta = -\frac{3}{\sqrt{13}} = -\frac{3\sqrt{13}}{13}$. Since $\frac{\pi}{2}<\theta<\pi$, $\sin\theta>0$ and thus:

$$\sin\theta = \sqrt{1-\cos^2\theta} = \sqrt{1-\left(-\frac{3\sqrt{13}}{13}\right)^2} = \sqrt{1-\tfrac{9}{13}} = \sqrt{\tfrac{4}{13}} = \frac{2}{\sqrt{13}} = \frac{2\sqrt{13}}{13}$$

Since $\csc\beta = 2$, $\sin\beta = \frac{1}{2}$. Since $0<\beta<\frac{\pi}{2}$, $\cos\beta>0$ and thus:

$$\cos\beta = \sqrt{1-\sin^2\beta} = \sqrt{1-\left(\tfrac{1}{2}\right)^2} = \sqrt{1-\tfrac{1}{4}} = \sqrt{\tfrac{3}{4}} = \frac{\sqrt{3}}{2}$$

Now using the addition formula for sin $(s + t)$:

$$\sin(\theta+\beta) = \sin\theta\cos\beta + \cos\theta\sin\beta = \frac{2\sqrt{13}}{13}\bullet\frac{\sqrt{3}}{2} - \frac{3\sqrt{13}}{13}\bullet\frac{1}{2} = \frac{2\sqrt{39}}{26} - \frac{3\sqrt{13}}{26} = \frac{2\sqrt{39}-3\sqrt{13}}{26}$$

Finally using the addition formula for cos $(s - t)$:

$$\cos(\beta-\theta) = \cos\beta\cos\theta + \sin\beta\sin\theta = \frac{\sqrt{3}}{2}\bullet\left(-\frac{3\sqrt{13}}{13}\right) + \frac{1}{2}\bullet\frac{2\sqrt{13}}{13} = -\frac{3\sqrt{39}}{26} + \frac{2\sqrt{13}}{26} = \frac{2\sqrt{13}-3\sqrt{39}}{26}$$

33. Using the addition formula for sin $(s + t)$: $\sin\left(t+\frac{\pi}{4}\right) = \sin t\cos\frac{\pi}{4} + \cos t\sin\frac{\pi}{4} = \frac{1}{\sqrt{2}}\sin t + \frac{1}{\sqrt{2}}\cos t = \frac{\sin t + \cos t}{\sqrt{2}}$

35. Using the results from Exercises 33 and 34: $\sin\left(t+\frac{\pi}{4}\right) + \cos\left(t+\frac{\pi}{4}\right) = \frac{\sin t + \cos t}{\sqrt{2}} + \frac{\cos t - \sin t}{\sqrt{2}} = \frac{2\cos t}{\sqrt{2}} = \sqrt{2}\cos t$

37. Using the addition formulas for tan $(s + t)$ and tan $(s - t)$:

$$\tan(s+t) = \frac{\tan s + \tan t}{1-\tan s\tan t} = \frac{2+3}{1-2\bullet 3} = \frac{5}{-5} = -1 \qquad \tan(s-t) = \frac{\tan s - \tan t}{1+\tan s\tan t} = \frac{2-3}{1+2\bullet 3} = \frac{-1}{7} = -\frac{1}{7}$$

39. Using the addition formulas for tan $(s + t)$ and tan $(s - t)$:

$$\tan(s+t) = \frac{\tan s + \tan t}{1-\tan s\tan t} = \frac{\tan\frac{3\pi}{4}+(-4)}{1-\tan\frac{3\pi}{4}\bullet(-4)} = \frac{-1-4}{1-(-1)(-4)} = \frac{-5}{-3} = \frac{5}{3}$$

$$\tan(s-t) = \frac{\tan s - \tan t}{1+\tan s\tan t} = \frac{\tan\frac{3\pi}{4}-(-4)}{1+\tan\frac{3\pi}{4}\bullet(-4)} = \frac{-1+4}{1+(-1)(-4)} = \frac{3}{5}$$

41. Since $\tan(s+t) = \frac{\tan s + \tan t}{1-\tan s\tan t}$, applying this formula with $s = t$ and $t = 2t$: $\frac{\tan t + \tan 2t}{1-\tan t\tan 2t} = \tan(t+2t) = \tan 3t$

43. Since $\tan(s-t) = \frac{\tan s - \tan t}{1+\tan s\tan t}$, applying this formula with $s = 70°$ and $t = 10°$:

$$\frac{\tan 70° - \tan 10°}{1+\tan 70°\tan 10°} = \tan(70°-10°) = \tan 60° = \sqrt{3}$$

45. Since $\tan(s+t) = \frac{\tan s + \tan t}{1-\tan s\tan t}$, applying this formula with $s = x - y$ and $t = y$:

$$\frac{\tan(x-y)+\tan y}{1-\tan(x-y)\tan y} = \tan(x-y+y) = \tan x$$

47. Using the addition formula for tan $(s + t)$ with $s = \frac{\pi}{3}$ and $t = \frac{\pi}{4}$:

$$\tan\frac{7\pi}{12} = \frac{\tan\frac{\pi}{3}+\tan\frac{\pi}{4}}{1-\tan\frac{\pi}{3}\bullet\tan\frac{\pi}{4}} = \frac{\sqrt{3}+1}{1-\sqrt{3}\bullet 1}\bullet\frac{1+\sqrt{3}}{1+\sqrt{3}} = \frac{1+2\sqrt{3}+3}{1-3} = \frac{4+2\sqrt{3}}{-2} = -2-\sqrt{3}$$

49. Using the addition formula for sin $(s + t)$: $\frac{\sin(s+t)}{\cos s\cos t} = \frac{\sin s\cos t + \cos s\sin t}{\cos s\cos t} = \frac{\sin s}{\cos s} + \frac{\sin t}{\cos t} = \tan s + \tan t$

51. Using the addition formulas for cos $(s - t)$ and cos $(s + t)$:

$$\begin{aligned}\cos(A-B) - \cos(A+B) &= (\cos A\cos B + \sin A\sin B) - (\cos A\cos B - \sin A\sin B)\\ &= \cos A\cos B + \sin A\sin B - \cos A\cos B + \sin A\sin B\\ &= 2\sin A\sin B\end{aligned}$$

53. Using the addition formulas for $\cos(s + t)$ and $\cos(s - t)$, as well as the identities $\sin^2 A = 1 - \cos^2 A$ and $\sin^2 B = 1 - \cos^2 B$:

$$\begin{aligned}\cos(A+B)\cos(A-B) &= (\cos A\cos B - \sin A\sin B)(\cos A\cos B + \sin A\sin B)\\ &= (\cos A\cos B)^2 - (\sin A\sin B)^2\\ &= \cos^2 A\cos^2 B - \sin^2 A\sin^2 B\\ &= \cos^2 A(1-\sin^2 B) - (1-\cos^2 A)\sin^2 B\\ &= \cos^2 A - \cos^2 A\sin^2 B - \sin^2 B + \cos^2 A\sin^2 B\\ &= \cos^2 A - \sin^2 B\end{aligned}$$

55. Using the addition formulas for $\cos(s + t)$ and $\sin(s + t)$, as well as the identity $\cos^2\beta + \sin^2\beta = 1$:

$$\begin{aligned}&\cos(\alpha+\beta)\cos\beta + \sin(\alpha+\beta)\sin\beta\\ &= (\cos\alpha\cos\beta - \sin\alpha\sin\beta)\cos\beta + (\sin\alpha\cos\beta + \cos\alpha\sin\beta)\sin\beta\\ &= \cos\alpha\cos^2\beta - \sin\alpha\sin\beta\cos\beta + \sin\alpha\sin\beta\cos\beta + \cos\alpha\sin^2\beta\\ &= \cos\alpha\cos^2\beta + \cos\alpha\sin^2\beta\\ &= \cos\alpha(\cos^2\beta + \sin^2\beta)\\ &= \cos\alpha\end{aligned}$$

57. Writing $2t = t + t$: $\tan 2t = \dfrac{\tan t + \tan t}{1 - \tan t \bullet \tan t} = \dfrac{2\tan t}{1 - \tan^2 t}$

59. The average rate of change is given by:

$$\begin{aligned}\frac{\Delta f}{\Delta x} &= \frac{f(x+h) - f(x)}{h}\\ &= \frac{\sin(x+h) - \sin x}{h}\\ &= \frac{\sin x\cos h + \cos x\sin h - \sin x}{h}\\ &= \frac{(\sin x)(\cos h - 1) + (\cos x)(\sin h)}{h}\\ &= (\sin x)\left(\frac{\cos h - 1}{h}\right) + (\cos x)\left(\frac{\sin h}{h}\right)\end{aligned}$$

61. The average rate of change is given by:

$$\begin{aligned}\frac{\Delta T}{\Delta x} &= \frac{T(x+h) - T(x)}{h}\\ &= \frac{\tan(x+h) - \tan x}{h}\\ &= \frac{\dfrac{\tan x + \tan h}{1 - \tan x\tan h} - \tan x}{h}\\ &= \frac{\tan x + \tan h - \tan x + \tan^2 x\tan h}{h(1 - \tan x\tan h)}\\ &= \frac{(\tan h)(1 + \tan^2 x)}{h(1 - \tan x\tan h)}\\ &= \frac{\tan h}{h} \bullet \frac{\sec^2 h}{1 - \tan x\tan h}\end{aligned}$$

63. **a.** Since θ is as pictured, $\sin\theta = \dfrac{b}{\sqrt{a^2+b^2}}$ and $\cos\theta = \dfrac{a}{\sqrt{a^2+b^2}}$. Using an addition formula:

$$\begin{aligned}\sqrt{a^2+b^2}\sin(x+\theta) &= \sqrt{a^2+b^2}(\sin x\cos\theta + \cos x\sin\theta)\\ &= \sqrt{a^2+b^2}\left(\frac{a}{\sqrt{a^2+b^2}}\sin x + \frac{b}{\sqrt{a^2+b^2}}\cos x\right)\\ &= a\sin x + b\cos x\end{aligned}$$

b. Since $\sin(x+\theta) \le 1$, a maximum value of $f(x) = a\sin x + b\cos x$ is $\sqrt{a^2+b^2}$.

65. **a.** Using an addition formula:

$$\sqrt{2}\cos\left(x-\tfrac{\pi}{4}\right)=\sqrt{2}\left(\cos x\cos\tfrac{\pi}{4}+\sin x\sin\tfrac{\pi}{4}\right)=\sqrt{2}\left(\frac{1}{\sqrt{2}}\cos x+\frac{1}{\sqrt{2}}\sin x\right)=\cos x+\sin x$$

b. Graphing $f(x)=\sqrt{2}\cos\left(x-\tfrac{\pi}{4}\right)=\cos x+\sin x$:

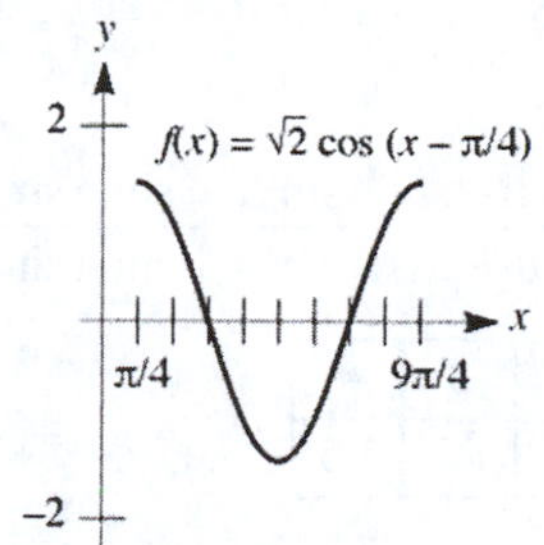

67. To prove this identity, instead show an alternate identity:

$$\begin{aligned}
&1-\cos^2 C-\cos^2 A-\cos^2 B-2\cos A\cos B\cos C\\
&=\sin^2 C-\cos^2 A-\cos^2 B-2\cos A\cos B\cos\left[\pi-(A+B)\right]\\
&=\sin^2\left[\pi-(A+B)\right]-\cos^2 A-\cos^2 B+2\cos A\cos B\cos(A+B)\\
&=\sin^2(A+B)-\cos^2 A-\cos^2 B+2\cos A\cos B(\cos A\cos B-\sin A\sin B)\\
&=(\sin A\cos B+\sin B\cos A)^2-\cos^2 A-\cos^2 B+2\cos^2 A\cos^2 B-2\cos A\cos B\sin A\sin B\\
&=\sin^2 A\cos^2 B+\sin^2 B\cos^2 A-\cos^2 A-\cos^2 B+2\cos^2 A\cos^2 B\\
&=\cos^2 B\left(\sin^2 A-1\right)+\cos^2 A\left(\sin^2 B-1\right)+2\cos^2 A\cos^2 B\\
&=-\cos^2 B\cos^2 A-\cos^2 A\cos^2 B+2\cos^2 A\cos^2 B\\
&=0
\end{aligned}$$

Hence $\cos^2 A+\cos^2 B+\cos^2 C+2\cos A\cos B\cos C=1$.

69. Since $a^2+b^2=1$ and $c^2+d^2=1$, there is some angle θ for which $a=\cos\theta$ and $b=\sin\theta$ and some angle ϕ for which $c=\cos\phi$ and $d=\sin\phi$. Thus: $|ac+bd|=|\cos\theta\cos\phi+\sin\theta\sin\phi|=|\cos(\theta-\phi)|\le 1$

71. Let $A=\frac{\pi}{3}-t$ and $B=\frac{\pi}{3}+t$, then: $\sin A\cos B+\cos A\sin B=\sin(A+B)$

But $A+B=\frac{2\pi}{3}$, so $\sin(A+B)=\sin\frac{2\pi}{3}=\dfrac{\sqrt{3}}{2}$.

73. If $\alpha+\beta=\frac{\pi}{4}$, then $\tan(\alpha+\beta)=\tan\frac{\pi}{4}=1$, so:

$$\begin{aligned}
1&=\tan(\alpha+\beta)\\
1&=\frac{\tan\alpha+\tan\beta}{1-\tan\alpha\tan\beta}\\
1-\tan\alpha\tan\beta&=\tan\alpha+\tan\beta\\
1&=\tan\alpha+\tan\beta+\tan\alpha\tan\beta
\end{aligned}$$

Working from the left-hand side: $(1+\tan\alpha)(1+\tan\beta)=1+\tan\alpha+\tan\beta+\tan\alpha\tan\beta=1+1=2$

75. **a.** Using ΔABH, $\cos(\alpha+\beta)=\dfrac{AB}{1}$, so $\cos(\alpha+\beta)=AB$.

b. Using ΔACF, $\cos\alpha=\dfrac{AC}{AF}=\dfrac{AC}{\cos\beta}$ from Exercise 74e, so $AC=\cos\alpha\cos\beta$.

c. Using ΔEFH, $\sin(\angle EHF)=\dfrac{EF}{HF}$. But $\angle EHF=\alpha$ from Exercise 74c, and $HF=\sin\beta$ from Exercise 74b, so $\sin\alpha=\dfrac{EF}{\sin\beta}$, and thus $EF=\sin\alpha\sin\beta$.

d. From part **a**, $\cos(\alpha+\beta) = AB = AC - BC$. But $AC = \cos\alpha\cos\beta$ from part **b**, and $BC = EF = \sin\alpha\sin\beta$ from part (c), so $\cos(\alpha+\beta) = \cos\alpha\cos\beta - \sin\alpha\sin\beta$.

77. Working from the left-hand side:

$$\frac{\sin(A+B)}{\sin(A-B)} = \frac{\sin A\cos B + \cos A\sin B}{\sin A\cos B - \cos A\sin B} \div \frac{\cos A\cos B}{\cos A\cos B} = \frac{\frac{\sin A}{\cos A} + \frac{\sin B}{\cos B}}{\frac{\sin A}{\cos A} - \frac{\sin B}{\cos B}} = \frac{\tan A + \tan B}{\tan A - \tan B}$$

79. Working from the left-hand side:

$$\cot(A+B) = \frac{\cos(A+B)}{\sin(A+B)} = \frac{\cos A\cos B - \sin A\sin B}{\sin A\cos B + \cos A\sin B} \div \frac{\sin A\sin B}{\sin A\sin B} = \frac{\frac{\cos A}{\sin A}\bullet\frac{\cos B}{\sin B} - 1}{\frac{\cos B}{\sin B} + \frac{\cos A}{\sin A}} = \frac{\cot A\cot B - 1}{\cot B + \cot A}$$

81. **a.** Complete the table:

t	1	2	3	4
$f(t)$	1.5	1.5	1.5	1.5

b. Conjecture: $f(t) = 1.5$

To prove this, first simplify the expressions:

$$\cos\left(t + \tfrac{2\pi}{3}\right) = \cos t\cos\tfrac{2\pi}{3} - \sin t\sin\tfrac{2\pi}{3} = -\tfrac{1}{2}\cos t - \tfrac{\sqrt{3}}{2}\sin t$$

$$\cos\left(t - \tfrac{2\pi}{3}\right) = \cos t\cos\tfrac{2\pi}{3} + \sin t\sin\tfrac{2\pi}{3} = -\tfrac{1}{2}\cos t + \tfrac{\sqrt{3}}{2}\sin t$$

Therefore:

$$\cos^2\left(t + \tfrac{2\pi}{3}\right) = \left(-\tfrac{1}{2}\cos t - \tfrac{\sqrt{3}}{2}\sin t\right)^2 = \tfrac{1}{4}\cos^2 t + \tfrac{\sqrt{3}}{2}\sin t\cos t + \tfrac{3}{4}\sin^2 t$$

$$\cos^2\left(t - \tfrac{2\pi}{3}\right) = \left(-\tfrac{1}{2}\cos t + \tfrac{\sqrt{3}}{2}\sin t\right)^2 = \tfrac{1}{4}\cos^2 t - \tfrac{\sqrt{3}}{2}\sin t\cos t + \tfrac{3}{4}\sin^2 t$$

Thus:

$$\begin{aligned} f(t) &= \cos^2 t + \left(\tfrac{1}{4}\cos^2 t + \tfrac{\sqrt{3}}{2}\sin t\cos t + \tfrac{3}{4}\sin^2 t\right) + \left(\tfrac{1}{4}\cos^2 t - \tfrac{\sqrt{3}}{2}\sin t\cos t + \tfrac{3}{4}\sin^2 t\right) \\ &= \tfrac{3}{2}\cos^2 t + \tfrac{3}{2}\sin^2 t \\ &= \tfrac{3}{2}\left(\cos^2 t + \sin^2 t\right) \\ &= \tfrac{3}{2} \end{aligned}$$

83. Working from the left-hand side:

$$\begin{aligned} \tan(A-B) &= \frac{\tan A - \dfrac{n\sin A\cos A}{1 - n\sin^2 A}}{1 + \tan A\bullet\dfrac{n\sin A\cos A}{1 - n\sin^2 A}} \\ &= \frac{\tan A\left(1 - n\sin^2 A\right) - n\sin A\cos A}{1 - n\sin^2 A + n\tan A\sin A\cos A} \\ &= \frac{\tan A - n\tan A\sin^2 A - n\sin A\cos A}{1 - n\sin^2 A + n\tan A\sin A\cos A} \\ &= \frac{\tan A - n\tan A\left(\sin^2 A + \cos^2 A\right)}{1 - n\sin^2 A + n\sin^2 A} \\ &= \frac{\tan A - n\tan A}{1} \\ &= (1-n)\tan A \end{aligned}$$

85. **a.** Using a calculator:

$\tan A + \tan B + \tan C = \tan 20° + \tan 50° + \tan 110° \approx -1.1918$

$\tan A \tan B \tan C = \tan 20° \tan 50° \tan 110° \approx -1.1918$

It appears that the two values are equal.

b. Using a calculator:

$\tan\alpha + \tan\beta + \tan\gamma = \tan\frac{\pi}{10} + \tan\frac{3\pi}{10} + \tan\frac{3\pi}{5} \approx -1.3764$

$\tan\alpha \tan\beta \tan\gamma = \tan\frac{\pi}{10} \bullet \tan\frac{3\pi}{10} \bullet \tan\frac{3\pi}{5} \approx -1.3764$

It appears that the two values are equal.

c. Since $A + B + C = \pi$, $\tan(A+B) = \tan(\pi - C) = -\tan C$. Since $\tan(A+B) = \dfrac{\tan A + \tan B}{1 - \tan A \tan B}$:

$$\frac{\tan A + \tan B}{1 - \tan A \tan B} = -\tan C$$
$$\tan A + \tan B = -\tan C + \tan A \tan B \tan C$$
$$\tan A + \tan B + \tan C = \tan A \tan B \tan C$$

8.2 The Double-Angle Formulas

1. Since $0° < \varphi < 90°$, $\sin\varphi > 0$ and thus: $\sin\varphi = \sqrt{1-\cos^2\varphi} = \sqrt{1-\left(\frac{7}{25}\right)^2} = \sqrt{1-\frac{49}{625}} = \sqrt{\frac{576}{625}} = \frac{24}{25}$

a. Using the double-angle identity for $\sin 2\varphi$: $\sin 2\varphi = 2\sin\varphi\cos\varphi = 2 \bullet \frac{24}{25} \bullet \frac{7}{25} = \frac{336}{625}$

b. Using the double-angle identity for $\cos 2\varphi$: $\cos 2\varphi = \cos^2\varphi - \sin^2\varphi = \left(\frac{7}{25}\right)^2 - \left(\frac{24}{25}\right)^2 = \frac{49}{625} - \frac{576}{625} = -\frac{527}{625}$

c. Using parts **a** and **b**: $\tan 2\varphi = \dfrac{\sin 2\varphi}{\cos 2\varphi} = \dfrac{\frac{336}{625}}{-\frac{527}{625}} = -\dfrac{336}{527}$

3. Since $\frac{3\pi}{2} < u < 2\pi$, $\sec u > 0$ and thus: $\sec u = \sqrt{1+\tan^2 u} = \sqrt{1+(-4)^2} = \sqrt{1+16} = \sqrt{17}$

Thus $\cos u = \frac{1}{\sqrt{17}}$. Since $\frac{3\pi}{2} < u < 2\pi$, $\sin u < 0$ and thus: $\sin u = -\sqrt{1-\cos^2 u} = -\sqrt{1-\frac{1}{17}} = -\sqrt{\frac{16}{17}} = -\frac{4}{\sqrt{17}}$

a. Using the double-angle identity for $\sin 2u$: $\sin 2u = 2\sin u\cos u = 2 \bullet \left(-\frac{4}{\sqrt{17}}\right) \bullet \left(\frac{1}{\sqrt{17}}\right) = -\frac{8}{17}$

b. Using the double-angle identity for $\cos 2u$: $\cos 2u = \cos^2 u - \sin^2 u = \left(\frac{1}{\sqrt{17}}\right)^2 - \left(-\frac{4}{\sqrt{17}}\right)^2 = \frac{1}{17} - \frac{16}{17} = -\frac{15}{17}$

c. Using parts **a** and **b**: $\tan 2u = \dfrac{\sin 2u}{\cos 2u} = \dfrac{-\frac{8}{17}}{-\frac{15}{17}} = \dfrac{8}{15}$

5. Since $0° < \alpha < 90°$, $\cos\alpha > 0$ and thus: $\cos\alpha = \sqrt{1-\sin^2\alpha} = \sqrt{1-\left(\frac{\sqrt{3}}{2}\right)^2} = \sqrt{1-\frac{3}{4}} = \sqrt{\frac{1}{4}} = \frac{1}{2}$

a. Since $0° < \alpha < 90°$, $0° < \frac{\alpha}{2} < 45°$ and thus $\sin\frac{\alpha}{2} > 0$. Therefore: $\sin\frac{\alpha}{2} = \sqrt{\dfrac{1-\cos\alpha}{2}} = \sqrt{\dfrac{1-\frac{1}{2}}{2}} = \sqrt{\frac{1}{4}} = \frac{1}{2}$

b. Since $0° < \alpha < 90°$, $0° < \frac{\alpha}{2} < 45°$ and thus $\cos\frac{\alpha}{2} > 0$. Therefore: $\cos\frac{\alpha}{2} = \sqrt{\dfrac{1+\cos\alpha}{2}} = \sqrt{\dfrac{1+\frac{1}{2}}{2}} = \sqrt{\frac{3}{4}} = \frac{\sqrt{3}}{2}$

c. Using parts **a** and **b**: $\tan\frac{\alpha}{2}=\frac{\sin\frac{\alpha}{2}}{\cos\frac{\alpha}{2}}=\frac{\frac{1}{2}}{\frac{\sqrt{3}}{2}}=\frac{1}{\sqrt{3}}=\frac{\sqrt{3}}{3}$

Notice that an alternate solution is to spot that since $\sin\alpha=\frac{\sqrt{3}}{2}$ and $0°<\alpha<90°$, $\alpha=60°$ and thus $\frac{\alpha}{2}=30°$. Thus $\sin\frac{\alpha}{2}=\frac{1}{2}$, $\cos\frac{\alpha}{2}=\frac{\sqrt{3}}{2}$ and $\tan\frac{\alpha}{2}=\frac{\sqrt{3}}{3}$.

7. a. Since $\frac{\pi}{2}<\theta<\pi$, $\frac{\pi}{4}<\frac{\theta}{2}<\frac{\pi}{2}$ and thus $\sin\frac{\theta}{2}>0$. Therefore: $\sin\frac{\theta}{2}=\sqrt{\frac{1-\cos\theta}{2}}=\sqrt{\frac{1+\frac{7}{9}}{2}}=\sqrt{\frac{8}{9}}=\frac{2\sqrt{2}}{3}$

b. Since $\frac{\pi}{2}<\theta<\pi$, $\frac{\pi}{4}<\frac{\theta}{2}<\frac{\pi}{2}$ and thus $\cos\frac{\theta}{2}>0$. Therefore: $\cos\frac{\theta}{2}=\sqrt{\frac{1+\cos\theta}{2}}=\sqrt{\frac{1-\frac{7}{9}}{2}}=\sqrt{\frac{1}{9}}=\frac{1}{3}$

c. Using parts **a** and **b**: $\tan\frac{\theta}{2}=\frac{\sin\frac{\theta}{2}}{\cos\frac{\theta}{2}}=\frac{\frac{2\sqrt{2}}{3}}{\frac{1}{3}}=2\sqrt{2}$

9. Since $\frac{\pi}{2}<\theta<\pi$, $\cos\theta<0$ and thus: $\cos\theta=-\sqrt{1-\sin^2\theta}=-\sqrt{1-\left(\frac{3}{4}\right)^2}=-\sqrt{1-\frac{9}{16}}=-\sqrt{\frac{7}{16}}=-\frac{\sqrt{7}}{4}$

a. Using the double-angle identity for $\sin 2\theta$: $\sin 2\theta=2\sin\theta\cos\theta=2\bullet\left(\frac{3}{4}\right)\bullet\left(-\frac{\sqrt{7}}{4}\right)=-\frac{3\sqrt{7}}{8}$

b. Using the double-angle identity for $\cos 2\theta$: $\cos 2\theta=\cos^2\theta-\sin^2\theta=\left(-\frac{\sqrt{7}}{4}\right)^2-\left(\frac{3}{4}\right)^2=\frac{7}{16}-\frac{9}{16}=-\frac{1}{8}$

c. Since $\frac{\pi}{4}<\frac{\theta}{2}<\frac{\pi}{2}$, $\sin\frac{\theta}{2}>0$ and thus: $\sin\frac{\theta}{2}=\sqrt{\frac{1-\cos\theta}{2}}=\sqrt{\frac{1-\left(-\frac{\sqrt{7}}{4}\right)}{2}}=\sqrt{\frac{4+\sqrt{7}}{8}}=\frac{\sqrt{8+2\sqrt{7}}}{4}$

d. Since $\frac{\pi}{4}<\frac{\theta}{2}<\frac{\pi}{2}$, $\cos\frac{\theta}{2}>0$ and thus: $\cos\frac{\theta}{2}=\sqrt{\frac{1+\cos\theta}{2}}=\sqrt{\frac{1-\frac{\sqrt{7}}{4}}{2}}=\sqrt{\frac{4-\sqrt{7}}{8}}=\frac{\sqrt{8-2\sqrt{7}}}{4}$

11. Since $180°<\theta<270°$, $\sin\theta<0$ and thus: $\sin\theta=-\sqrt{1-\cos^2\theta}=-\sqrt{1-\left(-\frac{1}{3}\right)^2}=-\sqrt{1-\frac{1}{9}}=-\sqrt{\frac{8}{9}}=-\frac{2\sqrt{2}}{3}$

a. Using the double-angle identity for $\sin 2\theta$: $\sin 2\theta=2\sin\theta\cos\theta=2\bullet\left(-\frac{2\sqrt{2}}{3}\right)\bullet\left(-\frac{1}{3}\right)=\frac{4\sqrt{2}}{9}$

b. Using the double-angle identity for $\cos 2\theta$: $\cos 2\theta=\cos^2\theta-\sin^2\theta=\left(-\frac{1}{3}\right)^2-\left(-\frac{2\sqrt{2}}{3}\right)^2=\frac{1}{9}-\frac{8}{9}=-\frac{7}{9}$

c. Since $90°<\frac{\theta}{2}<135°$, $\sin\frac{\theta}{2}>0$ and thus: $\sin\frac{\theta}{2}=\sqrt{\frac{1-\cos\theta}{2}}=\sqrt{\frac{1+\frac{1}{3}}{2}}=\sqrt{\frac{2}{3}}=\frac{\sqrt{6}}{3}$

d. Since $90°<\frac{\theta}{2}<135°$, $\cos\frac{\theta}{2}<0$ and thus: $\cos\frac{\theta}{2}=-\sqrt{\frac{1+\cos\theta}{2}}=-\sqrt{\frac{1-\frac{1}{3}}{2}}=-\sqrt{\frac{1}{3}}=-\frac{\sqrt{3}}{3}$

13. a. Since $0<\frac{\pi}{12}<\frac{\pi}{2}$, $\sin\frac{\pi}{12}>0$. Using the half-angle formula for $\sin\frac{1}{2}x$:

$$\sin\frac{\pi}{12}=\sqrt{\frac{1-\cos\frac{\pi}{6}}{2}}=\sqrt{\frac{1-\frac{\sqrt{3}}{2}}{2}}=\sqrt{\frac{2-\sqrt{3}}{4}}=\frac{\sqrt{2-\sqrt{3}}}{2}$$

b. Since $0<\frac{\pi}{12}<\frac{\pi}{2}$, $\cos\frac{\pi}{12}>0$. Using the half-angle formula for $\cos\frac{1}{2}x$:

$$\cos\frac{\pi}{12}=\sqrt{\frac{1+\cos\frac{\pi}{6}}{2}}=\sqrt{\frac{1+\frac{\sqrt{3}}{2}}{2}}=\frac{\sqrt{2+\sqrt{3}}}{2}$$

c. Using the half-angle formula for $\tan\frac{1}{2}x$: $\tan\frac{\pi}{12}=\dfrac{\sin\frac{\pi}{6}}{1+\cos\frac{\pi}{6}}=\dfrac{\frac{1}{2}}{1+\frac{\sqrt{3}}{2}}=\dfrac{1}{2+\sqrt{3}}\bullet\dfrac{2-\sqrt{3}}{2-\sqrt{3}}=\dfrac{2-\sqrt{3}}{4-3}=2-\sqrt{3}$

Note: We could also use parts **a** and **b**, as follows:

$$\begin{aligned}\tan\frac{\pi}{12}&=\frac{\sin\frac{\pi}{12}}{\cos\frac{\pi}{12}}\\&=\frac{\frac{\sqrt{2-\sqrt{3}}}{2}}{\frac{\sqrt{2+\sqrt{3}}}{2}}\\&=\frac{\sqrt{2-\sqrt{3}}}{\sqrt{2+\sqrt{3}}}\bullet\frac{\sqrt{2+\sqrt{3}}}{\sqrt{2+\sqrt{3}}}\\&=\frac{\sqrt{4-3}}{2+\sqrt{3}}\bullet\frac{2-\sqrt{3}}{2-\sqrt{3}}\\&=\frac{2-\sqrt{3}}{4-3}\\&=2-\sqrt{3}\end{aligned}$$

15. **a.** Since $90°<105°<180°$, $\sin 105°>0$. Using the half-angle formula for $\sin\frac{1}{2}x$:

$$\sin 105°=\sqrt{\frac{1-\cos 210°}{2}}=\sqrt{\frac{1+\frac{\sqrt{3}}{2}}{2}}=\sqrt{\frac{2+\sqrt{3}}{4}}=\frac{\sqrt{2+\sqrt{3}}}{2}$$

b. Since $90°<105°<180°$, $\cos 105°<0$. Using the half-angle formula for $\cos\frac{1}{2}x$:

$$\cos 105°=-\sqrt{\frac{1+\cos 210°}{2}}=-\sqrt{\frac{1-\frac{\sqrt{3}}{2}}{2}}=-\sqrt{\frac{2-\sqrt{3}}{4}}=-\frac{\sqrt{2-\sqrt{3}}}{2}$$

c. Using the half-angle formula for $\tan\frac{1}{2}x$:

$$\tan 105°=\frac{\sin 210°}{1+\cos 210°}=\frac{-\frac{1}{2}}{1-\frac{\sqrt{3}}{2}}=-\frac{1}{2-\sqrt{3}}\bullet\frac{2+\sqrt{3}}{2+\sqrt{3}}=-\frac{2+\sqrt{3}}{4-3}=-2-\sqrt{3}$$

17. From the first triangle we have $\sin\theta=\frac{3}{5}$, $\cos\theta=\frac{4}{5}$ and $\tan\theta=\frac{3}{4}$.

a. By the double-angle identity for $\sin 2\theta$: $\sin 2\theta=2\sin\theta\cos\theta=2\bullet\frac{3}{5}\bullet\frac{4}{5}=\frac{24}{25}$

b. By the double-angle identity for $\cos 2\theta$: $\cos 2\theta=\cos^2\theta-\sin^2\theta=\left(\frac{4}{5}\right)^2-\left(\frac{3}{5}\right)^2=\frac{16}{25}-\frac{9}{25}=\frac{7}{25}$

c. By the double-angle identity for $\tan 2\theta$: $\tan 2\theta=\dfrac{2\tan\theta}{1-\tan^2\theta}=\dfrac{2\bullet\frac{3}{4}}{1-\left(\frac{3}{4}\right)^2}=\dfrac{\frac{3}{2}}{\frac{7}{16}}=\dfrac{24}{7}$

Note: An easier approach, after doing **a** and **b**, would be to say: $\tan 2\theta=\dfrac{\sin 2\theta}{\cos 2\theta}=\dfrac{\frac{24}{25}}{\frac{7}{25}}=\dfrac{24}{7}$

19. From the first triangle we have $\sin\beta = \frac{4}{5}$ and $\cos\beta = \frac{3}{5}$.

a. By the double-angle identity for $\sin 2\beta$: $\sin 2\beta = 2\sin\beta\cos\beta = 2\bullet\frac{4}{5}\bullet\frac{3}{5} = \frac{24}{25}$

b. By the double-angle identity for $\cos 2\beta$: $\cos 2\beta = \cos^2\beta - \sin^2\beta = \left(\frac{3}{5}\right)^2 - \left(\frac{4}{5}\right)^2 = \frac{9}{25} - \frac{16}{25} = -\frac{7}{25}$

c. Using the identity for $\tan x$: $\tan 2\beta = \dfrac{\sin 2\beta}{\cos 2\beta} = \dfrac{\frac{24}{25}}{-\frac{7}{25}} = -\dfrac{24}{7}$

Note: We could also have used the double-angle formula for $\tan 2\beta$.

21. From the first triangle we have $\sin\theta = \frac{3}{5}$ and $\cos\theta = \frac{4}{5}$.

a. Since $\sin\frac{\theta}{2} > 0$, use the half-angle formula for $\sin\frac{\theta}{2}$ to obtain: $\sin\frac{\theta}{2} = \sqrt{\dfrac{1-\cos\theta}{2}} = \sqrt{\dfrac{1-\frac{4}{5}}{2}} = \sqrt{\frac{1}{10}} = \dfrac{\sqrt{10}}{10}$

b. Since $\cos\frac{\theta}{2} > 0$, use the half-angle formula for $\cos\frac{\theta}{2}$ to obtain: $\cos\frac{\theta}{2} = \sqrt{\dfrac{1+\cos\theta}{2}} = \sqrt{\dfrac{1+\frac{4}{5}}{2}} = \sqrt{\frac{9}{10}} = \dfrac{3\sqrt{10}}{10}$

c. Using the half-angle formula for $\tan\frac{\theta}{2}$ to obtain: $\tan\frac{\theta}{2} = \dfrac{\sin\theta}{1+\cos\theta} = \dfrac{\frac{3}{5}}{1+\frac{4}{5}} = \frac{3}{9} = \frac{1}{3}$

Note: We could also have computed this directly after parts **a** and **b** as: $\tan\frac{\theta}{2} = \dfrac{\sin\frac{\theta}{2}}{\cos\frac{\theta}{2}} = \dfrac{\frac{\sqrt{10}}{10}}{\frac{3\sqrt{10}}{10}} = \frac{1}{3}$

23. From the first triangle we have $\cos\beta = \frac{3}{5}$.

a. Since $\sin\frac{\beta}{2} > 0$, use the half-angle formula for $\sin\frac{\beta}{2}$ to obtain: $\sin\frac{\beta}{2} = \sqrt{\dfrac{1-\cos\beta}{2}} = \sqrt{\dfrac{1-\frac{3}{5}}{2}} = \sqrt{\frac{1}{5}} = \dfrac{\sqrt{5}}{5}$

b. Since $\cos\frac{\beta}{2} > 0$, use the half-angle formula for $\cos\frac{\beta}{2}$ to obtain: $\cos\frac{\beta}{2} = \sqrt{\dfrac{1+\cos\beta}{2}} = \sqrt{\dfrac{1+\frac{3}{5}}{2}} = \sqrt{\frac{4}{5}} = \dfrac{2\sqrt{5}}{5}$

c. Using the identity for $\tan x$ and parts **a** and **b**: $\tan\frac{\beta}{2} = \dfrac{\sin\frac{\beta}{2}}{\cos\frac{\beta}{2}} = \dfrac{\frac{\sqrt{5}}{5}}{\frac{2\sqrt{5}}{5}} = \frac{1}{2}$

25. Since $0 < \theta < \frac{\pi}{2}$, $\cos\theta > 0$. Since $\sin\theta = \frac{x}{5}$: $\cos\theta = \sqrt{1-\sin^2\theta} = \sqrt{1-\left(\dfrac{x}{5}\right)^2} = \sqrt{1-\dfrac{x^2}{25}} = \dfrac{\sqrt{25-x^2}}{5}$

Now apply the double angle formulas:

$$\sin 2\theta = 2\sin\theta\cos\theta = 2\bullet\frac{x}{5}\bullet\frac{\sqrt{25-x^2}}{5} = \frac{2x\sqrt{25-x^2}}{25}$$

$$\cos 2\theta = \cos^2\theta - \sin^2\theta = \left(\frac{\sqrt{25-x^2}}{5}\right)^2 - \left(\frac{x}{5}\right)^2 = \frac{25-x^2}{25} - \frac{x^2}{25} = \frac{25-2x^2}{25}$$

27. Since $0<\theta<\frac{\pi}{2}$, $\cos\theta>0$. Since $\sin\theta=\dfrac{x-1}{2}$:

$$\cos\theta=\sqrt{1-\sin^2\theta}=\sqrt{1-\left(\frac{x-1}{2}\right)^2}=\sqrt{1-\frac{x^2-2x+1}{4}}=\frac{\sqrt{3+2x-x^2}}{2}$$

Now apply the double-angle formulas:

$$\sin 2\theta=2\sin\theta\cos\theta=2\bullet\frac{x-1}{2}\bullet\frac{\sqrt{3+2x-x^2}}{2}=\frac{(x-1)\sqrt{3+2x-x^2}}{2}$$

$$\begin{aligned}\cos 2\theta&=\cos^2\theta-\sin^2\theta\\&=\left(\frac{\sqrt{3+2x-x^2}}{2}\right)^2-\left(\frac{x-1}{2}\right)^2\\&=\frac{3+2x-x^2}{4}-\frac{x^2-2x+1}{4}\\&=\frac{2+4x-2x^2}{4}\\&=\frac{1+2x-x^2}{2}\end{aligned}$$

29. Using the identity $\sin^2\theta=\dfrac{1-\cos 2\theta}{2}$:

$$\begin{aligned}\sin^4\theta&=\left(\sin^2\theta\right)^2\\&=\left(\frac{1-\cos 2\theta}{2}\right)^2\\&=\frac{1-2\cos 2\theta+(\cos 2\theta)^2}{4}\\&=\frac{1-2\cos 2\theta+\frac{1+\cos 4\theta}{2}}{4}\\&=\frac{2-4\cos 2\theta+1+\cos 4\theta}{8}\\&=\frac{3-4\cos 2\theta+\cos 4\theta}{8}\end{aligned}$$

31. Using the identities $\sin^2\frac{\theta}{2}=\dfrac{1-\cos\theta}{2}$ and $\cos^2\theta=\dfrac{1+\cos 2\theta}{2}$:

$$\begin{aligned}\sin^4\frac{\theta}{2}&=\left(\sin^2\frac{\theta}{2}\right)^2\\&=\left(\frac{1-\cos\theta}{2}\right)^2\\&=\frac{1-2\cos\theta+\cos^2\theta}{4}\\&=\frac{1-2\cos\theta+\frac{1+\cos 2\theta}{2}}{4}\\&=\frac{2-4\cos\theta+1+\cos 2\theta}{8}\\&=\frac{3-4\cos\theta+\cos 2\theta}{8}\end{aligned}$$

33. a. Replacing 2θ with $\theta + \theta$ and using the addition formula for cos $(s + t)$:

$$\cos 2\theta = \cos(\theta+\theta) = \cos\theta\cos\theta - \sin\theta\sin\theta = \cos^2\theta - \sin^2\theta$$

b. Replacing 2θ with $\theta + \theta$ and using the addition formula for tan $(s + t)$:

$$\tan 2\theta = \tan(\theta+\theta) = \frac{\tan\theta + \tan\theta}{1-\tan\theta\tan\theta} = \frac{2\tan\theta}{1-\tan^2\theta}$$

35. Working from the right-hand side: $\dfrac{1-\tan^2 s}{1+\tan^2 s} = \dfrac{1-\frac{\sin^2 s}{\cos^2 s}}{1+\frac{\sin^2 s}{\cos^2 s}} = \dfrac{\cos^2 s - \sin^2 s}{\cos^2 s + \sin^2 s} = \dfrac{\cos 2s}{1} = \cos 2s$

37. Writing $\theta = 2 \bullet \frac{\theta}{2}$, apply the double-angle formula:

$$\cos\theta = \cos\left(2\bullet\tfrac{\theta}{2}\right) = \cos^2\tfrac{\theta}{2} - \sin^2\tfrac{\theta}{2} = \cos^2\tfrac{\theta}{2} - \left(1-\cos^2\tfrac{\theta}{2}\right) = 2\cos^2\tfrac{\theta}{2} - 1$$

39. Using the identities $\sin^2\frac{\theta}{2} = \dfrac{1-\cos\theta}{2}$ and $\cos^2\theta = \dfrac{1+\cos 2\theta}{2}$:

$$\begin{aligned}
\sin^4\theta &= \left(\sin^2\theta\right)^2 \\
&= \left(\frac{1-\cos 2\theta}{2}\right)^2 \\
&= \frac{1-2\cos 2\theta + \cos^2 2\theta}{4} \\
&= \frac{1-2\cos 2\theta + \frac{1+\cos 4\theta}{2}}{4} \\
&= \frac{2-4\cos 2\theta + 1 + \cos 4\theta}{8} \\
&= \frac{3-4\cos 2\theta + \cos 4\theta}{8}
\end{aligned}$$

41. Working from the right-hand side: $\dfrac{2\tan\theta}{1+\tan^2\theta} = \dfrac{2\bullet\frac{\sin\theta}{\cos\theta}}{1+\frac{\sin^2\theta}{\cos^2\theta}} = \dfrac{2\sin\theta\cos\theta}{\cos^2\theta+\sin^2\theta} = \dfrac{\sin 2\theta}{1} = \sin 2\theta$

43. Working from the right-hand side: $2\sin^3\theta\cos\theta + 2\sin\theta\cos^3\theta = (2\sin\theta\cos\theta)\left(\sin^2\theta+\cos^2\theta\right) = \sin 2\theta \bullet 1 = \sin 2\theta$

45. Working from the left-hand side:

$$\frac{1+\tan\frac{\theta}{2}}{1-\tan\frac{\theta}{2}} = \frac{1+\dfrac{\sin\frac{\theta}{2}}{\cos\frac{\theta}{2}}}{1-\dfrac{\sin\frac{\theta}{2}}{\cos\frac{\theta}{2}}} = \frac{\cos\frac{\theta}{2}+\sin\frac{\theta}{2}}{\cos\frac{\theta}{2}-\sin\frac{\theta}{2}} = \frac{\left(\cos\frac{\theta}{2}+\sin\frac{\theta}{2}\right)^2}{\cos^2\frac{\theta}{2}-\sin^2\frac{\theta}{2}} = \frac{1+2\cos\frac{\theta}{2}\sin\frac{\theta}{2}}{\cos\theta} = \frac{1+\sin\theta}{\cos\theta} = \tan\theta + \sec\theta$$

47. Working from the left-hand side and using the addition formula for sin $(s - t)$:

$$\begin{aligned}
2\sin^2\left(45° - \theta\right) &= 2\left(\sin 45°\cos\theta - \cos 45°\sin\theta\right)^2 \\
&= 2\left(\frac{\cos\theta - \sin\theta}{\sqrt{2}}\right)^2 \\
&= 2\bullet\frac{\cos^2\theta - 2\sin\theta\cos\theta + \sin^2\theta}{2} \\
&= 1 - 2\cos\theta\sin\theta \\
&= 1 - \sin 2\theta
\end{aligned}$$

49. Simplifying the left-hand side:

$$1+\tan\theta\tan 2\theta = 1+\tan\theta \bullet \frac{2\tan\theta}{1-\tan^2\theta} = 1+\frac{2\tan^2\theta}{1-\tan^2\theta} = \frac{1-\tan^2\theta+2\tan^2\theta}{1-\tan^2\theta} = \frac{1+\tan^2\theta}{1-\tan^2\theta}$$

Simplifying the right-hand side:

$$\tan 2\theta\cot\theta - 1 = \frac{2\tan\theta}{1-\tan^2\theta} \bullet \frac{1}{\tan\theta} - 1 = \frac{2}{1-\tan^2\theta} - 1 = \frac{2-1+\tan^2\theta}{1-\tan^2\theta} = \frac{1+\tan^2\theta}{1-\tan^2\theta}$$

Since both sides simplify to the same quantity, the original equation is an identity.

51. Using the addition formula for $\tan(\alpha+\beta)$: $\tan(\alpha+\beta) = \dfrac{\tan\alpha+\tan\beta}{1-\tan\alpha\tan\beta} = \dfrac{\frac{1}{11}+\frac{5}{6}}{1-\frac{1}{11}\bullet\frac{5}{6}} = \dfrac{\frac{61}{66}}{\frac{61}{66}} = 1$

But if $\tan(\alpha+\beta) = 1$ and $0 < \alpha+\beta < \pi$, then $\alpha+\beta = \frac{\pi}{4}$.

53. **a.** Substituting $x = \cos 20° \approx 0.9397$ verifies that it is a root of the equation.

b. Substituting $\theta = 20°$ in the given identity:

$$\begin{aligned}\cos 60° &= 4\cos^3 20° - 3\cos 20°\\ \tfrac{1}{2} &= 4\cos^3 20° - 3\cos 20°\\ 1 &= 8\cos^3 20° - 6\cos 20°\\ 0 &= 8\cos^3 20° - 6\cos 20° - 1\end{aligned}$$

Thus $\cos 20°$ is a root of the cubic equation $8x^3 - 6x - 1 = 0$.

55. **a.** Using ΔODC and the fact that $OC = 1$, $OD = \cos\theta$ and $DC = \sin\theta$. Now using ΔADC and the fact that $AO = 1$: $\tan\frac{\theta}{2} = \dfrac{CD}{AO+OD} = \dfrac{\sin\theta}{1+\cos\theta}$

b. Using the formula from **a** and rationalizing denominators:

$$\tan 15° = \frac{\sin 30°}{1+\cos 30°} = \frac{\frac{1}{2}}{1+\frac{\sqrt{3}}{2}} = \frac{1}{2+\sqrt{3}} \bullet \frac{2-\sqrt{3}}{2-\sqrt{3}} = \frac{2-\sqrt{3}}{4-3} = 2-\sqrt{3}$$

Using the formula from **a** and rationalizing denominators:

$$\tan\frac{\pi}{8} = \frac{\sin\frac{\pi}{4}}{1+\cos\frac{\pi}{4}} = \frac{\frac{\sqrt{2}}{2}}{1+\frac{\sqrt{2}}{2}} = \frac{\sqrt{2}}{2+\sqrt{2}} \bullet \frac{2-\sqrt{2}}{2-\sqrt{2}} = \frac{2\sqrt{2}-2}{4-2} = \sqrt{2}-1$$

57. **a.** Following the suggestion and applying the double-angle formulas for $\sin 4\theta$ and $\sin 2\theta$:

$$\frac{\sin 4\theta}{4\sin\theta} = \frac{2\sin 2\theta\cos 2\theta}{4\sin\theta} = \frac{4\sin\theta\cos\theta\cos 2\theta}{4\sin\theta} = \cos\theta\cos 2\theta$$

b. Following the suggestion and applying the double-angle formulas for $\sin 8\theta$, $\sin 4\theta$ and $\sin 2\theta$:

$$\frac{\sin 8\theta}{8\sin\theta} = \frac{2\sin 4\theta\cos 4\theta}{8\sin\theta} = \frac{4\sin 2\theta\cos 2\theta\cos 4\theta}{8\sin\theta} = \frac{8\sin\theta\cos\theta\cos 2\theta\cos 4\theta}{8\sin\theta} = \cos\theta\cos 2\theta\cos 4\theta$$

c. Following the suggestion and applying the double-angle formulas for $\sin 16\theta$, $\sin 8\theta$, $\sin 4\theta$ and $\sin 2\theta$:

$$\begin{aligned}\frac{\sin 16\theta}{16\sin\theta} &= \frac{2\sin 8\theta\cos 8\theta}{16\sin\theta}\\ &= \frac{4\sin 4\theta\cos 4\theta\cos 8\theta}{16\sin\theta}\\ &= \frac{8\sin 2\theta\cos 2\theta\cos 4\theta\cos 8\theta}{16\sin\theta}\\ &= \frac{16\sin\theta\cos\theta\cos 2\theta\cos 4\theta\cos 8\theta}{16\sin\theta}\\ &= \cos\theta\cos 2\theta\cos 4\theta\cos 8\theta\end{aligned}$$

59. **a.** Using a calculator, $\cos 72° + \cos 144° = -0.5$.

b. Using the observation and the addition formulas for cosine:

$$\begin{aligned}\cos 72° + \cos 144° &= \cos(108° - 36°) + \cos(108° + 36°)\\ &= (\cos 108° \cos 36° + \sin 108° \sin 36°) + (\cos 108° \cos 36° - \sin 108° \sin 36°)\\ &= 2\cos 108° \cos 36°\end{aligned}$$

c. Since $108° = 180° - 72°$, $\cos 108° = -\cos 72°$. Similarly, since $36° = 180° - 144°$, $\cos 36° = -\cos 144°$. Therefore: $\cos 108° \cos 36° = (-\cos 72°)(-\cos 144°) = \cos 72° \cos 144°$

d. Using the above relationships: $\cos 72° + \cos 144° = 2\cos 108° \cos 36° = 2\cos 72° \cos 144° = 2\left(-\frac{1}{4}\right) = -\frac{1}{2}$

61. The coordinates of points A_1, A_2, and A_3 are:

$$A_1\ (1,0)$$

$$A_2\ (\cos 120°, \sin 120°) = A_2\left(-\tfrac{1}{2}, \tfrac{\sqrt{3}}{2}\right)$$

$$A_3\ (\cos 240°, \sin 240°) = A_3\left(-\tfrac{1}{2}, -\tfrac{\sqrt{3}}{2}\right)$$

Using the distance formula from the point $P(x, 0)$:

$$PA_1 = \sqrt{(x-1)^2 + (0-0)^2} = \sqrt{(x-1)^2} = |x-1| = 1 - x, \text{ since } x < 1$$

$$PA_2 = \sqrt{\left(x + \tfrac{1}{2}\right)^2 + \left(0 - \tfrac{\sqrt{3}}{2}\right)^2} = \sqrt{x^2 + x + \tfrac{1}{4} + \tfrac{3}{4}} = \sqrt{x^2 + x + 1}$$

$$PA_3 = \sqrt{\left(x + \tfrac{1}{2}\right)^2 + \left(0 + \tfrac{\sqrt{3}}{2}\right)^2} = \sqrt{x^2 + x + \tfrac{1}{4} + \tfrac{3}{4}} = \sqrt{x^2 + x + 1}$$

Now compute the product: $(PA_1)(PA_2)(PA_3) = (1-x)\sqrt{x^2+x+1}\sqrt{x^2+x+1} = (1-x)(1+x+x^2) = 1 - x^3$

63. **a.** The result is verified, since $\sin 18° \sin 54° = 0.25$.

b. (i) This is true because of the double-angle formula for sine, and the fact that $\cos 36° = \sin 54°$.

(ii) This is true because of the double-angle formula for sine, and the fact that $\cos 18° = \sin 72°$.

(iii) Dividing each side by $4 \sin 72°$ produces this result.

65. **a.** Using two addition formulas:

$$\begin{aligned}\cos(60° - \theta)\cos(60° + \theta) &= (\cos 60° \cos\theta + \sin 60° \sin\theta)(\cos 60° \cos\theta - \sin 60° \sin\theta)\\ &= \left(\tfrac{1}{2}\cos\theta + \tfrac{\sqrt{3}}{2}\sin\theta\right)\left(\tfrac{1}{2}\cos\theta - \tfrac{\sqrt{3}}{2}\sin\theta\right)\\ &= \tfrac{1}{4}\cos^2\theta - \tfrac{3}{4}\sin^2\theta\\ &= \tfrac{1}{4}\cos^2\theta - \tfrac{3}{4}(1 - \cos^2\theta)\\ &= \tfrac{1}{4}\cos^2\theta - \tfrac{3}{4} + \tfrac{3}{4}\cos^2\theta\\ &= \cos^2\theta - \tfrac{3}{4}\\ &= \frac{4\cos^2\theta - 3}{4}\end{aligned}$$

b. From part **a**: $\cos\theta \cos(60° - \theta)\cos(60° + \theta) = \cos\theta \bullet \dfrac{4\cos^2\theta - 3}{4} = \dfrac{4\cos^3\theta - 3\cos\theta}{4}$

From Example 3: $\dfrac{\cos 3\theta}{4} = \dfrac{4\cos^3\theta - 3\cos\theta}{4}$

The two expressions are equal.

c. Using $\theta = 20°$ in the identity:

$$\cos 20° \cos(60° - 20°)\cos(60° + 20°) = \frac{\cos 60°}{4}$$

$$\cos 20° \cos 40° \cos 80° = \frac{1/2}{4} = \frac{1}{8}$$

d. A calculator verifies the result.

67. **a.** Using the addition formulas and double-angle formulas:

$$\begin{aligned}\cos 3\theta &= \cos\theta\cos 2\theta - \sin\theta\sin 2\theta \\ &= \cos\theta\left(\cos^2\theta - \sin^2\theta\right) - \sin\theta(2\sin\theta\cos\theta) \\ &= \cos^3\theta - \cos\theta\left(1-\cos^2\theta\right) - 2\cos\theta\left(1-\cos^2\theta\right) \\ &= \cos^3\theta - \cos\theta + \cos^3\theta - 2\cos\theta + 2\cos^3\theta \\ &= 4\cos^3\theta - 3\cos\theta\end{aligned}$$

$$\begin{aligned}\cos 4\theta &= \cos^2 2\theta - \sin^2 2\theta \\ &= \left(\cos^2\theta - \sin^2\theta\right)^2 - 4\sin^2\theta\cos^2\theta \\ &= \cos^4\theta - 2\cos^2\theta\sin^2\theta + \sin^4\theta - 4\cos^2\theta\sin^2\theta \\ &= \cos^4\theta - 2\cos^2\theta\left(1-\cos^2\theta\right) + \left(1-\cos^2\theta\right)^2 - 4\cos^2\theta\left(1-\cos^2\theta\right) \\ &= \cos^4\theta - 2\cos^2\theta + 2\cos^4\theta + 1 - 2\cos^2\theta + \cos^4\theta - 4\cos^2\theta + 4\cos^4\theta \\ &= 8\cos^4\theta - 8\cos^2\theta + 1\end{aligned}$$

b. Since $\cos 3\theta = \cos\frac{6\pi}{7}$ and $\cos 4\theta = \cos\frac{8\pi}{7}$, both angles have a negative cosine with a reference angle of $\frac{\pi}{7}$. So $\cos\frac{6\pi}{7} = \cos\frac{8\pi}{7}$.

c. Since $\cos 4\theta = \cos 3\theta$:

$$\begin{aligned}8\cos^4\theta - 8\cos^2\theta + 1 &= 4\cos^3\theta - 3\cos\theta \\ 8\cos^4\theta - 4\cos^3\theta - 8\cos^2\theta + 3\cos\theta + 1 &= 0\end{aligned}$$

d. The factoring $(\cos\theta - 1)\left(8\cos^3\theta + 4\cos^2\theta - 4\cos\theta - 1\right) = 0$ checks, and since $\theta = \frac{2\pi}{7}$ is not a root of $\cos\theta - 1 = 0$, it must be a root to $8\cos^3\theta + 4\cos^2\theta - 4\cos\theta - 1 = 0$. Thus $\cos\frac{2\pi}{7}$ is a solution to the equation $8x^3 + 4x^2 - 4x - 1 = 0$.

e. The value $x = \cos\frac{2\pi}{7}$ checks in the equation.

8.3 The Product-to-Sum and Sum-to-Product Formulas

1. Using the product-to-sum formula for $\cos A\cos B$:

$$\cos 70°\cos 20° = \tfrac{1}{2}\left[\cos(70°-20°) + \cos(70°+20°)\right] = \tfrac{1}{2}\left[\cos 50° + \cos 90°\right] = \tfrac{1}{2}\left[\cos 50° + 0\right] = \tfrac{1}{2}\cos 50°$$

3. Using the product-to-sum formula for $\sin A\sin B$:

$$\sin 5°\sin 85° = \tfrac{1}{2}\left[\cos(5°-85°) - \cos(5°+85°)\right] = \tfrac{1}{2}\left[\cos(-80°) - \cos 90°\right] = \tfrac{1}{2}\left[\cos 80° - 0\right] = \tfrac{1}{2}\cos 80°$$

5. Using the product-to-sum formula for $\sin A\cos B$:

$$\sin 20°\cos 10° = \tfrac{1}{2}\left[\sin(20°-10°) + \sin(20°+10°)\right] = \tfrac{1}{2}\left[\sin 10° + \sin 30°\right] = \tfrac{1}{2}\left[\sin 10° + \tfrac{1}{2}\right] = \tfrac{1}{2}\sin 10° + \tfrac{1}{4}$$

7. Using the product-to-sum formula for $\cos A\cos B$:

$$\cos\tfrac{\pi}{5}\cos\tfrac{4\pi}{5} = \tfrac{1}{2}\left[\cos\left(\tfrac{\pi}{5} - \tfrac{4\pi}{5}\right) + \cos\left(\tfrac{\pi}{5} + \tfrac{4\pi}{5}\right)\right] = \tfrac{1}{2}\left[\cos\left(-\tfrac{3\pi}{5}\right) + \cos\pi\right] = \tfrac{1}{2}\left[\cos\tfrac{3\pi}{5} - 1\right] = \tfrac{1}{2}\cos\tfrac{3\pi}{5} - \tfrac{1}{2}$$

9. Using the product-to-sum formula for $\sin A\sin B$:

$$\sin\tfrac{2\pi}{7}\sin\tfrac{5\pi}{7} = \tfrac{1}{2}\left[\cos\left(\tfrac{2\pi}{7} - \tfrac{5\pi}{7}\right) - \cos\left(\tfrac{2\pi}{7} + \tfrac{5\pi}{7}\right)\right] = \tfrac{1}{2}\left[\cos\left(-\tfrac{3\pi}{7}\right) - \cos\pi\right] = \tfrac{1}{2}\left[\cos\tfrac{3\pi}{7} - (-1)\right] = \tfrac{1}{2}\cos\tfrac{3\pi}{7} + \tfrac{1}{2}$$

11. Using the product-to-sum formula for $\sin A\cos B$:

$$\sin\tfrac{7\pi}{12}\cos\tfrac{\pi}{12} = \tfrac{1}{2}\left[\sin\left(\tfrac{7\pi}{12} - \tfrac{\pi}{12}\right) + \sin\left(\tfrac{7\pi}{12} + \tfrac{\pi}{12}\right)\right] = \tfrac{1}{2}\left[\sin\tfrac{\pi}{2} + \sin\tfrac{2\pi}{3}\right] = \tfrac{1}{2}\left[1 + \tfrac{\sqrt{3}}{2}\right] = \tfrac{1}{2} + \tfrac{\sqrt{3}}{4}$$

13. Using the product-to-sum formula for $\sin A \sin B$:

$$\begin{aligned}\sin 3x \sin 4x &= \tfrac{1}{2}\left[\cos(3x-4x)-\cos(3x+4x)\right]\\ &= \tfrac{1}{2}\left[\cos(-x)-\cos(7x)\right]\\ &= \tfrac{1}{2}\left[\cos x-\cos 7x\right]\\ &= \tfrac{1}{2}\cos x-\tfrac{1}{2}\cos 7x\end{aligned}$$

15. Using the product-to-sum formula for $\sin A \cos B$:

$$\sin 6\theta \cos 5\theta = \tfrac{1}{2}\left[\sin(6\theta-5\theta)+\sin(6\theta+5\theta)\right]=\tfrac{1}{2}\left[\sin\theta+\sin 11\theta\right]=\tfrac{1}{2}\sin\theta+\tfrac{1}{2}\sin 11\theta$$

17. Using the product-to-sum formula for $\sin A \cos B$:

$$\sin\tfrac{3\theta}{2}\cos\tfrac{\theta}{2}=\tfrac{1}{2}\left[\sin\left(\tfrac{3\theta}{2}-\tfrac{\theta}{2}\right)+\sin\left(\tfrac{3\theta}{2}+\tfrac{\theta}{2}\right)\right]=\tfrac{1}{2}\left[\sin\theta+\sin 2\theta\right]=\tfrac{1}{2}\sin\theta+\tfrac{1}{2}\sin 2\theta$$

19. Using the product-to-sum formula for $\sin A \sin B$:

$$\begin{aligned}\sin(2x+y)\sin(2x-y) &= \tfrac{1}{2}\left[\cos(2x+y-2x+y)-\cos(2x+y+2x-y)\right]\\ &= \tfrac{1}{2}\left[\cos 2y-\cos 4x\right]\\ &= \tfrac{1}{2}\cos 2y-\tfrac{1}{2}\cos 4x\end{aligned}$$

21. Using the product-to-sum formula for $\sin A \cos B$:

$$\sin 2t\cos(s-t)=\tfrac{1}{2}\left[\sin(2t-s+t)+\sin(2t+s-t)\right]=\tfrac{1}{2}\left[\sin(3t-s)+\sin(t+s)\right]=\tfrac{1}{2}\sin(3t-s)+\tfrac{1}{2}\sin(t+s)$$

23. Using the sum-to-product formula for $\cos\alpha+\cos\beta$:

$$\cos 35^\circ+\cos 55^\circ=2\cos\frac{35^\circ+55^\circ}{2}\cos\frac{35^\circ-55^\circ}{2}=2\cos 45^\circ\cos(-10^\circ)=2\bullet\frac{\sqrt{2}}{2}\bullet\cos 10^\circ=\sqrt{2}\cos 10^\circ$$

25. Using the sum-to-product formula for $\sin\alpha-\sin\beta$:

$$\sin\tfrac{\pi}{5}-\sin\tfrac{3\pi}{10}=2\cos\frac{\frac{\pi}{5}+\frac{3\pi}{10}}{2}\sin\frac{\frac{\pi}{5}-\frac{3\pi}{10}}{2}=2\cos\tfrac{\pi}{4}\sin\left(-\tfrac{\pi}{20}\right)=-2\bullet\tfrac{\sqrt{2}}{2}\sin\tfrac{\pi}{20}=-\sqrt{2}\sin\tfrac{\pi}{20}$$

27. Using the sum-to-product formula for $\cos\alpha-\cos\beta$: $\cos 5\theta-\cos 3\theta=-2\sin\frac{5\theta+3\theta}{2}\sin\frac{5\theta-3\theta}{2}=-2\sin 4\theta\sin\theta$

29. Using the hint, $\cos 65^\circ=\sin(90^\circ-65^\circ)=\sin 25^\circ$. Now using the sum-to-product formula for $\sin\alpha+\sin\beta$:

$$\sin 35^\circ+\sin 25^\circ=2\sin\frac{35^\circ+25^\circ}{2}\cos\frac{35^\circ-25^\circ}{2}=2\sin 30^\circ\cos 5^\circ=2\bullet\tfrac{1}{2}\cos 5^\circ=\cos 5^\circ$$

31. Using the sum-to-product formula for $\sin\alpha-\sin\beta$:

$$\sin\left(\tfrac{\pi}{3}+2\theta\right)-\sin\left(\tfrac{\pi}{3}-2\theta\right)=2\cos\tfrac{2\pi/3}{2}\sin\tfrac{4\theta}{2}=2\cos\tfrac{\pi}{3}\sin 2\theta=2\bullet\tfrac{1}{2}\sin 2\theta=\sin 2\theta$$

33. Simplify the numerator and denominator separately. For the numerator, first use the identity $\cos\theta=\sin\left(\frac{\pi}{2}-\theta\right)$ so that $\cos\frac{5\pi}{12}=\sin\frac{\pi}{12}$. Now using the sum-to-product formula for $\sin\alpha+\sin\beta$:

$$\sin\tfrac{\pi}{12}+\sin\tfrac{5\pi}{12}=2\sin\frac{\frac{\pi}{12}+\frac{5\pi}{12}}{2}\cos\frac{\frac{\pi}{12}-\frac{5\pi}{12}}{2}=2\sin\tfrac{\pi}{4}\cos\left(-\tfrac{\pi}{6}\right)=2\bullet\tfrac{\sqrt{2}}{2}\bullet\tfrac{\sqrt{3}}{2}=\tfrac{\sqrt{6}}{2}$$

For the denominator, first use the identity $\cos\theta=\sin\left(\frac{\pi}{2}-\theta\right)$ so that $\cos\frac{\pi}{12}=\sin\frac{5\pi}{12}$. Now using the sum-to-product formula for $\sin\alpha-\sin\beta$: $\sin\frac{5\pi}{12}-\sin\frac{\pi}{12}=2\cos\frac{\frac{5\pi}{12}+\frac{\pi}{12}}{2}\sin\frac{\frac{5\pi}{12}-\frac{\pi}{12}}{2}=2\cos\frac{\pi}{4}\sin\frac{\pi}{6}=2\bullet\frac{\sqrt{2}}{2}\bullet\frac{1}{2}=\frac{\sqrt{2}}{2}$

Thus, the original problem becomes: $\dfrac{\cos\frac{5\pi}{12}+\sin\frac{5\pi}{12}}{\cos\frac{\pi}{12}-\sin\frac{\pi}{12}}=\dfrac{\frac{\sqrt{6}}{2}}{\frac{\sqrt{2}}{2}}=\dfrac{\sqrt{6}}{\sqrt{2}}=\sqrt{3}$

35. Using the sum-to-product formulas for $\sin s+\sin t$ and $\cos s+\cos t$:

$$\frac{\sin s+\sin t}{\cos s+\cos t}=\frac{2\sin\frac{s+t}{2}\cos\frac{s-t}{2}}{2\cos\frac{s+t}{2}\cos\frac{s-t}{2}}=\frac{\sin\frac{s+t}{2}}{\cos\frac{s+t}{2}}=\tan\frac{s+t}{2}$$

37. Using the sum-to-product formulas for sin 2x + sin 2y and cos 2x + cos 2y:

$$\frac{\sin 2x+\sin 2y}{\cos 2x+\cos 2y}=\frac{2\sin\frac{2x+2y}{2}\cos\frac{2x-2y}{2}}{2\cos\frac{2x+2y}{2}\cos\frac{2x-2y}{2}}=\frac{\sin(x+y)}{\cos(x+y)}=\tan(x+y)$$

Note that we can also use the result from Exercise 35 where $s = 2x$ and $t = 2y$.

39. Using the hint, use the sum-to-product formulas to obtain:

$$\cos 7\theta+\cos 5\theta=2\cos\frac{7\theta+5\theta}{2}\cos\frac{7\theta-5\theta}{2}=2\cos 6\theta\cos\theta$$
$$\cos 3\theta+\cos\theta=2\cos\frac{3\theta+\theta}{2}\cos\frac{3\theta-\theta}{2}=2\cos 2\theta\cos\theta$$

Thus:

$$\begin{aligned}\cos 7\theta+\cos 5\theta+\cos 3\theta+\cos\theta&=2\cos 6\theta\cos\theta+2\cos 2\theta\cos\theta\\&=(2\cos\theta)(\cos 6\theta+\cos 2\theta)\\&=(2\cos\theta)\left(2\cos\frac{6\theta+2\theta}{2}\cos\frac{6\theta-2\theta}{2}\right)\\&=4\cos\theta\cos 4\theta\cos 2\theta\end{aligned}$$

41. **a.** Since $\cos\frac{x}{2}=\sin\left(\frac{\pi}{2}-\frac{x}{2}\right)$, use the sum-to-product identity for $\sin\alpha+\sin\beta$ to obtain:

$$\begin{aligned}\sqrt{2}\left[\sin\frac{x}{2}+\cos\frac{x}{2}\right]&=\sqrt{2}\left[\sin\frac{x}{2}+\sin\left(\frac{\pi}{2}-\frac{x}{2}\right)\right]\\&=\sqrt{2}\left[2\sin\frac{\frac{\pi}{2}}{2}\cos\frac{x-\frac{\pi}{2}}{2}\right]\\&=\sqrt{2}\left[2\sin\frac{\pi}{4}\cos\left(\frac{x}{2}-\frac{\pi}{4}\right)\right]\\&=\sqrt{2}\cdot 2\cdot\frac{\sqrt{2}}{2}\cos\left(\frac{x}{2}-\frac{\pi}{4}\right)\\&=2\cos\left(\frac{x}{2}-\frac{\pi}{4}\right)\end{aligned}$$

b. Since $f(x)=2\cos\left(\frac{x}{2}-\frac{\pi}{4}\right)$, the amplitude is 2, the period is $\frac{2\pi}{1/2}=4\pi$, and the phase shift is $\frac{\pi/4}{1/2}=\frac{\pi}{2}$:

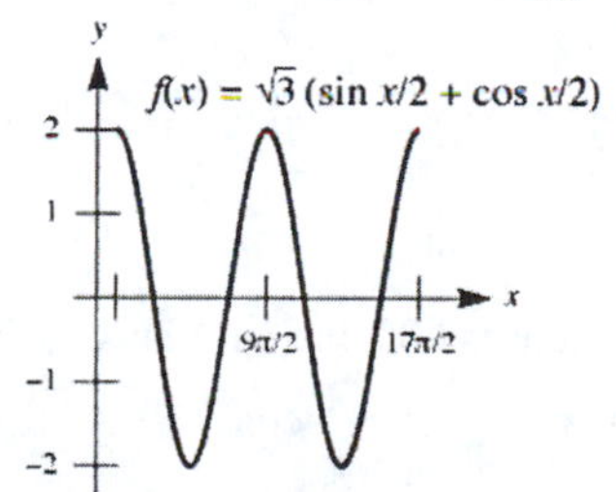

43. Begin with the product-to-sum formula: $\cos A\cos B=\frac{1}{2}\left[\cos(A-B)+\cos(A+B)\right]$

If we let $A + B = \alpha$ and $A - B = \beta$, then: $A=\frac{\alpha+\beta}{2}$ and $B=\frac{\alpha-\beta}{2}$

Substituting: $\cos\frac{\alpha+\beta}{2}\cos\frac{\alpha-\beta}{2}=\frac{1}{2}\left[\cos\beta+\cos\alpha\right]$

Multiplying by 2, we have the desired identity: $\cos\alpha+\cos\beta=2\cos\frac{\alpha+\beta}{2}\cos\frac{\alpha-\beta}{2}$

45. Using the identity for $\sin\alpha + \sin\beta$ (derived in Exercise 44), we replace β with $-\beta$ and note that $\sin(-\beta) = -\sin\beta$:

$$\sin\alpha+\sin(-\beta)=2\sin\frac{\alpha-\beta}{2}\cos\frac{\alpha+\beta}{2}$$
$$\sin\alpha-\sin\beta=2\cos\frac{\alpha+\beta}{2}\sin\frac{\alpha-\beta}{2}$$

47. Obtaining a common denominator in the expression, and noting that $\sin 10° = \cos(90° - 10°) = \cos 80°$:

$$\frac{1}{2\sin 10°} - 2\sin 70° = \frac{1-4\sin 10°\sin 70°}{2\sin 10°} = \frac{1-4\sin 10°\sin 70°}{2\cos 80°}$$

Now using the product-to-sum formula for $\sin A \sin B$:

$$\sin 10°\sin 70° = \tfrac{1}{2}\left[\cos(10°-70°)-\cos(10°+70°)\right] = \tfrac{1}{2}\left[\cos(-60°)-\cos 80°\right] = \tfrac{1}{2}\left[\tfrac{1}{2}-\cos 80°\right] = \tfrac{1}{4}-\tfrac{1}{2}\cos 80°$$

Thus: $$\frac{1}{2\sin 10°} - 2\sin 70° = \frac{1-4\sin 10°\sin 70°}{2\cos 80°} = \frac{1-4\left(\frac{1}{4}-\frac{1}{2}\cos 80°\right)}{2\cos 80°} = \frac{1-1+2\cos 80°}{2\cos 80°} = \frac{2\cos 80°}{2\cos 80°} = 1$$

49. First, note that $A + B + C = 180°$ implies that $A + B = 180° - C$, $B + C = 180° - A$ and $A + C = 180° - B$. Also, note the identities $\sin(180° - \alpha) = \sin\alpha$ and $\cos(180° - \alpha) = -\cos\alpha$. Working from the right-hand side, note that: $\sin\frac{A}{2}\sin\frac{B}{2} = \frac{1}{2}\cos\left(\frac{A}{2}-\frac{B}{2}\right) - \frac{1}{2}\cos\left(\frac{A}{2}+\frac{B}{2}\right)$. Thus:

$$\begin{aligned}
\sin\tfrac{A}{2}\sin\tfrac{B}{2}\sin\tfrac{C}{2} &= \tfrac{1}{2}\cos\left(\tfrac{A}{2}-\tfrac{B}{2}\right)\sin\tfrac{C}{2} - \tfrac{1}{2}\cos\left(\tfrac{A}{2}+\tfrac{B}{2}\right)\sin\tfrac{C}{2} \\
&= \tfrac{1}{2}\sin\left[90° - \tfrac{A}{2} + \tfrac{B}{2}\right]\sin\tfrac{C}{2} - \tfrac{1}{2}\sin\left[90° - \tfrac{A}{2} - \tfrac{B}{2}\right]\sin\tfrac{C}{2} \\
&= \tfrac{1}{4}\left[\cos\left(90° - \tfrac{A}{2} + \tfrac{B}{2} - \tfrac{C}{2}\right) - \cos\left(90° - \tfrac{A}{2} + \tfrac{B}{2} + \tfrac{C}{2}\right)\right] \\
&\quad - \tfrac{1}{4}\left[\cos\left(90° - \tfrac{A}{2} - \tfrac{B}{2} - \tfrac{C}{2}\right) - \cos\left(90° - \tfrac{A}{2} - \tfrac{B}{2} + \tfrac{C}{2}\right)\right] \\
&= \tfrac{1}{4}\left[\cos\frac{180°+B-(A+C)}{2} - \cos\frac{180°-A+(B+C)}{2} - \cos\frac{180°-(A+B+C)}{2} + \cos\frac{180°+C-(A+B)}{2}\right] \\
&= \tfrac{1}{4}\left[\cos\frac{180°+B-(180°-B)}{2} - \cos\frac{180°-A+(180°-A)}{2} - \cos\frac{180°-180°}{2} + \cos\frac{180°+C-(180°-C)}{2}\right] \\
&= \tfrac{1}{4}\left[\cos B - \cos(180° - A) - \cos 0 + \cos C\right] \\
&= \tfrac{1}{4}\left[\cos B + \cos A - 1 + \cos C\right]
\end{aligned}$$

Now prove the identity: $1 + 4\sin\frac{A}{2}\sin\frac{B}{2}\sin\frac{C}{2} = 1 + \cos B + \cos A - 1 + \cos C = \cos A + \cos B + \cos C$

51. **a.** Calculating the sum of the cosines for each triangle:

$$\cos 30° + \cos 70° + \cos 80° \approx 1.38$$
$$\cos 40° + \cos 25° + \cos 115° \approx 1.25$$
$$\cos 55° + \cos 55° + \cos 70° \approx 1.49$$

In each case the sum of the cosines is less than $\frac{3}{2}$.

b. Since each angle of an equilateral triangle is 60°, the sum is: $\cos 60° + \cos 60° + \cos 60° = \frac{1}{2} + \frac{1}{2} + \frac{1}{2} = \frac{3}{2}$

c. (i) This is the sum-to-product formula for $\cos A + \cos B$.

(ii) This is true since $\cos\dfrac{A-B}{2} \le 1$.

(iii) This is true since $A + B = 180° - C$.

(iv) This is just division by 2.

(v) The identities used are $\cos(90° - \theta) = \sin\theta$ and $\cos\theta = 1 - 2\sin^2\frac{\theta}{2}$.

(vi) Multiplying out this expression shows they are equal.

(vii) Since $2\left(\sin\frac{C}{2} - \frac{1}{2}\right)^2 \ge 0$, the expression is at most $\frac{3}{2}$.

8.4 Trigonometric Equations

1. For $\theta = \frac{\pi}{2}$, $2\cos^2\theta - 3\cos\theta = 2(0)^2 - 3(0) = 0$, so $\theta = \frac{\pi}{2}$ is a solution.

3. For $x = \frac{3\pi}{4}$, $\tan^2 x - 3\tan x + 2 = (-1)^2 + 3 + 2 = 6$, so $x = \frac{3\pi}{4}$ is not a solution.

5. Since $\sin\theta = \frac{\sqrt{3}}{2}$, $\theta = \frac{\pi}{3}$ and $\theta = \frac{2\pi}{3}$ are the primary solutions. All solutions will be of the form $\theta = \frac{\pi}{3} + 2\pi k$ or $\theta = \frac{2\pi}{3} + 2\pi k$, where k is any integer.

7. Since $\sin\theta = -\frac{1}{2}$, $\theta = \frac{7\pi}{6}$ and $\theta = \frac{11\pi}{6}$ are the primary solutions. All solutions will be of the form $\theta = \frac{7\pi}{6} + 2\pi k$ or $\theta = \frac{11\pi}{6} + 2\pi k$, where k is any integer.

9. Since $\cos\theta = -1$, $\theta = \pi$ is the primary solution. All solutions will be of the form $\theta = \pi + 2\pi k$, where k is any integer.

11. Since $\tan\theta = \sqrt{3}$, $\theta = \frac{\pi}{3}$ is the primary solution. All solutions will be of the form $\theta = \frac{\pi}{3} + \pi k$, where k is any integer.

13. Since $\tan x = 0$, $x = 0$ is the primary solution. All solutions will be of the form $x = 0 + \pi k = \pi k$, where k is any integer.

15. Since $2\cos^2\theta + \cos\theta = \cos\theta(2\cos\theta + 1) = 0$, the primary solutions are the solutions of $\cos\theta = 0$ or $\cos\theta = -\frac{1}{2}$, which are $\theta = \frac{\pi}{2}$, $\theta = \frac{3\pi}{2}$, $\theta = \frac{2\pi}{3}$ or $\theta = \frac{4\pi}{3}$. All solutions will be of the form $\theta = \frac{\pi}{2} + \pi k$, $\theta = \frac{2\pi}{3} + 2\pi k$ or $\theta = \frac{4\pi}{3} + 2\pi k$, where k is any integer.

17. Since $\cos^2 t\sin t - \sin t = \sin t(\cos^2 t - 1) = 0$, $\sin t = 0$ or $\cos t = \pm 1$. Therefore primary solutions are $t = 0$ or $t = \pi$. All solutions are of the form $t = \pi k$, where k is any integer.

19. Using the identity $\cos^2 x = 1 - \sin^2 x$:

$$2\cos^2 x - \sin x - 1 = 2\left(1 - \sin^2 x\right) - \sin x - 1 = -2\sin^2 x - \sin x + 1 = (-2\sin x + 1)(\sin x + 1)$$

So $\sin x = \frac{1}{2}$ or $\sin x = -1$. Thus the primary solutions are $x = \frac{\pi}{6}$, $x = \frac{5\pi}{6}$ or $x = \frac{3\pi}{2}$. All solutions are of the form $x = \frac{\pi}{6} + 2\pi k$, $x = \frac{5\pi}{6} + 2\pi k$ or $x = \frac{3\pi}{2} + 2\pi k$, where k is any integer.

21. Since $\sqrt{3}\sin t - \sqrt{1 + \sin^2 t} = 0$ is equivalent to $3\sin^2 t = 1 + \sin^2 t$ by squaring each side, $2\sin^2 t = 1$ and thus $\sin^2 t = \frac{1}{2}$ and $\sin t = \pm\frac{\sqrt{2}}{2}$. This would have primary solutions of $\frac{\pi}{4}, \frac{3\pi}{4}, \frac{5\pi}{4}$ or $\frac{7\pi}{4}$, but $\frac{5\pi}{4}$ and $\frac{7\pi}{4}$ do not work in the original equation. So the primary solutions are $t = \frac{\pi}{4}$ or $t = \frac{3\pi}{4}$. All solutions are of the form $t = \frac{\pi}{4} + 2\pi k$ or $t = \frac{3\pi}{4} + 2\pi k$, where k is any integer.

23. a. Graphing the equation:

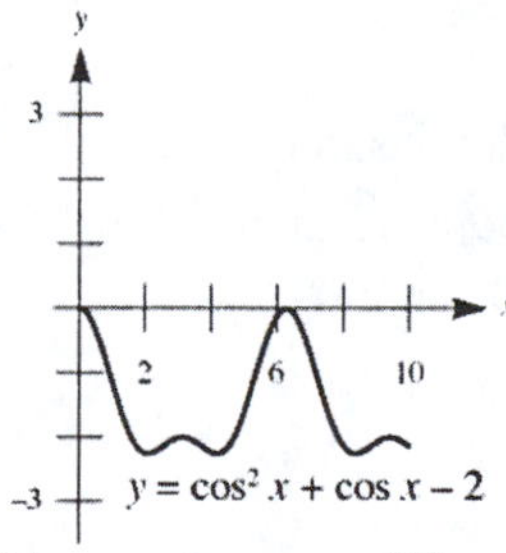

b. The x-intercept is approximately $x \approx 6.28$.

c. The exact value is $x = 2\pi$.

25. Since $\cos x = 0.184$, $x \approx 1.39$ or $x \approx 4.90$.

27. Since $\sin x = \frac{1}{\sqrt{5}}$, $x \approx 0.46$ or $x \approx 2.68$.

29. Since $\tan x = 6$, $x \approx 1.41$ or $x \approx 4.55$.

31. Dividing through by $\cos t$ results in $\tan t = 5$, thus $t \approx 1.37$ or $t \approx 4.51$.

33. Since $\sec t = 2.24$, $\cos t = \dfrac{1}{2.24} \approx 0.45$. Thus $t \approx 1.11$ or $t \approx 5.18$.

35. Factoring the equation:

$$\begin{aligned}\tan^2 x + \tan x - 12 &= 0\\ (\tan x + 4)(\tan x - 3) &= 0\\ \tan x &= -4 \text{ or } 3\end{aligned}$$

If $\tan x = -4$, then $x \approx 1.82$ or $x \approx 4.96$. If $\tan x = 3$, then $x \approx 1.25$ or $x \approx 4.39$. The solutions are approximately $x \approx 1.25, 1.82, 4.39, 4.96$.

37. Factoring the equation by grouping:

$$\begin{aligned}16\sin^3 x - 12\sin^2 x + 36\sin x - 27 &= 0\\ 4\sin^2 x(4\sin x - 3) + 9(4\sin x - 3) &= 0\\ (4\sin x - 3)\left(4\sin^2 x + 9\right) &= 0\\ \sin x &= \tfrac{3}{4}\end{aligned}$$

Note that $4\sin^2 x + 9 \neq 0$. If $\sin x = \frac{3}{4}$, then $x \approx 0.85$ or $x \approx 2.29$. The solutions are approximately $x \approx 0.85, 2.29$.

39. If $\sin\theta = \frac{1}{4}$, then $\theta \approx 14.5°, 165.5°$.

41. Factoring the equation:

$$\begin{aligned}9\tan^2\theta - 16 &= 0\\ (3\tan\theta + 4)(3\tan\theta - 4) &= 0\\ \tan\theta &= -\tfrac{4}{3}, \tfrac{4}{3}\end{aligned}$$

If $\tan\theta = -\frac{4}{3}$, then $\theta \approx 126.9°$ or $\theta \approx 306.9°$. If $\tan\theta = \frac{4}{3}$, then $\theta \approx 53.1°$ or $\theta \approx 233.1°$. The solutions are approximately $\theta \approx 53.1°, 126.9°, 233.1°, 306.9°$.

43. Using the quadratic formula: $\cos\theta = \dfrac{-(-1) \pm \sqrt{(-1)^2 - 4(1)(-1)}}{2(1)} = \dfrac{1 \pm \sqrt{1+4}}{2} = \dfrac{1 \pm \sqrt{5}}{2} \approx -0.6180, 1.6180$

Note that $\cos\theta = 1.6180$ is impossible. If $\cos\theta = -0.6180$, then $\theta \approx 128.2°$ or $\theta \approx 231.8°$. The solutions are approximately $\theta \approx 128.2°, 231.8°$.

45. Since $\cos 3\theta = 1$, $3\theta = 2\pi k$ for any integer k. So $\theta = \frac{2\pi k}{3}$. Thus the values of θ in the interval $[0°, 360°)$ are $0°$, $120°$ and $240°$.

47. Since $\sin 3\theta = -\frac{\sqrt{2}}{2}$, $3\theta = \frac{5\pi}{4} + 2\pi k$ or $\frac{7\pi}{4} + 2\pi k$. Thus $\theta = \frac{5\pi}{12} + \frac{2\pi k}{3}$ or $\frac{7\pi}{12} + \frac{2\pi k}{3}$. So the primary solutions are $75°$, $105°$, $195°$, $225°$, $315°$ or $345°$.

49. Using the double-angle identity for $\sin 2\theta$:

$$\begin{aligned}\sin 2\theta &= -2\cos\theta\\ 2\sin\theta\cos\theta &= -2\cos\theta\\ 2\sin\theta\cos\theta + 2\cos\theta &= 0\\ 2\cos\theta(\sin\theta + 1) &= 0\end{aligned}$$

Thus $\cos\theta = 0$ or $\sin\theta = -1$, which have solutions in the interval $[0°, 360°)$ of $\theta = 90°$ or $270°$.

51. Dividing each side by $\cos 2\theta$ results in $\tan 2\theta = \sqrt{3}$, thus $2\theta = 60° + 180°k$ and so $\theta = 30° + 90°k$. So the solutions in the interval $[0°, 360°)$ are $\theta = 30°$, $120°$, $210°$ or $300°$.

53. The x-coordinates of the intersection points are $x \approx 0.898, 5.385$:

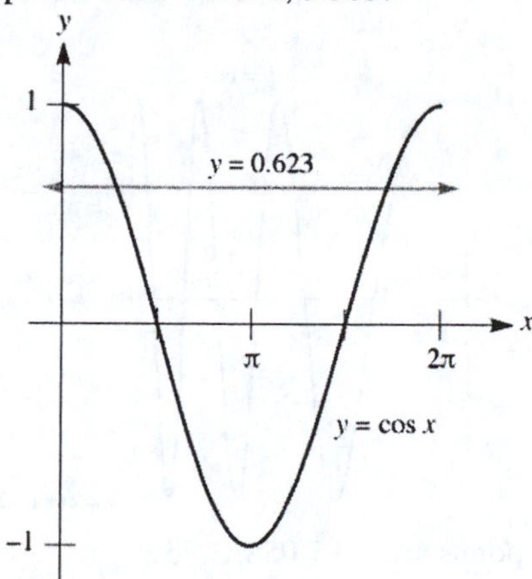

55. The x-coordinates of the intersection points are $x \approx 0.666, 2.475$:

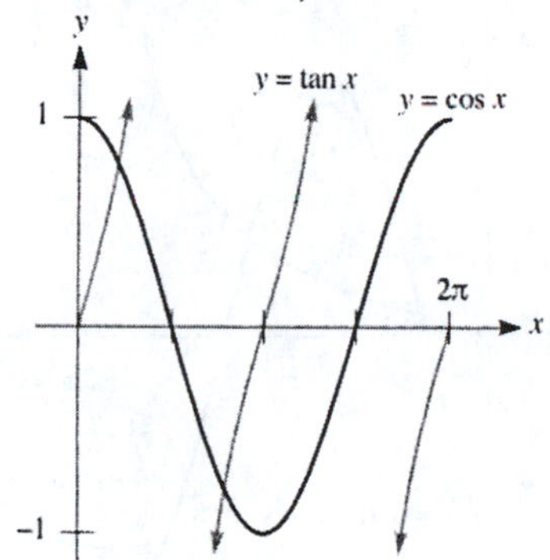

57. The x-coordinates of the intersection points are $x \approx 0.427, 2.715$:

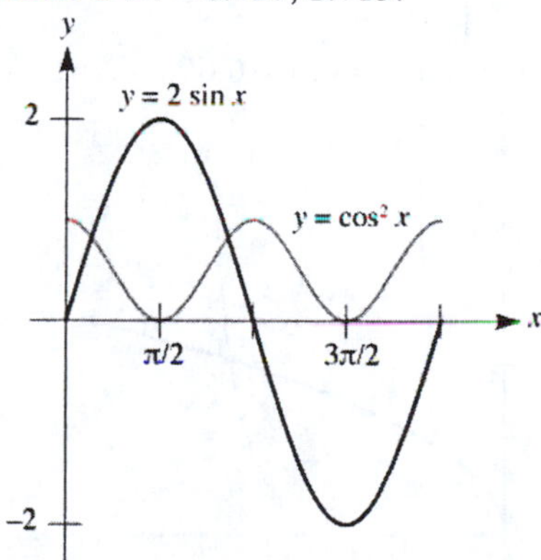

59. There are no solutions to the equation, since the two curves do not intersect:

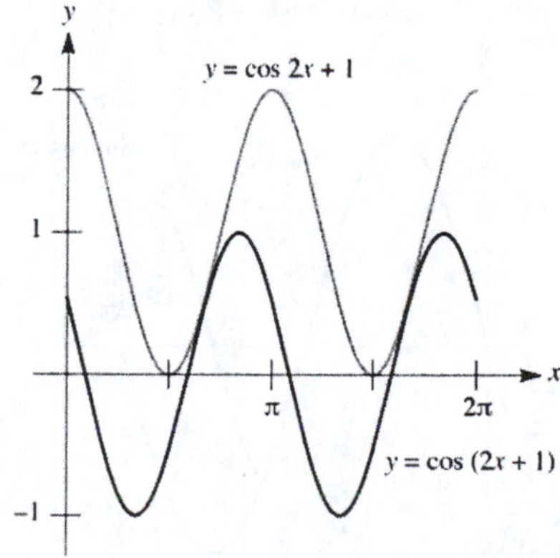

61. The x-coordinates of the intersection points are $x = 0, 1, x \approx 3.080, 4.080, 4.538, 5.538, 5.660$:

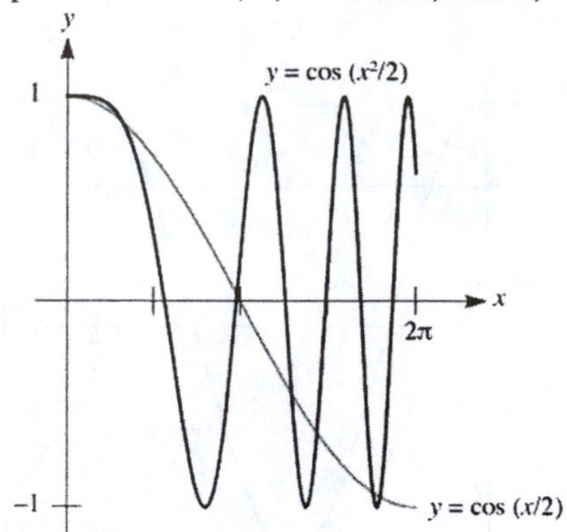

63. The x-coordinates of the intersection points are $x \approx 1.058, 3.739$:

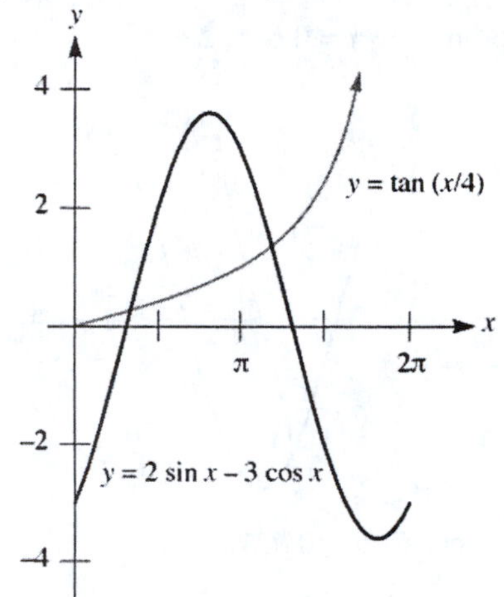

65. The x-coordinates of the intersection points are $x = 0, x \approx 0.695, 4.261$:

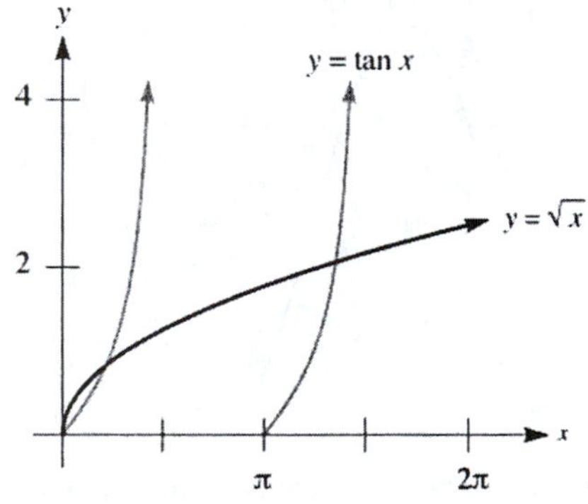

67. The x-coordinates of the intersection points are $x \approx 0.739, 3.881$:

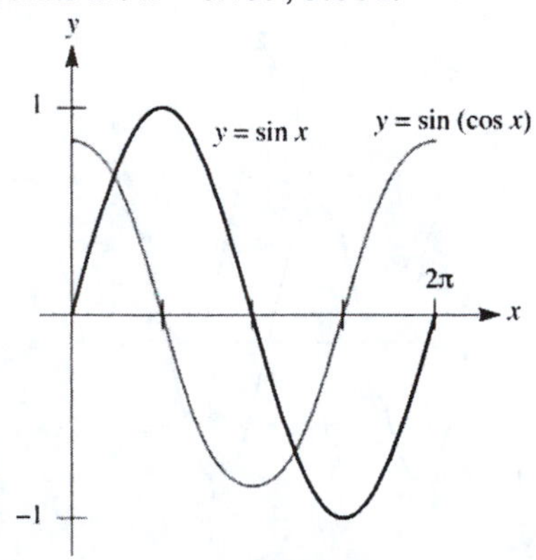

69. The x-coordinates of the intersection points are $x = 0$, $x \approx 4.493$:

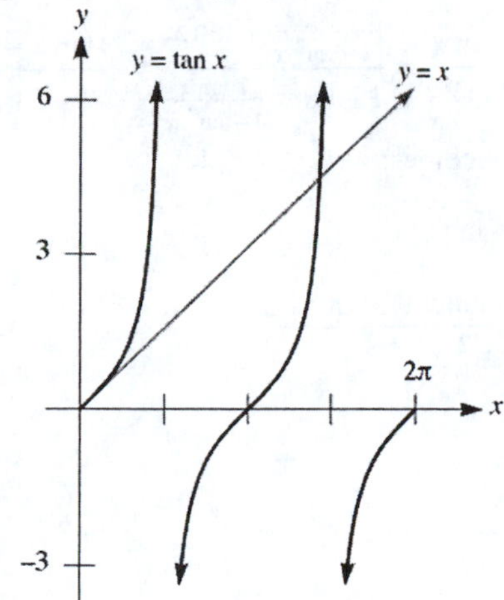

71. The x-coordinates of the intersection points are $x \approx 2.108, 5.746$:

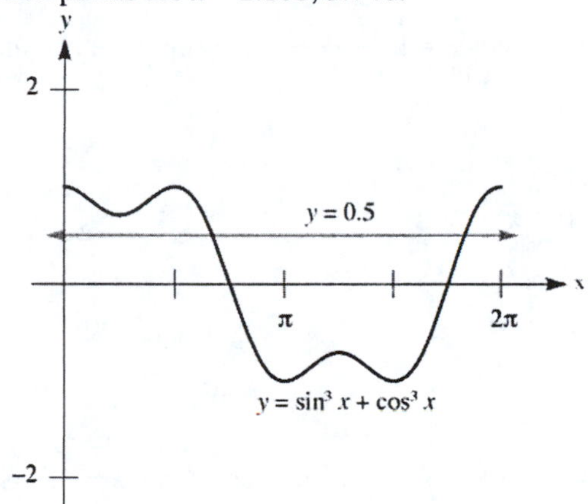

73. The x-coordinate of the intersection point is $x \approx 0.703$:

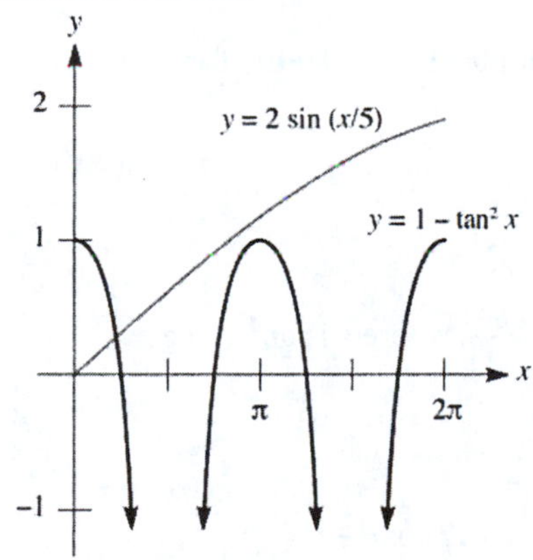

75. Following the hint, use the addition formula and double-angle formula for tangent to obtain:

$$\tan 3x = \tan(2x+x) = \frac{\tan 2x + \tan x}{1 - \tan x \tan 2x} = \frac{\frac{2\tan x}{1-\tan^2 x} + \tan x}{1 - \tan x \frac{2\tan x}{1-\tan^2 x}} = \frac{3\tan x - \tan^3 x}{1 - 3\tan^2 x}$$

So the left-hand side of the equation becomes:

$$\begin{aligned}\tan 3x - \tan x &= \frac{3\tan x - \tan^3 x}{1-3\tan^2 x} - \tan x \\ &= \frac{3\tan x - \tan^3 x - \tan x + 3\tan^3 x}{1 - 3\tan^2 x} \\ &= \frac{2\tan x + 2\tan^3 x}{1-3\tan^2 x} \\ &= \frac{2\tan x\left(1+\tan^2 x\right)}{1-3\tan^2 x}\end{aligned}$$

Since $1 + \tan^2 x \neq 0$, it must be that $\tan x = 0$ and thus $x = 0$ or $x = \pi$.

77. By the half-angle formula for cosine: $\cos\frac{x}{2} = \pm\sqrt{\frac{1+\cos x}{2}}$. Writing the original equation, then squaring:

$$\begin{aligned}\pm\sqrt{\frac{1+\cos x}{2}} &= 1 + \cos x \\ \frac{1+\cos x}{2} &= 1 + 2\cos x + \cos^2 x \\ 1 + \cos x &= 2 + 4\cos x + 2\cos^2 x \\ 0 &= 2\cos^2 x + 3\cos x + 1 \\ 0 &= (2\cos x + 1)(\cos x + 1) \\ \cos x = -\tfrac{1}{2} &\text{ or } \cos x = -1\end{aligned}$$

These equations have solutions of $x = \frac{2\pi}{3}, \frac{4\pi}{3}$ or π. Upon checking $x = \frac{4\pi}{3}$ is not a solution, and thus the solutions are $x = \frac{2\pi}{3}$ or $x = \pi$.

79. Using the identity for $\sec 4\theta$ and the double-angle identity for sine:

$$\begin{aligned}\sec 4\theta + 2\sin 4\theta &= 0 \\ \frac{1}{\cos 4\theta} + 2\sin 4\theta &= 0 \\ 1 + 2\sin 4\theta\cos 4\theta &= 0 \\ \sin 8\theta &= -1\end{aligned}$$

Thus $8\theta = \frac{3\pi}{2} + 2\pi k$, and so $\theta = \frac{3\pi}{16} + \frac{k\pi}{4}$, where k is any integer.

81. Following the hint:

$$\begin{aligned}4\sin\theta - 3\cos\theta &= 2 \\ 4\sin\theta &= 3\cos\theta + 2 \\ 16\sin^2\theta &= 9\cos^2\theta + 12\cos\theta + 4 \\ 16\left(1-\cos^2\theta\right) &= 9\cos^2\theta + 12\cos\theta + 4 \\ 16 - 16\cos^2\theta &= 9\cos^2\theta + 12\cos\theta + 4 \\ 25\cos^2\theta + 12\cos\theta - 12 &= 0\end{aligned}$$

This will not factor, so use the quadratic formula:

$$\cos\theta = \frac{-12 \pm \sqrt{(12)^2 - 4(25)(-12)}}{2(25)} = \frac{-12 \pm \sqrt{1344}}{50} = \frac{-12 \pm 8\sqrt{21}}{50} = \frac{-6 \pm 4\sqrt{21}}{25}$$

So $\cos\theta = 0.4932$ or $\cos\theta = -0.9732$, and thus $\theta = 60.45°$ (the other solution is not in the required interval).

83. **a.** Squaring each side:

$$\begin{aligned}\sin^2 x\cos^2 x &= 1\\ \sin^2 x\left(1-\sin^2 x\right) &= 1\\ \sin^2 x-\sin^4 x &= 1\\ \sin^4 x-\sin^2 x+1 &= 0\end{aligned}$$

b. Using the quadratic formula: $\sin^2 x=\dfrac{1\pm\sqrt{(-1)^2-4(1)(1)}}{2(1)}=\dfrac{1\pm\sqrt{1-4}}{2}=\dfrac{1\pm\sqrt{-3}}{2}$

Thus the original equation has no real-number solutions.

85. **a.** Since $\cos x=0.412$, $x\approx 1.146$ or $x\approx -1.146$. Now add multiples of 2π on to these values until we reach an x-value greater than 1000. Note that $\frac{1000}{2\pi}\approx 159$ so check $x = 1.146 + 159(2\pi)$ as a starting point. The first such value is at $x\approx 1000.173$.

b. Since $\cos x=-0.412$, $x\approx 1.995$ or $x\approx -1.995$. Again, add multiples of 2π on to these values until we reach an x-value greater than 1000. See the note from part **a**. The first such value is at $x\approx 1001.022$.

87. Solving the equation:

$$\begin{aligned}1+0.8\pi\cos(2\pi x) &= 0\\ 0.8\pi\cos(2\pi x) &= -1\\ \cos(2\pi x) &= -\frac{1}{0.8\pi}\\ 2\pi x &\approx 1.9800, 4.3032, 8.2632\\ x &\approx 0.315, 0.685, 1.315\end{aligned}$$

The x-coordinates of the turning points are approximately: P: 0.315, Q: 0.685, R: 1.315

89. **a.** Solving for α:

$$\begin{aligned}r &= \frac{v_0^2\sin 2\alpha}{g}\\ rg &= v_0^2\sin 2\alpha\\ \sin 2\alpha &= \frac{rg}{v_0^2}\\ 2\alpha &= \sin^{-1}\left(\frac{rg}{v_0^2}\right)\\ \alpha &= \tfrac{1}{2}\sin^{-1}\left(\frac{rg}{v_0^2}\right)\end{aligned}$$

b. Substituting $r = 100$ ft, $g=32\frac{\text{ft}}{\text{sec}^2}$, and $v_0=80\frac{\text{ft}}{\text{sec}}$: $\alpha=\frac{1}{2}\sin^{-1}\left(\frac{(100)(32)}{(80)^2}\right)=\frac{1}{2}\sin^{-1}(0.5)=\frac{1}{2}\left(\frac{\pi}{6}\right)=\frac{\pi}{12}$

So $\alpha=\frac{\pi}{12}=15°\approx 0.26$.

c. Working from the right-hand side: $\sin\left[2\left(\frac{\pi}{2}-\alpha\right)\right]=\sin(\pi-2\alpha)=\sin 2\alpha$

d. Using the result from part **c**:

$$\begin{aligned}2\left(\tfrac{\pi}{2}-\alpha\right) &= \tfrac{\pi}{6}\\ \tfrac{\pi}{2}-\alpha &= \tfrac{\pi}{12}\\ \alpha &= \tfrac{5\pi}{12}\end{aligned}$$

So $\alpha=\frac{5\pi}{12}=75°\approx 1.31$.

e. Substituting $r = 200$ ft, $g = 32\frac{\text{ft}}{\text{sec}^2}$, and $v_0 = 80\frac{\text{ft}}{\text{sec}}$: $\alpha = \frac{1}{2}\sin^{-1}\left(\frac{(200)(32)}{(80)^2}\right) = \frac{1}{2}\sin^{-1}(1) = \frac{1}{2}\left(\frac{\pi}{2}\right) = \frac{\pi}{4}$

So $\alpha = \frac{\pi}{4} = 45° \approx 0.79$. The other angle is given by:

$$2\left(\frac{\pi}{2} - \alpha\right) = \frac{\pi}{2}$$
$$\frac{\pi}{2} - \alpha = \frac{\pi}{4}$$
$$\alpha = \frac{\pi}{4}$$

Note that there is only one angle solution here.

91. a. Substituting $v_0 = 40\frac{\text{m}}{\text{sec}}$ and $g = 9.8\frac{\text{m}}{\text{sec}^2}$:

$$\frac{40^2\sin^2\alpha}{2(9.8)} = \frac{40^2\sin 2\alpha}{9.8}$$
$$\sin^2\alpha = 2\sin 2\alpha$$
$$\sin^2\alpha = 4\sin\alpha\cos\alpha$$
$$\frac{\sin\alpha}{\cos\alpha} = 4$$
$$\tan\alpha = 4$$
$$\alpha \approx 76.0°$$

b. Substituting $v_0 = 88\frac{\text{ft}}{\text{sec}}$ and $g = 32\frac{\text{ft}}{\text{sec}^2}$:

$$\frac{88^2\sin^2\alpha}{2(32)} = \frac{88^2\sin 2\alpha}{32}$$
$$\sin^2\alpha = 2\sin 2\alpha$$
$$\sin^2\alpha = 4\sin\alpha\cos\alpha$$
$$\frac{\sin\alpha}{\cos\alpha} = 4$$
$$\tan\alpha = 4$$
$$\alpha \approx 76.0°$$

c. The conclusion is that the initial velocity does not affect the solution. That is, an angle of 76.0° will result in the maximum height being equal to the range, regardless of the initial velocity.

93. Since P has coordinates $(x, \sin x)$, using the Pythagorean theorem:

$$x^2 + \sin^2 x = 2^2$$
$$\sin^2 x + x^2 - 4 = 0$$

Using a graphing calculator and rounding to two decimal places, the coordinates of P are $P(1.74, 0.99)$.

95. a. The coordinates of A are $(-1,0)$, the coordinates of B are $(\cos\theta, \sin\theta)$, and the coordinates of C are $(\cos(-\theta), \sin(-\theta)) = (\cos\theta, -\sin\theta)$. Thus $BC = 2\sin\theta$, and the height (distance from A to BC) is $1 + \cos\theta$. Thus the area is given by: $A(\theta) = \frac{1}{2} \bullet 2\sin\theta \bullet (1 + \cos\theta) = \sin\theta(1 + \cos\theta)$

b. The area of the unit circle is $\pi(1)^2 = \pi$, so find where $A(\theta) = 0.40\pi$: $\sin\theta(1 + \cos\theta) = 0.40\pi$
Using a graphing calculator, the solution is $\theta \approx 49.78°$.

c. Find where $A(\theta) = 0.42\pi$: $\sin\theta(1 + \cos\theta) = 0.42\pi$
No solution exists to this equation. The maximum area of the triangle will occur when $\theta \approx 1.05$ radian, and at this value the maximum area is approximately 1.30.
But $0.42\pi \approx 1.319$, thus the triangle area can never reach 42% of the unit circle area.

97. a. The volume of a right-circular cone is given by $V = \frac{1}{3}\pi r^2 h$. Finding r and h in terms of θ:

$$\cos\theta = \frac{h}{1}, \text{ so } h = \cos\theta \qquad \sin\theta = \frac{h}{1}, \text{ so } h = \sin\theta$$

Thus the volume is given as $V(\theta) = \frac{1}{3}\pi\sin^2\theta\cos\theta$.

b. Graphing the volume function:

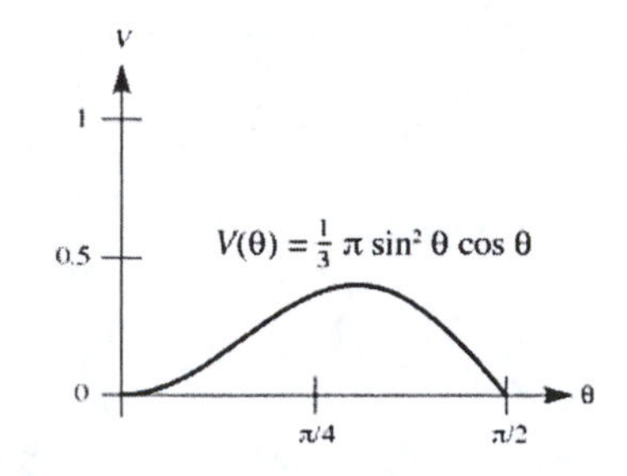

c. Finding where $V(\theta)=0.4$: $\frac{1}{3}\pi\sin^2\theta\cos\theta=0.4$

Using a graphing calculator, the solutions are $x\approx 0.905\approx 51.8°$ and $x\approx 1.005\approx 57.6°$.

d. The maximum occurs at the point (0.955,0.403), so the volume of the cone cannot equal 0.41 m^3.

99. a. Solving for α and using the hint:

$$r=\frac{2v_0^2\sin(\alpha-\beta)\cos\alpha}{g\cos^2\beta}$$

$$\frac{rg\cos^2\beta}{2v_0^2}=\sin(\alpha-\beta)\cos\alpha$$

$$\frac{rg\cos^2\beta}{2v_0^2}=\tfrac{1}{2}\left(\sin(\alpha-\beta-\alpha)+\sin(\alpha-\beta+\alpha)\right)$$

$$\frac{rg\cos^2\beta}{v_0^2}=\sin(-\beta)+\sin(2\alpha-\beta)$$

$$\frac{rg\cos^2\beta}{v_0^2}=\sin(2\alpha-\beta)-\sin\beta$$

$$\sin(2\alpha-\beta)=\frac{rg\cos^2\beta}{v_0^2}+\sin\beta$$

$$2\alpha-\beta=\sin^{-1}\left(\frac{rg\cos^2\beta}{v_0^2}+\sin\beta\right)$$

$$2\alpha=\beta+\sin^{-1}\left(\frac{rg\cos^2\beta}{v_0^2}+\sin\beta\right)$$

$$\alpha=\tfrac{1}{2}\left[\beta+\sin^{-1}\left(\frac{gr\cos^2\beta}{v_0^2}+\sin\beta\right)\right]$$

b. Substituting $\beta=\frac{\pi}{12}$, $v_0=20\frac{\text{ft}}{\text{sec}}$, $r=6$ ft, and $g=32\frac{\text{ft}}{\text{sec}^2}$:

$$\alpha=\tfrac{1}{2}\left[\tfrac{\pi}{12}+\sin^{-1}\left(\frac{32\bullet 6\bullet\cos^2\frac{\pi}{12}}{20^2}+\sin\tfrac{\pi}{12}\right)\right]\approx 30.0°\approx 0.52$$

c. Starting with the right-hand side:

$$\begin{aligned}f\left(\beta+\left(\tfrac{\pi}{2}-\alpha\right)\right)&=\sin\left(\beta+\tfrac{\pi}{2}-\alpha-\beta\right)\cos\left(\beta+\tfrac{\pi}{2}-\alpha\right)\\&=\sin\left(\tfrac{\pi}{2}-\alpha\right)\cos\left(\tfrac{\pi}{2}-(\alpha-\beta)\right)\\&=\cos\alpha\sin(\alpha-\beta)\\&=f(\alpha)\end{aligned}$$

d. Since $\sin(\alpha-\beta)\cos\alpha=\frac{rg\cos^2\beta}{2v_0^2}$, and $\alpha=0.10$ is a solution to this equation, then:

$$\begin{aligned}\beta+\tfrac{\pi}{2}-\alpha&=0.52\\\tfrac{\pi}{12}+\tfrac{\pi}{2}-\alpha&=0.52\\\alpha&=\tfrac{\pi}{12}+\tfrac{\pi}{2}-0.52\\\alpha&\approx 1.31\approx 75.0°\end{aligned}$$

101. **a.** Substituting $\beta = \frac{\pi}{12}$ and $v_0 = 20\frac{\text{ft}}{\text{sec}}$:

$$r = \frac{2(20)^2 \sin\left(\alpha - \frac{\pi}{12}\right)\cos\alpha}{g\cos^2\frac{\pi}{12}}$$
$$= \frac{800 \bullet \frac{1}{2}\left[\sin\left(\alpha - \frac{\pi}{12} - \alpha\right) + \sin\left(\alpha - \frac{\pi}{12} + \alpha\right)\right]}{g\cos^2\frac{\pi}{12}}$$
$$= \frac{400\left[\sin\left(-\frac{\pi}{12}\right) + \sin\left(2\alpha - \frac{\pi}{12}\right)\right]}{g\cos^2\frac{\pi}{12}}$$
$$= \frac{400\left[\sin\left(2\alpha - \frac{\pi}{12}\right) - \sin\frac{\pi}{12}\right]}{g\cos^2\frac{\pi}{12}}$$

b. Since $\sin\left(2\alpha - \frac{\pi}{12}\right) = 1$ when $2\alpha - \frac{\pi}{12} = \frac{\pi}{2}$, we have $2\alpha = \frac{7\pi}{12}$, so $\alpha = \frac{7\pi}{24}$. The maximum range is thus:

$$r_{max} = \frac{400\left(1 - \sin\frac{\pi}{12}\right)}{g\cos^2\frac{\pi}{12}}$$

c. Evaluating: $r_{max} = \dfrac{400\left(1 - \sin\frac{\pi}{12}\right)}{g\cos^2\frac{\pi}{12}} \approx 9.93$ ft

Note that r_{max} is less than 10 ft, as expected, and also $\dfrac{\frac{\pi}{12} + \frac{\pi}{2}}{2} = \dfrac{\frac{7\pi}{12}}{2} = \frac{7\pi}{24}$.

8.5 The Inverse Trigonometric Functions

1. We are asked to find the number x in the interval $\left[-\frac{\pi}{2}, \frac{\pi}{2}\right]$ such that $\sin x = \frac{\sqrt{3}}{2}$. Since $x = \frac{\pi}{3}$ is that number, $\sin^{-1}\left(\frac{\sqrt{3}}{2}\right) = \frac{\pi}{3}$.

3. We are asked to find the number x in the interval $\left(-\frac{\pi}{2}, \frac{\pi}{2}\right)$ such that $\tan x = \sqrt{3}$. Since $x = \frac{\pi}{3}$ is that number, $\tan^{-1}\sqrt{3} = \frac{\pi}{3}$.

5. We are asked to find the number x in the interval $\left(-\frac{\pi}{2}, \frac{\pi}{2}\right)$ such that $\tan x = -\frac{1}{\sqrt{3}}$. Since $x = -\frac{\pi}{6}$ is that number, $\arctan\left(-\frac{1}{\sqrt{3}}\right) = -\frac{\pi}{6}$.

7. We are asked to find the number x in the interval $\left(-\frac{\pi}{2}, \frac{\pi}{2}\right)$ such that $\tan x = 1$. Since $x = \frac{\pi}{4}$ is that number, $\tan^{-1} 1 = \frac{\pi}{4}$.

9. We are asked to find the number x in the interval $[0, \pi]$ such that $\cos x = 2\pi$. Since $\cos x \le 1$ for all x, this value for x does not exist, thus $\cos^{-1} 2\pi$ is undefined.

11. If $x = \sin^{-1}\left(\frac{1}{4}\right)$, then $\sin x = \frac{1}{4}$. Thus $\sin\left[\sin^{-1}\frac{1}{4}\right] = \frac{1}{4}$.

13. If $x = \cos^{-1}\left(\frac{3}{4}\right)$, then $\cos x = \frac{3}{4}$. Thus $\cos\left[\cos^{-1}\frac{3}{4}\right] = \frac{3}{4}$.

15. We are asked to find the number x in the interval $\left(-\frac{\pi}{2}, \frac{\pi}{2}\right)$ such that $\tan x = \tan\left(-\frac{\pi}{7}\right)$. Since $x = -\frac{\pi}{7}$ is that number, $\arctan\left[\tan\left(-\frac{\pi}{7}\right)\right] = -\frac{\pi}{7}$.

17. Since $\sin\frac{\pi}{2} = 1$, we must find arcsin 1. We are asked to find the number x in the interval $\left[-\frac{\pi}{2}, \frac{\pi}{2}\right]$ such that $\sin x = 1$. Since $x = \frac{\pi}{2}$ is that number, $\arcsin\left[\sin\frac{\pi}{2}\right] = \frac{\pi}{2}$.

19. Since $\cos 2\pi = 1$, we must find arccos 1. We are asked to find the number x in the interval $[0, \pi]$ such that $\cos x = 1$. Since $x = 0$ is that number, $\arccos(\cos 2\pi) = 0$.

21. **a.** The maximum value is approximately 1.57 (when $x = 1$), and the minimum value is approximately -1.57 (when $x = -1$):

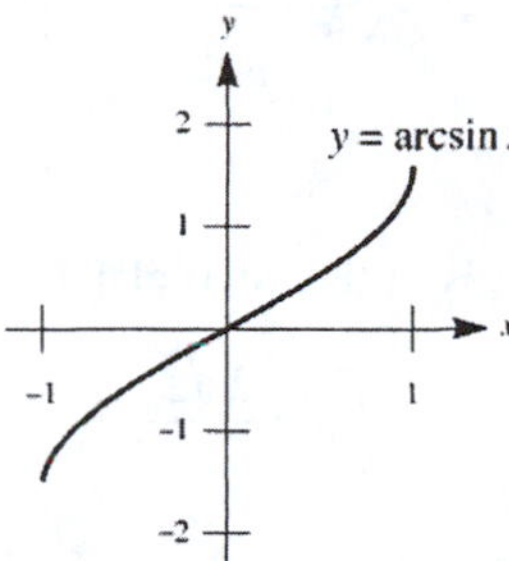

b. The maximum value is $\frac{\pi}{2}$ (when $x = 1$), and the minimum value is $-\frac{\pi}{2}$ (when $x = -1$).

23. **a.** Graphing the function:

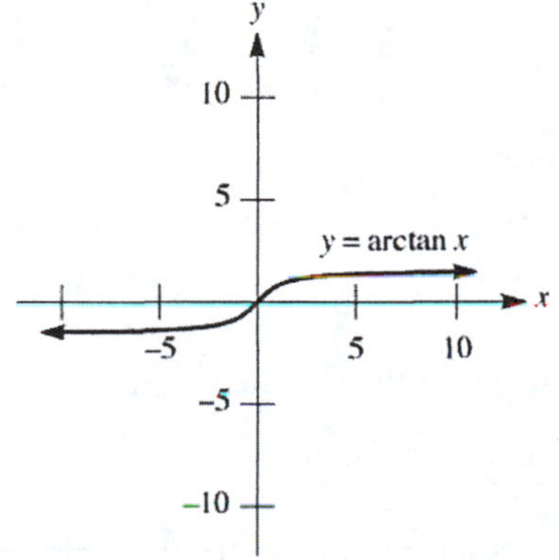

b. The two horizontal asymptotes are $y = -\frac{\pi}{2}$ and $y = \frac{\pi}{2}$. Graphing using the suggested viewing rectangle:

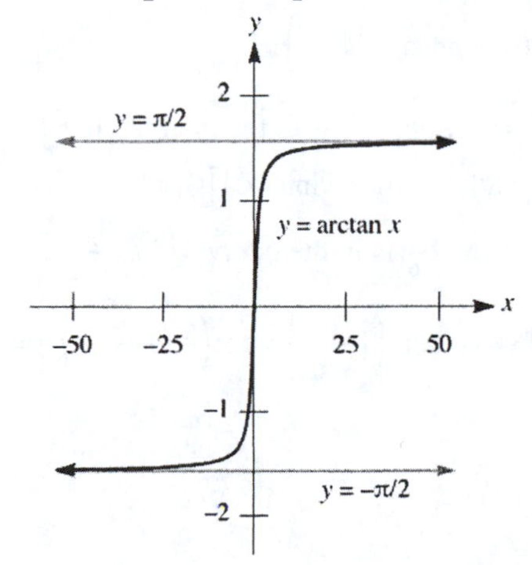

25. If $x = \arcsin\frac{2}{7}$, then $\sin x = \frac{2}{7}$ and since x is in the interval $\left[-\frac{\pi}{2}, \frac{\pi}{2}\right]$, x must lie in the first quadrant, so:

$$\cos x = \sqrt{1-\sin^2 x} = \sqrt{1-\left(\tfrac{2}{7}\right)^2} = \sqrt{1-\tfrac{4}{49}} = \sqrt{\tfrac{45}{49}} = \frac{3\sqrt{5}}{7}$$

Therefore: $\cos\left(\arcsin\frac{2}{7}\right) = \cos x = \dfrac{3\sqrt{5}}{7}$

27. Let $x = \tan^{-1}(-1)$, so $\tan x = -1$ and thus $x = -\frac{\pi}{4}$. Therefore: $\sin\left[\tan^{-1}(-1)\right] = \sin\left(-\frac{\pi}{4}\right) = -\frac{\sqrt{2}}{2}$

29. If $x = \sin^{-1}\left(\frac{2}{3}\right)$, then $\sin x = \frac{2}{3}$ and since x is in the interval $\left[-\frac{\pi}{2}, \frac{\pi}{2}\right]$, x must lie in the first quadrant, so:

$$\cos x = \sqrt{1-\sin^2 x} = \sqrt{1-\left(\tfrac{2}{3}\right)^2} = \sqrt{1-\tfrac{4}{9}} = \sqrt{\tfrac{5}{9}} = \frac{\sqrt{5}}{3}$$

Therefore: $\cos\left[\sin^{-1}\left(\frac{2}{3}\right)\right] = \cos x = \dfrac{\sqrt{5}}{3}$

31. If $x = \cos^{-1}\left(\frac{1}{3}\right)$, then $\sin x = \frac{1}{3}$ and since x is in the interval $[0, \pi]$, x must lie in the first quadrant, so:

$$\sin x = \sqrt{1-\cos^2 x} = \sqrt{1-\left(\tfrac{1}{3}\right)^2} = \sqrt{1-\tfrac{1}{9}} = \sqrt{\tfrac{8}{9}} = \frac{2\sqrt{2}}{3}$$

Therefore: $\sin\left[\cos^{-1}\left(\frac{1}{3}\right)\right] = \sin x = \dfrac{2\sqrt{2}}{3}$

33. If $x = \arcsin\frac{20}{29}$, then $\sin x = \frac{20}{29}$ and since x is in the interval $\left[-\frac{\pi}{2}, \frac{\pi}{2}\right]$, x must lie in the first quadrant, so:

$$\cos x = \sqrt{1-\sin^2 x} = \sqrt{1-\left(\tfrac{20}{29}\right)^2} = \sqrt{1-\tfrac{400}{841}} = \sqrt{\tfrac{441}{841}} = \tfrac{21}{29}$$

Therefore: $\tan\left(\arcsin\frac{20}{29}\right) = \tan x = \dfrac{\sin x}{\cos x} = \dfrac{\frac{20}{29}}{\frac{21}{29}} = \dfrac{20}{21}$

35. First compute $\sin^{-1}\left(\frac{1}{2}\right)$ and $\cos^{-1}\left(\frac{1}{2}\right)$:

$\sin^{-1}\left(\frac{1}{2}\right) = \frac{\pi}{6}$, since $\sin\frac{\pi}{6} = \frac{1}{2}$ and $\frac{\pi}{6}$ is in the interval $\left[-\frac{\pi}{2}, \frac{\pi}{2}\right]$

$\cos^{-1}\left(\frac{1}{2}\right) = \frac{\pi}{3}$, since $\cos\frac{\pi}{3} = \frac{1}{2}$ and $\frac{\pi}{3}$ is in the interval $[0, \pi]$

Therefore: $\csc\left[\sin^{-1}\left(\frac{1}{2}\right) - \cos^{-1}\left(\frac{1}{2}\right)\right] = \csc\left(\frac{\pi}{6} - \frac{\pi}{3}\right) = \csc\left(-\frac{\pi}{6}\right) = \dfrac{1}{\sin\left(-\frac{\pi}{6}\right)} = \dfrac{1}{-\frac{1}{2}} = -2$

37. First compute $\cos^{-1}\left(-\frac{1}{2}\right)$, $\cos^{-1}(0)$, and $\tan^{-1}\left(\frac{1}{\sqrt{3}}\right)$:

$\cos^{-1}\left(-\frac{1}{2}\right) = \frac{2\pi}{3}$, since $\cos\frac{2\pi}{3} = -\frac{1}{2}$ and $\frac{2\pi}{3}$ is in the interval $[0, \pi]$

$\cos^{-1}(0) = \frac{\pi}{2}$, since $\cos\frac{\pi}{2} = 0$ and $\frac{\pi}{2}$ is in the interval $[0, \pi]$

$\tan^{-1}\left(\frac{1}{\sqrt{3}}\right) = \frac{\pi}{6}$, since $\tan\frac{\pi}{6} = \frac{1}{\sqrt{3}}$ and $\frac{\pi}{6}$ is in the interval $\left(-\frac{\pi}{2}, \frac{\pi}{2}\right)$

Therefore: $\cot\left[\cos^{-1}\left(-\frac{1}{2}\right) + \cos^{-1}(0) + \tan^{-1}\left(\frac{1}{\sqrt{3}}\right)\right] = \cot\left(\frac{2\pi}{3} + \frac{\pi}{2} + \frac{\pi}{6}\right) = \cot\frac{4\pi}{3} = \frac{\sqrt{3}}{3}$

39. Note that the graphs appear to be identical:

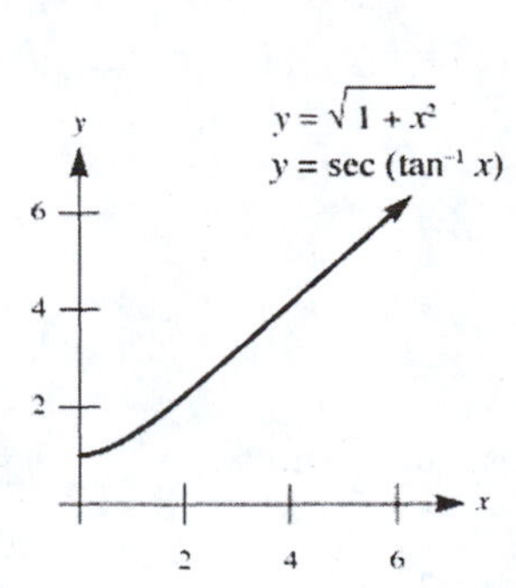

41. Note that the graphs appear to be identical:

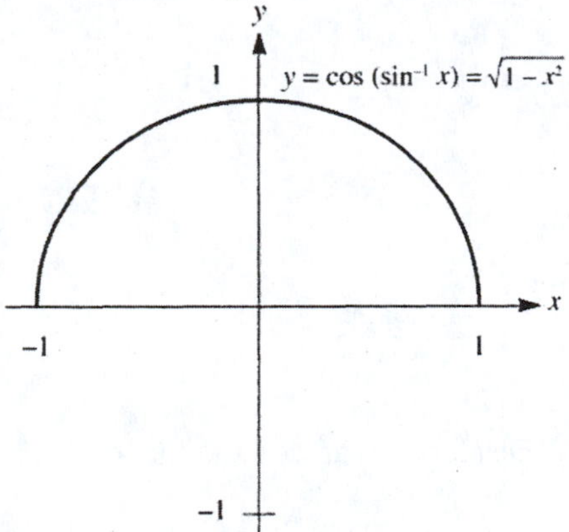

43. Since $\sin 2\theta = 2\sin\theta\cos\theta$ by the double-angle identity for sine, we must find $\cos\theta$. Since $0 < \theta < \frac{\pi}{2}$, $\cos\theta > 0$ and thus: $\cos\theta = \sqrt{1-\sin^2\theta} = \sqrt{1-\left(\frac{3x}{2}\right)^2} = \sqrt{1-\frac{9x^2}{4}} = \frac{\sqrt{4-9x^2}}{2}$. Since $\sin\theta = \frac{3x}{2}$, $\theta = \sin^{-1}\left(\frac{3x}{2}\right)$, and therefore:

$$\frac{\theta}{4} - \sin 2\theta = \frac{1}{4}\theta - 2\sin\theta\cos\theta = \frac{1}{4}\sin^{-1}\left(\frac{3x}{2}\right) - 2 \bullet \frac{3x}{2} \bullet \frac{\sqrt{4-9x^2}}{2} = \frac{1}{4}\sin^{-1}\left(\frac{3x}{2}\right) - \frac{3x\sqrt{4-9x^2}}{2}$$

45. Given $\tan\theta = \frac{x-1}{2}$, construct the triangle:

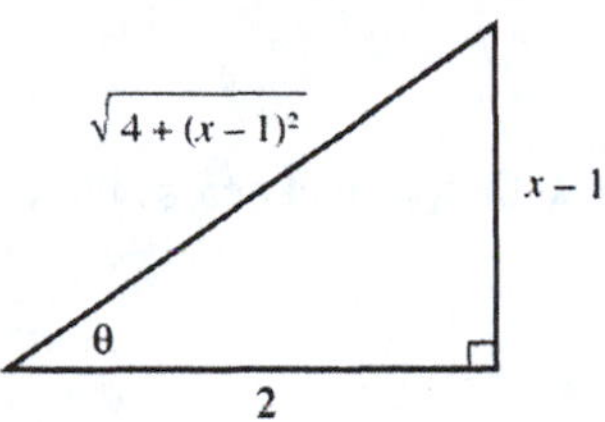

Therefore: $\theta - \cos\theta = \tan^{-1}\left(\frac{x-1}{2}\right) - \frac{2}{\sqrt{4+(x-1)^2}} = \tan^{-1}\left(\frac{x-1}{2}\right) - \frac{2}{\sqrt{5-2x+x^2}}$

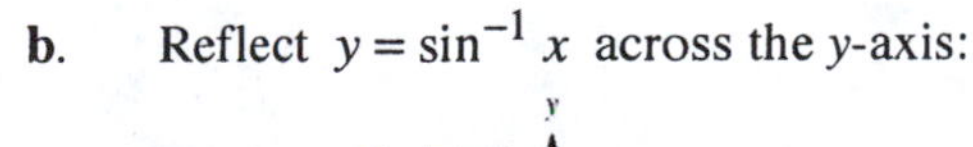

47. **a.** Reflect $y = \sin^{-1} x$ across the x-axis:

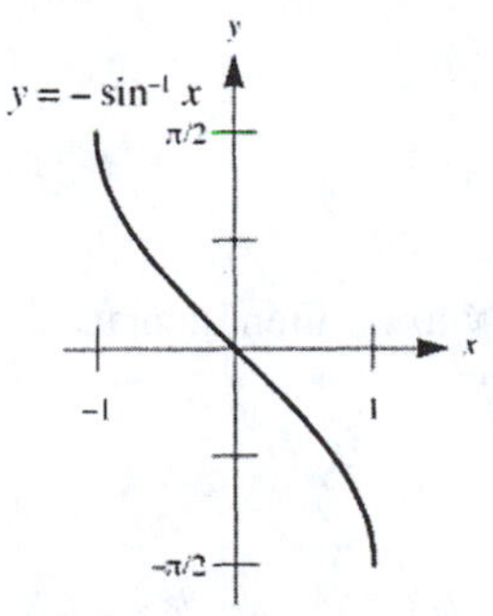

b. Reflect $y = \sin^{-1} x$ across the y-axis:

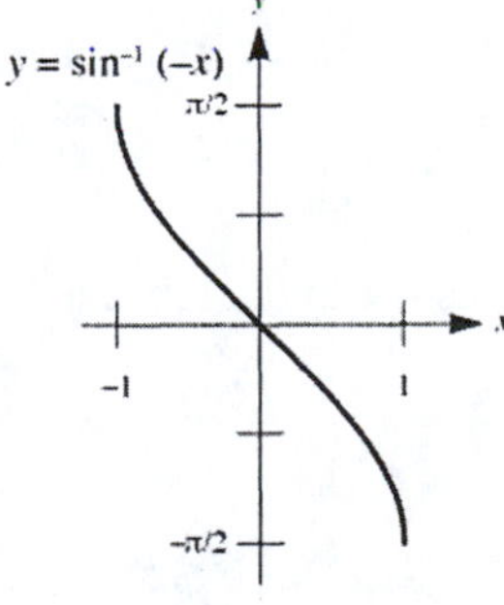

c. Reflect $y = \sin^{-1} x$ across both the x- and y-axes:

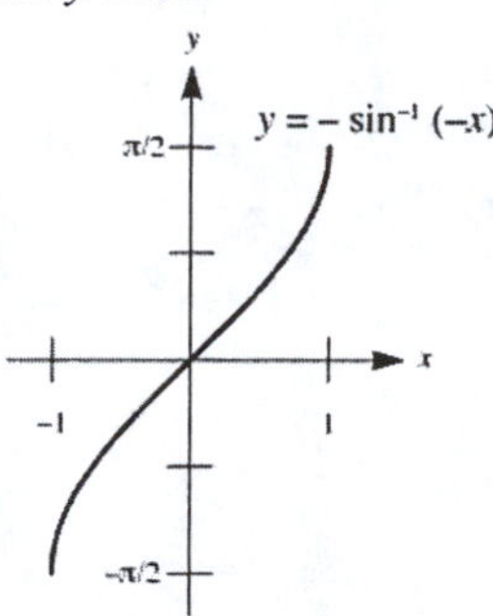

49. **a.** Displace $y = \arccos x$ to the left 1 unit:

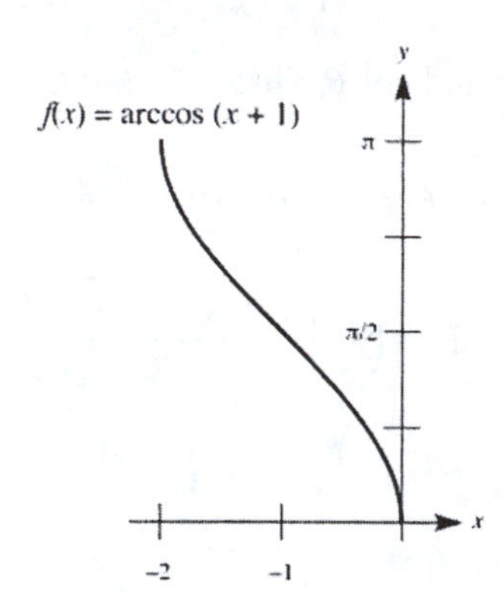

b. Displace $y = \arccos x$ up $\frac{\pi}{2}$ units:

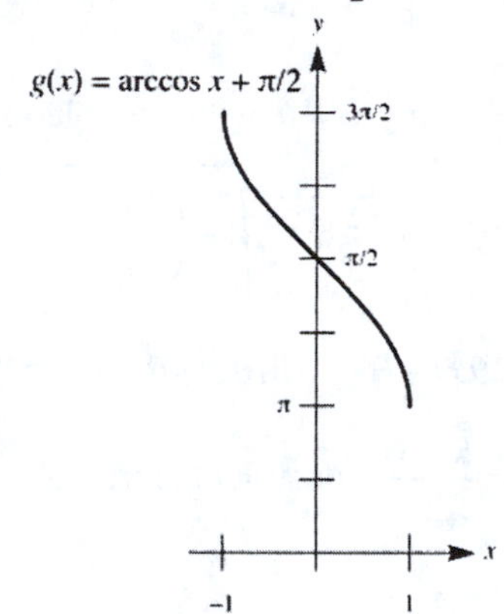

51. **a.** Displace $y = \arcsin x$ to the left 2 units, then reflect across the y-axis, then displace up $\frac{\pi}{2}$ units:

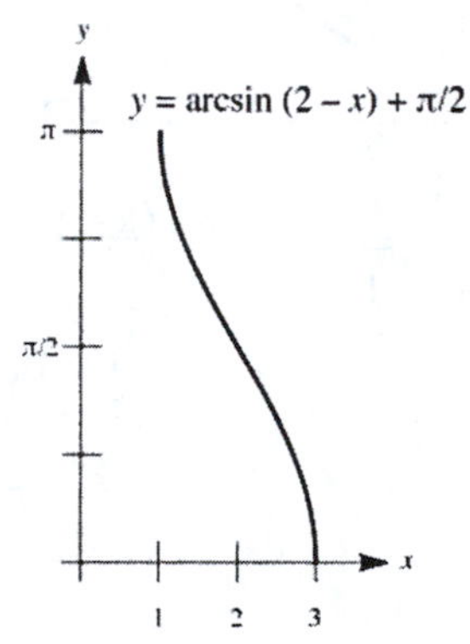

b. Displace $y = \arcsin x$ to the left 2 units, then reflect across both the x- and y-axes, then displace up $\frac{\pi}{2}$ units:

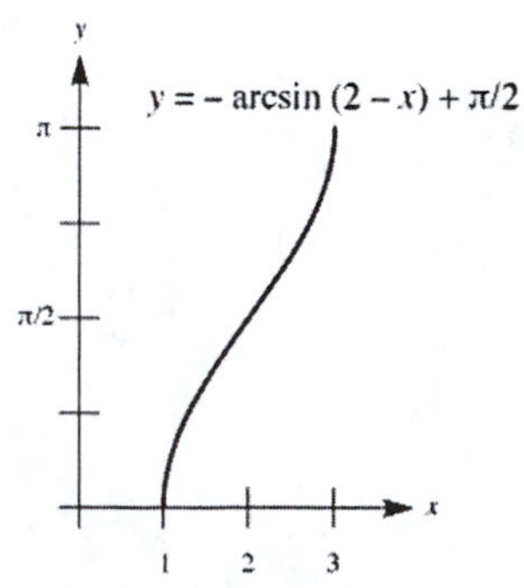

53. **a.** Reflect $y = \tan^{-1} x$ across the x-axis:

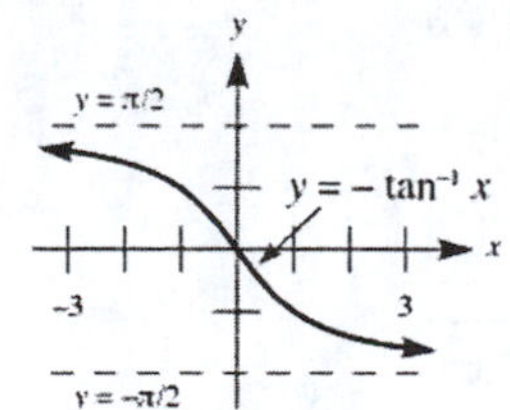

b. Reflect $y = \tan^{-1} x$ across the y-axis:

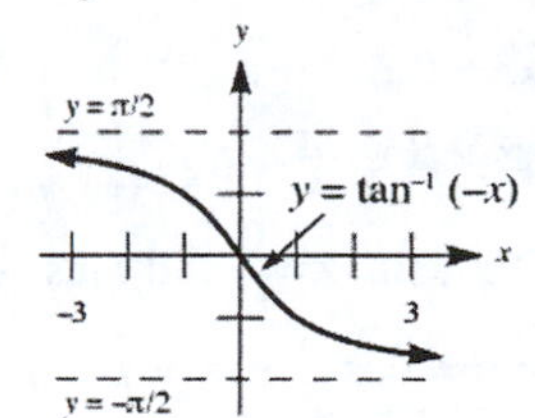

c. Reflect $y = \tan^{-1} x$ across both the x- and y-axes:

55. Displace $y = \arctan x$ to the left 1 unit, then reflect across both the x- and y-axes, then displace down $\frac{\pi}{2}$ units:

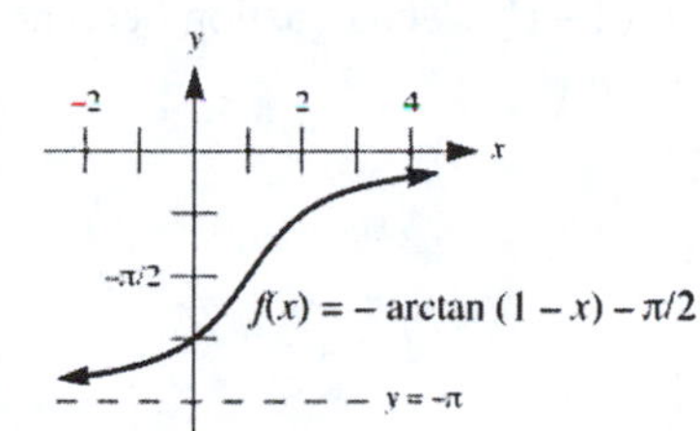

57. Let $\theta = \tan^{-1} 4$, so $\tan\theta = 4$. Draw the triangle:

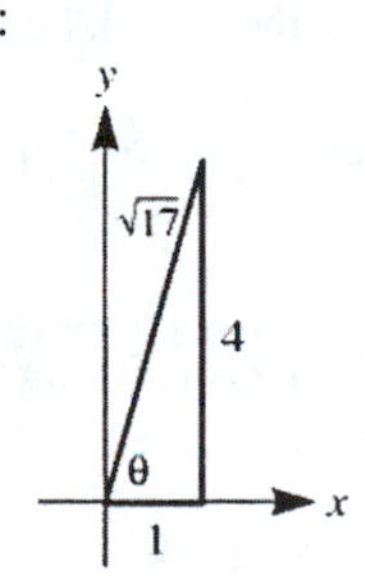

Using the double-angle formula for $\sin 2\theta$: $\sin\left(2\tan^{-1} 4\right) = \sin 2\theta = 2\sin\theta\cos\theta = 2 \bullet \frac{4}{\sqrt{17}} \bullet \frac{1}{\sqrt{17}} = \frac{8}{17}$

59. Let $s = \arccos\frac{3}{5}$, so $\cos s = \frac{3}{5}$. Drawing a triangle:

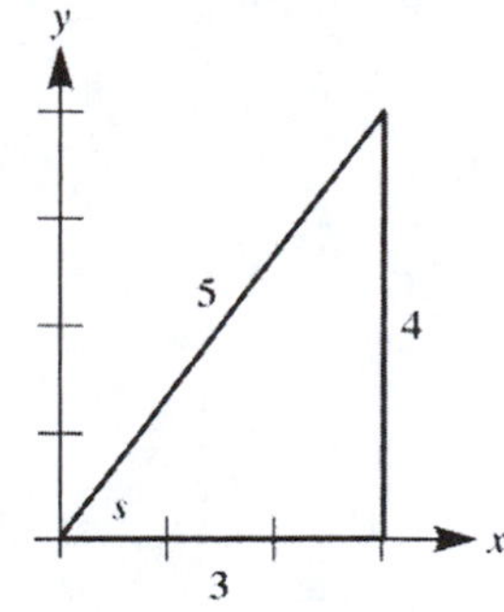

Let $t = \arctan\frac{7}{13}$, so $\tan t = \frac{7}{13}$. Drawing a triangle:

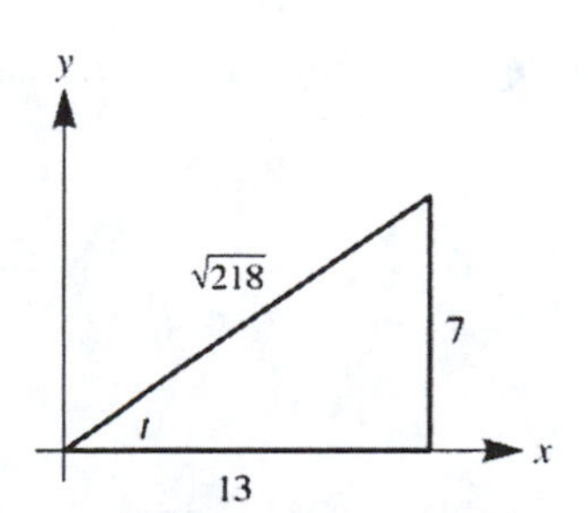

Using the addition formula for $\sin(s - t)$:

$$\sin\left(\arccos\tfrac{3}{5} - \arctan\tfrac{7}{13}\right) = \sin(s - t) = \sin s\cos t - \cos s\sin t = \tfrac{4}{5} \bullet \frac{13}{\sqrt{218}} - \tfrac{3}{5} \bullet \frac{7}{\sqrt{218}} = \frac{31}{5\sqrt{218}} = \frac{31\sqrt{218}}{1090}$$

61. a. Since $\alpha = \sin^{-1} x$, $-\frac{\pi}{2} \le \alpha \le \frac{\pi}{2}$ and since $\beta = \cos^{-1} x$, $0 \le \beta \le \pi$. Thus:

$$-\tfrac{\pi}{2} + 0 \le \alpha + \beta \le \tfrac{\pi}{2} + \pi$$
$$-\tfrac{\pi}{2} \le \alpha + \beta \le \tfrac{3\pi}{2}$$

b. Since $\alpha = \sin^{-1} x$, $\sin\alpha = x$ and thus: $\cos\alpha = \sqrt{1-\sin^2\alpha} = \sqrt{1-x^2}$

Similarly, since $\beta = \cos^{-1} x$, $\cos\beta = x$ and thus: $\sin\beta = \sqrt{1-\cos^2\beta} = \sqrt{1-x^2}$

Using the addition formula for $\sin(\alpha+\beta)$:

$$\sin(\alpha+\beta) = \sin\alpha\cos\beta + \cos\alpha\sin\beta = x \bullet x + \sqrt{1-x^2} \bullet \sqrt{1-x^2} = x^2 + 1 - x^2 = 1$$

But if $-\frac{\pi}{2} \le \alpha + \beta \le \frac{3\pi}{2}$ (from part **a**) and $\sin(\alpha+\beta) = 1$, then $\alpha + \beta = \frac{\pi}{2}$.

63. Using the hint:

$$\cos\left(\cos^{-1} t\right) = \cos\left(\sin^{-1} t\right)$$
$$t = \cos\left(\sin^{-1} t\right)$$

Let $u = \sin^{-1} t$, so $\sin u = t$. Then $\cos u = \sqrt{1-t^2}$. The equation becomes: $t = \sqrt{1-t^2}$. Squaring each side:

$$t^2 = 1 - t^2$$
$$2t^2 = 1$$
$$t^2 = \tfrac{1}{2}$$
$$t = \pm\frac{\sqrt{2}}{2}$$

Notice, however, that $t = -\frac{\sqrt{2}}{2}$ does not check in the original equation (remember, we squared each side of the equation which could introduce extraneous roots), since $\cos^{-1}\left(-\frac{\sqrt{2}}{2}\right) = \frac{3\pi}{4}$ while $\sin^{-1}\left(-\frac{\sqrt{2}}{2}\right) = -\frac{\pi}{4}$.

Thus $t = \frac{\sqrt{2}}{2}$ is the only solution.

65. Let $s = \arctan x$ and $t = \arctan y$, so $\tan s = x$ and $\tan t = y$. Using the addition formula for $\tan(s+t)$:

$$\tan(s+t) = \frac{\tan s + \tan t}{1 - \tan s \tan t} = \frac{x+y}{1-xy}$$

Then: $\arctan(\tan(s+t)) = \arctan\left(\frac{x+y}{1-xy}\right)$. So $s + t = \arctan x + \arctan y = \arctan\left(\frac{x+y}{1-xy}\right)$.

67. Following the hint, take the tangent of each side to obtain:

$$\tan\left(2\tan^{-1} x\right) = \tan\left[\tan^{-1}\left(\tfrac{1}{4x}\right)\right]$$
$$\tan\left(2\tan^{-1} x\right) = \tfrac{1}{4x}$$

Let $\alpha = \tan^{-1} x$, so using the double-angle formula for $\tan 2\alpha$:

$$\frac{2\tan\alpha}{1-\tan^2\alpha} = \frac{1}{4x}$$
$$\frac{2x}{1-x^2} = \frac{1}{4x}$$
$$8x^2 = 1 - x^2$$
$$9x^2 = 1$$
$$x^2 = \tfrac{1}{9}$$
$$x = \pm\tfrac{1}{3}$$

69. **a.** For $\cos^{-1} x$, x must lie in the interval $0 \le x \le \pi$, and for $\tan^{-1} x$, x must lie in the interval $-\frac{\pi}{2} < x < \frac{\pi}{2}$. Since $x = 0$ is not a solution, then $x > 0$.

b. Taking the tangent of each side of the equation:

$$\tan\left(\cos^{-1} x\right) = \tan\left(\tan^{-1} x\right)$$
$$\tan\left(\cos^{-1} x\right) = x$$

Let $\theta = \cos^{-1} x$, so $\cos \theta = x$. Draw the triangle:

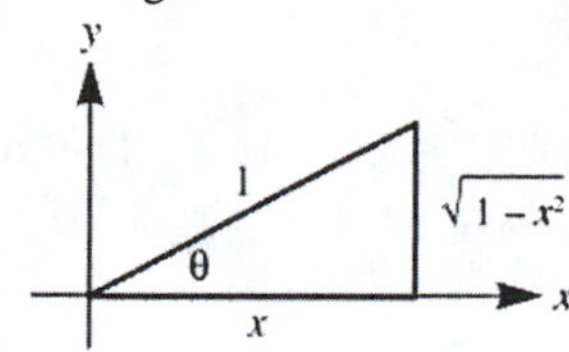

Therefore: $\tan\left(\cos^{-1} x\right) = \tan\theta = \dfrac{\sqrt{1-x^2}}{x}$. Substituting, solve the equation:

$$\frac{\sqrt{1-x^2}}{x} = x$$
$$\sqrt{1-x^2} = x^2$$
$$1-x^2 = x^4$$
$$0 = x^4 + x^2 - 1$$

Using the quadratic formula: $x^2 = \dfrac{-1 \pm \sqrt{1-4(-1)}}{2} = \dfrac{-1 \pm \sqrt{5}}{2}$

Since $x^2 \ge 0$ the negative root is discarded, and thus:

$$x^2 = \frac{\sqrt{5}-1}{2}, \text{ so } x = \sqrt{\frac{\sqrt{5}-1}{2}} \quad (x \ge 0 \text{ in the first quadrant})$$

c. Using a calculator, $x \approx 0.786$ and $y \approx \cos^{-1} 0.786 \approx 0.666$. The point of intersection is approximately $(0.786, 0.666)$.

71. The solution is $x \approx 0.74$:

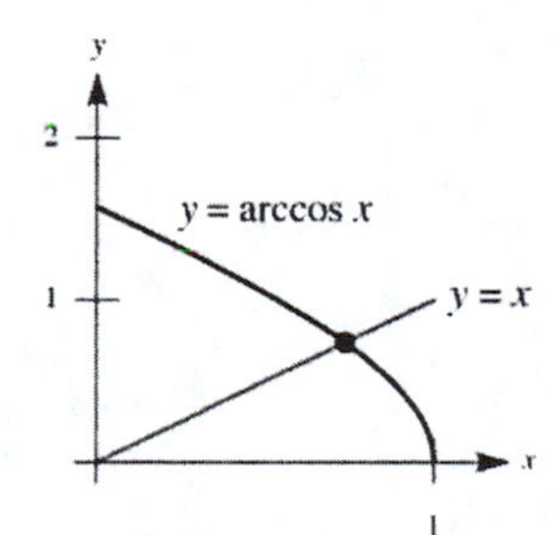

73. **a.** The solution is $x \approx 0.96$:

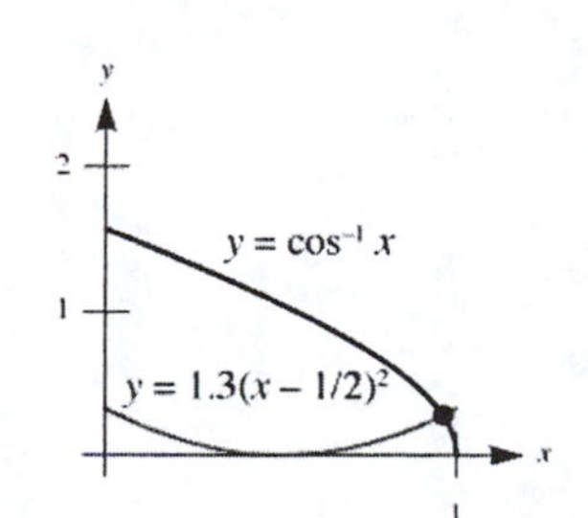

b. The solution is $x \approx 0.96$:

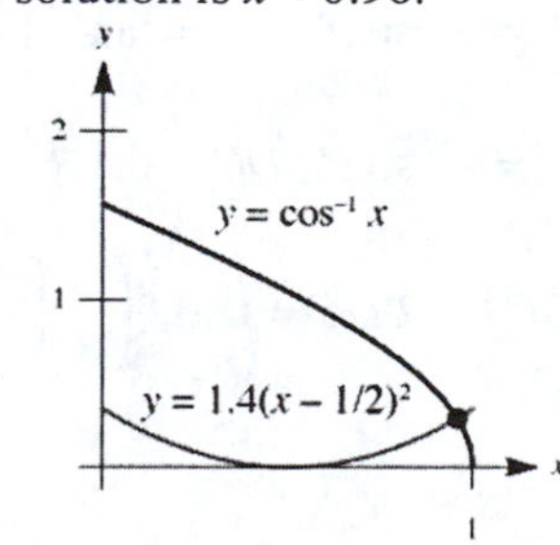

75. **a.** The solution is $x \approx 0.24$:

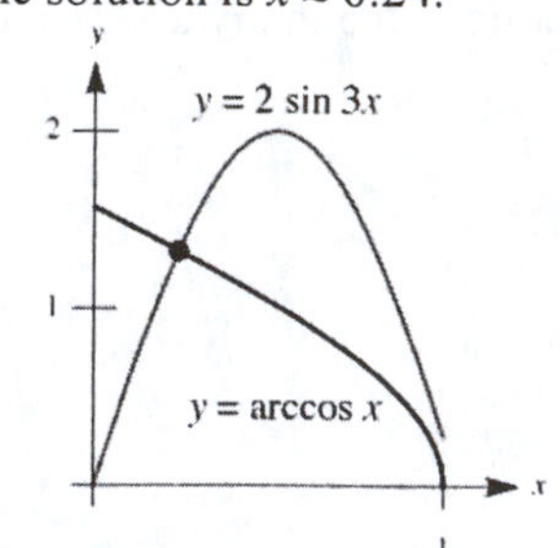

b. The solutions are $x \approx 0.19$ and $x \approx 0.68$:

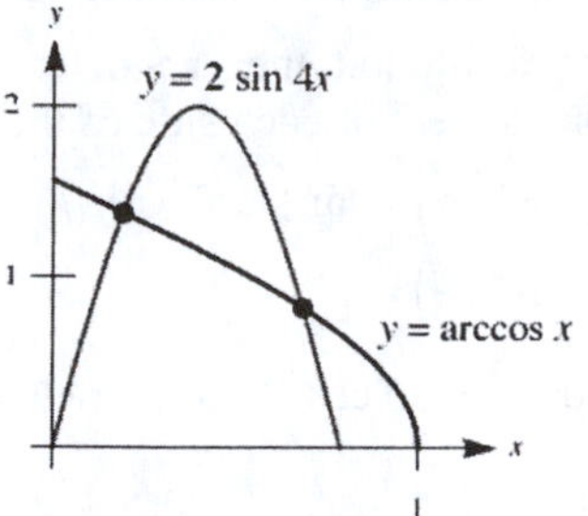

77. **a.** The solution is $x \approx 0.56$:

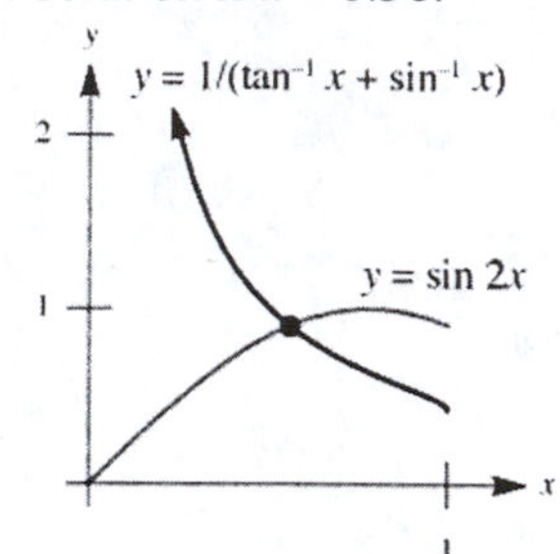

b. The solutions are $x \approx 0.51$ and $x \approx 0.84$:

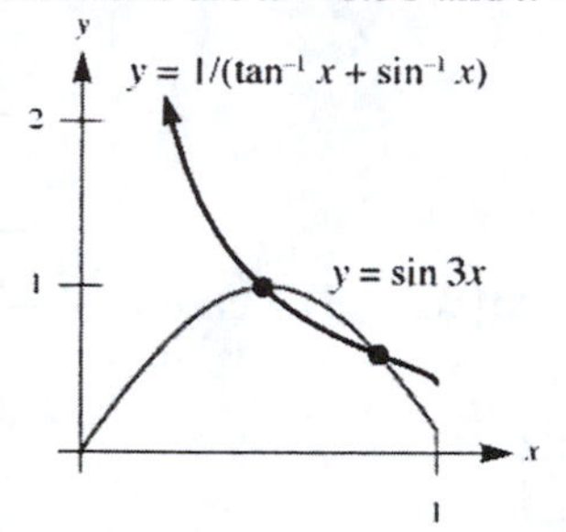

79. The solution is $x \approx 0.71$:

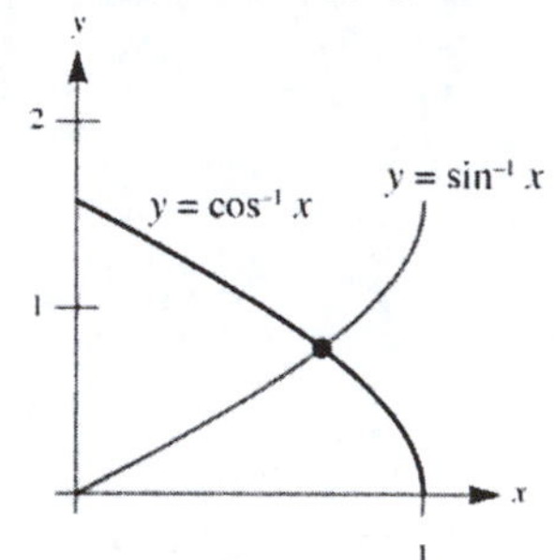

81. The solution is $x \approx 0.74$:

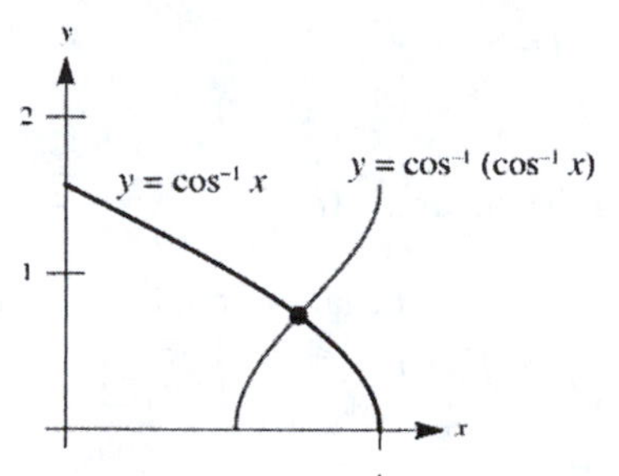

83. The solution is $x \approx 0.94$:

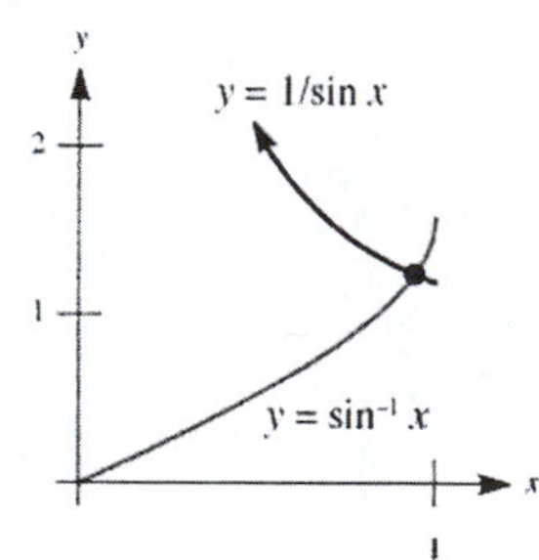

85. The solution is $x \approx 0.93$:

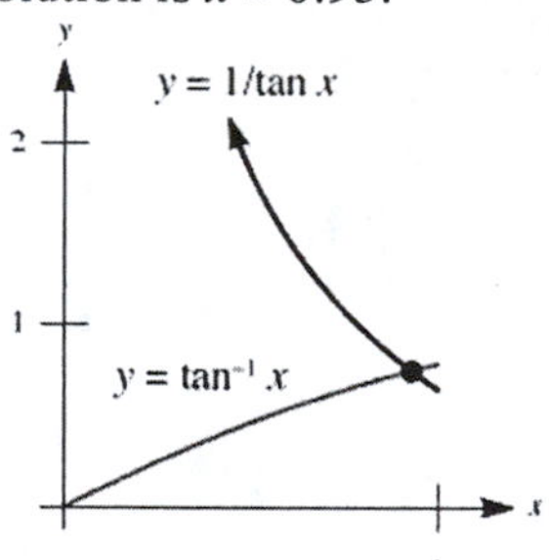

87. **a.** First find $\angle ADC$ and $\angle BDC$:

$$\tan \angle ADC = \frac{5}{x}, \text{ so } \angle ADC = \tan^{-1}\left(\frac{5}{x}\right)$$

$$\tan \angle BDC = \frac{3}{x}, \text{ so } \angle BDC = \tan^{-1}\left(\frac{3}{x}\right)$$

Thus $\theta = \angle ADC - \angle BDC = \tan^{-1}\left(\dfrac{5}{x}\right) - \tan^{-1}\left(\dfrac{3}{x}\right)$.

b. Graphing the curve:

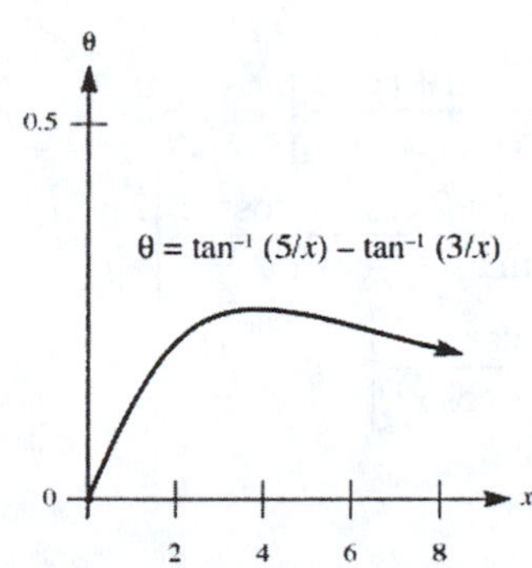

The maximum value of the curve occurs at the point (4,0.253). Thus, the maximum angle will occur when $x = 4$. Since this maximum angle corresponds with the best viewing of the picture, the person should stand 4 ft from the wall.

89. **a.** If A corresponds to $A(0,0)$, then the coordinates of the remaining points are $B(9,3)$, $C(12,0)$, $D(10,0)$ and $E(11,1)$. Following the hint, compute the slopes:

$$m_{\overline{DE}} = \frac{1-0}{11-10} = \frac{1}{1} = 1 \qquad m_{\overline{BC}} = \frac{0-3}{12-9} = \frac{-3}{3} = -1$$

$$m_{\overline{AB}} = \frac{3-0}{9-0} = \frac{3}{9} = \frac{1}{3} \qquad m_{\overline{BD}} = \frac{0-3}{10-9} = \frac{-3}{1} = -3$$

Since $m_{\overline{DE}} \bullet m_{\overline{BC}} = -1$, $\overline{DE}$ is perpendicular to $\overline{BC}$. Since $m_{\overline{AB}} \bullet m_{\overline{BD}} = -1$, $\overline{AB}$ is perpendicular to $\overline{BD}$.

b. Compute each distance using the coordinates specified in part **a** and the distance formula:

$$DE = \sqrt{(11-10)^2 + (1-0)^2} = \sqrt{1+1} = \sqrt{2} \qquad CE = \sqrt{(11-12)^2 + (1-0)^2} = \sqrt{1+1} = \sqrt{2}$$

$$BE = \sqrt{(11-9)^2 + (1-3)^2} = \sqrt{4+4} = 2\sqrt{2} \qquad AB = \sqrt{(9-0)^2 + (3-0)^2} = \sqrt{81+9} = 3\sqrt{10}$$

$$BD = \sqrt{(10-9)^2 + (0-3)^2} = \sqrt{1+9} = \sqrt{10}$$

c. Compute each tangent specified:

$$\tan\alpha = \frac{CE}{DE} = \frac{\sqrt{2}}{\sqrt{2}} = 1 \qquad \tan\beta = \frac{BE}{DE} = \frac{2\sqrt{2}}{\sqrt{2}} = 2 \qquad \tan\gamma = \frac{AB}{BD} = \frac{3\sqrt{10}}{\sqrt{10}} = 3$$

Thus $\alpha = \tan^{-1} 1$, $\beta = \tan^{-1} 2$ and $\gamma = \tan^{-1} 3$. Since $\alpha + \beta + \gamma = \pi$: $\tan^{-1} 1 + \tan^{-1} 2 + \tan^{-1} 3 = \pi$

Chapter 8 Review Exercises

1. Using the identity for $\cot x$ and the addition formula for $\tan(s + t)$:

$$\cot(x+y) = \frac{1}{\tan(x+y)} = \frac{1 - \tan x \tan y}{\tan x + \tan y} \bullet \frac{\cot x \cot y}{\cot x \cot y} = \frac{\cot x \cot y - 1}{\cot y + \cot x}$$

3. Working from the right-hand side: $\dfrac{2\tan x}{1+\tan^2 x} = \dfrac{\frac{2\sin x}{\cos x}}{1+\frac{\sin^2 x}{\cos^2 x}} = \dfrac{2\sin x\cos x}{\cos^2 x + \sin^2 x} = \dfrac{\sin 2x}{1} = \sin 2x$

5. Using the addition formulas for $\sin(x + y)$ and $\sin(x - y)$:

$$\begin{aligned}\frac{\sin(x+y)\sin(x-y)}{\cos^2 x\cos^2 y} &= \frac{(\sin x\cos y + \cos x\sin y)(\sin x\cos y - \cos x\sin y)}{\cos^2 x\cos^2 y} \\ &= \frac{\sin^2 x\cos^2 y - \cos^2 x\sin^2 y}{\cos^2 x\cos^2 y} \\ &= \frac{\sin^2 x}{\cos^2 x} - \frac{\sin^2 y}{\cos^2 y} \\ &= \tan^2 x - \tan^2 y\end{aligned}$$

7. Using the half-angle formula for $\tan \frac{x}{2}$:

$$\begin{aligned}\sin x\left(\tan \tfrac{x}{2}+\cot \tfrac{x}{2}\right)&=\sin x\left[\frac{\sin x}{1+\cos x}+\frac{1+\cos x}{\sin x}\right]\\&=\sin x\left[\frac{\sin^2 x+1+2\cos x+\cos^2 x}{(\sin x)(1+\cos x)}\right]\\&=\sin x\left[\frac{2+2\cos x}{\sin x(1+\cos x)}\right]\\&=\frac{2(1+\cos x)}{1+\cos x}\\&=2\end{aligned}$$

9. Using the addition formula for tan $(s-t)$: $\tan\left(\frac{\pi}{4}-x\right)=\dfrac{\tan\frac{\pi}{4}-\tan x}{1+\tan x\tan\frac{\pi}{4}}=\dfrac{1-\tan x}{1+\tan x}$

Now using the result of Exercise 8: $\tan\left(\frac{\pi}{4}+x\right)-\tan\left(\frac{\pi}{4}-x\right)=\dfrac{1+\tan x}{1-\tan x}-\dfrac{1-\tan x}{1+\tan x}=\dfrac{4\tan x}{1-\tan^2 x}=2\tan 2x$

11. Using the double-angle formula for $\sin 2\theta$: $2\sin\left(\frac{\pi}{4}-\frac{x}{2}\right)\cos\left(\frac{\pi}{4}-\frac{x}{2}\right)=\sin\left[2\bullet\left(\frac{\pi}{4}-\frac{x}{2}\right)\right]=\sin\left(\frac{\pi}{2}-x\right)=\cos x$

13. First simplify $\tan\left(\frac{\pi}{4}-t\right)$ using the addition formula for tan $(s-t)$: $\tan\left(\frac{\pi}{4}-t\right)=\dfrac{\tan\frac{\pi}{4}-\tan t}{1+\tan\frac{\pi}{4}\tan t}=\dfrac{1-\tan t}{1+\tan t}$

Therefore: $\dfrac{1-\tan\left(\frac{\pi}{4}-t\right)}{1+\tan\left(\frac{\pi}{4}-t\right)}=\dfrac{1-\frac{1-\tan t}{1+\tan t}}{1+\frac{1-\tan t}{1+\tan t}}=\dfrac{1+\tan t-1+\tan t}{1+\tan t+1-\tan t}=\dfrac{2\tan t}{2}=\tan t$

15. Using the addition formula for tan $(s+t)$: $\dfrac{\tan(\alpha-\beta)+\tan\beta}{1-\tan(\alpha-\beta)\tan\beta}=\tan\left[(\alpha-\beta)+\beta\right]=\tan\alpha$

17. Using the addition formula for tan $(s+t)$ and the double-angle formula for $\tan 2\theta$:

$$\begin{aligned}\tan 3\theta&=\tan(\theta+2\theta)\\&=\frac{\tan\theta+\tan 2\theta}{1-\tan\theta\tan 2\theta}\\&=\frac{\tan\theta+\frac{2\tan\theta}{1-\tan^2\theta}}{1-\tan\theta\bullet\frac{2\tan\theta}{1-\tan^2\theta}}\\&=\frac{\tan\theta\left(1-\tan^2\theta\right)+2\tan\theta}{1-\tan^2\theta-2\tan^2\theta}\\&=\frac{3\tan\theta-\tan^3\theta}{1-3\tan^2\theta}\\&=\frac{3t-t^3}{1-3t^2},\text{ where } t=\tan\theta\end{aligned}$$

19. Working from the right-hand side:

$$\frac{\cos x+\sin x}{\cos x-\sin x}=\frac{\cos x+\sin x}{\cos x-\sin x}\bullet\frac{\cos x+\sin x}{\cos x+\sin x}$$
$$=\frac{\cos^2 x+2\sin x\cos x+\sin^2 x}{\cos^2 x-\sin^2 x}$$
$$=\frac{1+\sin 2x}{\cos 2x}$$
$$=\frac{1}{\cos 2x}+\frac{\sin 2x}{\cos 2x}$$
$$=\tan 2x+\sec 2x$$

21. Using the double-angle formula for $\sin 2x$:

$$2\sin x+\sin 2x=2\sin x+2\sin x\cos x$$
$$=2\sin x(1+\cos x)$$
$$=2\sin x(1+\cos x)\bullet\frac{1-\cos x}{1-\cos x}$$
$$=\frac{2\sin x\left(1-\cos^2 x\right)}{1-\cos x}$$
$$=\frac{2\sin^3 x}{1-\cos x}$$

23. Working from the right-hand side:

$$\frac{1-\cos x+\sin x}{1+\cos x+\sin x}=\frac{1+\sin x-\cos x}{1+\sin x+\cos x}\bullet\frac{1+\sin x+\cos x}{1+\sin x+\cos x}$$
$$=\frac{(1+\sin x)^2-\cos^2 x}{\left[(1+\sin x)+\cos x\right]^2}$$
$$=\frac{1+2\sin x+\sin^2 x-\cos^2 x}{\left(1+2\sin x+\sin^2 x\right)+2\cos x(1+\sin x)+\cos^2 x}$$
$$=\frac{2\sin x+2\sin^2 x}{2+2\sin x+2\cos x(1+\sin x)}$$
$$=\frac{2\sin x(1+\sin x)}{2(1+\sin x)+2\cos x(1+\sin x)}$$
$$=\frac{2\sin x(1+\sin x)}{2(1+\sin x)(1+\cos x)}$$
$$=\frac{\sin x}{1+\cos x}$$
$$=\tan\tfrac{x}{2}$$

25. Using the addition formula for $\sin(s-t)$: $\sin(x+y)\cos y-\cos(x+y)\sin y=\sin\left[(x+y)-y\right]=\sin x$

27. Working from the left-hand side: $\dfrac{1-\tan^2\frac{x}{2}}{1+\tan^2\frac{x}{2}}=\dfrac{1-\dfrac{\sin^2\frac{x}{2}}{\cos^2\frac{x}{2}}}{1+\dfrac{\sin^2\frac{x}{2}}{\cos^2\frac{x}{2}}}=\dfrac{\cos^2\frac{x}{2}-\sin^2\frac{x}{2}}{\cos^2\frac{x}{2}+\sin^2\frac{x}{2}}=\cos^2\tfrac{x}{2}-\sin^2\tfrac{x}{2}=\cos\left(2\bullet\tfrac{x}{2}\right)=\cos x$

29. Using the double-angle formulas for $\sin 2\theta$ and $\cos 2\theta$:

$$\sin 4x=2\sin 2x\cos 2x$$
$$=2(2\sin x\cos x)\left(\cos^2 x-\sin^2 x\right)$$
$$=4\sin x\cos x\left(1-2\sin^2 x\right)$$
$$=4\sin x\cos x-8\sin^3 x\cos x$$

31. Using the addition formula for $\sin(s+t)$: $\sin 5x = \sin(4x+x) = \sin 4x\cos x + \cos 4x\sin x$

Simplifying each of these products using the double-angle formulas for $\sin 2\theta$ and $\cos 2\theta$:

$$\begin{aligned}\sin 4x\cos x &= 2\sin 2x\cos 2x\cos x\\ &= 2(2\sin x\cos x)(\cos^2 x-\sin^2 x)\cos x\\ &= 4\sin x\cos^2 x(1-2\sin^2 x)\\ &= 4\sin x(1-\sin^2 x)(1-2\sin^2 x)\\ &= 4\sin x-12\sin^3 x+8\sin^5 x\end{aligned}$$

$$\begin{aligned}\cos 4x\sin x &= (\cos^2 2x-\sin^2 2x)(\sin x)\\ &= \left[(\cos^2 x-\sin^2 x)^2-(2\sin x\cos x)^2\right]\sin x\\ &= \left[(1-2\sin^2 x)^2-4\sin^2 x\cos^2 x\right]\sin x\\ &= \left[1-4\sin^2 x+4\sin^4 x-4\sin^2 x(1-\sin^2 x)\right]\sin x\\ &= \left[1-8\sin^2 x+8\sin^4 x\right]\sin x\\ &= \sin x-8\sin^3 x+8\sin^5 x\end{aligned}$$

Therefore:

$$\begin{aligned}\sin 5x &= \sin 4x\cos x+\cos 4x\sin x\\ &= 4\sin x-12\sin^3 x+8\sin^5 x+\sin x-8\sin^3 x+8\sin^5 x\\ &= 16\sin^5 x-20\sin^3 x+5\sin x\end{aligned}$$

33. Using the sum-to-product formula for $\sin\alpha-\sin\beta$:

$$\sin 80°-\sin 20° = 2\cos\frac{80°+20°}{2}\sin\frac{80°-20°}{2} = 2\cos 50°\sin 30° = 2\cos 50°\bullet\tfrac{1}{2} = \cos 50°$$

35. Using the sum-to-product formulas for $\cos\alpha-\cos\beta$ and $\sin\alpha+\sin\beta$:

$$\cos x-\cos 3x = -2\sin\tfrac{x+3x}{2}\sin\tfrac{x-3x}{2} = -2\sin 2x\sin(-x) = 2\sin 2x\sin x$$

$$\sin x+\sin 3x = 2\sin\tfrac{x+3x}{2}\cos\tfrac{x-3x}{2} = 2\sin 2x\cos(-x) = 2\sin 2x\cos x$$

Therefore: $\dfrac{\cos x-\cos 3x}{\sin x+\sin 3x} = \dfrac{2\sin 2x\sin x}{2\sin 2x\cos x} = \dfrac{\sin x}{\cos x} = \tan x$

37. Using the sum-to-product formula for $\sin\alpha+\sin\beta$:

$$\sin\tfrac{5\pi}{12}+\sin\tfrac{\pi}{12} = 2\sin\frac{\frac{5\pi}{12}+\frac{\pi}{12}}{2}\cos\frac{\frac{5\pi}{12}-\frac{\pi}{12}}{2} = 2\sin\tfrac{\pi}{4}\cos\tfrac{\pi}{6} = 2\bullet\tfrac{\sqrt{2}}{2}\bullet\tfrac{\sqrt{3}}{2} = \tfrac{\sqrt{6}}{2}$$

39. Using the sum-to-product formulas for $\cos\alpha+\cos\beta$ and $\sin\alpha+\sin\beta$:

$$\cos 3y+\cos(2x-3y) = 2\cos\tfrac{2x}{2}\cos\tfrac{6y-2x}{2} = 2\cos x\cos(3y-x)$$

$$\sin 3y+\sin(2x-3y) = 2\sin\tfrac{2x}{2}\cos\tfrac{6y-2x}{2} = 2\sin x\cos(3y-x)$$

Therefore: $\dfrac{\cos 3y+\cos(2x-3y)}{\sin 3y+\sin(2x-3y)} = \dfrac{2\cos x\cos(3y-x)}{2\sin x\cos(3y-x)} = \dfrac{\cos x}{\sin x} = \cot x$

41. Using the sum-to-product formulas for $\sin\alpha-\sin\beta$ and $\cos\alpha-\cos\beta$:

$$\sin 40°-\sin 20° = 2\cos\frac{40°+20°}{2}\sin\frac{40°-20°}{2} = 2\cos 30°\sin 10° = 2\bullet\tfrac{\sqrt{3}}{2}\sin 10° = \sqrt{3}\sin 10°$$

$$\cos 20°-\cos 40° = -2\sin\frac{40°+20°}{2}\sin\frac{40°-20°}{2} = -2\sin 30°\sin(-10°) = 2\bullet\tfrac{1}{2}\sin 10° = \sin 10°$$

Therefore: $\dfrac{\sin 40°-\sin 20°}{\cos 20°-\cos 40°} = \dfrac{\sqrt{3}\sin 10°}{\sin 10°} = \sqrt{3}$. Using the result of Exercise 40, we have shown the required identity.

43. **a.** Since $a = 1$, the area is $\tan^{-1} 1 = \frac{\pi}{4}$.

b. (i) Since $\tan^{-1} a = 1.5$, $a = \tan 1.5 \approx 14$.

(ii) Since $\tan^{-1} a = 1.56$, $a = \tan 1.56 \approx 93$.

(iii) Since $\tan^{-1} a = 1.57$, $a = \tan 1.57 \approx 1256$.

45. The principal solution is $x = \tan^{-1} 4.26 \approx 1.34$. Since $\tan x$ is also positive in the third quadrant, the other solution in the interval $[0, 2\pi]$ is $1.34 + \pi \approx 4.48$.

47. Since $\csc x = 2.24$, $\sin x = \frac{1}{2.24} \approx 0.45$. The principal solution is $x = \sin^{-1} 0.45 \approx 0.46$. Since $\sin x$ is also positive in the second quadrant, the other solution in the interval $[0, 2\pi)$ is $\pi - 0.46 \approx 2.68$.

49. Given $\tan^2 x - 3 = 0$, so $\tan^2 x = 3$ and $\tan x = \pm\sqrt{3}$. If $\tan x = \sqrt{3}$ then $x = \frac{\pi}{3}$ or $\frac{4\pi}{3}$ while if $\tan x = -\sqrt{3}$ then $x = \frac{2\pi}{3}$ or $\frac{5\pi}{3}$. So the solutions in the interval $[0, 2\pi)$ are $x = \frac{\pi}{3}, \frac{2\pi}{3}, \frac{4\pi}{3}$ or $\frac{5\pi}{3}$.

51. Squaring each side of the equation:

$$
\begin{aligned}
(1+\sin x)^2 &= \cos^2 x \\
1+2\sin x+\sin^2 x &= 1-\sin^2 x \\
2\sin^2 x+2\sin x &= 0 \\
2\sin x(\sin x+1) &= 0
\end{aligned}
$$

So $\sin x = 0$ or $\sin x = -1$, thus $x = 0, \pi$ or $\frac{3\pi}{2}$. Upon checking we find that $x = \pi$ is not a solution (recall that squaring an equation can produce extraneous roots), so the solutions are $x = 0$ or $\frac{3\pi}{2}$.

53. Using the double-angle formula for $\cos 2x$:

$$
\begin{aligned}
\sin x-\left(\cos^2 x-\sin^2 x\right)+1 &= 0 \\
\sin x-\left(1-2\sin^2 x\right)+1 &= 0 \\
2\sin^2 x+\sin x &= 0 \\
\sin x(2\sin x+1) &= 0
\end{aligned}
$$

So $\sin x = 0$ or $\sin x = -\frac{1}{2}$. Thus the solutions are $x = 0, \pi, \frac{7\pi}{6}$ or $\frac{11\pi}{6}$.

55. Solving the equation:

$$
\begin{aligned}
3\csc x-4\sin x &= 0 \\
\frac{3}{\sin x} &= 4\sin x \\
\sin^2 x &= \frac{3}{4} \\
\sin x &= \pm\frac{\sqrt{3}}{2}
\end{aligned}
$$

So the solutions are $x = \frac{\pi}{3}, \frac{2\pi}{3}, \frac{4\pi}{3}$ or $\frac{5\pi}{3}$.

57. Factoring:

$$
\begin{aligned}
2\sin^4 x-3\sin^2 x+1 &= 0 \\
\left(2\sin^2 x-1\right)\left(\sin^2 x-1\right) &= 0 \\
\sin^2 x = \tfrac{1}{2} \quad &\text{or} \quad \sin^2 x = 1 \\
\sin x = \pm\tfrac{\sqrt{2}}{2} \quad &\text{or} \quad \sin x = \pm 1
\end{aligned}
$$

So the solutions are $x = \frac{\pi}{4}, \frac{\pi}{2}, \frac{3\pi}{4}, \frac{5\pi}{4}, \frac{3\pi}{2}$ or $\frac{7\pi}{4}$.

59. Using the identity $\sin^2 x = 1 - \cos^2 x$:

$$
\begin{aligned}
\left(1-\cos^2 x\right)^2+\cos^4 x &= \frac{5}{8} \\
1-2\cos^2 x+2\cos^4 x &= \frac{5}{8} \\
16\cos^4 x-16\cos^2 x+3 &= 0 \\
\left(4\cos^2 x-3\right)\left(4\cos^2 x-1\right) &= 0 \\
\cos^2 x = \tfrac{3}{4} \quad &\text{or} \quad \cos^2 x = \tfrac{1}{4} \\
\cos x = \pm\tfrac{\sqrt{3}}{2} \quad &\text{or} \quad \cos x = \pm\tfrac{1}{2}
\end{aligned}
$$

So the solutions are $x = \frac{\pi}{6}, \frac{\pi}{3}, \frac{2\pi}{3}, \frac{5\pi}{6}, \frac{7\pi}{6}, \frac{4\pi}{3}, \frac{5\pi}{3}$ or $\frac{11\pi}{6}$.

61. Using the suggestion, re-write the equation in terms of sines and cosines:

$$\cot x + \csc x + \sec x = \tan x$$

$$\frac{\cos x}{\sin x} + \frac{1}{\sin x} + \frac{1}{\cos x} = \frac{\sin x}{\cos x}$$

Multiplying each side of the equation by $\sin x \cos x$ yields:

$$\cos^2 x + \cos x + \sin x = \sin^2 x$$

$$\cos^2 x - \sin^2 x + \cos x + \sin x = 0$$

$$(\cos x + \sin x)(\cos x - \sin x + 1) = 0$$

$$\cos x + \sin x = 0 \quad \text{or} \quad \cos x - \sin x + 1 = 0$$

From the first equation $\sin x = -\cos x$, so $\tan x = -1$ and thus $x = \frac{3\pi}{4}$ or $\frac{7\pi}{4}$. From the second equation, isolate $\cos x$ and square each side:

$$\cos x = \sin x - 1$$

$$\cos^2 x = \sin^2 x - 2\sin x + 1$$

$$1 - \sin^2 x = \sin^2 x - 2\sin x + 1$$

$$0 = 2\sin^2 x - 2\sin x$$

$$0 = 2\sin x(\sin x - 1)$$

Now $\sin x = 0$ when $x = 0$ or π, but then $\csc x$ is undefined. Also $\sin x = 1$ when $x = \frac{\pi}{2}$, but then $\tan x$ is undefined. So the only solutions are $x = \frac{3\pi}{4}$ or $\frac{7\pi}{4}$.

63. Let $\theta = \tan^{-1}\left(\frac{\sqrt{2}}{2}\right)$, so $\tan\theta = \frac{\sqrt{2}}{2}$. Draw the triangle:

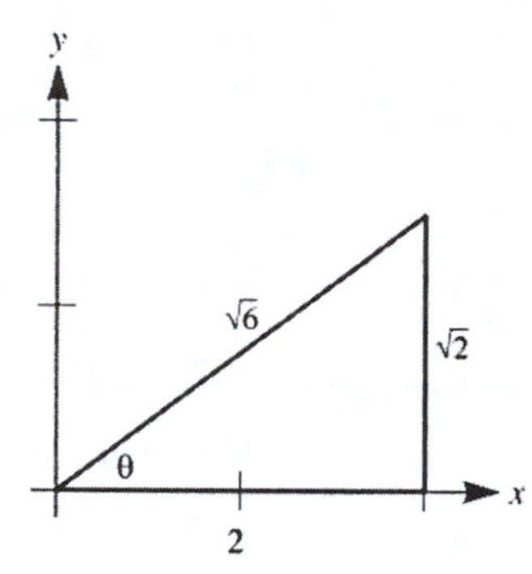

Thus $\sin\theta = \frac{\sqrt{2}}{\sqrt{6}} = \frac{\sqrt{3}}{3}$, so: $\cos\left\{\tan^{-1}\left[\sin\left(\tan^{-1}\left(\frac{\sqrt{2}}{2}\right)\right)\right]\right\} = \cos\left\{\tan^{-1}\left[\sin\theta\right]\right\} = \cos\left\{\tan^{-1}\left(\frac{\sqrt{3}}{3}\right)\right\} = \cos\frac{\pi}{6} = \frac{\sqrt{3}}{2}$

65. We are asked to find the number x in $\left(-\frac{\pi}{2}, \frac{\pi}{2}\right)$ such that $\tan x = \frac{\sqrt{3}}{3}$. Since $x = \frac{\pi}{6}$ is that number, $\arctan\frac{\sqrt{3}}{3} = \frac{\pi}{6}$.

67. We are asked to find the number x in $\left[-\frac{\pi}{2}, \frac{\pi}{2}\right]$ such that $\sin x = \frac{1}{2}$. Since $x = \frac{\pi}{6}$ is that number, $\arcsin\frac{1}{2} = \frac{\pi}{6}$.

69. We are asked to find the number x in $[0, \pi]$ such that $\cos x = \frac{1}{2}$. Since $x = \frac{\pi}{3}$ is that number, $\cos^{-1}\left(\frac{1}{2}\right) = \frac{\pi}{3}$.

71. We are asked to find the number x in $[0, \pi]$ such that $\cos x = -\frac{1}{2}$. Since $x = \frac{2\pi}{3}$ is that number, $\cos^{-1}\left(-\frac{1}{2}\right) = \frac{2\pi}{3}$.

73. Since $\cos\left(\cos^{-1} x\right) = x$ for every x in the interval $[-1, 1]$, $\cos\left[\cos^{-1}\left(\frac{2}{7}\right)\right] = \frac{2}{7}$.

75. Since 5 is in the interval $\left[\frac{3\pi}{2}, 2\pi\right]$, the value is $\cos^{-1}(\cos 5) = 2\pi - 5$.

77. Let $\theta = \tan^{-1}(-1)$, so $\tan\theta = -1$ and θ is in the interval $\left(-\frac{\pi}{2}, \frac{\pi}{2}\right)$, thus $\theta = -\frac{\pi}{4}$. Therefore:

$$\sin\left[\tan^{-1}(-1)\right] = \sin\left(-\frac{\pi}{4}\right) = -\frac{\sqrt{2}}{2}$$

79. Let $\theta = \cos^{-1}\left(\frac{\sqrt{2}}{3}\right)$, so $\cos\theta = \frac{\sqrt{2}}{3}$ and θ is in the interval $[0,\pi]$. Therefore:

$$\sec\left[\cos^{-1}\left(\frac{\sqrt{2}}{3}\right)\right] = \sec\theta = \frac{1}{\cos\theta} = \frac{1}{\frac{\sqrt{2}}{3}} = \frac{3}{\sqrt{2}} = \frac{3\sqrt{2}}{2}$$

81. Let $\theta = \sin^{-1}\left(\frac{5}{13}\right)$, so $\sin\theta = \frac{5}{13}$ and θ is in the interval $\left[-\frac{\pi}{2},\frac{\pi}{2}\right]$. Draw the triangle:

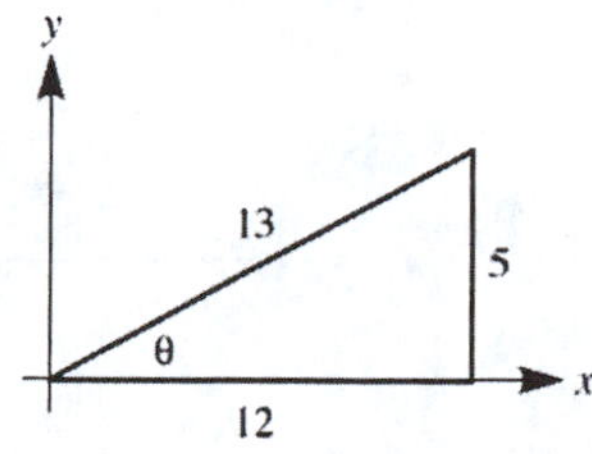

Using the addition formula for tan $(s+t)$: $\tan\left(\frac{\pi}{4}+\theta\right) = \dfrac{\tan\frac{\pi}{4}+\tan\theta}{1-\tan\frac{\pi}{4}\tan\theta} = \dfrac{1+\frac{5}{12}}{1-1\bullet\frac{5}{12}} = \dfrac{\frac{17}{12}}{\frac{7}{12}} = \frac{17}{7}$

83. Let $\theta = \tan^{-1} 2$, so $\tan\theta = 2$ and θ is in the interval $\left(-\frac{\pi}{2},\frac{\pi}{2}\right)$. Using the double-angle formula for $\tan 2\theta$:

$$\tan(2\theta) = \frac{2\tan\theta}{1-\tan^2\theta} = \frac{2\bullet 2}{1-(2)^2} = -\frac{4}{3}$$

85. Let $\theta = \cos^{-1}\left(\frac{4}{5}\right)$, so $\cos\theta = \frac{4}{5}$ and θ lies in the first quadrant. Thus $\frac{\theta}{2}$ lies in the first quadrant, so using the half-angle formula for $\cos\frac{\theta}{2}$: $\cos\frac{\theta}{2} = \sqrt{\dfrac{1+\cos\theta}{2}} = \sqrt{\dfrac{1+\frac{4}{5}}{2}} = \sqrt{\frac{9}{10}} = \dfrac{3}{\sqrt{10}} = \dfrac{3\sqrt{10}}{10}$

87. Let $\alpha = \tan^{-1} x$ and $\beta = \tan^{-1} y$, so $\tan\alpha = x$ and $\tan\beta = y$. Using the addition formula for tan $(s+t)$:

$$\tan\left(\tan^{-1} x + \tan^{-1} y\right) = \tan(\alpha+\beta) = \frac{\tan\alpha+\tan\beta}{1-\tan\alpha\tan\beta} = \frac{x+y}{1-xy}$$

89. Let $\theta = \arctan x$, so $\tan\theta = x$. Draw the triangle:

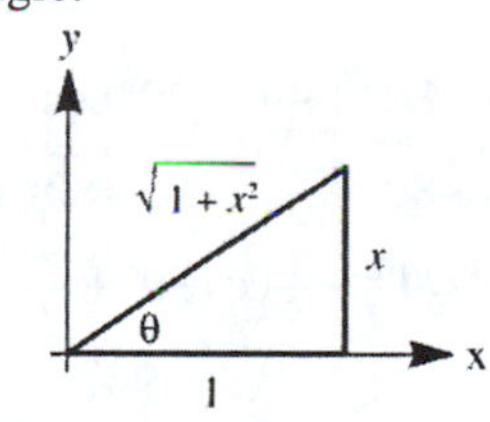

Using the double-angle formula for $\sin 2\theta$: $\sin(2\arctan x) = \sin 2\theta = 2\sin\theta\cos\theta = 2\bullet\dfrac{x}{\sqrt{x^2+1}}\bullet\dfrac{1}{\sqrt{x^2+1}} = \dfrac{2x}{x^2+1}$

91. Let $\theta = \sin^{-1}\left(x^2\right)$, so $\sin\theta = x^2$. Since $\sin\theta \ge 0$, θ must lie in the first quadrant. Draw the triangle:

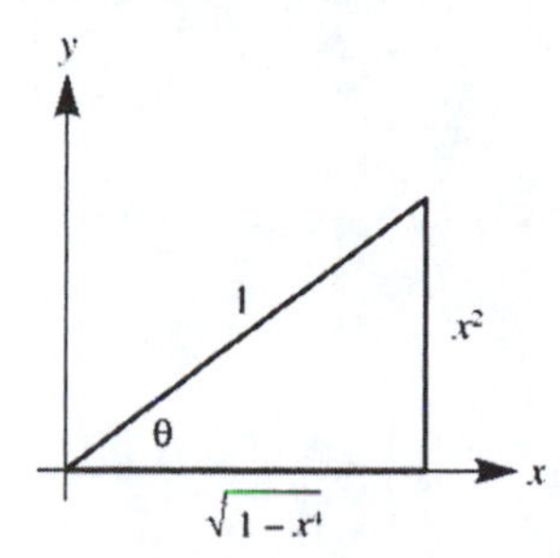

Noting that $\frac{\theta}{2}$ must also lie in the first quadrant, use the half-angle formula for $\sin\frac{\theta}{2}$ to obtain:

$$\sin\left[\frac{1}{2}\sin^{-1}\left(x^2\right)\right]=\sin\frac{\theta}{2}=\sqrt{\frac{1-\cos\theta}{2}}=\sqrt{\frac{1-\sqrt{1-x^4}}{2}}=\sqrt{\frac{1}{2}-\frac{1}{2}\sqrt{1-x^4}}$$

93. Let $\alpha=\arcsin\frac{4\sqrt{41}}{41}$ and $\beta=\arcsin\frac{\sqrt{82}}{82}$, so $\sin\alpha=\frac{4\sqrt{41}}{41}=\frac{4}{\sqrt{41}}$ and $\sin\beta=\frac{\sqrt{82}}{82}=\frac{1}{\sqrt{82}}$. Draw the triangles:

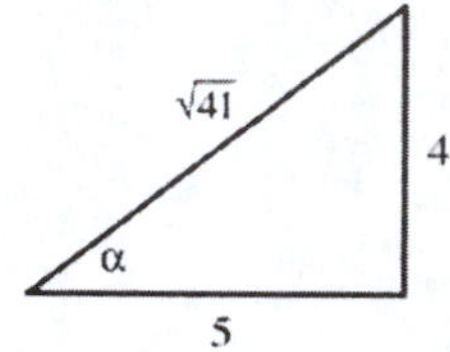

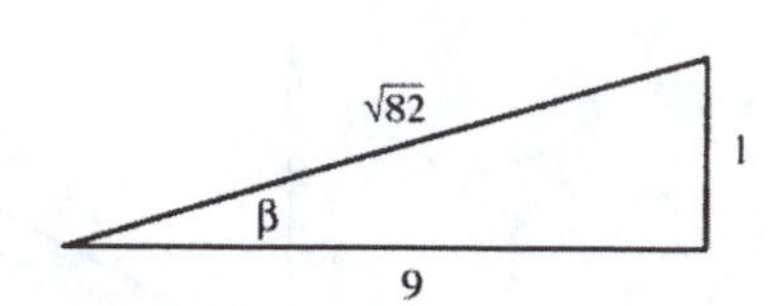

Now find tan ($\alpha+\beta$) by the addition formula for tan ($s + t$): $\tan(\alpha+\beta)=\frac{\tan\alpha+\tan\beta}{1-\tan\alpha\tan\beta}=\frac{\frac{4}{5}+\frac{1}{9}}{1-\frac{4}{5}\bullet\frac{1}{9}}=\frac{\frac{41}{45}}{\frac{41}{45}}=1$

Since $\tan(\alpha+\beta)=1$ and $0<\alpha+\beta<\pi$, $\alpha+\beta=\frac{\pi}{4}$. Therefore: $\arcsin\frac{4\sqrt{41}}{41}+\arcsin\frac{\sqrt{82}}{82}=\frac{\pi}{4}$

95. **a.** Using a calculator, $\cos 20°\cos 40°\cos 60°\cos 80° = 0.0625$.

b. Since $\cos 60°=\frac{1}{2}$: $\cos 20°\cos 40°\cos 60°\cos 80°=\frac{1}{2}\cos 20°\cos 40°\cos 80°$

Using the product-to-sum formula for $\cos A\cos B$:

$$\cos 20°\cos 40°=\frac{1}{2}\left[\cos 60°+\cos(-20°)\right]=\frac{1}{2}\left[\frac{1}{2}+\cos 20°\right]=\frac{1}{4}+\frac{1}{2}\cos 20°$$

Therefore:

$$\begin{aligned}\cos 20°\cos 40°\cos 60°\cos 80°&=\frac{1}{2}\cos 20°\cos 40°\cos 80°\\&=\frac{1}{2}\left(\frac{1}{4}+\frac{1}{2}\cos 20°\right)\cos 80°\\&=\frac{1}{8}\cos 80°+\frac{1}{4}\cos 20°\cos 80°\end{aligned}$$

Using the product-to-sum formula for $\cos A\cos B$, and the identity $\cos\theta=-\cos(180°-\theta)$:

$$\cos 20°\cos 80°=\frac{1}{2}\left[\cos 100°+\cos(-60°)\right]=\frac{1}{2}\left[-\cos 80°+\frac{1}{2}\right]=-\frac{1}{2}\cos 80°+\frac{1}{4}$$

Therefore:

$$\begin{aligned}\cos 20°\cos 40°\cos 60°\cos 80°&=\frac{1}{8}\cos 80°+\frac{1}{4}\cos 20°\cos 80°\\&=\frac{1}{8}\cos 80°+\frac{1}{4}\left(-\frac{1}{2}\cos 80°+\frac{1}{4}\right)\\&=\frac{1}{8}\cos 80°-\frac{1}{8}\cos 80°+\frac{1}{16}\\&=\frac{1}{16}\\&=0.0625\end{aligned}$$

97. Working from the right-hand side: $\frac{2}{\cot\theta+\tan\theta}=\frac{2}{\frac{\cos\theta}{\sin\theta}+\frac{\sin\theta}{\cos\theta}}=\frac{2}{\frac{\cos^2\theta+\sin^2\theta}{\sin\theta\cos\theta}}=\frac{2}{\frac{1}{\sin\theta\cos\theta}}=2\sin\theta\cos\theta=\sin 2\theta$

99. Using the double-angle formula for $\tan 2\theta$:

$$\frac{1}{1+\tan 2\theta\tan\theta}=\frac{1}{1+\frac{2\tan\theta}{1-\tan^2\theta}\bullet\tan\theta}$$
$$=\frac{1}{1+\frac{2\tan^2\theta}{1-\tan^2\theta}}$$
$$=\frac{1-\tan^2\theta}{\left(1-\tan^2\theta\right)+2\tan^2\theta}$$
$$=\frac{1-\tan^2\theta}{1+\tan^2\theta}$$
$$=\frac{1-\frac{\sin^2\theta}{\cos^2\theta}}{1+\frac{\sin^2\theta}{\cos^2\theta}}$$
$$=\frac{\cos^2\theta-\sin^2\theta}{\cos^2\theta+\sin^2\theta}$$
$$=\frac{\cos 2\theta}{1}$$
$$=\cos 2\theta$$

101. **a.** Working from the right-hand side: $\frac{2\tan\theta}{1+\tan^2\theta}=\frac{2\bullet\frac{\sin\theta}{\cos\theta}}{1+\frac{\sin^2\theta}{\cos^2\theta}}=\frac{2\sin\theta\cos\theta}{\cos^2\theta+\sin^2\theta}=\frac{\sin 2\theta}{1}=\sin 2\theta$

b. Working from the right-hand side: $\frac{1-\tan^2\theta}{1+\tan^2\theta}=\frac{1-\frac{\sin^2\theta}{\cos^2\theta}}{1+\frac{\sin^2\theta}{\cos^2\theta}}=\frac{\cos^2\theta-\sin^2\theta}{\cos^2\theta+\sin^2\theta}=\frac{\cos 2\theta}{1}=\cos 2\theta$

103. Using the product-to-sum formula for $\sin A\sin B$:

$$\sin\theta\sin\left[(n-1)\theta\right]=\tfrac{1}{2}\left[\cos(2\theta-n\theta)-\cos(n\theta)\right]=\tfrac{1}{2}\cos\left[(2-n)\theta\right]-\tfrac{1}{2}\cos n\theta=\tfrac{1}{2}\cos\left[(n-2)\theta\right]-\tfrac{1}{2}\cos n\theta$$

Therefore, the right-hand side of the original identity becomes:

$$\cos\left[(n-2)\theta\right]-2\sin\theta\sin\left[(n-1)\theta\right]=\cos\left[(n-2)\theta\right]-\cos\left[(n-2)\theta\right]+\cos n\theta=\cos n\theta$$

Chapter 8 Test

1. Using the addition formula for $\sin(s+t)$: $\sin\left(\theta+\frac{3\pi}{2}\right)=\sin\theta\cos\frac{3\pi}{2}+\cos\theta\sin\frac{3\pi}{2}=\sin\theta\bullet 0+\cos\theta\bullet(-1)=-\cos\theta$

2. Since $\frac{3\pi}{2}<t<2\pi$, draw the triangle:

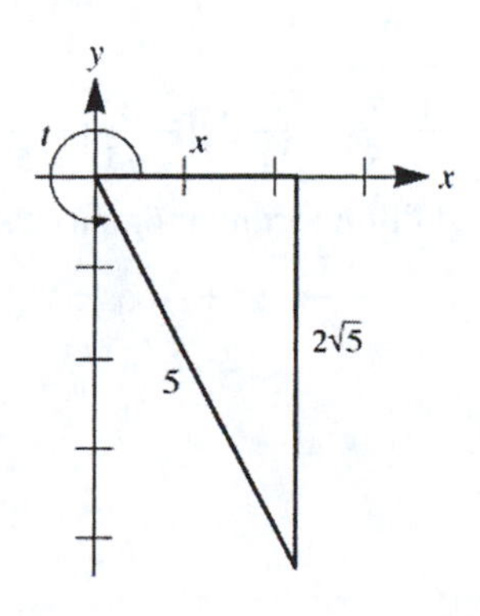

By the Pythagorean theorem:

$$\begin{aligned} x^2+\left(2\sqrt{5}\right)^2&=(5)^2\\ x^2+20&=25\\ x^2&=5\\ x&=\sqrt{5}\end{aligned}$$

So $\cos t=\frac{\sqrt{5}}{5}$, now use the double-angle formula for $\cos 2t$:

$$\cos 2t=\cos^2 t-\sin^2 t=\left(\frac{\sqrt{5}}{5}\right)^2-\left(-\frac{2\sqrt{5}}{5}\right)^2=\frac{1}{5}-\frac{4}{5}=-\frac{3}{5}$$

3. Since $\pi<\theta<\frac{3\pi}{2}$, draw the triangle:

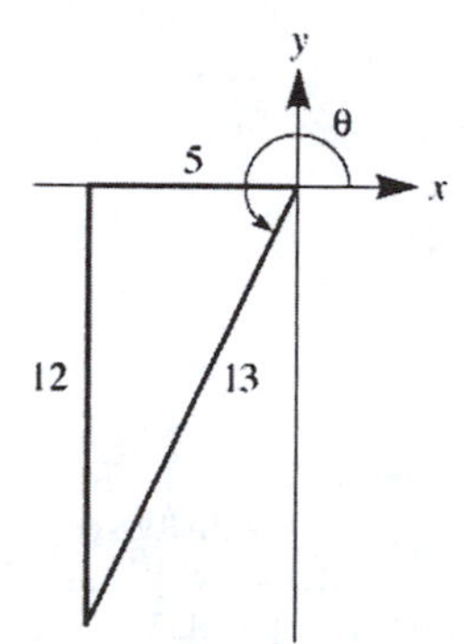

So $\sin\theta=-\frac{12}{13}$, now use the half-angle formula for $\tan\frac{\theta}{2}$: $\tan\frac{\theta}{2}=\dfrac{\sin\theta}{1+\cos\theta}=\dfrac{-\frac{12}{13}}{1-\frac{5}{13}}=\dfrac{-\frac{12}{13}}{\frac{8}{13}}=-\frac{3}{2}$

4. Dividing each side of the equation by $\cos x$ yields $\tan x=3$, which has a principal solution of $x=\tan^{-1}3\approx 1.25$. Since $\tan x$ is also positive in the third quadrant, the other solution in the interval $(0,2\pi)$ is $1.25+\pi\approx 4.39$.

5. Factoring:

$$\begin{aligned} 2\sin^2 x+7\sin x+3&=0\\ (2\sin x+1)(\sin x+3)&=0\\ \sin x=-\tfrac{1}{2} \text{ or } \sin x&=-3 \text{ (impossible)}\end{aligned}$$

So $x=\frac{7\pi}{6},\frac{11\pi}{6}$.

6. Since $\cos\alpha=\frac{2}{\sqrt{5}}$ and $\frac{3\pi}{2}<\alpha<2\pi$ (fourth quadrant): $\sin\alpha=-\sqrt{1-\cos^2\alpha}=-\sqrt{1-\frac{4}{5}}=-\frac{1}{\sqrt{5}}$

Similarly, since $\sin\beta=\frac{4}{5}$ and $\frac{\pi}{2}<\beta<\pi$ (second quadrant): $\cos\beta=-\sqrt{1-\sin^2\beta}=-\sqrt{1-\frac{16}{25}}=-\sqrt{\frac{9}{25}}=-\frac{3}{5}$

Using the addition formula for $\sin(\beta-\alpha)$:

$$\sin(\beta-\alpha)=\sin\beta\cos\alpha-\cos\beta\sin\alpha=\frac{4}{5}\bullet\frac{2}{\sqrt{5}}-\left(-\frac{3}{5}\right)\left(-\frac{1}{\sqrt{5}}\right)=\frac{8}{5\sqrt{5}}-\frac{3}{5\sqrt{5}}=\frac{5}{5\sqrt{5}}=\frac{\sqrt{5}}{5}$$

7. First, simplify the left-hand side using the addition formula for $\sin(\alpha+\beta)$:

$$\sin(x+30^\circ)=\sin x\cos 30^\circ+\cos x\sin 30^\circ=\frac{\sqrt{3}}{2}\sin x+\frac{1}{2}\cos x$$

Thus the equation is:

$$\begin{aligned}\frac{\sqrt{3}}{2}\sin x+\frac{1}{2}\cos x&=\sqrt{3}\sin x\\ \sqrt{3}\sin x+\cos x&=2\sqrt{3}\sin x\\ \cos x&=\sqrt{3}\sin x\\ \tan x&=\frac{1}{\sqrt{3}}\\ x&=30^\circ\end{aligned}$$

8. If $\csc\theta = -3$, then $\sin\theta = -\frac{1}{3}$. Since $\pi < \theta < \frac{3\pi}{2}$ (third quadrant): $\cos\theta = -\sqrt{1-\sin^2\theta} = -\sqrt{1-\frac{1}{9}} = -\sqrt{\frac{8}{9}} = -\frac{2\sqrt{2}}{3}$

Now $\frac{\pi}{2} < \frac{\theta}{2} < \frac{3\pi}{4}$ (second quadrant), so using the half-angle formula for $\sin\frac{\theta}{2}$:

$$\sin\frac{\theta}{2} = \sqrt{\frac{1-\cos\theta}{2}} = \sqrt{\frac{1+\frac{2\sqrt{2}}{3}}{2}} = \sqrt{\frac{3+2\sqrt{2}}{6}} = \frac{\sqrt{18+12\sqrt{2}}}{6}$$

9. For the restricted sine function, the domain is $\left[-\frac{\pi}{2},\frac{\pi}{2}\right]$ and the range is $[-1,1]$. For the function $y = \sin^{-1}x$, the domain is $[-1,1]$ and the range is $\left[-\frac{\pi}{2},\frac{\pi}{2}\right]$.

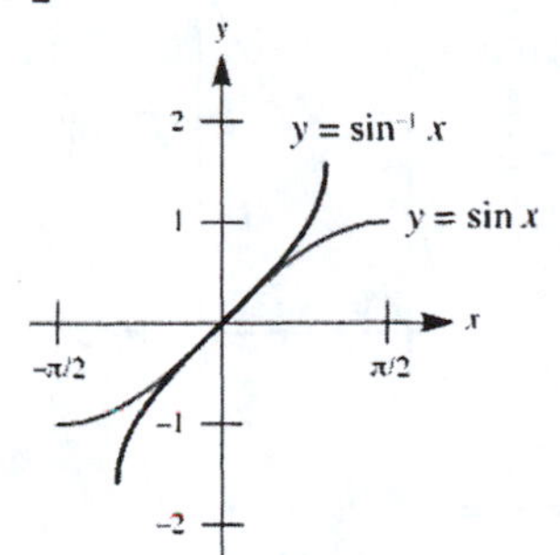

10. **a.** Since $\sin^{-1}(\sin x) = x$ for every x in the interval $\left[-\frac{\pi}{2},\frac{\pi}{2}\right]$, $\sin^{-1}\left(\sin\frac{\pi}{10}\right) = \frac{\pi}{10}$.

b. Since $\sin 2\pi = 0$, we are asked to find $x = \sin^{-1}0$. Then $\sin x = 0$ and x is in the interval $\left[-\frac{\pi}{2},\frac{\pi}{2}\right]$, thus $x = 0$. So $\sin^{-1}(\sin 2\pi) = 0$. Notice that we cannot use the identity $\sin^{-1}(\sin x) = x$, since 2π is not in the interval $\left[-\frac{\pi}{2},\frac{\pi}{2}\right]$.

11. Let $\theta = \arcsin\frac{3}{4}$, so $\sin\theta = \frac{3}{4}$ and θ is in the interval $\left[-\frac{\pi}{2},\frac{\pi}{2}\right]$. Draw the triangle:

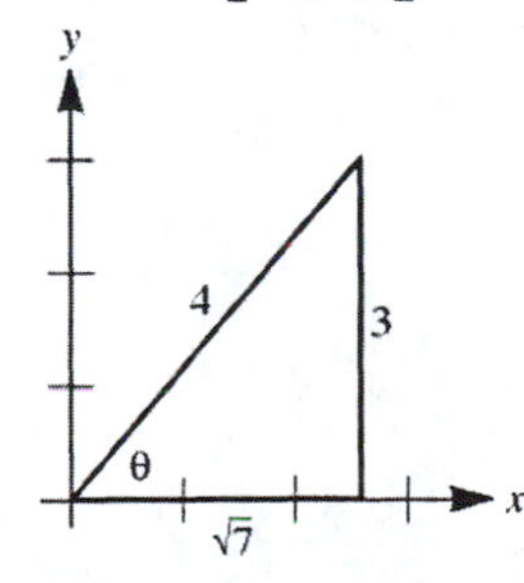

Therefore: $\cos\left(\arcsin\frac{3}{4}\right) = \cos\theta = \frac{\sqrt{7}}{4}$

12. Using the addition formula for $\tan(s+t)$: $\tan\left(\frac{\pi}{4}+\frac{\theta}{2}\right) = \dfrac{\tan\frac{\pi}{4}+\tan\frac{\theta}{2}}{1-\tan\frac{\pi}{4}\tan\frac{\theta}{2}} = \dfrac{1+\tan\frac{\theta}{2}}{1-\tan\frac{\theta}{2}}$

Now using the half-angle formula for $\tan\frac{\theta}{2}$: $\tan\left(\frac{\pi}{4}+\frac{\theta}{2}\right) = \dfrac{1+\tan\frac{\theta}{2}}{1-\tan\frac{\theta}{2}} = \dfrac{1+\frac{\sin\theta}{1+\cos\theta}}{1-\frac{\sin\theta}{1+\cos\theta}} = \dfrac{1+\cos\theta+\sin\theta}{1+\cos\theta-\sin\theta}$

13. Using the product-to-sum formula for $\sin A\cos B$:

$$\sin\frac{7\pi}{24}\cos\frac{\pi}{24} = \frac{1}{2}\left[\sin\left(\frac{7\pi}{24}+\frac{\pi}{24}\right)+\sin\left(\frac{7\pi}{24}-\frac{\pi}{24}\right)\right] = \frac{1}{2}\left[\sin\frac{\pi}{3}+\sin\frac{\pi}{4}\right] = \frac{1}{2}\left(\frac{\sqrt{3}}{2}+\frac{\sqrt{2}}{2}\right) = \frac{\sqrt{3}+\sqrt{2}}{4}$$

14. Using the sum-to-product formulas for $\sin\alpha+\sin\beta$ and $\cos\alpha+\cos\beta$:

$$\sin 3\theta+\sin 5\theta=2\sin\frac{3\theta+5\theta}{2}\cos\frac{3\theta-5\theta}{2}=2\sin 4\theta\cos(-\theta)=2\sin 4\theta\cos\theta$$

$$\cos 3\theta+\cos 5\theta=2\cos\frac{3\theta+5\theta}{2}\cos\frac{3\theta-5\theta}{2}=2\cos 4\theta\cos(-\theta)=2\cos 4\theta\cos\theta$$

Therefore: $\dfrac{\sin 3\theta+\sin 5\theta}{\cos 3\theta+\cos 5\theta}=\dfrac{2\sin 4\theta\cos\theta}{2\cos 4\theta\cos\theta}=\dfrac{\sin 4\theta}{\cos 4\theta}=\tan 4\theta$

15. Let $\theta=\arctan\sqrt{x^2-1}$, so $\tan\theta=\sqrt{x^2-1}$. Draw the triangle:

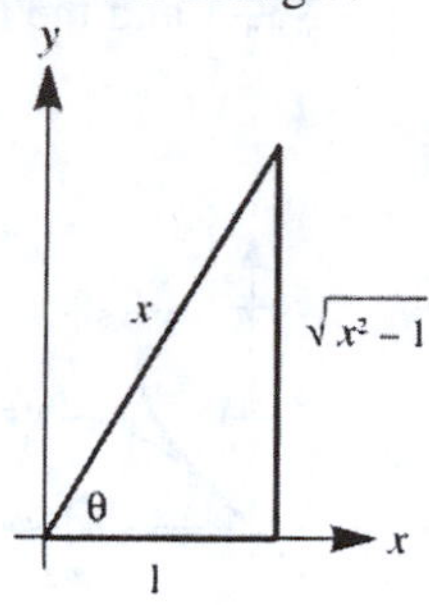

Therefore: $\sec\left[\arctan\sqrt{x^2-1}\right]=\sec\theta=\dfrac{x}{1}=x$

16. The domain of $y=\tan^{-1}x$ is $(-\infty,\infty)$, and the range is $\left(-\frac{\pi}{2},\frac{\pi}{2}\right)$.

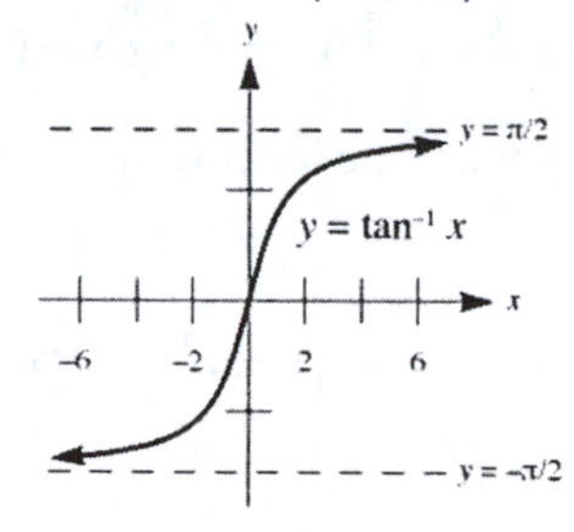

Chapter 9
Additional Topics in Trigonometry

9.1 Right-Triangle Applications

1. Draw the figure:

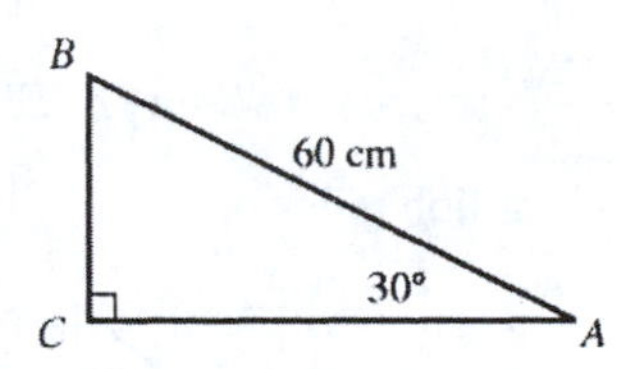

Since $\sin 30° = \frac{BC}{60}$: $BC = 60 \sin 30° = 60 \bullet \frac{1}{2} = 30$ cm

Since $\cos 30° = \frac{AC}{60}$: $AC = 60 \cos 30° = 60 \bullet \frac{\sqrt{3}}{2} = 30\sqrt{3}$ cm

3. Draw the figure:

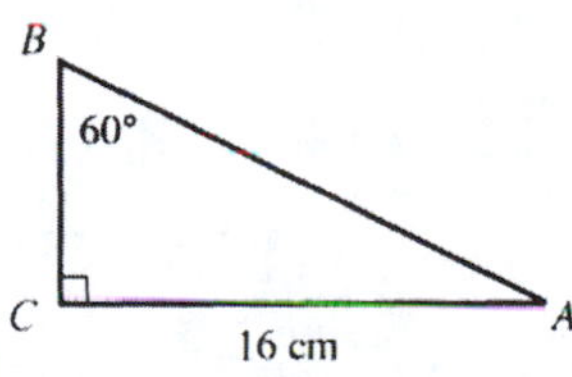

Since $\sin 60° = \frac{16}{AB}$: $AB = \frac{16}{\sin 60°} = \frac{16}{\frac{\sqrt{3}}{2}} = \frac{32}{\sqrt{3}} = \frac{32\sqrt{3}}{3}$ cm

Since $\tan 60° = \frac{16}{BC}$: $BC = \frac{16}{\tan 60°} = \frac{16}{\sqrt{3}} = \frac{16\sqrt{3}}{3}$ cm

5. Draw the figure:

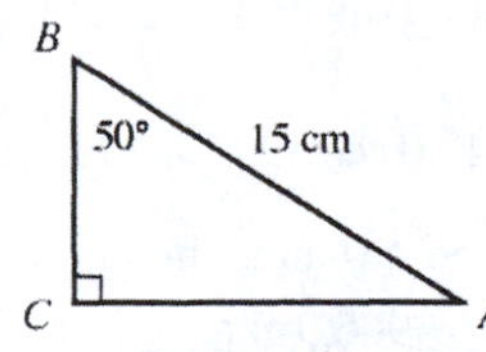

Since $\sin 50° = \frac{AC}{15}$: $AC = 15 \sin 50° \approx 11.5$ cm

Since $\cos 50° = \frac{BC}{15}$: $BC = 15 \cos 50° \approx 9.6$ cm

7. Draw a figure:

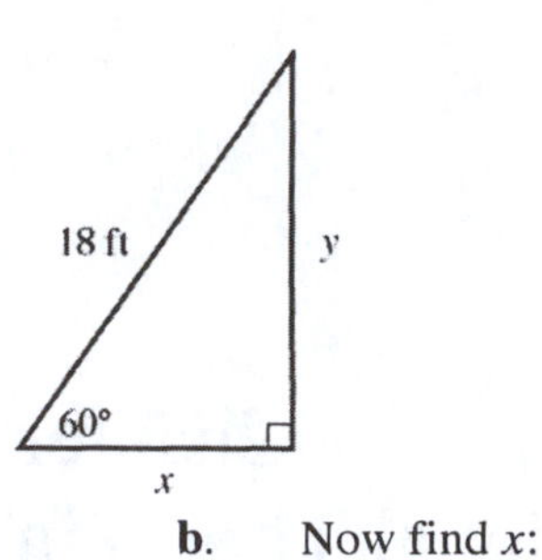

a. Now find y:

$$\sin 60° = \tfrac{y}{18}$$

$$y = 18\sin 60° = 18\left(\tfrac{\sqrt{3}}{2}\right) = 9\sqrt{3} \text{ ft}$$

Using a calculator, this is approximately 15.59 ft.

b. Now find x:

$$\cos 60° = \tfrac{x}{18}$$

$$x = 18\cos 60° = 18\left(\tfrac{1}{2}\right) = 9 \text{ ft}$$

9. **a.** Since $\sin\theta = \frac{5}{13}$, $\theta = \sin^{-1}\frac{5}{13} \approx 22.6°$.

b. Since $\cos\theta = \frac{12}{13}$, $\theta = \cos^{-1}\frac{12}{13} \approx 22.6°$.

c. Since $\tan\theta = \frac{5}{12}$, $\theta = \tan^{-1}\frac{5}{12} \approx 22.6°$.

11. Using the sine function, $\sin(\angle SEM) = \dfrac{MS}{SE}$, so: $MS = SE\sin(\angle SEM) = 93\sin 21.16° \approx 34$

Thus the distance MS is approximately 34 million miles.

13. Draw a figure:

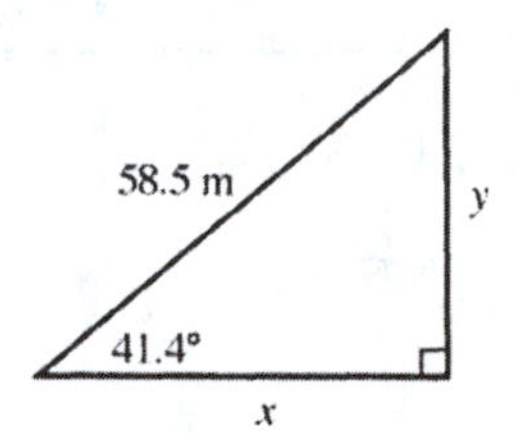

First find x and y:

$$\cos 41.4° = \frac{x}{58.5}, \text{ so } x = 58.5\cos 41.4° \approx 43.9 \text{ m} \qquad \sin 41.4° = \frac{y}{58.5}, \text{ so } y = 58.5\sin 41.4° \approx 38.7 \text{ m}$$

So the total length of fencing required is: $43.9 \text{ m} + 38.7 \text{ m} + 58.5 \text{ m} \approx 141.1 \text{ m}$

15. **a.** First draw the triangle:

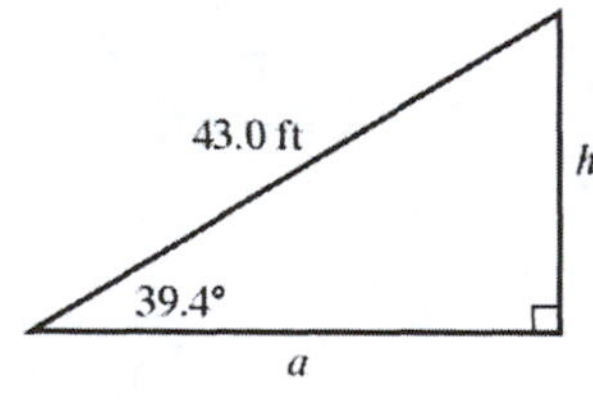

Now find h: $\sin 39.4° = \dfrac{h}{43.0}$, so $h = 43.0\sin 39.4° \approx 27.3$ ft

b. First find a: $\cos 39.4° = \dfrac{a}{43.0}$, so $a = 43.0\cos 39.4° \approx 33.2$ ft

Now the gable has a base of 2(33.2) = 66.4 ft and a height of 27.3 ft, so its area is given by:

$$\tfrac{1}{2}(\text{base})(\text{height}) \approx \tfrac{1}{2}(66.4)(27.3) \approx 906.9 \text{ ft}^2$$

Note: This calculation was done with the full calculator approximation, not the values rounded to one decimal place.

17. Using the formula $A = \frac{1}{2}ab\sin\theta$: $A = \frac{1}{2}(2)(3)\sin 30° = 1.5 \text{ in.}^2$

19. Start by finding the measure of a central angle of a triangle drawn from the center out to two adjacent vertices:

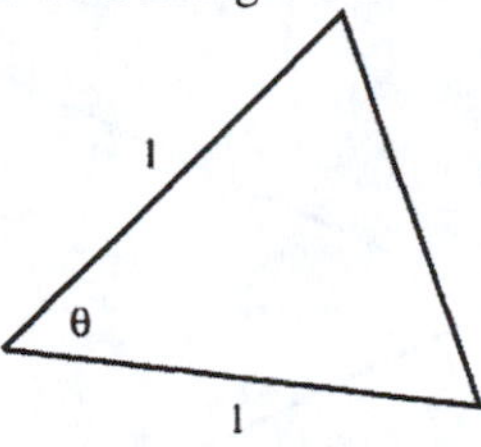

Since the polygon is 7-sided, $7\theta = 360°$, so $\theta = \frac{360°}{7}$. The area of this triangle is thus:

$$\tfrac{1}{2}ab\sin\theta = \tfrac{1}{2}(1)(1)\sin\tfrac{360°}{7} = \tfrac{1}{2}\sin\tfrac{360°}{7}$$

The area of the septagon (7-sided figure) is therefore: $7\left(\frac{1}{2}\sin\frac{360°}{7}\right) = \frac{7}{2}\sin\frac{360°}{7} \approx 2.736$ square units

21. Start by finding the measure of a central angle of a triangle drawn from the center out to two adjacent vertices:

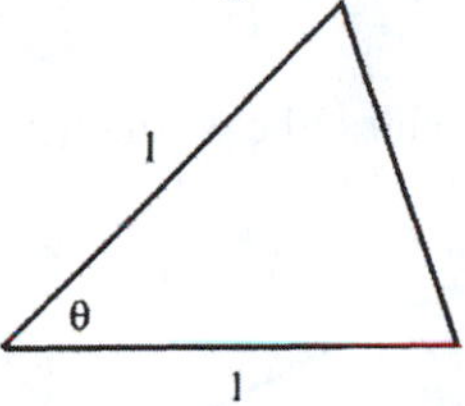

Since the polygon is 8-sided, $8\theta = 360°$, so $\theta = 45°$. The area of this triangle is thus:

$$\tfrac{1}{2}ab\sin\theta = \tfrac{1}{2}(1)(1)\sin 45° = \tfrac{\sqrt{2}}{4}$$

The area of the octagon (8-sided figure) is therefore: $8\left(\frac{\sqrt{2}}{4}\right) = 2\sqrt{2}$

The shaded area is obtained by subtracting this area from the circular area, therefore:

$$\text{Area} = \pi(1)^2 - 2\sqrt{2} = \pi - 2\sqrt{2} \approx 0.313 \text{ square units}$$

23. Start by finding the measure of a central angle of a triangle drawn from the center out to two adjacent vertices:

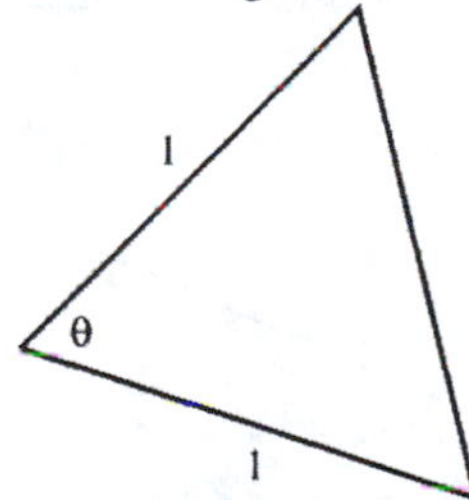

Since the polygon is 6-sided, $6\theta = 360°$, so $\theta = 60°$. The area of this triangle is thus: $\frac{1}{2}ab\sin\theta = \frac{1}{2}(1)(1)\sin 60° = \frac{\sqrt{3}}{4}$

Since the shaded area is comprised of four such triangles, the shaded area is: $4\left(\frac{\sqrt{3}}{4}\right) = \sqrt{3} \approx 1.732$ square units

25. The area of the sector is given by: $A_s = \frac{1}{2}(6 \text{ cm})^2(1.4) = 25.2 \text{ cm}^2$

The area of the triangle is given by: $A_t = \frac{1}{2}(6 \text{ cm})(6 \text{ cm})(\sin 1.4) = 18\sin 1.4 \text{ cm}^2$

Thus the shaded area is: $A_s - A_t = 25.2 \text{ cm}^2 - 18\sin 1.4 \text{ cm}^2 \approx 7.46 \text{ cm}^2$

27. Using the hint, draw the figure:

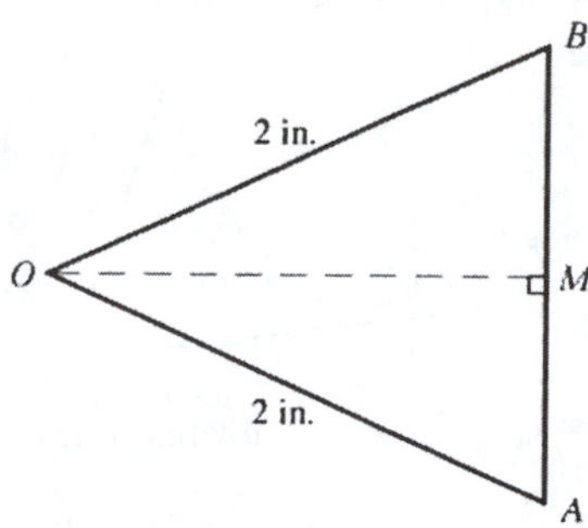

Now $\angle AOB = \frac{360°}{5} = 72°$, so $\angle AOM = 36°$. Now find AM:

$$\sin 36° = \frac{AM}{2}$$
$$AM = 2\sin 36°$$

Thus $AB = 4\sin 36°$, and since there are five sides, the perimeter is: $5(AB) = 5(4\sin 36°) = 20\sin 36°$ inches

29. Draw the figure:

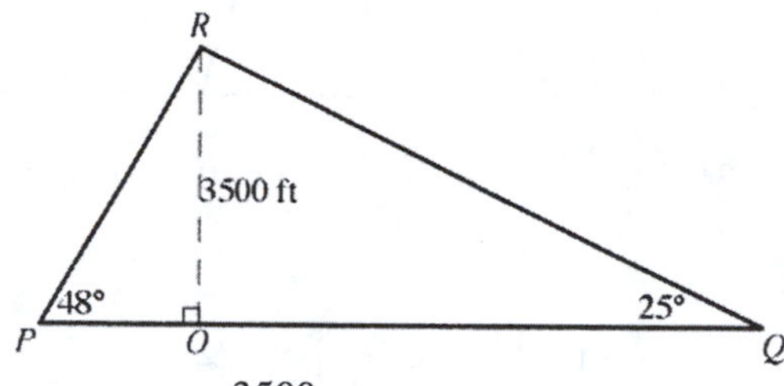

Now compute: $\tan 48° = \frac{3500}{PO}$ and $\tan 25° = \frac{3500}{OQ}$

So $PO = \frac{3500}{\tan 48°}$ and $OQ = \frac{3500}{\tan 25°}$. Thus $PQ = \frac{3500}{\tan 48°} + \frac{3500}{\tan 25°} \approx 10,660$ ft.

31. Draw a figure:

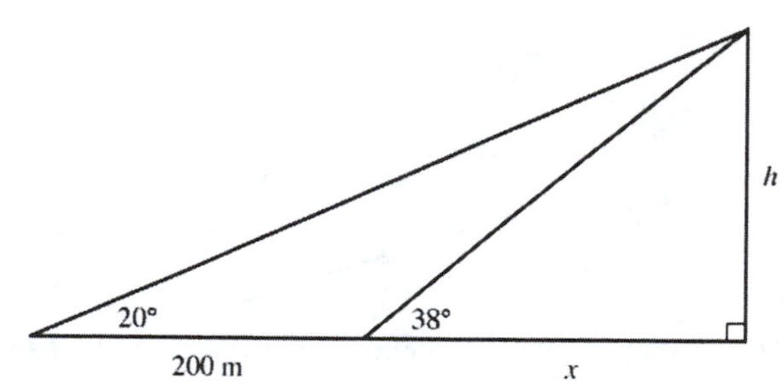

Now $\tan 38° = \frac{h}{x}$, so $x = \frac{h}{\tan 38°}$. Also $\tan 20° = \frac{h}{200 + x}$, so $h = (200 + x)\tan 20°$. Substituting:

$$h = \left(200 + \frac{h}{\tan 38°}\right)\tan 20°$$
$$h\tan 38° = 200\tan 38°\tan 20° + h\tan 20°$$
$$h(\tan 38° - \tan 20°) = 200\tan 38°\tan 20°$$
$$h = \frac{200\tan 38°\tan 20°}{\tan 38° - \tan 20°} \approx 136 \text{ m}$$

33. Draw the figure:

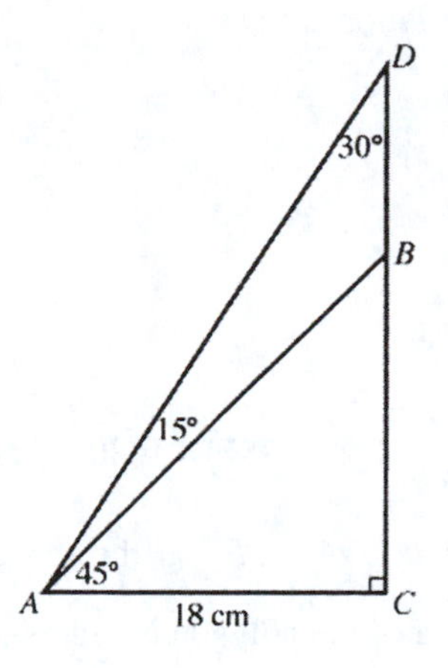

Now compute:

$\tan 45° = \frac{BC}{18}$, so $BC = 18 \tan 45°$ $\qquad$ $\tan 60° = \frac{CD}{18}$, so $CD = 18 \tan 60°$

Therefore: $BD = CD - BC = 18(\tan 60° - \tan 45°) = 18(\sqrt{3} - 1)$ cm

35. **a.** First, note that $\angle BOA = 90° - \theta$ since it forms a right angle with θ. Also, $\angle OAB = \theta$ since both angles are complementary to the same angle ($\angle AOB$). Now $\angle BAP = 90° - \theta$ since it forms a right angle with $\angle OAB$. Finally, $\angle BPA = \theta$ since $\angle BPA$ and $\angle OAB$ are both complementary to the same angle ($\angle BAP$).

b. Using ΔAOP:

$\sin\theta = \frac{AO}{OP} = \frac{AO}{1}$, so $AO = \sin\theta$ $\qquad$ $\cos\theta = \frac{AP}{OP} = \frac{AP}{1}$, so $AP = \cos\theta$

Using ΔAOB: $\sin\theta = \frac{OB}{OA} = \frac{OB}{\sin\theta}$, so $OB = \sin^2\theta$

Using ΔABP: $\cos\theta = \frac{BP}{AP} = \frac{BP}{\cos\theta}$, so $BP = \cos^2\theta$

37. **a.** Examine the similar triangles having $\overline{AB}$ and $\overline{BC}$ as hypotenuses, and notice that θ is also the angle at B in the smaller triangle. Thus: $\sin\theta = \frac{5}{BC}$, so $BC = \frac{5}{\sin\theta}$

b. Using the smaller triangle: $\cos\theta = \frac{4}{AB}$, so $AB = \frac{4}{\cos\theta}$

c. Since $AC = AB + BC$: $AC = \frac{4}{\cos\theta} + \frac{5}{\sin\theta} = 4\sec\theta + 5\csc\theta$

39. First observe that the figure $x^2 + y^2 = 1$ is a circle with a radius of one. Since OA, OD and OF all represent the radius, they are each equal to 1. In each case, look for a trigonometric relationship involving the required segment:

a. $\sin\theta = \frac{DE}{OD}$, so $DE = \sin\theta$ $\qquad$ **b.** $\cos\theta = \frac{OE}{OD}$, so $OE = \cos\theta$

c. $\tan\theta = \frac{CF}{OF}$, so $CF = \tan\theta$ $\qquad$ **d.** $\sec\theta = \frac{OC}{OF}$, so $OC = \sec\theta$

Going to ΔOAB, $\angle ABO = \theta$, thus:

e. $\cot\theta = \frac{AB}{OA}$, so $AB = \cot\theta$ $\qquad$ **f.** $\csc\theta = \frac{OB}{OA}$, so $OB = \csc\theta$

41. **a.** Use $\sin\theta$ to set up a trigonometric relationship:

$$\begin{aligned}\sin\theta &= \frac{r}{PS+r}\\ PS\sin\theta + r\sin\theta &= r\\ PS\sin\theta &= r - r\sin\theta\\ PS\sin\theta &= r(1-\sin\theta)\\ r &= \left(\frac{\sin\theta}{1-\sin\theta}\right)PS\end{aligned}$$

b. Using a calculator: $r = \left(\frac{\sin 0.257°}{1-\sin 0.257°}\right)(238{,}857) \approx 1080$ miles

43. **a.** The central angle $\angle AOB$ is $\frac{2\pi}{7}$. Since $OA = OB = 1$, the area of the triangle is: $\frac{1}{2}(1)(1)\sin\frac{2\pi}{7} = \frac{1}{2}\sin\frac{2\pi}{7}$

b. There are 7 such triangles, so the area of the polygon is $\frac{7}{2}\sin\frac{2\pi}{7}$.

c. There are n triangles within the polygon, each of which has an area of $\frac{1}{2}(1)(1)\sin\frac{2\pi}{n} = \frac{1}{2}\sin\frac{2\pi}{n}$, so the area of the entire polygon is $a_n = \frac{n}{2}\sin\frac{2\pi}{n}$.

d. Completing the table:

n	5	10	50	100	10^3	10^4	10^5
a_n	2.37764129	2.93892626	3.13333084	3.13952598	3.14157198	3.14159245	3.14159265

e. As n increases, the n-sided polygon begins to resemble the circle. With 10^5 sides, the polygon would be barely distinguishable from the circumscribed circle. As such, its area a_n should become closer and closer to the area of the circle, which is $\pi(1)^2 = \pi$.

45. Since $AD = AC = \sqrt{2}$ and the central angle is $90° = \frac{\pi}{2}$ radians, use the formula from Exercise 44**a**:

$$A_s = \frac{1}{2}\left(\sqrt{2}\right)^2\left(\frac{\pi}{2} - \sin\frac{\pi}{2}\right) = \frac{1}{2}(2)\left(\frac{\pi}{2} - 1\right) = \frac{\pi}{2} - 1$$

This represents the area of segment $ABCO$. Thus the shaded area is: $\frac{1}{2}(\pi)(1)^2 - \left(\frac{\pi}{2} - 1\right) = \frac{\pi}{2} - \frac{\pi}{2} + 1 = 1$

Since square $OCED$ has sides of length 1, the two areas are equal.

47. **a.** The coordinates of point A are $A(\cos\theta, \sin\theta)$.

b. Using the distance formula:

$$\begin{aligned}d &= \sqrt{(\cos\theta - 1)^2 + (\sin\theta - 0)^2}\\ &= \sqrt{\cos^2\theta - 2\cos\theta + 1 + \sin^2\theta}\\ &= \sqrt{\cos^2\theta + \sin^2\theta + 1 - 2\cos\theta}\\ &= \sqrt{1 + 1 - 2\cos\theta}\\ &= \sqrt{2 - 2\cos\theta}\end{aligned}$$

49. **a.** Using the chord-length formula to find CE: $CE = \sqrt{2 - 2\cos\theta}$

So the radius of semicircle CDE is $\frac{1}{2}\sqrt{2 - 2\cos\theta}$. Its area is given by:

$$\frac{1}{2} \bullet \pi\left(\frac{1}{2}\sqrt{2 - 2\cos\theta}\right)^2 = \frac{\pi}{8}(2 - 2\cos\theta) = \frac{\pi}{4}(1 - \cos\theta)$$

b. The area of segment CDE is given by: $\frac{1}{2}(1)^2(\theta - \sin\theta) = \frac{1}{2}(\theta - \sin\theta)$

Thus the area of lune CDE is given by: $\frac{\pi}{4}(1 - \cos\theta) - \frac{1}{2}(\theta - \sin\theta)$

c. Using the chord-length formula to find AC: $AC = \sqrt{2-2\cos(\pi-\theta)} = \sqrt{2+2\cos\theta}$

So the radius of semicircle ABC is $\frac{1}{2}\sqrt{2+2\cos\theta}$. Its area is then:

$$\frac{1}{2}\bullet\pi\left(\frac{1}{2}\sqrt{2+2\cos\theta}\right)^2 = \frac{\pi}{8}(2+2\cos\theta) = \frac{\pi}{4}(1+\cos\theta)$$

The area of segment ABC is given by: $\frac{1}{2}(1)^2(\pi-\theta-\sin(\pi-\theta)) = \frac{1}{2}(\pi-\theta-\sin\theta)$

Thus the area of lune ABC is given by: $\frac{\pi}{4}(1+\cos\theta)-\frac{1}{2}(\pi-\theta-\sin\theta)$

d. The base is $AE = 2$ and the height is $\sin\theta$, so the area of ΔACE is $\frac{1}{2}(2)\sin\theta = \sin\theta$.

e. The sum of the areas of the two lunes is:

$$\frac{\pi}{4}(1-\cos\theta)-\frac{1}{2}(\theta-\sin\theta)+\frac{\pi}{4}(1+\cos\theta)-\frac{1}{2}(\pi-\theta-\sin\theta) = \frac{\pi}{4}(2)-\frac{1}{2}(\pi-2\sin\theta) = \frac{\pi}{2}-\frac{\pi}{2}+\sin\theta = \sin\theta$$

Since this is the area of the triangle, the result is verified.

51. a. Draw a perpendicular from O to $\overline{AB}$, and call E the point of intersection. Since the central angle at θ has been bisected: $\sin\theta = \dfrac{AE}{AO} = \dfrac{AE}{1}$, so $AE = \sin\theta$. Thus $AB = 2AE = 2\sin\theta$.

b. $AB = \sin\theta$, since the y-coordinate at B is $\sin\theta$. By the same argument as in part **a**, $CB = 2\sin\theta$.

c. Since $OC = OB$, ΔOBC is isosceles. The sum of the angles of the triangle is 180°, so:

$$\angle OCB = \angle OBC = \frac{180° - 2\theta}{2} = 90° - \theta$$

d. Since $\angle OBC = 90° - \theta$: $\angle DBC = \angle OBC - \angle OBD = 90° - \theta - \theta = 90° - 2\theta$

e. Using ΔCDB:

$$\sin(90° - 2\theta) = \frac{CD}{2\sin\theta}$$
$$\cos 2\theta = \frac{CD}{2\sin\theta}$$
$$CD = 2\sin\theta\cos 2\theta$$

f. Since C lies on the unit circle at an angle of 3θ from the x-axis, its y-coordinate is $\sin 3\theta$. Since this is $AB + CD$: $\sin 3\theta = \sin\theta + 2\sin\theta\cos 2\theta$

g. Using a calculator, $\sin 10° + 2\sin 10°\cos 20° = 0.5$.

h. Using a calculator, $\sin\frac{\pi}{9} + 2\sin\frac{\pi}{9}\cos\frac{2\pi}{9} \approx 0.8660 \approx \dfrac{\sqrt{3}}{2}$.

53. First draw a figure:

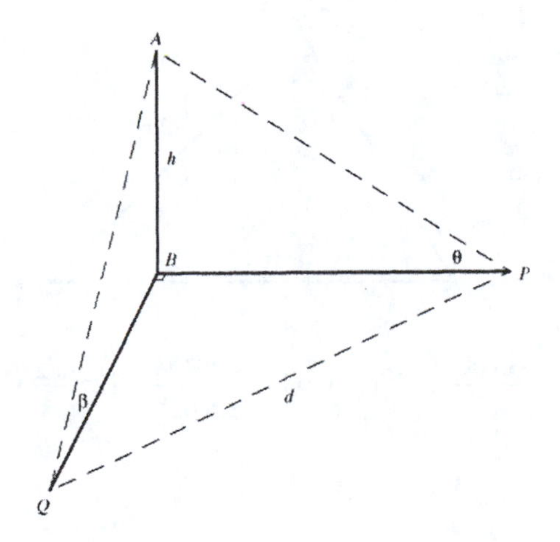

Denote the top of the tower by A, and the bottom by B. Then in ΔPBA we have $\cot\theta = \dfrac{BP}{h}$, and therefore $BP = h\cot\theta$. Similarly in ΔQBA, we have $\cot\beta = \dfrac{QB}{h}$, and therefore $QB = h\cot\beta$. Now apply the Pythagorean theorem in ΔQBP to obtain: $d^2 = (QB)^2 + (BP)^2 = h^2\cot^2\beta + h^2\cot^2\theta$. Solving for h:

$$\begin{aligned} h^2\cot^2\beta + h^2\cot^2\theta &= d^2 \\ h^2\left(\cot^2\beta + \cot^2\theta\right) &= d^2 \\ h^2 &= \frac{d^2}{\cot^2\beta + \cot^2\theta} \\ h &= \frac{d}{\sqrt{\cot^2\beta + \cot^2\theta}} \end{aligned}$$

55. **a.** The area below $\overline{AP}$ is composed of a sector and a triangle. The area of the sector is $\frac{1}{2}r^2\left(\frac{\pi}{2}-\theta\right)$, and the area of the triangle is given by: $\frac{1}{2}r^2\sin\left[\pi - \left(\frac{\pi}{2}-\theta\right)\right] = \frac{1}{2}r^2\sin\left(\frac{\pi}{2}+\theta\right) = \frac{1}{2}r^2\cos\theta$

Adding these two areas: $\frac{1}{2}r^2\left(\frac{\pi}{2}-\theta\right) + \frac{1}{2}r^2\cos\theta$

We want this to equal half the area of the semicircle, so:

$$\begin{aligned} \tfrac{1}{2}r^2\left(\tfrac{\pi}{2}-\theta\right) + \tfrac{1}{2}r^2\cos\theta &= \tfrac{1}{2}\left(\tfrac{1}{2}\pi r^2\right) \\ \tfrac{\pi}{2} - \theta + \cos\theta &= \tfrac{\pi}{2} \\ \cos\theta &= \theta \end{aligned}$$

b. Using the figure, $\theta \approx 0.75$.

c. The percentage error is given by: percentage error $= \dfrac{0.75 - 0.739}{0.739} \times 100 \approx 1.5\%$

Also, $\angle PCB = \frac{\pi}{2} - 0.739 \approx 0.832$, which is approximately 48°.

9.2 The Law of Sines and the Law of Cosines

1. Draw the triangle:

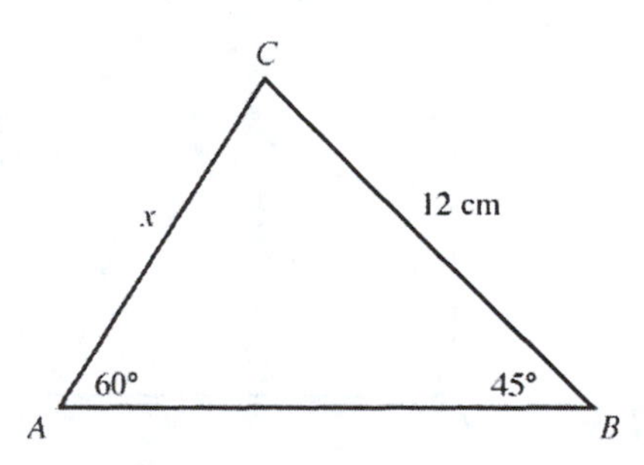

Using the law of sines: $\dfrac{\sin 45°}{x} = \dfrac{\sin 60°}{12}$. Thus: $x = 12 \bullet \dfrac{\sin 45°}{\sin 60°} = 12 \bullet \frac{\sqrt{2}}{2} \bullet \frac{2}{\sqrt{3}} = \frac{12\sqrt{2}}{\sqrt{3}} = 4\sqrt{6}$ cm

3. Draw the triangle:

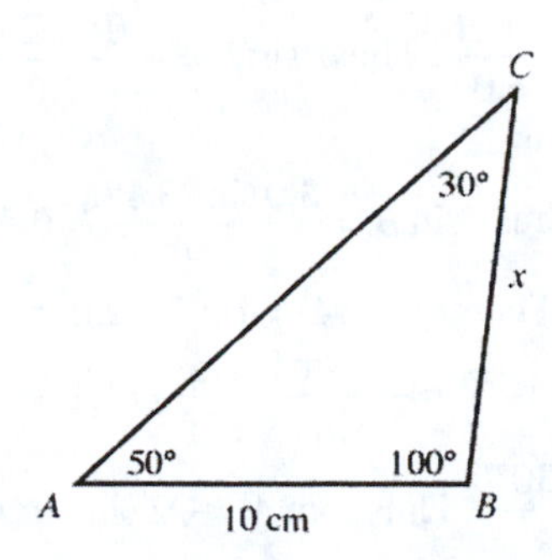

Using the law of sines: $\dfrac{\sin 50°}{x} = \dfrac{\sin 30°}{10}$. Thus: $x = \dfrac{10\sin 50°}{\sin 30°} = \dfrac{10\sin 50°}{\frac{1}{2}} = 20\sin 50°$ cm

5. Draw a triangle:

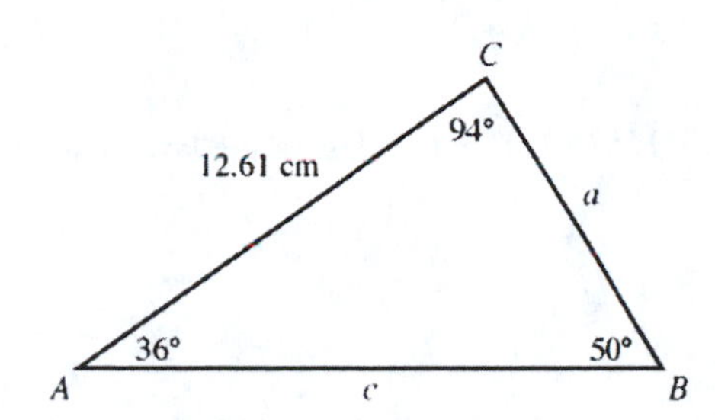

Using the law of sines: $\dfrac{\sin 36°}{a} = \dfrac{\sin 50°}{12.61}$. Thus: $a = \dfrac{12.61\sin 36°}{\sin 50°} \approx 9.7$ cm

Using the law of sines: $\dfrac{\sin 94°}{c} = \dfrac{\sin 50°}{12.61}$. Thus: $c = \dfrac{12.61\sin 94°}{\sin 50°} \approx 16.4$ cm

7. Draw a triangle:

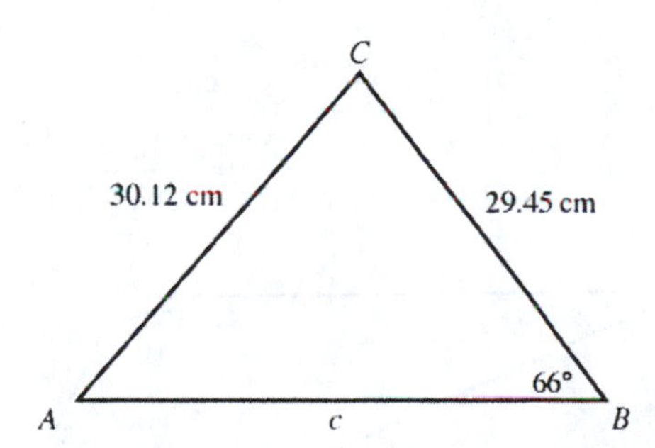

Using the law of sines: $\dfrac{\sin A}{29.45 \text{ cm}} = \dfrac{\sin 66°}{30.12 \text{ cm}}$. Thus: $\sin A = \dfrac{29.45\sin 66°}{30.12} \approx 0.8932$, so $A \approx 63.3°$

Then $C = 180° - 66° - 63.3° \approx 50.7°$. Using the law of sines: $\dfrac{\sin 50.7°}{c} = \dfrac{\sin 66°}{30.12}$.

Thus: $c = \dfrac{30.12\sin 50.7°}{\sin 66°} \approx 25.5$ cm

9. **a.** Since $\sin B = \frac{\sqrt{2}}{2}$, $\angle B = 45°$ or $\angle B = 135°$.

b. Since $\cos E = \frac{\sqrt{2}}{2}$, $\angle E = 45°$. Note that $\angle E \neq 135°$.

c. Since $\sin H = \frac{1}{4}$, $\angle H \approx 14.5°$ or $\angle H \approx 165.5°$.

d. Since $\cos K = -\frac{2}{3}$, $\angle K \approx 131.8°$.

11. **a.** Using the law of sines: $\frac{\sin 23.1°}{2.0} = \frac{\sin B}{6.0}$. Thus: $\sin B = \frac{6.0 \sin 23.1°}{2.0} \approx 1.18$

But $\sin B \le 1$, so no such triangle exists.

b. First find $\angle B$: $\frac{\sin 23.1°}{2.0} = \frac{\sin B}{3.0}$. Thus: $\sin B = \frac{3.0 \sin 23.1°}{2.0} \approx 0.5885$, so $B \approx 36.05°$ or $B \approx 143.95°$

Since $\angle B$ is obtuse, $\angle B \approx 143.95°$. Therefore $\angle C = 180° - 23.1° - 143.95° \approx 12.95°$, so by the law of sines: $\frac{\sin 12.95°}{c} = \frac{\sin 23.1°}{2.0}$. Thus: $c = \frac{2.0 \sin 12.95°}{\sin 23.1°} \approx 1.1$ feet

13. **a.** Using the law of sines: $\frac{\sin A}{\sqrt{2}} = \frac{\sin 30°}{1}$. Thus: $\sin A = \sqrt{2} \sin 30° = \sqrt{2} \bullet \frac{1}{2} = \frac{\sqrt{2}}{2}$

Therefore $\angle A = 45°$ or $\angle A = 135°$.

b. If $\angle A = 45°$, then $\angle C = 180° - 45° - 30° = 105°$. Using the law of sines: $\frac{\sin 105°}{c} = \frac{\sin 30°}{1}$

Thus: $c = \frac{\sin 105°}{\sin 30°} \approx 1.93$

c. If $\angle A = 135°$, then $\angle C = 180° - 135° - 30° = 15°$. Using the law of sines:

$$\frac{\sin 15°}{c} = \frac{\sin 30°}{1}$$

$$c = \frac{\sin 15°}{\sin 30°} \approx 0.52$$

d. Find the area of each triangle:

$$A = \tfrac{1}{2} ab \sin C = \tfrac{1}{2} \bullet \sqrt{2} \bullet 1 \sin 105° \approx 0.68 \qquad A = \tfrac{1}{2} ab \sin C = \tfrac{1}{2} \bullet \sqrt{2} \sin 15° \approx 0.18$$

15. Using the law of sines:

$$\frac{\sin 20°}{2} = \frac{\sin 100°}{a} = \frac{\sin 50°}{b} \qquad \frac{\sin 70°}{c} = \frac{\sin 95°}{b} = \frac{\sin 15°}{d}$$

Hence:

$$a = \frac{2 \sin 110°}{\sin 20°} \qquad b = \frac{2 \sin 50°}{\sin 20°} \text{ cm}$$

$$c = \frac{b \sin 70°}{\sin 95°} = \frac{2 \sin 50° \sin 70°}{\sin 20° \sin 95°} \text{ cm} \qquad d = \frac{b \sin 15°}{\sin 95°} = \frac{2 \sin 50° \sin 15°}{\sin 20° \sin 95°} \text{ cm}$$

17. Sketch the triangle:

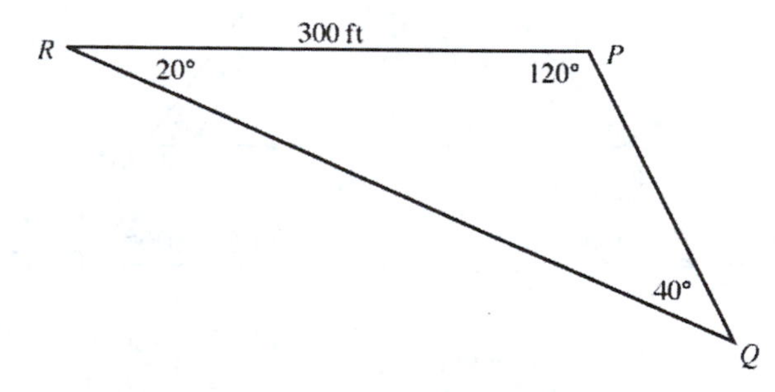

Applying the law of sines: $\frac{\sin 40°}{300} = \frac{\sin 20°}{PQ}$. So $PQ = \frac{300 \sin 20°}{\sin 40°} \approx 160$ ft.

19. **a.** Using the law of cosines: $x^2 = 5^2 + 8^2 - 2(5)(8)\cos 60° = 25 + 64 - 80 \bullet \frac{1}{2} = 49$. So $x = \sqrt{49} = 7$ cm.

b. Using the law of cosines: $x^2 = 5^2 + 8^2 - 2(5)(8)\cos 120° = 25 + 64 - 80 \bullet \left(-\frac{1}{2}\right) = 129$. So $x = \sqrt{129}$ cm.

21. **a.** Using the law of cosines:

$$\begin{aligned} x^2 &= (7.3)^2 + (11.5)^2 - 2(7.3)(11.5)\cos 40° \\ &= 53.29 + 132.25 - 167.9\cos 40° \\ &= 185.54 - 167.9\cos 40° \\ &\approx 56.92 \end{aligned}$$

So $x \approx \sqrt{56.92} \approx 7.5$ cm.

b. Using the law of cosines:

$$\begin{aligned} x^2 &= (7.3)^2 + (11.5)^2 - 2(7.3)(11.5)\cos 140° \\ &= 53.29 + 132.25 - 167.9\cos 140° \\ &= 185.54 - 167.9\cos 140° \\ &\approx 314.16 \end{aligned}$$

So $x \approx \sqrt{314.16} \approx 17.7$ cm.

23. This is incorrect because x is not the side opposite the 130° angle. The correct equation is:

$$6^2 = x^2 + 3^2 - 2(x)(3)\cos 130°$$

25. Using $a = 6$, $b = 7$, $c = 10$ and the law of cosines:

$$6^2 = 7^2 + 10^2 - 2(7)(10)\cos A, \text{ so } \cos A = \tfrac{113}{140}$$

$$7^2 = 6^2 + 10^2 - 2(6)(10)\cos B, \text{ so } \cos B = \tfrac{87}{120} = \tfrac{29}{40}$$

$$10^2 = 6^2 + 7^2 - 2(6)(7)\cos C, \text{ so } \cos C = -\tfrac{15}{84} = -\tfrac{5}{28}$$

27. Using the law of cosines to find angle A:

$$\begin{aligned} 7^2 &= 8^2 + 13^2 - 2(8)(13)\cos A \\ 49 &= 64 + 169 - 208\cos A \\ -184 &= -208\cos A \\ \cos A &= \tfrac{184}{208} \\ A &\approx 27.8° \end{aligned}$$

Now use the law of cosines to find angle B:

$$\begin{aligned} 8^2 &= 7^2 + 13^2 - 2(7)(13)\cos B \\ 64 &= 49 + 169 - 182\cos B \\ -154 &= -182\cos B \\ \cos B &= \tfrac{154}{182} \\ B &\approx 32.2° \end{aligned}$$

Since $A + B + C = 180°$: $C = 180° - A - B \approx 180° - 27.8° - 32.2° \approx 120°$

29. Using the law of cosines to find angle A:

$$\begin{aligned} \left(\tfrac{2}{\sqrt{3}}\right)^2 &= \left(\tfrac{2}{\sqrt{3}}\right)^2 + 2^2 - 2 \cdot \tfrac{2}{\sqrt{3}} \cdot 2\cos A \\ \tfrac{4}{3} &= \tfrac{4}{3} + 4 - \tfrac{8}{\sqrt{3}}\cos A \\ -4 &= -\tfrac{8}{\sqrt{3}}\cos A \\ \cos A &= \tfrac{\sqrt{3}}{2} \\ A &= 30° \end{aligned}$$

Since $a = b$, $B = 30°$. Since $A + B + C = 180°$: $C = 180° - A - B = 180° - 30° - 30° = 120°$

31. First draw a figure, noting that the central angle is $\frac{360°}{5} = 72°$:

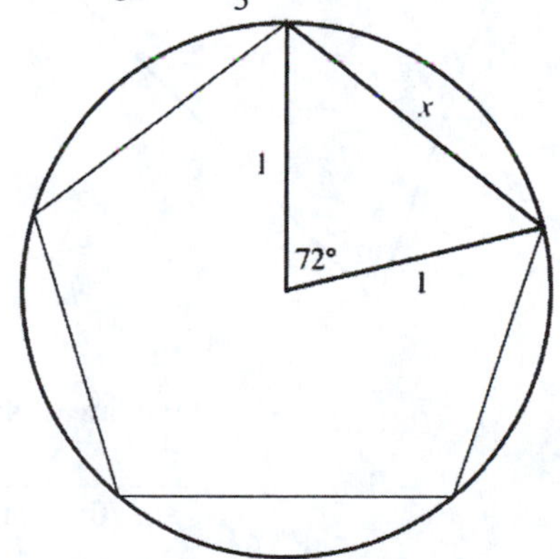

Apply the law of cosines to find x: $x^2 = 1^2 + 1^2 - 2(1)(1)\cos 72° = 2 - 2\cos 72° \approx 1.382$

So $x \approx \sqrt{1.382} \approx 1.18$, and thus the perimeter is $5(1.18) \approx 5.9$ units.

33. **a.** Using the law of cosines:

$$a^2 = (6.1)^2 + (3.2)^2 - 2(6.1)(3.2)\cos 40°$$
$$= 37.21 + 10.24 - 39.04\cos 40°$$
$$= 47.45 - 39.04\cos 40°$$
$$\approx 17.54$$

So $a \approx \sqrt{17.54} \approx 4.2$ cm.

b. Using the law of sines:

$$\frac{\sin C}{3.2} = \frac{\sin 40°}{4.2}$$
$$\sin C = \frac{3.2\sin 40°}{4.2}$$
$$\sin C \approx 0.49$$
$$C \approx 29.3°$$

c. Since $A + B + C = 180°$: $B = 180° - A - C \approx 180° - 40° - 29.3° \approx 110.7°$

35. First draw a figure indicating the relationship between Town A, Town B and Town C, where Town A is centered at the origin:

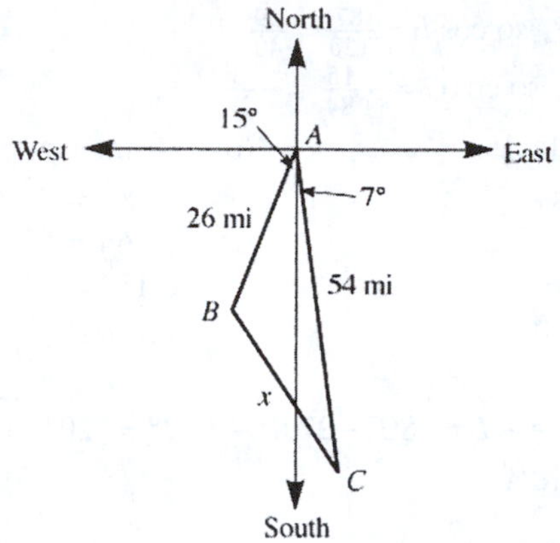

Using ΔABC and the law of cosines:

$$x^2 = 26^2 + 54^2 - 2(26)(54)\cos 22° = 676 + 2916 - 2808\cos 22° = 3592 - 2808\cos 22° \approx 988.5$$

So $x \approx \sqrt{988.5} \approx 31$. The distance between Towns B and C is approximately 31 miles.

37. Using P to denote the plane, note the following figure:

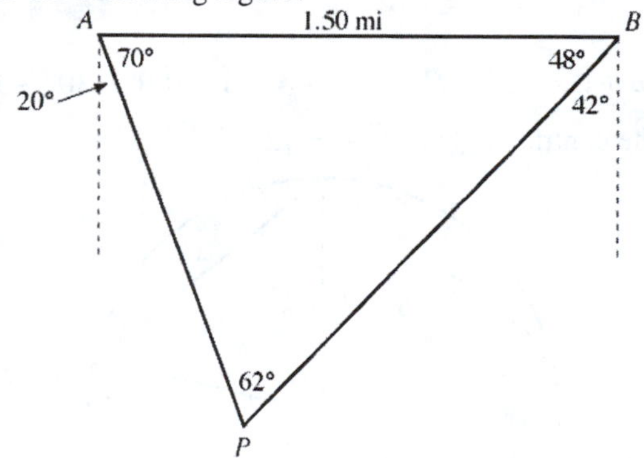

Using the law of sines to find AP:

$$\frac{AP}{\sin 48°} = \frac{1.5}{\sin 62°}$$
$$AP = \frac{1.5\sin 48°}{\sin 62°} \approx 1.26 \text{ mi}$$

Using the law of sines to find BP:

$$\frac{BP}{\sin 70°} = \frac{1.5}{\sin 62°}$$
$$BP = \frac{1.5\sin 70°}{\sin 62°} \approx 1.60 \text{ mi}$$

So the distance from the plane to lighthouse A is 1.26 miles and to lighthouse B is 1.60 miles.

39. Using the law of cosines:

$$\begin{aligned} D^2 &= d^2 + d^2 - 2(d)(d)\cos\tfrac{32}{60}^\circ \\ &= 2d^2 - 2d^2\cos\tfrac{32}{60}^\circ \\ &= 2d^2\left(1-\cos\tfrac{32}{60}^\circ\right) \\ &= 2(92{,}690{,}000)^2\left(1-\cos\tfrac{32}{60}^\circ\right) \\ &\approx 744{,}414{,}483{,}000 \end{aligned}$$

So $D \approx \sqrt{744{,}414{,}483{,}000} \approx 860{,}000$ miles. ***Note***: That is about 100 times the diameter of Earth!

41. **a.** Using the law of cosines: $\left(m^2+n^2+mn\right)^2 = \left(2mn+n^2\right)^2 + \left(m^2-n^2\right)^2 - 2\left(2mn+n^2\right)\left(m^2-n^2\right)\cos C$

After carrying out the indicated squaring operations, and then combining like terms, the equation becomes:

$$\begin{aligned} 2m^3n - 2mn^3 + m^2n^2 - n^4 &= -2\left(2mn+n^2\right)\left(m^2-n^2\right)\cos C \\ 2mn\left(m^2-n^2\right) + n^2\left(m^2-n^2\right) &= -2\left(2mn+n^2\right)\left(m^2-n^2\right)\cos C \\ \left(m^2-n^2\right)\left(2mn+n^2\right) &= -2\left(2mn+n^2\right)\left(m^2-n^2\right)\cos C \end{aligned}$$

Therefore $\cos C = -\frac{1}{2}$, and consequently $C = 120°$.

b. Let $m = 2$ and $n = 1$. Then by means of the expressions in part **a**, we obtain $a = 5$, $b = 3$, $c = 7$.

43. Draw the figure:

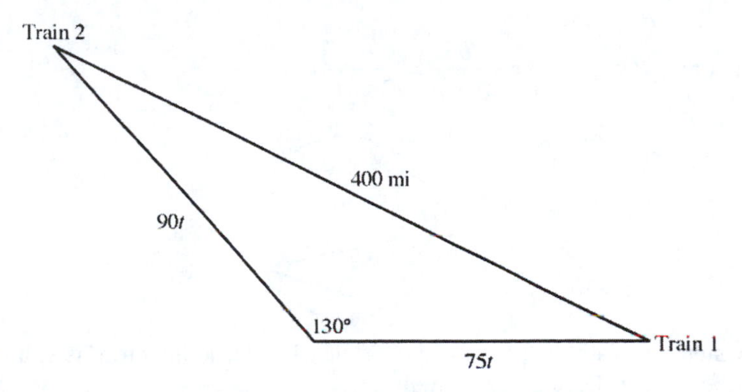

If t is the time of travel (in hours), then the trains have traveled $75t$ mi and $90t$ mi, as indicated in the figure. Using the law of cosines:

$$\begin{aligned} (400)^2 &= (90t)^2 + (75t)^2 - 2(90t)(75t)\cos 130° \\ 160000 &= 8100t^2 + 5625t^2 - 13500\cos 130° t^2 \\ t^2 &= \frac{160000}{13725 - 13500\cos 130°} \approx 7.142 \\ t &\approx 2.672 \text{ hr} \approx 160 \text{ minutes} \end{aligned}$$

The trains are 400 mi apart after 160 minutes, which occurs at 2:40 P.M.

45. Since $\angle AEB = 180° - 15° - 15° = 150°$, using the law of sines:

$$\begin{aligned} \frac{\sin\angle AEB}{AB} &= \frac{\sin 15°}{AE} \\ \frac{\sin 150°}{1} &= \frac{\sin 15°}{AE} \\ \tfrac{1}{2}AE &= \sin 15° \\ AE &= 2\sin 15° \end{aligned}$$

Now $\angle DAE = 90° - 15° = 75°$, so using the law of cosines:

$$\begin{aligned}(DE)^2 &= (AD)^2 + (AE)^2 - 2(AD)(AE)\cos 75° \\ &= 1 + (2\sin 15°)^2 - 2(1)(2\sin 15°)\cos 75° \\ &= 1 + 4\sin^2 15° - 4\sin 15° \cos 75° \\ &= 1 + 4\sin^2 15° - 4\sin 15° \sin 15° \\ &= 1\end{aligned}$$

So $DE = 1$, and since $CE = DE$ (congruent triangles), $CE = 1$. Because $DC = 1$, ΔCDE is equilateral.

47. **a.** The measure of the angle drawn from two vertices of a triangle to the center of the circumscribed circle is twice the measure of the remaining angle of the triangle at the third vertex.

b. Since $\overline{OT}$ is the perpendicular bisector of $\overline{AC}$, $AT = TC = \frac{b}{2}$.

c. By the side-angle-side postulate in geometry, the two triangles are congruent.

d. Since the two triangles are congruent, $\angle AOT = \angle COT$. Since $\angle AOC = 2\angle B$, then $\angle AOT = \angle COT = \angle B$.

e. Since $OC = R$ and $TC = b/2$:

$$\begin{aligned}\sin \angle TOC &= \frac{TC}{R} \\ \sin B &= \frac{b/2}{R} \\ R &= \frac{b}{2\sin B}\end{aligned}$$

From the law of sines: $R = \dfrac{a}{2\sin A} = \dfrac{b}{2\sin B} = \dfrac{c}{2\sin C}$

f. Consider the figure:

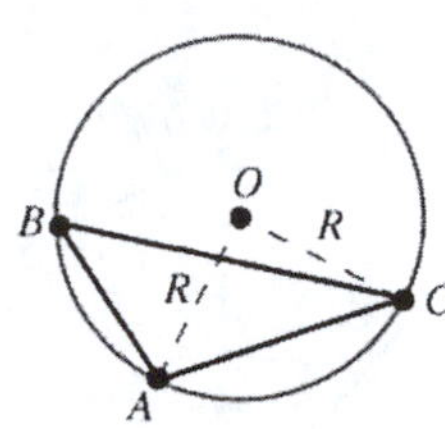

All of the above steps are valid if O lies outside of the triangle, so the proof is still valid.

g. Their circumscribed circles have the same radii.

49. **a.** The radial line drawn from the center of a circle is perpendicular to the tangent line at the point of tangency. Since r is the height of ΔAIC and b is its base, the area of ΔAIC is $\frac{1}{2}rb$.

b. Similarly, the area of ΔAIB is $\frac{1}{2}rc$ and the area of ΔBIC is $\frac{1}{2}ra$. Therefore:

$$A = A_{\Delta AIC} + A_{\Delta AIB} + A_{\Delta BIC} = \tfrac{1}{2}rb + \tfrac{1}{2}rc + \tfrac{1}{2}ra$$

c. Solving for r:

$$\begin{aligned}A &= \tfrac{1}{2}rb + \tfrac{1}{2}rc + \tfrac{1}{2}ra \\ 2A &= rb + rc + ra \\ 2A &= r(a+b+c) \\ r &= \frac{2A}{a+b+c}\end{aligned}$$

51. Using the results from the previous two exercises: $rR = \dfrac{2A}{a+b+c} \bullet \dfrac{abc}{4A} = \dfrac{abc}{2(a+b+c)}$

53. The hint that is given results in the two equations:

$$\lambda^2 = a^2 + d^2 - 2ad\cos(180° - \theta) \quad (1) \qquad \lambda^2 = b^2 + c^2 - 2bc\cos\theta \quad (2)$$

Equation (1) can be rewritten: $\lambda^2 = a^2 + d^2 + 2ad\cos\theta$. Solving this last equation for cos θ: $\cos\theta = \dfrac{\lambda^2 - a^2 - d^2}{2ad}$

Now use this expression for cos θ in equation (2) to obtain:

$$\lambda^2 = b^2 + c^2 - 2bc\left(\frac{\lambda^2 - a^2 - d^2}{2ad}\right)$$
$$\lambda^2 ad = b^2 ad + c^2 ad - bc\lambda^2 + a^2 bc + bcd^2$$
$$\lambda^2(ad + bc) = b^2 ad + c^2 ad + a^2 bc + d^2 bc$$
$$\lambda^2(ad + bc) = (c^2 ad + a^2 bc) + (b^2 ad + d^2 bc)$$
$$\lambda^2(ad + bc) = ac(cd + ab) + bd(ab + cd)$$
$$\lambda^2(ad + bc) = (ab + cd)(ac + bd)$$
$$\lambda^2 = \frac{(ab + cd)(ac + bd)}{ad + bc}$$

55. Using the hint:

$$a^4 + b^4 + c^4 = 2(a^2 + b^2)c^2$$
$$c^4 - 2(a^2 + b^2)c^2 + (a^4 + b^4) = 0$$

Using the quadratic formula:

$$c^2 = \frac{2(a^2 + b^2) \pm \sqrt{4(a^2 + b^2)^2 - 4(a^4 + b^4)}}{2}$$
$$= a^2 + b^2 \pm \sqrt{a^4 + 2a^2b^2 + b^4 - a^4 - b^4}$$
$$= a^2 + b^2 \pm \sqrt{2a^2b^2}$$
$$= a^2 + b^2 \pm ab\sqrt{2}$$

By the law of cosines: $c^2 = a^2 + b^2 - 2ab\cos C$. Setting these expressions equal:

$$a^2 + b^2 - 2ab\cos C = a^2 + b^2 \pm ab\sqrt{2}$$
$$-2ab\cos C = \pm ab\sqrt{2}$$
$$\cos C = \pm\frac{\sqrt{2}}{2}$$
$$C = 45° \text{ or } 135°$$

57. **a.** Simplifying each of the fractions:

$$\frac{\sin A}{a} = \frac{\frac{\sqrt{T}}{2bc}}{a} = \frac{\sqrt{T}}{2abc} \qquad \frac{\sin B}{b} = \frac{\frac{\sqrt{T}}{2ac}}{b} = \frac{\sqrt{T}}{2abc} \qquad \frac{\sin C}{c} = \frac{\frac{\sqrt{T}}{2ab}}{c} = \frac{\sqrt{T}}{2abc}$$

b. Since each of the fractions is equal to $\dfrac{\sqrt{T}}{2abc}$: $\dfrac{\sin A}{a} = \dfrac{\sin B}{b} = \dfrac{\sin C}{c}$

59. **a.** Following the hint, drop a perpendicular from O to $\overline{AD}$, and call the intersection point P. Using ΔAPO: $\cos\alpha = \frac{AP}{AO} = \frac{AP}{1} = AP$. Since $AD = 2 \bullet AP$, $AD = 2\cos\alpha$.

b. Following steps similar to part **a**, notice that ΔAOE is an isosceles triangle. Drop a perpendicular from O to $\overline{AE}$, and call the intersection point Q. Using ΔAQO: $\cos\beta = \frac{AQ}{AO} = \frac{AQ}{1} = AQ$. Since $AE = 2 \bullet AQ$, $AE = 2\cos\beta$.

c. Since $\angle AFC = 60°$, $\angle AFB = 180° - 60° = 120°$. Since the angles of ΔAFB must sum to 180°:

$$\begin{aligned} \alpha + 120° + \angle B &= 180° \\ \angle B &= 60° - \alpha \end{aligned}$$

Similarly, since the angles of ΔAFC must sum to 180°:

$$\begin{aligned} \beta + 60° + \angle C &= 180° \\ \angle C &= 120° - \beta \end{aligned}$$

d. Using the law of sines on ΔABC: $\dfrac{\sin(\angle A)}{BC} = \dfrac{\sin(\angle B)}{AC} = \dfrac{\sin(\angle C)}{AB}$

Since $\angle A = \alpha + \beta$, $\angle B = 60° - \alpha$ and $\angle C = 120° - \beta$:

$$\frac{\sin(\alpha+\beta)}{BC} = \frac{\sin(60° - \alpha)}{AC}, \text{ so } AC = \frac{BC \bullet \sin(60° - \alpha)}{\sin(\alpha+\beta)}$$

$$\frac{\sin(\alpha+\beta)}{BC} = \frac{\sin(120° - \beta)}{AB}, \text{ so } AB = \frac{BC \bullet \sin(120° - \beta)}{\sin(\alpha+\beta)}$$

e. Using the results in parts **a**, **b** and **d**:

$$\begin{aligned} AD \bullet AB - AE \bullet AC &= (2\cos\alpha) \bullet \frac{BC\sin(120° - \beta)}{\sin(\alpha+\beta)} - (2\cos\beta) \bullet \frac{BC\sin(60° - \alpha)}{\sin(\alpha+\beta)} \\ &= BC \bullet \left[\frac{2\cos\alpha\sin(120° - \beta) - 2\cos\beta\sin(60° - \alpha)}{\sin(\alpha+\beta)}\right] \end{aligned}$$

f. Using the addition formulas for sine:

$$\sin(120° - \beta) = \sin 120° \cos\beta - \cos 120° \sin\beta = \tfrac{\sqrt{3}}{2}\cos\beta + \tfrac{1}{2}\sin\beta$$

$$\sin(60° - \alpha) = \sin 60° \cos\alpha - \cos 60° \sin\alpha = \tfrac{\sqrt{3}}{2}\cos\alpha - \tfrac{1}{2}\sin\alpha$$

Therefore:

$$\begin{aligned} &2\cos\alpha\sin(120° - \beta) - 2\cos\beta\sin(60° - \alpha) \\ &\quad = 2\cos\alpha\left(\tfrac{\sqrt{3}}{2}\cos\beta + \tfrac{1}{2}\sin\beta\right) - 2\cos\beta\left(\tfrac{\sqrt{3}}{2}\cos\alpha - \tfrac{1}{2}\sin\alpha\right) \\ &\quad = \sqrt{3}\cos\alpha\cos\beta + \cos\alpha\sin\beta - \sqrt{3}\cos\alpha\cos\beta + \sin\alpha\cos\beta \\ &\quad = \sin\alpha\cos\beta + \cos\alpha\sin\beta \\ &\quad = \sin(\alpha+\beta) \end{aligned}$$

This completes the proof of the result.

61. a. Computing areas of each triangle:

$$\text{Area}_{\text{left}} = \tfrac{1}{2}(\text{base})(\text{height}) = \tfrac{1}{2}af\sin\tfrac{C}{2}$$

$$\text{Area}_{\text{right}} = \tfrac{1}{2}(\text{base})(\text{height}) = \tfrac{1}{2}bf\sin\tfrac{C}{2}$$

$$\text{Area}_{\text{entire}} = \tfrac{1}{2}(\text{base})(\text{height}) = \tfrac{1}{2}ab\sin C$$

Adding the left and right triangles: $\tfrac{1}{2}af\sin\tfrac{C}{2} + \tfrac{1}{2}bf\sin\tfrac{C}{2} = \tfrac{1}{2}ab\sin C$

b. Since $\sin C = 2\sin\frac{C}{2}\cos\frac{C}{2}$: $\tfrac{1}{2}af\sin\tfrac{C}{2} + \tfrac{1}{2}bf\sin\tfrac{C}{2} = ab\sin\tfrac{C}{2}\cos\tfrac{C}{2}$. Multiplying by $\frac{2}{\sin(C/2)}$:

$$\begin{aligned} af + bf &= 2ab\cos\tfrac{C}{2} \\ f(a+b) &= 2ab\cos\tfrac{C}{2} \\ f &= \frac{2ab\cos\frac{C}{2}}{a+b} \end{aligned}$$

c. Using the half-angle formula:

$$\cos\frac{C}{2}=\sqrt{\frac{1+\cos C}{2}}=\sqrt{\frac{1+\frac{a^2+b^2-c^2}{2ab}}{2}}=\sqrt{\frac{a^2+2ab+b^2-c^2}{4ab}}=\frac{1}{2}\sqrt{\frac{(a+b)^2-c^2}{ab}}=\frac{1}{2}\sqrt{\frac{(a+b-c)(a+b+c)}{ab}}$$

d. Combining our results from **b** and **c**:

$$f=\frac{2ab}{a+b}\cos\frac{C}{2}=\frac{ab}{a+b}\sqrt{\frac{(a+b-c)(a+b+c)}{ab}}=\frac{\sqrt{ab}}{a+b}\sqrt{(a+b-c)(a+b+c)}$$

63. From the law of sines, note that $a\sin B=b\sin A$. Now, using the double-angle identity for $\sin 2B$ and $\sin 2A$:

$$\begin{aligned}\frac{a^2\sin 2B+b^2\sin 2A}{4}&=\frac{2a^2\sin B\cos B+2b^2\sin A\cos A}{4}\\&=\frac{(a\sin B)(a\cos B)+(b\sin A)(b\cos A)}{2}\\&=\frac{(b\sin A)(a\cos B)+(a\sin B)(b\cos A)}{2}\\&=\frac{ab(\sin A\cos B+\cos A\sin B)}{2}\\&=\frac{ab\sin(A+B)}{2}\\&=\frac{ab\sin C}{2}\\&=\frac{1}{2}ab\sin C\end{aligned}$$

Note that the fact that $\sin(A+B)=\sin(180°-(A+B))=\sin C$ was used. Since this last expression is the area of ΔABC, the proof is complete.

9.3 Vectors in the Plane: A Geometric Approach

1. Graph the vector:

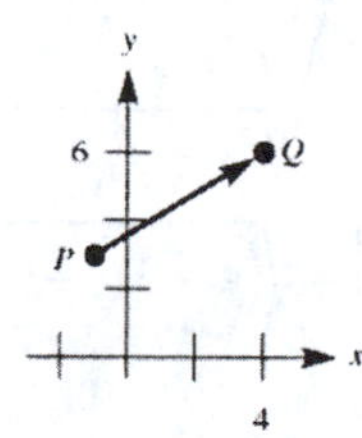

The magnitude is given by: $|\overrightarrow{PQ}|=\sqrt{(4-(-1))^2+(6-3)^2}=\sqrt{25+9}=\sqrt{34}$

3. Graph the vector:

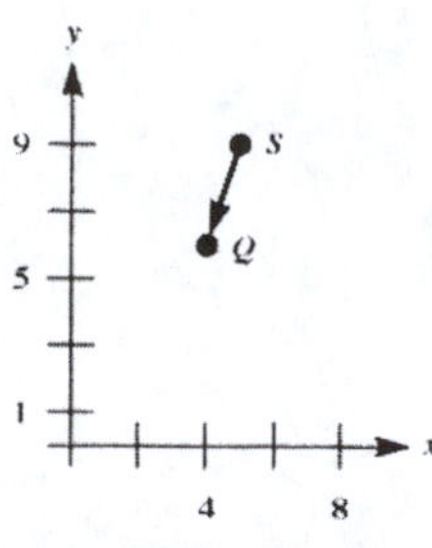

The magnitude is given by: $|\overrightarrow{SQ}| = \sqrt{(4-5)^2 + (6-9)^2} = \sqrt{1+9} = \sqrt{10}$

5. Graph the vector:

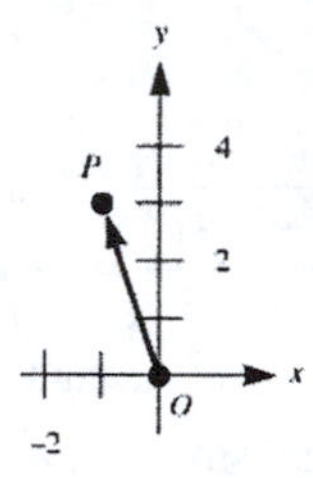

The magnitude is given by: $|\overrightarrow{OP}| = \sqrt{(-1-0)^2 + (3-0)^2} = \sqrt{1+9} = \sqrt{10}$

7. Graph the vector sum:

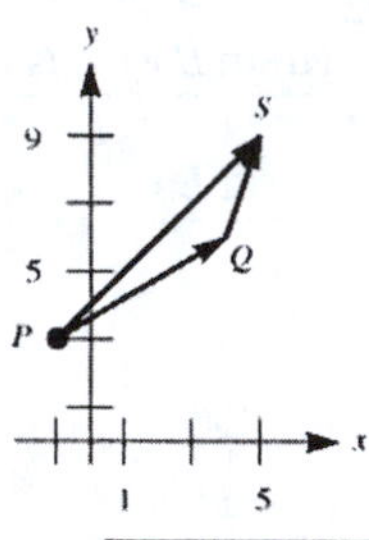

The magnitude is given by: $|\overrightarrow{PQ} + \overrightarrow{QS}| = |\overrightarrow{PS}| = \sqrt{(5-(-1))^2 + (9-3)^2} = \sqrt{36+36} = 6\sqrt{2}$

9. Graph the vector sum:

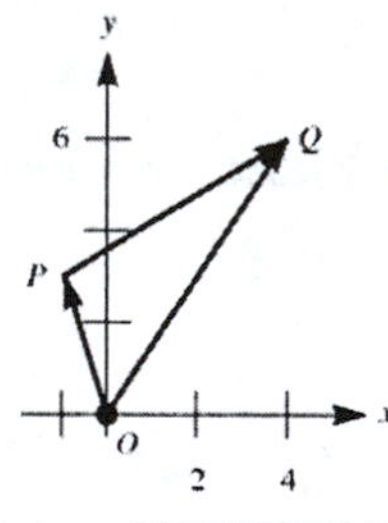

The magnitude is given by: $|\overrightarrow{OP} + \overrightarrow{PQ}| = |\overrightarrow{OQ}| = \sqrt{(4-0)^2 + (6-0)^2} = \sqrt{16+36} = 2\sqrt{13}$

11. Graph the vector sum:

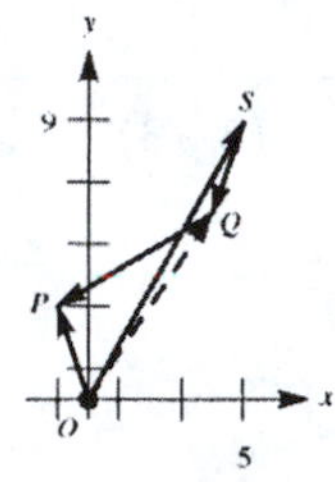

The magnitude is given by: $|\overrightarrow{OS} + \overrightarrow{SQ} + \overrightarrow{QP}| = |\overrightarrow{OQ} + \overrightarrow{QP}| = |\overrightarrow{OP}| = \sqrt{(-1-0)^2 + (3-0)^2} = \sqrt{1+9} = \sqrt{10}$

13. Graph the vector sum:

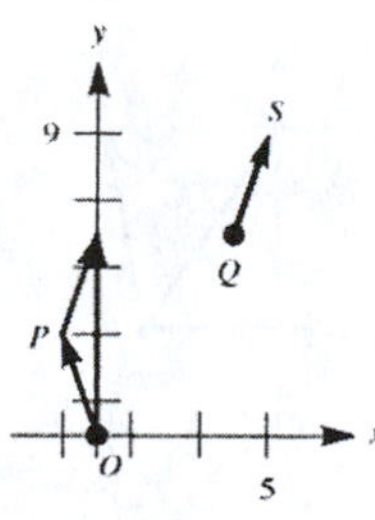

The magnitude is given by: $\left|\overrightarrow{OP}+\overrightarrow{QS}\right|=\sqrt{(0-0)^2+(6-0)^2}=\sqrt{0+36}=6$

15. Graph the vector sum:

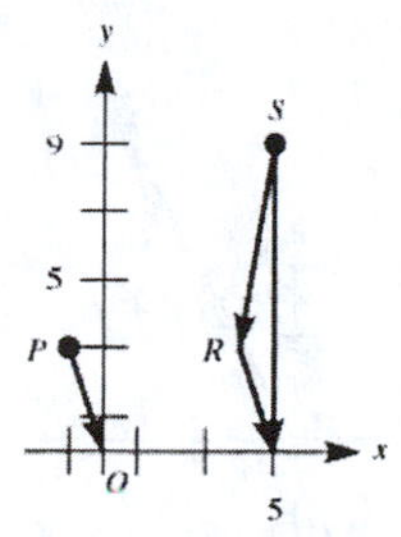

The magnitude is given by: $\left|\overrightarrow{SR}+\overrightarrow{PO}\right|=\sqrt{(5-5)^2+(9-0)^2}=\sqrt{0+81}=9$

17. Graph the vector sum:

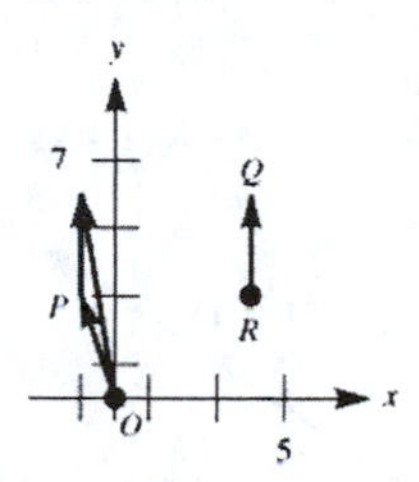

The magnitude is given by: $\left|\overrightarrow{OP}+\overrightarrow{RQ}\right|=\sqrt{(-1-0)^2+(6-0)^2}=\sqrt{1+36}=\sqrt{37}$

19. Graph the vector sum:

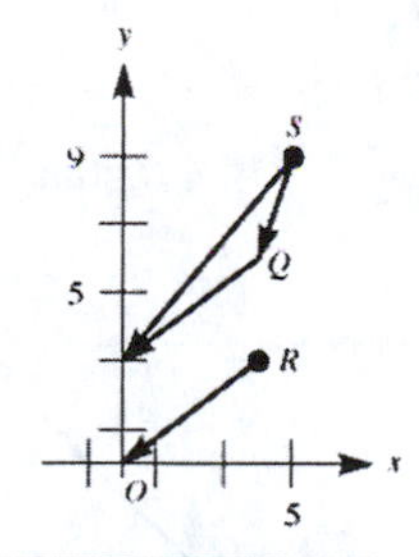

The magnitude is given by: $\left|\overrightarrow{SQ}+\overrightarrow{RO}\right|=\sqrt{(5-0)^2+(9-3)^2}=\sqrt{25+36}=\sqrt{61}$

21. Graph the vector sum:

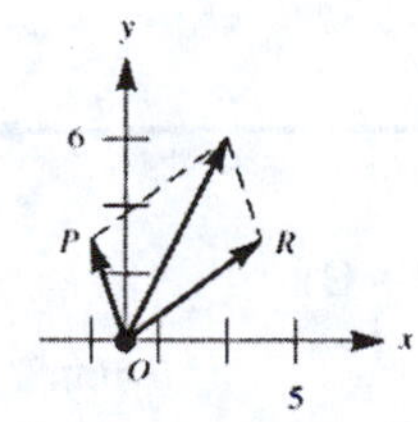

The magnitude is given by: $\left|\overrightarrow{OP}+\overrightarrow{OR}\right|=\sqrt{(3-0)^2+(6-0)^2}=\sqrt{9+36}=3\sqrt{5}$

23. Graph the vector sum:

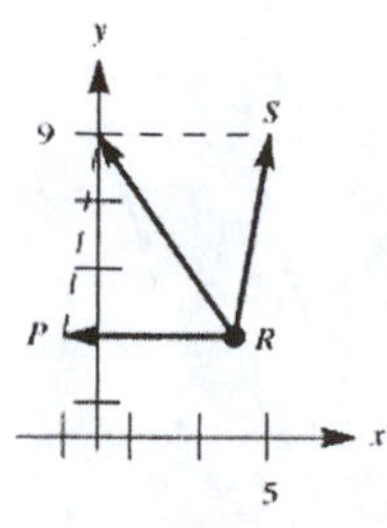

The magnitude is given by: $|\overrightarrow{RP}+\overrightarrow{RS}|=\sqrt{(0-4)^2+(9-3)^2}=\sqrt{16+36}=2\sqrt{13}$

25. Graph the vector sum:

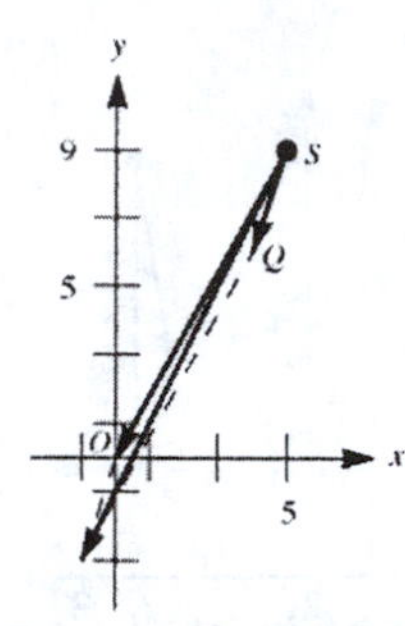

The magnitude is given by: $|\overrightarrow{SO}+\overrightarrow{SQ}|=\sqrt{(-1-5)^2+(-3-9)^2}=\sqrt{36+144}=6\sqrt{5}$

27. Draw the figure:

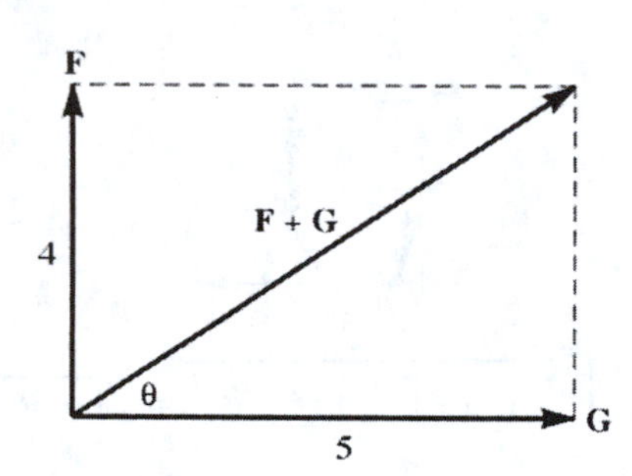

Now compute the magnitude and direction of **F** + **G**:

$|\mathbf{F}+\mathbf{G}|=\sqrt{4^2+5^2}=\sqrt{16+25}=\sqrt{41}$ N $\qquad \theta=\tan^{-1}\left(\frac{4}{5}\right)\approx 38.7°$

29. Draw the figure:

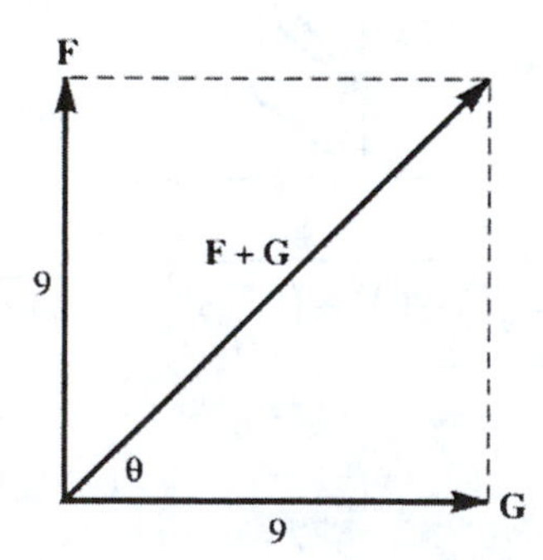

Now compute the magnitude and direction of **F** + **G**:

$|\mathbf{F}+\mathbf{G}|=\sqrt{9^2+9^2}=9\sqrt{2}$ N $\qquad \theta=\tan^{-1}\left(\frac{9}{9}\right)=\tan^{-1}1=45°$

31. Draw the figure:

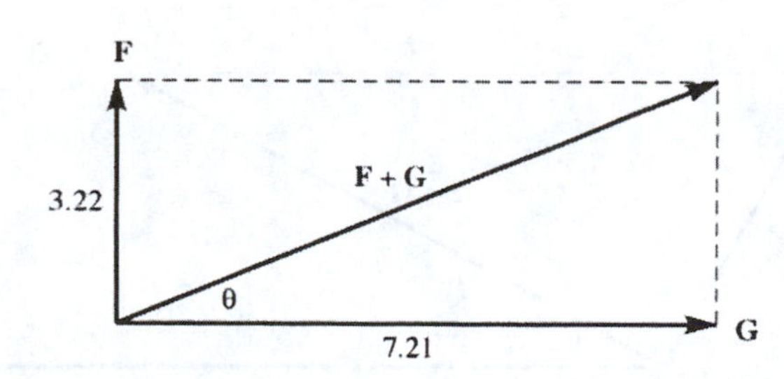

Now compute the magnitude and direction of **F** + **G**:

$$|\mathbf{F}+\mathbf{G}|=\sqrt{3.22^2+7.21^2}=\sqrt{62.3525}\approx 7.90 \text{ N} \qquad \theta=\tan^{-1}\left(\frac{3.22}{7.21}\right)\approx 24.1°$$

33. Draw the parallelogram:

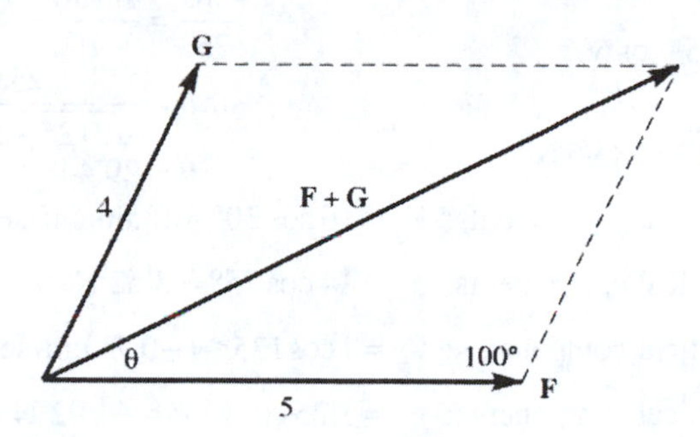

Let $d=|\mathbf{F}+\mathbf{G}|$. Then using the law of cosines:

$$d^2=5^2+4^2-2(5)(4)\cos 100°$$
$$d^2=41-40\cos 100°$$
$$d=\sqrt{41-40\cos 100°}\approx 6.92 \text{ N}$$

Find θ by the law of sines:

$$\frac{\sin\theta}{4}=\frac{\sin 100°}{d}$$
$$\sin\theta=\frac{4\sin 100°}{\sqrt{41-40\cos 100°}}\approx 0.5689$$
$$\theta\approx 34.67°$$

35. Draw the parallelogram:

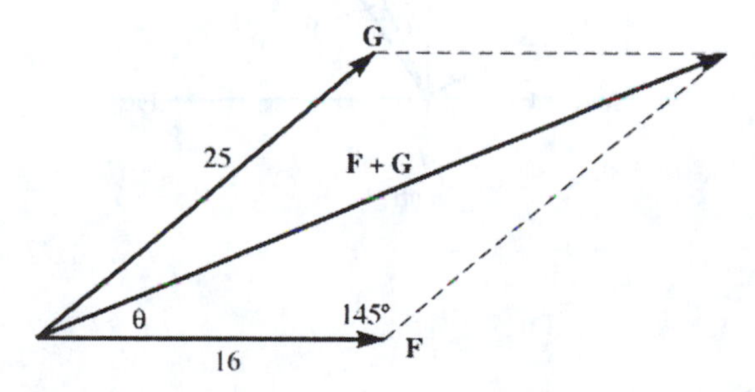

Let $d=|\mathbf{F}+\mathbf{G}|$. Then using the law of cosines:

$$d^2=16^2+25^2-2(16)(25)\cos 145°$$
$$d^2=881-800\cos 145°$$
$$d=\sqrt{881-800\cos 145°}\approx 39.20 \text{ N}$$

Find θ by the law of sines:

$$\frac{\sin\theta}{25}=\frac{\sin 145°}{d}$$
$$\sin\theta=\frac{25\sin 145°}{\sqrt{881-800\cos 145°}}\approx 0.3658$$
$$\theta\approx 21.46°$$

37. Draw the parallelogram:

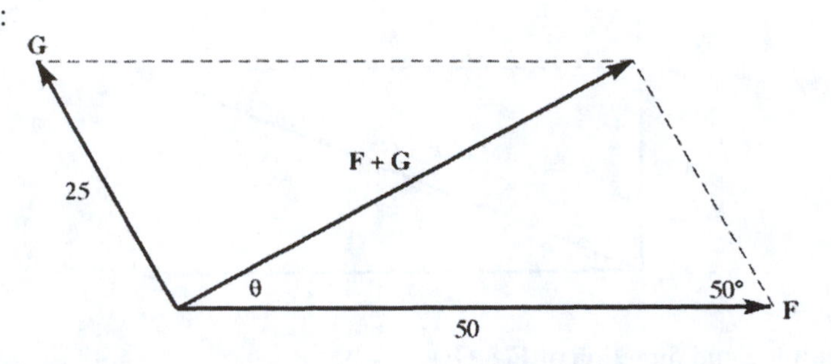

Let $d = |\mathbf{F} + \mathbf{G}|$. Then using the law of cosines:

$$d^2 = 50^2 + 25^2 - 2(50)(25)\cos 50°$$
$$d^2 = 3125 - 2500\cos 50°$$
$$d = \sqrt{3125 - 2500\cos 50°} \approx 38.96 \text{ N}$$

Find θ by the law of sines:

$$\frac{\sin\theta}{25} = \frac{\sin 50°}{d}$$
$$\sin\theta = \frac{25\sin 50°}{\sqrt{3125 - 2500\cos 50°}} \approx 0.4915$$
$$\theta \approx 29.44°$$

39. Compute the horizontal and vertical components: $V_x = 16\cos 30° \approx 13.86$ cm/sec $\quad V_y = 16\sin 30° = 8$ cm/sec

41. Compute the horizontal and vertical components: $F_x = 14\cos 75° \approx 3.62$ N $\quad F_y = 14\sin 75° \approx 13.52$ N

43. Compute the horizontal and vertical components: $V_x = 1\cos 135° \approx -0.71$ cm/sec $\quad V_y = 1\sin 135° \approx 0.71$ cm/sec

45. Compute the horizontal and vertical components: $F_x = 1.25\cos 145° \approx -1.02$ N $\quad F_y = 1.25\sin 145° \approx 0.72$ N

47. Draw the vectors:

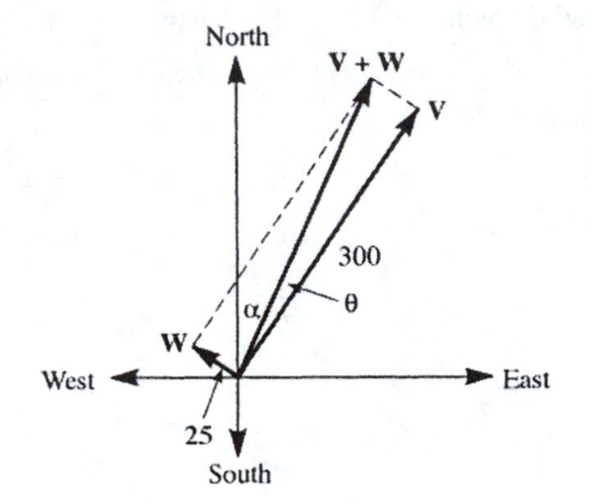

Let θ be the drift angle. Then:

$$\tan\theta = \frac{25}{300} = \frac{1}{12}$$
$$\theta = \tan^{-1}\left(\frac{1}{12}\right) \approx 4.76°$$

The ground speed is given by: $|\mathbf{V} + \mathbf{W}| = \sqrt{25^2 + 300^2} = \sqrt{90625} \approx 301.04$ mph

Let α be the course (bearing). Then: $\alpha = 30° - \theta \approx 30° - 4.76° \approx 25.24°$

49. Draw the vectors:

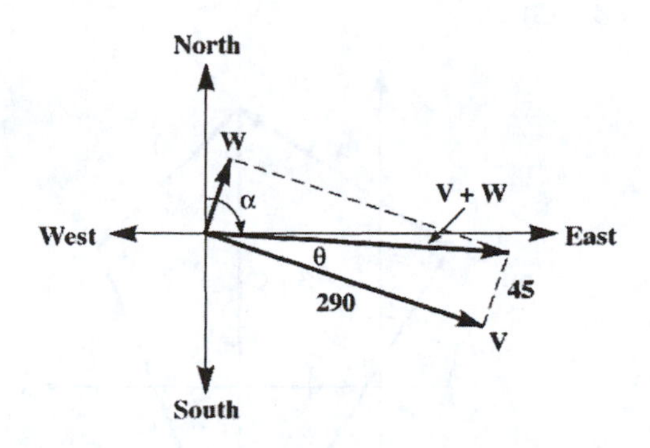

Let θ be the drift angle. Then:

$$\tan\theta = \frac{45}{290} = \frac{9}{58}$$

$$\theta = \tan^{-1}\left(\frac{9}{58}\right) \approx 8.82°$$

The ground speed is given by: $|\mathbf{V}+\mathbf{W}| = \sqrt{290^2 + 45^2} = \sqrt{86125} \approx 293.47$ mph

Let α be the course (bearing). Then: $\alpha = 100° - \theta \approx 100° - 8.82° \approx 91.18°$

51. Draw a figure:

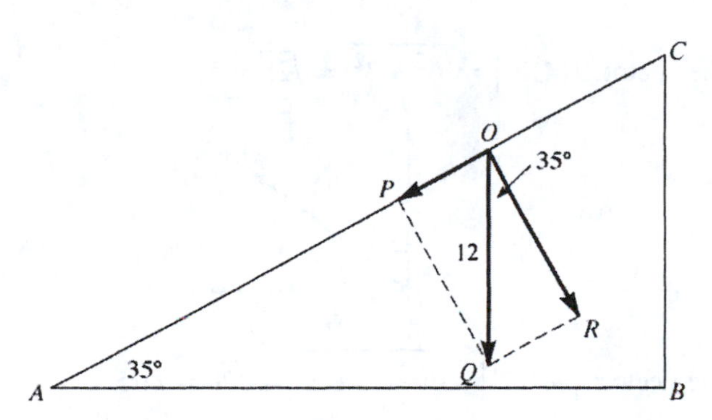

Notice that $\angle QOR = 35°$ since it is complementary to $\angle POQ$, but $\angle POQ = \angle ACB$. So the desired components are given by:

$$|\overrightarrow{OR}| = 12\cos 35° \approx 9.83 \text{ lb} \qquad |\overrightarrow{OP}| = 12\sin 35° \approx 6.88 \text{ lb}$$

53. Using the same approach as in Exercise 51:

perpendicular : $12\cos 10° \approx 11.82$ lb parallel : $12\sin 10° \approx 2.08$ lb

55. **a.** Draw the vector sum $(\mathbf{A}+\mathbf{B})+\mathbf{C}$:

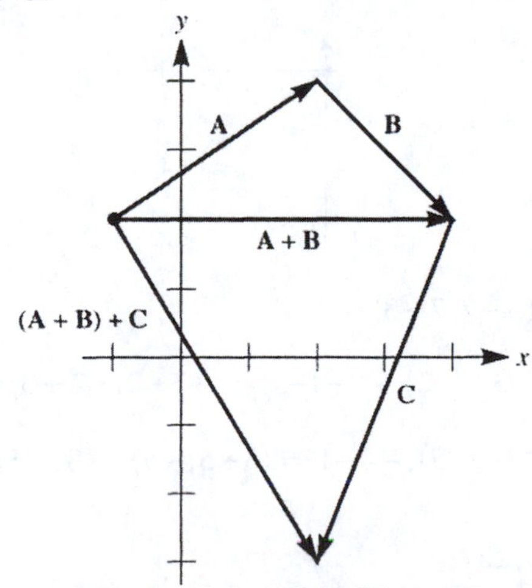

From the diagram, note that the initial point of $(\mathbf{A}+\mathbf{B})+\mathbf{C}$ is (–1,2) and the terminal point is (2,–3).

b. Draw the vector sum **A** + (**B** + **C**):

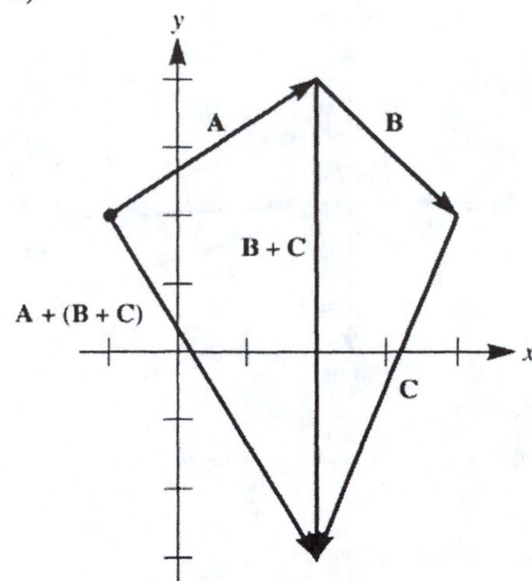

From the diagram, note that the initial point of **A** + (**B** + **C**) is (–1,2) and the terminal point is (2,–3).

9.4 Vectors in the Plane: An Algebraic Approach

1. Compute the length of the vector: $|\langle 4,3\rangle| = \sqrt{4^2+3^2} = \sqrt{25} = 5$

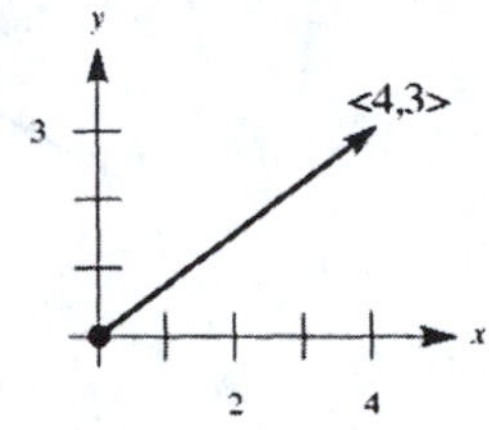

3. Compute the length of the vector: $|\langle -4,2\rangle| = \sqrt{(-4)^2+2^2} = \sqrt{20} = 2\sqrt{5}$

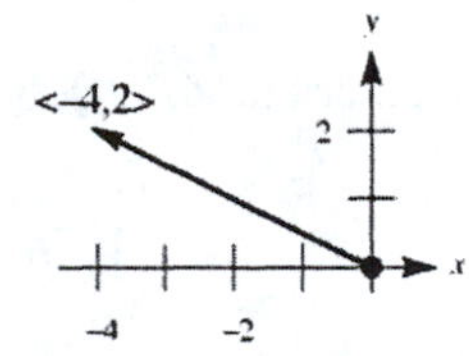

5. Compute the length of the vector: $\left|\left\langle \frac{3}{4}, -\frac{1}{2}\right\rangle\right| = \sqrt{\left(\frac{3}{4}\right)^2 + \left(-\frac{1}{2}\right)^2} = \sqrt{\frac{9}{16}+\frac{1}{4}} = \frac{\sqrt{13}}{4}$

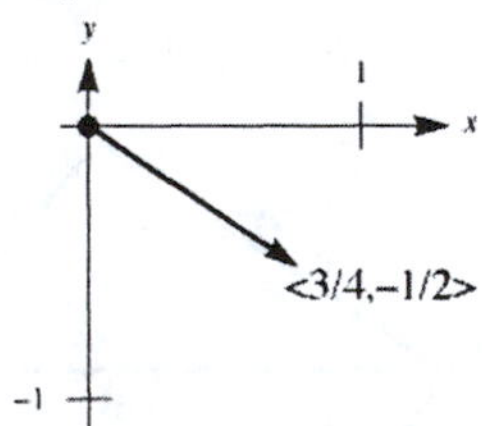

7. Subtracting components: $\overrightarrow{PQ} = \langle 3-2, 7-3\rangle = \langle 1,4\rangle$

9. Subtracting components: $\overrightarrow{PQ} = \langle -3-(-2), -2-(-3)\rangle = \langle -3+2, -2+3\rangle = \langle -1,1\rangle$

11. Subtracting components: $\overrightarrow{PQ} = \langle 3-(-5), -4-1\rangle = \langle 3+5, -5\rangle = \langle 8,-5\rangle$

13. $\mathbf{a}+\mathbf{b} = \langle 2+5, 3+4\rangle = \langle 7,7\rangle$

15. $2\mathbf{a}+4\mathbf{b} = \langle 4,6\rangle + \langle 20,16\rangle = \langle 24,22\rangle$

17. Since $\mathbf{b}+\mathbf{c}=\langle 5+6,4-1\rangle=\langle 11,3\rangle$: $|\mathbf{b}+\mathbf{c}|=\sqrt{11^2+3^2}=\sqrt{130}$

19. Since $\mathbf{a}+\mathbf{c}=\langle 2+6,3-1\rangle=\langle 8,2\rangle$:

$$|\mathbf{a}+\mathbf{c}|=\sqrt{8^2+2^2}=\sqrt{68}=2\sqrt{17} \qquad |\mathbf{a}|=\sqrt{2^2+3^2}=\sqrt{13} \qquad |\mathbf{c}|=\sqrt{6^2+(-1)^2}=\sqrt{37}$$

So $|\mathbf{a}+\mathbf{c}|-|\mathbf{a}|-|\mathbf{c}|=2\sqrt{17}-\sqrt{13}-\sqrt{37}$.

21. Since $\mathbf{b}+\mathbf{c}=\langle 5+6,4-1\rangle=\langle 11,3\rangle$: $\mathbf{a}+(\mathbf{b}+\mathbf{c})=\langle 2,3\rangle+\langle 11,3\rangle=\langle 13,6\rangle$

23. $3\mathbf{a}+4\mathbf{a}=\langle 6,9\rangle+\langle 8,12\rangle=\langle 14,21\rangle$

25. $\mathbf{a}-\mathbf{b}=\langle 2,3\rangle-\langle 5,4\rangle=\langle -3,-1\rangle$

27. $3\mathbf{b}-4\mathbf{d}=\langle 15,12\rangle-\langle -8,0\rangle=\langle 15+8,12-0\rangle=\langle 23,12\rangle$

29. Since $\mathbf{b}+\mathbf{c}=\langle 11,3\rangle$: $\mathbf{a}-(\mathbf{b}+\mathbf{c})=\langle 2,3\rangle-\langle 11,3\rangle=\langle -9,0\rangle$

31. Since $\mathbf{c}+\mathbf{d}=\langle 4,-1\rangle$ and $\mathbf{c}-\mathbf{d}=\langle 8,-1\rangle$: $|\mathbf{c}+\mathbf{d}|=\sqrt{16+1}=\sqrt{17}$ and $|\mathbf{c}-\mathbf{d}|=\sqrt{64+1}=\sqrt{65}$

Therefore: $|\mathbf{c}+\mathbf{d}|^2-|\mathbf{c}-\mathbf{d}|^2=17-65=-48$

33. Separating individual components: $\langle 3,8\rangle=\langle 3,0\rangle+\langle 0,8\rangle=3\mathbf{i}+8\mathbf{j}$

35. Separating individual components: $\langle -8,-6\rangle=\langle -8,0\rangle+\langle 0,-6\rangle=-8\mathbf{i}-6\mathbf{j}$

37. Separating individual components: $3\langle 5,3\rangle+2\langle 2,7\rangle=3(5\mathbf{i}+3\mathbf{j})+2(2\mathbf{i}+7\mathbf{j})=15\mathbf{i}+9\mathbf{j}+4\mathbf{i}+14\mathbf{j}=19\mathbf{i}+23\mathbf{j}$

39. Separating individual components: $\mathbf{i}+\mathbf{j}=\langle 1,1\rangle$

41. Separating individual components: $5\mathbf{i}-4\mathbf{j}=\langle 5,-4\rangle$

43. First compute the length of the vector: $|\langle 4,8\rangle|=\sqrt{4^2+8^2}=\sqrt{80}=4\sqrt{5}$

So a unit vector would be given by: $\frac{1}{4\sqrt{5}}\langle 4,8\rangle=\left\langle \frac{1}{\sqrt{5}},\frac{2}{\sqrt{5}}\right\rangle=\left\langle \frac{\sqrt{5}}{5},\frac{2\sqrt{5}}{5}\right\rangle$

45. First compute the length of the vector: $|\langle 6,-3\rangle|=\sqrt{6^2+(-3)^2}=\sqrt{45}=3\sqrt{5}$

So a unit vector would be given by: $\frac{1}{3\sqrt{5}}\langle 6,-3\rangle=\left\langle \frac{2}{\sqrt{5}},-\frac{1}{\sqrt{5}}\right\rangle=\left\langle \frac{2\sqrt{5}}{5},-\frac{\sqrt{5}}{5}\right\rangle$

47. First compute the length of the vector: $|8\mathbf{i}-9\mathbf{j}|=\sqrt{8^2+(-9)^2}=\sqrt{145}$

So a unit vector would be given by: $\frac{1}{\sqrt{145}}(8\mathbf{i}-9\mathbf{j})=\frac{8}{\sqrt{145}}\mathbf{i}-\frac{9}{\sqrt{145}}\mathbf{j}=\frac{8\sqrt{145}}{145}\mathbf{i}-\frac{9\sqrt{145}}{145}\mathbf{j}$

49. Compute the components u_1 and u_2: $u_1=\cos\frac{\pi}{6}=\frac{\sqrt{3}}{2}$ $\qquad u_2=\sin\frac{\pi}{6}=\frac{1}{2}$

51. Compute the components u_1 and u_2: $u_1=\cos\frac{2\pi}{3}=-\frac{1}{2}$ $\qquad u_2=\sin\frac{2\pi}{3}=\frac{\sqrt{3}}{2}$

53. Compute the components u_1 and u_2: $u_1=\cos\frac{5\pi}{6}=-\frac{\sqrt{3}}{2}$ $\qquad u_2=\sin\frac{5\pi}{6}=\frac{1}{2}$

55. Verify property 1:

$$\begin{aligned}\mathbf{u}+(\mathbf{v}+\mathbf{w})&=\langle u_1,u_2\rangle+(\langle v_1,v_2\rangle+\langle w_1,w_2\rangle)\\&=\langle u_1,u_2\rangle+\langle v_1+w_1,v_2+w_2\rangle\\&=\langle u_1+v_1+w_1,u_2+v_2+w_2\rangle\\&=\langle u_1+v_1,u_2+v_2\rangle+\langle w_1,w_2\rangle\\&=(\langle u_1,u_2\rangle+\langle v_1,v_2\rangle)+\langle w_1,w_2\rangle\\&=(\mathbf{u}+\mathbf{v})+\mathbf{w}\end{aligned}$$

Verify property 2:

$$\mathbf{0}+\mathbf{v}=\langle 0,0\rangle+\langle v_1,v_2\rangle=\langle 0+v_1,0+v_2\rangle=\langle v_1+0,v_2+0\rangle=\langle v_1,v_2\rangle+\langle 0,0\rangle=\mathbf{v}+\mathbf{0}$$
$$\mathbf{v}+\mathbf{0}=\langle v_1,v_2\rangle+\langle 0,0\rangle=\langle v_1+0,v_2+0\rangle=\langle v_1,v_2\rangle=\mathbf{v}$$

57. Verify property 5:

$$\begin{aligned} a(\mathbf{u}+\mathbf{v}) &= a(\langle u_1,u_2\rangle+\langle v_1,v_2\rangle) \\ &= a\langle u_1+v_1,u_2+v_2\rangle \\ &= \langle a(u_1+v_1),a(u_2+v_2)\rangle \\ &= \langle au_1+av_1,au_2+av_2\rangle \\ &= \langle au_1,au_2\rangle+\langle av_1,av_2\rangle \\ &= a\langle u_1,u_2\rangle+a\langle v_1,v_2\rangle \\ &= a\mathbf{u}+a\mathbf{v} \end{aligned}$$

Verify property 6:

$$\begin{aligned} (a+b)\mathbf{v} &= (a+b)\langle v_1,v_2\rangle \\ &= \langle (a+b)v_1,(a+b)v_2\rangle \\ &= \langle av_1+bv_1,av_2+bv_2\rangle \\ &= \langle av_1,av_2\rangle+\langle bv_1,bv_2\rangle \\ &= a\langle v_1,v_2\rangle+b\langle v_1,v_2\rangle \\ &= a\mathbf{v}+b\mathbf{v} \end{aligned}$$

59. Verify the property: $\mathbf{v}+(\mathbf{u}-\mathbf{v})=\mathbf{v}+[\mathbf{u}+(-\mathbf{v})]=\mathbf{v}+[(-\mathbf{v})+\mathbf{u}]=[\mathbf{v}+(-\mathbf{v})]+\mathbf{u}=\mathbf{0}+\mathbf{u}=\mathbf{u}$

61. **a.** Compute the dot products:

$$\mathbf{u}\bullet\mathbf{v}=\langle -4,5\rangle\bullet\langle 3,4\rangle=(-4)(3)+(5)(4)=-12+20=8$$
$$\mathbf{v}\bullet\mathbf{u}=\langle 3,4\rangle\bullet\langle -4,5\rangle=(3)(-4)+(4)(5)=-12+20=8$$

b. Compute the dot products:

$$\mathbf{v}\bullet\mathbf{w}=\langle 3,4\rangle\bullet\langle 2,-5\rangle=(3)(2)+(4)(-5)=6-20=-14$$
$$\mathbf{w}\bullet\mathbf{v}=\langle 2,-5\rangle\bullet\langle 3,4\rangle=(2)(3)+(-5)(4)=6-20=-14$$

c. Let $\mathbf{A}=\langle x_1,y_1\rangle$ and $\mathbf{B}=\langle x_2,y_2\rangle$. Compute each dot product:

$$\mathbf{A}\bullet\mathbf{B}=\langle x_1,y_1\rangle\bullet\langle x_2,y_2\rangle=x_1x_2+y_1y_2$$
$$\mathbf{B}\bullet\mathbf{A}=\langle x_2,y_2\rangle\bullet\langle x_1,y_1\rangle=x_2x_1+y_2y_1=x_1x_2+y_1y_2$$

Thus $\mathbf{A}\bullet\mathbf{B}=\mathbf{B}\bullet\mathbf{A}$.

63. **a.** Compute each quantity:

$$\mathbf{v}\bullet\mathbf{v}=\langle 3,4\rangle\bullet\langle 3,4\rangle=(3)(3)+(4)(4)=9+16=25$$
$$|\mathbf{v}|=\sqrt{3^2+4^2}=\sqrt{9+16}=\sqrt{25}=5,\ \text{so}\ |\mathbf{v}|^2=25$$

b. Compute each quantity:

$$\mathbf{w}\bullet\mathbf{w}=\langle 2,-5\rangle\bullet\langle 2,-5\rangle=(2)(2)+(-5)(-5)=4+25=29$$
$$|\mathbf{w}|=\sqrt{2^2+(-5)^2}=\sqrt{4+25}=\sqrt{29},\ \text{so}\ |\mathbf{w}|^2=29$$

65. Sketching the vectors:

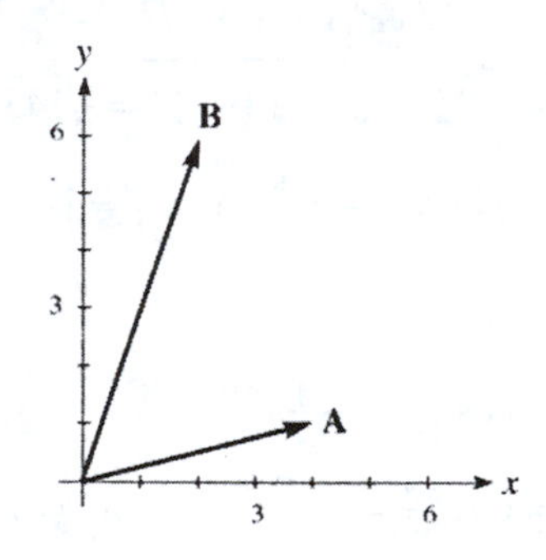

First do the computations:

$$|\mathbf{A}|=\sqrt{16+1}=\sqrt{17} \qquad |\mathbf{B}|=\sqrt{4+36}=\sqrt{40}=2\sqrt{10} \qquad \mathbf{A}\bullet\mathbf{B}=8+6=14$$

Thus: $\cos\theta=\dfrac{14}{\sqrt{17}\bullet 2\sqrt{10}}=\dfrac{7}{\sqrt{170}}$. So $\theta\approx 57.53^\circ$ or $\theta\approx 1.00$ radian.

67. Sketching the vectors:

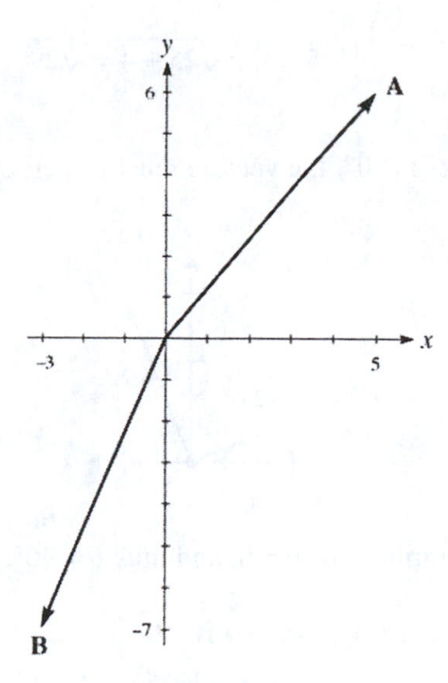

First do the computations:

$|\mathbf{A}| = \sqrt{25+36} = \sqrt{61}$ $\qquad |\mathbf{B}| = \sqrt{9+49} = \sqrt{58}$ $\qquad \mathbf{A} \cdot \mathbf{B} = -15-42 = -57$

Thus: $\cos\theta = \dfrac{-57}{\sqrt{61} \cdot \sqrt{58}} = \dfrac{-57}{\sqrt{3538}}$

So $\theta \approx 163.39°$ or $\theta \approx 2.85$ radians.

69. **a.** Sketching the vectors:

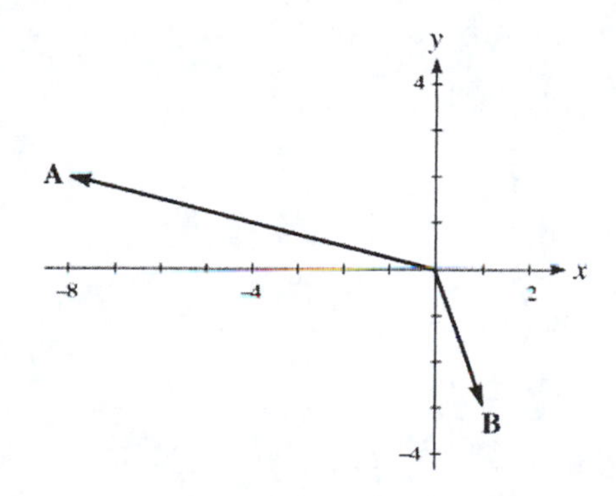

First do the computations:

$|\mathbf{A}| = \sqrt{64+4} = \sqrt{68} = 2\sqrt{17}$ $\qquad |\mathbf{B}| = \sqrt{1+9} = \sqrt{10}$ $\qquad \mathbf{A} \cdot \mathbf{B} = -8-6 = -14$

Thus: $\cos\theta = \dfrac{-14}{2\sqrt{17} \cdot \sqrt{10}} = \dfrac{-7}{\sqrt{170}}$. So $\theta \approx 122.47°$ or $\theta \approx 2.14$ radians.

b. Sketching the vectors:

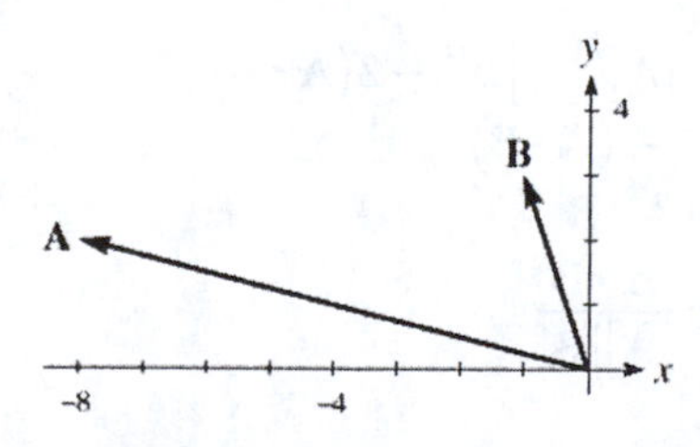

Again, do the computations:

$|\mathbf{A}| = \sqrt{64+4} = \sqrt{68} = 2\sqrt{17}$ $\qquad |\mathbf{B}| = \sqrt{1+9} = \sqrt{10}$ $\qquad \mathbf{A} \cdot \mathbf{B} = 8+6 = 14$

Thus: $\cos\theta = \dfrac{14}{2\sqrt{17} \cdot \sqrt{10}} = \dfrac{7}{\sqrt{170}}$. So $\theta \approx 57.53°$ or $\theta \approx 1.00$ radian.

71. **a.** First compute:

$|\langle 2,5\rangle| = \sqrt{4+25} = \sqrt{29}$ $\qquad |\langle -5,2\rangle| = \sqrt{25+4} = \sqrt{29}$ $\qquad \langle 2,5\rangle \bullet \langle -5,2\rangle = -10+10 = 0$

So $\cos\theta = \frac{0}{29} = 0$.

b. Since the angle between the vectors is 90°, the vectors must be perpendicular.

c. Draw the sketch:

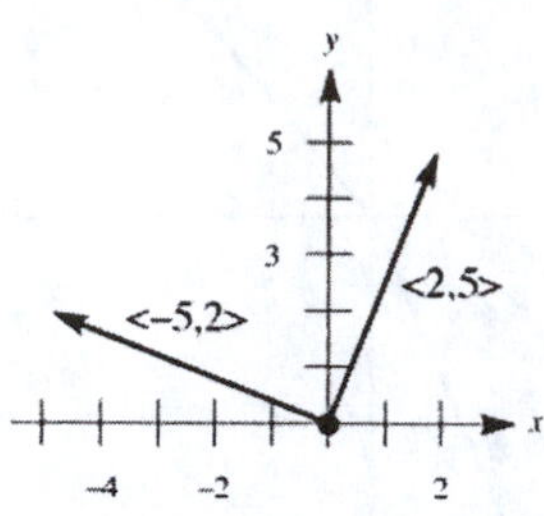

73. **a.** Since $\cos\theta = \dfrac{\mathbf{A} \bullet \mathbf{B}}{|\mathbf{A}||\mathbf{B}|}$, $\mathbf{A} \bullet \mathbf{B} = 0$ implies $\cos\theta = 0$, and thus $\theta = 90°$. So the vectors are perpendicular.

b. If **A** and **B** are perpendicular, then $\cos\theta = 0$, so $\mathbf{A} \bullet \mathbf{B} = 0$.

75. Call such a vector $\langle x,y\rangle$, so $x^2 + y^2 = 1$. Now $\langle x,y\rangle \bullet \langle -12,5\rangle = 0$, so:

$$\begin{aligned} -12x + 5y &= 0 \\ 5y &= 12x \\ y &= \tfrac{12}{5}x \end{aligned}$$

Substituting into $x^2 + y^2 = 1$:

$$\begin{aligned} x^2 + \tfrac{144}{25}x^2 &= 1 \\ \tfrac{169}{25}x^2 &= 1 \\ x^2 &= \tfrac{25}{169} \\ x &= \pm\tfrac{5}{13} \\ y &= \pm\tfrac{12}{13} \end{aligned}$$

So the two unit vectors are $\left\langle \frac{5}{13}, \frac{12}{13} \right\rangle$ and $\left\langle -\frac{5}{13}, -\frac{12}{13} \right\rangle$.

77. **a.** Using a vector argument:

$$\begin{aligned} |\mathbf{C}|^2 &= |\mathbf{A} - \mathbf{B}|^2 \\ &= (\mathbf{A} - \mathbf{B}) \bullet (\mathbf{A} - \mathbf{B}) \\ &= (\mathbf{A} - \mathbf{B}) \bullet \mathbf{A} - (\mathbf{A} - \mathbf{B}) \bullet \mathbf{B} \\ &= \mathbf{A} \bullet \mathbf{A} - \mathbf{B} \bullet \mathbf{A} - \mathbf{A} \bullet \mathbf{B} + \mathbf{B} \bullet \mathbf{B} \\ &= |\mathbf{A}|^2 - 2\mathbf{A} \bullet \mathbf{B} + |\mathbf{B}|^2 \end{aligned}$$

b. Using the suggestion given:

$$\begin{aligned} |\mathbf{A}|^2 + |\mathbf{B}|^2 - 2|\mathbf{A}||\mathbf{B}|\cos\theta &= |\mathbf{A}|^2 + |\mathbf{B}|^2 - 2(\mathbf{A} \bullet \mathbf{B}) \\ -2|\mathbf{A}||\mathbf{B}|\cos\theta &= -2(\mathbf{A} \bullet \mathbf{B}) \\ |\mathbf{A}||\mathbf{B}|\cos\theta &= \mathbf{A} \bullet \mathbf{B} \\ \cos\theta &= \frac{\mathbf{A} \bullet \mathbf{B}}{|\mathbf{A}||\mathbf{B}|} \end{aligned}$$

9.5 Parametric Equations

1. Find the x- and y-coordinates corresponding to $t = 0$:

$x = 2 - 4(0) = 2$ $\qquad$ $y = 3 - 5(0) = 3$

The point corresponding to $t = 0$ is $(2, 3)$.

3. Find the x- and y-coordinates corresponding to $t = \dfrac{\pi}{6}$:

$x = 5\cos\frac{\pi}{6} = 5 \bullet \dfrac{\sqrt{3}}{2} = \dfrac{5\sqrt{3}}{2}$ $\qquad$ $y = 2\sin\frac{\pi}{6} = 2 \bullet \frac{1}{2} = 1$

The point corresponding to $t = \dfrac{\pi}{6}$ is $\left(\dfrac{5\sqrt{3}}{2}, 1\right)$.

5. Find the x- and y-coordinates corresponding to $t = \dfrac{\pi}{4}$:

$x = 3\sin^3\frac{\pi}{4} = 3\left(\dfrac{\sqrt{2}}{2}\right)^3 = 3 \bullet \dfrac{\sqrt{2}}{4} = \dfrac{3\sqrt{2}}{4}$ $\qquad$ $y = 3\cos^3\frac{\pi}{4} = 3\left(\dfrac{\sqrt{2}}{2}\right)^3 = 3 \bullet \dfrac{\sqrt{2}}{4} = \dfrac{3\sqrt{2}}{4}$

The point corresponding to $t = \dfrac{\pi}{4}$ is $\left(\dfrac{3\sqrt{2}}{4}, \dfrac{3\sqrt{2}}{4}\right)$.

7. **a.** Graphing the equations for $0 \le t \le 1$:

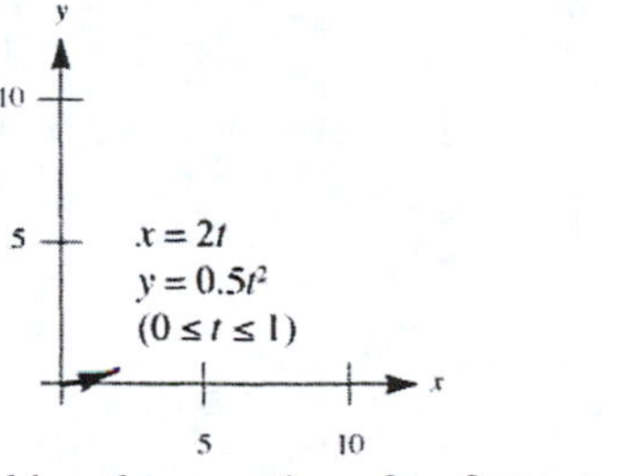

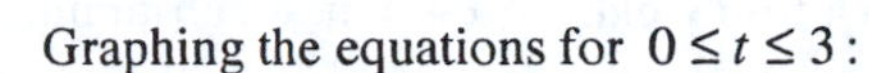
Graphing the equations for $0 \le t \le 3$:

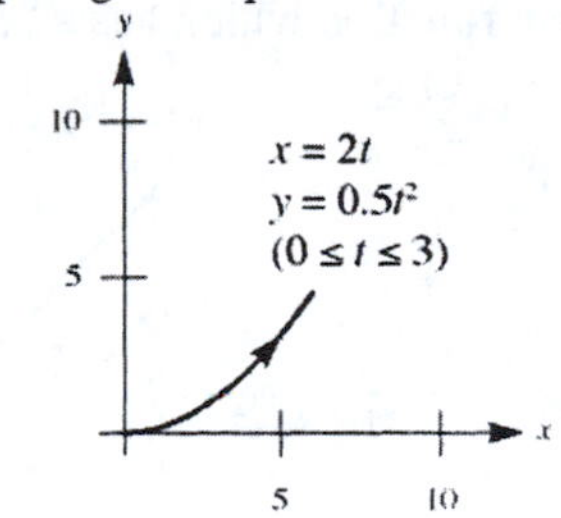

Graphing the equations for $0 \le t \le 4$:

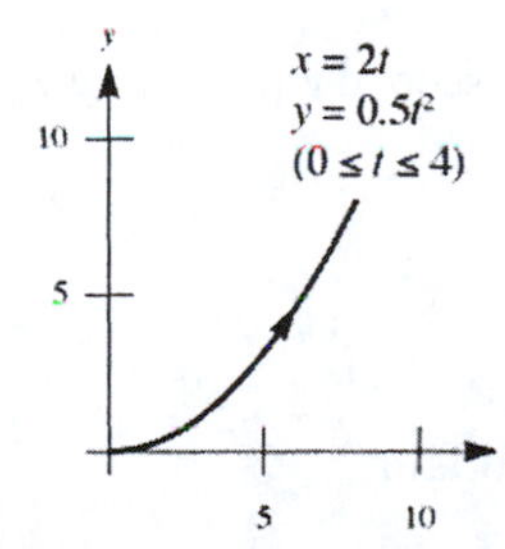

As the interval for t gets larger, the curve resembles a parabola.

b. Graphing the equations for $-5 \le t \le 5$:

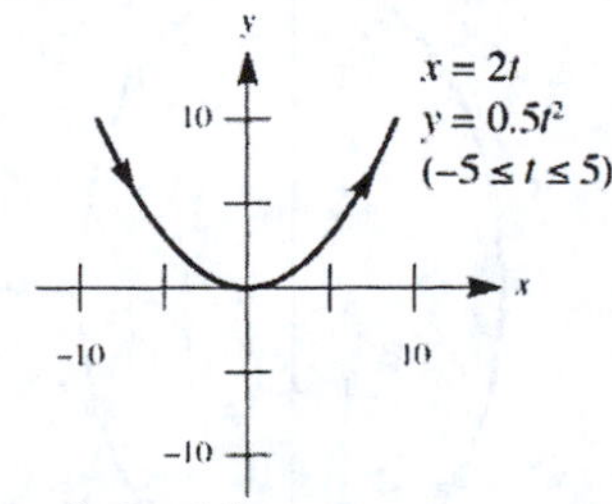

The restrictions on t in Figure 1(b) are $0 \le t \le 5$.

9. Solving $x = 2t - 1$ for t yields $t = \dfrac{x+1}{2}$, now substituting:

$$y = \left(\frac{x+1}{2}\right)^2 - 1 = \tfrac{1}{4}(x+1)^2 - 1$$
$$4(y+1) = (x+1)^2$$

Graph the parabola which has a vertex at (–1,–1):

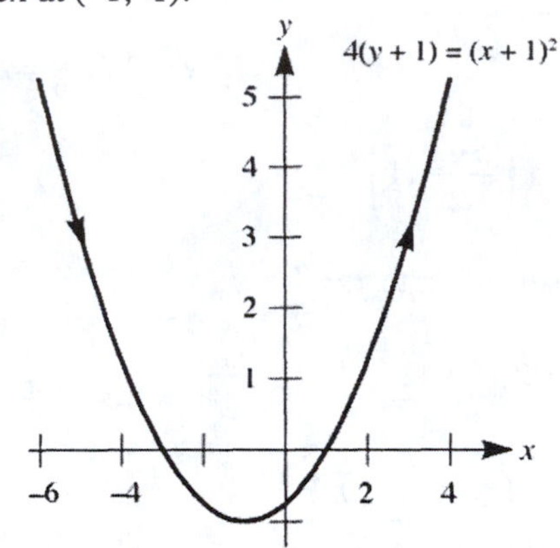

11. Solving $x = t - 4$ for t yields $t = x + 4$, now substituting: $y = |x+4|$

Graph the absolute value function which has a "corner" point at (–4,0):

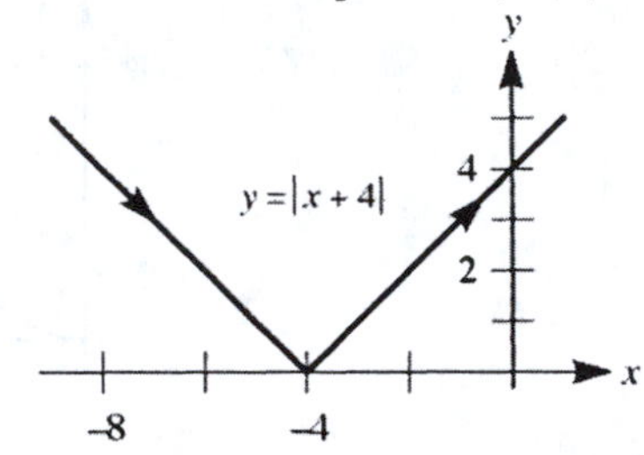

13. Multiplying the first equation by 3 and the second equation by 2 yields $3x = 6 \sin t$ and $2y = 6 \cos t$, so:

$$(3x)^2 + (2y)^2 = 36\sin^2 t + 36\cos^2 t$$
$$9x^2 + 4y^2 = 36$$
$$\frac{x^2}{4} + \frac{y^2}{9} = 1$$

Graph the ellipse which is centered at the origin:

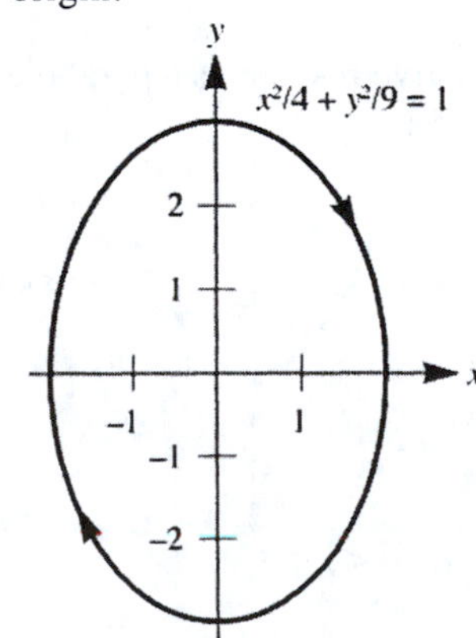

15. Multiplying the second equation by 2 yields $x = 2\cos\frac{t}{2}$ and $2y = 2\sin\frac{t}{2}$, so:

$$x^2 + (2y)^2 = 4\cos^2\tfrac{t}{2} + 4\sin^2\tfrac{t}{2}$$
$$x^2 + 4y^2 = 4$$
$$\frac{x^2}{4} + \frac{y^2}{1} = 1$$

Graph the ellipse which is centered at the origin:

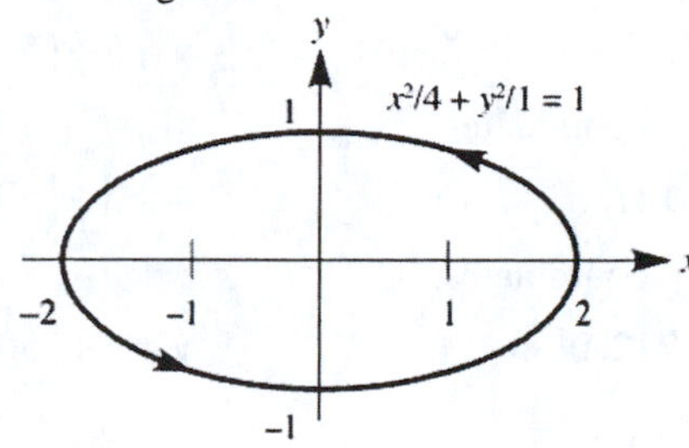

17. **a.** Compute:

$$x^2 + y^2 = 9\sin^2 t + 9\cos^2 t$$
$$x^2 + y^2 = 9$$

Graph the circle which is centered at the origin:

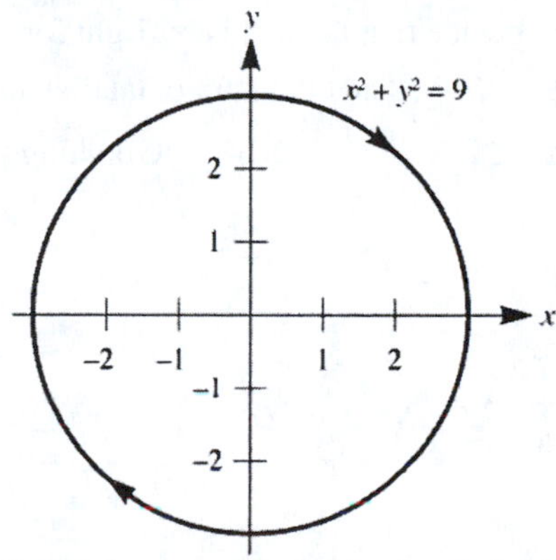

b. Multiplying the first equation by 3 and the second equation by 5 yields $3x = 15 \sin t$ and $5y = 15 \cos t$, so:

$$(3x)^2 + (5y)^2 = 225\sin^2 t + 225\cos^2 t$$
$$9x^2 + 25y^2 = 225$$
$$\frac{x^2}{25} + \frac{y^2}{9} = 1$$

Graph the ellipse which is centered at the origin:

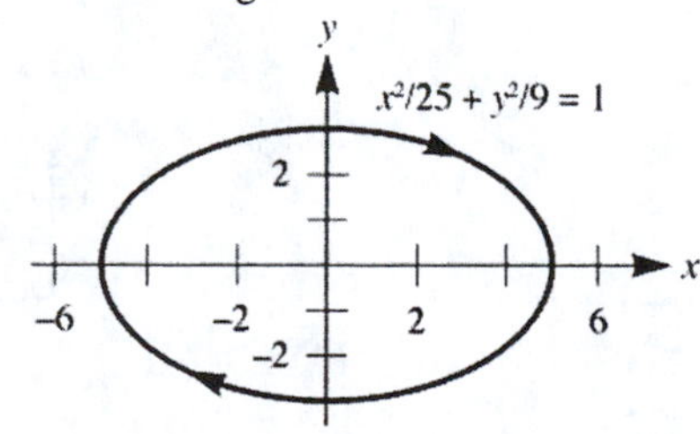

19. Solve for t:

$$6+(88\sin 35°)t-16t^2=0$$
$$16t^2-(88\sin 35°)-6=0$$

Using the quadratic formula: $t=\dfrac{88\sin 35° \pm \sqrt{7744\sin^2 35°+384}}{32} \approx 3.27,\,-0.11$. The solutions are verified.

21. **a.** When $t = 1$ compute the x- and y-coordinates:

$$x=\left(50\sqrt{2}\right)\bullet 1=50\sqrt{2}\approx 70.7 \qquad y=5+\left(50\sqrt{2}\right)\bullet 1-16(1)^2=50\sqrt{2}-11\approx 59.7$$

When $t = 2$ compute the x- and y-coordinates:

$$x=\left(50\sqrt{2}\right)\bullet 2=100\sqrt{2}\approx 141.4 \qquad y=5+\left(50\sqrt{2}\right)\bullet 2-16(2)^2=100\sqrt{2}-59\approx 82.4$$

When $t = 3$ compute the x- and y-coordinates:

$$x=\left(50\sqrt{2}\right)\bullet 3=150\sqrt{2}\approx 212.1 \qquad y=5+\left(50\sqrt{2}\right)\bullet 3-16(3)^2=150\sqrt{2}-139\approx 73.1$$

b. Find the value of t when $y = 0$:

$$5+\left(50\sqrt{2}\right)t-16t^2=0$$
$$16t^2-\left(50\sqrt{2}\right)t-5=0$$

Using the quadratic formula: $t=\dfrac{50\sqrt{2}\pm\sqrt{5000+320}}{32}=\dfrac{50\sqrt{2}\pm\sqrt{5320}}{32}\approx 4.49,\,-0.07$

Discard the negative value and conclude that the ball is in flight for approximately 4.49 seconds.

When $t \approx 4.49$: $x\approx\left(50\sqrt{2}\right)(4.49)\approx 317$. The total horizontal distance traveled is approximately 317 feet.

23. **a.** Graphing the equations for $-2 \le t \le 2$:

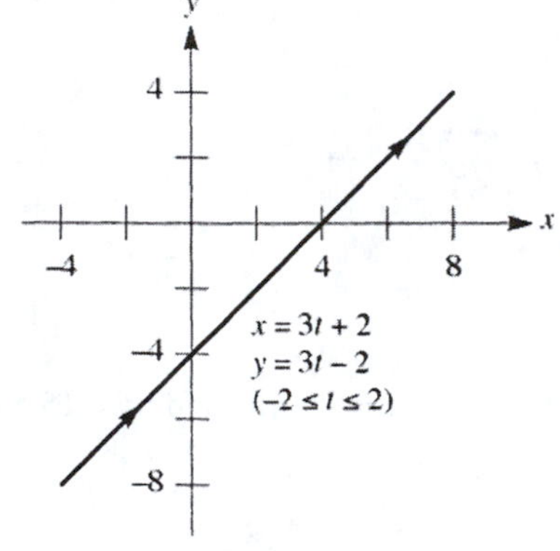

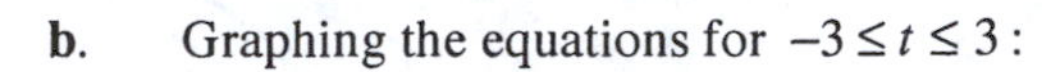

b. Graphing the equations for $-3 \le t \le 3$:

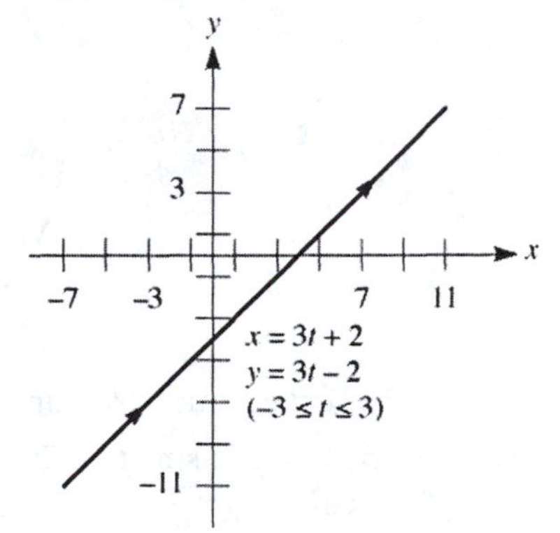

25. Graphing the equations for $0 \le t \le 2\pi$:

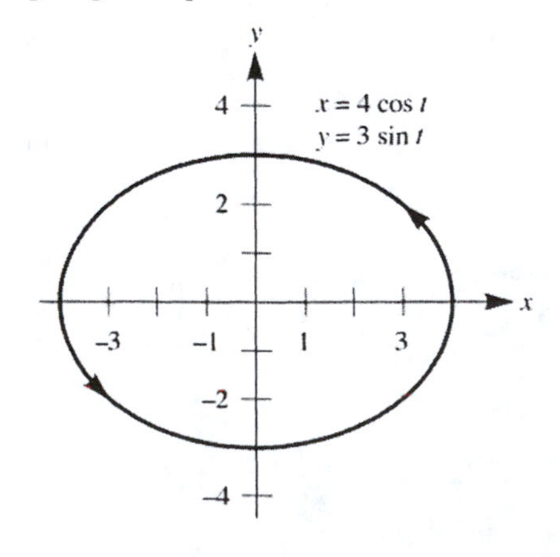

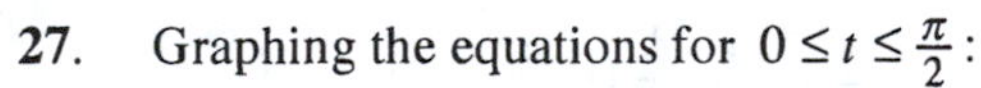

27. Graphing the equations for $0 \le t \le \frac{\pi}{2}$:

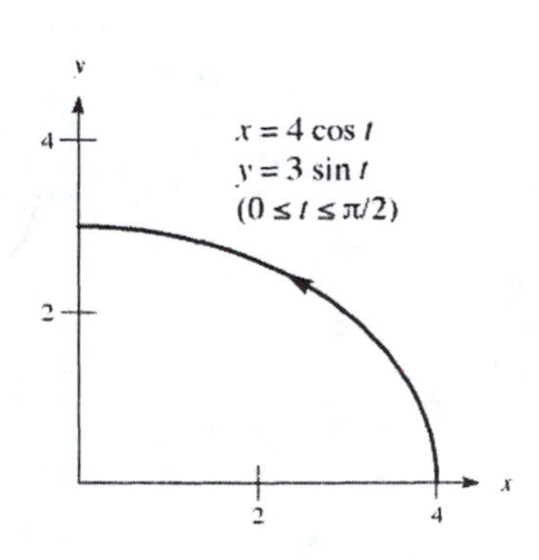

29. Graphing the equations for $0 \le t \le 2\pi$:

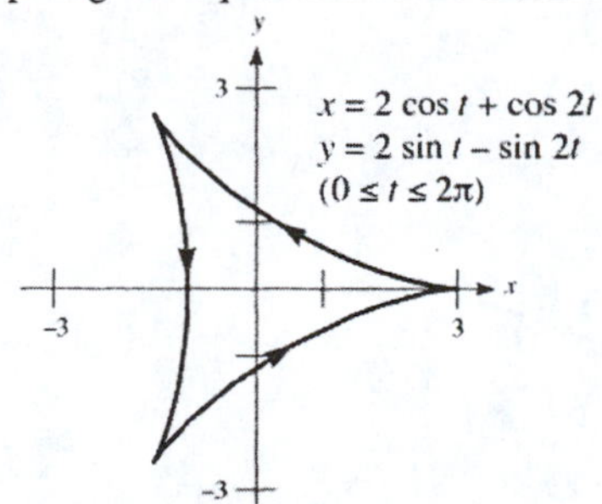

31. Graphing the equations for $0 \le t \le 2\pi$:

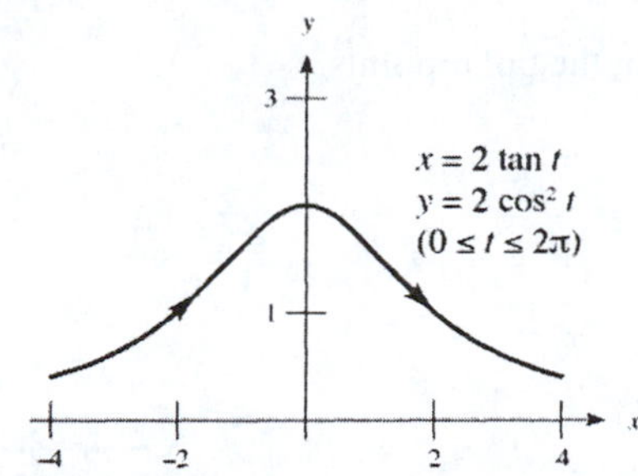

33. Graphing the equations for $-2 \le t \le 2$:

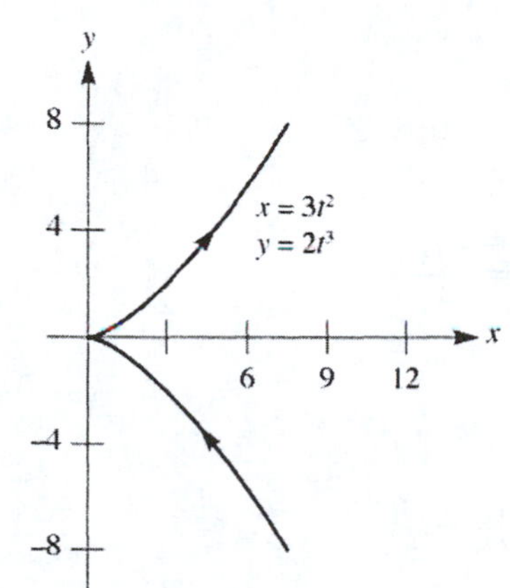

35. Graphing the equations for $-3 \le t \le 3$:

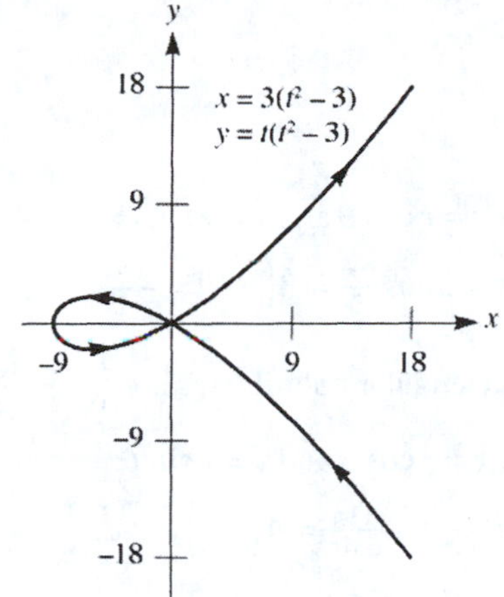

37. Graphing the equations for $0 \le t \le 2\pi$:

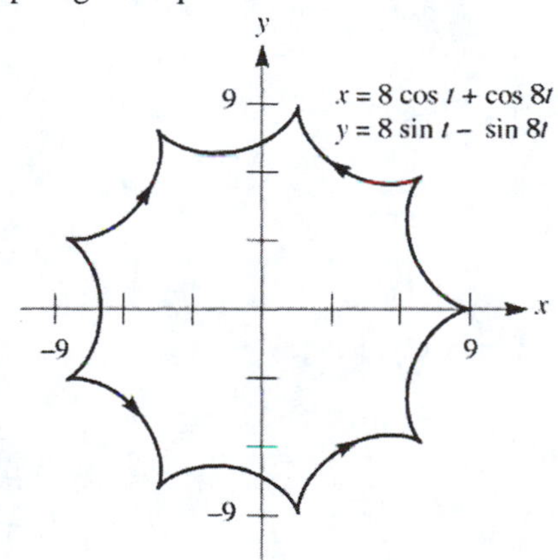

39. Graphing the equations for $0 \le t \le 10\pi$:

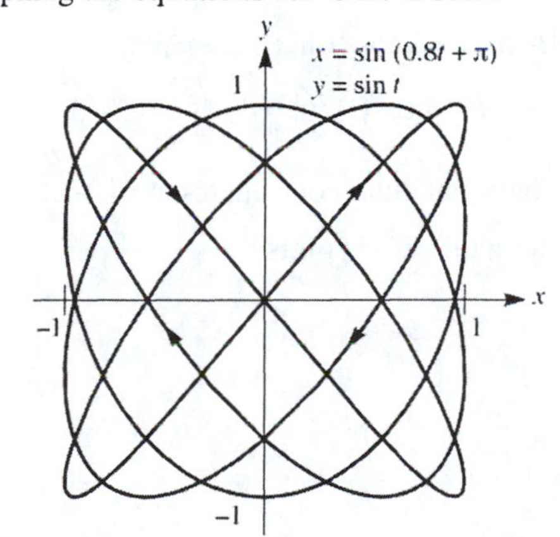

9.6 Introduction to Polar Coordinates

1. Sketching the polar points:

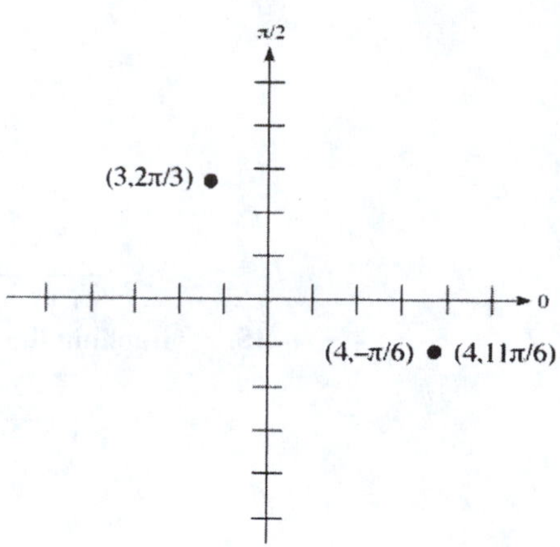

a. Using $x = r\cos\theta$ and $y = r\sin\theta$:

$$x = 3\cos\frac{2\pi}{3} = 3\left(-\frac{1}{2}\right) = -\frac{3}{2} \qquad y = 3\sin\frac{2\pi}{3} = 3\left(\frac{\sqrt{3}}{2}\right) = \frac{3\sqrt{3}}{2}$$

The rectangular coordinates are $\left(-\frac{3}{2}, \frac{3\sqrt{3}}{2}\right)$.

b. Using $x = r\cos\theta$ and $y = r\sin\theta$:

$$x = 4\cos\frac{11\pi}{6} = 4\left(\frac{\sqrt{3}}{2}\right) = 2\sqrt{3} \qquad y = 4\sin\frac{11\pi}{6} = 4\left(-\frac{1}{2}\right) = -2$$

The rectangular coordinates are $\left(2\sqrt{3}, -2\right)$.

c. Using $x = r\cos\theta$ and $y = r\sin\theta$:

$$x = 4\cos\left(-\frac{\pi}{6}\right) = 4\left(\frac{\sqrt{3}}{2}\right) = 2\sqrt{3} \qquad y = 4\sin\left(-\frac{\pi}{6}\right) = 4\left(-\frac{1}{2}\right) = -2$$

The rectangular coordinates are $\left(2\sqrt{3}, -2\right)$.

3. Sketching the polar points:

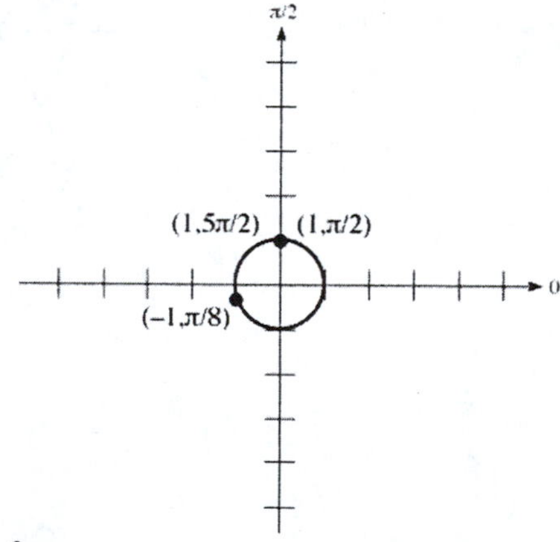

a. Using $x = r\cos\theta$ and $y = r\sin\theta$:

$$x = 1\cos\frac{\pi}{2} = 1(0) = 0 \qquad y = 1\sin\frac{\pi}{2} = 1(1) = 1$$

The rectangular coordinates are (0,1).

b. Using $x = r\cos\theta$ and $y = r\sin\theta$:

$$x = 1\cos\frac{5\pi}{2} = 1(0) = 0 \qquad y = 1\sin\frac{5\pi}{2} = 1(1) = 1$$

The rectangular coordinates are (0,1).

c. Using $x = r\cos\theta$ and $y = r\sin\theta$, and using the half-angle formulas for $\sin\theta$ and $\cos\theta$:

$$x = -1\cos\tfrac{\pi}{8} = -1\left(\sqrt{\frac{1+\frac{\sqrt{2}}{2}}{2}}\right) = -\sqrt{\frac{2+\sqrt{2}}{4}} = -\frac{\sqrt{2+\sqrt{2}}}{2}$$

$$y = -1\sin\tfrac{\pi}{8} = -1\left(\sqrt{\frac{1-\frac{\sqrt{2}}{2}}{2}}\right) = -\sqrt{\frac{2-\sqrt{2}}{4}} = -\frac{\sqrt{2-\sqrt{2}}}{2}$$

The rectangular coordinates are $\left(-\frac{\sqrt{2+\sqrt{2}}}{2}, -\frac{\sqrt{2-\sqrt{2}}}{2}\right)$.

5. Computing:

$r^2 = 1+1 = 2$, so $r = \sqrt{2}$

$\theta = \tan^{-1}\left(\frac{-1}{-1}\right) + \pi = \frac{\pi}{4} + \pi = \frac{5\pi}{4}$

The polar form is $\left(\sqrt{2}, \frac{5\pi}{4}\right)$.

7. If we first multiply by r:

$$\begin{aligned} r^2 &= 2r\cos\theta \\ x^2+y^2 &= 2x \\ x^2-2x+y^2 &= 0 \\ x^2-2x+1+y^2 &= 1 \\ (x-1)^2+y^2 &= 1 \end{aligned}$$

9. Substituting for r and $\tan\theta$:

$$\begin{aligned} \sqrt{x^2+y^2} &= \frac{y}{x} \\ x^2+y^2 &= \frac{y^2}{x^2} \\ x^4+x^2y^2 &= y^2 \\ x^4+x^2y^2-y^2 &= 0 \end{aligned}$$

11. Using the double-angle formula for $\cos 2\theta$: $r = 3\left(\cos^2\theta - \sin^2\theta\right)$. Multiplying by r^2:

$$\begin{aligned} r^3 &= 3\left(r^2\cos^2\theta - r^2\sin^2\theta\right) \\ \left(x^2+y^2\right)^{3/2} &= 3\left(x^2-y^2\right) \\ \left(x^2+y^2\right)^3 &= 9\left(x^2-y^2\right)^2 \end{aligned}$$

13. Multiplying each side by $2-\sin^2\theta$ yields:

$$\begin{aligned} 2r^2 - r^2\sin^2\theta &= 8 \\ 2\left(x^2+y^2\right) - y^2 &= 8 \\ 2x^2+2y^2-y^2 &= 8 \\ 2x^2+y^2 &= 8 \\ \frac{x^2}{4}+\frac{y^2}{8} &= 1 \end{aligned}$$

15. Using the addition formula for $\cos(s-t)$:

$$\begin{aligned} r\cos\left(\theta-\tfrac{\pi}{6}\right) &= 2 \\ r\left(\cos\theta\cos\tfrac{\pi}{6} + \sin\theta\sin\tfrac{\pi}{6}\right) &= 2 \\ \tfrac{\sqrt{3}}{2}r\cos\theta + \tfrac{1}{2}r\sin\theta &= 2 \\ \tfrac{\sqrt{3}}{2}x + \tfrac{1}{2}y &= 2 \\ \sqrt{3}x + y &= 4 \\ y &= -\sqrt{3}x+4 \end{aligned}$$

17. Substituting $x = r\cos\theta$ and $y = r\sin\theta$:

$$\begin{aligned} 3r\cos\theta - 4r\sin\theta &= 2 \\ r(3\cos\theta - 4\sin\theta) &= 2 \\ r &= \frac{2}{3\cos\theta - 4\sin\theta} \end{aligned}$$

19. Substituting $x = r\cos\theta$ and $y = r\sin\theta$:

$$\begin{aligned} r^2\sin^2\theta &= r^3\cos^3\theta \\ \sin^2\theta &= r\cos^3\theta \\ r &= \frac{\sin^2\theta}{\cos^3\theta} \\ r &= \tan^2\theta\sec\theta \end{aligned}$$

21. Substituting $x = r\cos\theta$ and $y = r\sin\theta$:

$$\begin{aligned} 2(r\cos\theta)(r\sin\theta) &= 1 \\ r^2(2\sin\theta\cos\theta) &= 1 \\ r^2\sin 2\theta &= 1 \\ r^2 &= \frac{1}{\sin 2\theta} \\ r^2 &= \csc 2\theta \end{aligned}$$

23. Substituting $x = r\cos\theta$ and $y = r\sin\theta$:

$$\begin{aligned} 9r^2\cos^2\theta + r^2\sin^2\theta &= 9 \\ r^2\left(9\cos^2\theta + \sin^2\theta\right) &= 9 \\ r^2 &= \frac{9}{9\cos^2\theta + \sin^2\theta} \end{aligned}$$

25. A: Since $\theta = \frac{\pi}{6}$: $r = \dfrac{4}{1+\sin\frac{\pi}{6}} = \dfrac{4}{1+\frac{1}{2}} = \dfrac{4}{\frac{3}{2}} = \frac{8}{3}$

B: Since $\theta = \frac{5\pi}{6}$: $r = \dfrac{4}{1+\sin\frac{5\pi}{6}} = \dfrac{4}{1+\frac{1}{2}} = \dfrac{4}{\frac{3}{2}} = \frac{8}{3}$

C: Since $\theta = \pi$: $r = \dfrac{4}{1+\sin\pi} = \dfrac{4}{1+0} = 4$

D: Since $\theta = \frac{7\pi}{6}$: $r = \dfrac{4}{1+\sin\frac{7\pi}{6}} = \dfrac{4}{1-\frac{1}{2}} = \dfrac{4}{\frac{1}{2}} = 8$

The coordinates of the points are $A\left(\frac{8}{3},\frac{\pi}{6}\right)$, $B\left(\frac{8}{3},\frac{5\pi}{6}\right)$, $C(4,\pi)$, and $D\left(8,\frac{7\pi}{6}\right)$.

27. A: Since $\theta = \frac{\pi}{6}$: $r = 2\cos\frac{\pi}{3} = 2\cdot\frac{1}{2} = 1$

B: Since $\theta = \frac{5\pi}{6}$: $r = 2\cos\frac{5\pi}{3} = 2\cdot\frac{1}{2} = 1$

C: Since $\theta = \frac{7\pi}{6}$: $r = 2\cos\frac{7\pi}{3} = 2\cdot\frac{1}{2} = 1$

D: Since $\theta = \frac{\pi}{2}$: $r = 2\cos\pi = 2\cdot(-1) = -2$

The coordinates of the points are $A\left(1,\frac{\pi}{6}\right)$, $B\left(1,\frac{5\pi}{6}\right)$, $C\left(1,\frac{7\pi}{6}\right)$, and $D\left(-2,\frac{\pi}{2}\right)$.

29. A: Since $\theta = 0$: $r = e^{0.6} = 1$

B: Since $\theta = \frac{\pi}{4}$: $r = e^{\pi/24} \approx 1.14$

C: Since $\theta = \frac{3\pi}{4}$: $r = e^{\pi/8} \approx 1.48$

D: Since $\theta = \pi$: $r = e^{\pi/6} \approx 1.69$

E: Since $\theta = \frac{5\pi}{4}$: $r = e^{5\pi/24} \approx 1.92$

F: Since $\theta = \frac{3\pi}{2}$: $r = e^{\pi/4} \approx 2.19$

G: Since $\theta = \frac{7\pi}{4}$: $r = e^{7\pi/24} \approx 2.50$

H: Since $\theta = 2\pi$: $r = e^{\pi/3} \approx 2.85$

I: Since $\theta = \frac{9\pi}{4}$: $r = e^{3\pi/8} \approx 3.25$

J: Since $\theta = \frac{11\pi}{4}$: $r = e^{11\pi/24} \approx 4.22$

K: Since $\theta = \frac{13\pi}{4}$: $r = e^{13\pi/24} \approx 5.48$

The coordinates of the points are $A(1,0)$, $B\left(1.14,\frac{\pi}{4}\right)$, $C\left(1.48,\frac{3\pi}{4}\right)$, $D(1.69,\pi)$, $E\left(1.92,\frac{5\pi}{4}\right)$, $F\left(2.19,\frac{3\pi}{2}\right)$, $G\left(2.50,\frac{7\pi}{4}\right)$, $H(2.85,2\pi)$, $I\left(3.25,\frac{9\pi}{4}\right)$, $J\left(4.22,\frac{11\pi}{4}\right)$, and $K\left(5.48,\frac{13\pi}{4}\right)$.

31. Use the distance formula, taking the points (r_1,θ_1) and (r_2,θ_2) to be $\left(2,\frac{2\pi}{3}\right)$ and $\left(4,\frac{\pi}{6}\right)$:

$$d^2 = r_1^2 + r_2^2 - 2r_1r_2\cos(\theta_2-\theta_1) = 2^2 + 4^2 - 2(2)(4)\cos\left(\tfrac{2\pi}{3}-\tfrac{\pi}{6}\right) = 20 - 16\cos\tfrac{\pi}{2} = 20 - 16\cdot 0 = 20$$

So $d = \sqrt{20} = 2\sqrt{5}$.

33. Use the distance formula, taking the points (r_1,θ_1) and (r_2,θ_2) to be $\left(4,\frac{4\pi}{3}\right)$ and $(1,0)$:

$$d^2 = r_1^2 + r_2^2 - 2r_1r_2\cos(\theta_2-\theta_1) = 4^2 + 1^2 - 2(4)(1)\cos\left(\tfrac{4\pi}{3}-0\right) = 17 - 8\cos\tfrac{4\pi}{3} = 17 - 8\cdot\left(-\tfrac{1}{2}\right) = 21$$

So $d = \sqrt{21}$.

35. **a.** Using the equation $r^2 + r_0^2 - 2rr_0\cos(\theta-\theta_0) = a^2$:

$$\begin{aligned} r^2 + 4^2 - 2(r)(4)\cos(\theta - 0) &= 2^2 \\ r^2 + 16 - 8r\cos\theta &= 4 \\ r^2 - 8r\cos\theta &= -12 \end{aligned}$$

b. Using the equation $r^2 + r_0^2 - 2rr_0 \cos(\theta - \theta_0) = a^2$:

$$r^2 + 4^2 - 2(r)(4)\cos\left(\theta - \tfrac{2\pi}{3}\right) = 2^2$$
$$r^2 + 16 - 8r\cos\left(\theta - \tfrac{2\pi}{3}\right) = 4$$
$$r^2 - 8r\cos\left(\theta - \tfrac{2\pi}{3}\right) = -12$$

c. Using the equation $r^2 + r_0^2 - 2rr_0 \cos(\theta - \theta_0) = a^2$:

$$r^2 + 0^2 - 2(r)(0)\cos(\theta - 0) = 2^2$$
$$r^2 = 4$$
$$r = 2$$

37. **a.** Using the equation $r^2 + r_0^2 - 2rr_0 \cos(\theta - \theta_0) = a^2$:

$$r^2 + 1^2 - 2(r)(1)\cos\left(\theta - \tfrac{3\pi}{2}\right) = 1^2$$
$$r^2 + 1 - 2r\cos\left(\theta - \tfrac{3\pi}{2}\right) = 1$$
$$r^2 = 2r\cos\left(\theta - \tfrac{3\pi}{2}\right)$$
$$r = 2\cos\left(\theta - \tfrac{3\pi}{2}\right)$$

b. Using the equation $r^2 + r_0^2 - 2rr_0 \cos(\theta - \theta_0) = a^2$:

$$r^2 + 1^2 - 2(r)(1)\cos\left(\theta - \tfrac{\pi}{4}\right) = 1^2$$
$$r^2 + 1 - 2r\cos\left(\theta - \tfrac{\pi}{4}\right) = 1$$
$$r^2 = 2r\cos\left(\theta - \tfrac{\pi}{4}\right)$$
$$r = 2\cos\left(\theta - \tfrac{\pi}{4}\right)$$

39. **a.** Since the line is of the form $r\cos(\theta - \alpha) = d$, $\alpha = \frac{\pi}{6}$ and $d = 2$. So the desired perpendicular distance is 2.

b. When $\theta = 0$:

$$r\cos\left(0 - \tfrac{\pi}{6}\right) = 2$$
$$r\left(\tfrac{\sqrt{3}}{2}\right) = 2$$
$$r = \frac{4}{\sqrt{3}} = \frac{4\sqrt{3}}{3}$$

When $\theta = \frac{\pi}{2}$:

$$r\cos\left(\tfrac{\pi}{2} - \tfrac{\pi}{6}\right) = 2$$
$$r \bullet \tfrac{1}{2} = 2$$
$$r = 4$$

The required polar points are $\left(\frac{4\sqrt{3}}{3}, 0\right)$ and $\left(4, \frac{\pi}{2}\right)$.

c. These coordinates are (d, α), which is the polar point $\left(2, \frac{\pi}{6}\right)$.

d. Sketch the line:

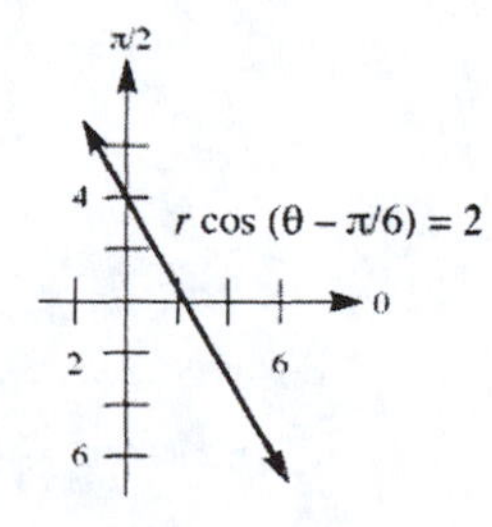

41. **a.** Since the line is of the form $r\cos(\theta-\alpha)=d$, $\alpha=-\frac{2\pi}{3}$ and $d=4$. So the desired perpendicular distance is 4.

b. When $\theta=0$:

$$r\cos\left(0+\tfrac{2\pi}{3}\right)=4$$
$$r\left(-\tfrac{1}{2}\right)=4$$
$$r=-8$$

When $\theta=\frac{\pi}{2}$:

$$r\cos\left(\tfrac{\pi}{2}+\tfrac{2\pi}{3}\right)=4$$
$$r\cos\tfrac{7\pi}{6}=4$$
$$r\bullet\left(-\tfrac{\sqrt{3}}{2}\right)=4$$
$$r=-\frac{8}{\sqrt{3}}=-\frac{8\sqrt{3}}{3}$$

The required polar points are $(-8,0)$ and $\left(-\frac{8}{3}\sqrt{3},\frac{\pi}{2}\right)$.

c. These coordinates are (d, α), which is the polar point $\left(4,-\frac{2\pi}{3}\right)$.

d. Sketch the line:

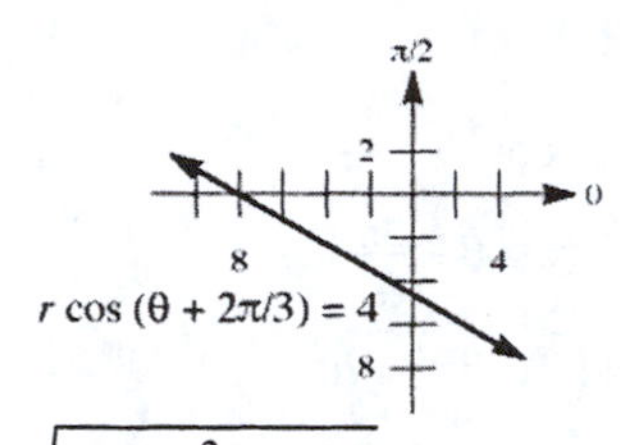

43. **a.** Since $x^2+y^2=r^2$: $r=\sqrt{x^2+y^2}=\sqrt{\left(-\sqrt{3}\right)^2+(1)^2}=\sqrt{4}=2$

Now using $x=r\cos\theta$ and $y=r\sin\theta$:

$2\cos\theta=-\sqrt{3}$, so $\cos\theta=-\frac{\sqrt{3}}{2}$ $\qquad$ $2\sin\theta=1$, so $\sin\theta=\frac{1}{2}$

Therefore a value of θ is $\frac{5\pi}{6}$. The polar coordinates of P are $\left(2,\frac{5\pi}{6}\right)$.

b. In the general equation $r\cos(\theta-\alpha)=d$, use the values $\alpha=\frac{5\pi}{6}$ and $d=2$ to obtain: $r\cos\left(\theta-\frac{5\pi}{6}\right)=2$

This is the polar equation for the tangent line.

c. The x-axis corresponds to $\theta=0$, therefore:

$$r\cos\left(0-\tfrac{5\pi}{6}\right)=2$$
$$r\left(-\tfrac{\sqrt{3}}{2}\right)=2$$
$$r=-\frac{4}{\sqrt{3}}=-\frac{4\sqrt{3}}{3}$$

So the x-intercept of the line is $-\frac{4\sqrt{3}}{3}$. The y-axis corresponds to $\theta=\frac{\pi}{2}$, so:

$$r\cos\left(\tfrac{\pi}{2}-\tfrac{5\pi}{6}\right)=2$$
$$r\cos\left(-\tfrac{\pi}{3}\right)=2$$
$$r\bullet\tfrac{1}{2}=2$$
$$r=4$$

So the y-intercept of the line is 4.

45. **a.** Since $r \cos \theta = x$, this is the graph of $x = 3$:

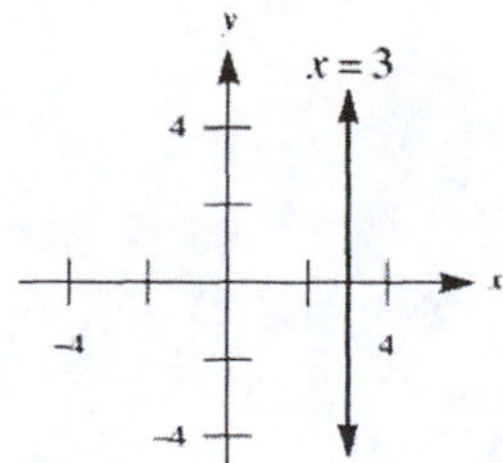

b. Since $r \sin \theta = y$, this is the graph of $y = 3$:

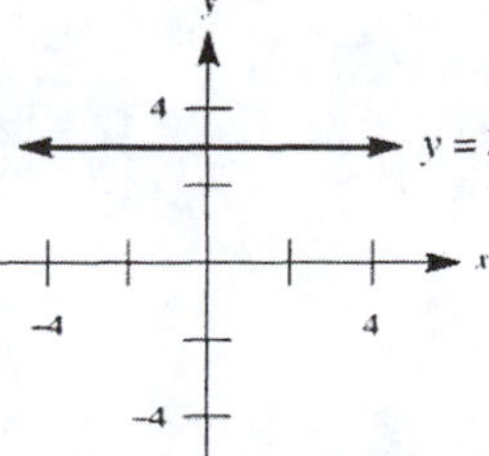

47. **a.** Since d is the perpendicular distance from the origin to L, and (d, α) is the foot of the perpendicular on L, the polar equation for L must be $r\cos(\theta - \alpha) = d$.

b. Using the formula $\cos(s - t) = \cos s \cos t + \sin s \sin t$:

$$\begin{aligned} r\cos(\theta - \alpha) &= d \\ r(\cos\theta\cos\alpha + \sin\theta\sin\alpha) &= d \\ (r\cos\theta)\cos\alpha + (r\sin\theta)\sin\alpha &= d \\ x\cos\alpha + y\sin\alpha &= d \end{aligned}$$

49. First establish an identity for $\sin 3\theta$:

$$\begin{aligned} \sin 3\theta &= \sin(2\theta + \theta) \\ &= \sin 2\theta\cos\theta + \cos 2\theta\sin\theta \\ &= (2\sin\theta\cos\theta)\cos\theta + \left(\cos^2\theta - \sin^2\theta\right)\sin\theta \\ &= 3\sin\theta\cos^2\theta - \sin^3\theta \end{aligned}$$

Now multiply the equation $r = a\sin 3\theta$ by r^3 and substitute:

$$\begin{aligned} r^4 &= a\left((3r\sin\theta)\left(r^2\cos^2\theta\right) - r^3\sin^3\theta\right) \\ \left(x^2 + y^2\right)^2 &= a\left(3yx^2 - y^3\right) \\ \left(x^2 + y^2\right)^2 &= ay\left(3x^2 - y^2\right) \end{aligned}$$

51. Substituting $x = r\cos\theta$ and $y = r\sin\theta$ results in the equation:

$$\frac{r^2\cos^2\theta}{a^2} - \frac{r^2\sin^2\theta}{b^2} = 1$$

$$\frac{r^2\left(b^2\cos^2\theta - a^2\sin^2\theta\right)}{a^2b^2} = 1$$

$$r^2 = \frac{a^2b^2}{b^2\cos^2\theta - a^2\sin^2\theta}$$

53. **a.** A is the point $(1, 0)$, B is the point $\left(1, \frac{2\pi}{3}\right)$, and C is the point $\left(1, \frac{4\pi}{3}\right)$.

b. Use the distance formula $d^2 = r_1^2 + r_2^2 - 2r_1r_2\cos\left(\theta_2 - \theta_1\right)$ to find each distance:

$$\begin{aligned} (PA)^2 &= r^2 + 1^2 - 2(r)(1)\cos(0 - 0) = r^2 + 1 - 2r \\ (PB)^2 &= r^2 + 1^2 - 2(r)(1)\cos\left(0 - \tfrac{2\pi}{3}\right) = r^2 + 1 + r \\ (PC)^2 &= r^2 + 1^2 - 2(r)(1)\cos\left(0 - \tfrac{4\pi}{3}\right) = r^2 + 1 + r \end{aligned}$$

Therefore: $(PA)^2(PB)^2(PC)^2 = \left(r^2 - 2r + 1\right)\left(r^2 + r + 1\right)^2 = (r-1)^2\left(r^2 + r + 1\right)^2 = \left(r^3 - 1\right)^2$

So: $(PA)(PB)(PC) = \sqrt{\left(r^3 - 1\right)^2} = 1 - r^3$, since $r < 1$

9.7 Curves in Polar Coordinates

1. Graphing the equations:

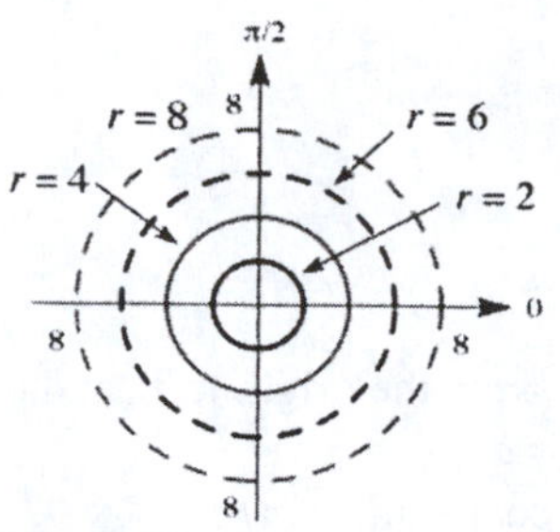

Notice the graphs are circles with radii 2, 4, 6, and 8.

3. Sketch the polar curve $r = \dfrac{\theta}{2\pi}$:

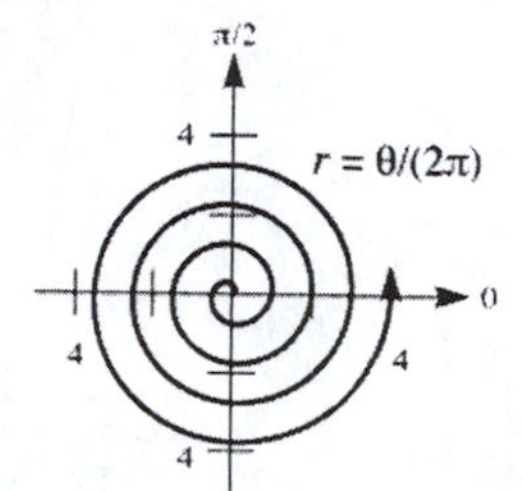

5. Sketch the polar curve $r = \ln\theta$:

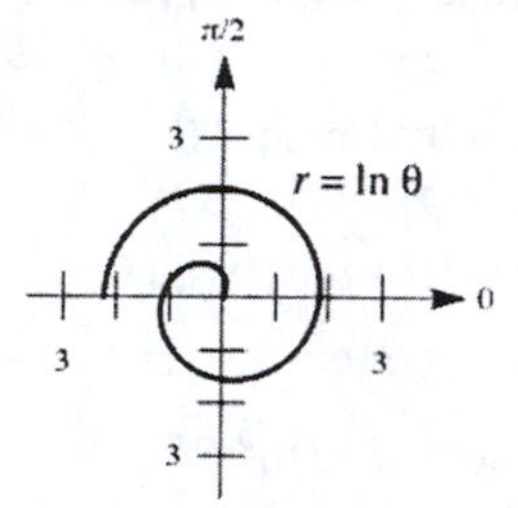

7. Sketch the polar curve $r = \theta$:

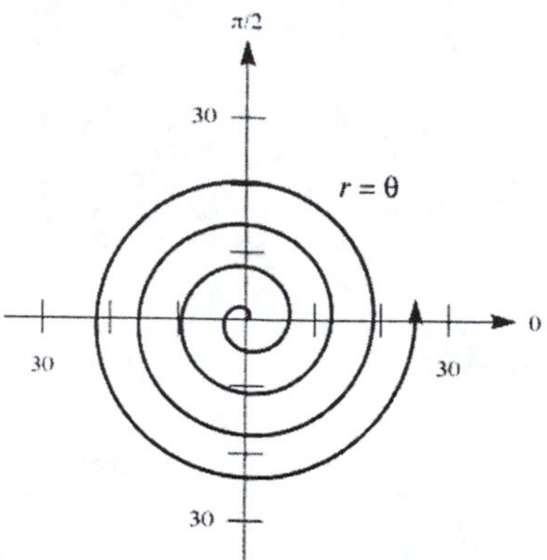

9. Sketch the polar curve $r = \dfrac{1}{\sqrt{\theta}}$:

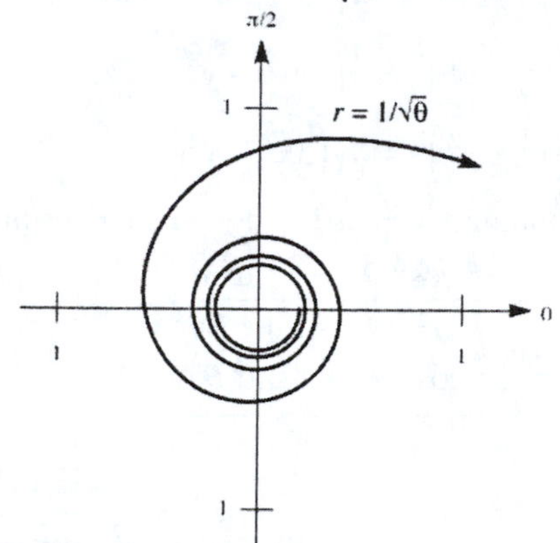

11. Sketch the polar curve $r = 1 + \cos\theta$:

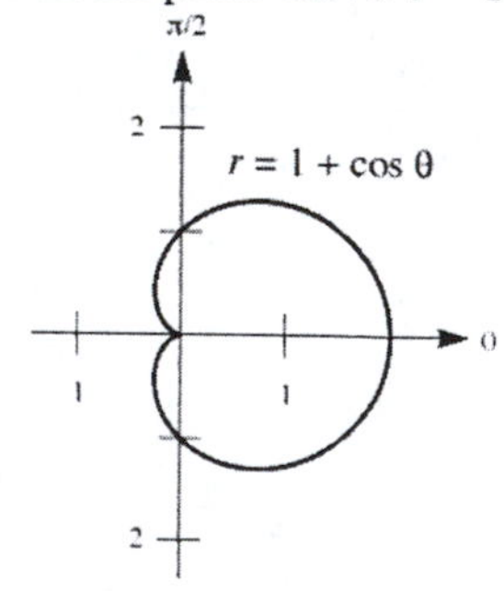

13. Sketch the polar curve $r = 2 - 2\sin\theta$:

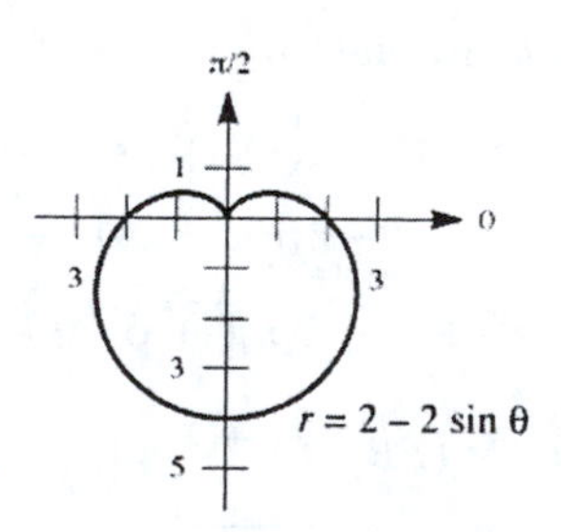

15. Sketch the polar curve $r = 1 - 2\sin\theta$:

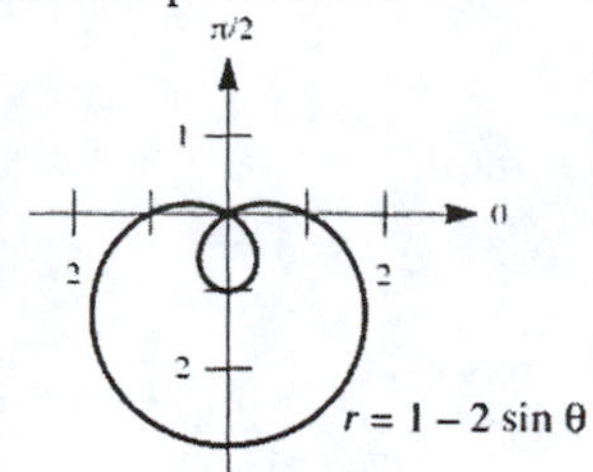

17. Sketch the polar curve $r = 2 + 4\cos\theta$:

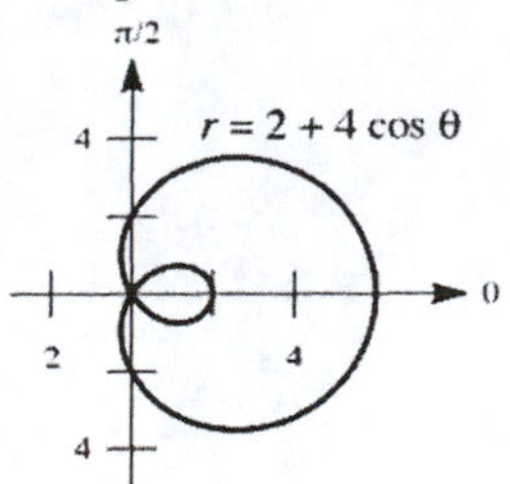

19. Sketch the polar curve $r^2 = 4\sin 2\theta$:

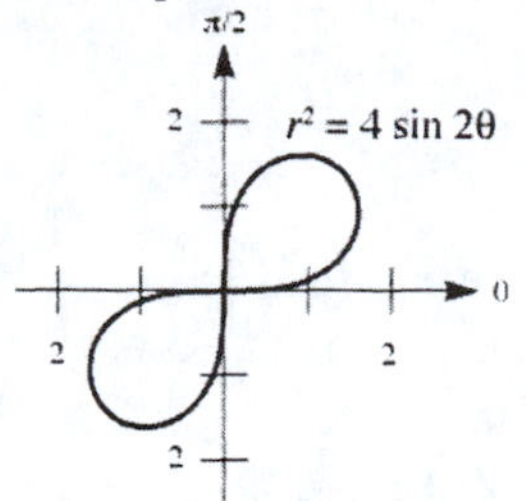

21. Sketch the polar curve $r^2 = \cos 4\theta$:

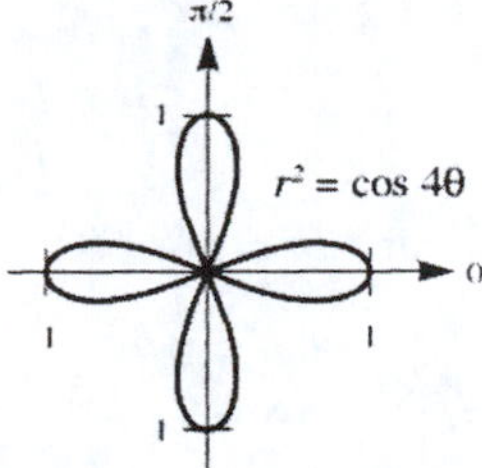

23. Sketch the polar curve $r = \cos 2\theta$:

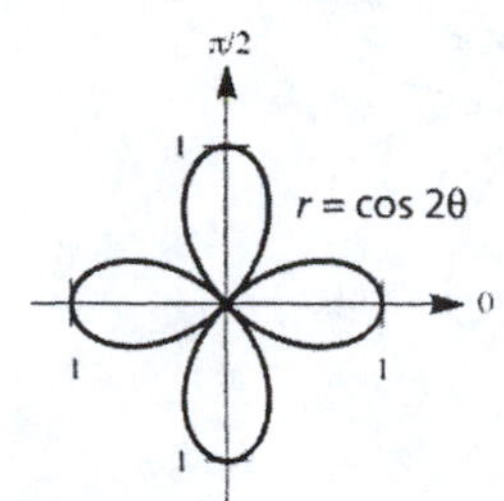

25. Sketch the polar curve $r = \sin 3\theta$:

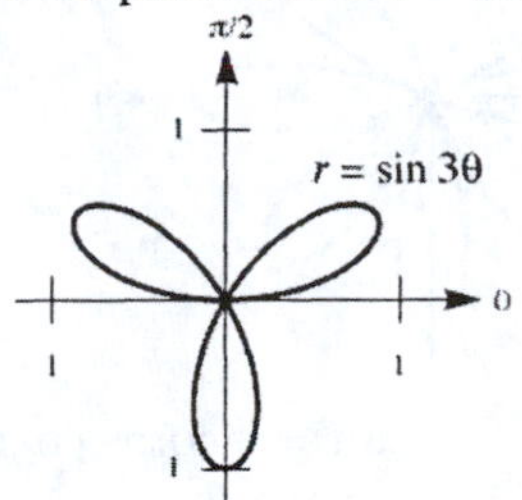

27. Sketch the polar curve $r = 4 + 2\sin\theta$:

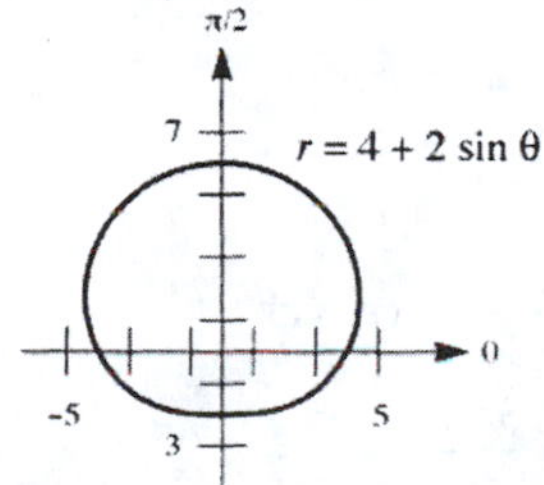

29. Sketch the polar curve $r = 8\tan\theta$:

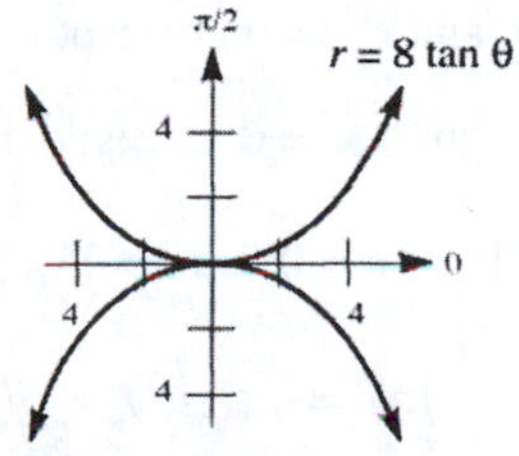

31. **a.** Replacing θ with $-\theta$: $r = \cos^2(-\theta) - 2\cos(-\theta) = \cos^2\theta - 2\cos\theta$
Since the equation is unchanged, the graph is symmetric about the x-axis.

b. Graphing the curve, note the symmetry about the x-axis:

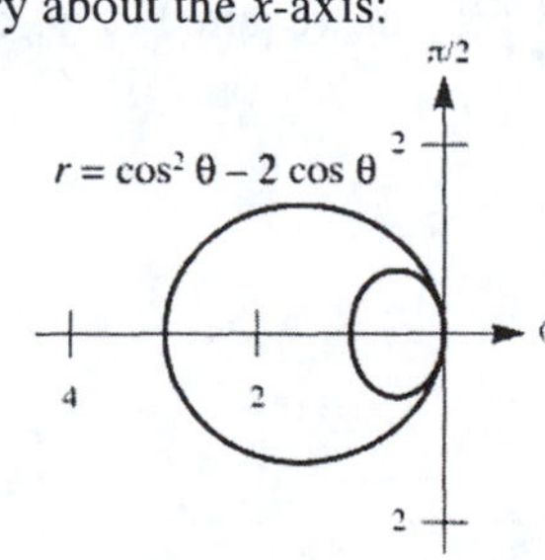

33. Sketch the polar curve $r = \cos 3\theta$:

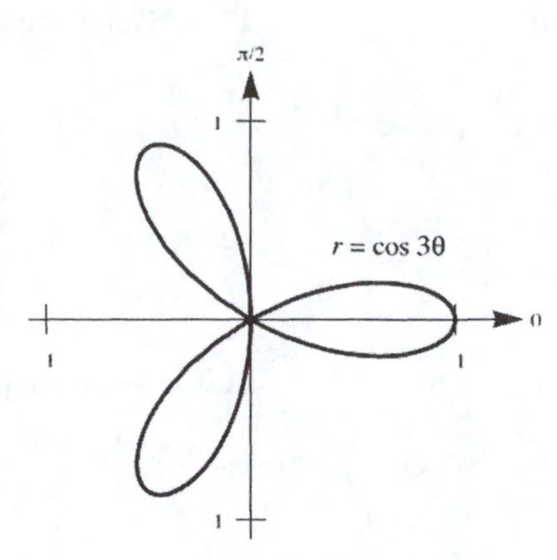

35. **a.** Sketch the polar curve $r = \cos 4\theta$:

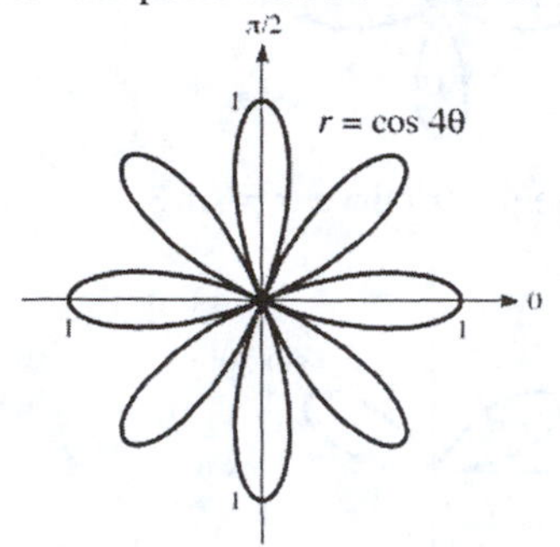

b. Sketch the polar curve $r = \sin 4\theta$:

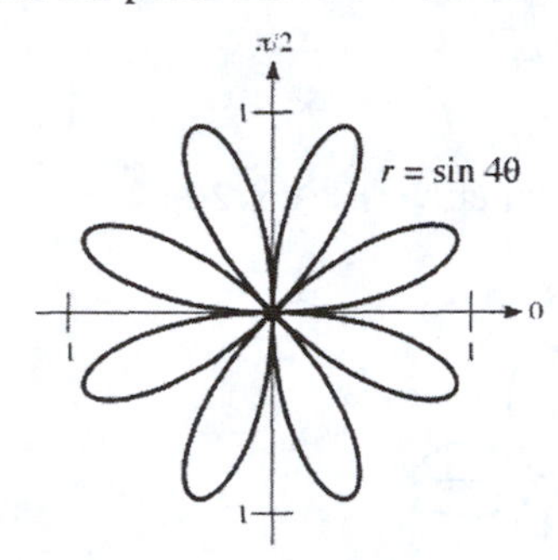

37. **a.** The graph is C, since $\theta = 0$ corresponds to $r = 5$ and $\theta = \frac{\pi}{2}$ corresponds to $r = 3$.

b. The graph is D, since $\theta = 0$ corresponds to $r = 5$ and $\theta = \frac{\pi}{2}$ corresponds to $r = 2$.

c. The graph is B, since $\theta = 0$ corresponds to $r = -1$ and $\theta = \frac{\pi}{2}$ corresponds to $r = 2$.

d. The graph is A, since $\theta = 0$ corresponds to $r = 1$ and $\theta = \frac{\pi}{2}$ corresponds to $r = 3$.

39. **a.** For A, let $\theta = 0$: $\ln r = 0$, so $r = 1$

For B, let $\theta = \frac{\pi}{2}$: $\ln r = \dfrac{a\pi}{2}$, so $r = e^{a\pi/2}$

For C, let $\theta = \pi$: $\ln r = a\pi$, so $r = e^{a\pi}$

For D, let $\theta = \frac{3\pi}{2}$: $\ln r = \dfrac{3a\pi}{2}$, so $r = e^{3a\pi/2}$

So the polar coordinates are $A(1, 0)$, $B\left(e^{a\pi/2}, \frac{\pi}{2}\right)$, $C\left(e^{a\pi}, \pi\right)$, and $D\left(e^{3a\pi/2}, \frac{3\pi}{2}\right)$. Since $OA = 1$, $OB = e^{a\pi/2}$, $OC = e^{a\pi}$, and $OD = e^{3a\pi/2}$, the required ratios are therefore:

$$\frac{OD}{OC} = \frac{e^{3a\pi/2}}{e^{a\pi}} = e^{3a\pi/2 - a\pi} = e^{a\pi/2}$$

$$\frac{OC}{OB} = \frac{e^{a\pi}}{e^{a\pi/2}} = e^{a\pi - a\pi/2} = e^{a\pi/2}$$

$$\frac{OB}{OA} = \frac{e^{a\pi/2}}{1} = e^{a\pi/2}$$

Thus: $\dfrac{OD}{OC} = \dfrac{OC}{OB} = \dfrac{OB}{OA} = e^{a\pi/2}$

b. Using the hint, the rectangular coordinates are $A(1,0)$, $B\left(0, e^{a\pi/2}\right)$, $C\left(-e^{a\pi}, 0\right)$, and $D\left(0, -e^{3a\pi/2}\right)$.

Now compute the slope of line segments $\overline{AB}$, $\overline{BC}$, and $\overline{CD}$:

$$m_{\overline{AB}} = \frac{e^{a\pi/2} - 0}{0-1} = -e^{a\pi/2}$$

$$m_{\overline{BC}} = \frac{0 - e^{a\pi/2}}{-e^{a\pi} - 0} = \frac{1}{e^{a\pi/2}}$$

$$m_{\overline{CD}} = \frac{-e^{3a\pi/2} - 0}{0 + e^{a\pi}} = -e^{a\pi/2}$$

Since $m_{\overline{AB}} \bullet m_{\overline{BC}} = -1$ and $m_{\overline{BC}} \bullet m_{\overline{CD}} = -1$, $\angle ABC$ and $\angle BCD$ are right angles.

41. a. Graph the curves. Note that the curves are identical.

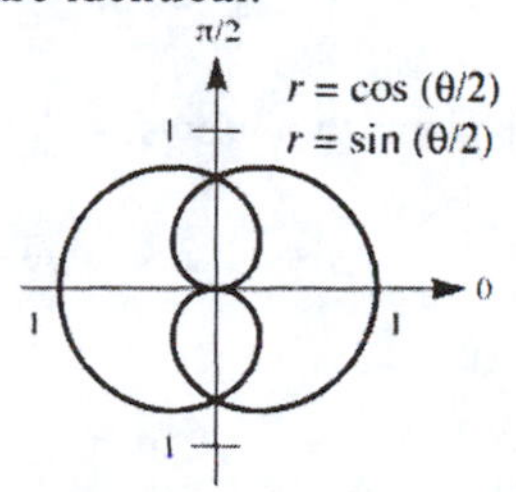

b. Square each side of the polar curve $r = \cos\frac{\theta}{2}$ and apply the half-angle identity:

$$r^2 = \cos^2\frac{\theta}{2}$$

$$r^2 = \frac{1+\cos\theta}{2}$$

$$2r^2 = 1 + \cos\theta$$

$$2\left(x^2+y^2\right) = 1 + \frac{x}{\sqrt{x^2+y^2}}$$

$$2\left(x^2+y^2\right)^{3/2} = \left(x^2+y^2\right)^{1/2} + x$$

$$\left(x^2+y^2\right)^{1/2}\left(2x^2+2y^2-1\right) = x$$

$$\left(x^2+y^2\right)\left(2x^2+2y^2-1\right)^2 = x^2$$

$$\left(x^2+y^2\right)\left[2\left(x^2+y^2\right)-1\right]^2 - x^2 = 0$$

Now square each side of the polar curve $r = \sin\frac{\theta}{2}$ and apply the half-angle identity:

$$r^2 = \sin^2\frac{\theta}{2}$$

$$r^2 = \frac{1-\cos\theta}{2}$$

$$2r^2 = 1 - \cos\theta$$

$$2\left(x^2+y^2\right) = 1 - \frac{x}{\sqrt{x^2+y^2}}$$

$$2\left(x^2+y^2\right)^{3/2} = \left(x^2+y^2\right)^{1/2} - x$$

$$\left(x^2+y^2\right)^{1/2}\left(2x^2+2y^2-1\right) = -x$$

$$\left(x^2+y^2\right)\left(2x^2+2y^2-1\right)^2 = x^2$$

$$\left(x^2+y^2\right)\left[2\left(x^2+y^2\right)-1\right]^2 - x^2 = 0$$

So both curves convert to the same rectangular equation.

43. Sketch the polar curve $r^2 = 4\cos 2\theta$:

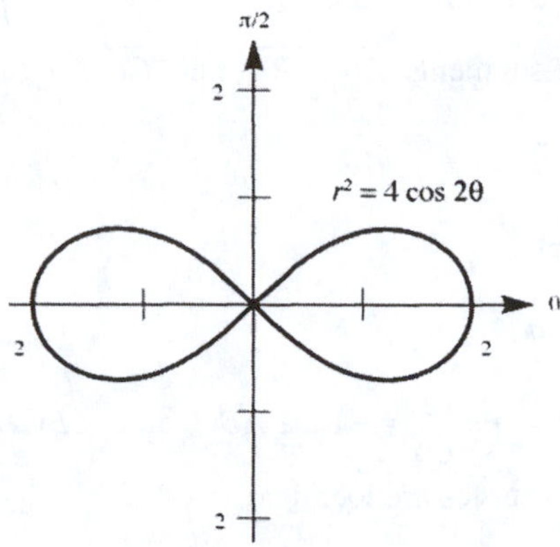

45. Graphing the polar curve $r = \cos^4(\theta/4)$ using two views:

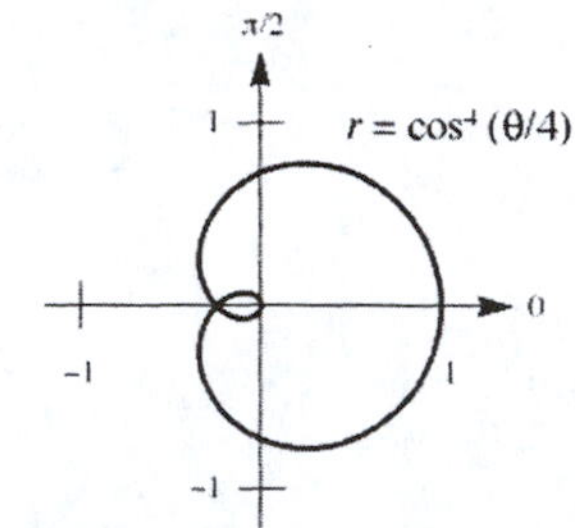

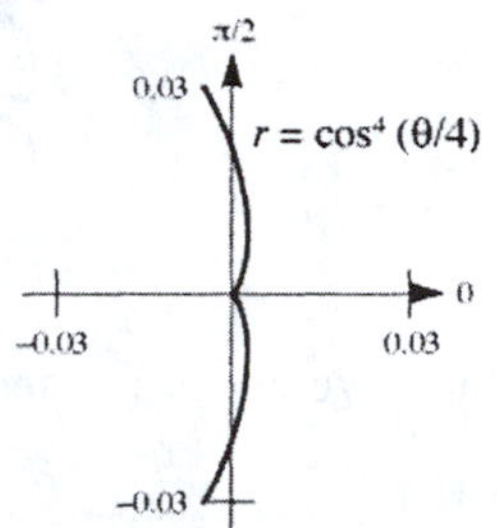

Zooming in several times, note that the inner loop near the origin is not simple, but rather a cardioid type shape which passes through both the first and fourth quadrants.

47. Graphing the polar curve $r = \dfrac{\sin\theta}{\theta}$:

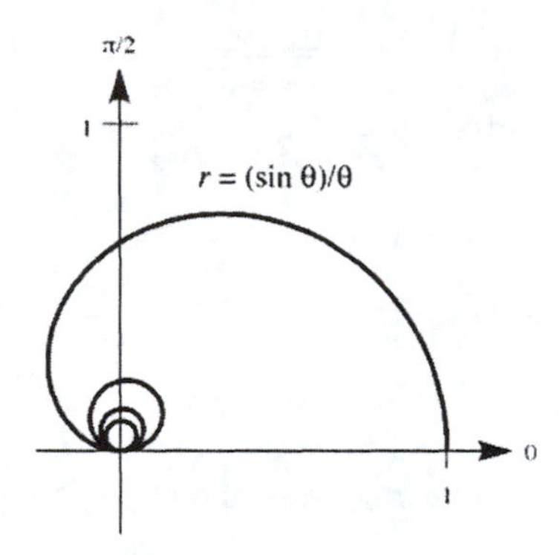

Chapter 9 Review Exercises

1. Using the identities for tan θ and cot θ:

$$\tan\theta + \cot\theta = 2$$

$$\frac{\sin\theta}{\cos\theta} + \frac{\cos\theta}{\sin\theta} = 2$$

$$\frac{\sin^2\theta + \cos^2\theta}{\sin\theta\cos\theta} = 2$$

$$1 = 2\sin\theta\cos\theta$$

Now $(\sin\theta + \cos\theta)^2 = \sin^2\theta + 2\sin\theta\cos\theta + \cos^2\theta = 1 + 1 = 2$, so $\sin\theta + \cos\theta = \sqrt{2}$ since $0° < \theta < 90°$.

3. Draw the figure:

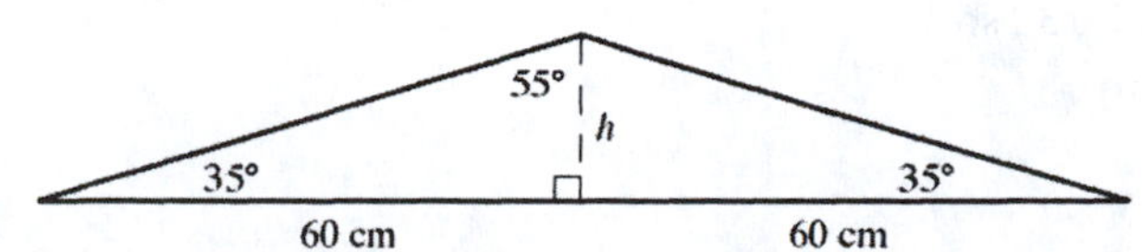

Note that the altitude of the triangle must bisect the upper vertex angle, which is 110° (180° – 35° – 35°).

Now find the height h: $\tan 35° = \dfrac{h}{60}$, so $h = 60\tan 35°$

Thus the area is given by: $\frac{1}{2}(\text{base})(\text{height}) = \frac{1}{2}(120)(60\tan 35°) = 3600\tan 35° \approx 2521\text{ cm}^2$

5. Draw the figure:

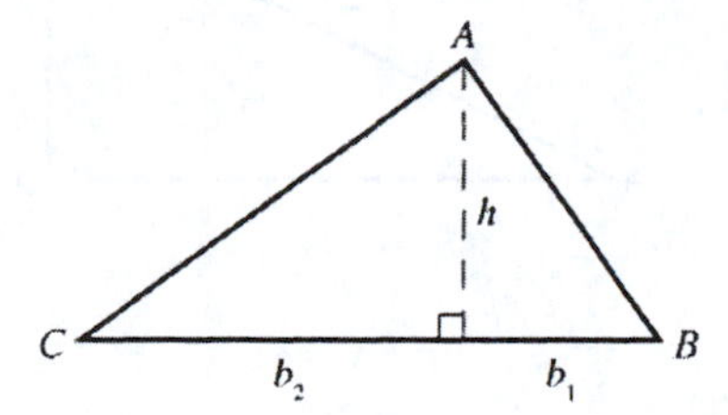

Now:

$$\cot B = \frac{b_1}{h}, \text{ so } b_1 = h\cot B \qquad \cot C = \frac{b_2}{h}, \text{ so } b_2 = h\cot C$$

So we have:

$$\begin{aligned} b_1 + b_2 &= a \\ h\cot B + h\cot C &= a \\ h(\cot B + \cot C) &= a \\ h &= \frac{a}{\cot B + \cot C} \end{aligned}$$

7. Draw the figure:

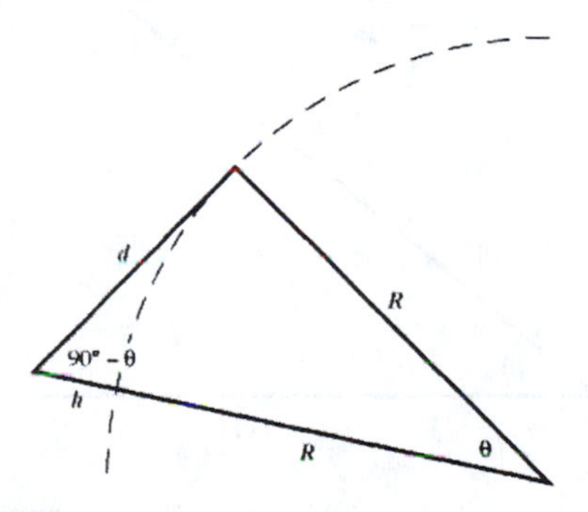

We have:

$$\begin{aligned} d^2 + R^2 &= (R+h)^2 \\ d^2 &= h^2 + 2Rh \\ d &= \sqrt{2Rh + h^2} \end{aligned}$$

Therefore: $\cot\theta = \dfrac{R}{d} = \dfrac{R}{\sqrt{2Rh + h^2}}$

9. Draw a triangle:

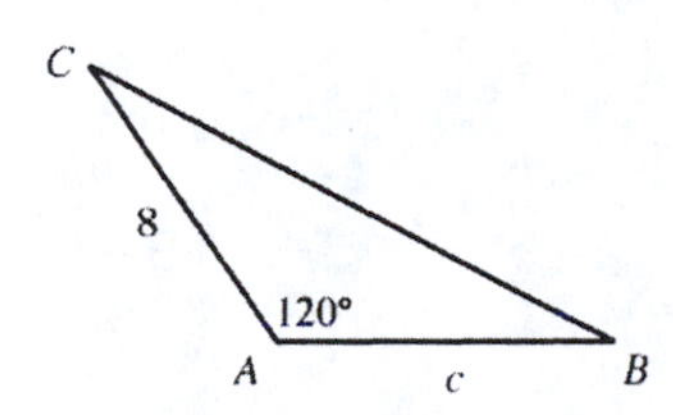

Now the area of $\Delta ABC = 12\sqrt{3}$, so:

$$12\sqrt{3} = \tfrac{1}{2}(c)(8)\sin 120°$$
$$12\sqrt{3} = 4c\left(\frac{\sqrt{3}}{2}\right)$$
$$12\sqrt{3} = 2\sqrt{3}c$$
$$6 = c$$

11. Draw a triangle:

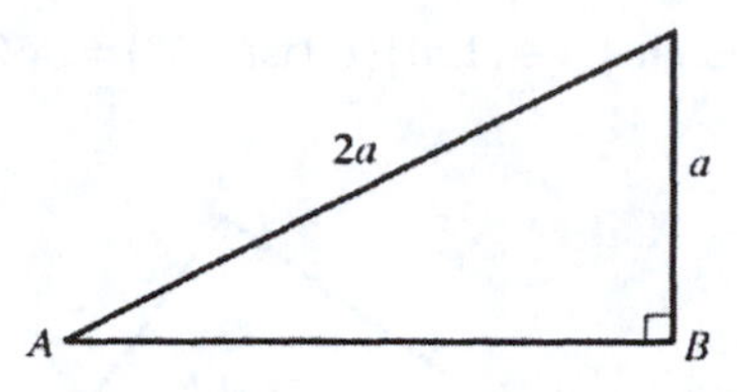

Since $\sin A = \dfrac{a}{2a} = \tfrac{1}{2}$, $A = 30°$.

13. Draw a triangle:

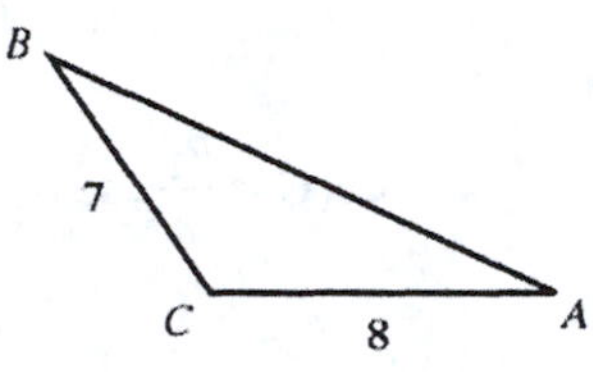

The area of ΔABC is given by: $\tfrac{1}{2}ab\sin\theta = \tfrac{1}{2}(8)(7)\sin C = 28 \bullet \tfrac{1}{4} = 7$ square units

15. Draw the triangle:

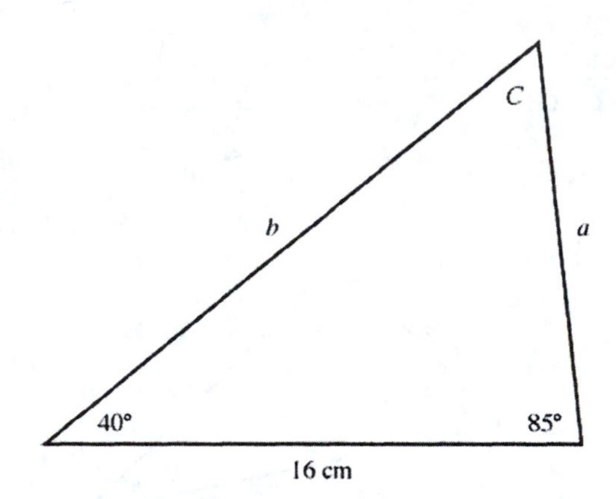

First note that $\angle C = 180° - 40° - 85° = 55°$. Using the law of sines: $\dfrac{\sin 40°}{a} = \dfrac{\sin 55°}{16}$, so $a = \dfrac{16\sin 40°}{\sin 55°} \approx 12.6$ cm

Using the law of sines: $\dfrac{\sin 85°}{b} = \dfrac{\sin 55°}{16}$, so $b = \dfrac{16\sin 85°}{\sin 55°} \approx 19.5$ cm

17. **a.** Draw the triangle:

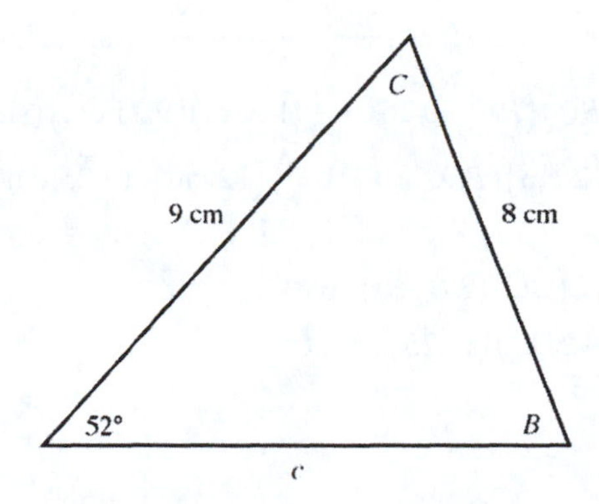

Using the law of sines: $\frac{\sin B}{9} = \frac{\sin 52°}{8}$, so $\sin B = \frac{9\sin 52°}{8} \approx 0.8865$

Thus $\angle B \approx 62.4°$. Now note that $\angle C \approx 180° - 52° - 62.4° \approx 65.6°$. Using the law of sines:

$$\frac{\sin 65.6°}{c} = \frac{\sin 52°}{8}, \text{ so } c = \frac{8\sin 65.6°}{\sin 52°} \approx 9.2 \text{ cm}$$

b. Note that $\sin B \approx 0.8865$, thus $\angle B \approx 117.6°$. Now note that $\angle C = 180° - 52° - 117.6° \approx 10.4°$. Using the law of sines: $\frac{\sin 10.4°}{c} = \frac{\sin 52°}{8}$, so $c = \frac{8\sin 10.4°}{\sin 52°} \approx 1.8$ cm

19. Draw the triangle:

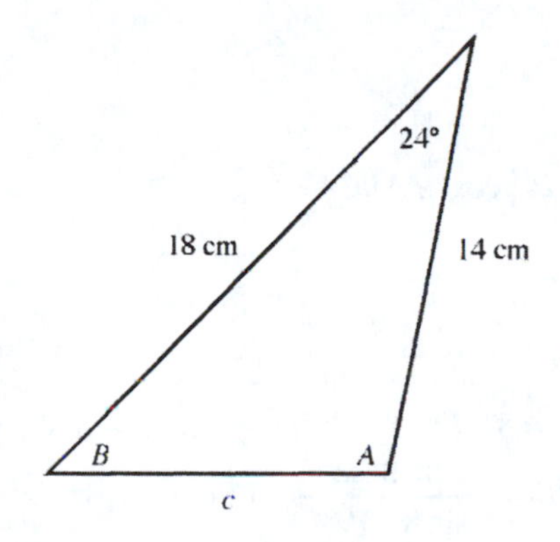

Using the law of cosines: $c^2 = 18^2 + 14^2 - 2(18)(14)\cos 24°$, so $c \approx 7.7$ cm

Using the law of sines: $\frac{\sin A}{18} = \frac{\sin 24°}{7.7}$, so $\sin A = \frac{18\sin 24°}{7.7} \approx 0.9486$

Then either $\angle A \approx 71.5°$ or $\angle A \approx 108.5°$. If $\angle A \approx 71.5°$, then $\angle B \approx 180° - 24° - 71.5° \approx 84.5°$. But this is impossible since $\angle B < \angle A$. If $\angle A \approx 108.5°$, then $\angle B \approx 180° - 24° - 108.5° \approx 47.5°$.

21. Draw the triangle:

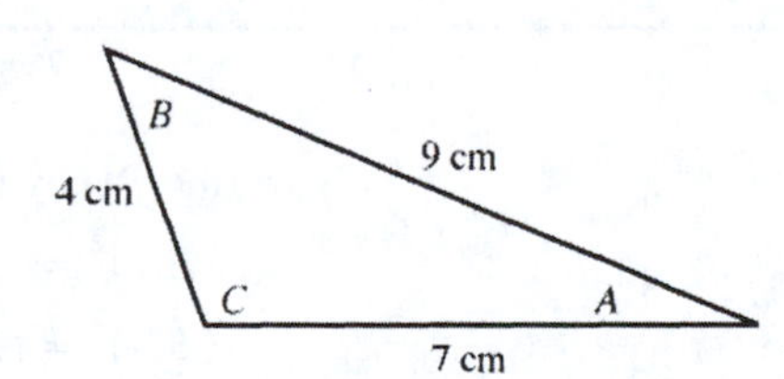

Using the law of cosines: $9^2 = 4^2 + 7^2 - 2(4)(7)\cos C$, so $\cos C \approx -0.2857$, thus $C \approx 106.6°$

Using the law of sines: $\frac{\sin 106.6°}{9} = \frac{\sin B}{7}$, so $\sin B = \frac{7\sin 106.6°}{9} \approx 0.7454$

Thus $B \approx 48.2°$. Therefore $\angle A \approx 180° - 106.6° - 48.2° \approx 25.2°$.

23. Note that $\angle AEB = 86°$. Using the law of sines: $\dfrac{\sin 50°}{BE} = \dfrac{\sin 86°}{12}$, so $BE = \dfrac{12\sin 50°}{\sin 86°} \approx 9.21$ cm

25. Using the area formula: $A = \frac{1}{2}(BC)(BE)\sin 36° = \frac{1}{2}(12\text{ cm})(9.21\text{ cm})(\sin 36°) \approx 32.48\text{ cm}^2$

27. Using the area formula: $A = \frac{1}{2}(12\text{ cm})(BD\sin 44°) = \frac{1}{2}(12\text{ cm})(13.25\text{ cm})(\sin 44°) \approx 55.23\text{ cm}^2$

29. Using the law of cosines:

$$\begin{aligned}(CD)^2 &= (BC)^2 + (BD)^2 - 2(BC)(BD)\cos 36° \\ &\approx (12)^2 + (13.25)^2 - 2(12)(13.25)\cos 36° \\ &\approx 319.56 - 318\cos 36° \\ &\approx 62.29\end{aligned}$$

So $CD \approx \sqrt{62.29} \approx 7.89$ cm.

31. Using the law of sines: $\dfrac{\sin 80°}{AC} = \dfrac{\sin 50°}{12}$, so $AC = \dfrac{12\sin 80°}{\sin 50°} \approx 15.43$ cm

33. Re-draw the figure:

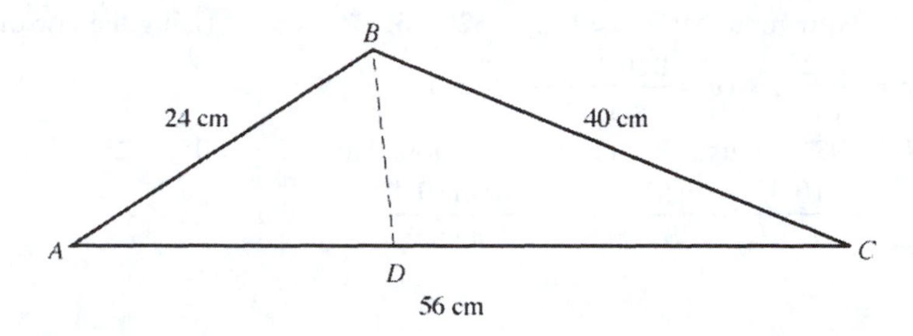

Use the law of cosines to find $\angle ABC$:

$$\begin{aligned}56^2 &= 40^2 + 24^2 - 2(40)(24)\cos(\angle ABC) \\ -0.5 &= \cos(\angle ABC) \\ \angle ABC &= 120°\end{aligned}$$

Now use the law of sines to find $\angle A$: $\dfrac{\sin A}{40} = \dfrac{\sin 120°}{56}$, so $\sin A \approx 0.6186$, thus $A \approx 38.21°$

Then $\angle ADB \approx 180° - 60° - 38.21° \approx 81.79°$. Find BD by the law of sines:

$$\frac{\sin 81.79°}{24} = \frac{\sin 38.21°}{BD}, \text{ so } BD = \frac{24\sin 38.21°}{\sin 81.79°} \approx 15 \text{ cm}$$

35. Re-draw the figure:

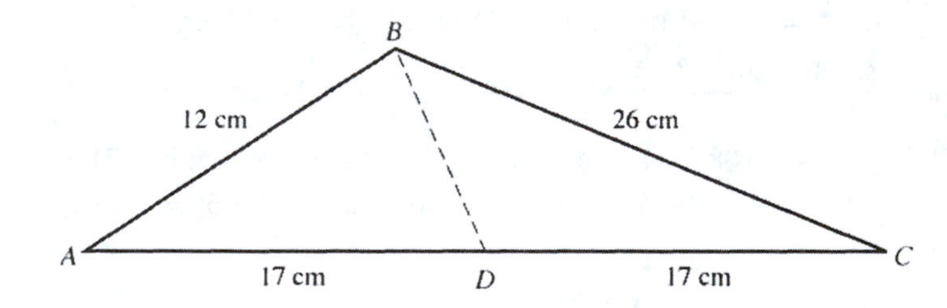

Using the law of cosines, find $\angle A$:

$$\begin{aligned}26^2 &= 12^2 + 34^2 - 2(12)(34)\cos A \\ \cos A &\approx 0.7647 \\ A &\approx 40.12°\end{aligned}$$

Now find BD by using the law of cosines (on ΔADB):

$$\begin{aligned}(BD)^2 &= 12^2 + 17^2 - 2(12)(17)\cos 40.12° \\ (BD)^2 &= 121 \\ BD &= 11 \text{ cm}\end{aligned}$$

37. **a.** Using the law of cosines:

$$a^2 = b^2 + c^2 - 2bc\cos A$$
$$4^2 = 5^2 + 6^2 - 2(5)(6)\cos A$$
$$16 = 25 + 36 - 60\cos A$$
$$-45 = -60\cos A$$
$$\tfrac{3}{4} = \cos A$$

Also using the law of cosines:

$$c^2 = a^2 + b^2 - 2ab\cos C$$
$$6^2 = 4^2 + 5^2 - 2(4)(5)\cos C$$
$$36 = 16 + 25 - 40\cos C$$
$$-5 = -40\cos C$$
$$\tfrac{1}{8} = \cos C$$

b. Since $\cos A = \frac{3}{4}$: $\sin A = \sqrt{1-\cos^2 A} = \sqrt{1-\frac{9}{16}} = \sqrt{\frac{7}{16}} = \frac{\sqrt{7}}{4}$

Therefore: $\cos 2A = \cos^2 A - \sin^2 A = \left(\frac{3}{4}\right)^2 - \left(\frac{\sqrt{7}}{4}\right)^2 = \frac{9}{16} - \frac{7}{16} = \frac{1}{8}$. But since $\cos C = \frac{1}{8}$, $C = 2A$.

39. Since $BE = BD$, ΔBDE is isosceles and thus its base angles are congruent. Therefore, calling $\angle BED = \theta$:

$$\theta + \theta + 108° = 180°$$
$$2\theta = 72°$$
$$\theta = 36°$$

So $\angle BED = 36°$.

41. Note that from Exercise 40 $\angle ABE = 36°$. Using the law of sines on ΔABE:

$$\frac{\sin \angle AEB}{16} = \frac{\sin 36°}{11}$$
$$\sin \angle AEB = \frac{16\sin 36°}{11} \approx 0.8550$$
$$\angle AEB \approx 58.76° \text{ or } 121.24°$$

Since the given figure indicates that $\angle AEB$ is an obtuse angle, $\angle AEB \approx 121.24°$.

43. Note that from Exercise 42 $\angle BAE \approx 22.76°$. Using the law of sines on ΔABE:

$$\frac{\sin 22.76°}{BE} = \frac{\sin 36°}{11}$$
$$BE = \frac{11\sin 22.76°}{\sin 36°} \approx 7.24 \text{ m}$$

45. **a.** First note that $\angle AOB = \frac{360°}{5} = 72°$. Draw a perpendicular from O to $\overline{AB}$, and call the intersection point P.

Then $\angle POB = 36°$, thus: $\sin 36° = \frac{PB}{OB} = \frac{PB}{1}$, so $PB = \sin 36°$. So $AB = 2PB = 2\sin 36°$.

b. Draw a perpendicular from O to $\overline{AD}$, and call the intersection point P. Then $\angle POD = 72°$, thus: $\sin 72° = \frac{PD}{OD} = \frac{PD}{1}$, so $PD = \sin 72°$. So $AD = 2PD = 2\sin 72°$.

c. Draw a perpendicular from V to $\overline{AB}$, and call the intersection point P. Then $AP = \sin 36°$ and $\angle AVP = \frac{108°}{2} = 54°$, so: $\sin 54° = \frac{AP}{AV} = \frac{\sin 36°}{AV}$, so $AV = \frac{\sin 36°}{\sin 54°} = \frac{\sin 36°}{\cos 36°} = \tan 36°$

d. Using the hint:

$$\begin{aligned} VW &= AC - 2AV \\ &= 2\sin 72° - 2\tan 36° \\ &= 4\sin 36°\cos 36° - \frac{2\sin 36°}{\cos 36°} \\ &= \frac{4\sin 36°\cos^2 36° - 2\sin 36°}{\cos 36°} \\ &= \frac{2\sin 36°}{\cos 36°}\left(2\cos^2 36° - 1\right) \\ &= 2\tan 36°\cos 72° \qquad \text{[using the double-angle identity]} \end{aligned}$$

e. Using the hint: $AW = AV + VW = \tan 36° + 2\tan 36°\cos 72° = \tan 36°\left(1 + 2\cos 72°\right)$

f. Simplifying the quotient: $\frac{AD}{AB} = \frac{2\sin 72°}{2\sin 36°} = \frac{4\sin 36°\cos 36°}{2\sin 36°} = 2\cos 36°$

g. Simplifying the quotient in (i):

$$\begin{aligned}
\frac{AV}{VW} &= \frac{\tan 36°}{2\tan 36°\cos 72°}\\
&= \frac{1}{2\cos 72°}\\
&= \frac{1}{2\left(2\cos^2 36° - 1\right)}\\
&= \frac{1}{4\bullet\frac{1}{16}\left(\sqrt{5}+1\right)^2 - 2}\\
&= \frac{1}{\frac{1}{4}\left(6+2\sqrt{5}\right) - 2}\\
&= \frac{4}{6+2\sqrt{5}-8}\\
&= \frac{2}{\sqrt{5}-1}\bullet\frac{\sqrt{5}+1}{\sqrt{5}+1}\\
&= \frac{2\left(\sqrt{5}+1\right)}{5-1}\\
&= \tfrac{1}{2}\left(\sqrt{5}+1\right)\\
&= 2\cos 36°
\end{aligned}$$

Simplifying the quotient in (ii):

$$\begin{aligned}
\frac{AW}{AV} &= \frac{\tan 36°\left(1+2\cos 72°\right)}{\tan 36°}\\
&= 1+2\cos 72°\\
&= 1+2\left(2\cos^2 36° - 1\right)\\
&= 4\cos^2 36° - 1\\
&= 4\bullet\tfrac{1}{16}\left(\sqrt{5}+1\right)^2 - 1\\
&= \tfrac{1}{4}\left(6+2\sqrt{5}\right) - 1\\
&= \tfrac{1}{2}+\tfrac{1}{2}\sqrt{5}\\
&= \tfrac{1}{2}\left(1+\sqrt{5}\right)\\
&= 2\cos 36°
\end{aligned}$$

Simplifying the quotient in (iii):

$$\begin{aligned}
\frac{AC}{AW} &= \frac{2\sin 72°}{\tan 36°\left(1+2\cos 72°\right)}\\
&= \frac{4\sin 36°\cos 36°}{\frac{\sin 36°}{\cos 36°}\left(1+4\cos^2 36° - 2\right)}\\
&= \frac{4\cos^2 36°}{4\cos^2 36° - 1}\\
&= \frac{4\bullet\frac{1}{16}\left(\sqrt{5}+1\right)^2}{4\bullet\frac{1}{16}\left(\sqrt{5}+1\right)^2 - 1}\\
&= \frac{\frac{1}{4}\left(6+2\sqrt{5}\right)}{\frac{1}{4}\left(6+2\sqrt{5}\right) - 1}\\
&= \frac{6+2\sqrt{5}}{2+2\sqrt{5}}\\
&= \frac{3+\sqrt{5}}{1+\sqrt{5}}\bullet\frac{1-\sqrt{5}}{1-\sqrt{5}}\\
&= \frac{-2-2\sqrt{5}}{-4}\\
&= \tfrac{1}{2}\left(1+\sqrt{5}\right)\\
&= 2\cos 36°
\end{aligned}$$

47. For the triangle, note the following figure:

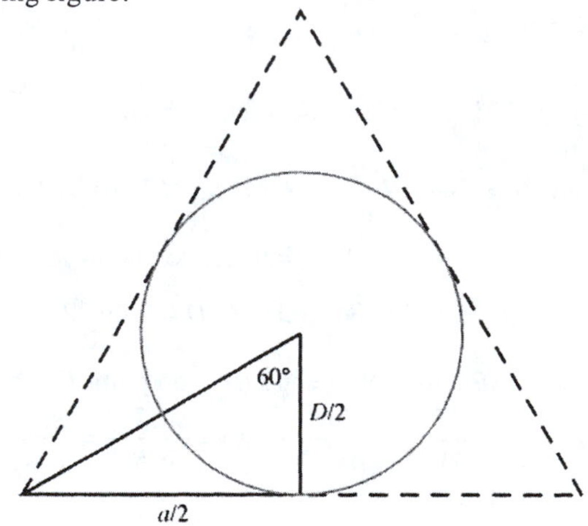

So $\tan 60° = \frac{a/2}{D/2}$, or $\sqrt{3} = \frac{a}{D}$, thus $a = D\sqrt{3}$. For the hexagon, note the following figure:

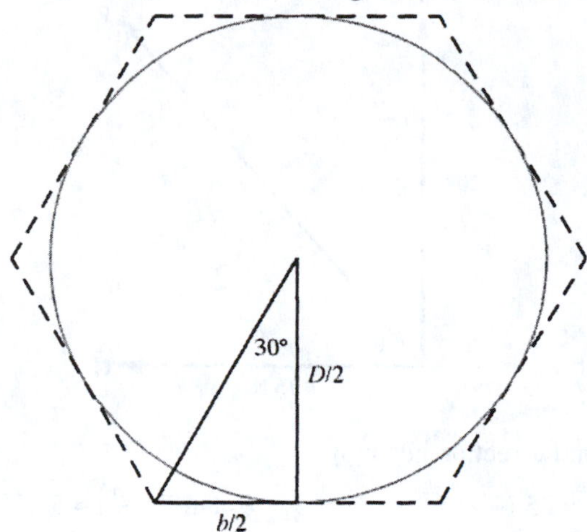

So $\tan 30° = \frac{b/2}{D/2}$, or $\frac{1}{\sqrt{3}} = \frac{b}{D}$, thus $b = \frac{D}{\sqrt{3}}$. Thus: $ab = \left(D\sqrt{3}\right)\left(\frac{D}{\sqrt{3}}\right) = D^2$

49. The area of the triangle is given by: $A = \frac{1}{2}(\text{base})(\text{height}) = \frac{1}{2}(a)(b\sin 60°) = \frac{\sqrt{3}}{4}ab$. Since the area is $10\sqrt{3}$ cm^2:

$$\frac{\sqrt{3}}{4}ab = 10\sqrt{3}$$
$$ab = 40$$

Find the third side d in terms of a and b using the law of cosines:

$$d^2 = a^2 + b^2 - 2ab\cos 60°$$
$$d^2 = a^2 + b^2 - 2(40)\left(\tfrac{1}{2}\right)$$
$$d = \sqrt{a^2 + b^2 - 40}$$

Since the perimeter of the triangle is 20 cm:

$$a + b + \sqrt{a^2 + b^2 - 40} = 20$$
$$\sqrt{a^2 + b^2 - 40} = 20 - (a + b)$$
$$a^2 + b^2 - 40 = 400 - 40(a + b) + (a + b)^2$$
$$a^2 + b^2 - 40 = 400 - 40a - 40b + a^2 + 2ab + b^2$$
$$-440 = -40a - 40b + 2(40)$$
$$-520 = -40a - 40b$$
$$13 = a + b$$

So, we have the system of equations:

$$a + b = 13$$
$$ab = 40$$

Solving the first equation for b yields $b = 13 - a$, now substitute:

$$a(13 - a) = 40$$
$$13a - a^2 = 40$$
$$a^2 - 13a + 40 = 0$$
$$(a - 8)(a - 5) = 0$$
$$a = 8, 5$$

Since a is the smaller of the two numbers, $a = 5$ and $b = 8$.

51. Draw a figure:

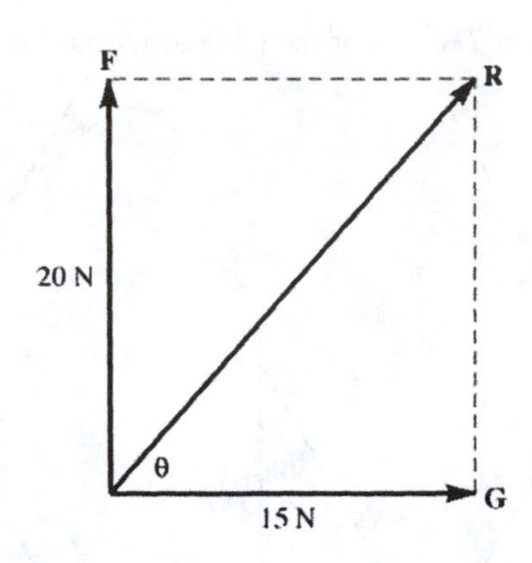

The resultant has a magnitude and direction given by:

$$|\mathbf{R}| = \sqrt{15^2 + 20^2} = \sqrt{625} = 25 \text{ N} \qquad \theta = \tan^{-1}\left(\tfrac{20}{15}\right) \approx 53.1°$$

53. The components are given by:

$$v_x = 50\cos 35° \approx 41.0 \text{ cm/sec} \qquad v_y = 50\sin 35° \approx 28.7 \text{ cm/sec}$$

55. Draw the figure:

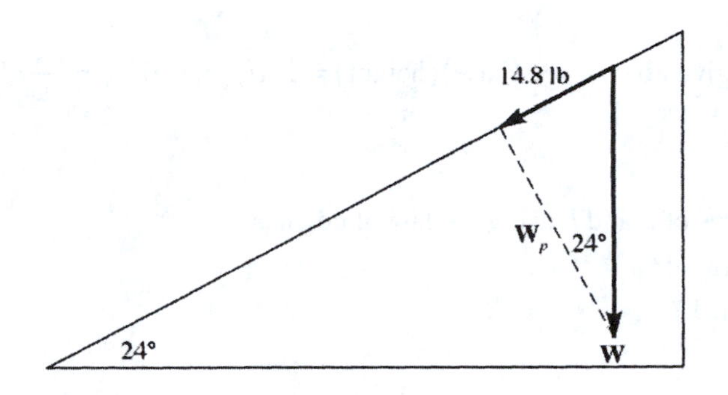

So the desired weights are given by:

$$\tan 24° = \frac{14.8}{|\mathbf{W}_p|}, \text{ so } |\mathbf{W}_p| = \frac{14.8}{\tan 24°} \approx 33.2 \text{ lb} \qquad \sin 24° = \frac{14.8}{|\mathbf{W}|}, \text{ so } |\mathbf{W}| = \frac{14.8}{\sin 24°} \approx 36.4 \text{ lb}$$

57. First find the lengths of $\langle 2,6\rangle$ and $\langle -5,b\rangle$:

$$|\langle 2,6\rangle| = \sqrt{2^2+6^2} = \sqrt{40} \qquad |\langle -5,b\rangle| = \sqrt{(-5)^2+b^2} = \sqrt{b^2+25}$$

Find b by solving the equation:

$$\begin{aligned}\sqrt{b^2+25} &= \sqrt{40}\\ b^2+25 &= 40\\ b^2 &= 15\\ b &= \pm\sqrt{15}\end{aligned}$$

59. $\mathbf{a}+\mathbf{b} = \langle 3,5\rangle + \langle 7,4\rangle = \langle 10,9\rangle$

61. Compute the quantity: $3\mathbf{c}+2\mathbf{a} = 3\langle 2,-1\rangle + 2\langle 3,5\rangle = \langle 6,-3\rangle + \langle 6,10\rangle = \langle 12,7\rangle$

63. Since $\mathbf{b}+\mathbf{d} = \langle 7,7\rangle$ and $\mathbf{b}-\mathbf{d} = \langle 7,1\rangle$: $|\mathbf{b}+\mathbf{d}|^2 - |\mathbf{b}-\mathbf{d}|^2 = \left(7^2+7^2\right) - \left(7^2+1^2\right) = 98-50 = 48$

65. $(\mathbf{a}+\mathbf{b})+\mathbf{c} = \langle 10,9\rangle + \langle 2,-1\rangle = \langle 12,8\rangle$

67. $(\mathbf{a}-\mathbf{b})-\mathbf{c} = \langle -4,1\rangle - \langle 2,-1\rangle = \langle -6,2\rangle$

69. $4\mathbf{c}+2\mathbf{a}-3\mathbf{b} = \langle 8,-4\rangle + \langle 6,10\rangle - \langle 21,12\rangle = \langle -7,-6\rangle$

71. $\langle 7,-6\rangle = 7\mathbf{i}-6\mathbf{j}$

73. First compute the length of $\langle 6,4\rangle$: $|\langle 6,4\rangle| = \sqrt{36+16} = \sqrt{52} = 2\sqrt{13}$

Therefore a unit vector in the same direction as $\langle 6,4\rangle$ would be: $\frac{1}{2\sqrt{13}}\langle 6,4\rangle = \left\langle \frac{3}{\sqrt{13}}, \frac{2}{\sqrt{13}}\right\rangle = \left\langle \frac{3\sqrt{13}}{13}, \frac{2\sqrt{13}}{13}\right\rangle$

75. Use the distance formula, taking the points (r_1, θ_1) and (r_2, θ_2) to be $\left(3, \frac{\pi}{12}\right)$ and $\left(2, \frac{17\pi}{18}\right)$:

$$d^2 = r_1^2 + r_2^2 - 2r_1r_2\cos(\theta_2 - \theta_1) = 3^2 + 2^2 - 2(3)(2)\cos\left(\frac{\pi}{12} - \frac{17\pi}{18}\right) = 13 - 12\cos\left(-\frac{31\pi}{36}\right) \approx 23.8757$$

So $d \approx \sqrt{23.8757} \approx 4.89$.

77. Using the equation $r^2 + r_0^2 - 2rr_0\cos(\theta - \theta_0) = a^2$:

$$r^2 + (5)^2 - 2(r)(5)\cos\left(\theta - \frac{\pi}{6}\right) = 3^2$$
$$r^2 + 25 - 10r\cos\left(\theta - \frac{\pi}{6}\right) = 9$$
$$r^2 - 10r\cos\left(\theta - \frac{\pi}{6}\right) = -16$$

79. **a.** Since the line is of the form $r\cos(\theta - \alpha) = d$, $\alpha = \frac{\pi}{3}$ and $d = 3$. So the desired perpendicular distance is 3.

b. When $\theta = 0$:

$$r\cos\left(0 - \frac{\pi}{3}\right) = 3$$
$$r\left(\frac{1}{2}\right) = 3$$
$$r = 6$$

When $\theta = \frac{\pi}{2}$:

$$r\cos\left(\frac{\pi}{2} - \frac{\pi}{3}\right) = 3$$
$$r\cos\frac{\pi}{6} = 3$$
$$r \bullet \frac{\sqrt{3}}{2} = 3$$
$$r = \frac{6}{\sqrt{3}} = 2\sqrt{3}$$

The required polar points are $(6, 0)$ and $\left(2\sqrt{3}, \frac{\pi}{2}\right)$.

c. These coordinates are (d, α), which is the polar point $\left(3, \frac{\pi}{3}\right)$.

d. Sketch the line:

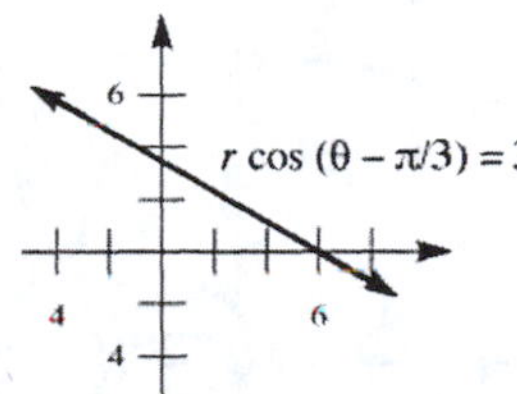

81. **a.** Sketch the polar curve $r = 2 - 2\cos\theta$:

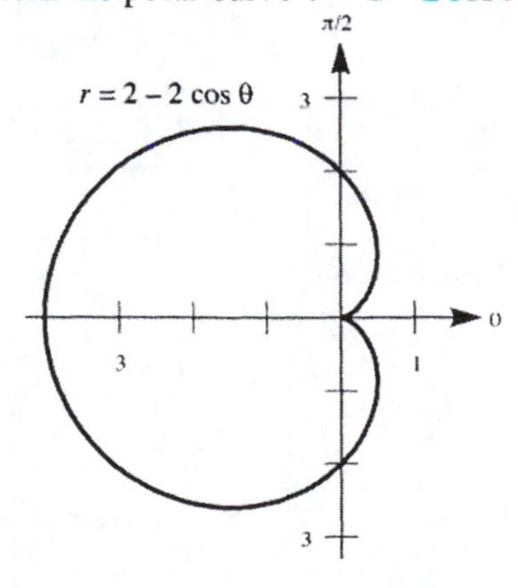

b. Sketch the polar curve $r = 2 - 2\sin\theta$:

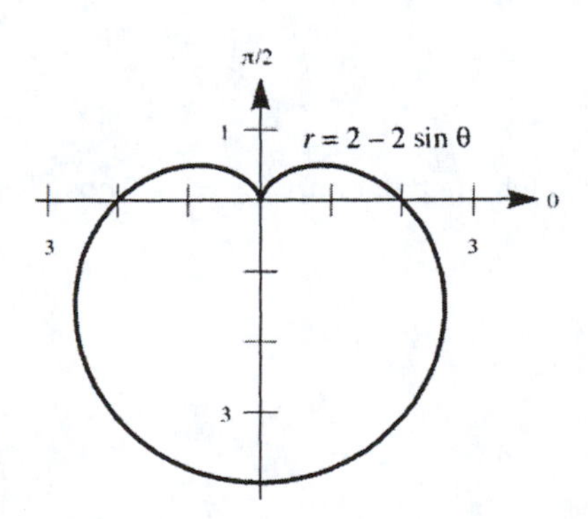

83. **a.** Sketch the polar curve $r = 2\cos\theta - 1$:

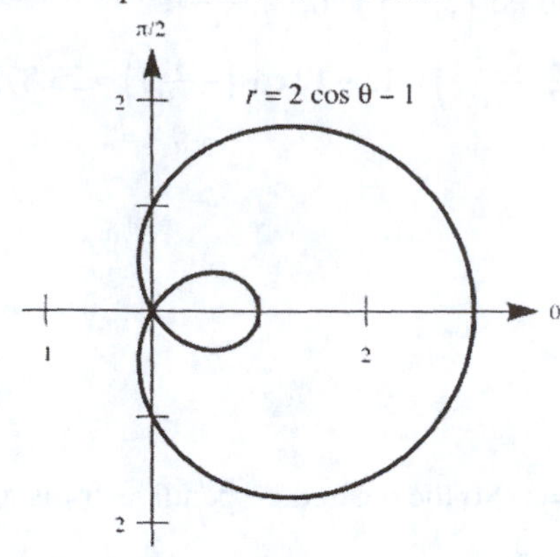

b. Sketch the polar curve $r = 2\sin\theta - 1$:

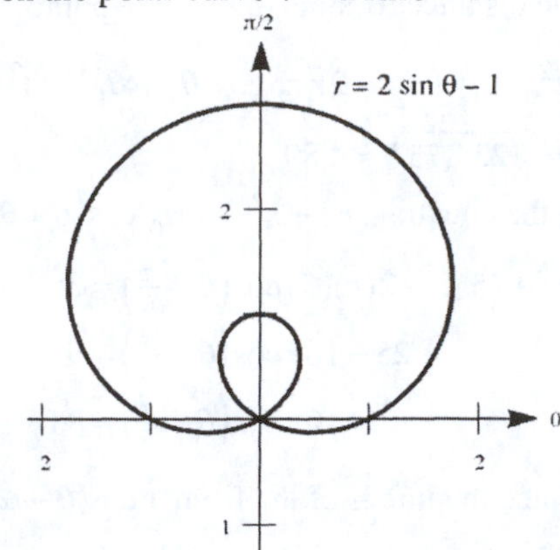

85. **a.** Sketch the polar curve $r = 4\sin 2\theta$:

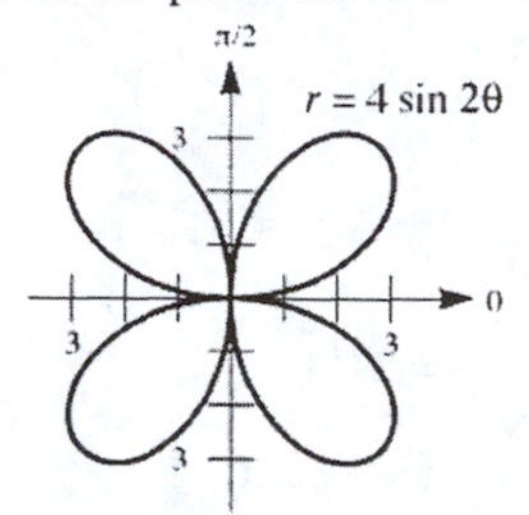

b. Sketch the polar curve $r = 4\cos 2\theta$:

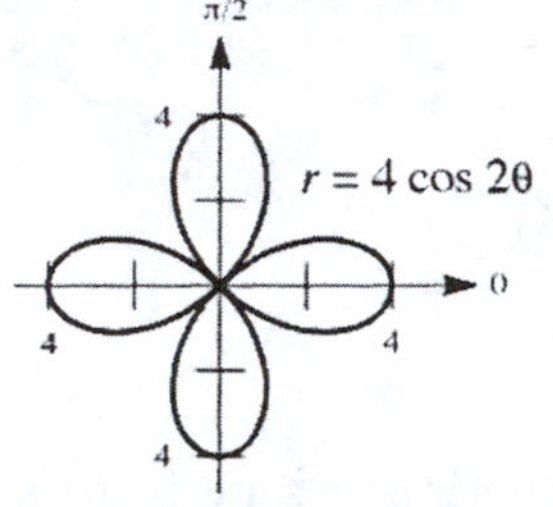

87. **a.** Sketch the polar curve $r = 1 + 2\sin(\theta / 2)$ (note that we must plot $0 \le \theta \le 4\pi$):

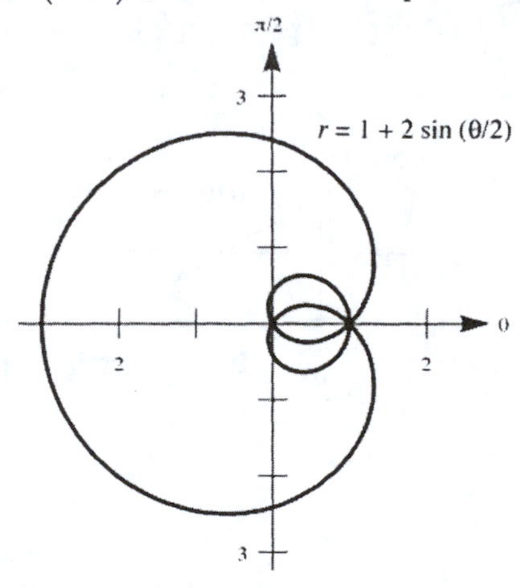

b. Sketch the polar curve $r = 1 - 2\cos(\theta / 2)$ (note that we must plot $0 \le \theta \le 4\pi$):

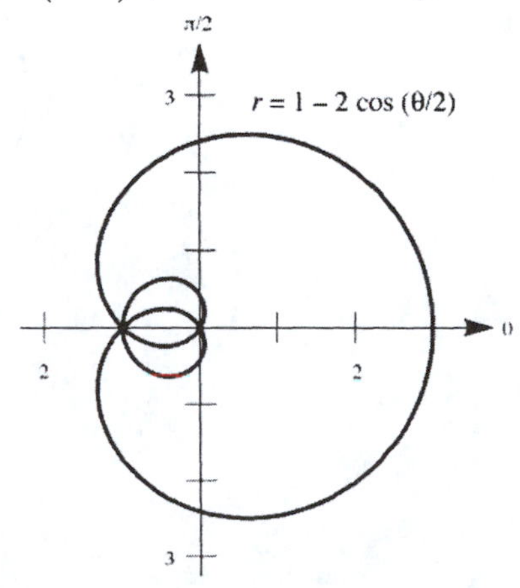

89. Solving $y = 1 + t$ for t yields $t = y - 1$, now substituting:

$$x = 3 - 5(y - 1)$$
$$x = 3 - 5y + 5$$
$$x + 5y = 8$$

So the given parametric equations determine a line.

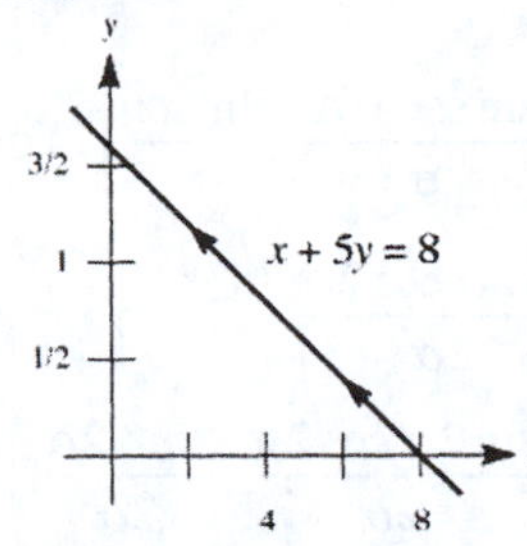

91. Multiplying the first equation by 2 yields $2x = 6\sin t$ and $y = 6\cos t$, so:

$$(2x)^2 + y^2 = 36\sin^2 t + 36\cos^2 t$$
$$4x^2 + y^2 = 36$$
$$\frac{x^2}{9} + \frac{y^2}{36} = 1$$

So the given parametric equations determine an ellipse.

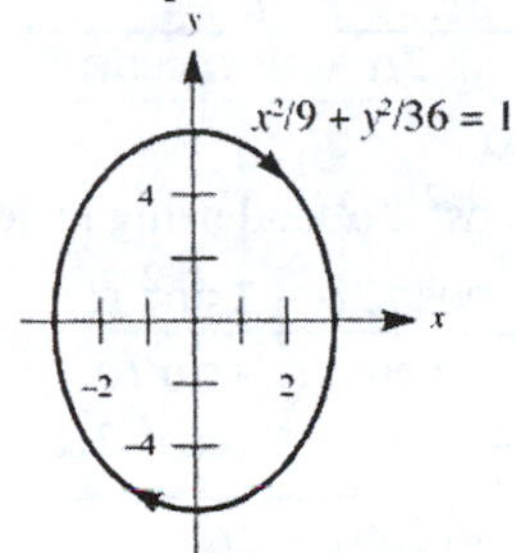

93. Multiplying the first equation by 3 and the second equation by 4 yields $3x = 12\sec t$ and $4y = 12\tan t$, so:

$$(3x)^2 - (4y)^2 = 144\sec^2 t - 144\tan^2 t$$
$$9x^2 - 16y^2 = 144$$
$$\frac{x^2}{16} - \frac{y^2}{9} = 1$$

So the given parametric equations determine a hyperbola.

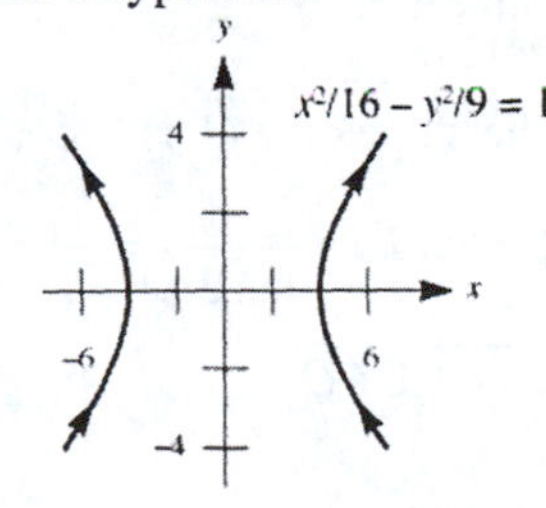

95. **a.** P can be written as $\left(\frac{\sin\alpha}{\alpha}, \alpha\right)$, while Q can be written as $\left(\frac{\sin 2\alpha}{2\alpha}, 2\alpha\right)$.

b. For point P:

$$x = r\cos\theta = \frac{\sin\alpha}{\alpha} \bullet \cos\alpha = \frac{\sin\alpha\cos\alpha}{\alpha} \qquad y = r\sin\theta = \frac{\sin\alpha}{\alpha} \bullet \sin\alpha = \frac{\sin^2\alpha}{\alpha}$$

So the rectangular coordinates are $P\left(\frac{\sin\alpha\cos\alpha}{\alpha}, \frac{\sin^2\alpha}{\alpha}\right)$. For point Q:

$$x = r\cos\theta = \frac{\sin 2\alpha}{2\alpha} \bullet \cos 2\alpha = \frac{\sin 2\alpha\cos 2\alpha}{2\alpha} \qquad y = r\sin\theta = \frac{\sin 2\alpha}{2\alpha} \bullet \sin 2\alpha = \frac{\sin^2 2\alpha}{2\alpha}$$

So the rectangular coordinates are $Q\left(\frac{\sin 2\alpha\cos 2\alpha}{2\alpha}, \frac{\sin^2 2\alpha}{2\alpha}\right)$.

c. The slope of $\overline{OQ}$ is given by: $\dfrac{\frac{\sin^2 2\alpha}{2\alpha} - 0}{\frac{\sin 2\alpha\cos 2\alpha}{2\alpha} - 0} = \dfrac{\sin^2 2\alpha}{\sin 2\alpha\cos 2\alpha} = \dfrac{\sin 2\alpha}{\cos 2\alpha} = \tan 2\alpha$

d. The slope of $\overline{PQ}$ is given by: $\dfrac{\frac{\sin^2 2\alpha}{2\alpha} - \frac{\sin^2\alpha}{\alpha}}{\frac{\sin 2\alpha\cos 2\alpha}{2\alpha} - \frac{\sin\alpha\cos\alpha}{\alpha}} = \dfrac{\sin^2 2\alpha - 2\sin^2\alpha}{\sin 2\alpha\cos 2\alpha - 2\sin\alpha\cos\alpha}$

e. Making the substitution $\sin^2 2\alpha = 1 - \cos^2 2\alpha$ and using the double-angle identities for sine and cosine:

$$\begin{aligned}
\frac{\sin^2 2\alpha - 2\sin^2\alpha}{\sin 2\alpha\cos 2\alpha - 2\sin\alpha\cos\alpha} &= \frac{1 - \cos^2 2\alpha - 2\sin^2\alpha}{\sin 2\alpha\cos 2\alpha - \sin 2\alpha} \\
&= \frac{\left(1 - 2\sin^2\alpha\right) - \cos^2 2\alpha}{\sin 2\alpha\left(\cos 2\alpha - 1\right)} \\
&= \frac{\cos 2\alpha - \cos^2 2\alpha}{\sin 2\alpha\left(\cos 2\alpha - 1\right)} \\
&= \frac{\cos 2\alpha\left(1 - \cos 2\alpha\right)}{\sin 2\alpha\left(\cos 2\alpha - 1\right)} \\
&= -\frac{\cos 2\alpha}{\sin 2\alpha} \\
&= -\cot 2\alpha
\end{aligned}$$

f. Since the slope of $\overline{OQ}$ is $\tan 2\alpha$ and the slope of $\overline{PQ}$ is $-\cot 2\alpha$:

$$m_{\overline{OQ}} \bullet m_{\overline{PQ}} = \left(\tan 2\alpha\right)\left(-\cot 2\alpha\right) = \tan 2\alpha \bullet \frac{-1}{\tan 2\alpha} = -1$$

Since the product of these slopes is -1, $\overline{OQ} \perp \overline{PQ}$.

97. **a.** If $\frac{a}{b}=\frac{x}{y}$, then: $\frac{a-b}{a+b}\div\frac{b}{b}=\frac{\frac{a}{b}-1}{\frac{a}{b}+1}=\frac{\frac{x}{y}-1}{\frac{x}{y}+1}\bullet\frac{y}{y}=\frac{x-y}{x+y}$

b. Since $\frac{a}{b}=\frac{\sin A}{\sin B}$, using the result in part **a** with $x=\sin A$ and $y=\sin B$: $\frac{a-b}{a+b}=\frac{x-y}{x+y}=\frac{\sin A-\sin B}{\sin A+\sin B}$

c. Using the sum-to-product formulas:

$$\sin A-\sin B=2\cos\tfrac{A+B}{2}\sin\tfrac{A-B}{2} \qquad \sin A+\sin B=2\sin\tfrac{A+B}{2}\cos\tfrac{A-B}{2}$$

Therefore our result from part **b** becomes:

$$\frac{a-b}{a+b}=\frac{2\cos\frac{A+B}{2}\sin\frac{A-B}{2}}{2\sin\frac{A+B}{2}\cos\frac{A-B}{2}}=\frac{\sin\frac{A-B}{2}}{\cos\frac{A-B}{2}}\div\frac{\sin\frac{A+B}{2}}{\cos\frac{A+B}{2}}=\frac{\tan\frac{1}{2}(A-B)}{\tan\frac{1}{2}(A+B)}$$

Chapter 9 Test

1. Draw the triangle:

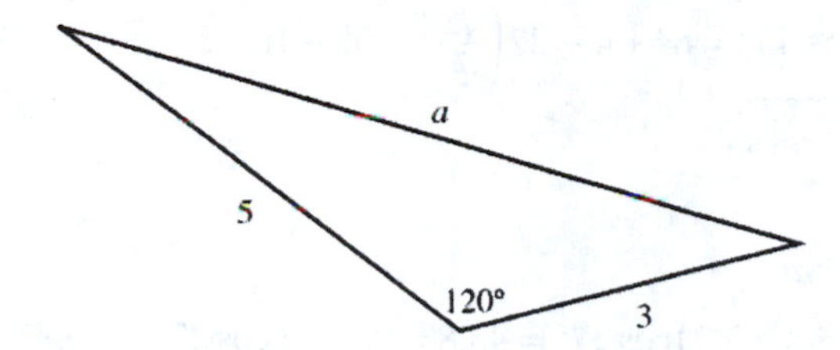

Using the law of cosines: $a^2=3^2+5^2-2(3)(5)\cos 120°=9+25-30\left(-\frac{1}{2}\right)=49$. So $a=7$ cm.

2. Draw the triangle:

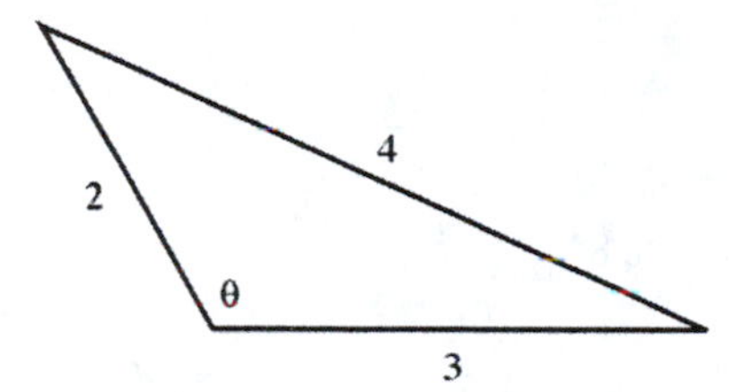

Using the law of cosines:

$$\begin{aligned}4^2&=2^2+3^2-2(2)(3)\cos\theta\\16&=4+9-12\cos\theta\\3&=-12\cos\theta\\\cos\theta&=-\tfrac{1}{4}\end{aligned}$$

The angle opposite the 4 cm side must be obtuse (not acute), since its cosine is negative.

3. Using the law of sines:

$$\frac{\sin 45°}{20\sqrt{2}}=\frac{\sin 30°}{x}$$

$$x=\frac{20\sqrt{2}\sin 30°}{\sin 45°}=\frac{20\sqrt{2}\bullet\frac{1}{2}}{\frac{\sqrt{2}}{2}}=20\text{ cm}$$

4. Let h be the vertical height we are trying to find. Then $\sin 60°=\frac{h}{10}$, so: $h=10\sin 60°=10\bullet\frac{\sqrt{3}}{2}=5\sqrt{3}$ feet

5. Using right-triangle trigonometry: $\tan A=\frac{5}{2}=2.5$, so $\angle A=\tan^{-1}2.5\approx 68.2°$

6. Since the coordinates of B are $(\cos\theta,\sin\theta)$, the base of the triangle is $1+\cos\theta$ and the height is $2\sin\theta$. Thus the area is given by: $A(\theta)=\frac{1}{2}(1+\cos\theta)(2\sin\theta)=\sin\theta(1+\cos\theta)$

7. Draw the figure:

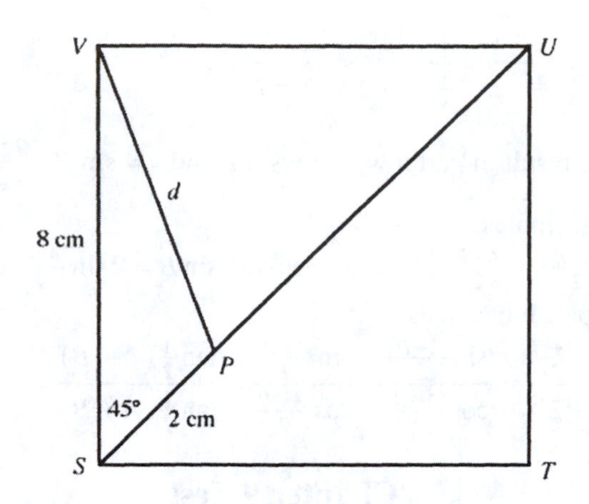

Note that the diagonals of a square bisect the vertex angles. Use the law of cosines with ΔSPV:

$$d^2 = 8^2 + 2^2 - 2(8)(2)\cos 45° = 64 + 4 - 32\left(\frac{\sqrt{2}}{2}\right) = 68 - 16\sqrt{2}$$

$$d = \sqrt{68 - 16\sqrt{2}} = 2\sqrt{17 - 4\sqrt{2}}$$

So $PV = 2\sqrt{17 - 4\sqrt{2}}$ cm.

8. Using the law of cosines to find a:

$$a^2 = (5.8)^2 + (3.2)^2 - 2(5.8)(3.2)\cos 27° = 43.88 - 37.12\cos 27° \approx 10.81$$

So $a \approx \sqrt{10.81} \approx 3.3$ cm. Now use the law of sines to find $\angle C$:

$$\frac{\sin C}{3.2} = \frac{\sin 27°}{3.3}$$

$$\sin C = \frac{3.2 \sin 27°}{3.3} \approx 0.442$$

$$C \approx 26.2°$$

Finally, find $B \approx 180° - 27° - 26.2° \approx 126.8°$.

9. Draw the figure:

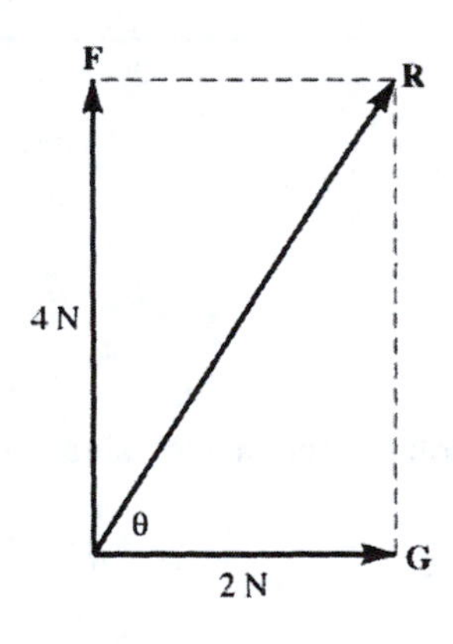

a. Find the magnitude of the resultant: $|\mathbf{R}| = \sqrt{2^2 + 4^2} = \sqrt{20} = 2\sqrt{5}$ N

b. $\tan\theta = \frac{4}{2} = 2$

10. Draw the figure:

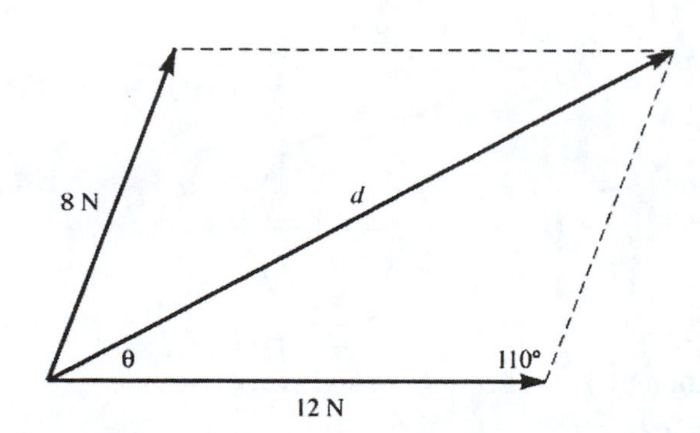

a. Using the law of cosines: $d^2 = 12^2 + 8^2 - 2(12)(8)\cos 110° = 208 - 192\cos 110°$

So $d = \sqrt{208 - 192\cos 110°} = 4\sqrt{13 - 12\cos 110°}$ N.

b. Using the law of sines:

$$\frac{\sin\theta}{8} = \frac{\sin 110°}{4\sqrt{13-12\cos 110°}}$$

$$\sin\theta = \frac{2\sin 110°}{\sqrt{13-12\cos 110°}}$$

11. Draw the figure:

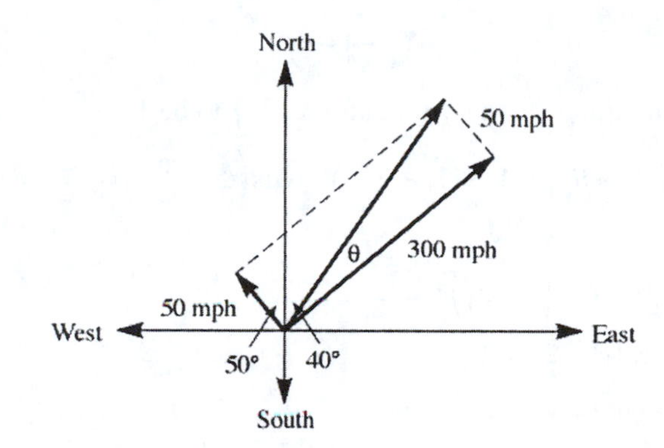

The heading vector and the wind vector are perpendicular to each other, so:

ground speed $= \sqrt{300^2 + 50^2} = 50\sqrt{37}$ mph $\qquad \tan\theta = \frac{50}{300} = \frac{1}{6}$

12. **a.** $2\mathbf{A} + 3\mathbf{B} = 2\langle 2,4\rangle + 3\langle 3,-1\rangle = \langle 13,5\rangle$

b. Compute the length: $|2\mathbf{A} + 3\mathbf{B}| = |\langle 13,5\rangle| = \sqrt{194}$

c. $\mathbf{C} - \mathbf{B} = \langle 4,-4\rangle - \langle 3,-1\rangle = \langle 1,-3\rangle = \mathbf{i} - 3\mathbf{j}$

13. Subtracting components: $\overrightarrow{PQ} = \langle -7,2\rangle - \langle 4,5\rangle = \langle -11,-3\rangle$

The required unit vector is: $\dfrac{\overrightarrow{PQ}}{|\overrightarrow{PQ}|} = \dfrac{1}{\sqrt{130}}\langle -11,-3\rangle = \left\langle \dfrac{-11\sqrt{130}}{130}, \dfrac{-3\sqrt{130}}{130} \right\rangle$

14. Using the double-angle formula for $\cos 2\theta$ and multiplying by r^2 yields:

$$r^2 = \cos^2\theta - \sin^2\theta$$

$$r^4 = r^2\cos^2\theta - r^2\sin^2\theta$$

$$\left(x^2 + y^2\right)^2 = x^2 - y^2$$

15. Plotting points in polar form:

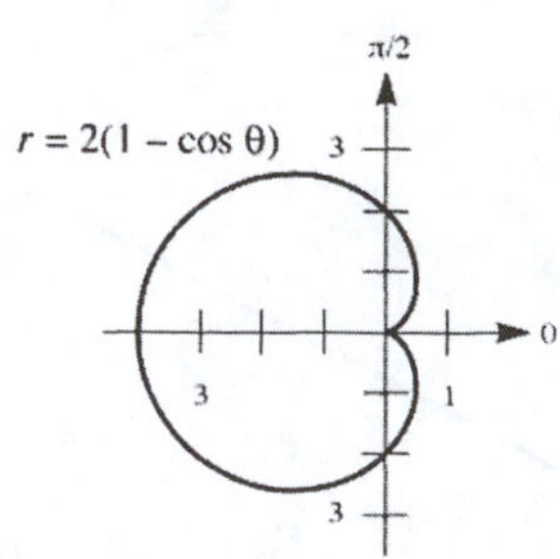

16. Multiplying the second equation by 2 yields $x = 4\sin t$ and $2y = 4\cos t$, so:

$$x^2 + (2y)^2 = 16\sin^2 t + 16\cos^2 t$$
$$x^2 + 4y^2 = 16$$
$$\frac{x^2}{16} + \frac{y^2}{4} = 1$$

So the given parametric equations determine an ellipse.

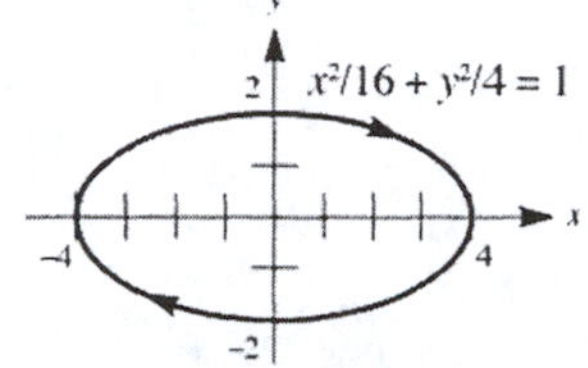

17. Use the distance formula, taking the points (r_1, θ_1) and (r_2, θ_2) to be $\left(4, \frac{10\pi}{21}\right)$ and $\left(1, \frac{\pi}{7}\right)$:

$$d^2 = r_1^2 + r_2^2 - 2r_1r_2\cos(\theta_2 - \theta_1) = 4^2 + 1^2 - 2(4)(1)\cos\left(\tfrac{\pi}{7} - \tfrac{10\pi}{21}\right) = 17 - 8\cos\left(-\tfrac{\pi}{3}\right) = 17 - 8 \bullet \tfrac{1}{2} = 13$$

So $d = \sqrt{13}$.

18. Using the equation $r^2 + r_0^2 - 2rr_0\cos(\theta - \theta_0) = a^2$:

$$r^2 + (5)^2 - 2(r)(5)\cos\left(\theta - \tfrac{\pi}{2}\right) = 2^2$$
$$r^2 + 25 - 10r\cos\left(\theta - \tfrac{\pi}{2}\right) = 4$$
$$r^2 - 10r\cos\left(\theta - \tfrac{\pi}{2}\right) = -21$$

Now substitute the polar point $\left(2, \frac{\pi}{6}\right)$ in the equation: $2^2 - 10(2)\cos\left(\frac{\pi}{6} - \frac{\pi}{2}\right) = 4 - 20\cos\left(-\frac{\pi}{3}\right) = 4 - 20 \bullet \frac{1}{2} = -6$

Since we did not obtain –21, the point does not lie on the circle.

19. **a.** The line will be of the form $r\cos(\theta - \alpha) = d$. Since $\alpha = \frac{5\pi}{6}$ and $d = 4$, the equation of the line is $r\cos\left(\theta - \frac{5\pi}{6}\right) = 4$.

b. Using the addition formula for cosine:

$$r\cos\left(\theta - \tfrac{5\pi}{6}\right) = 4$$
$$r\left(\cos\theta\cos\tfrac{5\pi}{6} + \sin\theta\sin\tfrac{5\pi}{6}\right) = 4$$
$$r\left[\cos\theta \bullet \left(-\tfrac{\sqrt{3}}{2}\right) + \sin\theta \bullet \tfrac{1}{2}\right] = 4$$
$$-r\sqrt{3}\cos\theta + r\sin\theta = 8$$
$$-r\sqrt{3}\cos\theta + r\sin\theta - 8 = 0$$

Chapter 10
Systems of Equations

10.1 Systems of Two Linear Equations in Two Unknowns

1. **a.** yes **b.** no--the xy term makes it non-linear
c. yes **d.** yes

3. Test (5,1) in the two equations and see if it produces true statements:

$2(5)-8(1)=2$
$10-8=2$ true

$3(5)+7(1)=22$
$15+7=22$ true

So (5,1) is a solution to the given system.

5. Test (0,–4) in the two equations and see if it produces true statements:

$\frac{1}{6}(0)+\frac{1}{2}(-4)=-2$
$0-2=-2$ true

$\frac{2}{3}(0)+\frac{3}{4}(-4)=2$
$0-3=2$ false

So (0,–4) is not a solution to the given system.

7. Graphing the two equations:

The system is consistent with exactly one solution.

9. Graphing the two equations:

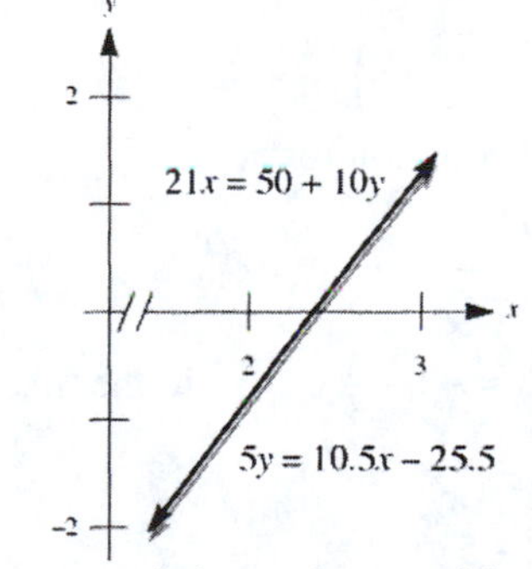

The system is inconsistent with no solution.

11. Solve the first equation for y:

$$\begin{aligned} 4x - y &= 7 \\ y &= 4x - 7 \end{aligned}$$

Now substitute $y = 4x - 7$ for y into the second equation, and solve for x:

$$\begin{aligned} -2x + 3y &= 9 \\ -2x + 3(4x - 7) &= 9 \\ -2x + 12x - 21 &= 9 \\ 10x - 21 &= 9 \\ 10x &= 30, \text{ so } x = 3 \end{aligned}$$

Substitute this back into the first equation: $y = 4x - 7 = 4(3) - 7 = 12 - 7 = 5$. The solution is (3,5).

13. Solve the first equation for x:

$$\begin{aligned} 6x - 2y &= -3 \\ 6x &= 2y - 3 \\ x &= \frac{2y-3}{6} \end{aligned}$$

Now substitute $x = \frac{2y-3}{6}$ for x in the second equation, and solve for y:

$$\begin{aligned} 5x + 3y &= 4 \\ 5\left(\frac{2y-3}{6}\right) + 3y &= 4 \end{aligned}$$

Multiply each side by 6:

$$\begin{aligned} 5(2y - 3) + 6(3y) &= 24 \\ 10y - 15 + 18y &= 24 \\ 28y - 15 &= 24 \\ 28y &= 39 \\ y &= \tfrac{39}{28} \end{aligned}$$

Substitute this back into the first equation: $x = \frac{2y-3}{6} = \frac{2\left(\frac{39}{28}\right) - 3}{6} = \frac{\frac{39}{14} - 3}{6} = \frac{39 - 42}{84} = -\frac{1}{28}$

The solution is $\left(-\frac{1}{28}, \frac{39}{28}\right)$.

15. Solve the second equation for x:

$$\begin{aligned} x + \tfrac{3}{4}y &= -1 \\ x &= -1 - \tfrac{3}{4}y \end{aligned}$$

Now substitute $x = -1 - \frac{3}{4}y$ for x in the first equation, and solve for y:

$$\begin{aligned} \tfrac{3}{2}x - 5y &= 1 \\ \tfrac{3}{2}\left(-1 - \tfrac{3}{4}y\right) - 5y &= 1 \\ -\tfrac{3}{2} - \tfrac{9}{8}y - 5y &= 1 \end{aligned}$$

Multiply by 8 to clear fractions:

$$\begin{aligned} -12 - 9y - 40y &= 8 \\ -12 - 49y &= 8 \\ -49y &= 20 \\ y &= -\tfrac{20}{49} \end{aligned}$$

Substitute this back into the second equation: $x = -1 - \frac{3}{4}y = -1 - \left(\frac{3}{4}\right)\left(-\frac{20}{49}\right) = -1 + \frac{15}{49} = -\frac{34}{49}$

The solution is $\left(-\frac{34}{49}, -\frac{20}{49}\right)$.

17. Solve the first equation for x:

$$4x+6y=3$$
$$4x=3-6y$$
$$x=\frac{3-6y}{4}$$

Now substitute $x=\frac{3-6y}{4}$ for x in the second equation, and solve for y:

$$-6x-9y=-\tfrac{9}{2}$$
$$-6\left(\frac{3-6y}{4}\right)-9y=-\tfrac{9}{2}$$

Multiply by 4 to clear fractions:

$$-6(3-6y)-36y=-18$$
$$-18+36y-36y=-18$$
$$-18=-18$$

The system is dependent. To represent the solution, solve the first equation for y:

$$4x+6y=3$$
$$6y=3-4x$$
$$y=\frac{3-4x}{6}$$

So any ordered pair of the form $\left(x,\frac{3-4x}{6}\right)$ satisfies the system. Another form of the answer is any ordered pair of the form $\left(\frac{3-6y}{4},y\right)$.

19. **a.** The solution is approximately (16.30,–24.5). Sketching the graph:

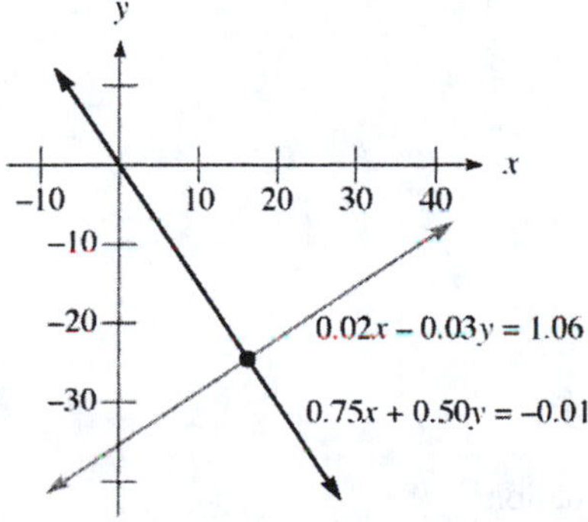

b. First multiply each equation by 100 to clear decimals:

$$100(0.02x-0.03y)=100(1.06)$$
$$100(0.75x+0.50y)=100(-0.01)$$

Therefore:

$$2x-3y=106$$
$$75x+50y=-1$$

Solve the first equation for x:

$$2x-3y=106$$
$$2x=3y+106$$
$$x=\frac{3y+106}{2}$$

Now substitute $x=\frac{3y+106}{2}$ for x in the second equation and solve for y:

$$75x+50y=-1$$
$$75\left(\frac{3y+106}{2}\right)+50y=-1$$

Multiply by 2 to clear fractions:

$$75(3y+106)+100y=-2$$
$$225y+7950+100y=-2$$
$$325y+7950=-2$$
$$325y=-7952$$
$$y=-\frac{7952}{325}$$

Substitute this back into the first equation:

$$x=\frac{3y+106}{2}=\frac{3\left(-\frac{7952}{325}\right)+106}{2}=\frac{3(-7952)+106(325)}{650}=\frac{-23856+34450}{650}=\frac{10594}{650}=\frac{5297}{325}$$

The solution is $\left(\frac{5297}{325},-\frac{7952}{325}\right)\approx(16.30,-24.47)$.

21. Multiply the second equation by 2:

$$5x+6y=4$$
$$4x-6y=-6$$

Adding:

$$9x=-2$$
$$x=-\frac{2}{9}$$

Substitute $x=-\frac{2}{9}$ into the first equation:

$$5x+6y=4$$
$$5\left(-\frac{2}{9}\right)+6y=4$$
$$-\frac{10}{9}+6y=4$$
$$6y=\frac{46}{9}$$
$$y=\frac{46}{54}=\frac{23}{27}$$

The solution is $\left(-\frac{2}{9},\frac{23}{27}\right)$.

23. Multiply the second equation by –2:

$$4x+\ 13y=-5$$
$$-4x+108y=2$$

Adding:

$$121y=-3$$
$$y=-\frac{3}{121}$$

Substitute $y=-\frac{3}{121}$ into the first equation:

$$4x+13y=-5$$
$$4x+13\left(-\frac{3}{121}\right)=-5$$
$$4x-\frac{39}{121}=-5$$
$$4x=-\frac{566}{121}$$
$$x=-\frac{283}{242}$$

The solution is $\left(-\frac{283}{242},-\frac{3}{121}\right)$.

25. Multiply the first equation by 12 and the second equation by 70 to clear fractions:

$$12\left(\frac{1}{4}x-\frac{1}{3}y\right)=12(4)$$
$$70\left(\frac{2}{7}x-\frac{1}{7}y\right)=70\left(\frac{1}{10}\right)$$

Therefore we have the system:

$$3x-\ \ 4y=48$$
$$20x-10y=7$$

Multiply the first equation by –5 and the second equation by 2:

$$-15x+20y=-240$$
$$40x-20y=14$$

Adding:

$$25x=-226$$
$$x=-\frac{226}{25}$$

Substitute $x=-\frac{226}{25}$ into the second equation:

$$20x-10y=7$$
$$20\left(-\frac{226}{25}\right)-10y=7$$
$$-\frac{904}{5}-10y=7$$
$$-10y=\frac{939}{5}$$
$$y=-\frac{939}{50}$$

The solution is $\left(-\frac{226}{25},-\frac{939}{50}\right)$.

27. Multiply the second equation by –4:

$$8x+16y=5$$
$$-8x-20y=-5$$

Adding: $-4y=0$, so $y=0$

Substitute $y=0$ into the first equation:

$$8x+16y=5$$
$$8x=5$$
$$x=\frac{5}{8}$$

The solution is $\left(\frac{5}{8},0\right)$.

29. The given points must satisfy the equation, so:

$$4=0^2=b\bullet 0+c \qquad\qquad 14=2^2+b\bullet 2+c$$
$$4=c \qquad\qquad 10=2b+c$$

This system, $c = 4$ and $2b + c = 10$, is easily solved by substitution:

$$2b+c=10$$
$$2b+4=10$$
$$2b=6$$
$$b=3$$
$$y=x^2+3x+4$$

So the equation of the parabola is $y=x^2+3x+4$.

31. Again, the points satisfy the equation:

$$Ax+By=2$$
$$A(-4)+B(5)=2$$
$$A(7)+B(-9)=2$$

So we have the system:

$$-4A+5B=2$$
$$7A-9B=2$$

Multiply the first equation by 7 and the second equation by 4:

$$-28A+35B=14$$
$$28A-36B=8$$
$$-B=22$$
$$B=-22$$

Substituting for B:

$$7A-9(-22)=2$$
$$7A+198=2$$
$$7A=-196$$
$$A=-28$$

33. **a.** Substituting $p = \$6$ in each equation:

supply: $q = 200 \bullet 6 = 1200$ demand: $q = 9600 - 400 \bullet 6 = 7200$

There will be a shortage of 6000 items. Substituting $p = \$12$ in each equation:

supply: $q = 200 \bullet 12 = 2400$ demand: $q = 9600 - 400 \bullet 12 = 4800$

There will be a shortage of 2400 items.

b. Substituting $p = \$20$ in each equation:

supply: $q = 200 \bullet 20 = 4000$ demand: $q = 9600 - 400 \bullet 20 = 1600$

There will be a surplus of 2400 items.

c. Setting the supply and demand equal:

$$200p=9600-400p$$
$$600p=9600$$
$$p=16$$

The equilibrium price is \$16 and the corresponding equilibrium quantity is 3200 items.

35. Let x = the amount of the 10% solution and y = the amount of the 35% solution. Then $x + y = 200$ cc. Also we know that the amount of acid in each separate solution, 10% of x and 35% of y, must add to equal the acid in the mixture, 25% of $(x + y)$. This second equation is usually in need of simplifying:

$$0.10x+0.35y=0.25(x+y)$$
$$10x+35y=25(x+y)$$
$$2x+7y=5x+5y$$
$$-3x+2y=0$$

This system is now solved by either method:

$$x+\ \ y=200$$
$$-3x+2y=0$$

Multiply the first equation by 3:

$$3x+3y=600$$
$$-3x+2y=0$$

Adding:

$$5y=600$$
$$y=120$$

So $x = 80$, and we need 80 cc of the 10% solution and 120 cc of the 35% solution.

37. Let x be the amount of \$5.20 coffee and y be the amount of \$5.80 coffee. Then $x + y = 16$ pounds. The total value of each bean is $5.20x$ and $5.80y$, respectively, and that of the mixture is $5.50(x + y)$, so we have:

$$5.20x+5.80y=5.50(x+y)$$
$$52x+58y=55(x+y)$$
$$52x+58y=55x+55y$$
$$-3x+3y=0$$
$$-x+y=0$$

Solve the system:

$$x+y=16$$
$$-x+y=0$$

Adding:

$$2y = 16$$
$$y = 8$$

Since $x = 16 - y$, $x = 16 - 8 = 8$. So we mix 8 pounds of \$5.20 coffee and 8 pounds of \$5.80 coffee.

39. **a.** Graphing the parabola $y = x^2 + \frac{5}{4}x - \frac{17}{4}$:

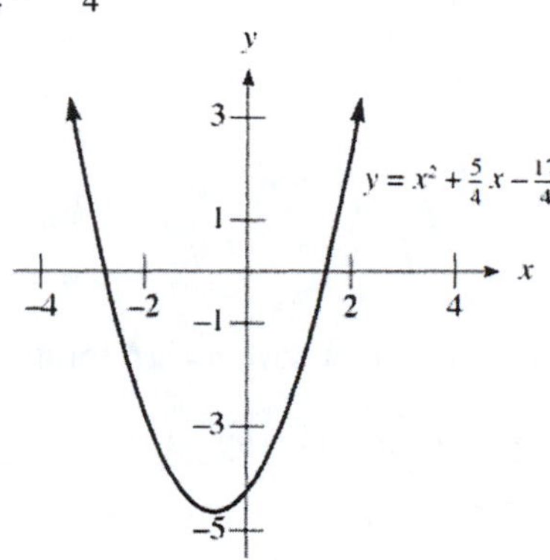

Note that it appears to pass through the given points.

b. Substituting each x-coordinate into the parabola:

$$y = (-3)^2 + \tfrac{5}{4}(-3) - \tfrac{17}{4} = 9 - \tfrac{15}{4} - \tfrac{17}{4} = 9 - 8 = 1$$
$$y = (1)^2 + \tfrac{5}{4}(1) - \tfrac{17}{4} = 1 + \tfrac{5}{4} - \tfrac{17}{4} = 1 - 3 = -2$$

41. Substituting the point (8,–7) results in the system of equations:

$$8a - 7b = 10$$
$$-7a + 8b = -5$$

Multiply the first equation by 7 and the second equation by 8:

$$56a - 49b = 70$$
$$-56a + 64b = -40$$

Adding yields:

$$15b = 30$$
$$b = 2$$

Substituting into the first equation:

$$8a - 14 = 10$$
$$8a = 24$$
$$a = 3$$

The constants are $a = 3$ and $b = 2$.

43. Eliminating fractions:

$$bx + ay = ab$$
$$ax + by = ab$$

Therefore:

$$abx + a^2 y = a^2 b$$
$$-abx - b^2 y = -ab^2$$

Adding:

$$\left(a^2 - b^2\right)y = a^2 b - ab^2$$
$$y = \frac{ab(a-b)}{a^2 - b^2}$$

If we factor and reduce, we have $y=\dfrac{ab}{a+b}$. Now substitute to solve for x:

$$\frac{x}{a}+\frac{\frac{ab}{a+b}}{b}=1$$
$$\frac{x}{a}+\frac{a}{a+b}=1$$
$$(a+b)x+a^2=a(a+b)$$
$$(a+b)x=a^2+ab-a^2$$
$$x=\frac{ab}{a+b}$$

The solution is $\left(\dfrac{ab}{a+b},\dfrac{ab}{a+b}\right)$. Note that we cannot have $a=\pm b$ for this solution.

45. Multiply the first equation by b and the second equation by $-a$:

$$abx+a^2by=b$$
$$-abx-ab^2y=-a$$

Adding:

$$\left(a^2b-ab^2\right)y=b-a$$
$$y=\frac{b-a}{a^2b-ab^2}=\frac{-1}{ab}$$

Substitute $y=\dfrac{-1}{ab}$ into the first equation:

$$ax+a^2y=1$$
$$ax-\frac{a}{b}=1$$
$$ax=\frac{a+b}{b}$$
$$x=\frac{a+b}{ab}$$

So the solution is $\left(\dfrac{a+b}{ab},\dfrac{-1}{ab}\right)$. Note that we cannot have $ab=0$, so $a\neq 0$ and $b\neq 0$.

47. Let $x=\dfrac{1}{s}$ and $y=\dfrac{1}{t}$. Then we have the system:

$$\tfrac{1}{2}x-\tfrac{1}{2}y=-10$$
$$2x+3y=5$$

Multiply the first equation by –4:

$$-2x+2y=40$$
$$2x+3y=5$$

Adding:

$$5y=45$$
$$y=9$$

Substitute $y=9$ into the second equation, and solve for x:

$$2x+3y=5$$
$$2x+3(9)=5$$
$$2x+27=5$$
$$2x=-22$$
$$x=-11$$

Since $s=\dfrac{1}{x}$ and $t=\dfrac{1}{y}$, then $s=-\tfrac{1}{11}$ and $t=\tfrac{1}{9}$. So the solution is $\left(-\tfrac{1}{11},\tfrac{1}{9}\right)$.

49. First clear fractions and put into a standard form: $\frac{2w-1}{3}+\frac{z+2}{4}=4$

Multiplying by 12 we have the equation:

$$\begin{aligned} 4(2w-1)+3(z+2) &= 12 \bullet 4 \\ 8w-4+3z+6 &= 48 \\ 8w+3z &= 46 \end{aligned}$$

The second equation will be $w + 2z = 9$. So we have the system:

$$\begin{aligned} 8w+3z &= 46 \\ w+2z &= 9 \end{aligned}$$

Multiply the second equation by –8:

$$\begin{aligned} 8w+\ \ 3z &= 46 \\ -8w-16z &= -72 \end{aligned}$$

Adding:

$$\begin{aligned} -13z &= -26 \\ z &= 2 \\ w &= 9-2(2) = 5 \end{aligned}$$

So the solution is $(w, z) = (5,2)$.

51. Letting $u = \ln x$ and $v = \ln y$, we have the system:

$$\begin{aligned} 2u-5v &= 11 \\ u+\ \ v &= -5 \end{aligned}$$

Using substitution, solve the second equation for v to obtain $v = -5 - u$. Now substitute into the first equation:

$$\begin{aligned} 2u-5(-5-u) &= 11 \\ 7u+25 &= 11 \\ 7u &= -14 \\ u &= -2 \\ v &= -5+2 = -3 \end{aligned}$$

So $\ln x = -2$ thus $x = e^{-2}$, and $\ln y = -3$ thus $y = e^{-3}$. The solution is (e^{-2}, e^{-3}), which is approximately (0.14,0.05).

53. Letting $u = e^x$ and $v = e^y$, we have the system:

$$\begin{aligned} u-3v &= 2 \\ 3u+\ \ v &= 16 \end{aligned}$$

Using substitution, solve the first equation for u to obtain $u = 2 + 3v$. Now substitute into the second equation:

$$\begin{aligned} 3(2+3v)+v &= 16 \\ 10v+6 &= 16 \\ 10v &= 10 \\ v &= 1 \\ u &= 5 \end{aligned}$$

So $e^x = 5$ thus $x = \ln 5$, and $e^y = 1$ thus $y = 0$. The solution is $(\ln 5, 0)$, which is approximately (1.61,0).

55. Letting $u=\sqrt{x^2-3x}$ and $v=\sqrt{y^2+6y}$, we have the system:

$$\begin{aligned} 4u-\ \ 3v &= -4 \\ \tfrac{1}{2}u+\tfrac{1}{2}v &= 3 \end{aligned}$$

Multiplying the second equation by 6, we have the system:

$$\begin{aligned} 4u-3v &= -4 \\ 3u+3v &= 18 \end{aligned}$$

Adding we obtain $7u = 14$, so $u = 2$. Substitute into the first equation:

$$\begin{aligned} 4(2)-3v &= -4 \\ 8-3v &= -4 \\ -3v &= -12 \\ v &= 4 \end{aligned}$$

So $u = 2$ and $v = 4$. Since $u = \sqrt{x^2 - 3x}$ and $v = \sqrt{y^2 + 6y}$, we have the two equations:

$$\begin{aligned} \sqrt{x^2 - 3x} &= 2 \\ x^2 - 3x &= 4 \\ x^2 - 3x - 4 &= 0 \\ (x-4)(x+1) &= 0 \\ x &= 4, -1 \end{aligned} \qquad \begin{aligned} \sqrt{y^2 + 6y} &= 4 \\ y^2 + 6y &= 16 \\ y^2 + 6y - 16 &= 0 \\ (y+8)(y-2) &= 0 \\ y &= -8, 2 \end{aligned}$$

So the solutions are (4,–8), (4,2), (–1,–8), and (–1,2).

57. If we take tu as our number, then it is important to distinguish between its value, $10t + u$, and the sum of its digits $t + u$. Here we are told that $t + u = 14$ and that $2t = u + 1$. We solve this system:

$$\begin{aligned} t + u &= 14 \\ 2t - u &= 1 \end{aligned}$$

Adding: $3t = 15$, so $t = 5$. Thus $u = 9$ and our original two-digit number was 59.

59. Let $u = \frac{1}{x}$ and $v = \frac{1}{y}$. Then:

$$\begin{aligned} \frac{a}{b}u + \frac{b}{a}v &= a + b \\ bu + av &= a^2 + b^2 \end{aligned}$$

Multiply the first equation by ab to clear the fractions:

$$\begin{aligned} a^2u + b^2v &= a^2b + ab^2 \\ bu + av &= a^2 + b^2 \end{aligned}$$

Multiply the first equation by $-b$ and the second equation by a^2:

$$\begin{aligned} -a^2bu - b^3v &= -a^2b^2 - ab^3 \\ a^2bu + a^3v &= a^4 + a^2b^2 \end{aligned}$$

Adding:

$$\begin{aligned} \left(a^3 - b^3\right)v &= a^4 - ab^3 \\ \left(a^3 - b^3\right)v &= a\left(a^3 - b^3\right) \\ v &= a \end{aligned}$$

Substituting $v = a$ into the second equation:

$$\begin{aligned} bu + av &= a^2 + b^2 \\ bu + a^2 &= a^2 + b^2 \\ bu &= b^2 \\ u &= b \end{aligned}$$

Since $x = \frac{1}{u}$ and $y = \frac{1}{v}$, then $x = \frac{1}{b}$ and $y = \frac{1}{a}$. So the solution is $\left(\frac{1}{b}, \frac{1}{a}\right)$.

61. **a.** Since the lines are concurrent, the following two systems must possess the same solution:

$$I.\quad \begin{aligned} 7x+5y&=4 \\ x+ky&=3 \end{aligned} \qquad\qquad II.\quad \begin{aligned} x+ky&=3 \\ 5x+\ y&=-k \end{aligned}$$

Using either method of this section, we find that the solution of system I is $\left(\frac{4k-15}{7k-5}, \frac{17}{7k-5}\right)$. Also the solution of system II is found to be $\left(\frac{-3-k^2}{5k-1}, \frac{k+15}{5k-1}\right)$. Since the two solutions are the same, the corresponding x- and y-coordinates must be equal. Equating the y-coordinates gives us: $\frac{17}{7k-5} = \frac{k+15}{5k-1}$

After clearing fractions and simplifying, this equation becomes $7k^2 + 15k - 58 = 0$. We factor to get $(7k + 29)(k - 2) = 0$ and therefore $k = -\frac{29}{7}$ or $k = 2$. These are the required values of k. If we equate the x-coordinates rather than the y-coordinates we obtain the equation $7k^3 + 15k^2 - 58k = 0$. We factor to get $k(7k + 29)(k - 2) = 0$, which has solutions $k = 0, -\frac{29}{7}$, and 2. The root $k = 0$ is extraneous, for in that case the two y-coordinates are not equal.

b. For $k = -\frac{29}{7}$ the graphs are: For $k = 2$ the graphs are:

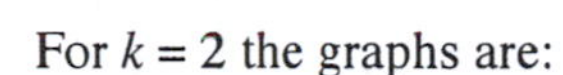

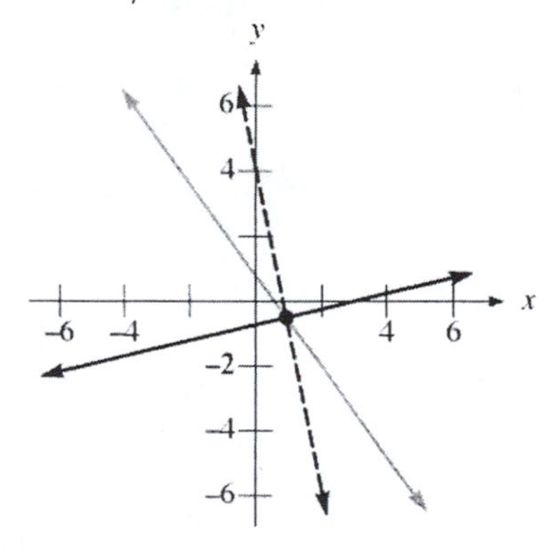

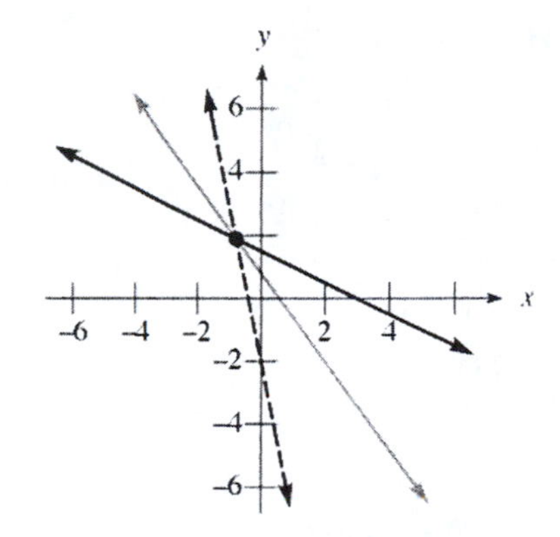

10.2 Gaussian Elimination

1. We are given the system:

$$\begin{aligned} 2x+y+\ z&=-9 \\ 3y-2z&=-4 \\ 8z&=-8 \end{aligned}$$

Solve for z in the third equation:

$$\begin{aligned} 8z&=-8 \\ z&=-1 \end{aligned}$$

Substitute into the second equation:

$$\begin{aligned} 3y-2z&=-4 \\ 3y-2(-1)&=-4 \\ 3y+2&=-4 \\ 3y&=-6 \\ y&=-2 \end{aligned}$$

Substitute into the first equation:

$$\begin{aligned} 2x+y+z&=-9 \\ 2x+(-2)+(-1)&=-9 \\ 2x-3&=-9 \\ 2x&=-6 \\ x&=-3 \end{aligned}$$

The solution is $(-3,-2,-1)$.

3. We are given the system:

$$8x+5y+3z=1$$
$$3y+4z=2$$
$$5z=3$$

Solve for z in the third equation:

$$5z=3$$
$$z=\frac{3}{5}$$

Substitute into the second equation:

$$3y+4z=2$$
$$3y+(4)\frac{3}{5}=2$$
$$3y+\frac{12}{5}=2$$
$$3y=-\frac{2}{5}$$
$$y=-\frac{2}{15}$$

Substitute into the first equation:

$$8x+5y+3z=1$$
$$8x+5\left(-\frac{2}{15}\right)+(3)\frac{3}{5}=1$$
$$8x-\frac{2}{3}+\frac{9}{5}=1$$
$$8x-\frac{10}{15}+\frac{27}{15}=1$$
$$8x=-\frac{2}{15}$$
$$x=-\frac{1}{60}$$

The solution is $\left(-\frac{1}{60},-\frac{2}{15},\frac{3}{5}\right)$.

5. We are given the system:

$$-4x+5y=0$$
$$3y+2z=1$$
$$3z=-1$$

Solve for z in the third equation:

$$3z=-1$$
$$z=-\frac{1}{3}$$

Substitute into the second equation:

$$3y+2z=1$$
$$3y+2\left(-\frac{1}{3}\right)=1$$
$$3y-\frac{2}{3}=1$$
$$3y=\frac{5}{3}$$
$$y=\frac{5}{9}$$

Substitute into the first equation:

$$-4x+5y=0$$
$$-4x+(5)\frac{5}{9}=0$$
$$-4x+\frac{25}{9}=0$$
$$-4x=-\frac{25}{9}$$
$$x=\frac{25}{36}$$

The solution is $\left(\frac{25}{36},\frac{5}{9},-\frac{1}{3}\right)$.

7. We are given the system:

$$\begin{aligned} -x+8y+3z&=0 \\ 2z&=0 \end{aligned}$$

Solve for z in the second equation:

$$\begin{aligned} 2z&=0 \\ z&=0 \end{aligned}$$

Substitute into the first equation:

$$\begin{aligned} -x+8y+3z&=0 \\ -x+8y+3(0)&=0 \\ -x+8y&=0 \\ 8y&=x \\ y&=\frac{x}{8} \end{aligned}$$

The solution is $\left(x,\frac{x}{8},0\right)$, where x is any real number.

9. We are given the system:

$$\begin{aligned} 2x+3y+\ z+\ \ w&=-6 \\ y+3z-4w&=23 \\ 6z-5w&=31 \\ -2w&=10 \end{aligned}$$

Solve for w in the fourth equation:

$$\begin{aligned} -2w&=10 \\ w&=-5 \end{aligned}$$

Substitute into the third equation:

$$\begin{aligned} 6z-5w&=31 \\ 6z-5(-5)&=31 \\ 6z+25&=31 \\ 6z&=6 \\ z&=1 \end{aligned}$$

Substitute into the second equation:

$$\begin{aligned} y+3z-4w&=23 \\ y+3(1)-4(-5)&=23 \\ y+23&=23 \\ y&=0 \end{aligned}$$

Substitute into the first equation:

$$\begin{aligned} 2x+3y+z+w&=-6 \\ 2x+3(0)+1+(-5)&=-6 \\ 2x-4&=-6 \\ 2x&=-2 \\ x&=-1 \end{aligned}$$

The solution is $(-1,0,1,-5)$.

11. We must arrange this according to echelon form. First add –2 times the first equation to the second one, and –3 times it to the third:

$$\begin{aligned} x+y+z&=12\\ -3y-3z&=-25\\ -y-2z&=-14 \end{aligned}$$

We want to eliminate the y term from the third equation so interchange the second and third equations:

$$\begin{aligned} x+y+z&=12\\ -y-2z&=-14\\ -3y-3z&=-25 \end{aligned}$$

Add –3 times the second equation to the third:

$$\begin{aligned} x+y+z&=12\\ -y-2z&=-14\\ 3z&=17 \end{aligned}$$

Solve for z in the third equation:

$$\begin{aligned} 3z&=17\\ z&=\tfrac{17}{3} \end{aligned}$$

Substitute into the second equation:

$$\begin{aligned} -y-2z&=-14\\ -y-(2)\tfrac{17}{3}&=-14\\ -y-\tfrac{34}{3}&=-14\\ -y&=-\tfrac{8}{3}\\ y&=\tfrac{8}{3} \end{aligned}$$

Substitute into the first equation:

$$\begin{aligned} x+y+z&=12\\ x+\tfrac{8}{3}+\tfrac{17}{3}&=12\\ x+\tfrac{25}{3}&=12\\ x&=\tfrac{11}{3} \end{aligned}$$

The solution is $\left(\frac{11}{3},\frac{8}{3},\frac{17}{3}\right)$.

13. We must arrange the system in echelon form. Multiply the first equation by –2 and add it to the second equation:

$$\begin{aligned} -4x+6y-4z&=-8\\ 4x+2y+3z&=7 \end{aligned}$$

Adding: $8y-z=-1$. Multiply the first equation by $-\frac{5}{2}$ and add it to the third equation:

$$\begin{aligned} -5x+\tfrac{15}{2}y-5z&=-10\\ 5x+4y+2z&=7 \end{aligned}$$

Adding:

$$\begin{aligned} \tfrac{23}{2}y-3z&=-3\\ 23y-6z&=-6 \end{aligned}$$

So we have the system:

$$\begin{aligned} 2x-3y+2z&=4\\ 8y-z&=-1\\ 23y-6z&=-6 \end{aligned}$$

Multiply the second equation by $-\frac{23}{8}$ and add it to the third equation:

$$\begin{aligned} -23y+\tfrac{23}{8}z&=\tfrac{23}{8}\\ 23y-6z&=-6 \end{aligned}$$

Adding: $-\frac{25}{8}z=-\frac{25}{8}$. So we have the system in echelon form:

$$\begin{aligned} 2x-3y+2z&=4 \\ 8y-z&=-1 \\ -\tfrac{25}{8}z&=-\tfrac{25}{8} \end{aligned}$$

Solve for z in the third equation:

$$\begin{aligned} -\tfrac{25}{8}z&=-\tfrac{25}{8} \\ z&=1 \end{aligned}$$

Substitute into the second equation:

$$\begin{aligned} 8y-z&=-1 \\ 8y-1&=-1 \\ 8y&=0 \\ y&=0 \end{aligned}$$

Substitute into the first equation:

$$\begin{aligned} 2x-3y+2z&=4 \\ 2x-3(0)+2(1)&=4 \\ 2x+2&=4 \\ 2x&=2 \\ x&=1 \end{aligned}$$

The solution is $(1,0,1)$.

15. We must arrange the system in echelon form. Multiply the first equation by -2 and add it to the second equation:

$$\begin{aligned} -6x-6y+4z&=-26 \\ 6x+2y-5z&=13 \end{aligned}$$

Adding: $-4y-z=-13$. Multiply the first equation by $-\frac{7}{3}$ and add it to the third equation:

$$\begin{aligned} -7x-7y+\tfrac{14}{3}z&=-\tfrac{91}{3} \\ 7x+5y-3z&=26 \end{aligned}$$

Adding:

$$\begin{aligned} -2y+\tfrac{5}{3}z&=-\tfrac{13}{3} \\ -6y+5z&=-13 \end{aligned}$$

So we have the system:

$$\begin{aligned} 3x+3y-2z&=13 \\ -4y-z&=-13 \\ -6y+5z&=-13 \end{aligned}$$

Multiply the second equation by $-\frac{3}{2}$ and add it to the third equation:

$$\begin{aligned} 6y+\tfrac{3}{2}z&=\tfrac{39}{2} \\ -6y+5z&=-13 \end{aligned}$$

Adding: $\frac{13}{2}z=\frac{13}{2}$. So we have the system in echelon form:

$$\begin{aligned} 3x+3y-2z&=13 \\ -4y-z&=-13 \\ \tfrac{13}{2}z&=\tfrac{13}{2} \end{aligned}$$

Solve for z in the third equation:

$$\begin{aligned} \tfrac{13}{2}z&=\tfrac{13}{2} \\ z&=1 \end{aligned}$$

Substitute into the second equation:

$$\begin{aligned} -4y-z&=-13 \\ -4y-1&=-13 \\ -4y&=-12 \\ y&=3 \end{aligned}$$

Substitute into the first equation:

$$\begin{aligned} 3x+3y-2z&=13\\ 3x+3(3)-2(1)&=13\\ 3x+7&=13\\ 3x&=6\\ x&=2 \end{aligned}$$

The solution is (2,3,1).

17. We must arrange the system in echelon form. Multiply the first equation by 2 and add it to the second equation:

$$\begin{aligned} 2x+2y+2z&=2\\ -2x+\ y+\ z&=-2 \end{aligned}$$

Adding:

$$\begin{aligned} 3y+3z&=0\\ y+\ z&=0 \end{aligned}$$

Multiply the first equation by –3 and add it to the third equation:

$$\begin{aligned} -3x-3y-3z&=-3\\ 3x+6y+6z&=5 \end{aligned}$$

Adding: $3y+3z=2$. So we have the system:

$$\begin{aligned} x+y+\ z&=1\\ y+\ z&=0\\ 3y+3z&=2 \end{aligned}$$

Multiply the second equation by –3 and add it to the third equation:

$$\begin{aligned} -3y-3z&=0\\ 3y+3z&=2 \end{aligned}$$

Adding: 0 = 2. Since this equation is false, there is no solution. This system is inconsistent.

19. First re-arrange the system as:

$$\begin{aligned} x+3y-2z&=2\\ 2x-\ y+\ z&=-1\\ -5x+6y-5z&=5 \end{aligned}$$

Multiply the first equation by –2 and add it to the second equation:

$$\begin{aligned} -2x-6y+4z&=-4\\ 2x-\ y+\ z&=-1 \end{aligned}$$

Adding: $-7y+5z=-5$. Multiply the first equation by 5 and add it to the third equation:

$$\begin{aligned} 5x+15y-10z&=10\\ -5x+\ 6y-\ 5z&=5 \end{aligned}$$

Adding:

$$\begin{aligned} 21y-15z&=15\\ 7y-\ 5z&=5 \end{aligned}$$

So we have the system:

$$\begin{aligned} x+3y-2z&=2\\ -7y+5z&=-5\\ 7y-5z&=5 \end{aligned}$$

Adding the second and third equations yields 0 = 0, so the system is dependent. Solving for x and y in terms of z yields the solution $\left(\dfrac{z+1}{-7},\dfrac{5(z+1)}{7},z\right)$, where z is any real number.

21. First re-arrange the system as:

$$\begin{aligned} x + 3y + 2z &= -1 \\ 2x - y + z &= 4 \\ 7x + 5z &= 11 \end{aligned}$$

Multiply the first equation by –2 and add it to the second equation:

$$\begin{aligned} -2x - 6y - 4z &= 2 \\ 2x - y + z &= 4 \end{aligned}$$

Adding: $-7y - 3z = 6$. Multiply the first equation by –7 and add it to the third equation:

$$\begin{aligned} -7x - 21y - 14z &= 7 \\ 7x + 5z &= 11 \end{aligned}$$

Adding:

$$\begin{aligned} -21y - 9z &= 18 \\ -7y - 3z &= 6 \end{aligned}$$

So we have the system:

$$\begin{aligned} x + 3y + 2z &= -1 \\ -7y - 3z &= 6 \\ -7y - 3z &= 6 \end{aligned}$$

Multiply the second equation by –1 and add it to the third equation:

$$\begin{aligned} 7y + 3z &= -6 \\ -7y - 3z &= 6 \end{aligned}$$

Adding: 0 = 0. So we have the system:

$$\begin{aligned} x + 3y + 2z &= -1 \\ -7y - 3z &= 6 \end{aligned}$$

Solve the second equation for y:

$$\begin{aligned} -7y - 3z &= 6 \\ -7y &= 3z + 6 \\ y &= \frac{-3z - 6}{7} \end{aligned}$$

Substitute into the first equation:

$$\begin{aligned} x + 3y + 2z &= -1 \\ x + 3\left(\frac{-3z - 6}{7}\right) + 2z &= -1 \\ x + \frac{-9z - 18 + 14z}{7} &= -1 \\ x + \frac{5z - 18}{7} &= -1 \\ x &= \frac{-7 - 5z + 18}{7} = \frac{-5z + 11}{7} \end{aligned}$$

The solution is $\left(\frac{11 - 5z}{7}, \frac{-3z - 6}{7}, z\right)$, where z is any real number.

23. Multiply the first equation by –1 and add it to the second equation:

$$\begin{aligned} -x - y - z - w &= -4 \\ x - 2y - z - w &= 3 \end{aligned}$$

Adding:

$$\begin{aligned} -3y - 2z - 2w &= -1 \\ 3y + 2z + 2w &= 1 \end{aligned}$$

Multiply the first equation by –2 and add it to the third equation:

$$\begin{aligned} -2x - 2y - 2z - 2w &= -8 \\ 2x - y + z - w &= 2 \end{aligned}$$

Adding:

$$\begin{aligned} -3y - z - 3w &= -6 \\ 3y + z + 3w &= 6 \end{aligned}$$

Multiply the first equation by –1 and add it to the fourth equation:

$$\begin{aligned} -x - y - z - w &= -4 \\ x - y + 2z - 2w &= -7 \end{aligned}$$

Adding:

$$\begin{aligned} -2y + z - 3w &= -11 \\ 2y - z + 3w &= 11 \end{aligned}$$

So we have the system:

$$\begin{aligned} x + y + z + w &= 4 \\ 3y + 2z + 2w &= 1 \\ 3y + z + 3w &= 6 \\ 2y - z + 3w &= 11 \end{aligned}$$

Multiply the second equation by –1 and add it to the third equation:

$$\begin{aligned} -3y - 2z - 2w &= -1 \\ 3y + z + 3w &= 6 \end{aligned}$$

Adding:

$$\begin{aligned} -z + w &= 5 \\ z - w &= -5 \end{aligned}$$

Multiply the second equation by $-\frac{2}{3}$ and add it to the fourth equation:

$$\begin{aligned} -2y - \tfrac{4}{3}z - \tfrac{4}{3}w &= -\tfrac{2}{3} \\ 2y - z + 3w &= 11 \end{aligned}$$

Adding:

$$\begin{aligned} -\tfrac{7}{3}z + \tfrac{5}{3}w &= \tfrac{31}{3} \\ -7z + 5w &= 31 \end{aligned}$$

So we have the system:

$$\begin{aligned} x + y + z + w &= 4 \\ 3y + 2z + 2w &= 1 \\ z - w &= -5 \\ -7z + 5w &= 31 \end{aligned}$$

Multiply the third equation by 7 and add it to the fourth equation:

$$\begin{aligned} 7z - 7w &= -35 \\ -7z + 5w &= 31 \end{aligned}$$

Adding: $-2w = -4$. So we have the system:

$$\begin{aligned} x + y + z + w &= 4 \\ 3y + 2z + 2w &= 1 \\ z - w &= -5 \\ -2w &= -4 \end{aligned}$$

Solve the fourth equation for w:

$$\begin{aligned} -2w &= -4 \\ w &= 2 \end{aligned}$$

Substitute into the third equation:

$$\begin{aligned} z - w &= -5 \\ z - 2 &= -5 \\ z &= -3 \end{aligned}$$

Substitute into the second equation:

$$\begin{aligned} 3y + 2z + 2w &= 1 \\ 3y + 2(-3) + 2(2) &= 1 \\ 3y - 2 &= 1 \\ 3y &= 3 \\ y &= 1 \end{aligned}$$

Substitute into the first equation:

$$\begin{aligned} x + y + z + w &= 4 \\ x + 1 - 3 + 2 &= 4 \\ x &= 4 \end{aligned}$$

The solution is (4,1,–3,2).

25. First re-arrange the system as:

$$\begin{aligned} x + 4y - 3z &= 1 \\ 2x + 3y + 2z &= 5 \end{aligned}$$

Multiply the first equation by –2 and add it to the second equation:

$$\begin{aligned} -2x - 8y + 6z &= -2 \\ 2x + 3y + 2z &= 5 \end{aligned}$$

Adding: $-5y + 8z = 3$. So we have the system:

$$\begin{aligned} x + 4y - 3z &= 1 \\ -5y + 8z &= 3 \end{aligned}$$

Solve the second equation for y:

$$\begin{aligned} -5y + 8z &= 3 \\ -5y &= 3 - 8z \\ y &= \frac{8z - 3}{5} \end{aligned}$$

Substitute into the first equation:

$$\begin{aligned} x + 4y - 3z &= 1 \\ x + 4\left(\frac{8z - 3}{5}\right) - 3z &= 1 \\ x + \frac{17z - 12}{5} &= 1 \\ x &= \frac{17 - 17z}{5} \end{aligned}$$

The solution is $\left(\frac{17 - 17z}{5}, \frac{8z - 3}{5}, z\right)$, where z is any real number.

27. Multiply the first equation by –3 and add it to the second equation:

$$\begin{aligned} -3x + 6y + 6z - 6w &= 30 \\ 3x + 4y - z - 3w &= 11 \end{aligned}$$

Adding: $10y + 5z - 9w = 41$. Multiply the first equation by 4 and add it to the third equation:

$$\begin{aligned} 4x - 8y - 8z + 8w &= -40 \\ -4x - 3y - 3z + 8w &= -21 \end{aligned}$$

Adding: $-11y - 11z + 16w = -61$. We have the system:

$$\begin{aligned} x - 2y - 2z + 2w &= -10 \\ 10y + 5z - 9w &= 41 \\ -11y - 11z + 16w &= -61 \end{aligned}$$

Add the second and third equations:

$$\begin{aligned} 10y + 5z - 9w &= 41 \\ -11y - 11z + 16w &= -61 \end{aligned}$$

Adding:

$$\begin{aligned} -y - 6z + 7w &= -20 \\ y + 6z - 7w &= 20 \end{aligned}$$

So we have the system:

$$\begin{aligned} x - 2y - 2z + 2w &= -10 \\ y + 6z - 7w &= 20 \\ -11y - 11z + 16w &= -61 \end{aligned}$$

Multiply the second equation by 11 and add it to the third equation:

$$\begin{aligned} 11y + 66z - 77w &= 220 \\ -11y - 11z + 16w &= -61 \end{aligned}$$

Adding: $55z - 61w = 159$. We have the system:

$$\begin{aligned} x - 2y - 2z + 2w &= -10 \\ y + 6z - 7w &= 20 \\ 55z - 61w &= 159 \end{aligned}$$

Solve the third equation for z:

$$\begin{aligned} 55z - 61w &= 159 \\ 55z &= 61w + 159 \\ z &= \frac{61w + 159}{55} \end{aligned}$$

Substitute into the second equation:

$$\begin{aligned} y + 6z - 7w &= 20 \\ y + 6\left(\frac{61w + 159}{55}\right) - 7w &= 20 \\ y + \frac{954 - 19w}{55} &= 20 \\ y &= \frac{146 + 19w}{55} \end{aligned}$$

Substitute into the first equation:

$$\begin{aligned} x - 2y - 2z + 2w &= -10 \\ x - 2\left(\frac{146 + 19w}{55}\right) - 2\left(\frac{61w + 159}{55}\right) + 2w &= -10 \\ x + \frac{-122 - 10w}{11} &= -10 \\ x &= \frac{12 + 10w}{11} \end{aligned}$$

The solution is $\left(\frac{12 + 10w}{11}, \frac{146 + 19w}{55}, \frac{61w + 159}{55}, w\right)$, where w is any real number.

29. Solving the second equation for y yields $y = \frac{2z+3}{3}$. Substitute this into the first equation to get $x = -\frac{5z}{12}$. The solution is $\left(-\frac{5z}{12}, \frac{2z+3}{3}, z\right)$, where z is any real number.

31. Let x represent the number of Type I jets, y represent the number of Type II jets, and z represent the number of Type III jets. The system of equations is:

$$\begin{aligned} 40x+80y+70z&=830 \\ 35x+75y+60z&=735 \\ 30x+44y+56z&=592 \end{aligned}$$

Dividing the first equation by 10, the second equation by 5, and the third equation by 2:

$$\begin{aligned} 4x+8y+7z&=83 \\ 7x+15y+12z&=147 \\ 15x+22y+28z&=296 \end{aligned}$$

Multiply the first equation by –7 and the second equation by 4:

$$\begin{aligned} -28x-56y-49z&=-581 \\ 28x+60y+48z&=588 \\ 4y-z&=7 \end{aligned}$$

Multiply the first equation by –15 and the third equation by 4:

$$\begin{aligned} -60x-120y-105z&=-1245 \\ 60x+88y+112z&=1184 \\ -32y+7z&=-61 \end{aligned}$$

So the system of equations becomes:

$$\begin{aligned} 4x+8y+7z&=83 \\ 4y-z&=7 \\ -32y+7z&=-61 \end{aligned}$$

Multiply the second equation by 8:

$$\begin{aligned} 32y-8z&=56 \\ -32y+7z&=-61 \end{aligned}$$

Adding yields:

$$\begin{aligned} -z&=-5 \\ z&=5 \end{aligned}$$

Substituting into the second equation:

$$\begin{aligned} 4y-5&=7 \\ 4y&=12 \\ y&=3 \end{aligned}$$

Substituting into the first equation:

$$\begin{aligned} 4x+24+35&=83 \\ 4x+59&=83 \\ 4x&=24 \\ x&=6 \end{aligned}$$

The company will need 6 Type I jets, 3 Type II jets, and 5 Type III jets.

33. **a.** Substituting the coordinates of each point, we have the system of equations:

$$\begin{aligned} a+b+c&=-2 \\ a-b+c&=0 \\ 4a+2b+c&=3 \end{aligned}$$

b. Adding –1 times the first equation to the second equation, and –4 times the first equation to the third equation:

$$\begin{aligned} a+b+c&=-2 \\ -2b&=2 \\ -2b-3c&=11 \end{aligned}$$

Dividing the second equation by –2, and adding 2 times the (new) second equation to the third equation:

$$\begin{aligned} a+b+c&=-2 \\ b&=-1 \\ -3c&=9 \end{aligned}$$

So $c=-3$ and $b=-1$. Substituting into the first equation yields $a-4=-2$, so $a=2$. The constants are $a=2$, $b=-1$, and $c=-3$.

c. Graphing the parabola:

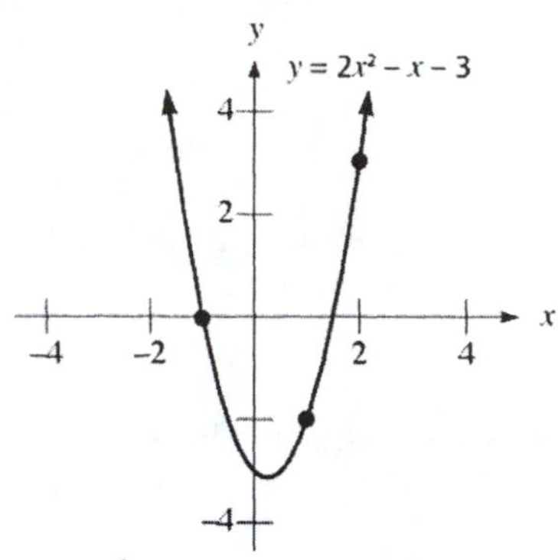

d. The quadratic regression model is $y = 2x^2 - x - 3$.

35. Let x represent the number of utility chairs, y represent the number of secretarial chairs, and z represent the number of managerial chairs. The system of equations is:

$$3x + 4y + 2z = 476$$
$$2x + 5y + 8z = 440$$
$$6x + 4y + z = 826$$

Multiply the first equation by –2 and the second equation by 3:

$$-6x - 8y - 4z = -952$$
$$6x + 15y + 24z = 1320$$
$$7y + 20z = 368$$

Multiply the first equation by –2:

$$-6x - 8y - 4z = -952$$
$$6x + 4y + z = 826$$
$$-4y - 3z = -126$$

The system of equations becomes:

$$3x + 4y + 2z = 476$$
$$7y + 20z = 368$$
$$-4y - 3z = -126$$

Multiplying the first equation by 4 and the second equation by 7:

$$28y + 80z = 1472$$
$$-28y - 21z = -882$$
$$59z = 590$$
$$z = 10$$

Substituting:

$$7y + 200 = 368$$
$$7y = 168$$
$$y = 24$$

Substituting into the first equation:

$$3x + 96 + 20 = 476$$
$$3x = 360$$
$$x = 120$$

The manufacturer should build 120 utility chairs, 24 secretarial chairs, and 10 managerial chairs.

37. **a.** Using $R(t) = at^2 + bt + c$. Substituting the three points:

$$\begin{aligned} 1 &= 0a + 0b + c \\ 3.7 &= 1a + 1b + c \\ 5.8 &= 4a + 2b + c \end{aligned}$$

Simplifying the system:

$$\begin{aligned} c &= 1 \\ a + b + c &= 3.7 \\ 4a + 2b + c &= 5.8 \end{aligned}$$

Substituting $c = 1$ simplifies the system:

$$\begin{aligned} a + b &= 2.7 \\ 4a + 2b &= 4.8 \end{aligned}$$

Multiplying the first equation by -2:

$$\begin{aligned} -2a - 2b &= -5.4 \\ 4a + 2b &= 4.8 \\ 2a &= -0.6 \\ a &= -0.3 \end{aligned}$$

Substituting to find b:

$$\begin{aligned} -0.3 + b &= 2.7 \\ b &= 3 \end{aligned}$$

So the parabola is $R(t) = -0.3t^2 + 3t + 1$.

b. Using the vertex formula: $t = \dfrac{-3}{-0.6} = 5$. The maximumum revenue occurs when $t = 5$, which is the month of December. Since $R(5) = 8.5$, the maximum revenue is \$850,000.

39. Using the hint, we let $A = e^x$, $B = e^y$, and $C = e^z$. Then we have the system:

$$\begin{aligned} A + B - 2C &= 2a \\ A + 2B - 4C &= 3a \\ \tfrac{1}{2}A - 3B + C &= -5a \end{aligned}$$

Multiply the first equation by -1 and add it to the second equation:

$$\begin{aligned} -A - B + 2C &= -2a \\ A + 2B - 4C &= 3a \end{aligned}$$

Adding: $B - 2C = a$. Multiply the first equation by -1 and the third equation by 2, then add the resulting equations:

$$\begin{aligned} -A - B + 2C &= -2a \\ A - 6B + 2C &= -10a \end{aligned}$$

Adding: $-7B + 4C = -12a$. We have the system:

$$\begin{aligned} A + B - 2C &= 2a \\ B - 2C &= a \\ -7B + 4C &= -12a \end{aligned}$$

Multiply the second equation by 7 and add it to the third equation:

$$\begin{aligned} 7B - 14C &= 7a \\ -7B + 4C &= -12a \end{aligned}$$

Adding:

$$\begin{aligned} -10C &= -5a \\ C &= \tfrac{a}{2} \end{aligned}$$

Substituting into the second equation:

$$\begin{aligned} B - 2C &= a \\ B - 2\left(\tfrac{a}{2}\right) &= a \\ B - a &= a \\ B &= 2a \end{aligned}$$

Substituting into the first equation:

$$\begin{aligned} A+B-2C &= 2a \\ A+2a-a &= 2a \\ A+a &= 2a \\ A &= a \end{aligned}$$

So $A = a$, $B = 2a$, and $C = \frac{a}{2}$. Since $A = e^x$, $B = e^y$, $C = e^z$, $x = \ln a$, $y = \ln 2a$ and $z = \ln \frac{a}{2}$.

41. The other two equations are:

$$T_2 = \frac{T_1 + 48 + T_3 + 42}{4} \qquad T_3 = \frac{T_2 + 50 + 35 + 32}{4}$$

Simplifying the three equations results in the system:

$$\begin{aligned} T_1 - 4T_2 + T_3 &= -90 \\ 4T_1 - T_2 &= 125 \\ T_2 - 4T_3 &= -117 \end{aligned}$$

Multiplying the first equation by –4:

$$\begin{aligned} -4T_1 + 16T_2 - 4T_3 &= 360 \\ 4T_1 - T_2 &= 125 \\ 15T_2 - 4T_3 &= 485 \end{aligned}$$

The system of equations becomes:

$$\begin{aligned} T_1 - 4T_2 + T_3 &= -90 \\ T_2 - 4T_3 &= -117 \\ 15T_2 - 4T_3 &= 485 \end{aligned}$$

Multiplying the second equation by –15:

$$\begin{aligned} -15T_2 + 60T_3 &= 1755 \\ 15T_2 - 4T_3 &= 485 \\ 56T_3 &= 2240 \\ T_3 &= 40 \end{aligned}$$

Substituting into the second equation:

$$\begin{aligned} T_2 - 160 &= -117 \\ T_2 &= 43 \end{aligned}$$

Substituting into the first equation:

$$\begin{aligned} T_1 - 172 + 40 &= -90 \\ T_1 &= 42 \end{aligned}$$

The temperatures are $T_1 = 42, T_2 = 43, T_3 = 40$.

43. Let α be the distance from A to B, β be the distance from B to C, and γ be the distance from C to A. The walking, riding and driving rates are, respectively, $\frac{1}{a}$ mi/min, $\frac{1}{b}$ mi/min and $\frac{1}{c}$ mi/min. From the statement of the problem we can obtain the following system of three equations:

$$\begin{aligned} \alpha a + \beta b + \gamma c &= 60(a+c-b) \\ \beta a + \gamma b + \alpha c &= 60(b+a-c) \\ \gamma a + \alpha b + \beta c &= 60(c+b-a) \end{aligned}$$

By adding all three equations: $\alpha(a+b+c) + \beta(a+b+c) + \gamma(a+b+c) = 60(a+b+c)$

Now divide through by the quantity $a + b + c$. This yields $\alpha + \beta + \gamma = 60$ miles.

10.3 Matrices

1. a. two by three (2×3) b. three by two (3×2)

3. five by four (5×4)

5. The coefficient matrix is $\begin{pmatrix} 2 & 3 & 4 \\ 5 & 6 & 7 \\ 8 & 9 & 10 \end{pmatrix}$ and the augmented matrix is $\begin{pmatrix} 2 & 3 & 4 & 10 \\ 5 & 6 & 7 & 9 \\ 8 & 9 & 10 & 8 \end{pmatrix}$.

7. The coefficient matrix is $\begin{pmatrix} 1 & 0 & 1 & 1 \\ 1 & 1 & 0 & 2 \\ 0 & 1 & 1 & 1 \\ 2 & -1 & -1 & 0 \end{pmatrix}$ and the augmented matrix is $\begin{pmatrix} 1 & 0 & 1 & 1 & -1 \\ 1 & 1 & 0 & 2 & 0 \\ 0 & 1 & 1 & 1 & 1 \\ 2 & -1 & -1 & 0 & 2 \end{pmatrix}$.

9. Form the augmented matrix: $\begin{pmatrix} 1 & -1 & 2 & 7 \\ 3 & 2 & -1 & -10 \\ -1 & 3 & 1 & -2 \end{pmatrix}$

Adding –3 times row 1 to row 2 and adding row 1 to row 3 yields: $\begin{pmatrix} 1 & -1 & 2 & 7 \\ 0 & 5 & -7 & -31 \\ 0 & 2 & 3 & 5 \end{pmatrix}$

Multiplying row 3 by –3 and adding to row 2 yields: $\begin{pmatrix} 1 & -1 & 2 & 7 \\ 0 & -1 & -16 & -46 \\ 0 & 2 & 3 & 5 \end{pmatrix}$

Multiplying row 2 by 2 and adding to row 3 yields: $\begin{pmatrix} 1 & -1 & 2 & 7 \\ 0 & -1 & -16 & -46 \\ 0 & 0 & -29 & -87 \end{pmatrix}$

So we have the system of equations:

$$\begin{aligned} x - y + 2z &= 7 \\ -y - 16z &= -46 \\ -29z &= -87 \end{aligned}$$

Solve equation three for z:

$$\begin{aligned} -29z &= -87 \\ z &= 3 \end{aligned}$$

Substitute into equation two:

$$\begin{aligned} -y - 48 &= -46 \\ -y &= 2 \\ y &= -2 \end{aligned}$$

Substitute into equation one:

$$\begin{aligned} x + 2 + 6 &= 7 \\ x &= -1 \end{aligned}$$

So the solution is $(-1, -2, 3)$.

11. Form the augmented matrix: $\begin{pmatrix} 1 & 0 & 1 & -2 \\ -3 & 2 & 0 & 17 \\ 1 & -1 & -1 & -9 \end{pmatrix}$

Adding 3 times row 1 to row 2, and -1 times row 1 to row 3 yields: $\begin{pmatrix} 1 & 0 & 1 & -2 \\ 0 & 2 & 3 & 11 \\ 0 & -1 & -2 & -7 \end{pmatrix}$

Switching row 2 and row 3: $\begin{pmatrix} 1 & 0 & 1 & -2 \\ 0 & -1 & -2 & -7 \\ 0 & 2 & 3 & 11 \end{pmatrix}$

Multiplying row 2 by 2 and adding it to row 3 yields: $\begin{pmatrix} 1 & 0 & 1 & -2 \\ 0 & -1 & -2 & -7 \\ 0 & 0 & -1 & -3 \end{pmatrix}$

So we have the system:

$$\begin{aligned} x \quad + z &= -2 \\ -y - 2z &= -7 \\ -z &= -3 \end{aligned}$$

Solve equation three for z:

$$\begin{aligned} -z &= -3 \\ z &= 3 \end{aligned}$$

Substitute into equation two:

$$\begin{aligned} -y - 6 &= -7 \\ -y &= -1 \\ y &= 1 \end{aligned}$$

Substitute into equation one:

$$\begin{aligned} x + 3 &= -2 \\ x &= -5 \end{aligned}$$

So the solution is $(-5,1,3)$.

13. Form the augmented matrix: $\begin{pmatrix} 1 & 1 & 1 & -4 \\ 2 & -3 & 1 & -1 \\ 4 & 2 & -3 & 33 \end{pmatrix}$

Adding -2 times row 1 to row 2, and -4 times row 1 to row 3 yields: $\begin{pmatrix} 1 & 1 & 1 & -4 \\ 0 & -5 & -1 & 7 \\ 0 & -2 & -7 & 49 \end{pmatrix}$

Adding -3 times row 3 to row 2 (to get a 1 entry) yields: $\begin{pmatrix} 1 & 1 & 1 & -4 \\ 0 & 1 & 20 & -140 \\ 0 & -2 & -7 & 49 \end{pmatrix}$

Adding 2 times row 2 to row 3 yields: $\begin{pmatrix} 1 & 1 & 1 & -4 \\ 0 & 1 & 20 & -140 \\ 0 & 0 & 33 & -231 \end{pmatrix}$

So we have the system of equations:

$$\begin{aligned} x + y + \quad z &= -4 \\ y + 20z &= -140 \\ 33z &= -231 \end{aligned}$$

Solving the last equation for z:

$$\begin{aligned} 33z &= -231 \\ z &= -7 \end{aligned}$$

Substituting into the second equation:

$$\begin{aligned} y+20z &= -140 \\ y-140 &= -140 \\ y &= 0 \end{aligned}$$

Substituting into the first equation:

$$\begin{aligned} x+y+z &= -4 \\ x+0-7 &= -4 \\ x &= 3 \end{aligned}$$

So the solution is (3,0,–7).

15. Form the augmented matrix: $\begin{pmatrix} 3 & -2 & 6 & 0 \\ 1 & 3 & 20 & 15 \\ 10 & -11 & -10 & -9 \end{pmatrix}$

Switching row 1 and row 2 yields: $\begin{pmatrix} 1 & 3 & 20 & 15 \\ 3 & -2 & 6 & 0 \\ 10 & -11 & -10 & -9 \end{pmatrix}$

Adding –3 times row 1 to row 2, and –10 times row 1 to row 3 yields: $\begin{pmatrix} 1 & 3 & 20 & 15 \\ 0 & -11 & -54 & -45 \\ 0 & -41 & -210 & -159 \end{pmatrix}$

Multiplying row 2 by –4 and adding it to row 3 yields: $\begin{pmatrix} 1 & 3 & 20 & 15 \\ 0 & -11 & -54 & -45 \\ 0 & 3 & 6 & 21 \end{pmatrix}$

Dividing row 3 by 3 and then switching it with row 2 yields: $\begin{pmatrix} 1 & 3 & 20 & 15 \\ 0 & 1 & 2 & 7 \\ 0 & -11 & -54 & -45 \end{pmatrix}$

Multiplying row 2 by 11 and adding it to row 3 yields: $\begin{pmatrix} 1 & 3 & 20 & 15 \\ 0 & 1 & 2 & 7 \\ 0 & 0 & -32 & 32 \end{pmatrix}$

So we have the system of equations:

$$\begin{aligned} x+3y+20z &= 15 \\ y+2z &= 7 \\ -32z &= 32 \end{aligned}$$

Solve equation three for z:

$$\begin{aligned} -32z &= 32 \\ z &= -1 \end{aligned}$$

Substitute into equation two:

$$\begin{aligned} y-2 &= 7 \\ y &= 9 \end{aligned}$$

Substitute into equation one:

$$\begin{aligned} x+27-20 &= 15 \\ x+7 &= 15 \\ x &= 8 \end{aligned}$$

So the solution is (8,9,–1).

17. Form the augmented matrix: $\begin{pmatrix} 4 & -3 & 3 & 2 \\ 5 & 1 & -4 & 1 \\ 9 & -2 & -1 & 3 \end{pmatrix}$

Subtracting row 1 from row 2 yields: $\begin{pmatrix} 4 & -3 & 3 & 2 \\ 1 & 4 & -7 & -1 \\ 9 & -2 & -1 & 3 \end{pmatrix}$

Switching row 1 and row 2 yields: $\begin{pmatrix} 1 & 4 & -7 & -1 \\ 4 & -3 & 3 & 2 \\ 9 & -2 & -1 & 3 \end{pmatrix}$

Adding -4 times row 1 to row 2, and -9 times row 1 to row 3 yields: $\begin{pmatrix} 1 & 4 & -7 & -1 \\ 0 & -19 & 31 & 6 \\ 0 & -38 & 62 & 12 \end{pmatrix}$

Multiplying row 2 by -2 and adding it to row 3 yields: $\begin{pmatrix} 1 & 4 & -7 & -1 \\ 0 & -19 & 31 & 6 \\ 0 & 0 & 0 & 0 \end{pmatrix}$

So we have the system of equations:

$$\begin{aligned} x + 4y - 7z &= -1 \\ -19y + 31z &= 6 \end{aligned}$$

Solve equation two for y:

$$-19y = 6 - 31z$$
$$y = \frac{31z - 6}{19}$$

Substitute into equation one:

$$x + 4\left(\frac{31z - 6}{19}\right) - 7z = -1$$
$$x + \frac{-9z - 24}{19} = -1$$
$$x = \frac{9z + 5}{19}$$

So the solution is $\left(\frac{9z+5}{19}, \frac{31z-6}{19}, z\right)$, for any real number z.

19. Form the augmented matrix: $\begin{pmatrix} 1 & -1 & 1 & 1 & 6 \\ 1 & 1 & -1 & 1 & 4 \\ 1 & 1 & 1 & -1 & -2 \\ -1 & 1 & 1 & 1 & 0 \end{pmatrix}$

Adding -1 times row 1 to both row 2 and row 3, and adding row 1 to row 4 yields: $\begin{pmatrix} 1 & -1 & 1 & 1 & 6 \\ 0 & 2 & -2 & 0 & -2 \\ 0 & 2 & 0 & -2 & -8 \\ 0 & 0 & 2 & 2 & 6 \end{pmatrix}$

Dividing rows 2, 3, and 4 by 2 and subtracting row 2 from row 3 yields: $\begin{pmatrix} 1 & -1 & 1 & 1 & 6 \\ 0 & 1 & -1 & 0 & -1 \\ 0 & 0 & 1 & -1 & -3 \\ 0 & 0 & 1 & 1 & 3 \end{pmatrix}$

Multiplying row 3 by –1 and adding it to row 4 yields: $\begin{pmatrix} 1 & -1 & 1 & 1 & 6 \\ 0 & 1 & -1 & 0 & -1 \\ 0 & 0 & 1 & -1 & -3 \\ 0 & 0 & 0 & 2 & 6 \end{pmatrix}$

So we have the system of equations:

$$\begin{aligned} x - y + z + w &= 6 \\ y - z \quad\ &= -1 \\ z - w &= -3 \\ 2w &= 6 \end{aligned}$$

Solve equation four for w:

$$\begin{aligned} 2w &= 6 \\ w &= 3 \end{aligned}$$

Substitute into equation three:

$$\begin{aligned} z - 3 &= -3 \\ z &= 0 \end{aligned}$$

Substitute into equation two:

$$\begin{aligned} y - 0 &= -1 \\ y &= -1 \end{aligned}$$

Substitute into equation one:

$$\begin{aligned} x + 1 + 0 + 3 &= 6 \\ x + 4 &= 6 \\ x &= 2 \end{aligned}$$

So the solution is (2,–1,0,3).

21. Form the augmented matrix: $\begin{pmatrix} 15 & 14 & 26 & 1 \\ 18 & 17 & 32 & -1 \\ 21 & 20 & 38 & 0 \end{pmatrix}$

Reduce coefficients by subtracting row 2 from row 3, and row 1 from row 2: $\begin{pmatrix} 15 & 14 & 26 & 1 \\ 3 & 3 & 6 & -2 \\ 3 & 3 & 6 & 1 \end{pmatrix}$

Subtracting row 2 from row 3 yields: $\begin{pmatrix} 15 & 14 & 26 & 1 \\ 3 & 3 & 6 & -2 \\ 0 & 0 & 0 & 3 \end{pmatrix}$

But this last row corresponds to the equation 0 = 3, which is false. So the system has no solution.

23. Adding the corresponding terms, $A + B$ will be: $\begin{pmatrix} 2 & 3 \\ -1 & 4 \end{pmatrix} + \begin{pmatrix} 1 & -1 \\ 3 & 0 \end{pmatrix} = \begin{pmatrix} 3 & 2 \\ 2 & 4 \end{pmatrix}$

25. Multiplying: $2A = \begin{pmatrix} 4 & 6 \\ -2 & 8 \end{pmatrix}$ and $2B = \begin{pmatrix} 2 & -2 \\ 6 & 0 \end{pmatrix}$. Therefore: $2A + 2B = \begin{pmatrix} 6 & 4 \\ 4 & 8 \end{pmatrix}$

27. The multiplication is defined, since # cols in A = # rows in B:

row 1, col 1: 2(1) + 3(3) = 2 + 9 = 11   row 1, col 2: 2(–1) + 3(0) = –2 + 0 = –2

row 2, col 1: –1(1) + 4(3) = –1 + 12 = 11   row 2, col 2: –1(–1) + 4(0) = 1 + 0 = 1

So $AB = \begin{pmatrix} 11 & -2 \\ 11 & 1 \end{pmatrix}$.

29. The multiplication is defined, since # cols in A = # rows in C:

row 1, col 1: 2(1) + 3(0) = 2 + 0 = 2   row 1, col 2: 2(0) + 3(1) = 0 + 3 = 3

row 2, col 1: –1(1) + 4(0) = –1 + 0 = –1   row 2, col 2: –1(0) + 4(1) = 0 + 4 = 4

So $AC = \begin{pmatrix} 2 & 3 \\ -1 & 4 \end{pmatrix}$.

31. This operation is not defined, since D and E do not have the same size.

33. Computing: $2F-3G=\begin{pmatrix} 10 & -2 \\ -8 & 0 \\ 4 & 6 \end{pmatrix}-\begin{pmatrix} 0 & 0 \\ 0 & 0 \\ 0 & 0 \end{pmatrix}=\begin{pmatrix} 10 & -2 \\ -8 & 0 \\ 4 & 6 \end{pmatrix}$

35. The multiplication is defined, since # cols of E = # rows of D.

row 1, col 1: $2(-1)+1(4)=-2+4=2$
row 1, col 2: $2(2)+1(0)=4+0=4$
row 1, col 3: $2(3)+1(5)=6+5=11$
row 2, col 1: $8(-1)-1(4)=-8-4=-12$
row 2, col 2: $8(2)-1(0)=16-0=16$
row 2, col 3: $8(3)-1(5)=24-5=19$
row 3, col 1: $6(-1)+5(4)=-6+20=14$
row 3, col 2: $6(2)+5(0)=12+0=12$
row 3, col 3: $6(3)+5(5)=18+25=43$

So $ED=\begin{pmatrix} 2 & 4 & 11 \\ -12 & 16 & 19 \\ 14 & 12 & 43 \end{pmatrix}$.

37. The multiplication is defined, since # cols of F = # rows of D:

row 1, col 1: $5(-1)-1(4)=-5-4=-9$
row 1, col 2: $5(2)-1(0)=10-0=10$
row 1, col 3: $5(3)-1(5)=15-5=10$
row 2, col 1: $-4(-1)+0(4)=4+0=4$
row 2, col 2: $-4(2)+0(0)=-8+0=-8$
row 2, col 3: $-4(3)+0(5)=-12+0=-12$
row 3, col 1: $2(-1)+3(4)=-2+12=10$
row 3, col 2: $2(2)+3(0)=4+0=4$
row 3, col 3: $2(3)+3(5)=6+15=21$

So $FD=\begin{pmatrix} -9 & 10 & 10 \\ 4 & -8 & -12 \\ 10 & 4 & 21 \end{pmatrix}$.

39. The operation is not defined, since G and A are not the same size.

41. The multiplication is defined, since # cols of G = # rows of D: $GD=\begin{pmatrix} 0 & 0 & 0 \\ 0 & 0 & 0 \\ 0 & 0 & 0 \end{pmatrix}$

43. Adding: $A+(B+C)=\begin{pmatrix} 2 & 3 \\ -1 & 4 \end{pmatrix}+\begin{pmatrix} 2 & -1 \\ 3 & 1 \end{pmatrix}=\begin{pmatrix} 4 & 2 \\ 2 & 5 \end{pmatrix}$

45. The multiplication is not defined, since # cols of $D \neq$ # rows of C.

47. The multiplication is defined, since # cols of A = # rows of A.

row 1, col 1: $2(2)+3(-1)=4-3=1$
row 1, col 2: $2(3)+3(4)=6+12=18$
row 2, col 1: $-1(2)+4(-1)=-2-4=-6$
row 2, col 2: $-1(3)+4(4)=-3+16=13$

So $A^2=\begin{pmatrix} 1 & 18 \\ -6 & 13 \end{pmatrix}$.

49. Multiplying: $AA^2=\begin{pmatrix} 2 & 3 \\ -1 & 4 \end{pmatrix}\begin{pmatrix} 1 & 18 \\ -6 & 13 \end{pmatrix}=\begin{pmatrix} -16 & 75 \\ -25 & 34 \end{pmatrix}$

51. **a.** Computing: $A(B+C)=\begin{pmatrix}-1&3&4\\3&2&-3\\9&1&6\end{pmatrix}\begin{pmatrix}11&6&2\\2&1&6\\-2&1&6\end{pmatrix}=\begin{pmatrix}-13&1&40\\43&17&0\\89&61&60\end{pmatrix}$

b. Computing:

$$\begin{aligned}AB+AC&=\begin{pmatrix}-1&3&4\\3&2&-3\\9&1&6\end{pmatrix}\begin{pmatrix}7&0&1\\0&0&3\\-1&2&4\end{pmatrix}+\begin{pmatrix}-1&3&4\\3&2&-3\\9&1&6\end{pmatrix}\begin{pmatrix}4&6&1\\2&1&3\\-1&-1&2\end{pmatrix}\\&=\begin{pmatrix}-11&8&24\\24&-6&-3\\57&12&36\end{pmatrix}+\begin{pmatrix}-2&-7&16\\19&23&3\\32&49&24\end{pmatrix}\\&=\begin{pmatrix}-13&1&40\\43&17&0\\89&61&60\end{pmatrix}\end{aligned}$$

c. Computing:

$$(AB)C=\left[\begin{pmatrix}-1&3&4\\3&2&-3\\9&1&6\end{pmatrix}\begin{pmatrix}7&0&1\\0&0&3\\-1&2&4\end{pmatrix}\right]\begin{pmatrix}4&6&1\\2&1&3\\-1&-1&2\end{pmatrix}=\begin{pmatrix}-11&8&24\\24&-6&-3\\57&12&36\end{pmatrix}\begin{pmatrix}4&6&1\\2&1&3\\-1&-1&2\end{pmatrix}=\begin{pmatrix}-52&-82&61\\87&141&0\\216&318&165\end{pmatrix}$$

d. Computing: $A(BC)=\begin{pmatrix}-1&3&4\\3&2&-3\\9&1&6\end{pmatrix}\begin{pmatrix}27&41&9\\-3&-3&6\\-4&-8&13\end{pmatrix}=\begin{pmatrix}-52&-82&61\\87&141&0\\216&318&165\end{pmatrix}$

53. Computing the product: $\begin{pmatrix}-32&14\\27&9\end{pmatrix}\begin{pmatrix}83&-19\\13&41\end{pmatrix}=\begin{pmatrix}-2474&1182\\2358&-144\end{pmatrix}$

55. Computing the product: $\begin{pmatrix}-6&9&-5&1\\9&-1&-5&2\\-5&-5&9&-3\\1&2&-3&1\end{pmatrix}\begin{pmatrix}0.5&1&1.5&2\\1&2&3.5&5.5\\1.5&3.5&7&12.5\\2&5.5&12.5&25\end{pmatrix}=\begin{pmatrix}0.5&0&0&0\\0&0.5&0&0\\0&0&0.5&0\\0&0&0&0.5\end{pmatrix}$

57. **a.** Compute A^2 and B^2: $A^2=\begin{pmatrix}3&5\\7&9\end{pmatrix}\begin{pmatrix}3&5\\7&9\end{pmatrix}=\begin{pmatrix}44&60\\84&116\end{pmatrix}$ $\quad B^2=\begin{pmatrix}2&4\\6&8\end{pmatrix}\begin{pmatrix}2&4\\6&8\end{pmatrix}=\begin{pmatrix}28&40\\60&88\end{pmatrix}$

So $A^2-B^2=\begin{pmatrix}16&20\\24&28\end{pmatrix}$.

b. We have $A-B=\begin{pmatrix}1&1\\1&1\end{pmatrix}$ and $A+B=\begin{pmatrix}5&9\\13&17\end{pmatrix}$, so: $(A-B)(A+B)=\begin{pmatrix}1&1\\1&1\end{pmatrix}\begin{pmatrix}5&9\\13&17\end{pmatrix}=\begin{pmatrix}18&26\\18&26\end{pmatrix}$

c. Compute the product: $(A+B)(A-B)=\begin{pmatrix}5&9\\13&17\end{pmatrix}\begin{pmatrix}1&1\\1&1\end{pmatrix}=\begin{pmatrix}14&14\\30&30\end{pmatrix}$

d. Compute AB and BA: $AB=\begin{pmatrix}3&5\\7&9\end{pmatrix}\begin{pmatrix}2&4\\6&8\end{pmatrix}=\begin{pmatrix}36&52\\68&100\end{pmatrix}$ $\quad BA=\begin{pmatrix}2&4\\6&8\end{pmatrix}\begin{pmatrix}3&5\\7&9\end{pmatrix}=\begin{pmatrix}34&46\\74&102\end{pmatrix}$

Therefore: $A^2+AB-BA-B^2=\begin{pmatrix}44&60\\84&116\end{pmatrix}+\begin{pmatrix}36&52\\68&100\end{pmatrix}-\begin{pmatrix}34&46\\74&102\end{pmatrix}-\begin{pmatrix}28&40\\60&88\end{pmatrix}=\begin{pmatrix}18&26\\18&26\end{pmatrix}$

59. **a.** The product AZ is given by: $AZ = \begin{pmatrix} 1 & 0 \\ 0 & -1 \end{pmatrix}\begin{pmatrix} x \\ y \end{pmatrix} = \begin{pmatrix} x \\ -y \end{pmatrix}$

b. The product BZ is given by: $BZ = \begin{pmatrix} -1 & 0 \\ 0 & 1 \end{pmatrix}\begin{pmatrix} x \\ y \end{pmatrix} = \begin{pmatrix} -x \\ y \end{pmatrix}$

c. The product AB is given by: $AB = \begin{pmatrix} 1 & 0 \\ 0 & -1 \end{pmatrix}\begin{pmatrix} -1 & 0 \\ 0 & 1 \end{pmatrix} = \begin{pmatrix} -1 & 0 \\ 0 & -1 \end{pmatrix}$

So $(AB)Z = \begin{pmatrix} -1 & 0 \\ 0 & -1 \end{pmatrix}\begin{pmatrix} x \\ y \end{pmatrix} = \begin{pmatrix} -x \\ -y \end{pmatrix}$. This would represent a reflection about the origin.

61. **a.** First compute the product: $AB = \begin{pmatrix} 1 & 2 \\ 3 & 4 \end{pmatrix}\begin{pmatrix} 3 & -1 \\ 5 & 8 \end{pmatrix} = \begin{pmatrix} 13 & 15 \\ 29 & 29 \end{pmatrix}$

Now compute the function values:

$f(A) = 1(4) - 2(3) = 4 - 6 = -2$
$f(B) = 3(8) - (-1)(5) = 24 + 5 = 29$
$f(AB) = 13(29) - 15(29) = -2(29) = -58$

So $f(A) \bullet f(B) = f(AB)$.

b. First compute the product: $AB = \begin{pmatrix} a & b \\ c & d \end{pmatrix}\begin{pmatrix} e & f \\ g & h \end{pmatrix} = \begin{pmatrix} ae+bg & af+bh \\ ce+dg & cf+dh \end{pmatrix}$

Now compute the function values:

$f(A) = ad - bc$
$f(B) = eh - fg$
$f(AB) = (ae+bg)(cf+dh) - (af+bh)(ce+dg)$
$= (acef + bcfg + adeh + bdgh) - (acef + bceh + adfg + bdgh)$
$= bcfg + adeh - bceh - adfg$

So $f(A) \bullet f(B) = f(AB)$.

63. **a.** Let $A = \begin{pmatrix} a & b \\ c & d \end{pmatrix}$ and $B = \begin{pmatrix} e & f \\ g & h \end{pmatrix}$. Then $A + B = \begin{pmatrix} a+e & b+f \\ c+g & d+h \end{pmatrix}$, so $(A+B)^T = \begin{pmatrix} a+e & c+g \\ b+f & d+h \end{pmatrix}$. Now

$A^T = \begin{pmatrix} a & c \\ b & d \end{pmatrix}$ and $B^T = \begin{pmatrix} e & g \\ f & h \end{pmatrix}$, so we have $A^T + B^T = \begin{pmatrix} a+e & c+g \\ b+f & d+h \end{pmatrix}$. Thus $(A+B)^T = A^T + B^T$.

b. Since $A^T = \begin{pmatrix} a & c \\ b & d \end{pmatrix}$, $\left(A^T\right)^T = \begin{pmatrix} a & b \\ c & d \end{pmatrix}$. Thus $\left(A^T\right)^T = A$.

c. Since $AB = \begin{pmatrix} ae+bg & af+bh \\ ce+dg & cf+dh \end{pmatrix}$, $(AB)^T = \begin{pmatrix} ae+bg & ce+dg \\ af+bh & cf+dh \end{pmatrix}$. Also

$B^T A^T = \begin{pmatrix} e & g \\ f & h \end{pmatrix}\begin{pmatrix} a & c \\ b & d \end{pmatrix} = \begin{pmatrix} ae+bg & ce+dg \\ af+bh & cf+dh \end{pmatrix}$. So we see that $(AB)^T = B^T A^T$.

10.4 The Inverse of a Square Matrix

1. Compute AI_2 and I_2A:

$$AI_2 = \begin{pmatrix} 4 & -1 \\ -5 & 2 \end{pmatrix}\begin{pmatrix} 1 & 0 \\ 0 & 1 \end{pmatrix} = \begin{pmatrix} 4 & -1 \\ -5 & 2 \end{pmatrix} = A \qquad I_2A = \begin{pmatrix} 1 & 0 \\ 0 & 1 \end{pmatrix}\begin{pmatrix} 4 & -1 \\ -5 & 2 \end{pmatrix} = \begin{pmatrix} 4 & -1 \\ -5 & 2 \end{pmatrix} = A$$

3. Compute CI_3 and I_3C:

$$CI_3 = \begin{pmatrix} 3 & 0 & -2 \\ 0 & 5 & 6 \\ 1 & 4 & -7 \end{pmatrix}\begin{pmatrix} 1 & 0 & 0 \\ 0 & 1 & 0 \\ 0 & 0 & 1 \end{pmatrix} = \begin{pmatrix} 3 & 0 & -2 \\ 0 & 5 & 6 \\ 1 & 4 & -7 \end{pmatrix} = C \qquad I_3C = \begin{pmatrix} 1 & 0 & 0 \\ 0 & 1 & 0 \\ 0 & 0 & 1 \end{pmatrix}\begin{pmatrix} 3 & 0 & -2 \\ 0 & 5 & 6 \\ 1 & 4 & -7 \end{pmatrix} = \begin{pmatrix} 3 & 0 & -2 \\ 0 & 5 & 6 \\ 1 & 4 & -7 \end{pmatrix} = C$$

5. We need to find numbers a, b, c and d such that: $\begin{pmatrix} 7 & 9 \\ 4 & 5 \end{pmatrix}\begin{pmatrix} a & b \\ c & d \end{pmatrix} = \begin{pmatrix} 1 & 0 \\ 0 & 1 \end{pmatrix}$ and $\begin{pmatrix} a & b \\ c & d \end{pmatrix}\begin{pmatrix} 7 & 9 \\ 4 & 5 \end{pmatrix} = \begin{pmatrix} 1 & 0 \\ 0 & 1 \end{pmatrix}$

Compute the left-hand product: $\begin{pmatrix} 7 & 9 \\ 4 & 5 \end{pmatrix}\begin{pmatrix} a & b \\ c & d \end{pmatrix} = \begin{pmatrix} 7a+9c & 7b+9d \\ 4a+5c & 4b+5d \end{pmatrix} = \begin{pmatrix} 1 & 0 \\ 0 & 1 \end{pmatrix}$

So we have the following systems of equations:

$$7a+9c=1 \qquad\qquad 7b+9d=0$$
$$4a+5c=0 \qquad\qquad 4b+5d=1$$

Multiplying the first equation by 5 and the second equation by –9 yields:

$$35a+45c=5 \qquad\qquad 35b+45d=0$$
$$-36a-45c=0 \qquad\qquad -36b-45d=-9$$

Adding:

$$-a=5 \qquad\qquad -b=-9$$
$$a=-5 \qquad\qquad b=9$$

Substituting for c and d, respectively:

$$4a+5c=0 \qquad\qquad 4b+5d=1$$
$$-20+5c=0 \qquad\qquad 36+5d=1$$
$$5c=20 \qquad\qquad 5d=-35$$
$$c=4 \qquad\qquad d=-7$$

So the inverse matrix is $A^{-1} = \begin{pmatrix} -5 & 9 \\ 4 & -7 \end{pmatrix}$.

7. We need to find numbers a, b, c, and d such that: $\begin{pmatrix} -3 & 1 \\ 5 & 6 \end{pmatrix}\begin{pmatrix} a & b \\ c & d \end{pmatrix} = \begin{pmatrix} 1 & 0 \\ 0 & 1 \end{pmatrix}$

Compute the left-hand product: $\begin{pmatrix} -3a+c & -3b+d \\ 5a+6c & 5b+6d \end{pmatrix} = \begin{pmatrix} 1 & 0 \\ 0 & 1 \end{pmatrix}$

So we have the following systems of equations:

$$-3a+\ c=1 \qquad\qquad -3b+\ d=0$$
$$5a+6c=0 \qquad\qquad 5b+6d=1$$

Solving for c and d respectively, then substituting:

$$-3a+c=1 \qquad\qquad -3b+d=0$$
$$c=3a+1 \qquad\qquad d=3b$$

Substituting:

$$5a+6(3a+1)=0 \qquad\qquad 5b+6(3b)=1$$
$$23a+6=0 \qquad\qquad 23b=1$$
$$23a=-6 \qquad\qquad b=\tfrac{1}{23}$$
$$a=-\tfrac{6}{23}$$

Now substituting to find c and d:

$$c = 3a + 1 \qquad d = 3b$$
$$c = -\frac{18}{23} + 1 \qquad d = \frac{3}{23}$$
$$c = \frac{5}{23}$$

So the inverse matrix is $A^{-1} = \begin{pmatrix} -\frac{6}{23} & \frac{1}{23} \\ \frac{5}{23} & \frac{3}{23} \end{pmatrix}$.

9. We need to find numbers a, b, c, and d such that: $\begin{pmatrix} -2 & 3 \\ -4 & 6 \end{pmatrix}\begin{pmatrix} a & b \\ c & d \end{pmatrix} = \begin{pmatrix} 1 & 0 \\ 0 & 1 \end{pmatrix}$

Compute the left-hand product: $\begin{pmatrix} -2a + 3c & -2b + 3d \\ -4a + 6c & -4b + 6d \end{pmatrix} = \begin{pmatrix} 1 & 0 \\ 0 & 1 \end{pmatrix}$

So we have the following systems of equations:

$$-2a + 3c = 1 \qquad -2b + 3d = 0$$
$$-4a + 6c = 0 \qquad -4b + 6d = 1$$

Multiplying the top equation by –2 yields:

$$4a - 6c = -2 \qquad 4b - 6d = 0$$
$$-4a + 6c = 0 \qquad -4b + 6d = 1$$

Adding:

$$0 = -2 \qquad 0 = 1$$

Since both of these equations are false, no A^{-1} exists.

11. We need to find numbers a, b, c, and d such that: $\begin{pmatrix} \frac{1}{3} & \frac{1}{3} \\ -\frac{1}{9} & \frac{2}{9} \end{pmatrix}\begin{pmatrix} a & b \\ c & d \end{pmatrix} = \begin{pmatrix} 1 & 0 \\ 0 & 1 \end{pmatrix}$

Compute the left-hand product: $\begin{pmatrix} \frac{1}{3}a + \frac{1}{3}c & \frac{1}{3}b + \frac{1}{3}d \\ -\frac{1}{9}a + \frac{2}{9}c & -\frac{1}{9}b + \frac{2}{9}d \end{pmatrix} = \begin{pmatrix} 1 & 0 \\ 0 & 1 \end{pmatrix}$

So we have the following systems of equations:

$$\frac{1}{3}a + \frac{1}{3}c = 1 \qquad \frac{1}{3}b + \frac{1}{3}d = 0$$
$$-\frac{1}{9}a + \frac{2}{9}c = 0 \qquad -\frac{1}{9}b + \frac{2}{9}d = 1$$

Multiplying the first equation by 3 and the second by 9 yields:

$$a + c = 3 \qquad b + d = 0$$
$$-a + 2c = 0 \qquad -b + 2d = 9$$

Adding:

$$3c = 3 \qquad 3d = 9$$
$$c = 1 \qquad d = 3$$

Now substituting to find a and b:

$$a + c = 3 \qquad b + d = 0$$
$$a + 1 = 3 \qquad b + 3 = 0$$
$$a = 2 \qquad b = -3$$

So the inverse is $A^{-1} = \begin{pmatrix} 2 & -3 \\ 1 & 3 \end{pmatrix}$.

13. Form the matrix: $\begin{pmatrix} 2 & 1 & 1 & 0 \\ 3 & 2 & 0 & 1 \end{pmatrix}$

Multiply row 1 by –1 and add to row 2: $\begin{pmatrix} 2 & 1 & 1 & 0 \\ 1 & 1 & -1 & 1 \end{pmatrix}$

Switch rows 1 and 2: $\begin{pmatrix} 1 & 1 & -1 & 1 \\ 2 & 1 & 1 & 0 \end{pmatrix}$

Multiply row 1 by –2 and add to row 2: $\begin{pmatrix} 1 & 1 & -1 & 1 \\ 0 & -1 & 3 & -2 \end{pmatrix}$

Add row 2 to row 1: $\begin{pmatrix} 1 & 0 & 2 & -1 \\ 0 & -1 & 3 & -2 \end{pmatrix}$

Multiply row 2 by –1: $\begin{pmatrix} 1 & 0 & 2 & -1 \\ 0 & 1 & -3 & 2 \end{pmatrix}$

So the inverse is $\begin{pmatrix} 2 & -1 \\ -3 & 2 \end{pmatrix}$.

15. Form the matrix: $\begin{pmatrix} 0 & -11 & 1 & 0 \\ 1 & 6 & 0 & 1 \end{pmatrix}$

Switch rows 1 and 2: $\begin{pmatrix} 1 & 6 & 0 & 1 \\ 0 & -11 & 1 & 0 \end{pmatrix}$

Multiply row 2 by $-\frac{1}{11}$: $\begin{pmatrix} 1 & 6 & 0 & 1 \\ 0 & 1 & -\frac{1}{11} & 0 \end{pmatrix}$

Multiply row 2 by –6 and add to row 1: $\begin{pmatrix} 1 & 0 & \frac{6}{11} & 1 \\ 0 & 1 & -\frac{1}{11} & 0 \end{pmatrix}$

So the inverse is $\begin{pmatrix} \frac{6}{11} & 1 \\ -\frac{1}{11} & 0 \end{pmatrix}$.

17. Form the matrix: $\begin{pmatrix} \frac{2}{3} & -\frac{1}{4} & 1 & 0 \\ -8 & 3 & 0 & 1 \end{pmatrix}$

Multiply row 1 by 12: $\begin{pmatrix} 8 & -3 & 12 & 0 \\ -8 & 3 & 0 & 1 \end{pmatrix}$

Add row 1 to row 2: $\begin{pmatrix} 8 & -3 & 12 & 0 \\ 0 & 0 & 12 & 1 \end{pmatrix}$

So the inverse does not exist.

19. Form the matrix: $\begin{pmatrix} -5 & 4 & -3 & 1 & 0 & 0 \\ 10 & -7 & 6 & 0 & 1 & 0 \\ 8 & -6 & 5 & 0 & 0 & 1 \end{pmatrix}$

Multiply row 1 by 2 and add to row 2: $\begin{pmatrix} -5 & 4 & -3 & 1 & 0 & 0 \\ 0 & 1 & 0 & 2 & 1 & 0 \\ 8 & -6 & 5 & 0 & 0 & 1 \end{pmatrix}$

Add row 3 to row 1 (to reduce the numbers): $\begin{pmatrix} 3 & -2 & 2 & 1 & 0 & 1 \\ 0 & 1 & 0 & 2 & 1 & 0 \\ 8 & -6 & 5 & 0 & 0 & 1 \end{pmatrix}$

Multiply row 2 by 2 and add to row 1, and also multiply row 2 by 6 and add to row 3: $\begin{pmatrix} 3 & 0 & 2 & 5 & 2 & 1 \\ 0 & 1 & 0 & 2 & 1 & 0 \\ 8 & 0 & 5 & 12 & 6 & 1 \end{pmatrix}$

Multiply row 1 by $-\frac{8}{3}$ and add to row 3: $\begin{pmatrix} 3 & 0 & 2 & 5 & 2 & 1 \\ 0 & 1 & 0 & 2 & 1 & 0 \\ 0 & 0 & -\frac{1}{3} & -\frac{4}{3} & \frac{2}{3} & -\frac{5}{3} \end{pmatrix}$

Multiply row 3 by –3: $\begin{pmatrix} 3 & 0 & 2 & 5 & 2 & 1 \\ 0 & 1 & 0 & 2 & 1 & 0 \\ 0 & 0 & 1 & 4 & -2 & 5 \end{pmatrix}$

Multiply row 3 by –2 and add to row 1: $\begin{pmatrix} 3 & 0 & 0 & -3 & 6 & -9 \\ 0 & 1 & 0 & 2 & 1 & 0 \\ 0 & 0 & 1 & 4 & -2 & 5 \end{pmatrix}$

Multiply row 1 by $\frac{1}{3}$: $\begin{pmatrix} 1 & 0 & 0 & -1 & 2 & -3 \\ 0 & 1 & 0 & 2 & 1 & 0 \\ 0 & 0 & 1 & 4 & -2 & 5 \end{pmatrix}$

So the inverse is $\begin{pmatrix} -1 & 2 & -3 \\ 2 & 1 & 0 \\ 4 & -2 & 5 \end{pmatrix}$.

21. Form the matrix: $\begin{pmatrix} 1 & 2 & -1 & 1 & 0 & 0 \\ 0 & 3 & 0 & 0 & 1 & 0 \\ -4 & 0 & 5 & 0 & 0 & 1 \end{pmatrix}$

Multiply row 1 by 4 and add to row 3: $\begin{pmatrix} 1 & 2 & -1 & 1 & 0 & 0 \\ 0 & 3 & 0 & 0 & 1 & 0 \\ 0 & 8 & 1 & 4 & 0 & 1 \end{pmatrix}$

Multiply row 2 by $\frac{1}{3}$: $\begin{pmatrix} 1 & 2 & -1 & 1 & 0 & 0 \\ 0 & 1 & 0 & 0 & \frac{1}{3} & 0 \\ 0 & 8 & 1 & 4 & 0 & 1 \end{pmatrix}$

Multiply row 2 by –2 and add to row 1, and multiply row 2 by –8 and add to row 3: $\begin{pmatrix} 1 & 0 & -1 & 1 & -\frac{2}{3} & 0 \\ 0 & 1 & 0 & 0 & \frac{1}{3} & 0 \\ 0 & 0 & 1 & 4 & -\frac{8}{3} & 1 \end{pmatrix}$

Add row 3 to row 1: $\begin{pmatrix} 1 & 0 & 0 & 5 & -\frac{10}{3} & 1 \\ 0 & 1 & 0 & 0 & \frac{1}{3} & 0 \\ 0 & 0 & 1 & 4 & -\frac{8}{3} & 1 \end{pmatrix}$

So the inverse is $\begin{pmatrix} 5 & -\frac{10}{3} & 1 \\ 0 & \frac{1}{3} & 0 \\ 4 & -\frac{8}{3} & 1 \end{pmatrix}$.

23. Form the matrix: $\begin{pmatrix} -7 & 5 & 3 & 1 & 0 & 0 \\ 3 & -2 & -2 & 0 & 1 & 0 \\ 3 & -2 & -1 & 0 & 0 & 1 \end{pmatrix}$

Multiply row 2 by –1 and add to row 3: $\begin{pmatrix} -7 & 5 & 3 & 1 & 0 & 0 \\ 3 & -2 & -2 & 0 & 1 & 0 \\ 0 & 0 & 1 & 0 & -1 & 1 \end{pmatrix}$

Multiply row 2 by 2 and add to row 1: $\begin{pmatrix} -1 & 1 & -1 & 1 & 2 & 0 \\ 3 & -2 & -2 & 0 & 1 & 0 \\ 0 & 0 & 1 & 0 & -1 & 1 \end{pmatrix}$

Multiply row 1 by 3 and add to row 2: $\begin{pmatrix} -1 & 1 & -1 & 1 & 2 & 0 \\ 0 & 1 & -5 & 3 & 7 & 0 \\ 0 & 0 & 1 & 0 & -1 & 1 \end{pmatrix}$

Multiply row 3 by 5 and add to row 2, and add row 3 to row 1: $\begin{pmatrix} -1 & 1 & 0 & 1 & 1 & 1 \\ 0 & 1 & 0 & 3 & 2 & 5 \\ 0 & 0 & 1 & 0 & -1 & 1 \end{pmatrix}$

Multiply row 1 by –1: $\begin{pmatrix} 1 & -1 & 0 & -1 & -1 & -1 \\ 0 & 1 & 0 & 3 & 2 & 5 \\ 0 & 0 & 1 & 0 & -1 & 1 \end{pmatrix}$

Add row 2 to row 1: $\begin{pmatrix} 1 & 0 & 0 & 2 & 1 & 4 \\ 0 & 1 & 0 & 3 & 2 & 5 \\ 0 & 0 & 1 & 0 & -1 & 1 \end{pmatrix}$

So the inverse is $\begin{pmatrix} 2 & 1 & 4 \\ 3 & 2 & 5 \\ 0 & -1 & 1 \end{pmatrix}$.

25. Form the matrix: $\begin{pmatrix} 1 & 2 & 3 & 1 & 0 & 0 \\ 4 & 5 & 6 & 0 & 1 & 0 \\ 7 & 8 & 9 & 0 & 0 & 1 \end{pmatrix}$

Multiply row 1 by –4 and add to row 2, and multiply row 1 by –7 and add to row 3: $\begin{pmatrix} 1 & 2 & 3 & 1 & 0 & 0 \\ 0 & -3 & -6 & -4 & 1 & 0 \\ 0 & -6 & -12 & -7 & 0 & 1 \end{pmatrix}$

Multiply row 2 by –2 and add to row 3: $\begin{pmatrix} 1 & 2 & 3 & 1 & 0 & 0 \\ 0 & -3 & -6 & -4 & 1 & 0 \\ 0 & 0 & 0 & 1 & -2 & 1 \end{pmatrix}$

So the inverse does not exist.

27. The inverse is $A^{-1}=\begin{bmatrix}-2.5 & 1.5\\ 2 & -1\end{bmatrix}$.

29. There is no inverse for D.

31. Note that $DE=\begin{bmatrix}-6 & 4 & 2\\ 16 & -10 & -3\\ 29 & -18 & -5\end{bmatrix}$, but there is no inverse for DE.

33. **a.** Since the system can be written as $A\bullet X=B$, where $X=\begin{pmatrix}x\\ y\end{pmatrix}$ and $B=\begin{pmatrix}5\\ 7\end{pmatrix}$, $X=A^{-1}\bullet B$.

So $X=\begin{pmatrix}11 & -8\\ -4 & 3\end{pmatrix}\begin{pmatrix}5\\ 7\end{pmatrix}=\begin{pmatrix}-1\\ 1\end{pmatrix}$, thus $x=-1$ and $y=1$.

b. Again this system can be written as $A\bullet X=B$, where $X=\begin{pmatrix}x\\ y\end{pmatrix}$ and $B=\begin{pmatrix}-12\\ 0\end{pmatrix}$, then $X=A^{-1}\bullet B$.

So $X=\begin{pmatrix}11 & -8\\ -4 & 3\end{pmatrix}\begin{pmatrix}-12\\ 0\end{pmatrix}=\begin{pmatrix}-132\\ 48\end{pmatrix}$, thus $x=-132$ and $y=48$.

35. Since the system can be written as $A\bullet X=B$, where $X=\begin{pmatrix}x\\ y\\ z\end{pmatrix}$ and $B=\begin{pmatrix}28\\ 9\\ 22\end{pmatrix}$, $X=A^{-1}\bullet B$.

So $X=\begin{pmatrix}1 & 2 & -2\\ -1 & 3 & 0\\ 0 & -2 & 1\end{pmatrix}\begin{pmatrix}28\\ 9\\ 22\end{pmatrix}=\begin{pmatrix}2\\ -1\\ 4\end{pmatrix}$, thus $x=2$, $y=-1$, $z=4$.

37. **a.** Enter $A=\begin{pmatrix}1 & 4\\ 2 & 7\end{pmatrix}$, so $A^{-1}=\begin{pmatrix}-7 & 4\\ 2 & -1\end{pmatrix}$.

b. Let $b=\begin{pmatrix}7\\ 12\end{pmatrix}$, so the solution is $A^{-1}b=\begin{pmatrix}-1\\ 2\end{pmatrix}$. The solution to the system is $(-1,2)$.

39. **a.** Enter $A=\begin{pmatrix}1 & 2 & 2\\ 3 & 1 & 0\\ 1 & 1 & 1\end{pmatrix}$, so $A^{-1}=\begin{pmatrix}-1 & 0 & 2\\ 3 & 1 & -6\\ -2 & -1 & 5\end{pmatrix}$.

b. Let $b=\begin{pmatrix}3\\ -1\\ 12\end{pmatrix}$, so the solution is $A^{-1}b=\begin{pmatrix}21\\ -64\\ 55\end{pmatrix}$. The solution to the system is $(21,-64,55)$.

41. **a.** Enter $A=\begin{pmatrix}2 & 3 & 1 & 1\\ 6 & 6 & -5 & -2\\ 1 & -1 & 1 & 1/6\\ 4 & 9 & 3 & 2\end{pmatrix}$, so $A^{-1}=\begin{pmatrix}\frac{31}{108} & \frac{2}{27} & \frac{7}{18} & -\frac{11}{108}\\ -\frac{53}{162} & \frac{1}{81} & -\frac{5}{27} & \frac{31}{162}\\ -\frac{55}{54} & -\frac{1}{27} & \frac{5}{9} & \frac{23}{54}\\ \frac{131}{54} & -\frac{4}{27} & -\frac{7}{9} & -\frac{43}{54}\end{pmatrix}$.

b. Let $b=\begin{pmatrix}3\\ 15\\ -3\\ -3\end{pmatrix}$, so the solution is $A^{-1}b=\begin{pmatrix}\frac{10}{9}\\ -\frac{22}{27}\\ -\frac{59}{9}\\ \frac{88}{9}\end{pmatrix}$. The solution to the system is $\left(\frac{10}{9},-\frac{22}{27},-\frac{59}{9},\frac{88}{9}\right)$.

43. We find $\begin{pmatrix} a & b \\ c & d \end{pmatrix}$ where $\begin{pmatrix} 2 & 5 \\ 6 & 15 \end{pmatrix}\begin{pmatrix} a & b \\ c & d \end{pmatrix} = \begin{pmatrix} 1 & 0 \\ 0 & 1 \end{pmatrix}$.

Carrying out the multiplication: $\begin{pmatrix} 2a+5c & 2b+5d \\ 6a+15c & 6b+15d \end{pmatrix} = \begin{pmatrix} 1 & 0 \\ 0 & 1 \end{pmatrix}$

So we have the following systems of equations:

$2a + 5c = 1$ $\qquad$ $2b + 5d = 0$

$6a + 15c = 0$ $\qquad$ $6b + 15d = 1$

Multiplying the first equation by –3 yields:

$-6a - 15c = -3$ $\qquad$ $-6b - 15d = 0$

$6a + 15c = 0$ $\qquad$ $6b + 15d = 1$

Adding these equations:

$0 = -3$ $\qquad$ $0 = 1$

Neither of these systems has a solution, thus the matrix has no inverse.

45. First convert the message TRY IT to the numerical sequence: 20, 18, 25, 0, 9, 20

Now encode each pair of values:

$$\begin{pmatrix} 9 & 8 \\ 8 & 7 \end{pmatrix}\begin{pmatrix} 20 \\ 18 \end{pmatrix} = \begin{pmatrix} 324 \\ 286 \end{pmatrix} \qquad \begin{pmatrix} 9 & 8 \\ 8 & 7 \end{pmatrix}\begin{pmatrix} 25 \\ 0 \end{pmatrix} = \begin{pmatrix} 225 \\ 200 \end{pmatrix} \qquad \begin{pmatrix} 9 & 8 \\ 8 & 7 \end{pmatrix}\begin{pmatrix} 9 \\ 20 \end{pmatrix} = \begin{pmatrix} 241 \\ 212 \end{pmatrix}$$

The encoded message is 324, 286, 225, 200, 241, 212.

47. First convert the message TURN NOW to the numerical sequence: 20, 21, 18, 14, 0, 14, 15, 23, 0

Now encode each trio of values:

$$\begin{pmatrix} 1 & 1 & 0 \\ 0 & -1 & 2 \\ 1 & 0 & 1 \end{pmatrix}\begin{pmatrix} 20 \\ 21 \\ 18 \end{pmatrix} = \begin{pmatrix} 41 \\ 15 \\ 38 \end{pmatrix} \qquad \begin{pmatrix} 1 & 1 & 0 \\ 0 & -1 & 2 \\ 1 & 0 & 1 \end{pmatrix}\begin{pmatrix} 14 \\ 0 \\ 14 \end{pmatrix} = \begin{pmatrix} 14 \\ 28 \\ 28 \end{pmatrix} \qquad \begin{pmatrix} 1 & 1 & 0 \\ 0 & -1 & 2 \\ 1 & 0 & 1 \end{pmatrix}\begin{pmatrix} 15 \\ 23 \\ 0 \end{pmatrix} = \begin{pmatrix} 38 \\ -23 \\ 15 \end{pmatrix}$$

The encoded message is 41, 15, 38, 14, 28, 28, 38, –23, 15.

49. First we need to find the inverse of the encoding message. Form the augmented matrix: $\begin{pmatrix} 12 & 47 & 1 & 0 \\ -1 & -4 & 0 & 1 \end{pmatrix}$

Multiply row 2 by –1, then switch rows 1 and 2: $\begin{pmatrix} 1 & 4 & 0 & -1 \\ 12 & 47 & 1 & 0 \end{pmatrix}$

Add –12 times row 1 to row 2: $\begin{pmatrix} 1 & 4 & 0 & -1 \\ 0 & -1 & 1 & 12 \end{pmatrix}$

Multiply row 2 by –1: $\begin{pmatrix} 1 & 4 & 0 & -1 \\ 0 & 1 & -1 & -12 \end{pmatrix}$

Add –4 times row 2 to row 1: $\begin{pmatrix} 1 & 0 & 4 & 47 \\ 0 & 1 & -1 & -12 \end{pmatrix}$

The inverse (decoding) matrix is $\begin{pmatrix} 4 & 47 \\ -1 & -12 \end{pmatrix}$. Now decode each pair of values:

$$\begin{pmatrix} 4 & 47 \\ -1 & -12 \end{pmatrix}\begin{pmatrix} 463 \\ -39 \end{pmatrix} = \begin{pmatrix} 19 \\ 5 \end{pmatrix} \qquad \begin{pmatrix} 4 & 47 \\ -1 & -12 \end{pmatrix}\begin{pmatrix} 60 \\ -5 \end{pmatrix} = \begin{pmatrix} 5 \\ 0 \end{pmatrix}$$

$$\begin{pmatrix} 4 & 47 \\ -1 & -12 \end{pmatrix}\begin{pmatrix} 825 \\ -70 \end{pmatrix} = \begin{pmatrix} 10 \\ 15 \end{pmatrix} \qquad \begin{pmatrix} 4 & 47 \\ -1 & -12 \end{pmatrix}\begin{pmatrix} 60 \\ -5 \end{pmatrix} = \begin{pmatrix} 5 \\ 0 \end{pmatrix}$$

The decoded message is 19, 5, 5, 0, 10, 15, 5, which is SEE JOE.

51. First we need to find the inverse of the encoding message. Form the augmented matrix: $\begin{pmatrix} 2 & -1 & -1 & 1 & 0 & 0 \\ -2 & 1 & 2 & 0 & 1 & 0 \\ -1 & 1 & 1 & 0 & 0 & 1 \end{pmatrix}$

Multiply row 3 by –1, then switch row 1 and row 3: $\begin{pmatrix} 1 & -1 & -1 & 0 & 0 & -1 \\ -2 & 1 & 2 & 0 & 1 & 0 \\ 2 & -1 & -1 & 1 & 0 & 0 \end{pmatrix}$

Add 2 times row 1 to row 2, and –2 times row 1 to row 3: $\begin{pmatrix} 1 & -1 & -1 & 0 & 0 & -1 \\ 0 & -1 & 0 & 0 & 1 & -2 \\ 0 & 1 & 1 & 1 & 0 & 2 \end{pmatrix}$

Add row 2 to row 3, and multiply row 2 by –1: $\begin{pmatrix} 1 & -1 & -1 & 0 & 0 & -1 \\ 0 & 1 & 0 & 0 & -1 & 2 \\ 0 & 0 & 1 & 1 & 1 & 0 \end{pmatrix}$

Add row 3 and row 2 to row 1: $\begin{pmatrix} 1 & 0 & 0 & 1 & 0 & 1 \\ 0 & 1 & 0 & 0 & -1 & 2 \\ 0 & 0 & 1 & 1 & 1 & 0 \end{pmatrix}$

The inverse (decoding) matrix is $\begin{pmatrix} 1 & 0 & 1 \\ 0 & -1 & 2 \\ 1 & 1 & 0 \end{pmatrix}$. Now decode each trio of values:

$$\begin{pmatrix} 1 & 0 & 1 \\ 0 & -1 & 2 \\ 1 & 1 & 0 \end{pmatrix}\begin{pmatrix} -3 \\ 24 \\ 22 \end{pmatrix} = \begin{pmatrix} 19 \\ 20 \\ 21 \end{pmatrix} \qquad \begin{pmatrix} 1 & 0 & 1 \\ 0 & -1 & 2 \\ 1 & 1 & 0 \end{pmatrix}\begin{pmatrix} -17 \\ 17 \\ 21 \end{pmatrix} = \begin{pmatrix} 4 \\ 25 \\ 0 \end{pmatrix}$$

The decoded message is 19, 20, 21, 4, 25, which is STUDY.

53. Form the augmented matrix: $\begin{pmatrix} 1 & 1 & 1 & 1 & 1 & 0 & 0 & 0 \\ 1 & 2 & 3 & 4 & 0 & 1 & 0 & 0 \\ 1 & 3 & 6 & 10 & 0 & 0 & 1 & 0 \\ 1 & 4 & 10 & 20 & 0 & 0 & 0 & 1 \end{pmatrix}$. We can keep the numbers small if we first subtract row 3 from row 4, row 2 from row 3, and row 1 from row 2: $\begin{pmatrix} 1 & 1 & 1 & 1 & 1 & 0 & 0 & 0 \\ 0 & 1 & 2 & 3 & -1 & 1 & 0 & 0 \\ 0 & 1 & 3 & 6 & 0 & -1 & 1 & 0 \\ 0 & 1 & 4 & 10 & 0 & 0 & -1 & 1 \end{pmatrix}$. Subtract row 3 from row 4, row 2 from row 3, and row 2 from row 1: $\begin{pmatrix} 1 & 0 & -1 & -2 & 2 & -1 & 0 & 0 \\ 0 & 1 & 2 & 3 & -1 & 1 & 0 & 0 \\ 0 & 0 & 1 & 3 & 1 & -2 & 1 & 0 \\ 0 & 0 & 1 & 4 & 0 & 1 & -2 & 1 \end{pmatrix}$. Now subtract row 3 from row 4, and multiply row 3 by –2 to add to row 2, and add row 3 to row 1 (we are clearing out the numbers in column 3): $\begin{pmatrix} 1 & 0 & 0 & 1 & 3 & -3 & 1 & 0 \\ 0 & 1 & 0 & -3 & -3 & 5 & -2 & 0 \\ 0 & 0 & 1 & 3 & 1 & -2 & 1 & 0 \\ 0 & 0 & 0 & 1 & -1 & 3 & -3 & 1 \end{pmatrix}$.

Finally, clear out column 4. Subtract row 4 from row 1, multiply row 4 by 3 and add to row 2, and multiply row 4 by –3 and add to row 3: $\begin{pmatrix} 1 & 0 & 0 & 0 & 4 & -6 & 4 & -1 \\ 0 & 1 & 0 & 0 & -6 & 14 & -11 & 3 \\ 0 & 0 & 1 & 0 & 4 & -11 & 10 & -3 \\ 0 & 0 & 0 & 1 & -1 & 3 & -3 & 1 \end{pmatrix}$. The inverse is $\begin{pmatrix} 4 & -6 & 4 & -1 \\ -6 & 14 & -11 & 3 \\ 4 & -11 & 10 & -3 \\ -1 & 3 & -3 & 1 \end{pmatrix}$.

55. **a.** Form the matrix: $\begin{pmatrix} 2 & 3 & 1 & 0 \\ 4 & 5 & 0 & 1 \end{pmatrix}$

Multiply row 1 by –2 and add to row 2: $\begin{pmatrix} 2 & 3 & 1 & 0 \\ 0 & -1 & -2 & 1 \end{pmatrix}$

Multiply row 2 by 3 and add to row 1: $\begin{pmatrix} 2 & 0 & -5 & 3 \\ 0 & -1 & -2 & 1 \end{pmatrix}$

Multiply row 1 by $\frac{1}{2}$ and row 2 by –1: $\begin{pmatrix} 1 & 0 & -\frac{5}{2} & \frac{3}{2} \\ 0 & 1 & 2 & -1 \end{pmatrix}$

So $A^{-1} = \begin{pmatrix} -\frac{5}{2} & \frac{3}{2} \\ 2 & -1 \end{pmatrix}$. Now form the matrix: $\begin{pmatrix} 7 & 8 & 1 & 0 \\ 6 & 7 & 0 & 1 \end{pmatrix}$

Subtract row 2 from row 1: $\begin{pmatrix} 1 & 1 & 1 & -1 \\ 6 & 7 & 0 & 1 \end{pmatrix}$

Multiply row 1 by – 6 and add to row 2: $\begin{pmatrix} 1 & 1 & 1 & -1 \\ 0 & 1 & -6 & 7 \end{pmatrix}$

Subtract row 2 from row 1: $\begin{pmatrix} 1 & 0 & 7 & -8 \\ 0 & 1 & -6 & 7 \end{pmatrix}$

So $B^{-1} = \begin{pmatrix} 7 & -8 \\ -6 & 7 \end{pmatrix}$. Then $B^{-1}A^{-1} = \begin{pmatrix} 7 & -8 \\ -6 & 7 \end{pmatrix}\begin{pmatrix} -\frac{5}{2} & \frac{3}{2} \\ 2 & -1 \end{pmatrix} = \begin{pmatrix} -\frac{67}{2} & \frac{37}{2} \\ 29 & -16 \end{pmatrix}$.

b. First find AB: $AB = \begin{pmatrix} 2 & 3 \\ 4 & 5 \end{pmatrix}\begin{pmatrix} 7 & 8 \\ 6 & 7 \end{pmatrix} = \begin{pmatrix} 32 & 37 \\ 58 & 67 \end{pmatrix}$. Now form the matrix: $\begin{pmatrix} 32 & 37 & 1 & 0 \\ 58 & 67 & 0 & 1 \end{pmatrix}$

Subtract row 1 from row 2 (to reduce numbers): $\begin{pmatrix} 32 & 37 & 1 & 0 \\ 26 & 30 & -1 & 1 \end{pmatrix}$

Subtract row 2 from row 1 (to reduce numbers): $\begin{pmatrix} 6 & 7 & 2 & -1 \\ 26 & 30 & -1 & 1 \end{pmatrix}$

Multiply row 1 by – 4 and add to row 2: $\begin{pmatrix} 6 & 7 & 2 & -1 \\ 2 & 2 & -9 & 5 \end{pmatrix}$

Multiply row 2 by –3 and add to row 1: $\begin{pmatrix} 0 & 1 & 29 & -16 \\ 2 & 2 & -9 & 5 \end{pmatrix}$

Multiply row 1 by –2 and add to row 2: $\begin{pmatrix} 0 & 1 & 29 & -16 \\ 2 & 0 & -67 & 37 \end{pmatrix}$

Multiply row 2 by $\frac{1}{2}$, then switch rows 1 and 2: $\begin{pmatrix} 1 & 0 & -\frac{67}{2} & \frac{37}{2} \\ 0 & 1 & 29 & -16 \end{pmatrix}$

So $(AB)^{-1} = \begin{pmatrix} -\frac{67}{2} & \frac{37}{2} \\ 29 & -16 \end{pmatrix}$, which is the same as $B^{-1}A^{-1}$ from part **a**.

57. a. First compute: $ad - bc = x(-x) - (1+x)(1-x) = -x^2 - 1 + x^2 = -1$

Thus the inverse of $\begin{pmatrix} a & b \\ c & d \end{pmatrix}$ is $\begin{pmatrix} \frac{-x}{-1} & \frac{-(1+x)}{-1} \\ \frac{-(1-x)}{-1} & \frac{x}{-1} \end{pmatrix} = \begin{pmatrix} x & 1+x \\ 1-x & -x \end{pmatrix}$. The inverse is the same as the original matrix.

b. With $x = 11$, the inverse is $\begin{pmatrix} 11 & 12 \\ -10 & -11 \end{pmatrix}$. With $x = \pi + 1$, the inverse is $\begin{pmatrix} \pi+1 & \pi+2 \\ -\pi & -1-\pi \end{pmatrix}$.

10.5 Determinants and Cramer's Rule

1. a. Evaluate the determinant: $\begin{vmatrix} 2 & -17 \\ 1 & 6 \end{vmatrix} = 2(6) - (-17)(1) = 12 + 17 = 29$

b. Evaluate the determinant: $\begin{vmatrix} 1 & 6 \\ 2 & -17 \end{vmatrix} = 1(-17) - (6)(2) = -17 - 12 = -29$

3. a. Evaluate the determinant: $\begin{vmatrix} 5 & 7 \\ 500 & 700 \end{vmatrix} = 100\begin{vmatrix} 5 & 7 \\ 5 & 7 \end{vmatrix} = 100[5(7) - 7(5)] = 100(35 - 35) = 0$

b. Evaluate the determinant: $\begin{vmatrix} 5 & 500 \\ 7 & 700 \end{vmatrix} = 100\begin{vmatrix} 5 & 5 \\ 7 & 7 \end{vmatrix} = 100[5(7) - 5(7)] = 100(35 - 35) = 0$

5. Evaluate the determinant: $\begin{vmatrix} \sqrt{2}-1 & \sqrt{2} \\ \sqrt{2} & \sqrt{2}+1 \end{vmatrix} = (\sqrt{2}-1)(\sqrt{2}+1) - \sqrt{2}(\sqrt{2}) = (2-1) - 2 = -1$

7. Evaluate the minor of 3: $\begin{vmatrix} 5 & 1 \\ 10 & -10 \end{vmatrix} = 10\begin{vmatrix} 5 & 1 \\ 1 & -1 \end{vmatrix} = 10[5(-1) - 1(1)] = 10(-6) = -60$

9. Evaluate the minor of –10: $\begin{vmatrix} -6 & 3 \\ 5 & -4 \end{vmatrix} = 3\begin{vmatrix} -2 & 1 \\ 5 & -4 \end{vmatrix} = 3[-2(-4) - 1(5)] = 3(3) = 9$

11. a. Compute: $-6\begin{vmatrix} -4 & 1 \\ 9 & -10 \end{vmatrix} + 3\begin{vmatrix} 5 & 1 \\ 10 & -10 \end{vmatrix} + 8\begin{vmatrix} 5 & -4 \\ 10 & 9 \end{vmatrix} = -6(31) + 3(-60) + 8(85) = 314$

b. Compute: $-6\begin{vmatrix} -4 & 1 \\ 9 & -10 \end{vmatrix} - 3\begin{vmatrix} 5 & 1 \\ 10 & -10 \end{vmatrix} + 8\begin{vmatrix} 5 & -4 \\ 10 & 9 \end{vmatrix} = -6(31) - 3(-60) + 8(85) = 674$

c. The answer in part **b** would be the determinant.

13. a. Expand along the second row: $-4\begin{vmatrix} 2 & 3 \\ 8 & 9 \end{vmatrix} + 5\begin{vmatrix} 1 & 3 \\ 7 & 9 \end{vmatrix} - 6\begin{vmatrix} 1 & 2 \\ 7 & 8 \end{vmatrix} = -4(-6) + 5(-12) - 6(-6) = 0$

b. Expand along the third row: $7\begin{vmatrix} 2 & 3 \\ 5 & 6 \end{vmatrix} - 8\begin{vmatrix} 1 & 3 \\ 4 & 6 \end{vmatrix} + 9\begin{vmatrix} 1 & 2 \\ 4 & 5 \end{vmatrix} = 7(-3) - 8(-6) + 9(-3) = 0$

c. Expand along the first column: $1\begin{vmatrix} 5 & 6 \\ 8 & 9 \end{vmatrix} - 4\begin{vmatrix} 2 & 3 \\ 8 & 9 \end{vmatrix} + 7\begin{vmatrix} 2 & 3 \\ 5 & 6 \end{vmatrix} = 1(-3) - 4(-6) + 7(-3) = 0$

d. Expand along the third column: $3\begin{vmatrix} 4 & 5 \\ 7 & 8 \end{vmatrix} - 6\begin{vmatrix} 1 & 2 \\ 7 & 8 \end{vmatrix} + 9\begin{vmatrix} 1 & 2 \\ 4 & 5 \end{vmatrix} = 3(-3) - 6(-6) + 9(-3) = 0$

15. Factoring 5 out of the first row: $\begin{vmatrix} 5 & 10 & 15 \\ 1 & 2 & 3 \\ -9 & 11 & 7 \end{vmatrix} = 5\begin{vmatrix} 1 & 2 & 3 \\ 1 & 2 & 3 \\ -9 & 11 & 7 \end{vmatrix} = 5\begin{vmatrix} 0 & 0 & 0 \\ 1 & 2 & 3 \\ -9 & 11 & 7 \end{vmatrix} = 0$

17. Expanding along the third row: $\begin{vmatrix} 1 & 2 & -3 \\ 4 & 5 & -9 \\ 0 & 0 & 1 \end{vmatrix} = 0\begin{vmatrix} 2 & -3 \\ 5 & -9 \end{vmatrix} - 0\begin{vmatrix} 1 & -3 \\ 4 & -9 \end{vmatrix} + 1\begin{vmatrix} 1 & 2 \\ 4 & 5 \end{vmatrix} = 1(-3) = -3$

19. Since the third column is a multiple of the first column, the determinant is 0.

21. Expanding along the first row: $\begin{vmatrix} 3 & 0 & 0 \\ 0 & 19 & 0 \\ 0 & 0 & 10 \end{vmatrix} = 3\begin{vmatrix} 19 & 0 \\ 0 & 10 \end{vmatrix} - 0\begin{vmatrix} 0 & 0 \\ 0 & 10 \end{vmatrix} + 0\begin{vmatrix} 0 & 19 \\ 0 & 0 \end{vmatrix} = 3(190) = 570$

23. Expanding along the second column: $\begin{vmatrix} 23 & 0 & 47 \\ -37 & 0 & 18 \\ 14 & 0 & 25 \end{vmatrix} = 0$

25. Using a graphing utility: $\begin{vmatrix} 15 & 14 & 26 \\ 18 & 17 & 32 \\ 21 & 20 & 42 \end{vmatrix} = 12$

27. **a.** The right-hand determinant should be 10 times the left-hand determinant.

b. Entering $A = \begin{pmatrix} 1 & 2 & 3 \\ -7 & -4 & 5 \\ 9 & 2 & 6 \end{pmatrix}$ and $B = \begin{pmatrix} 10 & 20 & 30 \\ -7 & -4 & 5 \\ 9 & 2 & 6 \end{pmatrix}$, we find det $A = 206$ and det $B = 2060$.

Thus det $B = 10$ x det A.

29. **a.** Adding column 2 to column 3: $\begin{vmatrix} 1 & -1 & -1 & 2 \\ 0 & 1 & 0 & 0 \\ 2 & 1 & 1 & -1 \\ -2 & 2 & 3 & 1 \end{vmatrix} = 1\begin{vmatrix} 1 & -1 & 2 \\ 2 & 1 & -1 \\ -2 & 3 & 1 \end{vmatrix}$

Adding twice column 3 to column 1: $\begin{vmatrix} 5 & -1 & 2 \\ 0 & 1 & -1 \\ 0 & 3 & 1 \end{vmatrix} = 5\begin{vmatrix} 1 & -1 \\ 3 & 1 \end{vmatrix} = 5(4) = 20$

b. Using a graphing utility: $\begin{vmatrix} 1 & -1 & -1 & 2 \\ 0 & 1 & 0 & 0 \\ 2 & 1 & 1 & -1 \\ -2 & 2 & 3 & 1 \end{vmatrix} = 20$

31. **a.** Expanding along the first row:

$$\begin{vmatrix} 2 & 7 & -1 & 9 \\ 4 & 0 & 3 & 6 \\ -8 & 5 & 1 & 3 \\ 11 & 2 & -6 & 1 \end{vmatrix} = 2\begin{vmatrix} 0 & 3 & 6 \\ 5 & 1 & 3 \\ 2 & -6 & 1 \end{vmatrix} - 7\begin{vmatrix} 4 & 3 & 6 \\ -8 & 1 & 3 \\ 11 & -6 & 1 \end{vmatrix} - 1\begin{vmatrix} 4 & 0 & 6 \\ -8 & 5 & 3 \\ 11 & 2 & 1 \end{vmatrix} - 9\begin{vmatrix} 4 & 0 & 3 \\ -8 & 5 & 1 \\ 11 & 2 & -6 \end{vmatrix}$$

$$= 2\left(0\begin{vmatrix} 1 & 3 \\ -6 & 1 \end{vmatrix} - 3\begin{vmatrix} 5 & 3 \\ 2 & 1 \end{vmatrix} + 6\begin{vmatrix} 5 & 1 \\ 2 & -6 \end{vmatrix}\right) - 7\left(4\begin{vmatrix} 1 & 3 \\ -6 & 1 \end{vmatrix} - 3\begin{vmatrix} -8 & 3 \\ 11 & 1 \end{vmatrix} + 6\begin{vmatrix} -8 & 1 \\ 11 & -6 \end{vmatrix}\right)$$

$$-1\left(4\begin{vmatrix} 5 & 3 \\ 2 & 1 \end{vmatrix} - 0\begin{vmatrix} -8 & 3 \\ 11 & 1 \end{vmatrix} + 6\begin{vmatrix} -8 & 5 \\ 11 & 2 \end{vmatrix}\right) - 9\left(4\begin{vmatrix} 5 & 1 \\ 2 & -6 \end{vmatrix} - 0\begin{vmatrix} -8 & 1 \\ 11 & -6 \end{vmatrix} + 3\begin{vmatrix} -8 & 5 \\ 11 & 2 \end{vmatrix}\right)$$

$$= 2(0+3-192) - 7(76+123+222) - 1(-4-0-426) - 9(-128-0-213)$$
$$= -378 - 2947 + 430 + 3069$$
$$= 174$$

b. Expanding along the second row:

$$\begin{vmatrix} 2 & 7 & -1 & 9 \\ 4 & 0 & 3 & 6 \\ -8 & 5 & 1 & 3 \\ 11 & 2 & -6 & 1 \end{vmatrix} = -4\begin{vmatrix} 7 & -1 & 9 \\ 5 & 1 & 3 \\ 2 & -6 & 1 \end{vmatrix} + 0\begin{vmatrix} 2 & -1 & 9 \\ -8 & 1 & 3 \\ 11 & -6 & 1 \end{vmatrix} - 3\begin{vmatrix} 2 & 7 & 9 \\ -8 & 5 & 3 \\ 11 & 2 & 1 \end{vmatrix} + 6\begin{vmatrix} 2 & 7 & -1 \\ -8 & 5 & 1 \\ 11 & 2 & -6 \end{vmatrix}$$

$$= -4\left(7\begin{vmatrix} 1 & 3 \\ -6 & 1 \end{vmatrix} + 1\begin{vmatrix} 5 & 3 \\ 2 & 1 \end{vmatrix} + 9\begin{vmatrix} 5 & 1 \\ 2 & -6 \end{vmatrix}\right) + 0$$

$$-3\left(2\begin{vmatrix} 5 & 3 \\ 2 & 1 \end{vmatrix} - 7\begin{vmatrix} -8 & 3 \\ 11 & 1 \end{vmatrix} + 9\begin{vmatrix} -8 & 5 \\ 11 & 2 \end{vmatrix}\right) + 6\left(2\begin{vmatrix} 5 & 1 \\ 2 & -6 \end{vmatrix} - 7\begin{vmatrix} -8 & 1 \\ 11 & -6 \end{vmatrix} - 1\begin{vmatrix} -8 & 5 \\ 11 & 2 \end{vmatrix}\right)$$

$$= -4(133 - 1 - 288) + 0 - 3(-2 + 287 - 639) + 6(-64 - 259 + 71)$$
$$= 624 + 0 + 1062 - 1512$$
$$= 174$$

33. Evaluate the determinant:

$$\begin{vmatrix} 1 & x & x^2 \\ 1 & y & y^2 \\ 1 & z & z^2 \end{vmatrix} = \begin{vmatrix} 1 & x & x^2 \\ 0 & y-x & y^2-x^2 \\ 0 & z-x & z^2-x^2 \end{vmatrix}$$
$$= \begin{vmatrix} y-x & y^2-x^2 \\ z-x & z^2-x^2 \end{vmatrix}$$
$$= \begin{vmatrix} y-x & (y+x)(y-x) \\ z-x & (z+x)(z-x) \end{vmatrix}$$
$$= (y-x)(z-x)\begin{vmatrix} 1 & y+x \\ 1 & z+x \end{vmatrix}$$
$$= (y-x)(z-x)[z+x-y-x]$$
$$= (y-x)(z-x)(z-y)$$

35. Subtracting row 1 from row 2 and row 1 from row 3: $\begin{vmatrix} 1 & 1 & 1 \\ 1 & 1+x & 1 \\ 1 & 1 & 1+y \end{vmatrix} = \begin{vmatrix} 1 & 1 & 1 \\ 0 & x & 0 \\ 0 & 0 & y \end{vmatrix} = 1\begin{vmatrix} x & 0 \\ 0 & y \end{vmatrix} = xy$

37. **a.** Subtract the first column from the second and add twice the first column to the third. The result is:

$$\begin{vmatrix} -20 & 22 & -43 \\ 6 & -10 & 13 \\ -1 & 0 & 0 \end{vmatrix} = -1\left[22(13) - (-43)(-10)\right] = 144$$

b. Multiply the second row by 2 and subtract it from the first row, and multiply the second row by 4 and subtract it from the third row. This yields: $\begin{vmatrix} 0 & -32 & -5 \\ 1 & 6 & 1 \\ 0 & -25 & -2 \end{vmatrix} = -1\left[(-32)(-2) - (-5)(-25)\right] = 61$

c. Subtract the first column from the second and add 10 times the first column to the third. This yields:

$$\begin{vmatrix} 2 & 0 & 0 \\ 1 & -5 & 16 \\ 4 & -5 & 39 \end{vmatrix} = 2\left[(-5)(39) - 16(-5)\right] = -230$$

39. Compute D: $D=\begin{vmatrix}3 & 4 & -1\\ 1 & -3 & 2\\ 5 & 0 & -6\end{vmatrix}=5\begin{vmatrix}4 & -1\\ -3 & 2\end{vmatrix}-6\begin{vmatrix}3 & 4\\ 1 & -3\end{vmatrix}=5(5)-6(-13)=103$

Now compute D_x: $D_x=\begin{vmatrix}5 & 4 & -1\\ 2 & -3 & 2\\ -7 & 0 & -6\end{vmatrix}=-7\begin{vmatrix}4 & -1\\ -3 & 2\end{vmatrix}-6\begin{vmatrix}5 & 4\\ 2 & -3\end{vmatrix}=-7(5)-6(-23)=103$

For D_y, adding –2 times column 1 to both column 2 and column 3 yields:

$$D_y=\begin{vmatrix}3 & 5 & -1\\ 1 & 2 & 2\\ 5 & -7 & -6\end{vmatrix}=\begin{vmatrix}3 & -1 & -7\\ 1 & 0 & 0\\ 5 & -17 & -16\end{vmatrix}=-1\begin{vmatrix}-1 & -7\\ -17 & -16\end{vmatrix}=-(-103)=103$$

For D_z, adding 3 times column 1 to column 2, and –2 times column 1 to column 3 yields:

$$D_z=\begin{vmatrix}3 & 4 & 5\\ 1 & -3 & 2\\ 5 & 0 & -7\end{vmatrix}=\begin{vmatrix}3 & 13 & -1\\ 1 & 0 & 0\\ 5 & 15 & -17\end{vmatrix}=-1\begin{vmatrix}13 & 1\\ 15 & -17\end{vmatrix}=-(-206)=206$$

So $x=\frac{D_x}{D}=1$, $y=\frac{D_y}{D}=1$, and $z=\frac{D_z}{D}=2$. So the solution is (1,1,2).

41. Subtracting column 1 from both column 2 and column 3 yields: $D=\begin{vmatrix}3 & 2 & -1\\ 2 & -3 & -4\\ 1 & 1 & 1\end{vmatrix}=\begin{vmatrix}3 & -1 & -4\\ 2 & -5 & -6\\ 1 & 0 & 0\end{vmatrix}=1\begin{vmatrix}-1 & -4\\ -5 & -6\end{vmatrix}=-14$

Adding –5 times column 3 to column 1, and subtracting column 3 from column 2 yields:

$$D_x=\begin{vmatrix}-6 & 2 & -1\\ -11 & -3 & -4\\ 5 & 1 & 1\end{vmatrix}=\begin{vmatrix}-1 & 3 & -1\\ 9 & 1 & -4\\ 0 & 0 & 1\end{vmatrix}=1\begin{vmatrix}-1 & 3\\ 9 & 1\end{vmatrix}=-28$$

Adding row 3 to row 1, and 4 times row 3 to row 2 yields: $D_y=\begin{vmatrix}3 & -6 & -1\\ 2 & -11 & -4\\ 1 & 5 & 1\end{vmatrix}=\begin{vmatrix}4 & -1 & 0\\ 6 & 9 & 0\\ 1 & 5 & 1\end{vmatrix}=1\begin{vmatrix}4 & -1\\ 6 & 9\end{vmatrix}=42$

Adding –2 times row 3 to row 1, and 3 times row 3 to row 2 yields:

$$D_z=\begin{vmatrix}3 & 2 & -6\\ 2 & 3 & -11\\ 1 & 1 & 5\end{vmatrix}=\begin{vmatrix}1 & 0 & -16\\ -1 & 0 & -26\\ 1 & 1 & 5\end{vmatrix}=-1\begin{vmatrix}1 & -16\\ -1 & -26\end{vmatrix}=-(-42)=42$$

So $x=\frac{D_x}{D}=2$, $y=\frac{D_y}{D}=-3$, and $z=\frac{D_z}{D}=-3$. So the solution is (2,–3,–3).

43. Adding –2 times row 3 to row 1, and –3 times row 3 to row 2 yields:

$$D=\begin{vmatrix}2 & 5 & 2\\ 3 & -1 & -4\\ 1 & 2 & -3\end{vmatrix}=\begin{vmatrix}0 & 1 & 8\\ 0 & -7 & 5\\ 1 & 2 & -3\end{vmatrix}=1\begin{vmatrix}1 & 8\\ -7 & 5\end{vmatrix}=61$$

Now compute:

$$D_x=\begin{vmatrix}0 & 5 & 2\\ 0 & -1 & -4\\ 0 & 2 & 3\end{vmatrix}=0 \qquad D_y=\begin{vmatrix}2 & 0 & 2\\ 3 & 0 & -4\\ 1 & 0 & -3\end{vmatrix}=0 \qquad D_z=\begin{vmatrix}2 & 5 & 0\\ 3 & -1 & 0\\ 1 & 2 & 0\end{vmatrix}=0$$

So $x=\frac{D_x}{D}=0$, $y=\frac{D_y}{D}=0$, and $z=\frac{D_z}{D}=0$. So the solution is (0,0,0).

45. Factoring 6 out of column 1 and 2 out of column 2: $D=\begin{vmatrix}12 & 0 & -11\\ 6 & 6 & -4\\ 6 & 2 & -5\end{vmatrix}=6\begin{vmatrix}2 & 0 & -11\\ 1 & 6 & -4\\ 1 & 2 & -5\end{vmatrix}=12\begin{vmatrix}2 & 0 & -11\\ 1 & 3 & -4\\ 1 & 1 & -5\end{vmatrix}$

Adding –3 times row 3 to row 2 yields: $D=12\begin{vmatrix}2 & 0 & -11\\ -2 & 0 & 11\\ 1 & 1 & -5\end{vmatrix}=12(-1)\begin{vmatrix}2 & -11\\ -2 & 11\end{vmatrix}=-12(0)=0$

So Cramer's Rule will not work. We form the augmented matrix (using equation two as row 1): $\begin{pmatrix}6 & 6 & -4 & 26\\ 12 & 0 & -11 & 13\\ 6 & 2 & -5 & 13\end{pmatrix}$

Adding –2 times row 1 to row 2 and –1 times row 1 to row 3 yields: $\begin{pmatrix}6 & 6 & -4 & 26\\ 0 & -12 & -3 & -39\\ 0 & -4 & -1 & -13\end{pmatrix}$

Multiply row 1 by $\frac{1}{2}$ and row 2 by $\frac{1}{3}$: $\begin{pmatrix}3 & 3 & -2 & 13\\ 0 & 4 & 1 & 13\\ 0 & -4 & -1 & -13\end{pmatrix}$

Adding row 2 to row 3 yields: $\begin{pmatrix}3 & 3 & -2 & 13\\ 0 & 4 & 1 & 13\\ 0 & 0 & 0 & 0\end{pmatrix}$

So we have the system:

$$3x+3y-2z=13$$
$$4y+z=13$$

Solve equation 2 for z:

$$4y+z=13$$
$$z=13-4y$$

Substitute into equation 1:

$$3x+3y-26+8y=13$$
$$3x=39-11y$$
$$x=13-\tfrac{11}{3}y$$

So the solution is $\left(13-\frac{11}{3}y, y, 13-4y\right)$, for any real number y.

47. Using a graphing utility, the following determinants are computed:

$$D=\begin{vmatrix}1 & 1 & 1 & 1\\ 1 & -1 & 1 & -1\\ 2 & -2 & -3 & -3\\ 3 & 2 & 1 & -1\end{vmatrix}=26 \qquad D_x=\begin{vmatrix}-7 & 1 & 1 & 1\\ -11 & -1 & 1 & -1\\ 26 & -2 & -3 & -3\\ -9 & 2 & 1 & -1\end{vmatrix}=-26 \qquad D_y=\begin{vmatrix}1 & -7 & 1 & 1\\ 1 & -11 & 1 & -1\\ 2 & 26 & -3 & -3\\ 3 & -9 & 1 & -1\end{vmatrix}=0$$

$$D_z=\begin{vmatrix}1 & 1 & -7 & 1\\ 1 & -1 & -11 & -1\\ 2 & -2 & 26 & -3\\ 3 & 2 & -9 & -1\end{vmatrix}=260 \qquad D_w=\begin{vmatrix}1 & 1 & 1 & -7\\ 1 & -1 & 1 & -11\\ 2 & -2 & -3 & 26\\ 3 & 2 & 1 & -9\end{vmatrix}=-52$$

So $x=\dfrac{D_x}{D}=1$, $y=\dfrac{D_y}{D}=0$, $z=\dfrac{D_z}{D}=-10$, and $w=\dfrac{D_w}{D}=2$. So the solution is (1,0,–10,2).

49. Expanding the determinant: $\begin{vmatrix} x-4 & 0 & 0 \\ 0 & x+4 & 0 \\ 0 & 0 & x+1 \end{vmatrix} = (x-4)\begin{vmatrix} x+4 & 0 \\ 0 & x+1 \end{vmatrix} = (x-4)(x+4)(x+1)$

This will equal 0 when $x = 4$, $x = -4$, or $x = -1$.

51. Expanding along the first column:

$$\begin{aligned}\begin{vmatrix} a & b & c \\ a & b & c \\ d & e & f \end{vmatrix} &= a\begin{vmatrix} b & c \\ e & f \end{vmatrix} - a\begin{vmatrix} b & c \\ e & f \end{vmatrix} + d\begin{vmatrix} b & c \\ b & c \end{vmatrix} \\ &= a(bf-ec) - a(bf-ec) + d(bc-bc) \\ &= abf - aec - abf + aec + 0 \\ &= 0\end{aligned}$$

53. Expanding the determinant:

$$\begin{aligned}\begin{vmatrix} a_1+A_1 & b_1 & c_1 \\ a_2+A_2 & b_2 & c_2 \\ a_3+A_3 & b_3 & c_3 \end{vmatrix} &= (a_1+A_1)\begin{vmatrix} b_2 & c_2 \\ b_3 & c_3 \end{vmatrix} - (a_2+A_2)\begin{vmatrix} b_1 & c_1 \\ b_3 & c_3 \end{vmatrix} + (a_3+A_3)\begin{vmatrix} b_1 & c_1 \\ b_2 & c_2 \end{vmatrix} \\ &= \left\{a_1\begin{vmatrix} b_2 & c_2 \\ b_3 & c_3 \end{vmatrix} - a_2\begin{vmatrix} b_1 & c_1 \\ b_3 & c_3 \end{vmatrix} + a_3\begin{vmatrix} b_1 & c_1 \\ b_2 & c_2 \end{vmatrix}\right\} + \left\{A_1\begin{vmatrix} b_2 & c_2 \\ b_3 & c_3 \end{vmatrix} - A_2\begin{vmatrix} b_1 & c_1 \\ b_3 & c_3 \end{vmatrix} + A_3\begin{vmatrix} b_1 & c_1 \\ b_2 & c_2 \end{vmatrix}\right\}\end{aligned}$$

Now observe that the expression in the first set of braces is: $\begin{vmatrix} a_1 & b_1 & c_1 \\ a_2 & b_2 & c_2 \\ a_3 & b_3 & c_3 \end{vmatrix}$

The expression in the second set of braces is: $\begin{vmatrix} A_1 & b_1 & c_1 \\ A_2 & b_2 & c_2 \\ A_3 & b_3 & c_3 \end{vmatrix}$

55. Expanding the determinant on the left-hand side of the given equation along its first row:

$$\begin{vmatrix} a_1 & b_1 & c_1 \\ a_2 & b_2 & c_2 \\ a_3 & b_3 & c_3 \end{vmatrix} = a_1\begin{vmatrix} b_2 & c_2 \\ b_3 & c_3 \end{vmatrix} - b_1\begin{vmatrix} a_2 & c_2 \\ a_3 & c_3 \end{vmatrix} + c_1\begin{vmatrix} a_2 & b_2 \\ a_3 & b_3 \end{vmatrix}$$

Next, expanding the determinant on the right-hand side of the given equation along its second row:

$$\begin{vmatrix} a_2 & b_2 & c_2 \\ a_1 & b_1 & c_1 \\ a_3 & b_3 & c_3 \end{vmatrix} = -a_1\begin{vmatrix} b_2 & c_2 \\ b_3 & c_3 \end{vmatrix} + b_1\begin{vmatrix} a_2 & c_2 \\ a_3 & c_3 \end{vmatrix} - c_1\begin{vmatrix} a_2 & b_2 \\ a_3 & b_3 \end{vmatrix}$$

By inspection now, we observe that the two expressions for the determinants are negatives of one another.

57. Subtract the fourth row from each of the other three rows. Therefore:

$$\begin{vmatrix} a & 0 & 0 & -d \\ 0 & b & 0 & -d \\ 0 & 0 & c & -d \\ 1 & 1 & 1 & 1+d \end{vmatrix} = abcd\begin{vmatrix} 1 & 0 & 0 & -1 \\ 0 & 1 & 0 & -1 \\ 0 & 0 & 1 & -1 \\ \frac{1}{a} & \frac{1}{b} & \frac{1}{c} & 1+\frac{1}{d} \end{vmatrix}$$

$$= abcd\begin{vmatrix} 1 & 0 & 0 & 0 \\ 0 & 1 & 0 & -1 \\ 0 & 0 & 1 & -1 \\ \frac{1}{a} & \frac{1}{b} & \frac{1}{c} & 1+\frac{1}{a}+\frac{1}{d} \end{vmatrix}$$

$$= abcd\begin{vmatrix} 1 & 0 & -1 \\ 0 & 1 & -1 \\ \frac{1}{b} & \frac{1}{c} & 1+\frac{1}{a}+\frac{1}{d} \end{vmatrix}$$

$$= abcd\begin{vmatrix} 1 & 0 & 0 \\ 0 & 1 & -1 \\ \frac{1}{b} & \frac{1}{c} & 1+\frac{1}{a}+\frac{1}{b}+\frac{1}{d} \end{vmatrix}$$

$$= abcd\begin{vmatrix} 1 & -1 \\ \frac{1}{c} & 1+\frac{1}{a}+\frac{1}{b}+\frac{1}{d} \end{vmatrix}$$

$$= abcd\left(1+\frac{1}{a}+\frac{1}{b}+\frac{1}{c}+\frac{1}{d}\right)$$

59. Subtracting row 1 from both row 2 and row 3: $\begin{vmatrix} 1 & bc & b+c \\ 1 & ca & c+a \\ 1 & ab & a+b \end{vmatrix} = \begin{vmatrix} 1 & bc & b+c \\ 0 & ca-bc & a-b \\ 0 & ab-bc & a-c \end{vmatrix} = 1\begin{vmatrix} c(a-b) & a-b \\ b(a-c) & a-c \end{vmatrix}$

Factoring $a-b$ out of the first row: $(a-b)\begin{vmatrix} c & 1 \\ b(a-c) & a-c \end{vmatrix}$

Factoring $a-c$ out of the second row: $(a-b)(a-c)\begin{vmatrix} c & 1 \\ b & 1 \end{vmatrix} = (a-b)(a-c)(c-b) = (a-b)(c-a)(b-c)$

61. Subtracting row 1 from each of row 2, row 3, and row 4 yields:

$$\begin{vmatrix} 1 & a & a & a \\ 1 & b & a & a \\ 1 & a & b & a \\ 1 & a & a & b \end{vmatrix} = \begin{vmatrix} 1 & a & a & a \\ 0 & b-a & 0 & 0 \\ 0 & 0 & b-a & 0 \\ 0 & 0 & 0 & b-a \end{vmatrix}$$

$$= 1\begin{vmatrix} b-a & 0 & 0 \\ 0 & b-a & 0 \\ 0 & 0 & b-a \end{vmatrix}$$

$$= (b-a)\begin{vmatrix} b-a & 0 \\ 0 & b-a \end{vmatrix}$$

$$= (b-a)(b-a)^2$$

$$= (b-a)^3$$

63. Form the augmented matrix: $\begin{pmatrix} a & b & c & k \\ a^2 & b^2 & c^2 & k^2 \\ a^3 & b^3 & c^3 & k^3 \end{pmatrix}$

Adding $-a$ times row 1 to row 2 and $-a^2$ times row 1 to row 3: $\begin{pmatrix} a & b & c & k \\ 0 & b^2-ab & c^2-ac & k^2-ak \\ 0 & b^3-a^2b & c^3-a^2c & k^3-a^2k \end{pmatrix}$

Factoring: $\begin{pmatrix} a & b & c & k \\ 0 & b(b-a) & c(c-a) & k(k-a) \\ 0 & b(b+a)(b-a) & c(c+a)(c-a) & k(k+a)(k-a) \end{pmatrix}$

Adding $-(b+a)$ times row 2 to row 3: $\begin{pmatrix} a & b & c & k \\ 0 & b(b-a) & c(c-a) & k(k-a) \\ 0 & 0 & c(c-a)(c-b) & k(k-a)(k-b) \end{pmatrix}$

So we have the system:

$$\begin{aligned} ax + by + cz &= k \\ b(b-a)y + c(c-a)z &= k(k-a) \\ c(c-a)(c-b)z &= k(k-a)(k-b) \end{aligned}$$

Solving the third equation for z:

$$\begin{aligned} c(c-a)(c-b)z &= k(k-a)(k-b) \\ z &= \frac{k(k-a)(k-b)}{c(c-a)(c-b)} \end{aligned}$$

Substitute into the second equation:

$$\begin{aligned} b(b-a)y + \frac{k(k-a)(k-b)}{c-b} &= k(k-a) \\ b(b-a)(c-b)y &= k(k-a)(c-b-k+b) \\ y &= \frac{k(k-a)(k-c)}{b(b-a)(b-c)} \end{aligned}$$

Substitute into the first equation:

$$\begin{aligned} ax + \frac{k(k-a)(k-c)}{(b-a)(b-c)} + \frac{k(k-a)(k-b)}{(c-a)(c-b)} &= k \\ ax &= \frac{k(k-b)(k-c)}{(b-a)(c-a)} \\ x &= \frac{k(k-b)(k-c)}{a(a-b)(a-c)} \end{aligned}$$

So the solution is $\left(\frac{k(k-b)(k-c)}{a(a-b)(a-c)}, \frac{k(k-a)(k-c)}{b(b-a)(b-c)}, \frac{k(k-a)(k-b)}{c(c-a)(c-b)}\right)$.

65. Evaluate the determinant by expanding along the first row:

$$\begin{vmatrix} x & y & 1 \\ -3 & -1 & 1 \\ 2 & 9 & 1 \end{vmatrix} = x(-1-9) - y(-3-2) + 1(-27+2) = -10x + 5y - 25$$

Setting this equal to 0:

$$\begin{aligned} -10x + 5y - 25 &= 0 \\ 5y &= 10x + 25 \\ y &= 2x + 5 \end{aligned}$$

67. Re-draw the figure:

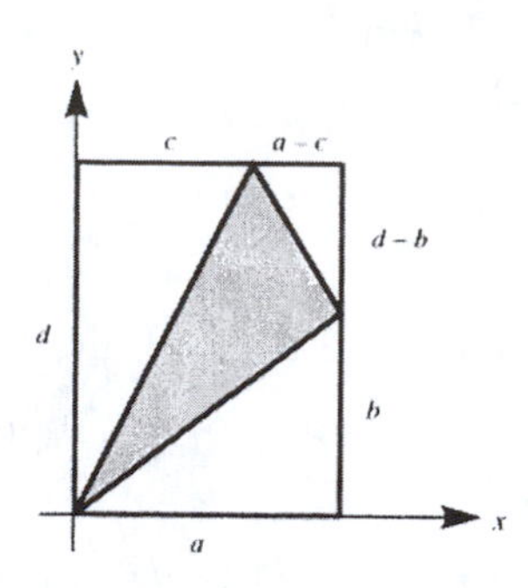

The area of the rectangle is *ad*, and the three triangles have areas of $\frac{1}{2}(ab)$, $\frac{1}{2}(a-c)(d-b)$, and $\frac{1}{2}(cd)$, so the area of the shaded triangle is: $ad-\left[\frac{1}{2}ab+\frac{1}{2}ad-\frac{1}{2}cd-\frac{1}{2}ab+\frac{1}{2}bc+\frac{1}{2}cd\right]=\frac{1}{2}ad-\frac{1}{2}bc=\frac{1}{2}(ad-bc)=\frac{1}{2}\begin{vmatrix} a & b \\ c & d \end{vmatrix}$

69. **a.** From the previous section, we found $A^{-1}=\begin{pmatrix} \frac{d}{ad-bc} & \frac{-b}{ad-bc} \\ \frac{-c}{ad-bc} & \frac{a}{ad-bc} \end{pmatrix}$. To check this, note that *A* times this matrix equals I_2. Since $D=ad-bc$, $A^{-1}=\frac{1}{D}\begin{pmatrix} d & -b \\ -c & a \end{pmatrix}$.

b. We have $D=-6(9)-7(1)=-54-7=-61$, so the inverse is: $-\frac{1}{61}\begin{pmatrix} 9 & -7 \\ -1 & -6 \end{pmatrix}=\begin{pmatrix} -\frac{9}{61} & \frac{7}{61} \\ \frac{1}{61} & \frac{6}{61} \end{pmatrix}$

10.6 Nonlinear Systems of Equations

1. Substituting:

$$x^2=3x$$
$$x^2-3x=0$$
$$x(x-3)=0$$

From which we get $x=0$ or $x=3$. If $x=0$, $y=3\bullet 0=0$ and if $x=3$, $y=9$. Our solutions are (0,0) and (3,9). Note that we have found the points of intersection of a line and a parabola.

3. Since $x^2=24y$, $x^2+y^2=25$ becomes:

$$24y+y^2=25$$
$$y^2+24y-25=0$$
$$(y+25)(y-1)=0$$
$$y=-25 \text{ and } y=1$$

When $y=1$ we get $x^2=24$ or $x=\pm 2\sqrt{6}$, but $y=-25$ means $x^2=-600$ which is impossible. Our two solutions are $\left(2\sqrt{6},1\right)$ and $\left(-2\sqrt{6},1\right)$.

5. Substitute into the first equation:

$$x\left(-x^2\right)=1$$
$$-x^3=1$$
$$x=-1$$

Since $y=\frac{1}{x}$, $y=-1$. So the only solution is $(-1,-1)$.

7. Multiply the first equation by –2:

$$-4x^2 - 2y^2 = -34$$
$$x^2 + 2y^2 = 22$$

Adding:

$$-3x^2 = -12$$
$$x^2 = 4$$
$$x = \pm 2$$

When $x = 2$:

$$2(4) + y^2 = 17$$
$$y^2 = 9$$
$$y = \pm 3$$

When $x = -2$:

$$2(4) + y^2 = 17$$
$$y^2 = 9$$
$$y = \pm 3$$

So the solutions are (2,3), (2,–3), (–2,3), and (–2,–3).

9. Substitute into the first equation:

$$x^2 - 1 = 1 - x^2$$
$$2x^2 = 2$$
$$x^2 = 1$$
$$x = \pm 1$$

So $y = 1 - 1 = 0$ for each value of x. So the solutions are (1,0) and (–1,0).

11. Substitute into the first equation:

$$x(4x+1) = 4$$
$$4x^2 + x - 4 = 0$$
$$x = \frac{-1 \pm \sqrt{1+64}}{8} = \frac{-1 \pm \sqrt{65}}{8}$$

When $x = \frac{-1+\sqrt{65}}{8}$: $y = \frac{-1+\sqrt{65}}{2} + 1 = \frac{1+\sqrt{65}}{2}$. When $x = \frac{-1-\sqrt{65}}{8}$: $y = \frac{-1-\sqrt{65}}{2} + 1 = \frac{1-\sqrt{65}}{2}$

So the solutions are $\left(\frac{-1+\sqrt{65}}{8}, \frac{1+\sqrt{65}}{2}\right)$ and $\left(\frac{-1-\sqrt{65}}{8}, \frac{1-\sqrt{65}}{2}\right)$.

13. Let $a = \frac{1}{x^2}$ and $b = \frac{1}{y^2}$, so:

$$a - 3b = 14$$
$$2a + b = 35$$

Multiply the first equation by –2:

$$-2a + 6b = -28$$
$$2a + b = 35$$

Adding:

$$7b = 7$$
$$b = 1$$

So $a - 3 = 14$, and $a = 17$. Since $a = \frac{1}{x^2}$ and $b = \frac{1}{y^2}$, we have $x^2 = \frac{1}{17}$ and $y^2 = 1$. So $x = \frac{\pm\sqrt{17}}{17}$ and $y = \pm 1$.

So the solutions are $\left(\frac{\sqrt{17}}{17}, 1\right), \left(\frac{\sqrt{17}}{17}, -1\right), \left(\frac{-\sqrt{17}}{17}, 1\right)$, and $\left(\frac{-\sqrt{17}}{17}, -1\right)$.

15. Substitute into the second equation:

$$(x-3)^2+\left(-\sqrt{x-1}\right)^2=4$$
$$x^2-6x+9+x-1=4$$
$$x^2-5x+4=0$$
$$(x-1)(x-4)=0$$
$$x=1 \text{ or } x=4$$

When $x=1,\ y=-\sqrt{1-1}=0$ and when $x=4,\ y=-\sqrt{4-1}=-\sqrt{3}$. So the solutions are (1,0) and $\left(4,-\sqrt{3}\right)$.

17. Since $y=2^{2x}-12=\left(2^x\right)^2-12$, substitute into the second equation:

$$y=y^2-12$$
$$0=y^2-y-12$$
$$0=(y-4)(y+3)$$
$$y=4 \text{ or } y=-3$$

When $y = 4$, we have $2^x = 4$, or $x = 2$. But $y = -3$ will not yield a solution. So the only solution is (2,4).

19. Let $u = \log_{10} x$ and $v = \log_{10} y$, so:

$$2u^2 - v^2 = -1$$
$$4u^2 - 3v^2 = -11$$

Multiply the first equation by –2:

$$-4u^2 + 2v^2 = 2$$
$$4u^2 - 3v^2 = -11$$

Adding:

$$-v^2 = -9$$
$$v^2 = 9$$
$$v = \pm 3$$

Substitute into the first equation:

$$2u^2 - 9 = -1$$
$$2u^2 = 8$$
$$u^2 = 4$$
$$u = \pm 2$$

Since $u = \log_{10} x,\ x = 10^{\pm 2}$. Similarly, $y = 10^{\pm 3}$. So the solutions are (100, 1000), $\left(100, \frac{1}{1000}\right)$, $\left(\frac{1}{100}, 1000\right)$, and $\left(\frac{1}{100}, \frac{1}{1000}\right)$.

21. First take the logarithm of each side of the first equation:

$$\ln\left(2^x 3^y\right) = \ln 4$$
$$\ln\left(2^x\right) + \ln\left(3^y\right) = \ln 2^2$$
$$(\ln 2)x + (\ln 3)y = 2\ln 2$$

Multiply the second equation by –ln 2:

$$(\ln 2)x + (\ln 3)y = 2\ln 2$$
$$(-\ln 2)x - (\ln 2)y = -5\ln 2$$

Adding:

$$(\ln 3 - \ln 2)y = -3\ln 2$$
$$y = \frac{3\ln 2}{\ln 2 - \ln 3}$$

Substitute into the second equation: $x = 5 - \frac{3\ln 2}{\ln 2 - \ln 3} = \frac{2\ln 2 - 5\ln 3}{\ln 2 - \ln 3}$. The solution is $\left(\frac{2\ln 2 - 5\ln 3}{\ln 2 - \ln 3}, \frac{3\ln 2}{\ln 2 - \ln 3}\right)$.

23. **a.** Graphing the two curves:

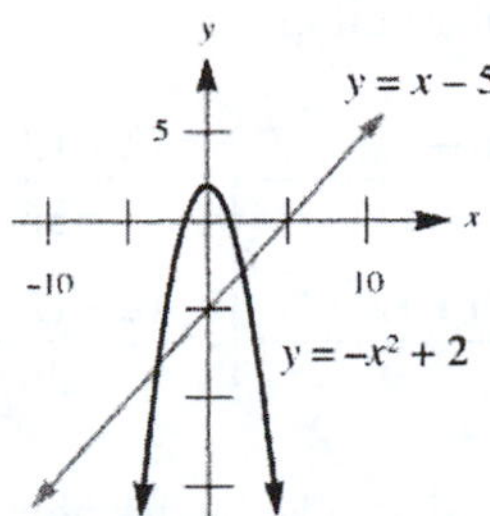

The intersection points are (2.19,–2.81) and (–3.19,–8.19).

b. Substituting $y = x - 5$:

$$x - 5 = -x^2 + 2$$
$$x^2 + x - 7 = 0$$

Using the quadratic formula: $x = \frac{-1 \pm \sqrt{(1)^2 - 4(1)(-7)}}{2(1)} = \frac{-1 \pm \sqrt{1+28}}{2} = \frac{-1 \pm \sqrt{29}}{2}$

If $x = \frac{-1+\sqrt{29}}{2}$, $y = \frac{-1+\sqrt{29}}{2} - 5 = \frac{-11+\sqrt{29}}{2}$. If $x = \frac{-1-\sqrt{29}}{2}$, $y = \frac{-1-\sqrt{29}}{2} - 5 = \frac{-11-\sqrt{29}}{2}$.

The intersection points are $\left(\frac{-1+\sqrt{29}}{2}, \frac{-11+\sqrt{29}}{2}\right) \approx (2.193, -2.807)$ and

$\left(\frac{-1-\sqrt{29}}{2}, \frac{-11-\sqrt{29}}{2}\right) \approx (-3.193, -8.193)$.

25. **a.** Graphing the two curves:

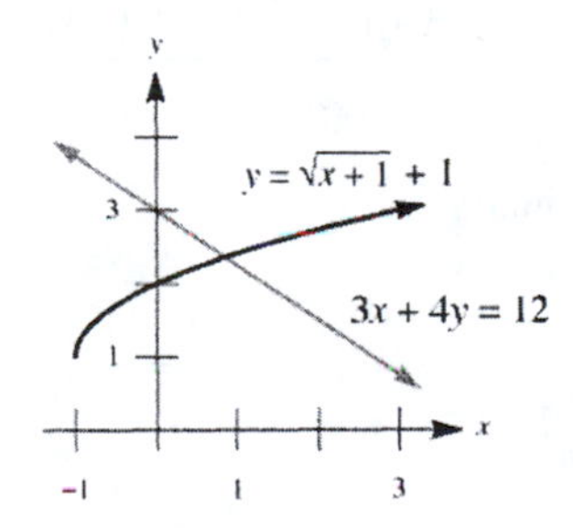

The intersection point is (0.85,2.36).

b. Substituting $y = \sqrt{x+1} + 1$:

$$3x + 4\left(\sqrt{x+1} + 1\right) = 12$$
$$3x + 4\sqrt{x+1} + 4 = 12$$
$$4\sqrt{x+1} = 8 - 3x$$
$$16(x+1) = 64 - 48x + 9x^2$$
$$16x + 16 = 64 - 48x + 9x^2$$
$$0 = 9x^2 - 64x + 48$$

Using the quadratic formula:

$$x = \frac{-(-64) \pm \sqrt{(-64)^2 - 4(9)(48)}}{2(9)} = \frac{64 \pm \sqrt{4096 - 1728}}{18} = \frac{64 \pm 8\sqrt{37}}{18} = \frac{32 \pm 4\sqrt{37}}{9}$$

Since $3x+4y=12$, $y=\dfrac{12-3x}{4}$. Substituting:

$$x=\frac{32-4\sqrt{37}}{9}:\quad y=\frac{1}{4}\left(12-\frac{96-12\sqrt{37}}{9}\right)=\frac{12+12\sqrt{37}}{36}=\frac{1+\sqrt{37}}{3}$$

$$x=\frac{32+4\sqrt{37}}{9}:\quad y=\frac{1}{4}\left(12-\frac{96+12\sqrt{37}}{9}\right)=\frac{12-12\sqrt{37}}{36}=\frac{1-\sqrt{37}}{3}$$

Since $y \geq 0$, this second point is discarded. The intersection point is $\left(\dfrac{32-4\sqrt{37}}{9},\dfrac{1+\sqrt{37}}{3}\right) \approx (0.852, 2.361)$.

27. **a.** Graphing the two curves:

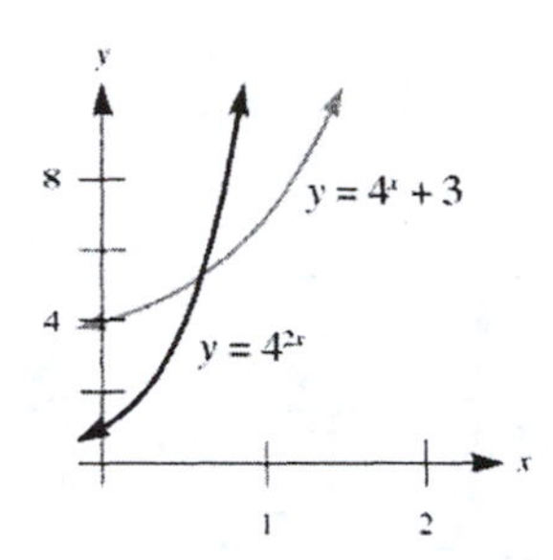

The intersection point is (0.60,5.30).

b. Setting the two equations equal: $4^{2x} = 4^x + 3$. Let $u = 4^x$. Then the equation becomes:

$$u^2 = u + 3$$
$$u^2 - u - 3 = 0$$

Using the quadratic formula: $u = \dfrac{-(-1) \pm \sqrt{(-1)^2 - 4(1)(-3)}}{2(1)} = \dfrac{1 \pm \sqrt{1+12}}{2} = \dfrac{1 \pm \sqrt{13}}{2}$

Since $u > 0$, choose $u = \dfrac{1+\sqrt{13}}{2}$, thus:

$$4^x = \frac{1+\sqrt{13}}{2}$$
$$x \ln 4 = \ln\left(\frac{1+\sqrt{13}}{2}\right)$$
$$x = \frac{\ln\left(\frac{1+\sqrt{13}}{2}\right)}{\ln 4}$$

Then $y = 4^x + 3 = \dfrac{1+\sqrt{13}}{2} + 3 = \dfrac{7+\sqrt{13}}{2}$. The solution is $\left(\dfrac{\ln\left(\frac{1+\sqrt{13}}{2}\right)}{\ln 4}, \dfrac{7+\sqrt{13}}{2}\right) \approx (0.602, 5.303)$.

29. Sketch the graphs:

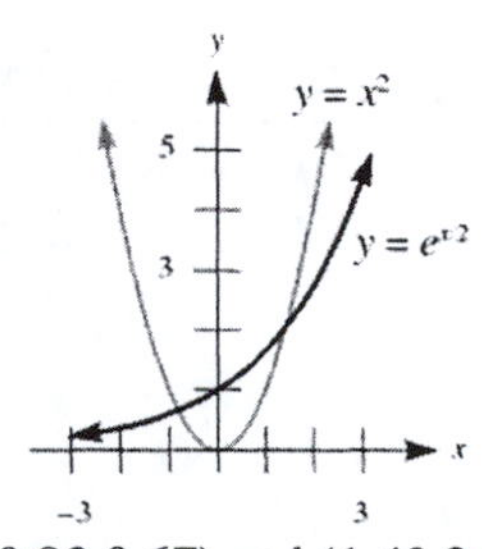

The intersection points are approximately (–0.82,0.67) and (1.43,2.04).

31. Sketch the graphs:

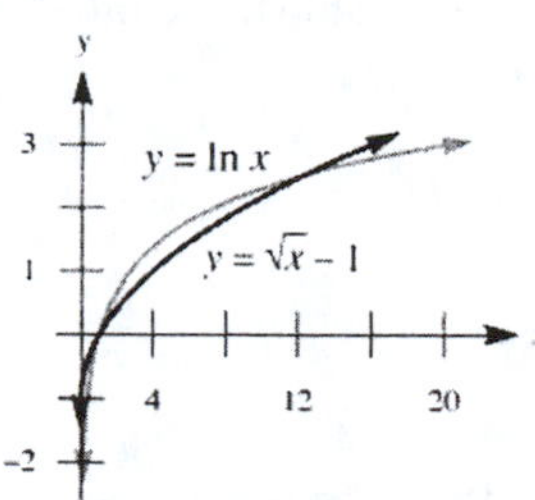

The intersection points are (1,0), which is exact, and approximately (12.34,2.51).

33. Sketch the graphs:

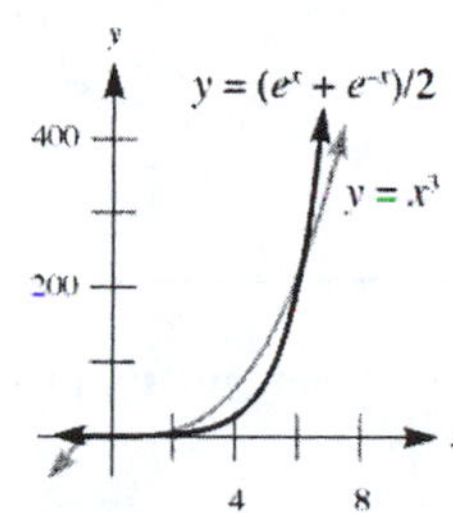

The intersection points are approximately (1.23,1.86) and (6.14,230.95).

35. **a.** Substituting x yields:

$$y = 3x + 1 = \frac{-3 + 3\sqrt{13}}{6} + 1 = \frac{3 + 3\sqrt{13}}{6} = \frac{1 + \sqrt{13}}{2}$$

$$y = \frac{1}{x} = \frac{6}{-1 + \sqrt{13}} \bullet \frac{-1 - \sqrt{13}}{-1 - \sqrt{13}} = \frac{-6\left(1 + \sqrt{13}\right)}{1 - 13} = \frac{-6\left(1 + \sqrt{13}\right)}{-12} = \frac{1 + \sqrt{13}}{2}$$

They both yield the same y-value.

b. Multiply the first equation by –3 and the second equation by 2, then add:

$$-6u + 9v = 18$$
$$6u + 8v = 118$$

Adding:

$$17v = 136$$
$$v = 8$$

Substituting into the first equation:

$$2u - 3v = -6$$
$$2u - 24 = -6$$
$$2u = 18$$
$$u = 9$$

So $u = 9$ and $v = 8$.

37. Since $ax + by = 2$, $by = 2 - ax$. Substitute into the second equation:

$$ax(by) = 1$$
$$ax(2 - ax) = 1$$
$$2ax - a^2x^2 = 1$$
$$a^2x^2 - 2ax + 1 = 0$$
$$(ax - 1)^2 = 0$$
$$ax = 1$$
$$x = \frac{1}{a}$$

When $ax = 1$, $by = 2 - 1 = 1$, so $y = \frac{1}{b}$. So the solution is $\left(\frac{1}{a}, \frac{1}{b}\right)$.

39. Solve the second equation for y to get $y = 23 - x$. Substitute into the first equation:

$$\begin{aligned} x^3 + (23-x)^3 &= 3473 \\ x^3 + 12167 - 1587x + 69x^2 - x^3 &= 3473 \\ 69x^2 - 1587x + 8694 &= 0 \\ x^2 - 23x + 126 &= 0 \\ (x-9)(x-14) &= 0 \\ x &= 9 \text{ or } x = 14 \end{aligned}$$

When $x = 9$, $y = 14$ and when $x = 14$, $y = 9$. So the solutions are (9,14) and (14,9).

41. First draw the rectangle:

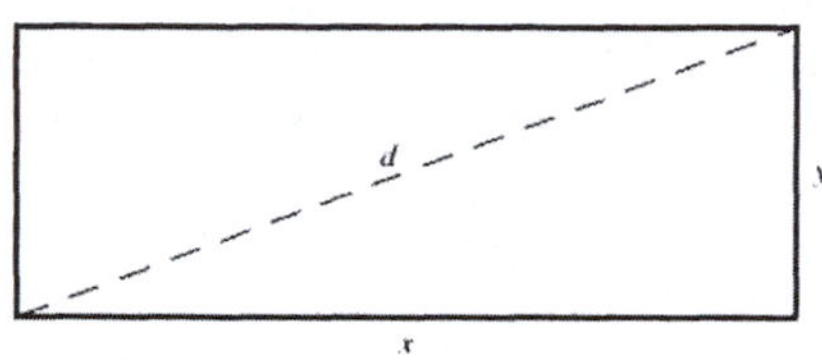

Now $x^2 + y^2 = d^2$ and $2x + 2y = 2p$. Solve the second equation for y:

$$\begin{aligned} 2x + 2y &= 2p \\ 2y &= 2p - 2x \\ y &= p - x \end{aligned}$$

Substitute into the first equation:

$$\begin{aligned} x^2 + (p-x)^2 &= d^2 \\ x^2 + p^2 - 2px + x^2 &= d^2 \\ 2x^2 - 2px + p^2 - d^2 &= 0 \end{aligned}$$

Using the quadratic formula: $x = \dfrac{2p \pm \sqrt{4p^2 - 8\left(p^2 - d^2\right)}}{4} = \dfrac{2p \pm 2\sqrt{2d^2 - p^2}}{4} = \dfrac{p \pm \sqrt{2d^2 - p^2}}{2}$

When $x = \dfrac{p + \sqrt{2d^2 - p^2}}{2}$: $y = p - \dfrac{p + \sqrt{2d^2 - p^2}}{2} = \dfrac{p - \sqrt{2d^2 - p^2}}{2}$

When $x = \dfrac{p - \sqrt{2d^2 - p^2}}{2}$: $y = p - \dfrac{p - \sqrt{2d^2 - p^2}}{2} = \dfrac{p + \sqrt{2d^2 - p^2}}{2}$

So the rectangle has dimensions $\dfrac{p - \sqrt{2d^2 - p^2}}{2}$ by $\dfrac{p - \sqrt{2d^2 - p^2}}{2}$.

43. **a.** Using the point (2,3), we have $3 = N_0 e^{2k}$. Using the point (8,24), we have $24 = N_0 e^{8k}$. Dividing these two yields:

$$\begin{aligned} \frac{24}{3} &= \frac{N_0 e^{8k}}{N_0 e^{2k}} \\ 8 &= e^{6k} \\ \ln 8 &= 6k \\ k &= \frac{\ln 8}{6} \end{aligned}$$

Substituting into the first equation:

$$\begin{aligned} 3 &= N_0 e^{\frac{1}{3}\ln 8} \\ 3 &= N_0 e^{\ln(8^{1/3})} \\ 3 &= N_0 \bullet 2 \\ N_0 &= \tfrac{3}{2} \end{aligned}$$

b. Using the point $\left(\frac{1}{2},1\right)$, we have $1=N_0e^{\frac{1}{2}k}$. Using the point (4,10), we have $10=N_0e^{4k}$. Dividing these two yields:

$$\frac{10}{1}=\frac{N_0e^{4k}}{N_0e^{\frac{1}{2}k}}$$
$$10=e^{\frac{7}{2}k}$$
$$\ln 10=\tfrac{7}{2}k$$
$$k=\tfrac{2}{7}\ln 10$$

Substituting into the first equation:

$$1=N_0e^{\frac{1}{7}\ln 10}$$
$$1=N_0e^{\ln(10^{1/7})}$$
$$1=N_0\bullet 10^{1/7}$$
$$N_0=10^{-1/7}$$

45. Let $w = x + y + z$, so:

$$xw=p^2$$
$$yw=q^2$$
$$zw=r^2$$

Adding the three equations:

$$(x+y+z)w=p^2+q^2+r^2$$
$$w^2=p^2+q^2+r^2$$
$$w=\pm\sqrt{p^2+q^2+r^2}$$

Substitute into the first equation: $x=\dfrac{p^2}{\pm\sqrt{p^2+q^2+r^2}}$. Similarly $y=\frac{q^2}{\pm A}$ and $z=\frac{r^2}{\pm A}$ where $A=\sqrt{p^2+q^2+r^2}$.

So the solutions are $\left(\frac{p^2}{A},\frac{q^2}{A},\frac{r^2}{A}\right)$ and $\left(\frac{-p^2}{A},\frac{-q^2}{A},\frac{-r^2}{A}\right)$, where $A=\sqrt{p^2+q^2+r^2}$.

47. The area is given by $A=\frac{1}{2}bh$, where b and h are the missing legs. Therefore: $180=\frac{1}{2}bh$, so $bh = 360$ thus $b=\frac{360}{h}$

Also from the Pythagorean theorem $b^2+h^2=41^2$, so:

$$\left(\frac{360}{h}\right)^2+h^2=41^2$$
$$\frac{129600}{h^2}+h^2=1681$$
$$129600+h^4=1681h^2$$
$$h^4-1681h^2+129600=0$$
$$\left(h^2-1600\right)\left(h^2-81\right)=0$$
$$h^2=1600 \text{ or } h^2=81$$
$$h=\pm 40 \text{ or } h=\pm 9$$

Since these must be the length of the sides, we can neglect the negative values. So $h = 40$ and $h = 9$ are solutions. Since $b=\frac{360}{h}$ we obtain $b = 9$ when $h = 40$, and $b = 40$ when $h = 9$. So the legs are 9 cm and 40 cm.

49. We have $lw = 60$ and $2l + 2w = 46$, so $l + w = 23$ and $l = 23 - w$. Substitute this into the first equation:

$$\begin{aligned} (23-w)(w) &= 60 \\ 23w - w^2 &= 60 \\ w^2 - 23w + 60 &= 0 \\ (w-20)(w-3) &= 0 \\ w = 20 \quad &\text{or} \quad w = 3 \end{aligned}$$

When $w = 20$ we have $l = 3$, and when $w = 3$ we have $l = 20$. So the rectangle must be 3 cm by 20 cm.

51. Solve $xy = 2$ to get $y = \frac{2}{x}$. Now substitute into the first equation:

$$\begin{aligned} x^2 + \frac{4}{x^2} &= 5 \\ x^4 + 4 &= 5x^2 \\ x^4 - 5x^2 + 4 &= 0 \\ \left(x^2 - 1\right)\left(x^2 - 4\right) &= 0 \\ x^2 = 1 \quad &\text{or} \quad x^2 = 4 \\ x = \pm 1 \quad &\quad\;\; x = \pm 2 \end{aligned}$$

When $x = 1$ we have $y = 2$, when $x = -1$ we have $y = -2$, when $x = 2$ we have $y = 1$, and when $x = -2$ we have $y = -1$. So the solutions are $(1,2)$, $(-1,-2)$, $(2,1)$, and $(-2,-1)$.

53. Multiply the second equation by 2: $2xy = 6$. Adding to the first equation:

$$\begin{aligned} x^2 + 2xy + y^2 &= 13 \\ (x+y)^2 &= 13 \\ x + y &= \pm\sqrt{13} \end{aligned}$$

Subtracting from the first equation:

$$\begin{aligned} x^2 - 2xy + y^2 &= 1 \\ (x-y)^2 &= 1 \\ x - y &= \pm 1 \end{aligned}$$

Solve the four systems of equations:

$$\begin{array}{llll} x+y=\sqrt{13} & x+y=\sqrt{13} & x+y=-\sqrt{13} & x+y=-\sqrt{13} \\ \underline{x-y=1} & \underline{x-y=-1} & \underline{x-y=1} & \underline{x-y=-1} \\ 2x=1+\sqrt{13} & 2x=-1+\sqrt{13} & 2x=1-\sqrt{13} & 2x=-1-\sqrt{13} \\ x=\dfrac{1+\sqrt{13}}{2} & x=\dfrac{-1+\sqrt{13}}{2} & x=\dfrac{1-\sqrt{13}}{2} & x=\dfrac{-1-\sqrt{13}}{2} \\ y=\dfrac{-1+\sqrt{13}}{2} & y=\dfrac{1+\sqrt{13}}{2} & y=\dfrac{-1-\sqrt{13}}{2} & y=\dfrac{1-\sqrt{13}}{2} \end{array}$$

So the solutions are $\left(\frac{1+\sqrt{13}}{2}, \frac{-1+\sqrt{13}}{2}\right)$, $\left(\frac{-1+\sqrt{13}}{2}, \frac{1+\sqrt{13}}{2}\right)$, $\left(\frac{1-\sqrt{13}}{2}, \frac{-1-\sqrt{13}}{2}\right)$, and $\left(\frac{-1-\sqrt{13}}{2}, \frac{1-\sqrt{13}}{2}\right)$.

55. Solving by factoring:

$$\begin{aligned} 2m^2 - 7m + 6 &= 0 \\ (2m-3)(m-2) &= 0 \\ m = \tfrac{3}{2} \quad &\text{or} \quad m = 2 \end{aligned}$$

So we have the equations:

$$\begin{array}{lcl} x^2\left(\frac{9}{2} - 4\right) = 2 & \text{or} & x^2(6-4) = 2 \\ x^2 = 4 & & x^2 = 1 \\ x = \pm 2 & & x = \pm 1 \\ y = \pm 3 & & y = \pm 2 \end{array}$$

So the solutions are $(2,3)$, $(-2,-3)$, $(1,2)$, and $(-1,-2)$.

57. Taking logs in the first equation:

$$\ln(x^4) = \ln(y^6)$$
$$4\ln x = 6\ln y$$
$$2\ln x = 3\ln y$$

The second equation is:

$$\ln x - \ln y = \frac{\ln x}{\ln y}$$
$$\ln x \ln y - (\ln y)^2 = \ln x$$

Let $u = \ln x$ and $v = \ln y$, so we have the equations $2u = 3v$ and $uv - v^2 = u$. Solving the first equation for u yields $u = \frac{3v}{2}$, and substituting into the second equation yields:

$$\left(\frac{3v}{2}\right)v - v^2 = \frac{3v}{2}$$
$$3v^2 - 2v^2 = 3v$$
$$v^2 - 3v = 0$$
$$v(v-3) = 0$$
$$v = 0 \text{ or } v = 3$$

When $v = 0$, $u = 0$ and when $v = 3$, $u = \frac{9}{2}$. Since $v = \ln y$, $v = 0$ cannot be a solution to the original second equation ($\ln y$ is the denominator). Thus $u = \ln x$ and $v = \ln y$ yields:

$\ln x = \frac{9}{2}$ so $x = e^{9/2}$ $\qquad\qquad$ $\ln y = 3$ so $y = e^3$

So the only solution is $\left(e^{9/2}, e^3\right)$.

10.7 Systems of Inequalities

1. **a.** Substitute the pair (1,2):

$$4(1) - 6(2) + 3 \geq 0$$
$$-5 \geq 0$$

Then (1,2) is not a solution.

b. Substitute the pair $\left(0, \frac{1}{2}\right)$:

$$4(0) - 6\left(\frac{1}{2}\right) + 3 \geq 0$$
$$0 \geq 0$$

Then $\left(0, \frac{1}{2}\right)$ is a solution.

3. Graphing the region:

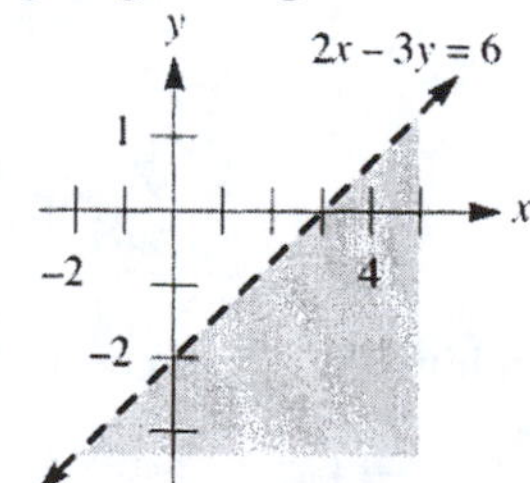

5. Graphing the region:

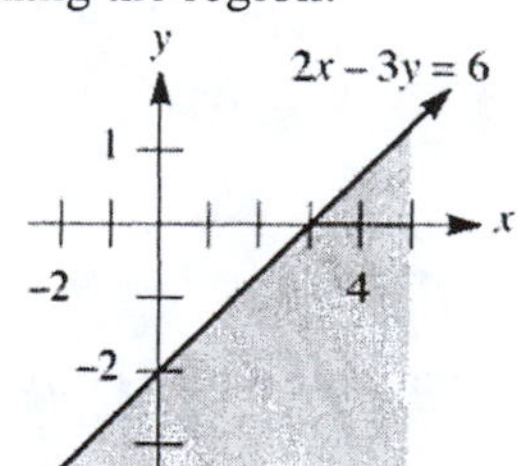

7. Graphing the region:

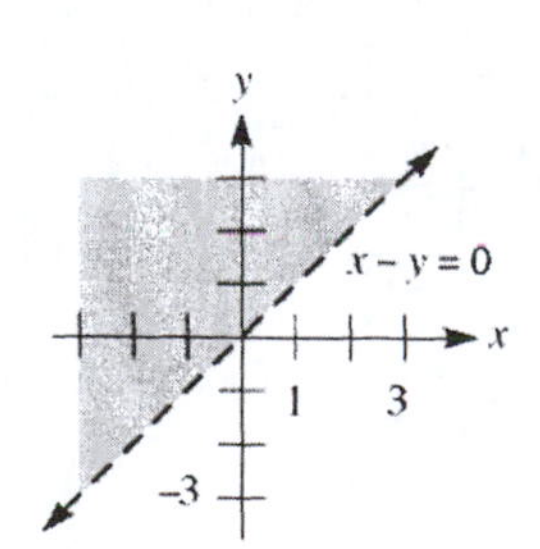

9. Graphing the region:

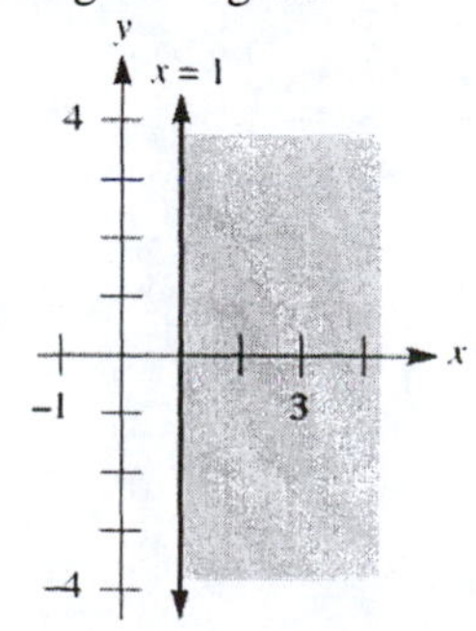

11. Graphing the region:

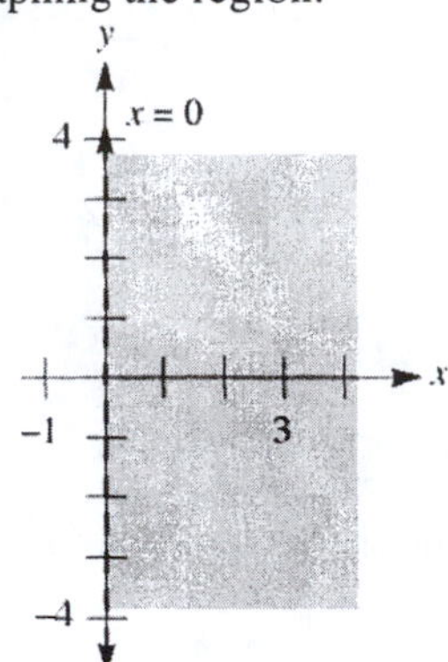

13. Graphing the region:

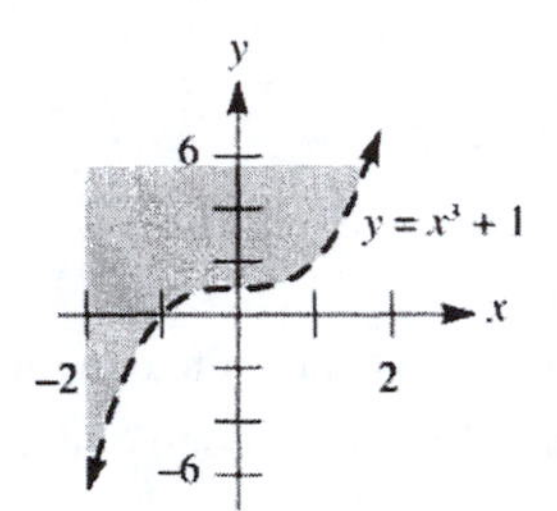

15. Graphing the region:

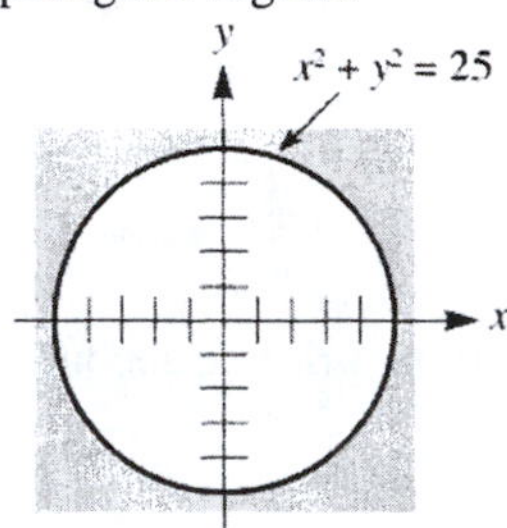

17. Graphing the region:

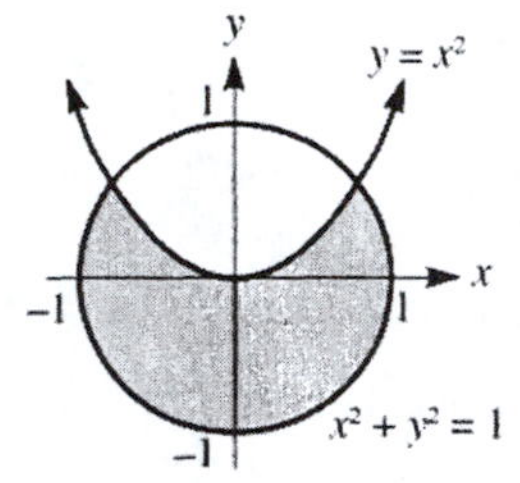

19. Graphing the system of inequalities:

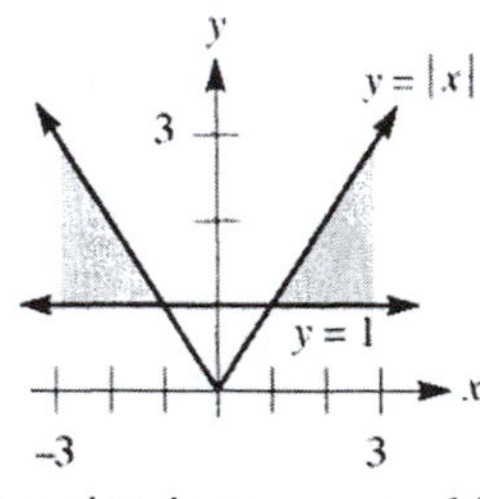

21. Graphing the system of inequalities:

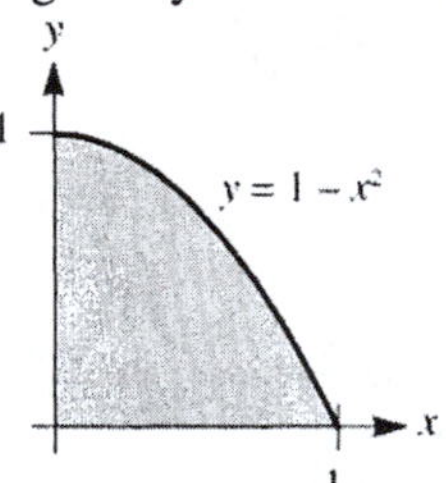

23. The region is convex and bounded. The vertices are (0,0), (7,0), (3,8), and (0,5). Graphing the system of inequalities:

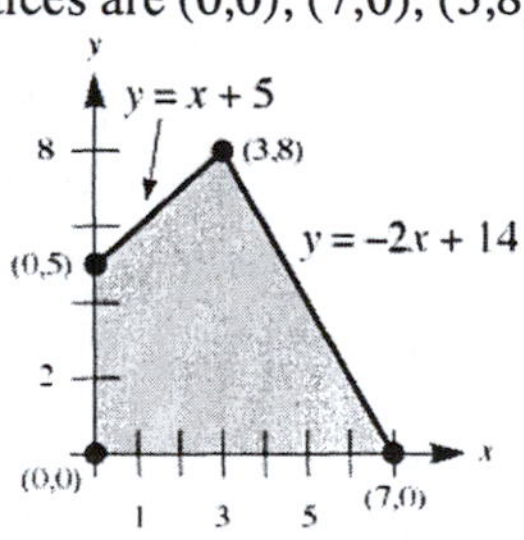

25. The region is convex and bounded. The vertices are (0,0), (0,4), (3,5), and (8,0). Graphing the system of inequalities:

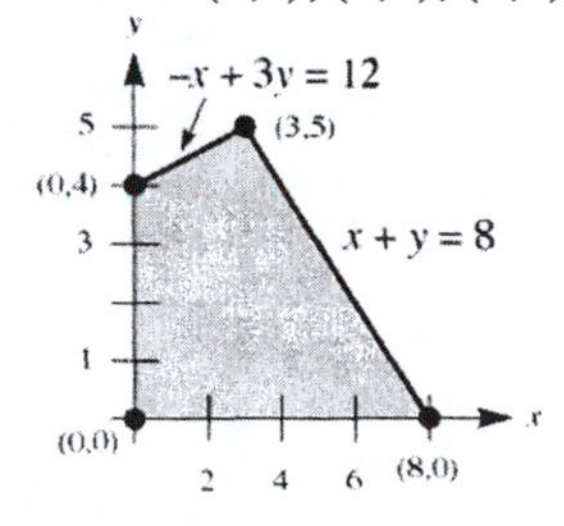

27. The region is convex but not bounded. The vertices are (2,7) and (8,5). Graphing the system of inequalities:

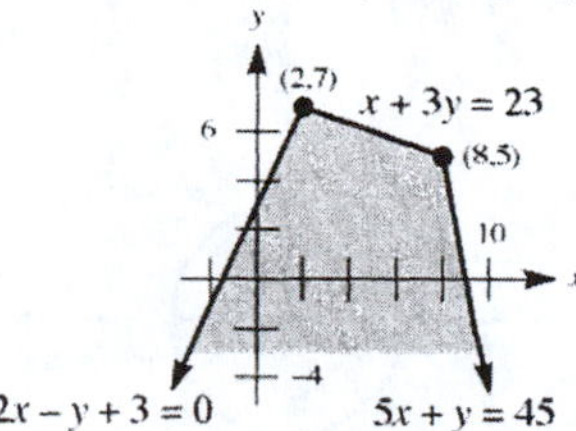

29. The region is convex but not bounded. The only vertex is (6,0). Graphing the system of inequalities:

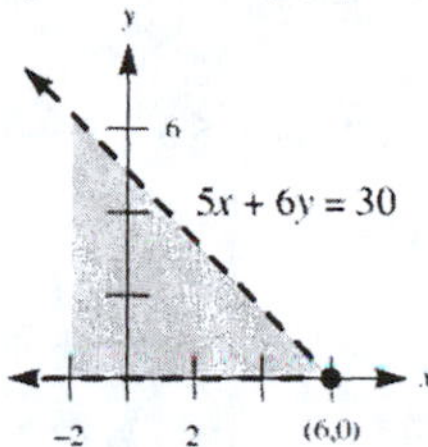

31. The region is convex and bounded. The vertices are (0,0), (0,5), and (6,0). Graphing the system of inequalities:

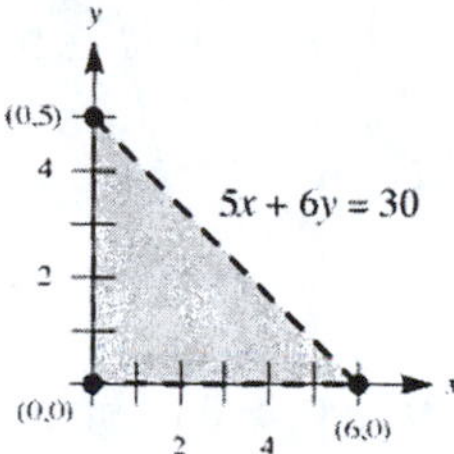

33. The region is convex and bounded. The vertices are (5,30), (10,30), (20,15), and (20,20). Graphing the system of inequalities:

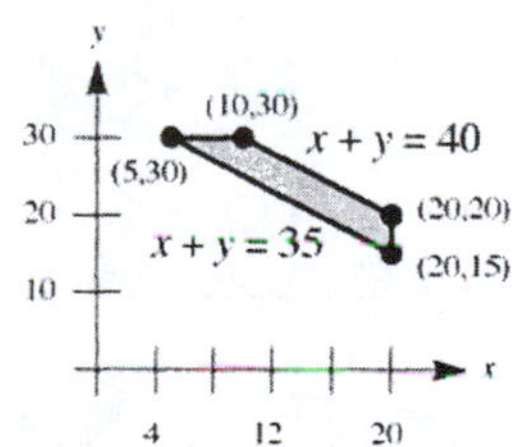

35. First find the intersection of $y = e^x$ and $y = e^{-x} + 1$:

$$e^x = e^{-x} + 1$$
$$\left(e^x\right)^2 = 1 + e^x$$
$$\left(e^x\right)^2 - e^x - 1 = 0$$

Using the quadratic formula: $e^x = \dfrac{1 \pm \sqrt{1-4(-1)}}{2} = \dfrac{1 \pm \sqrt{5}}{2}$

Since $e^x > 0$, discard the negative root to obtain $e^x = \frac{1+\sqrt{5}}{2}$, so $x = \ln\frac{1+\sqrt{5}}{2} \approx 0.48$. Since $y = 1 + e^{-x}$, $y = \frac{1+\sqrt{5}}{2} \approx 1.62$. The vertices are (0,1), (0,2) and $\left(\ln\frac{1+\sqrt{5}}{2}, \frac{1+\sqrt{5}}{2}\right)$. Graphing the inequalities:

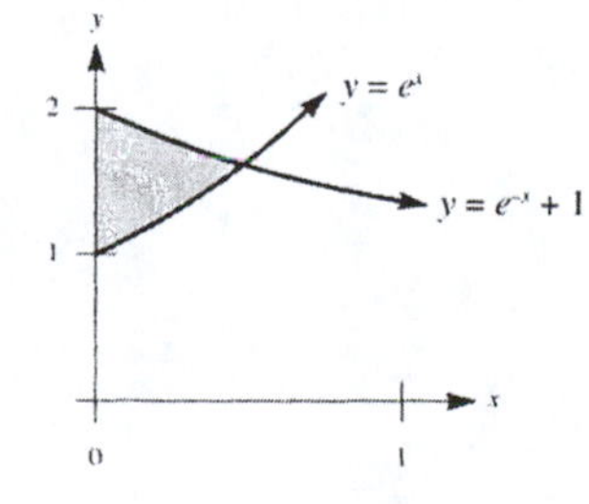

37. The domain results from the inequality $x^2 + y^2 - 1 \geq 0$, or $x^2 + y^2 \geq 1$. Sketching the graph:

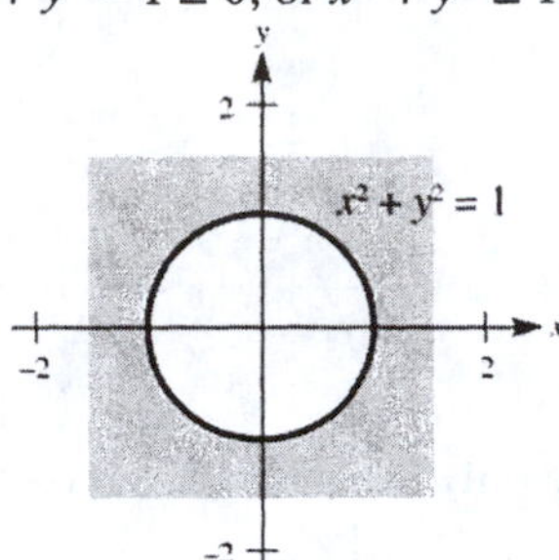

39. The domain results from the inequality $x^2 - y > 0$, so $y < x^2$. Sketching the graph:

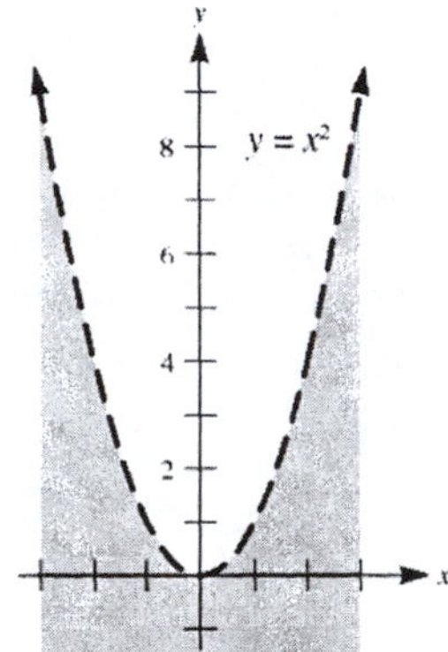

41. The domain results from two inequalities $x \geq 0$ and $y \geq 0$ (first quadrant and positive axes). Sketching the graph:

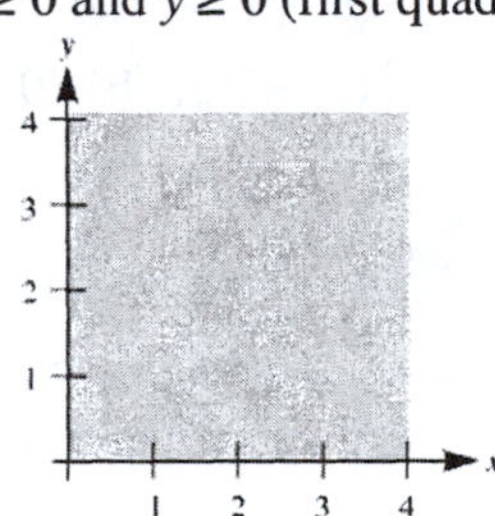

Chapter 10 Review Exercises

1. Adding the two equations yields:

$$2x = 6$$
$$x = 3$$

Substitute into equation 1:

$$3xy = -2$$
$$y = -5$$

So the solution is (3,–5).

3. Multiply the first equation by –2:

$$-4x - 2y = -4$$
$$x + 2y = 7$$

Adding:

$$-3x = 3$$
$$x = -1$$

Substitute into equation 1:

$$-2 + y = 2$$
$$y = 4$$

So the solution is (–1,4).

5. Multiply equation 1 by –5 and equation 2 by 2:

$$-35x - 10y = -45$$
$$8x + 10y = 126$$

Adding:

$$-27x = 81$$
$$x = -3$$

Substitute into equation 1:

$$-21 + 2y = 9$$
$$2y = 30$$
$$y = 15$$

So the solution is (–3,15).

7. Multiply the first equation by 2 and the second equation by 24 to clear fractions:

$$4x - y = -16$$
$$8x + 3y = -24$$

Multiply equation 1 by –2:

$$-8x + 2y = 32$$
$$8x + 3y = -24$$

Adding:

$$5y = 8$$
$$y = \frac{8}{5}$$

Substitute into equation 1:

$$4x - \frac{8}{5} = -16$$
$$4x = -\frac{72}{5}$$
$$x = -\frac{18}{5}$$

So the solution is $\left(-\frac{18}{5}, \frac{8}{5}\right)$.

9. Multiply the first equation by 2:

$$6x + 10y = 2$$
$$9x - 10y = 8$$

Adding:

$$15x = 10$$
$$x = \frac{2}{3}$$

Substitute into equation 1:

$$2 + 5y = 1$$
$$5y = -1$$
$$y = -\frac{1}{5}$$

So the solution is $\left(\frac{2}{3}, -\frac{1}{5}\right)$.

11. Multiply the first equation by 6 and the second equation by 2 to clear fractions:

$$4x + 3y = -72$$
$$x - 2y = 4$$

Multiply the second equation by –4:

$$4x + 3y = -72$$
$$-4x + 8y = -16$$

Adding:

$$11y = -88$$
$$y = -8$$

Substitute into equation 2:

$$x + 16 = 4$$
$$x = -12$$

So the solution is (–12,–8).

13. Let $a=\frac{1}{x}$ and $b=\frac{1}{y}$. So:

$$a+b=-1$$
$$2a+5b=-14$$

Multiply the first equation by –2:

$$-2a-2b=2$$
$$2a+5b=-14$$

Adding:

$$3b=-12$$
$$b=-4$$

Substitute into equation 1:

$$a-4=-1$$
$$a=3$$

Since $x=\frac{1}{a}$ and $y=\frac{1}{b}$, $x=\frac{1}{3}$ and $y=-\frac{1}{4}$. So the solution is $\left(\frac{1}{3},-\frac{1}{4}\right)$.

15. Multiply the second equation by $a-1$:

$$ax+(1-a)y=1$$
$$\left(-a^2+2a-1\right)x+(a-1)y=0$$

Adding:

$$\left(-a^2+3a-1\right)x=1$$
$$x=\frac{-1}{a^2-3a+1}$$

Substitute into equation 2:

$$\frac{a-1}{a^2-3a+1}+y=0$$
$$y=\frac{1-a}{a^2-3a+1}$$

So the solution is $\left(\frac{-1}{a^2-3a+1},\frac{1-a}{a^2-3a+1}\right)$. We must assume that $a^2-3a+1\neq 0$, or $a\neq\frac{3\pm\sqrt{5}}{2}$.

17. Multiply the first equation by 2:

$$4x-2y=6a^2-2$$
$$x+2y=-a^2+2$$

Adding:

$$5x=5a^2$$
$$x=a^2$$

Substitute into equation 2:

$$2y+a^2=2-a^2$$
$$2y=2-2a^2$$
$$y=1-a^2$$

So the solution is $\left(a^2,1-a^2\right)$.

19. Multiply the first equation by 3:

$$15x-3y=12a^2-18b^2$$
$$2x+3y=5a^2+b^2$$

Adding:

$$17x=17a^2-17b^2$$
$$x=a^2-b^2$$

Substitute into equation 1:

$$\begin{aligned} 5a^2-5b^2-y&=4a^2-6b^2\\ -y&=-a^2-b^2\\ y&=a^2+b^2 \end{aligned}$$

So the solution is $\left(a^2-b^2,a^2+b^2\right)$.

21. Multiply the first equation by p and the second equation by q:

$$\begin{aligned} p^2x-pqy&=pq^2\\ q^2x+pqy&=p^2q \end{aligned}$$

Adding:

$$\begin{aligned} \left(p^2+q^2\right)x&=pq(p+q)\\ x&=\frac{pq(p+q)}{p^2+q^2} \end{aligned}$$

Re-solve the system to find y. Multiply the first equation by $-q$ and the second equation by p:

$$\begin{aligned} -pqx+q^2y&=-q^3\\ pqx+p^2y&=p^3 \end{aligned}$$

Adding:

$$\begin{aligned} \left(p^2+q^2\right)y&=p^3-q^3\\ y&=\frac{p^3-q^3}{p^2+q^2} \end{aligned}$$

So the solution is $\left(\frac{pq(p+q)}{p^2+q^2},\frac{p^3-q^3}{p^2+q^2}\right)$. We must assume that p and q are not both 0.

23. Form the augmented matrix: $\begin{pmatrix} 1 & 1 & 1 & 9\\ 1 & -1 & -1 & -5\\ 2 & 1 & -2 & -1 \end{pmatrix}$

Add –1 times row 1 to row 2 and –2 times row 1 to row 3: $\begin{pmatrix} 1 & 1 & 1 & 9\\ 0 & -2 & -2 & -14\\ 0 & -1 & -4 & -19 \end{pmatrix}$

Switch row 2 and row 3, multiply each by –1: $\begin{pmatrix} 1 & 1 & 1 & 9\\ 0 & 1 & 4 & 19\\ 0 & 2 & 2 & 14 \end{pmatrix}$

Add –2 times row 2 to row 3: $\begin{pmatrix} 1 & 1 & 1 & 9\\ 0 & 1 & 4 & 19\\ 0 & 0 & -6 & -24 \end{pmatrix}$

So we have the system:

$$\begin{aligned} x+y+\ z&=9\\ y+4z&=19\\ -6z&=-24 \end{aligned}$$

Solve equation 3 for z:

$$\begin{aligned} -6z&=-24\\ z&=4 \end{aligned}$$

Substitute into equation 2:

$$\begin{aligned} y+16&=19\\ y&=3 \end{aligned}$$

Substitute into equation 1:

$$x+3+4=9$$
$$x=2$$

So the solution is (2,3,4).

25. Switching equations 1 and 3, form the augmented matrix: $\begin{pmatrix} 1 & 1 & 1 & -3 \\ 2 & 3 & 3 & -8 \\ 4 & -4 & 1 & 4 \end{pmatrix}$

Add –2 times row 1 to row 2 and –4 times row 1 to row 3: $\begin{pmatrix} 1 & 1 & 1 & -3 \\ 0 & 1 & 1 & -2 \\ 0 & -8 & -3 & 16 \end{pmatrix}$

Add 8 times row 2 to row 3: $\begin{pmatrix} 1 & 1 & 1 & -3 \\ 0 & 1 & 1 & -2 \\ 0 & 0 & 5 & 0 \end{pmatrix}$

So we have the system:

$$x+y+z=-3$$
$$y+z=-2$$
$$5z=0$$

Solve equation 3 for z:

$$5z=0$$
$$z=0$$

Substitute into equation 2:

$$y+0=-2$$
$$y=-2$$

Substitute into equation 1:

$$x-2+0=-3$$
$$x=-1$$

So the solution is (–1,–2,0).

27. Multiply the second equation by –2:

$$4x+2y-3z=15$$
$$-4x-2y-6z=-6$$

Adding:

$$-9z=9$$
$$z=-1$$

Substitute into the original equation:

$$2x+y-3=3$$
$$2x+y=6$$
$$y=6-2x$$

So the solution is $(x, 6-2x, -1)$, for any real number x.

29. Using equation 2 as row 1, form the augmented matrix: $\begin{pmatrix} -3 & 1 & -1 & 0 \\ 9 & 1 & 1 & 0 \\ 3 & -5 & 3 & 0 \end{pmatrix}$

Add 3 times row 1 to row 2, and add row 1 to row 3: $\begin{pmatrix} -3 & 1 & -1 & 0 \\ 0 & 4 & -2 & 0 \\ 0 & -4 & 2 & 0 \end{pmatrix}$

Add row 2 to row 3: $\begin{pmatrix} -3 & 1 & -1 & 0 \\ 0 & 4 & -2 & 0 \\ 0 & 0 & 0 & 0 \end{pmatrix}$

So we have the system:

$$\begin{aligned} -3x + y - z &= 0 \\ 4y - 2z &= 0 \end{aligned}$$

Solving equation 2 for y yields $y = \frac{1}{2}z$, and substituting into equation 1 yields $x = -\frac{1}{6}z$. The solution is $\left(-\frac{1}{6}z, \frac{1}{2}z, z\right)$, for any real number z.

31. Form the augmented matrix: $\begin{pmatrix} 1 & 1 & 1 & 1 & 8 \\ 3 & 3 & -1 & -1 & 20 \\ 4 & -1 & -1 & 2 & 18 \\ 2 & 5 & 5 & -5 & 8 \end{pmatrix}$

Add –3 times row 1 to row 2, –4 times row 1 to row 3, and –2 times row 1 to row 4: $\begin{pmatrix} 1 & 1 & 1 & 1 & 8 \\ 0 & 0 & -4 & -4 & -4 \\ 0 & -5 & -5 & -2 & -14 \\ 0 & 3 & 3 & -7 & -8 \end{pmatrix}$

Switch row 2 and row 3, and multiply row 3 by $-\frac{1}{4}$: $\begin{pmatrix} 1 & 1 & 1 & 1 & 8 \\ 0 & -5 & -5 & -2 & -14 \\ 0 & 0 & 1 & 1 & 1 \\ 0 & 3 & 3 & -7 & -8 \end{pmatrix}$

Add 2 times row 4 to row 2: $\begin{pmatrix} 1 & 1 & 1 & 1 & 8 \\ 0 & 1 & 1 & -16 & -30 \\ 0 & 0 & 1 & 1 & 1 \\ 0 & 3 & 3 & -7 & -8 \end{pmatrix}$

Add –3 times row 2 to row 4: $\begin{pmatrix} 1 & 1 & 1 & 1 & 8 \\ 0 & 1 & 1 & -16 & -30 \\ 0 & 0 & 1 & 1 & 1 \\ 0 & 0 & 0 & 41 & 82 \end{pmatrix}$

So we have the system:

$$\begin{aligned} x + y + z + w &= 8 \\ y + z - 16w &= -30 \\ z + w &= 1 \\ 41w &= 82 \end{aligned}$$

Solve equation 4 for w:

$$\begin{aligned} 41w &= 82 \\ w &= 2 \end{aligned}$$

Substitute into equation 3:

$$\begin{aligned} z + 2 &= 1 \\ z &= -1 \end{aligned}$$

Substitute into equation 2:

$$\begin{aligned} y - 1 - 32 &= -30 \\ y &= 3 \end{aligned}$$

Substitute into equation 1:

$$\begin{aligned} x + 3 - 1 + 2 &= 8 \\ x &= 4 \end{aligned}$$

So the solution is (4,3,–1,2).

33. Factoring 2 from row 1, 6 from row 2, and 5 from row 3: $\begin{vmatrix} 2 & 6 & 4 \\ 6 & 18 & 24 \\ 15 & 5 & -10 \end{vmatrix} = 60\begin{vmatrix} 1 & 3 & 2 \\ 1 & 3 & 4 \\ 3 & 1 & -2 \end{vmatrix}$

Subtracting row 2 from row 1 then expanding along the first row: $60\begin{vmatrix} 0 & 0 & -2 \\ 1 & 3 & 4 \\ 3 & 1 & -2 \end{vmatrix} = 60(-2)\begin{vmatrix} 1 & 3 \\ 3 & 1 \end{vmatrix} = -120(-8) = 960$

35. Evaluate the determinant: $\begin{vmatrix} 1 & 0 & 0 & 0 \\ 0 & 2 & 0 & 0 \\ 0 & 0 & 3 & 0 \\ 0 & 0 & 0 & 4 \end{vmatrix} = 1\begin{vmatrix} 2 & 0 & 0 \\ 0 & 3 & 0 \\ 0 & 0 & 4 \end{vmatrix} = 2\begin{vmatrix} 3 & 0 \\ 0 & 4 \end{vmatrix} = 2(12) = 24$

37. By expanding along column 1:

$$\begin{aligned} \begin{vmatrix} a & b & c \\ b & c & a \\ c & a & b \end{vmatrix} &= a\begin{vmatrix} c & a \\ a & b \end{vmatrix} - b\begin{vmatrix} b & c \\ a & b \end{vmatrix} + c\begin{vmatrix} b & c \\ c & a \end{vmatrix} \\ &= a\left(bc - a^2\right) - b\left(b^2 - ac\right) + c\left(ab - c^2\right) \\ &= abc - a^3 - b^3 + abc + abc - c^3 \\ &= 3abc - a^3 - b^3 - c^3 \end{aligned}$$

39. Substituting $(x, y) = (-2, 5)$ and $(x, y) = (2, 9)$:

$$5 = 4a - 2b - 1$$
$$9 = 4a + 2b - 1$$

Adding:

$$14 = 8a - 2$$
$$16 = 8a$$
$$a = 2$$

Substitute into the first equation:

$$5 = 8 - 2b - 1$$
$$-2 = -2b$$
$$b = 1$$

So $a = 2$ and $b = 1$.

41. See the figure:

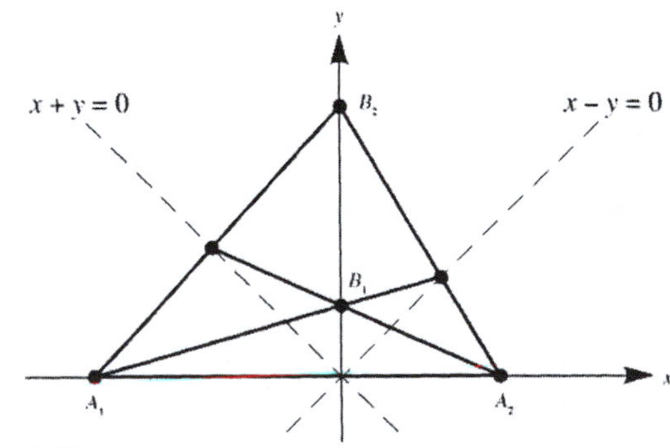

Start by finding the equation for A_2B_2:

$$m = -\frac{b_2}{a_2}$$
point: (a, a)

By the point-slope formula: $a_2y + b_2x = aa_2 + ab_2$. Since $(a_2, 0)$ lies on this curve, we have $a_2b_2 = aa_2 + ab_2$. (*)

Now find the equation for A_1B_1:

$$m = -\frac{b_1}{a_1}$$
$$\text{point}: (a, a)$$

By the point-slope formula: $a_1 y + b_1 x = aa_1 + ab_1$. Since $(a_1, 0)$ lies on this curve, we have $a_1 b_1 = aa_1 + ab_1$. (**)

Now look at A_1B_2:

$$m = -\frac{b_2}{a_1}$$
$$b_2 x + a_1 y = a_1 b_2$$

Also look at A_2B_1:

$$m = -\frac{b_1}{a_2}, \text{ so:}$$
$$b_1 x + a_2 y = a_2 b_1$$

Solve the system:

$$b_2 x + a_1 y = a_1 b_2$$
$$b_1 x + a_2 y = a_2 b_1$$

Multiply equation 1 by $-b_1$ and equation 2 by b_2:

$$-b_1 b_2 x - a_1 b_1 y = -a_1 b_1 b_2$$
$$b_1 b_2 x + a_2 b_2 y = a_2 b_1 b_2$$

Adding:

$$(a_2 b_2 - a_1 b_1) y = a_2 b_1 b_2 - a_1 b_1 b_2$$
$$y = \frac{b_1 b_2 (a_2 - a_1)}{a_2 b_2 - a_1 b_1}$$

Substituting into equation 1:

$$b_2 x + \frac{a_1 b_1 b_2 (a_2 - a_1)}{a_2 b_2 - a_1 b_1} = a_1 b_2$$
$$x = \frac{a_1 a_2 (b_2 - b_1)}{a_2 b_2 - a_1 b_1}$$

Now, we are asked to show that $x + y = 0$, so:

$$x + y = \frac{a_1 a_2 (b_2 - b_1) + b_1 b_2 (a_2 - a_1)}{a_2 b_2 - a_1 b_1} = \frac{a_1 a_2 b_2 - a_1 a_2 b_1 + a_2 b_1 b_2 - a_1 b_1 b_2}{a_2 b_2 - a_1 b_1}$$

Replacing $a_1 b_1$ and $a_2 b_2$ by equations (*) and (**):

$$x + y = \frac{a_1 (aa_2 + ab_2) - a_2 (aa_1 + ab_1) + b_1 (aa_2 + ab_2) - b_2 (aa_1 + ab_1)}{a_2 b_2 - a_1 b_1}$$
$$= \frac{a(a_1 a_2 + a_1 b_2 - a_1 a_2 - a_2 b_1 + a_2 b_1 + b_1 b_2 - a_1 b_2 - b_1 b_2)}{a_2 b_2 - a_1 b_1}$$
$$= 0$$

43. Find each intersection point:

$$\begin{aligned} y &= x - 1 \\ y &= -x - 2 \\ \hline 2y &= -3 \\ y &= -\tfrac{3}{2} \\ x &= -\tfrac{1}{2} \end{aligned} \qquad \left(-\tfrac{1}{2}, -\tfrac{3}{2}\right)$$

$$\begin{aligned} y &= x - 1 \\ y &= 2x + 3 \\ \hline x - 1 &= 2x + 3 \\ -4 &= x \\ y &= -5 \end{aligned} \qquad (-4, -5)$$

$$\begin{aligned} y &= -x - 2 \\ y &= 2x + 3 \\ \hline -x - 2 &= 2x + 3 \\ -5 &= 3x \\ x &= -\tfrac{5}{3} \\ y &= -\tfrac{1}{3} \end{aligned} \qquad \left(-\tfrac{5}{3}, -\tfrac{1}{3}\right)$$

Now substitute into the equation $(x-h)^2+(y-k)^2=r^2$:

$$\left(-\tfrac{1}{2}-h\right)^2+\left(-\tfrac{3}{2}-k\right)^2=r^2$$
$$(-4-h)^2+(-5-k)^2=r^2$$
$$\left(-\tfrac{5}{3}-h\right)^2+\left(-\tfrac{1}{3}-k\right)^2=r^2$$

These equations, when multiplied out, become:

$$\tfrac{5}{2}+h+h^2+3k+k^2=r^2$$
$$41+8h+h^2+10k+k^2=r^2$$
$$\tfrac{26}{9}+\tfrac{10}{3}h+h^2+\tfrac{2}{3}k+k^2=r^2$$

Subtracting the first equation from the other two yields:

$$\tfrac{77}{2}+7h+7k=0$$
$$\tfrac{7}{18}+\tfrac{7}{3}h-\tfrac{7}{3}k=0$$

These simplify to:

$$6h+6k=-33$$
$$6h-6k=-1$$

Adding:

$$12h=-34$$
$$h=-\tfrac{17}{6}$$

To find k, we subtract the two equations:

$$12k=-32$$
$$k=-\tfrac{8}{3}$$

Finally, find r: $r^2=(-4-h)^2+(-5-k)^2=\left(-\tfrac{7}{6}\right)^2+\left(-\tfrac{7}{3}\right)^2=\tfrac{245}{36}$

The equation of the circle is: $\left(x+\tfrac{17}{6}\right)^2+\left(y+\tfrac{8}{3}\right)^2=\tfrac{245}{36}$

45. Computing: $2A+2B=\begin{pmatrix}6&-4\\2&10\end{pmatrix}+\begin{pmatrix}4&2\\2&16\end{pmatrix}=\begin{pmatrix}10&-2\\4&26\end{pmatrix}$

47. Computing: $4B=\begin{pmatrix}8&4\\4&32\end{pmatrix}$

49. Computing: $AB=\begin{pmatrix}4&-13\\7&41\end{pmatrix}$

51. Computing: $AB-BA=\begin{pmatrix}7&-13\\7&41\end{pmatrix}-\begin{pmatrix}7&1\\11&38\end{pmatrix}=\begin{pmatrix}-3&-14\\-4&3\end{pmatrix}$

53. Computing: $B+C=\begin{pmatrix}1&1\\1&7\end{pmatrix}$

55. Computing: $AB+AC=\begin{pmatrix}4&-13\\7&41\end{pmatrix}+\begin{pmatrix}-3&2\\-1&-5\end{pmatrix}=\begin{pmatrix}1&-11\\6&36\end{pmatrix}$

57. Computing: $BA+CA=\begin{pmatrix}7&1\\11&38\end{pmatrix}+\begin{pmatrix}-3&2\\-1&-5\end{pmatrix}=\begin{pmatrix}4&3\\10&33\end{pmatrix}$

59. Computing: $A+(B+C)=\begin{pmatrix}3&-2\\1&5\end{pmatrix}+\begin{pmatrix}1&1\\1&7\end{pmatrix}=\begin{pmatrix}4&-1\\2&12\end{pmatrix}$

61. Computing: $A(BC)=\begin{pmatrix}3&-2\\1&5\end{pmatrix}+\begin{pmatrix}-2&-1\\-1&-8\end{pmatrix}=\begin{pmatrix}-4&13\\-7&-41\end{pmatrix}$

63. Compute A^2 and then A^3:

$$A^2 = \begin{pmatrix} 0 & 0 & 0 \\ a & 0 & 0 \\ b & c & 0 \end{pmatrix}\begin{pmatrix} 0 & 0 & 0 \\ a & 0 & 0 \\ b & c & 0 \end{pmatrix} = \begin{pmatrix} 0 & 0 & 0 \\ 0 & 0 & 0 \\ ac & 0 & 0 \end{pmatrix} \qquad A^3 = A \bullet A^2 = \begin{pmatrix} 0 & 0 & 0 \\ a & 0 & 0 \\ b & c & 0 \end{pmatrix}\begin{pmatrix} 0 & 0 & 0 \\ 0 & 0 & 0 \\ ac & 0 & 0 \end{pmatrix} = \begin{pmatrix} 0 & 0 & 0 \\ 0 & 0 & 0 \\ 0 & 0 & 0 \end{pmatrix}$$

65. **a.** Form the matrix: $\begin{pmatrix} 1 & -2 & 3 & 1 & 0 & 0 \\ 2 & -5 & 10 & 0 & 1 & 0 \\ -1 & 2 & -2 & 0 & 0 & 1 \end{pmatrix}$

Multiply row 1 by –2 and add to row 2, then add row 1 to row 3: $\begin{pmatrix} 1 & -2 & 3 & 1 & 0 & 0 \\ 0 & -1 & 4 & -2 & 1 & 0 \\ 0 & 0 & 1 & 1 & 0 & 1 \end{pmatrix}$

Multiply row 2 by –1: $\begin{pmatrix} 1 & -2 & 3 & 1 & 0 & 0 \\ 0 & 1 & -4 & 2 & -1 & 0 \\ 0 & 0 & 1 & 1 & 0 & 1 \end{pmatrix}$

Multiply row 2 by 2 and add to row 1: $\begin{pmatrix} 1 & 0 & -5 & 5 & -2 & 0 \\ 0 & 1 & -4 & 2 & -1 & 0 \\ 0 & 0 & 1 & 1 & 0 & 1 \end{pmatrix}$

Multiply row 3 by 4 and add to row 2, then multiply row 3 by 5 and add to row 1: $\begin{pmatrix} 1 & 0 & 0 & 10 & -2 & 5 \\ 0 & 1 & 0 & 6 & -1 & 4 \\ 0 & 0 & 1 & 1 & 0 & 1 \end{pmatrix}$

So the inverse is $\begin{pmatrix} 10 & -2 & 5 \\ 6 & -1 & 4 \\ 1 & 0 & 1 \end{pmatrix}$.

b. Since the system is equivalent to $A \bullet X = B$, where $A = \begin{pmatrix} 1 & -2 & 3 \\ 2 & -5 & 10 \\ -1 & 2 & -2 \end{pmatrix}$, $X = \begin{pmatrix} x \\ y \\ z \end{pmatrix}$, and $B = \begin{pmatrix} -2 \\ -3 \\ 6 \end{pmatrix}$:

$$X = A^{-1} \bullet B = \begin{pmatrix} 10 & -2 & 5 \\ 6 & -1 & 4 \\ 1 & 0 & 1 \end{pmatrix}\begin{pmatrix} -2 \\ -3 \\ 6 \end{pmatrix} = \begin{pmatrix} 16 \\ 15 \\ 4 \end{pmatrix}$$

So the solution is (16,15,4).

67. Adding 2 times column 2 to column 1 and column 2 to column 3 yields:

$$D = \begin{vmatrix} 2 & -1 & 1 \\ 3 & 2 & 2 \\ 1 & -5 & -3 \end{vmatrix} = \begin{vmatrix} 0 & -1 & 0 \\ 7 & 2 & 4 \\ -9 & -5 & -8 \end{vmatrix} = -(-1)\begin{vmatrix} 7 & 4 \\ -9 & -8 \end{vmatrix} = 1(-20) = -20$$

Adding 2 times row 1 to row 3 yields: $D_x = \begin{vmatrix} 1 & -1 & 1 \\ 0 & 2 & 2 \\ -2 & -5 & -3 \end{vmatrix} = \begin{vmatrix} 1 & -1 & 1 \\ 0 & 2 & 2 \\ 0 & -7 & -1 \end{vmatrix} = 1\begin{vmatrix} 2 & 2 \\ -7 & -1 \end{vmatrix} = 12$

Adding 2 times row 1 to row 3 yields: $D_y = \begin{vmatrix} 2 & 1 & 1 \\ 3 & 0 & 2 \\ 1 & -2 & -3 \end{vmatrix} = \begin{vmatrix} 2 & 1 & 1 \\ 3 & 0 & 2 \\ 5 & 0 & -1 \end{vmatrix} = -1\begin{vmatrix} 3 & 2 \\ 5 & -1 \end{vmatrix} = -(-13) = 13$

Adding 2 times row 1 to row 3 yields: $D_z = \begin{vmatrix} 2 & -1 & 1 \\ 3 & 2 & 0 \\ 1 & -5 & -2 \end{vmatrix} = \begin{vmatrix} 2 & -1 & 1 \\ 3 & 2 & 0 \\ 5 & -7 & 0 \end{vmatrix} = 1\begin{vmatrix} 3 & 2 \\ 5 & -7 \end{vmatrix} = -31$

So $x = \frac{D_x}{D} = -\frac{12}{20} = -\frac{3}{5}$, $y = \frac{D_y}{D} = -\frac{13}{20}$, and $z = \frac{D_z}{D} = \frac{31}{20}$. So the solution is $\left(-\frac{3}{5}, -\frac{13}{20}, \frac{31}{20}\right)$.

69. Subtracting column 2 from column 3 and column 1 from column 2 yields: $D = \begin{vmatrix} 1 & 2 & 3 \\ 4 & 5 & 6 \\ 7 & 8 & 9 \end{vmatrix} = \begin{vmatrix} 1 & 1 & 1 \\ 4 & 1 & 1 \\ 7 & 1 & 1 \end{vmatrix}$

Subtracting column 3 from column 2 yields: $D = \begin{vmatrix} 1 & 0 & 1 \\ 4 & 0 & 1 \\ 7 & 0 & 1 \end{vmatrix} = 0$

So Cramer's Rule will not work. Form the augmented matrix: $\begin{pmatrix} 1 & 2 & 3 & -1 \\ 4 & 5 & 6 & 2 \\ 7 & 8 & 9 & -3 \end{pmatrix}$

Add –4 times row 1 to row 2 and –7 times row 1 to row 3: $\begin{pmatrix} 1 & 2 & 3 & -1 \\ 0 & -3 & -6 & 6 \\ 0 & -6 & -12 & 4 \end{pmatrix}$

Add –2 times row 2 to row 3: $\begin{pmatrix} 1 & 2 & 3 & -1 \\ 0 & -3 & -6 & 6 \\ 0 & 0 & 0 & -8 \end{pmatrix}$

So 0 = –8, which is false. The system has no solution.

71. Since $D = 0$ from the previous exercise, form the augmented matrix: $\begin{pmatrix} 3 & 2 & -2 & 1 \\ 2 & 3 & -1 & -2 \\ 8 & 7 & -5 & 0 \end{pmatrix}$

Subtract row 2 from row 1: $\begin{pmatrix} 1 & -1 & -1 & 3 \\ 2 & 3 & -1 & -2 \\ 8 & 7 & -5 & 0 \end{pmatrix}$

Add –2 times row 1 to row 2 and –8 times row 1 to row 3: $\begin{pmatrix} 1 & -1 & -1 & 3 \\ 0 & 5 & 1 & -8 \\ 0 & 15 & 3 & -24 \end{pmatrix}$

Add –3 times row 2 to row 3: $\begin{pmatrix} 1 & -1 & -1 & 3 \\ 0 & 5 & 1 & -8 \\ 0 & 0 & 0 & 0 \end{pmatrix}$

So we have the system:

$$\begin{aligned} x - y - z &= 3 \\ 5y + z &= -8 \end{aligned}$$

Solve equation 2 for z: $z = -8 - 5y$. Substitute into equation 1:

$$\begin{aligned} x - y + 8 + 5y &= 3 \\ x &= -5 - 4y \end{aligned}$$

So the solution is $(-5 - 4y, y, -8 - 5y)$, for any real number y.

73. Adding 4 times column 2 to both column 1 and column 3 yields:

$$D=\begin{vmatrix} 2 & -1 & 1 & 3 \\ 1 & 2 & 0 & 2 \\ 0 & 3 & 3 & 4 \\ -4 & 1 & -4 & 0 \end{vmatrix}=\begin{vmatrix} -2 & -1 & -3 & 3 \\ 9 & 2 & 8 & 2 \\ 12 & 3 & 15 & 4 \\ 0 & 1 & 0 & 0 \end{vmatrix}=1\begin{vmatrix} -2 & -3 & 3 \\ 9 & 8 & 2 \\ 12 & 15 & 4 \end{vmatrix}$$

Subtracting row 2 from row 3 yields: $D=\begin{vmatrix} -2 & -3 & 3 \\ 9 & 8 & 2 \\ 3 & 7 & 2 \end{vmatrix}$

Adding $\frac{2}{3}$ times column 3 to column 1, and column 3 to column 2 yields:

$$D=\begin{vmatrix} 0 & 0 & 3 \\ \frac{31}{3} & 10 & 2 \\ \frac{13}{3} & 9 & 2 \end{vmatrix}=3\begin{vmatrix} \frac{31}{3} & 10 \\ \frac{13}{3} & 9 \end{vmatrix}=\begin{vmatrix} 31 & 10 \\ 13 & 9 \end{vmatrix}=149$$

Adding 11 times column 2 to column 1 and 4 times column 2 to column 3 yields:

$$D_x=\begin{vmatrix} 15 & -1 & 1 & 3 \\ 12 & 2 & 0 & 2 \\ 12 & 3 & 3 & 4 \\ -11 & 1 & -4 & 0 \end{vmatrix}=\begin{vmatrix} 4 & -1 & -3 & 3 \\ 34 & 2 & 8 & 2 \\ 45 & 3 & 15 & 4 \\ 0 & 1 & 0 & 0 \end{vmatrix}=1\begin{vmatrix} 4 & -3 & 3 \\ 34 & 8 & 2 \\ 45 & 15 & 4 \end{vmatrix}$$

Subtracting row 2 from row 3 yields: $D_x=\begin{vmatrix} 4 & -3 & 3 \\ 34 & 8 & 2 \\ 11 & 7 & 2 \end{vmatrix}$

Subtracting row 3 from row 2 yields: $D_x=\begin{vmatrix} 4 & -3 & 3 \\ 23 & 1 & 0 \\ 11 & 7 & 2 \end{vmatrix}$

Adding –23 times column 2 to column 1: $D_x=\begin{vmatrix} 73 & -3 & 3 \\ 0 & 1 & 0 \\ -150 & 7 & 2 \end{vmatrix}=\begin{vmatrix} 73 & 3 \\ -150 & 2 \end{vmatrix}=596$

Adding –2 times row 2 to row 1 and 4 times row 2 to row 4 yields:

$$D_y=\begin{vmatrix} 2 & 15 & 1 & 3 \\ 1 & 12 & 0 & 2 \\ 0 & 12 & 3 & 4 \\ -4 & -11 & -4 & 0 \end{vmatrix}=\begin{vmatrix} 0 & -9 & 1 & -1 \\ 1 & 12 & 0 & 2 \\ 0 & 12 & 3 & 4 \\ 0 & 37 & -4 & 8 \end{vmatrix}=-1\begin{vmatrix} -9 & 1 & -1 \\ 12 & 3 & 4 \\ 37 & -4 & 8 \end{vmatrix}$$

Adding 9 times column 2 to column 1 and column 2 to column 3 yields: $D_y=-1\begin{vmatrix} 0 & 1 & 0 \\ 39 & 3 & 7 \\ 1 & -4 & 4 \end{vmatrix}=1\begin{vmatrix} 39 & 7 \\ 1 & 4 \end{vmatrix}=149$

Adding –2 times row 2 and row 1 to 4 times row 2 to row 4 yields:

$$D_z=\begin{vmatrix} 2 & -1 & 15 & 3 \\ 1 & 2 & 12 & 2 \\ 0 & 3 & 12 & 4 \\ -4 & 1 & -11 & 0 \end{vmatrix}=\begin{vmatrix} 0 & -5 & -9 & -1 \\ 1 & 2 & 12 & 2 \\ 0 & 3 & 12 & 4 \\ 0 & 9 & 37 & 8 \end{vmatrix}=-1\begin{vmatrix} -5 & -9 & -1 \\ 3 & 12 & 4 \\ 9 & 37 & 8 \end{vmatrix}$$

Adding 4 times row 1 to row 2 and 8 times row 1 to row 3: $D_z = -1\begin{vmatrix} -5 & -9 & -1 \\ -17 & -24 & 0 \\ -31 & -25 & 0 \end{vmatrix} = 1\begin{vmatrix} -17 & -24 \\ -31 & -35 \end{vmatrix} = -149$

Adding –2 times row 2 to row 1 and 4 times row 2 to row 4:

$$D_w = \begin{vmatrix} 2 & -1 & 1 & 15 \\ 1 & 2 & 0 & 12 \\ 0 & 3 & 3 & 12 \\ -4 & 1 & -4 & -11 \end{vmatrix} = \begin{vmatrix} 0 & -5 & 1 & -9 \\ 1 & 2 & 0 & 12 \\ 0 & 3 & 3 & 12 \\ 0 & 9 & -4 & 37 \end{vmatrix} = -1\begin{vmatrix} -5 & 1 & -9 \\ 3 & 3 & 12 \\ 9 & -4 & 37 \end{vmatrix}$$

Factoring 3 out of row 2 yields: $D_w = -3\begin{vmatrix} -5 & 1 & -9 \\ 1 & 1 & 4 \\ 9 & -4 & 37 \end{vmatrix}$

Adding –1 times row 2 to row 1 and 4 times row 2 to row 3:

$$D_w = -3\begin{vmatrix} -6 & 0 & -13 \\ 1 & 1 & 4 \\ 13 & 0 & 53 \end{vmatrix} = -3\begin{vmatrix} -6 & -13 \\ 13 & 53 \end{vmatrix} = -3(-149) = 447$$

So $x = \frac{D_x}{D} = \frac{596}{149} = 4$, $y = \frac{D_y}{D} = \frac{149}{149} = 1$, $z = \frac{D_z}{D} = -\frac{149}{149} = -1$, and $w = \frac{D_w}{D} = \frac{447}{149} = 3$. So the solution is (4,1,–1,3).

75. Substitute to obtain:

$$\begin{aligned} x^2 &= 6x \\ x^2 - 6x &= 0 \\ x(x-6) &= 0 \\ x &= 0 \text{ or } x = 6 \end{aligned}$$

When $x = 0$, $y = 0$ and when $x = 6$, $y = 36$. So the solutions are (0,0) and (6,36).

77. Substitute to obtain:

$$\begin{aligned} x^2 - 9 &= 9 - x^2 \\ 2x^2 &= 18 \\ x^2 &= 9 \\ x &= \pm 3 \end{aligned}$$

When $x = \pm 3$, $y = 0$. So the solutions are (3,0) and (–3,0).

79. Adding the two equations yields:

$$\begin{aligned} 2x^2 &= 25 \\ x^2 &= \frac{25}{2} \\ x &= \frac{\pm 5\sqrt{2}}{2} \end{aligned}$$

Substituting for x yields:

$$\begin{aligned} \tfrac{25}{2} + y^2 &= 16 \\ y^2 &= \tfrac{7}{2} \\ y &= \frac{\pm\sqrt{14}}{2} \end{aligned}$$

So the solutions are $\left(\frac{5\sqrt{2}}{2}, \frac{\sqrt{14}}{2}\right), \left(-\frac{5\sqrt{2}}{2}, \frac{\sqrt{14}}{2}\right), \left(\frac{5\sqrt{2}}{2}, -\frac{\sqrt{14}}{2}\right)$, and $\left(-\frac{5\sqrt{2}}{2}, -\frac{\sqrt{14}}{2}\right)$.

81. Substitute to obtain:

$$x^2+x=1$$
$$x^2+x-1=0$$
$$x=\frac{-1\pm\sqrt{1+4}}{2}=\frac{-1\pm\sqrt{5}}{2}$$

Now $x=\frac{-1-\sqrt{5}}{2}$ is impossible, since $x\geq 0$.

So the only solution is $\left(\frac{-1+\sqrt{5}}{2},\sqrt{\frac{-1+\sqrt{5}}{2}}\right)$ or $\left(\frac{-1+\sqrt{5}}{2},\frac{\sqrt{-2+2\sqrt{5}}}{2}\right)$.

83. Multiply the second equation by –3:

$$-9x^2+3xy-3y^2=-54$$
$$x^2+2xy+3y^2=68$$

Adding:

$$-8x^2+5xy=14$$
$$-5xy=8x^2+14$$
$$y=\frac{8x^2+14}{5x}$$

Now substitute into the second equation: $3x^2-\frac{8x^2+14}{5}+\frac{\left(8x^2+14\right)^2}{25x^2}=18$

Multiply by $25x^2$:

$$75x^4-5x^2\left(8x^2+14\right)+\left(8x^2+14\right)^2=450x^2$$
$$75x^4-40x^4-70x^2+64x^4+224x^2+196=450x^2$$
$$99x^4-296x^2+196=0$$
$$\left(99x^2-98\right)\left(x^2-2\right)=0$$

Therefore we have the equations:

$$x^2=\frac{98}{99} \qquad \text{or} \qquad x^2=2$$
$$x=\frac{\pm 7\sqrt{22}}{33} \qquad \text{or} \qquad x=\pm\sqrt{2}$$

Now $y=\frac{8x^2+14}{5x}$, so:

When $x=\sqrt{2}$, $y=\frac{16+14}{5\sqrt{2}}=3\sqrt{2}$ When $x=-\sqrt{2}$, $y=\frac{16+14}{-5\sqrt{2}}=-3\sqrt{2}$

When $x=\frac{7\sqrt{22}}{33}$, $y=\frac{31\sqrt{22}}{33}$ When $x=\frac{-7\sqrt{22}}{33}$, $y=\frac{-31\sqrt{22}}{33}$

So the solutions are $\left(\sqrt{2},3\sqrt{2}\right),\left(-\sqrt{2},-3\sqrt{2}\right),\left(\frac{7\sqrt{22}}{33},\frac{31\sqrt{22}}{33}\right)$, and $\left(-\frac{7\sqrt{22}}{33},-\frac{31\sqrt{22}}{33}\right)$.

85. Let $u = x - 3$ and $v = y + 1$, so:

$$2u^2 - v^2 = -1$$
$$-3u^2 + 2v^2 = 6$$

Multiply the first equation by 2 and add:

$$u^2 = 4$$
$$u = \pm 2$$

Substituting:

$$8 - v^2 = -1$$
$$-v^2 = -9$$
$$v = \pm 3$$

So the solutions (u, v) are (2,3), (2,–3), (–2,3), and (–2,–3). Since $u = x - 3$ and $v = y + 1$, $x = u + 3$ and $y = v - 1$. So the solutions are (5,2), (5,–4), (1,2), and (1,–4).

87. Call the numbers x and y. So $x + y = s$ and $\frac{x}{y} = \frac{a}{b}$. So $y = s - x$, and substituting:

$$\frac{x}{s-x} = \frac{a}{b}$$
$$bx = as - ax$$
$$(a+b)x = as$$
$$x = \frac{as}{a+b}$$

Thus $y = s - \frac{as}{a+b} = \frac{bs}{a+b}$. So the two numbers are $\frac{as}{a+b}$ and $\frac{bs}{a+b}$.

89. Given the equations:

$$\tfrac{1}{2}x + \tfrac{1}{3}y + \tfrac{1}{4}z = 62$$
$$\tfrac{1}{3}x + \tfrac{1}{4}y + \tfrac{1}{5}z = 47$$
$$\tfrac{1}{4}x + \tfrac{1}{5}y + \tfrac{1}{6}z = 38$$

Multiply the first equation by 12, the second equation by 60, and the third equation by 60 to clear the fractions:

$$6x + 4y + 3z = 744$$
$$20x + 15y + 12z = 2820$$
$$15x + 12y + 10z = 2280$$

Use Cramer's Rule. Subtracting row 3 from row 2 yields: $D = \begin{vmatrix} 6 & 4 & 3 \\ 20 & 15 & 12 \\ 15 & 12 & 10 \end{vmatrix} = \begin{vmatrix} 6 & 4 & 3 \\ 5 & 3 & 2 \\ 15 & 12 & 10 \end{vmatrix}$

Subtracting row 2 from row 1 and adding –3 times row 2 to row 3 yields: $D = \begin{vmatrix} 1 & 1 & 1 \\ 5 & 3 & 2 \\ 0 & 3 & 4 \end{vmatrix}$

Subtracting column 1 from column 2 and column 3 yields: $D = \begin{vmatrix} 1 & 0 & 0 \\ 5 & -2 & -3 \\ 0 & 3 & 4 \end{vmatrix} = 1\begin{vmatrix} -2 & -3 \\ 3 & 4 \end{vmatrix} = 1(1) = 1$

Subtracting row 3 from row 2 yields: $D_x = \begin{vmatrix} 744 & 4 & 3 \\ 2820 & 15 & 12 \\ 2280 & 12 & 10 \end{vmatrix} = \begin{vmatrix} 744 & 4 & 3 \\ 540 & 3 & 2 \\ 2280 & 12 & 10 \end{vmatrix}$

Subtracting row 2 from row 1 and adding –4 times row 2 to row 3 yields: $D_x = \begin{vmatrix} 204 & 1 & 1 \\ 540 & 3 & 2 \\ 120 & 0 & 2 \end{vmatrix}$

Adding –3 times row 1 to row 2 yields: $D_x = \begin{vmatrix} 204 & 1 & 1 \\ -72 & 0 & -1 \\ 120 & 0 & 2 \end{vmatrix} = -1\begin{vmatrix} -72 & -1 \\ 120 & 2 \end{vmatrix} = -(-24) = 24$

Subtracting row 3 from row 2 yields: $D_y = \begin{vmatrix} 6 & 744 & 3 \\ 20 & 2820 & 12 \\ 15 & 2280 & 10 \end{vmatrix} = \begin{vmatrix} 6 & 744 & 3 \\ 5 & 540 & 2 \\ 15 & 2280 & 10 \end{vmatrix}$

Subtracting row 2 from row 1 and adding –3 times row 2 to row 3 yields: $D_y = \begin{vmatrix} 1 & 204 & 1 \\ 5 & 540 & 2 \\ 0 & 660 & 4 \end{vmatrix}$

Adding –5 times row 1 to row 2 yields: $D_y = \begin{vmatrix} 1 & 204 & 1 \\ 0 & -480 & -3 \\ 0 & 660 & 4 \end{vmatrix} = 1\begin{vmatrix} -480 & -3 \\ 660 & 4 \end{vmatrix} = 1(60) = 60$

Subtracting row 3 from row 2 yields: $D_z = \begin{vmatrix} 6 & 4 & 744 \\ 20 & 15 & 2820 \\ 15 & 12 & 2280 \end{vmatrix} = \begin{vmatrix} 6 & 4 & 744 \\ 5 & 3 & 540 \\ 15 & 12 & 2280 \end{vmatrix}$

Subtracting row 2 from row 1 and adding –3 times row 2 to row 3 yields: $D_z = \begin{vmatrix} 1 & 1 & 204 \\ 5 & 3 & 540 \\ 0 & 3 & 660 \end{vmatrix}$

Adding –5 times row 1 to row 2 yields: $D_z = \begin{vmatrix} 1 & 1 & 204 \\ 0 & -2 & -480 \\ 0 & 3 & 660 \end{vmatrix} = 1\begin{vmatrix} -2 & -480 \\ 3 & 660 \end{vmatrix} = 1(120) = 120$

So $x = \frac{D_x}{D} = 24$, $y = \frac{D_y}{D} = 60$, and $z = \frac{D_z}{D} = 120$. So the numbers are 24, 60, and 120.

91. Given the equations:

$$xy = m$$
$$x^2 + y^2 = n$$

Add twice the first equation to the second to obtain:

$$x^2 + 2xy + y^2 = n + 2m$$
$$(x+y)^2 = n + 2m$$
$$x + y = \pm\sqrt{n+2m}$$

Subtract twice the first equation from the second to obtain:

$$x^2 - 2xy + y^2 = n - 2m$$
$$(x-y)^2 = n - 2m$$
$$x - y = \pm\sqrt{n-2m}$$

Solve the systems:

$$x + y = \sqrt{n+2m}$$
$$x - y = \sqrt{n-2m}$$
$$2x = \sqrt{n+2m} + \sqrt{n-2m}$$
$$x = \frac{\sqrt{n+2m} + \sqrt{n-2m}}{2}$$
$$y = \frac{\sqrt{n+2m} - \sqrt{n-2m}}{2}$$

$$x + y = \sqrt{n+2m}$$
$$x - y = -\sqrt{n-2m}$$
$$2x = \sqrt{n+2m} - \sqrt{n-2m}$$
$$x = \frac{\sqrt{n+2m} - \sqrt{n-2m}}{2}$$
$$y = \frac{\sqrt{n+2m} + \sqrt{n-2m}}{2}$$

And also the systems:

$$\begin{aligned} x+y &= -\sqrt{n+2m} \\ x-y &= \sqrt{n-2m} \\ \hline 2x &= \sqrt{n-2m}-\sqrt{n+2m} \\ x &= \frac{\sqrt{n-2m}-\sqrt{n+2m}}{2} \\ y &= \frac{-\sqrt{n-2m}-\sqrt{n+2m}}{2} \end{aligned}$$

$$\begin{aligned} x+y &= -\sqrt{n+2m} \\ x-y &= -\sqrt{n-2m} \\ \hline 2x &= -\sqrt{n+2m}-\sqrt{n-2m} \\ x &= \frac{-\sqrt{n+2m}-\sqrt{n-2m}}{2} \\ y &= \frac{\sqrt{n-2m}-\sqrt{n+2m}}{2} \end{aligned}$$

So the possible pairs of numbers are:

$$\frac{\sqrt{n+2m}+\sqrt{n-2m}}{2} \text{ and } \frac{\sqrt{n+2m}-\sqrt{n-2m}}{2}, \quad \frac{\sqrt{n-2m}-\sqrt{n+2m}}{2} \text{ and } \frac{-\sqrt{n-2m}-\sqrt{n+2m}}{2}$$

93. The region is neither convex nor bounded. Graph the system of inequalities:

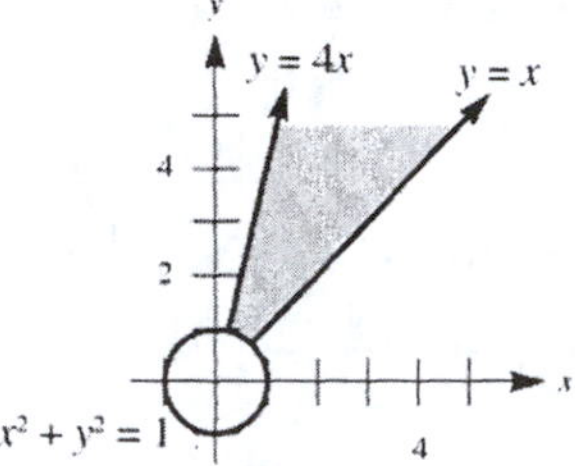

95. The region is convex and bounded. Graph the system of inequalities:

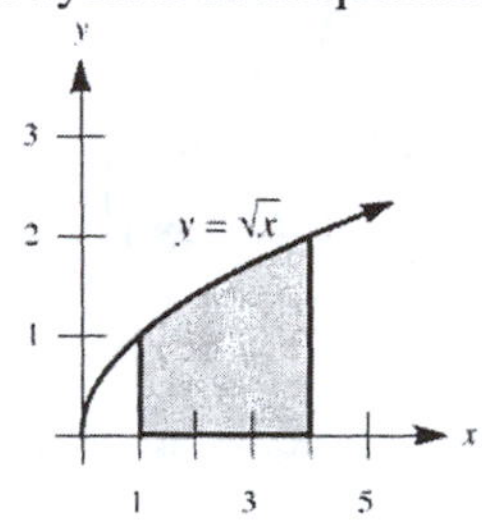

Chapter 10 Test

1. Substitute $y = x^2 + 2x + 3$ into the first equation:

$$\begin{aligned} 3x+4y &= 12 \\ 3x+4\left(x^2+2x+3\right) &= 12 \\ 3x+4x^2+8x+12 &= 12 \\ 4x^2+11x &= 0 \\ x(4x+11) &= 0 \\ x &= 0, -\tfrac{11}{4} \end{aligned}$$

When $x = 0$, $y = 3$ and when $x = -\frac{11}{4}$, $y = \frac{81}{16}$. So the solutions are (0,3) and $\left(-\frac{11}{4}, \frac{81}{16}\right)$.

2. Multiply the first equation by –3:

$$\begin{aligned}-3x+6y&=-39\\3x+5y&=-16\end{aligned}$$

Adding:

$$\begin{aligned}11y&=-55\\y&=-5\end{aligned}$$

Substitute into the first equation:

$$\begin{aligned}x-2y&=13\\x-2(-5)&=13\\x+10&=13\\x&=3\end{aligned}$$

So the solution is (3,–5).

3. **a.** Adding –3 times the first equation to the second equation, and also adding –4 times the first equation to the third equation results in the system:

$$\begin{aligned}x+4y-z&=0\\-11y+4z&=-1\\-20y+9z&=-7\end{aligned}$$

To reduce coefficients, multiply the second equation by –2 and add the third equation to the second equation:

$$\begin{aligned}x+4y-z&=0\\2y+z&=-5\\-20y+9z&=-7\end{aligned}$$

Multiply the second equation by 10 and add it to the third equation:

$$\begin{aligned}x+4y-z&=0\\2y+z&=-5\\19z&=-57\end{aligned}$$

Solving the third equation for z yields $z=-3$. Substitute into the second equation:

$$\begin{aligned}2y-3&=-5\\2y&=-2\\y&=-1\end{aligned}$$

Substitute into the first equation:

$$\begin{aligned}x+4(-1)-(-3)&=0\\x-4+3&=0\\x&=1\end{aligned}$$

So the solution is (1,–1,–3).

b. Adding row 1 to row 2, and 5 times row 1 to row 3: $D=\begin{vmatrix}1&4&-1\\3&1&1\\4&-4&5\end{vmatrix}=\begin{vmatrix}1&4&-1\\4&5&0\\9&16&0\end{vmatrix}$

Now expand along column 3: $D=-1\begin{vmatrix}4&5\\9&16\end{vmatrix}=-(64-45)=-19$

Adding column 1 to columns 2 and 3: $D_x=\begin{vmatrix}0&4&-1\\-1&1&1\\-7&-4&5\end{vmatrix}=\begin{vmatrix}0&4&-1\\-1&0&0\\-7&-11&-2\end{vmatrix}$

Now expand along row 2: $D_x=-(-1)\begin{vmatrix}4&-1\\-11&-2\end{vmatrix}=1(-8-11)=-19$

Adding column 1 to column 3: $D_y=\begin{vmatrix}1&0&-1\\3&-1&1\\4&-7&5\end{vmatrix}=\begin{vmatrix}1&0&0\\3&-1&4\\4&-7&9\end{vmatrix}$

Now expand along row 1: $D_y = 1\begin{vmatrix} -1 & 4 \\ -7 & 9 \end{vmatrix} = -9 + 28 = 19$

Adding –4 times column 1 to column 2: $D_z = \begin{vmatrix} 1 & 4 & 0 \\ 3 & 1 & -1 \\ 4 & -4 & -7 \end{vmatrix} = \begin{vmatrix} 1 & 0 & 0 \\ 3 & -11 & -1 \\ 4 & -20 & -7 \end{vmatrix}$

Now expand along row 1: $D_z = 1\begin{vmatrix} -11 & -1 \\ -20 & -7 \end{vmatrix} = 77 - 20 = 57$

So $x = \frac{D_x}{D} = \frac{-19}{-19} = 1$, $y = \frac{D_y}{D} = \frac{19}{-19} = -1$, and $z = \frac{D_z}{D} = \frac{57}{-19} = -3$. So the solution is (1,–1,–3), which verifies our answer from part **a**.

4. **a.** Computing: $2A - B = \begin{pmatrix} 2 & -6 \\ 4 & -2 \end{pmatrix} - \begin{pmatrix} 0 & 4 \\ 1 & 3 \end{pmatrix} = \begin{pmatrix} 2 & -10 \\ 3 & -5 \end{pmatrix}$

b. Computing: $BA = \begin{pmatrix} 0 & 4 \\ 1 & 3 \end{pmatrix}\begin{pmatrix} 1 & -3 \\ 2 & -1 \end{pmatrix} = \begin{pmatrix} 8 & -4 \\ 7 & -6 \end{pmatrix}$

5. Setting the two expressions equal:

$$24p + 180 = -30p + 1314$$
$$54p = 1134$$
$$p = 21$$

Finding q: $q = 24(21) + 180 = 684$

The equilibrium price is \$21 and the equilibrium quantity is 684 units.

6. Let $u = \frac{1}{x}$ and $v = \frac{1}{y}$, so we have the system:

$$\tfrac{1}{2}u + \tfrac{1}{3}v = 10$$
$$-5u - 4v = -4$$

Multiply the first equation by 10:

$$5u + \tfrac{10}{3}v = 100$$
$$-5u - 4v = -4$$

Adding:

$$-\tfrac{2}{3}v = 96$$
$$v = -144$$

Substitute into the first equation to find u:

$$\tfrac{1}{2}u + \tfrac{1}{3}(-144) = 10$$
$$\tfrac{1}{2}u - 48 = 10$$
$$\tfrac{1}{2}u = 58$$
$$u = 116$$

So $\frac{1}{x} = 116$ thus $x = \frac{1}{116}$, and $\frac{1}{y} = -144$ thus $y = -\frac{1}{144}$. So the solution is $\left(\frac{1}{116}, -\frac{1}{144}\right)$.

7. The coefficient matrix is $\begin{pmatrix} 1 & 1 & -1 \\ 2 & -1 & 2 \\ 1 & -2 & 1 \end{pmatrix}$, and the augmented matrix is $\begin{pmatrix} 1 & 1 & -1 & -1 \\ 2 & -1 & 2 & 11 \\ 1 & -2 & 1 & 10 \end{pmatrix}$.

8. Using the augmented matrix formed in the preceding exercise, add –2 times row 1 to row 2 and –1 times row 1 to row 3: $\begin{pmatrix} 1 & 1 & -1 & -1 \\ 0 & -3 & 4 & 13 \\ 0 & -3 & 2 & 11 \end{pmatrix}$. Multiply row 2 by –1 and add to row 3: $\begin{pmatrix} 1 & 1 & -1 & -1 \\ 0 & -3 & 4 & 13 \\ 0 & 0 & -2 & -2 \end{pmatrix}$

So we have the system of equations:

$$\begin{aligned} x + y - z &= -1 \\ -3y + 4z &= 13 \\ -2z &= -2 \end{aligned}$$

Solving the third equation for z yields $z = 1$. Substitute into the second equation:

$$\begin{aligned} -3y + 4(1) &= 13 \\ -3y &= 9 \\ y &= -3 \end{aligned}$$

Substitute into the first equation:

$$\begin{aligned} x - 3 - 1 &= -1 \\ x - 4 &= -1 \\ x &= 3 \end{aligned}$$

So the solution is (3,–3,1).

9. **a.** Writing the system of equations:

$$\begin{aligned} w + x + y + z &= 40 \\ x + y &= w + z \\ w &= \tfrac{1}{4}(x + y + z) \\ x + z &= \tfrac{2}{3}(w + y) \end{aligned}$$

Clearing fractions and writing the system in standard form:

$$\begin{aligned} w + x + y + z &= 40 \\ -w + x + y - z &= 0 \\ 4w - x - y - z &= 0 \\ -2w + 3x - 2y + 3z &= 0 \end{aligned}$$

b. The coefficient matrix is: $\begin{pmatrix} 1 & 1 & 1 & 1 \\ -1 & 1 & 1 & -1 \\ 4 & -1 & -1 & -1 \\ -2 & 3 & -2 & 3 \end{pmatrix}$

c. The inverse matrix is: $\begin{pmatrix} 0.2 & 0 & 0.2 & 0 \\ 0.1 & 0.5 & 0.2 & 0.2 \\ 0.4 & 0 & -0.2 & -0.2 \\ 0.3 & -0.5 & -0.2 & 0 \end{pmatrix}$

Therefore the solution is: $\begin{pmatrix} 0.2 & 0 & 0.2 & 0 \\ 0.1 & 0.5 & 0.2 & 0.2 \\ 0.4 & 0 & -0.2 & -0.2 \\ 0.3 & -0.5 & -0.2 & 0 \end{pmatrix} \begin{pmatrix} 40 \\ 0 \\ 0 \\ 0 \end{pmatrix} = \begin{pmatrix} 8 \\ 4 \\ 16 \\ 12 \end{pmatrix}$

The students spends 8 hours on math, 4 hours on English, 16 hours on chemistry, and 12 hours on economics.

10. **a.** The minor is $\begin{vmatrix} 2 & -1 \\ 0 & 4 \end{vmatrix} = 8$.

b. The cofactor is $-1\begin{vmatrix} 2 & -1 \\ 0 & 4 \end{vmatrix} = -1(8) = -8$.

11. Adding 2 times row 1 to row 2, and −4 times row 1 to row 3: $\begin{vmatrix} 4 & -5 & 0 \\ 0 & 0 & 7 \\ 0 & 40 & 14 \end{vmatrix}$

Expanding along column 1: $4\begin{vmatrix} 0 & 7 \\ 49 & 14 \end{vmatrix} = 4(0-280) = -1120$

12. If we multiply the second equation by 2 we have $2xy = 10$. Adding the two equations, we obtain:

$$x^2 + 2xy + y^2 = 25$$
$$(x+y)^2 = 25$$
$$x+y = \pm 5$$

Subtracting the two equations:

$$x^2 - 2xy + y^2 = 5$$
$$(x-y)^2 = 5$$
$$x-y = \pm\sqrt{5}$$

So we have the following four systems of equations:

$$\begin{aligned} x+y&=5 \\ x-y&=\sqrt{5} \end{aligned} \quad \begin{aligned} x+y&=5 \\ x-y&=-\sqrt{5} \end{aligned} \quad \begin{aligned} x+y&=-5 \\ x-y&=\sqrt{5} \end{aligned} \quad \begin{aligned} x+y&=-5 \\ x-y&=-\sqrt{5} \end{aligned}$$

Adding results in the four equations:

$$\begin{aligned} 2x&=5+\sqrt{5} \\ x&=\frac{5+\sqrt{5}}{2} \\ y&=\frac{5-\sqrt{5}}{2} \end{aligned} \quad \begin{aligned} 2x&=5-\sqrt{5} \\ x&=\frac{5-\sqrt{5}}{2} \\ y&=\frac{5+\sqrt{5}}{2} \end{aligned} \quad \begin{aligned} 2x&=-5+\sqrt{5} \\ x&=\frac{-5+\sqrt{5}}{2} \\ y&=\frac{-5-\sqrt{5}}{2} \end{aligned} \quad \begin{aligned} 2x&=-5-\sqrt{5} \\ x&=\frac{-5-\sqrt{5}}{2} \\ y&=\frac{-5+\sqrt{5}}{2} \end{aligned}$$

So the solutions are $\left(\frac{5+\sqrt{5}}{2},\frac{5-\sqrt{5}}{2}\right)$, $\left(\frac{5-\sqrt{5}}{2},\frac{5+\sqrt{5}}{2}\right)$, $\left(\frac{-5+\sqrt{5}}{2},\frac{-5-\sqrt{5}}{2}\right)$, and $\left(\frac{-5-\sqrt{5}}{2},\frac{-5+\sqrt{5}}{2}\right)$.

13. Multiply the first equation by –2:

$$-2A-4B-6C=-2$$
$$2A-B-C=2$$

Adding:

$$-5B-7C=0$$
$$-5B=7C$$
$$B=-\tfrac{7}{5}C$$

Substitute into the (original) first equation:

$$A+2B+3C=1$$
$$A-\tfrac{14}{5}C+3C=1$$
$$A+\tfrac{1}{5}C=1$$
$$A=1-\tfrac{1}{5}C$$

So the solution is $\left(1-\frac{1}{5}C,-\frac{7}{5}C,C\right)$, where C = any real number.

14. **a.** Form the matrix: $\begin{pmatrix} 10 & -2 & 5 & 1 & 0 & 0 \\ 6 & -1 & 4 & 0 & 1 & 0 \\ 1 & 0 & 1 & 0 & 0 & 1 \end{pmatrix}$. Switching rows: $\begin{pmatrix} 1 & 0 & 1 & 0 & 0 & 1 \\ 6 & -1 & 4 & 0 & 1 & 0 \\ 10 & -2 & 5 & 1 & 0 & 0 \end{pmatrix}$

Adding –6 times row 1 to row 2, and –10 times row 1 to row 3: $\begin{pmatrix} 1 & 0 & 1 & 0 & 0 & 1 \\ 0 & -1 & -2 & 0 & 1 & -6 \\ 0 & -2 & -5 & 1 & 0 & -10 \end{pmatrix}$

Multiply row 2 by –1, and adding 2 times this new row 2 to row 3: $\begin{pmatrix} 1 & 0 & 1 & 0 & 0 & 1 \\ 0 & 1 & 2 & 0 & -1 & 6 \\ 0 & 0 & -1 & 1 & -2 & 2 \end{pmatrix}$

Adding 2 times row 3 to row 2, row 3 to row 1, then multiplying row 3 by –1: $\begin{pmatrix} 1 & 0 & 0 & 1 & -2 & 3 \\ 0 & 1 & 0 & 2 & -5 & 10 \\ 0 & 0 & 1 & -1 & 2 & -2 \end{pmatrix}$

So the inverse is $\begin{pmatrix} 1 & -2 & 3 \\ 2 & -5 & 10 \\ -1 & 2 & -2 \end{pmatrix}$.

b. Calling A the coefficient matrix and $B = \begin{pmatrix} -1 \\ -2 \\ 3 \end{pmatrix}$, we know the solution will be given by $A^{-1}B$:

$$\begin{pmatrix} u \\ v \\ w \end{pmatrix} = \begin{pmatrix} 1 & -2 & 3 \\ 2 & -5 & 10 \\ -1 & 2 & -2 \end{pmatrix} \begin{pmatrix} -1 \\ -2 \\ 3 \end{pmatrix} = \begin{pmatrix} 12 \\ 38 \\ -9 \end{pmatrix}$$

So the solution is $(12, 38, -9)$.

15. Graph $5x - 6y \geq 30$:

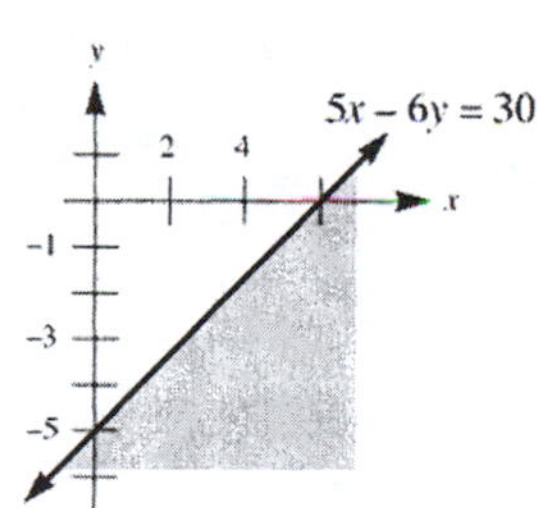

16. Substituting the two points into $y = Px^2 + Qx - 5$:

$$-1 = 4P - 2Q - 5$$
$$-2 = P - Q - 5$$

So we have the system:

$$4P - 2Q = 4$$
$$P - Q = 3$$

Divide the first equation by 2, and multiply the second equation by –1:

$$2P - Q = 2$$
$$-P + Q = -3$$

Adding, we obtain $P = -1$. Substitute into the second equation:

$$-1 - Q = 3$$
$$-Q = 4$$
$$Q = -4$$

So $P = -1$ and $Q = -4$.

17. Graph the inequality:

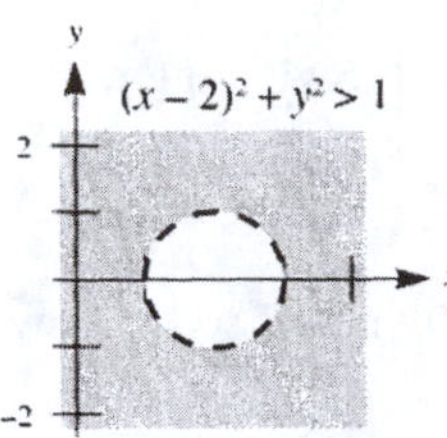

The solution set is neither bounded nor convex.

18. The vertices are (0,0), (0,7), (6,10), $\left(\frac{261}{26}, \frac{225}{26}\right)$, and (11,0). Sketch the graph:

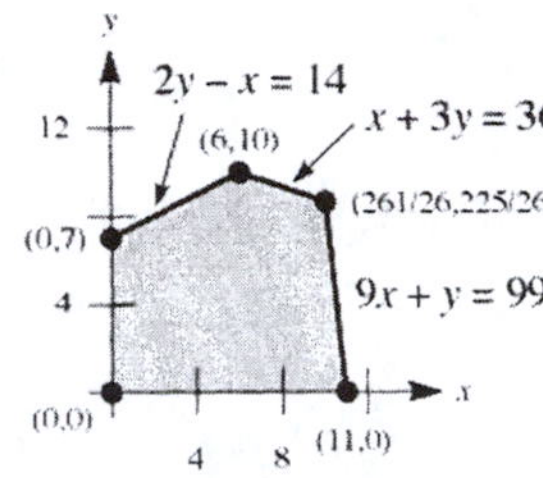

19. Since the solution is $A^{-1}b$ for $(\ln x, \ln y, \ln z)$: $A^{-1}b = \begin{pmatrix} -3 & 2 & -4 \\ -1 & 1 & -1 \\ 8 & -5 & 10 \end{pmatrix}\begin{pmatrix} 3 \\ -1 \\ 2 \end{pmatrix} = \begin{pmatrix} -19 \\ -6 \\ 49 \end{pmatrix}$

So $(\ln x, \ln y, \ln z) = (-19, -6, 49)$, and thus $(x, y, z) = \left(e^{-19}, e^{-6}, e^{49}\right)$.

Chapter 11
The Conic Sections

11.1 The Basic Equations

1. Using the distance formula: $d=\sqrt{(-5-3)^2+(-6+1)^2}=\sqrt{64+25}=\sqrt{89}$

3. Find the slope of the line:

$$\begin{aligned}4x-5y-20&=0\\-5y&=-4x+20\\y&=\tfrac{4}{5}x-4\end{aligned}$$

So the perpendicular slope is $m=-\frac{5}{4}$. Now find the y-intercept:

$$\begin{aligned}x-y+1&=0\\-y&=-x-1\\y&=x+1\end{aligned}$$

So $b = 1$, and the equation is $y=-\frac{5}{4}x+1$. Multiplying by 4: $4y=-5x+4$, or $5x+4y-4=0$

5. Find the slope of the line segment: $m=\frac{7-1}{6-2}=\frac{6}{4}=\frac{3}{2}$

So the perpendicular slope is $m=-\frac{2}{3}$. Find the midpoint: $M=\left(\frac{2+6}{2},\frac{1+7}{2}\right)=\left(\frac{8}{2},\frac{8}{2}\right)=(4,4)$

Now use the point-slope formula:

$$\begin{aligned}y-4&=-\tfrac{2}{3}(x-4)\\y-4&=-\tfrac{2}{3}x+\tfrac{8}{3}\\y&=-\tfrac{2}{3}x+\tfrac{20}{3}\end{aligned}$$

Multiplying by 3, the equation is $3y = -2x + 20$, or $2x + 3y - 20 = 0$.

7. Since the center is (1,0) and the radius is 5, the equation must be $(x-1)^2+y^2=25$. To find the x-intercepts, let $y = 0$ and solve the resulting equation for x:

$$\begin{aligned}y&=0\\(x-1)^2&=25\\x-1&=\pm5\\x&=6,-4\end{aligned}$$

To find the y-intercepts, let $x = 0$ and solve the resulting equation for y:

$$\begin{aligned} x &= 0 \\ (-1)^2 + y^2 &= 25 \\ y^2 &= 24 \\ y &= \pm\sqrt{24} = \pm 2\sqrt{6} \end{aligned}$$

The x-intercepts are 6 and -4, and the y-intercepts are $\pm 2\sqrt{6}$.

9. First find the midpoint of $\overline{AB}$: $M = \left(\frac{1+6}{2}, \frac{2+1}{2}\right) = \left(\frac{7}{2}, \frac{3}{2}\right)$

Now find the slope of the line: $m = \dfrac{8 - \frac{3}{2}}{7 - \frac{7}{2}} = \dfrac{\frac{13}{2}}{\frac{7}{2}} = \frac{13}{7}$

Using the point $(7, 8)$ in the point-slope formula:

$$\begin{aligned} y - 8 &= \tfrac{13}{7}(x - 7) \\ y - 8 &= \tfrac{13}{7}x - 13 \\ y &= \tfrac{13}{7}x - 5 \end{aligned}$$

Multiply by 7 to get $7y = 13x - 35$, or $13x - 7y - 35 = 0$.

11. Since C is the x-intercept, let $y = 0$ to obtain $\frac{x}{7} = 1$, or $x = 7$. So the coordinates of C are (7,0).

Similarly, since B is the y-intercept, let $x = 0$ to obtain $\frac{y}{5} = 1$, or $y = 5$. So the coordinates of B are (0,5). Find BC by the distance formula: $BC = \sqrt{(7-0)^2 + (0-5)^2} = \sqrt{49+25} = \sqrt{74}$

The perimeter is given by: $P = AB + BC + AC = 5 + \sqrt{74} + 7 = 12 + \sqrt{74}$

13. Since $\tan\theta = \sqrt{3}$, then $\theta = \frac{\pi}{3}$ or 60°.

15. **a.** Since $\tan\theta = 5$, then $\theta = 1.37$ or 78.69°.

b. Since $\tan\theta = -5$, then $\theta = 1.77$ or 101.31°.

17. **a.** Here $(x_0, y_0) = (1,4)$, $m = 1$, $b = -2$, so: $d = \dfrac{|1-2-4|}{\sqrt{1+1}} = \dfrac{5}{\sqrt{2}} = \dfrac{5\sqrt{2}}{2}$

b. Using $x - y - 2 = 0$, $A = 1$, $B = -1$, $C = -2$: $d = \dfrac{|1-4-2|}{\sqrt{1+1}} = \dfrac{5}{\sqrt{2}} = \dfrac{5\sqrt{2}}{2}$

19. **a.** Here $(x_0, y_0) = (-3,5)$. Convert to slope-intercept form:

$$\begin{aligned} 4x + 5y + 6 &= 0 \\ 5y &= -4x - 6 \\ y &= -\tfrac{4}{5}x - \tfrac{6}{5} \end{aligned}$$

So $m = -\frac{4}{5}$ and $b = -\frac{6}{5}$: $d = \dfrac{\left|\frac{12}{5} - \frac{6}{5} - 5\right|}{\sqrt{1 + \frac{16}{25}}} = \dfrac{\frac{19}{5}}{\frac{\sqrt{41}}{5}} = \dfrac{19}{\sqrt{41}} = \dfrac{19\sqrt{41}}{41}$

b. Given $A = 4$, $B = 5$, $C = 6$: $d = \dfrac{|-12+25+6|}{\sqrt{16+25}} = \dfrac{19}{\sqrt{41}} = \dfrac{19\sqrt{41}}{41}$

21. **a.** The radius of the circle is the distance from the point $(-2,-3)$ to the line $2x + 3y - 6 = 0$. Using $(x_0, y_0) = (-2,-3)$, $A = 2$, $B = 3$, $C = -6$: $r = \dfrac{|-4-9-6|}{\sqrt{4+9}} = \dfrac{19}{\sqrt{13}}$

So the equation of the circle is $(x+2)^2 + (y+3)^2 = \frac{361}{13}$.

b. Since the radius is the distance from the point (1,3) to the line $y=\frac{1}{2}x+5$, $(x_0, y_0)=(1,3)$,

$$m=\frac{1}{2}, b=5:\ r=\frac{\left|\frac{1}{2}+5-3\right|}{\sqrt{1+\frac{1}{4}}}=\frac{\frac{5}{2}}{\frac{\sqrt{5}}{2}}=\sqrt{5}$$

23. Using the suggestion, work with ΔABC and ΔCDA. For ΔABC, find the base using the distance formula: $AB=\sqrt{(8-0)^2+(2-0)^2}=\sqrt{64+4}=\sqrt{68}=2\sqrt{17}$

Now find the equation of the line through A and B. Find the slope: $m=\frac{2-0}{8-0}=\frac{1}{4}$

So the equation is $y=\frac{1}{4}x$. Find the distance from $C(4,7)$ to this line: $h=\frac{|1+0-7|}{\sqrt{1+\frac{1}{16}}}=\frac{6}{\frac{\sqrt{17}}{4}}=\frac{24}{\sqrt{17}}$

So ΔABC will have an area of: $\text{Area}_{\Delta ABC}=\frac{1}{2}(\text{base})(\text{height})=\frac{1}{2}\left(2\sqrt{17}\right)\left(\frac{24}{\sqrt{17}}\right)=24$

For ΔCDA, find the base using the distance formula: $CD=\sqrt{(1-4)^2+(6-7)^2}=\sqrt{9+1}=\sqrt{10}$

Now find the equation of the line through C and D. Find the slope: $m=\frac{6-7}{1-4}=\frac{-1}{-3}=\frac{1}{3}$

Using $C(4,7)$, in the point-slope formula:

$$y-7=\frac{1}{3}(x-4)$$
$$y-7=\frac{1}{3}x-\frac{4}{3}$$
$$y=\frac{1}{3}x+\frac{17}{3}$$
$$3y=x+17 \text{ or } x-3y+17=0$$

Find the distance from $A(0,0)$ to this line: $h=\frac{|0+0+17|}{\sqrt{1+9}}=\frac{17}{\sqrt{10}}$

So ΔCDA will have an area of : $\text{Area}_{\Delta CDA}=\frac{1}{2}(\text{base})(\text{height})=\frac{1}{2}\left(\sqrt{10}\right)\left(\frac{17}{\sqrt{10}}\right)=\frac{17}{2}$

Thus, the total combined area of quadrilateral $ABCD$ is: $24+\frac{17}{2}=\frac{65}{2}$

25. Using the point (0,–5) in the point-slope formula:

$$y+5=m(x-0)$$
$$y=mx-5$$

Now, since the distance from the center (3,0) to this line is 2: $2=\frac{|3m-5-0|}{\sqrt{1+m^2}}=\frac{|3m-5|}{\sqrt{1+m^2}}$

Squaring each side:

$$4=\frac{9m^2-30m+25}{1+m^2}$$
$$4+4m^2=9m^2-30m+25$$
$$0=5m^2-30m+21$$

Using the quadratic formula: $m=\frac{30\pm\sqrt{900-420}}{10}=\frac{30\pm\sqrt{480}}{10}=\frac{30\pm4\sqrt{30}}{10}=\frac{15\pm2\sqrt{30}}{5}$

27. Use the same approach as in the preceding exercise. Let d_1 and d_2 represent the distances from (0,0) to the lines $3x + 4y - 12 = 0$ and $3x + 4y - 24 = 0$, respectively. For d_1, $(x_0, y_0) = (0, 0)$, $A = 3$, $B = 4$, $C = -12$: $d_1 = \dfrac{|0+0-12|}{\sqrt{9+16}} = \dfrac{12}{\sqrt{25}} = \frac{12}{5}$

For d_2, $(x_0, y_0) = (0,0)$, $A = 3$, $B = 4$, $C = -24$: $d_2 = \dfrac{|0+0-24|}{\sqrt{9+16}} = \dfrac{24}{\sqrt{25}} = \frac{24}{5}$

So the distance between the lines is: $d_2 - d_1 = \frac{24}{5} - \frac{12}{5} = \frac{12}{5}$

29. For the described line segment to be bisected by the point (2,6), then (2,6) must be the midpoint of the x- and y-intercepts (as points) of the line. Using the point-slope formula:

$$y - 6 = m(x-2)$$
$$y - 6 = mx - 2m$$
$$y = mx - 2m + 6$$

To find the x-intercept, let $y = 0$ and solve the resulting equation for x:

$$y = 0$$
$$mx - 2m + 6 = 0$$
$$mx = 2m - 6$$
$$x = \frac{2m-6}{m}$$

To find the y-intercept, let $x = 0$ and solve the resulting equation for y:

$$x = 0$$
$$y = -2m + 6$$

Since (2,6) must be the midpoint of $\left(\dfrac{2m-6}{m}, 0\right)$ and $(0, -2m + 6)$, we have the two sets of equations:

$$\frac{2m-6}{2m} = 2$$
$$2m - 6 = 4m$$
$$-6 = 2m$$
$$-3 = m$$

$$\frac{-2m+6}{2} = 6$$
$$-2m + 6 = 12$$
$$-2m = 6$$
$$m = -3$$

Now use the point-slope formula with the point (2,6):

$$y - 6 = -3(x-2)$$
$$y - 6 = -3x + 6$$
$$y = -3x + 12$$

Here is an alternate solution. Let the intercepts be $(a,0)$ and $(0,b)$. Since (2,6) is their midpoint:

$2 = \dfrac{a+0}{2}$, so $a = 4$ $6 = \dfrac{b+0}{2}$, so $b = 12$

Thus the slope is given by: $m = \dfrac{12-0}{0-4} = -3$. Thus the equation is $y = -3x + 12$.

Special thanks to John Fay for this alternate solution.

31. Using the hint, let d_1 and d_2 represent the distances from (x, y) to $x - y + 1 = 0$ and $x + 7y - 49 = 0$, respectively. Then:

$$d_1 = \frac{|x-y+1|}{\sqrt{1+1}} = \frac{|x-y+1|}{\sqrt{2}} \qquad d_2 = \frac{|x+7y-49|}{\sqrt{1+49}} = \frac{|x+7y-49|}{\sqrt{50}}$$

Since $d_1 = d_2$: $\dfrac{|x-y+1|}{\sqrt{2}} = \dfrac{|x+7y-49|}{\sqrt{50}}$. Multiplying each side by $\sqrt{2}$:

$$|x-y+1| = \frac{|x+7y-49|}{5}$$
$$5|x-y+1| = |x+7y-49|$$

Rather than squaring each side, note that one of the two equations must hold:

$$\begin{aligned} 5(x-y+1) &= x+7y-49 \\ 5x-5y+5 &= x+7y-49 \\ 4x-12y+54 &= 0 \\ 2x-6y+27 &= 0 \\ y &= \tfrac{1}{3}x+\tfrac{9}{2} \end{aligned}$$

$$\begin{aligned} 5(x-y+1) &= -x-7y+49 \\ 5x-5y+5 &= -x-7y+49 \\ 6x+2y-44 &= 0 \\ 3x+y-22 &= 0 \\ y &= -3x+22 \end{aligned}$$

Note that the first of these lines, $y=\frac{1}{3}x+\frac{9}{2}$, is the solution as indicated in the figure. The second line we found is the second angle bisector passing through the same intersection point but bisecting the larger vertical angles.

33. **a.** The standard form for a circle is $(x-h)^2+(y-k)^2=r^2$. Substitute each point for (x,y):

$$\begin{aligned} (-12-h)^2+(1-k)^2 &= r^2 \\ (2-h)^2+(1-k)^2 &= r^2 \\ (0-h)^2+(7-k)^2 &= r^2 \end{aligned}$$

Setting the first two equations equal:

$$\begin{aligned} (-12-h)^2+(1-k)^2 &= (2-h)^2+(1-k)^2 \\ 144+24h+h^2 &= 4-4h+h^2 \\ 28h &= -140 \\ h &= -5 \end{aligned}$$

Setting the second two equations equal:

$$\begin{aligned} (2-h)^2+(1-k)^2 &= (0-h)^2+(7-k)^2 \\ 4-4h+h^2+1-2k+k^2 &= h^2+49-14k+k^2 \\ -4h+12k &= 44 \\ h-3k &= -11 \end{aligned}$$

Substituting $h=-5$:

$$\begin{aligned} -5-3k &= -11 \\ -3k &= -6 \\ k &= 2 \end{aligned}$$

Now substitute into the second equation to find r:

$$\begin{aligned} (7)^2+(-1)^2 &= r^2 \\ 50 &= r^2 \\ 5\sqrt{2} &= r \end{aligned}$$

So the center is $(-5,2)$ and the radius is $5\sqrt{2}$.

b. First draw the sketch:

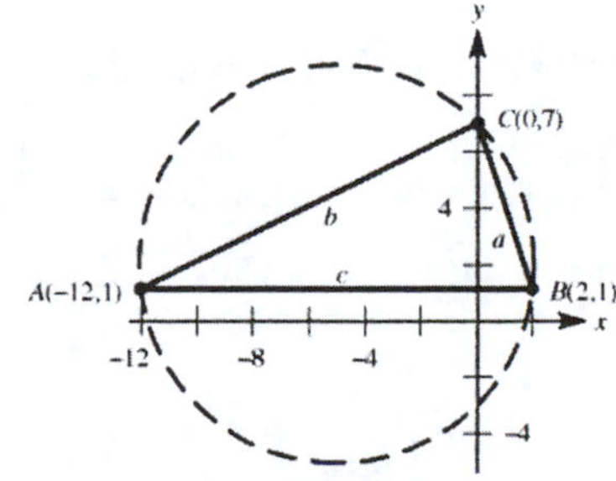

Find the lengths of three sides (using the distance formula):

$$\begin{aligned} a &= \sqrt{(2-0)^2+(1-7)^2} = \sqrt{4+36} = \sqrt{40} = 2\sqrt{10} \\ b &= \sqrt{(-12-0)^2+(1-7)^2} = \sqrt{144+36} = \sqrt{180} = 6\sqrt{5} \\ c &= \sqrt{(-12-2)^2+(1-1)^2} = \sqrt{196+0} = \sqrt{196} = 14 \end{aligned}$$

Using $R = 5\sqrt{2}$ from part **a**: $\dfrac{abc}{4R} = \dfrac{(2\sqrt{10})(6\sqrt{5})(14)}{4(5\sqrt{2})} = \dfrac{42\sqrt{50}}{5\sqrt{2}} = 42$

For the area of the triangle, note that the base = 14 and the height = 6, so:

$$\text{Area} = \tfrac{1}{2}(\text{base})(\text{height}) = \tfrac{1}{2}(14)(6) = 42$$

This verifies the result.

35. First find points P and Q. Solving for x, we have $x = 7y - 44$. Now substitute:

$$\begin{aligned}(7y-44)^2 - 4(7y-44) + y^2 - 6y &= 12\\ 49y^2 - 616y + 1936 - 28y + 176 + y^2 - 6y &= 12\\ 50y^2 - 650y + 2100 &= 0\\ 50(y^2 - 13y + 42) &= 0\\ 50(y-7)(y-6) &= 0\\ y &= 7, 6\end{aligned}$$

When $y = 7$, $x = 49 - 44 = 5$. When $y = 6$, $x = 42 - 44 = -2$. So the two points are $P(5,7)$ and $Q(-2,6)$. Now use the distance formula: $PQ = \sqrt{(-2-5)^2 + (6-7)^2} = \sqrt{49+1} = \sqrt{50} = 5\sqrt{2}$

37. Let d_1 be the distance from $(0, c)$ to $ax + y = 0$: $d_1 = \dfrac{|0+c+0|}{\sqrt{a^2+1}} = \dfrac{|c|}{\sqrt{a^2+1}}$

Let d_2 be the distance from $(0, c)$ to $x + by = 0$: $d_2 = \dfrac{|0+bc+0|}{\sqrt{1+b^2}} = \dfrac{|bc|}{\sqrt{1+b^2}}$

So the product of these distances is given by: $d_1 d_2 = \dfrac{|c|}{\sqrt{a^2+1}} \bullet \dfrac{|bc|}{\sqrt{1+b^2}} = \dfrac{|bc^2|}{\sqrt{a^2 + a^2b^2 + b^2 + 1}}$

39. **a.** Draw the triangle:

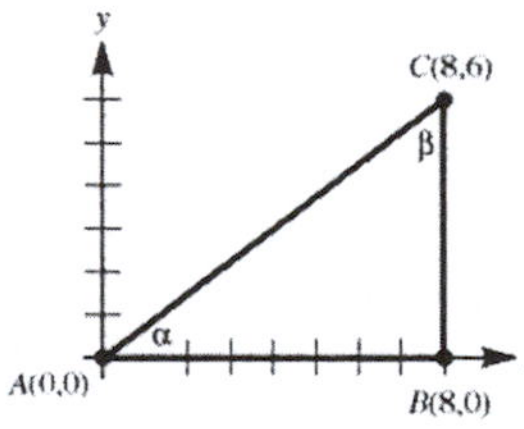

Since the triangle is a right triangle, note that $\sin\alpha = \frac{6}{10} = \frac{3}{5}$, $\cos\alpha = \frac{8}{10} = \frac{4}{5}$, $\sin\beta = \frac{8}{10} = \frac{4}{5}$ and $\cos\beta = \frac{6}{10} = \frac{3}{5}$. To find the bisector for A, note that: $m = \tan\frac{\alpha}{2} = \dfrac{\sin\alpha}{1+\cos\alpha} = \dfrac{\frac{3}{5}}{1+\frac{4}{5}} = \frac{3}{9} = \frac{1}{3}$

Since this line passes through the point (0,0), its equation is $y = \frac{1}{3}x$. To find the bisector for B, note that its slope is -1 and that it passes through (8,0), so:

$$\begin{aligned}y - 0 &= -1(x-8)\\ y &= -x + 8\end{aligned}$$

To find the bisector for C, first draw the figure:

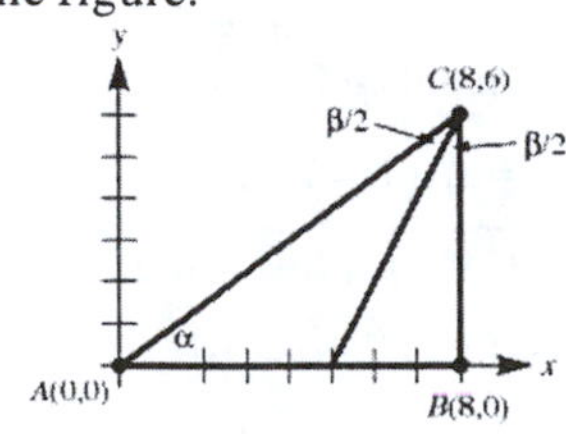

Now note that: $m=\tan\left(90°-\frac{\beta}{2}\right)=\cot\frac{\beta}{2}=\frac{1+\cos\beta}{\sin\beta}=\frac{1+\frac{3}{5}}{\frac{4}{5}}=2$

With the point (8,6), use the point-slope formula:

$$y-6=2(x-8)$$
$$y-6=2x-16$$
$$y=2x-10$$

So the bisectors at each vertex are: $A: y=\frac{1}{3}x$ $\quad B: y=-x+8$ $\quad C: y=2x-10$

b. Setting A and B equal:

$$\frac{1}{3}x=-x+8$$
$$\frac{4}{3}x=8$$
$$x=6$$
$$y=2$$

Setting B and C equal:

$$-x+8=2x-10$$
$$18=3x$$
$$6=x$$
$$2=y$$

Setting A and C equal:

$$\frac{1}{3}x=2x-10$$
$$-\frac{5}{3}x=-10$$
$$x=6$$
$$y=2$$

So the angle bisectors are concurrent at the point (6,2).

41. The distance from the point (x,y) to $\left(0,\frac{1}{4}\right)$ is: $d=\sqrt{(x-0)^2+\left(y-\frac{1}{4}\right)^2}=\sqrt{x^2+y^2-\frac{1}{2}y+\frac{1}{16}}$

The distance from the point (x,y) to $y=-\frac{1}{4}$ is: $d=\frac{\left|0-\frac{1}{4}-y\right|}{\sqrt{1+0}}=\left|y+\frac{1}{4}\right|$

Setting these distances equal: $\left|y+\frac{1}{4}\right|=\sqrt{x^2+y^2-\frac{1}{2}y+\frac{1}{16}}$

Squaring:

$$y^2+\frac{1}{2}y+\frac{1}{16}=x^2+y^2-\frac{1}{2}y+\frac{1}{16}$$
$$y=x^2$$

Thus, the points satisfy the equation $y=x^2$.

43. **a.** Write the line in slope-intercept form:

$$Ax+By+C=0$$
$$By=-Ax-C$$
$$y=-\frac{A}{B}x-\frac{C}{B}$$

So the slope is $-\frac{A}{B}$ and the y-intercept is $-\frac{C}{B}$.

b. Using the given formula: $d=\frac{\left|mx_0+b-y_0\right|}{\sqrt{1+m^2}}=\frac{\left|-\frac{Ax_0+By_0+C}{B}\right|}{\sqrt{1+\frac{A^2}{B^2}}}=\frac{\frac{\left|Ax_0+By_0+C\right|}{|B|}}{\frac{\sqrt{A^2+B^2}}{|B|}}=\frac{\left|Ax_0+By_0+C\right|}{\sqrt{A^2+B^2}}$

45. First find the slope of the line passing through $\left(x_2,y_2\right)$ and $\left(x_3,y_3\right)$: $m=\frac{y_2-y_3}{x_2-x_3}$

Using the point-slope formula:

$$y-y_2=\frac{y_2-y_3}{x_2-x_3}\left(x-x_2\right)$$
$$\left(x_2-x_3\right)y-x_2y_2+x_3y_2=\left(y_2-y_3\right)x-x_2y_2+x_2y_3$$
$$\left(y_3-y_2\right)x+\left(x_2-x_3\right)y+\left(x_3y_2-x_2y_3\right)=0$$

Now find the distance from $\left(x_1, y_1\right)$ to the line:

$$d=\frac{\left|x_1\left(y_3-y_2\right)+y_1\left(x_2-x_3\right)+x_3y_2-x_2y_3\right|}{\sqrt{\left(y_3-y_2\right)^2+\left(x_2-x_3\right)^2}}=\frac{\left|x_1y_3-x_1y_2+x_2y_1-x_3y_1+x_3y_2-x_2y_3\right|}{\sqrt{\left(x_2-x_3\right)^2+\left(y_2-y_3\right)^2}}$$

Expanding D along the third column, we find that its terms exactly match the numerator of d.

47. Since the center can be written as $(x, 2x-7)$, find the distance to each point. Let d_1 represent the distance to (6,3) and d_2 represent the distance to (–4,–3), then:

$$d_1=\sqrt{(x-6)^2+(2x-7-3)^2}=\sqrt{x^2-12x+36+4x^2-40x+100}=\sqrt{5x^2-52x+136}$$

$$d_2=\sqrt{(x+4)^2+(2x-7+3)^2}=\sqrt{x^2+8x+16+4x^2-16x+16}=\sqrt{5x^2-8x+32}$$

Since $d_1=d_2$ (they both represent the radius of the circle):

$$\begin{aligned}\sqrt{5x^2-52x+136}&=\sqrt{5x^2-8x+32}\\5x^2-52x+136&=5x^2-8x+32\\-44x&=-104\\x&=\tfrac{26}{11}\\y&=2\left(\tfrac{26}{11}\right)-7=\tfrac{52}{11}-\tfrac{77}{11}=-\tfrac{25}{11}\end{aligned}$$

So the center is $\left(\frac{26}{11},-\frac{25}{11}\right)$. Find the radius using the point (6,3) in the distance formula:

$$r=\sqrt{\left(6-\tfrac{26}{11}\right)^2+\left(3+\tfrac{25}{11}\right)^2}=\sqrt{\frac{1600}{121}+\frac{3364}{121}}=\frac{\sqrt{4964}}{11}$$

So the equation of the circle is $\left(x-\frac{26}{11}\right)^2+\left(y+\frac{25}{11}\right)^2=\frac{4964}{121}$.

49. Since the line $2x+3y=26$ has a slope of $-\frac{2}{3}$, the radial line will have a slope of $\frac{3}{2}$. Call (h,k) the center of the circle:

$$\begin{aligned}\frac{k-6}{h-4}&=\frac{3}{2}\\2k-12&=3h-12\\2k&=3h\\k&=\tfrac{3}{2}h\end{aligned}$$

The circle has an equation of $(x-h)^2+(y-k)^2=25$, so using the point (4,6):

$$\begin{aligned}(4-h)^2+(6-k)^2&=25\\16-8h+h^2+36-12k+k^2&=25\\h^2-8h+k^2-12k&=-27\end{aligned}$$

Substituting $k=\frac{3}{2}h$:

$$\begin{aligned}h^2-8h+\tfrac{9}{4}h^2-18h&=-27\\4h^2-32h+9h^2-72h&=-108\\13h^2-104h+108&=0\\h&=\frac{104\pm\sqrt{5200}}{26}=4\pm\frac{10\sqrt{13}}{13}\\k&=6\pm\frac{15\sqrt{13}}{13}\end{aligned}$$

So the two circles are $\left(x-4-\frac{10\sqrt{13}}{13}\right)^2+\left(y-6-\frac{15\sqrt{13}}{13}\right)^2=25$ and $\left(x-4+\frac{10\sqrt{13}}{13}\right)^2+\left(y-6+\frac{15\sqrt{13}}{13}\right)^2=25$.

51. **a.** Replacing y with tx in the equation of the folium yields:

$$x^3+t^3x^3=6x(tx)$$
$$x^3+t^3x^3=6x^2t \text{ (since } x\neq 0 \text{ at point } Q)$$
$$x+t^3x=6t$$
$$x\left(1+t^3\right)=6t$$
$$x=\frac{6t}{1+t^3}$$
$$y=tx=\frac{6t^2}{1+t^3}$$

So the coordinates of point Q are $x=\dfrac{6t}{1+t^3}$ and $y=\dfrac{6t^2}{1+t^3}$.

b. The slope of $\overline{PQ}$ is given by: $\dfrac{3-\dfrac{6t^2}{1+t^3}}{3-\dfrac{6t}{1+t^3}}=\dfrac{1-\dfrac{2t^2}{1+t^3}}{1-\dfrac{2t}{1+t^3}}\bullet\dfrac{1+t^3}{1+t^3}=\dfrac{1+t^3-2t^2}{1+t^3-2t}=\dfrac{t^3-2t^2+1}{t^3-2t+1}$

We can factor both the numerator and denominator of this expression. For the numerator, add and subtract the term t^2 as follows:

$$\begin{aligned}t^3-2t^2+1&=t^3-2t^2+t^2-t^2+1\\&=\left(t^3-t^2\right)-\left(t^2-1\right)\\&=t^2(t-1)-(t+1)(t-1)\\&=(t-1)\left(t^2-t-1\right)\end{aligned}$$

For the denominator, add and subtract the term t^2 as follows:

$$\begin{aligned}t^3-2t+1&=t^3-t^2+t^2-2t+1\\&=\left(t^3-t^2\right)+\left(t^2-2t+1\right)\\&=t^2(t-1)+(t-1)^2\\&=(t-1)\left(t^2+t-1\right)\end{aligned}$$

So the slope of $\overline{PQ}$ is given by: $\dfrac{t^3-2t^2+1}{t^3-2t+1}=\dfrac{(t-1)\left(t^2-t-1\right)}{(t-1)\left(t^2+t-1\right)}=\dfrac{t^2-t-1}{t^2+t-1}$

c. Cross-multiplying and combining like terms:

$$\left(t^2-t-1\right)\left(u^2+u-1\right)=\left(t^2+t-1\right)\left(u^2-u-1\right)$$
$$t^2u^2+t^2u-t^2-tu^2-tu+t-u^2-u+1=t^2u^2-t^2u-t^2+tu^2-tu-t-u^2+u+1$$
$$2t^2u-2tu^2+2t-2u=0$$
$$t^2u-tu^2+t-u=0$$
$$tu(t-u)+1(t-u)=0$$
$$(t-u)(tu+1)=0$$
$$t=u \text{ or } tu=-1$$

Since points R, O and Q are not collinear, $t\neq u$, and thus $tu=-1$. Therefore $\overline{OR}\perp\overline{OQ}$, as required.

11.2 The Parabola

1. Note that $4p = 4$, so $p = 1$. So the focus is $(0,1)$, the directrix is $y = -1$, and the focal width is 4.

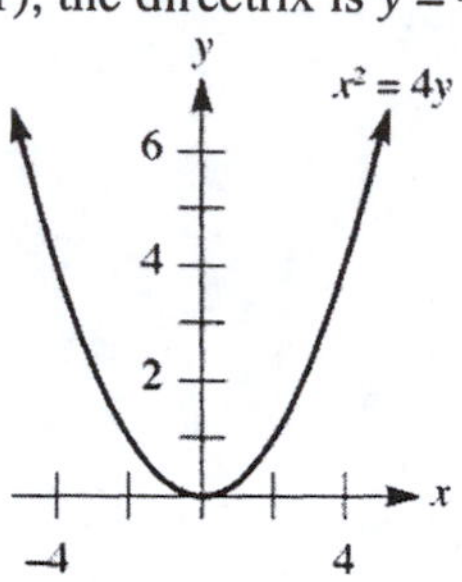

3. Note that $4p = 8$, so $p = 2$. So the focus is $(-2,0)$, the directrix is $x = 2$, and the focal width is 8.

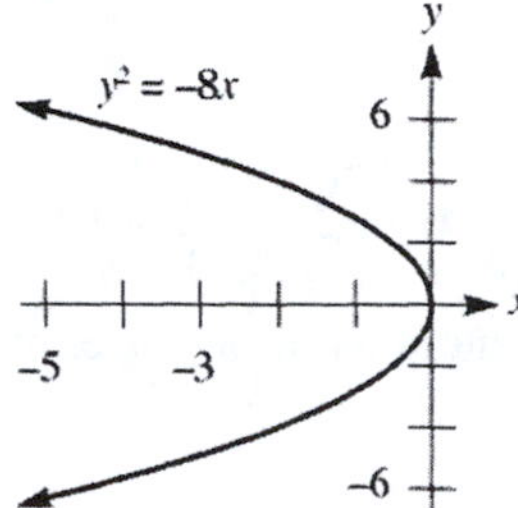

5. Note that $4p = 20$, so $p = 5$. So the focus is $(0,-5)$, the directrix is $y = 5$, and the focal width is 20.

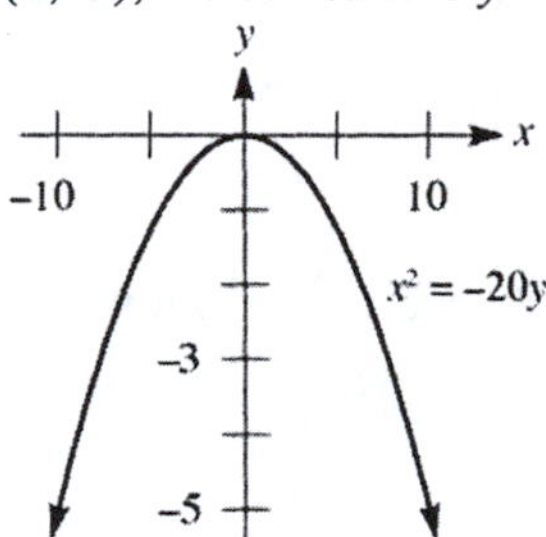

7. Since $y^2 = -28x$, then $4p = 28$ so $p = 7$. So the focus is $(-7,0)$, the directrix is $x = 7$, and the focal width is 28.

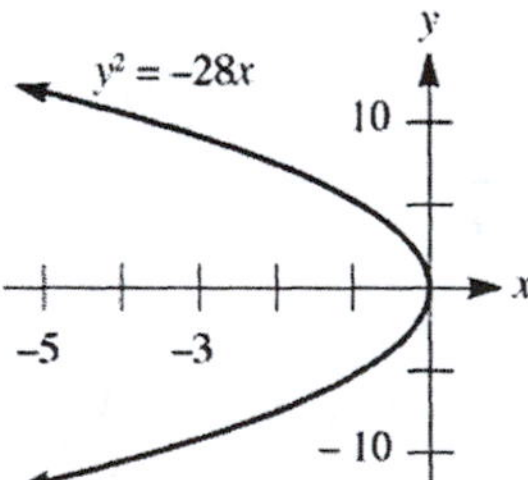

9. Note that $4p = 6$, so $p = \frac{3}{2}$. So the focus is $\left(0, \frac{3}{2}\right)$, the directrix is $y = -\frac{3}{2}$, and the focal width is 6.

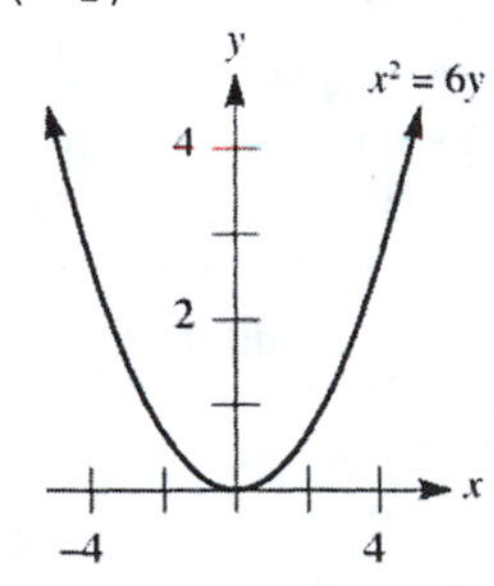

11. Since $x^2 = \frac{7}{4}y$, then $4p = \frac{7}{4}$ so $p = \frac{7}{16}$. So the focus is $\left(0, \frac{7}{16}\right)$, the directrix is $y = -\frac{7}{16}$, and the focal width is $\frac{7}{4}$.

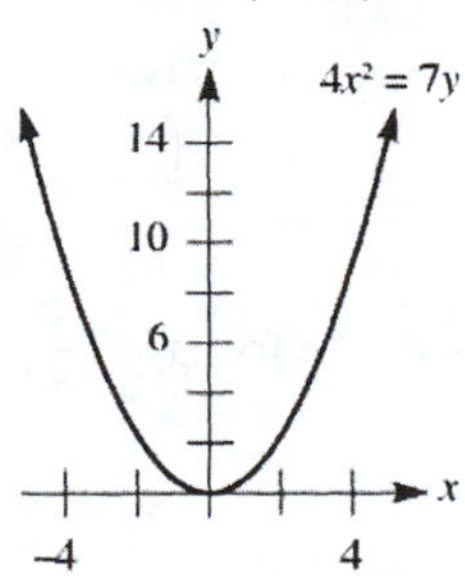

13. First complete the square:

$$y^2 - 6y - 4x + 17 = 0$$
$$y^2 - 6y = 4x - 17$$
$$y^2 - 6y + 9 = 4x - 17 + 9$$
$$(y-3)^2 = 4x - 8 = 4(x-2)$$

Note that $4p = 4$, so $p = 1$. The vertex is (2,3), the focus is (3,3), the directrix is $x = 1$, and the focal width is 4.

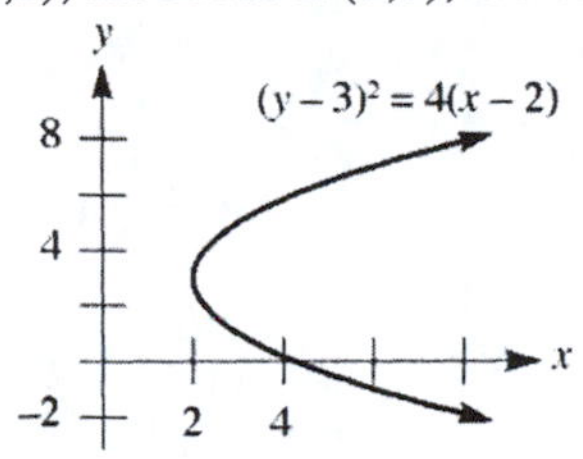

15. First complete the square:

$$x^2 - 8x - y + 18 = 0$$
$$x^2 - 8x = y - 18$$
$$x^2 - 8x + 16 = y - 18 + 16$$
$$(x-4)^2 = y - 2$$

Note that $4p = 1$, so $p = \frac{1}{4}$. The vertex is (4,2), the focus is $\left(4, \frac{9}{4}\right)$, the directrix is $y = \frac{7}{4}$, and the focal width is 1.

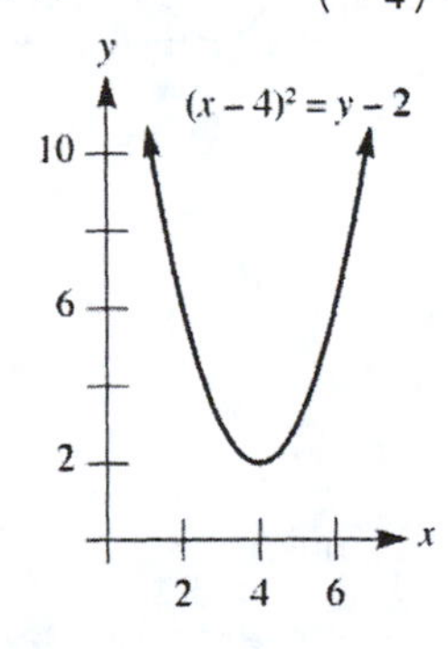

17. First complete the square:

$$\begin{aligned} y^2+2y-x+1&=0 \\ y^2+2y&=x-1 \\ y^2+2y+1&=x-1+1 \\ (y+1)^2&=x \end{aligned}$$

Note that $4p=1$, so $p=\frac{1}{4}$. The vertex is (0,–1), the focus is $\left(\frac{1}{4},-1\right)$, the directrix is $x=-\frac{1}{4}$, and the focal width is 1.

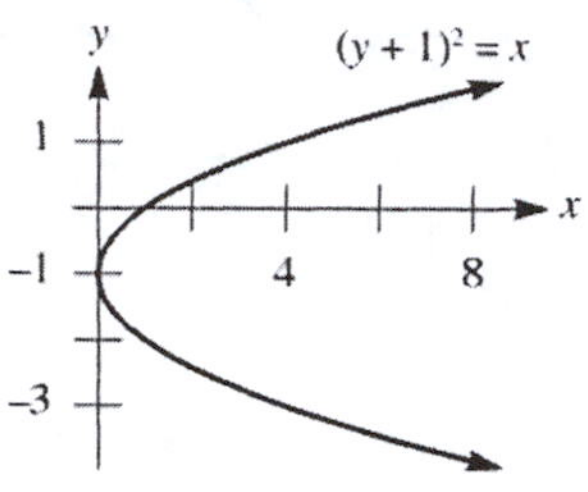

19. First complete the square:

$$\begin{aligned} 2x^2-12x-y+18&=0 \\ 2x^2-12x&=y-18 \\ x^2-6x&=\tfrac{1}{2}y-9 \\ x^2-6x+9&=\tfrac{1}{2}y-9+9 \\ (x-3)^2&=\tfrac{1}{2}y \end{aligned}$$

Note that $4p=\frac{1}{2}$, so $p=\frac{1}{8}$. The vertex is (3,0), the focus is $\left(3,\frac{1}{8}\right)$, the directrix is $y=-\frac{1}{8}$, and the focal width is $\frac{1}{2}$.

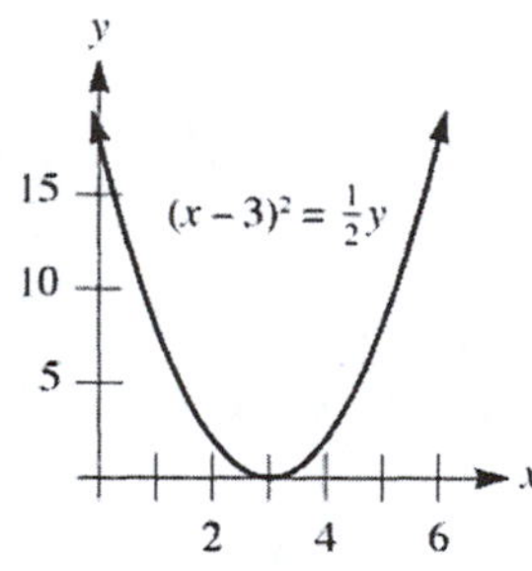

21. First complete the square:

$$\begin{aligned} 2x^2-16x-y+33&=0 \\ 2x^2-16x&=y-33 \\ 2(x-4)^2&=y-1 \\ (x-4)^2&=\tfrac{1}{2}(y-1) \end{aligned}$$

Note that $4p=\frac{1}{2}$, so $p=\frac{1}{8}$. The vertex is (4,1), the focus is $\left(4,\frac{9}{8}\right)$, the directrix is $y=\frac{7}{8}$, and the focal width is $\frac{1}{2}$.

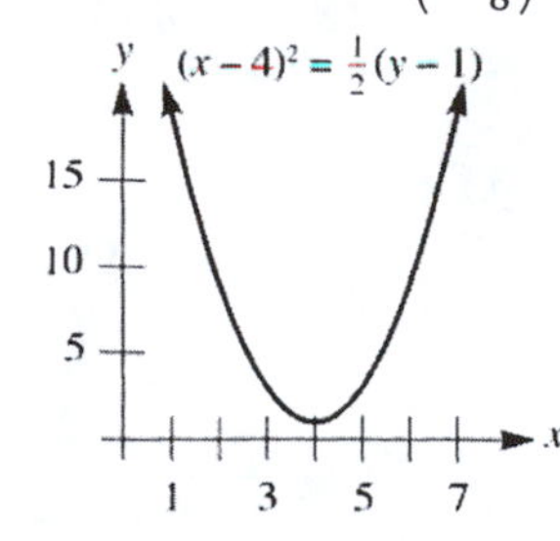

23. **a.** From the figure, the point (38.1,15.847) lies on the parabola $x^2 = 4py$. Substituting:

$$(38.1)^2 = 4p(15.847)$$
$$p = \frac{(38.1)^2}{4(15.847)} \approx 22.9 \text{ m}$$

b. The diameter of the dish is 76.2 m, so the focal ratio is: $\frac{p}{d} = \frac{22.9}{76.2} \approx 0.3$

25. **a.** Since the point (20,–15) lies on the parabola $x^2 = -4py$:

$$(20)^2 = -4p(-15)$$
$$400 = 60p$$
$$p = \frac{20}{3}$$

Thus the equation of the arch is $x^2 = -\frac{80}{3}y$.

b. Substituting $x = \pm 5$:

$$(\pm 5)^2 = -\frac{80}{3}y$$
$$25 = -\frac{80}{3}y$$
$$y = -\frac{15}{16}$$

The height of the arch is $14\frac{1}{16}$ feet.

27. The line of symmetry has equation $y = 1$. The figure is graphed as:

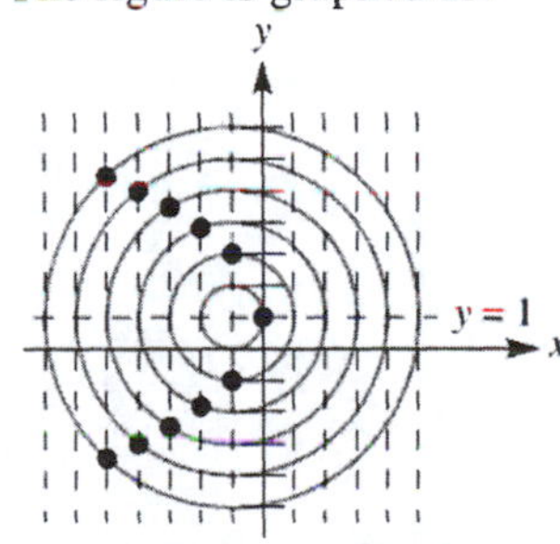

29. Given that the focus is at (0,3), we see that $p = 3$ and thus $4p = 12$. Since the parabola opens up, it has the $x^2 = 4py$ form, and its equation must be $x^2 = 12y$.

31. The directrix is $x = -32$, so (32,0) is the focus and the parabola opens to the right. So the parabola is of the form $y^2 = 4px$, where $p = 32$. Thus the equation is $y^2 = 128x$.

33. Since the parabola is symmetric about the x-axis, then $y^2 = \pm 4px$ is the form of the parabola. Since the x-coordinate of the focus is negative, then $y^2 = -4px$. Finally we are given $4p = 9$, so the equation is $y^2 = -9x$.

35. **a.** We have $4p = 2$, so $p = \frac{1}{2}$, and thus the focus is $\left(0, \frac{1}{2}\right)$. To find A', find the slope of the line containing A and the focus: $m = \frac{8 - \frac{1}{2}}{4 - 0} = \frac{\frac{15}{2}}{4} = \frac{15}{8}$. So $y = \frac{15}{8}x + \frac{1}{2}$ is its equation. Now substitute into the parabola:

$$x^2 = 2\left(\frac{15}{8}x + \frac{1}{2}\right)$$
$$x^2 = \frac{15}{4}x + 1$$
$$4x^2 = 15x + 4$$
$$4x^2 - 15x - 4 = 0$$
$$x = -\frac{1}{4}, 4$$
$$y = \frac{1}{32}, 8$$

So $A' = \left(-\frac{1}{4}, \frac{1}{32}\right)$. To find B', find the slope of the line containing B and the focus: $m = \dfrac{2-\frac{1}{2}}{-2-0} = \dfrac{\frac{3}{2}}{-2} = -\frac{3}{4}$

So $y = -\frac{3}{4}x + \frac{1}{2}$ is its equation. Now substitute into the parabola:

$$x^2 = 2\left(-\tfrac{3}{4}x + \tfrac{1}{2}\right)$$
$$x^2 = -\tfrac{3}{2}x + 1$$
$$2x^2 + 3x - 2 = 0$$
$$x = \tfrac{1}{2}, -2$$
$$y = \tfrac{1}{8}, 2$$

So $B' = \left(\frac{1}{2}, \frac{1}{8}\right)$. Now find the slope through A and B': $m = \dfrac{8-\frac{1}{8}}{4-\frac{1}{2}} = \dfrac{\frac{63}{8}}{\frac{7}{2}} = \frac{9}{4}$

Using (4,8) in the point-slope formula:

$$y - 8 = \tfrac{9}{4}(x-4)$$
$$y - 8 = \tfrac{9}{4}x - 9$$
$$y = \tfrac{9}{4}x - 1$$

b. Find the slope: $m = \dfrac{2-\frac{1}{32}}{-2+\frac{1}{4}} = \dfrac{\frac{63}{32}}{-\frac{7}{4}} = -\frac{9}{8}$. Using (–2,2) in the point-slope formula:

$$y - 2 = -\tfrac{9}{8}(x+2)$$
$$y - 2 = -\tfrac{9}{8}x - \tfrac{9}{4}$$
$$y = -\tfrac{9}{8}x - \tfrac{1}{4}$$

c. Setting the y-coordinates equal:

$$\tfrac{9}{4}x - 1 = -\tfrac{9}{8}x - \tfrac{1}{4}$$
$$18x - 8 = -9x - 2$$
$$27x = 6$$
$$x = \tfrac{2}{9}$$
$$y = \tfrac{1}{2} - 1 = -\tfrac{1}{2}$$

So the lines intersect at the point $\left(\frac{2}{9}, -\frac{1}{2}\right)$. Since the directrix is $y = -\frac{1}{2}$, then this point lies on the directrix.

37. a. Since $4p = 1$, $p = \frac{1}{4}$, thus the focus is $\left(0, \frac{1}{4}\right)$. Find the slope of the focal chord from P: $m = \dfrac{4-\frac{1}{4}}{2-0} = \dfrac{\frac{15}{4}}{2} = \frac{15}{8}$

So its equation is $y = \frac{15}{8}x + \frac{1}{4}$. Now find the intersection of this line with the parabola $y = x^2$:

$$x^2 = \tfrac{15}{8}x + \tfrac{1}{4}$$
$$8x^2 = 15x + 2$$
$$8x^2 - 15x - 2 = 0$$
$$(8x+1)(x-2) = 0$$
$$x = -\tfrac{1}{8}, 2$$
$$y = \tfrac{1}{64}, 4$$

So the coordinates of Q are $\left(-\frac{1}{8}, \frac{1}{64}\right)$.

b. Find the midpoint of $\overline{PQ}$: $M = \left(\dfrac{2-\frac{1}{8}}{2}, \dfrac{4+\frac{1}{64}}{2}\right) = \left(\frac{15}{16}, \frac{257}{128}\right)$

c. The coordinates of S are $\left(0, \frac{257}{128}\right)$. We can find T, since the slope of this line (which is perpendicular to $\overline{PQ}$) is $-\frac{8}{15}$, and thus by the point-slope formula:

$$y - \tfrac{257}{128} = -\tfrac{8}{15}\left(x - \tfrac{15}{16}\right)$$
$$y - \tfrac{257}{128} = -\tfrac{8}{15}x + \tfrac{1}{2}$$
$$y = -\tfrac{8}{15}x + \tfrac{321}{128}$$

When $x = 0$, $y = \frac{321}{128}$, so the coordinates of T are $\left(0, \frac{321}{128}\right)$. Thus the length of $\overline{ST}$ is $\frac{321}{128} - \frac{257}{128} = \frac{64}{128} = \frac{1}{2}$. Since the focal width is $4p = 1$, this verifies that ST is one-half the focal width.

39. The sketch in the book will have the following coordinates: $O(0,0)$, $B\left(x, \dfrac{x^2}{4p}\right)$, $A\left(-x, \dfrac{x^2}{4p}\right)$

We require that $AB = OB$. So: $2x = \sqrt{(x-0)^2 + \left(\dfrac{x^2}{4p} - 0\right)^2} = \sqrt{x^2 + \dfrac{x^4}{16p^2}}$. Squaring:

$$4x^2 = x^2 + \frac{x^4}{16p^2}$$
$$0 = -3x^2 + \frac{x^4}{16p^2}$$
$$0 = -48x^2p^2 + x^4$$
$$0 = x^2\left(-48p^2 + x^2\right)$$

So $x^2 = 0$ or $x = 0$ is one root and $x^2 = 48p^2$ or $x = \pm 4\sqrt{3}p$ is the other. The distance from A to B is therefore:

$2\left(4\sqrt{3}p\right) = 8\sqrt{3}p$ units

To find the area we note that this is an equilateral triangle. Half of it is a 30°- 60°- 90° triangle whose height will be $\sqrt{3}$ times its base. The height is therefore $\sqrt{3} \bullet 4\sqrt{3}p = 12p$. The area is: $\frac{1}{2}\left(8\sqrt{3}p\right)(12p) = 48\sqrt{3}p^2$ sq. units

41. **a.** The parabola passes through the points (2,4.5) and (4,0), so:

$$2^2 = -4p(4.5 - k) \qquad 4^2 = -4p(0 - k)$$
$$4 = -4p(4.5 - k) \qquad 16 = 4pk$$
$$p = -\frac{1}{4.5 - k} \qquad p = \frac{4}{k}$$

Setting these expressions for p equal:

$$-\frac{1}{4.5 - k} = \frac{4}{k}$$
$$-k = 18 - 4k$$
$$3k = 18$$
$$k = 6$$

The height of the arch is 6 m.

b. Since R is the focus of the parabola, and $p = \dfrac{4}{6} = \dfrac{2}{3}$, the focus is at the point $\left(0, 5\frac{1}{3}\right)$. The distance from R to the base is $5\frac{1}{3}$ m.

43. Since the focus is $(p,0)$, the slope of the focal chord is: $m = \dfrac{y_0 - 0}{\frac{y_0^2}{4p} - p} = \dfrac{4py_0}{y_0^2 - 4p^2}$

Using the point-slope formula, the equation of this focal chord is: $y - y_0 = m\left(x - \dfrac{y_0^2}{4p}\right)$

This line intersects the parabola when $x = \dfrac{y^2}{4p}$, so:

$$y - y_0 = m\left(\frac{y^2}{4p} - \frac{y_0^2}{4p}\right)$$
$$4p(y - y_0) = m(y + y_0)(y - y_0)$$
$$4p = m(y + y_0)$$

Substitute $m = \dfrac{4py_0}{y_0^2 - 4p^2}$:

$$4p = \frac{4py_0}{y_0^2 - 4p^2}(y + y_0)$$
$$\frac{y_0^2 - 4p^2}{y_0} = y + y_0$$
$$y = -\frac{4p^2}{y_0}$$

Therefore: $x = \dfrac{1}{4p}\left(-\dfrac{4p^2}{y_0}\right)^2 = \dfrac{1}{4p}\left(\dfrac{16p^4}{y_0^2}\right) = \dfrac{4p^3}{y_0^2}$

But $y_0^2 = 4px_0$, so: $x = \dfrac{4p^3}{4px_0} = \dfrac{p^2}{x_0}$. Thus the coordinates of Q are $\left(\dfrac{p^2}{x_0}, -\dfrac{4p^2}{y_0}\right)$.

45. Let P have coordinates (x_0, y_0). Then, from Exercise 43, Q has coordinates $\left(\dfrac{p^2}{x_0}, -\dfrac{4p^2}{y_0}\right)$.

The following results summarize the calculations required for this problem. Note: At each step we have replaced the quantity y_0^2 by $4px_0$. This simplifies matters a great deal.

Coordinates of M: $\left(\dfrac{x_0^2 + p^2}{2x_0}, \dfrac{2p(x_0 - p)}{y_0}\right)$ Coordinates of S: $\left(\dfrac{p^2 + x_0^2}{2x_0}, 0\right)$

Slope of $\overline{PQ}$: $\dfrac{4px_0}{y_0(x_0 - p)}$ Slope of perpendicular: $\dfrac{y_0(p - x_0)}{4px_0}$

Since $-\dfrac{ST}{MS}$ is the slope, m, of the line through M and T, we have:

$$ST = -(MS)m = -\frac{2p(x_0 - p)}{y_0} \bullet \frac{4px_0}{y_0(p - x_0)} = \frac{8p^2x_0}{y_0^2} = 2p$$

Thus ST is one-half of the focal width.

47. **a.** Since the focus is $(0, p)$, then the slope of the focal chord is: $m = \dfrac{\frac{x_0^2}{4p} - p}{x_0 - 0} = \dfrac{x_0^2 - 4p^2}{4px_0}$

The equation of the focal chord must be $y - p = mx$. This line intersects the parabola when $y = \dfrac{x^2}{4p}$, so we have:

$$\frac{x^2}{4p} - p = mx$$
$$\frac{x^2}{4p} - p = \frac{x_0^2 - 4p^2}{4px_0} \bullet x$$
$$x^2 x_0 - 4p^2 x_0 = xx_0^2 - 4p^2 x$$
$$xx_0\left(x - x_0\right) = -4p^2\left(x - x_0\right)$$
$$xx_0 = -4p^2$$
$$x = -\frac{4p^2}{x_0}$$

Therefore: $y = \dfrac{1}{4p}\left(-\dfrac{4p^2}{x_0}\right)^2 = \dfrac{1}{4p}\left(\dfrac{16p^4}{x_0^2}\right) = \dfrac{4p^3}{x_0^2}$

But $x_0^2 = 4py_0$, so: $y = \dfrac{4p^3}{4py_0} = \dfrac{p^2}{y_0}$. Thus the coordinates of Q are $\left(-\dfrac{4p^2}{x_0}, \dfrac{p^2}{y_0}\right)$.

b. By the midpoint formula: Midpoint of $PQ = \left(\dfrac{x_0 - \frac{4p^2}{x_0}}{2}, \dfrac{y_0 + \frac{p^2}{y_0}}{2}\right) = \left(\dfrac{x_0^2 - 4p^2}{2x_0}, \dfrac{y_0^2 + p^2}{2y_0}\right)$

c. The distance from P to the directrix is $y_0 + p$. Therefore $PF = y_0 + p$. The distance from Q to the directrix is $\dfrac{p^2}{y_0} + p$. So we have $QF = \dfrac{p^2}{y_0} + p$ and consequently:

$$PQ = QF + PF = \frac{p^2}{y_0} + p + y_0 + p = \frac{p^2 + 2py_0 + y_0^2}{y_0} = \frac{\left(p + y_0\right)^2}{y_0}$$

d. The distance from the center of the circle to the directrix is found by adding p to the y-coordinate of the midpoint of $\overline{PQ}$. This yields: $\dfrac{y_0^2 + p^2}{2y_0} + p = \dfrac{y_0^2 + p^2 + 2py_0}{2y_0} = \dfrac{\left(y_0 + p\right)^2}{2y_0}$

Comparing this result with the expression for PQ determined in part **c**, we conclude that the distance from the center of the circle to the directrix equals the radius of the circle. This implies that the circle is tangent to the directrix, as we wished to show.

11.3 Tangents to Parabolas (Optional Section)

1. Call m the slope of the ta ngent line, so: $y-4=m(x-2)$. Now substitute $y=x^2$:

$$x^2-4=m(x-2)$$
$$(x+2)(x-2)=m(x-2)$$
$$(x-2)(x+2-m)=0$$

So the solutions are $x = 2$ and $x = m - 2$. But since the line is tangent to the parabola, these two x-values must be equal:

$$m-2=2$$
$$m=4$$

Thus the tangent line is given by:

$$y-4=4(x-2)$$
$$y-4=4x-8$$
$$y=4x-4$$

Sketch the parabola and tangent line:

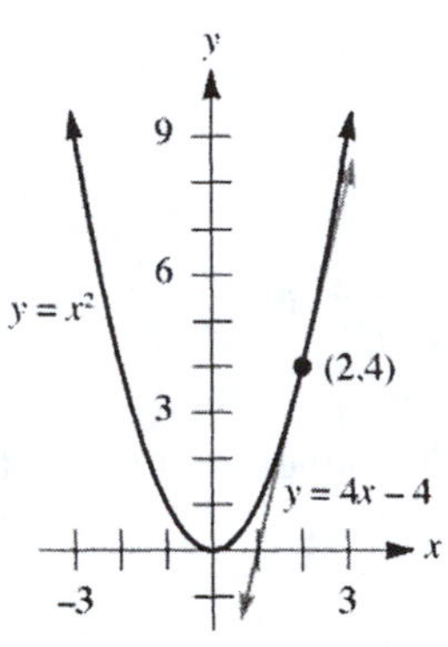

3. Call m the slope of the tangent line, so: $y-2=m(x-4)$. Solving $x^2=8y$ for y yields $y=\frac{x^2}{8}$, now substitute:

$$\frac{x^2}{8}-2=m(x-4)$$
$$x^2-16=8m(x-4)$$
$$(x+4)(x-4)=8m(x-4)$$
$$(x-4)(x+4-8m)=0$$

So the solutions are $x = 4$ and $x = 8m - 4$. But since the line is tangent to the parabola, these two x-values must be equal:

$$8m-4=4$$
$$8m=8$$
$$m=1$$

Thus the tangent line is given by:

$$y-2=1(x-4)$$
$$y-2=x-4$$
$$y=x-2$$

Sketch the parabola and tangent line:

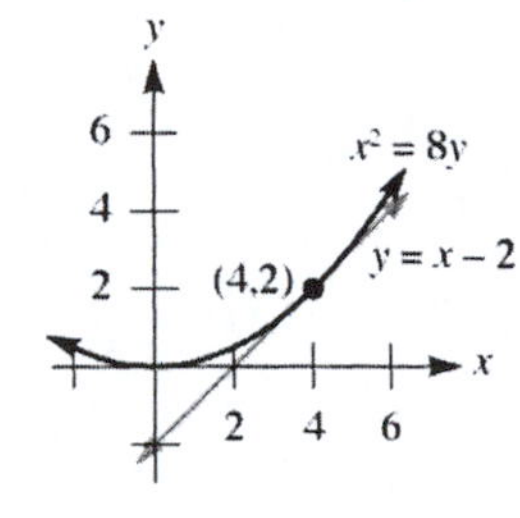

5. Call m the slope of the tangent line, so: $y+9=m(x+3)$. Solving $x^2=-y$ for y yields $y=-x^2$, now substitute:

$$\begin{aligned} -x^2+9&=m(x+3)\\ x^2-9&=-m(x+3)\\ (x+3)(x-3)&=-m(x+3)\\ (x+3)(x-3+m)&=0 \end{aligned}$$

So the solutions are $x=-3$ and $x=3-m$. But since the line is tangent to the parabola, these two x-values must be equal:

$$\begin{aligned} 3-m&=-3\\ -m&=-6\\ m&=6 \end{aligned}$$

Thus the tangent line is given by:

$$\begin{aligned} y+9&=6(x+3)\\ y+9&=6x+18\\ y&=6x+9 \end{aligned}$$

Sketch the parabola and tangent line:

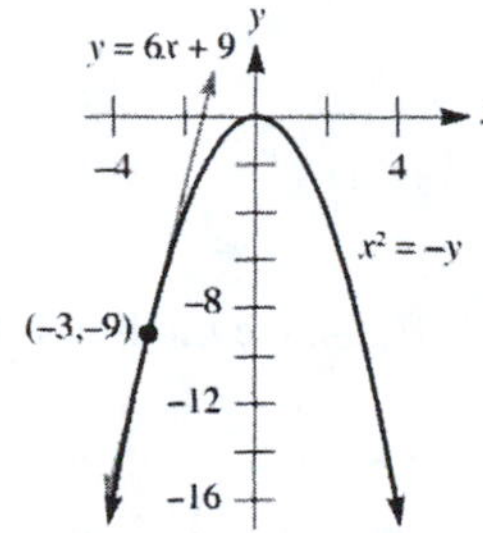

7. Call m the slope of the tangent line, so: $y-2=m(x-1)$. Solving $y^2=4x$ for x yields $x=\dfrac{y^2}{4}$, now substitute:

$$\begin{aligned} y-2&=m\left(\frac{y^2}{4}-1\right)\\ 4(y-2)&=m\left(y^2-4\right)\\ 4(y-2)&=m(y+2)(y-2)\\ (y-2)[4-m(y+2)]&=0 \end{aligned}$$

So the solutions are $y=2$ and $y=\dfrac{4}{m}-2$. But since the line is tangent to the parabola, these two x-values must be equal:

$$\begin{aligned} \frac{4}{m}-2&=2\\ \frac{4}{m}&=4\\ 4&=4m\\ 1&=m \end{aligned}$$

Thus the tangent line is given by:

$$\begin{aligned} y-2&=1(x-1)\\ y-2&=x-1\\ y&=x+1 \end{aligned}$$

Sketch the parabola and tangent line:

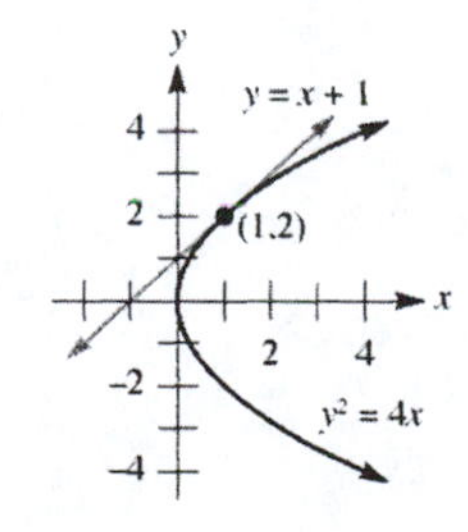

9. Let m be the slope, so $y - 2 = m(x - 4)$. Since $y = \sqrt{x}$, then $x = y^2$:

$$y-2=m\left(y^2-4\right)$$
$$y-2=m(y+2)(y-2)$$
$$0=(y-2)\left[m(y+2)-1\right]$$

So the solutions are $y = 2$ and $y=\frac{1}{m}-2$. But since the line is tangent to the parabola, these two x-values must be equal:

$$\frac{1}{m}-2=2$$
$$\frac{1}{m}=4$$
$$m=\tfrac{1}{4}$$

11. Let m be the slope, so $y - 8 = m(x - 2)$. Substitute $y = x^3$:

$$x^3-8=m(x-2)$$
$$(x-2)\left(x^2+2x+4\right)=m(x-2)$$
$$(x-2)\left(x^2+2x+4-m\right)=0$$

So the solutions are $x = 2$ and $m = x^2 + 2x + 4$. But since the line is tangent to the parabola, we can substitute $x = 2$ to obtain: $m = 2^2 + 4 + 4 = 12$

13. Let the equation of the tangent line be $y-y_0=m\left(x-x_0\right)$. Replacing y and y_0 by $\frac{x^2}{4p}$ and $\frac{x_0^2}{4p}$, respectively, we obtain:

$$\frac{x^2}{4p}-\frac{x_0^2}{4p}=m\left(x-x_0\right)$$
$$x^2-x_0^2-4pm\left(x-x_0\right)=0$$
$$\left(x-x_0\right)\left[\left(x+x_0\right)-4pm\right]=0$$
$$\left(x-x_0\right)\left(x+x_0-4pm\right)=0$$

Setting the second factor equal to zero yields $x=4pm-x_0$. But since $x=x_0$ is known to be the unique solution, we have $x_0=4pm-x_0$. Thus $m=\frac{2x_0}{4p}=\frac{x_0}{2p}$. The equation of the tangent line is therefore:

$$y-y_0=\frac{x_0}{2p}\left(x-x_0\right)$$
$$y=\frac{x_0}{2p}x-\frac{x_0^2}{2p}+y_0$$

Now if in this last equation we replace x_0^2 by $4py_0$, we obtain $y=\frac{x_0}{2p}x-y_0$, as required.

15. **a.** Let the coordinates of P be (a, b), so that $a^2 = 4pb$. Then according to Exercise 42 of the previous section, the coordinates of Q are $\left(-\frac{4p^2}{a}, \frac{p^2}{b}\right)$. The slope of the tangent at P is $\frac{a}{2p}$. The slope of the tangent at Q is $\frac{-\frac{4p^2}{a}}{2p} = -\frac{2p}{a}$. Thus the two slopes are negative reciprocals, from which it follows that the tangents are perpendicular to one another.

b. The equation of the tangent at P is $y = \frac{a}{2p}x - b$. The equation of the tangent at Q is found to be $y = -\frac{2p}{a}x - \frac{p^2}{b}$. By solving this system of two equations, we find that the coordinates of the point of intersection are $x = \frac{a(b-p)}{2b}$ and $y = -p$. Since the y-value here is $-p$, we conclude that the two tangents intersect at a point on the directrix.

c. The coordinates of the intersection point D are given in the solution to part **b**. Using those coordinates, the slope of the line from D to the focus is found to be $\frac{4bp}{a(p-b)}$. On the other hand, the slope of $\overline{PQ}$ is found to be $\frac{a(b-p)}{4bp}$. (***Note***: In computing the slope of $\overline{PQ}$, the relationship $a^2 = 4pb$ helps in simplifying the expressions that arise.) Now by inspection we see that the two expressions for slope are negative reciprocals, thus the lines are perpendicular.

17. Summarizing our calculations, we have:

Equation of normal: $y - y_0 = -\frac{y_0}{2p}(x - x_0)$ Coordinates of A: $(2p + x_0, 0)$

Equation of line $\overline{AZ}$: $y = \frac{p - x_0}{y_0}(x - 2p - x_0)$ Equation of line $\overline{PF}$: $y = \frac{y_0}{x_0 - p}(x - p)$

Coordinates of Z: $x = \frac{2p^3 + p^2x_0 + x_0^3}{(x_0 + p)^2} = \frac{2p^2 - px_0 + x_0^2}{x_0 + p}$, $y = \frac{y_0(x_0 - p)}{x_0 + p}$

With these coordinates for Z, the distance ZP can be determined by using the distance formula. The result is $ZP = 2p$, as indicated in the problem.

19. **a.** The x-intercept of the tangent line is $-x_0$. Therefore $FA = x_0 + p$. Next, according to the definition of a parabola, FP is equal to the distance from P to the directrix. But the distance from P to the directrix is $x_0 + p$. Consequently $FP = FA$ because both equal $x_0 + p$.

b. Since $AF = FP$, ΔAFP is isosceles, and consequently $\alpha = \gamma$. Also, since HP is parallel to the x-axis, we have $\beta = \gamma$.

c. We have $\alpha = \gamma = \beta$ and therefore $\alpha = \beta$, as required.

11.4 The Ellipse

1. Dividing by 36, the standard form is $\frac{x^2}{9}+\frac{y^2}{4}=1$. The length of the major axis is 6 and the length of the minor axis is 4. Using $c^2=a^2-b^2$: $c=\sqrt{a^2-b^2}=\sqrt{9-4}=\sqrt{5}$. So the foci are $\left(\pm\sqrt{5},0\right)$ and the eccentricity is $\frac{\sqrt{5}}{3}$.

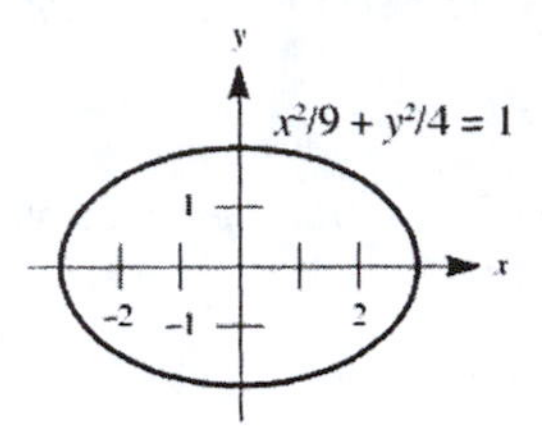

3. Dividing by 16, the standard form is $\frac{x^2}{16}+\frac{y^2}{1}=1$. The length of the major axis is 8 and the length of the minor axis is 2. Using $c^2=a^2-b^2$: $c=\sqrt{a^2-b^2}=\sqrt{16-1}=\sqrt{15}$. So the foci are $\left(\pm\sqrt{15},0\right)$ and the eccentricity is $\frac{\sqrt{15}}{4}$.

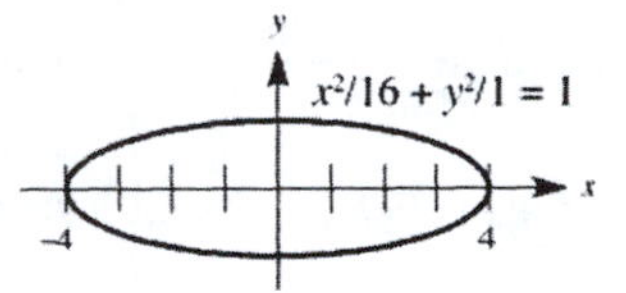

5. Dividing by 2, the standard form is $\frac{x^2}{2}+\frac{y^2}{1}=1$. The length of the major axis is $2\sqrt{2}$ and the length of the minor axis is 2. Using $c^2=a^2-b^2$: $c=\sqrt{a^2-b^2}=\sqrt{2-1}=1$. So the foci are $(\pm 1,0)$ and the eccentricity is $\frac{\sqrt{2}}{2}$.

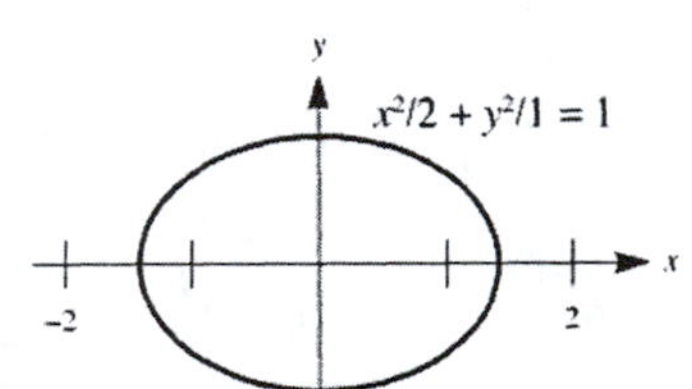

7. Dividing by 144, the standard form is $\frac{x^2}{9}+\frac{y^2}{16}=1$. The length of the major axis is 8 and the length of the minor axis is 6. Using $c^2=a^2-b^2$: $c=\sqrt{a^2-b^2}=\sqrt{16-9}=\sqrt{7}$. So the foci are $\left(0,\pm\sqrt{7}\right)$ and the eccentricity is $\frac{\sqrt{7}}{4}$.

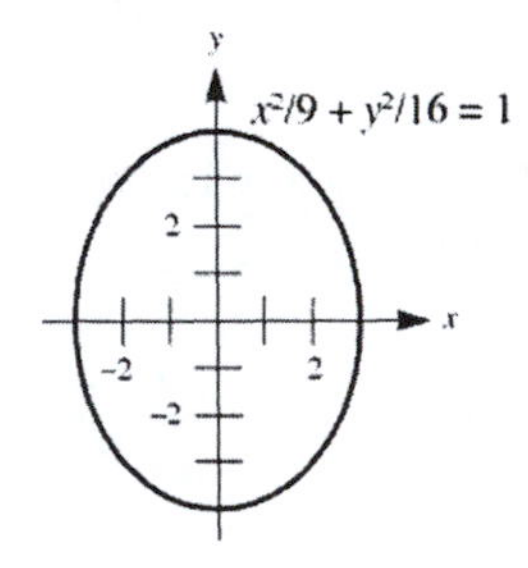

9. Dividing by 5, the standard form is $\frac{x^2}{1/3}+\frac{y^2}{5/3}=1$. The lengths of the major and minor axes are given by:

major axis : $2\sqrt{\frac{5}{3}}=\frac{2\sqrt{5}}{\sqrt{3}}=\frac{2\sqrt{15}}{3}$ minor axis : $2\sqrt{\frac{1}{3}}=\frac{2}{\sqrt{3}}=\frac{2\sqrt{3}}{3}$

Using $c^2=a^2-b^2$: $c=\sqrt{\frac{5}{3}-\frac{1}{3}}=\sqrt{\frac{4}{3}}=\frac{2}{\sqrt{3}}=\frac{2\sqrt{3}}{3}$. So the foci are $\left(0,\pm\frac{2\sqrt{3}}{3}\right)$.

The eccentricity is given by: $e=\frac{\frac{2\sqrt{3}}{3}}{\frac{\sqrt{15}}{3}}=\frac{2\sqrt{3}}{\sqrt{15}}=\frac{2}{\sqrt{5}}=\frac{2\sqrt{5}}{5}$

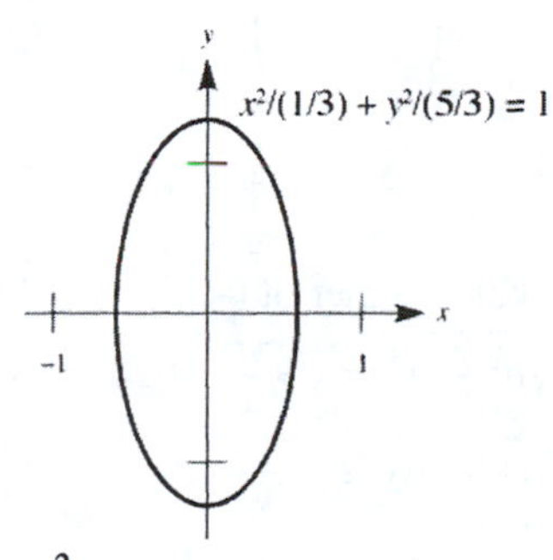

11. Dividing by 4, the standard form is $\frac{x^2}{2}+\frac{y^2}{4}=1$. The length of the major axis is 4 and the length of the minor axis is $2\sqrt{2}$. Using $c^2=a^2-b^2$: $c=\sqrt{a^2-b^2}=\sqrt{4-2}=\sqrt{2}$. So the foci are $\left(0,\pm\sqrt{2}\right)$ and the eccentricity is $\frac{\sqrt{2}}{2}$.

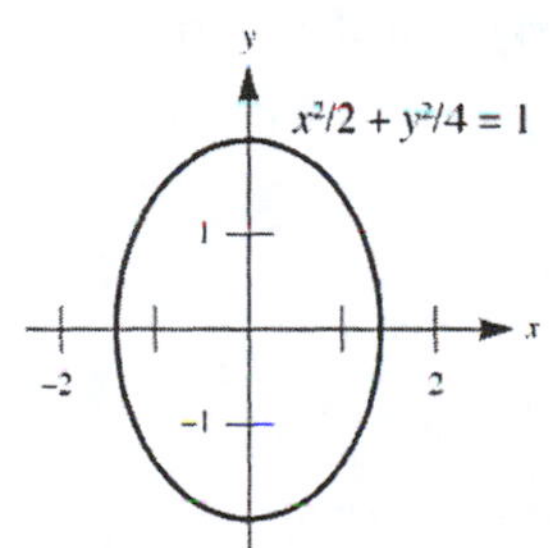

13. The equation is already in standard form with a center of (5,–1). The length of the major axis is 10 and the length of the minor axis is 6. Using $c^2=a^2-b^2$: $c=\sqrt{a^2-b^2}=\sqrt{25-9}=\sqrt{16}=4$

So the foci are $(5+4,-1)=(9,-1)$ and $(5-4,-1)=(1,-1)$, and the eccentricity is $\frac{4}{5}$.

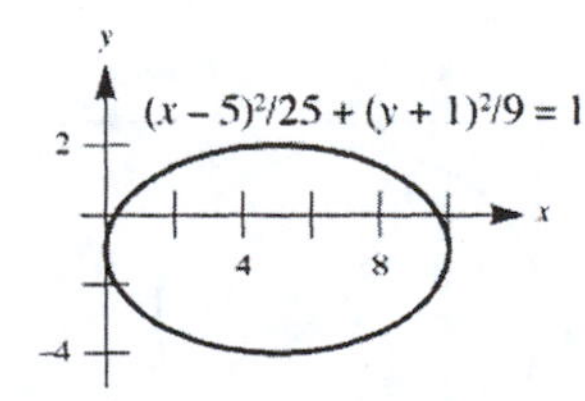

15. The equation is already in standard form with a center of (1,2). The length of the major axis is 4 and the length of the minor axis is 2. Using $c^2 = a^2 - b^2$: $c = \sqrt{a^2 - b^2} = \sqrt{4-1} = \sqrt{3}$

So the foci are $\left(1, 2 \pm \sqrt{3}\right)$ and the eccentricity is $\frac{\sqrt{3}}{2}$.

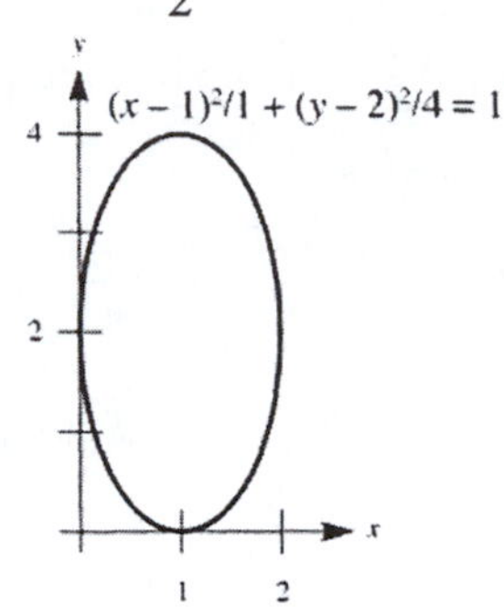

17. The equation is already in standard form with a center of (–3,0). The length of the major axis is 6 and the length of the minor axis is 2. Using $c^2 = a^2 - b^2$: $c = \sqrt{a^2 - b^2} = \sqrt{9-1} = \sqrt{8} = 2\sqrt{2}$

So the foci are $\left(-3 \pm 2\sqrt{2}, 0\right)$ and the eccentricity is $\frac{2\sqrt{2}}{3}$.

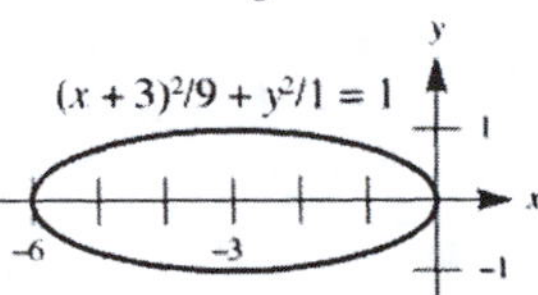

19. Complete the square to convert the equation to standard form:

$$3x^2 + 4y^2 - 6x + 16y + 7 = 0$$
$$3\left(x^2 - 2x\right) + 4\left(y^2 + 4y\right) = -7$$
$$3\left(x^2 - 2x + 1\right) + 4\left(y^2 + 4y + 4\right) = -7 + 3 + 16$$
$$3(x-1)^2 + 4(y+2)^2 = 12$$
$$\frac{(x-1)^2}{4} + \frac{(y+2)^2}{3} = 1$$

The center is (1,–2), the length of the major axis is 4, and the length of the minor axis is $2\sqrt{3}$. Using $c^2 = a^2 - b^2$, we find: $c = \sqrt{a^2 - b^2} = \sqrt{4-3} = 1$. So the foci are (1 + 1,–2) = (2,–2) and (1 – 1,–2) = (0,–2), and the eccentricity is $\frac{1}{2}$.

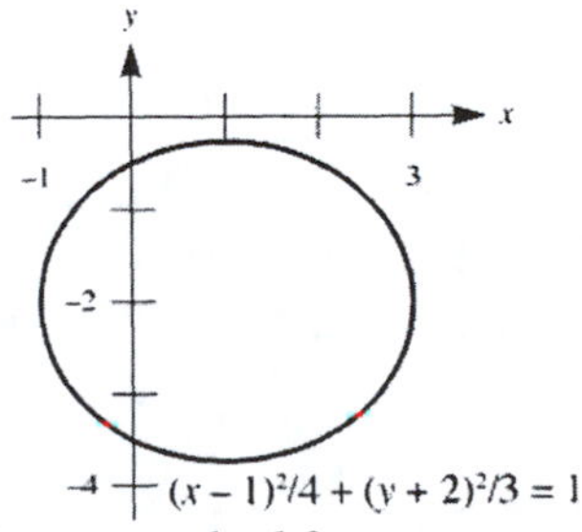

21. Complete the square to convert the equation to standard form:

$$5x^2 + 3y^2 - 40x - 36y + 188 = 0$$
$$5\left(x^2 - 8x\right) + 3\left(y^2 - 12y\right) = -188$$
$$5\left(x^2 - 8x + 16\right) + 3\left(y^2 - 12y + 36\right) = -188 + 80 + 108$$
$$5(x-4)^2 + 3(y-6)^2 = 0$$

Notice that the only solution to this is the center (4,6). This is called a degenerate ellipse, or, more commonly, a point!

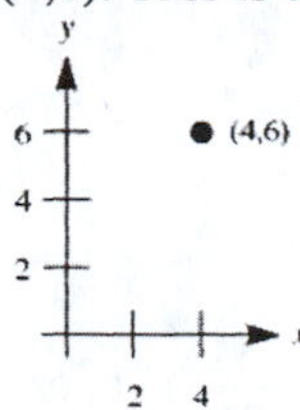

23. Complete the square to convert the equation to standard form:

$$16x^2+25y^2-64x-100y+564=0$$
$$16(x^2-4x)+25(y^2-4y)=-564$$
$$16(x^2-4x+4)+25(y^2-4y+4)=-564+64+100$$
$$16(x-2)^2+25(y-2)^2=-400$$

Notice that there is no solution to this equation, since the left-hand side is non-negative. So there is no graph.

25. We are given $c=3$ and $a=5$, so we have the equation in the form: $\frac{x^2}{5^2}+\frac{y^2}{b^2}=1$. Since $c^2=a^2-b^2$, we find b:

$$9=25-b^2$$
$$b^2=16$$
$$b=4$$

So the equation is $\frac{x^2}{25}+\frac{y^2}{16}=1$, or $16x^2+25y^2=400$.

27. We are given $a=4$, so the equation has a form of: $\frac{x^2}{16}+\frac{y^2}{b^2}=1$. Now $\frac{c}{a}=\frac{1}{4}$, so $\frac{c}{4}=\frac{1}{4}$ and thus $c=1$. We find b:

$$c^2=a^2-b^2$$
$$1=16-b^2$$
$$b^2=15$$

So the equation is $\frac{x^2}{16}+\frac{y^2}{15}=1$, or $15x^2+16y^2=240$.

29. We have $c=2$ and $a=5$, so the equation has a form of: $\frac{x^2}{b^2}+\frac{y^2}{25}=1$. Now find b:

$$c^2=a^2-b^2$$
$$4=25-b^2$$
$$b^2=21$$

So the equation is $\frac{x^2}{21}+\frac{y^2}{25}=1$, or $25x^2+21y^2=525$.

31. We know $a=2b$ and that the equation has a form of: $\frac{x^2}{a^2}+\frac{y^2}{b^2}=1$. Using the point $(1,\sqrt{2})$ and $a=2b$:

$$\frac{(1)^2}{(2b)^2}+\frac{(\sqrt{2})^2}{b^2}=1$$
$$\frac{1}{4b^2}+\frac{2}{b^2}=1$$

Multiply by $4b^2$:

$$1+8=4b^2$$
$$b^2=\frac{9}{4}$$
$$b=\frac{3}{2}$$

Since $a=2b$, then $a=3$. So the equation is $\frac{x^2}{9}+\frac{y^2}{9/4}=1$, or $x^2+4y^2=9$.

33. **a.** Using the equation $\frac{x_1x}{a^2}+\frac{y_1y}{b^2}=1$, where $(x_1,y_1)=(8,2)$, $a^2=76$, $b^2=\frac{76}{3}$:

$$\frac{8x}{76}+\frac{2y}{76/3}=1$$
$$8x+6y=76$$
$$6y=-8x+76$$
$$y=-\frac{4}{3}x+\frac{38}{3}$$

b. Here $(x_1,y_1)=(-7,3)$, $a^2=76$, $b^2=\frac{76}{3}$:

$$-\frac{7x}{76}+\frac{3y}{76/3}=1$$
$$-7x+9y=76$$
$$9y=7x+76$$
$$y=\frac{7}{9}x+\frac{76}{9}$$

c. Here $(x_1,y_1)=(1,-5)$, $a^2=76$, $b^2=\frac{76}{3}$:

$$\frac{x}{76}-\frac{5y}{76/3}=1$$
$$x-15y=76$$
$$-15y=-x+76$$
$$y=\frac{1}{15}x-\frac{76}{15}$$

35. **a.** Using the tangent formula where $(x_1,y_1)=(4,2)$, $a^2=\frac{52}{3}$, $b^2=52$:

$$\frac{4x}{52/3}+\frac{2y}{52}=1$$
$$12x+2y=52$$
$$2y=-12x+52$$
$$y=-6x+26$$

b. The y-intercept is 26 and the x-intercept is $\frac{13}{3}$, so the area is: $\frac{1}{2}(\text{base})(\text{height})=\frac{1}{2}\bullet\frac{13}{3}\bullet 26=\frac{169}{3}$

37. First find c: $c=ae=(39.44)(0.2484)\approx 9.80$

At perihelion, the distance is: $a-c=39.44-9.80=29.64$ AU

At aphelion, the distance is: $a+c=39.44+9.80=49.24$ AU

39. Using the expressions for perihelion and aphelion, the system of equations is:

$$a+c=2.288$$
$$a-c=2.132$$

Adding yields:

$$2a=4.42$$
$$a=2.21$$

Subtracting yields:

$$2c=0.156$$
$$c=0.078$$

Now find b: $b=\sqrt{a^2-c^2}=\sqrt{2.21^2-0.078^2}\approx 2.209$

The semimajor axis has length $2b\approx 4.42$ AU and the eccentricity is $e=\frac{0.078}{2.21}\approx 0.04$.

41. **a.** Substituting $(-x, y)$ and $(x, -y)$ both result in the same equation, to the graph is symmetric about both the x-axis and the y-axis.

b. Solving for y:

$$\frac{x^2}{9} + \frac{y^2}{4} = 1$$
$$\frac{y^2}{4} = 1 - \frac{x^2}{9}$$
$$\frac{y^2}{4} = \frac{9 - x^2}{9}$$
$$y^2 = \frac{36 - 4x^2}{9}$$
$$y = \pm\sqrt{\frac{36 - 4x^2}{9}} = \pm\frac{2}{3}\sqrt{9 - x^2}$$

c. For the domain, we must solve:

$$36 - 4x^2 \geq 0$$
$$9 - x^2 \geq 0$$
$$(3 + x)(3 - x) \geq 0$$

Using a sign chart, the domain is the interval [–3,3].

d. Use a calculator to complete the table:

x	0	0.5	1.0	1.5	2.0	2.5	3.0
y	±2	±1.97	±1.89	±1.73	±1.49	±1.11	0

e. Sketching the graph:

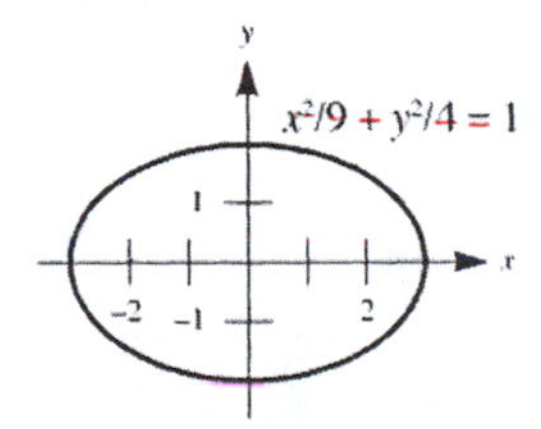

43. The focus of $y^2 = 4x$ is (1,0), which is the center of the circle. Since the circle passes through the origin, $r = 1$, so the equation of the circle is $(x-1)^2 + y^2 = 1$. For the ellipse, its equation has the form $\frac{(x-h)^2}{a^2} + \frac{y^2}{b^2} = 1$. Since the ellipse passes through the origin, $h = a$, so the equation has the form $\frac{(x-a)^2}{a^2} + \frac{y^2}{b^2} = 1$. Since $e = \frac{c}{a} = \frac{3}{4}$, $c = \frac{3}{4}a$. Also, since (1,0) is a focus of the ellipse, $a - c = 1$, so:

$$a - \frac{3}{4}a = 1$$
$$\frac{1}{4}a = 1$$
$$a = 4, c = 3$$

Finally, find b: $b = \sqrt{a^2 - c^2} = \sqrt{16 - 9} = \sqrt{7}$. So the equation of the ellipse is $\frac{(x-4)^2}{16} + \frac{y^2}{7} = 1$.

45. From Figure 4, we have $F_1P + F_2P > F_1F_2$. But by definition, $F_1P + F_2P = 2a$, while $F_1F_2 = 2c$. Thus $2a > 2c$, or $a > c$, so $a^2 > c^2$ (since a and c are positive), and consequently $a^2 - c^2 > 0$.

47. The tangent at (x_1, y_1) has equation $\frac{x_1 x}{a^2} + \frac{y_1 y}{b^2} = 1$, which has slope $\frac{-x_1 b^2}{y_1 a^2}$. Thus the normal has slope $\frac{y_1 a^2}{x_1 b^2}$ and equation:

$$y - y_1 = \frac{y_1 a^2}{x_1 b^2}(x - x_1)$$
$$b^2 x_1 y - b^2 x_1 y_1 = y_1 a^2 x - a^2 x_1 y_1$$
$$a^2 y_1 x - b^2 x_1 y = (a^2 - b^2) x_1 y_1$$

49. If we first multiply each equation by $a^2 b^2$:

$$b^2 x^2 + a^2 y^2 = a^2 b^2$$
$$a^2 x^2 + b^2 y^2 = a^2 b^2$$

Multiplying the first equation by $-a^2$ and the second equation by b^2 yields:

$$-a^2 b^2 x^2 - a^4 y^2 = -a^4 b^2$$
$$a^2 b^2 x^2 + b^4 y^2 = a^2 b^4$$

Adding:

$$(b^4 - a^4) y^2 = a^2 b^2 (b^2 - a^2)$$
$$y^2 = \frac{a^2 b^2}{b^2 + a^2}$$
$$y = \pm \frac{ab}{\sqrt{a^2 + b^2}}$$

Substituting into the first equation:

$$b^2 x^2 + \frac{a^4 b^2}{a^2 + b^2} = a^2 b^2$$
$$b^2 x^2 = \frac{a^2 b^4}{a^2 + b^2}$$
$$x^2 = \frac{a^2 b^2}{a^2 + b^2}$$
$$x = \pm \frac{ab}{\sqrt{a^2 + b^2}}$$

So there are four intersection points: $\left(\frac{ab}{A}, \frac{ab}{A}\right), \left(\frac{ab}{A}, -\frac{ab}{A}\right), \left(-\frac{ab}{A}, \frac{ab}{A}\right), \left(-\frac{ab}{A}, -\frac{ab}{A}\right)$, where $A = \sqrt{a^2 + b^2}$

Sketch the intersection:

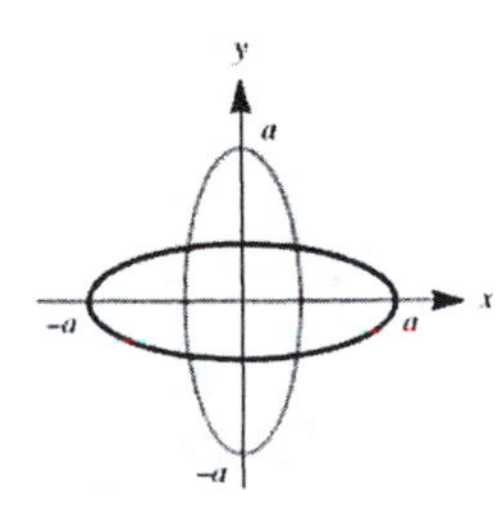

51. Using Exercise 48, we find that N is the point $\left(\frac{(a^2-b^2)x_1}{a^2},0\right)$. Thus:

$$FN=\frac{(a^2-b^2)x_1}{a^2}+c=\frac{c^2}{a^2}x_1+c=e^2x_1+c$$

It can be shown using $\frac{x_1^2}{a^2}+\frac{y_1^2}{b^2}=1$ that $y_1=a^2-c^2-x_1^2+e^2x_1^2$. Now:

$$FP=\sqrt{(x_1+c)^2+y_1^2}=\sqrt{x_1^2+2x_1c+c^2+a^2-c^2-x_1^2+e^2x_1^2}=\sqrt{e^2x_1^2+2aex_1+a^2}=ex_1+a$$

Thus: $\frac{FN}{FP}=\frac{e^2x_1+c}{ex_1+a}=\frac{e(ex_1+a)}{ex_1+a}=e$

53. **a.** Substitute the points:

$$5^2+3(1)^2=25+3=28$$
$$4^2+3(-2)^2=16+12=28$$
$$(-1)^2+3(3)^2=1+27=28$$

b. $\overline{AB}$ has slope $\frac{-2-1}{4-5}=3$. The line containing C parallel to $\overline{AB}$ has equation:

$$y-3=3(x+1)$$
$$y=3x+6$$

The intersections of this line with the ellipse have x-coordinates such that:

$$x^2+3(3x+6)^2=28$$
$$x^2+27x^2+108x+108=28$$
$$7x^2+27x+20=0$$
$$(7x+20)(x+1)=0$$
$$x=-\tfrac{20}{7},-1$$

Thus the point D we want is $\left(-\frac{20}{7},-\frac{18}{7}\right)$.

c. The point O is (0,0). Use the suggested formula:

$$\text{Area of }\Delta OAC=\tfrac{1}{2}\left|0-0+5\bullet 3-(-1)(1)+0-0\right|=8$$
$$\text{Area of }\Delta OBD=\tfrac{1}{2}\left|0-0+4\left(-\tfrac{18}{7}\right)-(-2)\left(-\tfrac{20}{7}\right)+0-0\right|=8$$

Thus the areas are equal.

55. **a.** We have $a\sqrt{(x+c)^2+y^2}=a^2+xc$. Dividing by a and noting that $e=\frac{c}{a}$: $\sqrt{(x+c)^2+y^2}=a+xe$

But the radical is F_1P by the distance formula, so: $F_1P=a+xe$

b. We have:

$$F_1P+F_2P=2a$$
$$(a+xe)+F_2P=2a$$
$$F_2P=a-xe$$

57. **a.** Use the tangent formula, and multiply by a^2b^2:

$$\frac{x_1x}{a^2}+\frac{y_1y}{b^2}=1$$
$$b^2x_1x+a^2y_1y=a^2b^2$$

b. Since (h, k) lies on this line, replacing (x, y) with (h, k) results in: $b^2x_1h+a^2y_1k=a^2b^2$

c. Repeating part **a**, we have $b^2x_2x+a^2y_2y=a^2b^2$. Now substitute (h, k) to obtain $b^2x_2h+a^2y_2k=a^2b^2$.

d. Replacing (x, y) with (x_1, y_1) and (x_2, y_2), respectively, results in the equations we have proved in parts **b** and **c**. Thus this line must pass through the points (x_1, y_1) and (x_2, y_2).

59. **a.** Dividing by 12, graph the ellipse $\frac{x^2}{12}+\frac{y^2}{4}=1$:

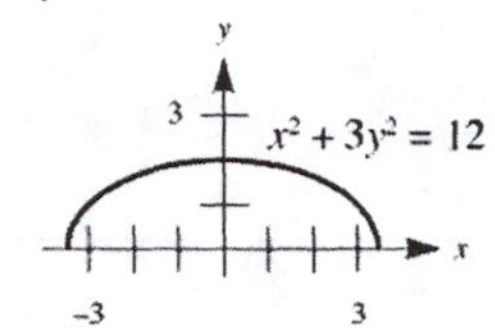

b. Since $a^2 = 12$ and $b^2 = 4$, $a = 2\sqrt{3}$ and $b = 2$. Therefore: $c^2 = a^2 - b^2 = 12 - 4 = 8$, so $c = \sqrt{8} = 2\sqrt{2}$

c. The auxiliary circle has an equation $x^2 + y^2 = 12$. Graphing the ellipse and auxiliary circle:

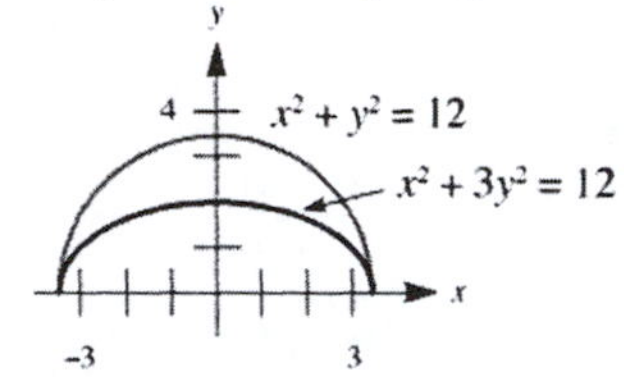

d. Verify that the point $P(3,1)$ lies on the ellipse: $(3)^2 + 3(1)^2 = 9 + 3 = 12$
Using $(x_0, y_0) = (3,1)$, find the equation of the tangent line to the ellipse at P:

$$\frac{3x}{12}+\frac{1y}{4}=1$$
$$\frac{x}{4}+\frac{y}{4}=1$$
$$x+y=4$$

e. Graph the upper halves of the ellipse and circle, and the line $y = -x + 4$:

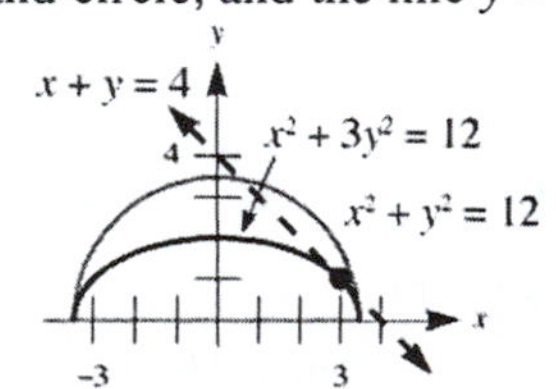

f. The perpendicular to the tangent at $P(3,1)$ has slope $m = 1$. Using the point $(-2\sqrt{2}, 0)$ and the point-slope formula:

$$y-0=1\left(x+2\sqrt{2}\right)$$
$$y=x+2\sqrt{2}$$

Now graph the upper halves of the ellipse and circle, as well as the lines $y = -x + 4$ and $y = x + 2\sqrt{2}$:

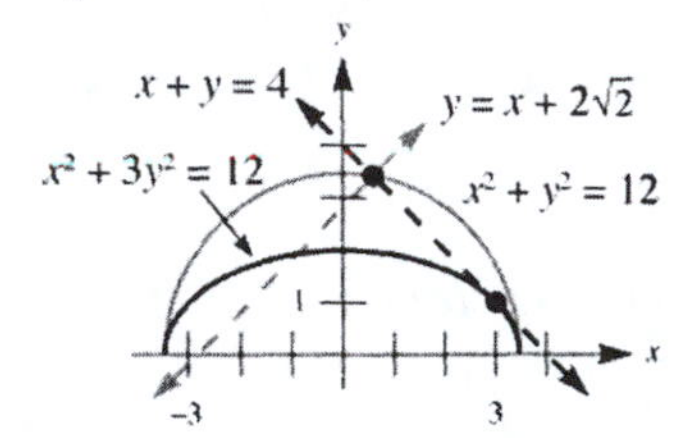

61. a. Multiplying by a^2b^2: $b^2x^2 + a^2y^2 = a^2b^2$

Substituting (x_1, y_1) for (x, y) yields: $b^2x_1^2 + a^2y_1^2 = a^2b^2$

b. Combining b^2 and a^2 terms results in: $b^2(x^2 - x_1^2) + a^2(y^2 - y_1^2) = 0$

c. We know (x_1, y_1) satisfies equation (4), and since it also satisfies equation (5), (x_1, y_1) is a solution to the system. Since (5) represents a tangent line, by definition there must be only one solution.

d. Solving for y yields $y = m(x - x_1) + y_1$, now substitute:

$$b^2(x^2 - x_1^2) + a^2\left[m^2(x - x_1)^2 + 2my_1(x - x_1) + y_1^2 - y_1^2\right] = 0$$
$$b^2(x^2 - x_1^2) + a^2m^2(x - x_1)^2 + 2a^2my_1(x - x_1) = 0$$

e. Factoring out $x - x_1$: $(x - x_1)\left[b^2(x + x_1) + a^2m^2(x - x_1) + 2a^2my_1\right] = 0$

f. Replacing x with x_1 in the brackets:

$$b^2(x_1 + x_1) + a^2m^2(x_1 - x_1) + 2a^2my_1 = 0$$
$$2b^2x_1 + 2a^2my_1 = 0$$
$$2a^2my_1 = -2b^2x_1$$
$$m = -\frac{b^2x_1}{a^2y_1}$$

g. Equation (5) becomes:

$$y - y_1 = -\frac{b^2x_1}{a^2y_1}(x - x_1)$$
$$a^2y_1y - a^2y_1^2 = -b^2x_1x + b^2x_1^2$$
$$b^2x_1x + a^2y_1y = b^2x_1^2 + a^2y_1^2$$

h. Using equation (2): $b^2x_1x + a^2y_1y = a^2b^2$

Dividing by a^2b^2: $\frac{x_1x}{a^2} + \frac{y_1y}{b^2} = 1$

11.5 The Hyperbola

1. Dividing by 4, the standard form is $\frac{x^2}{4} - \frac{y^2}{1} = 1$. The vertices are $(\pm 2, 0)$, the length of the transverse axis is 4, the length of the conjugate axis is 2, and the asymptotes are $y = \pm\frac{1}{2}x$. Using $c^2 = a^2 + b^2$: $c = \sqrt{a^2 + b^2} = \sqrt{4+1} = \sqrt{5}$

The foci are $(\pm\sqrt{5}, 0)$ and the eccentricity is $\frac{\sqrt{5}}{2}$.

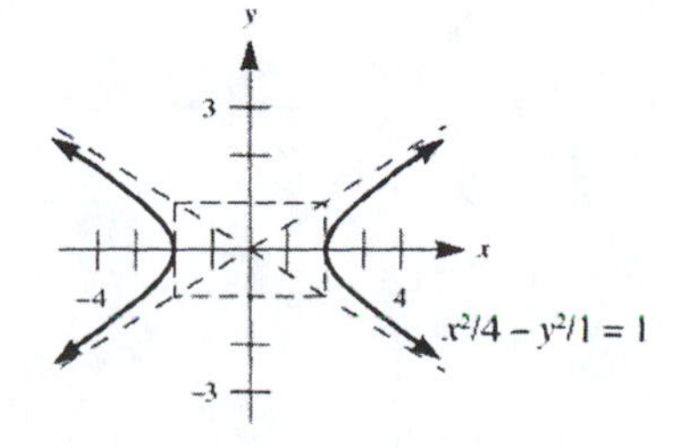

3. Dividing by 4, the standard form is $\frac{y^2}{4} - \frac{x^2}{1} = 1$. The vertices are $(0,\pm 2)$, the length of the transverse axis is 4, the length of the conjugate axis is 2, and the asymptotes are $y = \pm 2x$. Using $c^2 = a^2 + b^2$: $c = \sqrt{a^2 + b^2} = \sqrt{4+1} = \sqrt{5}$

The foci are $\left(0, \pm\sqrt{5}\right)$ and the eccentricity is $\frac{\sqrt{5}}{2}$.

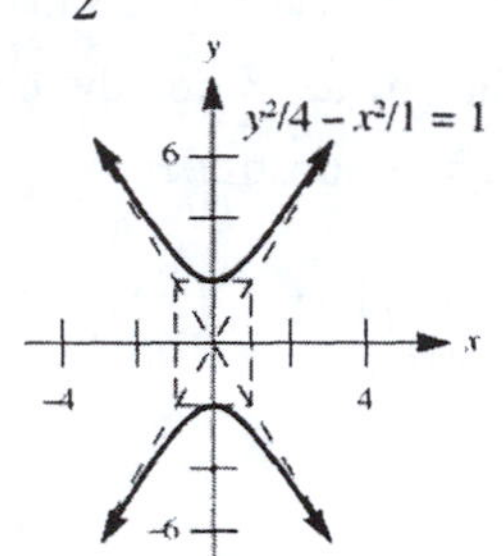

5. Dividing by 400, the standard form is $\frac{x^2}{25} - \frac{y^2}{16} = 1$. The vertices are $(\pm 5,0)$, the length of the transverse axis is 10, the length of the conjugate axis is 8, and the asymptotes are $y = \pm\frac{4}{5}x$. Using $c^2 = a^2 + b^2$:

$$c = \sqrt{a^2 + b^2} = \sqrt{25+16} = \sqrt{41}$$

The foci are $\left(\pm\sqrt{41}, 0\right)$ and the eccentricity is $\frac{\sqrt{41}}{5}$.

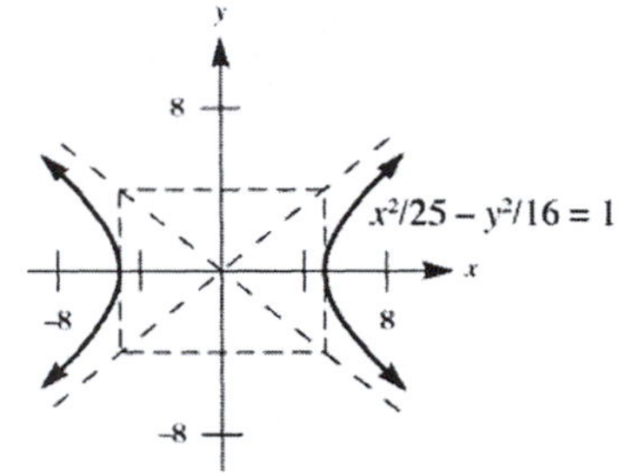

7. Rewriting the equation, the standard form is $\frac{y^2}{1/2} - \frac{x^2}{1/3} = 1$. The vertices are $\left(0, \pm\sqrt{\frac{1}{2}}\right) = \left(0, \pm\frac{\sqrt{2}}{2}\right)$, the length of the transverse axis is $\sqrt{2}$, the length of the conjugate axis is $2\sqrt{\frac{1}{3}} = \frac{2\sqrt{3}}{3}$, and the asymptotes are $y = \pm\sqrt{\frac{3}{2}}x = \pm\frac{\sqrt{6}}{2}x$.
Using $c^2 = a^2 + b^2$: $c = \sqrt{a^2 + b^2} = \sqrt{\frac{1}{2} + \frac{1}{3}} = \sqrt{\frac{5}{6}} = \frac{\sqrt{5}}{\sqrt{6}} = \frac{\sqrt{30}}{6}$

The foci are $\left(0, \pm\frac{\sqrt{30}}{6}\right)$ and the eccentricity is: $e = \frac{\frac{\sqrt{30}}{6}}{\frac{\sqrt{2}}{2}} = \frac{\sqrt{30}}{3\sqrt{2}} = \frac{\sqrt{15}}{3}$

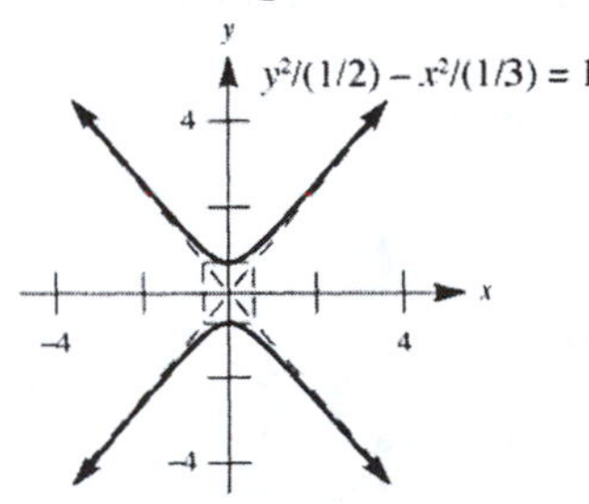

9. Dividing by 100, the standard form is $\frac{y^2}{25} - \frac{x^2}{4} = 1$. The vertices are (0,±5), the length of the transverse axis is 10, the length of the conjugate axis is 4, and the asymptotes are $y = \pm\frac{5}{2}x$. Using $c^2 = a^2 + b^2$:

$$c = \sqrt{a^2 + b^2} = \sqrt{25+4} = \sqrt{29}$$

The foci are $\left(0, \pm\sqrt{29}\right)$ and the eccentricity is $\frac{\sqrt{29}}{5}$.

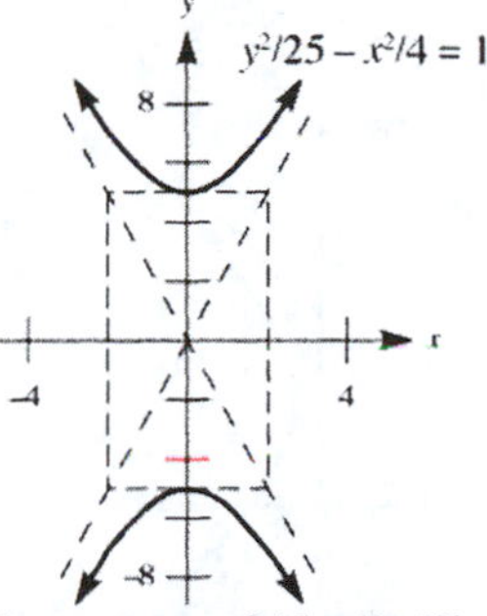

11. The equation is already in standard form with a center of (5,–1). The vertices are (5 + 5, –1) = (10,–1) and (5 – 5, –1) = (0, –1), the length of the transverse axis is 10, and the length of the conjugate axis is 6. The asymptotes have slopes of $\pm\frac{3}{5}$, so using the point-slope formula:

$$y+1 = \tfrac{3}{5}(x-5) \qquad y+1 = -\tfrac{3}{5}(x-5)$$
$$y+1 = \tfrac{3}{5}x - 3 \qquad y+1 = -\tfrac{3}{5}x + 3$$
$$y = \tfrac{3}{5}x - 4 \qquad y = -\tfrac{3}{5}x + 2$$

Using $c^2 = a^2 + b^2$: $c = \sqrt{a^2+b^2} = \sqrt{25+9} = \sqrt{34}$. The foci are $\left(5 \pm \sqrt{34}, -1\right)$ and the eccentricity is $\frac{\sqrt{34}}{5}$.

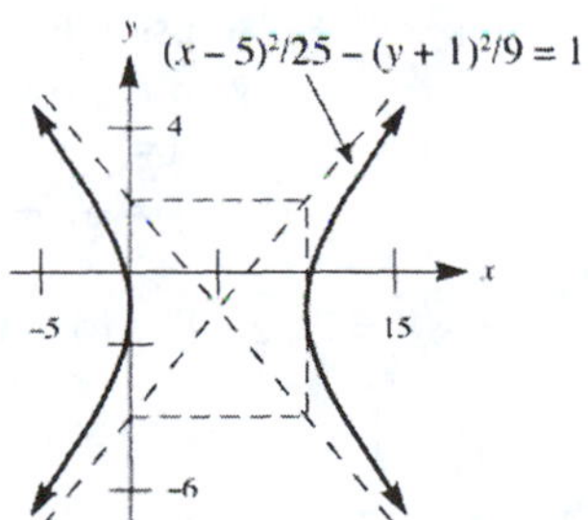

13. The equation is already in standard form with a center of (1,2). The vertices are (1, 2 + 2) = (1,4) and (1, 2 – 2) = (1,0), the length of the transverse axis is 4, and the length of the conjugate axis is 2. The asymptotes have slopes of ±2, so using the point-slope formula:

$$y-2 = 2(x-1) \qquad y-2 = -2(x-1)$$
$$y-2 = 2x-2 \qquad y-2 = -2x+2$$
$$y = 2x \qquad y = -2x+4$$

Using $c^2 = a^2 + b^2$: $c = \sqrt{a^2+b^2} = \sqrt{4+1} = \sqrt{5}$. The foci are $\left(1, 2 \pm \sqrt{5}\right)$ and the eccentricity is $\frac{\sqrt{5}}{2}$.

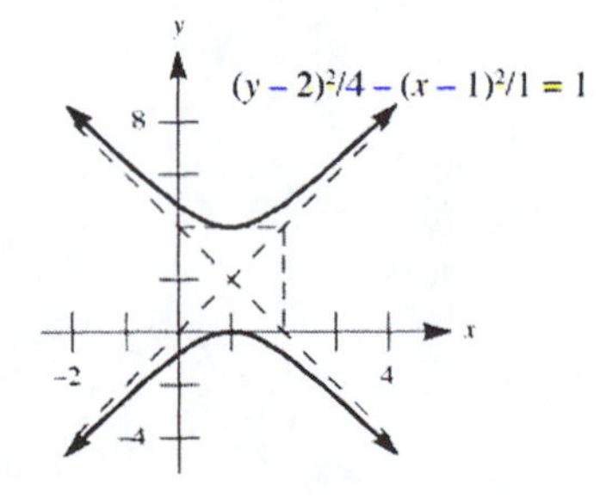

15. The equation is already in standard form with a center of (–3,4). The vertices are (–3 + 4, 4) = (1,4) and (–3 – 4, 4) = (–7,4), the length of the transverse axis is 8, and the length of the conjugate axis is 8. The asymptotes have slopes of ±1, so using the point-slope formula:

$$y-4=1(x+3) \qquad y-4=-1(x+3)$$
$$y-4=x+3 \qquad y-4=-x-3$$
$$y=x+7 \qquad y=-x+1$$

Using $c^2=a^2+b^2$: $c=\sqrt{a^2+b^2}=\sqrt{16+16}=\sqrt{32}=4\sqrt{2}$

The foci are $\left(-3\pm4\sqrt{2},4\right)$ and the eccentricity is $\dfrac{4\sqrt{2}}{4}=\sqrt{2}$.

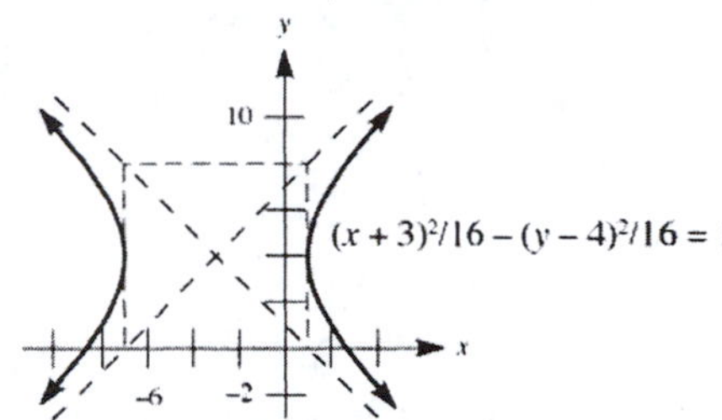

17. Complete the square to convert the equation to standard form:

$$x^2-y^2+2y-5=0$$
$$x^2-\left(y^2-2y\right)=5$$
$$x^2-\left(y^2-2y+1\right)=5-1$$
$$x^2-(y-1)^2=4$$
$$\frac{x^2}{4}-\frac{(y-1)^2}{4}=1$$

The center is (0,1), the vertices are (0 + 2, 1) = (2,1) and (0 – 2, 1) = (–2,1), and the lengths of both the transverse and conjugate axes are 4. The asymptotes have slopes of ±1, so using the point-slope formula:

$$y-1=1(x-0) \qquad y-1=-1(x-0)$$
$$y-1=x \qquad y-1=-x$$
$$y=x+1 \qquad y=-x+1$$

Using $c^2=a^2+b^2$: $c=\sqrt{a^2+b^2}=\sqrt{4+4}=\sqrt{8}=2\sqrt{2}$. The foci are $\left(\pm2\sqrt{2},1\right)$ and the eccentricity is $\dfrac{2\sqrt{2}}{2}=\sqrt{2}$.

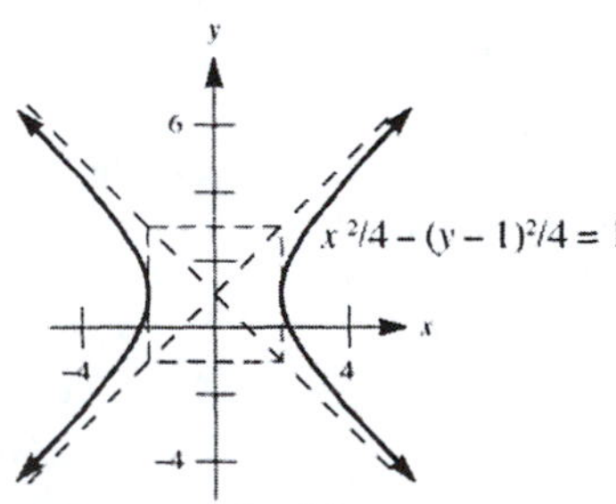

19. Complete the square to convert the equation to standard form:

$$x^2-y^2-4x+2y-6=0$$
$$\left(x^2-4x\right)-\left(y^2-2y\right)=6$$
$$\left(x^2-4x+4\right)-\left(y^2-2y+1\right)=6+4-1$$
$$(x-2)^2-(y-1)^2=9$$
$$\frac{(x-2)^2}{9}-\frac{(y-1)^2}{9}=1$$

The center is (2,1), the vertices are (2 + 3, 1) = (5,1) and (2 − 3, 1) = (−1,1), and the lengths of both the transverse and conjugate axes are 6. The asymptotes have slopes of ±1, so using the point-slope formula:

$$y-1=1(x-2) \qquad y-1=-1(x-2)$$
$$y-1=x-2 \qquad y-1=-x+2$$
$$y=x-1 \qquad y=-x+3$$

Using $c^2=a^2+b^2$: $c=\sqrt{a^2+b^2}=\sqrt{9+9}=\sqrt{18}=3\sqrt{2}$

The foci are $\left(2\pm 3\sqrt{2},1\right)$ and the eccentricity is $\dfrac{3\sqrt{2}}{3}=\sqrt{2}$.

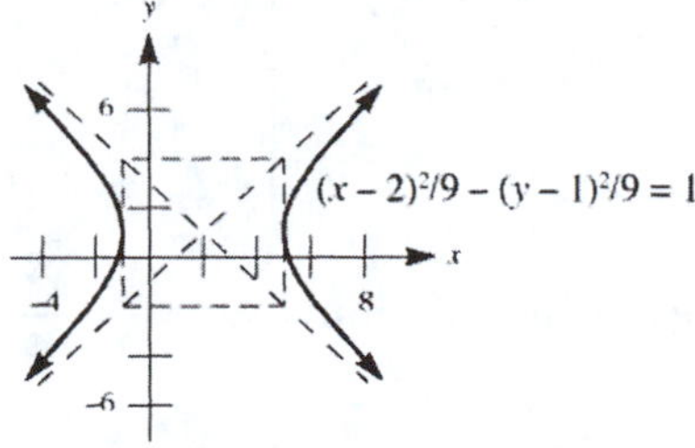

21. Complete the square to convert the equation to standard form:

$$y^2-25x^2+8y-9=0$$
$$\left(y^2+8y\right)-25x^2=9$$
$$\left(y^2+8y+16\right)-25x^2=9+16$$
$$(y+4)^2-25x^2=25$$
$$\frac{(y+4)^2}{25}-\frac{x^2}{1}=1$$

The center is (0,−4), the vertices are (0, −4 + 5) = (0,1) and (0, −4 − 5) = (0,−9), the length of the transverse axis is 10, and the length of the conjugate axis is 2. The asymptotes have slopes of ±5, so using the point-slope formula:

$$y+4=5(x-0) \qquad y+4=-5(x-0)$$
$$y+4=5x \qquad y+4=-5x$$
$$y=5x-4 \qquad y=-5x-4$$

Using $c^2=a^2+b^2$: $c=\sqrt{a^2+b^2}=\sqrt{25+1}=\sqrt{26}$. The foci are $\left(0,-4\pm\sqrt{26}\right)$ and the eccentricity is $\dfrac{\sqrt{26}}{5}$.

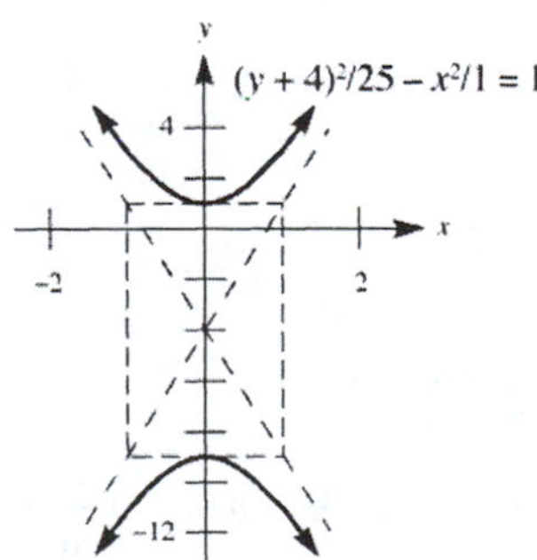

23. Complete the square to convert the equation to standard form:

$$x^2+7x-y^2-y+12=0$$
$$\left(x^2+7x\right)-\left(y^2+y\right)=-12$$
$$\left(x^2+7x+\tfrac{49}{4}\right)-\left(y^2+y+\tfrac{1}{4}\right)=-12+\tfrac{49}{4}-\tfrac{1}{4}$$
$$\left(x+\tfrac{7}{2}\right)^2-\left(y+\tfrac{1}{2}\right)^2=0$$

Notice that this is a degenerate hyperbola, and the graph consists of the "would-be" asymptotes with slopes ±1:

$$y+\tfrac{1}{2}=1\left(x+\tfrac{7}{2}\right) \qquad y+\tfrac{1}{2}=-1\left(x+\tfrac{7}{2}\right)$$
$$y+\tfrac{1}{2}=x+\tfrac{7}{2} \qquad y+\tfrac{1}{2}=-x-\tfrac{7}{2}$$
$$y=x+3 \qquad y=-x-4$$

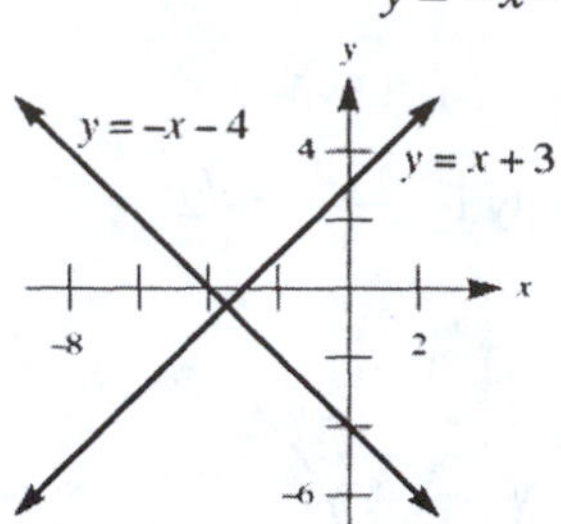

25. Since $P(x, y)$ lies on $\frac{x^2}{4}-\frac{y^2}{1}=1$, we can find y in terms of x: $y^2=\frac{x^2}{4}-1=\frac{x^2-4}{4}$

Taking roots, we have $y=\frac{\sqrt{x^2-4}}{2}$, since $P(x, y)$ lies in the first quadrant. So the coordinates of P are $\left(x, \frac{\sqrt{x^2-4}}{2}\right)$. Since $Q(x, y)$ lies in the first quadrant on the asymptote, find the equation of the asymptote:

$$y-0=\tfrac{1}{2}(x-0)$$
$$y=\tfrac{1}{2}x$$

So the coordinates of Q are $\left(x, \tfrac{1}{2}x\right)$. Since P and Q have the same x-coordinate PQ is the difference between their y-coordinates: $PQ=\frac{x}{2}-\frac{\sqrt{x^2-4}}{2}=\frac{x-\sqrt{x^2-4}}{2}$. The order of subtraction is because the asymptote lies above the hyperbola in the first quadrant, and thus $\frac{x}{2}$ is larger than $\frac{\sqrt{x^2-4}}{2}$. This proves the desired result.

27. Since the foci are (±4,0) and the vertices are (±1,0), then $c = 4$, $a = 1$, and the hyperbola has the form: $\frac{x^2}{1}-\frac{y^2}{b^2}=1$

Find b:

$$c^2=a^2+b^2$$
$$16=1+b^2$$
$$b^2=15$$

So the equation is $\frac{x^2}{1}-\frac{y^2}{15}=1$, or $15x^2-y^2=15$.

29. The slope of the asymptotes is $\pm\frac{1}{2}$, which tells us that the ratio $\frac{b}{a}=\frac{1}{2}$ in this hyperbola. Also, since the vertices are (±2,0) then $a = 2$. The required ratio is therefore: $\frac{b}{2}=\frac{1}{2}$ and $b=1$. The equation is $\frac{x^2}{4}-\frac{y^2}{1}=1$, or $x^2-4y^2=4$.

31. Since the asymptotes are $y=\pm\frac{\sqrt{10}}{5}x$, then $\frac{b}{a}=\frac{\sqrt{10}}{5}$, so $b=\frac{\sqrt{10}}{5}a$. Now, since the foci are $(\pm\sqrt{7},0)$, then $c=\sqrt{7}$ and the hyperbola has the form: $\frac{x^2}{a^2}-\frac{y^2}{b^2}=1$. Since $b=\frac{\sqrt{10}}{5}a$ and $c=\sqrt{7}$, we have:

$$\begin{aligned}c^2&=a^2+b^2\\7&=a^2+\left(\frac{\sqrt{10}}{5}a\right)^2\\7&=a^2+\tfrac{2}{5}a^2\\7a^2&=35\\a^2&=5\\b^2&=\tfrac{2}{5}a^2=\tfrac{2}{5}(5)=2\end{aligned}$$

So the equation is $\frac{x^2}{5}-\frac{y^2}{2}=1$, or $2x^2-5y^2=10$.

33. The vertices are at $(0,\pm7)$ so we know it is a "vertical" hyperbola. Its equation will be $\frac{y^2}{49}-\frac{x^2}{b^2}=1$, but we also know that (1,9) is a point satisfying the equation. Use it to find b:

$$\begin{aligned}\tfrac{81}{49}-\frac{1}{b^2}&=1\\81b^2-49&=49b^2\\32b^2&=49\\b^2&=\tfrac{49}{32}\end{aligned}$$

So the equation is $\frac{y^2}{49}-\frac{x^2}{49/32}=1$, or $y^2-32x^2=49$.

35. We have $2a=6$, so $a=3$. Also $2b=2$, so $b=1$. Since the foci are on the y-axis, the hyperbola will have the form:

$$\frac{y^2}{a^2}-\frac{x^2}{b^2}=1$$

So the equation is $\frac{y^2}{9}-\frac{x^2}{1}=1$, or $y^2-9x^2=9$.

37. Writing the equation as $\frac{x^2}{16}-\frac{y^2}{16}=1$, we have $a=b=4$. So the slopes of the asymptotes are $\pm\frac{b}{a}=\pm\frac{4}{4}=\pm1$. But these are negative reciprocals of each other, so the asymptotes are perpendicular to each other.

39. **a.** Substituting $P(5,6)$ into $5y^2-4x^2=80$, we have: $5(6)^2-4(5)^2=5(36)-4(25)=180-100=80$
So $P(5,6)$ lies on the hyperbola.

b. Dividing by 80 yields $\frac{y^2}{16}-\frac{x^2}{20}=1$, so $a=4$ and $b=2\sqrt{5}$. Using $c^2=a^2+b^2$:

$$c=\sqrt{a^2+b^2}=\sqrt{16+20}=\sqrt{36}=6$$

So $c=6$ and the foci are $(0,\pm6)$.

c. Compute the distances:

$$\begin{aligned}F_1P&=\sqrt{(5-0)^2+(6-6)^2}=5\\F_2P&=\sqrt{(5-0)^2+(6-(-6))^2}=\sqrt{25+144}=13\end{aligned}$$

d. Verify the result: $|F_1P-F_2P|=|5-13|=|-8|=8=2(4)=2a$

41. a. Solving for y:

$$\frac{(y-3)^2}{25}-\frac{(x-4)^2}{9}=1$$
$$9(y-3)^2-25(x-4)^2=225$$
$$9(y-3)^2=25(x-4)^2+225$$
$$(y-3)^2=\tfrac{25}{9}(x-4)^2+25$$
$$y-3=\pm 5\sqrt{1+\tfrac{1}{9}(x-4)^2}$$
$$y=3\pm 5\sqrt{1+\tfrac{1}{9}(x-4)^2}$$

b. Graphing the hyperbola in the standard viewing rectangle:

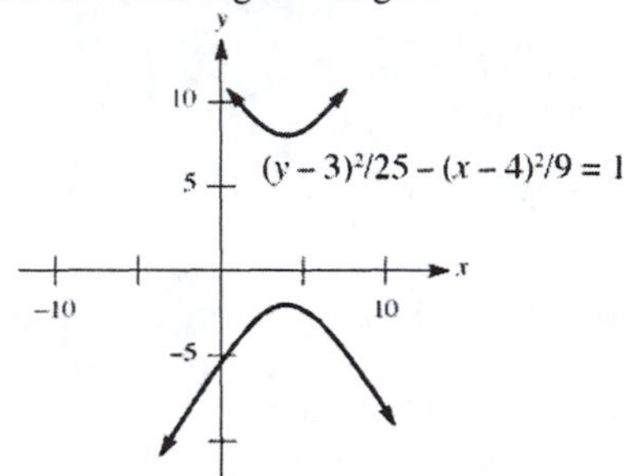

Using the suggested range settings, graph the hyperbola:

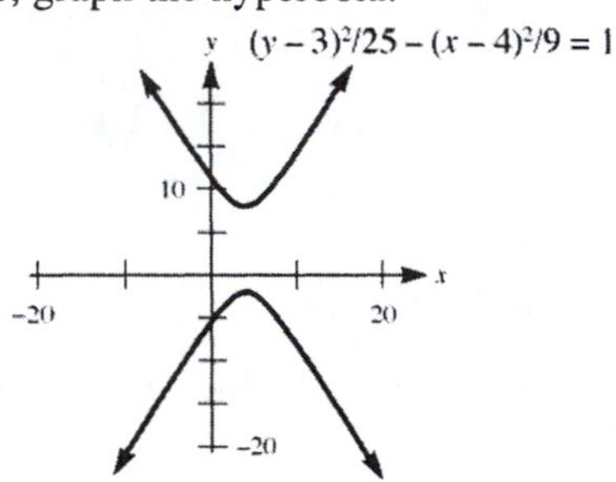

c. Since $a^2=25$ and $b^2=9$, $a=5$ and $b=3$, thus the slopes of the asymptotes are $\pm\frac{5}{3}$. Using the point (4,3) in the point-slope formula: $y-3=\pm\frac{5}{3}(x-4)$, so $y=\pm\frac{5}{3}(x-4)+3$. Now graph the hyperbola and the two asymptotes:

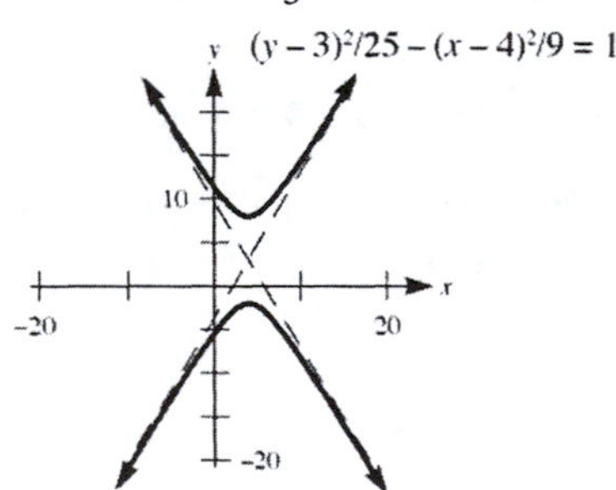

43. Using the quadratic formula to solve $-2y^2 + xy + \left(x^2 - 1\right) = 0$:

$$y = \frac{-x \pm \sqrt{x^2 - 4(-2)\left(x^2 - 1\right)}}{2(-2)} = \frac{-x \pm \sqrt{x^2 + 8x^2 - 8}}{-4} = \frac{-x \pm \sqrt{9x^2 - 8}}{-4} = \tfrac{1}{4}x \pm \tfrac{1}{4}\sqrt{9x^2 - 8}$$

Graphing the hyperbola using the suggested range settings:

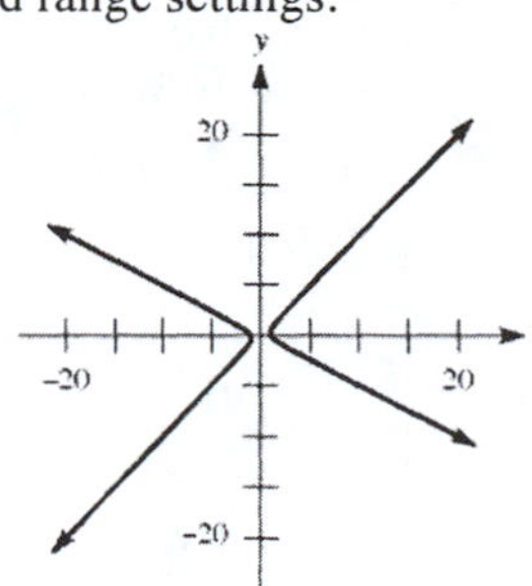

45. **a.** Since the asymptotes must have slopes of $\pm\frac{b}{a}$, then:

$$-\frac{b}{a} = \frac{-1}{\frac{b}{a}}$$

$$\frac{b^2}{a^2} = 1$$

$$b^2 = a^2$$

$$b = a$$

Now, since $c^2 = a^2 + b^2 = 2a^2$, then the eccentricity is: $\frac{c}{a} = \frac{\sqrt{2a^2}}{a} = \frac{\sqrt{2}a}{a} = \sqrt{2}$

b. The slopes of the asymptotes are $\pm\frac{a}{a} = \pm 1$, so the hyperbola will have perpendicular asymptotes. The eccentricity is $\frac{\sqrt{2}a}{a} = \sqrt{2}$.

47. **a.** This equation is just $F_1P - F_2P = 2a$, the defining relation of a hyperbola. Since P is on the right-hand branch, $F_1P > F_2P$.

b. Squaring each side of the equation:

$$\sqrt{(x+c)^2 + y^2} = 2a + \sqrt{(x-c)^2 + y^2}$$

$$(x+c)^2 + y^2 = 4a^2 + 4a\sqrt{(x-c)^2 + y^2} + (x-c)^2 + y^2$$

$$4xc = 4a^2 + 4a\sqrt{(x-c)^2 + y^2}$$

$$xc - a^2 = a\sqrt{(x-c)^2 + y^2}$$

$$xc - a^2 = a\left(F_2P\right)$$

c. Dividing by a:

$$\frac{xc}{a} - a = F_2P$$

$$xe - a = F_2P$$

49. For this hyperbola we find $a^2 = b^2 = k^2$, and $e = \sqrt{2}$. Also $d^2 = x^2 + y^2 = x^2 + (x^2 - k^2) = 2x^2 - k^2$. Thus we want to show that $F_1P \bullet F_2P = 2x^2 - k^2$. Using the formulas for F_1P and F_2P developed in Exercises 47 and 48:

$$F_1P \bullet F_2P = (xe + a)(xe - a) = x^2e^2 - a^2 = x^2(2) - k^2 = 2x^2 - k^2 = d^2$$

51. Draw the sketch:

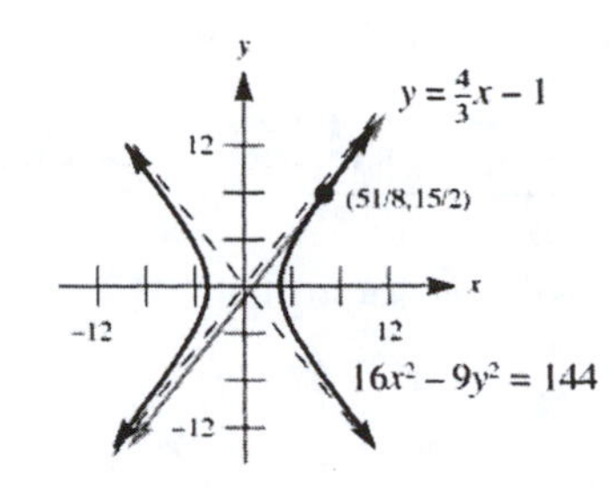

By substitution:

$$16x^2 - 9\left(\tfrac{4}{3}x - 1\right)^2 = 144$$
$$16x^2 - 9\left(\tfrac{16}{9}x^2 - \tfrac{8}{3}x + 1\right) = 144$$
$$16x^2 - 16x^2 + 24x - 9 = 144$$
$$24x = 153$$
$$x = \tfrac{51}{8}$$
$$y = \tfrac{4}{3} \bullet \tfrac{51}{8} - 1 = \tfrac{17}{2} - 1 = \tfrac{15}{2}$$

Therefore $\left(\frac{51}{8}, \frac{15}{2}\right)$ is the intersection point.

53. We have $(x_1, y_1) = (4,6)$, $a^2 = 4$, $b^2 = 12$:

$$\frac{4x}{4} - \frac{6y}{12} = 1$$
$$x - \frac{y}{2} = 1$$
$$2x - y = 2$$
$$-y = -2x + 2$$
$$y = 2x - 2$$

55. Assume (x, y) satisfy equation (2) and let $b^2 = c^2 - a^2$, so:

$$\frac{x^2}{a^2} - \frac{y^2}{c^2 - a^2} = 1$$
$$\left(c^2 - a^2\right)x^2 - a^2y^2 = a^2\left(c^2 - a^2\right)$$
$$a^2y^2 = \left(c^2 - a^2\right)\left(x^2 - a^2\right)$$
$$y^2 = \frac{\left(c^2 - a^2\right)\left(x^2 - a^2\right)}{a^2}$$

The difference of the distances from (x,y) to $(-c,0)$ and $(c,0)$ is given by:

$$\sqrt{(x+c)^2+y^2}-\sqrt{(x-c)^2+y^2}$$
$$=\sqrt{(x+c)^2+\frac{(c^2-a^2)(x^2-a^2)}{a^2}}-\sqrt{(x-c)^2+\frac{(c^2-a^2)(x^2-a^2)}{a^2}}$$
$$=\sqrt{\frac{a^2(x+c)^2+(c^2-a^2)(x^2-a^2)}{a^2}}-\sqrt{\frac{a^2(x-c)^2+(c^2-a^2)(x^2-a^2)}{a^2}}$$
$$=\sqrt{\frac{a^2x^2+2a^2cx+a^2c^2+a^4-a^2c^2-a^2x^2+c^2x^2}{a^2}}-\sqrt{\frac{a^2x^2-2a^2cx+a^2c^2+a^4-a^2c^2-a^2x^2+c^2x^2}{a^2}}$$
$$=\sqrt{\frac{a^4+2a^2cx+c^2x^2}{a^2}}-\sqrt{\frac{a^4-2a^2cx+c^2x^2}{a^2}}$$
$$=\sqrt{\frac{(a^2+cx)^2}{a^2}}-\sqrt{\frac{(a^2-cx)^2}{a^2}}$$
$$=\left|\frac{a^2+cx}{a}\right|-\left|\frac{a^2-cx}{a}\right|$$

If $x \geq a$ and $c > a$, then $a^2+cx \geq 0$ and $a^2-cx \leq 0$, so this difference is:

$$\frac{a^2+cx}{a}-\left(-\frac{a^2-cx}{a}\right)=\frac{a^2+cx+a^2-cx}{a}=\frac{2a^2}{a}=2a$$

If $x \leq -a$ and $c < a$, then $a^2+cx \leq 0$ and $a^2-cx \geq 0$, so this difference is:

$$-\frac{a^2+cx}{a}-\frac{a^2-cx}{a}=\frac{-a^2-cx-a^2+cx}{a}=\frac{-2a^2}{a}=-2a$$

In either case, the absolute value of the difference is $2a$, thus the point (x,y) satisfies the definition of the hyperbola.

57. The slope of the tangent line at P is $\frac{b^2x_1}{a^2y_1}$, so the slope of the normal is $\frac{-a^2y_1}{b^2x_1}$ and its equation is:

$$y-y_1=\frac{-a^2y_1}{b^2x_1}(x-x_1)$$
$$b^2x_1y-b^2x_1y_1=-a^2y_1x+a^2y_1x_1$$
$$a^2y_1x+b^2x_1y=x_1y_1(a^2+b^2)$$

59. The coordinates of S and T are found to be: $S\left(a,\frac{b^2(x_1-a)}{ay_1}\right)$ and $T\left(-a,\frac{-b^2(x_1+a)}{ay_1}\right)$

We shall show that the circle with diameter ST passes through the focus $F_2(c,0)$. (The calculations for the other focus are similar.) In order to show that the circle passes through F_2, it is enough to establish that $\angle TF_2S$ is a right angle. This can be done by comparing slopes. The slopes of $\overline{SF_2}$ and $\overline{TF_2}$ are easy to calculate using the coordinates for S and T given above. The results are:

$$\text{slope of } SF_2=\frac{b^2(x_1-a)}{ay_1(a-c)} \qquad \text{slope of } TF_2=\frac{b^2(x_1+a)}{ay_1(a+c)}$$

The product of these two slopes is:

$$\frac{b^4\left(x_1^2-a^2\right)}{a^2y_1^2\left(a^2-c^2\right)}=\frac{b^4\left(x_1^2-a^2\right)}{a^2y_1^2\left(-b^2\right)}=\frac{-b^2x_1^2+a^2b^2}{a^2y_1^2}=\frac{-b^2x_1^2+b^2x_1^2-a^2y_1^2}{a^2y_1^2}=\frac{-a^2y_1^2}{a^2y_1^2}=-1$$

This shows that $\overline{SF_2}$ is perpendicular to $\overline{TF_2}$. Therefore $\angle TF_2S$ is a right angle, and consequently, F_2 lies on the circle with diameter ST.

11.6 The Focus-Directrix Property of Conics

1. In order to use the formulas for the focal radii, we must find a and e. Dividing by 76, the standard form is $\frac{x^2}{76}+\frac{y^2}{76/3}=1$. So $a=\sqrt{76}=2\sqrt{19}$. To find the eccentricity, first find c:

$$c=\sqrt{a^2-b^2}=\sqrt{76-\tfrac{76}{3}}=\sqrt{\tfrac{152}{3}}=\frac{2\sqrt{38}}{\sqrt{3}}=\frac{2\sqrt{114}}{3}$$

The eccentricity is: $e=\frac{c}{a}=\frac{\frac{2\sqrt{114}}{3}}{2\sqrt{19}}=\frac{\sqrt{6}}{3}$. The focal radii are given by:

$$F_1P=a+ex=2\sqrt{19}+\frac{\sqrt{6}}{3}(-8)=2\sqrt{19}-\frac{8\sqrt{6}}{3}=\frac{6\sqrt{19}-8\sqrt{6}}{3}$$

$$F_2P=a-ex=2\sqrt{19}-\frac{\sqrt{6}}{3}(-8)=2\sqrt{19}+\frac{8\sqrt{6}}{3}=\frac{6\sqrt{19}+8\sqrt{6}}{3}$$

3. The equation is already in standard form with $a=15$ and $b=5$. Finding c:

$$c=\sqrt{a^2-b^2}=\sqrt{225-25}=\sqrt{200}=10\sqrt{2}$$

The eccentricity is: $e=\frac{c}{a}=\frac{10\sqrt{2}}{15}=\frac{2\sqrt{2}}{3}$. The focal radii are given by:

$$F_1P=a+ex=15+\frac{2\sqrt{2}}{3}(9)=15+6\sqrt{2} \qquad F_2P=a-ex=15-\frac{2\sqrt{2}}{3}(9)=15-6\sqrt{2}$$

5. a. The ellipse is already in standard form with $a=4$ and $b=3$. Finding c: $c=\sqrt{a^2-b^2}=\sqrt{16-9}=\sqrt{7}$

The foci are $\left(\pm\sqrt{7},0\right)$ and the eccentricity is $\frac{\sqrt{7}}{4}$. The directrices are given by:

$$x=\pm\frac{a}{e}=\pm\frac{4}{\frac{\sqrt{7}}{4}}=\pm\frac{16}{\sqrt{7}}=\pm\frac{16\sqrt{7}}{7}$$

b. The hyperbola is already in standard form with $a=4$ and $b=3$. Finding c: $c=\sqrt{a^2+b^2}=\sqrt{16+9}=\sqrt{25}=5$

The foci are $(\pm5,0)$ and the eccentricity is $\frac{5}{4}$. The directrices are given by: $x=\pm\frac{a}{e}=\pm\frac{4}{\frac{5}{4}}=\pm\frac{16}{5}$

7. a. Dividing by 156, the standard form for the ellipse is $\frac{x^2}{13}+\frac{y^2}{12}=1$, and so $a=\sqrt{13}$ and $b=\sqrt{12}=2\sqrt{3}$.

Finding c: $c=\sqrt{a^2-b^2}=\sqrt{13-12}=\sqrt{1}=1$

The foci are $(\pm1,0)$ and the eccentricity is $\frac{1}{\sqrt{13}}=\frac{\sqrt{13}}{13}$. The directrices are given by: $x=\pm\frac{a}{e}=\pm\frac{\sqrt{13}}{\frac{\sqrt{13}}{13}}=\pm13$

b. Dividing by 156, the standard form for the hyperbola is $\frac{x^2}{13} - \frac{y^2}{12} = 1$, and so $a = \sqrt{13}$ and $b = \sqrt{12} = 2\sqrt{3}$.

Finding c: $c = \sqrt{a^2 + b^2} = \sqrt{13+12} = \sqrt{25} = 5$

The foci are $(\pm 5, 0)$ and the eccentricity is $\frac{5}{\sqrt{13}} = \frac{5\sqrt{13}}{13}$. The directrices are given by: $x = \pm\frac{a}{e} = \pm\frac{\sqrt{13}}{\frac{5\sqrt{13}}{13}} = \pm\frac{13}{5}$

9. **a.** Dividing by 900, the standard form for the ellipse is $\frac{x^2}{36} + \frac{y^2}{25} = 1$, and so $a = 6$ and $b = 5$. Finding c:

$$c = \sqrt{a^2 - b^2} = \sqrt{36 - 25} = \sqrt{11}$$

The foci are $\left(\pm\sqrt{11}, 0\right)$ and the eccentricity is $\frac{\sqrt{11}}{6}$. The directrices are given by:

$$x = \pm\frac{a}{e} = \pm\frac{6}{\frac{\sqrt{11}}{6}} = \pm\frac{36}{\sqrt{11}} = \pm\frac{36\sqrt{11}}{11}$$

b. Dividing by 900, the standard form for the hyperbola is $\frac{x^2}{36} - \frac{y^2}{25} = 1$, and so $a = 6$ and $b = 5$. Finding c:

$$c = \sqrt{a^2 + b^2} = \sqrt{36 + 25} = \sqrt{61}$$

The foci are $\left(\pm\sqrt{61}, 0\right)$ and the eccentricity is $\frac{\sqrt{61}}{6}$. The directrices are given by:

$$x = \pm\frac{a}{e} = \pm\frac{6}{\frac{\sqrt{61}}{6}} = \pm\frac{36}{\sqrt{61}} = \pm\frac{36\sqrt{61}}{61}$$

11. Since the foci are $(\pm 1, 0)$, $c = 1$. Since the directrices are $x = \pm 4$, $\frac{a}{e} = 4$, so $a = 4e$. But since $e = \frac{c}{a} = \frac{1}{a}$:

$$a = 4 \bullet \frac{1}{a}$$
$$a^2 = 4$$

Since $a^2 - b^2 = c^2$, we can find b^2:

$$2^2 - b^2 = 1^2$$
$$-b^2 = -3$$
$$b^2 = 3$$

The equation of the ellipse is $\frac{x^2}{4} + \frac{y^2}{3} = 1$, or $3x^2 + 4y^2 = 12$.

13. Since the foci are $(\pm 2, 0)$, $c = 2$. Since the directrices are $x = \pm 1$, $\frac{a}{e} = 1$, so $a = e$. But since $e = \frac{c}{a} = \frac{2}{a}$:

$$a = \frac{2}{a}$$
$$a^2 = 2$$

Since $a^2 + b^2 = c^2$, we can find b^2:

$$2 + b^2 = 2^2$$
$$b^2 = 2$$

So the equation of the hyperbola is $\frac{x^2}{2} - \frac{y^2}{2} = 1$, or $x^2 - y^2 = 2$.

15. a. By the distance formula:

$$d_1 = \sqrt{(x+c)^2+(y-0)^2} = \sqrt{(x+c)^2+y^2} \qquad d_2 = \sqrt{(x-c)^2+(y-0)^2} = \sqrt{(x-c)^2+y^2}$$

Squaring:

$$d_1^2 = (x+c)^2+y^2 \qquad d_2^2 = (x-c)^2+y^2$$

b. Working from the left-hand side: $d_1^2 - d_2^2 = (x+c)^2-(x-c)^2 = x^2+2cx+c^2-x^2+2cx-c^2 = 4cx$

c. Since d_1 and d_2 represent the distances from the foci to a point on the ellipse, $d_1+d_2 = 2a$ by the definition of an ellipse.

d. Factoring:

$$\begin{aligned} d_1^2 - d_2^2 &= 4cx \\ (d_1+d_2)(d_1-d_2) &= 4cx \\ 2a(d_1-d_2) &= 4cx \\ d_1-d_2 &= \frac{2cx}{a} \end{aligned}$$

e. Adding the two equations:

$$\begin{aligned} 2d_1 &= 2a+\frac{2cx}{a} \\ d_1 &= a+\frac{c}{a}x = a+ex \end{aligned}$$

f. Substituting the result from part **e**:

$$\begin{aligned} a+ex+d_2 &= 2a \\ d_2 &= a-ex \end{aligned}$$

11.7 The Conics in Polar Coordinates

1. a. Comparing the given equation with the four basic types, it appears this is the type associated with Figure 2.

Divide both numerator and denominator by 3 to obtain: $r = \dfrac{2}{1+\frac{2}{3}\cos\theta} = \dfrac{\frac{2}{3}\cdot 3}{1+\frac{2}{3}\cos\theta}$

Therefore $e = \frac{2}{3}$ and $d = 3$. Since $e < 1$, this confirms the given conic is an ellipse. The eccentricity is $\frac{2}{3}$ and the directrix is $x = 3$. Computing the values of r when $\theta = 0, \frac{\pi}{2}, \pi$ and $\frac{3\pi}{2}$:

θ	0	$\frac{\pi}{2}$	π	$\frac{3\pi}{2}$
r	$\frac{6}{5}$	2	6	2

Since the major axis of this ellipse lies along the x-axis, the length of the major axis is:

$$2a = \tfrac{6}{5}+6 = \tfrac{36}{5}, \text{ so } a = \tfrac{18}{5}$$

The endpoints of the major axis are at $\left(\frac{6}{5},0\right)$ and $(-6,0)$, so the x-coordinate of the center is:

$$\tfrac{1}{2}\left(-6+\tfrac{6}{5}\right) = \tfrac{1}{2}\left(-\tfrac{24}{5}\right) = -\tfrac{12}{5}$$

So the center is $\left(-\frac{12}{5},0\right)$. Finally, calculate b: $b = a\sqrt{1-e^2} = \frac{18}{5}\sqrt{1-\frac{4}{9}} = \frac{18}{5}\sqrt{\frac{5}{9}} = \frac{18\sqrt{5}}{15} = \frac{6\sqrt{5}}{5}$

So the endpoints of the minor axis are $\left(-\frac{12}{5}, \pm\frac{6\sqrt{5}}{5}\right)$. Graph the ellipse:

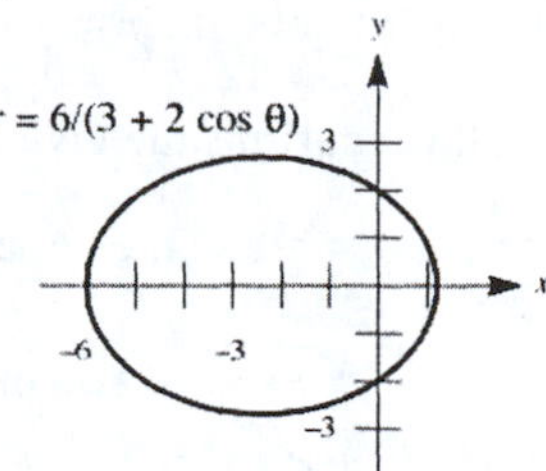

b. Comparing the given equation with the four basic types, it appears this is the type associated with Figure 3.

Divide both numerator and denominator by 3 to obtain: $r = \dfrac{2}{1-\frac{2}{3}\cos\theta} = \dfrac{\frac{2}{3}\bullet 3}{1-\frac{2}{3}\cos\theta}$

Therefore $e = \frac{2}{3}$ and $d = 3$. Since $e < 1$, this confirms the given conic is an ellipse. The eccentricity is $\frac{2}{3}$ and the directrix is $x = -3$. Computing the values of r when $\theta = 0, \frac{\pi}{2}, \pi$ and $\frac{3\pi}{2}$:

θ	0	$\frac{\pi}{2}$	π	$\frac{3\pi}{2}$
r	6	2	$\frac{6}{5}$	2

Since the major axis of this ellipse lies along the x-axis, the length of the major axis is:

$2a = \frac{6}{5} + 6 = \frac{36}{5}$, so $a = \frac{18}{5}$

The endpoints of the major axis are at $(6, 0)$ and $\left(-\frac{6}{5}, 0\right)$, so the x-coordinate of the center is:

$\frac{1}{2}\left(6 - \frac{6}{5}\right) = \frac{1}{2}\left(\frac{24}{5}\right) = \frac{12}{5}$

So the center is $\left(\frac{12}{5}, 0\right)$. Finally, calculate b: $b = a\sqrt{1-e^2} = \frac{18}{5}\sqrt{1-\frac{4}{9}} = \frac{18}{5}\sqrt{\frac{5}{9}} = \frac{18\sqrt{5}}{15} = \frac{6\sqrt{5}}{5}$

So the endpoints of the minor axis are $\left(\frac{12}{5}, \pm\frac{6\sqrt{5}}{5}\right)$. Graph the ellipse:

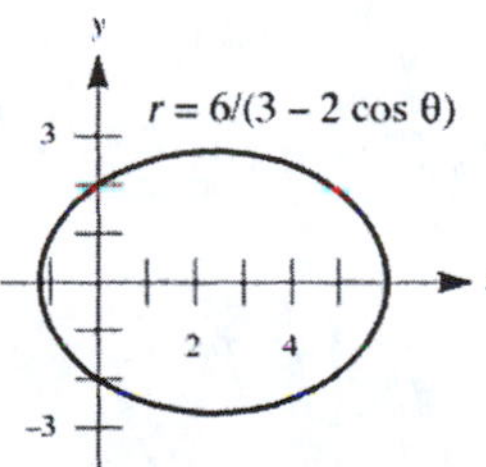

3. **a.** Comparing the given equation with the four basic types, it appears this is the type associated with Figure 2.

Divide both numerator and denominator by 2 to obtain: $r = \dfrac{\frac{5}{2}}{1+\cos\theta} = \dfrac{1\bullet\frac{5}{2}}{1+\cos\theta}$

Therefore $e = 1$ and $d = \frac{5}{2}$. Since $e = 1$, this confirms the given conic is a parabola. The directrix is $x = \frac{5}{2}$.

Computing the value of r when $\theta = 0$ yields $r = \frac{5}{4}$, so the vertex is $\left(\frac{5}{4}, 0\right)$. Graph the parabola:

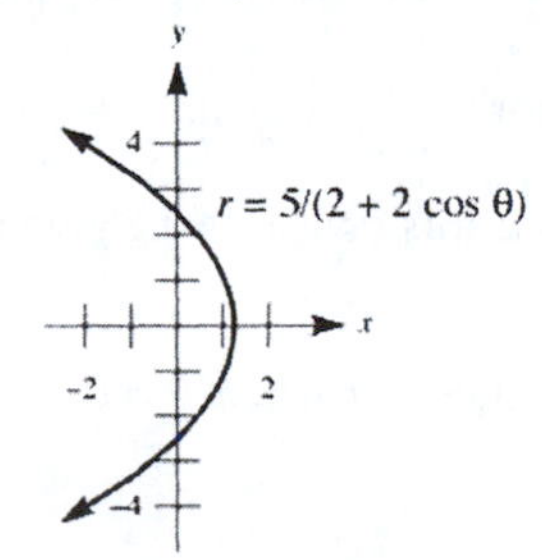

b. Comparing the given equation with the four basic types, it appears this is the type associated with Figure 3.

Divide both numerator and denominator by 2 to obtain: $r = \dfrac{\frac{5}{2}}{1-\cos\theta} = \dfrac{1 \bullet \frac{5}{2}}{1-\cos\theta}$

Therefore $e = 1$ and $d = \frac{5}{2}$. Since $e = 1$, this confirms the given conic is a parabola. The directrix is $x = -\frac{5}{2}$.

Computing the value of r when $\theta = \pi$ yields $r = \frac{5}{4}$, so the vertex is $\left(-\frac{5}{4}, 0\right)$. Graph the parabola:

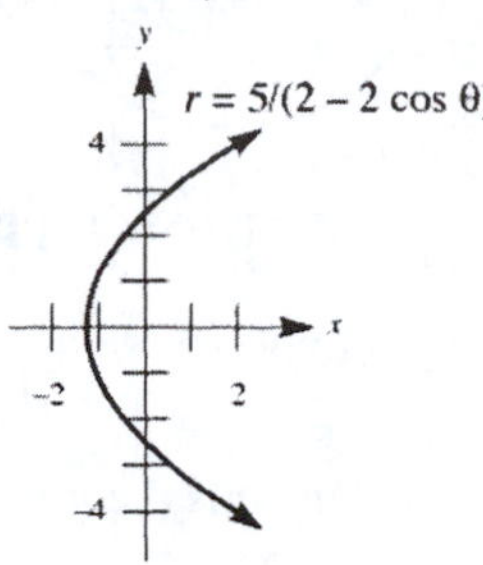

5. **a.** Comparing the given equation with the four basic types, it appears this is the type associated with Figure 2.

Divide both numerator and denominator by 2 to obtain: $r = \dfrac{3}{2+4\cos\theta} = \dfrac{\frac{3}{2}}{1+2\cos\theta} = \dfrac{2 \bullet \frac{3}{4}}{1+2\cos\theta}$

Therefore $e = 2$ and $d = \frac{3}{4}$. Since $e > 1$, this confirms the given conic is a hyperbola. The eccentricity is 2 and the directrix is $x = \frac{3}{4}$. Computing the values of r when $\theta = 0, \frac{\pi}{2}, \pi$ and $\frac{3\pi}{2}$:

θ	0	$\frac{\pi}{2}$	π	$\frac{3\pi}{2}$
r	$\frac{1}{2}$	$\frac{3}{2}$	$-\frac{3}{2}$	$\frac{3}{2}$

Since the two vertices $\left(\frac{1}{2}, 0\right)$ and $\left(\frac{3}{2}, 0\right)$ lie on the transverse axis, then:

$2a = \frac{3}{2} - \frac{1}{2} = 1$, so $a = \frac{1}{2}$

The center of the hyperbola is the midpoint of these two vertices, which is $(1, 0)$. Since a focus is $(0, 0)$, then $c = 1$.

Finally, find b: $b = \sqrt{c^2 - a^2} = \sqrt{1 - \frac{1}{4}} = \sqrt{\frac{3}{4}} = \frac{1}{2}\sqrt{3}$

Notice that we could also find b from the eccentricity: $b = a\sqrt{e^2 - 1} = \frac{1}{2}\sqrt{4-1} = \frac{1}{2}\sqrt{3}$

Graph the hyperbola:

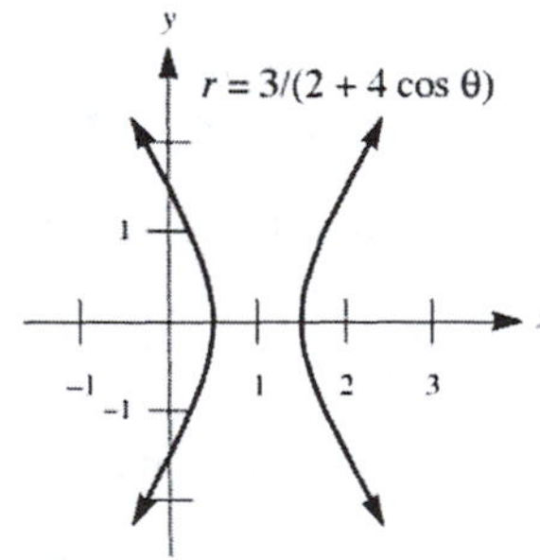

b. Comparing the given equation with the four basic types, it appears this is the type associated with Figure 3.

Divide both numerator and denominator by 2 to obtain: $r = \dfrac{3}{2-4\cos\theta} = \dfrac{\frac{3}{2}}{1-2\cos\theta} = \dfrac{2 \bullet \frac{3}{4}}{1-2\cos\theta}$

Therefore $e = 2$ and $d = \frac{3}{4}$. Since $e > 1$, this confirms the given conic is a hyperbola. The eccentricity is 2 and the directrix is $x = -\frac{3}{4}$. Computing the values of r when $\theta = 0, \frac{\pi}{2}, \pi$ and $\frac{3\pi}{2}$:

θ	0	$\frac{\pi}{2}$	π	$\frac{3\pi}{2}$
r	$-\frac{3}{2}$	$\frac{3}{2}$	$\frac{1}{2}$	$\frac{3}{2}$

Since the two vertices $\left(-\frac{3}{2},0\right)$ and $\left(-\frac{1}{2},0\right)$ lie on the transverse axis, then:

$$2a=-\tfrac{1}{2}+\tfrac{3}{2}=1, \text{ so } a=\tfrac{1}{2}$$

The center of the hyperbola is the midpoint of these two vertices, which is $(-1,0)$. Since a focus is $(0,0)$, then $c=1$. Finally, find b: $b=\sqrt{c^2-a^2}=\sqrt{1-\frac{1}{4}}=\sqrt{\frac{3}{4}}=\frac{1}{2}\sqrt{3}$

Notice that we could also find b from the eccentricity: $b=a\sqrt{e^2-1}=\frac{1}{2}\sqrt{4-1}=\frac{1}{2}\sqrt{3}$

Graph the hyperbola:

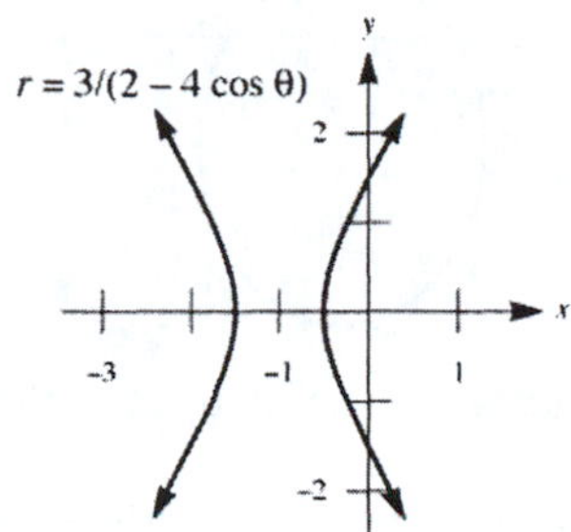

7. Comparing the given equation with the four basic types, it appears this is the type associated with Figure 3.

Divide both numerator and denominator by 2 to obtain: $r=\dfrac{24}{2-3\cos\theta}=\dfrac{12}{1-\frac{3}{2}\cos\theta}=\dfrac{\frac{3}{2}\bullet 8}{1-\frac{3}{2}\cos\theta}$

Since the eccentricity is $e=\frac{3}{2}>1$, this conic is a hyperbola. Computing the values of r when $\theta=0,\frac{\pi}{2},\pi$ and $\frac{3\pi}{2}$:

θ	0	$\frac{\pi}{2}$	π	$\frac{3\pi}{2}$
r	-24	12	$\frac{24}{5}$	12

Since the two vertices $(-24,0)$ and $\left(-\frac{24}{5},0\right)$ lie on the transverse axis, its length must be:

$$2a=-\tfrac{24}{5}+24=\tfrac{96}{5}, \text{ so } a=\tfrac{48}{5}$$

The center of the hyperbola is the midpoint of these two vertices, which is $\left(-\frac{72}{5},0\right)$. Since a focus is $(0,0)$, $c=\frac{72}{5}$ and thus: $b=\sqrt{c^2-a^2}=\sqrt{\frac{5184}{25}-\frac{2304}{25}}=\sqrt{\frac{2880}{25}}=\frac{24\sqrt{5}}{5}$

So the length of the conjugate axis is $2b=\frac{48\sqrt{5}}{5}$. Graph the hyperbola:

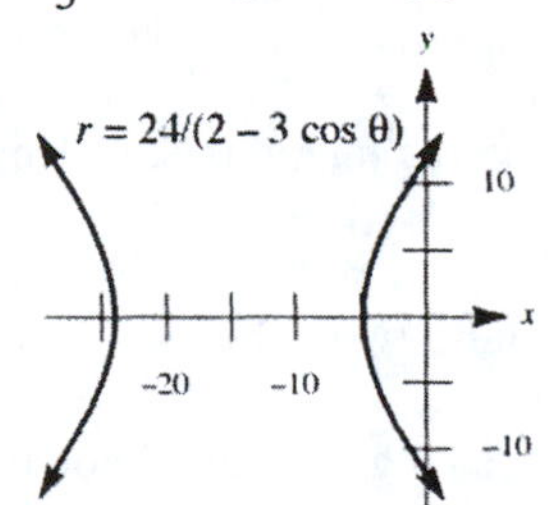

9. Comparing the given equation with the four basic types, it appears this is the type associated with Figure 4.

Divide both numerator and denominator by 5 to obtain: $r=\dfrac{8}{5+3\sin\theta}=\dfrac{\frac{8}{5}}{1+\frac{3}{5}\sin\theta}=\dfrac{\frac{3}{5}\bullet\frac{8}{3}}{1+\frac{3}{5}\sin\theta}$

Since the eccentricity is $e=\frac{3}{5}<1$, this conic is an ellipse. Computing the values of r when $\theta=0,\frac{\pi}{2},\pi$ and $\frac{3\pi}{2}$:

θ	0	$\frac{\pi}{2}$	π	$\frac{3\pi}{2}$
r	$\frac{8}{5}$	1	$\frac{8}{5}$	4

Since the two vertices (0, 1) and (0, –4) lie on the major axis, its length must be:

$2a = 1 + 4 = 5$, so $a = \frac{5}{2}$

The center of the ellipse is the midpoint of these two vertices, which is $\left(0, -\frac{3}{2}\right)$. Since a focus is $(0,0)$, $c = \frac{3}{2}$ and thus: $b = \sqrt{a^2 - c^2} = \sqrt{\frac{25}{4} - \frac{9}{4}} = \sqrt{\frac{16}{4}} = 2$. So the length of the minor axis is $2b = 4$. Graph the ellipse:

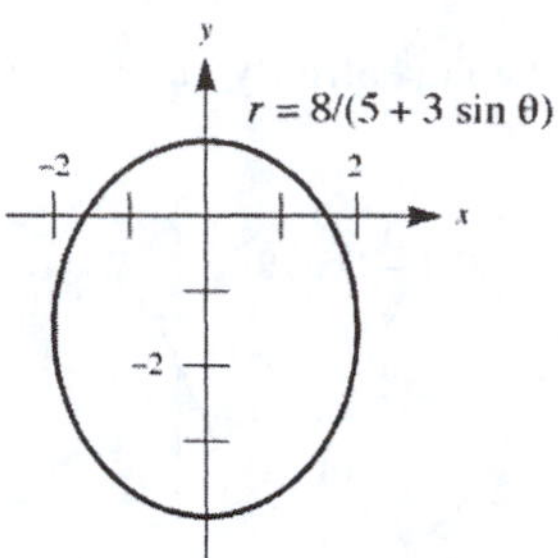

11. Comparing the given equation with the four basic types, it appears this is the type associated with Figure 5.

Divide both numerator and denominator by 5 to obtain: $r = \dfrac{12}{5 - 5\sin\theta} = \dfrac{\frac{12}{5}}{1 - \sin\theta} = \dfrac{1 \bullet \frac{12}{5}}{1 - \sin\theta}$

Since the eccentricity is $e = 1$, this conic is a parabola with directrix $y = -\frac{12}{5}$. Since the focus is (0,0), the vertex must be the midpoint of $(0, 0)$ and $\left(0, -\frac{12}{5}\right)$, which is $\left(0, -\frac{6}{5}\right)$. Graph the parabola:

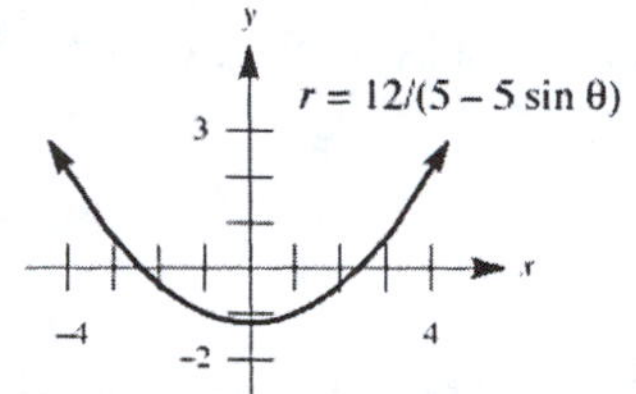

13. Comparing the given equation with the four basic types, it appears this is the type associated with Figure 2.

Divide both numerator and denominator by 7 to obtain: $r = \dfrac{12}{7 + 5\cos\theta} = \dfrac{\frac{12}{7}}{1 + \frac{5}{7}\cos\theta} = \dfrac{\frac{5}{7} \bullet \frac{12}{5}}{1 + \frac{5}{7}\cos\theta}$

Since the eccentricity is $e = \frac{5}{7} < 1$, this conic is an ellipse. Computing the values of r when $\theta = 0, \frac{\pi}{2}, \pi$ and $\frac{3\pi}{2}$:

θ	0	$\frac{\pi}{2}$	π	$\frac{3\pi}{2}$
r	1	$\frac{12}{7}$	6	$\frac{12}{7}$

Since the two vertices (1, 0) and (–6, 0) lie on the major axis, its length must be:

$2a = 1 + 6 = 7$, so $a = \frac{7}{2}$

The center of the ellipse is the midpoint of these two vertices, which is $\left(-\frac{5}{2}, 0\right)$. Since a focus is $(0,0)$, then $c = \frac{5}{2}$ and thus: $b = \sqrt{a^2 - c^2} = \sqrt{\frac{49}{4} - \frac{25}{4}} = \sqrt{6}$. So the length of the minor axis is $2b = 2\sqrt{6}$. Graph the ellipse:

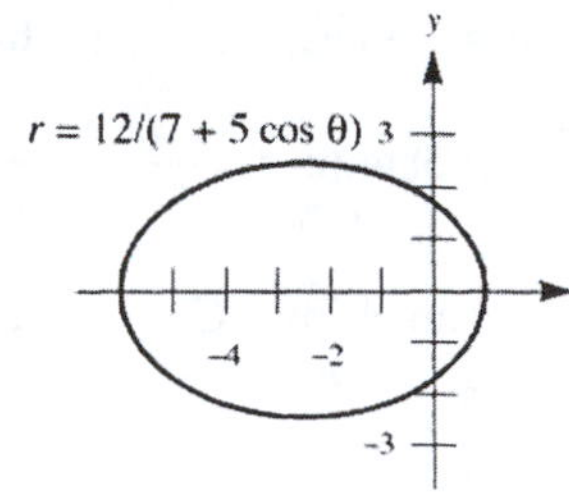

15. Comparing the given equation with the four basic types, it appears this is the type associated with Figure 4.

Divide both numerator and denominator by 5 to obtain: $r = \dfrac{4}{5+5\sin\theta} = \dfrac{\frac{4}{5}}{1+\sin\theta} = \dfrac{1\bullet\frac{4}{5}}{1+\sin\theta}$

Since the eccentricity is $e = 1$, this conic is a parabola with directrix $y = \frac{4}{5}$. Since the focus is (0,0), the vertex must be the midpoint of (0,0) and $\left(0,\frac{4}{5}\right)$, which is $\left(0,\frac{2}{5}\right)$. Graph the parabola:

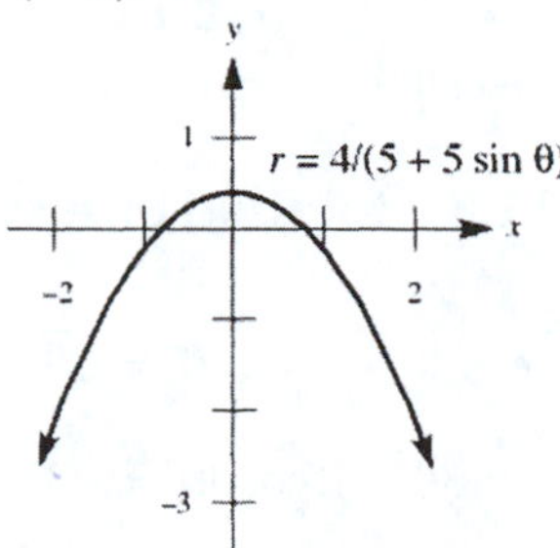

17. Comparing the given equation with the four basic types, it appears this is the type associated with Figure 3.
The equation is already in standard form with eccentricity $e = 2 > 1$, so this conic is a hyperbola. Computing the values of r when $\theta = 0, \frac{\pi}{2}, \pi$ and $\frac{3\pi}{2}$:

θ	0	$\frac{\pi}{2}$	π	$\frac{3\pi}{2}$
r	–9	9	3	9

Since the two vertices (–9,0) and (–3,0) lie on the transverse axis, its length must be:

$2a = -3+9 = 6$, so $a = 3$

The center of the hyperbola is the midpoint of these two vertices, which is (–6,0). Since a focus is (0, 0), $c = 6$ and thus: $b = \sqrt{c^2 - a^2} = \sqrt{36-9} = \sqrt{27} = 3\sqrt{3}$. So the length of the conjugate axis is $2b = 6\sqrt{3}$. Graph the hyperbola:

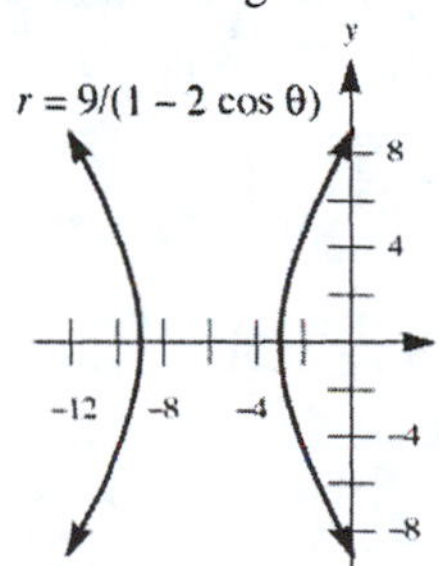

19. Since the coordinates of P are (r, θ), then the coordinates of Q are $(r, \theta + \pi)$. Now find FP and FQ, noting that $\cos(\theta + \pi) = -\cos\theta$:

$$FP = r = \frac{ed}{1-e\cos\theta} \qquad FQ = r = \frac{ed}{1-e\cos(\theta+\pi)} = \frac{ed}{1+e\cos\theta}$$

Therefore: $\dfrac{1}{FP} + \dfrac{1}{FQ} = \dfrac{1-e\cos\theta}{ed} + \dfrac{1+e\cos\theta}{ed} = \dfrac{2}{ed}$

This is remarkable in that $\dfrac{2}{ed}$ is a constant, even though P is a variable point.

21. Draw a focal chord $\overline{AB}$, with A representing the endpoint on the "left". Since $\overline{AB}$ is a 90° rotation from $\overline{PQ}$, then A corresponds to the polar coordinates $A\left(r,\theta+\frac{\pi}{2}\right)$ and B corresponds to the coordinates $B\left(r,\theta+\frac{3\pi}{2}\right)$.

Using the identities $\cos\left(\theta+\frac{\pi}{2}\right) = -\sin\theta$ and $\cos\left(\theta+\frac{3\pi}{2}\right) = \sin\theta$:

$$AF = r = \frac{ed}{1-e\cos\left(\theta+\frac{\pi}{2}\right)} = \frac{ed}{1+e\sin\theta} \qquad FB = r = \frac{ed}{1-e\cos\left(\theta+\frac{3\pi}{2}\right)} = \frac{ed}{1-e\sin\theta}$$

Since $AB = AF + FB$:

$$AB = \frac{ed}{1+e\sin\theta} + \frac{ed}{1-e\sin\theta} = \frac{ed(1-e\sin\theta)+ed(1+e\sin\theta)}{(1+e\sin\theta)(1-e\sin\theta)} = \frac{ed-e^2d\sin\theta+ed+e^2d\sin\theta}{1-e^2\sin^2\theta} = \frac{2ed}{1-e^2\sin^2\theta}$$

Using the result from Exercise 10, we show the required sum is constant:

$$\frac{1}{PQ}+\frac{1}{AB} = \frac{1-e^2\cos^2\theta}{2ed}+\frac{1-e^2\sin^2\theta}{2ed} = \frac{2-e^2\left(\sin^2\theta+\cos^2\theta\right)}{2ed} = \frac{2-e^2}{2ed}$$

But since e and d are constants, we have proven the desired result.

11.8 Rotation of Axes

1. For x, we have: $x = x'\cos\theta - y'\sin\theta = \sqrt{3}\cos 30° - 2\sin 30° = \sqrt{3}\cdot\frac{\sqrt{3}}{2} - 2\cdot\frac{1}{2} = \frac{3}{2} - 1 = \frac{1}{2}$

For y, we have: $y = x'\sin\theta + y'\cos\theta = \sqrt{3}\sin 30° + 2\cos 30° = \sqrt{3}\cdot\frac{1}{2} + 2\cdot\frac{\sqrt{3}}{2} = \frac{\sqrt{3}}{2} + \sqrt{3} = \frac{3\sqrt{3}}{2}$

So the coordinates in the x-y system are $\left(\frac{1}{2}, \frac{3\sqrt{3}}{2}\right)$.

3. For x, we have: $x = x'\cos\theta - y'\sin\theta = \sqrt{2}\cos 45° + \sqrt{2}\sin 45° = \sqrt{2}\cdot\frac{1}{\sqrt{2}} + \sqrt{2}\cdot\frac{1}{\sqrt{2}} = 1 + 1 = 2$

For y, we have: $y = x'\sin\theta + y'\cos\theta = \sqrt{2}\sin 45° - \sqrt{2}\cos 45° = \sqrt{2}\cdot\frac{1}{\sqrt{2}} - \sqrt{2}\cdot\frac{1}{\sqrt{2}} = 1 - 1 = 0$

So the coordinates in the x-y system are $(2, 0)$.

5. For x', we have:

$$x' = x\cos\theta + y\sin\theta = -3\cos\left[\sin^{-1}\left(\tfrac{5}{13}\right)\right] + 1\sin\left[\sin^{-1}\left(\tfrac{5}{13}\right)\right] = -3\cdot\tfrac{12}{13} + 1\cdot\tfrac{5}{13} = -\tfrac{31}{13}$$

For y', we have: $y' = -x\sin\theta + y\cos\theta = 3\sin\left[\sin^{-1}\left(\frac{5}{13}\right)\right] + 1\cos\left[\sin^{-1}\left(\frac{5}{13}\right)\right] = 3\cdot\frac{5}{13} + 1\cdot\frac{12}{13} = \frac{27}{13}$

So the coordinates in the x'-y' system are $\left(-\frac{31}{13}, \frac{27}{13}\right)$.

7. We have:

$$\cot 2\theta = \frac{A-C}{B} = \frac{25-18}{-24} = -\tfrac{7}{24}, \text{ so } \tan 2\theta = -\tfrac{24}{7} \qquad \sec^2 2\theta = 1+\tan^2 2\theta = 1+\left(-\tfrac{24}{7}\right)^2 = \tfrac{625}{49}$$

So $\sec 2\theta = -\frac{25}{7}$ (second quadrant, since $\cot 2\theta < 0$), and thus $\cos 2\theta = -\frac{7}{25}$. Since θ is in the first quadrant:

$$\sin\theta = \sqrt{\frac{1-\cos 2\theta}{2}} = \sqrt{\frac{1+\frac{7}{25}}{2}} = \sqrt{\frac{16}{25}} = \frac{4}{5} \qquad \cos\theta = \sqrt{\frac{1+\cos 2\theta}{2}} = \sqrt{\frac{1-\frac{7}{25}}{2}} = \sqrt{\frac{9}{25}} = \frac{3}{5}$$

9. We have:

$$\cot 2\theta = \frac{1-8}{-24} = \frac{-7}{-24} = \tfrac{7}{24}, \text{ so } \tan 2\theta = \tfrac{24}{7} \qquad \sec^2 2\theta = 1+\tan^2 2\theta = 1+\left(\tfrac{24}{7}\right)^2 = \tfrac{625}{49}$$

So $\sec 2\theta = \frac{25}{7}$ (first quadrant, since $\cot 2\theta > 0$), and thus $\cos 2\theta = \frac{7}{25}$. Since θ is in the first quadrant:

$$\sin\theta = \sqrt{\frac{1-\cos 2\theta}{2}} = \sqrt{\frac{1-\frac{7}{25}}{2}} = \sqrt{\frac{9}{25}} = \frac{3}{5} \qquad \cos\theta = \sqrt{\frac{1+\cos 2\theta}{2}} = \sqrt{\frac{1+\frac{7}{25}}{2}} = \sqrt{\frac{16}{25}} = \frac{4}{5}$$

11. We have: $\cot 2\theta = \frac{A-C}{B} = \frac{1-(-1)}{-2\sqrt{3}} = -\frac{1}{\sqrt{3}}$, so $\tan 2\theta = -\sqrt{3}$

Therefore $2\theta = 120°$, and thus $\theta = 60°$. So:

$$\sin\theta = \sin 60° = \frac{\sqrt{3}}{2} \qquad \cos\theta = \cos 60° = \frac{1}{2}$$

13. We have:

$$\cot 2\theta = \frac{A-C}{B} = \frac{0-(-240)}{161} = \frac{240}{161}, \text{ so } \tan 2\theta = \frac{161}{240} \quad \sec^2 2\theta = 1+\tan^2 2\theta = 1+\left(\frac{161}{240}\right)^2 = \frac{83521}{57600}$$

So $\sec 2\theta = \frac{289}{240}$ (first quadrant, since $\cot 2\theta > 0$), and thus $\cos 2\theta = \frac{240}{289}$. Since θ is in the first quadrant:

$$\sin\theta = \sqrt{\frac{1-\cos 2\theta}{2}} = \sqrt{\frac{1-\frac{240}{289}}{2}} = \sqrt{\frac{49}{578}} = \frac{7\sqrt{2}}{34} \qquad \cos\theta = \sqrt{\frac{1+\cos 2\theta}{2}} = \sqrt{\frac{1+\frac{240}{289}}{2}} = \sqrt{\frac{529}{578}} = \frac{23\sqrt{2}}{34}$$

15. Using the rotation equations:

$$x = x'\cos\theta - y'\sin\theta = x'\cos 45° - y'\sin 45° = \frac{\sqrt{2}}{2}x' - \frac{\sqrt{2}}{2}y'$$
$$y = x'\sin\theta + y'\cos\theta = x'\sin 45° + y'\cos 45° = \frac{\sqrt{2}}{2}x' + \frac{\sqrt{2}}{2}y'$$

So the equation $2xy = 9$ becomes:

$$2\left(\frac{\sqrt{2}}{2}x' - \frac{\sqrt{2}}{2}y'\right)\left(\frac{\sqrt{2}}{2}x' + \frac{\sqrt{2}}{2}y'\right) = 9$$
$$2\left(\frac{1}{2}x'^2 - \frac{1}{2}y'^2\right) = 9$$
$$x'^2 - y'^2 = 9$$

Graph the equation:

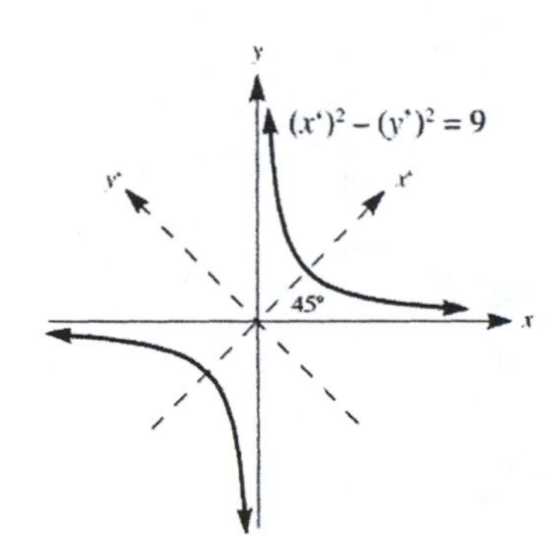

17. We find:

$$\cot 2\theta = \frac{7-1}{8} = \frac{3}{4}, \text{ so } \tan 2\theta = \frac{4}{3}$$
$$\sec^2 2\theta = 1+\tan^2 2\theta = 1+\left(\frac{4}{3}\right)^2 = \frac{25}{9}, \text{ so } \sec 2\theta = \frac{5}{3} \quad (\text{since } 2\theta < 90°)$$

Thus $\cos 2\theta = \frac{3}{5}$, and therefore:

$$\sin\theta = \sqrt{\frac{1-\cos 2\theta}{2}} = \sqrt{\frac{1-\frac{3}{5}}{2}} = \sqrt{\frac{1}{5}} = \frac{\sqrt{5}}{5} \qquad \cos\theta = \sqrt{\frac{1+\cos 2\theta}{2}} = \sqrt{\frac{1+\frac{3}{5}}{2}} = \sqrt{\frac{4}{5}} = \frac{2\sqrt{5}}{5}$$

Thus $\theta = \sin^{-1}\left(\frac{\sqrt{5}}{5}\right) \approx 26.6°$. Now:

$$x = x'\cos\theta - y'\sin\theta = \frac{2\sqrt{5}}{5}x' - \frac{\sqrt{5}}{5}y' \qquad y = x'\sin\theta + y'\cos\theta = \frac{\sqrt{5}}{5}x' + \frac{2\sqrt{5}}{5}y'$$

Making the substitutions into $7x^2 + 8xy + y^2 - 1 = 0$:

$$7\left(\frac{4}{5}x'^2 - \frac{4}{5}x'y' + \frac{1}{5}y'^2\right) + 8\left(\frac{2}{5}x'^2 + \frac{3}{5}x'y' - \frac{2}{5}y'^2\right) + \left(\frac{1}{5}x'^2 + \frac{4}{5}x'y' + \frac{4}{5}y'^2\right) - 1 = 0$$
$$\frac{28}{5}x'^2 - \frac{28}{5}x'y' + \frac{7}{5}y'^2 + \frac{16}{5}x'^2 + \frac{24}{5}x'y' - \frac{16}{5}y'^2 + \frac{1}{5}x'^2 + \frac{4}{5}x'y' + \frac{4}{5}y'^2 - 1 = 0$$
$$9x'^2 - y'^2 = 1$$
$$\frac{x'^2}{\frac{1}{9}} - \frac{y'^2}{1} = 1$$

Rotating $\theta = 26.6°$, sketch the hyperbola:

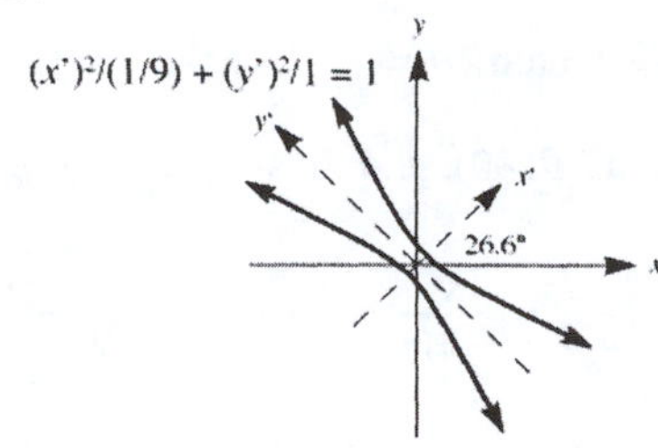

19. Use an alternate approach here:

$$x^2 + 4xy + 4y^2 = 1$$
$$(x + 2y)^2 = 1$$
$$x + 2y = 1$$
$$y = -\tfrac{1}{2}x + \tfrac{1}{2}$$

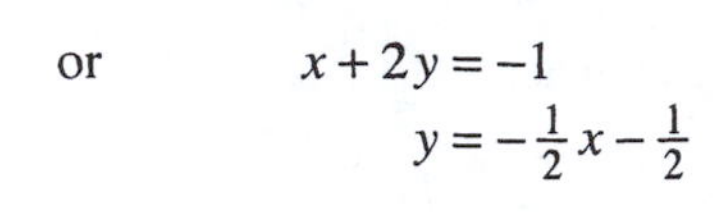

or

$$x + 2y = -1$$
$$y = -\tfrac{1}{2}x - \tfrac{1}{2}$$

The graph consists of two lines:

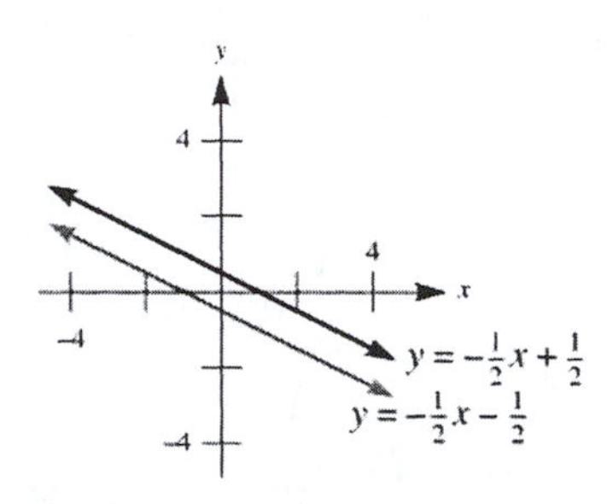

21. We find:

$$\cot 2\theta = \frac{9-16}{-24} = \frac{7}{24}, \text{ so } \tan 2\theta = \frac{24}{7}$$

$$\sec^2 2\theta = 1 + \tan^2 2\theta = 1 + \left(\frac{24}{7}\right)^2 = \frac{625}{49}, \text{ so } \sec 2\theta = \frac{25}{7} \quad (\text{since } 2\theta < 90°)$$

Thus $\cos 2\theta = \frac{7}{25}$, and therefore:

$$\sin\theta = \sqrt{\frac{1-\cos 2\theta}{2}} = \sqrt{\frac{1-\frac{7}{25}}{2}} = \sqrt{\frac{9}{25}} = \frac{3}{5} \qquad \cos\theta = \sqrt{\frac{1+\cos 2\theta}{2}} = \sqrt{\frac{1+\frac{7}{25}}{2}} = \sqrt{\frac{16}{25}} = \frac{4}{5}$$

Thus $\theta = \sin^{-1}\left(\frac{3}{5}\right) \approx 36.9°$. Now:

$$x = x'\cos\theta - y'\sin\theta = \tfrac{4}{5}x' - \tfrac{3}{5}y' \qquad y = x'\sin\theta + y'\cos\theta = \tfrac{3}{5}x' + \tfrac{4}{5}y'$$

Making the substitutions into $9x^2 - 24xy + 16y^2 - 400x - 300y = 0$ and collecting like terms:

$$25y'^2 - 500x' = 0$$
$$y'^2 = 20x'$$

Rotating 36.9°, sketch the parabola:

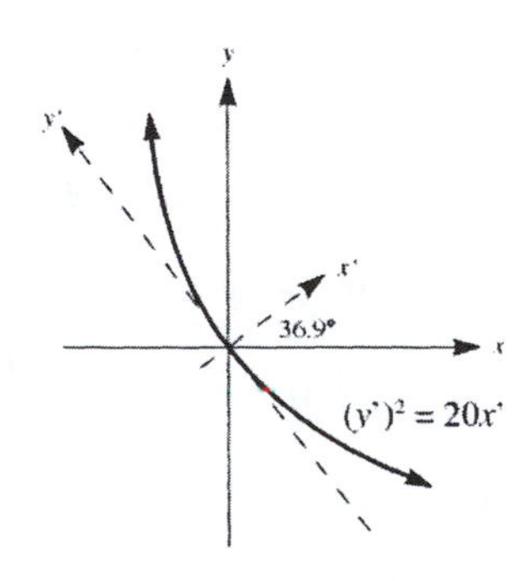

23. We find:

$$\cot 2\theta = \frac{0-3}{4} = -\tfrac{3}{4}, \text{ so } \tan 2\theta = -\tfrac{4}{3}$$

$$\sec^2 2\theta = 1 + \tan^2 2\theta = 1 + \left(-\tfrac{4}{3}\right)^2 = \tfrac{25}{9}$$

$$\sec 2\theta = -\tfrac{5}{3} \quad (\text{since } 2\theta > 90°)$$

Thus $\cos 2\theta = -\frac{3}{5}$, and therefore:

$$\sin\theta = \sqrt{\frac{1-\cos 2\theta}{2}} = \sqrt{\frac{1+\frac{3}{5}}{2}} = \sqrt{\tfrac{4}{5}} = \frac{2\sqrt{5}}{5} \qquad \cos\theta = \sqrt{\frac{1+\cos 2\theta}{2}} = \sqrt{\frac{1-\frac{3}{5}}{2}} = \sqrt{\tfrac{1}{5}} = \frac{\sqrt{5}}{5}$$

Thus $\theta = \cos^{-1}\left(\frac{\sqrt{5}}{5}\right) \approx 63.4°$. Now:

$$x = x'\cos\theta - y'\sin\theta = \tfrac{\sqrt{5}}{5}x' - \tfrac{2\sqrt{5}}{5}y' \qquad y = x'\sin\theta + y'\cos\theta = \tfrac{2\sqrt{5}}{5}x' + \tfrac{\sqrt{5}}{5}y'$$

Making the substitutions into $4xy + 3y^2 + 4x + 6y = 1$ and completing the square on x' and y' terms:

$$\frac{\left(x' + \frac{2\sqrt{5}}{5}\right)^2}{1} - \frac{\left(y' + \frac{\sqrt{5}}{5}\right)^2}{4} = 1$$

Rotating 63.4°, sketch the hyperbola:

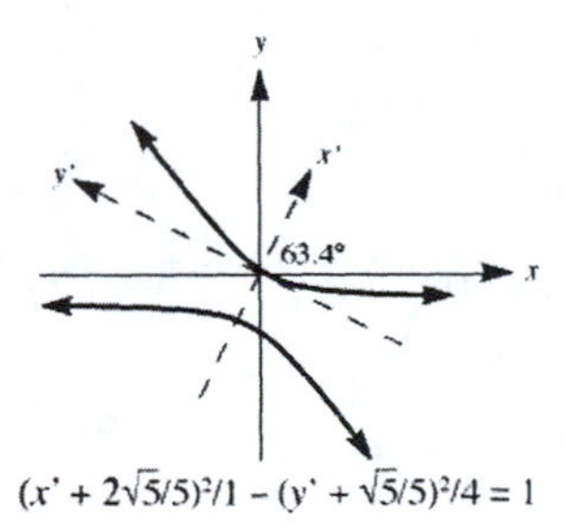

25. We find: $\cot 2\theta = \dfrac{3-3}{-2} = 0$, so $2\theta = 90°$ and thus $\theta = 45°$. Therefore:

$$x = x'\cos\theta - y'\sin\theta = x'\cos 45° - y'\sin 45° = \tfrac{\sqrt{2}}{2}x' - \tfrac{\sqrt{2}}{2}y'$$

$$y = x'\sin\theta + y'\cos\theta = x'\sin 45° + y'\cos 45° = \tfrac{\sqrt{2}}{2}x' + \tfrac{\sqrt{2}}{2}y'$$

Making the substitutions into $3x^2 - 2xy + 3y^2 - 6\sqrt{2}\,x + 2\sqrt{2}\,y + 4 = 0$ and completing the square on x' and y' terms: $\dfrac{(x'-1)^2}{1} + \dfrac{(y'+1)^2}{\frac{1}{2}} = 1$. Rotating 45°, sketch the ellipse:

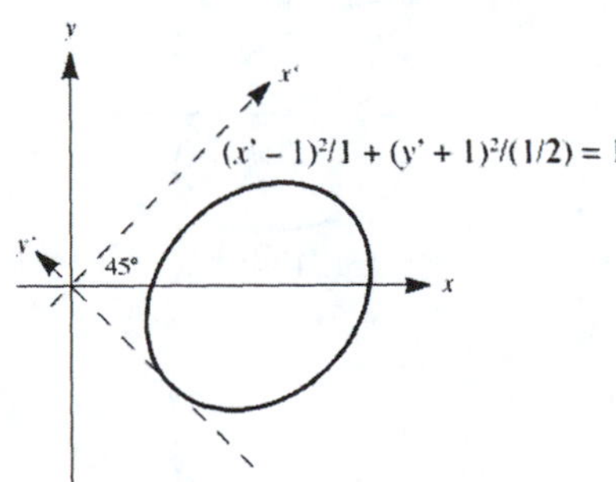

27. First multiply out to get:

$$x^2 - 2xy + y^2 = 8y - 48$$
$$x^2 - 2xy + y^2 - 8y + 48 = 0$$

Now: $\cot 2\theta = \frac{1-1}{-2} = 0$, so $2\theta = 90°$ and thus $\theta = 45°$. Therefore:

$$x = x'\cos\theta - y'\sin\theta = \tfrac{\sqrt{2}}{2}x' - \tfrac{\sqrt{2}}{2}y' \qquad y = x'\sin\theta + y'\cos\theta = \tfrac{\sqrt{2}}{2}x' + \tfrac{\sqrt{2}}{2}y'$$

Making the substitutions and completing the square on x' and y' terms yields: $\left(y' - \sqrt{2}\right)^2 = 2\sqrt{2}\left(x' - \frac{11\sqrt{2}}{2}\right)$

Rotating 45°, sketch the parabola:

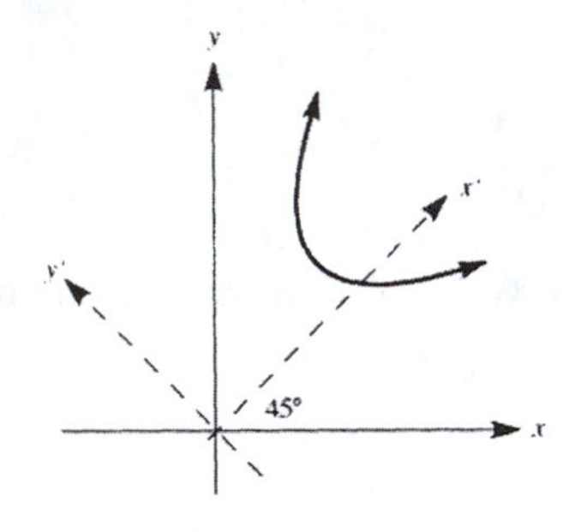

29. We find:

$$\cot 2\theta = \frac{3-6}{4} = -\tfrac{3}{4}, \text{ so } \tan 2\theta = -\tfrac{4}{3}$$
$$\sec^2 2\theta = 1 + \tan^2 2\theta = 1 + \left(-\tfrac{4}{3}\right)^2 = \tfrac{25}{9}$$

So $\sec 2\theta = -\frac{5}{3}$ (since $2\theta > 90°$), and thus $\cos 2\theta = -\frac{3}{5}$. Therefore:

$$\sin\theta = \sqrt{\frac{1-\cos 2\theta}{2}} = \sqrt{\frac{1+\frac{3}{5}}{2}} = \frac{2\sqrt{5}}{5} \qquad \cos\theta = \sqrt{\frac{1+\cos 2\theta}{2}} = \sqrt{\frac{1-\frac{3}{5}}{2}} = \frac{\sqrt{5}}{5}$$

Thus $\theta = \cos^{-1}\left(\frac{\sqrt{5}}{5}\right) \approx 63.4°$. Now:

$$x = x'\cos\theta - y'\sin\theta = \tfrac{\sqrt{5}}{5}x' - \tfrac{2\sqrt{5}}{5}y' \qquad y = x'\sin\theta + y'\cos\theta = \tfrac{2\sqrt{5}}{5}x' + \tfrac{\sqrt{5}}{5}y'$$

Making the substitutions into $3x^2 + 4xy + 6y^2 = 7$: $\frac{x'^2}{1} + \frac{y'^2}{\frac{7}{2}} = 1$

Rotating 63.4°, sketch the ellipse:

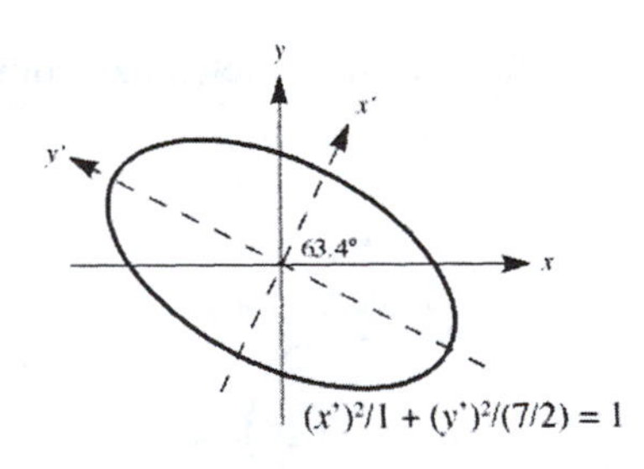

31. We find: $\cot 2\theta = \dfrac{17-8}{-12} = -\frac{3}{4}$, so $\tan 2\theta = -\frac{4}{3}$

As with Exercise 29 we find $\sin\theta = \frac{2\sqrt{5}}{5}$ and $\cos\theta = \frac{\sqrt{5}}{5}$, so:

$$x = \frac{\sqrt{5}}{5}x' - \frac{2\sqrt{5}}{5}y' \qquad\qquad y = \frac{2\sqrt{5}}{5}x' + \frac{\sqrt{5}}{5}y'$$

Substituting into $17x^2 - 12xy + 8y^2 - 80 = 0$ yields: $\dfrac{x'^2}{16} + \dfrac{y'^2}{4} = 1$. Rotating 63.4°, sketch the ellipse:

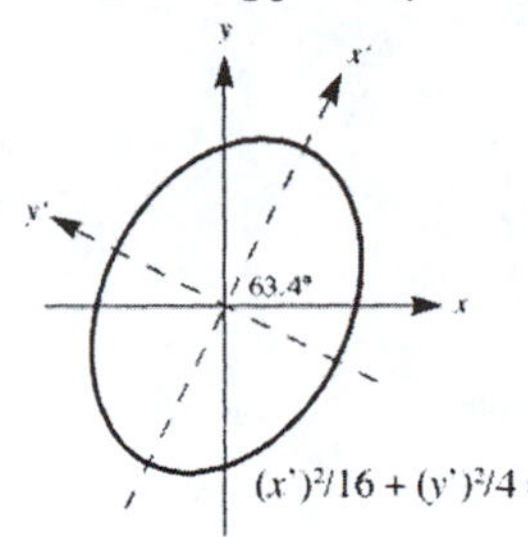

33. We find:

$$\cot 2\theta = \frac{0+4}{3} = \frac{4}{3}, \text{ so } \tan 2\theta = \frac{3}{4}$$

$$\sec^2 2\theta = 1 + \tan^2 2\theta = 1 + \frac{9}{16} = \frac{25}{16}, \text{ so } \sec 2\theta = \frac{5}{4}$$

Then $\cos 2\theta = \frac{4}{5}$, and thus:

$$\sin\theta = \sqrt{\frac{1-\cos 2\theta}{2}} = \sqrt{\frac{1-\frac{4}{5}}{2}} = \frac{\sqrt{10}}{10} \qquad\qquad \cos\theta = \sqrt{\frac{1+\cos 2\theta}{2}} = \sqrt{\frac{1+\frac{4}{5}}{2}} = \frac{3\sqrt{10}}{10}$$

Then $\theta = \sin^{-1}\left(\frac{\sqrt{10}}{10}\right) \approx 18.4°$, and:

$$x = \frac{3\sqrt{10}}{10}x' - \frac{\sqrt{10}}{10}y' \qquad\qquad y = \frac{\sqrt{10}}{10}x' + \frac{3\sqrt{10}}{10}y'$$

Substituting into $3xy - 4y^2 + 18 = 0$ results in: $\dfrac{y'^2}{4} - \dfrac{x'^2}{36} = 1$. Rotating 18.4°, sketch the hyperbola:

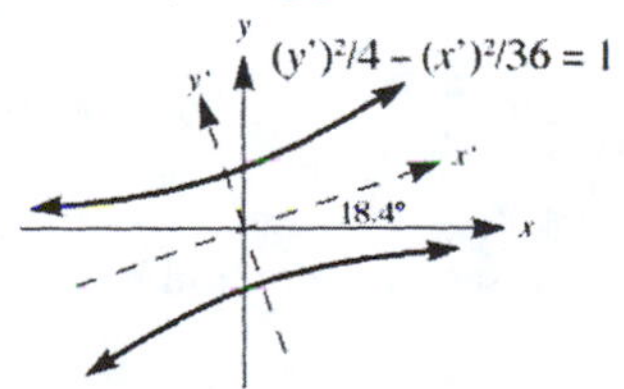

35. First multiply out terms to obtain: $x^2 + 2xy + y^2 + 4\sqrt{2}x - 4\sqrt{2}y = 0$

We find: $\cot 2\theta = \dfrac{1-1}{2} = 0$, so $2\theta = 90°$ and thus $\theta = 45°$

Now:

$$x = \frac{\sqrt{2}}{2}x' - \frac{\sqrt{2}}{2}y' \qquad\qquad y = \frac{\sqrt{2}}{2}x' + \frac{\sqrt{2}}{2}y'$$

Substituting into $x^2 + 2xy + y^2 + 4\sqrt{2}\,x - 4\sqrt{2}\,y = 0$ results in: $x'^2 = 4y'$. Rotating 45°, sketch the parabola:

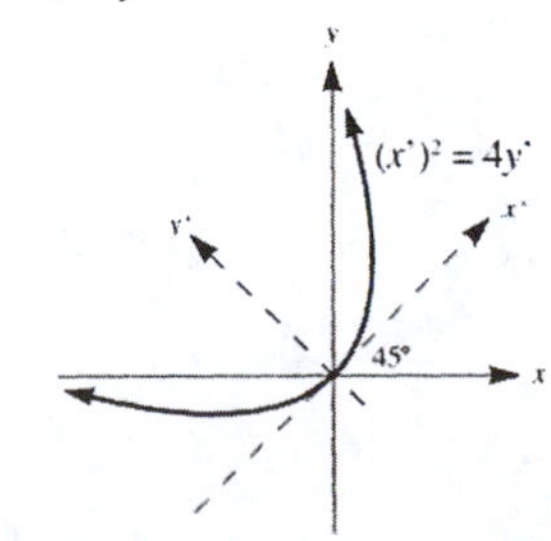

37. We find:

$$\cot 2\theta = \frac{3-2}{-\sqrt{15}} = -\frac{1}{\sqrt{15}}, \text{ so } \tan 2\theta = -\sqrt{15}$$

$$\sec^2 2\theta = 1 + \tan^2 2\theta = 1 + 15 = 16, \text{ so } \sec 2\theta = -4 \quad (\text{since } 2\theta > 90°)$$

Thus $\cos 2\theta = -\frac{1}{4}$ and we find:

$$\sin\theta = \sqrt{\frac{1-\cos 2\theta}{2}} = \sqrt{\frac{1+\frac{1}{4}}{2}} = \sqrt{\frac{5}{8}} = \frac{\sqrt{10}}{4} \qquad \cos\theta = \sqrt{\frac{1+\cos 2\theta}{2}} = \sqrt{\frac{1-\frac{1}{4}}{2}} = \sqrt{\frac{3}{8}} = \frac{\sqrt{6}}{4}$$

Then $\theta = \cos^{-1}\left(\frac{\sqrt{6}}{4}\right) \approx 52.2°$, and:

$$x = \frac{\sqrt{6}}{4}x' - \frac{\sqrt{10}}{4}y' \qquad y = \frac{\sqrt{10}}{4}x' + \frac{\sqrt{6}}{4}y'$$

Substituting into $3x^2 - \sqrt{15}\,xy + 2y^2 = 3$ results in: $\frac{x'^2}{6} + \frac{y'^2}{\frac{2}{3}} = 1$. Rotating 52.2°, sketch the ellipse:

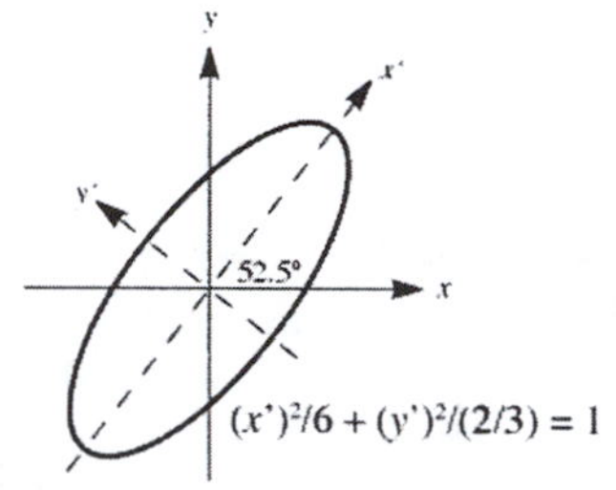

39. We find: $\cot 2\theta = \frac{3-3}{-2} = 0$, so $2\theta = 90°$ and thus $\theta = 45°$. Now:

$$x = \frac{\sqrt{2}}{2}x' - \frac{\sqrt{2}}{2}y' \qquad y = \frac{\sqrt{2}}{2}x' + \frac{\sqrt{2}}{2}y'$$

Substituting into $3x^2 - 2xy + 3y^2 + 2 = 0$ results in: $x'^2 + 2y'^2 = -1$

But clearly this is impossible, so there is no graph.

41. Since $\frac{A-C}{B}$ can be any real number, and if $0° < \theta < 90°$, then $0° < 2\theta < 180°$, which is the principal period for the cotangent function, then there is always an angle 2θ such that $\cot 2\theta = \frac{A-C}{B}$. Recall that the range of the cotangent function is any real number.

43. Multiplying the first equation by $\sin\theta$ and the second equation by $\cos\theta$ yields:

$$(\sin\theta\cos\theta)x + (\sin^2\theta)y = x'\sin\theta$$
$$(-\sin\theta\cos\theta)x + (\cos^2\theta)y = y'\cos\theta$$

Adding:

$$(\sin^2\theta + \cos^2\theta)y = x'\sin\theta + y'\cos\theta$$
$$y = x'\sin\theta + y'\cos\theta$$

Multiplying the first equation by $\cos\theta$ and the second equation by $-\sin\theta$ yields:

$$(\cos^2\theta)x + (\sin\theta\cos\theta)y = x'\cos\theta$$
$$(\sin^2\theta)x - (\sin\theta\cos\theta)y = -y'\sin\theta$$

Adding:

$$(\sin^2\theta + \cos^2\theta)x = x'\cos\theta - y'\sin\theta$$
$$x = x'\cos\theta - y'\sin\theta$$

45. Making the substitutions for x and y:

$$A(x'\cos\theta - y'\sin\theta)^2 + B(x'\cos\theta - y'\sin\theta)(x'\sin\theta + y'\cos\theta)$$
$$+C(x'\sin\theta + y'\cos\theta)^2 + D(x'\cos\theta - y'\sin\theta) + E(x'\sin\theta + y'\cos\theta) + F = 0$$
$$A\cos^2\theta x'^2 - 2A\sin\theta\cos\theta x'y' + A\sin^2\theta y'^2 + B\sin\theta\cos\theta x'^2$$
$$+B\left(\cos^2\theta - \sin^2\theta\right)x'y' - B\sin\theta\cos\theta y'^2 + C\sin^2\theta x'^2 + 2C\sin\theta\cos\theta x'y'$$
$$+C\cos^2\theta y'^2 + D\cos\theta x' - D\sin\theta y' + E\sin\theta x' + E\cos\theta y' + F = 0$$
$$\left(A\cos^2\theta + B\sin\theta\cos\theta + C\sin^2\theta\right)x'^2 + \left[2(C-A)\sin\theta\cos\theta\right.$$
$$\left.+B\left(\cos^2\theta - \sin^2\theta\right)\right]x'y' + \left(A\sin^2\theta - B\sin\theta\cos\theta + C\cos^2\theta\right)y'^2$$
$$+(D\cos\theta + E\sin\theta)x' + (E\cos\theta - D\sin\theta)y' + F = 0$$
$$A'x'^2 + B'x'y' + C'y'^2 + D'x' + E'y' + F' = 0$$

47. **a.** Substituting for A' and C':

$$\begin{aligned} A' - C' &= A\cos^2\theta + B\sin\theta\cos\theta + C\sin^2\theta - A\sin^2\theta + B\sin\theta\cos\theta - C\cos^2\theta \\ &= A\left(\cos^2\theta - \sin^2\theta\right) - C\left(\cos^2\theta - \sin^2\theta\right) + 2B\sin\theta\cos\theta \\ &= A\cos 2\theta - C\cos 2\theta + B\sin 2\theta \\ &= (A-C)\cos 2\theta + B\sin 2\theta \end{aligned}$$

b. Substituting for B':

$$B' = (C-A)(2\sin\theta\cos\theta) + B\left(\cos^2\theta - \sin^2\theta\right) = (C-A)\sin 2\theta + B\cos 2\theta = B\cos 2\theta - (A-C)\sin 2\theta$$

c. First compute the powers:

$$\begin{aligned} (A'-C')^2 &= (A-C)^2\cos^2 2\theta + 2B(A-C)\sin 2\theta\cos 2\theta + B^2\sin^2 2\theta \\ B'^2 &= B^2\cos^2 2\theta - 2B(A-C)\sin 2\theta + (A-C)^2\sin^2 2\theta \end{aligned}$$

Adding these two equations:

$$(A'-C')^2 + B'^2 = (A-C)^2\left(\cos^2 2\theta + \sin^2 2\theta\right) + B^2\left(\sin^2 2\theta + \cos^2 2\theta\right) = (A-C)^2 + B^2$$

d. Squaring the equation $A' + C' = A + C$ we obtain:

$$A'^2 + 2A'C' + C'^2 = A^2 + 2AC + C^2$$

By subtracting the equation in part **c**:

$$\begin{aligned} -4A'C' + B'^2 &= -4AC + B^2 \\ B'^2 - 4A'C' &= B^2 - 4AC \end{aligned}$$

Chapter 11 Review Exercises

1. The x-axis contains $\overline{AB}$ and has equation $y = 0$. The slope of the line containing $\overline{BC}$ is $-\frac{1}{b}$ and the equation is:

$$\begin{aligned} y &= -\tfrac{1}{b}(x - 6b) \\ by &= 6b - x \\ x + by &= 6b \end{aligned}$$

The slope of the line containing $\overline{AC}$ is $-\frac{1}{a}$ and the equation is:

$$\begin{aligned} y &= -\tfrac{1}{a}(x - 6a) \\ x + ay &= 6a \end{aligned}$$

3. Using the equations from Exercise 2, we show $G(2a + 2b, 2)$ lies on all three medians:

$$2(2a+2b) + (a+b)(2) = 4a + 4b + 2a + 2b = 6(a+b)$$
$$2a + 2b - (b - 2a)(2) = 2a + 2b - 2b + 4a = 6a$$
$$2a + 2b - (a - 2b)(2) = 2a + 2b - 2a + 4b = 6b$$

5. Using the equations from Exercise 4, we show $H(0,-6ab)$ lies on each altitude:

$$x = 0$$
$$-6ab = b \cdot 0 - 6ab = -6ab$$
$$-6ab = a \cdot 0 - 6ab = -6ab$$

7. Using the equations from Exercise 6, we show $O(3a + 3b, 3ab + 3)$ lies on each perpendicular bisector:

$$x = 3a + 3b$$
$$b(3a+3b) - (3ab+3) = 3ab + 3b^2 - 3ab - 3 = 3b^2 - 3$$
$$a(3a+3b) - (3ab+3) = 3a^2 + 3ab - 3ab - 3 = 3a^2 - 3$$

9. Find each distance and use the result from Exercise 8:

$$p = \sqrt{(6b)^2 + 6^2} = 6\sqrt{b^2+1} \qquad q = \sqrt{(6a)^2 + 6^2} = 6\sqrt{a^2+1}$$
$$r = 6(a-b) \qquad R = 3\sqrt{(a^2+1)(b^2+1)}$$

Thus:

$$\frac{pqr}{4R} = \frac{\left(6\sqrt{b^2+1}\right)\left(6\sqrt{a^2+1}\right)\left(6(a-b)\right)}{4 \cdot 3\sqrt{\left(a^2+1\right)\left(b^2+1\right)}} = 18(a-b)$$

$$\text{Area of } \Delta ABC = \tfrac{1}{2}\left(6(a-b)\right)(6) = 18(a-b)$$

So the area is $\frac{pqr}{4R}$, as required.

11. Compute each side of the identity:

$$\begin{aligned} OH^2 &= (3a+3b-0)^2 + (3ab+3+6ab)^2 \\ &= 9\left(a^2+2ab+b^2\right) + 9(3ab+1)^2 \\ &= 9\left(a^2+2ab+b^2+9a^2b^2+6ab+1\right) \\ &= 81a^2b^2 + 9a^2 + 9b^2 + 72ab + 9 \end{aligned}$$

$$\begin{aligned} 9R^2 - \left(p^2+q^2+r^2\right) & \\ &= 9\left[9\left(a^2+1\right)\left(b^2+1\right)\right] - 36\left(b^2+1\right) - 36\left(a^2+1\right) - 36\left(a^2-2ab+b^2\right) \\ &= 81a^2b^2 + 81a^2 + 81b^2 + 81 - 36b^2 - 36 - 36a^2 - 36 - 36a^2 + 72ab - 36b^2 \\ &= 81a^2b^2 + 9a^2 + 9b^2 + 72ab + 9 \end{aligned}$$

Thus $OH^2 = 9R^2 - (p^2 + q^2 + r^2)$.

13. First compute the squares:

$$HA^2 = (6a)^2 + (-6ab)^2 = 36a^2 + 36a^2b^2$$
$$HB^2 = (6b)^2 + (-6ab)^2 = 36b^2 + 36a^2b^2$$
$$HC^2 = 0^2 + (6+6ab)^2 = 36 + 72ab + 36a^2b^2$$

So $HA^2 + HB^2 + HC^2 = 108a^2b^2 + 36a^2 + 36b^2 + 72ab + 36$. Now compute the right-hand side:

$$\begin{aligned} 12R^2 - \left(p^2+q^2+r^2\right) &= 12\left[9\left(a^2+1\right)\left(b^2+1\right)\right] - \left[36\left(b^2+1\right) + 36\left(a^2+1\right) + 36\left(a^2-2ab+b^2\right)\right] \\ &= 108a^2b^2 + 108a^2 + 108b^2 + 108 - 36b^2 - 36 - 36a^2 - 36 - 36a^2 + 72ab - 36b^2 \\ &= 108a^2b^2 + 36a^2 + 36b^2 + 72ab + 36 \end{aligned}$$

Thus $HA^2 + HB^2 + HC^2 = 12R^2 - (p^2 + q^2 + r^2)$.

15. In Exercise 12 we saw that: $GH^2 = 4\left(9a^2b^2 + a^2 + b^2 + 8ab + 1\right)$

Therefore: $GH = 2\sqrt{9a^2b^2 + a^2 + b^2 + 8ab + 1}$

Now:

$$\begin{aligned} 2GO &= 2\sqrt{(2a+2b-3a-3b)^2 + (2-3ab-3)^2} \\ &= 2\sqrt{a^2 + 2ab + b^2 + 1 + 6ab + 9a^2b^2} \\ &= 2\sqrt{9a^2b^2 + a^2 + b^2 + 8ab + 1} \end{aligned}$$

Thus $GH = 2GO$.

17. Since $\tan\theta = -\frac{2}{3}$, then $\theta \approx 146.3°$.

19. Given $(x_0, y_0) = (-1,-3)$, $A = 5$, $B = 6$ and $C = -30$, so using the distance formula from a point to a line yields:

$$d = \frac{|5(-1)+6(-3)-30|}{\sqrt{5^2+6^2}} = \frac{53}{\sqrt{61}} = \frac{53\sqrt{61}}{61}$$

21. Label $A(-6,0)$, $B(6,0)$ and $C\left(0,6\sqrt{3}\right)$. The height of the triangle is $6\sqrt{3}$. Since the line $\overline{AB}$ is the x-axis, the distance from (1,2) to $\overline{AB}$ is 2. The line containing $\overline{AC}$ has slope $\sqrt{3}$ and equation:

$$\begin{aligned} y - 0 &= \sqrt{3}(x+6) \\ y &= \sqrt{3}x + 6\sqrt{3} \end{aligned}$$

The distance from (1,2) to $\overline{AC}$ is thus: $\dfrac{\left|\sqrt{3}(1)+6\sqrt{3}-2\right|}{\sqrt{1^2+\left(-\sqrt{3}\right)^2}} = \dfrac{7\sqrt{3}-2}{2}$

The line containing $\overline{BC}$ has slope $-\sqrt{3}$ and equation:

$$\begin{aligned} y - 0 &= -\sqrt{3}(x-6) \\ y &= -\sqrt{3}x + 6\sqrt{3} \end{aligned}$$

The distance from (1,2) to $\overline{BC}$ is thus: $\dfrac{\left|-\sqrt{3}(1)+6\sqrt{3}-2\right|}{\sqrt{1^2+\left(\sqrt{3}\right)^2}} = \dfrac{5\sqrt{3}-2}{2}$

The sum of these distances is $6\sqrt{3}$, which is also the height.

23. **a.** The form is $y^2 = 4px$, where $p = 4$. Thus the equation is $y^2 = 16x$.

b. The form is $x^2 = 4py$, where $p = 4$. Thus the equation is $x^2 = 16y$.

25. Since the parabola is symmetric about the positive y-axis, its equation must be of the form $x^2 = 4py$, where $p > 0$. Now the focal width is 12, so $4p = 12$. Thus the equation is $x^2 = 12y$.

27. We have $c = 2$ and $a = 8$, and the ellipse must have the form: $\dfrac{x^2}{8^2} + \dfrac{y^2}{b^2} = 1$

Since $c^2 = a^2 - b^2$, find b:

$$\begin{aligned} 4 &= 64 - b^2 \\ b^2 &= 60 \end{aligned}$$

The equation is $\dfrac{x^2}{64} + \dfrac{y^2}{60} = 1$, or $15x^2 + 16y^2 = 960$.

29. Since one end of the minor axis is (–6,0), $b = 6$ and the ellipse has a form of: $\frac{x^2}{36}+\frac{y^2}{a^2}=1$

Now $\frac{c}{a}=\frac{4}{5}$, so $c=\frac{4}{5}a$. Finding a:

$$c^2 = a^2 - b^2$$
$$\left(\tfrac{4}{5}a\right)^2 = a^2 - 36$$
$$\tfrac{16}{25}a^2 = a^2 - 36$$
$$36 = \tfrac{9}{25}a^2$$
$$100 = a^2$$

The equation is $\frac{x^2}{36}+\frac{y^2}{100}=1$, or $25x^2+9y^2=900$.

31. Since the foci are (±2,0), $c = 2$. Since the directrices are $x = \pm 5$, $\frac{a}{e}=5$, so $a = 5e$. But since $e=\frac{c}{a}=\frac{2}{a}$:

$$a = 5\bullet\frac{2}{a}$$
$$a^2 = 10$$

Since $a^2 - b^2 = c^2$, we can find b^2:

$$10 - b^2 = 2^2$$
$$-b^2 = -6$$
$$b^2 = 6$$

The equation of the ellipse is $\frac{x^2}{10}+\frac{y^2}{6}=1$, or $3x^2+5y^2=30$.

33. Since the foci are (0,±3), $c = 3$ and the equation has the form: $\frac{y^2}{a^2}-\frac{x^2}{b^2}=1$

Now $\frac{a}{b}=2$, so $a = 2b$. Substitute:

$$c^2 = a^2 + b^2$$
$$9 = (2b)^2 + b^2$$
$$9 = 5b^2$$
$$\tfrac{9}{5} = b^2$$
$$\tfrac{36}{5} = a^2$$

The equation is $\frac{y^2}{36/5}-\frac{x^2}{9/5}=1$, or $5y^2-20x^2=36$.

35. Here $2a = 3$, so $a=\frac{3}{2}$. We can also find c:

$$\frac{c}{a}=\frac{5}{4}$$
$$\frac{c}{\frac{3}{2}}=\frac{5}{4}$$
$$c=\frac{15}{8}$$

Now find b by substitution:

$$a^2 + b^2 = c^2$$
$$\tfrac{9}{4} + b^2 = \tfrac{225}{64}$$
$$144 + 64b^2 = 225$$
$$64b^2 = 81$$
$$b^2 = \tfrac{81}{64}$$

Since the focal axis is horizontal, the equation must have the form: $\dfrac{x^2}{a^2} - \dfrac{y^2}{b^2} = 1$

The equation is $\dfrac{x^2}{9/4} - \dfrac{y^2}{81/64} = 1$, or $36x^2 - 64y^2 = 81$.

37. Note that $4p = 10$, so $p = \tfrac{5}{2}$. The vertex is (0,0), the focus is $\left(0, \tfrac{5}{2}\right)$, the directrix is $y = -\tfrac{5}{2}$, and the focal width is 10.

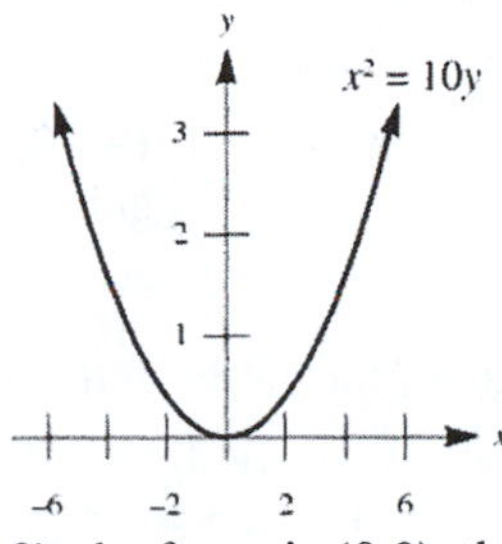

39. Note that $4p = 12$, so $p = 3$. The vertex is (0,3), the focus is (0,0), the directrix is $y = 6$, and the focal width is 12.

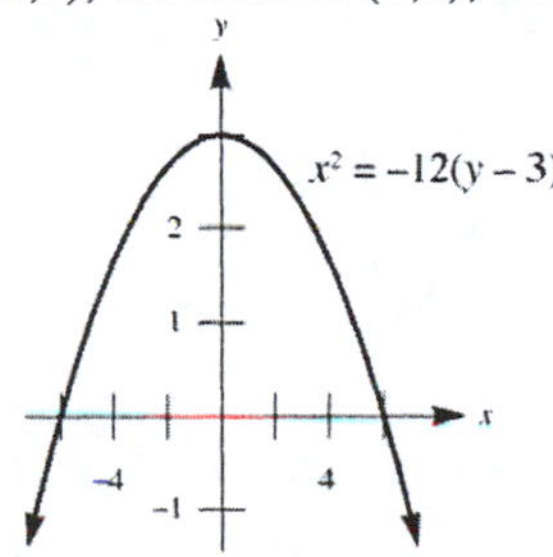

41. Note that $4p = 4$, so $p = 1$. The vertex is (1,1), the focus is (0,1), the directrix is $x = 2$, and the focal width is 4.

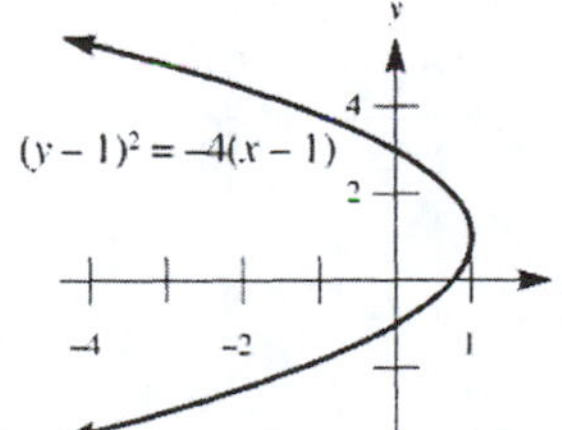

43. Dividing by 144, the standard form is $\dfrac{x^2}{36} + \dfrac{y^2}{16} = 1$. The center is (0,0), the length of the major axis is 12, and the length of the minor axis is 8. Using $c^2 = a^2 - b^2$: $c = \sqrt{a^2 - b^2} = \sqrt{36 - 16} = \sqrt{20} = 2\sqrt{5}$

The foci are $\left(\pm 2\sqrt{5}, 0\right)$ and the eccentricity is $\dfrac{2\sqrt{5}}{6} = \dfrac{\sqrt{5}}{3}$.

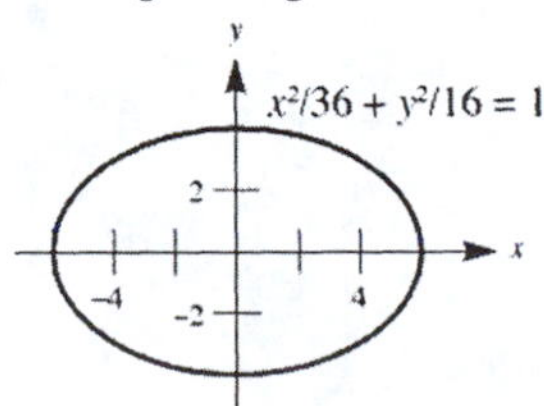

45. Dividing by 9, the standard form is $\frac{x^2}{1}+\frac{y^2}{9}=1$. The center is (0,0), the length of the major axis is 6, and the length of the minor axis is 2. Using $c^2=a^2-b^2$: $c=\sqrt{a^2-b^2}=\sqrt{9-1}=\sqrt{8}=2\sqrt{2}$

The foci are $\left(0,\pm 2\sqrt{2}\right)$ and the eccentricity is $\frac{2\sqrt{2}}{3}$.

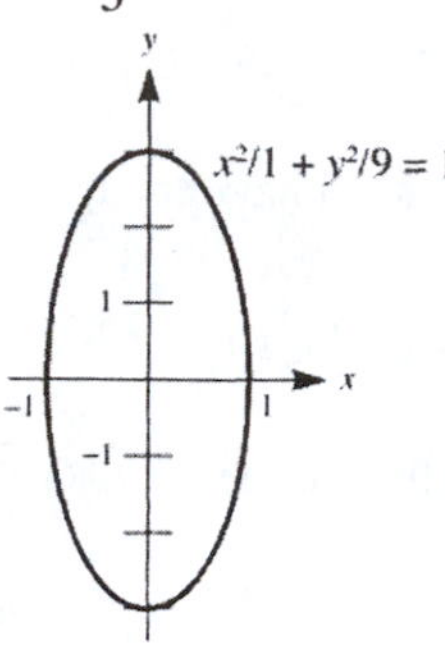

47. The equation is already in standard form where the center is (–3, 0), and the lengths of the major and minor axes are 6. Since the lengths of the major and minor axes are equal this ellipse is actually a circle.

Using $c^2=a^2-b^2$: $c=\sqrt{a^2-b^2}=\sqrt{9-9}=0$

The only focus is at (–3,0), which is the center of the circle, and the eccentricity is $\frac{0}{3}=0$.

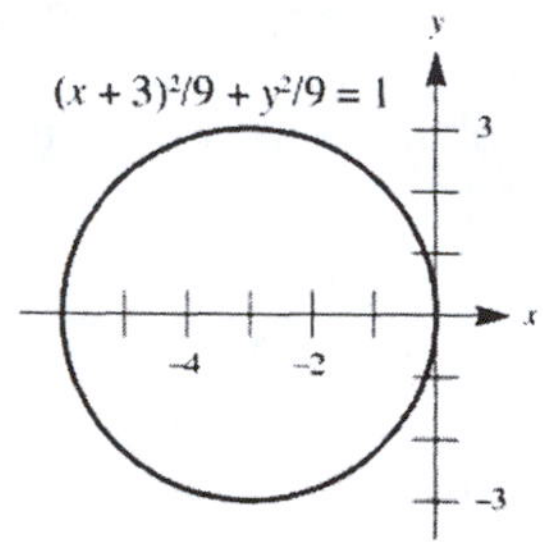

49. Dividing by 144, the standard form is $\frac{x^2}{36}-\frac{y^2}{16}=1$. The center is (0,0), the vertices are (±6,0), and the asymptotes are $y=\pm\frac{4}{6}x=\pm\frac{2}{3}x$. Using $c^2=a^2+b^2$: $c=\sqrt{a^2+b^2}=\sqrt{36+16}=\sqrt{52}=2\sqrt{13}$

The foci are $\left(\pm 2\sqrt{13},0\right)$ and the eccentricity is $\frac{2\sqrt{13}}{6}=\frac{\sqrt{13}}{3}$.

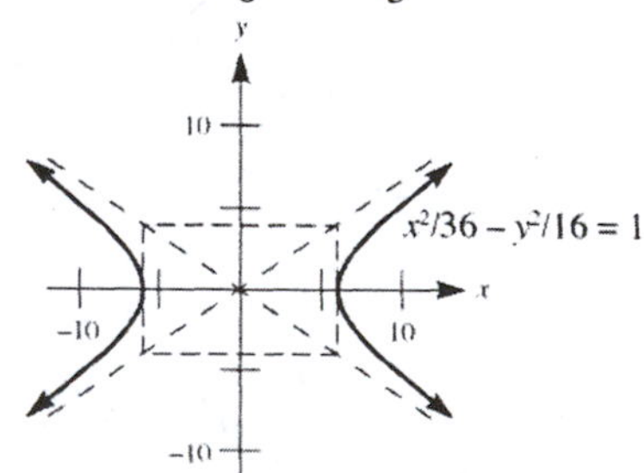

51. Dividing by 9, the standard form is $\frac{y^2}{1}-\frac{x^2}{9}=1$. The center is (0,0), the vertices are (0,±1), and the asymptotes are $y=\pm\frac{1}{3}x$. Using $c^2=a^2+b^2$: $c=\sqrt{a^2+b^2}=\sqrt{1+9}=\sqrt{10}$

The foci are $\left(0,\pm\sqrt{10}\right)$ and the eccentricity is $\frac{\sqrt{10}}{1}=\sqrt{10}$.

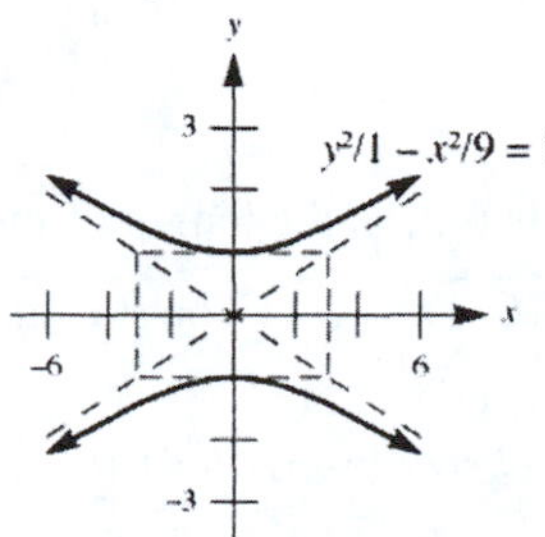

53. The equation is already in standard form where the center is (0,–3) and the vertices are (0, –3 + 3) = (0,0) and (0, –3 – 3) = (0,– 6). The asymptotes have slopes of $\pm\frac{3}{3}=\pm1$, so using the point-slope formula:

$$\begin{aligned} y-(-3)&=1(x-0) \\ y+3&=x \\ y&=x-3 \end{aligned} \qquad \begin{aligned} y-(-3)&=-1(x-0) \\ y+3&=-x \\ y&=-x-3 \end{aligned}$$

Using $c^2=a^2+b^2$: $c=\sqrt{a^2+b^2}=\sqrt{9+9}=\sqrt{18}=3\sqrt{2}$

The foci are $\left(0,-3\pm3\sqrt{2}\right)$ and the eccentricity is $\frac{3\sqrt{2}}{3}=\sqrt{2}$.

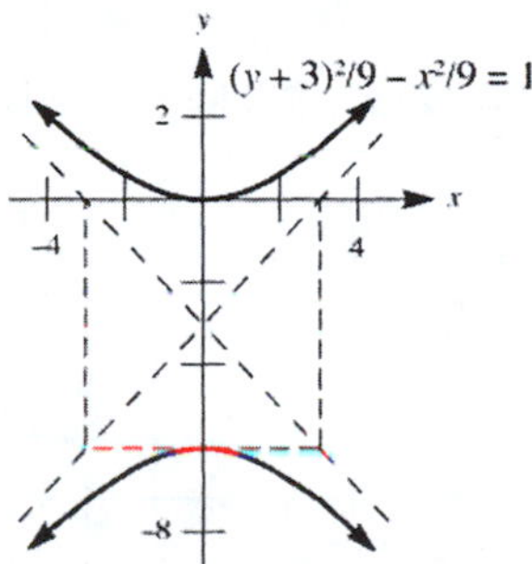

55. Complete the square to convert the equation to standard form:

$$\begin{aligned} y^2-8y&=16x-80 \\ y^2-8y+16&=16x-64 \\ (y-4)^2&=16(x-4) \end{aligned}$$

This is the equation of a parabola. Since $4p$ = 16, p = 4. The vertex is (4,4), the axis of symmetry is y = 4, the focus is (8,4), and the directrix is x = 0.

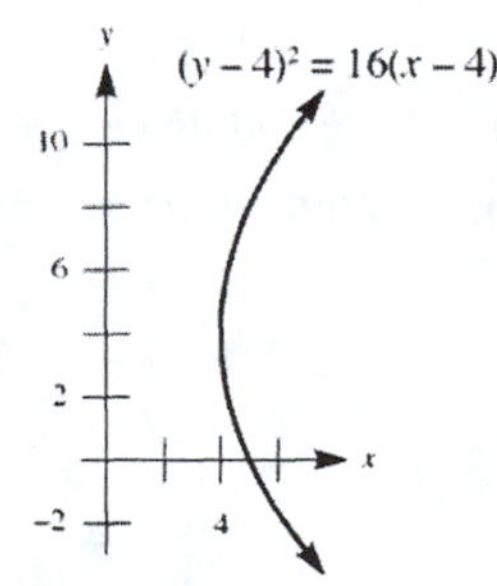

57. Complete the square to convert the equation to standard form:

$$16x^2 + 64x + 9y^2 - 54y = -1$$
$$16(x^2 + 4x) + 9(y^2 - 6y) = -1$$
$$16(x^2 + 4x + 4) + 9(y^2 - 6y + 9) = -1 + 64 + 81$$
$$16(x+2)^2 + 9(y-3)^2 = 144$$
$$\frac{(x+2)^2}{9} + \frac{(y-3)^2}{16} = 1$$

This is the equation of an ellipse. Its center is (–2,3), the length of the major axis is 8, and the length of the minor axis is 6. Using $c^2 = a^2 - b^2$: $c = \sqrt{a^2 - b^2} = \sqrt{16-9} = \sqrt{7}$. The foci are $\left(-2, 3 \pm \sqrt{7}\right)$.

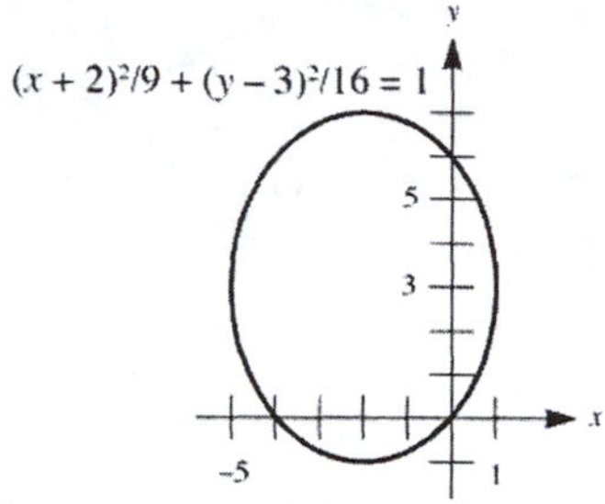

59. Complete the square to convert the equation to standard form:

$$x^2 + 6x = 12y - 33$$
$$(x+3)^2 = 12y - 24$$
$$(x+3)^2 = 12(y-2)$$

This is the equation of a parabola. Since $4p = 12$, $p = 3$. The vertex is (–3,2), the axis of symmetry is $x = -3$, the focus is (–3,5), and the directrix is $y = -1$.

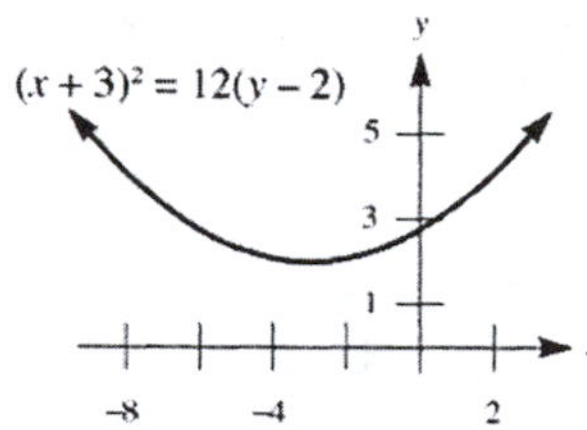

61. Complete the square to convert the equation to standard form:

$$x^2 - 4x - y^2 + 2y = 6$$
$$(x^2 - 4x + 4) - (y^2 - 2y + 1) = 6 + 4 - 1$$
$$(x-2)^2 - (y-1)^2 = 9$$
$$\frac{(x-2)^2}{9} - \frac{(y-1)^2}{9} = 1$$

This is the equation of a hyperbola. Its center is (2,1) and its vertices are (2 + 3, 1) = (5,1) and (2 – 3, 1) = (–1,1). The asymptotes have slopes of $\pm\frac{3}{3} = \pm 1$, so using the point-slope formula:

$$y - 1 = 1(x-2) \qquad y - 1 = -1(x-2)$$
$$y - 1 = x - 2 \qquad y - 1 = -x + 2$$
$$y = x - 1 \qquad y = -x + 3$$

Using $c^2 = a^2 + b^2$: $c = \sqrt{a^2 + b^2} = \sqrt{9+9} = \sqrt{18} = 3\sqrt{2}$. The foci are $\left(2 \pm 3\sqrt{2}, 1\right)$.

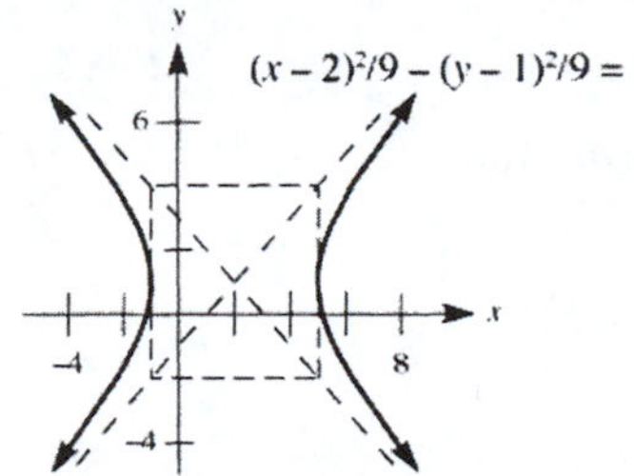

63. Complete the square to convert the equation to standard form:

$$x^2 + 2y - 12 = 0$$
$$x^2 = -2y + 12$$
$$x^2 = -2(y - 6)$$

This is the equation of a parabola. Since $4p = -2$, $p = -\frac{1}{2}$. The vertex is (0,6), the axis of symmetry is $x = 0$, the focus is $\left(0, \frac{11}{2}\right)$, and the directrix is $y = \frac{13}{2}$.

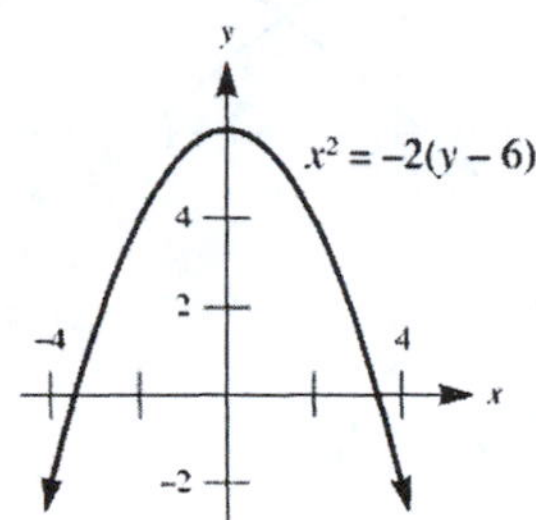

65. Complete the square to convert the equation to standard form:

$$x^2 + 16y^2 - 160y = -384$$
$$x^2 + 16\left(y^2 - 10y\right) = -384$$
$$x^2 + 16(y-5)^2 = -384 + 400$$
$$x^2 + 16(y-5)^2 = 16$$
$$\frac{x^2}{16} + \frac{(y-5)^2}{1} = 1$$

This is the equation of an ellipse. Its center is (0,5), the length of the major axis is 8, and the length of the minor axis is 2. Using $c^2 = a^2 - b^2$: $c = \sqrt{a^2 - b^2} = \sqrt{16-1} = \sqrt{15}$. The foci are $\left(\pm\sqrt{15}, 5\right)$.

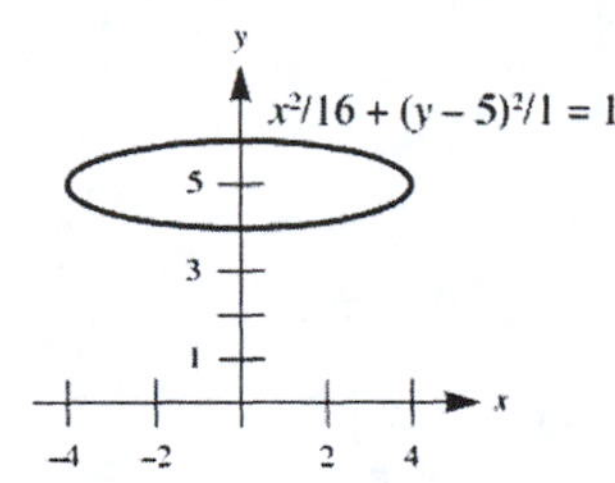

67. Complete the square to convert the equation to standard form:

$$16x^2-64x-25y^2+100y=36$$
$$16\left(x^2-4x\right)-25\left(y^2-4y\right)=36$$
$$16(x-2)^2-25(y-2)^2=36+64-100$$
$$16(x-2)^2-25(y-2)^2=0$$
$$16(x-2)^2=25(y-2)^2$$
$$\pm4(x-2)=5(y-2)$$

The graph of this equation consists of two lines:

$$4(x-2)=5(y-2) \qquad -4(x-2)=5(y-2)$$
$$4x-8=5y-10 \qquad -4x+8=5y-10$$
$$4x-5y=-2 \qquad 4x+5y=18$$

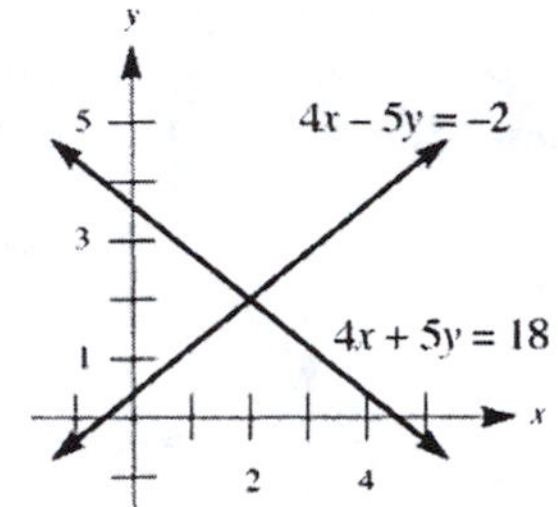

69. Complete the square:

$$Ax^2+Dx+Ey+F=0$$
$$Ey=-A\left(x^2+\frac{D}{A}x\right)-F$$
$$y=-\frac{A}{E}\left(x^2+\frac{D}{A}x+\frac{D^2}{4A^2}\right)+\frac{D^2}{4AE}-\frac{F}{E}$$
$$y=-\frac{A}{E}\left(x+\frac{D}{2A}\right)^2+\frac{D^2-4AF}{4AE}$$

Thus the vertex has x-coordinate $-\frac{D}{2A}$ and y-coordinate $\frac{D^2-4AF}{4AE}$.

71. Since the distance VF is equal to p, we wish to show the intersection point has a y-coordinate of $\frac{p}{2}$.

The parabola has equation $x^2=4py$ and the circle has equation $x^2+y^2=\left(\frac{3p}{2}\right)^2$. Substitute:

$$4py+y^2=\frac{9p^2}{4}$$
$$16py+4y^2=9p^2$$
$$4y^2+16py-9p^2=0$$
$$(2y-p)(2y+9p)=0$$
$$2y=p \quad \text{or} \quad 2y=-9p$$
$$y=\frac{p}{2} \qquad y=-\frac{9p}{2}$$

Clearly $y=-\frac{9p}{2}$ indicates a parabola opening downward, so $y=\frac{p}{2}$. Thus $\overline{VF}$ is bisected by the indicated chord.

73. The given line will be tangent to the circle at (a,b) if and only if (a,b) lies on the line and the perpendicular distance from the center of the circle (h,k) to the line is the radius r. Since (a,b) is on the given circle:

$$(a-h)(a-h)+(b-k)(b-k)=(a-h)^2+(b-k)^2=r^2$$

So (a,b) lies on the line. Using the formula $d=\dfrac{|Ax_0+By_0+C|}{\sqrt{A^2+B^2}}$ we obtain:

$$d=\frac{\left|(a-h)h+(b-k)k-ah+h^2-bk+k^2-r^2\right|}{\sqrt{(a-h)^2+(b-k)^2}}=\frac{\left|-r^2\right|}{\sqrt{r^2}}=r$$

The perpendicular distance from (h,k) to the line is r. Thus the given line is the tangent line at (a,b).

75. The coordinates of A are $(r_1,0)$ and for some y_1 the coordinates of V are $\left(\dfrac{r_1+r_2}{2},y_1\right)$. Then:

$$VO^2=\left(\frac{r_1+r_2}{2}\right)^2+y_1^2$$

$$VA^2=\left(\frac{r_1+r_2}{2}-r_1\right)^2+y_1^2=\left(\frac{r_2-r_1}{2}\right)^2+y_1^2$$

Thus $VO^2-VA^2=\dfrac{r_1^2+2r_1r_2+r_2^2-r_2^2+2r_1r_2-r_1^2}{4}=r_1r_2$.

Chapter 11 Test

1. Since $4p=12$, then $p=3$. The focus is $(-3,0)$ and the directrix is $x=3$.

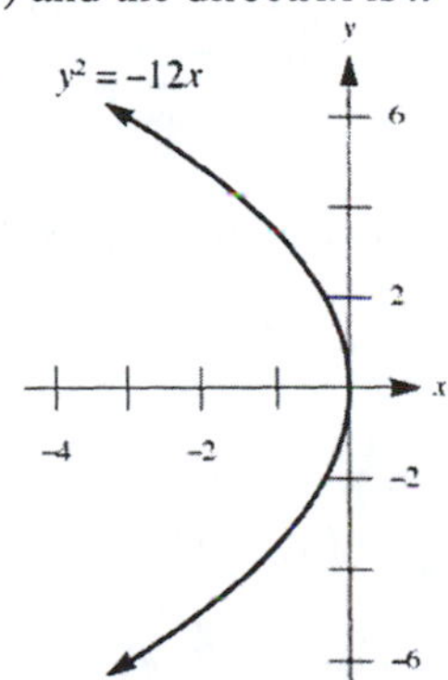

2. Dividing by 4, the standard form is $\dfrac{x^2}{4}-\dfrac{y^2}{1}=1$. The asymptotes are $y=\pm\frac{1}{2}x$. Using $c^2=a^2+b^2$:

$$c=\sqrt{a^2+b^2}=\sqrt{4+1}=\sqrt{5}$$

The foci are $\left(\pm\sqrt{5},0\right)$.

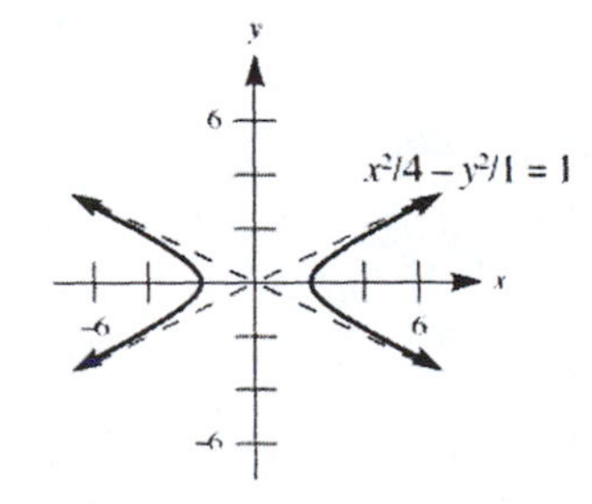

3. Using the expressions for perihelion and aphelion, the system of equations is:

$$a + c = 9.5447$$
$$a - c = 9.5329$$

Adding yields:

$$2a = 19.0776$$
$$a = 9.5388$$

Subtracting yields:

$$2c = 0.0118$$
$$c = 0.0059$$

Now find b: $b = \sqrt{a^2 - c^2} = \sqrt{9.5388^2 - 0.0059^2} \approx 9.539$

The semimajor axis has length $2b \approx 19.078$ AU and the eccentricity is $e = \dfrac{0.0059}{9.5388} \approx 0.001$.

4. **a.** We have: $\cot 2\theta = \dfrac{A - C}{B} = \dfrac{1 - 3}{2\sqrt{3}} = -\dfrac{1}{\sqrt{3}}$, so $\tan 2\theta = -\sqrt{3}$

Thus $2\theta = 120°$, or $\theta = 60°$.

b. Applying the rotation formulas:

$$x = x' \cos 60° - y' \sin 60° = \tfrac{1}{2}x' - \tfrac{\sqrt{3}}{2}y' \qquad y = x' \sin 60° + y' \cos 60° = \tfrac{\sqrt{3}}{2}x' + \tfrac{1}{2}y'$$

Substituting into $x^2 + 2\sqrt{3}\,xy + 3y^2 - 12\sqrt{3}\,x + 12y = 0$ yields: $x'^2 = -6y'$

Rotating 60°, graph the parabola:

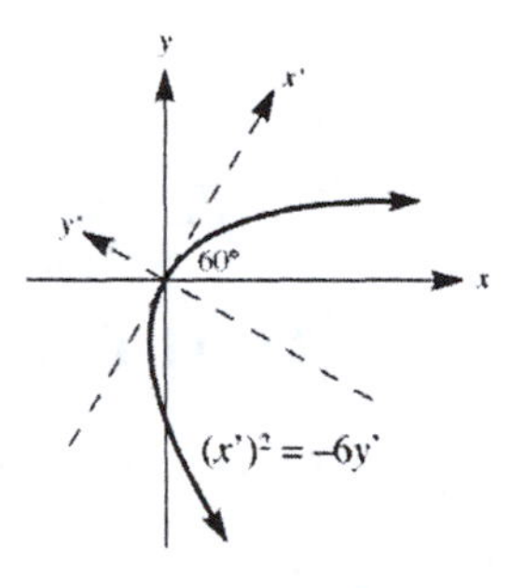

5. Since $\tan\theta = \frac{1}{\sqrt{3}}$, then $\theta = 30°$.

6. Since $e = \dfrac{c}{a}$, then $\dfrac{c}{a} = \frac{1}{2}$, so $a = 2c$. Since the foci are $(0, \pm 2)$, $c = 2$ and thus $a = 4$. Now find b^2:

$$c^2 = a^2 - b^2$$
$$4 = 16 - b^2$$
$$b^2 = 12$$

The equation of the ellipse is $\dfrac{x^2}{12} + \dfrac{y^2}{16} = 1$.

7. Call m the slope of the tangents, so:

$$y - 0 = m(x+4)$$
$$y = mx + 4m$$

Now find the distance from the center of the circle (0,0) to this line: $r = \frac{|0+0-4m|}{\sqrt{1+m^2}} = \frac{|4m|}{\sqrt{1+m^2}}$

Since this radius is 1: $1 = \frac{|4m|}{\sqrt{1+m^2}}$. Squaring:

$$1 = \frac{16m^2}{1+m^2}$$
$$1 + m^2 = 16m^2$$
$$1 = 15m^2$$
$$m^2 = \tfrac{1}{15}$$
$$m = \pm\frac{\sqrt{15}}{15}$$

8. The slope is given by $m = \tan 60° = \sqrt{3}$. Using the point (2,0) in the point-slope formula:

$$y - 0 = \sqrt{3}(x-2)$$
$$y = \sqrt{3}x - 2\sqrt{3}$$
$$\sqrt{3}x - y - 2\sqrt{3} = 0$$

9. Since the foci are $(\pm 2,0)$, $c = 2$. Also $\frac{b}{a} = \frac{1}{\sqrt{3}}$, so $a = b\sqrt{3}$. So:

$$c^2 = a^2 + b^2$$
$$4 = \left(b\sqrt{3}\right)^2 + b^2$$
$$4 = 4b^2$$
$$1 = b^2$$
$$1 = b$$

So $a = 1\sqrt{3} = \sqrt{3}$, thus the equation is $\frac{x^2}{3} - \frac{y^2}{1} = 1$.

10. **a.** Substituting $P(6,5)$ into the equation $5x^2 - 4y^2 = 80$: $5(6)^2 - 4(5)^2 = 5(36) - 4(25) = 180 - 100 = 80$
So $P(6,5)$ lies on the hyperbola.

b. The quantity $(F_1P - F_2P)^2$ can be computed without determining the coordinates of F_1 and F_2. By definition we have $|F_1P - F_2P| = 2a$ for any point P on the hyperbola. Squaring both sides yields $(F_1P - F_2P)^2 = 4a^2$. Now to compute a, convert the equation $5x^2 - 4y^2 = 80$ to standard form. The result is $\frac{x^2}{16} - \frac{y^2}{20} = 1$. Therefore $a = 4$ and we obtain $(F_1P - F_2P)^2 = 4a^2 = 4(16) = 64$.

11. Dividing by 100 yields the standard form $\frac{x^2}{25}+\frac{y^2}{4}=1$. The length of the major axis is 10 and the length of the minor axis is 4. Using $c^2=a^2-b^2$: $c=\sqrt{a^2-b^2}=\sqrt{25-4}=\sqrt{21}$. The foci are $\left(\pm\sqrt{21},0\right)$.

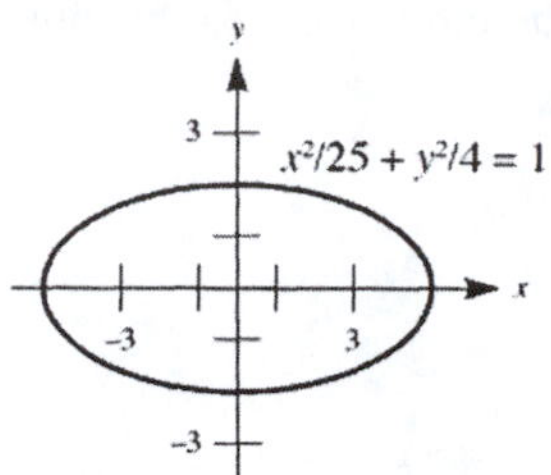

12. Here $(x_0, y_0)=(-1,0)$, $A=2$, $B=-1$ and $C=-1$, so using the distance formula from a point to a line yields:

$$d=\frac{\left|Ax_0+By_0+C\right|}{\sqrt{A^2+B^2}}=\frac{\left|-2+0-1\right|}{\sqrt{4+1}}=\frac{3}{\sqrt{5}}=\frac{3\sqrt{5}}{5}$$

13. Complete the square to convert the equation to standard form:

$$16x^2+y^2-64x+2y+65=0$$
$$16\left(x^2-4x\right)+\left(y^2+2y\right)=-65$$
$$16\left(x^2-4x+4\right)+\left(y^2+2y+1\right)=-65+64+1$$
$$16(x-2)^2+(y+1)^2=0$$

Since the only solution to this equation is the point (2,–1), the graph consists of a single point.

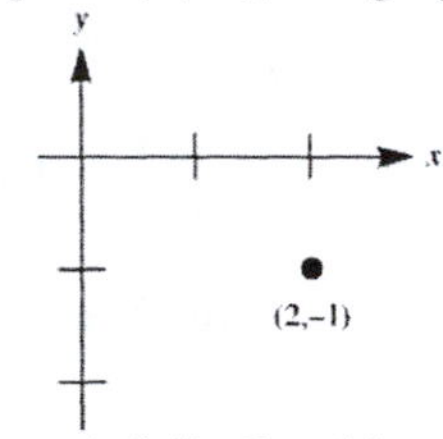

14. The equation represents a hyperbola with center (–4,4). Graphing the hyperbola:

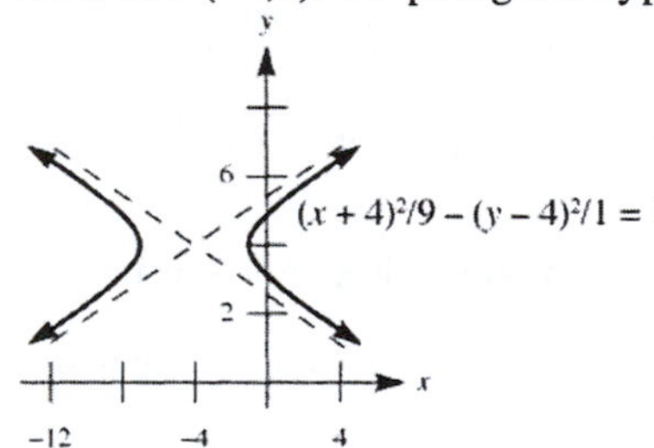

15. Comparing the given equation with the four basic types, it appears this is the type associated with Figure 3 (Section 11.7). Divide both numerator and denominator by 5 to obtain: $r = \dfrac{9}{5-4\cos\theta} = \dfrac{\frac{9}{5}}{1-\frac{4}{5}\cos\theta}$

Since the eccentricity is $e = \frac{4}{5} < 1$, this conic is an ellipse. Computing the values of r when $\theta = 0, \frac{\pi}{2}, \pi$ and $\frac{3\pi}{2}$:

θ	0	$\frac{\pi}{2}$	π	$\frac{3\pi}{2}$
r	9	$\frac{9}{5}$	1	$\frac{9}{5}$

So the vertices on the major axis are (9,0) and (–1,0), and thus the center is (4,0) and the length of the major axis is $2a = 10$, so $a = 5$. Since a focus is (0,0), then $c = 4$ and thus: $b = \sqrt{a^2 - c^2} = \sqrt{25-16} = \sqrt{9} = 3$

Graph the ellipse $r = \dfrac{9}{5-4\cos\theta}$:

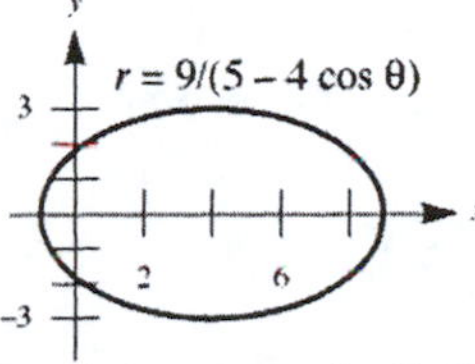

16. Since $4p = 8$, then $p = 2$. The focal width is 8 and the vertex is (1,2).

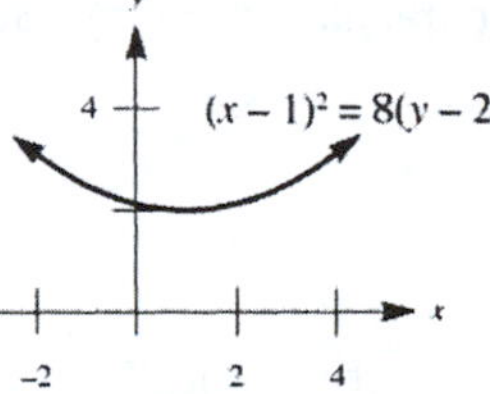

17. **a.** Since $a = 6$ and $b = 5$, find c: $c = \sqrt{a^2 - b^2} = \sqrt{36-25} = \sqrt{11}$

So the eccentricity is $e = \dfrac{c}{a} = \dfrac{\sqrt{11}}{6}$. Thus the directrices are given by: $x = \pm\dfrac{a}{e} = \pm\dfrac{6}{\frac{\sqrt{11}}{6}} = \pm\dfrac{36}{\sqrt{11}} = \pm\dfrac{36\sqrt{11}}{11}$

b. The focal radii are given by:

$$F_1P = a + ex = 6 + \frac{\sqrt{11}}{6}(3) = 6 + \frac{\sqrt{11}}{2} = \frac{12+\sqrt{11}}{2}$$
$$F_2P = a - ex = 6 - \frac{\sqrt{11}}{6}(3) = 6 - \frac{\sqrt{11}}{2} = \frac{12-\sqrt{11}}{2}$$

18. The equation of the tangent line is $\dfrac{x_1x}{a^2} + \dfrac{y_1y}{b^2} = 1$. With $x_1 = -2$ and $y_1 = 4$, this becomes $\dfrac{-2x}{a^2} + \dfrac{4y}{b^2} = 1$. To determine a^2 and b^2, divide both sides of the equation $x^2 + 3y^2 = 52$ by 52. This yields $\dfrac{x^2}{52} + \dfrac{y^2}{52/3} = 1$. Thus $a^2 = 52$ and $b^2 = \frac{52}{3}$, and the equation of the tangent becomes $\dfrac{-2x}{52} + \dfrac{4y}{52/3} = 1$. When we simplify and solve for y, the result is $y = \frac{1}{6}x + \frac{13}{3}$. This is the equation of the tangent line, as required.

19. Let m denote the slope for the required tangent line. Then the equation of the line is $y-8=m(x-4)$. Because this is the tangent line, the system $\begin{aligned} y-8 &= m(x-4) \\ x^2 &= 2y \end{aligned}$ must have exactly one solution, namely (4,8). Solving the second equation yields $y=\frac{x^2}{2}$. Substitute into the first equation:

$$\begin{aligned} \frac{x^2}{2}-8 &= m(x-4) \\ x^2-16 &= 2m(x-4) \\ (x-4)(x+4) &= 2m(x-4) \\ (x-4)(x+4-2m) &= 0 \end{aligned}$$

The solutions are $x = 4$ and $x = 2m - 4$. But since the line is tangent to the parabola, these two x-values must be equal:

$$\begin{aligned} 2m-4 &= 4 \\ 2m &= 8 \\ m &= 4 \end{aligned}$$

With this value for m, the equation $y-8=m(x-4)$ becomes:

$$\begin{aligned} y-8 &= 4(x-4) \\ y-8 &= 4x-16 \\ y &= 4x-8 \end{aligned}$$

This is the required tangent line.

20. From the figure, the point (16,5) lies on the parabola $x^2=4py$. Substituting:

$$\begin{aligned} 16^2 &= 4p(5) \\ p &= \frac{16^2}{4(5)} = 12.8 \text{ m} \end{aligned}$$

The focal length is 12.8 m. The diameter of the reflector is 32 m, so the focal ratio is: $\frac{p}{d}=\frac{12.8}{32}=0.4$

Chapter 12
Roots of Polynomial Equations

12.1 The Complex Number System

1. Complete the table:

i^2	i^3	i^4	i^5	i^6	i^7	i^8
-1	$-i$	1	i	-1	$-i$	1

3. **a.** The real part is 4 and the imaginary part is 5. **b.** The real part is 4 and the imaginary part is –5.
c. The real part is $\frac{1}{2}$ and the imaginary part is –1. **d.** The real part is 0 and the imaginary part is 16.

5. Equating the real parts gives $2c = 8$, and therefore $c = 4$. Similarly, equating the imaginary parts yields $d = -3$.

7. **a.** $(5-6i)+(9+2i)=(5+9)+(-6+2)i=14-4i$ **b.** $(5-6i)-(9+2i)=(5-9)+(-6-2)i=-4-8i$

9. **a.** $(3-4i)(5+i)=15-17i-4i^2=19-17i$ **b.** $(5+i)(3-4i)=19-17i$, from part **a.**

c. Compute the quotient: $\dfrac{3-4i}{5+i}\bullet\dfrac{5-i}{5-i}=\dfrac{15-23i+4i^2}{25-i^2}=\dfrac{11-23i}{26}=\frac{11}{26}-\frac{23}{26}i$

d. Compute the quotient: $\dfrac{5+i}{3-4i}\bullet\dfrac{3+4i}{3+4i}=\dfrac{15+23i+4i^2}{9-16i^2}=\dfrac{11+23i}{25}=\frac{11}{25}+\frac{23}{25}i$

11. **a.** Computing: $z+w=(2+3i)+(9-4i)=11-i$ **b.** Computing: $\overline{z}+w=(2-3i)+(9-4i)=11-7i$
c. Computing: $z+\overline{z}=(2+3i)+(2-3i)=4$

13. Evaluate the expression: $(z+w)+w_1=\left[(2+3i)+(9-4i)\right]+(-7-i)=(11-i)+(-7-i)=4-2i$

15. Compute the product: $zw=(2+3i)(9-4i)=18+19i-12i^2=30+19i$

17. Computing: $z\overline{z}=(2+3i)(2-3i)=4-9i^2=13$

19. First compute the product: $ww_1=(9-4i)(-7-i)=-63+19i+4i^2=-67+19i$
Therefore: $z(ww_1)=(2+3i)(-67+19i)=-134-163i+57i^2=-191-163i$

21. First compute the sum: $w+w_1=(9-4i)+(-7-i)=2-5i$
Therefore: $z(w+w_1)=(2+3i)(2-5i)=4-4i-15i^2=19-4i$

23. First compute the powers:

$$z^2 = (2+3i)(2+3i) = 4+12i+9i^2 = -5+12i$$
$$w^2 = (9-4i)(9-4i) = 81-72i+16i^2 = 65-72i$$

Therefore: $z^2 - w^2 = (-5+12i)-(65-72i) = -70+84i$

25. Since $zw = 30 + 19i$ (from Exercise 15): $(zw)^2 = (30+19i)(30+19i) = 900+1140i+361i^2 = 539+1140i$

27. Since $z^2 = -5 + 12i$ (from Exercise 23), we have: $z^3 = z \bullet z^2 = (2+3i)(-5+12i) = -10+9i+36i^2 = -46+9i$

29. Compute the quotient: $\dfrac{z}{w} = \dfrac{2+3i}{9-4i} \bullet \dfrac{9+4i}{9+4i} = \dfrac{18+35i+12i^2}{81-16i^2} = \dfrac{6+35i}{97} = \frac{6}{97} + \frac{35}{97}i$

31. Using the values of $\overline{z}$ and $\overline{w}$: $\dfrac{\overline{z}}{\overline{w}} = \dfrac{2-3i}{9+4i} \bullet \dfrac{9-4i}{9-4i} = \dfrac{18-35i+12i^2}{81-16i^2} = \dfrac{6-35i}{97} = \frac{6}{97} - \frac{35}{97}i$

33. Using the value of $\overline{z}$: $\dfrac{z}{\overline{z}} = \dfrac{2+3i}{2-3i} \bullet \dfrac{2+3i}{2+3i} = \dfrac{4+12i+9i^2}{4-9i^2} = \dfrac{-5+12i}{13} = -\frac{5}{13} + \frac{12}{13}i$

35. Since $w - \overline{w} = (9-4i)-(9+4i) = -8i$: $\dfrac{w-\overline{w}}{2i} = \dfrac{-8i}{2i} = -4$

37. Compute the quotient: $\dfrac{i}{5+i} \bullet \dfrac{5-i}{5-i} = \dfrac{5i-i^2}{25-i^2} = \dfrac{1+5i}{26} = \frac{1}{26} + \frac{5}{26}i$

39. Compute the quotient: $\dfrac{1}{i} \bullet \dfrac{i}{i} = \dfrac{i}{i^2} = \dfrac{i}{-1} = -i$

41. Simplifying: $i^{17} = i^{16} \bullet i = \left(i^2\right)^8 \bullet i = (-1)^8 \bullet i = i$

43. Simplifying: $i^{26} = \left(i^2\right)^{13} = (-1)^{13} = -1$

45. Writing the numbers in complex form: $\sqrt{-49} + \sqrt{-9} + \sqrt{-4} = 7i + 3i + 2i = 12i$

47. Writing the numbers in complex form:

$$\sqrt{-20} - 3\sqrt{-45} + \sqrt{-80} = \sqrt{4}\sqrt{-5} - 3\sqrt{9}\sqrt{-5} + \sqrt{16}\sqrt{-5} = 2\sqrt{5}i - 9\sqrt{5}i + 4\sqrt{5}i = -3\sqrt{5}i$$

49. Writing the numbers in complex form: $1+\sqrt{-36}\sqrt{-36} = 1+(6i)(6i) = 1+36i^2 = -35$

51. Writing the numbers in complex form:

$$3\sqrt{-128} - 4\sqrt{-18} = 3\sqrt{-64}\sqrt{2} - 4\sqrt{-9}\sqrt{2} = 3(8i)\sqrt{2} - 4(3i)\sqrt{2} = 24\sqrt{2}i - 12\sqrt{2}i = 12\sqrt{2}i$$

53. **a.** Computing the discriminant: $b^2 - 4ac = (-1)^2 - 4(1)(1) = 1-4 = -3$

Since the discriminant is negative, the quadratic will have complex roots.

b. Using the quadratic formula: $x = \dfrac{-(-1) \pm \sqrt{-3}}{2(1)} = \dfrac{1 \pm i\sqrt{3}}{2} = \dfrac{1}{2} \pm \dfrac{\sqrt{3}}{2}i$

55. **a.** Computing the discriminant: $b^2 - 4ac = (2)^2 - 4(5)(2) = 4-40 = -36$

Since the discriminant is negative, the quadratic will have complex roots.

b. Using the quadratic formula: $z = \dfrac{-2 \pm \sqrt{-36}}{2(5)} = \dfrac{-2 \pm 6i}{10} = -\frac{1}{5} \pm \frac{3}{5}i$

57. **a.** Computing the discriminant: $b^2 - 4ac = (3)^2 - 4(2)(4) = 9-32 = -23$

Since the discriminant is negative, the quadratic will have complex roots.

b. Using the quadratic formula: $z = \dfrac{-3 \pm \sqrt{-23}}{2(2)} = \dfrac{-3 \pm i\sqrt{23}}{4} = -\dfrac{3}{4} \pm \dfrac{\sqrt{23}}{4}i$

59. **a.** Computing the discriminant: $b^2-4ac=\left(-\frac{1}{4}\right)^2-4\left(\frac{1}{6}\right)(1)=\frac{1}{16}-\frac{2}{3}=-\frac{29}{48}$

Since the discriminant is negative, the quadratic will have complex roots.

b. Using the quadratic formula: $z=\dfrac{-(-3)\pm\sqrt{-87}}{2(2)}=\dfrac{3\pm i\sqrt{87}}{4}=\dfrac{3}{4}\pm\dfrac{\sqrt{87}}{4}i$

61. **a.** Substituting $x=2-i\sqrt{2}$: $x^2-4x+6=\left(2-i\sqrt{2}\right)^2-4\left(2-i\sqrt{2}\right)+6=4-4i\sqrt{2}-2-8+4i\sqrt{2}+6=0$

b. This verifies the solution found in Example 6.

63. **a.** Notice that the result does agree with the definition: $z+w=(a+bi)+(c+di)=(a+c)+(b+d)i$

b. Notice that the result does agree with the definition: $z-w=(a+bi)-(c+di)=(a-c)+(b-d)i$

c. Notice that the result does agree with the definition:

$$zw=(a+bi)(c+di)=ac+bci+adi+bdi^2=(ac-bd)+(bc+ad)i$$

d. Notice that the result does agree with the definition:

$$\frac{z}{w}=\frac{a+bi}{c+di}\bullet\frac{c-di}{c-di}=\frac{ac+bci-adi-bdi^2}{c^2-d^2i^2}=\frac{ac+bd}{c^2+d^2}+\frac{bc-ad}{c^2+d^2}i$$

65. **a.** Compute each power:

$$\begin{aligned}z^3&=\left(\frac{-1+\sqrt{3}i}{2}\right)^3\\&=\frac{\left(-1+\sqrt{3}i\right)^3}{8}\\&=\frac{\left(-1+\sqrt{3}i\right)\left(-1+\sqrt{3}i\right)^2}{8}\\&=\frac{\left(-1+\sqrt{3}i\right)\left(1-2\sqrt{3}i+3i^2\right)}{8}\\&=\frac{\left(-1+\sqrt{3}i\right)\left(-2-2\sqrt{3}i\right)}{8}\\&=\frac{2-2\sqrt{3}i+2\sqrt{3}i-6i^2}{8}\\&=\frac{8}{8}\\&=1\end{aligned}$$

$$\begin{aligned}w^3&=\left(\frac{-1-\sqrt{3}i}{2}\right)^3\\&=\frac{\left(-1-\sqrt{3}i\right)^3}{8}\\&=\frac{\left(-1-\sqrt{3}i\right)\left(-1-\sqrt{3}i\right)^2}{8}\\&=\frac{\left(-1-\sqrt{3}i\right)\left(1+2\sqrt{3}i+3i^2\right)}{8}\\&=\frac{\left(-1-\sqrt{3}i\right)\left(-2+2\sqrt{3}i\right)}{8}\\&=\frac{2+2\sqrt{3}i-2\sqrt{3}i-6i^2}{8}\\&=\frac{8}{8}\\&=1\end{aligned}$$

b. Compute the product: $zw=\left(\dfrac{-1+\sqrt{3}i}{2}\right)\left(\dfrac{-1-\sqrt{3}i}{2}\right)=\dfrac{1-\sqrt{3}i+\sqrt{3}i-3i^2}{4}=\dfrac{4}{4}=1$

c. Compute each power:

$$w^2=\left(\frac{-1-\sqrt{3}i}{2}\right)^2=\frac{1+2\sqrt{3}i+3i^2}{4}=\frac{-2+2\sqrt{3}i}{4}=\frac{-1+\sqrt{3}i}{2}=z$$

$$z^2=\left(\frac{-1+\sqrt{3}i}{2}\right)^2=\frac{1-2\sqrt{3}i+3i^2}{4}=\frac{-2-2\sqrt{3}i}{4}=\frac{-1-\sqrt{3}i}{2}=w$$

Note: Another approach (using parts **a** and **b**) is to recognize:

$$z^3=zw,\text{ so }z^2=\frac{z^3}{z}=\frac{zw}{z}=w \qquad w^3=zw,\text{ so }w^2=\frac{w^3}{w}=\frac{zw}{w}=z$$

d. Compute the product:

$$\begin{aligned}\left(1-z+z^2\right)\left(1+z-z^2\right)&=(1-z+w)(1+z-w)\\&=\left(\frac{2}{2}-\frac{-1+\sqrt{3}i}{2}+\frac{-1-\sqrt{3}i}{2}\right)\left(\frac{2}{2}+\frac{-1+\sqrt{3}i}{2}-\frac{-1-\sqrt{3}i}{2}\right)\\&=\left(\frac{2-2\sqrt{3}i}{2}\right)\left(\frac{2+2\sqrt{3}i}{2}\right)\\&=\left(1-\sqrt{3}i\right)\left(1+\sqrt{3}i\right)\\&=1-3i^2\\&=4\end{aligned}$$

67. **a.** Let $z = a + bi$, then:

$$\begin{aligned}0+z&=(0+0i)+(a+bi)=a+bi=z\\z+0&=(a+bi)+(0+0i)=a+bi=z\end{aligned}$$

b. Let $z = a + bi$, then:

$$\begin{aligned}0\bullet z&=(0+0i)(a+bi)=0+0i=0\\z\bullet 0&=(a+bi)(0+0i)=0\end{aligned}$$

69. **a.** Compute each sum:

$$\begin{aligned}z+w&=(a+bi)+(c+di)=(a+c)+(b+d)i\\w+z&=(c+di)+(a+bi)=(c+a)+(d+b)i=z+w\end{aligned}$$

b. Compute each product:

$$\begin{aligned}zw&=(a+bi)(c+di)=ac+bci+adi+bdi^2=(ac-bd)+(bc+ad)i\\wz&=(c+di)(a+bi)=ac+adi+bci+bdi^2=(ac-bd)+(bc+ad)i=zw\end{aligned}$$

71. Factoring by grouping:

$$\begin{aligned}x^3-3x^2+4x-12&=0\\x^2(x-3)+4(x-3)&=0\\(x-3)\left(x^2+4\right)&=0\end{aligned}$$

So $x = 3$ or $x^2+4=0$, thus $x^2=-4$ and $x=\pm 2i$. The roots are 3, $\pm 2i$.

73. Factoring by grouping:

$$\begin{aligned}x^6-9x^4+16x^2-144&=0\\x^4\left(x^2-9\right)+16\left(x^2-9\right)&=0\\\left(x^2-9\right)\left(x^4+16\right)&=0\end{aligned}$$

So $x^2=9$, or $x=\pm 3$, or $x^4=-16$, so $x^2=\pm 4i$, so $x=\pm\sqrt{4i}=\pm 2\sqrt{i}$ or $x=\pm\sqrt{-4i}=\pm 2i\sqrt{i}$. The roots are ± 3, $\pm 2\sqrt{i}$, and $\pm 2i\sqrt{i}$.

75. Compute each quotient then add the complex numbers:

$$\begin{aligned}\frac{a+bi}{a-bi}+\frac{a-bi}{a+bi}&=\frac{a+bi}{a-bi}\bullet\frac{a+bi}{a+bi}+\frac{a-bi}{a+bi}\bullet\frac{a-bi}{a-bi}\\&=\frac{(a+bi)^2}{a^2+b^2}+\frac{(a-bi)^2}{a^2+b^2}\\&=\frac{a^2+2abi-b^2+a^2-2abi-b^2}{a^2+b^2}\\&=\frac{2a^2-2b^2}{a^2+b^2}\end{aligned}$$

Thus the real part is $\dfrac{2a^2-2b^2}{a^2+b^2}$, and the imaginary part is 0.

77. Compute each quotient then add the complex numbers:

$$\frac{(a+bi)^2}{a-bi}-\frac{(a-bi)^2}{a+bi}=\frac{(a+bi)^3-(a-bi)^3}{(a-bi)(a+bi)}$$
$$=\frac{\left(a^3+3a^2bi+3ab^2i^2+b^3i^3\right)-\left(a^3-3a^2bi+3ab^2i^2-b^3i^3\right)}{a^2-b^2i^2}$$
$$=\frac{\left(a^3-3ab^2\right)+\left(3a^2b-b^3\right)i-\left(a^3-3ab^2\right)+\left(3a^2b-b^3\right)}{a^2+b^2}$$
$$=\frac{6a^2b-2b^3}{a^2+b^2}i$$

Thus the real part is 0.

79. First compute zw: $zw=(a+bi)(c+di)=ac+bci+adi+bdi^2=(ac-bd)+(bc+ad)i$

Since $zw=0$, $ac-bd=0$ and $bc+ad=0$. Now assuming $a\neq 0$, we can solve the first equation for c to obtain $c=\frac{bd}{a}$.

Substituting into the second equation yields:

$$b\left(\tfrac{bd}{a}\right)+ad=0$$
$$b^2d+a^2d=0$$
$$\left(b^2+a^2\right)d=0$$

But if $a\neq 0$, then $b^2+a^2\neq 0$ and thus $d=0$. Since $c=\frac{bd}{a}$, $c=0$. Thus $w=0$.

12.2 Division of Polynomials

1. Using long division:

$$\begin{array}{r} x-5 \\ x-3\overline{)x^2-8x+4} \\ \underline{x^2-3x} \\ -5x+4 \\ \underline{-5x+15} \\ -11 \end{array}$$

The quotient is $x-5$ and the remainder is -11. Write the equation: $x^2-8x+4=(x-3)(x-5)-11$

3. Using long division:

$$\begin{array}{r} x-11 \\ x+5\overline{)x^2-6x-2} \\ \underline{x^2+5x} \\ -11x-2 \\ \underline{-11x-55} \\ 53 \end{array}$$

The quotient is $x-11$ and the remainder is 53. Write the equation: $x^2-6x-2=(x+5)(x-11)+53$

5. Using long division:

$$\begin{array}{r}
3x^2 - \frac{3}{2}x - \frac{1}{4} \\
2x+1 \overline{\big)\, 6x^3 + 0x^2 - 2x + 3} \\
\underline{6x^3 + 3x^2} \qquad\qquad \\
-3x^2 - 2x \qquad \\
\underline{-3x^2 - \frac{3}{2}x} \qquad \\
-\frac{1}{2}x + 3 \\
\underline{-\frac{1}{2}x - \frac{1}{4}} \\
\frac{13}{4}
\end{array}$$

The quotient is $3x^2 - \frac{3}{2}x - \frac{1}{4}$ and the remainder is $\frac{13}{4}$. Write the equation:

$$6x^3 - 2x + 3 = (2x+1)\left(3x^2 - \frac{3}{2}x - \frac{1}{4}\right) + \frac{13}{4}$$

7. Using long division:

$$\begin{array}{r}
x^4 - 3x^3 + 9x^2 - 27x + 81 \\
x+3 \overline{\big)\, x^5 + 0x^4 + 0x^3 + 0x^2 + 0x + 2} \\
\underline{x^5 + 3x^4} \qquad\qquad\qquad\qquad\qquad \\
-3x^4 + 0x^3 \qquad\qquad\qquad\qquad \\
\underline{-3x^4 - 9x^3} \qquad\qquad\qquad\qquad \\
9x^3 + 0x^2 \qquad\qquad\qquad \\
\underline{9x^3 + 27x^2} \qquad\qquad\qquad \\
-27x^2 + 0x \qquad\quad \\
\underline{-27x^2 - 81x} \qquad\quad \\
81x + 2 \\
\underline{81x + 243} \\
-241
\end{array}$$

The quotient is $x^4 - 3x^3 + 9x^2 - 27x + 81$ and the remainder is -241. Write the equation:

$$x^5 + 2 = (x+3)\left(x^4 - 3x^3 + 9x^2 - 27x + 81\right) - 241$$

9. Using long division:

$$\begin{array}{r}
x^5 + 2x^4 + 4x^3 + 8x^2 + 16x + 32 \\
x-2 \overline{\big)\, x^6 + 0x^5 + 0x^4 + 0x^3 + 0x^2 + 0x - 64} \\
\underline{x^6 - 2x^5} \qquad\qquad\qquad\qquad\qquad\qquad \\
2x^5 + 0x^4 \qquad\qquad\qquad\qquad\qquad \\
\underline{2x^5 - 4x^4} \qquad\qquad\qquad\qquad\qquad \\
4x^4 + 0x^3 \qquad\qquad\qquad\qquad \\
\underline{4x^4 - 8x^3} \qquad\qquad\qquad\qquad \\
8x^3 + 0x^2 \qquad\qquad\qquad \\
\underline{8x^3 - 16x^2} \qquad\qquad\qquad \\
16x^2 + 0x \qquad\quad \\
\underline{16x^2 - 32x} \qquad\quad \\
32x - 64 \\
\underline{32x - 64} \\
0
\end{array}$$

The quotient is $x^5 + 2x^4 + 4x^3 + 8x^2 + 16x + 32$ and the remainder is 0. Write the equation:

$$x^6 - 64 = (x-2)\left(x^5 + 2x^4 + 4x^3 + 8x^2 + 16x + 32\right) + 0$$

11. Using long division:

$$\begin{array}{r} 5x^2+15x+17 \\ x^2-3x+5 \overline{\big) 5x^4+0x^3-3x^2+0x+2} \\ \underline{5x^4-15x^3+25x^2} \\ 15x^3-28x^2+0x \\ \underline{15x^3-45x^2+75x} \\ 17x^2-75x+2 \\ \underline{17x^2-51x+85} \\ -24x-83 \end{array}$$

The quotient is $5x^2 + 15x + 17$ and the remainder is $-24x - 83$. Write the equation:

$$5x^4-3x^2+2=\left(x^2-3x+5\right)\left(5x^2+15x+17\right)+(-24x-83)$$

13. Using long division:

$$\begin{array}{r} 3y-19 \\ y^2+5y+2 \overline{\big) 3y^3-4y^2+0y-3} \\ \underline{3y^3+15y^2+6y} \\ -19y^2-6y-3 \\ \underline{-19y^2-95y-38} \\ 89y+35 \end{array}$$

The quotient is $3y - 19$ and the remainder is $89y + 35$. Write the equation:

$$3y^3-4y^2-3=\left(y^2+5y+2\right)(3y-19)+(89y+35)$$

15. Using long division:

$$\begin{array}{r} t^2-2t-4 \\ t^2-2t+4 \overline{\big) t^4-4t^3+4t^2+0t-16} \\ \underline{t^4-2t^3+4t^2} \\ -2t^3+0t^2+0t \\ \underline{-2t^3+4t^2-8t} \\ -4t^2+8t-16 \\ \underline{-4t^2+8t-16} \\ 0 \end{array}$$

The quotient is $t^2 - 2t - 4$ and the remainder is 0. Write the equation:

$$t^4-4t^3+4t^2-16=\left(t^2-2t+4\right)\left(t^2-2t-4\right)+0$$

17. Using long division:

$$\begin{array}{r} z^4+z^3+z^2+z+1 \\ z-1\overline{)z^5+0z^4+0z^3+0z^2+0z-1} \\ \underline{z^5-1z^4} \\ z^4+0z^3 \\ \underline{z^4-1z^3} \\ z^3+0z^2 \\ \underline{z^3-1z^2} \\ z^2+0z \\ \underline{z^2-1z} \\ z-1 \\ \underline{z-1} \\ 0 \end{array}$$

The quotient is $z^4+z^3+z^2+z+1$ and the remainder is 0. Write the equation:

$$z^5-1=(z-1)\left(z^4+z^3+z^2+z+1\right)+0$$

19. Using long division:

$$\begin{array}{r} ax+(b+ar) \\ x-r\overline{)ax^2+bx+c} \\ \underline{ax^2-arx} \\ (b+ar)x+c \\ \underline{(b+ar)x-r(b+ar)} \\ c+r(b+ar) \end{array}$$

The quotient is $ax+(b+ar)$ and the remainder is $c+r(b+ar)=ar^2+br+c$. Write the equation:

$$ax^2+bx+c=(x-r)(ax+(b+ar))+\left(ar^2+br+c\right)$$

21. Using synthetic division:

$$\begin{array}{r|rrr} 5 & 1 & -6 & -2 \\ & & 5 & -5 \\ \hline & 1 & -1 & -7 \end{array}$$

The quotient is $x-1$ and the remainder is -7. So $x^2-6x-2=(x-5)(x-1)-7$.

23. Using synthetic division:

$$\begin{array}{r|rrr} -1 & 4 & -1 & -5 \\ & & -4 & 5 \\ \hline & 4 & -5 & 0 \end{array}$$

The quotient is $4x-5$ and the remainder is 0. So $4x^2-x-5=(x+1)(4x-5)+0$.

25. Using synthetic division:

$$\begin{array}{r|rrrr} 4 & 6 & -5 & 2 & 1 \\ & & 24 & 76 & 312 \\ \hline & 6 & 19 & 78 & 313 \end{array}$$

The quotient is $6x^2+19x+78$ and the remainder is 313. So $6x^3-5x^2+2x+1=(x-4)(6x^2+19x+78)+313$.

27. Using synthetic division:

2	1	0	0	–1
		2	4	8
	1	2	4	7

The quotient is $x^2 + 2x + 4$ and the remainder is 7. So $x^3 - 1 = (x - 2)(x^2 + 2x + 4) + 7$.

29. Using synthetic division:

–2	1	0	0	0	0	–1
		–2	4	–8	16	–32
	1	–2	4	–8	16	–33

The quotient is $x^4 - 2x^3 + 4x^2 - 8x + 16$ and the remainder is –33. So $x^5 - 1 = (x + 2)(x^4 - 2x^3 + 4x^2 - 8x + 16) - 33$.

31. Using synthetic division:

–4	1	–6	0	0	2
		–4	40	–160	640
	1	–10	40	–160	642

The quotient is $x^3 - 10x^2 + 40x - 160$ and the remainder is 642.
So $x^4 - 6x^3 + 2 = (x + 4)(x^3 - 10x^2 + 40x - 160) + 642$.

33. Using synthetic division:

10	1	–4	–3	6
		10	60	570
	1	6	57	576

The quotient is $x^2 + 6x + 57$ and the remainder is 576. So $x^3 - 4x^2 - 3x + 6 = (x - 10)(x^2 + 6x + 57) + 576$.

35. Using synthetic division:

–5	1	–1	0	0
		–5	30	–150
	1	–6	30	–150

The quotient is $x^2 - 6x + 30$ and the remainder is –150. So $x^3 - x^2 = (x + 5)(x^2 - 6x + 30) - 150$.

37. Using synthetic division:

2 / 3	54	–27	–27	14
		36	6	–14
	54	9	–21	0

The quotient is $54x^2 + 9x - 21$ and the remainder is 0. So $54x^3 - 27x^2 - 27x + 14 = \left(x - \frac{2}{3}\right)\left(54x^2 + 9x - 21\right) + 0$.

39. Using synthetic division:

3	1	0	3	0	12
		3	9	36	108
	1	3	12	36	120

The quotient is $x^3 + 3x^2 + 12x + 36$ and the remainder is 120. So $x^4 + 3x^2 + 12 = (x - 3)(x^3 + 3x^2 + 12x + 36) + 120$.

41. Since $\sqrt[5]{32} = 2$ is a root, use synthetic division:

2	1	0	0	0	0	–32
		2	4	8	16	32
	1	2	4	8	16	0

So $x^5 - 32 = (x - 2)\left(x^4 + 2x^3 + 4x^2 + 8x + 16\right)$.

43. Since $\sqrt[4]{81}=3$ is a root, use synthetic division:

$$\begin{array}{r|rrrrr} 3 & 1 & 0 & 0 & 0 & -81 \\ & & 3 & 9 & 27 & 81 \\ \hline & 1 & 3 & 9 & 27 & 0 \end{array}$$

So $z^4-81=(z-3)(z^3+3z^2+9z+27)$.

45. Since the two quotients are the same, we can now perform synthetic division:

$$\begin{array}{r|rrr} 4/3 & 2 & -8/3 & 1/3 \\ & & 8/3 & 0 \\ \hline & 2 & 0 & 1/3 \end{array}$$

The quotient is $2x$ and the remainder is $\frac{1}{3}$. So the original problem has a quotient of $2x$ and a remainder of $3\left(\frac{1}{3}\right)=1$.

47. Adapting the hint from Exercise 45, first divide the numerator and denominator by 2 to form the quotient:

$$\frac{3x^3+\frac{1}{2}}{x+\frac{1}{2}}$$

Using synthetic division:

$$\begin{array}{r|rrrr} -1/2 & 3 & 0 & 0 & 1/2 \\ & & -3/2 & 3/4 & -3/8 \\ \hline & 3 & -3/2 & 3/4 & 1/8 \end{array}$$

The quotient is $3x^2-\frac{3}{2}x+\frac{3}{4}$ and the remainder is $\frac{1}{8}$. So the original problem has a quotient of $3x^2-\frac{3}{2}x+\frac{3}{4}$ and a remainder of $2\left(\frac{1}{8}\right)=\frac{1}{4}$.

49. Using synthetic division:

$$\begin{array}{r|rrrr} -1 & 1 & 0 & k & 1 \\ & & -1 & 1 & -k-1 \\ \hline & 1 & -1 & k+1 & -k \end{array}$$

So $-k=-4$, and thus $k=4$.

51. Using synthetic division:

$$\begin{array}{r|rrr} p & 1 & 2p & -3q^2 \\ & & p & 3p^2 \\ \hline & 1 & 3p & 3p^2-3q^2 \end{array}$$

Since the remainder is 0:

$$\begin{aligned} 3p^2-3q^2&=0 \\ 3p^2&=3q^2 \\ p^2&=q^2 \end{aligned}$$

53. Using synthetic division:

$$\begin{array}{r|rrr} i & 1 & -4 & 1 \\ & & i & -1-4i \\ \hline & 1 & -4+i & -4i \end{array}$$

The quotient is $x+(-4+i)$ and the remainder is $-4i$.

55. Using synthetic division:

$$\begin{array}{r|rrr} 1+i & 1 & -2 & 2 \\ & & 1+i & -2 \\ \hline & 1 & -1+i & 0 \end{array}$$

The quotient is $x+(-1+i)$ and the remainder is 0.

57. We know that: $f(x)=d(x)\bullet q(x)+R(x)$. Since $d(\sqrt{3})=3-3=0$, we have: $f(\sqrt{3})=R(\sqrt{3})=-6\sqrt{3}+57$

59. Using synthetic division:

$$\begin{array}{r|rrrrrr} a & 1 & 0 & 0 & 0 & -5a^4 & 4a^5 \\ & & a & a^2 & a^3 & a^4 & -4a^5 \\ \hline & 1 & a & a^2 & a^3 & 4a^4 & 0 \end{array}$$

The remainder is 0.

12.3 The Remainder Theorem and the Factor Theorem

1. Substitute $x = 10$:

$$12(10) - 8 = 112$$
$$120 - 8 = 112$$

Yes, it is a root.

3. Substitute $x = 1 - \sqrt{5}$:

$$\left(1-\sqrt{5}\right)^2 - 2\left(1-\sqrt{5}\right) - 4 = 0$$
$$1 - 2\sqrt{5} + 5 - 2 + 2\sqrt{5} - 4 = 0$$
$$4 - 4 = 0$$

Yes, it is a root.

5. Substitute $x = \frac{1}{2}$:

$$2\left(\tfrac{1}{2}\right)^2 - 3\left(\tfrac{1}{2}\right) + 1 = 0$$
$$\tfrac{1}{2} - \tfrac{3}{2} + 1 = 0$$
$$-1 + 1 = 0$$

Yes, it is a root.

7. Compute $f\left(\frac{2}{3}\right)$: $f\left(\frac{2}{3}\right) = 3\left(\frac{2}{3}\right) - 2 = 2 - 2 = 0$. So $x = \frac{2}{3}$ is a zero of $f(x)$.

9. Compute $h(-1)$: $h(-1) = 5(-1)^3 - (-1)^2 + 2(-1) + 8 = -5 - 1 - 2 + 8 = 0$. So $x = -1$ is a zero of $h(x)$.

11. Compute $f(2)$: $f(2) = 1 + 2(2) + (2)^3 - (2)^5 = 1 + 4 + 8 - 32 = -19$. So $t = 2$ is not a zero of $f(t)$.

13. **a.** Compute $f\left(\frac{\sqrt{3}-1}{2}\right)$: $f\left(\frac{\sqrt{3}-1}{2}\right) = 2\left(\frac{\sqrt{3}-1}{2}\right)^3 - 3\left(\frac{\sqrt{3}-1}{2}\right) + 1 = \frac{3\sqrt{3}-5}{2} - \frac{3\sqrt{3}-3}{2} + 1 = -1 + 1 = 0$

So $x = \frac{\sqrt{3}-1}{2}$ is a zero of $f(x)$.

b. Compute $f\left(\frac{\sqrt{3}+1}{2}\right)$: $f\left(\frac{\sqrt{3}+1}{2}\right) = 2\left(\frac{\sqrt{3}+1}{2}\right)^3 - 3\left(\frac{\sqrt{3}+1}{2}\right) + 1 = \frac{5+3\sqrt{3}}{2} - \frac{3\sqrt{3}-3}{2} + 1 = 4 + 1 = 5$

So $x = \frac{\sqrt{3}+1}{2}$ is not a zero of $f(x)$.

15. The roots are 1, 2 (multiplicity 3), and 3.

17. **a.** The multiplicity at –1 is 2:

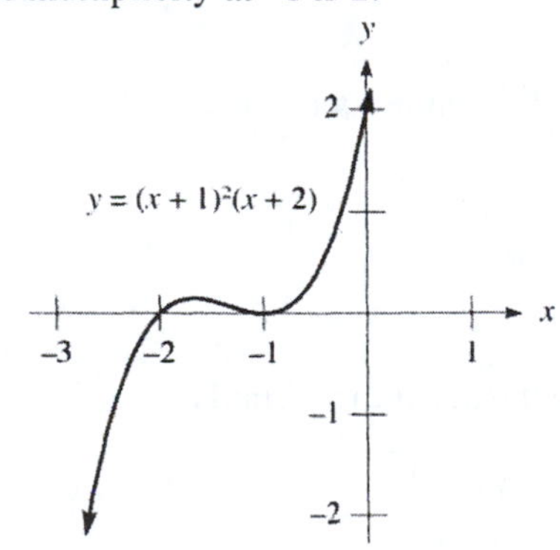

b. The multiplicity at –2 is 3:

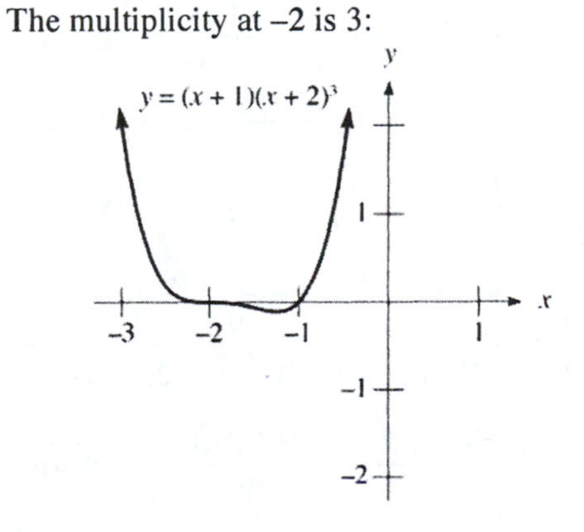

c. The multiplicity at –1 is 2 and the multiplicity at –2 is 3:

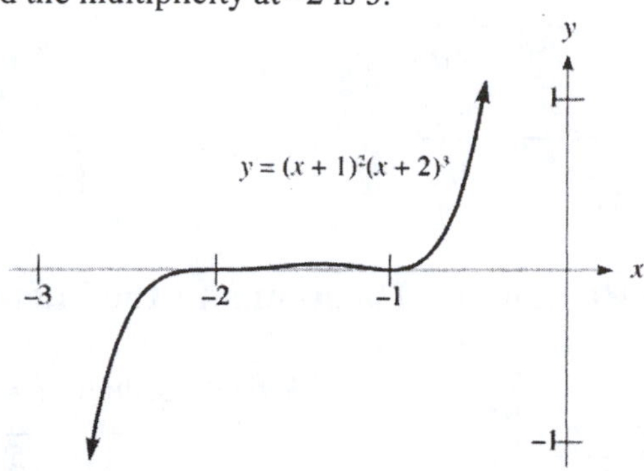

19. Using synthetic division:

–3	4	–6	1	–5
		–12	54	–165
	4	–18	55	–170

So $f(-3) = -170$.

21. Using synthetic division:

1/2	6	5	–8	–10	–3
		3	4	–2	–6
	6	8	–4	–12	–9

So $f\left(\frac{1}{2}\right) = -9$.

23. Using synthetic division:

$-\sqrt{2}$	1	3	–4
		$-\sqrt{2}$	$-3\sqrt{2}+2$
	1	$3-\sqrt{2}$	$-3\sqrt{2}-2$

So $f\left(-\sqrt{2}\right) = -3\sqrt{2} - 2$.

25. Since $x = -1$ is a root, $x + 1$ will be a factor. Using synthetic division:

–1	2	0	–1	1
		–2	2	–1
	2	–2	1	0

So the equation factors as $(x+1)\left(2x^2 - 2x + 1\right) = 0$. Using the quadratic formula:

$$x = \frac{2 \pm \sqrt{(-2)^2 - 4(2)(1)}}{2(2)} = \frac{2 \pm \sqrt{4-8}}{4} = \frac{2 \pm 2i}{4} = \frac{1 \pm i}{2}$$

The roots are $x = -1, \dfrac{1 \pm i}{2}$.

27. Using synthetic division:

–3	1	–4	–9	36
		–3	21	–36
	1	–7	12	0

So $x^3 - 4x^2 - 9x + 36 = (x+3)\left(x^2 - 7x + 12\right) = (x+3)(x-4)(x-3)$. So the roots are ±3, 4.

29. Using synthetic division:

1	1	1	–7	5
		1	2	–5
	1	2	–5	0

So $x^3 + x^2 - 7x + 5 = (x-1)\left(x^2 + 2x - 5\right)$. So $x = 1$ is a root. Using the quadratic formula:

$$x = \frac{-2 \pm \sqrt{4+20}}{2} = \frac{-2 \pm 2\sqrt{6}}{2} = -1 \pm \sqrt{6}$$

So the roots are $1, -1 \pm \sqrt{6}$.

31. Using synthetic division:

-2	3	-5	-16	12
		-6	22	-12
	3	-11	6	0

So $3x^3-5x^2-16x+12=(x+2)(3x^2-11x+6)=(x+2)(3x-2)(x-3)$. So the roots are $-2, \frac{2}{3}$, and 3.

33. Using synthetic division:

$-3/2$	2	1	-5	-3
		-3	3	3
	2	-2	-2	0

So $2x^3+x^2-5x-3=\left(x+\frac{3}{2}\right)(2x^2-2x-2)=2\left(x+\frac{3}{2}\right)(x^2-x-1)$. So $x=-\frac{3}{2}$ is a root. Use the quadratic formula: $x=\frac{1\pm\sqrt{1+4}}{2}=\frac{1\pm\sqrt{5}}{2}$. So the roots are $-\frac{3}{2}, \frac{1\pm\sqrt{5}}{2}$.

35. Using synthetic division:

5	1	-15	75	-125	0
		5	-50	125	0
	1	-10	25	0	0

Therefore: $x^4-15x^3+75x^2-125x=(x-5)(x^3-10x^2+25x)=x(x-5)(x^2-10x+25)=x(x-5)^3$

So the roots are 0 and 5.

37. Using synthetic division:

-4	1	2	-23	-24	144
		-4	8	60	-144
	1	-2	-15	36	0

Use synthetic division again:

3	1	-2	-15	36
		3	3	-36
	1	1	-12	0

So $x^4+2x^3-23x^2-24x+144=(x+4)(x-3)(x^2+x-12)=(x+4)^2(x-3)^2$. So the roots are -4 and 3.

39. Sketching the graph:

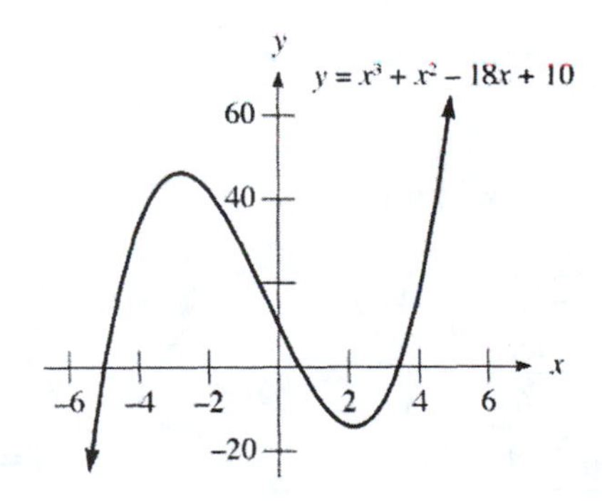

Using synthetic division:

-5	1	1	-18	10
		-5	20	-10
	1	-4	2	0

So $x=-5$ is a root. For the remaining factor $x^2-4x+2=0$, use the quadratic formula to find the remaining roots:

$$x=\frac{4\pm\sqrt{(-4)^2-4(1)(2)}}{2(1)}=\frac{4\pm\sqrt{8}}{2}=\frac{4\pm2\sqrt{2}}{2}=2\pm\sqrt{2}\approx0.59,3.41$$

The roots are -5, $2-\sqrt{2}\approx0.59$, and $2+\sqrt{2}\approx3.41$.

41. Sketching the graph:

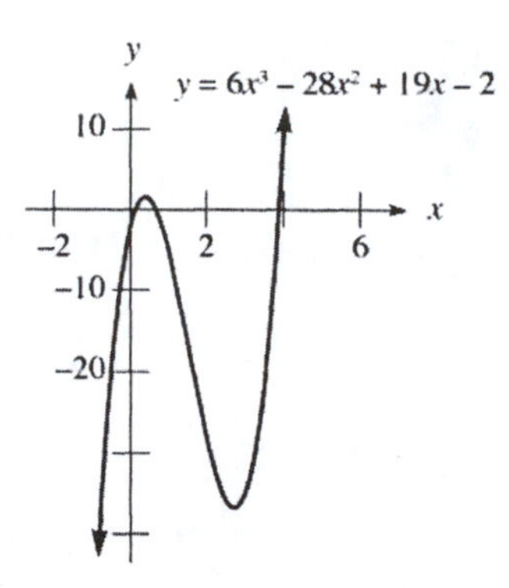

Using synthetic division:

$$\begin{array}{r|rrrr} 2/3 & 6 & -28 & 19 & -2 \\ & & 4 & -16 & 2 \\ \hline & 6 & -24 & 3 & 0 \end{array}$$

So $x = 2/3$ is a root. For the remaining factor $6x^2 - 24x + 3 = 0$, or $2x^2 - 8x + 1 = 0$, use the quadratic formula:

$$x = \frac{8 \pm \sqrt{(-8)^2 - 4(2)(1)}}{2(2)} = \frac{8 \pm \sqrt{56}}{4} = \frac{8 \pm 2\sqrt{14}}{4} = \frac{4 \pm \sqrt{14}}{2} \approx 0.13, 3.87$$

The roots are 2/3, $\frac{4-\sqrt{14}}{2} \approx 0.13$, and $\frac{4+\sqrt{14}}{2} \approx 3.87$.

43. Sketching the graph:

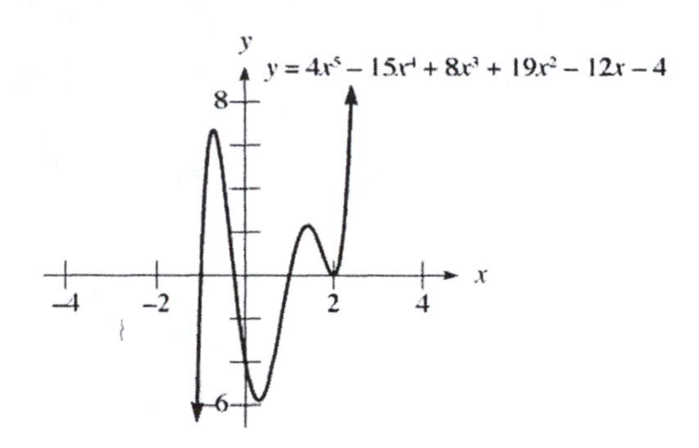

Using synthetic division:

$$\begin{array}{r|rrrrrr} 1 & 4 & -15 & 8 & 19 & -12 & -4 \\ & & 4 & -11 & -3 & 16 & 4 \\ \hline & 4 & -11 & -3 & 16 & 4 & 0 \end{array}$$

So $x = 1$ is a root. For the remaining factor $4x^4 - 11x^3 - 3x^2 + 16x + 4 = 0$, use synthetic division:

$$\begin{array}{r|rrrrr} -1 & 4 & -11 & -3 & 16 & 4 \\ & & -4 & 15 & -12 & -4 \\ \hline & 4 & -15 & 12 & 4 & 0 \end{array}$$

So $x = -1$ is a root. For the remaining factor $4x^3 - 15x^2 + 12x + 4 = 0$, use synthetic division:

$$\begin{array}{r|rrrr} -1/4 & 4 & -15 & 12 & 4 \\ & & -1 & 4 & -4 \\ \hline & 4 & -16 & 16 & 0 \end{array}$$

For the remaining factor:

$$\begin{aligned} 4x^2 - 16x + 16 &= 0 \\ x^2 - 4x + 4 &= 0 \\ (x-2)^2 &= 0 \\ x &= 2 \end{aligned}$$

The roots are 1, −1, −1/4, and 2.

45. **a.** $R\left(\frac{1}{2}\right)=f\left(\frac{1}{2}\right)=1.125$

b. $R(1.25)=f(1.25)=-0.046875$

c. Since $f(1)=0$, $t-1$ is a linear factor of $f(t)$.

d. Since 1 is a solution, use synthetic division:

$$\begin{array}{r|rrrr} 1 & 1 & 0 & -4 & 3 \\ & & 1 & 1 & -3 \\ \hline & 1 & 1 & -3 & 0 \end{array}$$

Using the quadratic formula to solve $t^2+t-3=0$: $t=\dfrac{-1\pm\sqrt{1+12}}{2}=\dfrac{-1\pm\sqrt{13}}{2}$

So the solutions are $1, \dfrac{-1\pm\sqrt{13}}{2}$.

47. We must have $f(x)=(x-3)(x-5)(x+4)=0$. Multiplying out, we have $x^3-4x^2-17x+60=0$.

49. We must have $f(x)=(x+1)^2(x+6)=0$. Multiplying out, we have $x^3+8x^2+13x+6=0$.

51. Since $\frac{1}{2}$ is not a root of x^2-3x-4, we know that such an equation must have $f(x)=\left(x^2-3x-4\right)\left(x-\frac{1}{2}\right)^3$, which has degree 5. Thus no such polynomial of degree 4 exists.

53. **a.** $g(r)=ar^3+br^2+cr+d$

b. Using synthetic division:

$$\begin{array}{r|cccc} r & a & b & c & d \\ & & ar & ar^2+br & ar^3+br^2+cr \\ \hline & a & ar+b & ar^2+br+c & ar^3+br^2+cr+d \end{array}$$

55. Note that this is just a restatement of one root of the quadratic formula. So we know it is a zero.

57. Using synthetic division:

$$\begin{array}{r|cccc} 1 & 1 & 1 & a & b \\ & & 1 & 2 & a+2 \\ \hline & 1 & 2 & a+2 & a+b+2 \end{array}$$

Using synthetic division again:

$$\begin{array}{r|cccc} 1 & 1 & -1 & -a & b \\ & & 1 & 0 & -a \\ \hline & 1 & 0 & -a & -a+b \end{array}$$

If 1 is a root, then:

$$a+b+2=0$$
$$-a+b=0$$

Adding these, we get $2b+2=0$, so $b=-1$. Substituting we get $a=-1$. So $a=-1$ and $b=-1$.

59. Let r_1 and r_2 be the roots, so $r_2=2r_1$. Then: $x^2+bx+1=\left(x-r_1\right)\left(x-2r_1\right)=x^2-3r_1x+2r_1^2$

Since r_1 is a constant, $-3r_1=b$ and $2r_1^2=1$, so $r_1^2=\frac{1}{2}$ and thus $r_1=\pm\dfrac{\sqrt{2}}{2}$. So $b=\pm\dfrac{3\sqrt{2}}{2}$.

61. Let r denote the root with multiplicity 2. Using synthetic division:

$$\begin{array}{r|cccc} r & 1 & 0 & -12 & 16 \\ & & r & r^2 & r^3-12r \\ \hline & 1 & r & r^2-12 & r^3-12r+16 \end{array}$$

Since r is a double root, the equation $x^2+rx+(r^2-12)=0$ must also have r as a root. Substituting $x=r$:

$$\begin{aligned} r^2+r^2+r^2-12&=0 \\ 3r^2&=12 \\ r^2&=4 \\ r&=\pm 2 \end{aligned}$$

But $r=-2$ does not check in the original equation, thus $r=2$. Thus the quadratic equation becomes $x^2+2x-8=0$, which factors to $(x+4)(x-2)=0$, and thus $x=-4$ is also a solution. So the solutions are 2 (multiplicity 2) and -4.

12.4 The Fundamental Theorem of Algebra

1. a. yes
   b. yes
   c. yes
   d. no--not a polynomial equation
3. The equation does not have any real-number roots:

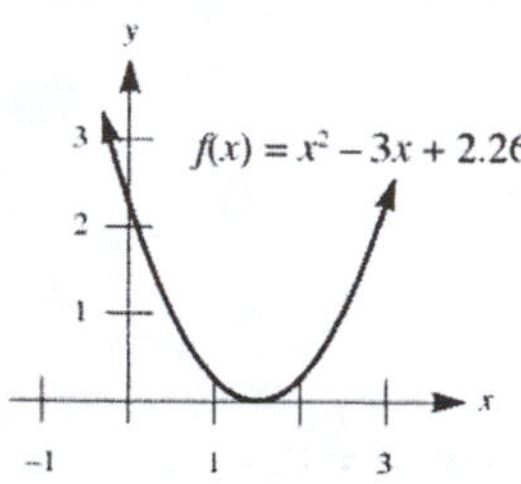

5. The equation has three real-number roots:

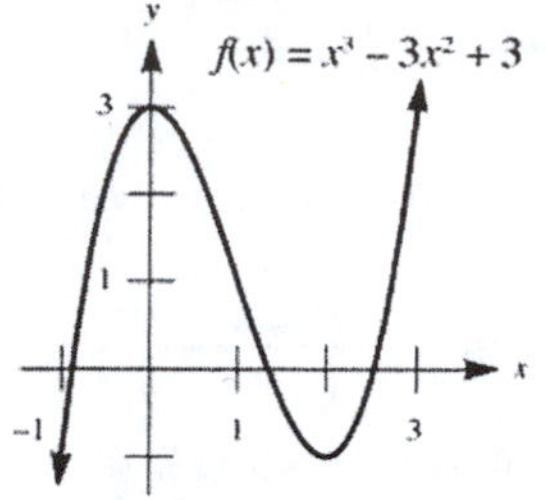

7. The equation does not have any real-number roots:

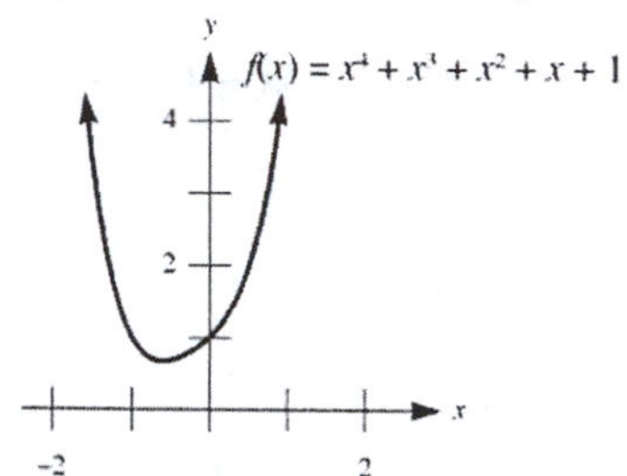

9. The equation has one real-number root:

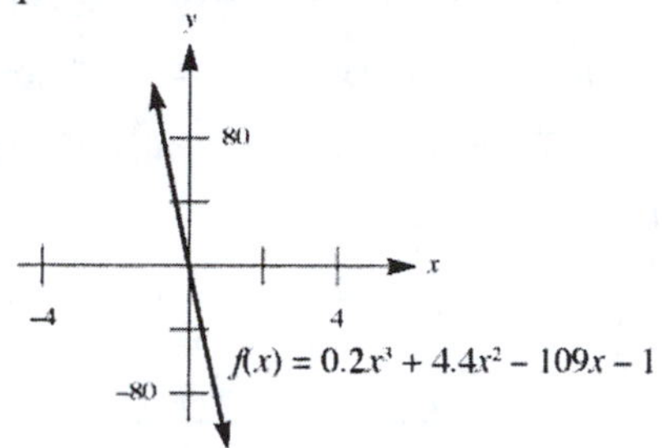

11. Factoring: $x^2 - 2x - 3 = (x+1)(x-3) = [x-(-1)](x-3)$

13. Factoring: $4x^2 + 23x - 6 = (4x-1)(x+6) = 4\left(x - \frac{1}{4}\right)[x-(-6)]$

15. Factoring: $x^2 - 5 = \left(x+\sqrt{5}\right)\left(x-\sqrt{5}\right) = \left[x-\left(-\sqrt{5}\right)\right]\left(x-\sqrt{5}\right)$

17. Factoring:

$$\begin{aligned} x^5 - 7x^3 - 18x &= x\left(x^4 - 7x^2 - 18\right) \\ &= x\left(x^2 - 9\right)\left(x^2 + 2\right) \\ &= x(x+3)(x-3)\left(x+\sqrt{2}i\right)\left(x-\sqrt{2}i\right) \\ &= (x-0)(x-(-3))(x-3)\left(x-\left(-\sqrt{2}i\right)\right)\left(x-\sqrt{2}i\right) \end{aligned}$$

19. Writing the polynomial as: $f(x) = (x-1)^2(x+3) = x^3 + x^2 - 5x + 3$

21. Writing the polynomial as: $f(x) = (x-2)(x+2)(x-2i)(x+2i) = \left(x^2-4\right)\left(x^2+4\right) = x^4 - 16$

23. Writing the polynomial as: $f(x) = \left(x-\sqrt{3}\right)^2\left[x-\left(-\sqrt{3}\right)\right]^2(x-4i)[x-(-4i)] = x^6 + 10x^4 - 87x^2 + 144$

25. **a.** Writing the polynomial as: $f(x)=x^2(x-1)(x-3)=x^2(x^2-4x+3)=x^4-4x^3+3x^2$

b. Sketching the graph:

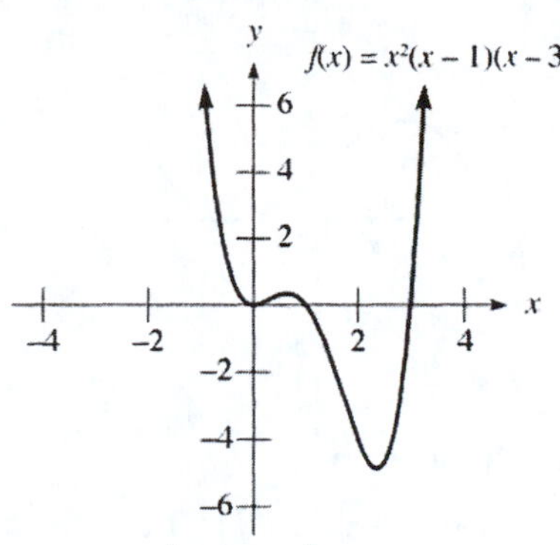

27. **a.** Writing the polynomial as: $f(x)=x(x-1)^2(x-3)^2$

b. Sketching the graph:

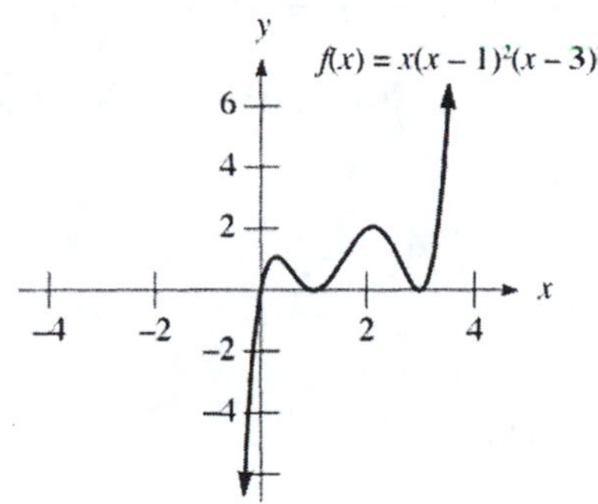

29. **a.** Writing the polynomial as: $f(x)=x^2(x-1)(x-3)^3$

b. Sketching the graph:

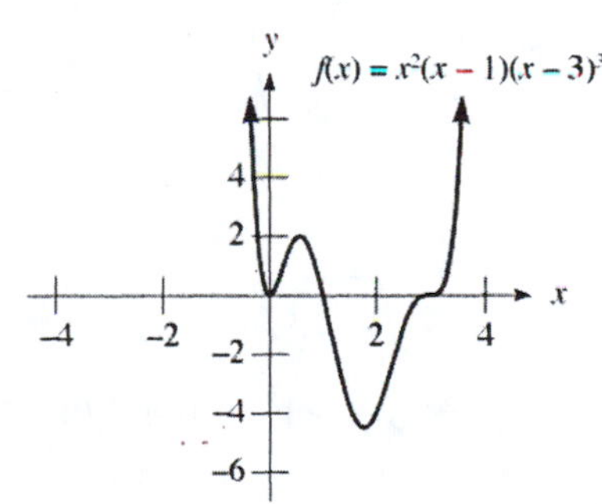

31. We can write $f(x)$ as $f(x)=a(x+4)(x-9)$, for some constant a. Since $f(3)=5$:

$$5=a(7)(-6)$$
$$a=-\tfrac{5}{42}$$

So $f(x)=-\frac{5}{42}(x^2-5x-36)=-\frac{5}{42}x^2+\frac{25}{42}x+\frac{30}{7}$.

33. **a.** We can write $f(x)$ as $f(x)=a(x+5)(x-2)(x-3)$, for some constant a. Since $f(0)=1$:

$$1=a(5)(-2)(-3)$$
$$a=\tfrac{1}{30}$$

So $f(x)=\frac{1}{30}(x^3-19x+30)=\frac{1}{30}x^3-\frac{19}{30}x+1$.

b. Graphing the function:

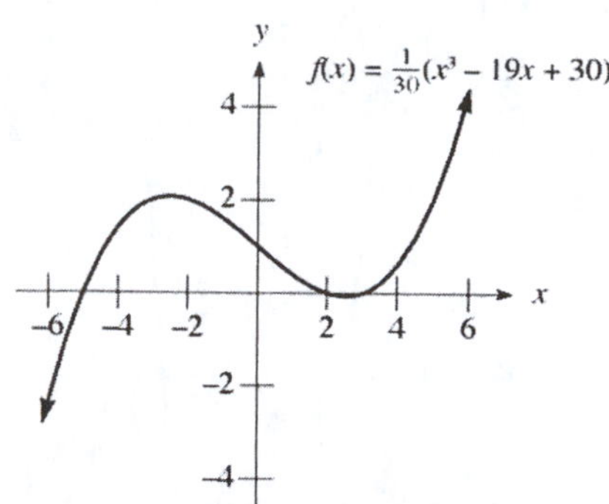

The graph confirms the information from part **a**.

35. We have $-b = -i - \sqrt{3}$, so $b = i + \sqrt{3}$. Also $c = -i\left(-\sqrt{3}\right) = i\sqrt{3}$. So $x^2 + \left(i + \sqrt{3}\right)x + i\sqrt{3} = 0$.

37. We have $-b = 3$, so $b = -3$. Also $c = (9)(-6) = -54$. So $x^2 - 3x - 54 = 0$.

39. We have $-b = 2$, so $b = -2$. Also $c = \left(1+\sqrt{5}\right)\left(1-\sqrt{5}\right) = -4$. So $x^2 - 2x - 4 = 0$.

41. We have $-B = 2a$, so $B = -2a$. Also $C = \left(a+\sqrt{b}\right)\left(a-\sqrt{b}\right) = a^2 - b$. So $x^2 - 2ax + a^2 - b = 0$.

43. **a.** Using $a = 3$ and $b = 76$:

$$\begin{aligned} x &= \sqrt[3]{\frac{76}{2} + \sqrt{\frac{76^2}{4} + \frac{3^3}{27}}} - \sqrt[3]{-\frac{76}{2} + \sqrt{\frac{76^2}{4} + \frac{3^3}{27}}} \\ &= \sqrt[3]{38 + \sqrt{1444 + 1}} - \sqrt[3]{-38 + \sqrt{1444 + 1}} \\ &= \sqrt[3]{38 + \sqrt{1445}} - \sqrt[3]{-38 + \sqrt{1445}} \\ &= 4 \end{aligned}$$

b. Graphing the function in the suggested viewing rectangle:

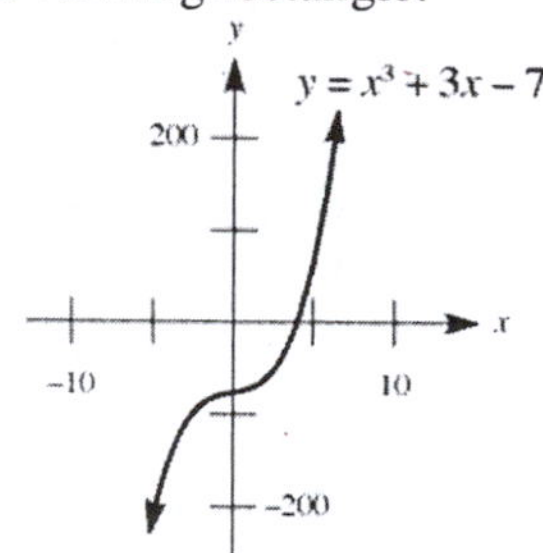

Note the x-intercept is 4. Substituting $x = 4$: $y = (4)^3 + 3(4) - 76 = 64 + 12 - 76 = 0$

45. Using the hint:

$$\begin{aligned} x^4 + 64 &= x^4 + 16x^2 + 64 - 16x^2 \\ &= \left(x^2 + 8\right)^2 - (4x)^2 \\ &= \left(x^2 + 8 + 4x\right)\left(x^2 + 8 - 4x\right) \\ &= \left(x^2 + 4x + 8\right)\left(x^2 - 4x + 8\right) \end{aligned}$$

Now $x^2 + 4x + 8 = 0$ when $x = \dfrac{-4 \pm \sqrt{16 - 32}}{2} = \dfrac{-4 \pm 4i}{2} = -2 \pm 2i$. Also $x^2 - 4x + 8 = 0$ when $x = \dfrac{4 \pm \sqrt{16 - 32}}{2} = \dfrac{4 \pm 4i}{2} = 2 \pm 2i$. Therefore:

$$\begin{aligned} x^4 + 64 &= \left(x^2 + 4x + 8\right)\left(x^2 - 4x + 8\right) \\ &= \left[x - (-2 + 2i)\right]\left[x - (-2 - 2i)\right]\left[x - (2 + 2i)\right]\left[x - (2 - 2i)\right] \\ &= (x + 2 - 2i)(x + 2 + 2i)(x - 2 - 2i)(x - 2 + 2i) \end{aligned}$$

47. We know: $x^3+bx^2+cx+d=(x-r_1)(x-r_2)(x-r_3)=x^3-(r_1+r_2+r_3)x^2+(r_1r_2+r_1r_3+r_2r_3)x-r_1r_2r_3$

Therefore:

$$\begin{aligned} r_1+r_2+r_3&=-b\\ r_1r_2+r_1r_3+r_2r_3&=c\\ r_1r_2r_3&=-d \end{aligned}$$

49. Let r_1, r_2 and r_3 be the roots, and assume $r_1=r_2$. The identities from Exercise 47 give us:

$$\begin{aligned} r_1+r_2+r_3&=0\\ 2r_1+r_3&=0\\ r_3&=-2r_1 \end{aligned}$$

Therefore:

$$\begin{aligned} r_1r_2r_3&=-250\\ r_1(r_1)(-2r_1)&=-250\\ r_1^3&=125\\ r_1&=5 \end{aligned}$$

So $r_2=5$ and $r_3=-10$. So the roots are 5 (multiplicity 2) and –10.

51. **a.** Using $r_1=\cos 72°$ and $r_2=\cos 144°$, we are given that $r_1+r_2=-\frac{1}{2}$ and $r_1r_2=-\frac{1}{4}$, so $b=\frac{1}{2}$ and $c=-\frac{1}{4}$. Thus a quadratic equation is $x^2+\frac{1}{2}x-\frac{1}{4}=0$. A better form (with integer coefficients) is $4x^2+2x-1=0$.

b. Solving the equation $4x^2+2x-1=0$ using the quadratic formula:

$$x=\frac{-2\pm\sqrt{4-4(4)(-1)}}{2(4)}=\frac{-2\pm\sqrt{20}}{8}=\frac{-2\pm2\sqrt{5}}{8}=\frac{1}{4}\left(-1\pm\sqrt{5}\right)$$

The smaller of these two roots is cos 144° (since it is negative), thus $\cos 144°=\frac{1}{4}\left(-1-\sqrt{5}\right)$ and $\cos 72°=\frac{1}{4}\left(-1+\sqrt{5}\right)$.

c. Using a calculator verifies that $\cos 72°\approx 0.3090$ and $\cos 144°\approx -0.8090$.

53. **a.** Since r_1, r_2, r_3 and r_4 are real roots, the equation can be written as $(x-r_1)(x-r_2)(x-r_3)(x-r_4)=0$. When expanded, the x^3 term has a coefficient of $-(r_1+r_2+r_3+r_4)$, so this value must be zero. Thus $r_1+r_2+r_3+r_4=0$.

b. The circle has an equation $(x-h)^2+(y-k)^2=r^2$. Substituting $y=x^2$:

$$\begin{aligned} (x-h)^2+(x^2-k)^2&=r^2\\ x^2-2hx+h^2+x^4-2kx^2+k^2&=r^2\\ x^4+(1-2k)x^2-2hx+(h^2+k^2-r^2)&=0 \end{aligned}$$

Since x_1, x_2, x_3 and x_4 are roots to this equation, and since the equation has no x^3 term, from part **a** we have $x_1+x_2+x_3+x_4=0$.

55. Proceed as in the preceding exercise. Let:

$$f(x)=\frac{(x-a)(x-b)c^2}{(c-a)(c-b)}+\frac{(x-b)(x-c)a^2}{(a-b)(a-c)}+\frac{(x-c)(x-a)b^2}{(b-c)(b-a)}-x^2$$

$$f(a)=0+\frac{(a-b)(a-c)a^2}{(a-b)(a-c)}+0-a^2=a^2-a^2=0$$

$$f(b)=0+0+\frac{(b-c)(b-a)b^2}{(b-c)(b-a)}-b^2=b^2-b^2=0$$

$$f(c)=\frac{(c-a)(c-b)c^2}{(c-a)(c-b)}+0+0-c^2=c^2-c^2=0$$

Since a, b, c are distinct (for denominators to be non-zero), and all three are roots, by Exercise 54a, $f(x) = 0$ for all values of x. Thus the equation is an identity.

12.5 Rational and Irrational Roots

1. a. See textbook.
   b. The possible rational roots are $\pm 1, \pm 11$.
3. a. The possible rational roots are $\pm 1, \pm 5$, so there cannot be a rational root larger than 5. No, 5 is not a root, since substituting $x = 5$ into the left side of the equation results in 3050, rather than 0.
   b. The possible rational roots are $\pm 1, \pm 5$, so 2 cannot be a root. Substituting $x = 2$ into the left side of the equation results in -1, rather than 0.
   c. Graphing the function using the viewing rectangle [1.8,2.2] x [–1,1]:

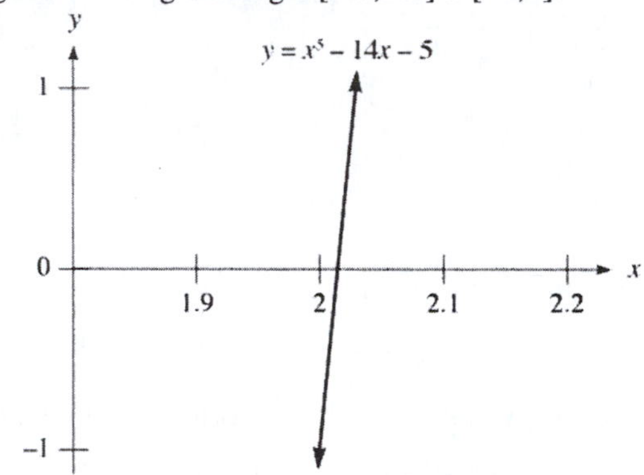

   Note that 2 is not a root of the equation $x^5 - 14x - 5 = 0$.

5. p-factors: $\pm 1, \pm 3$ q-factors: $\pm 1, \pm 2, \pm 4$

   possible rational roots $\frac{p}{q}$: $\pm 1, \pm \frac{1}{2}, \pm \frac{1}{4}, \pm 3, \pm \frac{3}{2}, \pm \frac{3}{4}$

7. p-factors: $\pm 1, \pm 3, \pm 9$ q-factors: $\pm 1, \pm 2, \pm 4, \pm 8$

   possible rational roots $\frac{p}{q}$: $\pm 1, \pm \frac{1}{2}, \pm \frac{1}{4}, \pm \frac{1}{8}, \pm 3, \pm \frac{3}{2}, \pm \frac{3}{4}, \pm \frac{3}{8}, \pm 9, \pm \frac{9}{2}, \pm \frac{9}{4}, \pm \frac{9}{8}$

9. First multiply by 3 (we need integer coefficients) to get: $2x^3 - 3x^2 - 15x + 6 = 0$
   p-factors: $\pm 1, \pm 2, \pm 3, \pm 6$ q-factors: $\pm 1, \pm 2$

   possible rational roots $\frac{p}{q}$: $\pm 1, \pm \frac{1}{2}, \pm 2, \pm 3, \pm \frac{3}{2}, \pm 6$

11. The possible rational roots are ± 1. Using synthetic division:

$$\begin{array}{r|rrrr} 1 & 1 & 0 & -3 & 1 \\ & & 1 & 1 & -2 \\ \hline & 1 & 1 & -2 & -1 \end{array}$$

   So $x = 1$ is not a root.

$$\begin{array}{r|rrrr} -1 & 1 & 0 & -3 & 1 \\ & & -1 & 1 & 2 \\ \hline & 1 & -1 & -2 & 3 \end{array}$$

   So $x = -1$ is not a root. There are no rational roots.

13. The possible rational roots are ± 1. Using synthetic division:

$$\begin{array}{r|rrrr} 1 & 1 & 1 & -1 & 1 \\ & & 1 & 2 & 1 \\ \hline & 1 & 2 & 1 & 2 \end{array}$$

So $x = 1$ is not a root.

$$\begin{array}{r|rrrr} -1 & 1 & 1 & -1 & 1 \\ & & -1 & 0 & 1 \\ \hline & 1 & 0 & -1 & 2 \end{array}$$

So $x = -1$ is not a root. There are no rational roots.

15. The possible rational roots are $\pm 1, \pm \frac{1}{2}, \pm \frac{1}{3}, \pm \frac{1}{4}, \pm \frac{1}{6}, \pm \frac{1}{12}, \pm 2, \pm \frac{2}{3}, \pm 3, \pm \frac{3}{2}, \pm \frac{3}{4}, \pm 6$. Using synthetic division:

$$\begin{array}{r|rrrrr} 1 & 12 & 0 & -1 & 0 & -6 \\ & & 12 & 12 & 11 & 11 \\ \hline & 12 & 12 & 11 & 11 & 5 \end{array}$$

So $x = 1$ is not a root. Since this row is all positive, we can exclude $x = 2$, $x = 3$, $x = \frac{3}{2}$ and $x = 6$.

$$\begin{array}{r|rrrrr} -1 & 12 & 0 & -1 & 0 & -6 \\ & & -12 & 12 & -11 & 11 \\ \hline & 12 & -12 & 11 & -11 & 5 \end{array}$$

So $x = -1$ is not a root. Since this row alternates signs, we can exclude $x = -2$, $x = -3$, $x = -\frac{3}{2}$ and $x = -6$.

$$\begin{array}{r|rrrrr} 1/4 & 12 & 0 & -1 & 0 & -6 \\ & & 3 & 3/4 & -1/16 & -1/64 \\ \hline & 12 & 3 & -1/4 & -1/16 & -385/64 \end{array}$$

So $x = \frac{1}{4}$ is not a root.

$$\begin{array}{r|rrrrr} 1/3 & 12 & 0 & -1 & 0 & -6 \\ & & 4 & 4/3 & 1/9 & 1/27 \\ \hline & 12 & 4 & 1/3 & 1/9 & -161/27 \end{array}$$

Proceeding in a similar fashion, we find that none of the candidates are rational roots.

17. The possible rational roots are $\pm 1, \pm 3$. Using synthetic division:

$$\begin{array}{r|rrrr} 1 & 1 & 3 & -1 & -3 \\ & & 1 & 4 & 3 \\ \hline & 1 & 4 & 3 & 0 \end{array}$$

So $x^3 + 3x^2 - x - 3 = (x - 1)(x^2 + 4x + 3) = (x - 1)(x + 1)(x + 3)$. So the roots are 1, -1, and -3.

19. The possible rational roots are $\pm 1, \pm \frac{1}{2}, \pm \frac{1}{4}, \pm 5, \pm \frac{5}{2}, \pm \frac{5}{4}$. Using synthetic division:

$$\begin{array}{r|rrrr} -1/4 & 4 & 1 & -20 & -5 \\ & & -1 & 0 & 5 \\ \hline & 4 & 0 & -20 & 0 \end{array}$$

Therefore: $4x^3 + x^2 - 20x - 5 = \left(x + \frac{1}{4}\right)\left(4x^2 - 20\right) = 4\left(x + \frac{1}{4}\right)\left(x^2 - 5\right) = 4\left(x + \frac{1}{4}\right)\left(x + \sqrt{5}\right)\left(x - \sqrt{5}\right)$

So the roots are $-\frac{1}{4}, -\sqrt{5}$ and $\sqrt{5}$.

21. The possible rational roots are $\pm 1, \pm\frac{1}{3}, \pm\frac{1}{9}, \pm 2, \pm\frac{2}{3}, \pm\frac{2}{9}$. Using synthetic division:

$$\begin{array}{r|rrrr} 1 & 9 & 18 & 11 & 2 \\ & & 9 & 27 & 38 \\ \hline & 9 & 27 & 38 & 40 \end{array}$$

So $x = 2$ is excluded also (the row is positive).

$$\begin{array}{r|rrrr} -1 & 9 & 18 & 11 & 2 \\ & & -9 & -9 & -2 \\ \hline & 9 & 9 & 2 & 0 \end{array}$$

So $9x^3+18x^2+11x+2=(x+1)(9x^2+9x+2)=(x+1)(3x+2)(3x+1)$. So the roots are $-1, -\frac{2}{3}$ and $-\frac{1}{3}$.

23. The possible rational roots are ±1, ±2, ±3, ±4, ±6, ±8, ±12, ±24. Using synthetic division:

$$\begin{array}{r|rrrrr} 1 & 1 & 1 & -25 & -1 & 24 \\ & & 1 & 2 & -23 & -24 \\ \hline & 1 & 2 & -23 & -24 & 0 \end{array}$$

So $x = 1$ is a root. Using synthetic division again:

$$\begin{array}{r|rrrr} -1 & 1 & 2 & -23 & -24 \\ & & -1 & -1 & 24 \\ \hline & 1 & 1 & -24 & 0 \end{array}$$

So $x=-1$ is also a root. So $x^4+x^3-25x^2-x+24=(x-1)(x+1)(x^2+x-24)$. Use the quadratic formula:

$$x=\frac{-1\pm\sqrt{1+96}}{2}=\frac{-1\pm\sqrt{97}}{2}$$

So the roots are $1, -1, \dfrac{-1+\sqrt{97}}{2}$ and $\dfrac{-1-\sqrt{97}}{2}$.

25. The possible rational roots are ±1. Using synthetic division:

$$\begin{array}{r|rrrrr} 1 & 1 & -4 & 6 & -4 & 1 \\ & & 1 & -3 & 3 & -1 \\ \hline & 1 & -3 & 3 & -1 & 0 \end{array}$$

Using synthetic division again:

$$\begin{array}{r|rrrr} 1 & 1 & -3 & 3 & -1 \\ & & 1 & -2 & 1 \\ \hline & 1 & -2 & 1 & 0 \end{array}$$

So $x^4-4x^3+6x^2-4x+1=(x-1)^2(x^2-2x+1)=(x-1)^4$. So the only root is 1 (with multiplicity 4).

27. First multiply by 3 to obtain integer coefficients: $3x^3-17x^2-10x+24=0$

The possible rational roots are $\pm 1, \pm\frac{1}{3}, \pm 2, \pm\frac{2}{3}, \pm 3, \pm 4, \pm\frac{4}{3}, \pm 6, \pm 8, \pm\frac{8}{3}, \pm 12, \pm 24$. Using synthetic division:

$$\begin{array}{r|rrrr} 1 & 3 & -17 & -10 & 24 \\ & & 3 & -14 & -24 \\ \hline & 3 & -14 & -24 & 0 \end{array}$$

So $3x^3-17x^2-10x+24=(x-1)(3x^2-14x-24)=(x-1)(3x+4)(x-6)$. So the roots are $1, -\frac{4}{3}$ and 6.

29. a. The possible rational roots are $\pm 1, \pm\frac{1}{5}, \pm 2, \pm\frac{2}{5}, \pm 3, \pm\frac{3}{5}, \pm 4, \pm\frac{4}{5}, \pm 6, \pm\frac{6}{5}, \pm 12, \pm\frac{12}{5}$. Using synthetic division:

$$\begin{array}{r|rrrrr} 2 & 5 & 0 & 0 & -10 & -12 \\ & & 10 & 20 & 40 & 60 \\ \hline & 5 & 10 & 20 & 30 & 48 \end{array}$$

Since this row is all positive, $x = 2$ is an upper bound.

-1	5	0	0	-10	-12
		-5	5	-5	15
	5	-5	5	-15	3

Since this row alternates signs, $x = -1$ is a lower bound.

b. The possible rational roots are $\pm 1, \pm \frac{1}{3}, \pm 2, \pm \frac{2}{3}, \pm 4, \pm \frac{4}{3}$. Using synthetic division:

4/3	3	-4	5	-2	-4
		4	0	20/3	56/9
	3	0	5	14/3	20/9

So $x = \frac{4}{3}$ is the upper bound.

-2/3	3	-4	5	-2	-4
		-2	4	-6	16/3
	3	-6	9	-8	4/3

So $x = -\frac{2}{3}$ is the lower bound. Since the question asked for the integral upper and lower bounds, the upper bound is 2 and the lower bound is -1.

c. The possible rational roots are $\pm 1, \pm \frac{1}{2}, \pm 2, \pm 3, \pm \frac{3}{2}, \pm 4, \pm 6, \pm 12$. Using synthetic division:

6	2	-7	-5	28	-12
		12	30	150	1068
	2	5	25	178	1056

So $x = 6$ is an upper bound.

-2	2	-7	-5	28	-12
		-4	22	-34	12
	2	-11	17	-6	0

Actually -2 is a root, but also a lower bound.

31. The root lies between 0 and 1.

$f(0.5) = -0.375$ $f(0.7) = 0.043$ $f(0.6) = -0.373$

So the root lies between 0.6 and 0.7.

$f(0.65) = -0.075$ $f(0.66) = -0.052$ $f(0.67) = -0.029$

$f(0.68) = -0.005$ $f(0.69) = 0.018$

So the root lies between 0.68 and 0.69.

33. Let $f(x) = x^5 - 200$.

$f(2) = -168$ $f(3) = 43$

So the root lies between 2 and 3.

$f(2.5) = -102$ $f(2.8) = -27.9$ $f(2.9) = 5.11$

So the root lies between 2.8 and 2.9.

$f(2.87) = -5.28$ $f(2.88) = -1.86$ $f(2.89) = 1.60$

So the root lies between 2.88 and 2.89.

35. Let $f(x) = x^3 - 8x^2 + 21x - 22$.

$f(4) = -2$ $f(5) = 8$

So the root lies between 4 and 5.

$f(4.5) = 1.63$ $f(4.4) = 0.704$ $f(4.3) = -0.113$

So the root lies between 4.3 and 4.4.

$f(4.33) = 0.12$ $f(4.32) = 0.04$ $f(4.31) = -0.03$

So the root lies between 4.31 and 4.32.

37. **a.** Sketching the graph:

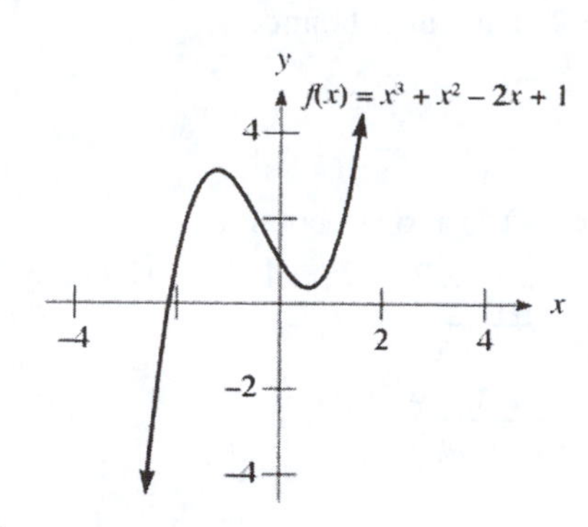

The root lies between –3 and –2.

b. Let $f(x) = x^3 + x^2 - 2x + 1$.

$f(-2) = 1$ $\quad$ $f(-3) = -11$

So the root lies between –3 and –2.

$f(-2.1) \approx 0.35$ $\quad$ $f(-2.2) \approx -0.41$

So the root lies between –2.2 and –2.1.

$f(-2.15) \approx -0.02$ $\quad$ $f(-2.14) \approx 0.05$

So the root lies between –2.15 and –2.14.

$f(-2.148) \approx -0.001$ $\quad$ $f(-2.147) \approx 0.01$

So the root lies between –2.148 and –2.147.

39. **a.** Sketching the graph:

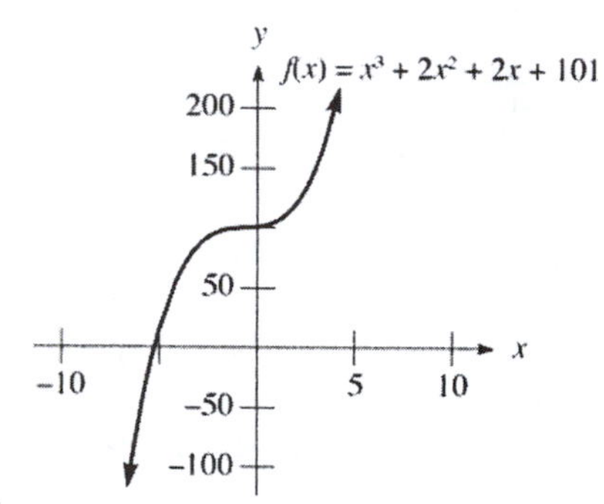

The root lies between –6 and –5.

b. Let $f(x) = x^3 + 2x^2 + 2x + 101$.

$f(-5) = 16$ $\quad$ $f(-6) = -55$

So the root lies between –6 and –5.

$f(-5.3) \approx -2.30$ $\quad$ $f(-5.2) \approx 4.07$

So the root lies between –5.3 and –5.2.

$f(-5.27) \approx -0.36$ $\quad$ $f(-5.26) \approx 0.28$

So the root lies between –5.27 and –5.26.

$f(-5.265) \approx -0.04$ $\quad$ $f(-5.264) \approx 0.03$

So the root lies between –5.265 and –5.264.

41. **a.** Since 2 is a factor of $8 \bullet 5 = 40$, and 2 is not a factor of 5, 2 must be a factor of 8, which is true.

b. The result guarantees only that A is a factor of B in the case where A and C have no factor in common. Here A and C have a common factor of 5, so the result does not apply.

c. Since $x = \frac{p}{q}$ is a root of the equation, this statement must be true.

d. Subtract a_0 and multiply by q^n to get: $a_n p^n + a_{n-1} q p^{n-1} + a_{n-2} q^2 p^{n-2} + \ldots + a_1 q^{n-1} p = -a_0 q^n$

Therefore: $p\left(a_n p^{n-1} + a_{n-1} q p^{n-2} + a_{n-2} q^2 p^{n-3} + \ldots + a_1 q^{n-1}\right) = -a_0 q^n$

43. Sketching the graph:

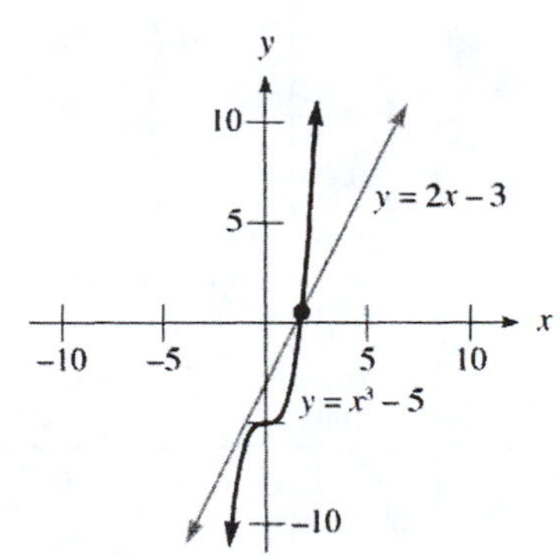

To use successive approximations, we find where $f(x)=(x^3-5)-(2x-3)=x^3-2x-2=0$.

$f(1)=-3$ $\quad f(2)=2$

So the root lies between 1 and 2.

$f(1.7)=-0.487$ $\quad f(1.8)=0.232$

So the root lies between 1.7 and 1.8.

$f(1.76)\approx -0.068$ $\quad f(1.77)\approx 0.005$

So the root lies between 1.76 and 1.77.

$f(1.769)\approx -0.0022$ $\quad f(1.770)\approx 0.0052$

The x-coordinate is approximately 1.769.

45. Sketching the graph:

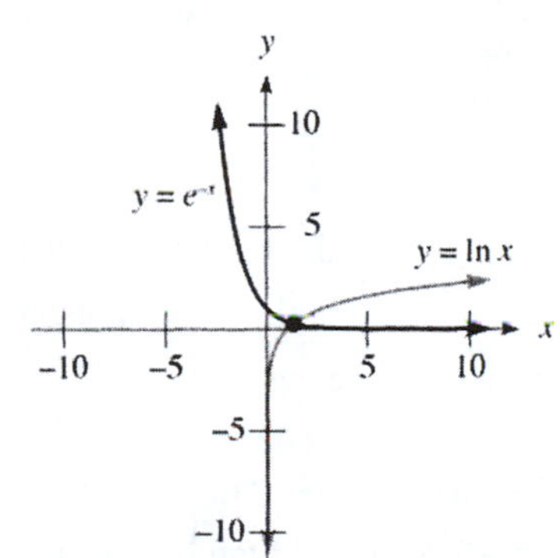

To use successive approximations, we find where $f(x)=e^{-x}-\ln x=0$.

$f(1)\approx 0.3679$ $\quad f(2)\approx -0.5578$

So the root lies between 1 and 2.

$f(1.3)\approx 0.0102$ $\quad f(1.4)\approx -0.0899$

So the root lies between 1.3 and 1.4.

$f(1.30)\approx 0.0102$ $\quad f(1.31)\approx -0.0002$

So the root lies between 1.30 and 1.31.

$f(1.309)\approx 0.0008$ $\quad f(1.310)\approx -0.0002$

The x-coordinate is approximately 1.310.

47. Sketching the graph:

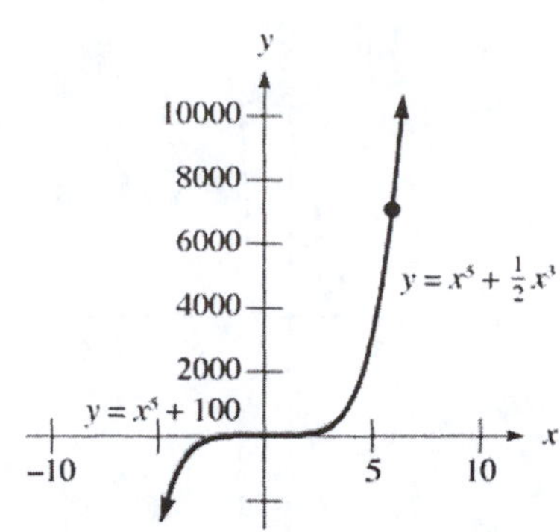

To use successive approximations, we find where $f(x)=\left(x^5+100\right)-\left(x^5+\frac{1}{2}x^3\right)=100-\frac{1}{2}x^3=0$.

$f(5)=37.5$ $\qquad$ $f(6)=-8$

So the root lies between 5 and 6.

$f(5.8)=2.444$ $\qquad$ $f(5.9)=-2.6895$

So the root lies between 5.8 and 5.9.

$f(5.84)\approx 0.4116$ $\qquad$ $f(5.85)\approx -0.1008$

So the root lies between 5.84 and 5.85.

$f(5.848)\approx 0.0018$ $\qquad$ $f(5.849)\approx -0.0495$

The x-coordinate is approximately 5.848. Solving the equation:

$$\begin{aligned} x^5+100&=x^5+\tfrac{1}{2}x^3\\ \tfrac{1}{2}x^3&=100\\ x^3&=200\\ x&=\sqrt[3]{200}\approx 5.848 \end{aligned}$$

49. **a.** Here $a_n=9$, $a_0=27$ and $f(1)=29$. Since these three numbers are odd, the equation has no rational roots.

b. Here $a_n=5$, $a_0=-25$ and $f(1)=-29$. Since these three numbers are odd, the equation has no rational roots.

51. **a.** Using $\tan 9°\approx 0.1584$ verifies it is a root of the equation.

b. Substituting $x=\tan 9°$ into the equation:

$$\begin{aligned} &\tan^5 9°-5\tan^4 9°-10\tan^3 9°+10\tan^2 9°+5\tan 9°-1\\ &=\left(\tan^5 9°-10\tan^3 9°+5\tan 9°\right)-\left(5\tan^4 9°-10\tan^2 9°+1\right)\\ &=\left(\tan 45°\right)\left(5\tan^4 9°-10\tan^2 9°+1\right)-\left(5\tan^4 9°-10\tan^2 9°+1\right)\\ &=\left(5\tan^4 9°-10\tan^2 9°+1\right)\left(\tan 45°-1\right)\\ &=\left(5\tan^4 9°-10\tan^2 9°+1\right)(1-1)\\ &=0 \end{aligned}$$

c. The possible rational roots are −1 and 1. Synthetic division with −1 does not produce a root. Using synthetic division with 1, the equation factors as $(x-1)\left(x^4-4x^3-14x^2-4x+1\right)=0$.

The reduced equation is $x^4-4x^3-14x^2-4x+1=0$.

d. Since $\tan 9°\neq 1$, it must be a root of $x^4-4x^3-14x^2-4x+1=0$. Also, since 1 and −1 were the only possible rational roots, then $\tan 9°$ must be irrational.

53. **a.** Since $\cos 3\theta = \cos\frac{6\pi}{7}$ and $\cos 4\theta = \cos\frac{8\pi}{7}$, both angles have the same reference angle of $\frac{\pi}{7}$.
Since both cosines are negative, it must be the case that $\cos 3\theta = \cos 4\theta$.

b. Since $\cos 3\theta = \cos 4\theta$:

$$4\cos^3\frac{2\pi}{7} - 3\cos\frac{2\pi}{7} = 8\cos^4\frac{2\pi}{7} - 8\cos^2\frac{2\pi}{7} + 1$$
$$0 = 8\cos^4\frac{2\pi}{7} - 4\cos^3\frac{2\pi}{7} - 8\cos^2\frac{2\pi}{7} + 3\cos\frac{2\pi}{7} + 1$$

Thus $x = \cos\frac{2\pi}{7}$ is a root of the equation $8x^4 - 4x^3 - 8x^2 + 3x + 1 = 0$.

c. The possibilities for rational roots are $\pm 1, \pm\frac{1}{2}, \pm\frac{1}{4}, \pm\frac{1}{8}$. Of these, only $x = 1$ is a rational root.

Using synthetic division, the equation factors as $(x-1)\left(8x^3 + 4x^2 - 4x - 1\right) = 0$. Since

$\cos\frac{2\pi}{7} \neq 1$, it must be a root of the reduced equation $8x^3 + 4x^2 - 4x - 1 = 0$.

d. First write the reduced equation as $x^3 + \frac{1}{2}x^2 - \frac{1}{2}x - \frac{1}{8} = 0$. Using Table 2:

$$\cos\frac{2\pi}{7}\cos\frac{4\pi}{7}\cos\frac{6\pi}{7} = r_1 r_2 r_3 = -\left(-\frac{1}{8}\right) = \frac{1}{8}$$
$$\cos\frac{2\pi}{7} + \cos\frac{4\pi}{7} + \cos\frac{6\pi}{7} = r_1 + r_2 + r_3 = -\frac{1}{2}$$

55. Since p is a prime number, the only possible rational roots are $\pm 1, \pm p$.
Let $f(x) = x^3 + x^2 + x - p$

$f(1) = 1 + 1 + 1 - p = 3 - p$

So 1 is a root only if $p = 3$.

$f(-1) = -1 + 1 - 1 - p = -1 - p$

So -1 is a root only if $p = -1$, but -1 is not prime.

$$f(p) = p^3 + p^2 + p - p = p^3 + p^2 = p^2(p+1)$$

So p is a root only if $p = 0$ or $p = -1$, neither of which is prime.

$$f(-p) = -p^3 + p^2 - p - p = -p^3 + p^2 - 2p = -p\left(p^2 - p + 2\right)$$

So $-p$ is a root only if $p = 0$ or $p = \dfrac{1 \pm \sqrt{1-8}}{2}$, neither of which is prime. Thus $p = 3$ is the only prime number such that $f(x) = 0$ will have rational roots. When $p = 3$, we have $x^3 + x^2 + x - 3 = 0$. We know $x = 1$ is a root. Now using synthetic division:

1) 1	1	1	−3
	1	2	3
1	2	3	0

Since $x^2 + 2x + 3 = 0$ has no real roots, $x = 1$ is the only root.

57. **a.** If $x = 1$ is a root, then $1 + p - q = 0$, thus $q - p = 1$. But the only two prime numbers which are 1 unit apart would be $q = 3$ and $p = 2$. The remaining quadratic is then $x^2 + x + 3 = 0$, which we solve using the quadratic formula:

$$x = \frac{-1 \pm \sqrt{1-12}}{2} = \frac{-1 \pm i\sqrt{11}}{2}$$

So the remaining roots are $\dfrac{-1 \pm i\sqrt{11}}{2}$.

b. Suppose -1 is a root. Then $-1 - p - q = 0$, so $p + q = -1$. Clearly this is impossible for prime numbers p and q.
Suppose q is a root. If $x = q$:

$$q^3 + pq - q = 0$$
$$q\left(q^2 + p - 1\right) = 0$$

Clearly $q \neq 0$, so $q^2 + p = 1$. But since $q > 1$ and $p > 1$, this is also impossible. Suppose $-q$ is a root. If $x = -q$:

$$-q^3 - pq - q = 0$$
$$-q\left(q^2 + p + 1\right) = 0$$

Clearly $q \neq 0$, so $q^2 + p = -1$. But since $q > 1$ and $p > 1$, this is also impossible.

59. The possible rational roots are $\pm 1, \pm 2, \pm 4$. Using synthetic division:

$$\begin{array}{r|rrrr} 1 & 1 & -b^2 & 3b & -4 \\ & & 1 & -b^2+1 & -b^2+3b+1 \\ \hline & 1 & -b^2+1 & -b^2+3b+1 & -b^2+3b-3 \end{array}$$

Setting the remainder equal to 0:

$$-3 + 3b - b^2 = 0$$
$$b^2 - 3b + 3 = 0$$
$$b = \frac{3 \pm \sqrt{9-12}}{2}, \text{ which is not an integer}$$

Use synthetic division again:

$$\begin{array}{r|rrrr} -1 & 1 & -b^2 & 3b & -4 \\ & & -1 & b^2+1 & -b^2-3b-1 \\ \hline & 1 & -b^2-1 & b^2+3b+1 & -b^2-3b-5 \end{array}$$

Setting the remainder equal to 0:

$$-b^2 - 3b - 5 = 0$$
$$b^2 + 3b + 5 = 0$$
$$b = \frac{-3 \pm \sqrt{9-20}}{2}, \text{ which is not an integer}$$

Using synthetic division again:

$$\begin{array}{r|rrrr} 2 & 1 & -b^2 & 3b & -4 \\ & & 2 & -2b^2+4 & -4b^2+6b+8 \\ \hline & 1 & -b^2+2 & -2b^2+3b+4 & -4b^2+6b+4 \end{array}$$

Setting the remainder equal to 0:

$$4 + 6b - 4b^2 = 0$$
$$2b^2 - 3b - 2 = 0$$
$$(2b+1)(b-2) = 0$$
$$b = -\tfrac{1}{2} \text{ or } b = 2$$

Hence, if $b = 2$, the given equation has $x = 2$ as a rational root. Using synthetic division again:

$$\begin{array}{r|rrrr} -2 & 1 & -b^2 & 3b & -4 \\ & & -2 & 2b^2+4 & -4b^2-6b-8 \\ \hline & 1 & -b^2-2 & 2b^2+3b+4 & -4b^2-6b-12 \end{array}$$

Setting the remainder equal to 0:

$$-4b^2 - 6b - 12 = 0$$
$$2b^2 + 3b + 6 = 0$$
$$b = \frac{-3 \pm \sqrt{9-48}}{4}, \text{ which is not an integer}$$

Using synthetic division again:

$$\begin{array}{r|llll} 4 & 1 & -b^2 & 3b & -4 \\ & & 4 & -4b^2+16 & -16b^2+12b+64 \\ \hline & 1 & -b^2+4 & -4b^2+3b+16 & -16b^2+12b+60 \end{array}$$

Setting the remainder equal to 0:

$$60+12b-16b^2=0$$
$$4b^2-3b-15=0$$
$$b=\frac{3\pm\sqrt{9+240}}{8}, \text{ which is not an integer}$$

Using synthetic division again:

$$\begin{array}{r|llll} -4 & 1 & -b^2 & 3b & -4 \\ & & -4 & 4b^2+16 & -16b^2-12b-64 \\ \hline & 1 & -b^2-4 & 4b^2+3b+16 & -16b^2-12b-68 \end{array}$$

Setting the remainder equal to 0:

$$-16b^2-12b-68=0$$
$$4b^2+3b+17=0$$
$$b=\frac{-3\pm\sqrt{9-272}}{8}, \text{ which is not an integer}$$

So $b = 2$ is the only integral value.

12.6 Conjugate Roots and Descartes's Rule of Signs

1. The other root must be $7 + 2i$.

3. One other root must be $5 - 2i$. So $[x-(5-2i)][x-(5+2i)]$ are factors, or $x^2 - 10x + 29$. Using long division:

$$\begin{array}{r|l} & x-3 \\ x^2-10x+29 & x^3-13x^2+59x-87 \\ & \underline{x^3-10x^2+29x} \\ & -3x^2+30x-87 \\ & \underline{-3x^2+30x-87} \\ & 0 \end{array}$$

Since $x - 3$ is the other factor, 3 is the other root. So the other two roots are $5 - 2i$ and 3.

5. One other root must be $-2 - i$, so $[x-(-2+i)][x-(-2-i)]$ are factors, or $x^2 + 4x + 5$. Using long division:

$$\begin{array}{r|l} & x^2+6x+9 \\ x^2+4x+5 & x^4+10x^3+38x^2+66x+45 \\ & \underline{x^4+4x^3+5x^2} \\ & 6x^3+33x^2+66x \\ & \underline{6x^3+24x^2+30x} \\ & 9x^2+36x+45 \\ & \underline{9x^2+36x+45} \\ & 0 \end{array}$$

Since $x^2 + 6x + 9 = (x + 3)^2$, the other root is -3. So the remaining roots are $-2 - i$ and -3 (multiplicity 2).

7. One other root must be $6 + 5i$, so $[x-(6-5i)][x-(6+5i)]$ are factors, or $x^2 - 12x + 61$. Using long division:

$$\begin{array}{r} 4x+1 \\ x^2-12x+61\overline{)4x^3-47x^2+232x+61} \\ \underline{4x^3-48x^2+244x} \\ x^2-12x+61 \\ \underline{x^2-12x+61} \\ 0 \end{array}$$

Since $4x + 1$ is the other factor, $-\frac{1}{4}$ is the other root. So the remaining roots are $6 + 5i$ and $-\frac{1}{4}$.

9. One other root must be $4-\sqrt{2}i$, so $\left[x-\left(4+\sqrt{2}i\right)\right]\left[x-\left(4-\sqrt{2}i\right)\right]$ are factors, or $x^2 - 8x + 18$. Using long division:

$$\begin{array}{r} 4x^2+9 \\ x^2-8x+18\overline{)4x^4-32x^3+81x^2-72x+162} \\ \underline{4x^4-32x^3+72x^2} \\ 9x^2-72x+162 \\ \underline{9x^2-72x+162} \\ 0 \end{array}$$

Since $4x^2 + 9 = (2x + 3i)(2x - 3i)$, the other roots are $-\frac{3i}{2}$ and $\frac{3i}{2}$. So the remaining roots are $4-\sqrt{2}i, -\frac{3i}{2}$ and $\frac{3i}{2}$.

11. One other root must be $10 - 2i$, so $[x-(10+2i)][x-(10-2i)]$ are factors, or $x^2 - 20x + 104$. Using long division:

$$\begin{array}{r} x^2-2x-4 \\ x^2-20x+104\overline{)x^4-22x^3+140x^2-128x-416} \\ \underline{x^4-20x^3+104x^2} \\ -2x^3+36x^2-128x \\ \underline{-2x^3+40x^2-208x} \\ -4x^2+80x-416 \\ \underline{-4x^2+80x-416} \\ 0 \end{array}$$

So the other factor is $x^2 - 2x - 4$. We use the quadratic formula: $x = \dfrac{2\pm\sqrt{4+16}}{2} = \dfrac{2\pm 2\sqrt{5}}{2} = 1\pm\sqrt{5}$

So the remaining roots are $10 - 2i$, $1+\sqrt{5}$ and $1-\sqrt{5}$.

13. One other root must be $\frac{1-\sqrt{2}i}{3}$ so $\left(x-\frac{1+\sqrt{2}i}{3}\right)\left(x-\frac{1-\sqrt{2}i}{3}\right)$ are factors, or $x^2-\frac{2}{3}x+\frac{1}{3}$. Using long division:

$$\begin{array}{r} 15x-6 \\ x^2-\frac{2}{3}x+\frac{1}{3}\overline{)15x^3-16x^2+9x-2} \\ \underline{15x^3-10x^2+5x} \\ -6x^2+4x-2 \\ \underline{-6x^2+4x-2} \\ 0 \end{array}$$

So the other factor is $15x-6$, so $x=\frac{2}{5}$ is a root. The remaining roots are $\frac{1-i\sqrt{2}}{3}$ and $\frac{2}{5}$.

15. We know $3 + 2i$ is a root, so $[x-(3-2i)][x-(3+2i)]$ are factors, which is $x^2 - 6x + 13$. Using long division:

$$
\begin{array}{r}
x^5 + 3x^4 + x^3 - 3x^2 - 4x + 2 \\
x^2 - 6x + 13 \overline{\smash{\big)}\, x^7 - 3x^6 - 4x^5 + 30x^4 + 27x^3 - 13x^2 - 64x + 26} \\
\underline{x^7 - 6x^6 + 13x^5} \\
3x^6 - 17x^5 + 30x^4 \\
\underline{3x^6 - 18x^5 + 39x^4} \\
x^5 - 9x^4 + 27x^3 \\
\underline{x^5 - 6x^4 + 13x^3} \\
-3x^4 + 14x^3 - 13x^2 \\
\underline{-3x^4 + 18x^3 - 39x^2} \\
-4x^3 + 26x^2 - 64x \\
\underline{-4x^3 + 24x^2 - 52x} \\
2x^2 - 12x + 26 \\
\underline{2x^2 - 12x + 26} \\
0
\end{array}
$$

We know $x = -1 - i$ will be a root, so $[x-(-1+i)][x-(-1-i)]$ are factors, which is $x^2 + 2x + 2$. Using long division:

$$
\begin{array}{r}
x^3 + x^2 - 3x + 1 \\
x^2 + 2x + 2 \overline{\smash{\big)}\, x^5 + 3x^4 + x^3 - 3x^2 - 4x + 2} \\
\underline{x^5 + 2x^4 + 2x^3} \\
x^4 - x^3 - 3x^2 \\
\underline{x^4 + 2x^3 + 2x^2} \\
-3x^3 - 5x^2 - 4x \\
\underline{-3x^3 - 6x^2 - 6x} \\
x^2 + 2x + 2 \\
\underline{x^2 + 2x + 2} \\
0
\end{array}
$$

Finally, we know $x = 1$ is a root, so synthetically divide:

$$
\begin{array}{r|rrrr}
1 & 1 & 1 & -3 & 1 \\
 & & 1 & 2 & -1 \\
\hline
 & 1 & 2 & -1 & 0
\end{array}
$$

So we are left with $x^2 + 2x - 1 = 0$. Using the quadratic formula: $x = \frac{-2 \pm \sqrt{4+4}}{2} = \frac{-2 \pm 2\sqrt{2}}{2} = -1 \pm \sqrt{2}$

So the remaining roots are $3 + 2i$, $-1 - i$, $-1+\sqrt{2}$, and $-1-\sqrt{2}$.

17. Here $r_1 = 1+\sqrt{6}$ and $r_2 = 1-\sqrt{6}$, so $\left[x-\left(1+\sqrt{6}\right)\right]\left[x-\left(1-\sqrt{6}\right)\right] = 0$. Multiplied out, we obtain $x^2 - 2x - 5 = 0$.

19. Here $r_1 = \frac{2+\sqrt{10}}{3}$ and $r_2 = \frac{2-\sqrt{10}}{3}$, so $\left(x - \frac{2+\sqrt{10}}{3}\right)\left(x - \frac{2-\sqrt{10}}{3}\right) = 0$. Multiplied out, we obtain $x^2 - \frac{4}{3}x - \frac{2}{3} = 0$.

21. a. Computing $f(-x)$: $f(-x)=2(-x)^4-3(-x)^3+12(-x)^2+22(-x)-60=2x^4+3x^3+12x^2-22x-60$
Since the resulting function has only one sign change, there can only be one negative real root.

b. Since $f(x)=2x^4-3x^3+12x^2+22x-60$ has three sign changes, there can be either one or three positive real roots.

c. Graphing the function:

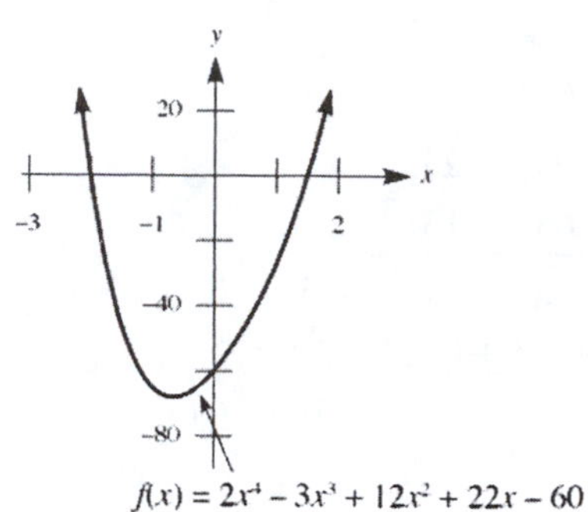

From the graph it appears there is one positive real root.

d. The roots are –2.000 and 1.500, which agrees with the roots $-2, \frac{3}{2}$ found in the text.

23. a. Since $f(x)=x^3+8x+5$ has no sign changes, there are no positive real roots. Compute:

$$f(-x)=(-x)^3+8(-x)+5=-x^3-8x+5$$

Since there is one sign change, there is exactly one negative real root.

b. Graphing the function:

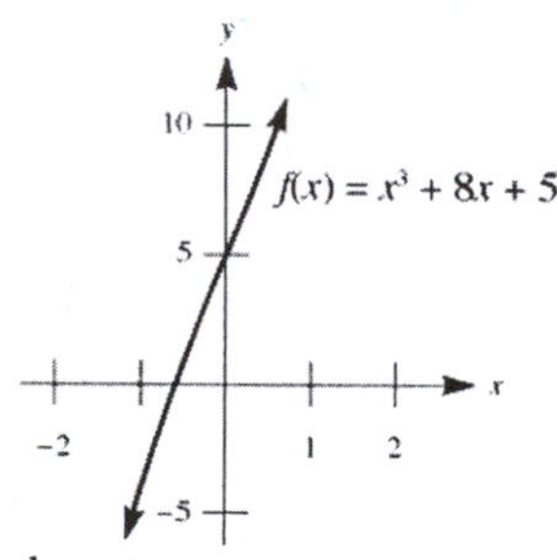

Note that there is only one negative real root.

c. The root is approximately –0.5982.

d. Write the equation as $x^3+8x=-5$. Now use the formula with $a=8$ and $b=-5$:

$$\begin{aligned} x &= \sqrt[3]{\frac{-5}{2}+\sqrt{\frac{(-5)^2}{4}+\frac{8^3}{27}}}-\sqrt[3]{-\frac{-5}{2}+\sqrt{\frac{(-5)^2}{4}+\frac{8^3}{27}}} \\ &= \sqrt[3]{-2.5+\sqrt{\frac{25}{4}+\frac{512}{27}}}-\sqrt[3]{2.5+\sqrt{\frac{25}{4}+\frac{512}{27}}} \\ &= \sqrt[3]{-2.5+\sqrt{\frac{2723}{108}}}-\sqrt[3]{2.5+\sqrt{\frac{2723}{108}}} \\ &\approx -0.5982 \end{aligned}$$

Note that our value agrees with that found in part **c**.

25. Let $f(x)=x^3+5$. Since there are no sign changes, there will be no positive roots. Now $f(-x)=-x^3+5$. Since there is one sign change, there is 1 negative root. So the equation has 2 complex roots and 1 negative real root.

27. Let $f(x)=2x^5+3x+4$. Since there are no sign changes, there will be no positive roots. Now $f(-x)=-2x^5-3x+4$. Since there is one sign change, there is 1 negative root. So the equation has 4 complex roots and 1 negative real root.

29. Let $f(x)=5x^4+2x-7$. Since there is one sign change, there will be 1 positive root. Now $f(-x)=5x^4-2x-7$. Since there is one sign change, there will be 1 negative root. So the equation has 2 complex roots, 1 positive real root and 1 negative real root.

31. Let $f(x) = x^3 - 4x^2 - x - 1$. Since there is one sign change, there will be 1 positive root. Now $f(-x) = -x^3 - 4x^2 + x - 1$. Since there are two sign changes, there will be 0 or 2 negative roots. So the equation has either 1 positive real root and 2 negative real roots, or 1 positive real root and 2 complex roots.

33. Let $f(x) = 3x^8 + x^6 - 2x^2 - 4$. Since there is one sign change, there is 1 positive root. Now $f(-x) = 3x^8 + x^6 - 2x^2 - 4$. Since there is one sign change, there is 1 negative root. So the equation has 1 positive real root, 1 negative real root and 6 complex roots.

35. Let $f(x) = x^9 - 2$. Since there is one sign change, there is 1 positive root. Now $f(-x) = -x^9 - 2$. Since there are no sign changes, there are no negative roots. So the equation has 1 positive real root and 8 complex roots.

37. Let $f(x) = x^8 - 2$. Since there is one sign change, there is 1 positive root. Now $f(-x) = x^8 - 2$. Since there is one sign change, there is 1 negative root. So the equation has 1 positive real root, 1 negative real root and 6 complex roots.

39. Let $f(x) = x^6 + x^2 - x - 1$. Since there is one sign change, there is 1 positive root. Now $f(-x) = x^6 + x^2 + x - 1$. Since there is one sign change, there is 1 negative root. So the equation has 1 positive real root, 1 negative real root and 4 complex roots.

41. Let $f(x) = x^4 + cx^2 + dx - e$. Since there is one sign change, there is 1 positive root. Now $f(-x) = x^4 + cx^2 - dx - e$. Since there is one sign change, there is 1 negative root. So the equation has 1 positive real root, 1 negative real root and 2 complex roots.

43. **a.** The two roots $\sqrt{3}+2i$ and $\sqrt{3}-2i$ must be included, so:

$$f(x)=\left[x-\left(\sqrt{3}+2i\right)\right]\left[x-\left(\sqrt{3}-2i\right)\right]=x^2-2\sqrt{3}x+7=0$$

Unfortunately, not all coefficients are rational: $x^2+7=2\sqrt{3}x$

Squaring:

$$x^4+14x^2+49=12x^2$$
$$x^4+2x^2+49=0$$

So $f(x) = x^4 + 2x^2 + 49$ is the desired polynomial.

b. Use long division to find the other factor:

$$\begin{array}{r}
x^2+2\sqrt{3}x+7 \\
x^2-2\sqrt{3}x+7\overline{\big)\,x^4+0x^3+2x^2+0x+49} \\
\underline{x^4-2\sqrt{3}x^3+7x^2} \\
2\sqrt{3}x^3-5x^2+0x \\
\underline{2\sqrt{3}x^3-12x^2+14\sqrt{3}x} \\
7x^2-14\sqrt{3}x+49 \\
\underline{7x^2-14\sqrt{3}x+49} \\
0
\end{array}$$

So the other factor is $x^2+2\sqrt{3}x+7$. Now solving $x^2+2\sqrt{3}x+7=0$ using the quadratic formula:

$$x=\frac{-2\sqrt{3}\pm\sqrt{\left(2\sqrt{3}\right)^2-4(1)(7)}}{2}=\frac{-2\sqrt{3}\pm\sqrt{12-28}}{2}=\frac{-2\sqrt{3}\pm\sqrt{-16}}{2}=\frac{-2\sqrt{3}\pm 4i}{2}=-\sqrt{3}\pm 2i$$

The other roots are $-\sqrt{3}\pm 2i$.

45. **a.** If $b = 0$, then $a+b\sqrt{c}=a=a-b\sqrt{c}$, so $a-b\sqrt{c}$ is also a root.

b. Compute $d\left(a+b\sqrt{c}\right)$: $d\left(a+b\sqrt{c}\right)=\left[a+b\sqrt{c}-\left(a+b\sqrt{c}\right)\right]\left[a+b\sqrt{c}-\left(a-b\sqrt{c}\right)\right]=0\left(2b\sqrt{c}\right)=0$

c. Factor $d(x) = x^2 - 2ax + a^2 - b^2c = (x - a)^2 - b^2c$.

d. If $x = a + b\sqrt{c}$:

$$f\left(a+b\sqrt{c}\right) = d\left(a+b\sqrt{c}\right)Q\left(a+b\sqrt{c}\right) + C\left(a+b\sqrt{c}\right) + D$$
$$0 = 0 \bullet Q\left(a+b\sqrt{c}\right) + Ca + D + bC\sqrt{c}$$
$$0 = Ca + D + bC\sqrt{c}$$

But if C and D are rational, then $C = 0$ (otherwise $\sqrt{c}$ will not be "cancelled" out). If $C = 0$, then $D = 0$. So $C = D = 0$.

e. Compute $f\left(a-b\sqrt{c}\right)$:

$$f\left(a-b\sqrt{c}\right) = \left[a-b\sqrt{c}-\left(a+b\sqrt{c}\right)\right]\left[a-b\sqrt{c}-\left(a-b\sqrt{c}\right)\right]Q(x) = \left(-2b\sqrt{c}\right)(0)Q(x) = 0$$

So $x = a - b\sqrt{c}$ is also a root of $f(x) = 0$.

12.7 Introduction to Partial Fractions

1. a. Clearing fractions: $7x - 6 = A(x+2) + B(x-2) = (A+B)x + (2A-2B)$

Equating coefficients:

$$A + B = 7$$
$$2A - 2B = -6$$

Multiplying the first equation by 2:

$$2A + 2B = 14$$
$$2A - 2B = -6$$

Adding yields $4A = 8$, so $A = 2$ and $B = 5$.

b. Substituting $x = 2$ into $7x - 6 = A(x+2) + B(x-2)$:

$$14 - 6 = A(2+2) + B(2-2)$$
$$8 = 4A$$
$$2 = A$$

Substituting $x = -2$ into $7x - 6 = A(x+2) + B(x-2)$:

$$-14 - 6 = A(-2+2) + B(-2-2)$$
$$-20 = -4B$$
$$5 = B$$

So $A = 2$ and $B = 5$.

3. a. Clearing fractions: $6x - 25 = A(2x-5) + B(2x+5) = (2A+2B)x + (-5A+5B)$

Equating coefficients:

$$2A + 2B = 6$$
$$-5A + 5B = -25$$

Dividing the first equation by 2 and the second equation by –5:

$$A + B = 3$$
$$A - B = 5$$

Adding yields $2A = 8$, so $A = 4$ and $B = -1$.

b. Substituting $x = \frac{5}{2}$ into $6x - 25 = A(2x-5) + B(2x+5)$:

$$15 - 25 = A(5-5) + B(5+5)$$
$$-10 = 10B$$
$$-1 = B$$

Substituting $x = -\frac{5}{2}$ into $6x - 25 = A(2x-5) + B(2x+5)$:

$$-15 - 25 = A(-5-5) + B(-5+5)$$
$$-40 = -10A$$
$$4 = A$$

So $A = 4$ and $B = -1$.

5. **a.** Clearing fractions: $1 = A(3x-1)+B(x+1) = (3A+B)x+(-A+B)$

Equating coefficients:

$$3A+B=0$$
$$-A+B=1$$

Multiplying the second equation by –1:

$$3A+B=0$$
$$A-B=-1$$

Adding yields $4A=-1$, so $A=-\frac{1}{4}$. Substituting into $-A+B=1$:

$$\frac{1}{4}+B=1$$
$$B=\frac{3}{4}$$

So $A=-\frac{1}{4}$ and $B=\frac{3}{4}$.

b. Substituting $x=\frac{1}{3}$ into $1=A(3x-1)+B(x+1)$:

$$1=A(1-1)+B\left(\frac{1}{3}+1\right)$$
$$1=\frac{4}{3}B$$
$$\frac{3}{4}=B$$

Substituting $x=-1$ into $1=A(3x-1)+B(x+1)$:

$$1=A(-3-1)+B(-1+1)$$
$$1=-4A$$
$$-\frac{1}{4}=A$$

So $A=-\frac{1}{4}$ and $B=\frac{3}{4}$.

7. Clearing fractions: $8x+3=A(x+3)+B=Ax+(3A+B)$

Equating coefficients:

$$A=8$$
$$3A+B=3$$

Substituting $A=8$ into $3A+B=3$:

$$24+B=3$$
$$B=-21$$

So $A=8$ and $B=-21$.

9. Clearing fractions: $6-x=A(5x+4)+B=5Ax+(4A+B)$

Equating coefficients:

$$5A=-1$$
$$4A+B=6$$

So $A=-\frac{1}{5}$. Substituting $A=-\frac{1}{5}$ into $4A+B=6$:

$$-\frac{4}{5}+B=6$$
$$B=\frac{34}{5}$$

So $A=-\frac{1}{5}$ and $B=\frac{34}{5}$.

11. Clearing fractions:

$$3x^2+7x-2=A\left(x^2+1\right)+(Bx+C)(x-1)=Ax^2+A+Bx^2+Cx-Bx-C=(A+B)x^2+(-B+C)x+(A-C)$$

Equating coefficients:

$$A+B=3$$
$$-B+C=7$$
$$A-C=-2$$

Adding all three equations yields $2A = 8$, so $A = 4$, $B = -1$, and $C = 6$.

13. Clearing fractions:

$$x^2+1=A\left(x^2+4\right)+(Bx+C)(x+1)=Ax^2+4A+Bx^2+Cx+Bx+C=(A+B)x^2+(B+C)x+(4A+C)$$

Equating coefficients:

$$A+B=1$$
$$B+C=0$$
$$4A+C=1$$

Form the augmented matrix: $\begin{pmatrix} 1 & 1 & 0 & 1 \\ 0 & 1 & 1 & 0 \\ 4 & 0 & 1 & 1 \end{pmatrix}$

Adding –4 times row 1 to row 3: $\begin{pmatrix} 1 & 1 & 0 & 1 \\ 0 & 1 & 1 & 0 \\ 0 & -4 & 1 & -3 \end{pmatrix}$

Adding 4 times row 2 to row 3: $\begin{pmatrix} 1 & 1 & 0 & 1 \\ 0 & 1 & 1 & 0 \\ 0 & 0 & 5 & -3 \end{pmatrix}$

Dividing row 3 by 5, and adding –1 times row 3 to row 2: $\begin{pmatrix} 1 & 1 & 0 & 1 \\ 0 & 1 & 0 & \frac{3}{5} \\ 0 & 0 & 1 & -\frac{3}{5} \end{pmatrix}$

Adding –1 times row 2 to row 1: $\begin{pmatrix} 1 & 0 & 0 & \frac{2}{5} \\ 0 & 1 & 0 & \frac{3}{5} \\ 0 & 0 & 1 & -\frac{3}{5} \end{pmatrix}$

So $A=\frac{2}{5}$, $B=\frac{3}{5}$, and $C=-\frac{3}{5}$.

15. Clearing fractions: $1=A\left(x^2-x+1\right)+(Bx+C)x=Ax^2-Ax+A+Bx^2+Cx=(A+B)x^2+(-A+C)x+A$

Equating coefficients:

$$A+B=0$$
$$-A+C=0$$
$$A=1$$

So $A=1$, $B=-1$, and $C=1$.

17. Clearing fractions: $3x^2-2=A(x+1)(x-1)+B(x-2)(x-1)+C(x-2)(x+1)$

Substituting $x=-1$:

$$3-2=A(0)(-2)+B(-3)(-2)+C(-3)(0)$$
$$1=6B$$
$$B=\frac{1}{6}$$

Substituting $x=1$:

$$3-2=A(2)(0)+B(-1)(0)+C(-1)(-2)$$
$$1=-2C$$
$$C=-\frac{1}{2}$$

Substituting $x=2$:

$$12-2=A(3)(1)+B(0)(1)+C(0)(3)$$
$$10=3A$$
$$A=\frac{10}{3}$$

So $A=\frac{10}{3}$, $B=\frac{1}{6}$, and $C=-\frac{1}{2}$.

19. Clearing fractions:

$$\begin{aligned} 4x^2-47x+133 &= A(x-6)^2+B(x-6)+C \\ &= Ax^2-12Ax+36A+Bx-6B+C \\ &= Ax^2+(-12A+B)x+(36A-6B+C) \end{aligned}$$

Equating coefficients:

$$A=4$$
$$-12A+B=-47$$
$$36A-6B+C=133$$

Substituting $A=4$ into $-12A+B=-47$:

$$-48+B=-47$$
$$B=1$$

Substituting into $36A-6B+C=133$:

$$\begin{aligned}144-6+C&=133\\138+C&=133\\C&=-5\end{aligned}$$

So $A=4$, $B=1$, and $C=-5$.

21. Clearing fractions:

$$x^2-2=(Ax+B)\left(x^2+2\right)+(Cx+D)=Ax^3+Bx^2+2Ax+2B+Cx+D=Ax^3+Bx^2+(2A+C)x+(2B+D)$$

Equating coefficients:

$$\begin{aligned}A&=0\\B&=1\\2A+C&=0\\2B+D&=-2\end{aligned}$$

Substituting $A=0$ into $2A+C=0$ results in $C=0$. Substituting $B=1$ into $2B+D=-2$:

$$\begin{aligned}2+D&=-2\\C&=-4\end{aligned}$$

So $A=0$, $B=1$, $C=0$, and $D=-4$.

23. **a.** Note that the two graphs are not the same:

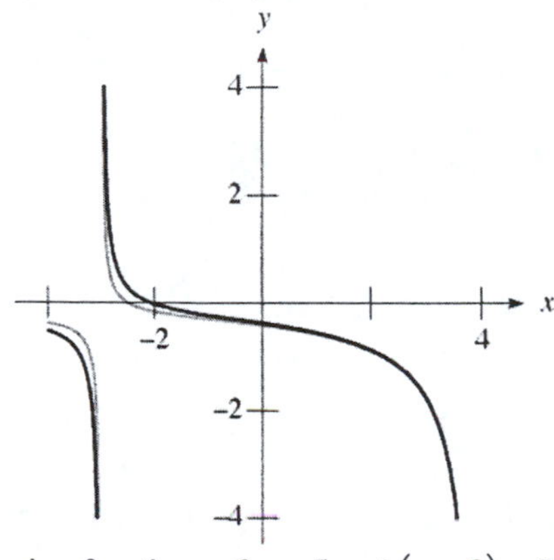

b. Again note the two graphs are not the same:

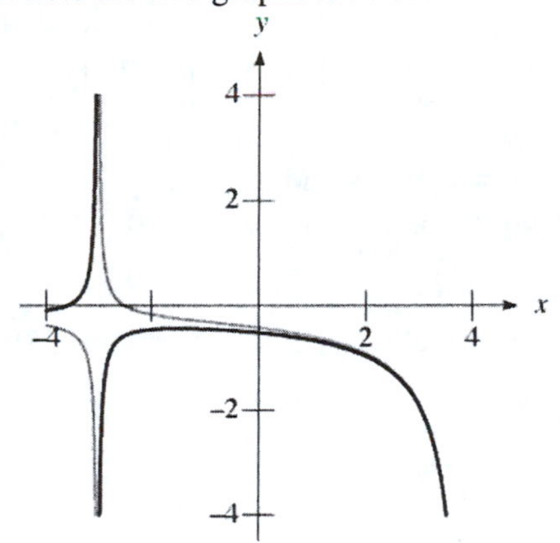

c. Clearing fractions: $2x+5=A(x+3)+B(x-4)=Ax+3A+Bx-4B=(A+B)x+(3A-4B)$

Equating coefficients:

$$\begin{aligned}A+B&=2\\3A-4B&=5\end{aligned}$$

Multiply the first equation by 4:

$$\begin{aligned}4A+4B&=8\\3A-4B&=5\end{aligned}$$

Adding yields $7A=13$, so $A=\frac{13}{7}$. Substituting into the first equation yields $B=\frac{1}{7}$. The partial fraction decomposition is $\dfrac{2x+5}{(x-4)(x+3)}=\dfrac{13/7}{x-4}+\dfrac{1/7}{x+3}$.

25. **a.** Form the augmented matrix: $\begin{pmatrix} 1 & 1 & 0 & 7 \\ 0 & -2 & 1 & -9 \\ 9 & 0 & -2 & 29 \end{pmatrix}$

Adding –9 times row 1 to row 3: $\begin{pmatrix} 1 & 1 & 0 & 7 \\ 0 & -2 & 1 & -9 \\ 0 & -9 & -2 & -34 \end{pmatrix}$

Dividing row 2 by –2, and adding 9 times row 2 to row 3: $\begin{pmatrix} 1 & 1 & 0 & 7 \\ 0 & 1 & -\frac{1}{2} & \frac{9}{2} \\ 0 & 0 & -\frac{13}{2} & \frac{13}{2} \end{pmatrix}$

Dividing row 3 by $-\frac{13}{2}$, and adding $\frac{1}{2}$ times row 3 to row 2: $\begin{pmatrix} 1 & 1 & 0 & 7 \\ 0 & 1 & 0 & 4 \\ 0 & 0 & 1 & -1 \end{pmatrix}$

Adding –1 times row 2 to row 1: $\begin{pmatrix} 1 & 0 & 0 & 3 \\ 0 & 1 & 0 & 4 \\ 0 & 0 & 1 & -1 \end{pmatrix}$

So $A = 3$, $B = 4$, and $C = -1$.

b. Multiplying the second equation by 2:

$$-9B - 2C = -34$$
$$-4B + 2C = -18$$

Adding yields $-13B = -52$, so $B = 4$. Substituting into $-2B + C = -9$:

$$-8 + C = -9$$
$$C = -1$$

So $B = 4$ and $C = -1$.

27. **a.** The possible rational roots are –1 and 1. Since $f(-1) = 5$ and $f(1) = 1$, the equation has no rational roots.

b. Graphing the function:

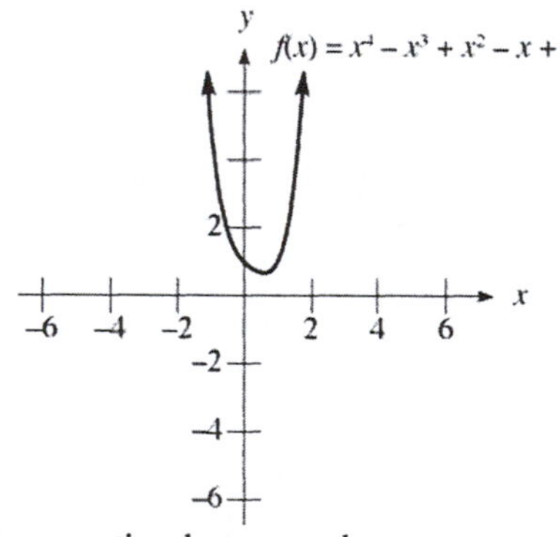

Since there are no x-intercepts, the equation has no real roots.

c. Multiplying out the factors:

$$(x^2 + bx + 1)(x^2 + cx + 1) = x^4 + cx^3 + x^2 + bx^3 + bcx^2 + bx + x^2 + cx + 1$$
$$= x^4 + (b + c)x^3 + (2 + bc)x^2 + (b + c)x + 1$$

Equating coefficients:

$$b + c = -1$$
$$2 + bc = 1$$

Since $b+c=-1$, $c=-1-b$. Substituting:

$$2+b(-1-b)=1$$
$$2-b-b^2=1$$
$$0=b^2+b-1$$

Using the quadratic formula: $b=\dfrac{-1\pm\sqrt{1-4(1)(-1)}}{2(1)}=\dfrac{-1\pm\sqrt{1+4}}{2}=\dfrac{-1\pm\sqrt{5}}{2}$

If $b=\dfrac{-1+\sqrt{5}}{2}$: $c=-1-\dfrac{-1+\sqrt{5}}{2}=\dfrac{-2+1-\sqrt{5}}{2}=\dfrac{-1-\sqrt{5}}{2}$

If $b=\dfrac{-1-\sqrt{5}}{2}$: $c=-1-\dfrac{-1-\sqrt{5}}{2}=\dfrac{-2+1+\sqrt{5}}{2}=\dfrac{-1+\sqrt{5}}{2}$

Thus one factorization is: $\left(x^2+\dfrac{-1+\sqrt{5}}{2}x+1\right)\left(x^2+\dfrac{-1-\sqrt{5}}{2}x+1\right)$

d. For each factor, use the quadratic formula:

$$x=\frac{-\frac{-1+\sqrt{5}}{2}\pm\sqrt{\left(\frac{-1+\sqrt{5}}{2}\right)^2-4}}{2}=\frac{\frac{1-\sqrt{5}}{2}\pm\sqrt{\frac{6-2\sqrt{5}}{4}-4}}{2}=\frac{1-\sqrt{5}\pm\sqrt{6-2\sqrt{5}-16}}{4}=\frac{1-\sqrt{5}\pm\sqrt{-10-2\sqrt{5}}}{4}=\frac{1-\sqrt{5}\pm i\sqrt{10+2\sqrt{5}}}{4}$$

$$x=\frac{-\frac{-1-\sqrt{5}}{2}\pm\sqrt{\left(\frac{-1-\sqrt{5}}{2}\right)^2-4}}{2}=\frac{\frac{1+\sqrt{5}}{2}\pm\sqrt{\frac{6+2\sqrt{5}}{4}-4}}{2}=\frac{1+\sqrt{5}\pm\sqrt{6+2\sqrt{5}-16}}{4}=\frac{1+\sqrt{5}\pm\sqrt{-10+2\sqrt{5}}}{4}=\frac{1+\sqrt{5}\pm i\sqrt{10-2\sqrt{5}}}{4}$$

29. **a.** Graphing the two functions:

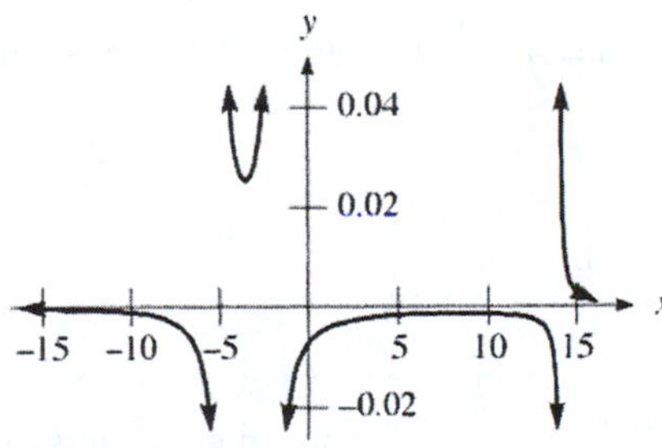

Note that the two graphs appear identical.

b. Using the viewing rectangle [–14.005,–14] x [–0.000331,–0.000330]:

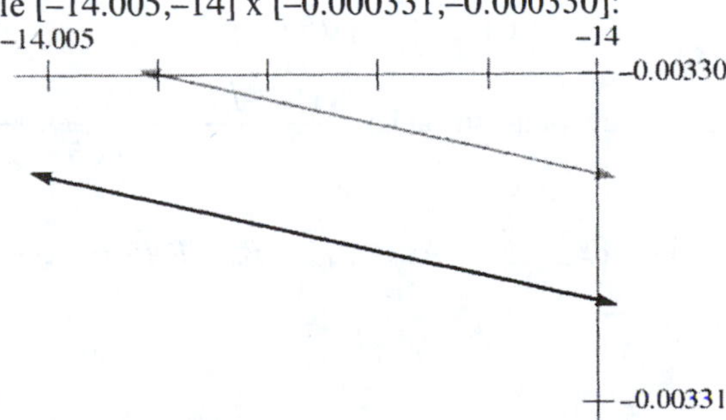

Note that the two graphs do not appear the same.

c. Clearing fractions: $1 = A(x+5)(x-14) + B(x+2)(x-14) + C(x+2)(x+5)$

Substituting $x = -5$:

$$1 = B(-3)(-19)$$
$$B = \tfrac{1}{57}$$

Substituting $x = 14$:

$$1 = C(16)(19)$$
$$C = \tfrac{1}{304}$$

Substituting $x = -2$:

$$1 = A(3)(-16)$$
$$A = -\tfrac{1}{48}$$

The partial fractions decomposition is $\dfrac{1}{(x+2)(x+5)(x-14)} = \dfrac{-1/48}{x+2} + \dfrac{1/57}{x+5} + \dfrac{1/304}{x-14}$.

12.8 More About Partial Fractions

1. a. No, this is reducible, since:

$$x^2 - 16 = 0$$
$$(x+4)(x-4) = 0$$
$$x = -4, 4$$

Thus the polynomial equation $f(x) = 0$ has two real roots.

b. Yes, this is irreducible, since $x^2 + 16 = 0$ has no real roots.

3. a. No, this is reducible, since:

$$x^2 + 3x - 4 = 0$$
$$(x+4)(x-1) = 0$$
$$x = -4, 1$$

Thus the polynomial equation $f(x) = 0$ has two real roots.

b. Yes, this is irreducible, since $x^2 + 3x + 4 = 0$ has no real roots.

5. a. The denominator factors as $x^2 - 100 = (x+10)(x-10)$.

b. The form of the partial fraction decomposition is: $\dfrac{11x+30}{x^2-100} = \dfrac{A}{x+10} + \dfrac{B}{x-10}$

c. Clearing fractions yields: $11x + 30 = A(x-10) + B(x+10) = Ax - 10A + Bx + 10B = (A+B)x + (-10A + 10B)$

Equating coefficients:

$$A + B = 11$$
$$-10A + 10B = 30$$

Dividing the second equation by 10:

$$A + B = 11$$
$$-A + B = 3$$

Adding yields $2B = 14$, so $B = 7$ and $A = 4$. The decomposition is: $\dfrac{11x+30}{x^2-100} = \dfrac{4}{x+10} + \dfrac{7}{x-10}$

7. a. The denominator factors as $x^2 - 5 = \left(x+\sqrt{5}\right)\left(x-\sqrt{5}\right)$.

b. The form of the partial fraction decomposition is: $\dfrac{8x-2\sqrt{5}}{x^2-5} = \dfrac{A}{x+\sqrt{5}} + \dfrac{B}{x-\sqrt{5}}$

c. Clearing fractions yields:

$$8x - 2\sqrt{5} = A\left(x-\sqrt{5}\right) + B\left(x+\sqrt{5}\right) = Ax - A\sqrt{5} + Bx + B\sqrt{5} = (A+B)x + (-A+B)\sqrt{5}$$

Equating coefficients:

$$A + B = 8$$
$$-A + B = -2$$

Adding yields $2B = 6$, so $B = 3$ and $A = 5$. The decomposition is: $\dfrac{8x-2\sqrt{5}}{x^2-5} = \dfrac{5}{x+\sqrt{5}} + \dfrac{3}{x-\sqrt{5}}$

9. **a.** The denominator factors as $x^2 - x - 6 = (x-3)(x+2)$.

b. The form of the partial fraction decomposition is: $\dfrac{7x+39}{x^2-x-6} = \dfrac{A}{x+2} + \dfrac{B}{x-3}$

c. Clearing fractions yields: $7x + 39 = A(x-3) + B(x+2) = Ax - 3A + Bx + 2B = (A+B)x + (-3A+2B)$

Equating coefficients:

$$A + B = 7$$
$$-3A + 2B = 39$$

Multiplying the first equation by 3:

$$3A + 3B = 21$$
$$-3A + 2B = 39$$

Adding yields $5B = 60$, so $B = 12$ and $A = -5$. The decomposition is: $\dfrac{7x+39}{x^2-x-6} = \dfrac{-5}{x+2} + \dfrac{12}{x-3}$

11. **a.** The denominator factors as: $x^3 - 3x^2 - 4x + 12 = x^2(x-3) - 4(x-3) = (x-3)\left(x^2-4\right) = (x-3)(x-2)(x+2)$

b. The form of the partial fraction decomposition is: $\dfrac{3x^2+17x-38}{x^3-3x^2-4x+12} = \dfrac{A}{x-3} + \dfrac{B}{x-2} + \dfrac{C}{x+2}$

c. Clearing fractions yields:

$$\begin{aligned} 3x^2 + 17x - 38 &= A(x-2)(x+2) + B(x-3)(x+2) + C(x-3)(x-2) \\ &= A\left(x^2-4\right) + B\left(x^2-x-6\right) + C\left(x^2-5x+6\right) \\ &= Ax^2 - 4A + Bx^2 - Bx - 6B + Cx^2 - 5Cx + 6C \\ &= (A+B+C)x^2 + (-B-5C)x + (-4A-6B+6C) \end{aligned}$$

Equating coefficients:

$$A + B + C = 3$$
$$-B - 5C = 17$$
$$-4A - 6B + 6C = -38$$

Multiplying the first equation by 4:

$$4A + 4B + 4C = 12$$
$$-4A - 6B + 6C = -38$$

Adding yields $-2B + 10C = -26$, so $B - 5C = 13$. We have:

$$B - 5C = 13$$
$$-B - 5C = 17$$

Adding yields $-10C = 30$, so $C = -3$. Substituting:

$$B - 5(-3) = 13$$
$$B + 15 = 13$$
$$B = -2$$

Substituting:

$$A - 2 - 3 = 3$$
$$A = 8$$

So $A = 8$, $B = -2$, and $C = -3$. The decomposition is: $\dfrac{3x^2+17x-38}{x^3-3x^2-4x+12} = \dfrac{8}{x-3} + \dfrac{-2}{x-2} + \dfrac{-3}{x+2}$

13. **a.** The denominator factors as: $x^3+x^2+x=x\left(x^2+x+1\right)$

b. The form of the partial fraction decomposition is: $\dfrac{5x^2+2x+5}{x^3+x^2+x}=\dfrac{A}{x}+\dfrac{Bx+C}{x^2+x+1}$

c. Clearing fractions yields:

$$5x^2+2x+5=A\left(x^2+x+1\right)+(Bx+C)x=Ax^2+Ax+A+Bx^2+Cx=(A+B)x^2+(A+C)x+A$$

Equating coefficients:

$$A+B=5$$
$$A+C=2$$
$$A=5$$

So $A=5, B=0$, and $C=-3$. The decomposition is: $\dfrac{5x^2+2x+5}{x^3+x^2+x}=\dfrac{5}{x}+\dfrac{-3}{x^2+x+1}$

15. **a.** The denominator factors as $x^4+2x^2+1=\left(x^2+1\right)^2$.

b. The form of the partial fraction decomposition is: $\dfrac{2x^3+5x-4}{x^4+2x^2+1}=\dfrac{Ax+B}{x^2+1}+\dfrac{Cx+D}{\left(x^2+1\right)^2}$

c. Clearing fractions yields:

$$\begin{aligned}2x^3+5x-4&=(Ax+B)\left(x^2+1\right)+Cx+D\\&=Ax^3+Bx^2+Ax+B+Cx+D\\&=Ax^3+Bx^2+(A+C)x+(B+D)\end{aligned}$$

Equating coefficients:

$$A=2$$
$$B=0$$
$$A+C=5$$
$$B+D=-4$$

So $A=2, B=0, C=3$, and $D=-4$. The decomposition is: $\dfrac{2x^3+5x-4}{x^4+2x^2+1}=\dfrac{2x}{x^2+1}+\dfrac{3x-4}{\left(x^2+1\right)^2}$

17. Since $x^3-3x^2-16x-12=(x-6)(x+1)(x+2)$, the form of the decomposition is:

$$\frac{x^2+2}{x^3-3x^2-16x-12}=\frac{A}{x-6}+\frac{B}{x+1}+\frac{C}{x+2}$$

Clearing fractions yields: $x^2+2=A(x+1)(x+2)+B(x-6)(x+2)+C(x-6)(x+1)$

Substituting $x=-1$:

$$1+2=A(0)+B(-7)(1)+C(0)$$
$$3=-7B$$
$$B=-\tfrac{3}{7}$$

Substituting $x=-2$:

$$4+2=A(0)+B(0)+C(-8)(-1)$$
$$6=8C$$
$$C=\tfrac{3}{4}$$

Substituting $x=6$:

$$36+2=A(7)(8)+B(0)+C(0)$$
$$38=56A$$
$$A=\tfrac{19}{28}$$

The partial fraction decomposition is: $\dfrac{x^2+2}{x^3-3x^2-16x-12}=\dfrac{\frac{19}{28}}{x-6}+\dfrac{-\frac{3}{7}}{x+1}+\dfrac{\frac{3}{4}}{x+2}$

19. Since $6x^2-19x+15=(3x-5)(2x-3)$, the form of the decomposition is: $\dfrac{5-x}{6x^2-19x+15}=\dfrac{A}{3x-5}+\dfrac{B}{2x-3}$

Clearing fractions yields: $5-x=A(2x-3)+B(3x-5)=2Ax-3A+3Bx-5B=(2A+3B)x+(-3A-5B)$

Equating coefficients:

$$2A+3B=-1$$
$$-3A-5B=5$$

Multiplying the first equation by 3 and the second equation by 2:

$$6A+9B=-3$$
$$-6A-10B=10$$

Adding yields $-B=7$, so $B=-7$ and $A=10$. The partial fraction decomposition is:

$$\frac{5-x}{6x^2-19x+15}=\frac{10}{3x-5}+\frac{-7}{2x-3}$$

21. Since $x^3-5x=x\left(x^2-5\right)=x\left(x+\sqrt{5}\right)\left(x-\sqrt{5}\right)$, the form of the decomposition is: $\dfrac{2x+1}{x^3-5x}=\dfrac{A}{x}+\dfrac{B}{x+\sqrt{5}}+\dfrac{C}{x-\sqrt{5}}$

Clearing fractions yields:

$$\begin{aligned}2x+1&=A\left(x+\sqrt{5}\right)\left(x-\sqrt{5}\right)+Bx\left(x-\sqrt{5}\right)+Cx\left(x+\sqrt{5}\right)\\&=Ax^2-5A+Bx^2-B\sqrt{5}x+Cx^2+C\sqrt{5}x\\&=(A+B+C)x^2+\left(-\sqrt{5}B+\sqrt{5}C\right)x-5A\end{aligned}$$

Equating coefficients:

$$A+B+C=0$$
$$-\sqrt{5}B+\sqrt{5}C=2$$
$$-5A=1$$

Solving the third equation yields $A=-\frac{1}{5}$. Substituting into the first equation yields:

$$-\tfrac{1}{5}+B+C=0$$
$$-\sqrt{5}B+\sqrt{5}C=2$$

Multiplying the first equation by 5 and the second equation by $\sqrt{5}$:

$$5B+5C=1$$
$$-5B+5C=2\sqrt{5}$$

Adding yields $10C=1+2\sqrt{5}$, so $C=\dfrac{1+2\sqrt{5}}{10}$. Subtracting yields $10B=1-2\sqrt{5}$, so $B=\dfrac{1-2\sqrt{5}}{10}$. The partial fraction decomposition is: $\dfrac{2x+1}{x^3-5x}=\dfrac{-1/5}{x}+\dfrac{\left(1-2\sqrt{5}\right)/10}{x+\sqrt{5}}+\dfrac{\left(1+2\sqrt{5}\right)/10}{x-\sqrt{5}}$

23. Since $x^4+8x^2+16=\left(x^2+4\right)^2$, the form of the decomposition is: $\dfrac{x^3+2}{x^4+8x^2+16}=\dfrac{Ax+B}{x^2+4}+\dfrac{Cx+D}{\left(x^2+4\right)^2}$

Clearing fractions yields:

$$x^3+2=(Ax+B)\left(x^2+4\right)+(Cx+D)=Ax^3+Bx^2+4Ax+4B+Cx+D=Ax^3+Bx^2+(4A+C)x+(4B+D)$$

Equating coefficients:

$$A=1$$
$$B=0$$
$$4A+C=0$$
$$4B+D=2$$

Since $4(1)+C=0$, $C=-4$, and since $4(0)+D=2$, $D=2$. The partial fraction decomposition is:

$$\frac{x^3+2}{x^4+8x^2+16}=\frac{x}{x^2+4}+\frac{-4x+2}{\left(x^2+4\right)^2}$$

25. Factoring the denominator: $x^4 - 15x^3 + 75x^2 - 125x = x\left(x^3 - 15x^2 + 75x - 125\right) = x(x-5)^3$

The form of the partial fraction decomposition is: $\dfrac{x^3 + x - 3}{x(x-5)^3} = \dfrac{A}{x} + \dfrac{B}{x-5} + \dfrac{C}{(x-5)^2} + \dfrac{D}{(x-5)^3}$

Clearing fractions yields: $x^3 + x - 3 = A(x-5)^3 + Bx(x-5)^2 + Cx(x-5) + Dx$

Substituting $x = 0$:

$$-3 = A(-5)^3$$
$$-3 = -125A$$
$$A = \tfrac{3}{125}$$

Substituting $x = 5$:

$$127 = 5D$$
$$D = \tfrac{127}{5}$$

Substituting $x = 4$:

$$64 + 4 - 3 = -A + 4B - 4C + 4D$$
$$65 = -\tfrac{3}{125} + 4B - 4C + \tfrac{508}{5}$$
$$65 = \tfrac{12697}{125} + 4B - 4C$$
$$-\tfrac{4572}{125} = 4B - 4C$$
$$-\tfrac{1143}{125} = B - C$$

Substituting $x = 6$:

$$216 + 6 - 3 = A + 6B + 6C + 6D$$
$$219 = \tfrac{3}{125} + 6B + 6C + \tfrac{762}{5}$$
$$219 = \tfrac{19053}{125} + 6B + 6C$$
$$\tfrac{8322}{125} = 6B + 6C$$
$$\tfrac{1387}{125} = B + C$$

So we have the system of equations:

$$B + C = \tfrac{1387}{125}$$
$$B - C = -\tfrac{1143}{125}$$

Adding yields $2B = \frac{244}{125}$, so $B = \frac{122}{125}$. Subtracting yields $2C = \frac{2530}{125}$, so $C = \frac{1265}{125} = \frac{253}{25}$. The partial fraction decomposition is: $\dfrac{x^3 + x - 3}{x^4 - 15x^3 + 75x^2 - 125x} = \dfrac{\frac{3}{125}}{x} + \dfrac{\frac{122}{125}}{x-5} + \dfrac{\frac{253}{25}}{(x-5)^2} + \dfrac{\frac{127}{5}}{(x-5)^3}$

27. Since $x^3 - 1 = (x-1)\left(x^2 + x + 1\right)$, the form of the decomposition is: $\dfrac{1}{x^3 - 1} = \dfrac{A}{x-1} + \dfrac{Bx + C}{x^2 + x + 1}$

Clearing fractions yields:

$$1 = A\left(x^2 + x + 1\right) + (Bx + C)(x-1) = Ax^2 + Ax + A + Bx^2 + Cx - Bx - C = (A+B)x^2 + (A - B + C)x + (A - C)$$

Equating coefficients:

$$A + B = 0$$
$$A - B + C = 0$$
$$A - C = 1$$

Adding the first two equations results in the system:

$$2A + C = 0$$
$$A - C = 1$$

Adding yields $3A = 1$, so $A = \frac{1}{3}$. Thus $B = -\frac{1}{3}$ and $C = -\frac{2}{3}$. The partial fraction decomposition is:

$$\frac{1}{x^3 - 1} = \frac{\frac{1}{3}}{x-1} + \frac{-\frac{1}{3}x - \frac{2}{3}}{x^2 + x + 1}$$

29. Factoring the denominator: $x^4 + 2x^3 + x^2 = x^2\left(x^2 + 2x + 1\right) = x^2(x+1)^2$

The form of the partial fraction decomposition is: $\dfrac{7x^3 + 11x^2 - x - 2}{x^2(x+1)^2} = \dfrac{A}{x} + \dfrac{B}{x^2} + \dfrac{C}{x+1} + \dfrac{D}{(x+1)^2}$

Clearing fractions yields: $7x^3 + 11x^2 - x - 2 = Ax(x+1)^2 + B(x+1)^2 + Cx^2(x+1) + Dx^2$

Substituting $x = 0$:

$$-2 = B$$

Substituting $x = -1$:

$$-7 + 11 + 1 - 2 = D$$
$$3 = D$$

Substituting $x=1$:

$$\begin{aligned} 7+11-1-2 &= 4A+4B+2C+D \\ 15 &= 4A-8+2C+3 \\ 20 &= 4A+2C \\ 10 &= 2A+C \end{aligned}$$

Substituting $x=-2$:

$$\begin{aligned} -56+44+2-2 &= -2A+B-4C+4D \\ -12 &= -2A-2-4C+12 \\ -22 &= -2A-4C \\ 11 &= A+2C \end{aligned}$$

So we have the system of equations:

$$\begin{aligned} 2A+C &= 10 \\ A+2C &= 11 \end{aligned}$$

Multiplying the first equation by –2:

$$\begin{aligned} -4A-2C &= -20 \\ A+2C &= 11 \end{aligned}$$

Adding yields $-3A=-9$, so $A=3$. Substituting:

$$\begin{aligned} 3+2C &= 11 \\ 2C &= 8 \\ C &= 4 \end{aligned}$$

The partial fraction decomposition is: $\dfrac{7x^3+11x^2-x-2}{x^4+2x^3+x^2}=\dfrac{3}{x}+\dfrac{-2}{x^2}+\dfrac{4}{x+1}+\dfrac{3}{(x+1)^2}$

31. Factoring the denominator: $x^4-81=\left(x^2+9\right)\left(x^2-9\right)=\left(x^2+9\right)(x+3)(x-3)$

The form of the partial fraction decomposition is: $\dfrac{x^3-5}{x^4-81}=\dfrac{A}{x+3}+\dfrac{B}{x-3}+\dfrac{Cx+D}{x^2+9}$

Clearing fractions yields: $x^3-5=A(x-3)\left(x^2+9\right)+B(x+3)\left(x^2+9\right)+(Cx+D)(x+3)(x-3)$

Substituting $x=3$:

$$\begin{aligned} 27-5 &= 6(9+9)B \\ 22 &= 108B \\ B &= \tfrac{11}{54} \end{aligned}$$

Substituting $x=-3$:

$$\begin{aligned} -27-5 &= (-6)(9+9)A \\ -32 &= -108A \\ A &= \tfrac{8}{27} \end{aligned}$$

Substituting $x=0$:

$$\begin{aligned} -5 &= -27A+27B-9D \\ -5 &= -8+\tfrac{11}{2}-9D \\ -\tfrac{5}{2} &= -9D \\ D &= \tfrac{5}{18} \end{aligned}$$

Substituting $x=1$:

$$\begin{aligned} 1-5 &= -20A+40B-8C-8D \\ 1 &= 5A-10B+2C+2D \\ 1 &= \tfrac{40}{27}-\tfrac{55}{27}+2C+\tfrac{5}{9} \\ 1 &= 2C \\ C &= \tfrac{1}{2} \end{aligned}$$

The partial fraction decomposition is: $\dfrac{x^3-5}{x^4-81}=\dfrac{\frac{8}{27}}{x+3}+\dfrac{\frac{11}{54}}{x-3}+\dfrac{\frac{1}{2}x+\frac{5}{18}}{x^2+9}$

33. Following the hint, factor the denominator:

$$x^4+x^3+2x^2+x+1=\left(x^4+x^3+x^2\right)+\left(x^2+x+1\right)=x^2\left(x^2+x+1\right)+\left(x^2+x+1\right)=\left(x^2+x+1\right)\left(x^2+1\right)$$

The form of the partial fraction decomposition is: $\dfrac{1}{\left(x^2+x+1\right)\left(x^2+1\right)}=\dfrac{Ax+B}{x^2+1}+\dfrac{Cx+D}{x^2+x+1}$

Clearing fractions yields:

$$\begin{aligned} 1 &= (Ax+B)\left(x^2+x+1\right)+(Cx+D)\left(x^2+1\right) \\ &= Ax^3+Ax^2+Ax+Bx^2+Bx+B+Cx^3+Dx^2+Cx+D \\ &= (A+C)x^3+(A+B+D)x^2+(A+B+C)x+(B+D) \end{aligned}$$

Equating coefficients:

$$A+C=0$$
$$A+B+D=0$$
$$A+B+C=0$$
$$B+D=1$$

Subtracting the first equation from the third equation yields $B=0$, so $D=1$, $A=-1$, and $C=1$. The partial fraction decomposition is: $\frac{1}{x^4+x^3+2x^2+x+1}=\frac{-x}{x^2+1}+\frac{x+1}{x^2+x+1}$

35. Using synthetic division with $x=2$:

2) 1	2	–5	–6
	2	8	6
1	4	3	0

Thus the equation factors as:

$$x^3+2x^2-5x-6=0$$
$$(x-2)\left(x^2+4x+3\right)=0$$
$$(x-2)(x+1)(x+3)=0$$
$$x=-3,-1,2$$

The roots of the equation are $x=-3,-1,2$.

37. **a.** Simplifying identity (6):

$$\begin{aligned}3x^3-x^2+7x-3&=(Ax+B)\left(x^2+3\right)+(Cx+D)\\&=Ax^3+Bx^2+3Ax+3B+Cx+D\\&=Ax^3+Bx^2+(3A+C)x+(3B+D)\end{aligned}$$

Equating coefficients:

$$A=3$$
$$B=-1$$
$$3A+C=7$$
$$3B+D=-3$$

b. Substituting $A=3$:

$$3(3)+C=7$$
$$9+C=7$$
$$C=-2$$

Substituting $B=-1$:

$$3(-1)+D=-3$$
$$-3+D=-3$$
$$D=0$$

These agree with the values given in the text.

39. Using long division:

$$\begin{array}{r} 6x+8 \\ x^2-4x+3\overline{)6x^3-16x^2-13x+25} \\ \underline{6x^3-24x^2+18x} \\ 8x^2-31x+25 \\ \underline{8x^2-32x+24} \\ x+1 \end{array}$$

The expression can be written as $6x+8+\frac{x+1}{x^2-4x+3}$.

Since $x^2-4x+3=(x-3)(x-1)$, the form of the decomposition is: $\frac{x+1}{x^2-4x+3}=\frac{A}{x-3}+\frac{B}{x-1}$

Clearing fractions yields: $x+1=A(x-1)+B(x-3)=Ax-A+Bx-3B=(A+B)x+(-A-3B)$

Equating coefficients:

$$A+B=1$$
$$-A-3B=1$$

Adding yields $-2B=2$, so $B=-1$ and $A=2$. Thus we can write: $\dfrac{6x^3-16x^2-13x+25}{x^2-4x+3}=6x+8+\dfrac{2}{x-3}+\dfrac{-1}{x-1}$

41. Using long division:

$$\begin{array}{r} x-4 \\ x^4-6x^3+12x^2-8x\,\big)\overline{x^5-10x^4+36x^3-55x^2+32x+1} \\ \underline{x^5-6x^4+12x^3-8x^2} \\ -4x^4+24x^3-47x^2+32x \\ \underline{-4x^4+24x^3-48x^2+32x} \\ x^2+1 \end{array}$$

The expression can be written as $x-4+\dfrac{x^2+1}{x^4-6x^3+12x^2-8x}$. Now factor the denominator:

$$x^4-6x^3+12x^2-8x=x\left(x^3-6x^2+12x-8\right)=x(x-2)^3$$

Thus the form of the decomposition is: $\dfrac{x^2+1}{x(x-2)^3}=\dfrac{A}{x}+\dfrac{B}{x-2}+\dfrac{C}{(x-2)^2}+\dfrac{D}{(x-2)^3}$

Clearing fractions yields: $x^2+1=A(x-2)^3+Bx(x-2)^2+Cx(x-2)+Dx$

Substituting $x=0$:

$$1=-8A$$
$$A=-\tfrac{1}{8}$$

Substituting $x=2$:

$$5=2D$$
$$D=\tfrac{5}{2}$$

Substituting $x=3$:

$$9+1=A+3B+3C+3D$$
$$10=-\tfrac{1}{8}+3B+3C+\tfrac{15}{2}$$
$$10=\tfrac{59}{8}+3B+3C$$
$$\tfrac{21}{8}=3B+3C$$
$$\tfrac{7}{8}=B+C$$

Substituting $x=1$:

$$1+1=-A+B-C+D$$
$$2=\tfrac{1}{8}+B-C+\tfrac{5}{2}$$
$$2=\tfrac{21}{8}+B-C$$
$$-\tfrac{5}{8}=B-C$$

Thus the system of equations is:

$$B+C=\tfrac{7}{8}$$
$$B-C=-\tfrac{5}{8}$$

Adding yields $2B=\tfrac{1}{4}$, so $B=\tfrac{1}{8}$ and $C=\tfrac{3}{4}$. Thus we can write:

$$\frac{x^5-10x^4+36x^3-55x^2+32x+1}{x^4-6x^3+12x^2-8x}=x-4+\frac{-\frac{1}{8}}{x}+\frac{\frac{1}{8}}{x-2}+\frac{\frac{3}{4}}{(x-2)^2}+\frac{\frac{5}{2}}{(x-2)^3}$$

43. Using long division:

$$\begin{array}{r|l} & x^2+2x+5 \\ \hline x^4-1 & x^6+2x^5+5x^4-x^2-2x-4 \\ & x^6 \qquad\quad -x^2 \\ \hline & 2x^5+5x^4 \qquad -2x-4 \\ & 2x^5 \qquad\qquad -2x \\ \hline & 5x^4 \qquad\qquad -4 \\ & 5x^4 \qquad\qquad -5 \\ \hline & 1 \end{array}$$

The expression can be written as $x^2+2x+5+\frac{1}{x^4-1}$. Now factor the denominator:

$$x^4-1=\left(x^2+1\right)\left(x^2-1\right)=\left(x^2+1\right)(x+1)(x-1)$$

Thus the form of the decomposition is: $\frac{1}{x^4-1}=\frac{A}{x+1}+\frac{B}{x-1}+\frac{Cx+D}{x^2+1}$

Clearing fractions yields: $1=A(x-1)\left(x^2+1\right)+B(x+1)\left(x^2+1\right)+(Cx+D)(x+1)(x-1)$

Substituting $x=1$:

$$1=4B$$
$$B=\tfrac{1}{4}$$

Substituting $x=-1$:

$$1=-4A$$
$$A=-\tfrac{1}{4}$$

Substituting $x=0$:

$$1=-A+B-D$$
$$1=\tfrac{1}{4}+\tfrac{1}{4}-D$$
$$\tfrac{1}{2}=-D$$
$$-\tfrac{1}{2}=D$$

Substituting $x=2$:

$$1=5A+15B+6C+3D$$
$$1=-\tfrac{5}{4}+\tfrac{15}{4}+6C-\tfrac{3}{2}$$
$$0=6C$$
$$0=C$$

Thus we can write: $\frac{x^6+2x^5+5x^4-x^2-2x-4}{x^4-1}=x^2+2x+5+\frac{-\frac{1}{4}}{x+1}+\frac{\frac{1}{4}}{x-1}+\frac{-\frac{1}{2}}{x^2+1}$

45. Rewriting the quadratic polynomial: $x^2-2ax+a^2+b^2=(x-a)^2+b^2$

Since this is a sum of two squares, it is irreducible.

47. The form of the partial fraction decomposition is: $\frac{px+q}{(x-a)(x-b)}=\frac{A}{x-a}+\frac{B}{x-b}$

Clearing fractions yields: $px+q=A(x-b)+B(x-a)=Ax-Ab+Bx-Ba=(A+B)x+(-Ab-Ba)$

Equating coefficients:

$$A+B=p$$
$$-Ab-Ba=q$$

Multiplying the first equation by a:

$$Aa+Ba=pa$$
$$-Ab-Ba=q$$

Adding yields $A(a-b)=pa+q$, so $A=\frac{pa+q}{a-b}$. Multiplying the first equation by b:

$$Ab+Bb=pb$$
$$-Ab-Ba=q$$

Adding yields $B(b-a)=pb+q$, so $B=\frac{pb+q}{b-a}$. The partial fraction decomposition is:

$$\frac{px+q}{(x-a)(x-b)}=\frac{(pa+q)/(a-b)}{x-a}+\frac{(pb+q)/(b-a)}{x-b}$$

49. The form of the partial fraction decomposition is: $\frac{px+q}{(x-a)(x+a)} = \frac{A}{x-a} + \frac{B}{x+a}$

Clearing fractions yields: $px+q = A(x+a)+B(x-a) = Ax+Aa+Bx-Ba = (A+B)x+(Aa-Ba)$

Equating coefficients:

$$\begin{aligned} A+B &= p \\ Aa-Ba &= q \end{aligned}$$

Multiplying the first equation by a:

$$\begin{aligned} Aa+Ba &= pa \\ Aa-Ba &= q \end{aligned}$$

Adding yields $2Aa = pa+q$, so $A = \frac{pa+q}{2a}$. Multiplying the first equation by $-a$:

$$\begin{aligned} -Aa-Ba &= -pa \\ Aa-Ba &= q \end{aligned}$$

Adding yields $-2Ba = -pa+q$, so $B = \frac{pa-q}{2a}$. The partial fraction decomposition is:

$$\frac{px+q}{(x-a)(x+a)} = \frac{(pa+q)/2a}{x-a} + \frac{(pa-q)/2a}{x+a}$$

51. The form of the partial fraction decomposition is: $\frac{1}{(1-ax)(1-bx)(1-cx)} = \frac{A}{1-ax} + \frac{B}{1-bx} + \frac{C}{1-cx}$

Clearing fractions yields: $1 = A(1-bx)(1-cx)+B(1-ax)(1-cx)+C(1-ax)(1-bx)$

Substituting $x = \frac{1}{a}$:

$$\begin{aligned} 1 &= A\left(1-\frac{b}{a}\right)\left(1-\frac{c}{a}\right) \\ 1 &= A \bullet \frac{(a-b)(a-c)}{a^2} \\ A &= \frac{a^2}{(a-b)(a-c)} \end{aligned}$$

Substituting $x = \frac{1}{b}$:

$$\begin{aligned} 1 &= B\left(1-\frac{a}{b}\right)\left(1-\frac{c}{b}\right) \\ 1 &= B \bullet \frac{(b-a)(b-c)}{b^2} \\ B &= \frac{b^2}{(b-a)(b-c)} \end{aligned}$$

Substituting $x = \frac{1}{c}$:

$$\begin{aligned} 1 &= C\left(1-\frac{a}{c}\right)\left(1-\frac{b}{c}\right) \\ 1 &= C \bullet \frac{(c-a)(c-b)}{c^2} \\ C &= \frac{c^2}{(c-a)(c-b)} \end{aligned}$$

The partial fraction decomposition is: $\frac{1}{(1-ax)(1-bx)(1-cx)} = \frac{\frac{a^2}{(a-b)(a-c)}}{1-ax} + \frac{\frac{b^2}{(b-a)(b-c)}}{1-bx} + \frac{\frac{c^2}{(c-a)(c-b)}}{1-cx}$

53. Factor the denominator by adding and subtracting $2x^2$:

$$x^4+1=x^4+2x^2+1-2x^2=\left(x^2+1\right)^2-2x^2=\left(x^2+\sqrt{2}x+1\right)\left(x^2-\sqrt{2}x+1\right)$$

The form of the partial fraction decomposition is: $\dfrac{1}{x^4+1}=\dfrac{Ax+B}{x^2+\sqrt{2}x+1}+\dfrac{Cx+D}{x^2-\sqrt{2}x+1}$

Clearing fractions yields:

$$\begin{aligned}1&=(Ax+B)\left(x^2-\sqrt{2}x+1\right)+(Cx+D)\left(x^2+\sqrt{2}x+1\right)\\&=Ax^3-\sqrt{2}Ax^2+Ax+Bx^2-\sqrt{2}Bx+B+Cx^3+\sqrt{2}Cx^2+Cx+Dx^2+\sqrt{2}Dx+D\\&=(A+C)x^3+\left(-\sqrt{2}A+B+\sqrt{2}C+D\right)x^2+\left(A-\sqrt{2}B+C+\sqrt{2}D\right)x+(B+D)\end{aligned}$$

Equating coefficients:

$$\begin{aligned}A+C&=0\\-\sqrt{2}A+B+\sqrt{2}C+D&=0\\A-\sqrt{2}B+C+\sqrt{2}D&=0\\B+D&=1\end{aligned}$$

Substitute $C=-A$ and $D=1-B$ into the second equation:

$$\begin{aligned}-\sqrt{2}A+B+\sqrt{2}(-A)+1-B&=0\\-2\sqrt{2}A+1&=0\\1&=2\sqrt{2}A\\A&=\frac{1}{2\sqrt{2}}=\frac{\sqrt{2}}{4}\\C&=-\frac{\sqrt{2}}{4}\end{aligned}$$

Substituting into the third equation:

$$\begin{aligned}\frac{\sqrt{2}}{4}-\sqrt{2}B-\frac{\sqrt{2}}{4}+\sqrt{2}(1-B)&=0\\-\sqrt{2}B+\sqrt{2}-\sqrt{2}B&=0\\-2\sqrt{2}B&=-\sqrt{2}\\B&=\tfrac{1}{2}\\D&=1-\tfrac{1}{2}=\tfrac{1}{2}\end{aligned}$$

The partial fraction decomposition is: $\dfrac{1}{x^4+1}=\dfrac{\frac{\sqrt{2}}{4}x+\frac{1}{2}}{x^2+\sqrt{2}x+1}+\dfrac{-\frac{\sqrt{2}}{4}x+\frac{1}{2}}{x^2-\sqrt{2}x+1}$

Chapter 12 Review Exercises

1. Using synthetic division:

-2	1	3	-1	-5	1
		-2	-2	6	-2
	1	1	-3	1	-1

So $q(x)=x^3+x^2-3x+1$ and $R(x)=-1$.

3. Using synthetic division:

3	1	0	-2	0	8
		3	9	21	63
	1	3	7	21	71

So the quotient is $x^3+3x^2+7x+21$ and the remainder is 71.

5. Using synthetic division:

-4	2	-5	-6	-3
		-8	52	-184
	2	-13	46	-187

So the quotient is $2x^2 - 13x + 46$ and the remainder is -187.

7. Using synthetic division:

-0.2	5	-19	-4
		-1	4
	5	-20	0

So the quotient is $5x - 20$ and the remainder is 0.

9. Using synthetic division:

10	1	0	0	0	-10	4
		10	100	1000	10000	99900
	1	10	100	1000	9990	99904

So $f(10) = 99{,}904$.

11. Using synthetic division:

$1/10$	1	-10	1	-1
		$1/10$	$-99/100$	$1/1000$
	1	$-99/10$	$1/100$	$-999/1000$

So $f\left(\frac{1}{10}\right) = -\frac{999}{1000}$.

13. Using synthetic division:

$a-1$	1	3	3	1
		$a-1$	a^2+a-2	a^3-1
	1	$a+2$	a^2+a+1	a^3

So $f(a-1) = a^3$.

15. **a.** Using synthetic division:

-0.3	1	4	-6	-8	-2
		-0.3	-1.11	2.133	1.7601
	1	3.7	-7.11	-5.867	-0.24

So $f(-0.3) \approx -0.24$.

b. Using synthetic division:

-0.39	1	4	-6	-8	-2
		-0.39	-1.4079	2.8891	1.9933
	1	3.61	-7.4079	-5.1109	-0.007

So $f(-0.39) \approx -0.007$.

c. Using synthetic division:

-0.394	1	4	-6	-8	-2
		-0.394	-1.420764	2.92378	2.00003
	1	3.606	-7.420764	-5.07622	0.00003

So $f(-0.394) \approx 0.00003$

17. Using synthetic division:

3	1	-4	$-a$	-6
		3	-3	$-3a-9$
	1	-1	$-a-3$	$-3a-15$

If 3 is a root, then:

$$-3a - 15 = 0$$
$$-3a = 15$$
$$a = -5$$

19. Using synthetic division:

1	a^2	$3a$	0	2
		a^2	a^2+3a	a^2+3a
	a^2	a^2+3a	a^2+3a	a^2+3a+2

If $x - 1$ is a factor, then:

$$a^2 + 3a + 2 = 0$$
$$(a+1)(a+2) = 0$$
$$a = -1, -2$$

So $a = -1$ or $a = -2$.

21. a. Let $f(x+h)=0$, and let $x=r-h$. Then $f(r-h+h)=f(r)=0$, since r is a root of $f(x)=0$.

b. Let $f(-x)=0$, and let $x=-r$. Then $f(-(-r))=f(r)=0$, since r is a root of $f(x)=0$.

c. Let $f\left(\frac{x}{k}\right)=0$, and let $x=kr$. Then $f\left(\frac{kr}{k}\right)=f(r)=0$, since r is a root of $f(x)=0$.

23. p-factors: $\pm1, \pm2, \pm3, \pm6, \pm9, \pm18$ q-factors: ±1

possible rational roots: $\pm1, \pm2, \pm3, \pm6, \pm9, \pm18$

25. p-factors: $\pm1, \pm2, \pm4, \pm8$ q-factors: $\pm1, \pm2$

possible rational roots: $\pm1, \pm\frac{1}{2}, \pm2, \pm4, \pm8$

27. p-factors: $\pm p, \pm1$ q-factors: ±1

possible rational roots: $\pm p, \pm1$

29. The possible rational roots are $\pm1, \pm\frac{1}{2}, \pm2, \pm3, \pm\frac{3}{2}, \pm6$. Using synthetic division:

$$\begin{array}{r|rrrr} 2 & 2 & 1 & -7 & -6 \\ & & 4 & 10 & 6 \\ \hline & 2 & 5 & 3 & 0 \end{array}$$

So $2x^3+x^2-7x-6=(x-2)(2x^2+5x+3)=(x-2)(2x+3)(x+1)$. So the roots are $2, -\frac{3}{2}$ and -1.

31. The possible rational roots are $\pm1, \pm\frac{1}{2}, \pm2, \pm5, \pm\frac{5}{2}, \pm10$. Using synthetic division:

$$\begin{array}{r|rrrr} 5/2 & 2 & -1 & -14 & 10 \\ & & 5 & 10 & -10 \\ \hline & 2 & 4 & -4 & 0 \end{array}$$

So $2x^3-x^2-14x+10=\left(x-\frac{5}{2}\right)(2x^2+4x-4)=2\left(x-\frac{5}{2}\right)(x^2+2x-2)$.

Using the quadratic formula: $x=\dfrac{-2\pm\sqrt{4+8}}{2}=\dfrac{-2\pm2\sqrt{3}}{2}=-1\pm\sqrt{3}$. So the roots are $\frac{5}{2}, -1+\sqrt{3}$ and $-1-\sqrt{3}$.

33. First multiply by 2 to obtain $3x^3+x^2+x-2=0$. The possible rational roots are $\pm1, \pm\frac{1}{3}, \pm2, \pm\frac{2}{3}$. Using synthetic division:

$$\begin{array}{r|rrrr} 2/3 & 3 & 1 & 1 & -2 \\ & & 2 & 2 & 2 \\ \hline & 3 & 3 & 3 & 0 \end{array}$$

So $3x^3+x^2+x-2=\left(x-\frac{2}{3}\right)(3x^2+3x+3)=3\left(x-\frac{2}{3}\right)(x^2+x+1)$.

Using the quadratic formula: $x=\dfrac{-1\pm\sqrt{1-4}}{2}=\dfrac{-1\pm i\sqrt{3}}{2}$

So the roots are $\frac{2}{3}, \frac{-1+i\sqrt{3}}{2}$, and $\frac{-1-i\sqrt{3}}{2}$.

35. The possible rational roots are $\pm1, \pm7, \pm49$. Using synthetic division:

$$\begin{array}{r|rrrrrr} -1 & 1 & 1 & -14 & -14 & 49 & 49 \\ & & -1 & 0 & 14 & 0 & -49 \\ \hline & 1 & 0 & -14 & 0 & 49 & 0 \end{array}$$

Therefore:

$$x^5+x^4-14x^3-14x^2+49x+49=(x+1)(x^4-14x^2+49)=(x+1)(x^2-7)^2=(x+1)(x+\sqrt{7})^2(x-\sqrt{7})^2$$

So the roots are -1, $-\sqrt{7}$ (multiplicity 2), and $\sqrt{7}$ (multiplicity 2).

37. Since $x^3 - 9x^2 + 24x - 20 = 0$ has a root r with multiplicity 2, we can use synthetic division by r twice, each time the remainder must be 0:

$$\begin{array}{r|llll} r & 1 & -9 & 24 & -20 \\ & & r & r^2 - 9r & r^3 - 9r^2 + 24r \\ \hline & 1 & r - 9 & r^2 - 9r + 24 & r^3 - 9r^2 + 24r - 20 \end{array}$$

Continue the division:

$$\begin{array}{r|lll} r & 1 & r - 9 & r^2 - 9r + 24 \\ & & r & 2r^2 - 9r \\ \hline & 1 & 2r - 9 & 3r^2 - 18r + 24 \end{array}$$

So $3r^2 - 18r + 24 = 3(r^2 - 6r + 8) = 3(r - 4)(r - 2) = 0$, thus $r = 4$ or $r = 2$. We check these in $r^3 - 9r^2 + 24r - 20 = 0$:

$$r = 4:\ (4)^3 - 9(4)^2 + 24(4) - 20 = 64 - 144 + 96 - 20 = -4 \neq 0$$

So $r = 4$ cannot be a root.

$$r = 2:\ (2)^3 - 9(2)^2 + 24(2) - 20 = 8 - 36 + 48 - 20 = 0$$

So $r = 2$ is the root with multiplicity 2. Since the synthetic division resulted in $x + (2r - 9) = 0$, $x + (-5) = 0$ and thus $x = 5$. So the roots are 2 (multiplicity 2) and 5.

Note: Actually, an easier (and more direct) approach is to find the roots directly. The possible rational roots are $\pm 1, \pm 2, \pm 4, \pm 5, \pm 10, \pm 20$. Note that there are 3 sign changes, so there can be 1 or 3 positive real roots. Since $f(-x)$ has no sign changes, all 3 roots must be positive real numbers for one of them to have multiplicity of 2 (note that it cannot have a radical or be complex–why?). We use synthetic division:

$$\begin{array}{r|llll} 2 & 1 & -9 & 24 & -20 \\ & & 2 & -14 & 20 \\ \hline & 1 & -7 & 10 & 0 \end{array}$$

So $x^3 - 9x^2 + 24x - 20 = (x - 2)(x^2 - 7x + 10) = (x - 2)^2 (x - 5)$. So the roots are 2 (with multiplicity 2) and 5.

39. **a.** Let $p(x)$ and $d(x)$ be the polynomials where $d(x) \neq 0$. Then there are unique polynomials $q(x)$ and $R(x)$ such that

$$p(x) = d(x) \bullet q(x) + R(x)$$

where either $R(x) = 0$ or the degree of $R(x)$ is less than the degree of $d(x)$.

b. When a polynomial $f(x)$ is divided by $x - r$, the remainder is $f(r)$.

c. Let $f(x)$ be a polynomial. If $f(r) = 0$, then $x - r$ is a factor of $f(x)$. Conversely, if $x - r$ is a factor of $f(x)$, then $f(r) = 0$.

d. Every polynomial equation of the form $a_n x^n + a_{n-1} x^{n-1} + \ldots + a_1 x + a_0 = 0$ $(n \geq 1, a_n \neq 0)$ has at least one root among the complex numbers. (This root may be a real number.)

41. Factor: $6x^2 + 7x - 20 = (3x - 4)(2x + 5) = 6\left(x - \frac{4}{3}\right)\left[x - \left(-\frac{5}{2}\right)\right]$

43. Factor: $x^4 - 4x^3 + 5x - 20 = x^3(x - 4) + 5(x - 4) = (x - 4)(x^3 + 5) = (x - 4)(x + \sqrt[3]{5})(x^2 - \sqrt[3]{5}x + \sqrt[3]{25})$

Solve $x^2 - \sqrt[3]{5}x + \sqrt[3]{25} = 0$ using the quadratic formula, which yields: $x = \dfrac{\sqrt[3]{5} \pm i\sqrt{3\sqrt[3]{25}}}{2}$

So $x^4 - 4x^3 + 3x - 20 = (x - 4)\left(x - \left(-\sqrt[3]{5}\right)\right)\left(x - \dfrac{\sqrt[3]{5} + i\sqrt{3\sqrt[3]{25}}}{2}\right)\left(x - \dfrac{\sqrt[3]{5} - i\sqrt{3\sqrt[3]{25}}}{2}\right)$.

45. One other root is $2 + 3i$, so $[x-(2-3i)][x-(2+3i)]$ are factors, which is $x^2 - 4x + 13$. Using long division:

$$\begin{array}{r} x-3 \\ x^2-4x+13\,\overline{)\,x^3-7x^2+25x-39} \\ \underline{x^3-4x^2+13x} \\ -3x^2+12x-39 \\ \underline{-3x^2+12x-39} \\ 0 \end{array}$$

So $x - 3$ is the other factor. So the roots are $2 - 3i$, $2 + 3i$, and 3.

47. One other root is $1-i\sqrt{2}$, so $\left[x-\left(1+i\sqrt{2}\right)\right]\left[x-\left(1-i\sqrt{2}\right)\right]$ are factors, which is $x^2 - 2x + 3$. Using long division:

$$\begin{array}{r} x^2-7 \\ x^2-2x+3\,\overline{)\,x^4-2x^3-4x^2+14x-21} \\ \underline{x^4-2x^3+3x^2} \\ -7x^2+14x-21 \\ \underline{-7x^2+14x-21} \\ 0 \end{array}$$

So $x^2-7=\left(x+\sqrt{7}\right)\left(x-\sqrt{7}\right)$ is the other factor. So the roots are $1+i\sqrt{2}$, $1-i\sqrt{2}$, $\sqrt{7}$, and $-\sqrt{7}$.

49. Let $f(x) = x^3 + 8x - 7$. There is one sign change, so there is 1 positive root. Now $f(-x) = -x^3 - 8x - 7$. Since there are no sign changes, there are no negative roots. So the equation has 1 positive real root and 2 complex roots.

51. Let $f(x) = x^3 + 3x + 1$. There are no sign changes, so there are no positive roots. Now $f(-x) = -x^3 - 3x + 1$. Since there is one sign change, there is 1 negative root. So the equation has 1 negative real root and 2 complex roots.

53. Let $f(x) = x^4 - 10$. There is one sign change, so there is 1 positive root. Since $f(-x) = f(x)$, there is also 1 negative root. So the equation has 1 positive real root, 1 negative real root and 2 complex roots.

55. **a.** Let $f(x) = x^3 + x^2 + x + 1$, so $f(-x) = -x^3 + x^2 - x + 1$. There are three sign changes, so there are either 1 or 3 negative roots.

b. Using the hint, we have $(x-1)\left(x^3+x^2+x+1\right)=x^4-1$. Let $g(x) = x^4 - 1$. Then $g(-x) = x^4 - 1$, so this can only have 1 negative root. Since $x = 1$ is a positive root, $x^3 + x^2 + x + 1 = 0$ can only have 1 negative root also.

c. Factoring, we have $x^3+x^2+x+1=x^2(x+1)+(x+1)=(x+1)\left(x^2+1\right)$. So $x = -1$ is a root. Also:

$$\begin{aligned} x^2+1&=0 \\ x^2&=-1 \\ x&=\pm i \end{aligned}$$

So the roots are -1, i, and $-i$.

57. Substituting $y = x^3$, we have $x^2+\left(x^3\right)^2=1$, so $x^6 + x^2 - 1 = 0$. Let $f(x) = x^6 + x^2 - 1$.

$f(0) = -1$ $\qquad$ $f(1) = 1$

So there is a root between 0 and 1.

$f(0.8) = -0.09$ $\qquad$ $f(0.9) = 0.34$

So there is a root between 0.8 and 0.9.

$f(0.82) = -0.02$ $\qquad$ $f(0.83) = 0.01$

So there is a root between 0.82 and 0.83. So the x-coordinate lies between 0.82 and 0.83.

59. **a.** Let $f(x) = x^3 - 36x - 84$. Since there is only one sign change, there is 1 positive real root.

b. Using synthetic division:

$$\begin{array}{r|rrrr} 7 & 1 & 0 & -36 & -84 \\ & & 7 & 49 & 91 \\ \hline & 1 & 7 & 13 & 7 \end{array}$$

Since the row is all positive, there are no real roots to the equation greater than 7.

c. Here $f(6) = -84$ and $f(7) = 7$, so there is a root between 6 and 7. $f(6.8) = -14.4$ and $f(6.9) = -3.89$, so there is a root between 6.9 and 7.0. $f(6.93) = -0.67$ and $f(6.94) = 0.42$, so there is a root between 6.93 and 6.94.

61. Another root will be $4+\sqrt{5}$, so:

$$\left[x-\left(4-\sqrt{5}\right)\right]\left[x-\left(4+\sqrt{5}\right)\right]=0$$
$$x^2-8x+11=0$$

63. Other roots will be $6 + 2i$ and $-\sqrt{5}$:

$$\left(x-\sqrt{5}\right)\left[x-\left(-\sqrt{5}\right)\right]\left[x-(6-2i)\right]\left[x-(6+2i)\right]=0$$
$$\left(x^2-5\right)\left(x^2-12x+40\right)=0$$
$$x^4-12x^3+35x^2+60x-200=0$$

65. Write $x-1=\sqrt{2}+\sqrt{3}$. Squaring both sides:

$$x^2-2x+1=5+2\sqrt{6}$$
$$x^2-2x-4=2\sqrt{6}$$

Squaring both sides again:

$$x^4-4x^3-4x^2+16x+16=24$$
$$x^4-4x^3-4x^2+16x-8=0$$

67. Factoring, we have $y=x^3-2x^2-3x=x\left(x^2-2x-3\right)=x(x-3)(x+1)$. The zeros are 0, 3, and -1.

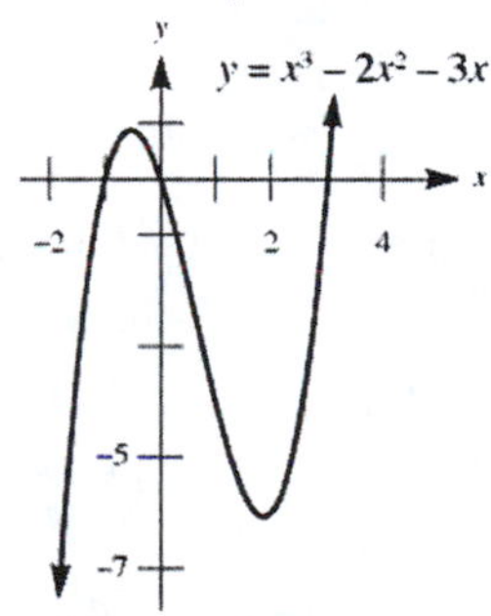

69. Factoring, we have $y=x^4-4x^2=x^2\left(x^2-4\right)=x^2(x+2)(x-2)$. The zeros are 0, -2, and 2.

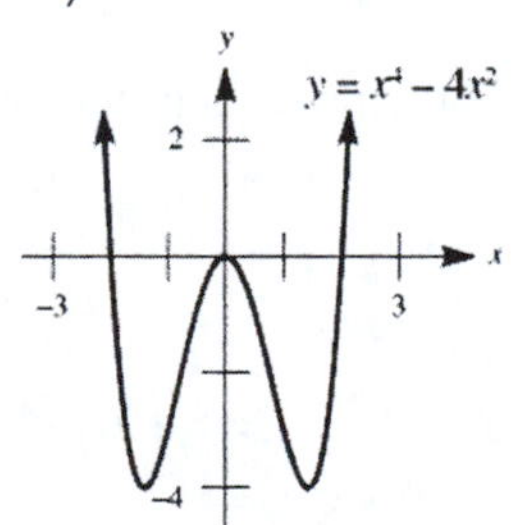

71. Performing the indicated operations yields:

$$(3-2i)(3+2i)+(1+3i)^2=9-4i^2+1+6i+9i^2=9+4+1+6i-9=5+6i$$

73. Performing the indicated operations yields: $\left(1+i\sqrt{2}\right)\left(1-i\sqrt{2}\right)+\left(\sqrt{2}+i\right)\left(\sqrt{2}-i\right)=1-2i^2+2-i^2=1+2+2+1=6$

75. Performing the indicated operations yields: $\dfrac{3-i\sqrt{3}}{3+i\sqrt{3}}\bullet\dfrac{3-i\sqrt{3}}{3-i\sqrt{3}}=\dfrac{9-6i\sqrt{3}-3}{9+3}=\dfrac{6-6i\sqrt{3}}{12}=\dfrac{1-i\sqrt{3}}{2}=\frac{1}{2}-\frac{\sqrt{3}}{2}i$

77. Writing in complex form: $-\sqrt{-2}\sqrt{-9}+\sqrt{-8}-\sqrt{-72}=-\left(i\sqrt{2}\right)(3i)+2i\sqrt{2}-6i\sqrt{2}=3\sqrt{2}-4i\sqrt{2}=3\sqrt{2}-4\sqrt{2}i$

79. Let $z=a+bi$, so $\overline{z}=a-bi$. Therefore: $\dfrac{z+\overline{z}}{2}=\dfrac{a+bi+a-bi}{2}=\dfrac{2a}{2}=a=\text{Re}(z)$

81. **a.** Compute the absolute values:

$$|6+2i|=\sqrt{6^2+2^2}=\sqrt{36+4}=\sqrt{40}=2\sqrt{10}$$
$$|6-2i|=\sqrt{6^2+(-2)^2}=\sqrt{36+4}=\sqrt{40}=2\sqrt{10}$$

b. Using the new definition: $|-3|=|-3+0i|=\sqrt{(-3)^2+0^2}=\sqrt{9}=3$

c. Let $z=a+bi$, so $\overline{z}=a-bi$. We now compute:

$$z\overline{z}=(a+bi)(a-bi)=a^2-b^2i^2=a^2+b^2$$
$$|z|^2=\left(\sqrt{a^2+b^2}\right)^2=a^2+b^2$$

Thus $z\overline{z}=|z|^2$.

83. Combining the fractions: $\dfrac{1}{a-bi}-\dfrac{1}{a+bi}=\dfrac{(a+bi)-(a-bi)}{(a-bi)(a+bi)}=\dfrac{2bi}{a^2-b^2i^2}=\dfrac{2bi}{a^2+b^2}$

85. Combining the fractions: $\dfrac{a+bi}{a-bi}-\dfrac{a-bi}{a+bi}=\dfrac{(a+bi)^2-(a-bi)^2}{(a-bi)(a+bi)}=\dfrac{a^2+2abi-b^2-a^2+2abi+b^2}{a^2+b^2}=\dfrac{4abi}{a^2+b^2}$

87. Since $100-x^2=(10+x)(10-x)$, the form of the decomposition is: $\dfrac{2x-1}{100-x^2}=\dfrac{A}{10+x}+\dfrac{B}{10-x}$

Clearing fractions yields: $2x-1=A(10-x)+B(10+x)=(-A+B)x+(10A+10B)$

Equating coefficients:

$$-A+B=2$$
$$10A+10B=-1$$

Multiplying the first equation by 10:

$$-10A+10B=20$$
$$10A+10B=-1$$

Adding yields $20B=19$, so $B=\frac{19}{20}$. Substituting into $-A+B=2$:

$$-A+\tfrac{19}{20}=2$$
$$-A=\tfrac{21}{20}$$
$$A=-\tfrac{21}{20}$$

The partial fraction decomposition is: $\dfrac{2x-1}{100-x^2}=\dfrac{-\frac{21}{20}}{10+x}+\dfrac{\frac{19}{20}}{10-x}$

89. Since $x^3+2x^2+x=x\left(x^2+2x+1\right)=x(x+1)^2$, the form of the decomposition is:

$$\frac{1}{x^3+2x^2+x}=\frac{A}{x}+\frac{B}{x+1}+\frac{C}{(x+1)^2}$$

Clearing fractions yields:

$$1=A(x+1)^2+Bx(x+1)+Cx=Ax^2+2Ax+A+Bx^2+Bx+Cx=(A+B)x^2+(2A+B+C)x+A$$

Equating coefficients:

$$A+B=0$$
$$2A+B+C=0$$
$$A=1$$

So $B=-1$. Substituting into $2A+B+C=0$:

$$2-1+C=0$$
$$1+C=0$$
$$C=-1$$

The partial fraction decomposition is: $\dfrac{1}{x^3+2x^2+x}=\dfrac{1}{x}+\dfrac{-1}{x+1}+\dfrac{-1}{(x+1)^2}$

91. Since $x^4+6x^2+9=\left(x^2+3\right)^2$, the form of the decomposition is: $\dfrac{x^3+2}{x^4+6x^2+9}=\dfrac{Ax+B}{x^2+3}+\dfrac{Cx+D}{\left(x^2+3\right)^2}$

Clearing fractions yields:

$$x^3+2=(Ax+B)\left(x^2+3\right)+(Cx+D)=Ax^3+Bx^2+3Ax+3B+Cx+D=Ax^3+Bx^2+(3A+C)x+(3B+D)$$

Equating coefficients:

$$A=1$$
$$B=0$$
$$3A+C=0$$
$$3B+D=2$$

So $C=-3$ and $D=2$. The partial fraction decomposition is: $\dfrac{x^3+2}{x^4+6x^2+9}=\dfrac{x}{x^2+3}+\dfrac{-3x+2}{\left(x^2+3\right)^2}$

93. **a.** Multiplying out factors: $\left(x^2+bx+1\right)\left(x^2+cx-1\right)=x^4+(b+c)x^3+bcx^2+(-b+c)x-1$

Equating coefficients:

$$b+c=-2$$
$$bc=1$$
$$-b+c=0$$

Adding the first and third equations yields $2c=-2$, so $c=-1$ and $b=-1$. Thus the factorization is given by:

$$x^4-2x^3+x^2-1=\left(x^2-x+1\right)\left(x^2-x-1\right)$$

b. The form of the decomposition is: $\dfrac{x^3}{x^4-2x^3+x^2-1}=\dfrac{Ax+B}{x^2-x+1}+\dfrac{Cx+D}{x^2-x-1}$

Clearing fractions yields:

$$\begin{aligned}x^3&=(Ax+B)\left(x^2-x-1\right)+(Cx+D)\left(x^2-x+1\right)\\&=Ax^3-Ax^2-Ax+Bx^2-Bx-B+Cx^3-Cx^2+Cx+Dx^2-Dx+D\\&=(A+C)x^3+(-A+B-C+D)x^2+(-A-B+C-D)x+(-B+D)\end{aligned}$$

Equating coefficients:

$$A+C=1$$
$$-A+B-C+D=0$$
$$-A-B+C-D=0$$
$$-B+D=0$$

Adding the second and third equations yields $-2A=0$, so $A=0$ and $C=1$. Using the second and fourth equations:

$$B-1+D=0$$
$$-B+D=0$$

Adding yields $-1+2D=0$, so $D=\frac{1}{2}$ and $B=\frac{1}{2}$. The partial fraction decomposition is:

$$\frac{x^3}{x^4-2x^3+x^2-1}=\frac{\frac{1}{2}}{x^2-x+1}+\frac{x+\frac{1}{2}}{x^2-x-1}$$

Chapter 12 Test

1. Since $f\left(\frac{1}{2}\right)$ will be the remainder after synthetic division, synthetically divide by $\frac{1}{2}$:

1/2	6	–5	7	–2	–2
		3	–1	3	1/2
	6	–2	6	1	–3/2

So $f\left(\frac{1}{2}\right)=-\frac{3}{2}$.

2. Using synthetic division:

–3	1	1	–11	–15
		–3	6	15
	1	–2	–5	0

So $x^3+x^2-11x-15=(x+3)\left(x^2-2x-5\right)$. Using the quadratic formula: $x=\frac{2\pm\sqrt{4+20}}{2}=\frac{2\pm2\sqrt{6}}{2}=1\pm\sqrt{6}$

So the roots of the equation are –3, $1+\sqrt{6}$, and $1-\sqrt{6}$.

3. The possible rational roots are: $\pm1,\pm\frac{1}{2},\pm2,\pm3,\pm\frac{3}{2},\pm6$

4. Such a function would be $y = a(x-1)(x+8)$. Now $y = -24$ when $x = 0$, so $-24 = a(-1)(8)$, so $a = 3$. So the function is $y = 3(x-1)(x+8)$, or $f(x) = 3x^2 + 21x - 24$.

5. Using synthetic division:

–1	4	1	–8	3
		–4	3	5
	4	–3	–5	8

The quotient is $4x^2-3x-5$ and the remainder is 8.

6. a. Let $f(x)$ be a polynomial. If $f(r) = 0$, then $x-r$ is a factor of $f(x)$. Conversely, if $x-r$ is a factor of $f(x)$, then $f(r) = 0$.

   b. Every polynomial of the form $a_nx^n+a_{n-1}x^{n-1}+\ldots+a_1x+a_0=0$ $(n\geq1, a_n\neq0)$ has at least one root among the complex numbers. (This root may be a real number.)

   c. Let $f(x)$ be a polynomial, all of whose coefficients are real numbers. Then $f(x)$ can be factored (over the real numbers) into a product of linear and/or irreducible quadratic factors.

7. a. Using synthetic division with 1:

1	1	–2	0	–1
		1	–1	–1
	1	–1	–1	–2

Now use 2:

2	1	–2	0	–1
		2	0	0
	1	0	0	–1

Now use 3:

3	1	–2	0	–1
		3	3	9
	1	1	3	8

So 3 is an upper bound for the roots.

   b. Since $f(2.2) = -0.032$ and $f(2.3) = 0.587$, the root lies between 2.2 and 2.3.

8. Two other roots are $1 - i$ and $3 + 2i$, so $x^2 - 2x + 2$ and $x^2 - 6x + 13$ are factors. Using long division:

$$\begin{array}{r}
x^3 - 4x^2 + x + 26 \\
x^2 - 2x + 2 \,\big)\, \overline{x^5 - 6x^4 + 11x^3 + 16x^2 - 50x + 52} \\
\underline{x^5 - 2x^4 + 2x^3} \qquad\qquad\qquad\qquad \\
-4x^4 + 9x^3 + 16x^2 \qquad\qquad\quad \\
\underline{-4x^4 + 8x^3 - 8x^2} \qquad\qquad\quad \\
x^3 + 24x^2 - 50x \qquad \\
\underline{x^3 - 2x^2 + 2x} \qquad \\
26x^2 - 52x + 52 \\
\underline{26x^2 - 52x + 52} \\
0
\end{array}$$

Using long division again:

$$\begin{array}{r}
x + 2 \\
x^2 - 6x + 13 \,\big)\, \overline{x^3 - 4x^2 + x + 26} \\
\underline{x^3 - 6x^2 + 13x} \qquad \\
2x^2 - 12x + 26 \\
\underline{2x^2 - 12x + 26} \\
0
\end{array}$$

So the roots are $1 \pm i$, $3 \pm 2i$, and -2.

9. Using long division:

$$\begin{array}{r}
x^2 + 2x - 1 \\
x^2 + 1 \,\big)\, \overline{x^4 + 2x^3 + 0x^2 - x + 6} \\
\underline{x^4 + \qquad\quad x^2} \qquad\qquad \\
2x^3 - x^2 - x \qquad \\
\underline{2x^3 \qquad + 2x} \qquad \\
-x^2 - 3x + 6 \\
\underline{-x^2 \qquad\quad - 1} \\
-3x + 7
\end{array}$$

So $q(x) = x^2 + 2x - 1$ and $R(x) = -3x + 7$.

10. First find the roots by the quadratic formula: $x = \dfrac{6 \pm \sqrt{36 - 40}}{4} = \dfrac{6 \pm 2i}{4} = \frac{3}{2} \pm \frac{1}{2}i$

So $2x^2 - 6x + 5 = 2\left(x^2 - 3x + \frac{5}{2}\right) = 2\left[x - \left(\frac{3}{2} + \frac{1}{2}i\right)\right]\left[x - \left(\frac{3}{2} - \frac{1}{2}i\right)\right]$.

11.
a. The possible rational roots are $\pm 1, \pm 2, \pm 3, \pm 4, \pm 6, \pm 8, \pm 12$, and ± 24.

b. Using synthetic division:

$$\begin{array}{r|rrrrr}
2 & 1 & -1 & 0 & 0 & 24 \\
 & & 2 & 2 & 4 & 8 \\
\hline
 & 1 & 1 & 2 & 4 & 32
\end{array}$$

Since this last row consists of all positive numbers, we know that 2 is an upper bound for the roots of this equation.

c. Only $x = 1$, since $x = 2$ is an upper bound and not a root.

d. Using synthetic division:

$$\begin{array}{r|rrrrr}
1 & 1 & -1 & 0 & 0 & 24 \\
 & & 1 & 0 & 0 & 0 \\
\hline
 & 1 & 0 & 0 & 0 & 24
\end{array}$$

Since $x = 1$ was the only possibility, there are no positive rational roots.

12. a. The possible rational roots are $\pm 1, \pm\frac{1}{2}, \pm 3, \pm\frac{3}{2}$. Using synthetic division:

$$\begin{array}{r|rrrr} 3/2 & 2 & -1 & -1 & -3 \\ & & 3 & 3 & 3 \\ \hline & 2 & 2 & 2 & 0 \end{array}$$

So $2x^3 - x^2 - x - 3 = \left(x - \frac{3}{2}\right)\left(2x^2 + 2x + 2\right) = 2\left(x - \frac{3}{2}\right)\left(x^2 + x + 1\right)$. Using the quadratic equation, we have:

$$x = \frac{-1 \pm \sqrt{1-4}}{2} = \frac{-1 \pm i\sqrt{3}}{2}$$

So the only rational root is $\frac{3}{2}$.

b. All solutions are $\frac{3}{2}$, $\frac{-1+i\sqrt{3}}{2}$, and $\frac{-1-i\sqrt{3}}{2}$.

13. Let $f(x) = 3x^4 + x^2 - 5x - 1$. Since there is one sign change, there is 1 positive root. Now $f(-x) = 3x^4 + x^2 + 5x - 1$. Since there is one sign change, there is 1 negative root. So the equation has 1 positive real root, 1 negative real root, and 2 complex roots.

14. The factors are $x + 2$, $x - (1 - 3i)$, and $x - (1 + 3i)$, so:

$$\left(x+2\right)\left[x-\left(1-3i\right)\right]\left[x-\left(1+3i\right)\right] = 0$$
$$\left(x+2\right)\left(x^2 - 2x + 10\right) = 0$$
$$x^3 + 6x + 20 = 0$$

15. Here $f(x) = \left(x-2\right)\left(x-3i\right)^3\left[x - \left(1+\sqrt{2}\right)\right]^2$. Note that if a restriction of rational coefficients been added, we would also have $-3i$ (multiplicity 3) and $1-\sqrt{2}$ (multiplicity 2) as roots.

16. Simplify the expression: $(3+2i)(5-3i) + \sqrt{-3} = 15 + 10i - 9i - (-6) + i\sqrt{3} = 21 + \left(1+\sqrt{3}\right)i$

17. Rationalizing: $\frac{3+i}{1-4i} \bullet \frac{1+4i}{1+4i} = \frac{3+i+12i-4}{1+16} = \frac{-1+13i}{17} = -\frac{1}{17} + \frac{13}{17}i$

18. Since $x^3 - 16x = x\left(x^2 - 16\right) = x(x+4)(x-4)$, the decomposition will have the form: $\frac{3x-1}{x^3-16x} = \frac{A}{x} + \frac{B}{x+4} + \frac{C}{x-4}$

Clearing fractions yields: $3x - 1 = A(x+4)(x-4) + Bx(x-4) + Cx(x+4)$

Substituting $x = -4$:

$$-12 - 1 = A(0)(-8) + B(-4)(-8) + C(-4)(0)$$
$$-13 = 32B$$
$$B = -\frac{13}{32}$$

Substituting $x = 4$:

$$12 - 1 = A(8)(0) + B(4)(0) + C(4)(8)$$
$$11 = 32C$$
$$C = \frac{11}{32}$$

Substituting $x = 0$:

$$0 - 1 = A(4)(-4) + B(0)(-4) + C(0)(4)$$
$$-1 = -16A$$
$$A = \frac{1}{16}$$

So $A = \frac{1}{16}$, $B = -\frac{13}{32}$, and $C = \frac{11}{32}$. The partial fraction decomposition is: $\frac{3x-1}{x^3-16x} = \frac{\frac{1}{16}}{x} + \frac{-\frac{13}{32}}{x+4} + \frac{\frac{11}{32}}{x-4}$

19. Factoring the denominator by grouping: $x^3 - x^2 + 3x - 3 = x^2(x-1) + 3(x-1) = (x-1)(x^2+3)$

The decomposition will have the form: $\dfrac{1}{x^3 - x^2 + 3x - 3} = \dfrac{A}{x-1} + \dfrac{Bx+C}{x^2+3}$

Clearing fractions yields:

$$1 = A(x^2+3) + (Bx+C)(x-1) = Ax^2 + 3A + Bx^2 + Cx - Bx - C = (A+B)x^2 + (-B+C)x + (3A-C)$$

Equating coefficients:

$$A + B = 0$$
$$-B + C = 0$$
$$3A - C = 1$$

Form the augmented matrix: $\begin{pmatrix} 1 & 1 & 0 & 0 \\ 0 & -1 & 1 & 0 \\ 3 & 0 & -1 & 1 \end{pmatrix}$

Adding –3 times row 1 to row 3 and multiplying row 2 by –1: $\begin{pmatrix} 1 & 1 & 0 & 0 \\ 0 & 1 & -1 & 0 \\ 0 & -3 & -1 & 1 \end{pmatrix}$

Adding 3 times row 2 to row 3: $\begin{pmatrix} 1 & 1 & 0 & 0 \\ 0 & 1 & -1 & 0 \\ 0 & 0 & -4 & 1 \end{pmatrix}$

Dividing row 3 by –4, and adding it to row 2: $\begin{pmatrix} 1 & 1 & 0 & 0 \\ 0 & 1 & 0 & -\frac{1}{4} \\ 0 & 0 & 1 & -\frac{1}{4} \end{pmatrix}$

Adding –1 times row 2 to row 1: $\begin{pmatrix} 1 & 0 & 0 & \frac{1}{4} \\ 0 & 1 & 0 & -\frac{1}{4} \\ 0 & 0 & 1 & -\frac{1}{4} \end{pmatrix}$

So $A = \frac{1}{4}$, $B = -\frac{1}{4}$, and $C = -\frac{1}{4}$. The partial fraction decomposition is: $\dfrac{1}{x^3 - x^2 + 3x - 3} = \dfrac{\frac{1}{4}}{x-1} + \dfrac{-\frac{1}{4}x - \frac{1}{4}}{x^2+3}$

20. Since $x^3 - 4x^2 + 4x = x\left(x^2 - 4x + 4\right) = x(x-2)^2$, the decomposition will have the form:

$$\frac{4x^2 - 15x + 20}{x^3 - 4x^2 + 4x} = \frac{A}{x} + \frac{B}{x-2} + \frac{C}{(x-2)^2}$$

Clearing fractions yields:

$$\begin{aligned} 4x^2 - 15x + 20 &= A(x-2)^2 + Bx(x-2) + Cx \\ &= Ax^2 - 4Ax + 4A + Bx^2 - 2Bx + Cx \\ &= (A+B)x^2 + (-4A - 2B + C)x + 4A \end{aligned}$$

Equating coefficients:

$$\begin{aligned} A + B &= 4 \\ -4A - 2B + C &= -15 \\ 4A &= 20 \end{aligned}$$

So $A = 5$ and $B = -1$. Substituting into $-4A - 2B + C = -15$:

$$\begin{aligned} -4(5) - 2(-1) + C &= -15 \\ -20 + 2 + C &= -15 \\ C &= 3 \end{aligned}$$

The partial fraction decomposition is: $\dfrac{4x^2 - 15x + 20}{x^3 - 4x^2 + 4x} = \dfrac{5}{x} + \dfrac{-1}{x-2} + \dfrac{3}{(x-2)^2}$

Chapter 13
Additional Topics in Algebra

13.1 Mathematical Induction

1. Here P_n denotes the statement: $1+2+3+...+n=\dfrac{n(n+1)}{2}$

Since $1=\frac{1(1+1)}{2}=1$, P_1 is true. It remains to show that P_k implies P_{k+1}. Assume P_k is true: $1+2+...+k=\dfrac{k(k+1)}{2}$

Thus: $1+2+...+k+k+1=\dfrac{k(k+1)}{2}+k+1=(k+1)\left(\dfrac{k}{2}+1\right)=\dfrac{(k+1)(k+2)}{2}=\dfrac{(k+1)[(k+1)+1]}{2}$

So P_{k+1} is true and the induction is complete.

3. Here P_n denotes the statement: $1+4+7+...+(3n-2)=\dfrac{n(3n-1)}{2}$

Since $1=\frac{1(3(1)-1)}{2}=1$, P_1 is true. Assume P_k is true: $1+4+...+(3k-2)=\dfrac{k(3k-1)}{2}$. Thus:

$$\begin{aligned}1+4+...+(3k-2)+[3(k+1)-2]&=\frac{k(3k-1)}{2}+3(k+1)-2\\&=\tfrac{1}{2}\left[k(3k-1)+6(k+1)-4\right]\\&=\tfrac{1}{2}\left(3k^2-k+6k+2\right)\\&=\tfrac{1}{2}\left(3k^2+5k+2\right)\\&=\frac{(k+1)(3k+2)}{2}\\&=\frac{(k+1)\left[3(k+1)-1\right]}{2}\end{aligned}$$

So P_{k+1} is true and the induction is complete.

5. Here P_n denotes: $1^2+2^2+3^2+\ldots+n^2=\frac{n(n+1)(2n+1)}{6}$

Since $1^2=1=\frac{1(1+1)[2(1)+1]}{6}=\frac{1(2)(3)}{6}=1$, P_1 is true. Assume P_k is true: $1^2+2^2+\ldots+k^2=\frac{k(k+1)(2k+1)}{6}$

Thus:

$$\begin{aligned}1^2+2^2+\ldots+k^2+(k+1)^2&=\frac{k(k+1)(2k+1)}{6}+(k+1)^2\\&=\frac{(k+1)[k(2k+1]+6(k+1)^2}{6}\\&=\frac{(k+1)\left(2k^2+k+6k+6\right)}{6}\\&=\frac{(k+1)\left(2k^2+7k+6\right)}{6}\\&=\frac{(k+1)(k+2)(2k+3)}{6}\\&=\frac{(k+1)[(k+1)+1][2(k+1)+1]}{6}\end{aligned}$$

So P_{k+1} is true and the induction is complete.

7. Here P_n denotes: $1^2+3^2+5^2+\ldots+(2n-1)^2=\frac{n(2n-1)(2n+1)}{3}$

As $1^2=1=\frac{1[2(1)-1][2(1)+1]}{3}=1$, P_1 is true. Assume P_k is true: $1^2+3^2+\ldots+(2k-1)^2=\frac{k(2k-1)(2k+1)}{3}$

Thus:

$$\begin{aligned}1^2+3^2+\ldots+(2k-1)^2+[2(k+1)-1]^2&=\frac{k(2k-1)(2k+1)}{3}+[2(k+1)-1]^2\\&=\frac{k(2k-1)(2k+1)}{3}+\frac{3(2k+1)^2}{3}\\&=\frac{2k+1}{3}[k(2k-1)+3(2k+1)]\\&=\frac{[2(k+1)-1]}{3}\left(2k^2+5k+3\right)\\&=\frac{(k+1)[2(k+1)-1](2k+3)}{3}\\&=\frac{(k+1)[2(k+1)-1][2(k+1)+1]}{3}\end{aligned}$$

So P_{k+1} is true and the induction is complete.

9. Here P_n denotes: $3+3^2+3^3+\ldots+3^n=\frac{1}{2}\left(3^{n+1}-3\right)$

Since $3=\frac{1}{2}\left(3^2-3\right)=\frac{1}{2}(6)=3$, P_1 is true. Assume P_k is true: $3+3^2+\ldots+3^k=\frac{1}{2}\left(3^{k+1}-3\right)$

Thus:

$$3+3^2+\ldots+3^k+3^{k+1}=\frac{1}{2}\left(3^{k+1}-3\right)+3^{k+1}=\frac{3}{2}\left(3^{k+1}\right)-\frac{3}{2}=\frac{1}{2}\left[3\left(3^{k+1}\right)-3\right]=\frac{1}{2}\left(3^{k+2}-3\right)=\frac{1}{2}\left[3^{(k+1)+1}-3\right]$$

So P_{k+1} is true and the induction is complete.

11. Here P_n denotes: $1^3+2^3+3^3+\ldots+n^3=\left[\frac{n(n+1)}{2}\right]^2$

Since $1^3=1=\left[\frac{1(2)}{2}\right]^2=1$, P_1 is true. Assume P_k is true: $1^3+2^3+\ldots+k^3=\left[\frac{k(k+1)}{2}\right]^2$. Thus:

$$\begin{aligned}1^3+2^3+\ldots+k^3+(k+1)^3&=\left[\frac{k(k+1)}{2}\right]^2+(k+1)^3\\&=(k+1)^2\left[\frac{k^2}{4}+(k+1)\right]\\&=\frac{(k+1)^2}{4}\left[k^2+4k+4\right]\\&=\frac{(k+1)^2(k+2)^2}{4}\\&=\frac{(k+1)^2\left[(k+1)+1\right]^2}{2^2}\\&=\left(\frac{(k+1)\left[(k+1)+1\right]}{2}\right)^2\end{aligned}$$

So P_{k+1} is true and the induction is complete.

13. Here P_n denotes: $1^3+3^3+5^3+\ldots+(2n-1)^3=n^2\left(2n^2-1\right)$

Since $1^3=1(1)^2\left[2(1)^2-1\right]=(1)(1)=1$, P_1 is true. Assume P_k is true: $1^3+3^3+\ldots+(2k-1)^3=k^2\left(2k^2-1\right)$. Thus:

$$\begin{aligned}1^3+3^3+\ldots+(2k-1)^3+\left[2(k+1)-1\right]^3&=k^2\left(2k^2-1\right)+(2k+1)^3\\&=2k^4-k^2+8k^3+12k^2+6k+1\\&=2k^4+8k^3+11k^2+6k+1\\&=(k+1)^2\left(2k^2+4k+1\right)\\&=(k+1)^2\left[2\left(k^2+2k+1\right)-1\right]\\&=(k+1)^2\left[2(k+1)^2-1\right]\end{aligned}$$

So P_{k+1} is true and the induction is complete.

15. Here P_n denotes: $1\times3+3\times5+5\times7+\ldots+(2n-1)(2n+1)=\frac{n\left(4n^2+6n-1\right)}{3}$

Since $1\times3=3=\frac{(1)\left[4(1)^2+6(1)-1\right]}{3}=\frac{9}{3}=3$, P_1 is true. Assume P_k is true:

$$1\times3+3\times5+\ldots+(2k-1)(2k+1)=\frac{k\left(4k^2+6k-1\right)}{3}$$

Thus (noting that –1 is a root):

$$1\times 3+3\times 5+...+(2k-1)(2k+1)+[2(k+1)-1][2(k+1)+1]=\frac{k(4k^2+6k-1)}{3}+(2k+1)(2k+3)$$
$$=\frac{4k^3+6k^2-k+12k^2+24k+9}{3}$$
$$=\frac{4k^3+18k^2+23k+9}{3}$$
$$=\frac{(k+1)(4k^2+14k+9)}{3}$$
$$=\frac{(k+1)\left[4(k+1)^2+6(k+1)-1\right]}{3}$$

So P_{k+1} is true and the induction is complete.

17. Here P_n denotes: $1+\frac{3}{2}+\frac{5}{2^2}+...+\frac{2n-1}{2^{n-1}}=6-\frac{2n+3}{2^{n-1}}$

Since $1=6-\frac{[2(1)+3]}{2^0}=6-5=1$, P_1 is true. Assume P_k is true: $1+\frac{3}{2}+...+\frac{2k-1}{2^{k-1}}=6-\frac{2k+3}{2^{k-1}}$

Thus: $1+\frac{3}{2}+...+\frac{2k-1}{2^{k-1}}+\frac{2(k+1)-1}{2^k}=6-\frac{2k+3}{2^{k-1}}+\frac{2k+1}{2^k}=6-\frac{4k+6-2k-1}{2^k}=6-\frac{2k+5}{2^k}=6-\frac{2(k+1)+3}{2^{(k+1)-1}}$

So P_{k+1} is true and the induction is complete.

19. Let P_n denote the statement: $n\le 2^{n-1}$

Since $2^{1-1}=2^0=1\ge 1$, P_1 is true. Assume P_k is true: $k\le 2^{k-1}$. Hence: $k+1\le 2^{k-1}+1$

Since $1\le 2^{k-1}$ (use induction to prove this), we have: $k+1\le 2^{k-1}+1\le 2^{k-1}+2^{k-1}=2\left(2^{k-1}\right)=2^k=2^{(k+1)-1}$

So P_{k+1} is true and the induction is complete.

21. For $n=2$: $2^2+4=8<(2+1)^2$

Assuming: $k^2+4<(k+1)^2$. Then:

$$k^2+2k+4<(k+1)^2+2k$$
$$(k+1)^2+3<k^2+4k+1$$
$$(k+1)^2+4<k^2+4k+2$$
$$(k+1)^2+4<k^2+4k+4$$
$$(k+1)^2+4<(k+2)^2$$

This completes the induction for $n\ge 2$.

23. For $n=7$, we have: $(1.5)^7\approx 17.09>2(7)$, so P_1 is true

Assume P_k is true, so: $(1.5)^k>2k$. Then for $k\ge 7$:

$$(1.5)^{k+1}=1.5(1.5)^k$$
$$>1.5(2k)$$
$$=3k$$
$$=2k+k$$
$$>2k+2\quad(\text{since } k>2)$$
$$=2(k+1)$$

So P_{k+1} is true and the induction is complete.

25. For $n = 39$, we have: $(1.1)^{39} \approx 41.1 > 39$, so P_{39} is true

Assume P_k is true for some $k \geq 39$, so: $(1.1)^k > k$

Then: $(1.1)^{k+1} = 1.1(1.1)^k > 1.1k = k + 0.1k > k+1$ (since $k > 10$)

So P_{k+1} is true and the induction is complete.

27. **a.** Complete the table:

n	1	2	3	4	5
$f(n)$	$\frac{1}{2}$	$\frac{2}{3}$	$\frac{3}{4}$	$\frac{4}{5}$	$\frac{5}{6}$

b. $\frac{6}{7}$, $f(6) = f(5) + \frac{1}{42} = \frac{5}{6} + \frac{1}{42} = \frac{36}{42} = \frac{6}{7}$

c. $f(n) = \dfrac{1}{1\times 2} + \dfrac{1}{2\times 3} + \ldots + \dfrac{1}{n(n+1)} = \dfrac{n}{n+1}$

$f(1) = \frac{1}{2} = \dfrac{1}{1+1} = \frac{1}{2}$. Assuming $f(k) = \dfrac{k}{k+1}$, then:

$$\begin{aligned} f(k+1) &= f(k) + \frac{1}{(k+1)(k+2)} \\ &= \frac{k}{k+1} + \frac{1}{(k+1)(k+2)} \\ &= \frac{k(k+2)+1}{(k+1)(k+2)} \\ &= \frac{k^2+2k+1}{(k+1)(k+2)} \\ &= \frac{(k+1)^2}{(k+1)(k+2)} \\ &= \frac{k+1}{k+2} \end{aligned}$$

This completes the induction.

29. **a.** Complete the table:

n	1	2	3	4	5
$f(n)$	1	4	9	16	25

b. 36, $f(6) = f(5) + 2\sqrt{f(5)} + 1 = 25 + 2\sqrt{25} + 1 = 36$

c. Let $f(n) = n^2$, so $f(1) = 1 = 1^2$. Assume that for some $k \geq 1$ that $f(k) = k^2$. Then we have:

$$f(k+1) = f(k) + 2\sqrt{f(k)} + 1 = k^2 + 2\sqrt{k^2} + 1 = k^2 + 2k + 1 = (k+1)^2$$

The induction is complete.

31. For $n = 1$: 13 is prime and P_1 is true

For $n = 2$: 17 is prime and P_2 is true

For $n = 3$: 23 is prime and P_3 is true

For $n = 4$: 31 is prime and P_4 is true

For $n = 5$: 41 is prime and P_5 is true

For $n = 6$: 53 is prime and P_6 is true

For $n = 7$: 67 is prime and P_7 is true

For $n = 8$: 83 is prime and P_8 is true

For $n = 9$: 101 is prime and P_9 is true

However, for $n = 10$, we have $(10)^2 + 10 + 11 = 121 = (11)(11)$ and so P_{10} is false.

33. Let P_n denote: $1 + r + r^2 + \ldots + r^{n-1} = \dfrac{r^n - 1}{r-1}$ when $r \neq 1$

Then since $1 = \dfrac{r-1}{r-1} = 1$, P_1 is true. Assuming P_k is true, then: $1 + r + \ldots + r^{k-1} = \dfrac{r^k - 1}{r-1}$

Thus: $1 + r + \ldots + r^{k-1} + r^k = \dfrac{r^k - 1}{r-1} + r^k = \dfrac{r^k - 1 + r^k(r-1)}{r-1} = \dfrac{r^{k+1} - 1}{r-1}$

So P_{k+1} is true and the induction is complete.

35. Let P_n denote: $n^5 - n = 5R$ for some natural number R. Since $2^5 - 2 = 32 - 2 = 30 = 5(6)$, P_2 is true. Since:

$$(k+1)^5 - (k+1) = k^5 + 5k^4 + 10k^3 + 10k^2 + 4k = k^5 - k + 5\left(k^4 + 2k^3 + 2k^2 + k\right)$$

Assuming P_k is true results in: $(k+1)^5 - (k+1) = 5(R + k^4 + 2k^3 + 2k^2 + k)$

So P_{k+1} is true and the induction is complete.

37. For $n = 0$, we have: $2^1 + 3^1 = 5 = 5(1)$, so P_0 is true

Assume P_k is true, so: $2^{2k+1} + 3^{2k+1} = 5A$, for some natural number A. Then:

$$\begin{aligned}2^{2k+3} + 3^{2k+3} &= 4 \bullet 2^{2k+1} + 9 \bullet 3^{2k+1} \\ &= 4 \bullet 2^{2k+1} + 9\left(5A - 2^{2k+1}\right) \\ &= 4 \bullet 2^{2k+1} + 45A - 9 \bullet 2^{2k+1} \\ &= 45A - 5 \bullet 2^{2k+1} \\ &= 5\left(9A - 2^{2k+1}\right)\end{aligned}$$

Since 5 is a factor of this expression, P_{k+1} is true and the induction is complete.

39. For $n = 0$, we have: $2^1 + (-1)^0 = 2 + 1 = 3 = 3(1)$, so P_0 is true

Assume P_k is true, so: $2^{k+1} + (-1)^k = 3A$, for some natural number A. Then:

$$\begin{aligned}2^{k+2} + (-1)^{k+1} &= 2 \bullet 2^{k+1} - (-1)^k \\ &= 2\left[3A - (-1)^k\right] - (-1)^k \\ &= 6A - 2(-1)^k - (-1)^k \\ &= 6A - 3(-1)^k \\ &= 3\left[2A - (-1)^k\right]\end{aligned}$$

Since 3 is a factor of this expression, P_{k+1} is true and the induction is complete.

41. Let P_n denote: $x^n - y^n = (x-y)\left(Q(x,y)\right)$ where $Q(x,y)$ is a polynomial in x and y.

Then: $x - y = (x-y)(1)$, so P_1 is true

While using the suggested identity: $x^{k+1} - y^{k+1} = x^k(x-y) + (x^k - y^k)y$

So assuming P_k, we have $x^k - y^k = (x-y)Q(x,y)$: $x^{k+1} - y^{k+1} = \left(x^k + yQ(x,y)\right)(x-y)$

So P_{k+1} is true and the induction is complete.

43. For $n = 1$: $(1+p) \geq (1+p)$, so P_1 is true. When $p > -1$ then $p + 1 > 0$. Assuming $(1+p)^k \geq 1 + kp$ then:

$$(1+p)^{k+1} \geq (1+kp)(1+p) \geq 1 + p + kp + kp^2 = 1 + (k+1)p + kp^2 \geq 1 + (k+1)p$$

The induction is complete.

13.2 The Binomial Theorem

1. **a.** Compute: $(a+b)^2 = (a+b)(a+b) = a^2 + ab + ba + b^2 = a^2 + 2ab + b^2$

b. Compute:

$$\begin{aligned}(a+b)^3 &= (a+b)(a+b)^2 \\ &= (a+b)\left(a^2 + 2ab + b^2\right) \\ &= a^3 + 2a^2b + ab^2 + ba^2 + 2ab^2 + b^3 \\ &= a^3 + 3a^2b + 3ab^2 + b^3\end{aligned}$$

3. $5! = (5)(4)(3)(2)(1) = 120$

5. Compute: $\binom{7}{3}\binom{3}{2} = \frac{7!}{3!(4!)} \bullet \frac{3!}{2!(1!)} = \frac{7 \bullet 6 \bullet 5}{3 \bullet 2} \bullet 3 = 105$

7. **a.** Compute: $\binom{5}{3} = \frac{5!}{3!(5-3)!} = \frac{5(4)}{2} = 10$ **b.** Compute: $\binom{5}{4} = \frac{5!}{4!(5-4)!} = 5$

9. Simplify: $\frac{(n+2)!}{n!} = \frac{(n+2)(n+1)n!}{n!} = n^2 + 3n + 2$

11. Compute: $\binom{6}{4} + \binom{6}{3} - \binom{7}{4} = \frac{6!}{4!2!} + \frac{6!}{3!3!} - \frac{7!}{4!3!} = \frac{(6)(5)}{2} + \frac{(6)(5)(4)}{6} - \frac{(7)(6)(5)}{6} = 15 + 20 - 35 = 0$

13. Expand using the binomial theorem:

$$(a+b)^9 = \binom{9}{0}a^9 + \binom{9}{1}a^8 b + \binom{9}{2}a^7 b^2 + \binom{9}{3}a^6 b^3 + \binom{9}{4}a^5 b^4 + \binom{9}{5}a^4 b^5 + \binom{9}{6}a^3 b^6 + \binom{9}{7}a^2 b^7 + \binom{9}{8}ab^8 + \binom{9}{9}b^9$$
$$= a^9 + 9a^8 b + 36a^7 b^2 + 84a^6 b^3 + 126a^5 b^4 + 126a^4 b^5 + 84a^3 b^6 + 36a^2 b^7 + 9ab^8 + b^9$$

15. Expand using the binomial theorem:

$$(2A+B)^3 = \binom{3}{0}(2A)^3 + \binom{3}{1}(2A)^2 B + \binom{3}{2}(2A)B^2 + \binom{3}{3}B^3 = 8A^3 + 12A^2 B + 6AB^2 + B^3$$

17. Expand using the binomial theorem:

$$(1-2x)^6 = \binom{6}{0}1^6 - \binom{6}{1}1^5(2x) + \binom{6}{2}1^4(2x)^2 - \binom{6}{3}1^3(2x)^3 + \binom{6}{4}1^2(2x)^4 - \binom{6}{5}1(2x)^5 + \binom{6}{6}(2x)^6$$
$$= 1 - 12x + 60x^2 - 160x^3 + 240x^4 - 192x^5 + 64x^6$$

19. Expand using the binomial theorem:

$$\left(\sqrt{x}+\sqrt{y}\right)^4 = \binom{4}{0}\left(\sqrt{x}\right)^4 + \binom{4}{1}\left(\sqrt{x}\right)^3\left(\sqrt{y}\right) + \binom{4}{2}\left(\sqrt{x}\right)^2\left(\sqrt{y}\right)^2 + \binom{4}{3}\left(\sqrt{x}\right)\left(\sqrt{y}\right)^3 + \binom{4}{4}\left(\sqrt{y}\right)^4$$
$$= x^2 + 4x\sqrt{xy} + 6xy + 4y\sqrt{xy} + y^2$$

21. Expand using the binomial theorem:

$$\left(x^2+y^2\right)^5 = \binom{5}{0}\left(x^2\right)^5 + \binom{5}{1}\left(x^2\right)^4 y^2 + \binom{5}{2}\left(x^2\right)^3\left(y^2\right)^2 + \binom{5}{3}\left(x^2\right)^2\left(y^2\right)^3 + \binom{5}{4}\left(x^2\right)\left(y^2\right)^4 + \binom{5}{5}\left(y^2\right)^5$$
$$= x^{10} + 5x^8 y^2 + 10x^6 y^4 + 10x^4 y^6 + 5x^2 y^8 + y^{10}$$

23. Expand using the binomial theorem:

$$\left(1-\frac{1}{x}\right)^6 = \binom{6}{0} - \binom{6}{1}\frac{1}{x} + \binom{6}{2}\left(\frac{1}{x}\right)^2 - \binom{6}{3}\left(\frac{1}{x}\right)^3 + \binom{6}{4}\left(\frac{1}{x}\right)^4 - \binom{6}{5}\left(\frac{1}{x}\right)^5 + \binom{6}{6}\left(\frac{1}{x}\right)^6$$
$$= 1 - \frac{6}{x} + \frac{15}{x^2} - \frac{20}{x^3} + \frac{15}{x^4} - \frac{6}{x^5} + \frac{1}{x^6}$$

25. Expand using the binomial theorem:

$$\left(\frac{x}{2}-\frac{y}{3}\right)^3 = \binom{3}{0}\left(\frac{x}{2}\right)^3 - \binom{3}{1}\left(\frac{x}{2}\right)^2\left(\frac{y}{3}\right) + \binom{3}{2}\left(\frac{x}{2}\right)\left(\frac{y}{3}\right)^2 - \binom{3}{3}\left(\frac{y}{3}\right)^3 = \frac{x^3}{8} - \frac{x^2 y}{4} + \frac{xy^2}{6} - \frac{y^3}{27}$$

27. Expand using the binomial theorem:

$$\left(ab^2+c\right)^7 = \binom{7}{0}\left(ab^2\right)^7 + \binom{7}{1}\left(ab^2\right)^6 c + \binom{7}{2}\left(ab^2\right)^5 c^2 + \binom{7}{3}\left(ab^2\right)^4 c^3 + \binom{7}{4}\left(ab^2\right)^3 c^4 + \binom{7}{5}\left(ab^2\right)^2 c^5 + \binom{7}{6}\left(ab^2\right)c^6 + \binom{7}{7}c^7$$
$$= a^7 b^{14} + 7a^6 b^{12} c + 21a^5 b^{10} c^2 + 35a^4 b^8 c^3 + 35a^3 b^6 c^4 + 21a^2 b^4 c^5 + 7ab^2 c^6 + c^7$$

29. Expand using the binomial theorem:

$$\left(x+\sqrt{2}\right)^8 = \binom{8}{0}x^8 + \binom{8}{1}x^7\sqrt{2} + \binom{8}{2}x^6\left(\sqrt{2}\right)^2 + \binom{8}{3}x^5\left(\sqrt{2}\right)^3 + \binom{8}{4}x^4\left(\sqrt{2}\right)^4$$
$$+\binom{8}{5}x^3\left(\sqrt{2}\right)^5 + \binom{8}{6}x^2\left(\sqrt{2}\right)^6 + \binom{8}{7}x\left(\sqrt{2}\right)^7 + \binom{8}{8}\left(\sqrt{2}\right)^8$$
$$= x^8 + 8\sqrt{2}x^7 + 56x^6 + 112\sqrt{2}x^5 + 280x^4 + 224\sqrt{2}x^3 + 224x^2 + 64\sqrt{2}x + 16$$

31. Expand using the binomial theorem:

$$\left(\sqrt{2}-1\right)^3 = \binom{3}{0}\left(\sqrt{2}\right)^3 - \binom{3}{1}\left(\sqrt{2}\right)^2 + \binom{3}{2}\left(\sqrt{2}\right) - \binom{3}{3} = 2\sqrt{2} - 6 + 3\sqrt{2} - 1 = 5\sqrt{2} - 7$$

33. Expand using the binomial theorem:

$$\left(\sqrt{2}+\sqrt{3}\right)^5 = \binom{5}{0}\left(\sqrt{2}\right)^5 + \binom{5}{1}\left(\sqrt{2}\right)^4\sqrt{3} + \binom{5}{2}\left(\sqrt{2}\right)^3\left(\sqrt{3}\right)^2 + \binom{5}{3}\left(\sqrt{2}\right)^2\left(\sqrt{3}\right)^3 + \binom{5}{4}\left(\sqrt{2}\right)\left(\sqrt{3}\right)^4 + \binom{5}{5}\left(\sqrt{3}\right)^5$$
$$= 4\sqrt{2} + 20\sqrt{3} + 60\sqrt{2} + 60\sqrt{3} + 45\sqrt{2} + 9\sqrt{3}$$
$$= 89\sqrt{3} + 109\sqrt{2}$$

35. Expand using the binomial theorem:

$$\left(2\sqrt[3]{2}-\sqrt[3]{4}\right)^3 = \binom{3}{0}\left(2\sqrt[3]{2}\right)^3 - \binom{3}{1}\left(2\sqrt[3]{2}\right)^2\left(\sqrt[3]{4}\right) + \binom{3}{2}\left(2\sqrt[3]{2}\right)\left(\sqrt[3]{4}\right)^2 - \binom{3}{3}\left(\sqrt[3]{4}\right)^3$$
$$= 16 - 12\sqrt[3]{16} + 6\sqrt[3]{32} - 4$$
$$= 12 - 24\sqrt[3]{2} + 12\sqrt[3]{4}$$

37. Expand using the binomial theorem:

$$\left[x^2-(2x+1)\right]^5$$
$$= \binom{5}{0}\left(x^2\right)^5 - \binom{5}{1}\left(x^2\right)^4(2x+1) + \binom{5}{2}\left(x^2\right)^3(2x+1)^2 - \binom{5}{3}\left(x^2\right)^2(2x+1)^3$$
$$+\binom{5}{4}\left(x^2\right)(2x+1)^4 - \binom{5}{5}(2x+1)^5$$
$$= x^{10} - \left[5x^8(2x+1)\right] + \left[10x^6\left(4x^2+4x+1\right)\right] - \left[10x^4\left(8x^3+12x^2+6x+1\right)\right]$$
$$+\left[5x^2\left(16x^4+32x^3+24x^2+8x+1\right)\right] - \left(32x^5+80x^4+80x^3+40x^2+10x+1\right)$$
$$= x^{10} - \left(10x^9+5x^8\right) + \left(40x^8+40x^7+10x^6\right) - \left(80x^7+120x^6+60x^5+10x^4\right)$$
$$+\left(80x^6+160x^5+120x^4+40x^3+5x^2\right) - \left(32x^5+80x^4+80x^3+40x^2+10x+1\right)$$
$$= x^{10} - 10x^9 + 35x^8 - 40x^7 - 30x^6 + 68x^5 + 30x^4 - 40x^3 - 35x^2 - 10x - 1$$

39. The fifteenth term will be given by: $\binom{16}{14}a^2b^{14} = 120a^2b^{14}$

41. The one-hundredth term will be given by: $\binom{100}{99}x^{99} = 100x^{99}$

43. Here $n - r + 1 = 8$, so $10 - r + 1 = 8$, so $r = 3$. Hence the coefficient is: $\binom{10}{2} = 45$

45. Here $r - 1 = 8$, thus $r = 9$. Hence the coefficient is: $\frac{1}{2}(-4)^8\binom{9}{8} = 9 \bullet \frac{1}{2}(65536) = 294912$

47. Here $r - 1 = 6$. Hence $r = 7$, and the coefficient is: $\binom{8}{6} = \frac{8(7)}{2} = 28$

49. Here $(12 - r + 1)(-1) + 2(r - 1) = 0$, so $-12 + r - 1 + 2r - 2 = 0$, then $3r = 15$, so $r = 5$ and the coefficient is:

$$(3)^4\binom{12}{4} = \frac{(12)(11)(10)(9)}{(4)(3)(2)}(3)^4 = 495(81) = 40095$$

51. Here $r - 1 = n$, so $r = n + 1$, so the coefficient is: $\binom{2n}{n} = \dfrac{(2n)!}{n!n!} = \dfrac{(2n)!}{(n!)^2}$

53. **a.** Complete the table:

k	0	1	2	3	4	5	6	7	8
$\binom{8}{k}$	1	8	28	56	70	56	28	8	1

b. We have $1 + 8 + 28 + 56 + 70 + 56 + 28 + 8 + 1 = 256$, while $2^8 = 256$.

c. Since a^L and b^L equal 1 for any L because $a = b = 1$, from the binomial theorem:

$$2^n = (1+1)^n = \binom{n}{0} + \binom{n}{1} + \ldots + \binom{n}{n}$$

55. **a.** Simplify: $(1+x)^n\left(1+\dfrac{1}{x}\right)^n = (1+x)^n\left(\dfrac{x+1}{x}\right)^n = (1+x)^n \bullet \dfrac{(1+x)^n}{x^n} = \dfrac{(1+x)^{2n}}{x^n}$

b. The nth term of expansion on the right is: $\dfrac{\binom{2n}{n}(1)^n(x)^n}{x^n} = \binom{2n}{n}$

This verifies the result.

c. Using the binomial theorem:

$$(1+x)^n = \binom{n}{0} + \binom{n}{1}x + \binom{n}{2}x^2 + \binom{n}{3}x^3 + \ldots + \binom{n}{n}x^n$$

$$\left(1+\frac{1}{x}\right)^n = \binom{n}{0} + \binom{n}{1}x^{-1} + \binom{n}{2}x^{-2} + \binom{n}{3}x^{-3} + \ldots + \binom{n}{n}x^{-n}$$

When multiplied, the terms that do not contain x come from terms with corresponding positive and negative exponents: $\binom{n}{1}x \bullet \binom{n}{1}x^{-1} + \binom{n}{2}x^2 \bullet \binom{n}{2}x^{-2} + \ldots + \binom{n}{n}x^n \bullet \binom{n}{n}x^{-n} = \binom{n}{1}^2 + \binom{n}{2}^2 + \ldots + \binom{n}{n}^2$

This verifies the identity.

13.3 Introduction to Sequences and Series

1. Since $a_n = \dfrac{n}{n+1}$, we can write out the first four terms by setting n equal to the natural numbers 1 through 4:

$$a_1 = \frac{1}{1+1} = \frac{1}{2},\quad a_2 = \frac{2}{2+1} = \frac{2}{3},\quad a_3 = \frac{3}{3+1} = \frac{3}{4},\quad a_4 = \frac{4}{4+1} = \frac{4}{5}$$

So the first four terms are $\frac{1}{2}, \frac{2}{3}, \frac{3}{4}, \frac{4}{5}$.

3. Here $b_n = (-1)^n$, so $b_1 = -1$, $b_2 = 1$, $b_3 = -1$, and $b_4 = 1$. The first four terms are $-1, 1, -1, 1$.

5. Here $c_n = 2^{-n}$, so: $c_1 = 2^{-1} = \frac{1}{2}$, $c_2 = 2^{-2} = \frac{1}{4}$, $c_3 = 2^{-3} = \frac{1}{8}$, $c_4 = 2^{-4} = \frac{1}{16}$

The first four terms are $\frac{1}{2}, \frac{1}{4}, \frac{1}{8}, \frac{1}{16}$.

7. Here $x_n = 3n$, so: $x_1 = 3(1) = 3$, $x_2 = 3(2) = 6$, $x_3 = 3(3) = 9$, $x_4 = 3(4) = 12$

The first four terms are 3, 6, 9, 12.

9. Here $b_n = \left(1+\frac{1}{n}\right)^n$, so: $b_1 = \left(1+\frac{1}{1}\right)^1 = 2$, $b_2 = \left(1+\frac{1}{2}\right)^2 = \frac{9}{4}$, $b_3 = \left(1+\frac{1}{3}\right)^3 = \frac{64}{27}$, $b_4 = \left(1+\frac{1}{4}\right)^4 = \frac{625}{256}$

The first four terms are $2, \frac{9}{4}, \frac{64}{27}, \frac{625}{256}$.

11. Here $a_n = \frac{n-1}{n+1}$, so: $a_0 = \frac{0-1}{0+1} = -1$, $a_1 = \frac{1-1}{1+1} = 0$, $a_2 = \frac{2-1}{2+1} = \frac{1}{3}$, $a_3 = \frac{3-1}{3+1} = \frac{1}{2}$

The first four terms are $-1, 0, \frac{1}{3}, \frac{1}{2}$.

13. Here $b_n = \frac{(-2)^{n+1}}{(n+1)^2}$, so: $b_0 = \frac{(-2)^{0+1}}{(0+1)^2} = -2$, $b_1 = \frac{(-2)^{1+1}}{(1+1)^2} = 1$, $b_2 = \frac{(-2)^{2+1}}{(2+1)^2} = -\frac{8}{9}$, $b_3 = \frac{(-2)^{3+1}}{(3+1)^2} = 1$

The first four terms are $-2, 1, -\frac{8}{9}, 1$.

15. Here $a_1 = 1$ and $a_n = \left(1 + a_{n-1}\right)^2$ for $n \ge 2$, so:

$$a_1 = 1 \qquad a_2 = \left(1 + a_1\right)^2 = 2^2 = 4$$
$$a_3 = \left(1 + a_2\right)^2 = 5^2 = 25 \qquad a_4 = \left(1 + a_3\right)^2 = 26^2 = 676$$
$$a_5 = \left(1 + a_4\right)^2 = 677^2 = 458329$$

The first five terms are 1, 4, 25, 676, 458329.

17. Here $a_1 = 2$, $a_2 = 2$, and $a_n = a_{n-1}a_{n-2}$ for $n \ge 3$, so $a_1 = 2$, $a_2 = 2$, $a_3 = 4$, $a_4 = 8$, and $a_5 = 32$. The first five terms are 2, 2, 4, 8, 32.

19. Here $a_1 = 1$, $a_{n+1} = na_n$ for $n \ge 1$, so $a_1 = 1$, $a_2 = 1$, $a_3 = 2$, $a_4 = 6$, and $a_5 = 24$. The first five terms are 1, 1, 2, 6, 24.

21. Here $a_1 = 0$, $a_n = 2^{a_{n-1}}$ for $n \ge 2$, so $a_1 = 0$, $a_2 = 1$, $a_3 = 2$, $a_4 = 4$, and $a_5 = 16$. The first five terms are 0, 1, 2, 4, 16.

23. First compute the required values of a_n as shown in the table:

n	0	1	2	3	4
a_n	1	$1\frac{1}{2}$	2	$2\frac{1}{2}$	3

Now graph these values:

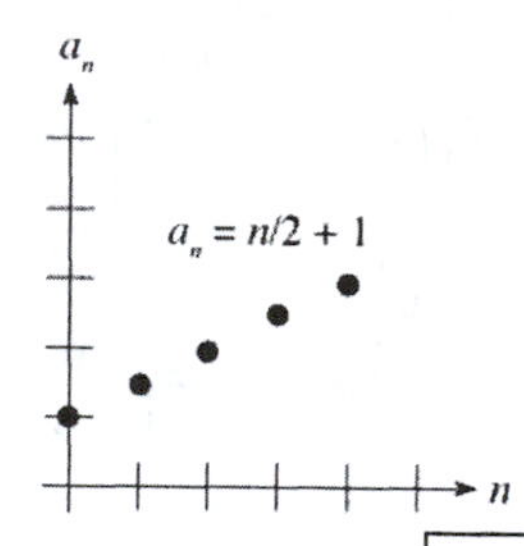

25. First compute the required values of c_n as shown in the table:

n	1	2	3	4	5
c_n	5	$2\frac{1}{2}$	$1\frac{2}{3}$	$1\frac{1}{4}$	1

Now graph these values:

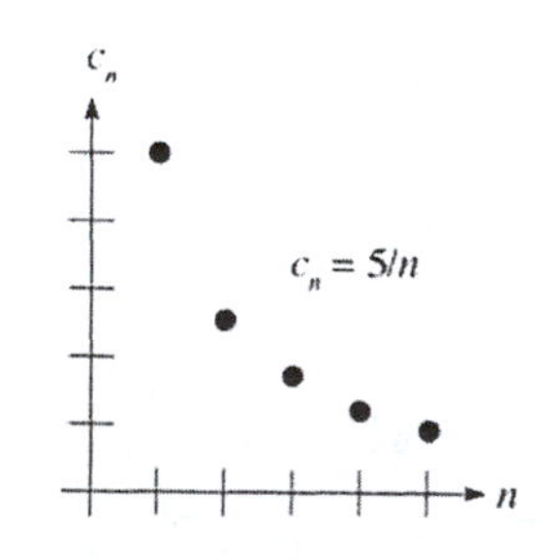

27. First compute the required values of a_n as shown in the table:

n	1	2	3	4	5
a_n	1	0	0	0	0

Now graph these values:

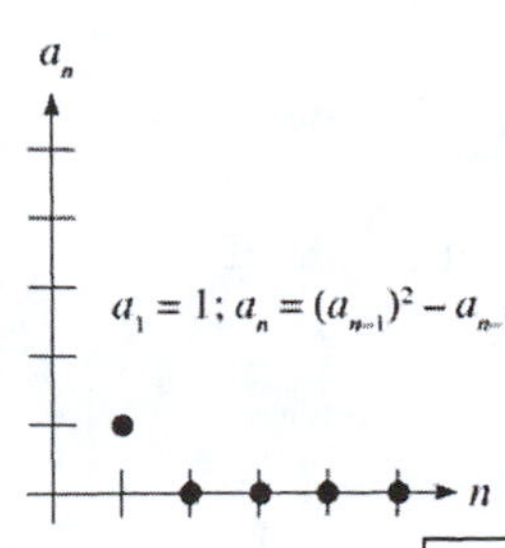

29. First compute the required values of b_n as shown in the table:

n	0	1	2	3	4
b_n	2	−1	2	−1	2

Now graph these values:

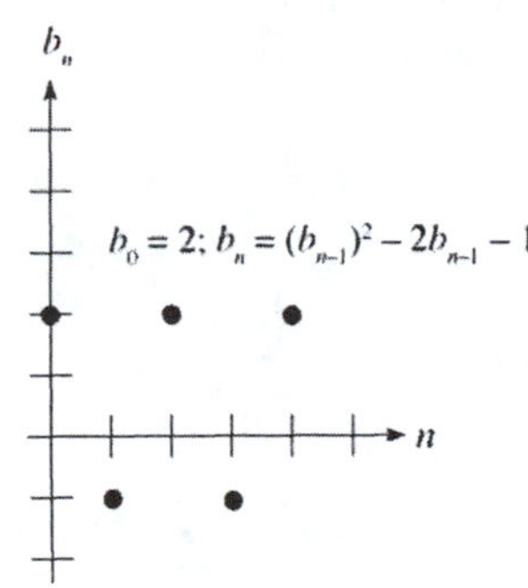

31. First compute the required values of A_n as shown in the table:

n	0	1	2	3	4
A_n	0	−3	3	0	−3

Now graph these values:

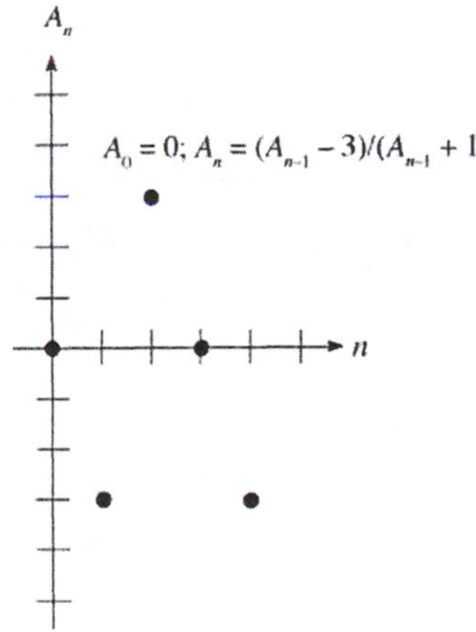

33. Since $a_n = 2^n$, the sum is $2 + 4 + 8 + 16 + 32 = 62$.

35. Since $a_n = n^2 - n$, the sum is $0 + 2 + 6 + 12 + 20 = 40$.

37. Since $a_n = \dfrac{(-1)^n}{n!}$, the sum is $-1 + \frac{1}{2} - \frac{1}{6} + \frac{1}{24} - \frac{1}{120} = -\frac{19}{30}$.

39. Here $a_1 = 1$, $a_2 = 2$, and $a_n = a_{n-1}^2 + a_{n-2}^2$ for $n \geq 3$, so the sum is $1 + 2 + 5 + 29 + 866 = 903$.

41. Expand the sum: $\sum_{k=1}^{3}(k-1) = 0 + 1 + 2 = 3$

43. Expand the sum: $\sum_{k=4}^{5} k^2 = 16 + 25 = 41$

45. Expand the sum: $\sum_{n=1}^{3} x^n = x + x^2 + x^3$

47. Expand the sum: $\sum_{n=1}^{4} \frac{1}{n} = 1 + \frac{1}{2} + \frac{1}{3} + \frac{1}{4} = \frac{25}{12}$

49. Expand the sum:

$$\begin{aligned}\sum_{j=1}^{9} \log_{10} \frac{j}{j+1} &= \log_{10} \tfrac{1}{2} + \log_{10} \tfrac{2}{3} + \log_{10} \tfrac{3}{4} + \log_{10} \tfrac{4}{5} + \log_{10} \tfrac{5}{6} + \log_{10} \tfrac{6}{7} + \log_{10} \tfrac{7}{8} + \log_{10} \tfrac{8}{9} + \log_{10} \tfrac{9}{10} \\ &= \log_{10} \left(\tfrac{1}{2} \bullet \tfrac{2}{3} \bullet \tfrac{3}{4} \bullet \tfrac{4}{5} \bullet \tfrac{5}{6} \bullet \tfrac{6}{7} \bullet \tfrac{7}{8} \bullet \tfrac{8}{9} \bullet \tfrac{9}{10} \right) \\ &= \log_{10} \tfrac{1}{10} \\ &= -1\end{aligned}$$

51. Expand the sum: $\sum_{j=1}^{6} \left(\frac{1}{j} - \frac{1}{j+1} \right) = \left(1 - \frac{1}{2}\right) + \left(\frac{1}{2} - \frac{1}{3}\right) + \left(\frac{1}{3} - \frac{1}{4}\right) + \left(\frac{1}{4} - \frac{1}{5}\right) + \left(\frac{1}{5} - \frac{1}{6}\right) + \left(\frac{1}{6} - \frac{1}{7}\right) = 1 - \frac{1}{7} = \frac{6}{7}$

53. Write the sum in sigma notation: $5 + 5^2 + 5^3 + 5^4 = \sum_{j=1}^{4} 5^j$

55. Write the sum in sigma notation: $x + x^2 + x^3 + x^4 + x^5 + x^6 = \sum_{j=1}^{6} x^j$

57. Write the sum in sigma notation: $1 + \frac{1}{2} + \frac{1}{3} + \ldots + \frac{1}{12} = \sum_{k=1}^{12} \frac{1}{k}$

59. Write the sum in sigma notation: $2 - 2^2 + 2^3 - 2^4 + 2^5 = \sum_{j=1}^{5} (-1)^{j+1} 2^j$

61. Write the sum in sigma notation: $1 - 2 + 3 - 4 + 5 = \sum_{j=1}^{5} (-1)^{j+1} j$

63. **a.** The first six terms are given by:

$s_1 = 0.7$ $\qquad$ $s_2 = (s_1)^2 = (0.7)^2 = 0.49$

$s_3 = (s_2)^2 = (0.49)^2 = 0.2401$ $\qquad$ $s_4 = (s_3)^2 = (0.2401)^2 \approx 0.05765$

$s_5 = (s_4)^2 \approx (0.05765)^2 \approx 0.00332$ $\qquad$ $s_6 = (s_5)^2 \approx (0.00332)^2 \approx 0.00001$

The answers appear to be approaching 0.

b. $s_{10} \approx 4.90 \times 10^{-80}$

c. The value of s_{100} would be virtually identical to 0.

65. **a.** First compute the required values of P_t as shown in the table:

t	0	1	2	3	4	5
P_t	300	1111	1829	1468	1691	1559

Sketching the graph:

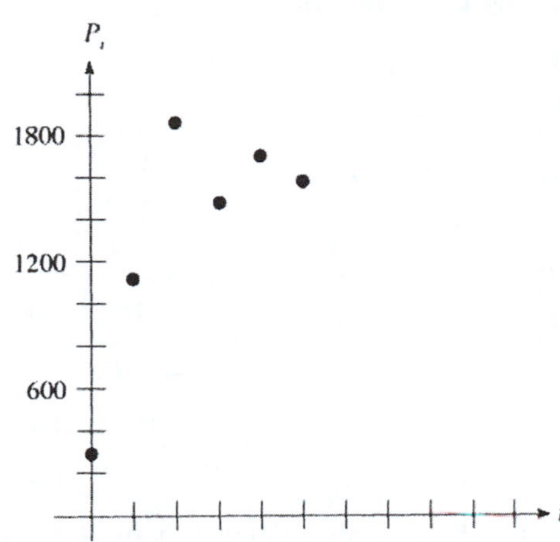

The population seems to be oscillating closer to a value near 1600.

b. Compute the required values of P_t as shown in the tables:

t	0	1	2	3	4	5	6	7	8	9	10
P_t	300	1111	1829	1468	1691	1559	1640	1591	1621	1602	1614

t	10	11	12	13	14	15	16	17	18	19	20
P_t	1614	1607	1611	1608	1610	1609	1610	1609	1610	1609	1610

Sketching the graph:

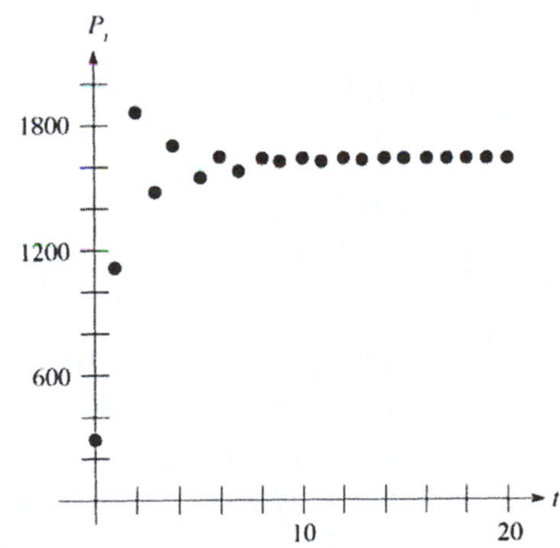

The population seems to be approaching an equilibrium population of 1609-1610.

c. Solving the equation:

$$P_{t-1} = 5P_{t-1}e^{-P_{t-1}/1000}$$
$$1 = 5e^{-P_{t-1}/1000}$$
$$0.2 = e^{-P_{t-1}/1000}$$
$$-P_{t-1}/1000 = \ln 0.2$$
$$-P_{t-1} = 1000\ln 0.2$$
$$P_{t-1} = -1000\ln 0.2 \approx 1609$$

67. **a.** Completing the table:

F_1	F_2	F_3	F_4	F_5	F_6	F_7	F_8	F_9	F_{10}
1	1	2	3	5	8	13	21	34	55

b. Since $F_{22} = F_{20} + F_{21}$, $F_{22} = 6,765 + 10,946 = 17,711$. Also:

$$\begin{aligned} F_{19} + F_{20} &= F_{21} \\ F_{19} + 6765 &= 10946 \\ F_{19} &= 4181 \end{aligned}$$

c. We have:

$$\begin{aligned} F_{29} + F_{30} &= F_{31} \\ 514,229 + F_{30} &= 1,346,269 \\ F_{30} &= 832,040 \end{aligned}$$

69. **a.** For $n = 1$: $F_1 = 1$ $\qquad F_3 - 1 = 2 - 1 = 1$

For $n = 2$: $F_1 + F_2 = 1 + 1 = 2$ $\qquad F_4 - 1 = 3 - 1 = 2$

For $n = 3$: $F_1 + F_2 + F_3 = 1 + 1 + 2 = 4$ $\qquad F_5 - 1 = 5 - 1 = 4$

b. The values completed in part **a** show the hypothesis to be true for $n = 1$. Assuming: $F_1 + F_2 + \ldots + F_k = F_{k+2} - 1$

Therefore: $F_1 + F_2 + \ldots + F_k + F_{k+1} = F_{k+2} - 1 + F_{k+1} = F_{k+3} - 1 = F_{(k+1)+2} - 1$

This completes the induction.

71. **a.** For $n = 1$: $F_2^2 = 1^2 = 1$ $\qquad F_1F_3 + (-1)^1 = 1 \bullet 2 - 1 = 1$

For $n = 2$: $F_3^2 = 2^2 = 4$ $\qquad F_2F_4 + (-1)^2 = 1 \bullet 3 + 1 = 4$

For $n = 3$: $F_4^2 = 3^2 = 9$ $\qquad F_3F_5 + (-1)^3 = 2 \bullet 5 - 1 = 9$

b. Since $F_2^2 = 1^2 = 1 = F_1F_3 + (-1)^1 = 2 - 1 = 1$, the hypothesis is true for $n = 1$. Assume: $F_{k+1}^2 = F_kF_{k+2} + (-1)^k$

Then:

$$\begin{aligned} F_{k+1}F_{k+2} + F_{k+1}^2 &= F_kF_{k+2} + F_{k+1}F_{k+2} + (-1)^k \\ F_{k+1}\left(F_{k+2} + F_{k+1}\right) &= F_{k+2}\left(F_k + F_{k+1}\right) + (-1)^k \\ F_{k+1}F_{k+3} &= F_{k+2}^2 + (-1)^k \\ F_{k+2}^2 &= F_{k+1}F_{k+3} + (-1)^{k+1} \end{aligned}$$

This completes the induction.

73. **a.** For $n = 1$: $F_1 = \dfrac{\left(1+\sqrt{5}\right)^1 - \left(1-\sqrt{5}\right)^1}{2^1\sqrt{5}} = \dfrac{1+\sqrt{5}-1+\sqrt{5}}{2\sqrt{5}} = \dfrac{2\sqrt{5}}{2\sqrt{5}} = 1$

For $n = 2$: $F_2 = \dfrac{\left(1+\sqrt{5}\right)^2 - \left(1-\sqrt{5}\right)^2}{2^2\sqrt{5}} = \dfrac{\left(6+2\sqrt{5}\right)-\left(6-2\sqrt{5}\right)}{4\sqrt{5}} = \dfrac{4\sqrt{5}}{4\sqrt{5}} = 1$

b. Substituting $n = 24$ and $n = 25$:

$$F_{24} = \frac{\left(1+\sqrt{5}\right)^{24} - \left(1-\sqrt{5}\right)^{24}}{2^{24}\sqrt{5}} = 46,368 \qquad F_{25} = \frac{\left(1+\sqrt{5}\right)^{25} - \left(1-\sqrt{5}\right)^{25}}{2^{25}\sqrt{5}} = 75,025$$

c. Using the formula: $F_{26} = \dfrac{\left(1+\sqrt{5}\right)^{26} - \left(1-\sqrt{5}\right)^{26}}{2^{26}\sqrt{5}} = 121,393$

Checking using the results from part **b**: $F_{24} + F_{25} = 46,368 + 75,025 = 121,393$

75. **a.** Subtracting the two equations:

$$\alpha^n - \beta^n = (F_n\alpha + F_{n-1}) - (F_n\beta + F_{n-1})$$
$$\alpha^n - \beta^n = F_n\alpha - F_n\beta$$
$$\alpha^n - \beta^n = F_n(\alpha - \beta)$$
$$F_n = \frac{\alpha^n - \beta^n}{\alpha - \beta}$$

b. Using the quadratic formula to solve $x^2 - x - 1 = 0$: $x = \frac{-(-1) \pm \sqrt{(-1)^2 - 4(1)(-1)}}{2(1)} = \frac{1 \pm \sqrt{1+4}}{2} = \frac{1 \pm \sqrt{5}}{2}$

Choosing α to be the larger root yields $\alpha = \frac{1+\sqrt{5}}{2}$ and $\beta = \frac{1-\sqrt{5}}{2}$.

c. Substituting these values for α and β:

$$F_n = \frac{\alpha^n - \beta^n}{\alpha - \beta} = \frac{\left(\frac{1+\sqrt{5}}{2}\right)^n - \left(\frac{1-\sqrt{5}}{2}\right)^n}{\frac{1+\sqrt{5}}{2} - \frac{1-\sqrt{5}}{2}} = \frac{\frac{(1+\sqrt{5})^n}{2^n} - \frac{(1-\sqrt{5})^n}{2^n}}{\frac{1+\sqrt{5}-1+\sqrt{5}}{2}} = \frac{(1+\sqrt{5})^n - (1-\sqrt{5})^n}{2^n\sqrt{5}}$$

d. For $n = 1$, the equation states: $F_1 = \frac{(1+\sqrt{5})^1 - (1-\sqrt{5})^1}{2^1\sqrt{5}} = \frac{1+\sqrt{5}-1+\sqrt{5}}{2\sqrt{5}} = \frac{2\sqrt{5}}{2\sqrt{5}} = 1$

Thus the equation holds for $n \geq 1$.

13.4 Arithmetic Sequences and Series

1. **a.** To find the common difference d, subtract any term from the succeeding term. Here that can be:

$3 - 1 = 2 \qquad 5 - 3 = 2 \qquad 7 - 5 = 2$

So 2 is the required common difference.

b. Again subtract terms: $6 - 10 = -4 \qquad 2 - 5 = -4 \qquad -2 - 2 = -4$

The common difference is −4. It is not necessary to try all three pairs, but this helps to verify that we have an arithmetic sequence.

c. Subtract: $1 - \frac{2}{3} = \frac{4}{3} - 1 = \frac{5}{3} - \frac{4}{3} = \frac{1}{3}$. So $\frac{1}{3}$ is the common difference.

d. Subtract: $1 + \sqrt{2} - 1 = \sqrt{2}$ or $(1 + 2\sqrt{2}) - (1 + \sqrt{2}) = \sqrt{2}$. Here $\sqrt{2}$ is the common difference.

3. Since $a = 10$ and $d = 11$, using $a_n = a + (n-1)d$ results in: $a_{12} = 10 + (12-1)11 = 131$

5. Finding the required term: $a_{100} = 6 + (100 - 1)(5) = 6 + 495 = 501$

7. Finding the required term: $a_{1000} = -1 + (1000 - 1)(1) = 998$

9. Here $a_4 = -6$ and $a_{10} = 5$, thus:

$$-6 = a + 3d$$
$$5 = a + 9d$$

Subtracting the second equation from the first yields:

$$-6d = -11$$
$$d = \frac{11}{6}$$

Therefore: $a = 5 - 9\left(\frac{11}{6}\right) = 5 - \frac{33}{2} = -\frac{23}{2}$

11. Here $a_{60} = 105$ and $d = 5$, so:

$$105 = a + 59(5)$$
$$105 = a + 295$$
$$a = -190$$

13. Since $a_{15} = a + 14d$ and $a_7 = a + 6d$:

$$a_{15} - a_7 = 8d$$
$$-1 = 8d$$
$$d = -\frac{1}{8}$$

15. Since $a = 1$ and $d = 1$: $S_{1000} = \frac{1000}{2}(2 + 999) = 500500$

17. For $\frac{\pi}{3}+\frac{2\pi}{3}+\pi+\frac{4\pi}{3}+\ldots+\frac{13\pi}{3}$, $a=\frac{\pi}{3}$, $d=\frac{\pi}{3}$, so the sum is: $S_{13}=\frac{13}{2}\left(\frac{2\pi}{3}+\frac{12\pi}{3}\right)=\frac{14(13)(\pi)}{6}=\frac{91\pi}{3}$

19. Since $d=5$ and $S_{38}=3534$:

$$\begin{aligned}3534&=\tfrac{38}{2}[2a+37(5)]\\3534&=38a+3515\\19&=38a\\a&=\tfrac{1}{2}\end{aligned}$$

21. Here $a=4$, $a_{16}=-100$, therefore: $S_{16}=16\left(\frac{4-100}{2}\right)=-768$. Therefore:

$$\begin{aligned}-768&=\tfrac{16}{2}(2(4)+15d)\\-768&=64+120d\\-832&=120d\\d&=-\tfrac{104}{15}\end{aligned}$$

23. Since $a_8=5$ and $S_{10}=20$:

$$\begin{aligned}5&=a+7d\\20&=5(2a+9d)\\4&=2a+9d\end{aligned}$$

Solve the system:

$$\begin{aligned}a+7d&=5\\2a+9d&=4\end{aligned}$$

Multiplying the first equation by –2 and adding to the second equation:

$$\begin{aligned}-2a-14d&=-10\\2a+9d&=4\\-5d&=-6\\d&=\tfrac{6}{5}\end{aligned}$$

Therefore: $a=5-7\left(\frac{6}{5}\right)=-\frac{17}{5}$

25. Here $S=S_{20}=\sum_{k=1}^{20}(4k+3)$, thus $a=7$ and $a_{20}=83$. So $S=20\left(\frac{7+83}{2}\right)=900$.

27. Let x denote the middle term, so the three terms are $x-d, x, x+d$:

$$\begin{aligned}x-d+x+x+d&=30\\3x&=30\\x&=10\end{aligned}$$

Therefore:

$$\begin{aligned}x(x-d)(x+d)&=360\\10\left(100-d^2\right)&=360\\100-d^2&=36\\-d^2&=-64\\d&=\pm 8\end{aligned}$$

The terms are 2, 10, 18 or 18, 10, 2.

29. Using equations as in Exercise 27: $3x=6$, so $x=2$

$$\begin{aligned}(x-d)^3+x^3+(x+d)^3&=132\\(2-d)^3+8+(2+d)^3&=132\\(2-d)^3+(2+d)^3&=124\\16+12d^2&=124\\d^2&=9\\d&=\pm 3\end{aligned}$$

The terms are –1, 2, 5 or 5, 2, –1.

31. **a.** Subtracting yields:

$$a_2 - a_1 = -1 - \frac{1}{1+\sqrt{2}} = \frac{-2-\sqrt{2}}{1+\sqrt{2}} = \frac{\left(1-\sqrt{2}\right)\left(-2-\sqrt{2}\right)}{-1} = -\sqrt{2}$$

$$a_3 - a_2 = \frac{1}{1-\sqrt{2}} + 1 = \frac{2-\sqrt{2}}{1-\sqrt{2}} = \frac{\left(2-\sqrt{2}\right)\left(1+\sqrt{2}\right)}{-1} = -\sqrt{2}$$

So $a_2 - a_1 = a_3 - a_2 = -\sqrt{2}$.

b. Using $a = \frac{1}{1+\sqrt{2}}$ and $d = -\sqrt{2}$ results in:

$$\begin{aligned}
S_6 &= 3\left[2\left(\frac{1}{1+\sqrt{2}}\right) + 5\left(-\sqrt{2}\right)\right] \\
&= 3\left(\frac{2}{1+\sqrt{2}} - \frac{10+5\sqrt{2}}{1+\sqrt{2}}\right) \\
&= 3\left(\frac{-8-5\sqrt{2}}{1+\sqrt{2}}\right) \\
&= -\frac{24+15\sqrt{2}}{1+\sqrt{2}} \\
&= \left(24+15\sqrt{2}\right)\left(1-\sqrt{2}\right) \\
&= 24 - 30 - 9\sqrt{2} \\
&= -6 - 9\sqrt{2}
\end{aligned}$$

33. Using $a = \dfrac{1}{1+\sqrt{b}}$ results in: $d = \dfrac{1}{2}\left(\dfrac{1}{1-\sqrt{b}} - \dfrac{1}{1+\sqrt{b}}\right) = \dfrac{1}{2}\left(\dfrac{2\sqrt{b}}{1-b}\right) = \dfrac{\sqrt{b}}{1-b}$

Therefore: $S_n = \dfrac{n}{2}\left[\dfrac{2}{1+\sqrt{b}} + (n-1)\dfrac{\sqrt{b}}{1-b}\right] = \dfrac{n}{2}\left[\dfrac{\left(2-2\sqrt{b}\right)+(n-1)\sqrt{b}}{1-b}\right] = \dfrac{n}{2(1-b)}\left[2+(n-3)\sqrt{b}\right]$

35. Using the given ratio:

$$\begin{aligned}
\frac{n^2}{m^2} &= \frac{\frac{n}{2}[2a+(n-1)d]}{\frac{m}{2}[2a+(m-1)d]} \\
\frac{2a+(n-1)d}{2a+(m-1)d} &= \frac{n}{m} \\
m[2a+(n-1)d] &= n[2a+(m-1)d] \\
2am + m(n-1)d &= 2an + n(m-1)d \\
2a(m-n) &= [n(m-1) - m(n-1)]d \\
d &= \frac{2a(m-n)}{m-n} \\
d &= 2a \quad (\text{assuming } m \neq n)
\end{aligned}$$

Consequently: $\dfrac{a_n}{a_m} = \dfrac{a+(n-1)(2a)}{a+(m-1)(2a)} = \dfrac{1+2n-2}{1+2m-2} = \dfrac{2n-1}{2m-1}$

37. Let the three consecutive terms be $x - d$, x, and $x + d$, then using the Pythagorean theorem:

$$\begin{aligned}
(x-d)^2 + x^2 &= (x+d)^2 \\
x^2 - 2xd + d^2 + x^2 &= x^2 + 2xd + d^2 \\
x^2 - 4xd &= 0 \\
x = 0 \quad &\text{or} \quad x = 4d
\end{aligned}$$

Since the side of the triangle cannot have zero length, $x = 4d$ and the three terms are $3d$, $4d$, and $5d$, which is clearly similar to a 3-4-5 right triangle.

39. We have $\frac{1}{b}-\frac{1}{a}=\frac{1}{c}-\frac{1}{b}$, from which it follows that $b=\frac{2ac}{a+c}$. Therefore:

$$\begin{aligned}\ln(a+c)+\ln(a-2b+c)&=\ln(a+c)+\ln\left(a-\frac{4ac}{a+c}+c\right)\\&=\ln(a+c)+\ln\left(\frac{a^2+ac-4ac+ac+c^2}{a+c}\right)\\&=\ln(a+c)+\ln\frac{(a-c)^2}{a+c}\\&=\ln(a-c)^2\\&=2\ln(a-c)\end{aligned}$$

13.5 Geometric Sequences and Series

1. The first three terms have a common ratio, hence: $\frac{x}{9}=\frac{4}{x}$, therefore $x^2=36$ and $x=\pm 6$

Since the ratio is positive the second term is 6.

3. Letting the common ratio be r, we have: $4, 4r,$ and $4r^2$ so $64r^3=8000$

Therefore, $r^3=125$, so $r=5$, and the second and third terms are 20 and 100, respectively.

5. Here $a_1=-1$ and $r=-1$, so $a_{100}=-1(-1)^{99}=1$.

7. Here $a_1=\frac{2}{3}$ and $r=\frac{2}{3}$, so $a_8=\frac{2}{3}\left(\frac{2}{3}\right)^7=\frac{256}{6561}$.

9. Here $4096=a_7=r^6$, so $r=\pm\sqrt[6]{4096}=\pm 4$.

11. Compute the sum: $S_{10}=\frac{7\left(1-2^{10}\right)}{1-2}=-7(1-1024)=7161$

13. Compute the sum: $1+\sqrt{2}+2+\ldots+32=S_{11}=\frac{1\left[1-\left(\sqrt{2}\right)^{11}\right]}{1-\sqrt{2}}=\frac{1-32\sqrt{2}}{1-\sqrt{2}}=63+31\sqrt{2}$

15. Compute the sum: $\sum_{k=1}^{6}\left(\frac{3}{2}\right)^k=S_6=\frac{\frac{3}{2}\left[1-\left(\frac{3}{2}\right)^6\right]}{1-\frac{3}{2}}=-3\left(1-\frac{729}{64}\right)=\frac{1995}{64}$

17. Compute the sum:

$$\sum_{k=2}^{6}\left(\frac{1}{10}\right)^k=S_6-a_1=\frac{\frac{1}{10}\left[1-\left(\frac{1}{10}\right)^6\right]}{1-\frac{1}{10}}-\frac{1}{10}=\frac{1}{9}\left(1-\frac{1}{1000000}\right)-\frac{1}{10}=\frac{1}{9}\bullet\frac{999999}{1000000}-\frac{1}{10}=\frac{111111}{1000000}-\frac{1}{10}=\frac{11111}{1000000}$$

This is equivalent is 0.011111.

19. Compute the sum: $\frac{2}{3}-\frac{4}{9}+\frac{8}{27}-\ldots=\frac{\frac{2}{3}}{1+\frac{2}{3}}=\frac{2}{5}$

21. Compute the sum: $1+\frac{1}{1.01}+\frac{1}{(1.01)^2}+\ldots=\frac{1}{1-\frac{1}{1.01}}=101$

23. Writing as a geometric sum: $0.555\ldots=\frac{5}{10}+\frac{5}{100}+\ldots=\frac{\frac{5}{10}}{1-\frac{1}{10}}=\frac{5}{9}$

25. Writing as a geometric sum: $0.12323\ldots=\frac{1}{10}+\frac{23}{1000}+\frac{23}{100000}+\ldots=\frac{1}{10}+\frac{\frac{23}{1000}}{1-\frac{1}{100}}=\frac{1}{10}+\frac{23}{990}=\frac{122}{990}=\frac{61}{495}$

27. Writing as a geometric sum: $0.432\ldots = \frac{432}{1000} + \frac{432}{1000000} + \ldots = \dfrac{\frac{432}{1000}}{1 - \frac{1}{1000}} = \frac{432}{999} = \frac{16}{37}$

29. For $\frac{a}{r}$, a, ar:

$$\frac{a}{r}(a)(ar) = -1000$$
$$a^3 = -1000$$
$$a = -10$$

So:

$$\frac{a}{r} + a + ar = 15$$
$$\frac{-10}{r} - 10 - 10r = 15$$
$$2r^2 + 5r + 2 = 0$$
$$(2r+1)(r+2) = 0$$
$$r = -\tfrac{1}{2}, -2$$

31. Compute the sum: $\dfrac{\sqrt{3}}{\sqrt{3}+1} + \dfrac{\sqrt{3}}{\sqrt{3}+3} + \ldots = \dfrac{\frac{\sqrt{3}}{\sqrt{3}+1}}{1 - \frac{\sqrt{3}+1}{\sqrt{3}+3}} = \dfrac{\frac{\sqrt{3}}{\sqrt{3}+1}}{\frac{2}{\sqrt{3}+3}} = \dfrac{\sqrt{3}\left(\sqrt{3}+3\right)}{2\left(\sqrt{3}+1\right)} = \dfrac{3+3\sqrt{3}}{2\left(1+\sqrt{3}\right)} = \frac{3}{2}$

33. For $a_1, a_2, \ldots$ a geometric sequence, with ratio r: $S = a_1 + a_2 + \ldots + a_n = \dfrac{a_1\left(1-r^n\right)}{1-r}$

Also: $T = \dfrac{1}{a_1} + \dfrac{1}{a_2} + \ldots + \dfrac{1}{a_n} = \dfrac{\frac{1}{a_1}\left(1 - \frac{1}{r^n}\right)}{1 - \frac{1}{r}}$

Therefore: $\dfrac{S}{T} = \dfrac{\frac{a_1\left(1-r^n\right)}{1-r}}{\frac{\frac{1}{a_1}\left(1-\frac{1}{r^n}\right)}{\frac{r-1}{r}}} = \dfrac{a_1\left(1-r^n\right)a_1(r-1)r^n}{(1-r)r\left(r^n-1\right)} = a_1^2 r^{n-1} = a_1\left(a_1 r^{n-1}\right) = a_1 a_n$

35. The total distance traveled by the ball consists of two infinite geometric series:

down: $6 + \frac{1}{3}(6) + \frac{1}{9}(6) + \ldots$

up: $2 + \frac{1}{3}(2) + \frac{1}{9}(2) + \ldots$

Now use the formula $S = \frac{a}{1-r}$ to find each sum:

down: $a = 6, r = \frac{1}{3}$

$$S = \frac{6}{1-\frac{1}{3}} = \frac{6}{\frac{2}{3}} = 9 \text{ ft.}$$

up: $a = 1, r = \frac{1}{3}$

$$S = \frac{2}{1-\frac{1}{3}} = \frac{2}{\frac{2}{3}} = 3 \text{ ft.}$$

So the total distance traveled is 12 ft.

13.6 DeMoivre's Theorem

1. The complex number $4 + 2i$ is identified with the point (4,2):

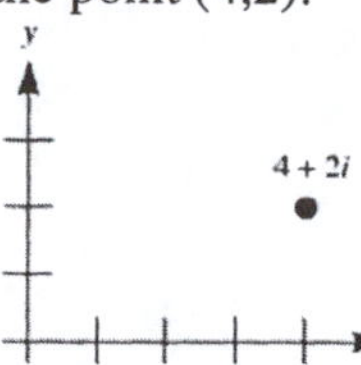

3. The complex number $-5 + i$ is identified with the point (–5,1):

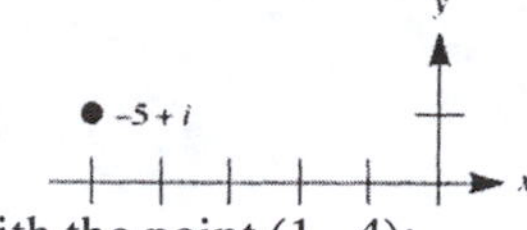

5. The complex number $1 - 4i$ is identified with the point (1,–4):

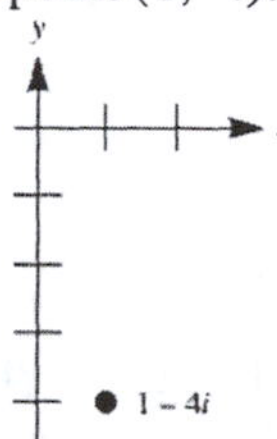

7. The complex number $-i$, or $0 - 1i$, is identified with the point (0,–1):

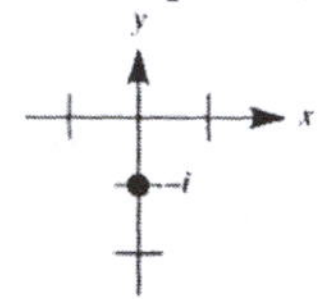

9. Convert the complex number to rectangular form: $2\left[\cos\frac{\pi}{4} + i\sin\frac{\pi}{4}\right] = 2\left[\frac{\sqrt{2}}{2} + \frac{\sqrt{2}}{2}i\right] = \sqrt{2} + \sqrt{2}i$

11. Convert the complex number to rectangular form: $4\left[\cos\frac{5\pi}{6} + i\sin\frac{5\pi}{6}\right] = 4\left[-\frac{\sqrt{3}}{2} + \frac{1}{2}i\right] = -2\sqrt{3} + 2i$

13. Convert the complex number to rectangular form: $\sqrt{2}\left[\cos 225° + i\sin 225°\right] = \sqrt{2}\left[-\frac{\sqrt{2}}{2} - \frac{\sqrt{2}}{2}i\right] = -1 - i$

15. Convert the complex number to rectangular form: $\sqrt{3}\left[\cos\frac{\pi}{2} + i\sin\frac{\pi}{2}\right] = \sqrt{3}(0 + 1i) = \sqrt{3}i$

17. Use the hint to find $\cos 75°$ and $\sin 75°$:

$$\cos 75° = \cos(30° + 45°) = \cos 30°\cos 45° - \sin 30°\sin 45° = \frac{\sqrt{3}}{2} \bullet \frac{\sqrt{2}}{2} - \frac{1}{2} \bullet \frac{\sqrt{2}}{2} = \frac{\sqrt{6}-\sqrt{2}}{4}$$

$$\sin 75° = \sin(30° + 45°) = \sin 30°\cos 45° + \cos 30°\sin 45° = \frac{1}{2} \bullet \frac{\sqrt{2}}{2} + \frac{\sqrt{3}}{2} \bullet \frac{\sqrt{2}}{2} = \frac{\sqrt{6}+\sqrt{2}}{4}$$

Therefore: $4\left[\cos 75° + i\sin 75°\right] = 4\left[\frac{\sqrt{6}-\sqrt{2}}{4} + \frac{\sqrt{6}+\sqrt{2}}{4}i\right] = \left(\sqrt{6} - \sqrt{2}\right) + \left(\sqrt{6} + \sqrt{2}\right)i$

19. Here $a = \frac{\sqrt{3}}{2}$ and $b = \frac{1}{2}$, so: $r = \sqrt{a^2 + b^2} = \sqrt{\frac{3}{4} + \frac{1}{4}} = 1$

Now find θ such that $\cos\theta = \frac{a}{r} = \frac{\sqrt{3}}{2}$ and $\sin\theta = \frac{b}{r} = \frac{1}{2}$. Such a θ is $\theta = \frac{\pi}{6}$. Thus: $\frac{\sqrt{3}}{2} + \frac{1}{2}i = \cos\frac{\pi}{6} + i\sin\frac{\pi}{6}$

21. Here $a = -1$ and $b = \sqrt{3}$, so: $r = \sqrt{a^2 + b^2} = \sqrt{1 + 3} = \sqrt{4} = 2$

Now find θ such that $\cos\theta = \frac{a}{r} = -\frac{1}{2}$ and $\sin\theta = \frac{b}{r} = \frac{\sqrt{3}}{2}$. Such a θ is $\theta = \frac{2\pi}{3}$. Thus: $-1 + \sqrt{3}i = 2\left[\cos\frac{2\pi}{3} + i\sin\frac{2\pi}{3}\right]$

23. Here $a = -2\sqrt{3}$ and $b = -2$, so: $r = \sqrt{a^2 + b^2} = \sqrt{12 + 4} = \sqrt{16} = 4$

Now find θ such that $\cos\theta = \frac{a}{r} = -\frac{2\sqrt{3}}{4} = -\frac{\sqrt{3}}{2}$ and $\sin\theta = \frac{b}{r} = -\frac{2}{4} = -\frac{1}{2}$. Such a θ is $\theta = \frac{7\pi}{6}$. Thus:

$$-2\sqrt{3} - 2i = 4\left[\cos\frac{7\pi}{6} + i\sin\frac{7\pi}{6}\right]$$

25. Here $a = 0$ and $b = -6$, so: $r = \sqrt{a^2 + b^2} = \sqrt{0+36} = 6$

Now find θ such that $\cos\theta = \frac{a}{r} = \frac{0}{6} = 0$ and $\sin\theta = \frac{b}{r} = -\frac{6}{6} = -1$. Such a θ is $\theta = \frac{3\pi}{2}$.

Thus: $-6i = 6\left[\cos\frac{3\pi}{2} + i\sin\frac{3\pi}{2}\right]$

27. Here $a = \frac{\sqrt{3}}{4}$ and $b = -\frac{1}{4}$, so: $r = \sqrt{a^2 + b^2} = \sqrt{\frac{3}{16} + \frac{1}{16}} = \sqrt{\frac{1}{4}} = \frac{1}{2}$

Now find θ such that $\cos\theta = \frac{a}{r} = \frac{\sqrt{3}/4}{1/2} = \frac{\sqrt{3}}{2}$ and $\sin\theta = \frac{b}{r} = \frac{-1/4}{1/2} = -\frac{1}{2}$. Such a θ is $\theta = \frac{11\pi}{6}$.

Thus: $\frac{\sqrt{3}}{4} - \frac{1}{4}i = \frac{1}{2}\left[\cos\frac{11\pi}{6} + i\sin\frac{11\pi}{6}\right]$

29. Performing the multiplication and adding angles:

$$2\left[\cos 22° + i\sin 22°\right] \bullet 3\left[\cos 38° + i\sin 38°\right] = 6\left[\cos 60° + i\sin 60°\right] = 6\left(\frac{1}{2} + \frac{\sqrt{3}}{2}i\right) = 3 + 3\sqrt{3}i$$

31. Performing the multiplication and adding angles:

$$\sqrt{2}\left[\cos\frac{\pi}{3} + i\sin\frac{\pi}{3}\right] \bullet \sqrt{2}\left[\cos\frac{4\pi}{3} + i\sin\frac{4\pi}{3}\right] = 2\left[\cos\frac{5\pi}{3} + i\sin\frac{5\pi}{3}\right] = 2\left(\frac{1}{2} - \frac{\sqrt{3}}{2}i\right) = 1 - \sqrt{3}i$$

33. Performing the multiplication and adding angles:

$$3\left[\cos\frac{\pi}{7} + i\sin\frac{\pi}{7}\right] \bullet \sqrt{2}\left[\cos\frac{\pi}{7} + i\sin\frac{\pi}{7}\right] = 3\sqrt{2}\left[\cos\frac{2\pi}{7} + i\sin\frac{2\pi}{7}\right] = 3\sqrt{2}\cos\frac{2\pi}{7} + \left(3\sqrt{2}\sin\frac{2\pi}{7}\right)i$$

35. Performing the division and subtracting angles:

$$6\left[\cos 50° + i\sin 50°\right] \div 2\left[\cos 5° + i\sin 5°\right] = 3\left[\cos 45° + i\sin 45°\right] = 3\left(\frac{\sqrt{2}}{2} + \frac{\sqrt{2}}{2}i\right) = \frac{3\sqrt{2}}{2} + \frac{3\sqrt{2}}{2}i$$

37. Performing the division and subtracting angles:

$$2^{4/3}\left[\cos\frac{5\pi}{12} + i\sin\frac{5\pi}{12}\right] \div 2^{1/3}\left[\cos\frac{\pi}{4} + i\sin\frac{\pi}{4}\right] = 2\left[\cos\frac{\pi}{6} + i\sin\frac{\pi}{6}\right] = 2\left(\frac{\sqrt{3}}{2} + \frac{1}{2}i\right) = \sqrt{3} + i$$

39. Performing the division and subtracting angles: $\left[\cos\frac{2\pi}{5} + i\sin\frac{2\pi}{5}\right] \div \left[\cos\frac{2\pi}{5} + i\sin\frac{2\pi}{5}\right] = \cos 0 + i\sin 0 = 1 + 0i = 1$

41. Using DeMoivre's theorem: $\left[3\left(\cos\frac{\pi}{3} + i\sin\frac{\pi}{3}\right)\right]^5 = 3^5\left(\cos\frac{5\pi}{3} + i\sin\frac{5\pi}{3}\right) = 243\left(\frac{1}{2} - \frac{\sqrt{3}}{2}i\right) = \frac{243}{2} - \frac{243\sqrt{3}}{2}i$

43. Using DeMoivre's theorem: $\left[\frac{1}{2}\left(\cos\frac{\pi}{24} + i\sin\frac{\pi}{24}\right)\right]^6 = \left(\frac{1}{2}\right)^6\left(\cos\frac{\pi}{4} + i\sin\frac{\pi}{4}\right) = \frac{1}{64}\left(\frac{\sqrt{2}}{2} + \frac{\sqrt{2}}{2}i\right) = \frac{\sqrt{2}}{128} + \frac{\sqrt{2}}{128}i$

45. Using DeMoivre's theorem: $\left[2^{1/5}\left(\cos 63° + i\sin 63°\right)\right]^{10} = \left(2^{1/5}\right)^{10}\left(\cos 630° + i\sin 630°\right) = 4(0 - 1i) = -4i$

47. Performing the multiplication and adding angles:

$$\begin{aligned} 2(\cos 100° + i\sin 200°) \times \sqrt{2}(\cos 20° + i\sin 20°) \times \tfrac{1}{2}(\cos 5° + i\sin 5°) &= \sqrt{2}(\cos 225° + i\sin 225°) \\ &= \sqrt{2}\left(-\frac{\sqrt{2}}{2} - \frac{\sqrt{2}}{2}i\right) \\ &= -1 - i \end{aligned}$$

49. First write $\frac{1}{2} - \frac{\sqrt{3}}{2}i$ in trigonometric form: $\frac{1}{2} - \frac{\sqrt{3}}{2}i = \cos\frac{5\pi}{3} + i\sin\frac{5\pi}{3}$

Now using DeMoivre's theorem: $\left(\frac{1}{2} - \frac{\sqrt{3}}{2}i\right)^5 = \cos\frac{25\pi}{3} + i\sin\frac{25\pi}{3} = \frac{1}{2} + \frac{\sqrt{3}}{2}i$

51. First write $-2 - 2i$ in trigonometric form: $-2 - 2i = 2\sqrt{2}\left(-\frac{\sqrt{2}}{2} - \frac{\sqrt{2}}{2}i\right) = 2\sqrt{2}\left(\cos\frac{5\pi}{4} + i\sin\frac{5\pi}{4}\right)$

Now using DeMoivre's theorem: $(-2 - 2i)^5 = \left(2\sqrt{2}\right)^5\left(\cos\frac{25\pi}{4} + i\sin\frac{25\pi}{4}\right) = 128\sqrt{2}\left(\frac{\sqrt{2}}{2} + \frac{\sqrt{2}}{2}i\right) = 128 + 128i$

53. First write $-2\sqrt{3} - 2i$ in trigonometric form: $-2\sqrt{3} - 2i = 4\left(-\frac{\sqrt{3}}{2} - \frac{1}{2}i\right) = 4\left(\cos\frac{7\pi}{6} + i\sin\frac{7\pi}{6}\right)$

Now using DeMoivre's theorem: $\left(-2\sqrt{3} - 2i\right)^4 = 4^4\left[\cos\frac{14\pi}{3} + i\sin\frac{14\pi}{3}\right] = 256\left(-\frac{1}{2} + \frac{\sqrt{3}}{2}i\right) = -128 + 128\sqrt{3}i$

55. Begin by writing $-27i$ in trigonometric form: $-27i = 27(0-1i) = 27\left[\cos\frac{3\pi}{2} + i\sin\frac{3\pi}{2}\right]$

Now let $z = r(\cos\theta + i\sin\theta)$ denote a cube root of $-27i$. Then: $z^3 = r^3(\cos 3\theta + i\sin 3\theta) = 27\left[\cos\frac{3\pi}{2} + i\sin\frac{3\pi}{2}\right]$

Then $r^3 = 27$ so $r = 3$, and $3\theta = \frac{3\pi}{2} + 2\pi k$, so $\theta = \frac{\pi}{2} + \frac{2\pi}{3}k$.

When $k = 0$: $z_1 = 3\left[\cos\frac{\pi}{2} + i\sin\frac{\pi}{2}\right] = 3(0+i) = 3i$

When $k = 1$: $z_2 = 3\left[\cos\frac{7\pi}{6} + i\sin\frac{7\pi}{6}\right] = 3\left(-\frac{\sqrt{3}}{2} - \frac{1}{2}i\right) = -\frac{3\sqrt{3}}{2} - \frac{3}{2}i$

When $k = 2$: $z_3 = 3\left[\cos\frac{11\pi}{6} + i\sin\frac{11\pi}{6}\right] = 3\left(\frac{\sqrt{3}}{2} - \frac{1}{2}i\right) = \frac{3\sqrt{3}}{2} - \frac{3}{2}i$

So the cube roots of $-27i$ are $3i$, $-\frac{3\sqrt{3}}{2} - \frac{3}{2}i$ and $\frac{3\sqrt{3}}{2} - \frac{3}{2}i$.

57. Begin by writing 1 in trigonometric form: $1 = 1(1+0i) = 1[\cos 0 + i\sin 0]$

Now let $z = r(\cos\theta + i\sin\theta)$ denote an eighth root of 1. Then: $z^8 = r^8(\cos 8\theta + i\sin 8\theta) = 1[\cos 0 + i\sin 0]$

Then $r^8 = 1$ so $r = 1$, and $8\theta = 0 + 2\pi k$, so $\theta = \frac{\pi}{4}k$.

When $k = 0$: $z_1 = 1[\cos 0 + i\sin 0] = 1 + 0i = 1$

When $k = 1$: $z_2 = 1\left[\cos\frac{\pi}{4} + i\sin\frac{\pi}{4}\right] = \frac{\sqrt{2}}{2} + \frac{\sqrt{2}}{2}i$

When $k = 2$: $z_3 = 1\left[\cos\frac{\pi}{2} + i\sin\frac{\pi}{2}\right] = 0 + 1i = i$

When $k = 3$: $z_4 = 1\left[\cos\frac{3\pi}{4} + i\sin\frac{3\pi}{4}\right] = -\frac{\sqrt{2}}{2} + \frac{\sqrt{2}}{2}i$

When $k = 4$: $z_5 = 1[\cos\pi + i\sin\pi] = -1 + 0i = -1$

When $k = 5$: $z_6 = 1\left[\cos\frac{5\pi}{4} + i\sin\frac{5\pi}{4}\right] = -\frac{\sqrt{2}}{2} - \frac{\sqrt{2}}{2}i$

When $k = 6$: $z_7 = 1\left[\cos\frac{3\pi}{2} + i\sin\frac{3\pi}{2}\right] = 0 - 1i = -i$

When $k = 7$: $z_8 = 1\left[\cos\frac{7\pi}{4} + i\sin\frac{7\pi}{4}\right] = \frac{\sqrt{2}}{2} - \frac{\sqrt{2}}{2}i$

So the eighth roots of 1 are 1, $\frac{\sqrt{2}}{2} + \frac{\sqrt{2}}{2}i$, i, $-\frac{\sqrt{2}}{2} + \frac{\sqrt{2}}{2}i$, -1, $-\frac{\sqrt{2}}{2} - \frac{\sqrt{2}}{2}i$, $-i$, and $\frac{\sqrt{2}}{2} - \frac{\sqrt{2}}{2}i$. These final results were obtained using the half-angle formulas for sine and cosine.

59. Begin by writing 64 in trigonometric form: $64 = 64(1+0i) = 64[\cos 0 + i\sin 0]$

Now let $z = r(\cos\theta + i\sin\theta)$ denote a cube root of 64. Then: $z^3 = r^3(\cos 3\theta + i\sin 3\theta) = 64[\cos 0 + i\sin 0]$

Then $r^3 = 64$ so $r = 4$, and $3\theta = 0 + 2\pi k$, so $\theta = 0 + \frac{2\pi}{3}k$.

When $k = 0$: $z_1 = 4[\cos 0 + i\sin 0] = 4(1+0i) = 4$

When $k = 1$: $z_2 = 4\left[\cos\frac{2\pi}{3} + i\sin\frac{2\pi}{3}\right] = 4\left(-\frac{1}{2} + \frac{\sqrt{3}}{2}i\right) = -2 + 2\sqrt{3}i$

When $k = 2$: $z_3 = 4\left[\cos\frac{4\pi}{3} + i\sin\frac{4\pi}{3}\right] = 4\left(-\frac{1}{2} - \frac{\sqrt{3}}{2}i\right) = -2 - 2\sqrt{3}i$

So the cube roots of 64 are 4, $-2 + 2\sqrt{3}i$, and $-2 - 2\sqrt{3}i$.

61. First write 729 in trigonometric form: $729 = 729(1+0i) = 729[\cos 0 + i\sin 0]$

Now let $z = r(\cos\theta + i\sin\theta)$ denote a sixth root of 729. Then: $z^6 = r^6[\cos 6\theta + i\sin 6\theta] = 729[\cos 0 + i\sin 0]$

Then $r^6 = 729$ so $r = 3$, and $6\theta = 0 + 2\pi k$, so $\theta = 0 + \frac{\pi}{3}k$.

When $k = 0$: $z_1 = 3[\cos 0 + i\sin 0] = 3(1+0i) = 3$

When $k = 1$: $z_2 = 3\left[\cos\frac{\pi}{3} + i\sin\frac{\pi}{3}\right] = 3\left(\frac{1}{2} + \frac{\sqrt{3}}{2}i\right) = \frac{3}{2} + \frac{3\sqrt{3}}{2}i$

When $k = 2$: $z_3 = 3\left[\cos\frac{2\pi}{3} + i\sin\frac{2\pi}{3}\right] = 3\left(-\frac{1}{2} + \frac{\sqrt{3}}{2}i\right) = -\frac{3}{2} + \frac{3\sqrt{3}}{2}i$

When $k = 3$: $z_4 = 3[\cos\pi + i\sin\pi] = 3(-1+0i) = -3$

When $k=4$: $z_5=3\left[\cos\frac{4\pi}{3}+i\sin\frac{4\pi}{3}\right]=3\left(-\frac{1}{2}-\frac{\sqrt{3}}{2}i\right)=-\frac{3}{2}-\frac{3\sqrt{3}}{2}i$

When $k=5$: $z_6=3\left[\cos\frac{5\pi}{3}+i\sin\frac{5\pi}{3}\right]=3\left(\frac{1}{2}-\frac{\sqrt{3}}{2}i\right)=\frac{3}{2}-\frac{3\sqrt{3}}{2}i$

So the sixth roots of 729 are $3, \frac{3}{2}+\frac{3\sqrt{3}}{2}i, -\frac{3}{2}+\frac{3\sqrt{3}}{2}i, -3, -\frac{3}{2}-\frac{3\sqrt{3}}{2}i$ and $\frac{3}{2}-\frac{3\sqrt{3}}{2}i$.

63. First write $7-7i$ in trigonometric form: $7-7i=7\sqrt{2}\left(\frac{1}{\sqrt{2}}-\frac{1}{\sqrt{2}}i\right)=7\sqrt{2}\left[\cos\frac{7\pi}{4}+i\sin\frac{7\pi}{4}\right]$

Using DeMoivre's theorem: $(7-7i)^8=\left(7\sqrt{2}\right)^8\left[\cos 14\pi+i\sin 14\pi\right]=\left(7^8\right)\left(2^4\right)(1+0i)=92{,}236{,}816$

65. Begin by writing i in trigonometric form: $i=1(0+1i)=1\left[\cos\frac{\pi}{2}+i\sin\frac{\pi}{2}\right]$

Now let z be a fifth root of i, where $z=r(\cos\theta+i\sin\theta)$. Then: $z^5=r^5\left[\cos 5\theta+i\sin 5\theta\right]=1\left[\cos\frac{\pi}{2}+i\sin\frac{\pi}{2}\right]$

So $r^5=1$ thus $r=1$, and $5\theta=\frac{\pi}{2}+2\pi k=90°+360°k$, so $\theta=18°+72°k$.

When $k=0$: $z_1=1\left[\cos 18°+i\sin 18°\right]=0.95+0.31i$
When $k=1$: $z_2=1\left[\cos 90°+i\sin 90°\right]=i$
When $k=2$: $z_3=1\left[\cos 162°+i\sin 162°\right]=-0.95+0.31i$
When $k=3$: $z_4=1\left[\cos 234°+i\sin 234°\right]=-0.59-0.81i$
When $k=4$: $z_5=1\left[\cos 306°+i\sin 306°\right]=0.59-0.81i$

So the fifth roots of i are $0.95+0.31i$, i, $-0.95+0.31i$, $-0.59-0.81i$ and $0.59-0.81i$. Each of the real and imaginary parts here has been rounded off to two decimal places.

67. Begin by writing $8-8\sqrt{3}i$ in trigonometric form: $8-8\sqrt{3}i=16\left(\frac{1}{2}-\frac{\sqrt{3}}{2}i\right)=16\left[\cos\frac{5\pi}{3}+i\sin\frac{5\pi}{3}\right]$

Now let $z=r(\cos\theta+i\sin\theta)$ denote a fourth root of $8-8\sqrt{3}i$.
Then: $z^4=r^4\left[\cos 4\theta+i\sin 4\theta\right]=16\left[\cos\frac{5\pi}{3}+i\sin\frac{5\pi}{3}\right]$
Then $r^4=16$ so $r=2$, and $4\theta=\frac{5\pi}{3}+2\pi k$, so $\theta=\frac{5\pi}{12}+\frac{\pi}{2}k$.

When $k=0$: $z_1=2\left[\cos\frac{5\pi}{12}+i\sin\frac{5\pi}{12}\right]=2\left[\frac{\sqrt{6}-\sqrt{2}}{4}+\frac{\sqrt{6}+\sqrt{2}}{4}i\right]=\frac{\sqrt{6}-\sqrt{2}}{2}+\frac{\sqrt{6}+\sqrt{2}}{2}i$

When $k=1$: $z_2=2\left[\cos\frac{11\pi}{12}+i\sin\frac{11\pi}{12}\right]=2\left[\frac{-\sqrt{2}-\sqrt{6}}{4}+\frac{\sqrt{6}-\sqrt{2}}{4}i\right]=\frac{-\sqrt{2}-\sqrt{6}}{2}+\frac{\sqrt{6}-\sqrt{2}}{2}i$

When $k=2$: $z_3=2\left[\cos\frac{17\pi}{12}+i\sin\frac{17\pi}{12}\right]=2\left[\frac{\sqrt{2}-\sqrt{6}}{4}+\frac{-\sqrt{2}-\sqrt{6}}{4}i\right]=\frac{\sqrt{2}-\sqrt{6}}{2}-\frac{\sqrt{2}+\sqrt{6}}{2}i$

When $k=3$: $z_4=2\left[\cos\frac{23\pi}{12}+i\sin\frac{23\pi}{12}\right]=2\left[\frac{\sqrt{2}+\sqrt{6}}{4}+\frac{\sqrt{2}-\sqrt{6}}{4}i\right]=\frac{\sqrt{2}+\sqrt{6}}{2}+\frac{\sqrt{2}-\sqrt{6}}{2}i$

So the fourth roots of $8-8\sqrt{3}i$ are $\frac{\sqrt{6}-\sqrt{2}}{2}+\frac{\sqrt{6}+\sqrt{2}}{2}i$, $\frac{-\sqrt{2}-\sqrt{6}}{2}+\frac{\sqrt{6}-\sqrt{2}}{2}i$, $\frac{\sqrt{2}-\sqrt{6}}{2}+\frac{-\sqrt{2}-\sqrt{6}}{2}i$ and $\frac{\sqrt{2}+\sqrt{6}}{2}+\frac{\sqrt{2}-\sqrt{6}}{2}i$. These final results were obtained using the addition formulas for sine and cosine.
Note: If you used half-angle formulas, your answers, though identical in value, may "look" vastly different. Those answers (which are correct) are $\sqrt{2-\sqrt{3}}+\sqrt{2+\sqrt{3}}i$, $-\sqrt{2+\sqrt{3}}+\sqrt{2-\sqrt{3}}i$, $-\sqrt{2-\sqrt{3}}-\sqrt{2+\sqrt{3}}i$ and $\sqrt{2+\sqrt{3}}-\sqrt{2-\sqrt{3}}i$.

69. a. Let $z = r(\cos\theta + i\sin\theta)$, so $z^3 = r^3(\cos 3\theta + i\sin 3\theta)$. Since $1 = 1(\cos 0 + i\sin 0)$, we have $r^3 = 1$ so $r = 1$, and $3\theta = 0 + 2\pi k$, so $\theta = 0 + \frac{2\pi}{3}k$.

When $k = 0$: $z_1 = 1[\cos 0 + i\sin 0] = 1(1 + 0i) = 1$

When $k = 1$: $z_2 = 1\left[\cos\frac{2\pi}{3} + i\sin\frac{2\pi}{3}\right] = 1\left(-\frac{1}{2} + \frac{\sqrt{3}}{2}i\right) = -\frac{1}{2} + \frac{\sqrt{3}}{2}i$

When $k = 2$: $z_3 = 1\left[\cos\frac{4\pi}{3} + i\sin\frac{4\pi}{3}\right] = 1\left(-\frac{1}{2} - \frac{\sqrt{3}}{2}i\right) = -\frac{1}{2} - \frac{\sqrt{3}}{2}i$

So the cube roots of 1 are $1, -\frac{1}{2} + \frac{\sqrt{3}}{2}i$ and $-\frac{1}{2} - \frac{\sqrt{3}}{2}i$.

b. Compute the sum: $z_1 + z_2 + z_3 = 1 + \left(-\frac{1}{2} + \frac{\sqrt{3}}{2}i\right) + \left(-\frac{1}{2} - \frac{\sqrt{3}}{2}i\right) = 0 + 0i = 0$

Now compute the products:

$z_1 z_2 = 1\left[\cos\frac{2\pi}{3} + i\sin\frac{2\pi}{3}\right] = -\frac{1}{2} + \frac{\sqrt{3}}{2}i$ (Note this is z_2)

$z_2 z_3 = 1[\cos 2\pi + i\sin 2\pi] = 1$ (Note this is z_1)

$z_3 z_1 = 1\left[\cos\frac{4\pi}{3} + i\sin\frac{4\pi}{3}\right] = -\frac{1}{2} - \frac{\sqrt{3}}{2}i$ (Note this is z_3)

So $z_1z_2 + z_2z_3 + z_3z_1 = z_1 + z_2 + z_3 = 0$.

71. Using trigonometric forms and DeMoivre's theorem, compute the powers:

$$\left[\frac{-1+i\sqrt{3}}{2}\right]^5 = \left[-\frac{1}{2} + \frac{\sqrt{3}}{2}i\right]^5 = \left[\cos\frac{2\pi}{3} + i\sin\frac{2\pi}{3}\right]^5 = \cos\frac{10\pi}{3} + i\sin\frac{10\pi}{3} = -\frac{1}{2} - \frac{\sqrt{3}}{2}i$$

$$\left[\frac{-1-i\sqrt{3}}{2}\right]^5 = \left[-\frac{1}{2} - \frac{\sqrt{3}}{2}i\right]^5 = \left[\cos\frac{4\pi}{3} + i\sin\frac{4\pi}{3}\right]^5 = \cos\frac{20\pi}{3} + i\sin\frac{20\pi}{3} = -\frac{1}{2} + \frac{\sqrt{3}}{2}i$$

Now compute the sum: $\left[\frac{-1+i\sqrt{3}}{2}\right]^5 + \left[\frac{-1-i\sqrt{3}}{2}\right]^5 = \left(-\frac{1}{2} - \frac{\sqrt{3}}{2}i\right) + \left(-\frac{1}{2} + \frac{\sqrt{3}}{2}i\right) = -1$

73. Since the multiplication results in a difference of squares:

$$\begin{aligned}(\cos\theta + i\sin\theta)(\cos\theta - i\sin\theta) &= \cos^2\theta - i^2\sin^2\theta \\ &= \cos^2\theta + \sin^2\theta \quad (\text{since } i^2 = -1) \\ &= 1\end{aligned}$$

75. Using the hint:

$$\begin{aligned}\frac{r(\cos\alpha + i\sin\alpha)}{R(\cos\beta + i\sin\beta)} \bullet \frac{(\cos\beta - i\sin\beta)}{(\cos\beta - i\sin\beta)} &= \frac{r\left(\cos\alpha\cos\beta + i\sin\alpha\cos\beta - i\cos\alpha\sin\beta - i^2\sin\alpha\sin\beta\right)}{R\left(\cos^2\beta - i^2\sin^2\beta\right)} \\ &= \frac{r\left[(\cos\alpha\cos\beta + \sin\alpha\sin\beta) + i(\sin\alpha\cos\beta - \cos\alpha\sin\beta)\right]}{R\left(\cos^2\beta + \sin^2\beta\right)}\end{aligned}$$

Using the difference identities for sine and cosine: $\frac{r[\cos(\alpha-\beta) + i\sin(\alpha-\beta)]}{R} = \frac{r}{R}[\cos(\alpha-\beta) + i\sin(\alpha-\beta)]$

77. Using the hint and the identities $\cos(-\theta) = \cos\theta$ and $\sin(-\theta) = -\sin\theta$:

$$\begin{aligned}\frac{1}{z} &= \frac{1(\cos 0 + i\sin 0)}{r(\cos\theta + i\sin\theta)} \\ &= \frac{1}{r}[\cos(0-\theta) + i\sin(0-\theta)] \quad (\text{by Exercise 75}) \\ &= \frac{1}{r}[\cos(-\theta) + i\sin(-\theta)] \\ &= \frac{1}{r}(\cos\theta - i\sin\theta)\end{aligned}$$

79. For convenience, denote sin θ and cos θ by S and C, respectively. Then, following the hint in the text:

$$\frac{(1+S)+iC}{(1+S)-iC}\bullet\frac{(1+S)+iC}{(1+S)+iC}=\frac{(1+S)^2+2iC(1+S)-C^2}{(1+S)^2+C^2}$$
$$=\frac{(1+S)^2+2iC(1+S)-\left(1-S^2\right)}{(1+S)^2+\left(1-S^2\right)}$$
$$=\frac{(1+S)^2+2iC(1+S)-(1-S)(1+S)}{(1+S)^2+(1-S)(1+S)}$$

Notice that every term in the numerator and in the denominator of this last expression contains the common factor (1 + S), and this factor is nonzero since $\theta \neq \frac{3\pi}{2}k$ and thus sin $\theta \neq -1$. Dividing out this factor, we see that the expression becomes: $\frac{(1+S)+2iC-(1-S)}{(1+S)+(1-S)}=\frac{2S+2iC}{2}=S+iC$

Since this expression represents sin θ + i cos θ, we have proven the result.

81. From Exercise 80, the solutions are $w=\cos\theta+i\sin\theta$ and $w=\cos\theta-i\sin\theta$. If $w=\cos\theta+i\sin\theta$, then:

$$w^n+\frac{1}{w^n}=\cos n\theta+i\sin n\theta+\frac{1}{\cos n\theta+i\sin n\theta}=\cos n\theta+i\sin n\theta+\cos n\theta-i\sin n\theta=2\cos n\theta$$

If $w=\cos\theta-i\sin\theta$, then:

$$w^n+\frac{1}{w^n}=\cos n\theta-i\sin n\theta+\frac{1}{\cos n\theta-i\sin n\theta}=\cos n\theta-i\sin n\theta+\cos n\theta+i\sin n\theta=2\cos n\theta$$

In either case, the result is proved.

Chapter 13 Review Exercises

1. When n = 1: $5=\frac{5}{2}(1)(1+1)=5$. Assuming: $5+10+\ldots+5k=\frac{5}{2}k(k+1)$

Then: $5+10+\ldots+5k+5(k+1)=\frac{5}{2}k(k+1)+5(k+1)=(k+1)\left(\frac{5}{2}k+5\right)=\frac{5}{2}(k+1)(k+2)$

This completes the induction.

3. When n = 1: $1\bullet 2=2=\frac{1}{3}(1)(1+1)(1+2)=2$. Assuming: $1\bullet 2+2\bullet 3+\ldots+k(k+1)=\frac{1}{3}k(k+1)(k+2)$. Then:

$$1\bullet 2+2\bullet 3+\ldots+k(k+1)+(k+1)(k+2)=\frac{1}{3}k(k+1)(k+2)+(k+1)(k+2)$$
$$=\left(\frac{1}{3}k+1\right)(k+1)(k+2)$$
$$=\frac{1}{3}(k+1)(k+2)(k+3)$$

This completes the induction.

5. For n = 1: $1=3+(2-3)2=3-2=1$. Assuming: $1+3\bullet 2+5\bullet 2^2+\ldots+(2k-1)\bullet 2^{k-1}=3+(2k-3)\bullet 2^k$. Then:

$$1+3\bullet 2+\ldots+(2k-1)2^{k-1}+(2k+1)2^k=3+(2k-3)2^k+(2k+1)2^k$$
$$=3+(4k-2)2^k$$
$$=3+(2k-1)2^{k+1}$$
$$=3+\left(2(k+1)-3\right)\bullet 2^{k+1}$$

The induction is complete.

7. For $n = 1$: $1=\left(1^2-2+3\right)2-3=1$. Assuming: $1+2^2\bullet 2+3^2\bullet 2^2+4^2\bullet 2^3+\ldots+k^2\bullet 2^{k-1}=\left(k^2-2k+3\right)2^k-3$

Then:

$$\begin{aligned}1+2^2\bullet 2+\ldots+(k+1)^2 2^k &=\left(k^2-2k+3\right)2^k-3+(k+1)^2 2^k\\ &=2^k\left[k^2-2k+3+(k+1)^2\right]-3\\ &=2^k\left(2k^2+4\right)-3\\ &=2^{k+1}\left(k^2+2\right)-3\\ &=2^{k+1}\left[(k+1)^2-2k+1\right]-3\\ &=2^{k+1}\left[(k+1)^2-2(k+1)+3\right]-3\end{aligned}$$

The induction is complete.

9. For $n = 1$: $7^1-1=6=3(2)$. Assuming: $7^k-1=3L$

Then: $7^{k+1}-1=3L+7^{k+1}-7^k=3L+7^k(7-1)=3L+6\left(7^k\right)=3\left(L+2\bullet 7^k\right)$

The induction is complete.

11. Expand using the binomial theorem:

$$\begin{aligned}\left(3a+b^2\right)^4 &=\binom{4}{0}(3a)^4+\binom{4}{1}(3a)^3\left(b^2\right)+\binom{4}{2}(3a)^2\left(b^2\right)^2+\binom{4}{3}(3a)\left(b^2\right)^3+\binom{4}{4}\left(b^2\right)^4\\ &=81a^4+108a^3b^2+54a^2b^4+12ab^6+b^8\end{aligned}$$

13. Expand using the binomial theorem:

$$\begin{aligned}\left(x+\sqrt{x}\right)^4 &=\binom{4}{0}x^4+\binom{4}{1}x^3\left(\sqrt{x}\right)+\binom{4}{2}x^2\left(\sqrt{x}\right)^2+\binom{4}{3}x\left(\sqrt{x}\right)^3+\binom{4}{4}\left(\sqrt{x}\right)^4\\ &=x^4+4x^3\sqrt{x}+6x^3+4x^2\sqrt{x}+x^2\end{aligned}$$

15. Expand using the binomial theorem:

$$\begin{aligned}\left(x^2-2y^2\right)^5 &=\binom{5}{0}\left(x^2\right)^5-\binom{5}{1}\left(x^2\right)^4\left(2y^2\right)+\binom{5}{2}\left(x^2\right)^3\left(2y^2\right)^2-\binom{5}{3}\left(x^2\right)^2\left(2y^2\right)^3\\ &\qquad+\binom{5}{4}\left(x^2\right)\left(2y^2\right)^4-\binom{5}{5}\left(2y^2\right)^5\\ &=x^{10}-10x^8y^2+40x^6y^4-80x^4y^6+80x^2y^8-32y^{10}\end{aligned}$$

17. Expand using the binomial theorem:

$$\left(1+\frac{1}{x}\right)^5=\binom{5}{0}+\binom{5}{1}\frac{1}{x}+\binom{5}{2}\left(\frac{1}{x}\right)^2+\binom{5}{3}\left(\frac{1}{x}\right)^3+\binom{5}{4}\left(\frac{1}{x}\right)^4+\binom{5}{5}\left(\frac{1}{x}\right)^5=1+\frac{5}{x}+\frac{10}{x^2}+\frac{10}{x^3}+\frac{5}{x^4}+\frac{1}{x^5}$$

19. Expand using the binomial theorem:

$$\begin{aligned}\left(a\sqrt{b}-b\sqrt{a}\right)^4 &=a^2b^2\left(\sqrt{a}-\sqrt{b}\right)^4\\ &=a^2b^2\left[\left(\sqrt{a}\right)^4-4\left(\sqrt{a}\right)^3\sqrt{b}+6\left(\sqrt{a}\right)^2\left(\sqrt{b}\right)^2-4\left(\sqrt{a}\right)\left(\sqrt{b}\right)^3+\left(\sqrt{b}\right)^4\right]\\ &=a^2b^2\left(a^2-4a\sqrt{ab}+6ab-4b\sqrt{ab}+b^2\right)\\ &=a^4b^2-4a^3b^2\sqrt{ab}+6a^3b^3-4a^2b^3\sqrt{ab}+a^2b^4\end{aligned}$$

21. The fifth term is given by: $\binom{5}{5-1}(3x)\left(y^2\right)^4=5(3x)y^8=15xy^8$

23. Here $7 - r + 1 = 5$, so $r = 3$, and the coefficient of the third term is: $\binom{7}{2}(2)^2=(4)\dfrac{7!}{5!2!}=84$

25. Here $r - 1 = 6$, so $r = 7$, and the coefficient of the seventh term is: $\binom{8}{6}=\dfrac{8!}{6!2!}=28$

27. Computing each side:

$$\binom{2}{0}^2 + \binom{2}{1}^2 + \binom{2}{2}^2 = (1)^2 + (2)^2 + (1)^2 = 6 \qquad \binom{4}{2} = \frac{4!}{2!2!} = 6$$

29. Computing each side:

$$\binom{4}{0}^2 + \binom{4}{1}^2 + \binom{4}{2}^2 + \binom{4}{3}^2 + \binom{4}{4}^2 = 1^2 + 4^2 + 6^2 + 4^2 + 1^2 = 70$$

$$\binom{8}{4} = \frac{8!}{4!4!} = 70$$

31. Computing values: $\binom{3}{0}+\binom{3}{1}+\binom{3}{2}+\binom{3}{3} = 1+3+3+1 = 8 = 2^3$

33. Computing values: $a_1 = \frac{2}{1+1} = 1,\quad a_2 = \frac{4}{2+1} = \frac{4}{3},\quad a_3 = \frac{6}{3+1} = \frac{3}{2},\quad a_4 = \frac{8}{4+1} = \frac{8}{5}$

The first four terms are $1, \frac{4}{3}, \frac{3}{2}, \frac{8}{5}$. Graph these points:

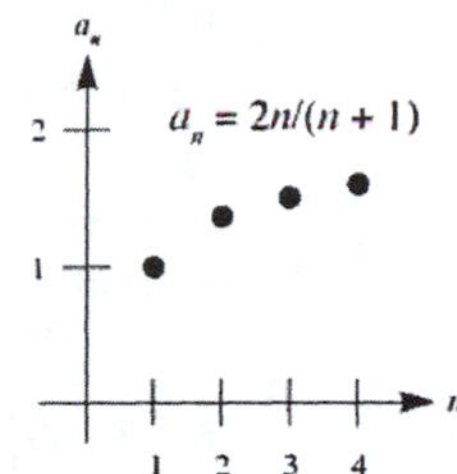

35. Computing values: $a_1 = (-1)\left(1-\frac{1}{2}\right) = -\frac{1}{2},\quad a_2 = (1)\left(1-\frac{1}{3}\right) = \frac{2}{3},\quad a_3 = (-1)\left(1-\frac{1}{4}\right) = -\frac{3}{4},\quad a_4 = (1)\left(1-\frac{1}{5}\right) = \frac{4}{5}$

The first four terms are $-\frac{1}{2}, \frac{2}{3}, -\frac{3}{4}, \frac{4}{5}$. Graph these points:

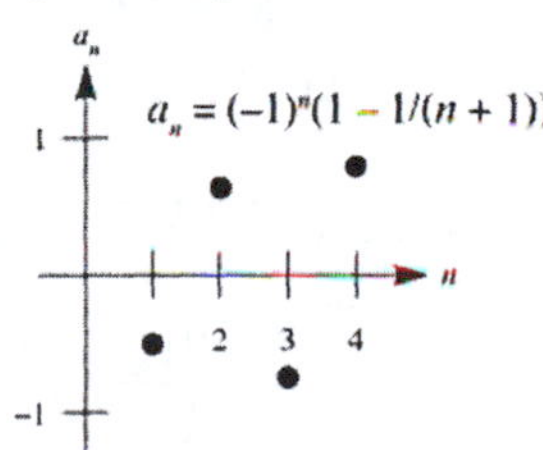

37. Computing values: $a_0 = -3,\quad a_1 = -12,\quad a_2 = -48,\quad a_3 = -192$

The first four terms are –3, –12, –48, –192.

39. **a.** Expanding the sum: $\sum_{k=1}^{3} (-1)^k (2k+1) = -3+5-7 = -5$

b. Expanding the sum:

$$\sum_{k=0}^{8}\left(\frac{1}{k+1} - \frac{1}{k+2}\right) = \left(1-\frac{1}{2}\right)+\left(\frac{1}{2}-\frac{1}{3}\right)+\left(\frac{1}{3}-\frac{1}{4}\right)+\left(\frac{1}{4}-\frac{1}{5}\right)+\left(\frac{1}{5}-\frac{1}{6}\right)+\left(\frac{1}{6}-\frac{1}{7}\right)+\left(\frac{1}{7}-\frac{1}{8}\right)+\left(\frac{1}{8}-\frac{1}{9}\right)+\left(\frac{1}{9}-\frac{1}{10}\right)$$

$$= 1 - \frac{1}{10}$$

$$= \frac{9}{10}$$

41. The sum can be written as: $\frac{5}{3} + \frac{5}{3^2} + \frac{5}{3^3} + \frac{5}{3^4} + \frac{5}{3^5} = \sum_{k=1}^{5} \frac{5}{3k}$

43. Here $a_n = 1 + 4n$, so $a_{18} = 1 + 4(18) = 73$.

45. Here $a_n = \frac{10}{2^{n-1}}$, so $a_{12} = \frac{10}{2^{11}} = \frac{5}{1024}$.

47. The sum is given by: $S_{12} = (12)\left(\dfrac{8+\frac{43}{2}}{2}\right) = \frac{59}{2}(6) = 177$

49. By inspection $S_1 = 7, S_2 = 77, \ldots$ so $S_{10} = 7{,}777{,}777{,}777$.

51. Compute:

$$a_3 = 4$$
$$a_5 = 10$$
$$\frac{a_5}{a_3} = \frac{10}{4} = \frac{5}{2} = r^2 \text{ and } r = \pm\sqrt{\frac{5}{2}}$$

Since r is known to be negative, $r = -\sqrt{\frac{5}{2}}$ and $a_6 = -\sqrt{\frac{5}{2}}(10) = -5\sqrt{10}$.

53. The sum is given by: $S = \dfrac{\frac{3}{5}}{1-\frac{1}{5}} = \frac{3}{4}$

55. The sum is given by: $S = \dfrac{\frac{1}{9}}{1+\frac{1}{9}} = \frac{1}{10}$

57. We have $0.\overline{45} = 0.45 + 0.0045 + 0.000045 + \ldots$. This is a geometric sequence with $a = 0.45$, $r = \frac{1}{100}$, so:

$$0.\overline{45} = \frac{0.45}{1-\frac{1}{100}} = \frac{45}{99} = \frac{15}{33} = \frac{5}{11}$$

59. The sum is given by: $S_n = \frac{n}{2}\left[2(1)+(n-1)^1\right] = \frac{n}{2}(2+n-1) = \frac{n}{2}(1+n) = \frac{n^2}{2}+\frac{n}{2} = n+\frac{n^2}{2}-\frac{n}{2} = n+\frac{n(n-1)}{2}$

61. The sum is given by: $S_n = \frac{n}{2}[2(1)+(n-1)3] = \frac{n}{2}(2+3n-3) = \frac{n}{2}(3n-1) = \frac{3n^2}{2}-\frac{n}{2} = n+\frac{3n^2}{2}-\frac{3n}{2} = n+\frac{3n(n-1)}{2}$

63. **a.** Using formula (1), we have: $1^2+2^2+\ldots+50^2 \approx \dfrac{\left(50+\frac{1}{2}\right)^{2+1}}{2+1} = \dfrac{\left(\frac{101}{2}\right)^3}{3} = \dfrac{1030301}{24} \approx 42929$

b. The exact sum is given by: $\dfrac{50(51)(101)}{6} = 42925$. The percent error is: $100 \bullet \frac{4}{42925} \approx 0.00932\%$

c. Using formula (1), we have: $1^4+2^4+\ldots+200^4 \approx \dfrac{(200.5)^5}{5} \approx 6.48040\times10^{10}$

d. Compute the sum: $1^4+2^4+\ldots+200^4 \approx \dfrac{200(201)(401)(120599)}{30} \approx 6.48027\times10^{10}$

The percent error is: $100 \bullet \dfrac{0.00013\times10^{10}}{6.48027\times10^{10}} = 2\times10^{-3}\ \%$

65. Call the three terms x, xr, xr^2. Then:

$$x + xr + xr^2 = 13$$
$$\frac{1}{x}+\frac{1}{xr}+\frac{1}{xr^2} = \frac{13}{9}$$

Combining fractions in the second equation: $\dfrac{r^2+r+1}{xr^2} = \frac{13}{9}$

The first equation can be written as:

$$x\left(1+r+r^2\right) = 13$$
$$1+r+r^2 = \frac{13}{x}$$

Substituting:

$$\frac{\frac{13}{x}}{xr^2} = \frac{13}{9}$$
$$\frac{1}{x^2r^2} = \frac{1}{9}$$
$$x^2r^2 = 9$$
$$xr = \pm 3$$

If $xr = 3$, then $x = \frac{3}{r}$, and the first equation becomes:

$$\frac{3}{r} + 3 + 3r = 13$$
$$3r^2 - 10r + 3 = 0$$
$$(3r-1)(r-3) = 0$$
$$r = \frac{1}{3}, 3$$

If $xr = -3$, then $x = -\frac{3}{r}$, and the first equation becomes:

$$-\frac{3}{r} - 3 - 3r = 13$$
$$3r^2 + 16r + 3 = 0$$
$$r = \frac{-16 \pm \sqrt{220}}{6} = \frac{-8 \pm \sqrt{55}}{3}$$

So the possible values of r are $\frac{1}{3}$, 3, $\frac{-8 \pm \sqrt{55}}{3}$.

67. Let $b = ra$ and $c = r^2a$. Since a, $2b$, c are consecutive terms in an arithmetic sequence, their common difference is constant:

$$c - 2b = 2b - a$$
$$a + c = 4b$$
$$a + r^2a = 4ra$$

Since $a \neq 0$, we can divide by a:

$$1 + r^2 = 4r$$
$$r^2 - 4r + 1 = 0$$
$$r = \frac{4 \pm \sqrt{16-4}}{2} = \frac{4 \pm 2\sqrt{3}}{2} = 2 \pm \sqrt{3}$$

69. Let $b = ra$ and $c = r^2a$. We wish to show that: $\frac{1}{2b} - \frac{1}{a+b} = \frac{1}{c+b} - \frac{1}{2b}$

Substitute with each side of the equation:

$$\frac{1}{2b} - \frac{1}{a+b} = \frac{1}{2ra} - \frac{1}{a+ra} = \frac{a+ra-2ra}{2ra(a+ra)} = \frac{a-ra}{2ra^2(1+r)} = \frac{1-r}{2ar(1+r)}$$
$$\frac{1}{c+b} - \frac{1}{2b} = \frac{1}{r^2a+ra} - \frac{1}{2ra} = \frac{2ra - r^2a - ra}{2ra\left(r^2a+ra\right)} = \frac{ra(1-r)}{2r^2a^2(r+1)} = \frac{1-r}{2ra(r+1)}$$

These two sides are the same, which proves the desired result.

71. Start with the given equation:

$$\ln(A+C) + \ln(A+C-2B) = 2\ln(A-C)$$
$$\ln\left[(A+C)(A+C-2B)\right] = \ln(A-C)^2$$
$$A^2 + 2AC + C^2 - 2AB - 2BC = A^2 - 2AC + C^2$$
$$4AC - 2AB - 2BC = 0$$
$$2AC = AB + BC$$

We wish to prove that $\frac{1}{A}, \frac{1}{B}, \frac{1}{C}$ are consecutive terms in an arithmetic sequence, or: $\frac{1}{B}-\frac{1}{A}=\frac{1}{C}-\frac{1}{B}$

Multiply by ABC:

$$AC-BC=AB-AC$$
$$2AC=AB+BC$$

This is what we have already shown.

73. **a.** Here $a_1=1, r=2, a_{14}=14$, and $d=1$, so: $S=\frac{1-2^{14}\bullet 14}{1-2}+\frac{1\left(2-2^{14}\right)}{(1-2)^2}=-1+229376+2-16384=212993$

b. Here $a_1=2, r=4, a_7=20$, and $d=3$, so: $S=\frac{2-4^7\bullet 20}{1-4}+\frac{3\left(4-4^7\right)}{(1-4)^2}=109226-5460=103766$

c. Here $a_1=3, r=-\frac{1}{2}, a_{10}=21$, and $d=2$, so:

$$\begin{aligned}S&=\frac{3-\left(-\frac{1}{2}\right)^{10}\bullet 21}{1-\left(-\frac{1}{2}\right)}+\frac{2\left[-\frac{1}{2}-\left(-\frac{1}{2}\right)^{10}\right]}{\left[1-\left(-\frac{1}{2}\right)\right]^2}\\&=\frac{3-\frac{21}{1024}}{\frac{3}{2}}+\frac{-1-\frac{1}{512}}{\frac{9}{4}}\\&=2-\frac{7}{512}-\frac{4}{9}-\frac{1}{1152}\\&=\frac{9216-63-2048-4}{4608}\\&=\frac{7101}{4608}\\&=\frac{789}{512}\text{ or }1.541015625\end{aligned}$$

75. Convert to rectangular form: $\cos\frac{\pi}{6}+i\sin\frac{\pi}{6}=\frac{\sqrt{3}}{2}+\frac{1}{2}i$

77. Convert to rectangular form: $5\left[\cos\left(-\frac{\pi}{4}\right)+i\sin\left(-\frac{\pi}{4}\right)\right]=5\left[\frac{\sqrt{2}}{2}-\frac{\sqrt{2}}{2}i\right]=\frac{5\sqrt{2}}{2}-\frac{5\sqrt{2}}{2}i$

79. Convert to trigonometric form: $3i=3(0+1i)=3\left[\cos\frac{\pi}{2}+i\sin\frac{\pi}{2}\right]$

81. Convert to trigonometric form: $2\sqrt{3}-2i=4\left[\frac{\sqrt{3}}{2}-\frac{1}{2}i\right]=4\left[\cos\frac{11\pi}{6}+i\sin\frac{11\pi}{6}\right]$

83. Compute the product: $4\left[\cos\frac{\pi}{12}+i\sin\frac{\pi}{12}\right]\bullet 3\left[\cos\frac{\pi}{12}+i\sin\frac{\pi}{12}\right]=12\left[\cos\frac{\pi}{6}+i\sin\frac{\pi}{6}\right]=12\left[\frac{\sqrt{3}}{2}+\frac{1}{2}i\right]=6\sqrt{3}+6i$

85. Compute the quotient:

$$4[\cos 32^\circ+i\sin 32^\circ]\div\sqrt{2}[\cos 2^\circ+i\sin 2^\circ]=\frac{4}{\sqrt{2}}[\cos 30^\circ+i\sin 30^\circ]=2\sqrt{2}\left[\frac{\sqrt{3}}{2}+\frac{1}{2}i\right]=\sqrt{6}+\sqrt{2}i$$

87. Compute the power using DeMoivre's theorem:

$$\left[2\left(\cos\frac{2\pi}{15}+i\sin\frac{2\pi}{15}\right)\right]^5=2^5\left[\cos\frac{2\pi}{3}+i\sin\frac{2\pi}{3}\right]=32\left[-\frac{1}{2}+\frac{\sqrt{3}}{2}i\right]=-16+16\sqrt{3}i$$

89. First write $\sqrt{2}-\sqrt{2}\,i$ in trigonometric form: $\sqrt{2}-\sqrt{2}i=2\left[\frac{\sqrt{2}}{2}-\frac{\sqrt{2}}{2}i\right]=2\left[\cos\frac{7\pi}{4}+i\sin\frac{7\pi}{4}\right]$

Now compute the power using DeMoivre's theorem:

$$\begin{aligned}\left(\sqrt{2}-\sqrt{2}\,i\right)^{15}&=\left[2\left(\cos\tfrac{7\pi}{4}+i\sin\tfrac{7\pi}{4}\right)\right]^{15}\\&=2^{15}\left[\cos\tfrac{105\pi}{4}+i\sin\tfrac{105\pi}{4}\right]\\&=2^{15}\left[\cos\tfrac{\pi}{4}+i\sin\tfrac{\pi}{4}\right]\\&=2^{15}\left[\tfrac{\sqrt{2}}{2}+\tfrac{\sqrt{2}}{2}i\right]\\&=2^{14}\sqrt{2}+2^{14}\sqrt{2}\,i\\&=16384\sqrt{2}+16384\sqrt{2}\,i\end{aligned}$$

91. Let $z=r(\cos\theta+i\sin\theta)$ and $-64i=64(0-i)=64\left[\cos\frac{3\pi}{2}+i\sin\frac{3\pi}{2}\right]$. Then

$z^3=r^3(\cos 3\theta+i\sin 3\theta)=64\left[\cos\frac{3\pi}{2}+i\sin\frac{3\pi}{2}\right]$. So $r=4$ and $3\theta=\frac{3\pi}{2}+2\pi k$, thus $\theta=\frac{\pi}{2}+\frac{2\pi}{3}k$.

When $k=0$: $z_1=4\left[\cos\frac{\pi}{2}+i\sin\frac{\pi}{2}\right]=4(0+i)=4i$

When $k=1$: $z_2=4\left[\cos\frac{7\pi}{6}+i\sin\frac{7\pi}{6}\right]=4\left[-\frac{\sqrt{3}}{2}-\frac{1}{2}i\right]=-2\sqrt{3}-2i$

When $k=2$: $z_3=4\left[\cos\frac{11\pi}{6}+i\sin\frac{11\pi}{6}\right]=4\left[\frac{\sqrt{3}}{2}-\frac{1}{2}i\right]=2\sqrt{3}-2i$

So the cube roots of $-64i$ are $4i$, $-2\sqrt{3}-2i$ and $2\sqrt{3}-2i$.

93. Let $z=r(\cos\theta+i\sin\theta)$ and $1+\sqrt{3}i=2\left[\frac{1}{2}+\frac{\sqrt{3}}{2}i\right]=2\left[\cos\frac{\pi}{3}+i\sin\frac{\pi}{3}\right]$.

Then: $z^4=r^4(\cos 4\theta+i\sin 4\theta)=2\left[\cos\frac{\pi}{3}+i\sin\frac{\pi}{3}\right]$

So $r^4=2$ and $r=\sqrt[4]{2}$, and also $4\theta=\frac{\pi}{3}+2\pi k$, so $\theta=\frac{\pi}{12}+\frac{\pi}{2}k$.

When $k=0$: $z_1=\sqrt[4]{2}\left[\cos\frac{\pi}{12}+i\sin\frac{\pi}{12}\right]=\frac{\sqrt{2\sqrt{2}}+\sqrt{6\sqrt{2}}}{4}+\frac{\sqrt{6\sqrt{2}}-\sqrt{2\sqrt{2}}}{4}i$

When $k=1$: $z_2=\sqrt[4]{2}\left[\cos\frac{7\pi}{12}+i\sin\frac{7\pi}{12}\right]=\frac{\sqrt{2\sqrt{2}}-\sqrt{6\sqrt{2}}}{4}+\frac{\sqrt{6\sqrt{2}}+\sqrt{2\sqrt{2}}}{4}i$

When $k=2$: $z_3=\sqrt[4]{2}\left[\cos\frac{13\pi}{12}+i\sin\frac{13\pi}{12}\right]=\frac{-\sqrt{6\sqrt{2}}-\sqrt{2\sqrt{2}}}{4}+\frac{\sqrt{2\sqrt{2}}-\sqrt{6\sqrt{2}}}{4}i$

When $k=3$: $z_4=\sqrt[4]{2}\left[\cos\frac{19\pi}{12}+i\sin\frac{19\pi}{12}\right]=\frac{\sqrt{6\sqrt{2}}-\sqrt{2\sqrt{2}}}{4}-\frac{\sqrt{6\sqrt{2}}+\sqrt{2\sqrt{2}}}{4}i$

These are the fourth roots of $1+\sqrt{3}\,i$.

95. Label the vertices A, B, C. Then, by hypothesis:

$$\begin{aligned}\cot B-\cot A&=\cot C-\cot B\\2\cot B-\cot A&=\cot C\\\frac{2\cos B}{\sin B}-\frac{\cos A}{\sin A}&=\frac{\cos C}{\sin C}\end{aligned}$$

Using the formulas given in the text, substitute:

$$\frac{2(c^2+a^2-b^2)/2ca}{\sqrt{T}/2ac}-\frac{(b^2+c^2-a^2)/2bc}{\sqrt{T}/2bc}=\frac{(a^2+b^2-c^2)/2ab}{\sqrt{T}/2ab}$$

$$\begin{aligned}2(c^2+a^2-b^2)-(b^2+c^2-a^2)&=a^2+b^2-c^2\\2c^2+2a^2-2b^2-b^2-c^2+a^2&=a^2+b^2-c^2\\2a^2-2b^2&=2b^2-2c^2\\b^2-a^2&=c^2-b^2\end{aligned}$$

Therefore a^2, b^2, and c^2 are consecutive terms in an arithmetic sequence.

Chapter 13 Test

1. For $n = 1$, we have: $1^2 = \frac{1(1+1)(2+1)}{6} = \frac{6}{6} = 1$. Assume P_k is true, so: $1^2 + 2^2 + 3^2 + \ldots + k^2 = \frac{k(k+1)(2k+1)}{6}$

Then:

$$\begin{aligned}1^2 + 2^2 + \ldots + k^2 + (k+1)^2 &= \frac{k(k+1)(2k+1)}{6} + (k+1)^2\\ &= (k+1)\left[\frac{k(2k+1)}{6} + k + 1\right]\\ &= (k+1)\left(\frac{2k^2 + k + 6k + 6}{6}\right)\\ &= \frac{(k+1)(2k^2 + 7k + 6)}{6}\\ &= \frac{(k+1)(k+2)(2k+3)}{6}\\ &= \frac{(k+1)(k+2)[2(k+1)+1]}{6}\end{aligned}$$

So P_{k+1} is true and the induction is complete.

2. a. Expanding the sum: $\sum_{k=0}^{2}(10k-1) = -1 + 9 + 19 = 27$

   b. Expanding the sum: $\sum_{k=1}^{3}(-1)^k k^2 = -1 + 4 - 9 = -6$

3. a. $S_n = \frac{a(1-r^n)}{1-r}$

   b. Here $a = \frac{3}{2}$, $r = \frac{3}{2}$, and $n = 10$, so: $S_{10} = \frac{\frac{3}{2}\left[1-\left(\frac{3}{2}\right)^{10}\right]}{1-\frac{3}{2}} = \frac{\frac{3}{2}\left(1-\frac{59049}{1024}\right)}{-\frac{1}{2}} = -3\left(-\frac{58025}{1024}\right) = \frac{174075}{1024}$

4. a. Here $11 - r + 1 = 3$, so $r = 9$, and the coefficient is: $\binom{11}{8}(-2)^8 = \frac{(11)(10)(9)}{(3)(2)}(256) = 42240$

   b. The fifth term is: $\binom{11}{4}(a)^7(-2b^3)^4 = \frac{(11)(10)(9)(8)}{(4)(3)(2)}a^7(16b^{12}) = 5280a^7b^{12}$

5. Expand using the binomial theorem:

$$\begin{aligned}(3x^2+y^3)^5 &= \binom{5}{0}(3x^2)^5 + \binom{5}{1}(3x^2)^4(y^3) + \binom{5}{2}(3x^2)^3(y^3)^2 + \binom{5}{3}(3x^2)^2(y^3)^3\\ &\quad + \binom{5}{4}(3x^2)(y^3)^4 + \binom{5}{5}(y^3)^5\\ &= 243x^{10} + 405x^8y^3 + 270x^6y^6 + 90x^4y^9 + 15x^2y^{12} + y^{15}\end{aligned}$$

6. The sum is: $S_{12} = 12\left(\frac{8+\frac{43}{2}}{2}\right) = \frac{59}{2}(6) = 177$

7. The sum is: $S = \frac{a}{1-r} = \frac{\frac{7}{10}}{1-\frac{1}{10}} = \frac{7}{9}$

8. Here $a_1 = 1$ and $a_2 = 1$, so: $a_3 = a_2^2 + a_1 = 1 + 1 = 2$, $a_4 = a_3^2 + a_2 = 2^2 + 1 = 5$, $a_5 = a_4^2 + a_3 = 5^2 + 2 = 27$

The fourth and fifth terms are 5 and 27, respectively.

9. Since $a_3 = 4$ and $a_5 = 10$, $a_5 = a_3 r^2$, so:

$$10 = 4r^2$$
$$\frac{5}{2} = r^2$$
$$r = -\sqrt{\frac{5}{2}} = -\frac{\sqrt{10}}{2}$$

Then $a_6 = ra_5 = -\frac{\sqrt{10}}{2}(10) = -5\sqrt{10}$.

10. Here $a_n = -61 + 15(n-1)$, so: $a_{20} = -61 + 15(19) = 224$

11. a. The first four terms are given in the table:

n	0	1	2	3
a_n	0	0	1	3

Graphing these terms:

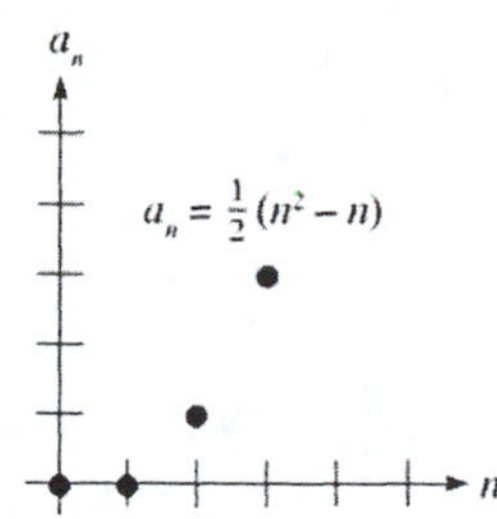

b. The first four terms are given in the table:

n	1	2	3	4
b_n	1	−1	4	0

Graphing these terms:

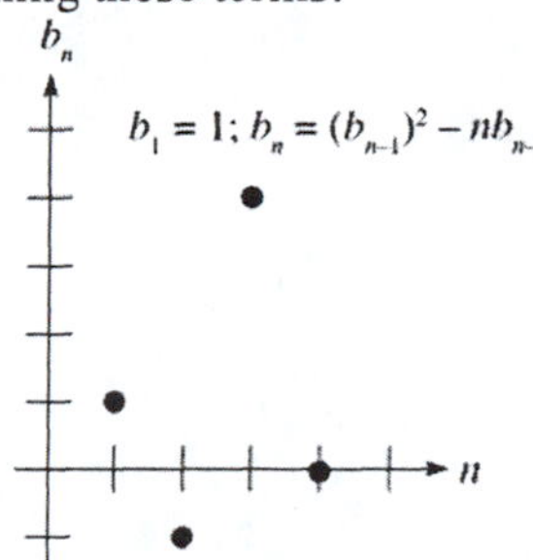

12. Convert to rectangular form: $z = 2\left(\cos\frac{2\pi}{3} + i\sin\frac{2\pi}{3}\right) = 2\left(-\frac{1}{2} + \frac{\sqrt{3}}{2}i\right) = -1 + \sqrt{3}i$

13. Convert to trigonometric form: $\sqrt{2} - \sqrt{2}i = 2\left(\frac{\sqrt{2}}{2} - \frac{\sqrt{2}}{2}i\right) = 2\left(\cos\frac{7\pi}{4} + i\sin\frac{7\pi}{4}\right)$

14. Compute the product: $zw = 3\left(\cos\frac{2\pi}{9} + i\sin\frac{2\pi}{9}\right) \cdot 5\left(\cos\frac{\pi}{9} + i\sin\frac{\pi}{9}\right) = 15\left(\cos\frac{\pi}{3} + i\sin\frac{\pi}{3}\right) = 15\left(\frac{1}{2} + \frac{\sqrt{3}}{2}i\right) = \frac{15}{2} + \frac{15\sqrt{3}}{2}i$

15. Let $z = r(\cos\theta + i\sin\theta)$ be a cube root of $64i$. Write $64i$ in trigonometric form: $64i = 64(0 + 1i) = 64\left(\cos\frac{\pi}{2} + i\sin\frac{\pi}{2}\right)$

Thus: $z^3 = r^3(\cos 3\theta + i\sin 3\theta) = 64\left(\cos\frac{\pi}{2} + i\sin\frac{\pi}{2}\right)$

So $r^3 = 64$ and thus $r = 4$, and $3\theta = \frac{\pi}{2} + 2\pi k$ thus $\theta = \frac{\pi}{6} + \frac{2\pi}{3}k$.

When $k = 0$: $z_1 = 4\left(\cos\frac{\pi}{6} + i\sin\frac{\pi}{6}\right) = 4\left(\frac{\sqrt{3}}{2} + \frac{1}{2}i\right) = 2\sqrt{3} + 2i$

When $k = 1$: $z_2 = 4\left(\cos\frac{5\pi}{6} + i\sin\frac{5\pi}{6}\right) = 4\left(-\frac{\sqrt{3}}{2} + \frac{1}{2}i\right) = -2\sqrt{3} + 2i$

When $k = 2$: $z_3 = 4\left(\cos\frac{3\pi}{2} + i\sin\frac{3\pi}{2}\right) = 4(0 - i) = -4i$

So the cube roots of $64i$ are $2\sqrt{3} + 2i$, $-2\sqrt{3} + 2i$, and $-4i$.

Appendix B

B.1 Review of Integer Exponents

1. Substituting $x=-2$: $2(-2)^3-(-2)+4=2(-8)+2+4=-10$

3. Substituting $x=-\frac{1}{2}$: $\dfrac{1-2\left(-\frac{1}{2}\right)^2}{1+2\left(-\frac{1}{2}\right)^3}=\dfrac{1-2\left(\frac{1}{4}\right)}{1+2\left(-\frac{1}{8}\right)}=\dfrac{1-\frac{1}{2}}{1-\frac{1}{4}}=\dfrac{\frac{1}{2}}{\frac{3}{4}}=\dfrac{2}{3}$

5. **a.** Simplifying: $a^3a^{12}=a^{3+12}=a^{15}$ **b.** Simplifying: $(a+1)^3(a+1)^{12}=(a+1)^{15}$

c. Simplifying: $(a+1)^{12}(a+1)^3=(a+1)^{15}$

7. **a.** Simplifying: $yy^2y^8=y^{1+2+8}=y^{11}$ **b.** Simplifying: $(y+1)(y+1)^2(y+1)^8=(y+1)^{11}$

c. Simplifying: $\left[(y+1)(y+1)^8\right]^2=\left[(y+1)^9\right]^2=(y+1)^{18}$

9. **a.** Using properties of exponents: $\dfrac{\left(x^2+3\right)^{10}}{\left(x^2+3\right)^9}=\left(x^2+3\right)^{10-9}=x^2+3$

b. Using properties of exponents: $\dfrac{\left(x^2+3\right)^9}{\left(x^2+3\right)^{10}}=\left(x^2+3\right)^{9-10}=\left(x^2+3\right)^{-1}=\dfrac{1}{x^2+3}$

c. Using properties of exponents: $\dfrac{12^{10}}{12^9}=12^{10-9}=12$

11. **a.** Using properties of exponents: $\dfrac{t^{15}}{t^9}=t^{15-9}=t^6$ **b.** Using properties of exponents: $\dfrac{t^9}{t^{15}}=t^{9-15}=t^{-6}=\dfrac{1}{t^6}$

c. Using properties of exponents: $\dfrac{\left(t^2+3\right)^{15}}{\left(t^2+3\right)^9}=\left(t^2+3\right)^{15-9}=\left(t^2+3\right)^6$

13. a. Using properties of exponents: $\dfrac{x^6y^{15}}{x^2y^{20}}=\dfrac{x^{6-2}}{y^{20-15}}=\dfrac{x^4}{y^5}$

b. Using properties of exponents: $\dfrac{x^2y^{20}}{x^6y^{15}}=\dfrac{y^{20-15}}{x^{6-2}}=\dfrac{y^5}{x^4}$

c. Using properties of exponents: $\left(\dfrac{x^2y^{20}}{x^6y^{15}}\right)^2=\left(\dfrac{y^{20-15}}{x^{6-2}}\right)^2=\left(\dfrac{y^5}{x^4}\right)^2=\dfrac{y^{10}}{x^8}$

15. a. Simplifying: $4\left(x^3\right)^2=4x^6$

b. Simplifying: $\left(4x^3\right)^2=4^2x^6=16x^6$

c. Using properties of exponents: $\dfrac{\left(4x^2\right)^3}{\left(4x^3\right)^2}=\dfrac{4^3x^6}{4^2x^6}=4^{3-2}=4$

17. a. Since $x^0=1$ for any value of $x\neq0$, $64^0=1$.

b. Since $x^0=1$ for any value of $x\neq0$, $\left(64^3\right)^0=1$.

c. Since $x^0=1$ for any value of $x\neq0$, $\left(64^0\right)^3=(1)^3=1$.

19. a. Evaluating negative exponents: $10^{-1}+10^{-2}=\frac{1}{10}+\frac{1}{100}=\frac{10}{100}+\frac{1}{100}=\frac{11}{100}$

b. Using the result from part (a): $\left(10^{-1}+10^{-2}\right)^{-1}=\left(\frac{11}{100}\right)^{-1}=\frac{100}{11}$

c. Evaluating negative exponents: $\left[\left(10^{-1}\right)\left(10^{-2}\right)\right]^{-1}=\left[\frac{1}{10}\bullet\frac{1}{100}\right]^{-1}=\left(\frac{1}{1000}\right)^{-1}=1000$

21. Evaluating negative exponents: $\left(\frac{1}{3}\right)^{-1}+\left(\frac{1}{4}\right)^{-1}=3+4=7$

23. a. Evaluating negative exponents: $5^{-2}+10^{-2}=\frac{1}{25}+\frac{1}{100}=\frac{4}{100}+\frac{1}{100}=\frac{5}{100}=\frac{1}{20}$

b. Evaluating negative exponents: $(5+10)^{-2}=(15)^{-2}=\frac{1}{225}$

25. Using properties of exponents: $\left(a^2bc^0\right)^{-3}=\dfrac{1}{\left(a^2b\right)^3}=\dfrac{1}{a^6b^3}$

27. Using properties of exponents: $\left(a^{-2}b^{-1}c^3\right)^{-2}=a^4b^2c^{-6}=\dfrac{a^4b^2}{c^6}$

29. Using properties of exponents: $\left(\dfrac{x^3y^{-2}z}{xy^2z^{-3}}\right)^{-3}=\dfrac{x^{-9}y^6z^{-3}}{x^{-3}y^{-6}z^9}=\dfrac{y^{6-(-6)}}{x^{-3+9}z^{9+3}}=\dfrac{y^{12}}{x^6z^{12}}$

31. Using properties of exponents: $\left(\dfrac{x^4y^{-8}z^2}{xy^2z^{-6}}\right)^{-2}=\dfrac{x^{-8}y^{16}z^{-4}}{x^{-2}y^{-4}z^{12}}=\dfrac{y^{16+4}}{x^{-2+8}z^{12+4}}=\dfrac{y^{20}}{x^6z^{16}}$

33. Using properties of exponents: $\dfrac{x^2}{y^{-3}}\div\dfrac{x^2}{y^3}=\dfrac{x^2}{y^{-3}}\bullet\dfrac{y^3}{x^2}=\dfrac{x^2y^3}{x^2y^{-3}}=y^6$

35. Subtracting exponents: $\dfrac{b^{p+1}}{b^p}=b^{p+1-p}=b$

37. Adding exponents: $\left(x^p\right)\left(x^p\right)=x^{p+p}=x^{2p}$

39. Factoring $9=3^2$ and $12=2^2\bullet3$: $\dfrac{2^8\bullet3^{15}}{9\bullet3^{10}\bullet12}=\dfrac{2^8\bullet3^{15}}{3^2\bullet3^{10}\bullet2^2\bullet3}=\dfrac{2^8\bullet3^{15}}{2^2\bullet3^{13}}=2^6\bullet3^2=576$

41. Factoring $24 = 2^3 \bullet 3$, $32 = 2^5$, and $12 = 2^2 \bullet 3$: $\dfrac{24^5}{32 \bullet 12^4} = \dfrac{\left(2^3 \bullet 3\right)^5}{2^5 \bullet \left(2^2 \bullet 3\right)^4} = \dfrac{2^{15} \bullet 3^5}{2^5 \bullet 2^8 \bullet 3^4} = \dfrac{2^{15} \bullet 3^5}{2^{13} \bullet 3^4} = 2^2 \bullet 3 = 12$

43. Written in scientific notation: 9.29×10^7 miles

45. Written in scientific notation: 6.68×10^4 mph

47. Written in scientific notation: 2.5×10^{19} miles

49. a. Written in scientific notation: 1.0×10^{-9} seconds
 b. Written in scientific notation: 1.0×10^{-18} seconds
 c. Written in scientific notation: 1.0×10^{-24} seconds

B.2 Review of *n*th Roots

1. false

3. true

5. true

7. true

9. a. $\sqrt[3]{-64} = -4$
 b. $\sqrt[4]{-64}$ is not a real number

11. a. $\sqrt[3]{\frac{8}{125}} = \frac{2}{5}$
 b. $\sqrt[3]{-\frac{8}{125}} = -\frac{2}{5}$

13. a. $\sqrt{-16}$ is not a real number
 b. $\sqrt[4]{-16}$ is not a real number

15. a. $\sqrt[4]{\frac{256}{81}} = \frac{4}{3}$
 b. $\sqrt[3]{-\frac{27}{125}} = -\frac{3}{5}$

17. a. $\sqrt[5]{-32} = -2$
 b. $-\sqrt[5]{-32} = -(-2) = 2$

19. a. Factoring $18 = 9 \bullet 2$: $\sqrt{18} = \sqrt{9}\sqrt{2} = 3\sqrt{2}$
 b. Factoring $54 = 27 \bullet 2$: $\sqrt[3]{54} = \sqrt[3]{27}\sqrt[3]{2} = 3\sqrt[3]{2}$

21. a. Factoring $98 = 49 \bullet 2$: $\sqrt{98} = \sqrt{49}\sqrt{2} = 7\sqrt{2}$
 b. Factoring $-64 = -32 \bullet 2$: $\sqrt[5]{-64} = \sqrt[5]{-32}\sqrt[5]{2} = -2\sqrt[5]{2}$

23. a. $\sqrt{\frac{25}{4}} = \frac{5}{2}$
 b. $\sqrt[4]{\frac{16}{625}} = \frac{2}{5}$

25. a. Factoring $8 = 4 \bullet 2$: $\sqrt{2} + \sqrt{8} = \sqrt{2} + \sqrt{4}\sqrt{2} = \sqrt{2} + 2\sqrt{2} = 3\sqrt{2}$
 b. Factoring $16 = 8 \bullet 2$: $\sqrt[3]{2} + \sqrt[3]{16} = \sqrt[3]{2} + \sqrt[3]{8}\sqrt[3]{2} = \sqrt[3]{2} + 2\sqrt[3]{2} = 3\sqrt[3]{2}$

27. a. Factoring $50 = 25 \bullet 2$ and $128 = 64 \bullet 2$: $4\sqrt{50} - 3\sqrt{128} = 4\sqrt{25}\sqrt{2} - 3\sqrt{64}\sqrt{2} = 20\sqrt{2} - 24\sqrt{2} = -4\sqrt{2}$
 b. Factoring $32 = 16 \bullet 2$ and $162 = 81 \bullet 2$: $\sqrt[4]{32} + \sqrt[4]{162} = \sqrt[4]{16}\sqrt[4]{2} + \sqrt[4]{81}\sqrt[4]{2} = 2\sqrt[4]{2} + 3\sqrt[4]{2} = 5\sqrt[4]{2}$

29. a. The value is 0.3 [because $(0.3)^2 = 0.09$].
 b. The value is 0.2 [because $(0.2)^3 = 0.008$].

31. Simplifying each radical yields: $4\sqrt{24} - 8\sqrt{54} + 2\sqrt{6} = 4\sqrt{4}\sqrt{6} - 8\sqrt{9}\sqrt{6} + 2\sqrt{6} = 8\sqrt{6} - 24\sqrt{6} + 2\sqrt{6} = -14\sqrt{6}$

33. Simplifying: $\sqrt{\sqrt{64}} = \sqrt{8} = \sqrt{4}\sqrt{2} = 2\sqrt{2}$

35. a. Simplifying: $\sqrt{36x^2} = 6x$, since $x > 0$
 b. Simplifying: $\sqrt{36y^2} = -6y$, since $y < 0$

37. a. Multiplying radicals yields: $\sqrt{ab^2}\sqrt{a^2b} = \sqrt{a^3b^3} = ab\sqrt{ab}$
 b. Multiplying radicals yields: $\sqrt{ab^3}\sqrt{a^3b} = \sqrt{a^4b^4} = a^2b^2$

39. Simplifying the radical yields: $\sqrt{72a^3b^4c^5} = \sqrt{36a^2b^4c^4}\sqrt{2ac} = 6ab^2c^2\sqrt{2ac}$

41. Simplifying the radical yields: $\sqrt[4]{16a^4b^5} = \sqrt[4]{16a^4b^4} \bullet \sqrt[4]{b} = 2ab\sqrt[4]{b}$

43. Simplifying the radical yields: $\sqrt[3]{\dfrac{16a^{12}b^2}{c^9}} = \dfrac{\sqrt[3]{8a^{12}}\sqrt[3]{2b^2}}{\sqrt[3]{c^9}} = \dfrac{2a^4\sqrt[3]{2b^2}}{c^3}$

45. Rationalizing the denominator: $\dfrac{4}{\sqrt{7}} \bullet \dfrac{\sqrt{7}}{\sqrt{7}} = \dfrac{4\sqrt{7}}{7}$

47. Rationalizing the denominator: $\dfrac{1}{\sqrt{8}} \bullet \dfrac{\sqrt{2}}{\sqrt{2}} = \dfrac{\sqrt{2}}{\sqrt{16}} = \dfrac{\sqrt{2}}{4}$

49. Rationalizing the denominator: $\dfrac{1}{1+\sqrt{5}} \bullet \dfrac{1-\sqrt{5}}{1-\sqrt{5}} = \dfrac{1-\sqrt{5}}{1-5} = \dfrac{1-\sqrt{5}}{-4} = \dfrac{\sqrt{5}-1}{4}$

51. Rationalizing the denominator: $\frac{1+\sqrt{3}}{1-\sqrt{3}} \bullet \frac{1+\sqrt{3}}{1+\sqrt{3}} = \frac{1+\sqrt{3}+\sqrt{3}+3}{1-3} = \frac{4+2\sqrt{3}}{-2} = -2-\sqrt{3}$

53. Rationalizing the denominator and simplifying:

$$\frac{1}{\sqrt{5}} + 4\sqrt{45} = \frac{1}{\sqrt{5}} \bullet \frac{\sqrt{5}}{\sqrt{5}} + 4\sqrt{9}\sqrt{5} = \frac{\sqrt{5}}{5} + 12\sqrt{5} = \frac{\sqrt{5}}{5} + \frac{60\sqrt{5}}{5} = \frac{61\sqrt{5}}{5}$$

55. Rationalizing the denominator: $\frac{1}{\sqrt[3]{25}} \bullet \frac{\sqrt[3]{5}}{\sqrt[3]{5}} = \frac{\sqrt[3]{5}}{\sqrt[3]{125}} = \frac{\sqrt[3]{5}}{5}$

57. Rationalizing the denominator: $\frac{3}{\sqrt[4]{3}} \bullet \frac{\sqrt[4]{3^3}}{\sqrt[4]{3^3}} = \frac{3\sqrt[4]{27}}{3} = \sqrt[4]{27}$

59. Rationalizing the denominator: $\frac{3}{\sqrt[5]{16a^4b^9}} \bullet \frac{\sqrt[5]{2ab}}{\sqrt[5]{2ab}} = \frac{3\sqrt[5]{2ab}}{\sqrt[5]{32a^5b^{10}}} = \frac{3\sqrt[5]{2ab}}{2ab^2}$

61. a. Rationalizing the denominator: $\frac{2}{\sqrt{3}+1} \bullet \frac{\sqrt{3}-1}{\sqrt{3}-1} = \frac{2\left(\sqrt{3}-1\right)}{3-1} = \frac{2\left(\sqrt{3}-1\right)}{2} = \sqrt{3}-1$

b. Rationalizing the denominator: $\frac{2}{\sqrt{3}-1} \bullet \frac{\sqrt{3}+1}{\sqrt{3}+1} = \frac{2\left(\sqrt{3}+1\right)}{3-1} = \frac{2\left(\sqrt{3}+1\right)}{2} = \sqrt{3}+1$

63. a. Rationalizing the denominator: $\frac{x}{\sqrt{x}-2} \bullet \frac{\sqrt{x}+2}{\sqrt{x}+2} = \frac{x\left(\sqrt{x}+2\right)}{x-4}$

b. Rationalizing the denominator: $\frac{x}{\sqrt{x}-y} \bullet \frac{\sqrt{x}+y}{\sqrt{x}+y} = \frac{x\left(\sqrt{x}+y\right)}{x-y^2}$

65. a. Rationalizing the denominator: $\frac{\sqrt{x}+\sqrt{2}}{\sqrt{x}-\sqrt{2}} \bullet \frac{\sqrt{x}+\sqrt{2}}{\sqrt{x}+\sqrt{2}} = \frac{x+2\sqrt{2x}+2}{x-2}$

b. Rationalizing the denominator: $\frac{\sqrt{x}+\sqrt{a}}{\sqrt{x}-\sqrt{a}} \bullet \frac{\sqrt{x}+\sqrt{a}}{\sqrt{x}+\sqrt{a}} = \frac{x+2\sqrt{ax}+a}{x-a}$

67. Rationalizing the denominator: $\frac{-2}{\sqrt{x+h}-\sqrt{x}} \bullet \frac{\sqrt{x+h}+\sqrt{x}}{\sqrt{x+h}+\sqrt{x}} = \frac{-2\left(\sqrt{x+h}+\sqrt{x}\right)}{x+h-x} = \frac{-2\left(\sqrt{x+h}+\sqrt{x}\right)}{h}$

69. Rationalizing the denominator: $\frac{1}{\sqrt[3]{a}-1} \bullet \frac{\left(\sqrt[3]{a}\right)^2+\sqrt[3]{a}+1}{\left(\sqrt[3]{a}\right)^2+\sqrt[3]{a}+1} = \frac{\sqrt[3]{a^2}+\sqrt[3]{a}+1}{\left(\sqrt[3]{a}\right)^3-1} = \frac{\sqrt[3]{a^2}+\sqrt[3]{a}+1}{a-1}$

71. Rationalizing the numerator: $\frac{\sqrt{x}-\sqrt{5}}{x-5} \bullet \frac{\sqrt{x}+\sqrt{5}}{\sqrt{x}+\sqrt{5}} = \frac{x-5}{(x-5)\left(\sqrt{x}+\sqrt{5}\right)} = \frac{1}{\sqrt{x}+\sqrt{5}}$

73. Rationalizing the numerator: $\frac{\sqrt{2+h}+\sqrt{2}}{h} \bullet \frac{\sqrt{2+h}-\sqrt{2}}{\sqrt{2+h}-\sqrt{2}} = \frac{2+h-2}{h\left(\sqrt{2+h}-\sqrt{2}\right)} = \frac{h}{h\left(\sqrt{2+h}-\sqrt{2}\right)} = \frac{1}{\sqrt{2+h}-\sqrt{2}}$

75. Rationalizing the numerator: $\frac{\sqrt{x+h}-\sqrt{x}}{h} \bullet \frac{\sqrt{x+h}+\sqrt{x}}{\sqrt{x+h}+\sqrt{x}} = \frac{x+h-x}{h\left(\sqrt{x+h}+\sqrt{x}\right)} = \frac{h}{h\left(\sqrt{x+h}+\sqrt{x}\right)} = \frac{1}{\sqrt{x+h}+\sqrt{x}}$

77. Let $a=9$ and $b=16$, then:

$\sqrt{a+b} = \sqrt{9+16} = \sqrt{25} = 5$ $\qquad$ $\sqrt{a}+\sqrt{b} = \sqrt{9}+\sqrt{16} = 3+4 = 7$

So the formula is not valid.

79. Let $u = 1$ and $v = 8$, then:

$\sqrt[3]{u+v} = \sqrt[3]{1+8} = \sqrt[3]{9} \approx 2.08$ $\qquad \sqrt[3]{u} + \sqrt[3]{v} = \sqrt[3]{1} + \sqrt[3]{8} = 1+2 = 3$

So the formula is not valid.

81. **a.** To six decimal places, both values are 1.645751.

b. Following the hint, square each side: $\left(\sqrt{7}-1\right)^2 = 7 - 2\sqrt{7} + 1 = 8 - 2\sqrt{7}$

B.3 Review of Rational Exponents

1. Writing in radical form: $a^{3/5} = \sqrt[5]{a^3} = \left(\sqrt[5]{a}\right)^3$

3. Writing in radical form: $5^{2/3} = \sqrt[3]{5^2} = \left(\sqrt[3]{5}\right)^2$

5. Writing in radical form: $\left(x^2+1\right)^{3/4} = \sqrt[4]{\left(x^2+1\right)^3} = \left(\sqrt[4]{x^2+1}\right)^3$

7. Writing in radical form: $2^{xy/3} = \sqrt[3]{2^{xy}} = \left(\sqrt[3]{2}\right)^{xy}$

9. Writing in exponential form: $\sqrt[3]{p^2} = p^{2/3}$

11. Writing in exponential form: $\sqrt[7]{(1+u)^4} = (1+u)^{4/7}$

13. Writing in exponential form: $\sqrt[p]{\left(a^2+b^2\right)^3} = \left(a^2+b^2\right)^{3/p}$

15. Simplifying: $16^{1/2} = \sqrt{16} = 4$

17. Simplifying: $\left(\frac{1}{36}\right)^{1/2} = \sqrt{\frac{1}{36}} = \frac{1}{6}$

19. Simplifying: $(-16)^{1/2} = \sqrt{-16}$, which is not a real number

21. Simplifying: $625^{1/4} = \sqrt[4]{625} = 5$

23. Simplifying: $8^{1/3} = \sqrt[3]{8} = 2$

25. Simplifying: $8^{2/3} = \left(\sqrt[3]{8}\right)^2 = (2)^2 = 4$

27. Simplifying: $(-32)^{1/5} = \sqrt[5]{-32} = -2$

29. Simplifying: $(-1000)^{1/3} = \sqrt[3]{-1000} = -10$

31. Simplifying: $49^{-1/2} = \left(\sqrt{49}\right)^{-1} = 7^{-1} = \frac{1}{7}$

33. Simplifying: $(-49)^{-1/2} = \left(\sqrt{-49}\right)^{-1}$, which is not a real number

35. Simplifying: $(36)^{-3/2} = \left(\sqrt{36}\right)^{-3} = 6^{-3} = \frac{1}{6^3} = \frac{1}{216}$

37. Simplifying: $125^{2/3} = \left(\sqrt[3]{125}\right)^2 = (5)^2 = 25$

39. Simplifying: $(-1)^{3/5} = \left(\sqrt[5]{-1}\right)^3 = (-1)^3 = -1$

41. Simplifying: $32^{4/5} - 32^{-4/5} = \left(\sqrt[5]{32}\right)^4 - \left(\sqrt[5]{32}\right)^{-4} = 2^4 - 2^{-4} = 16 - \frac{1}{16} = \frac{255}{16}$

43. Simplifying:

$$\left(\frac{9}{16}\right)^{-5/2} - \left(\frac{1000}{27}\right)^{4/3} = \left(\frac{16}{9}\right)^{5/2} - \left(\frac{1000}{27}\right)^{4/3} = \left(\sqrt{\frac{16}{9}}\right)^5 - \left(\sqrt[3]{\frac{1000}{27}}\right)^4 = \left(\frac{4}{3}\right)^5 - \left(\frac{10}{3}\right)^4 = \frac{1024}{243} - \frac{10000}{81} = -\frac{28976}{243}$$

45. Simplifying: $\left(2a^{1/3}\right)\left(3a^{1/4}\right) = 6a^{7/12}$, since $\frac{1}{3} + \frac{1}{4} = \frac{4}{12} + \frac{3}{12} = \frac{7}{12}$

47. Simplifying: $\sqrt[4]{\frac{64a^{2/3}}{a^{1/3}}} = \sqrt[4]{64a^{1/3}} = \left(2^6 a^{1/3}\right)^{1/4} = 2^{3/2} a^{1/12}$

49. Subtracting exponents yields: $\dfrac{\left(x^2+1\right)^{3/4}}{\left(x^2+1\right)^{-1/4}} = \left(x^2+1\right)^{4/4} = x^2+1$

51. **a.** Using rational exponents: $\sqrt{3}\sqrt[3]{6} = 3^{1/2} 6^{1/3} = 3^{1/2} 2^{1/3} 3^{1/3} = 2^{1/3} 3^{5/6}$

b. Using rational exponents: $3^{1/2} 6^{1/3} = 3^{3/6} 6^{2/6} = \sqrt[6]{3^3 6^2} = \sqrt[6]{972}$

53. a. Using rational exponents: $\sqrt[3]{6}\,\sqrt[4]{2} = 6^{1/3}\,2^{1/4} = 2^{1/3}3^{1/3}2^{1/4} = 2^{7/12}\,3^{1/3}$

b. Using rational exponents: $6^{1/3}\,2^{1/4} = 6^{4/12}\,2^{3/12} = \sqrt[12]{6^4\,2^3} = \sqrt[12]{10368}$

55. a. Using rational exponents: $\sqrt[3]{x^2}\,\sqrt[5]{y^4} = x^{2/3}\,y^{4/5}$

b. Using rational exponents: $x^{2/3}\,y^{4/5} = x^{10/15}\,y^{12/15} = \sqrt[15]{x^{10}\,y^{12}}$

57. a. Using rational exponents: $\sqrt[4]{x^a}\,\sqrt[3]{x^b}\,\sqrt{x^{a/6}} = x^{a/4}\,x^{b/3}\,x^{a/12} = x^{3a/12}\,x^{4b/12}\,x^{a/12} = x^{(4a+4b)/12} = x^{(a+b)/3}$

b. Using rational exponents: $x^{(a+b)/3} = \sqrt[3]{x^{a+b}}$

59. Rewriting: $\sqrt[3]{(x+1)^2} = (x+1)^{2/3}$

61. Rewriting: $\left(\sqrt[5]{x+y}\right)^2 = (x+y)^{2/5}$

63. Using rational exponents: $\sqrt[3]{\sqrt{x}} + \sqrt{\sqrt[3]{x}} = \left(x^{1/2}\right)^{1/3} + \left(x^{1/3}\right)^{1/2} = x^{1/6} + x^{1/6} = 2x^{1/6}$

65. Using rational exponents: $\sqrt{\sqrt[3]{x}\,\sqrt[4]{y}} = \left(x^{1/3}\,y^{1/4}\right)^{1/2} = x^{1/6}\,y^{1/8}$

67. Compute $9^{10/9} \approx 11.5$ and $10^{9/10} \approx 7.9$. Thus $9^{10/9}$ is larger.

69. Using $a = 9$ and $b = 16$:

$(a+b)^{1/2} = (9+16)^{1/2} = 25^{1/2} = 5$ $\qquad a^{1/2} + b^{1/2} = 9^{1/2} + 16^{1/2} = 3 + 4 = 7$

So the formula is not valid.

71. Using $u = 1$ and $v = 8$:

$(u+v)^{1/3} = (1+8)^{1/3} = 9^{1/3} \approx 2.08$ $\qquad u^{1/3} + v^{1/3} = 1^{1/3} + 8^{1/3} = 1 + 2 = 3$

So the formula is not valid.

73. Using $x = 4$ and $m = 2$:

$x^{1/m} = 4^{1/2} = 2$ $\qquad \dfrac{1}{x^m} = \dfrac{1}{4^2} = \dfrac{1}{16}$

So the formula is not valid.

B.4 Review of Factoring

1. a. Using a difference of squares: $x^2 - 64 = (x+8)(x-8)$

b. Using a common factor of $7x^2$: $7x^4 + 14x^2 = 7x^2\left(x^2+2\right)$

c. Using a common factor of z then a difference of squares: $121z - z^3 = z\left(121 - z^2\right) = z(11+z)(11-z)$

d. Using a difference of squares: $a^2b^2 - c^2 = (ab)^2 - (c)^2 = (ab+c)(ab-c)$

3. a. Factoring by trial and error: $x^2 + 2x - 3 = (x+3)(x-1)$

b. Factoring by trial and error: $x^2 - 2x - 3 = (x-3)(x+1)$

c. Factoring by trial and error, we find that $x^2 - 2x + 3$ is irreducible.

d. Factoring by trial and error: $-x^2 + 2x + 3 = (-x+3)(x+1)$

Note that we could also use a common factor of –1, then use trial and error:

$$-x^2 + 2x + 3 = -1\left(x^2 - 2x - 3\right) = -(x-3)(x+1)$$

5. a. By the sum of cubes: $x^3 + 1 = (x+1)\left(x^2 - x + 1\right)$

b. Since $6^3 = 216$, use the sum of cubes to obtain: $x^3 + 216 = x^3 + 6^3 = (x+6)\left(x^2 - 6x + 36\right)$

c. Using a common factor of 8 then a difference of cubes:

$$1000 - 8x^6 = 8\left(125 - x^6\right) = 8\left(5^3 - x^6\right) = 8\left(5 - x^2\right)\left(25 + 5x^2 + x^4\right)$$

d. By the difference of cubes: $64a^3x^3 - 125 = (4ax)^3 - 5^3 = (4ax - 5)\left(16a^2x^2 + 20ax + 25\right)$

7. **a.** Using a difference of squares: $144 - x^2 = (12 + x)(12 - x)$

b. Since this is a sum of squares, the expression $144 + x^2$ is irreducible.

c. Using a difference of squares: $144 - (y-3)^2 = (12 + y - 3)(12 - y + 3) = (9 + y)(15 - y)$

9. **a.** Using a common factor of h^3 then a difference of squares: $h^3 - h^5 = h^3\left(1 - h^2\right) = h^3(1+h)(1-h)$

b. Using a common factor of h^3 then a difference of squares: $100h^3 - h^5 = h^3\left(100 - h^2\right) = h^3(10+h)(10-h)$

c. Using a common factor of $(h+1)^3$ then a difference of squares:

$$100(h+1)^3 - (h+1)^5 = (h+1)^3\left[100 - (h+1)^2\right] = (h+1)^3(10+h+1)(10-h-1) = (h+1)^3(11+h)(9-h)$$

11. **a.** Factoring by trial and error: $x^2 - 13x + 40 = (x-8)(x-5)$

b. Factoring by trial and error, we find that $x^2 - 13x - 40$ is irreducible.

13. **a.** Factoring by trial and error: $x^2 + 5x - 36 = (x+9)(x-4)$

b. Factoring by trial and error: $x^2 - 13x + 36 = (x-9)(x-4)$

15. **a.** Factoring by trial and error: $3x^2 - 22x - 16 = (3x+2)(x-8)$

b. Factoring by trial and error, we find that $3x^2 - x - 16$ is irreducible.

17. **a.** Factoring by trial and error: $6x^2 + 13x - 5 = (3x-1)(2x+5)$

b. Factoring by trial and error: $6x^2 - x - 5 = (6x+5)(x-1)$

19. **a.** Using the special product for $(A+B)^2$: $t^4 + 2t^2 + 1 = \left(t^2+1\right)^2$

b. Using the special product for $(A-B)^2$ and the difference of squares: $t^4 - 2t^2 + 1 = \left(t^2-1\right)^2 = (t+1)^2(t-1)^2$

c. By trial and error, we find that $t^4 - 2t^2 - 1$ is irreducible.

21. **a.** Using the common factor x: $4x^3 - 20x^2 - 25x = x\left(4x^2 - 20x - 25\right)$

b. Using the common factor x and the special product for $(A-B)^2$:

$$4x^3 - 20x^2 + 25x = x\left(4x^2 - 20x + 25\right) = x(2x-5)^2$$

23. **a.** Using grouping: $ab - bc + a^2 - ac = b(a-c) + a(a-c) = (a-c)(b+a)$

b. Using grouping: $(u+v)x - xy + (u+v)^2 - (u+v)y = x(u+v-y) + (u+v)(u+v-y) = (u+v-y)(x+u+v)$

25. Using grouping: $x^2z^2 + xzt + xyz + yt = xz(xz+t) + y(xz+t) = (xz+t)(xz+y)$

27. Using the special product for $(A-B)^2$: $a^4 - 4a^2b^2c^2 + 4b^4c^4 = \left(a^2 - 2b^2c^2\right)^2$

29. Since this is a sum of squares, $A^2 + B^2$ is irreducible.

31. Since $4^3 = 64$, using the sum of cubes factoring yields: $x^3 + 64 = (x+4)\left(x^2 - 4x + 16\right)$

33. Using the difference of cubes factoring:

$$(x+y)^3 - y^3 = (x+y-y)\left[(x+y)^2 + (x+y)(y) + y^2\right] = x\left(x^2 + 2xy + y^2 + xy + y^2 + y^2\right) = x\left(x^2 + 3xy + 3y^2\right)$$

35. Using the difference of cubes factoring, then grouping:

$$x^3 - y^3 + x - y = (x-y)\left(x^2 + xy + y^2\right) + (x-y)(1) = (x-y)\left(x^2 + xy + y^2 + 1\right)$$

37. **a.** Using the difference of squares factoring (twice): $p^4 - 1 = \left(p^2+1\right)\left(p^2-1\right) = \left(p^2+1\right)(p+1)(p-1)$

b. Using the difference of squares factoring (three times):

$$p^8 - 1 = \left(p^4+1\right)\left(p^4-1\right) = \left(p^4+1\right)\left(p^2+1\right)\left(p^2-1\right) = \left(p^4+1\right)\left(p^2+1\right)(p+1)(p-1)$$

39. Using the special product for $(A+B)^3$: $x^3+3x^2+3x+1=(x+1)^3$

Note: If you do not recognize the special product, you can also use grouping:

$$\begin{aligned} x^3+3x^2+3x+1 &= (x^3+1)+(3x^2+3x) \\ &= (x+1)(x^2-x+1)+3x(x+1) \\ &= (x+1)(x^2-x+1+3x) \\ &= (x+1)(x^2+2x+1) \\ &= (x+1)(x+1)^2 \\ &= (x+1)^3 \end{aligned}$$

41. Since this is a sum of squares, x^2+16y^2 is irreducible.

43. Using the difference of squares factoring: $\frac{25}{16}-c^2=\left(\frac{5}{4}+c\right)\left(\frac{5}{4}-c\right)$

45. Using the difference of squares factoring (twice): $z^4-\frac{81}{16}=\left(z^2+\frac{9}{4}\right)\left(z^2-\frac{9}{4}\right)=\left(z^2+\frac{9}{4}\right)\left(z+\frac{3}{2}\right)\left(z-\frac{3}{2}\right)$

47. Using the difference of cubes factoring: $\dfrac{125}{m^3n^3}-1=\left(\dfrac{5}{mn}-1\right)\left(\dfrac{25}{m^2n^2}+\dfrac{5}{mn}+1\right)$

49. Using the special product for $(A+B)^2$: $\frac{1}{4}x^2+xy+y^2=\left(\frac{1}{2}x+y\right)^2$

51. Using the common factor $x-a$, then a difference of squares:

$$\begin{aligned} 64(x-a)^3-x+a &= (x-a)\left[64(x-a)^2-1\right] \\ &= (x-a)[8(x-a)+1][8(x-a)-1] \\ &= (x-a)(8x-8a+1)(8x-8a-1) \end{aligned}$$

53. Using grouping: $x^2-a^2+y^2-2xy=\left(x^2-2xy+y^2\right)-a^2=(x-y)^2-a^2=(x-y-a)(x-y+a)$

55. Using the common factor x, then trial and error: $21x^3+82x^2-39x=x\left(21x^2+82x-39\right)=x(7x-3)(3x+13)$

57. Using grouping, the special product for $(A-B)^2$, and the difference of squares factoring:

$$\begin{aligned} 12xy+25-4x^2-9y^2 &= 5^2-\left(4x^2-12xy+9y^2\right) \\ &= 5^2-(2x-3y)^2 \\ &= [5-(2x-3y)][5+(2x-3y)] \\ &= (5-2x+3y)(5+2x-3y) \end{aligned}$$

59. Factoring by trial and error: $ax^2+(a+b)x+b=(ax+b)(x+1)$

61. Using the common factor $(x+1)^{1/2}$:

$$(x+1)^{1/2}-(x+1)^{3/2}=(x+1)^{1/2}\left[1-(x+1)^{2/2}\right]=(x+1)^{1/2}(1-x-1)=(x+1)^{1/2}(-x)=-x(x+1)^{1/2}$$

63. Using the common factor $(x+1)^{-3/2}$:

$$(x+1)^{-1/2}-(x+1)^{-3/2}=(x+1)^{-3/2}\left[(x+1)^1-1\right]=(x+1)^{-3/2}(x)=x(x+1)^{-3/2}$$

65. Using the common factor $\frac{1}{3}(2x+3)^{1/2}$:

$$\begin{aligned}(2x+3)^{1/2}-\tfrac{1}{3}(2x+3)^{3/2}&=\tfrac{1}{3}(2x+3)^{1/2}\left[3-(2x+3)^1\right]\\&=\tfrac{1}{3}(2x+3)^{1/2}(3-2x-3)\\&=\tfrac{1}{3}(2x+3)^{1/2}(-2x)\\&=-\tfrac{2}{3}x(2x+3)^{1/2}\end{aligned}$$

67. a. Using the difference of squares factoring: $100^2-99^2=(100+99)(100-99)=(199)(1)=199$

b. Using the difference of cubes factoring: $8^3-6^3=(8-6)\left(8^2+8\cdot 6+6^2\right)=2(64+48+36)=2(148)=296$

c. Using the difference of squares factoring: $1000^2-999^2=(1000+999)(1000-999)=(1999)(1)=1999$

69. Factoring the sum of cubes and then grouping:

$$\begin{aligned}A^3+B^3+3AB(A+B)&=(A+B)\left(A^2-AB+B^2\right)+(A+B)(3AB)\\&=(A+B)\left(A^2-AB+B^2+3AB\right)\\&=(A+B)\left(A^2+2AB+B^2\right)\\&=(A+B)(A+B)^2\\&=(A+B)^3\end{aligned}$$

71. Using the common factor $x\left(a^2+x^2\right)^{-3/2}$:

$$\begin{aligned}2x\left(a^2+x^2\right)^{-1/2}-x^3\left(a^2+x^2\right)^{-3/2}&=x\left(a^2+x^2\right)^{-3/2}\left[2\left(a^2+x^2\right)-x^2\right]\\&=x\left(a^2+x^2\right)^{-3/2}\left(2a^2+2x^2-x^2\right)\\&=x\left(a^2+x^2\right)^{-3/2}\left(2a^2+x^2\right)\end{aligned}$$

73. Using grouping (first with fourth term, second with third term):

$$\begin{aligned}y^4-p^2q^2-(p+q)y^3+pq(p+q)y&=\left(y^2+pq\right)\left(y^2-pq\right)-(p+q)(y)\left(y^2-pq\right)\\&=\left(y^2-pq\right)\left(y^2+pq-py-qy\right)\\&=\left(y^2-pq\right)\left(y^2-(p+q)y+pq\right)\\&=\left(y^2-pq\right)(y-p)(y-q)\end{aligned}$$

B.5 Review of Fractional Expressions

1. Factor and reduce: $\dfrac{x^2-9}{x+3}=\dfrac{(x-3)(x+3)}{x+3}=x-3$

3. Factor and reduce: $\dfrac{x+2}{x^4-16}=\dfrac{x+2}{\left(x^2-4\right)\left(x^2+4\right)}=\dfrac{x+2}{(x-2)(x+2)\left(x^2+4\right)}=\dfrac{1}{(x-2)\left(x^2+4\right)}$

5. Factor and reduce: $\dfrac{x^2+2x+4}{x^3-8}=\dfrac{x^2+2x+4}{(x-2)\left(x^2+2x+4\right)}=\dfrac{1}{x-2}$

7. Factor and reduce: $\dfrac{9ab-12b^2}{6a^2-8ab}=\dfrac{3b(3a-4b)}{2a(3a-4b)}=\dfrac{3b}{2a}$

9. Factor and reduce: $\dfrac{a^3+a^2+a+1}{a^2-1}=\dfrac{a^2(a+1)+(a+1)}{(a-1)(a+1)}=\dfrac{(a+1)\left(a^2+1\right)}{(a-1)(a+1)}=\dfrac{a^2+1}{a-1}$

11. Factor and reduce: $\dfrac{x^3-y^3}{(x-y)^3}=\dfrac{(x-y)(x^2+xy+y^2)}{(x-y)^3}=\dfrac{x^2+xy+y^2}{(x-y)^2}$

13. Multiplying and factoring: $\dfrac{2}{x-2}\bullet\dfrac{x^2-4}{x+2}=\dfrac{2(x-2)(x+2)}{(x-2)(x+2)}=2$

15. Multiplying and factoring: $\dfrac{x^2-x-2}{x^2+x-12}\bullet\dfrac{x^2-3x}{x^2-4x+4}=\dfrac{(x-2)(x+1)}{(x+4)(x-3)}\bullet\dfrac{x(x-3)}{(x-2)^2}=\dfrac{x(x+1)}{(x+4)(x-2)}=\dfrac{x^2+x}{(x+4)(x-2)}$

17. Dividing and factoring: $\dfrac{x^3+y^3}{x^2-4xy+3y^2}\div\dfrac{(x+y)^3}{x^2-2xy-3y^2}=\dfrac{(x+y)(x^2-xy+y^2)}{(x-3y)(x-y)}\bullet\dfrac{(x+y)(x-3y)}{(x+y)^3}=\dfrac{x^2-xy+y^2}{(x-y)(x+y)}$

19. Obtaining common denominators and subtracting yields: $\dfrac{4}{x}-\dfrac{2}{x^2}=\dfrac{4x}{x^2}-\dfrac{2}{x^2}=\dfrac{4x-2}{x^2}$

21. Obtaining common denominators and subtracting yields: $\dfrac{6}{a}-\dfrac{a}{6}=\dfrac{36}{6a}-\dfrac{a^2}{6a}=\dfrac{36-a^2}{6a}$

23. Obtaining common denominators and adding yields:

$$\frac{1}{x+3}+\frac{3}{x+2}=\frac{x+2}{(x+3)(x+2)}+\frac{3(x+3)}{(x+3)(x+2)}=\frac{x+2+3(x+3)}{(x+3)(x+2)}=\frac{4x+11}{(x+3)(x+2)}$$

25. Obtaining common denominators and subtracting yields:

$$\frac{3x}{x-2}-\frac{6}{x^2-4}=\frac{3x}{x-2}-\frac{6}{(x-2)(x+2)}=\frac{3x(x+2)-6}{(x-2)(x+2)}=\frac{3x^2+6x-6}{(x-2)(x+2)}$$

27. Obtaining common denominators and adding yields:

$$\begin{aligned}\frac{a}{x-1}+\frac{2ax}{(x-1)^2}+\frac{3ax^2}{(x-1)^3}&=\frac{a(x-1)^2+2ax(x-1)+3ax^2}{(x-1)^3}\\&=\frac{ax^2-2ax+a+2ax^2-2ax+3ax^2}{(x-1)^3}\\&=\frac{6ax^2-4ax+a}{(x-1)^3}\end{aligned}$$

29. Factoring out –1 yields: $\dfrac{4}{x-5}-\dfrac{4}{5-x}=\dfrac{4}{x-5}+\dfrac{4}{x-5}=\dfrac{8}{x-5}$

31. Obtaining common denominators and combining yields:

$$\begin{aligned}\frac{a^2+b^2}{a^2-b^2}+\frac{a}{a+b}+\frac{b}{b-a}&=\frac{a^2+b^2}{(a-b)(a+b)}+\frac{a}{a+b}-\frac{b}{a-b}\\&=\frac{a^2+b^2+a(a-b)-b(a+b)}{(a-b)(a+b)}\\&=\frac{a^2+b^2+a^2-ab-ab-b^2}{(a-b)(a+b)}\\&=\frac{2a^2-2ab}{(a-b)(a+b)}\\&=\frac{2a(a-b)}{(a-b)(a+b)}\\&=\frac{2a}{a+b}\end{aligned}$$

33. Obtaining common denominators and subtracting yields:

$$\frac{1}{x^2+x-20}-\frac{1}{x^2-8x+16}=\frac{1}{(x+5)(x-4)}-\frac{1}{(x-4)^2}=\frac{x-4-(x+5)}{(x+5)(x-4)^2}=\frac{-9}{(x+5)(x-4)^2}$$

35. Obtaining common denominators and subtracting yields:

$$\frac{2q+p}{2p^2-9pq-5q^2}-\frac{p+q}{p^2-5pq}=\frac{2q+p}{(2p+q)(p-5q)}-\frac{p+q}{p(p-5q)}$$
$$=\frac{(2q+p)p-(p+q)(2p+q)}{p(2p+q)(p-5q)}$$
$$=\frac{2pq+p^2-2p^2-3pq-q^2}{p(2p+q)(p-5q)}$$
$$=\frac{-p^2-pq-q^2}{p(2p+q)(p-5q)}$$

37. Multiplying by $\frac{x}{x}$ yields: $\dfrac{\frac{1}{x}+1}{\frac{1}{x}-1}\bullet\frac{x}{x}=\frac{1+x}{1-x}$

39. Multiplying by $\frac{xa}{xa}$ yields: $\dfrac{\frac{1}{x}-\frac{1}{a}}{x-a}\bullet\frac{xa}{xa}=\frac{a-x}{xa(x-a)}=-\frac{1}{ax}$

41. Multiplying by $\frac{a}{a}$ yields: $\dfrac{a-\frac{1}{a}}{1+\frac{1}{a}}\bullet\frac{a}{a}=\frac{a^2-1}{a+1}=\frac{(a+1)(a-1)}{a+1}=a-1$

43. Multiplying by $\frac{2(2+h)}{2(2+h)}$ yields: $\dfrac{\frac{1}{2+h}-\frac{1}{2}}{h}\bullet\frac{2(2+h)}{2(2+h)}=\frac{2-2-h}{2h(2+h)}=\frac{-h}{2h(2+h)}=-\frac{1}{4+2h}$

45. Multiplying by $\frac{a^2x^2}{a^2x^2}$ yields: $\dfrac{\frac{a}{x^2}+\frac{x}{a^2}}{a^2-ax+x^2}\bullet\frac{a^2x^2}{a^2x^2}=\frac{a^3+x^3}{a^2x^2\left(a^2-ax+x^2\right)}=\frac{(a+x)\left(a^2-ax+x^2\right)}{a^2x^2\left(a^2-ax+x^2\right)}=\frac{a+x}{a^2x^2}$

47. Rewriting the expression without negative exponents: $\left(x^{-1}+2\right)^{-1}=\frac{1}{x^{-1}+2}=\dfrac{1}{\frac{1}{x}+2}\bullet\frac{x}{x}=\frac{x}{1+2x}$

49. Rewriting the expression without negative exponents: $\left(\frac{1}{a^{-1}}+\frac{1}{a^{-2}}\right)^{-1}=\left(a+a^2\right)^{-1}=\frac{1}{a+a^2}$

51. Rewriting the expression without negative exponents:

$$x(x+y)^{-1}+y(x-y)^{-1}=\frac{x}{x+y}+\frac{y}{x-y}=\frac{x(x-y)+y(x+y)}{x^2-y^2}=\frac{x^2+y^2}{x^2-y^2}$$

53. Multiplying by $\frac{(a-b)(a+b)}{(a-b)(a+b)}$ yields:

$$\frac{(a-b)(a+b)}{(a-b)(a+b)} \bullet \frac{\frac{a+b}{a-b}+\frac{a-b}{a+b}}{\frac{a-b}{a+b}-\frac{a+b}{a-b}} \bullet \frac{ab\left(b^2-a^2\right)}{a^2+b^2} = \frac{(a+b)^2+(a-b)^2}{(a-b)^2-(a+b)^2} \bullet \frac{ab(b-a)(b+a)}{a^2+b^2}$$

$$= \frac{2a^2+2b^2}{-4ab} \bullet \frac{ab(b-a)(b+a)}{a^2+b^2}$$

$$= \frac{2\left(a^2+b^2\right)}{-4} \bullet \frac{(b-a)(b+a)}{a^2+b^2}$$

$$= \frac{(b-a)(b+a)}{-2}$$

$$= \frac{b^2-a^2}{-2}$$

$$= \frac{a^2-b^2}{2}$$

55. Simplifying the complex fractions:

$$\frac{\left(a+\frac{1}{b}\right)^a\left(a-\frac{1}{b}\right)^b}{\left(b+\frac{1}{a}\right)^a\left(b-\frac{1}{a}\right)^b} = \frac{\left(\frac{ab+1}{b}\right)^a\left(\frac{ab-1}{b}\right)^b}{\left(\frac{ab+1}{a}\right)^a\left(\frac{ab-1}{a}\right)^b} = \left(\frac{\frac{ab+1}{b}}{\frac{ab+1}{a}}\right)^a\left(\frac{\frac{ab-1}{b}}{\frac{ab-1}{a}}\right)^b = \left(\frac{a}{b}\right)^a\left(\frac{a}{b}\right)^b = \left(\frac{a}{b}\right)^{a+b}$$